Dictionary of Plant Breeding

Dictionary of Plant Breeding

Third Edition

Rolf H. J. Schlegel

CRC Press
Taylor & Francis Group
Boca Raton London New York

CRC Press is an imprint of the
Taylor & Francis Group, an **informa** business

CRC Press
Taylor & Francis Group
6000 Broken Sound Parkway NW,
Suite 300, Boca Raton, FL 33487

© 2020 by Taylor & Francis Group, LLC

CRC Press is an imprint of Taylor & Francis Group, an Informa business

No claim to original U.S. Government works

ISBN 13: 978-0-367-49216-8 (hbk)

**Visit the Taylor & Francis Web site at
http://www.taylorandfrancis.com**

**and the CRC Press Web site at
http://www.crcpress.com**

Contents

List of Figures vii

List of Tables xi

Preface xiii

About the Author xv

User's Guide 1

Abbreviations and Acronyms 3

Glossary 9

Important Crop Plants, Weeds, Ornamental, Herb, Industrial, Woody, and Other Plants of the World 475

Figures 625

Tables 687

Bibliography 731

List of Figures

Figure 1   Basic scheme of selection in mutation breeding 625

Figure 2   Main steps in the establishment of a hybrid variety of wheat................................. 626

Figure 3   Development of allopolyploid (amphidiploid) hybrids, for example, a wheat–rye hybrid (octoploid triticale) 627

Figure 4   Residue seed method of breeding: half-sib progeny selection or method of overstored seeds... 628

Figure 5   Single plant selection, including testing of progeny, in autogamous plants 629

Figure 6   Inheritance and segregation patterns in subsequent generations of flower color from a cross of a red-flowered with a white-flowered plant, and with dominant inheritance of red flowers 630

Figure 7   Pedigree breeding (cross-combination breeding) in autogamous plants.................. 631

Figure 8   Some diploid and allopolyploid species of the genus *Brassica,* having agricultural and breeding importance.. 632

Figure 9   Experimental field design of a Latin square considering four variants and four replications (four blocks, four columns) and a Latin rectangle considering ten variants (five blocks, five columns) 633

Figure 10   The phylogeny of wheat (*Triticum* spp.) 634

Figure 11   Types of chromosome and/or centromere constrictions..................................... 634

Figure 12   The karyogram of diploid rye, *Secale cereale* L.. 635

Figure 13   Pairing failure of meiotic chromosomes in an interlocked configuration................ 635

Figure 14   Different types of trisomics in plants 636

Figure 15   A diagrammatic representation of different meiotic chromosome configurations observed at diakinesis and metaphase, including the minimum chiasmata.............. 637

Figure 16   Combination breeding, using the bulk method 638

Figure 17   A comparison of the genetic segregation patterns of sexually derived and doubled-haploid-derived F_2 progenies from F_1 heterozygotes................................... 639

Figure 18   Schematic interpretation of the "dominance hypothesis," considering inbred lines and the F_1 hybrid... 640

Figure 19   Schematic drawing of a topcross design, including progeny testing for general combining ability.. 641

Figure 20   Schematic drawing of a recurrent selection design.. 642

Figure 21   Schematic drawing of a reciprocal recurrent selection design 643

Figure 22   Schematic drawing of F_1 hybrid seed production in maize.................................. 644

Figure 23   Schematic drawing of F_1 hybrid seed production, after utilization of cytoplasmic male-sterility 645

Figure 24   Schematic drawing of unequal crossing-over... 645

Figure 25   Schematic drawing of embryo sac and pollen formation 646

Figure 26   Breeding scheme using doubled haploids 647

Figure 27   Schematic drawing of *Agrobacterium*-mediated gene transfer............................. 648

Figure 28   Different methods of asexual reproduction in plants 649

Figure 29 F_1 hybrid seed production in allogamous rye 650

Figure 30   Designs of spreader nurseries... 651

Figure 31   Basic crossing schemes in plant breeding 652

Figure 32   Schematic drawing of honeycomb and grid designs 653

Figure 33   Several types of hill plots 654

Figure 34   Basic shapes and types of spikes in wheat... 654

Figure 35   Schematic drawing of sexual organs of a plant flower 655

Figure 36   Types of duplicated chromosome segments 656

Figure 37   Schematic drawing of aneuploid types in wheat, applicable to other diploids and polyploids; it describes the chromosome number, meiotic configuration, and name of the aberration; in isosomics, 1^{II} means a ring-like structure) 657

Figure 38   Estimation of heritability, h^2, based on selection advantage, "R", selection coefficient, "S", and selection intensity, "i" 658

Figure 39   Schematic drawing of mass selection and single-plant selection 659

Figure 40   Schematic drawing of positive mass selection 660

Figure 41   Couple method of breeding in allogamous plants 661

Figure 42   A general breeding scheme called "the variety machine." (Modified after N. F. Jensen, 1988. *Plant Breeding Methodology*. John Wiley & Sons, New York, pp. 676.) 661

Figure 43   A general scheme of integration of genomic resources for crop improvement. (Modified after Varshney, R. K. et al., 2005. *Trends in Plant Science* 10: 621–630.) 662

Figure 44   Schematic drawing of presumed pedigree of sugar beet. (Modified after Zhukovski,1971.) 663

Figure 45   Molecular procedure of chromosome walking 664

Figure 46   Schematic drawing of gene tagging, using transposons 665

Figure 47   (A) Scheme of F_1 hybrid seed production in rapeseed (*Brassica napus* ssp. *napus*), strip cropping as ratio 2:1; and (B) scheme of line production by using cytoplasmic male-sterile mutants 666

Figure 48   Schematic drawing of restriction fragment length polymorphism (RFLP) analysis, using gel electrophoresis and a Southern transfer technique 667

Figure 49   Experimental field design of quasi-complete Latin square, considering five variants and five replications (five blocks, five columns) 668

Figure 50   Breeding schemes for clonal varieties 669

Figure 51   Breeding schemes for synthetic varieties, using inbred lines 670

Figure 52   Schematic drawing of a labeling reaction with labeled dNTP for *in situ* hybridization experiments 671

Figure 53   Schematic drawing of oligolabeling reaction with unlabeled and labeled dNTP for *in situ* hybridization experiments 672

Figure 54   Schematic drawing of the nick translation reaction with unlabeled and labeled dNTP for *in situ* hybridization experiments 673

Figure 55   Schematic drawing of F_1 hybrid seed production in barley, using the BTT system 674

Figure 56   Types of periclinal chimera 674

Figure 57   Schematic drawing of isolation and regeneration of protoplasts 675

Figure 58   Schematic drawing of haploid production by androgenesis (anther and pollen culture) 676

Figure 59   General scheme of marker-assisted selection 677

Figure 60   Diagramatic scheme of phenotype vs. genotype selection. The height of the lighter portion of each column reflects the variability caused by the genetic constitution, while the darker portion gives the phenotypic variation, i.e. non-inherited characters. The total height of each column reflects the total expression of the character. When 20 individuals of cross progeny are arranged by their genotypic background, then five superior individuals can be selected, based on a specific quality limit (dotted horizontal line). When it is selected by genotypic variation (as revealed by progeny testing), then just four individuals would need to be considered (beyond the dotted vertical line). Therefore, genotypic selection is more efficient when applied from the very beginning 678

Figure 61 Genomic relationships among citrus fruits...679

Figure 62 Schematic drawing of common plant diseases..680

Figure 63 Schematic drawing of SSR and SNP markers. (Source: R. Schlegel.) 680

Figure 64 Phylogenetic relationships among rye species (*Secale* spp.) and their evolutionary development. (Source: R. Schlegel.)......................................681

Figure 65 Main steps of sex chromosome evolution in plants. (Modified after Ming et al. 2011.)..682

Figure 66 Polytene chromosome of bean (*Phaseolus vulgaris*) antipodal cells in comparison to the standard metaphase chromosomes. (Used with permission of E. Badaeva, 2012.)..683

Figure 67 Schematic drawing of leaf anatomy in C_3 and C_4 plants. In C_4 plants, a wreath-like arrangement of cells acts to concentrate carbon dioxide. A ring of mesophyll cells captures the carbon dioxide, that is conveyed to an inner ring of bundle-sheath cells. It is called Kranz ("wreath") anatomy 684

Figure 68 Scheme of reduction of variability during the process of domestication, i.e. from wild plants to advanced varieties..684

Figure 69 Simplified phylogeny of the green plant lineage focusing on the occurrence of events. Estimates for the age of angiosperms have suggested a range of 167–199 million years ago (Mya); then rapid radiations responsible for the extant angiosperm diversity occurred after the early diversification of the Mesangiospermae 139–156 Mya, with a burst of diversification specific to the Cretaceous Period, <125 Mya. (Modified after Alix et al. (2017), http://molcyt.org/2017/07/13/polyploidy-and-interspecific-hybridisation-partners-for-adaptation-speciation-and-evolution-in-plants/.)......................................685

Figure 70 Schematic drawing of a monocentric and a holocentric chromosome. (Source: R. Schlegel.) 685

Figure 71 Schematic drawing of genomic compartments of a plant species. (Source: R. Schlegel.)..686

List of Tables

Table 1   Classification of Wheats (*Triticum* spp.) 687

Table 2   Number of Gametes, Genotypes, and Phenotypes, Considering One and Multifactorial Heterozygosity in F_1 and F_2 Generations 688

Table 3   Segregation of Recessive Nulliplex Genotypes from Triplex, Duplex, and Simplex Genotypes, Considering Selfing, Random Chromosome Distribution, and Complete Dominance 688

Table 4   Expected F_2 Segregations of Trisomic F_1 Plants from a Critical Cross of Trisomic by Disomic, Excluding Any Selection, Male Transmission of *n + 1* Gametes and Abnormal Chromosome Segregation 689

Table 5   Basic Methods of Plant Breeding 689

Table 6   Genotypic and Phenotypic Segregation in F_2 Populations, Considering Two Genes and Interacting in Different Manners 690

Table 7   Frequencies and Ratios of Completely Recessive Plants in F_2 Progeny of Doubled Haploids, Diploids and Tetraploids 690

Table 8   Genome Relationships between Embryo, Endosperm, and Ovary, after Crossing Parents with Different Ploidy Levels 691

Table 9   Genetic Segregation Patterns Depending on the Number of Genes Involved 691

Table 10   Frequencies of Homozygotes and Heterozygotes in a Progeny of a Heterozygous Individual after Subsequent Self-Pollinations 691

Table 11   Examples of Seed Conditioning in Some Crop Plants 692

Table 12   Taxonomic Classification System in Plants 692

Table 13   Decimal Code for Plant Growth in Cereals 693

Table 14   Genome Relationships of Oats (*Avena* spp.) 694

Table 15   The Approximate Protein Composition (%) in Some Cereals 695

Table 16   Food Reserves of Some Crop Plants 695

Table 17   Taxonomic Relationships of Some Tuberous *Solanum* spp. 696

Table 18   Types of Flowers in Higher Plants 697

Table 19   Segregation of a Single Gene and/or Alleles in Subsequent Generations 697

Table 20   Phenotypic Relations of Homozygotes and Heterozygotes, Depending on Different Dominance Levels in Diploids 697

Table 21   Phenotypic Ratios in the F_2 Generation for Two Unlinked Genes, Depending on the Degree of Dominance at Each Locus and Epistasis between Loci 698

Table 22   Calculation of Recombination Frequency between Two Loci from a Cross between *AaBb* x *aaaa* 698

Table 23   Test Crosses with Monosomics to Determine the Location of a Dominant Allele 699

Table 24   Possible Planting Arrangements for a Diallel Crossing (Six Parents, No Reciprocal Cross, No Self-Pollination) 699

Table 25   A Randomized Complete-Block Design for Five Entries and Ten Replications 700

Table 26   Lattice Design (42 Entries, Three Replications, No Blocks within the Replication, Entries Assigned at Random to the 42 Plots) 700

Table 27   Recovering of Genes from the Recurrent Parent During Backcrossing 701

Table 28   Scheme of Seed Purification and Increase 701

Table 29   Types and Characteristics of Several Markers in Breeding and Genetics 702

Table 30   Typical Characteristics for Identification of Wheat, Barley, or Wild Oat 703

Table 31   Taxonomy of the Genus *Helianthus* 703

Table 32   About the Evolution of Triticeae 704

Table 33   Inheritance and Variation of Several Breeding Characteristics 704

Table 34   Some Examples of Heritability of Breeding Characteristics 704

Table 35   Proposed Breeding Schemes Depending on the Crop, Reproduction System and
Basic Population Features (+ Applicable) 705

Table 36   The Size Reduction of Populations Depending on Various Numbers of
Overdominance Loci and Different Selective Disadvantages for Homozygotes 705

Table 37   Types of Hybrids and/or Hybridization 706

Table 38   The Genetic Code 706

Table 39   Cleaving Sites of Some Restriction Enzymes 707

Table 40   Guidelines for Selection of an Experimental Design 707

Table 41   Genome Relationships of Barley (*Hordeum* sp.) 708

Table 42   Developmental Stages in Plants 709

Table 43   Types of Chromatin in Chromosomes and Their Properties 710

Table 44   Scheme of Soil Classification (According to the Guidelines for Soil Description,
FAO 1990) 711

Table 45   Species of Rice (*Oryza* spp.), Their Karyological Composition and Distribution 712

Table 46   Toxic Plants and Their Toxins 713

Table 47   Different Techniques of Plant Tissue Culture and Their Applications in Plant
Improvement 716

Table 48   Total World Area of Major Cereal Crops and World Production in 2005, As Well
As Comparative Values of Energy Provided, Total Content Of Protein, and Lipid
in 100 g of Cereal Grains 718

Table 49   General Issues of Variety Development and Maintenance 718

Table 50   Examples of Monosomic Series Developed in Europe 719

Table 51   Genetically Localized Genes of Mildew Resistance (*Blumeria graminis* f. sp.
tritici) in Wheat 720

Table 52   Genetically Localized Genes of Brown Rust (*Puccinia triticina*) in Wheat 721

Table 53   Phylogenetic Relationships in Wheat Considering Genome-Plasmon
Constitutions 722

Table 54   Phylogenetic Relationships in Barley Considering Genome-Plasmon
Constitutions 723

Table 55   Major Food Crops of the World, Ranked According to Total Tonnage Produced
Annually 724

Table 56   Estimated Domestication of Main Crop Plants and Their Origin 725

Table 57   Scheme of a Conventional Breeding Program in Cereals, Including Subsequent
Performance Testing 726

Table 58   Comparision of Association and QTL Mapping 727

Table 59   Types of Repetitive DNA 727

Table 60   *Aegilops* Species Used in Introgression Experiments of Wheat and Their
Genomic Designations 728

Table 61   Genotypes Derived from a Single-Locus Heterozygote and Their Frequencies in
Subsequent Selfing Generations 728

Table 62   Applied Methods of Micropropagation, from a Stock Plant to Final Plantlet
Regeneration 729

Preface

Modern plant breeding has improved the livelihoods of hundreds of millions of people in the world. Plant breeding began in Neolithic times, when the people collected and later sowed seed of plants they liked. But it was not until the encounter with Mendel's discoveries thousands of years later that the genetic basis of breeding was understood. Developments since then have provided further insight into how genes acting alone, or in concert with other genes and the environment, result in a particular phenotype.

As a complex subject, plant breeding makes use of many scientific and technological disciplines, such as agronomy, horticulture, forestry, seed production, genetics, molecular genetics, biotechnology, biochemistry, mutagenesis, population genetics, biometry, botany, cytology, cytogenetics, ecology, plant evolution or genetic conservation, and laboratory technologies. Often it is difficult to know the precise meaning of many terms and to accurately interpret specific concepts of plant breeding. Most dictionaries and glossaries available are highly specific or fragmentary. Until now, no attempt has been made to provide a comprehensive compilation of plant breeding terms. This dictionary includes the specific terms of plant breeding and terms that are adjusted from other disciplines.

Moreover, the intent was to create a book that includes not only present terms, but also some terms used during the long history of plant breeding. In addition, this book offers most of the words used by a plant breeder and seed producer together in one source. In order to also serve students, teachers, and research workers, the book is supplemented with breeding schemes, tables, examples, and a list of crop plants, including a few details.

Completeness could not be achieved. The modern subjects, such as biotechnology, molecular genetics, cytogenetics, and genetic engineering, were included as far as possible. Selection was made on the basis of the author's experience and his knowledge of the breeder's requirements. This is a modest effort to serve the scientific community.

Plant breeding is a rapidly developing subject. The recent integration of advances in biotechnology, genomic research, and molecular marker applications with conventional plant breeding practices has created the foundation for molecular plant breeding, an interdisciplinary science that is revolutionizing 21st-century crop improvement. Techniques such as cutting and modifying the genome during the repair process by zinc finger nuclease; TALENs, and CRISPR/Cas, genome editing to introduce changes to just a few base pairs (by oligonucleotide-directed mutagenesis), transferring genes from an identical or closely related species (by cisgenesis), adding in a reshuffled set of regulatory instructions from same species (by intragenesis), deploying processes that alter gene activity without altering the DNA itself (by epigenetic methods), or grafting of unaltered plant onto a genetically modified rootstock are considered in this edition.

Therefore, the definitions or descriptions given here may subsequently be modified. The author, therefore, accepts no responsibility for the legal validity, accuracy, adequacy, or interpretation of the terms given in this book. Suggestions or errors brought to his attention will be considered whenever possible.

This third edition of the dictionary has again undergone extensive revision and supplements since the first and second successful publications in 2003 as well as 2010, respectively. A good handbook is never completed! It provides an update on the research worldwide. In addition, several new technical terms are included, including improvement as well as supplements of descriptions. This totally revised and expanded reference provides comprehensive coverage of the latest discoveries in cytogenetics, molecular genetics, marker-assisted selection, experimental gene transfer, seed sciences, crop physiology, and genetically modified crops.

The book offers an encyclopedic list of nearly 10,000 contemporary references, over 40% more than the second edition, and over 50 useful tables and 65 drawings and/or figures. The third edition is an up-to-date resource for plant and crop physiologists; plant, crop, soil, and environmental scientists; botanists; agronomists; agriculturists; horticulturists; biochemists; foresters; plant culturists; and upper-level undergraduate and graduate students in these disciplines.

About the Author

Rolf H. J. Schlegel, Ph.D., is Professor of Cytogenetics and Applied Genetics, with over 50 years of experience in research and teaching of advanced genetics and plant breeding in Germany and Bulgaria. Prof. Schlegel is the author of more than 160 research papers, monographs, and scientific contributions, co-coordinator of international research projects, and has been a scientific consultant at the Bulgarian Academy of Agricultural Sciences for several years. He received his Master's degree in Agriculture and Plant Breeding, and his Ph.D. and D.Sc. in Genetics and Cytogenetics from the Martin-Luther University Halle/S., Germany. Later he became Head of Laboratory of Chromosome Manipulation and the Department of Applied Genetics and Genetic Resources at the Institute of Plant Genetics and Crop Plant Research, Gatersleben, Germany, and the Head of the Genebank at the Institute of Wheat and Sunflower Research, General Toshevo/Varna, as well as at the AgroBioInstitute, Sofia, Bulgaria. He is now emeritus of the institute.

User's Guide

This dictionary provides a representative selection of technical terms from the huge vocabulary of plant breeders, seed producers, and all those who work in related fields. Different terms included in this book have been arranged alphabetically on a word-by-word basis. When Greek letters were necessary in association with some words, they were translated into English and also arranged according to alphabetical order.

Names of scientists and/or family names are in capital letters. However, when used adjectivally (e.g., Mendelian) or as a unit of measure (e.g., Ångström), only the first letter is capitalized.

Explanations of a given term may be more or less extensive. Several definitions have simply been separated by a semicolon. Terms that are either very general in nature or self-explanatory have been avoided. Alternate names of terms have been given within the definition.

Cross-references have been provided wherever necessary for economizing space, demonstrating interrelationships, and organizing the material in a clear manner. Cross-referenced terms are indicated by the symbol >>>.

Several terms are glossed and supplemented by tables, figures, and illustrations. Cross-reference figures and tables are also indicated by the symbol >>>, together with the number of the figure or table.

Since plant breeders and specialized scientists make use of different meanings for identical terms or vice versa, the terms were categorized according to the scheme given under abbreviations. Whenever it was necessary or possible, the term was associated with one or more abbreviations of a scientific field and/or category.

Figures, tables, and a list of crops, weeds, and some other important plants are included in separate sections. The latter shows the common names and the scientific descriptions, together with the chromosome number, genome constitution, DNA content, and other details.

Abbreviations – Acronyms

ABA	abscisic acid
ABRE	abscisic acid-response element
ACC	1-amino-cyclopropane-1-carboxylic acid
ACCase	acetyl-CoA-carboxylase
ACMV	African cassava mosaic virus
ACP	acyl carrier protein
adm	administration
AdoMetDC	S-adenoysl-t-methionine decarboxylase
AFLP	amplified fragment length polymorphism
agr	agriculture
Al	aluminum
AlMV	alfalfa mosaic virus
ALS	acetolactate synthase
AMPA	aminomethylphosphonic acid
anat	anatomy
ANOVA	analysis of variance
A-PAGE	acid polyacrylamide gel electrophoresis
aqu	aquaculture
ArMV	*Arabidopsis* mosaic virus
ASI	anthesis-silking interval
AUDPC	area under the disease progress curve
AuxRE	auxin-response element
BADH	betaine aldehyde dehydrogenase
BAP	6-benzylaminopurine
BILs	backcross inbred lines
bio	biology
biot	biotechnology
BMV	barley mosaic virus
BNYVV	beet necrotic yellow vein virus
bot	botany
bp	base pair(s)
Bq	becquerel
BrdU	bromodeoxyuridine
bree	breeding
Bt	*Bacillus thuringensis*
bu	bushels
BYDV	barley yellow dwarf virus
bZIP	basic leucine zipper
°C	degree Celsius
ca	approximately, about
CaMV	cauliflower mosaic virus
CAT	chloramphenicol acetyltransferase
cc	cubic centimeter
CCN	cereal cyst nematode, **Heterodera avenae**
CDK	cyclin-dependent kinase
cDNA	complementary DNA
CE	coupling element

cf	confer, compare
CHA	chemical hybridizing agent
chem	chemistry
cm	centimeter
cm³	cubic centimeter
cM	centimorgan
CMO	choline monooxygenase
CMS	cytoplasmic male sterile
CMV	cucumber mosaic virus
CoA	coenzyme A
COR	cold-responsive
CP	coat protein
CPMV	cowpea mosaic virus
CSSLs	chromosome segment substitution lines
cyto	cytology
2,4 D	2,4-dichlorophenoxyacetic acid
da	dalton
DAPI	4',6-diamidino-2-phenylindole
dATP	deoxyadenosine triphosphate
DCL	dicer-like protein
dCTP	deoxycytidine triphosphate
dGTP	deoxyguanine triphosphate
DH	doubled haploid
Dnase	deoxyribonuclease
dNTP	deoxyribonucleotide
ds	double-stranded
DTF	days to flowering
DTH	days to heading
DTT	dithiothreitel
dUTP	deoxuridine triphosphate
eco	ecology
EDTA	ethylenediamine tetra acetic acid
EGMS	environment-induced genic male sterility
ELISA	enzyme-linked immunosorbent assay
EM	electron microscope
EMS	ethyl methanesulfonate
env	environment
EST	expressed sequence tag
eth	ethology
EU	European Union
evol	evolution
FEA	formalin-ethanol-acetic acid
FHB	**Fusarium** head blight
FNP	functional nucleotide polymorphism
foo	food science and production
fore	forestry
f.sp.	forma specialis
f.w.	formular weight
GCA	general combining ability
GD	genetic distance
gen	general issues

gene	genetics
GIS	geographic-information system
GM	genetically modified
Gpb	giga base pairs
GUS	ß-glucoronidase
h	hour
HMW	high molecular weight (glutenin)
hort	horticulture
HPLC	high-performance liquid chromatography
HPV	human papilloma virus
Hz	hertz
IR	infrared
IRAP	inter-retrotransposon amplified polymorphism
ISSR	inter-simple sequence repeat
ITSR	internal transcribed spacer region
IUCN	International Union of Conservation of Nature and Natural Resources, Gland, Switzerland
k	kilo
kb	kilobases
kda	kilodalton
kg	kilogram
km	kilometer
l	liter
Lat	Latin
LD_{50}	lethal dose/fifty LD50
LMW	low molecular weight (glutenins)
log	logarithm
μ	micro
m	meter
M	mega
M	molar concentration
M	morgan
MAS	marker-assisted selection
Mb	mega basepairs
me	milli-equivalents
meth	methods, methodology
micr	microscopy
min	minute
ml	milliliter
μm	micrometer
mM	millimol
mm	millimeter
mmt	million metric tons
mol	mole **chem**
mRNA	messenger RNA
m.w.	molecular weight
mya	million years ago
n	nano
N	normal **chem**
NBT	nitroblue tetrazolium salt
NGO	non-governmental organization

NILs	near-isogenic lines
NIR	near infrared
nm	nanometer
nt	nucleotide
org	organization
PAGE	polyacrylamide gel electrophoresis
PBS	phosphate-buffered saline
PCR	polymerase chain reaction
PEG	polyethylene glycol
PFG	paraformaldehyde-glutaraldehyde
PFGE	pulsed-field gel electrophoresis
pg	picogram
phy	physics
phys	physiology
phyt	phytopathology
pl	plural
PMCs	pollen mother cells
ppm	parts per million
PPO	polyphenol oxidase
prep	preparation, techniques
Q	quintals
QTL	quantitative trait loci
R	roentgen
RAPD	random amplified polymorphic DNA
RCA-RCA	restriction and circularization-aided rolling circle amplification
RCB	randomized-complete block
REMAP	retrotransposon-microsatellite amplified polymorphism
RFLP	restriction fragment length polymorphism
RILs	recombinant inbred lines
RNase	ribonuclease
rpm	rounds per minute
RT-PCR	real-time polymerase-chain reaction
RWA	Russian wheat aphid
SAMPL	selective amplification of microsatellite polymorphic loci
SAUDPC	standardized area under the disease progress curve
SBMV	soilborne mosaic virus
SCAR	sequence-characterized amplified region
SD	standard deviation
SDS	sodium dodecyl sulfate
SDS-PAGE	sodium dodecyl sulphate polyacrylamide gel electrophoresis
sec	second
seed	seed science, seed production
sero	serology
SINGER	Systemwide Information Network for Genetic Resources
SLB	**Septoria** leaf blotch
SNP	single nucleotide repeat
sp.	species
SSC	standard saline citrate buffer
SSD	single seed descent
SSLP	simple sequence length polymorphism
ssp.	subspecies

SSR	single sequence repeat
stat	statistics, experimental planning
STMS	sequence tagged microsatellite site
STS	sequence tagged site
syn	synonymous
tax	taxonomy
TBS	tris-buffered saline
TE	tris-EDTA buffer
tech	technology
TET	transiently expressed transposase
TKW	thousand-kernel weight
T_m	melting temperature of DNA
TRIM	terminal-repeat retrotransposon in miniature
tya	thousand years ago
U.K.	United Kingdom
UNEP	United Nations Environment Program, Nairobi, Kenya
URL	uniform resource locator
U.S.	term used in the United States
U.S.A.	United States of America
UV	ultraviolet
V	volt
ver	verb
VIGS	virus-induced gene silencing
v/v	volume to volume
WDF	wheat dwarf mosaic
WGA	whole genome amplification
WSBMV	wheat soilborne mosaic virus
WSMV	wheat streak mosaic virus
WSSMV	wheat spindle streak mosaic virus
WTO	World Trade Organization, Geneva, Switzerland
w/v	weight per volume
WYMV	wheat yellow mosaic virus
YAC	yeast artificial chromosome
zoo	zoology
~	about, adequate
>>>	see

Glossary

A

AI, AII: The first and second anaphase, respectively, in meiosis *cyto* >>> anaphase

AARS: >>> All-America Rose Selections Inc.

ABA: >>> abscisic acid

abasic site: A gap in the nucleic acid sequence that originates from the loss of a base *gene*

abaxial: The surface of a leaf facing away from the axis or stem of a plant, as opposed to adaxial *bot*

aberration: Variation of chromosome structure caused by induced or spontaneous mutations; in general, a non-typical form or function cyto; in microscopy, failure of an optical or electro-optical lens to produce exact geometric (and chromatic) correspondence between an object and its image *micr*

aberration rate: The portion of chromosomal changes as compared to normal chromosomes *cyto*

aberrant: Having uncommon characteristics or not strictly true to the phenotype *gene*

ab initio gene prediction: The prediction of >>> exon structure within a gene using computational algorithms; exons are ligated to form the structural protein that is encoded by the gene; comparison of sequence differences between coding and non-coding regions has encouraged the development of prediction methods based on the probabilistic modeling of DNA sequences that helps to overcome the limitations of homology-based methods; therefore, ab initio gene-finding programs recognize signals of compositional features in an input genomic sequence by pattern matching or statistical means *biot meth*

abiogenesis: Spontaneous generation of living organisms from non-living matter *evol*

abiogeny: >>> abiogenesis

abiotic: Factors or processes of the non-living environment (climate, geology, and atmosphere) eco

abiotic stress-related trait: Response by the plant in terms of resistivity or sensitivity to abiotic stress *phys eco*

abjection: The separating of a spore from a sporophore or sterigma by a fungus *bot*

abjunction: The cutting off of a spore from a hypha by a septum *bot*

ablastous: Without germ or bud *bot*

abnormal: Not normal; unusual variance from the natural habit; describes a state, condition, or behavior that is unusual or different from what is considered normal *gene bot*

abort: To fail in the early stages of formation; the collapse or disappearance of seeds or cells bot

abortive: Defective or barren; development arrested while incomplete or imperfect *bot*

abortive mitosis model: The hypothesis that cells undergo a programmed death when induced to mitosis at an inappropriate phase of the cell cycle *bio*

abortive infection: When pathogenic microorganisms fail to become established in the tissue of the host *phyt*

abortive transduction: An event in which transducing DNA fails to be incorporated into the recipient >>> chromosome *biot*

aboveground biomass: >>> biomass

abrasion injury: Observed on a plant that was caused by sand, hail, or wind resulting in moderate to severe tissue loss on stem, leave, flower, or fruit *hort agr*

abrasive polisher: A machine that uses abrasive action between the kernel and the emery stone to remove the >>> bran from the >>> rice or other cereal kernel *agr seed*

abrupt speciation: The result of a sudden change in chromosome number or constitution, which produces an almost irreversible barrier to crossing populations or species *eco evol*

abscise: Separate by abscission, as a leaf from a stem *bot*

abscisic acid (ABA): A growth regulator or plant hormone such as >>> auxins, >>> gibberellins, or >>> cytokinins; it occurs in various tissues and seeds; the substance is thought to play an important role in the regulation of certain aspects of seed growth and development, as well as being involved in fruit growth, rejection of plant organs such as leaves and fruits, and certain other physiological phenomena; regulatory mechanisms of ABA-dependent gene expression have been studied, for example, using a vp1 (viviparous1) mutant of >>> maize and abi (ABA-insensitive) mutants of >>> thale cress (Arabidopsis thaliana), both of which showed reduced levels of seed dormancy and sensitivity to exogenous ABA for inhibition of germination; ABA biosynthesis is required for seed maturation and >>> dormancy during seed development, while in vegetative tissues ABA is synthesized de novo mainly in response to drought and high salinity stresses; many genes as components of stress signaling pathways are induced by exogenous ABA in thale cress and rice; ABA regulatory pathways are not completely independent; several Cor/Lea genes are in fact responsive to exogenous ABA, and their promoter sequences commonly contain ABA responsive element (ABRE); information on roles of ABA in the regulation of ABRE-containing genes under LT conditions is still limited in wheat and its related species; an ABA-insensitive, non-dormant line of common wheat, "EH47-1" was derived from an ABA-sensitive and dormant line, "Kitakei-1354" a single dominant mutant by EMS (ethylmethan sulfonate) mutagenesis; embryos of the mutant line lose sensitivity to ABA during the later process of seed maturation, while embryos of the parental line maintain the sensitivity even after maturity; comparative studies of freezing tolerance after cold acclimation and Cor/Lea gene expression between "Kitakei-1354" and "EH47-1" suggest that ABA sensitivity contributes to determining the basal level of freezing tolerance in wheat; additional results indicate that the two mutations of ABA sensitivity mainly affect developing seeds similar to Arabidopsis abi3 mutation; Arabidopsis abi3, which is an ortholog of a maize vp1, has >>> pleiotropic effects on seed maturation, regulation of sensitivity to ABA inhibition of germination, expression of some seed-specific genes, acquisition of desiccation tolerance, and dormancy *phys*

abscisin(e): >>> abscisic acid

abscissa: In monovariable distributions the abscissa (x-axis) is used for plotting the trait, while the ordinate (y-axis) gives the frequencies *stat*

abscission: Rejection of plant organs (e.g., of leaves in autumn) bot phys >>> abscisic acid

abscission layer: Cells of >>> abscission zone disintegrate, causing the separation of plant organs, e.g., leaf, fruit, pod, or branch *bot*

abscission zone: Zone at base of leaf, flower, fruit, or other plant part that contains an >>> abscission and/or separation layer and a protective layer, both involved in the abscission of the plant part phys *bot*

absorbency: A measure of the loss of intensity of radiation passing through an absorbing medium meth *phy*

absorption: Uptake of substances, usually nutrients, water, or light, by plant cells or tissue; in soil science, the physical uptake of water or ions by a (soil) substance phys; in microscopy, the interaction of light with matter, resulting in decreased intensity across entire spectrum or loss of intensity from a portion of the spectrum *micr*

absorption spectrum: A graph that shows the percentage of each wavelength of light absorbed by a pigment (e.g., chlorophyll) *phys*

absorptive: The state or process of being absorbed

abundance: The estimated number of individuals of a species in an area or population *phyt*

abzyme: An >>> antibody engineered to carry out an enzymatic reaction (i.e., a catalytic antibody); typically the antibody is raised against a transition state analog for the reaction to be catalyzed *biot*

acantha: A spine, thorn, or prickle *bot*

acanthocarpous: A fruit showing prickles *bot*

acaricide: A pesticide used to kill or control mites or ticks *phyt*

acarides: Related to the spider animals; more than 10,000 species are known; many species parasitically live on plants *zoo phyt*

acarpous: Describes a plant that is sterile *bot* >>> sterile

acaudate: Not having a tail *phyt*

acaulescent: Without a visible stem or culm *bot*

acauline: >>> acaulescent

accelerated aging test: >>> aging test

acceleration: The time rate of change of velocity with respect to magnitude or direction *phy*; in genetics, the speeding-up of the time of action of a gene so that the character it controls develops earlier in the life cycle than it did in ancestral forms *gene*

accent plant: Plant used in a formal bed or border to emphasize contrasts of color, height, or texture *hort*

acceptable daily intake: A measure of the chemical that is believed to be consumed on a daily basis over a lifetime without harm; in U.S. to establish safe levels of potentially hazardous substances in food *bio meth*

acceptor: An atom that receives a pair of electrons to form a chemical bond *chem phys*

accession: A distinct sample of germplasm (cultivar, breeding line, population) that is maintained in a genebank for conservation and evaluation; in order to represent the genetic variation of a sample, ideally 4,000 seeds are needed for genetically homogeneous lines and about 12,000 seeds for heterogeneous lines *seed* >>> Figure 51

accession number: A unique identifier assigned to each accession when it is registered within a genebank *seed*

accessory bud: Buds that are at or near the nodes but not in the axils of leaves *bot*

accessory chromosome: A chromosome that is present in addition to the normal chromosome complement *cyto* >>> B chromosome

accessory DNA: Surplus >>> DNA present in certain cells or cell stages due to gene amplification *gene*

accessory genome: >>> dispensable genome >>> pan genome

accessory plate: The supplementary metaphase plate, originating from bivalents remaining outside the equatorial plate because of non-congression or a lack of centromere orientation *cyto*

accidental evolution: Evolution which confers no selective advantages and which owes its origin to mutations of more or less neutral effect *evol* >>> Figure 68

accidental host: That type of host on which the pathogen or parasite lives only for a limited time; it has no particular importance for the reproduction of the pathogen or parasite *phyt* >>> host

accidental sample: A sampling technique that makes no attempt to achieve representativeness, but chooses subjects based on convenience and accessibility *stat*

acclimation: >>> acclimatization

acclimatization: Changes involving the synthesis of proteins, membranes, and metabolites that occur in a plant in response to chilling or freezing temperatures that protect tissues or confer tolerance to the cold *phys* >>> low-temperature acclimatization

acclimatized A state of physiological adjustment by plants to changed environmental or stress conditions *phys* >>> www.plantstress.com

accommodation: The act of adjusting the eye to bring objects that are closer to the eye into focus *micr*

ACC oxidase: >>> 1-aminocyclopropane-1-carboxylic acid oxidase

ACC synthase: >>> 1-aminocyclopropane-1-carboxylic acid synthase

accumbent: Used to describe the first sprouts of an embryo when they lie against the body of the seed *bot*

accumulation center: An area where a great deal of variation of a given species or crop plant may be found, but which is not considered a center of origin *eco gene*

accuracy: A term used for measurement or an estimate of measurement; it is the degree to which a measurement represents the true value *stat*

acellular: Describing tissue or organisms that are not made up of separate cells but often have more than one nucleus *bot*

acentric: Chromosome, chromosome segments, or chromatids that show no >>> centromere; also applied to a chromosome fragment formed during cell division that lacks a centromere; this fragment is unable to follow the rest of the chromosomes in migration toward one or the other pole, as it has lost its point of attachment to the cell spindle *cyto* >>> Figure 11

acentric-dicentric translocation: >>> aneucentric translocation

acentric fragment: A piece of chromosome that is the result of a chromosome breakage; it does not contain a centromere; it got lost during cell division *cyto*

acentric inversion: A segment of chromosome which does not involve a centromere *cyto* >>> paracentric inversion

acephalous: Not having a head *phyt*

acephate: A systemic insecticide that is used to control pests (e.g., aphids, blackflies, caterpillars, fungus gnats, mushroom flies, scale, sciarid flies, symphylans, thrips, and whiteflies) *phyt*

acerose: Needlelike and stiff, like pine needles *bot*

acervate: Growth in heaps or groups *bot*

acetabuliform: Saucer-like in form *bot*

acetaldehyde: A simple aldehyde that is a bridge product of alcoholic fermentation *chem*

acetanilide: An active ingredient in a class of selective herbicides used predominantly in >>> maize *phyt*

acetate: A salt or ester of acetic acid *chem*

acetic acid (ethanoic acid): A carboxylic acid, CH_3COOH, and simple fatty acid; final product of several fermentation, oxidation, or rot processes; it plays a crucial role in energy metabolism *chem phys*

aceto-carmine: staining Used as a dye for staining of chromosomes; usually cells or tissues are pretreated (fixation) for 12–24 hours with a mixture of alcohol and acetic acid (3:1) or alcohol, acetic acid, and chloroform after Carnoy; prior to squashing the material it is stained with aceto-carmine *cyto micr* >>> opuntia >>> Carnoys

acetohydroxy acid synthase (AHAS): >>> acetolactate synthase

acetolactate synthase (ALS): An enzyme essential for amino acid production; it is the target of many herbicides including the imidazolinones and sulfonylureas; this enzyme has mutated in some species to become herbicide-resistant *phyt biot phys*

acetone: A simple but most important >>> ketone that is often used as a lipid solvent *chem*

aceto-orcein: A fluid consisting of the dye >>> orcein dissolved in acetic acid that is used in >>> chromosome staining *cyto micr*

acetylation: Introduction of an acetyl group into a chemical compound; acetylation of proteins may affect how they act in the organism *chem phys*

acetyl group: A small molecule made of two carbon, three hydrogen, and one oxygen atoms; acetyl groups are added to or removed from other molecules and may affect how the molecules act in the organism; it is a chemical group that is added and taken off of histones; acetylation of histones allows transcription to occur, and deacetylation inhibits transcription *chem biot*

achene: A small, usually single-seeded, dry, indehiscent fruit formed from a single carpel (e.g., the feathery achene of *Clematis* ssp. or buttercup); variants of the achene include caryopsis, >>> cypsela, nut, and samara *bot*

achiasmate: Meiosis and/or chromosome pairing without crossing over and chiasma formation *cyto* >>> Figure 15

achiasmatic: >>> achiasmate

achiral: A compound that may be superimposed on its mirror image; achiral molecules do not display topological handedness *biot chem*

achlamydeous (of flowers): Lacking petals and sepals, i.e., calyx nor corolla (e.g., in willows [*Salix* ssp.]) bot

achlorophyllous: A plant or leaf without chlorophyll bot

achromat: A microscope objective corrected for axial chromatic aberration *micr* >>> achromatic lens

achromatic: Parts of the nucleus not stainable by common chromosome dyes; in general, without color *cyto*

achromatic aplanatic condenser: A well-adapted microscope condenser lens that is corrected for chromatic and spherical aberrations *micr*

achromatic figure: The nuclear spindle during mitosis or meiosis *cyto*

achromatic lens (achromat): A lens cluster whose foci and power are made the same for two wavelengths; the simplest achromat is a doublet that combines two single lenses with different dispersions and curvatures to achromatize the combination *micr*

achromatin: That part of the nucleus that does not stain with basic dye *cyto*

A chromosome: Any of the standard chromosomes of a given genome *cyto*

achromycin: The trade name for tetracycline; an antibacterial antibiotic from *Streptomyces* ssp. *biot*

acicular: Pointed or needle shaped *bot*

acicular leaf: A pointed or needle-shaped leaf (e.g., in conifers) bot

acidic dye: An organic anion that stains positively charged macromolecules and acts on protoplasm *cyto*

acid phosphatase (Acph): An enzyme that is a member of hydrolases; because of its variability it is sometimes used as a biochemical marker in genetic studies *chem gene*

acid soil: Specifically, a soil with pH value <7.0, which is caused by the presence of active hydrogen and/or aluminum ions; the pH value decreases as the activity of these ions increases *agr* >>> Table 44

acid tolerance: The weathering of soils can result in acidification, limiting the types of plant or microbes that will grow; repeated fertilization with ammonia can also lead to acidification; to overcome such problems the soil must be limed, or plant species and microbes used that are tolerant of acidity *agr phys*

acid(ic) dye: An organic anion that stains positively charged macromolecules and acts on >>> protoplasm *micr*

acidity: In soil science, a measure of the activity of the hydrogen and aluminum ions in wet soil, usually expressed as pH value; crop plants show specific requirements; some of them grow reasonably well on acid soils (e.g., cowpea pH 5.0–7.0; oats, rye pH 5.5–7.5; maize, sorghum, wheat pH 6.0–7.5; barley pH 6.5–7.5; alfalfa pH 7.0–8.0) *agr* >>> Table 44

acidophilic: Having an affinity for acid stains (eosinophilic) cyto; thriving in or requiring an acid environment *eco*

acidophilous: >>> acidophilic

acinaceous: Consisting of or full of kernels *bot*

acinaciform: Shaped like a scimitar (e.g., the shape of the pods of some beans) *bot*

aconitase (Aco): An enzyme (dehydratase) that catalyzes the production of isocitric acid from citric acid; because of its variability it is sometimes used as a biochemical marker in genetic studies chem *gene*

acorn: The non-splitting, one-seeded fruit of, for example, an oak tree *bot*

acquired character: A non-heritable modification of structures or functions impressed on an individual by environmental influences during its development *gene* >>> modification

acquired immunity: An immunity that may be induced by means of preimmunization *phyt*

acquired mutation: A non-heritable genetic change occurring within a somatic cell gene >>> somatic mutation

acquired resistance: Plant resistance to a disease activated after inoculation of the plant with certain microorganisms or treatment with certain chemical compounds *phyt*

acre: 1 acre = 0.4047 hectare = 4,840 square yards = 10 square chains; 640 acres = 1 square mile (also called a section)

acreable: In terms of an acre or per acre *agr*

acreage: Extent or area in acres *agr* >>> Table 48

acridine: A chemical that is capable of causing >>> frameshift mutations in the DNA sequence; several derivatives of acridine, such as >>> acridine orange, are used as dyes or biological stains *chem gene cyto* >>> fluorescence staining >>> mutagen

acridine orange: An acridine dye that functions as both a fluorochrome and a mutagen *micr cyto gene* >>> fluorescence staining >>> mutagen

acriflavine: An acridine dye that produces frameshift mutations *gene meth* >>> mutation

acrocarpic: Fruits and/or seeds are formed on the top of a stem of a plant *bot*

acrocarpous: >>> acrocarpic

acrocentric: The >>> centromere is present on the end or close to the end of a >>> chromosome *cyto* >>> Figure 11

acropetal: Toward the >>> apex; the opposite of basipetal *bot*

acrosyndesis (*syn* **telosyndesis**): Incomplete end-to-end >>> chromosome pairing during meiosis; it is thought to be an artifact *cyto*

acrotrisomic: An individual with a normal chromosome complement plus an extra >>> acrocentric chromosome *cyto*

acrylamide: Chemical with the basic formula C_3H_5NO that can be chemically induced to form a gel matrix; this matrix can be used for electrophoretic separation of proteins and nucleic acids *chem meth*

actinomorp(hic): Radially symmetrical (rotate) of a flower *bot*

actinomycin: An antibiotic produced by Streptomyces chrysomallus that prevents the transcription of mRNA *micr cyto* >>> tetracycline

activation tagging: Insertion of activation >>> vectors carrying strong >>> promoters causing >>> mutagenesis; novel phenotypes can be generated *biot*

activator: In enzymology, a protein-like substance that is able to stimulate developmental processes phys; in molecular biology, a protein upstream from a gene on which the DNA binds; it activates the transcription of the gene *gene*

activator-dissociator system: A group of two interacting transposable elements "Ac" and "Ds" in maize; "Ac" is a 4.6 kb autonomous element carrying a transposase gene; "Ds" is often a derivative of "Ac" element that no longer produces a functional transposase; "Ac" and "Ds" loci are recognized and mapped by their action on neighboring genes *gene*

active chromatin: Mostly euchromatic regions of the nucleus that support transcription of the underlying genes *gene cyto*

active collection: A collection of >>> germplasm used for regeneration, multiplication, distribution, characterization, and evaluation; ideally, germplasm should be maintained in sufficient quantity to be available on request; it is commonly duplicated in a >>> base collection and is often stored under medium- to long-term storage conditions *meth*

active immunity: All means and reactions that enable a plant to prevent an interaction with a pathogen *phyt* >>> immunity

active ingredient: In any pesticide product, the component that kills or controls target pests *phyt* >>> active substance

active resistance: Resistance resulting from host reactions occurring in response to the presence of the pathogen or its metabolites *phyt*

active site: That portion of an >>> enzyme where the substrate molecules combine and are transformed into their reaction products *phys*

active substance: Pesticides or herbicides are usually mixtures of different substances; among them the active substance is the most important one as it attacks the pathogen *phyt* >>> active ingredient

active transport: The passage of substances across a cell membrane against a concentration gradient that requires energy *phys*

activity rhythm: An individual's daily pattern of physiological activity *phys*

aculeate: Armed with prickles *bot*

acuminate: Gradually tapering to a sharp point; point is drawn out *bot*

acute: Sharply pointed, but less tapering than acuminate; angle 90° or less *bot*; in plant pathology, pertaining to symptoms that develop suddenly *phyt*

acute toxicity: Ability of a single dose of a compound to poison, as opposed to chronic toxicity *phyt*

acyclic: Not cyclic; an acyclic flower *bot*; of or pertaining to a chemical compound not containing a closed chain or ring of atoms *chem*

adaptability: The potentiality for >>> adaptation; the ability of an individual or taxon to cope with environmental stress; the range and extent of reaction is genetically determined *phys* >>> www.plantstress.com

adaptable: Capable of being adapted or able to adjust oneself readily to different environmental conditions *phys* >>> adaptability

adaptation: The process of changes of an individual's structure, morphology, and function that makes it better suited to survive in a given environment *phys*

adapted race: >>> physiological race

adaptedness: The state of being adapted *phys*

adapter: Synthetic double-stranded oligonucleotide; a specific type of linker; it is applied to attach sticky ends to a blunt-ended DNA molecule *biot*

adaptive: Changes of a plant that act to preserve its full development *phys*

adaptive capacity: The genetically set range or flexibility of reactions of a plant and/or population enabling it to respond in different ways to differing conditions *eco gene*

adaptive character: A functional or structural characteristic of an organism that enables or enhances the probability of survival and reproduction *gene*

adaptive radiation: The diversification from a common ancestry, of population, groups, or lines, which leads to a variety of types, which may subsequently evolve to become a new species; the process happens over a range of similar environments, usually over a relatively short period and by natural selection *eco evol*

adaptive reaction: >>> adaptive capacity

adaptive selection: The evolution of comparable forms in separate but ecologically similar areas *eco gene evol*

adaptive trait: >>> adaptive character

adaptive trials: Multilocational coordinated field experiments across the crop-growing regions in one or more countries that test adaptation of varieties or breeding strains under specific ecological conditions *meth*

adaptive value: A measure of the reproductive efficiency of an organism or genotype compared with other organisms or genotypes *gene* >>> selective value

adaptiveness: >>> adaptedness

adaptivity: >>> adaptability

adaxial: Toward the axis syn ventral *bot* >>> abaxial

addendum (addenda pl): An item or a constituent substance to be added in formulation of tissue culture media *biot*

addition line: A cell line or line of individuals carrying chromosomes or chromosome arms in addition the normal standard chromosome set *cyto*

addition rule: The probability of either of two separate events occurring equal to the sum of the probabilities or of events occurring independently *stat*

additive effects (of genes): Gene action in which the effects on a genetic trait are enhanced by each additional gene, either an >>> allele at the same locus or genes at different loci *gene* >>> Tables 6, 20, 21

additive genes: Gene interaction without dominance (if allele), or without epistasis (if non-allele); the expression of any genetic trait is enhanced to the simple sum of the individual genetic or allelic effects contributing to that character *gene stat* >>> Tables 6, 20, 21

additive main-effect and multiplicative interaction (AMMI) analysis: The statistical model combines an additive model (analysis of variance, AOV) for main effects and a multiplicative model (principal components analysis, PCA) for interaction effects; it is applicable to trials with at least three genotypes and three environments; AMMI has proven useful for (1) understanding and visualizing complex GE interactions, and (2) gaining accuracy and thereby improving selections; AMMI is implemented by MATMODEL; for replicated data, MATMODEL can calculate the AMMI gain factor, which is the effective number of replications with AMMI analysis divided by the actual number of replications; e.g., when ten replications without AMMI are required to match the accuracy of four replications with AMMI, the AMMI gain factor is 10/4 or 2.5; for a wide range of yield trials, MATMODEL typically achieves a gain factor of 2 to 4; YTDESIGN can calculate the optimal number of replications both with and without AMMI analysis; clearly, this optimal number is usually different and smaller when MATMODEL is used, i.e., the design and the analysis of an experiment interact, so the best design depends on the efficiency of the analysis; efficient experiments require good designs and good analyses, appropriately integrated (GAUCH 1992) *stat* >>> biplot analysis

additive resistance: Resistance governed by more than one gene, each of which can be expressed independently, but which is reinforced by the expression of each of the additional genes *phyt gene*

additive (genetic) variance (VA): The proportion of the genetic variance due to additive effects *stat*

adduct: The covalent complex formed when a chemical binds a biomolecule, such as DNA or a protein *biot*

adelphogamy: Sib pollination or pollination involving a stigma and pollen belonging to two different individuals that are vegetatively derived from the same mother *gene*

adenine (A): A white crystalline purine base that occurs in both >>> DNA and >>> RNA, and nucleotides such as >>> adenosine diphosphate and >>> adenosine triphosphate; ($C_5H_5N_5$; formulary weight [f.w.] = 135.14); it belongs to B-group vitamin (B4), generally available as $C_5H_5N_5.3H_2O$; (m.w. 189.13); it is added to some tissue culture media, as adenine sulfate, to promote shoot formation and for its weak cytokinin effect; it is present in plant tissues combined with aminoamide, phosphoric acids, and D-ribose *chem gene biot* >>> Table 38

adenine sulfate: A growth factor used in some tissue culture media *biot* >>> adenine

adenose: Having glands or gland-like organs *bot*

adenosine ($C_{10}H_{13}N_5O_4$): The nucleoside formed when adenine is linked to ribose sugar *chem gene* >>> Table 38

adenosine diphosphate (ADP): High-energy phosphoric ester (nucleotide) of the nucleoside adenosine that functions as the principal energy-carrying compound in the living cell *phys*

adenosine triphosphate (ATP): High-energy phosphoric ester (nucleotide) of the nucleoside adenosine that functions as the principal energy-carrying compound in the living cell *phys* >>> mitochondrion

adhere: To stick to *bot*

adherent: Sticking or clinging *bot*

adhesion: The molecular attraction between substances causing their surfaces to remain in contact *meth*

adjacent distribution: The orientation and distribution of adjacent chromosomes in the ring or chain configuration of >>> translocation heterozygotes *cyto* >>> Figure 15

adjacent segregation: A reciprocal >>> translocation heterozygote in which during >>> meiosis the segregation of a translocated and a normal chromosome happens together, giving unbalanced gametes with duplications and deficiencies leading to non-viable zygotes; adjacent segregation is of two kinds depending on whether non-homologous (adjacent-1) or homologous (adjacent-2) centromeres segregate together; adjacent-1 segregation is the usual type of adjacent segregation and adjacent-2 segregation is rare *cyto* >>> Figure 15

adjacent-1 segregation: Segregation of non-homologous centromeres during meiosis in a reciprocal translocation heterozygote such that unbalanced gametes with duplications and deficiencies are produced, as opposed to >>> alternate segregation and adjacent-2 segregation *cyto* >>> Figure 15

adjacent-2 segregation: Segregation of homologous centromeres during meiosis in a reciprocal translocation heterozygote such that unbalanced gametes with duplications and deficiencies are produced, as opposed to >>> alternate segregation and adjacent-1 segregation *cyto*

A-DNA: The dehydrated form of right-handed helical >>> DNA obtained under non-physiological conditions *gene*

adnate: Fusion of unlike parts (e.g., fusion of palea to the caryopsis in Bromus grass) *bot*

adosculation: The fertilization of plants by pollen falling on the pistils *bot*

adpressed: Lying flat against (e.g., the rachilla against the palea in the grain of barley or oats) bot

adsorption: The physical binding of a particle of a particular substance to the surface of another by adhesion or penetration phy; attachment of phage to host bacterium (e.g., phage lamda adsorbs to a >>> maltose-binding protein) *biot* >>> lamda phage

adsorption complex: The various substances in the soil that are capable of adsorption (e.g., clay or humus) *agr* >>> Table 44

adspersed: To have a wide distribution; scattered *bot*

adult: Having attained full size and strength; mature *phys*

adult resistance: Resistance not expressed at the seedling stage; it increases with plant maturity (mature-plant resistance or age resistance); generally attributed to >>> horizontal resistance *phyt*

adult stage: >>> adult

adulthood: >>> adult

aduncate: Hooked, crooked, and bent *bot*

advance crop: >>> forecrop

advanced: Opposite of primitive; in phylogeny, a plant or character further removed from an evolutionary divergence than a more primitive one *evol* >>> Figure 68 >>> advanced character

advanced character: A feature that shows a real deviation from the ancestral trait or type (e.g., an agronomic trait such as brittle rachis) >>> advanced

advanced generations synthetic variety (beyond Syn1): Derives from an initial intercrossing of a specific set of clones or seed-propagated lines; usually stable for only a limited number of generations, for example, the varieties "Ranger" and "Moapa" in alfalfa, "Saratoga" in bromegrass, or "Pennlate" orchardgrass *seed* >>> Figures 50, 51

adventitious: Growing from an unusual position (e.g., roots from a leaf or stem) *bot* >>> Figure 28

adventitious bud: A bud appearing in an unusual place (e.g., a bud on leaves) *bot*

adventitious embryony: A condition in a seed in which the embryo arises from somatic rather than reproductive tissue; the development of a diploid >>> embryo from nucellary or integumentary tissue (sporophyte tissue); common in certain grasses and often results in multiple embryos *bot* >>> twin seedling >>> Figure 28

adventitious plant: An individual that arises from somatic rather than reproductive tissue *bot*

adventitious root: Arising from any structure other than a >>> root, for example, from a node of a stem or from a leaf; the phenomenon is quite common in >>> triticale and wheat; the adventitious root development is probably the response to accumulation of auxins in the base of the plant; sometimes this accumulation occurs because the plants are waterlogged and the existing roots system is no longer a sink for the >>> auxins due to lack of oxygen and/or dying off of the roots *bot* >>> Figure 48

adventive: A plant that has been introduced but is not yet naturalized *agr*

aecial host: Host plant on which heteroecious rust fungus produces >>> aeciospores and pycniospores, as opposed to telial host *phyt*

aeciospore: Dikaryotic spore of a rust fungus produced in an aecium; in heteroecios rusts, a spore stage that infects the >>> alternate host *phyt* >>> aecial host

aeolian soil: A type of soil that is transported from one place to another by the wind *agr* >>> Table 44

aerating tissue: >>> aerenchyma

aeration: Bringing air into a substance, tissue, or soil (e.g., by earthworms or digging and turning the soil to loosen) *phys agr*

aerator: Any implement that is used for breaking up compacted soil to facilitate air and gas exchange *agr*

aerenchyma: Plant tissue of thin-walled cells with large, air-filled intercellular spaces, found in the cortex of roots and stems of some aquatic and marsh plants *bot*

aerial pathogens: Antagonistic microorganisms that inhibit numerous fungal pathogens of aerial plant parts (e.g., *Tilletiopsis* ssp. parasitize the cucumber powdery mildew fungus Spaerotheca fuligena); present in crop soils and exert a certain degree of biological control over one or many plant pathogens *phyt* >>> biological control

aerial pest control: Pest control by utilization of aircraft and helicopters in order to be more efficient, to prevent damages of the crop, or to cope with difficult landscape and soil conditions *phyt meth*

aerial root: In some epiphytic orchids the leaves and the shoot axis are reduced or missing; then flatted and green roots take over the fixation of the plant and the function of the leaves (>>> photosynthesis) *bot*

aerial shoots: Shoots growing high above the ground (e.g., trees, bushes, etc.) *bot*

aerobe: An organism needing free oxygen for growth *bot*

aerobic rice: Rice varieties that would normally be grown flooded shallowly, instead being grown in soil without flooding, i.e., with irrigation merely to maintain a moist but aerobic soil; alternatively, wetting and drying methods are applied in order to minimize the use of irrigation by flooding the field followed by drainage until the soil is at field capacity and then reflooding *agr meth*

aerobium: >>> aerobe

aerosol: A colloidal substance that is suspended in the air *phyt meth*

aesculin: >>> glucoside

aestivation: Condition in which an organism may pass an unfavorable hot or dry season and in which its normal activities are greatly curtailed or temporarily suspended (McFarland 2006) *phys* >>> hibernation

afflux: The act of flowing to or toward some point or organ in a plant *phys*

afforest: Converting bare or cultivated land into forest *eco fore agr*

afforestation: The establishment of forest by natural succession or by the planting of trees on land where they formerly did not grow *eco*

affymetrix genechip arrays: High-density oligonucleotide microarrays; they are manufactured using a photolithographic technology with up to 500,000 oligos being synthesized on a single chip; the system allows measurement of the relative concentration of a >>> DNA or

>>> RNA sequence in a complex mixture of nucleic acids; they can be used (1) to monitor global >>> mRNA abundance in a range of species; (2) to identify genes that are uniquely expressed in samples of normal or mutated tissue, (3) to investigate changes in gene expression associated with environmental or other changes, (4) to identify putative functions for uncharacterized >>> ESTs, or (5) to identify gene expression "fingerprints" of efficiency and toxicity *biot* >>> Figure 48

aflatoxin: One of a group of mycotoxins ($C_{17}H_{10}O_6$) produced by fungi of the genus *Aspergillus* (molds) that bind to >>> DNA and prevent >>> replication and >>> transcription; a known carcinogen *phys*

AFLP: >>> amplified fragment length polymorphisms >>> random amplified polymorphic DNA (RAPD) technique

African cassava mosaic virus (ACMV): This disease continues to be the major constraint to both the commercial and subsistence production of >>> cassava across central and southern Africa, causing up to 100% crop losses; the causative agent of the disease has been shown to be a >>> geminivirus of the Begomovirus *phyt*

Africa Rice Center (WARDA): An autonomous intergovernmental research association of African member states; one of the 15 international agricultural research centers supported by the Consultative Group on International Agricultural Research (>>> CGIAR); it was created in 1971 by 11 African countries with the assistance of the United Nations Development Programme (UNDP), the Food and Agriculture Organization of the United Nations (>>> FAO), and the Economic Commission for Africa (ECA); until end September 2007, it comprised 17 member states: Benin, Burkina Faso, Cameroon, Chad, Côte d'Ivoire, Gambia, Ghana, Guinea, Guinea Bissau, Liberia, Mali, Mauritania, Niger, Nigeria, Senegal, Sierra Leone, and Togo; in September 2007, four more countries joined WARDA: Central African Republic, Democratic Republic of Congo, Republic of Congo, and Uganda; Egypt joined WARDA in June 2008; with this the total number of WARDA member countries has gone up to 22, including countries from West, East, Central, and North Africa; WARDA was constituted as the West Africa Rice Development Association – a name that it carried until 2003 when it was designated by the WARDA Council of Ministers as the "Africa Rice Center" in recognition of its increasing and leading role in rice research; its mission is to contribute to poverty alleviation and food security in Africa, through research, development, and partnership activities aimed at increasing the productivity and profitability of the rice sector in ways that ensure the sustainability of the farming environment *org* >>> www.warda.org

after-harvest cultivation: Any discing or plowing of the land after harvesting that will incorporate crop seed from the crop residue into the soil so that subsequent deterioration will prevent the seed from >>> volunteering in the following crop *agr meth*

aftermath: Second crop of grass cut from a field in a season *agr*

after-ripening: A term for the collective changes that occur in a dormant seed that make it capable of >>> germination; it is usually considered to denote physiological changes *seed*

agameon: A plant species reproducing exclusively by >>> apomixis *bot*

agamete: Any non-reproducing germ cell *bot* >>> apomixis

agamic: Reproducing asexually *bot* >>> apomixis

agamic complex: Refers to hybrids or their derivatives that are partially or entirely reproduced by asexual seed formation *bot* >>> apomixis

agamobium: The asexual generation of an organism having alternation of generations *evol gene*

agamogenesis: Any reproduction without the male gametes *bot* >>> apomixis

agamogony: A type of >>> apomixis in which cells undergo abnormal >>> meiosis during >>> megasporogenesis, resulting in a diploid >>> embryo sac rather than the normal haploid embryo sac *bot* >>> Figure 28

agamont: An asexual individual in whose agametangia the agametes are formed *bot* >>> apomixis

agamospecies: Populations morphologically differentiated from one another and reproducing apo-mictically *bot* >>> Figure 28 >>> apomixis

agamospermy: All types of >>> apomixis in which seeds are formed by asexual means; it does not include vegetative reproduction *bot*

agamous: >>> agamic

agar: A complex >>> polysaccharide obtained from certain types of seaweed (red algae); when it is heated with water and subsequently cooled to about <45°C, it forms a gel *prep*

agar culture: Cells, organs, tissue, or embryos artificially grown on a solid medium composed of agar together with certain nutrients, hormones, etc. *biot*

agar gel: Gels for electrophoresis that were produced from >>> agar *prep*

agar medium: >>> agar culture

agar-agar (Malay): >>> agar

agarose (starch) gel: An inert matrix used in >>> electrophoresis for the separation of nucleic acids based on their size or conformation; the molecules are visualized in the gel by ultra-violet fluorescence of >>> ethidium bromide, which is either included in the gel or in the running buffer, or used to stain the gel after electrophoresis *prep*

AG complex: The complete set of factors assumed to be responsible for the formation of sexual organs *bot*

agenesis: The absence of development *bio phys*

agent A natural force, object, or substance producing or used for obtaining specific results *prep* >>> www.plantstress.com

age of stand: The number of years during which a perennial >>> crop may be offered for pedi-gree from one planting; the first seed crop is considered the first year in which a seed crop would normally be harvested, irrespective of time or method of planting; each calendar year thereafter is considered a seed crop year *agr seed*

age resistance: >>> adult resistance

agglomerate: In biotechnology, a mass of cells clustered together *biot*; in soil science, rock com-posed of rounded or angular volcanic fragments *agr* >>> Table 44.

agglutinate: Fixed together as if with glue *prep*

agglutination: The clumping of cellular components *prep*

agglutinin: Any >>> antibody capable of causing clumping of types of cells *prep*

agglutinogen: An >>> antigen that causes the production of agglutinins *chem biot*

aggregate: In soil science, a cluster of soil particles forming a pad *agr* >>> Table 44

aggregate fruit: A fruit development from several pistils in one flower, as in strawberry or black-berry *bot* >>> composite fruit

aggressiveness: The ability of a >>> pathogen to infect a plant, to break its resistance, to become a >>> parasite on a plant, or to use a host plant for reproduction; the degree of aggressive-ness can be estimated only when a pathogen meets a resistant host; when the aggressive-ness copes with the resistance of the host, it is termed >>> virulence *phyt*

aging test: A method to originally evaluate seed storability; it subjects unimbibed seeds to condi-tions of high temperature (+41°C) and relative humidity (~100%) for short periods (3–4 days); the seeds are then removed from the stress conditions and placed under optimum germination conditions; the two environmental variables cause rapid seed deterioration; high-vigor seed lots will withstand these extreme stress conditions and deteriorate at a slower rate than low-vigor seeds *seed* >>> storability of seeds

agnation: Relationship through the male line *gene*

agonisis: >>> certation

AGPase: A glucose-1-phosphate adenylyltransferase is an enzyme that catalyzes the chemical reaction; two substrates of this enzyme are >>> ATP and alpha-D-glucose 1-phosphate, whereas its two products are diphosphate and ADP-glucose; transcriptional and allosteric regulation of ADP-glucose pyrophosphorylase (AGPase) plays a major role in the regulation of starch synthesis; it has a significant role in crop productivity; thermotolerant variants of AGPase in cereals may be used for developing cultivars, which may enhance productivity under heat stress *phys*

agribusiness: A combination of the producing operations of a farm, the manufacture and distribution of farm equipment and supplies, and the processing, storage, and distribution of farm commodities *agr*

agriculture: The science of transforming sunlight energy into plant and animal products that can be utilized by humans; the selective breeding of crop and farm animals has had an enormous impact on productivity in agriculture; modern varieties of crop plants have increased nutritional value and resistance to disease; recent developments in genetic engineering have enabled the potential use of transgenic organisms in agriculture to be explored *agr*

agricultural chemical: A broad term used to cover pesticides, adjuvants, conditioning agents, and other chemical tools used in improving agricultural production, protecting crops, or controlling pests, diseases, and physiological conditions of crop plants *phyt agr*

agri-genomics: Study of the make-up of and interaction between genes in crops and combinatorial chemistry *biot*

agritourism: Tourism based on attracting visitors to farm operations; it is comprised of businesses such as crop and animal farms, U-pick operations, wineries, aquaculture and for-fee fishing operations, Christmas tree farms, herb farms and greenhouses, maple syrup and cheese producers, and farm stands or exhibition of old crops, breeding strains, and races *agr*

Agrobacterium: A genus of bacteria that includes several plant pathogenic species, causing tumor-like symptoms *bio* >>> Figure 27 >>> *Agrobacterium rhizogenes* >>> *Agrobacterium tumefaciens* >>> crown gall >>> hairy root culture >>> Ri plasmid >>>Ti plasmid >>> www.pflanzengallen.de

Agrobacterium rhizogenes: A species of Gram-negative, rod-shaped soil bacteria, often harboring large plasmids, called Ri plasmids; it can cause a tumorous growth known as hairy root disease in certain plants *bot biot* >>> Figure 27 >>> *Agrobacterium tumefaciens* >>> Agrobacterium-mediated transformation

Agrobacterium tumefaciens: A bacterium that causes crown gall disease in some plants; it infects a wound, and injects a short stretch of >>> DNA into some of the cells around the wound; the DNA comes from a large plasmid – the tumor induction (Ti) plasmid – a short region of which (called transferred DNA [T-DNA]) is transferred to the plant cell, where it causes the cell to grow into a tumor-like structure; the T-DNA contains genes that inter alia allow the infected plant cells to make two unusual compounds, >>> nopaline and >>> octopine, both characteristic of transformed cells; the cells form a gall that hosts the bacterium; this DNA-transfer mechanism is exploited in the genetic engineering of plants; the Ti plasmid is modified so that a foreign gene is transferred into the plant cell along with, or instead of, the nopaline synthesis genes; when the bacterium is cultured with isolated plant cells or with wounded plant tissues, the novel gene is injected into the cells and ends up integrated into the chromosomes of the plant *bot biot* >>> Figure 27 >>> www.pflanzengallen.de

Agrobacterium-mediated transformation: Agrobacterium is the generic name of a soil bacterium that frequently causes crown gall in many plant species; besides *A. rhizogenes*, *A. tumefaciens* is one species that is most used in DNA transfer by manipulating the tumor-inducing (Ti) plasmid that is harbored by these bacteria *biot* >>> Figure 27 >>> *Agrobacterium rhizogenes* >>> *Agrobacterium tumefaciens* >>> www.pflanzengallen.de

agrobiodiversity: The variety and variability of animals, plants, and microorganisms used directly or indirectly for food and agriculture (crops, livestock, forestry, and fisheries); it comprises the diversity of genetic resources (varieties, breeds, etc.) and species used for food, fuel, fodder, fiber, and pharmaceuticals *eco*

agrobiology: The scientific study of plant life in relation to agriculture, especially with regard to plant genetics, cultivation, and crop yield *agr bio*

agrobiotechnology: Modern biological knowledge and methods that can be applied to human goals in agriculture *biot meth agr*

agro-ecotype: A group of >>> agrotypes all having similar environmental preferences *eco*

agroenergy crop (plants): In narrow sense, renewable crop plants used for production of motor fuels from biomass and/or food, such as sweet potato (*Ipomoea batatas*), cotton (*Gossypium hirsutum*), cashew nut (*Anacardium occidentale*), peanut (*Arachis hypogaea*), jojoba (*Simmondsia chinensis*), cocos palm (*Cocos nucifera*), false flax (*Camelina sativa*), cassava (*Manihot esculenta*), maize (*Zea mays*), (>>> *Elaeis oleifera*, >>> *E. oleifera, Attalea maripa*), pine tree (*Pinus ssp.*), pongamia (*Pongamia pinnata*), physic nut (*Jatropha curcas*), rice (*Oriza sativa*), castor (*Ricinus communis*), rapeseed (*Brassica napus oleifera*), rye (*Secale cereale*), sesame (*Sesamum indicum*), sunflower (*Helianthus anuus*), soybean (*Glycine max*), Jerusalem artichoke (*Heliathus tuberosus*), wheat (*Triticum aestivum*), sorghum (*Sorghum bicolor*), sugarcane (*Saccharum officinarum*), sugarbeet (*Beta vulgaris*); in a broad sense, crop plants that are used for methane gaze, oil, and ethanol production, heating, etc., and/or food; it is part of national or international medium- and long-term strategies in order to achieve the following benefits: reducing fossil fuel use, broadening biofuel production and consumption; protecting the environment and contributing to social inclusion *agr*

agroforestry: A land-use system in which woody perennials are grown with agricultural crops (together with other land uses, like animal production) *agr*

agroinfection: Infection of plants via soilborne pathogens *phyt*

agronomist: A specialist in farm management and the production of field crops *agr*

agronomy: The area of agriculture devoted to the production of crops and soil management; the scientific utilization of agricultural land *agr*

agrostology: The branch of systematic botany that encompasses grasses (Gramineae) *bot tax*

agrotain (NBPT): A product that inhibits conversion of urea to ammonium carbonate, thereby reducing the potential for ammonia volatilization from urea materials; similar to "N-Serve", it reduces potential nitrogen losses in seasons when cultivation or rain does not incorporate the urea into the soil soon after application; it is most useful when urea is applied without incorporation to the surface of fields with high levels of crop residue, such as in no-till situations, or fields with high pH levels at the surface *agr phys*

agroytype: An agricultural race *agr eco* >>> agro-ecotype

a-helix: >>> alpha-helix

air checking: In potato breeding, shallow splits resembling fingernail cuts in the skin of a potato tuber; a phenomenon that occurs when highly hydrated tubers are harvested under cool temperatures as well as low humidity *breed agr*

air layering: A method of plant propagation in which roots are induced to form around a stem; a very narrow strip of bark is removed from around the branch or stem; a sliver of wood can be inserted into the cut to keep it open; a bundle of moist sphagnum moss is tied securely around the cut area; the moss must remain moist and the roots of the plant somewhat dry; new roots will sprout from the incision; the new plant is then cut off below the moss, potted, and kept in a humid atmosphere until it is established *hort meth*

airlock: An airtight chamber permitting passage to or from a space *seed*

air plant: >>> epiphytic

air-screen cleaner: The basic piece of equipment for cleaning seed, utilizing airflow and perforated screens *seed*

akaryote: >>> akaryotic

akaryotic: Without a nucleus; a stage in the nuclear cycle before meiosis in which no or little chromatin is seen in the nucleus *cyto*

akinete: >>> akinetic

akinetic: A non-motile reproductive structure (e.g., a resting cell) *cyto*

alanine (ala): An amino acid present in almost all proteins chem >>> Table 38

alate: Winged *bot*

albido: The white tissue beneath the peel of >>> citrus *bot hort*

albinism: In plants, a deficiency of chromoplasts *bot*

albino: A plant lacking chromoplasts *bot*

albumen: Starchy and other nutritive material in a seed, stored as endosperm inside the embryo sac, or as perisperm in the surrounding nucellary cells; in general, any deposit of nutritive material accompanying the embryo *phys* >>> endosperm >>> albumin(e)

albumin(e): Any of certain proteins soluble in distilled water at neutral or slightly acid pH and in dilute aqueous salt solution; they coagulate by heat (e.g., leucosins in cereal grains, ricin in rice, or legumelins in pulse seeds, which are mainly enzymes); they function as soluble, enzymatic, and metabolic proteins *chem* >>> Table 15

albuminoid: Containing or resembling albumen or albumin *bot*

albuminous seed: A seed having a well-developed >>> endosperm or >>> perisperm *seed* >>> Table 15

alepidote: Having no scales of scurf, smooth *bot*

aleurodid: >>> whitefly

aleurone: A granulated protein that forms the outermost layer of a cereal grain *bot* >>> aleurone grain >>> Table 15

aleurone grain: Small protein grains present in cells of storage tissue *bot*

aleurone layer: A layer of cells below the testa of some seeds (e.g., cereals), which contains hydrolytic enzymes (e.g., amylases and proteases) for the digestion of the food stored in the endosperm; the production of enzymes is activated by gibberellins when the seed is soaked in water prior to germination *bot*

aleuroplast: A leucoplast in which protein granules are present as a main storage product *bot*

alien addition line: A line (strain) of plants with one or more extra chromosomes of an alien species *cyto*

alien chromosome: A chromosome from a more or less related species transferred to a crop plant *cyto*

alien chromosome transfer: Cytogenetic methods that facilitate the transfer of individual chromosomes from one species to another *meth cyto*

alien gene transfer: The transfer of genes between species or genera by different means *gene biot* >>> Figure 59

alien germplasm: Genes introduced from a wild relative or non-adapted species *cyto gene* >>> Figure 59

alien species: An organism that has invaded or been introduced by man and is growing in a new region *eco*; in cytogenetics, a species that serves as donor of genomes, chromosomes, or chromosome segments to be transferred to a recipient species or genotype *cyto*

alien substitution line: A line of plants in which one or more alien chromosomes from a certain donor species replace one or more chromosomes of a recipient species *cyto*

A line: The seed-bearing parent line used to produce hybrid seed that is male sterile; in wheat hybrid seed production, a male-sterile parent line used to produce hybrid seeds and hence the seed-producing parental line *seed meth* >>> B line >>> Figure 2

aliquot: A part, such as a representative sample, that divides the whole without a remainder; two is an aliquot of six because it is contained exactly three times; loosely, it is used for any fraction or portion *prep*

alkali: A substance capable of furnishing hydroxyl-OH ions to its solution; the most important alkali metals are potassium and sodium *chem*

alkaline phosphatase: An enzyme that is a member of hydrolases; it cleaves the 5'-terminal phosphate group from linear DNA or RNA molecules; dephosphorylated 5' DNA or RNA ends cannot be joined by ligase to 3′ ends; polynucleotide kinase reverses the reaction; sometimes used as a biochemical marker in genetic studies *phys biot gene*

alkaline soil: Specifically, a soil with pH value >7.0 caused by the presence of carbonates of calcium, magnesium, potassium, and sodium; commonly used for soils showing a pH value of 8.5 *agr* >>> Table 44

alkaloid: One of a group of basic, nitrogenous, normally heterocyclic compounds of a complex nature; alkaloids occur in several plants (e.g., coniine in >>> hemlock, morphine in >>> poppy fruits, strychnine in seeds of *Strychnos nux-vomica*, atropine in nightshades, colchicine in >>> meadow saffron, caffeine in >>> coffee and tea, nicotine in >>> tobacco leaves, theobromine in >>> cacao) *chem phys* >>> Table 46

alkaloidity: The alkaloid content of cell, tissue, organs, or individuals of plants *phys*

alkylating agent: A chemical agent that can add alkyl groups (e.g., ethyl or methyl groups to another molecule; many mutagens act through alkylation) *chem*

alkylresorcinols: Phenolic, amphiphilic lipids present at levels of up to 0.15% of whole grain of, e.g., >>> wheat and >>> rye; little is known about their presence in food, absorption in animals and humans, or their *in vivo* biological effects; because alkylresorcinols are present in the human diet in significant amounts only in products containing whole grain wheat or rye, they have potential to be >>> biomarkers of whole grain wheat and rye intake; a rapid gas chromatographic method is available to analyze alkylresorcinols in whole cereal grains; wheat, rye, and triticale all contain moderate to high amounts of alkylresorcinols (300–1500 µg/g dry matter), while barley contains low amounts (~50 µg/g dry matter); in these cereals, alkylresorcinols are present in the bran fraction; all other cereals (rice, oats, maize, sorghum, and millet) do not contain any detectable amounts of alkylresorcinols; an extraction method using hot propanol:water was able to recover all alkylresorcinols from experimental breads, indicating that alkylresorcinols are not destroyed during baking; the absorption of alkylresorcinols in rats, pigs, and humans was determined, with values for absorption ranging from 34 to 79% *phys* >>> biomarkers

All-America Rose Selections Inc. (AARS): An association of U.S. commercial rose growers that tests and approves new rose varieties for commercial use *org hort*

allautogamia: The state of having a facultative method of pollination in addition to a normal method *bot meth*

allele: One of two or more alternate forms of a gene occupying the same locus on a particular chromosome; currently, different alleles of a given gene are usually recognized by phenotype rather than by comparison of their nucleotide sequences gene; in molecular biology, an allele can be defined as variant of a DNA sequence; it should have exactly the same genetic properties as a phenotypically defined allele; mutations within protein coding sequences define molecular allelism at many levels and degrees of severity (identical sequences – the baseline against which mutations are measured; silent mutation – mutation in "degenerate" position in a codon, so that no amino acid replacement occurs; missense mutation – mutation leading to an amino acid substitution; many, and perhaps most missense mutations have little or no effect on the function of the protein; evidence for this comes from the fact that comparison of amino acid sequences from homologous proteins from many species often shows that some regions of a protein are highly conserved, while other regions appear to tolerate a great deal of amino acid substitution; non-sense mutation – mutation which converts a codon into a >>> stop codon; occurrence of a stop codon within a protein coding sequence results in a truncated protein; insertions/deletions/rearrangements – insertions and deletions within a protein coding sequence can interrupt protein function; as well, insertions or deletions whose length is not a multiple of "3" will cause a frameshift mutation, meaning that incorrect amino acids will be inserted downstream from the mutation; rearrangements in a protein coding sequence will probably result in loss of function); mutations within genes are only a small percentage of the total number of mutations occurring in the genome because only a small percentage of eukaryotic genomes codes for proteins, most mutations will occur within non-coding DNA, and will usually be selectively neutral; therefore, molecular alleles should segregate by the same Mendelian principles as phenotypic alleles; in most cases, molecular alleles are selectively neutral; all of the rules of population genetics apply to molecular alleles; practically, there are an infinite number of possible molecular alleles; a given phenotypic allele observed in the population could, in principle, consist of a set of different molecularly defined alleles; because most mutations are selectively neutral and can occur anywhere in the genome, the molecular alleles can be used as markers for >>> chromosome mapping, that is, RFLPs are molecular alleles *biot* >>> allelism >>> Figure 43

allele extinction: Disappearance of an allele from the gene pool *gene*

allele diversification: The evolution of multiple alleles from a common ancestor *gene* >>> Figure 68

allele frequency: A measure of the commonness of an >>> allele in a population of alleles *gene*

allele mining: An approach to access new and useful genetic variation in >>> crop plant collections; it focuses on the detection of allelic variation in important genes and/or traits within a germplasm collection; if the targeted DNA (either a gene of known function or a given sequence) is known, then the allelic variation (usually point mutations) in a collection can be identified *biot meth* >>> association mapping

allele shift: A modification of allele frequency in a population due to either natural or artificial selection *gene*

allele-specific amplification (ASA): Use of >>> polymerase chain reaction (PCR) at a sufficiently high stringency that only a primer with exactly the same sequence as the target DNA will be amplified; a powerful means of genotyping for single-locus disorders that have been characterized at the molecular level *biot*

allele-specific associated primer(s) (ASAP): A >>> PCR variant in which the sequence of the decamer oligo is derived from normal >>> RAPD, which generated an absence and/or presence >>> polymorphism; these polymorphisms do not require electrophoretic separation of the sample; the presence of an >>> amplification product is detected by measuring fluorescence of ethidium-bromide stained DNA (GU et al. 1995) *biot* >>> Figure 43

allele-specific expression: The transcription of only one allele or both of a genetic locus to different extents; it may be detected by allele-specific amplification or >>> PCR *gene biot*

allele-specific methylation: The methylation of cytosyl residues in either the paternal or maternal allele of the gene; it regulates the expression of imprinted genes in a wide range of eukaryotes *gene phys*

allele-specific oligo(nucleotide) (ASO): A special kind of oligo for an >>> allele-specific PCR; the sequence of the oligo is designed in such a way to allow and/or inhibit hybridization at the spot where the mutant (resistant) allele differs from the wild-type (susceptible) allele *biot*

allele-specific oligo(nucleotide) ligation: A technique for the detection of >>> SNPs in genomic DNA; the target DNA is first amplified by >>> PCR; then allele-specific oligonucleotides complementary to the target sequence are annealed to the >>> amplicon just adjacent to the polymorphic site; only if the oligonucleotide fully matches the target will it be ligated; no ligation happens if there is mismatching *biot meth*

allele-specific PCR (AS-PCR): Refers to amplification of specific alleles or DNA sequence variants at the same locus; specificity is achieved by designing one or both >>> PCR primers so that they partially overlap the site of sequence difference between the amplified alleles; variants of this technique have been described under different names such as >>> RAPD, >>> AP-PCR, or >>> DNA amplification finger printing (DAF); they are based on polymerase chain reaction (PCR), which involves randomly synthesized short oligo-nucleotide sequences as primers; these primers are specific to the ends of a given sequence in the

DNA of a plant; genetic differences between individuals can be detected when the size of the segment of DNA bracketed by the primers is different; the bracketed region can be isolated and purified, so that restriction mapping and sequencing is possible; these markers are known as randomly amplified polymorphic DNA (RAPD) markers that do not involve >>> SOUTHERN analysis and radioactive labeling; randomly amplified polymorphic DNA markers are, therefore, cheaper and easier to use than RFLP and >>> AFLP markers; however, RAPDs are very sensitive to laboratory conditions, are usually dominant markers, are potentially population specific, and show a low level of polymorphism with PCR amplification using a single random primer *biot* >>> Figure 43

allele-specific RNA interference: The >>> silencing of only one allele of a specific gene by an allele-specific interference of RNA; it starts with the cloning of the target gene into a plasmid vector; its mutation *in vitro* creates a variant of the gene, whose >>> mRNA is not subject to the RNA interference, caused by the >>> dsRNA that is able to knockdown the endogenous protein *biot meth*

allele trend: A directed change in allele frequency of populations per time unit *gene*

allele turnover: The replacement in a population of old alleles by new ones over time *gene*

allelic complementation: The production of a non-mutant phenotype when two independent mutations at the same gene locus, but on different homologous chromosomes, are introduced *gene*

allelic deletion: The deletion of one complete allele of a gene or an exon; it can be detected by comparative and quantitative real-time >>> PCR *gene biot*

allelic diversification: The evolution of multiple alleles from a common ancestor *gene evo*

allelic exclusion: The expression of genes from the maternal or paternal chromosome but not both, due to chromosomal inactivation *gene*

allelic imbalance: The presence of one non-functional allele at a specific locus, or the complete loss of an allele, i.e., loss of heterozygosity *gene*

allelic mining: >>> allele mining

allelic recombination: A >>> recombination event occurring between sequences located at similar or identical positions on homologous chromosomes or sister chromatids *cyto*

allelic turnover: The replacement in a population of old alleles by new ones over time *gene evo*

allelism: The common shortening of the term "allelomorphism"; one of the two or more forms of a gene arising by mutation and occupying the same relative position (locus) on homologous chromosomes *gene* >>> allele

allelobrachial: Changes of chromosome structure in which the arms of homologous chromosomes are included *cyto*

allelogenous: Females that produce only males or only females in different progenies *bot*

allelomorph: A term that is commonly shortened to "allele" *gene* >>> allele

allelomorphism: >>> alternative inheritance >>> allelopmorph

allelopathy: Refers to the release into the environment by an organism of a chemical substance that acts as a germination or growth inhibitor to another organism; economic and environmental constraints of crop production systems have stimulated interest in alternative weed management strategies; allelopathy offers potential for selective biological weed management through the production and release of allelochemicals from leaves, flowers, seeds, stems, and roots of living decomposing plant materials; actually, the term allelopathy refers to biochemical interactions among plants, including those mediated by microorganisms; this broad definition of allelopathy is appropriate because considerable research has indicated the involvement of microorganisms and lower plants in the production of phytotoxins; a variety of allelochemicals have been identified, including the phenolic acids, coumarins, terpenoids, flavonoids, alkaloids, glycosides, and glucosinolates; allelopathic inhibition typically results from the combined action of a group of allelochemicals which, collectively, interfere with severe physiological processes; allelopathy is strongly coupled with inherent stresses of the crop environment, including insects and disease, temperature extremes, nutrient and moisture variables, radiation, and herbicides; these stress conditions often enhance allelochemical production, thus increasing the potential for allelopathic interference (Lincoln and others 1998) *eco phyt* >>> www.plantstress.com

allelotype: The genetic composition (i.e., allele frequency) of a breeding population *gene*

alleome: A collection of different allotypes or allelic protein variants, a new type of protein library *biot*

allergenicity: The tendency of a substance to cause allergic reactions *bio*

all-exon array: A >>> microarray onto which synthetic >>> oligonucleotides are immobilized that span all exons of the genome *biot meth*

alliaceous: Onion-like in smell or form *bot*

alliance: A ranked category in vegetation classification, comprising one or more closely related associations (Lincoln et al. 1998) *bot*

allocarpy: The production of fruit following cross-fertilization *bot*

allocompetition (intergenotype competition): Cultivation at high plant density implies the presence of strong interplant competition eco; the individual plants, clones, lines, or families are evaluated when being subjected to intergenotypic competition; also called intergenotype competition *stat* >>> Figure 50

allocycly: Differences in chromosome coiling caused by environmental or genotypic effects *cyto*

allodiploid: Cells or individuals in which one or more chromosome pairs are exchanged for one or more pairs from another species *cyto*

allogamous: Cross-fertilizing in plants; as opposed to >>> autogamous *bot* >>> Table 35

allogamy: Cross-fertilization; as opposed to >>> autogamy *bot* >>> Table 35

allogene: >>> recessive allele (as opposed to >>> protogene)

allogeneic: Two genetically dissimilar individuals of the same species *gene*

allogenetic: Cells or tissues related but sufficiently dissimilar in genotype to interact antigenically *phys*

allogenic: Applied to successional change due to a change in abiotic environments *eco*

allogenous flora (plant): Relic plants of an earlier prevailing flora and environment *eco*

allograft: A graft of tissue from a donor of one genotype to a host of a different genotype but of the same species *hort*

allohaploid: A haploid cell or individual derived from an >>> allopolyploid and composed of two or more different chromosome sets *cyto* >>> Imperata cylindrica procedure >>> Figure 3

alloheteroploid: Heteroploid individuals or cells whose chromosomes derive from various genomes *cyto* >>> autoheteroploid >>> Figure 3

alloiobiogenesis: An alternation of a sexual with an asexual form *gene*; in cytology, the alternation of a haploid with a diploid stage, i.e., alternation of generations *cyto*

alloiogenesis: Growth of a part of an organism in relation to the growth of the whole organism or some other part of it *phys*

allometric: Growth in which the growth rate of one part of the plant differs from that of another part or of the rest of the plant; it is of common use in morphogenesis studies, where organ dimensions and growth rates frequently can be fit by the relation *bot*; in statistics, it also can be explained as the relationship between two variables of the form $y = a \times b$ where a and b are constants; b is termed the allometric coefficient *stat*

allometry: The relation between the growth rate of a part of an individual and the growth rye of the whole or of another part; or the relationship between growth rates of different groups, races, genera, etc. >>> alloiogenesis

allopatric: Applied to species that occupy separate habitats and that do not occur together in nature (cf parapatric and sympatric) *eco*

allopatric hybridization: Hybridization between incompletely differentiated species, in a border zone, owing to the premature breakdown of a geographic barrier, as opposed to >>> sympatric hybridization *eco*

allophene: A phenotype not due to the mutant genetic constitution of the cell or the tissue in question; such a cell or tissue will develop a normal phenotype if it is transplanted to a wild-type host *gene*

allophenic: Characteristics that arise by intercellular >>> gene action; sometimes used as chimeric, i.e., composed of cells of two different genotypes (also called hybrid) *gene* >>> chimera >>> hybrid

alloplasm: Cytoplasm from an alien species that has been transferred by backcrossing into a cultivated species; common and readily observable effects are >>> male sterility, female sterility, or reduced plant vigor; if the alien cytoplasm does not manifest itself, the cytoplasm is not considered alien *gene meth*

alloplasmic: An individual having the common nucleus, but an alien cytoplasm (e.g., alloplasmic rye containing a wheat cytoplasm); usually leads to meiotic disturbances and >>> sterility *gene*

alloploid (syn allopolyploid): A plant that arises after natural or experimental crossing of two or more species or genera; they may contain genomes of the parents in one or more copies *cyto* >>> amphiploid >>> Figures 3, 45

alloploidy: >>> alloploid >>> allopolyploid

allopolyploid: Plants with more than two sets of chromosomes that originate from two or more parents; the sets contain at least some non-homologous chromosomes *cyto* >>> Figure 3 >>> Table 17

allopolyploidy: >>> allopolyploid

allosome: A chromosome deviating in size, form, or behavior from the other chromosomes (>>> autosomes), such as the >>> sex chromosome or >>> B chromosome *cyto* >>> heterochromosome >>> marker chromosome >>> sSMC

allosomal inheritance: The inheritance of characters governed by genes located in an >>> allosome *gene*

allosteric: An enzyme whose activity is altered when its structure is distorted by an organic compound at a non-substrate site *phys chem*

allosteric effect: The binding of a ligand to one site on a protein molecule in such a way that the properties of another site on the same protein are affected *chem*

allosteric transition: A change from one conformation of a >>> protein to another conformation *chem*

allosubstitution: The replacement of a chromosome or chromosome arm by an >>> alien chromosome or chromosome arm *cyto*

allosynapsis: >>> allosyndesis

allosyndesis: Chromosome pairing of completely or partially >>> homologous (>>> homoeologous) chromosomes *cyto*

allotetraploid (syn amphidiploid): A plant that is diploid for two genomes, each from a different species *cyto* >>> Figure 8

allotetraploidy: >>> allotetraploid

allotopic: A type specimen of the sex opposite that of the holotype *bot*; in immunology, an antibody that acts as an antigen to other antibodies of the same species that have variant molecular sites *meth*

allotropous flower: Shaped so that nectar is easy to access for insects *bot*

allotype: A classification of immunoglobulin molecules according to the antigenicity of the constant regions; a variation that is determined by a single allele *biot*

allozygosity: Homozygosity in which the two alleles are alike but unrelated *gene*

allozygote: A zygote heterozygous for different mutant alleles *gene*

allozyme: Isoenzymes of protein nature whose synthesis is usually controlled by codominant alleles and inherited by monogenic ratios; they show a specific banding pattern if separated by >>> electrophoresis *phys* >>> Table 29

alluvial soil: Soils developed on fairly recent alluvium (a sediment deposited by streams and varying widely in particle size); usually they show no horizon development *agr* >>> Table 44

alpha-amylase: An enzyme that breaks down starch in germination grains; it splits starch molecules at random points, forming smaller molecules of widely varying size, for example, the so-called dextrinizing enzyme of malt *chem* >>> falling number

alpha-amylase activity: Catalysis of the endohydrolysis of 1,4-α-D-glucosidic linkages in polysaccharides containing three or more 1,4-α-linked D-glucose units *phys* >>> falling number

alpha-amylase/trypsin inhibitor (ATI): A molecule contributing to pest resistance in wheat, that is a strong activator of innate immune response in monocytes, macrophages, and dendritic cells of humans; this protein in wheat may influence the inflammation of chronic health conditions *phys*

alpha-bromonaphthalene: A chemical agent that is used for artificial >>> chromosome condensation; for several hours root tips are treated with a saturated water solution prior to staining *cyto*

alpha complementation (of beta-galactosidase): pUC18 and similar molecular vectors contain only a small part of the whole gene for beta-galactosidase; this small part gives rise to a truncated protein that forms an enzymatically active hetero dimer with a specific mutant beta-galactosidase *biot*

alpha helix: The right-handed, or less commonly left-handed, coil-like configuration of a >>> polypeptide chain that represents the secondary three-dimensional structure of some protein molecules in which the linear sequence of amino acids is folded into a spiral that is stabilized by hydrogen bonds between the carboxyl oxygen of each peptide bond *chem bio gene*

alpha level: >>> significance level

alpha lattice design: One of a class of incomplete >>> block designs that cater to a wider range of numbers of entries than >>> "square" or >>> "rectangular" lattices; alpha lattice design (after PATTERSON and WILLIAMS, 1976) is an incomplete block design; the blocks are unbalanced in the sense that not all genotypes are represented in this block, and that treatment pair does not occur once in a block, so not all entry pairs are compared with the same accuracy; if the aim would be to have a high accuracy each treatment pair had to be one in each block which implies that the number of replications would be: $r = k + 1$, k is the block size (i.e., 300, in each block 30 entries); this is not feasible with a large genotype set; within the alpha lattice design, the block size can be chosen as well as the number of blocks; as such the alpha lattice design is the most flexible design; the entries are assigned randomly to the blocks *stat*

ALS-AHAS: >>> Acetolactate synthase or acetohydroxy acid synthase

Alternaria: A genus of fungi; it forms yellowish-brown conidia that are divided by transverse and longitudinal septa; there are many species, including important plant pathogens (early blight of potato, Alternaria solani; black rot of carrot and rape, A. radicina and A. brassica; Alternaria disease of wheat) *phyt*

alternate: Not opposite to each other on the axis, but borne at regular intervals at different levels (e.g., of leaves) *bot*

alternate dominance: Sex determination which supposes all individuals to be heterozygous for sex but that the male determiners are dominant in male offspring and the female determiners dominant in female offspring *gene*

alternate host: >>> alternative host

alternate segregation: At meiosis in a reciprocal translocation heterozygote, the segregation of both normal chromosomes to one pole and both translocated chromosomes to the other pole, giving genetically balanced gametes, or segregation of centromeres during meiosis in a reciprocal translocation heterozygote such that genetically balanced gametes are produced *cyto*

alternating dominance: A change of dominance from one >>> allele to the other (*A1a2 a1A2*) of a pair of alleles during ontogenetic development of a heterozygous hybrid; the phenotypic expression of the alleles acts one after another *gene*

alternation: The population density alternates between high and low values in successive generations if the key factor is density-dependent and strongly overcompensates for a change in population density *eco gene*

alternation of generation: The alternation of two or more generations, reproducing themselves in different ways (i.e., alternation of gametophyte [sexual reproductive] and sporophyte [asexual reproductive] stages in the life cycle of a plant) *bot*

alternative disjunction: The distribution of >>> interchange chromosomes at >>> anaphase I of meiosis is determined by their >>> centromere orientation; in the case of alternative disjunction, chromosomes located alternatively in the pairing configuration are distributed to the same >>> spindle pole; as opposed to adjacent disjunction *cyto* >>> adjacent disjunction >>> translocation

alternative drip irrigation: >>> irrigation

alternative host: A host that harbors a pest or disease while the primary host is absent or out of season *phyt* >>> host

alternative inheritance: Relationship between two or more factors such that they are of necessity separated into sister gametes in germ cell formation *gene*

alternative splicing: Formation of diverse mRNAs through differential splicing of the same RNA precursor; it may result in proteins with different compositions of amino acids, or it may involve just the length of 3′ UTR; a reason for alternative/differential splicing is base modification during RNA editing causing a change in slice sites *gene biot*

altimeter: >>> hypsometer

ALTMANN's granules: >>> mitochondria

altricial: Offspring that show a marked delay in attainment of independent self-maintenance *bio*

Alu element: A repetitive DNA element approximately 300 bp long that is abundantly dispersed throughout the >>> genome of organisms; the name derived from the AluI >>> restriction enzyme cleavage site that is within most Alu elements *biot* >>> Table 39

aluminum (Al): Has no specific importance in the metabolism of higher plants; small amounts of uptake favor the imbibition of the cytoplasm; higher concentrations in the soil may cause severe inhibition of plant growth; aluminum tolerance is a main task of plant breeding in several regions of the world *chem phys agr* >>> aluminum toxicity >>> Tables 33, 44

aluminum toxicity: High levels of aluminum or manganese are common in acid soils (>>> Table 44) and can be toxic to both the plant and soil microorganisms; aluminum is the most abundant metal on Earth, is highly toxic to plant growth, and is found in about 2.5 billion ha of acid soils worldwide; many of the world's farmers living on acid soils do not have the management options required to improve cereal production; therefore, the development of cereal/crop cultivars capable of improved production on acid soils is needed; >>> rye, the world's most acid soil-tolerant cereal, contains an enormous diversity in organization of gene complexes that play a role in controlling aluminum tolerance; the presence of malate transporter (ALMT) genes contributes to the up-regulation of aluminum tolerance expression in rye; there is variation in copy number and chromosome location of ALMT genes in rye; at least two ALMT gene complexes are characterized that vary in gene copy number and expression of rye aluminum tolerance *agr phys* >>> aluminum >>> www.rye-gene -map.de/

alvar: Plant community that occurs on shallow, alkaline limestone soils; it is generally dominated by mosses and herbs; woody plants are usually dwarfed, with trees generally absent or at least forming a discontinuous canopy *eco*

alveograph: Dough-testing instrument for measuring the extensibility and resistance to stretching of dough, thereby providing a prediction of >>> baking quality; as a standard disc of dough is blown into a bubble, pressure change and bursting pressure are charted versus time *meth*

alveolate: Of a surface or structure shaped like a honeycomb, for example, pollen or seed surface *bot*

AMBA: >>> American Malting Barley Association

amber codon: Amber suppressor mutation that changes anticodon of amino acid-carrying tRNA to UAG *biot*

amber mutant: It synthesizes >>> mRNA containing the codon UAG as a consequence of a point mutation in a corresponding gene *gene*

ambisexual: A plant that has the reproductive organs of both sexes *bot* >>> bisexual

ambivalent gene: Genes with both advantageous and disadvantageous effects *gene*

ambosexual: >>> ambisexual

ameiosis: The failure of meiosis and its replacement by nuclear division without reduction of the chromosome number *cyto*

ameiotic parthenogenesis: Parthenogenesis in which meiosis has been entirely suppressed *gene cyto* >>> parthenogenesis

amendment: An alteration or addition to soil to correct a problem; commonly, any substance (such as sand, calcined clay, peat, or sawdust) added to soil for the purpose of altering physical conditions *agr* >>> Table 44

ament(um): >>> catkin

American Seed Trade Association (ASTA): Founded in 1883, it is one of the oldest trade organizations in the U.S.; its membership consists of ~850 companies involved in seed production and distribution, plant breeding, and related industries in North America; as an authority on plant germplasm, ASTA advocates science and policy issues of industry-wide importance *org*

AMES test: A biological assay to assess the mutagenic potential of chemical compounds; a positive test indicates that the chemical might act as a carcinogen/mutagen although a number of false-positives and false-negatives are known; named after B. AMES (Berkeley, California), who described in a series of papers from the early 1970s the procedure and effects of testing *meth gene*

amidase: An >>> enzyme that catalyzes the hydrolysis of an acid amide *chem phys*

amide: A compound derived from ammonia by replacement of one or more of the hydrogens with organic acid groups *chem*

amine: An organic base derived from ammonia by replacement of one or more of the hydrogens with organic radical groups *chem*

aminoacetic acid: >>> glycine

aminoacyl-tRNA: A >>> tRNA molecule covalently bound to an amino acid via an acyl bond between the carboxyl group of the amino acid and the 3'-OH of the tRNA *chem*

aminoacyl-tRNA ligase: An enzyme that synthesizes a specific aminoacyl-tRNA molecule, employing a specific amino acid, for example, >>> alanine, its cognate tRNA, and >>> ATP to form, for example, alanyl-tRNA alanine *chem phys biot* >>> Table 38

amino acid: An organic compound containing an acidic carboxyl group (–COOH) and a basic amino group (NH2); amino acid molecules combine to form proteins; according to the side group R, they are subdivided into polar or hydrophilic (>>> serine, >>> threonine, >>> tyrosine, >>> asparagine, and >>> glutamine); non-polar or hydrophobic (>>> glycine, alanine, >>> valine, >>> leucine, >>> isoleucine, >>> proline, >>> phenylalanine, >>> tryptophan, and >>> cysteine); acidic (aspartic acid and glutamic acid) and basics (>>> lysine, >>> arginine, >>> histidine); the sequence of amino acids determines the shape, properties, and biological role of a protein; they are the fundamental constituent of living matter; they are synthesized by autotrophic organisms, such as green plants *chem bio* >>> Table 38

amino acid sequence: The sequence of amino acid residues in a >>> polypeptide chain that represents the primary structure of a protein; the sequence is unique to each protein and influences the protein structure (secondary, tertiary, quaternary) *gene*

1-aminocyclopropane-1-carboxylic acid (ACC) oxidase: The enzyme in the ethylene biosynthesis pathway converting ACC to ethylene *phys biot chem*

1-aminocyclopropane-1-carboxylic acid (ACC) synthase: The enzyme in the ethylene biosynthesis pathway converting SAM (S-adenosyl-methionine) to ACC *phys biot chem*

aminopeptidase (Amp): An enzyme that catalyzes the hydrolysis of amino acids in a >>> polypeptide chain by acting on the peptide bond adjacent to the essential free amino group *chem phys*

amitosis: Nuclear division by a process other than mitosis *cyto*

amixia: Cross sterility *bot*

amixis: Reproduction in which the essential events of sexual reproduction are absent *bot*

AMMI analysis: >>> additive main-effect and multiplicative interaction

ammonia (NH_3): Compound that plays an important role in the natural nitrogen pathways *chem agr*

ammonification: Conversion of organic matter during decay by bacteria, fungi, and other organisms into >>> ammonia and ultimately ammonium, NH_4, which can be absorbed by plants *phys*

ammonium fixation: In soil science, adsorption of ammonium ions by clay minerals rendering them insoluble and non-exchangeable agr >>> Table 44

ammonium phosphate [$(NH_4)_3PO_4$]: Used as a mineral fertilizer *chem agr*

ammonium sulfate [$(NH_4)_2SO_4$]: Used as a mineral fertilizer with nitrogen content of about 21% *chem agr*

amorph: A gene that is inactive – an amorphic gene; in botany, sometimes used to refer to something that lacks a discernable shape and can thus be described as "amorphous" *gene bot*

amorphous: >>> amorph

amorphous wax: Smooth or glasslike wax occurring on leaf surfaces and is generally easy to wet *bot* >>> waxy

AMOVA: >>> Analysis of molecular variance

amphiapomict: Reproducing both sexually and apomictically *bot gene*

amphibiotic: Applied to an organism that can be either parasitic on or symbiotic with a particular host organism *phyt*

amphibivalent: A ring-like >>> interchange configuration of four chromosomes *cyto*

amphicarpous: Plants producing two classes of fruit that differ either in form or in time of ripening *bot*

amphidiploid: >>> allotetraploid >>> didiploid

amphihaploid: >>> allohaploid

amphikaryon: The nucleus of the zygote produced after fertilization *cyto*

amphimict: A species or individual that is reproduced by fusion of nuclei during sexual reproduction *bot*

amphimictic: >>> amphimict

amphimixis: >>> amphimict

amphiplasty (nucleolar dominance): Originally, the loss of a satellite from a chromosome; later, morphological changes of chromosomes after >>> interspecific hybridization; occurs when the genomes of the parental species are spatially separated in the hybrid nucleus in a concentric fashion; the genome occupying the central position has an active >>> nucleolus organizer region (NOR), while the NOR of the peripherical genome is suppressed (e.g., in hexaploid wheat, NORs on chromosome 1B and 6B function preferentially, although 1A and 5D also carry NORs; however, when the NOR on 1B is deleted, then the NORs on 5D and/or 1A are used to a greater extent) *cyto*

amphiploid: >>> alloploid

amphiploidy: >>> alloploidy >>> allopolyploidy

amphitene: >>> zygotene

amphitoky: In some insects (e.g., butterfly), progeny of both sexes may develop from unfertilized eggs *zoo*

amphitropous ovule: A type of >>> ovule arrangement in which the ovule is slightly curved so the >>> micropyle is near the funicular attachment *bot*

amphogenic: Producing offspring consisting of approximately equal proportions of either sex *gene*

ampicillin (beta-lactamase): An antibiotic substance; the resistance to ampicillin is sometimes used as a screening marker in genetic experiments with bacteria (e.g., the cloning vector pBR322 carries a gene for ampicillin resistance; it interferes with bacterial cell wall synthesis; the bla gene is used as a selective marker in many vectors, including pBR322 and pUC18) *biot* >>> Figure 46

amplexicaul: A leaf whose base wholly or partly surrounds the stem *bot*

amplicon: A piece of DNA formed by >>> PCR amplified DNA sequence *biot*

amplification: The intrachromosomal or extrachromosomal production of many DNA copies from a certain region of DNA; it can happen spontaneously or it can be done by molecular techniques (e.g., PCR) *cyto biot*

amplified fragment length polymorphisms (AFLPs): Polymorphic DNA fragments are amplified through >>> PCR procedure; their differences are used for genotype identification and linkage studies *biot* >>> differential AFLP

amplify: >>> amplification

amplimer: PCR-amplified segment of the genome (including >>> STSs and >>> ESTs) *biot*

ampoule: A bottle with a bulbous body and narrow neck *prep*

ampulliform: Flask-like in form *bot*

amygdaliform: Almond-shaped *bot*

amygdalin: >>> glucoside

amylaceous: >>> amyliferous

amylase: A member of a group of enzymes that hydrolyze starch or glycogen by splitting of glucosidic bonds, giving rise to sugars, glucose, dextrin, or maltose; they occur particularly in germinating seeds in which the amylase mobilizes food reserves for the growth of the seedling *chem phys* >>> alpha amylase

amyliferous: Containing starch *bot*

amylograph: An instrument used to determine change of viscosity with time in a heated mixture of water and a starchy material such as flour *meth*

amylolytic enzymes: Those enzymes that hydrolyze starches and similar glucose polymers *chem*

amylopectin: Larger, highly branched chains of glucose molecules *phys*

amyloplast (*syn* anaplast *syn* anaplastid): A plastid that synthesizes and stores >>> starch to the exclusion of other activities *bot*

amylose: Relatively short, unbranched chains of >>> glucose molecules; a polysaccharide consisting of linear chains of between 100 and 1,000 linked glucose molecules; a constituent of starch; in water, amylose reacts with iodine to give a characteristic blue color *chem phys*

anabiose: The situation in the life cycle of some plants in which there is no visible metabolic activity (resting period) *bot*

anabolic: Pertaining to an enzymatic reaction leading to the synthesis of a more complex biological molecule from a less complex one *chem phys*

anachromasis: The change which takes place in a nucleus during prophase *cyto*

anandrous: Having no stames *bot*

ananthous: Having no inflorescence *bot*

anaerobe: >>> anaerobic

anaerobic: An organism able to grow without free oxygen *phys*

anagenesis: A mode of evolution characterized by cumulative changes in an evolutionary lineage *eco* >>> Figure 68

analysis of molecular variance (AMOVA): A method for studying molecular variation within a species meth *biot*

analysis of variance (ANOVA): A statistical method that allows the partitioning of the total variation observed in an experiment among several statistically independent possible causes of the variation; among such causes are treatment effects, grouping effects, and experimental errors; the statistical test of the hypothesis that the treatment had no effect is the >>> F-test or variance-ratio test; if the ratio of the mean square for treatments to the mean square for error exceeds a certain constant that depends on the respective >>> degrees of freedom of the two mean squares at a chosen >>> significance level, then the treatments are inferred to have been effective; the analysis of variance of data from a randomized block design is correct only if all of the treatments are equally replicated and if all blocks have the same number of observations; this introduces complications if observations are missing from the data; if only one observation is missing, the procedure is to calculate a substitute by means of a special formula *stat*

anandrous: Free of anthers (male sexual organs) *bot*

anaphase (A): A stage that occurs once in mitosis and twice in meiosis and that involves the separation of chromosomal material to a given two sets of chromosomes, which will eventually form part of new cell nuclei; the separation is controlled by the >>> spindle; in anaphase of mitosis and anaphase II of meiosis, the >>> centromere becomes functionally double and daughter chromosomes separate from the equator, moving toward the opposite poles of the cell; the spindle then elongates and pushes the two groups of chromosomes further apart; in anaphase II of meiosis the centromere does not divide *cyto*

anaphase movement: The movement of chromosomes and/or chromatids toward the >>> cell poles during mitotic or meiotic anaphase *cyto* >>> Figure 15

anaphase separation: The disjunction of the chromatids of each chromosome during >>> mitosis and anaphase II of meiosis or the separation of chromosomes in anaphase I of >>> meiosis *cyto*

anaphragmic: Mutations that lead to increased enzyme activity by removal of inhibitors *phys gene*

anaschistic: Bivalents that are longitudinally split at first meiotic division; these are bivalents with chiasmata close to the spindle attachment (as opposed to diaschistic) *cyto*

anastomsis: Natural grafting that can occur in either stems or roots, e.g., mango seeds contain both a nucellary embryo and a normal embryo that is the result of open-pollination; trees growing from casually discarded seeds often consist of two trunks joined at the base by anastomosis; one trunk is the nucellary seedling and is identical to the maternal parent, while the other is an open-pollinated variant and is visibly different in many characteristics, including fruit quality and resistance to parasites *bot*

anatropous ovule: A type of ovule arrangement in which the ovule is completely inverted, having a long >>> funiculus with the >>> micropyle adjacent to the base of the funiculus *bot*

ancestor: The form or stock from which an organism has descended or the actual or assumed earlier type from which a species or other >>> taxon evolved *gene*

ancestry: A series of ancestors; ancestral descent; lineage *gene*

anchorage dependence: Describes the normal eukaryotic cell's need for a surface to attach to in order to grow in culture *biot*

anchor gene: A gene that has been positioned on both the physical map and the linkage map of a chromosome *biot*

ancient: Dating from a remote period; very old; aged

ancient clones: The importance of ancient clones is that they provide proof of the durability of horizontal resistance; such clones may date from centuries, even millennia, ago; they are common in figs, olives, date palms, >>> citrus, horseradish, garlic, ginger, turmeric, saffron, rhubarb, etc. *hort* >>> Figure 50

androdioecious: Describes a species having male and hermaphroditic (>>> perfect) flowers on separate individuals *bot*

androdioecy: Where male and hermaphrodite >>> genets coexist *bot*

androecious: Applied to a plant that possesses only male flowers *bot*

androecium: A collection of >>> stamens that form the male reproductive organs of flowering plants *bot*

androecy: Applied to a plant that possesses only male flowers *bot*

androgenesis: Development of a haploid embryo from a male nucleus *bot*

androgenetic embryo: An embryo that contains two paternally derived sets of chromosomes and no maternally derived chromosomes *biot*

androgenetic haploid: Plant having chromosomes from the male parent only *cyto* >>> haploid

androgenous: Producing only male offspring *bot*

androgynary: Having flowers whose stames and pistils are petaloid *bot*

androgynism: Possessing both stames and pistils >>> bisexuality >>> hermaphroditism

androgynophore: The stipe or column on which >>> stamens and >>> carpels are borne *bot*

andromonoecious: A species having male and >>> hermaphroditic (>>> perfect) flowers on the same individual *bot*

andromonoecism: Plants with staminate and perfect >>> hermaphroditic flowers *bot*

andromonoecy: Where a >>> hermaphrodite bears both male and hermaphrodite flowers *bot*

androphore: The stalk or column supporting the >>> stamens of certain flowers; usually formed by a union of the filaments (e.g., in Leguminosae) *bot*

androsome: Any chromosome exclusively present in the nucleus of the male *cyto*

anemochory: Distribution of seeds by the wind *bot* >>> Table 35

anemophilous: Windborne pollen *bot* >>> Table 35

anemophily: Pollination of a flower in which the pollen is carried by the wind (e.g., in grasses) *bot* >>> anemophilous >>> Table 35

aneucentric: Applied to an aberrant >>> chromosome possessing more than one >>> centromere *cyto*

aneucentric translocation: A translocation involving the centromere so that an acentric chromosome and a dicentric chromosome result *cyto*

aneuhaploid: When the chromosome number deviates from the haploid standard chromosome number of the species or individual *cyto*

aneuploid: A cell or organism whose nuclei possess a chromosome number that is greater or smaller by a certain number than the normal chromosome number of that species; an aneuploid results from non-disjunction of one or more pairs of homologous chromosomes; beside nullisomics, monosomics, trisomics, tetrasomics, telosomics in allopolyploids, mono-, di-, or trisome-autotetraploids were proposed (NAKAJIMA 1979) *cyto* >>> Table 50

aneuploid reduction: Reduction of the genetic variability by decreasing the chromosome number *cyto*

aneuploidy: >>> aneuploid

aneusomatic: Individuals whose cells exhibit variable numbers of chromosomes, e.g., euploid and aneuploidy cells *cyto*

aneusomic: Individuals or cells that contain unequal sets of the individual (homologous) chromosomes *cyto*

angiocarpous: Having a fruit enclosed within a distinct covering (e.g., a filbert within its husk) *bot*

angiosperm: Seeds formed within an ovary *bot*

angiospermous: >>> angiosperms

angiosperms: Any vascular plant of the phylum, having the seeds enclosed in a fruit, grain, pod, or >>> capsule, and comprising all flowering plants; they produce seeds enclosed in fruit (an ovary); the dominant type of plant today; there are over 250,000 species; flowers are used in reproduction; they evolved about 145 mya, during the late Jurassic period, and were eaten by dinosaurs; they became the dominant land plants about 100 mya (edging out conifers, a type of gymnosperm); angiosperms are divided into the monocots (grasses) and dicots (beet) *bot* >>> Table 32

Ångstrom (Å): A unit of length equal to 10^{-10} m; formerly used to measure wavelengths and intermolecular distances but has now been replaced by the nanometer (1 Å = 0.1 nm); the unit is named after Swedish pioneer of spectroscopy A. J. ÅNGSTRÖM (1814–1874)

angustifoliate: With small leaves *bot*

anhydride: A compound formed by removing water from a more complex compound *chem*

anion: An ion that carries a negative electrical charge *chem*

anisogamete: A >>> gamete that differs in size, appearance, structure, or >>> sex chromosome content from the gamete of the opposite sex *bot*

anisogamy: Unequal gametes fusing during fertilization *bot*

anisomeric: Non-equivalent genes that interact to produce particular phenotypes *gene*

anisoplethy: The departure of ratios of >>> genet morphs in population from the expectation of unity (e.g., one to one, of males and females, pins and thrums) *bot*

anisoploid: An individual with an odd number of chromosome sets in somatic cells *cyto*

anisoploid seeds: A mixture of seeds of different ploidy levels (e.g., in sugarbeet varieties = 2x, 3x, 4x) *seed cyto*

anisotrisomic: A mixture of seeds or individuals that are not only >>> trisomic *cyto*

ANITA: The basal angiosperms are the flowering plants which diverged from the lineage leading to most flowering plants; in particular, the most basal angiosperms were called the ANITA grade which is made up of *Amborella* (a single species of shrub from New Caledonia), *Nymphaeales* (water lilies, together with some other aquatic plants), and *Austrobaileyales* (woody aromatic plants including star anise); thus ANITA stands for *Amborella, Nymphaeales and Illiciales, Trimeniaceae within the Austrobaileyales*; the basal angiosperms are only a few hundred species, compared with hundreds of thousands of species of eudicots, monocots, and magnoliids; they diverged from the ancestral angiosperm lineage before the five groups comprising the mesangiosperms diverged from each other *tax bot*

anneal: In molecular genetics, heating that results in the separation of the individual strands of any double-stranded nucleic acid helix, and cooling that leads to the pairing of any molecules that have segments with complementary base pairs; synonymously used for hybridization *biot*

annotation: Analysis and commentary added to sequence data in databases; annotation provides in-context information about coding and non-coding sequence within genes, patterning and motifs, similarities, known or predicted protein structure and function, as well as links to external data, such as morphological observations *biot*

annouline: The fluorescent protein pigment exudated by the roots of, for example, ryegrass seedlings; the fluorescent nature of this material is useful in distinguishing annual and perennial >>> ryegrass *seed*

annual: A plant that completes its life cycle within a single growing season *bot* >>> biennial >>> perennial

annual ring: In woody plants, the layer of wood produced each year that can be seen when the wood is cut into a cross section; the number of rings equals the age of the tree *fore*

annulus: Applied to any of a number of ring-shaped nuclear pores *bot cyto*

anonymity: A situation in which no one, including the researcher, can link individuals' identities to their responses or behaviors that serve as research data *stat*

anonymous DNA marker: A DNA marker detectable by virtue of variation in its sequence, irrespective of whether or not it actually occurs in or near a coding sequence; >>> microsatellites are typical anonymous DNA markers *biot*

anonymous single nucleotide polymorphism: One of the most frequent occurring >>> SNPs that has no known effect on the function of the gene *gene biot* >>> SNP

anorthogenesis: Adaptive changes of evolutionary significance based on preadaptation *evol* >>> Figure 68

ANOVA: >>> analysis of variance

Antennaria type: Mitotic >>> diplospory where the megaspore mother cell does not enter meiosis but proceeds directly into the first mitosis; the megaspore mother cell thus functions as an unreduced gamete *bot*

antephase: >>> prophase

anther: The terminal portion of a >>> stamen of a flowering plant; the pollen sacs containing pollen are borne on the anther; the number of anthers in a flower varies from three to ten among most species *bot* >>> Figure 58

anther culture: Culturing of anthers containing pollen or of single pollen grains; the method is used for the production of haploid plants, for the production of >>> doubled haploids (which are homozygous) after spontaneous or induced rediploidization, or for breeding on a lower ploidy level *biot* >>> microspore culture >>> speargrass >>> Figures 17, 26, 58 >>> Tables 7, 47

antheridial mother cell: A cell within the pollen grain of flowering plants which divides to form the male gametes *bot* >>> generative nucleus

antheridium (*pl* antheridia): The male sex organ or >>> gametangium within which male gametes are formed, e.g., in mosses and liverworts *bot*

anther lobe: The half of anther containing two pollen sacs or male sporangia *bot* >>> theca

antherozoid: A motile male gamete, e.g., in Pteridophyta or Cycadopsida *bot*

anthesis: The time of flowering in a plant, the opening of a flower bud, or the time when the >>> stigma is ready to receive the pollen *bot*

anthocyan: >>> anthocyanin

anthocyanin: Water-soluble, nitrogenous pigments that contribute to the autumnal colors of the leaves of temperate-climate plants; widely distributed in leaves, stems, and flowers; red- and purple-colored matter found in various parts of the plants (e.g., in the auricles, awns, nodes, and coleoptiles of many cereals); anthocyanins contribute to plants' defense against a number of abiotic and biotic stress agents; e.g., the anthocyanin pigmentation of the barley leaf sheath is genetically determined by *Ant1*, a gene which maps to a region of chromosome 7HS delimited by the microsatellite loci *Xgbms0226* and *Xgbms0240*; the sequence of the maize gene *C1* (encoding an R2R3 MYB factor regulating anthocyanin synthesis) is homologous to the barley sequence *chem phys*

anthracnose: A general term for any of several plant diseases in which symptoms include the formation of dark and often sunken spots on leaves, fruits, etc. (e.g., caused by *Colletotrichum lindemuthianmum* in dwarf bean, *Elsinoe ampelina* in grape, *Kabtiella caulivora* in clover, *Gloeosporium ribis glossulariae* in gooseberry, *Colletotrichum linicolum* in linseed, *C. oligochaetum* in cucumber, *C. graminicolum* in rye and maize, or *C. orbiculare* in melons) *phyt*

anthropochore: A wild plant or species dispersed as a result of accidental human activity *eco*

anthropochory: Dispersal of wild plants, such as seeds or fruits, as a result of accidental human activity *eco*

antiauxin: A chemical that interferes with the auxin response; it may or may not involve prevention of auxin transport or movement in plants; some antiauxins are said to promote morphogenesis *in vitro* and *in vivo*, such as 2,3,5-tri-iodobenzoate, or 2,4,5-trichlorophenoxy-acetate, which stimulate the growth of some *in vitro* cultures *chem biot*

antibiosis: The phenomenon whereby a natural organic substance secreted by one organism has an injurious effect on normal growth and development of another organism when the two organisms are brought together *eco*

antibiotic: A substance that is produced by some organisms; it may inhibit the growth of another organism or even kill it; several antibiotics are produced by fungi; they act against bacteria *phys*

antibiotic resistance gene: A gene that encodes an >>> enzyme that degrades or excretes an antibiotic, thus conferring resistance *gene phyt*

antibody: A protein, usually found in serum, whose presence can be shown by its specific reactivity with an antigen; it binds with high affinity to antigens and thereby destroys them; in molecular biology, it is used in >>> Western blots; a normal immuneserum contains different antibodies, which recognize all the many different antigens an animal normally produces; monospecific antiserum is purified to exclusively contain antibodies that recognize epitopes of a single macromolecule; >>> monocolonal antibodies contain only one particular type of antibody that is specific for a single epitope; they are produced by hybridoma cell cultures *phys biot*

anticlinal: Referring to a layer of cells running orthogonal to the surface of a plant part *bot*

anticline: >>> anticlinal

anticodon: A triplet sequence of nucleotides in >>> tRNA that during protein synthesis binds by base pairing to a complementary sequence, the codon, in >>> mRNA attached to a >>> ribosome *gene*

antigen: A molecule, normally of a protein although sometimes of a polysaccharide, usually found on the surface of a cell, whose shape causes the production in the invaded organism of antibodies that will bind specifically to the antigen *phys biot*

antigenicity: >>> antigen

antimetabolite: A substance that resembles in chemical structure some naturally occurring compounds and which specifically antagonizes the biological action of such compounds *phys*

antimitotic: Substances that may lead to the cessation of >>> mitosis *cyto*

antimorph: A mutant allele that acts in a direction opposite to the normal allele *gene* >>> amorph >>> hypermorph >>> neomorph

antimutagenic: Substances that can reduce the rate of mutations *gene*

antimutator gene: Mutant genes that decrease the mutation rates; most probably, its product increases the efficiency of the proofreading and/or editing function of the DNA polymerase during replication *gene*

antinutritive character: A crop product containing substances causing either diseases or negative influences in humans and animals (e.g., the >>> pentosans of rye and triticale) *agr*

antioxidant: A substance that delays the onset or slows down the rate of oxidation of oxidizable substrates; for example, whole cereal grains contain a number of antioxidants such as vitamins (e.g., vitamins E and beta-carotene) and trace elements essential for enzymes performing antioxidant functions (e.g., Se, Mn, Mg), and non-nutrients such as phenolic compounds (e.g., lignans) and antinutrients (e.g., >>> phytic acid) *chem phys*

antiparallel orientation: The normal arrangement of the two strands of a DNA molecule, and of other nucleic-acid duplexes (DNA–RNA, RNA–RNA), in which the two strands are oriented in opposite directions so that the 5′-phosphate end of one strand is aligned with the 3′-hydroxyl end of the complementary strand *chem biot*

antipodal cells: Three haploid nuclei that are formed during >>> megasporogenesis in plants; all are located opposite the >>> micropylar end of an ovule *bot* >>> Figures 25, 35

antipodal nuclei: Three of the eight nuclei that result from the >>> megaspore by mitotic cell divisions within the developing megagametophyte (>>> embryo sac); they are usually located at the base of the embryo sac and have no apparent function in most species *bot* >>> Figure 25

antipodes: >>> antipodal cells

antisense DNA: Non-coding DNA of one of the double-stranded DNA, as opposed to >>> sense strand DNA, which is the coding DNA (i.e., which is transcribed as mRNA) *gene biot*

antisense gene: A gene construct placed in inverted orientation relative to a promoter; when it is transcribed it produces a transcript complementary to the >>> mRNA transcribed from the normal orientation of the gene *biot* >>> antisense technique

antisense RNA: A complementary RNA sequence that binds to a naturally occurring (sense) mRNA molecule; in this way it thus blocks its translation *biot* >>> antisense technique

antisense technique: A method of inhibiting the activity of a certain gene; the genetic information encoded by genes is conveyed with the help of the messenger RNA (>>> mRNA); messenger RNA moves to ribosome sites of protein production; by antisense technique plants are transformed with reversed or "antisense" genes; when antisense genes are expressed, an mRNA molecule is produced that is a mirror image of the targeted gene; the two, opposite mRNAs bind to one another, disrupting their function and making protein synthesis impossible; in effect, the targeted gene is blocked; for example, the antisense strategy is used by plant breeders to block the synthesis of >>> amylose starch in potatoes *biot meth*

antisense therapy: *In vivo* treatment of a genetic disease by blocking translation of a protein with a DNA or an RNA sequence that is complementary to a specific mRNA *meth biot* >>> antisense technique

antiserum: A serum that contains >>> antibodies *meth*

antitrypsin factors: Chemicals found in certain grains and other food raw materials that hinder the digestive action of the enzyme trypsin; known to be present in >>> rye *chem*

antixenosis: A non-host preference by the pest; estimated by the degree of colonization on the plants *phyt*

antrorse: Directed upward or forward toward the apex; the opposite of retrorse *bot*

anucleolate: Without a nucleus *cyto*

AOSA: Association of Official Seed Analysts

AOSCA: >>> Association of Official Seed Certifying Agencies

apatite: An important inorganic phosphate of soil, but only of limited use for plant nutrition *chem agr*

aperture: In an optical instrument, the opening of a lens *micr*; a depressed region in the pollen wall in which thick intine is covered by thin >>> exine; the pollen grain emerges through the aperture *bot*

aperture diaphragm: An adjustable diaphragm in illumination path that regulates amount of excitation intensity (numerical aperture of excitation light) *micr*

aperture plane: In a microscope adjusted for >>> Koehler illumination, the conjugate planes that include the light source, the condenser iris diaphragm, the objective lens back aperture, and the eye point *micr*

apetalous: A plant showing no flowers *bot*

apex (pl apices): Extreme point or distal end *bot*

apex culture: >>> shoot-tip culture

apex of the leaf: >>> leaf tip

apex separator: A wheat-cleaning machine using a travelling continuous mesh *meth*

aphid vector: Any of numerous tiny soft-bodied insects of the family Aphididae transferring viruses *phyt*

aphids: Any of numerous tiny soft-bodied insects of the family Aphididae that suck the sap from the stems and leaves of various plants; they are also called plant louses; aphids are major agricultural pests that cause significant yield losses of crop plants each year; excessive dependence on insecticides for aphid control is undesirable because of the development of insecticide resistance, the potential negative effects on non-target organisms, and environmental pollution; transgenic plants engineered for resistance to aphids via a non-toxic mode of action could be an efficient alternative strategy; although the transgenic plants developed through expressing aphid-resistant genes, manipulating plant secondary metabolism and plant-mediated >>> RNAi strategy have been demonstrated to confer improved aphid resistance to some degree; so far, no aphid-resistant transgenic crop plants have ever been commercialized *zoo phyt*

aphyllous: Leafless; applied to flowering plants that are naturally leafless (e.g., many species of cactus) *bot*

aphylly: Leafnessness *bot* >>> aphyllous

apical: At the end; related to the apex or tip bot

apical cell: A meristematic initial in the apical meristem of shoots or roots of plants; as this cell divides, new tissue is formed *bot*

apical dominance: A condition in plants where the stem apex prevents the development of lateral branches near the apex *phys*

apical meristem: An area of actively dividing cells at the tip of a shoot, branch, or root; it is also the precursor of the primary tissue of root or shoots *bot*

apical placentation: A type of precentral placentation in fruit where the seeds are attached near the top of the central ovary axis *bot*

apical segment (of spike): The uppermost segment of the rachis *bot*

apical spikelet: The spikelet occurring at the apex (tip) of the ear or panicle *bot*

apiculate: Ending in a short, flexible point *bot*

apiculture: Beekeeping for the sale of honey and for pollination of crop plants *seed*

aplanate: Lying in a plane; leaves may be displayed on the twigs of a plant to give aplanate foliage; flattened *bot*

apoamphimict: A plant that reproduces predominantly by >>> apomixis but also sexually *bot*

apocarp: >>> apocarpy

apocarpy: The condition in which the female reproductive organs (carpels) of a flower are not joined to each other (e.g., in buttercup, *Ranunculus* ssp.) *bot* >>> syncarpy

apochromat: A microscope objective corrected for spherical and chromatic aberration *micr*

apoenzyme: The portion of a conjugated enzyme that is a protein *phys*

apogamety: Autonomous development of a vegetative cell, not the egg cell, into an embryo in an apomict *bot*

apogamy: A type of >>> apomixis involving the suppression of gametophyte formation so that seeds are formed directly from somatic cells of the parent tissue *bot* >>> apogamety >>> Figure 28

apolegamy: Selective breeding *meth*

apomeiosis: Sporogenesis without reduction of chromosome number during meiosis and giving rise to apomixis *bot*

apomictic: Plants that form asexual progenies *bot* >>> Figure 28

apomixis: Asexual reproduction in plants without fertilization or meiosis, for example, *Poa* ssp.; these seeds are produced from flowers, just as regular seeds are, but no pollen is involved; the plants grown from such seeds are perfect clones of the original plant; e.g., >>> mangosteen, a tropical fruit produces apomictic seeds within some of the segments of the fruit; the common dandelion (*Taraxacum* ssp.) produces many apomictic seeds as well; there are a lot efforts in order to transfer apomictic growth behavior to hybrid crops; the Saharan cypress (*Cupressus dupreziana*) exhibits a unique form of natural male apomixis, in which the seeds develop from the genetic content of pollen, not from ovules; this means that the entire genome is derived from the male side with no contribution from the female *bot* >>> Figure 28

apomorphy: A derived state of a character that has changed in state in relation to its predecessor *bot*

apophase: The period of postmeiotic reconstruction of the cell *cyto*

apoplast: Areas of a plant that lie outside the >>> plasmalemma, such as cell walls and dead tissue of the >>> xylem; the apoplast may represent one of the main pathways of water through the plant *bot*

apoplastic path: Water moves in the cell wall or non-living region between cells without crossing any membranes; apoplastic means non-living tissue *phys*

apoptosis: The process by which a cell dies in a programmed way, or in other words, kills itself; it is the most common form of physiological (as opposed to pathological) cell death; it is an active process requiring metabolic activity by the dying cell; often characterized by shrinkage of the cell and cleavage of the DNA into fragments *bot phys*

apospory: The development of a diploid embryo sac in some plants by the somatic division of a nucellus or integument cell without meiosis; a sort of >>> agamospermy in which a seed is produced without fertilization; it occurs in the genera of Beta, Brachiaria, Eragrostis, Hieracium, Hypericum, Malus, Panicum, Purthenium, Puspalurn, Puspalurn, Pennisetum, Poa, Potentilla, Ranunculu, and Rubus *bot* >>> apomixis >>> Figure 28

apostatic selection: A selection of very rare genotypes or phenotypes that stand out from the norm (e.g., a macromutation) *gene* >>> Figure 60

apothecium: A roughly cup-shaped or dish-like ascocarp, in which the asci line the inner surface and are thus exposed to the atmosphere *bot*

appeal: Refers to the process whereby a seed grower may request an official organization to reconsider the status of an inspected crop based on factors not given on the report of inspections; for example, in Canada, an appeals committee reviews the case and recommends a decision to the official organization; all cases brought to the appeals committee are considered anonymously and without bias *seed*

appendage: A process of outgrowth of any sort *bot*

apple maggot: A North American native fruit fly (*R. pomonella*) that lays its eggs inside the fruit of the hawthorn tree; this agricultural pest began plaguing U.S. apple growers in the 1850s, likely after undergoing extensive and genome-wide changes in a single generation; in the 1850s, a splinter group of *Rhagoletis* began laying eggs in apples in upstate New York, a move that required the flies to adapt the timing of their annual egg-laying cycle to match the fruiting time of their new hosts; today, there are two forms of the fruit fly, the ancestral form that times its life cycle to the hawthorn tree and a derived form that is timed to apple trees, which fruit about three to four weeks earlier; both forms have evolved very distinct differences and are on the path to evolving into two new species; the differences happened in no more than 170 generations; the two populations are also known to interbreed; in a given year, about 5% move between populations (*Rhagoletis pomonella*) *phyt*

apple scab (*Venturia inaequalis*): A common disease of apple trees in which the most obvious symptom is the appearance of superficial, dark, corky scabs on the fruit; it is one of the most damaging diseases affecting commercial apple production; some wild Malus species possess resistance against apple scab; one gene, HcrVf2, from a cluster of three genes derived from the wild apple Malus floribunda, clone 821, has recently been shown to confer resistance to apple scab when transferred into a scab-susceptible variety *phyt* >>> Figure 50

applied genetics: Studying genetic factors and processes for utilization in agriculture and other branches of science *gene*

applied research: Designed with a practical outcome in mind and with the assumption that some group or society as a whole will gain specific benefits from the research *meth bio*

appressed: Lying close and flat against *bot*

appressorium: In certain parasitic fungi, an attachment organ consisting of a flattened hypha that presses closely to the tissue of the host as a preliminary stage in the infection process *bot*

apurinic site: Gap in a nucleic acid molecule created by the removal of purine *gene*

aquaporins: Cell membrane proteins that facilitate water movement across bio-membranes; they are suggested to mediate not only water but also other molecules transports; aquaporin genes are known as major intrinsic protein (MIP) genes; e.g., in barley >>> EST database, putative 24 MIPs (contigs) were identified and 11 genes of plasma membrane type aquaporin (PIPs) were detected by >>> PCR; expression of these 11 PIPs was investigated under salt (NaCl), osmotic (manitol), heavy metals ($CuCl_2$ and $CdCl_2$), and oxidative (H_2O_2) stresses; one of them, HvPIP2-1, was most abundant and its protein expression was also analyzed; it was confirmed that HvPIP2-1, encoded water channel activity in Xenopus laevis oocytes injected with HvPIP2-1 cRNA; transcripts and proteins of HvPIP2-1 were reduced in barely roots under salt stress; over-expression of HvPIP2-1 increased the shoot/root ratio and raised salt sensitivity in transgenic rice plants, indicating HvPIP2-1 is involved in the cellular mechanism of >>> salt tolerance; overexpression of the HvPIP2-1 also increased internal CO_2 conductance and CO_2 assimilation in the leaves of transgenic rice plants, suggesting that HvPIP2-1 permeates CO_2 in addition to H_2O; recent reports suggested that aquaporins were involved in the flood-induced reduction of root water uptake, chilling-induced decrease of root water permeability, and other many physiological functions in plants *phys biot*

aquatic: Living in water *bot*

aquifer: An area in which groundwater accumulates *agr*

arabinose (Ara): An aldose that contains five carbon atoms; a member of pentoses (e.g., it occurs in sugarbeet) *chem phys*

arabinoxylan: A polysaccharide composed of xylose and arabinose, is part of the soluble and insoluble fiber in cereals; for example, rye is a good source of arabinoxylan (pentosan) *chem phys*

arable: Capable of producing crops by plowing or tillage *agr*

arable land: Land capable of producing crops by plowing or tillage *agr*

arachin: >>> globulin

arachnoid: Covered with hairs or fibers or formed of hairs or fibers *bot*

arborescent: Treelike shape *bot*

arboretum: A park or garden where trees and shrubs are grown for educational and/or scientific uses *bot*

archegonium (*pl* archegonia): Female sex (reproductive) organ of liverworts, mosses, ferns, and most gymnosperms; usually a flask-like organ, comprising a swollen base or center containing a single egg cell and a slender elongated neck containing one or more layers of cells *bot*

archesporial: The differentiated cell situated in the nucellary tissue of the ovule that is destined to undergo meiosis and give rise to the haploid generation *bot*

archesporium: Cells formed by mitosis of the micro- and macrospore mother cells *bot*

arcuate: Curved or arched *bot*

areola (*pl* areolae): A small, well-defined area on the surface of a cactus bearing spines or flowers *bot*

area sampling: A multistage sampling technique that involves moving from larger clusters of units to smaller and smaller ones until the unit of analysis *stat meth* >>> cluster sampling

arginine (Arg): An aliphatic, basic, polar amino acid that contains the guanido group *chem phys* >>> Table 38

arid: Dry climates or dry regions (<250 mm rainfall in temperate climates and <350 mm rainfall in tropical climates); commonly applied to a region or a climate in which precipitation is too low to support crop production *eco*

aril(lus): A usually fleshy and often brightly colored outgrowth from the >>> funicle or >>> hilum of a seed (outer seed covering), a third integument; arils probably often aid seed dispersal by drawing attention to the seed after the fruit has dehisced and by providing food as an attractant and reward to the disperser *bot*

arista: >>> awn

aristate: Showing awns; bearded; bristle-tipped *bot* >>> Figure 34

arithmetic mean: An average; the number found by dividing the sum of a series by the number of items in the series *stat*

armyworm: Causal agent is *Pseudaletia unipuncta*, *Spodoptera mauritia*, and *S. praefica*; the swarming caterpillars cause severe damage to >>> rice plants; the worms appear suddenly in masses and move like an army from field to field so that seedbeds or the direct seeded fields look as if grazed by cattle; generally, a transplanted crop is not severely affected; damage by armyworms is most serious during periods of stem elongation and grain formation; larvae defoliate plants, typically by chewing angular pieces off leaves; they may also feed on the panicle near the developing kernels, causing these kernels to dry before filling; this feeding causes all or parts of the panicle to turn white; if the entire panicle is white, the damage may also be due to stem rot or feeding by rats; significant yield reduction can occur if defoliation is greater than 25% at 2 to 3 weeks before heading *phyt*

arnautka: Strains of durum wheat [*Triticum durum*, 2n = 4x = 28 (BBA^{u}A^u)] *agr* >>> Table 1

aromatic: Having an odor, fragrant or otherwise; bearing volatile essential oils *bot*

aromatic compounds: Compounds that contain at least one benzene ring in their structure, or compounds that have a fragrance or smell, usually restricted to those with pleasant odors *chem*

arrect: Stiffly upright *bot*

arrhenogenic: Producing offspring consisting entirely, or almost entirely, of males *gene*

arrhenotoky: In insects (e.g., bees), the development of male progenies from non-fertilized eggs, i.e., parthnogenesis *zoo*

arrow: The inflorescence of >>> sugarcane *agr*

arsenic (As): May be found in plants in very low dosage; higher concentrations in soil may be toxic for plants *chem agr*

articulate: Jointed; having a node or joint *bot*

artifact: A human-made object; something observed that usually is not present but that has arisen as a result of the process of observation or investigation *meth*

artificial germplasm: >>> synthetic germplasm

artificial chromosome: A chromosome experimentally created and constituted, in addition to genetically coding DNA sequences, by ligating origin of replication, autonomous replicating sequences, and telomeric and centromeric sequences *biot*

artificial light: Light other than sunlight; often from fluorescent tubes; used to grow plants in greenhouses or growing chambers usually out of season *hort agr*

artificial seeds: Based on embryogenic suspension cultures, embryos may be coated with water-soluble hygrogels and other substances in order to guarantee a proper germination even under field conditions *seed biot* >>> synthetic seeds

artificial selection: Plant selection by human or agronomic means; it is the practice of choosing individuals from a population for reproduction, usually because these individuals possess one or more desirable traits *meth agr* >>> breeding >>> plant breeding >>> genetics >>> Figure 60

artioploid: Refers to >>> polyploids with even sets of genomes *cyto*

ASAP: >>> allele-specific associated primers

ascendent: >>> ancestor

ascending: Rising somewhat obliquely, or curved upward *bot*

ascogonium: The female gametangium of some fungi *bot*

ascorbic acid: The water-soluble vitamin C ($C_6H_8O_6$) that occurs in large quantities in fruits and vegetables *chem phys*

ascospore: A sexually produced, haploid spore formed within an >>> ascus by some fungi *bot*

ascus: A minute, baglike structure within which ascospores develop in some fungi *bot*

aseptic: Free from living microorganisms or pathogens; sterile *pre*

asexual: Any reproductive process that does not involve the fusion of gametes *bot* >>> Figure 28

asexual reproduction: A propagation without formation of zygotes by sexual organs and genetic recombination; in plants, there are two types of asexual reproduction: vegetative propagation (by stolons, rhizomes, tubers, tillers, >>> bulbs, bulbils, or corms) and apomixis (by vegetative proliferation or agamospermy) *bot* >>> Figure 28

ash content: The amount of inorganic minerals in food samples; the ash content of flour indicates how much of the outer layers of the kernel have been ground to flour; for example, in white wheat flour (ash content 0.7% or less) about 30% of the outer layers of the kernel have been removed; the ash content of whole meal rye flour is about 2% *meth phys chem*

Asian rice gall midge (*Orseolia oryzae*, Diptera: Cecidomyiidae): A major pest of rice in several South and Southeast Asian countries causing severe yield losses; the maggots feed internally on the growing tips of the tillers and transform them into tubular galls, onion-leaf like structures called "silver shoots"; host resistance has been suggested as the most logical and economical approach for control *phyt* >>> www.pflanzengallen.de

Asian soybean rust: Caused by the fungal pathogen *Phakopsora pachyrhizi*; it was first identified in Brazil in 2001 and quickly infected soybean areas in several countries in South America; primary efforts to combat this disease involve the development of resistant cultivars; five distinct genes that confer resistance against ASR have been reported: *Rpp1*, *Rpp2*, *Rpp3*, *Rpp4*, and *Rpp5*; however, no cultivar carrying any of those resistance loci has been released so far; however, the application of molecular breeding recently improved soybean rust resistance and minimized the adverse effects from overuse of fungicides; a backcross breeding program introgressed the *Rpp5* gene of SBR resistance into "HL203", an elite Vietnamese soybean variety, by using molecular markers *phyt*

ASO: >>> allele-specific oligo

AS-PCR: >>> allele-specific PCR

asparagine (Asn): An amino acid ($C_4H_8N_2O_3$) found in storage proteins of plants (e.g., in peas and beans); the designation derived from the presence in asparagus *chem phys* >>> Table 38

asparagus knife: A tool for prying and pulling out long-rooted plants syn dandelion weeder *syn* fishtail weeder *hort*

aspartic acid (Asp): An aliphatic, acidic, polar alpha-amino acid ($HO_2CCH[NH_2]CH_2CO_2H$) *chem phys* >>> Table 38

aspirator: An air-blast separator; a seed conditioning (cleaning) machine that uses air to separate according to specific gravity (weight) and resistance to air flow *seed* >>> Table 11

assignment test: A test that determines whether a locus is on a specific chromosome by observation of the concordance of the locus and the specific chromosome in hybrid cell lines *gene meth*

assimilation: The production of an organic substance from anorganic elements and compounds via photosynthesis *phys* >>> carbon dioxide

association: In the vegetation classification hierarchy, the level naming the dominant overstory species plus the dominant species in subordinate layers (midstory, understory, and/or ground layer); subordinate layers are indicated by hyphens; sometimes used to refer to large assemblages of organisms in a particular area or to a group of plants growing together and forming a small unit of natural vegetation *bot eco* >>> faciation

association analysis: >>> association mapping >>> association genetics

association genetics: An attractive approach to identifying markers very tightly linked to desirable QTL alleles, when a trait is affected by a large number of genes each of small effect *meth gene*

Association of Official Seed Certifying Agencies (AOSCA): Established in 1919 in the United States; it sets minimum standards for genetic purity and identity and recommends minimum standards for seed quality; its goal is to standardize certification regulations and procedures internationally so companies compete less than one set of standards *org* >>> www.aosca.org/

association mapping: A population-based survey of molecular marker analysis in order to identify trait-marker relationships based on linkage disequilibrium; the association between a pair of linked markers is also called >>> linkage disequilibrium (LD) or, less frequently, gametic disequilibrium; however, association has a broader meaning that includes combinations of three or more linked markers, at least some of which are in LD; these combinations are called haplotypes if specified for a single chromosome; the >>> sex chromosomes in males and certain chromosomal aberrations are >>> monosomic, with each individual carrying only one >>> haplotype; with these exceptions (and some chromosomal aberrations) haplotypes occur in pairs, called diplotypes, consisting of one haplotype from each parent; the two haplotypes of a diplotype cannot be ascertained with certainty if two or more markers are heterozygous, except in special cases that include family studies, physical separation of chromosomes, and zero frequency of alternative haplotypes; association mapping depends on the choice of map taken to represent LD; >>> physical maps specify distance in the DNA sequence, ideally measured in bp; the closest approach to this ideal is by the DNA sequence nominally finished, although errors in many relatively small areas remain and, of course, polymorphisms affecting the DNA sequence are represented by one arbitrary allele; for association mapping, it is convenient to represent location in kb to three decimal places, retaining full precision in the finished maps; two physical maps at lower resolution are derived from chromosome breakage in radiation hybrids, the utility of which is limited to organisms without a finished DNA sequence, and chromosome bands that project cytogenetic assignments to the physical map; at all levels of resolution, physical maps have the additivity that defines a linear map; thus, if the distance in the ith interval between two adjacent markers is d_j and if the intervals are mutually exclusive and jointly exhaustive, the distance between any two markers is Σd_j, just as in any road map; >>> genetic maps also have additivity, but the distance in the ith interval is proportional to $\varepsilon_j d_j$, where ε_j is not a constant but an interval-specific scaling factor such that $\varepsilon_j \geq 0$, and not all ε_j are equal; the distance between any two markers if the intervals are mutually exclusive and jointly exhaustive is proportional to $\Sigma \varepsilon_j d_j$; there are two types of genetic maps; >>> linkage maps long antedated physical maps; their development began in 1913, when A. H. STURTEVANT (1891–1970) elaborated the concept of linear arrangement of genes separated by crossing-over; in 1919, J. B. S. HALDANE (1892–1964) introduced the Morgan unit (w) as the length of a chromatid that on average has experienced one crossover event per meiosis, thereby taking $w_j = \varepsilon_j d_j / t$ as his measurement of distance, where t is the number of generations observed; until recently, linkage maps have been estimated directly from recombination, since values of $\varepsilon_j = w_j / d_j$ could not be determined with accuracy until the physical map was finished; in contrast, LD maps determine distance not from recombination, but from LD, and so distance in the ith interval is expected to be $\varepsilon_j d_j = w_j t$; the number of generations is large and can be reliably determined only from a population genetics model that allows ε_j to be estimated directly; whereas >>> haplotype inference from diplotypes complicates association mapping, the two types of data provide virtually identical LD maps *meth biot* >>> Figure 43 >>> Table 58 >>> QTL mapping

assortative mating: Non-random mating based on the characteristics of partners; it occurs if the plants mating resemble each other, with regard to some traits; usually used for positive assertive mating, whereby the traits of mating partners are more similar than random pairing; negative assertive mating occurs when the traits of mating partners are less similar than expected from random pairing *gene* >>> disassortive mating

assortment: The separation of genes, or of chromosomes, of maternal and paternal origin at meiosis *cyto gene*

ASTA: >>> American Seed Trade Association

asymmetrical: Not having planes that divide the structure into mirror-image halves *bot*

asymmetric cell division: Division of a cell into two cells of different chromosome numbers *cyto*

asymmetric fusion: A cell formed by the fusion of dissimilar cells; referred to as a >>> hetero-karyon *biot* >>> cell fusion

asynapsis: Chromosomes of meiosis I in which pairing either fails or is incomplete *cyto*

asynaptic: >>> asynapsis

asyndesis: >>> asynapsis

atavism: The reappearance of a character after several generations; the reversion to an ancestral or earlier type of character; the character being the expression of a recessive gene or of complementary genes *bio gene*

atelomitic: Applied to chromosomes with non-terminal spindle attachments *cyto*

ATI: >>> alpha-amylase/trypsin inhibitor

a-tomatine: The main saponin in tomato (*Lycopersicon esculentum*), which shows very strong antifungal activity *phys* >>> tomatinase

atomic force microscopy (AFM): A form of scanning probe microscopy that provides atom-level information about molecules; the surface of a molecule is scanned with a microprobe in an x–y grid, and the force encountered is measured with piezoelectric sensors *micr*

atomic mass unit (Amu/amu or Dalton/dalton): The basic unit of mass on an atomic scale; 1 dalton is 1/12 the mass of a carbon 12 atom (i.e., the mass of a hydrogen atom is $1.66 \times 10{-24}$ g; therefore, there are 6.023×1023 dalton in one gram) *chem*

atomizer: Machine for adding water to grain in finely divided spray *meth*

atrazine: A selective herbicide, widely used in >>> maize cropping; it is environmentally significant, since it was the second most commonly detected pesticide residue in a U.S. Environmental Protection Agency (EPA) survey of drinking water wells conducted during 1988–1990; due to concerns about groundwater contamination and worker exposure, the EPA is conducting a special review of atrazine registration *phyt*

atrophy: Reduced or diminished organ size, shape, or function, usually a deteriorate change *bot phys*

atropine: A poisonous crystalline alkaloid, $C_{17}H_{23}NO_3$, it can be extracted from deadly nightshade and other solanaceous plants; used in medicine to treat colic, to reduce secretion, and to dilate the pupil of the eye *chem phys*

attar: A term used for a perfume from flowers (an essential oil, e.g., the attar of roses) *hort*

attractant: A chemical or agent that lures insects or other pests by stimulating their sense of smell; attractants are a non-toxic technique for luring insects into traps and are heavily used in orchard crops; though distinct from toxic baits, attractants are regulated as pesticides *phyt*

att site: Loci on a phage and the bacterial chromosome at which site-specific recombination takes place *biot*

attenuate: Gradually narrowing to a pointed apex or base; sharper than acute *bot*

attenuator: A nucleotide sequence, located in the leader region between the promoter and the structural genes of some >>> operons, that causes RNA polymerase to cease >>> transcription in the leader region before transcribing the structural genes of the operon *gene biot*

atypical: Having no distinct typical character; not typical; not conformable to the type *meth*

auger: A tool for boring a hole in the soil; used in planting or transplanting seeds, seedlings, or >>> bulbs or in fertilizing shrubs, trees, and ground covers *prep hort*; a long pipe with a twist inside to carry the grain up into a granary *seed*

auricle: Claw-like outgrowth arising at the junction of the leaf blade and sheath (e.g., present in wheat and barley, absent in oats) *bot* >>> Table 30

auriculate: Furnished with auricles *bot* >>> auricle

aurofusarin: An orange-yellow pigment of the fungus Fusarium culmorum *phyt*

Australian Wheat Board (AWB): A statutory marketing agency that handles Australia's domestic marketing of wheat and export marketing of wheat and flour; under the Australian system, farmers take their wheat to elevators designated as official handling agents for the board; following delivery, farmers receive an initial payment, then over a period of time they receive additional payments until the full price has been paid; AWB is a grower-owned and controlled company operating under Australian corporation laws *org*

authorship claim to a variety: >>> seed breeder's rights

autoallopolyploid: Cells or individuals whose genomes show characteristics of both auto- and alloploidy *cyto*

autoallopolyploidy: >>> autoallopolyploid

autobivalent: A bivalent of meiosis I that is formed from two structurally and genetically completely identical >>> sister chromosomes *cyto* >>> Figure 15

autochthon(e): One of the indigenous plants of a region *eco*

autochthonous: Applied to material that originated in its present position *eco* >>> indigenous

autoclave: An apparatus in which media, glassware, etc., are sterilized by steam and/or pressure *meth*

autoecious: The converse of heteroecious, which means that a rust fungus or an aphid is obliged to change its species of host in order to complete its life cycle; an autoecious rust fungus is one that completes its entire life cycle on one species of host; entomologists use the term "monoecious" in place of autoecious when describing aphids; in botany, monoecious means that separate male or female flowers occur on a single plant *phyt* >>> dioecious >>> hermaphrodite >>> Table 52

autofertility: >>> autogamy >>> self-fertility

autofluorescence: Fluorescence from objects in a microscope sample other than from fluorophores *micr*

autogamous: >>> autogamy

autogamy: Obligatory self-fertilization, for example, pea, peanut, flax, barley, oats, lupine, rice, soybean, or wheat *bot* >>> gametocide >>> Table 35

autogenomatic: Genomes that are completely homologous and pair normally in meiosis *cyto*

autogenous control: The action of a gene product that either inhibits (negative autogenous control) or activates (positive autogenous control) expression of its gene coding *gene biot*

autograft: A graft of tissue from a donor of one genotype to a host of the same genotype; the graft usually takes place from one part to another part of the same individual *hort*

autoicous: Having staminate and pistillate flowers on the one plant *bot*

autolysis: The destruction of a cell or some of its components through the action of its own hydrolic enzymes *phys*

automatic selection: >>> unconscious selection

automixis: Obligatory self-fertilization or fusion of embryo sac nuclei *bot*

automorphy: Unique derived characteristic; a trait present in only one member of a lineage or in only one lineage among many *gene*

automutagen: Any >>> mutagen formed by the organism itself that may induce mutations *gene*

automutation: A mutation that arises without exogenic application of mutagens *gene* >>> spontaneous mutation

autonomic movement: Plant movement as a result of internal stimuli *phys*

autonomous apomixis: Agamic seed formation that does not depend on pollination *bot* >>> diploid parthenogenesis >>> parthenogenesis

autonomous transposon: Any one of a class of transposons that encode a set of proteins catalyzing its transposon, of the transposition of non-autonomous elements *gene*

autoorientation: The positioning of >>> multivalents at >>> diakinesis and >>> metaphase I of >>> meiosis in a >>> polyploid in such a way that a regular segregation of >>> centromeres is achieved and fertile spores result *cyto*

autophene: A genetically controlled character that is manifested by the cell's own genotype and which shows special behavior in transplants and explants *gene*

autophilous: >>> autogamous

autophyte: Any organism that synthesizes its own food, such as a >>> photosynthetic plant; as opposed to >>> heterophyte *bot*

autoploid: A cell or individual with genomes characteristic of the species itself *cyto*

autopolyploid: A polyploid organism that originates by the multiplication of a single genome of the same species *cyto*

autopolyploidization: The occurrence of doublings of chromosome number by failure of chromosomes to divide equationally in a mitosis following chromosome replication; plants seem to have commonly used autopolyploidization as an evolutionary tool *cyto* >>> Figure 68 >>> C mitosis

autopolyploidy: >>> autopolyploid

autoradiography: A method of determination of amounts and distributions of radioactive substances using photographic material that is blackened when it is exposed to radiation; usually tritium, the radioactive isotope of hydrogen, or radioactive phosphorus are incorporated into molecules instead of hydrogen or common phosphorus; in this way, certain compounds can be traced *meth biot*

autoreduplication: Biological systems that generate the template for their own reproduction and duplicate themselves *bio*

autoregression: The generation of a series of observations whereby the value of each observation is partly dependent on the values of those that have immediately preceded it *stat*

autoregulation: A regulatory system of gene expression in which the product of a structural gene modulates its own expression *gene*

autosegregation: The occurrence of changes in the chromosome complement during the formation of the egg cell *cyto*

autosomal gene: A gene located on an autosome (i.e., a chromosome that is not a sex-determining chromosome) *gene*

autosome: Any chromosome in the cell nucleus other than a >>> sex chromosome *cyto*

autosyndesis: The pairing of complete or partial homologues of chromosomes *cyto*

autotetraploid An >>> autopolyploid with four similar genomes; if a given gene exists in two allelic forms A and a, then five genotypic classes can be formed: AAAA (quadruplex), AAAa (triplex), AAaa (duplex), Aaaa (simplex), and aaaa (nulliplex) *cyto gene* >>> Table 3

autotetraploidy: >>> autotetraploid

autotroph: >>> autotrophic

autotrophic: Cells or organisms that synthesize cell components from simple chemical substances *phys*

autotropism: The ability of plants to self-regulate a stimulated crooking of an organ in a way that the previous shape is reestablished *bot*

autozygosity: Homozygosity in which the two alleles are identical by descent (e.g., they are copies of an ancestral gene) *gene*

autozygote: A diploid individual in which the two genes of a locus are identical by descent from an ancestral gene *gene*

autumnal: Growth form for late summer and fall *bot*

autumn wood: >>> late wood

auxesis: The chemical or physical induction of cell division *bot phys*

auxetic: Any substance which induces, or increases, cell division *phys*

auxin: A hormone that promotes longitudinal growth in the cell of higher plants; in combination with cytokinin, auxin is required for the sustained proliferation of many cultured plant tissues; synthetic auxins such as 2,4-dichlorophenoxy acetic acid (2,4-D) and 2,4,5-trichlorophenoxyacetic acid (2,4,5-T) stimulate uncontrolled DNA and protein synthesis and are effective herbicides; 2,4-D is still used as a herbicide, but use of 2,4,5-T has been banned because it tends to be contaminated with carcinogenic dioxin; broad-leaved weeds like dandelions are much more susceptible to auxins than narrow-leafed plants like grass and cereal crops; at extremely low concentrations, 2,4-D is used to promote growth of orchid seedlings or *in vitro* cell cultures *chem phys phyt* >>> www.pan-uk.org/pestnews/Actives/24d.htm

auxin–cytokinin ratio: The relative proportion of auxin to cytokinin present in plant-tissue-culture media; varying the relative amounts of these two hormone groups in tissue culture formulae affects the proportional growth of shoots and roots *in vitro*; as the ratio is increased (increased auxin or decreased cytokinin), roots are more likely to be produced, and as the ratio is decreased root growth declines and shoot initiation and growth are promoted; it was first recognized by C. O. MILLER and F. SKOOG in the 1950s *biot*

auxotroph(ic): Fails to grow on a medium containing the minimum nutrients essential for the growth of the wild type *phys*

availability: Describes the amount of a nutrient and/or water in fertilizer or the soil, respectively, that a plant can immediately absorb; it can be different from the actual amount of the nutrient present *agr phys* >>> available water capacity >>> available water >>> Table 44

available water: That part of the water in the soil that can be taken up by plant roots *agr phys* >>> Table 44

available water capacity: The weight percentage of water that a soil can store in a form available to plants; it is about equal to the moisture content at >>> field capacity minus that at the >>> wilting point *agr phys*

avenacin: >>> saponin

average: A quantity or rating that represents or approximates an arithmetic mean *stat*

average effect of a gene: The change in >>> mean value of the >>> population produced by combining a gene with a random sample of gametes from the original population *gene*

avidin: A glycoprotein component of egg white that binds strongly to the vitamin >>> biotin; proteins and nucleic acids can be linked to biotin (biotinylated) and the avidin–biotin reaction can then be used in a number of assay methods, such as antigen–antibody reactions or >>> DNA hybridization (e.g., enzymes conjugated with avidin can be used to bind to biotinylated antibodies) *chem cyto micr* >>> chromosome staining

avirulence: The inability of a pathogen to infect *phyt*

avirulent: A strain of a parasite unable to infect and cause disease in a host plant *phyt*

avirulent gene: A gene that does not contribute to parasitic ability *phyt gene*

avoidance: Ability of an organism to prevent an injurious stress, pathogen, or predator from penetrating its tissues (e.g., drought avoidance may be achieved through restriction of water loss or by expansion of the root system to a greater supply of water) *phys phyt*

AVRDC: Asian Vegetable Research and Development Center

awn: The bristle-like projection arising from the top of the glume and lemma (e.g., in barley, the top of the lemma in wheat, and from the back of the lemma in oats) *bot* >>> germination >>> Figure 34

awned: >>> aristate

axenic: A pure culture of one species; it implies that cultures are free of microorganisms *biot* >>> aseptic

axil: The angle between the upper surface of a leaf and the stem that bears it *bot*

axil placentation: The type of ovule attachment within a fruit in which the seeds are attached along the central axis at the junction of the septa *bot*

axillary: In or related to the axis *bot*

axillary bud: Develops in the axil of a leaf; the presence of axillary buds distinguishes a leaf from a leaflet *bot*

axillary bud proliferation: Propagation in culture using protocols and media which promote axillary (lateral shoot) growth; a technique for mass production (>>> micropropagation) of plantlets in culture, achieved primarily through hormonal inhibition of apical dominance and stimulation of lateral branching *biot*

axillary shoot: Shoot formed axillary to the primary shoot meristem *bot phys*

axillary tiller: A tiller may form a bud located at the coleoptilar node (coleoptilar tiller) and at each crown node (axillary tiller); the coleoptilar tiller can emerge at any time, independent of the number of leaves on the main stem; axillary tillers usually begin to emerge when the plant has three leaves; rarely are more than five axillary tillers formed on a cereal plant *bot*

axis (of cereal plants): The stem or central column upon which other parts are borne; in general, the central part of a longitudinal support on which organs or parts are arranged *bot*

azacytidine (5-azacytidine): A drug ($C_8H_{12}N_4O_5$) that may activate the expression of rRNA genes by reduction of their methylation level *biot* >>> Table 43

azalea pot: Container that is half as tall as it is wide *hort meth*

azotobacter: Bacteria living in soil and water that are able to bind and incorporate atmospheric nitrogen into their cells *bio agr*

B

B1, B2, B3 …: >>> BC1, BC2, BC3 …

baby: A plantlet or seedling *hort*

baby potato: Any of various red, white, yellow, blue-purple potato varieties harvested in the early stages of growth, before maturity in order to keep them small (<3cm) and tender *breed agr* >>> www.tuckertaters.com/potato-glossary.pdf

BAC: >>> bacterial artificial chromosomes

bacciferous: Berry load-bearing, producing berries *bot*

bacciform: Berry-shaped *bot*

BAC clone(s): Useful tools as chromosome markers; in BAC-FISH karyotyping they can be even specific for chromosome arms *cyto* >>> bacterial artificial chromosome >>> BAC fingerprinting

BAC fingerprinting: A variant of the >>> restriction fragment length polymorphism technique, which allows discrimination between different clones of a bacterial artificial chromosome (BAC) library by digestion of the clones with a set of >>> restriction endonucleases *biot meth* >>> bacterial artificial chromosome >>> BAC clone

*Bacillus thuringiensis***:** A bacterium that kills insects; a major component of the microbial pesticide industry and a subject in biotechnology *biot*

backbulb: A dormant water-storing thickened stem that looks like a bulb, for example, in Orchidaceae; it grows actively as a pseudobulb the first year, then becomes dormant when the next year's pseudobulb takes over *bot hort*

backcross: A cross of an F1 hybrid or heterozygote with an individual of genotype identical to that on one or the other of the two parental individuals; matings involving a hybrid genotype are used in genetic analyses to determine linkage and crossing-over values; the portion of donor genome is reduced by subsequent backcrossing from 50% (BC0) to 25% (BC1), 12.5% (BC2), 6.25% (BC3), 3.125% (BC4) etc. *meth* >>> design of experiment >>> Figures 2, 31, 60 >>> Tables 27, 35, 53

backcross-assisted selection (BCAS): A method that allows the selection of plants carrying a favorable recessive allele at each generation, limiting the need for a progeny test, which is common in traditional backcrossing; in cases where the traditional means of selection are limited by environmental conditions (e.g., the presence of an abiotic or a biotic stress such as drought) this selection strategy is superior to conventional ones; particularly in genetic transformation approaches, where the transgenes can be used as markers, BCAS may show a considerable advantage *meth biot* >>> Tables 35, 53 >>> www.plantstress.com

backcross breeding: A system of breeding whereby recurrent backcrosses are made to one of the parents of a hybrid, accompanied by selection for a specific character *meth* >>> design of experiment >>> Figure 31 >>> Tables 27, 35, 53

backcross (donor) parent: That parent of a hybrid with which it is again crossed or with which it is repeatedly crossed; backcrosses may involve individuals of genotype identical to the parent rather than the parent itself *meth* >>> Figures 2, 31, 60 >>> Tables 27, 35, 53

backcross method: >>> backcross breeding

backcross population: In mapping studies, to analyze specific DNA fragments derived from parent A in the background of parent B, a hybrid F1 plant is backcrossed to parent B; in this situation, parent A is the >>> donor of DNA fragments and parent B is the >>> recipient; the latter is also called the >>> recurrent parent; during this process two goals are achieved: unlinked donor fragments are separated by segregation and linked donor fragments are minimized due to recombination with the recurrent parent; to reduce the number and size of donor fragments, backcrossing is repeated and, as a result, so-called advanced backcross lines are generated; with each cycle of backcrossing, the proportion of the donor genome is reduced by 50%; molecular markers help to monitor this process and to speed it up; in an analysis of the chromosomal segments retained around the *Tm2* locus of tomato, it was estimated that marker-assisted selection reduced the number of required backcrosses from 100 – in the case of no marker selection – to two; the progeny of each backcross is later screened for the trait introduced by the donor; in the case of dominant traits, the progeny can be screened directly; in the case of recessive traits, the selfed progeny of each backcross plant has to be assessed >>> mapping population >>> Tables 27, 35, 53

backcross ratio: The proportion of heterozygotes to >>> bottom recessives expected in a backcross is given by calculating "2^n" where "n" is the number of factors involved, e.g., with 2 factors $2^n = 4$ and the expected ratio is 3:1 with 3 factors $2^n = 8$ and the expected ratio is 7:1, etc. *gene* >>> Tables 27, 35, 53

backcrossing: >>> backcross

backfill: Filling in around roots in a planting hole with a soil mix for better establishing the plant *meth*

backhoe: A shovel mounted on the rear of a tractor, hydraulically operated to dig trenches or pits in soil *agr*

backmutation: A reverse mutation in which a mutant gene reverts to the original standard form and/or wild type; it is rare to forward mutations, but often strongly selected for; the >>> AMES test relies on backmutation for the detection of mutagens *gene* >>> chimera >>> Figure 56

backward selection: Selection of parent plants based on results from a progeny test *meth*

BAC library: >>> bacterial artificial chromosome library

BAC map: The ordered alignment of >>> BAC clones such that a physical map of the genome is constructed *biot meth*

BAC vector: An Escherichia coli vector for >>> DNA fragments; larger than >>> cosmids; alternative to >>> YAC vectors *biot*

bacterial artificial chromosome (BAC): Pieces of plant DNA that have been cloned inside living bacteria; they can be used as probes to detect complementary DNA sequences within large pieces of DNA via hybridization techniques, or for marker-assisted selection by faster selection of segregant-bearing genes for a particular trait and to develop future crop varieties faster *biot* >>> BAC fingerprinting

bacterial artificial chromosome library (BAC library): One of the molecular tools in plant science to break complex genomes into manageable pieces for study; most living organisms have large numbers of genes in their genomes; bacteria have between 5,000 and 8,000, humans 35,000, and wheat 150,000; the large collection of genes of an entire genome can be broken into small fragments, with one or two genes per fragment, and inserted into pieces of DNA; when the bacterial artificial chromosome is placed again into bacteria they make many hundreds of copies of themselves, including the foreign genes introduced; building a library of genes in this way is the first step to studying the structure of individual genes, understanding what it is they do in the plant and how they interact with other genes to control plant development and metabolism; the BAC library consists of 1.2 million BAC clones with a size of between 100 and 200 kb; the genome of wheat is about five times larger than that of humans and includes a total of 150,000 genes *biot* >>> BAC fingerprinting

bacterial blight disease: A plant disease spread by bacteria that causes death of leaves, spotting of bean pods, and discoloration of seed; for example, in rice, the causal agent is *Xanthomonas oryzae* var. *oryzae*; symptoms are lesions usually starting near the leaf tips or leaf margins or both, and extending down the outer edge; young lesions are pale to grayish green, later turning yellow to gray (dead) with time; in very susceptible varieties, lesions may extend the entire leaf length into the leaf sheath *phyt*

bacterial diseases: Diseases caused by specialized bacteria *phyt* >>> disease

bacterial ring rot (disease of potato tubers): It is caused by *Corynebacterium sepedonicum*, that forms characteristic breakdown of the vascular ring within the tuber; this ring often appears as a creamy-yellow to light-brown, cheesy rot; the outer surface of severely diseased tubers may show slightly sunken, dry, and cracked areas; symptoms can be less obvious than described above, appearing as only a broken, sporadically appearing dark line or as a continuous, yellowish discoloration; in the past, bacterial ring rot was a huge problem in potato production *phyt*

bacterial soft rot (of potato): A disease caused by the bacterium *Erwinia carotovora* subsp. *carotovora*; this bacterium has an extensive host range, including most fleshy vegetables; it survives readily in soil and surface waters; these bacteria are capable of multiplying and persisting in the root zones of many host and non-host crop and weed species; bacterial soft rot occurs on a wide range of crops and is one of the most severe postharvest diseases of potatoes worldwide; loss may occur during storage, transport, or marketing; all potatoes varieties are susceptible; contamination of potato tubers occurs anytime they come into contact with the bacterium; it invades the potato tuber chiefly through wounds; most of the soft rot infections are in tissues that have been weakened, invaded, or killed by pathogens or by mechanical means; soft rot in tubers is favored by immaturity, wounding, invasion by other pathogens, warm tuber and storage temperatures, free water, and low oxygen conditions *phyt*

bactericidal: Killing or hampering bacteria *phyt*

bacteriocide: A chemical compound that kills bacteria *phyt*

bacteriocins: Bactericidal substances produced by certain strains of bacteria and active against some other strains of the same or closely related species *phyt*

bacterioid: Bacteria cells that are not normal-shaped, usually found in root nodules of legumes *agr*

bacteriology: The branch of science for bacteria *phyt*

bacteriolysis: The lysis of bacterial cells, usually induced by antibodies formed by the host organism *phyt*

bacteriophage: A virus that infects bacteria; consists of a polyhedral head containing DNA or RNA enclosed in a protein coat (e.g., the bacteriophages T4, M13, P1, and PS8 are used in genetic engineering) *biot*

bacteriosis: >>> bacterial diseases

bacteriostatic: A chemical or physical agent that prevents multiplication of bacteria without killing them *phyt*

bagasse: The crushed stalks of sorghum and sugarcane after extracting the juice by heavy grooved rollers – that is, the fine organic material removed from vacuum drums during processing;

because it is of high organic matter and sugar content, bagasse makes a good inoculant carrier, but bagasse must first be sterilized to stop the growth of fungi and other contaminant organisms *agr phyt*

bagging: >>> caging

bag storage: Storing of grain in bags, usually made of jute (gunny) or polyethylene *seed meth*

bait: A material used to lure insects; it is often added to pesticides (e.g., against snails) *meth*

Bakanae disease: Seedling disease of rice caused by fungus-producing gibberellins *phyt* >>> gibberellin

Baker's mealy bugs: Sometimes called grape mealy bugs; a common species of mealy bug known to feed on some ornamental plants, Pseudococcus maritimus *phyt hort*

baking potato: A widely used category of tablestock potatoes; in the U.S., such potatoes will normally have russeted skin and white flesh, but they may also be found with white or yellow skin or with yellow flesh. The important factor is not skin or flesh color, but the high specific gravity which makes them high in starch and gives them a dry, mealy texture; the flesh becomes light and fluffy when cooked; they are ideal for baking, mashing, and French fries; popular varieties for baking are "Russet Burbank", "Russet Norkotah", "Russet Arcadia", "Norgold Russe", "Goldrush, White Rose", etc. *agr*

baking test: A procedure for estimating the overall quality of a flour or meal, with respect to its performance in the bakery and the characteristics of bread made from it *meth*

balanced design: An experimental design in which all treatment combinations have the same number of observations; there are various different kinds of balance (variance balance, efficiency balance, etc.), none of which are equivalent to the combinatorialist's pairwise balance, but become equivalent to it if some extra conditions are satisfied (e.g., in binary equireplicate uniform block designs) *stat* >>> design of experiment

balanced diallelic: The genotype involving a multiple allelic locus in an autotetraploid where two different alleles are represented an equal number of times (e.g., *a1a1a2a2*) gene >>> design of experiment

balanced fertilizer: Fertilizer which supplies approximately equal percentages of the primary elements *phys agr*

balanced incomplete block design (BIB): A design in which one constant value for the residual variance of the difference between candidates for all pairs of candidates is indicated *stat meth* >>> design of experiment

balanced lattice: A special group of balanced incomplete block; allows incomplete blocks to be combined into one or more separate complete replicates *stat meth* >>> design of experiment

balanced lethals: Recessive lethals at different loci, so that each homologous chromosome carries at least one lethal, and associated with inversions, so that no recombination occurs between the homologous chromosomes *gene*

balanced lighting: Refers to a light source that provides both red and blue light; balanced lighting is important because some plants require both red and blue light to grow and flower *phys hort*

balanced polymorphism: A genetic polymorphism that is stable, and is maintained in a population by natural selection, because the heterozygotes for particular alleles have a higher adaptive fitness than either homozygote; it is referred to as overdominance, as opposed to underdominance, where the heterozygotes have a lower fitness, giving rise to unstable equilibrium *gene*

balanced tertiary trisomic (BTT): A specific interchange trisomic spontaneously selected or experimentally designed in a way that it is heterozygous (Aaa); its trisomic progeny after selfing is genetically similar to the parent; the dominant allele is present on the translocated chromosome linked to the break point; BTTs were thought to be used for hybrid seed production in barley; the American variety "Hembar" was the first commercial hybrid barley; it was first grown in the U.S. in 1969–1970 and produced by the BBT system from

RAMAGE (1965), crossing a male sterile diploid with the cultivar "Arivat"; the line 63-j-18-17 used in the hybrid was a BTT 27d msg2, that is, the extra chromosome is the translocated chromosome 27d with the centromere of chromosome 2; usually, it was difficult to increase trisomics because the BTTs were weak; if the BTT plants were vigorous, then it was difficult to eliminate them in the F1 crossing block; the progeny of the BTTs segregate for trisomics and sterile diploids, which would appear in the F1 hybrid; with the occurrence of male sterile plants there is a greater risk for >>> ergot; these problems appeared with some of the BTTs that were produced in Sweden and Germany from new combinations of translocations and male steriles; in order to try to eliminate the rouging of diploids in trisomic propagations, vigorous BTTs were treated with mutagens; the objectives were twofold: (1) the first was to induce a seedling lethal that was closely linked to a male sterile gene, (2) the second objective was the induction of a pollen lethal on the extra chromosome to eliminate its transmission through the pollen; linkages of male steriles with seedling lethals and decreased transmission of the extra chromosome were found; unfortunately, none of these linkages were considered to be sufficiently close enough for large scale use; although the genetic system was available – allogamy can be selected among barley accessions – the problem of the heterotic effect of combining ability was the greatest limitation (e.g., as compared to maize); hybrids produced by this system were last marketed in 1978; the hybrids were not able to compete with the new high-yielding short stiff-strawed varieties *cyto meth* >>> Figures 14, 51, 55

balancing selection: Selection involving opposing forces in which selective advantages and disadvantages cancel each other out; heterozygote advantage (or overdominant selection) is an example in which an allele selected against in the homozygous state is retained because of the superiority of heterozygotes; other balanced states may occur including when: (1) an allele is favored at one developmental stage and is selected against at another (antagonistic pleiotropy), (2) an allele is favored in one sex and selected against in another (sexual antagonism), and (3) an allele is favored when it is rare and selected against when it is common (negative frequency dependent selection) *gene* >>> Table 36

bale: A standard bundle of agricultural merchandise such as straw, cotton, or paper usually pressed or bound; the volume and weight of a bale depends on the commodity and locality *agr*

baler: A machine that picks up dry hay or straw after harvest and bundles it into big rectangular or round bales; the bales are tied together with baler twine *agr*

ballistic fruits: >>> catapult fruits

ball metaphase: A form of mitosis with characteristically clumped chromosomes *cyto*

balm: >>> balsam

Bal31 nuclease: An exonuclease that degrades both strands of a DNA molecule at the same time *biot* >>> endonuclease >>> exonuclease >>> Table 39

balsam: A mixture of resins and ethereal oils of sticky consistency, secreted by some plants *micr*

band: Specific heterochromatic regions along a chromosome that can be stained by different banding methods *cyto*

band application: The spreading of fertilizer or other chemicals over, or next to, each row of plants in a field, as opposed to broadcast application *meth agr*

banding: A special staining technique for chromosomes, which results in a longitudinal differentiation (e.g., Giemsa staining, which is a complex of stains specific for the phosphate groups of DNA); C-bands represent constitutive heterochromatin; G-banding patterns on mitotic chromosomes correspond very closely to the chromomere patterns of meiotic chromosomes bivalent at pachytene, that is, the chromatin that is densely packed enough to see; chromomeres appear to be the focus of chromatin condensation along the chromosome, and may be the sites where chromatin condensation is initiated; G-bands may be the corresponding initiation sites for chromosome condensation during mitosis; chromomeres may be more apparent in meiosis because of the relative degree of condensation of the chromosomes; meiotic chromosomes are very extended and the homologues are paired which enhances chromomere patterns; mitotic chromosomes are overall more condensed, masking chromomeres and requiring banding pretreatment to reveal them; on mitotic prophase chromosomes, up to 2,000 or more bands can be observed; the bands on these high-resolution chromosomes correspond to meiosis chromomeres; as the chromosomes condense in the prophase to metaphase transition, there is a progressive coalescence of bands such that each band and its component subbands retain the same relative location and staining intensity; the bands observed at metaphase are made up of a collection of smaller subbands; in plant mitotic chromosomes, C-banding but no G-banding has been produced, although pachytene chromosomes of plant meiotic cells demonstrate chromomeres; one suggestion is that this is due to the much greater quantity of DNA per unit length in plant chromosomes compared to those in animals; the chromomeres are too tightly packed to be resolved by G-banding; the reason is still not clear as compaction seems to be roughly equivalent *cyto*; in agriculture, placing fertilizer in continuous narrow bands and then covering it with soil *agr*; in horticulture, encircling part of a plant (e.g., a trunk) or a portion of a garden with some type of material that traps, kills, or keeps out pests (e.g., poisonous baits or copper stripping) *hort meth* >>> C banding >>> G banding >>> N banding >>> Table 43

banding pattern: The linear pattern of deeply stained bands and weakly staining interbands that results from more or less defined local differences in the degree of DNA compaction along the chromosome *cyto* >>> C banding >>> G banding >>> N banding >>> Table 43

band-pass: A microscopic filter that passes light of a certain restricted range of wavelength *micr*

band-seeding: Placing forage crop seed in rows directly above but not in contact with a band of fertilizer *agr*

barb: A stiff bristle or hair terminating an awn or prickle *bot*

barbate: Having one or more groups of hairs; bearded *bot*

barbed: Having sharp points or hooks such as those on awns of wheat or barley *bot*

BARC: >>> Bioinformatics Applications Research Center

bar chart: >>> histogram

bar code: A pattern of light and dark lines on labels that can be read by a light pen for direct entry into a computer; used for tagging and labeling of plants, seed accessions, etc. *meth* >>> DNA bar coding

barcoding: >>> DNA bar coding

bar diagram: >>> histogram

bare root: A plant that is sold or shipped dormant with no soil surrounding its roots *hort*

bare root transplant: A soilless young plant lifted from a seedbed for transplanting, as opposed to a module or potted plant *hort meth*

bar gene: A gene from Streptomyces hygroscopices that encodes the enzyme phosphinothricin acetyltransferase; it confers resistance to "Bialaphos" herbicide; used in genetic transformation studies as a marker gene for selection of successful transformants *biot*

bark: The outer skin of a tree trunk, outside the secondary, vascular cambium; it is composed of phloem tissue, which occurs as living inner and dead outer zones; the outer zone is penetrated by the cork layers formed from the cork cambia *bot*

bark ringing: A method used for forcing fruit trees to flower; a complete ring is cut around the trunk below the lowest branch and another ring is cut right below the first; the bark between the rings is removed; the scar should be covered with grafting wax *meth hort*

barleycorn: A traditional English unit of length equal to one third of an >>> inch; use of seeds as average units of weight or length was common in societies based on agriculture; in the Anglo-Saxon era, barley was an important subsistence crop and barleycorns were used as units of length; three barleycorns "smooth and round laid end to end" were equal to the

"Saxon once" (inch); barleycorns, or grains, became the basis of all English weight systems *meth agr*

barley yellow dwarf virus (BYDV): Infects all cereal species but barley and oats are usually more severely affected than wheat; plants are most vulnerable to infection early in growth; infection results in stunting, discoloration, and substantial yield loss; the virus is transmitted by several species of cereal aphids (mainly bird-cherry aphid, *Rhopaosiphum padi*, or grain aphid, *Sitobion avenae*) *phyt* >>> aphid vector

barn: A building for storing straw, hay, grain, etc. *agr*

barnase gene: >>> gametocide

barren: Sterile or otherwise incapable of reproducing *bot*

barren glume: >>> spikelet glume

basal: Referring to the base or located there; the lower portion of a structure *bot*

basal area: Total area of ground covered by trees measured at breast height, or actual surface area of soil covered by a plant measured close to the ground *eco*

basal cell: The lowermost cell formed after the first division of the zygote *bot*

basal stem cutting: Cutting taken from the base of (a usually herbaceous) plant, mostly in spring *hort*

basal node: The node or joint at the base of the stem *bot*

basal placentation: A type of free-central placentation in which the seeds are attached at the bottom of the central ovary axis *bot*

basal plate: The bottom of a >>> bulb; new bulbs or bulblets can grow from the basal plate *hort*

basal rosette: In some plants, a cluster of leaves around the stem on or near the ground *bot*

basal treatment: An application to the stems of plants at or just above the level of the soil *phyt agr*

base: A chemical compound that reacts with an acid to give water (and a salt); a base that dissolves in water to produce hydroxide ions is called an alkali chem; in biology, a cyclic, nitrogen-containing compound that is one of the essential components of nucleic acids; it exists in five main forms (>>> adenine, A; >>> guanine, G; >>> thymine, T; >>> cytosine, C; >>> uracil, U); A and G have a similar structure and are called purines; T, C, and U have a similar structure and are called pyrimidines; a base joined to a ribose sugar joined to a phosphate group is a nucleotide – the building block of nucleic acids *bio* >>> Table 38

base grade: A selected grade of cotton used by cotton merchants as a basis for contracts, premiums, and discounts *agr meth*

base analogues: A purine or pyrimidine base that differs slightly in structure from a normal base, but that because of its similarity to that base may act as a mutagen when incorporated into DNA (e.g., uracil, 5-bromouracil, 5-fluorouracil, 5-methylcytosin, 5-bromocytosin, hypoxanthin) *chem gene* >>> Table 38

base collection: A collection of germplasm that is kept for long-term, secure conservation and is not to be used as a routine distribution source; seeds are usually stored at subzero temperatures and low moisture content *meth*

base pair (bp): The nitrogenous bases (adenine-thymine/uracil; guanine-cytosine) that pair in double-stranded DNA or RNA molecules; 1,000 bp = 1 kb *gene* >>> Table 38

base pair map: A physical map of a genome, or parts of it, for which the sequence is known base by base *meth biot*

base pairing: A complementary binding by means of hydrogen bonds of a purine to a pyrimidine base in nucleic acids *gene* >>> base

base population: The initial set of genotypes from which selections will be taken to establish a breeding population, for example, the wild forest; it is also sometimes referred to in the same meaning as recruitment population (in the first generation they are actually the same, and it can be referred to as one of the populations in a stratified tiered structure) *meth* >>> Figures 51, 60

base seed: Particularly valuable seeds, usually derived from highly productive single plants (elite plants), which are used for seed production of commercially grown material; seed stock produced from breeder's seed by, or under the control of, an appropriate agricultural authority; the source of certified seed, either directly or as registered seed *seed* >>> Table 28

base sequence analysis: A technique, often automated, for determining the base sequence of DNA, RNA, etc. *meth biot*

base substitution: A form of mutation in which one of the bases in DNA replaces another *gene*

basic chromosome set: The standard chromosome number of a given species *gene cyto*

basic form: >>> primitive form

basic local alignment search tool (BLAST): A bioinformatics tool to find matches between a DNA sequence and known sequences stored in databases *biot meth*

basic number (of chromosomes): The haploid number of chromosomes in diploid ancestors of polyploids, represented by "x" *gene cyto* >>> basic chromosome set >>> genome

basic research: Research conducted for the purpose of advancing knowledge about biological behavior with little concern for the immediate or practical benefits that might result *meth bio*

basic seed: >>> base seed >>> elite >>> super-elite >>> Table 57

basidiospore: From a basidium of Basidiomycetes-produced haploid spore that is formed after meiosis and exogenously laced up from a steringma *bot phyt*

basidium: A stand-like cell, mostly club-shaped, from which exogenously laced up haploid spores after >>> karyogamy *bot*

basifixed: Anthers attached to the filaments by their bases *bot*

basmati rice: A variety of long grain rice, famous for its fragrance and delicate flavor; its name means "the fragrant one" in Sanskrit, but it can also mean the "soft rice"; India and Pakistan are the largest cultivators and exporters of this rice, primarily grown through paddy field farming in the Punjab region *seed* >>> rice

bast: Any of several strong, woody fibers, such as flax, hemp, ramie, or jute, obtained from phloem tissue *bot* >>> phloem

bast plant: Crop plants used for fiber production, such as flax or hemp *agr*

bastard: The product of crossing two sperm cells of genetically different constitution gene >>> hybrid

batch culture: A cell suspension grown in liquid medium of a set volume; inocula of successive subcultures are of similar size and cultures contain about the same cell mass at the end of each passage; cultures commonly exhibit five distinct phases per passage (a lag phase follows inoculation, then an exponential growth phase, a linear growth phase, a deceleration phase, and finally a stationary phase) *biot*

batch drying: Drying seeds in relatively small quantities held in a stationary position (as opposed to drying in a continuous moving line) *meth seed*

Batesian mimicry: A form of mimicry typified by a situation where a harmless species has evolved to imitate the warning signals of a harmful species directed at a common predator; it is named after the English naturalist H. W. BATES (1825–1892); mimicry may broadly be defined as imitation or copying of an action or image; in biological systems, mimicry specifically refers to the fascinating resemblance of an organism, called the "mimic", to another somewhat distantly related organism, called the "model"; the set of mimic and model species involved is often referred to as a mimicry complex; usually through escape from predation, the mimicry of a trait or traits helps the mimic to survive; this, coupled with the fact that the resemblance traits are genetically based, implies that mimicry complexes have been shaped by natural selection; there are two major types of mimicry, Batesian and Müllerian *eco* >>> Vavilovian mimicry

B–A translocation: Spontaneous or induced interchanges between A chromosomes and B chromosomes, e.g., found in rye *cyto*

Bayesian analysis: A statistical approach for constructing phylogenetic trees related to maximum likelihood that operates on a priori weighting factors and probabilities; in Bayesian mapping everything is treated as an unknown variable with a prior distribution; a variable can be classified into one of two classes: observables and unobservables; the observables include data (phenotypic values, marker scores, pedigrees, etc.), the unobservables include parameters; in mapping studies, the Bayesian shrinkage estimation method or the penalized maximum likelihood method can be applied *stat meth*

Bayoud disease (of date palm): A vascular wilt caused by the fungus *Fusarium oxysporum f. sp. albedinis* in the dioecious monocotyledonous date palm *phyt* >>> Figure 62

BC1, BC2, BC3, etc.: Symbols indicating the first, second, third, etc., backcross or backcross generation *meth*

BCAS: >>> backcross-assisted selection

B chromosome: Any chromosome of a heterogeneous group of chromosomes present in several plant species, which differ in their morphology, numerical variation, meiotic pairing, and mitotic behavior from normal A chromosomes; they are also called supernumerary chromosomes, accessory chromosomes, or extra chromosomes; a B chromosome derives from the A chromosome complement by aberrant division processes and subsequent modifications; up to 12 and more B chromosomes have been observed in addition to the diploid A chromosome complement (e.g., in rye) *cyto*

beak: The extension of the keel at the tip of the glume or lemma in wheat *bot*

bean golden mosaic virus (BGMV): The causal agent of bean golden mosaic of common beans; a transgenic bean line that has been developed based on RNA interference to silence the BGMV *rep* gene shows immunity to the virus; crosses were done between a transgenic line and six bean cultivars followed by four backcrosses to the commercial cultivars "Pérola" and "BRS Pontal"; the transgene locus was consistently inherited from the crosses analyzed in a Mendelian fashion in the segregating populations; the disease resistance reaction co-inherits with the transgene; nevertheless, the expression of disease resistance displayed a dosage effect phenomenon in the F_1 generation; the analysis of the homozygous near-isogenic lines in field conditions, under high BGMV disease incidence, indicates that the transgenic lines were completely resistant *phyt biot* >>> gene silencing

bean pod weevil (Apion godmani): A serious insect pest of common beans (*Phaseolus vulgaris*) grown in Mexico and Central America that is best controlled by host-plant resistance, e.g., available in the varieties "Durango" or "Jalisco" *hort phyt*

bean rust: A disease of beans caused by the fungus *Uromyces appendiculatus phyt*

bean dwarf mosaic virus (BDMV): A single-stranded DNA virus (genus Begomovirus, family Geminiviridae) that infects common beans (*Phaseolus vulgaris*) and causes stunted plant growth, and mosaic and mottle symptoms in leaves; BDMV shows differential pathogenicity in common beans, infecting germplasm of the Andean gene pool (e.g., the snap bean cultivar "Topcrop"), but not that of the Middle American gene pool (e.g., the pinto bean cultivar "Othello"); for example, resistance to BDMV in "Othello" is associated with development of a hypersensitive response in vascular (phloem) tissues *phyt*

bean yellow dwarf virus (BYDV): A disease occurring in French beans that can cause up to 90% losses in yield *phyt*

bearded: Bearing long, stiff hairs, for example, at culm nodes of *Dichanthium bot*

beat(ting) up: Restocking failed areas in a crop or stand by further sowings or plantings; there are several other terms in use, for example, "blanking", "filling", "gapping", "infilling", "recruiting", "reinforcement planting"; in forestry, it means to replace dead trees with new ones, especially during the early years of the establishment or reestablishment of a plantation *fore*

Beaumont period: Period of weather required for potato blight *Phytophthora infestans* infection *phyt*

Becquerel (Bq): The SI unit of radioactivity; the unit is named after the discoverer of radioactivity, A. H. BECQUEREL *phy*

bed: An area within a garden or lawn in which plants are grown; in agriculture, a narrow flat-topped ridge on which crops are grown with a furrow on each side for drainage of water or an area in which seedlings or sprouts are grown before transplanting *hort agr*

bedding plant: A plant grown for its flowers or foliage that is suited by habit for growing in beds or masses *hort*

beet: Any of various biennial plants of the genus Beta, of the goosefoot family, especially *B. vulgaris*, having a fleshy red or white root and dark-green red-veined leaves; sugarbeet derived from *B. vulgaris* by selection for high sugar content *bot hort*

beet yellows virus (BYV): A virus disease that can dramatically reduce yield from sugarbeet crops *phyt*

behavior flexibility: All means of plant behavior permitting temporary adaptation to environmental conditions *eco*

behavior genetics: A branch of genetics dealing with the inheritance of different types and/or forms of behavior *gene*

bell: A bloom type; it describes a single flower that has an elongated corolla tube giving the flower a bell-shaped appearance *hort*

belowground biomass: >>> biomass

belt conveyor: Endless band used for conveying grain or products in a continuous stream *meth*

bench setting: Height that the cutting plane of a mower is set above a hard, level surface *agr*

beneficial insect: An insect that serves the best interest of man – for example, insect pest predators and parasitoids, and pollinating insects phyt *agr hort*

Benlate: Popular trademark for a fungicide containing benomyl, a systemic fungicide; it is used to treat Botrytis, powdery mildew, Rhizoctonia, and root rot; benomyl is classified for general use by the U.S. Environmental Protection Agency (EPA) *phyt hort* >>> Figure 62 >>> Table 51 >>> www.epa.gov/agriculture/lfra.html

BERG micromethod: A method for characterization of milling quality in cereals – for example, 10 g of wheat of normal water content is ground to a meal, which is then sifted for 10 min on a sieving machine with three bolting cloths; No 9, 12, and 15 ("silk cloth XX") representing a mesh width of 150, 105, and 75, respectively; the flour fraction in g under cloth 15 gives the unit of BERG and has proved very useful for determination of the grittiness; it was used to a great extent in the wheat breeding work at Weibullsholm, Sweden, during the 1950–1960s *meth*

berry: A simple, fleshy, or pulpy and usually many-seeded fruit that has two or more compartments and does not burst open to release its seeds when ripe (e.g., banana, tomato, potato, grape) *bot*

berry-bearing: >>> bacciferous

berry-shaped: >>> bacciform

best linear prediction (BLP): A statistical method that utilizes matrix algebra to predict the breeding values for any trait or selection index; in BLP fixed effects are assumed to be known; BLP is especially suited for analyses of messy or unbalanced data *meth stat*

best linear unbiased prediction (BLUP): A statistical method that predicts breeding values for any trait or selection index; this method was originally developed in animal breeding for the estimation of breeding values and is now widely used in many areas of research; it has become the most widely used method for genetic assessment of perennial species, and it is potentially relevant for annual crops; based on breeding values predicted by BLUP analysis, the most efficient selection procedure is mass selection; it does not, however, seem to have gained the same popularity in plant breeding and variety testing as it has in animal breeding; in plants, application of mixed models with random genetic effects has up until recently been mainly restricted to the estimation of genetic and non-genetic components of variance, whereas estimation of genotypic values is mostly based on a model with fixed effects; the latest experiments demonstrate that BLUP has good predictive accuracy compared to other procedures; while pedigree information is often included via the so-called numerator relationship matrix (A), it is suggested that it is frequently straightforward to exploit the same information by a simple mixed model without explicit reference to the A-matrix *meth stat*

beta-amylase activity: Catalysis of the hydrolysis of 1,4-alpha-glucosidic linkages in polysaccharides to remove successive maltose units from the non-reducing ends of the chains phys

beta-glucan: A component of soluble fiber in cereals; oat is very rich in beta-glucan, it lowers serum cholesterol in humans; development of (cereal) cultivars with greater groat (caryopsis) beta-glucan content may increase the nutritional and economic value of the crop *phys agr*

betacyanin: >>> table beet

beta-DNA: The normal form of DNA found in biological systems, which exists as a right-handed helix *gene*

betalain: >>> table beet

betaxanthin: >>> table beet

bevel (of lemma): A depression variable in depth in the base of the lemma, rounded in barley, transverse in oats *bot*

BGMV: >>> bean golden mosaic virus

biallelic map: >>> diallelic map

bias: A consistent departure of the expected value of a statistic from its parameter *stat*

biased: >>> bias

Bibliomics: A subset of high-quality and rare information, retrieved and organized by systematic literature-searching tools from existing databases, and related to a subset of genes functioning together *biot*

bi-cropping: A method of growing cereals in a leguminous living mulch; it could potentially reduce the need for synthetic inputs to cereal production while preventing losses of nutrients and increasing soil biological activity; also a low-input production system for cereals *agr meth*

bidirectional replication: When two >>> replication forks move away from the same origin in different directions *cyto*

biennial: A plant that lives for two years; during the first season food may be stored for use during the flower and seed production in the second year *bot* >>> annual >>> perennial

biennial crop: >>> biennial

bifid: Apex with one cleft or having two teeth *bot*

bifloral: Showing two flowers *bot*

bifoliate: Showing two leaves *bot*

BIFS: >>> biologically integrated farming systems

bifurcate: Split in two; forked *bot*

bigeneric hybrid: Offspring derived from crossing two different genera *meth bree*

bigerm: Having two seeds *bot*

bilateral sexual polyploidization: Formation of polyploid progeny where >>> unreduced gametes are produced by both parents *cyto bot* >>> unilateral sexual polyploidization

bilateral symmetry: Composed of two corresponding halves, each a mirror image of the other; it is usually used to describe flowers *bot hort*

biliabate: Being divided into two upper lobes, e.g., the flower of *Anthirrinum majus bot*

bimitosis: The simultaneous occurrence of two mitoses in binucleate cells *cyto*

bimodal distribution: A statistical distribution having two modes *stat*

binary: This term has several meanings, associated with the number 2 or the set (0,1) in some way; in statistical design theory, a >>> block design is binary if no treatment occurs more than once in a block (that is, a part of the treatment partition and a part of the block partition meet in at most one plot); if this holds, then a block can be thought of as a set (rather than a multiset) of treatments, and the block design can be represented as an incidence structure *stat*

binary fission: Mode of reproduction not involving any sex but division of a parent cell into two equally sized offspring *gene*

binary scale: A scale for scoring data where there are only two possible responses *meth*

binary vector system: A two-plasmid system in *Agrobacterium tumefaciens* for transferring into plant cells a segment of T-DNA that carries cloned genes; one plasmid contains the virulence gene (responsible for transfer of the T-DNA), and another plasmid contains the T-DNA borders, the selectable marker, and the DNA to be transferred *biot*

binder: A machine which cuts and binds a crop in bundles *agr*

binemic: Chromosomes that contain two DNA helices per metaphase chromatid *cyto*

binomial distribution: A calculation that measures the likelihood of events taking place where the probability is measured between 0 (the event will certainly not occur) and 1 (the event is absolutely certain) *stat*

binomial expansion: The probability that an event will occur 0, 1, 2, ..., n times out of n is given by the successive terms of the expression $(p + q)^n$, where p is the probability of the event occurring, and $q = 1 - p$ *stat*

binomial nomenclature: The system of naming organisms using a two-part Latinized (or scientific) name that was devised by the Swedish botanist C. LINNAEUS (1707–1778); the first part is the generic name (genus), the second is the specific epithet or name (species); the Latin name is usually printed in italics, starting with a capital *tax bot*

binucleate: Cells with two nuclei *cyto*

bioactive molecule: Showing a direct or measurable effect on living tissues and cells *biot*

bioassay: The use of living cells or organisms to make quantitative and qualitative measurements; in cell biology and molecular genetics, determination of the effectiveness of a compound by measuring its effect on plants in comparison with a standard preparation *meth biot*

biocatalysis: >>> biotechnology

biocatalyst: A biological substance used to cause a particular chemical or biochemical reaction *phys chem*

biochemical genetics: A branch of genetics dealing with the chemical nature of hereditary determinants *gene*

biochemical profiling: The global analysis of cellular metabolites and other low m.w. organic molecules produced by cells and organs; the methods that are employed to separate and identify these molecules are liquid chromatography, gas chromatography, and mass spectrum analysis; analyses enable understanding of the impact of various interventions (genes, drugs) on cellular function *biot*

biochemical trait: The traits associated with assay of either the affected metabolite profile or enzymatic activity or biological processes *phys*

biochemistry: The chemistry of life; the branch of chemistry that is concerned with biological processes *chem*

biochip: An electronic device that uses organic molecules to form a semiconductor; first used with regard to an electronic device that utilizes biological molecules as the "framework" for other molecules that act as semiconductors, and functions as an integrated circuit; during the 1990s, this term also became commonly used for various screening procedures, e.g., to analyze very small samples of DNA, to assess the impact of specific cellular receptors (ligand-receptor response of cell), to size and sort DNA fragments (genes) via the (proportional) fluorescence of dyes intercalated in the DNA molecules, to detect presence of a specific DNA fragment (gene) via hybridization to a probe (that was fabricated onto the "chip"), to size and sort protein molecules, to screen for compounds that act against a disease (e.g., by applying antibodies linked to fluorescent-molecules, then measuring electronically the fluorescence that is triggered by antibody-binding), to conduct gene expression analysis by measuring fluorescence of messenger RNA (specific to which particular gene is "turned on") when that mRNA hybridizes with DNA (from genome) on the hybridization surface on chip *biot*

biocide: A natural or synthetic substance toxic to living organisms *phyt*

biocoenosis: A community of organisms and its interaction with abiotic factors of habitat *eco*

biocontainment: >>> biological containment

biocontrol: >>> biological control

biodegradable plastics: Plastics that can be decomposed by microorganisms *biot*

biodiversity: The existence of a wide variety of species (species diversity), other taxa of plants, or other organisms in a natural environment or habitat, or communities within a particular environment (ecological diversity), or of genetic variation within a species (genetic diversity); genetic diversity provides resources for genetic resistance to pests and diseases; in agriculture, biodiversity is a production system characterized by the presence of multiple plant and/or animal species, as contrasted with the genetic specialization of monoculture; the importance of shrinking biodiversity was demonstrated in a genome-wide examination of 75 Canadian hard red spring wheat (*Triticum aestivum*) cultivars released from 1845 to 2004; using 370 >>> simple sequence repeat (or >>> SSR) markers that were widely distributed over all 21 wheat chromosomes, a total of 2,280 SSR alleles were detected; allelic reduction occurred in every part of the wheat genome, and a majority of the reduced alleles resided in only a few early cultivars; significant allelic reduction started in the 1930s; considering 2,010 SSR alleles detected in the 20 earliest cultivars, 38% of them were retained, 18% were new, and 44% were lost in the 20 most recent cultivars; the net reduction of the total SSR variation in 20 recent cultivars was 17%; this clear-cut evidence not only supports the contention that modern plant breeding reduces the genetic diversity of wheat, but also underlies the need for conserving germplasm and introducing genetic diversity into breeding (cf. FU & SOMERS 2009) *evol eco agr* >>> Figure 63

Biodiversity International: In effect from December 1, 2006, IPGRI and INIBAP operate under the name Bioversity International, "Bioversity" for short; this new name echoes the strategy, which focuses on improving people's lives through biodiversity research *org* >>> www.bioversityinternational.org

bioeconomy: Refers to all economic activity derived from scientific and research activity focused on understanding mechanisms and processes at the genetic and molecular levels and its application to industrial process; it is often used interchangeably with biotechonomy; the term is widely used by regional development agencies, international organizations, and biotechnology companies; it is closely linked to the evolution of the biotechnology industry; the ability to study, understand, and manipulate genetic material has been possible due to scientific breakthroughs and technological progress; the evolution of the biotechnology industry and its application to agriculture is a classic example of bioeconomic activity; the term was first defined at the Genomics Seminar in the American Academy of Arts and Sciences (AAAS) meeting by J. ENRIQUEZ and R. MARTINEZ in 1997 (ENRIQUEZ 1998) *biot meth*

bioethics: A field of study and counsel concerned with the implications of certain genetic and medical procedures, such as organ transplants, genetic engineering, and care of the terminally ill *bio*

bioenrichment: A bioremediation strategy that involves adding nutrients or oxygen, thereby bolstering the activity of microbes as they break down pollutants *biot phys*

bioflavonoid: >>> catechin

biofortification: The process of breeding food crops that are rich in bioavailable micronutrients; these crops fortify themselves – that is, they load high levels of minerals and vitamins in their seeds and roots, which are then harvested and eaten by animals or humans; through biofortification, breeders can provide farmers with crop varieties that naturally reduce anemia, cognitive impairment, and other nutritionally related health problems, potentially for use in Third-World countries *agr*

bio-fuels: Non-fossil fuels, produced from agriculture sources, residues, and waste; for example, bio-ethanol refers to ethanol produced from crops (maize-ethanol, sugar-ethanol) *agr*

biogenesis: The production of living organisms from other living organisms *bio*

bioinformatics: The application of quantitative and analytical computational techniques to model biological systems or developing analytical tools to discover knowledge in the data is the second, and more scientific, aspect of bioinformatics; the term is often used to describe computational molecular biology, for example, the use of computers to store, search, and characterize the genetic code of genes, the proteins linked to each gene, and their associated functions *meth biot*

Bioinformatics Applications Research Center (BARC): A collaboration of North Queensland (NQ) (Australia) researchers conducting R&D in the broad area of biotechnology applications for NQ and other tropical regions of the world; BARC is built on the strengths of many component bioscience research programs; thus, many projects are strongly cross-disciplinary; it consolidates complementary biotechnological applications research programs into a cohesive broad-based endeavor, underpinned by the key enabling technologies of super-computing and advanced networking; it uses advanced software development expertise to create novel software methods and tools for the biotechnology industry *biot stat*

biolistic gene gun: "Biolistic" derived from a contraction of the words "biological" and "ballistic"; it refers to a projectile fired from a gun; used to shoot pellets that are loaded with genes into plant seeds or tissues, in order for them to become integrated and/or expressed in the foreign background; the gun uses an actual explosive to propel the material; compressed air or steam may also be used as the propellant *biot*

biological assay: >>> bioassay

biological containment: A technique by which the genetic constitution of an organism is altered in order to minimize its ability to grow outside the laboratory; precautions are taken to prevent the spread of recombinant DNA molecules in the natural environment; disabled host organisms (e.g., with stable auxotrophic requirements or defective cell walls), together with non-transmissible cloning vectors are used; biological containment is especially important when toxin genes from pathogens are expressed in *Escherichia coli* or other vectors *biot* >>> gene containment

biological control: The practice of using beneficial natural organisms to attack and control harmful plants, animal pests, and weeds is called biological control, or biocontrol; this can include introducing predators, parasites, and disease organisms, or releasing sterilized individuals; biocontrol methods may be an alternative or complement to chemical and gene-engineered pest control methods *phyt* >>> Bt >>> Bt gene

biological determinant: A biological factor such as crop species, variety, weeds, insect pests, or disease that determines the crop configuration and performance of a cropping pattern at a given site or area *agr*

biologically integrated farming systems (BIFS): Integrated farming systems that have been proven to economically reduce the use of farm chemicals by utilizing a combination of the best practices from organic, integrated pest management, and conventional farming systems *agr eco*

biological nitrogen fixation: Capture of gaseous >>> nitrogen from the atmosphere into biological compounds, carried out by certain >>> bacteria either living freely in soil or in symbiotic association with a green plant *bio* >>> nitrogen fixation

biological pesticide: A chemical which is derived from plants, fungi, bacteria, or other natural synthesis and which can be used for pest control *phyt*

biological species concept: A system in which organisms are classified in the same species if they are potentially capable of interbreeding and producing fertile offspring *evol*

biological yield: The total yield of plant material (i.e., the total biomass including the economic yield – for example, the grain yield); the larger the biological yield, the greater the >>> photosynthetic efficiency *phys*

biomarker: It is not always possible to establish a quantitative measurement of the benefits or risks between diet and health and disease; a biomarker in the diet is a compound that can be measured, and its presence is related to the potential benefits or risks to health *phys meth*

biomass: The total weight of organic material in a given area or volume; it can be divided into aboveground and belowground biomass; modern techniques allow easy determination of cell biomass, e.g., during fermentation; radio frequency impedance spectroscopy (RFIS) is a robust method for non-invasive in-line monitoring of the passive electrical properties of cells in suspension and can distinguish between living and dead cells based on their distinct behavior in an applied radio frequency field; RFIS is a robust method for the high resolution of the in-line measurement, revealing subtle changes in cell growth which were not accessible using conventional methods *phys*

biome: Interactive groups of individuals of one or more species occupying a major terrestrial region; they are created and maintained by climate *eco*

biometrical genetics: >>> quantitative genetics

biometrics: >>> biometry

biometry: Mathematical statistics applied to biological investigations *stat* >>> bioinformatics >>> statistics

bionanotechnology: A branch of >>> nanotechnology that uses biological starting materials, utilizes biological design or fabrication principles, or is applied in medicine or biotechnology; it has become an exciting field of research and an area of technology development, especially since the length scale that nanotechnology can access more and more coincides with the length scale of basic biological structures and fundamental biological components *biot meth*

biopesticide: >>> biological pesticide >>> Bt

biopharming: The production of pharmaceutical proteins in genetically engineered plants; pharmaceuticals can be made in plants at a significantly reduced cost compared to current production methods; major concerns with biopharming are that food or feed crops may become contaminated with pharmaceutical products, and that the products may have negative effects on natural ecosystems; in the U.S., biopharm crops are regulated by two federal agencies (USDA and FDA) and by state departments of agriculture; manufacturing pharmaceutical products in crops has been one of the promised benefits of plant genetic engineering for the past 20 years; this use of >>> biotechnology, sometimes known as "pharming", "biopharming", or "molecular farming", has migrated from speculation to the application phase in fields and greenhouses across several countries; biopharming promises more plentiful and cheaper supplies of pharmaceutical drugs, including vaccines for infectious diseases and therapeutic proteins for the treatment of such things as cancer and heart disease; plant-made pharmaceuticals are produced by genetically engineering plants to produce specific compounds, generally proteins, which are extracted and purified after harvest; as used here, the term "biopharming" does not include naturally occurring plant products or nutritionally enhanced foods *biot* >>> pharm crop

biopiracy: The collecting and patenting of plants and other biological material, formerly held in common, and their exploitation for profit *biot* >>> Intellectual Property Rights

bioreactor: A culture vessel used for experimental or large-scale bioprocessing *biot*

bioremediation: The use of biological systems, usually microorganisms, to decompose or sequester toxic and unwanted substances in the environment *meth*

bioseeds: Seeds produced via genetic engineering of existing plants *biot seed*

biosensor: A device, especially an electrochemical device, that detects, quantitatively and in real time, the presence of an analyte or some biological events (e.g., respiration, enzymic activity, binding to an antibody) and converts it into an electrical signal *biot meth*

biosome: Any autonomous cell constituent multiplying by autoreduplication *gene*

biostatistics: The application of statistics to biological data *stat* >>> biometry

biosynthesis: The synthesis of the chemical components of the cell from simple precursors *phys*

biotechnology: Any technique (e.g., recombinant DNA methods, protein engineering, cell fusion, nucleotide synthesis, biocatalysis, fermentation, cell cultures, cell manipulations, etc.) that uses living organisms or parts of them to make or modify products, to improve organisms, or to make them available for specific uses; more practically for plant breeding, applications include anther culture for haploid production, embryo/ovule culture after interspecific hybridization, genetic engineering (transformation), >>> *in vitro* selection, *in vitro* germplasm conservation and exchange, >>> micropropagation, cell and organ culture, >>> somaclonal variation, somatic cell hybridization (protoplast fusion), or somatic embryogenesis; the term was first coined by the Hungarian engineer, K. EREKY, in 1909; the origin of biotechnology can be traced back to prehistoric times when microorganisms were already used for processes like fermentation, preparation of yoghurt and cheese from milk, vinegar from molasses, and production of butanol and acetone from starch by *Clostridium acetobutylicum*; remarkable contributions to the development of biotechnology were: the first attempts of plant tissue culture by HABERLANDT in 1902, embryo culture in some crucifers by HANNIG in 1904, asymbiotic *in vitro* germination of orchid seeds by KNUDSON in 1922, *in vitro* culture of root tips by ROBBINS in 1922, embryo culture in interspecific hybrids of linseed by KAIBACH in 1925, successful culture of tomato roots by WHITE in 1934, successful establishment of continuously growing callus culture by GAUTHERET et al. in 1939, the use of coconut milk containing a cell division factor for *Datura* by OVERBEEK in 1941, *in vitro* adventitious shoot formation in tobacco by SKOOG in 1944, raising of whole plants by shoot tip culture in *Lupinus* and *Tropaeolum* by BALL in 1950, first application of >>> micrografting by MOREL et al. in 1952, discovery of kinetin as a cell division hormone by MILLER et al. in 1955, *in vitro* regeneration of somatic embryos in *Citrus* by MAHESHWARI et al. in 1958, regeneration of *Daucus* embryos from cell suspensions by REINERT in 1959, first test-tube fertilization in *Papaver* by KANTA in 1960, enzymatic degradation of cell walls for production of protoplasts by COCKING in 1960, first haploid plant production from *Daucus* pollen grains by GUHA in 1964, first achievement of protoplast fusion by POWER in 1970, discovery of the first restriction endonuclease, >>> HindII, in 1970 by SMITH, first plant regeneration from protoplasts by TAKEBE in 1971, first interspecific protoplast fusion in tobacco by CARLSON in 1972, etc. *biot* >>> native trait recovery >>> Figure 58

biotechonomy: >>> bioeconomy

biotin(e): Functions as a coenzyme; is a part of the vitamin B complex; also called vitamin H; present in all living cells, bound to polypeptides or proteins; important in fat, protein, and carbohydrate metabolism; a common addition to plant tissue culture media *biot phys*

biotinylated probe: A DNA sequence in which biotinylated dUTP is incorporated and labeled with biotin; it is used in >>> DNA-DNA hybridization experiments, such as >>> SOUTHERN transfer or *in situ* hybridization, with chromosomes; the detection of hybrid molecules is realized by a complex of streptavidin, biotin, and horseradish-peroxidase; if there is a hybridization, then the complex shows a green fluorescence color *micr cyto biot* >>> Figures 48, 52, 53, 54

biotope: A portion of a habitat characterized by uniformity in climate and distribution of biotic and abiotic components *eco*

biotrophic pathogen: A parasitic organism that obtains its nutrient supply only from living host tissue regardless of whether or not it can be artificially cultured *phyt*

biotrophy: Obtains nutrients from living cells; biotrophs are typically obligate parasites; the term can also cover the phase in the infection process where a necrotroph does not destroy the host (hemibiotrophic) *phyt*

biotype: A group of genetically identical individuals; sometimes, a physiologic race or population; it may be homozygous or heterozygous *gene*

bioversity: >>> Biodiversity International

biparental inheritance: Plant zygotes that show traits indicating chloroplast chromosomes from both parents are present and active *gene*

biparental populations: The progeny derived after crossing two genotypes as male and female parents; such populations include F2 genotypes generated from F1 progeny, lines generated after doubling the haploids (DHs, obtained from F1 plants through anther, egg cell or ovule culture or distant hybridization), or recombinant inbred lines (RILs), which are derived by single-seed descent for at least five or more generations by repeated selfing or >>> sibling mating *gene meth* >>> Figure 51

bipartitioning: Showing normal meiosis *cyto*

biplot analysis: A statistical test used for studying genotype × environment interaction or any two-way tables; its descriptive and visualization capabilities along with the availability of user-friendly software have enabled plant scientists to examine any two-way data by a click on a computer button; the additive main effects and multiplicative interaction model and the genotype main effects and genotype × environment interaction effects model have been the two most commonly used models for the biplot analysis; despite widespread use, the validity and limitations of biplot analysis have not been completely examined (YANG et al. 2009) *stat*

bird cherry-oat aphid: >>> barley yellow dwarf virus

bird netting: Different types of mesh used as a drape to keep birds out of fruit trees, berry patches, vegetable gardens, or field experiments *meth hort agr*

bird pollination: >>> ornithophily

bird resistance: A characteristic of a genotype or individual plant in which it is avoided by birds until other food sources are exhausted, or until the plant is weathered *phyt*

birdscare: Scarecrow

birimose: Opening by two slits (e.g., anthers of plants) *bot*

bisexual: Species comprises individuals of both sexes or a hermaphrodite organism in which an individual plant possesses both stamens and pistils in the flower *bot* >>> co-sexual >>> Table 18

bisulfite genomic sequencing: A procedure in which bisulfite is used to deaminate >>> cytosine to >>> uracil in genomic DNA; conditions are chosen so that 5-methylcytosine is not changed; >>> PCR amplification and subsequent DNA sequencing reveal that cytosines are methylated in genomic DNA *biot* >>> Table 38

bisulfite mutagenesis: A variant of chemical mutagenesis of >>> ssDNA molecules that uses sodium bisulfite ($NaHSO_3$) for the deamination of cytosine residues to yield uracil; it is a substitution mutagenesis *gene meth* >>> TAL effector nuclease

bitter pit: A physiological disorder believed to be induced by calcium deficiency in apple fruits; the incidence of bitter pit usually occurs during storage, but in some cases it can also develop at harvest; the fact that total calcium in the fruit is not able to accurately predict bitter pit incidence has puzzled many scientists for a long time; the high correlation with no predictive accuracy between calcium and bitter pit makes the development of this disorder one of the most complex and challenging mechanisms present in plants; the most effective way of prevention is spraying the trees (usually apple or pear) with calcium chloride or calcium nitrate *hort phys*

bivalency: The property of an immunoglobulin G molecule, and some other immunoglobulins, of having two antigen-binding sites *chem*

bivalent: Two homologous chromosomes when they are paired during prophase-metaphase of the first meiotic division *cyto* >>> Figure 15

bivalent formation: The association of two homologous chromosomes as a ring or rod configuration depending on chiasma formation *cyto* >>> Figure 15

bivalent interlocking: >>> interlocking

bivariate statistics: Statistics that describe the relationship between two variables stat meth

bla gene: Beta-lactamase gene conferring resistance to ampicillin; commonly used as selective marker for plasmid vectors *biot*

black leg (of beets): A number of diseases (e.g., caused by Pythium debaryanum) of which symptoms include blackening of the base of the stem, often followed by the collapse of the stem *phyt*

black leg (of potato): A bacterial disease (*Erwinia carotovora ssp. atroseptica*) causing severe yield loss, particularly in wet conditions *phyt*

black rot (of rapeseed and cabbage): Productivity and quality of rapeseed is heavily affected by the disease, caused by the bacterium *Xanthomonas campestris* pv. *campestris*; several races have been described, the race 6 being the most frequent in rapeseed and races 1 and 4 the most frequent in cabbage crops; the control of the disease can be aided by the employment of resistant varieties; resistance can be evaluated in landraces *phyt*

black sigatoka (in banana): A leaf spot disease, also known as black leaf streak, causing significant reductions in leaf area, yield losses of 50% or more, and premature ripening, a serious defect in exported fruit; it is more damaging and difficult to control than the related yellow sigatoka disease, and has a wider host range that includes the plantains and dessert triploid ABB cooking bananas that are usually not affected by yellow sigatoka; black sigatoka was first recognized in the Sigatoka Valley of Fiji in 1963, but was probably widespread in Southeast Asia and the South Pacific by that time; in the western hemisphere, it first appeared in 1972 in Honduras and now occurs on the mainland from central Mexico south to Bolivia and northwestern Brazil, and in the Caribbean basin in Cuba, Jamaica, the Dominican Republic, and southern Florida; in Africa, the disease was first recorded in Zambia in 1973 and has since spread throughout the sub-Saharan portions of that continent; in most areas, black sigatoka has now replaced yellow sigatoka to become the predominant leaf spot disease of banana; the disease is caused by the ascomycete, *Mycosphaerella fijiensis*, anamorph: *Paracercospora fijiensis* (a variant of the pathogen, *M. fijiensis var. difformis*, that was previously reported in tropical America, is no longer recognized); the pathogen produces conidia and ascospores, both of which are infective; they are formed under high moisture conditions, and are disseminated by wind, and in the case of conidia, also by rain and irrigation water; due to their greater abundance and small size, ascospores are more important than conidia in spreading the disease within plants and plantations; in contrast, infected planting material and leaves, which are used often in the developing world as packing materials, are usually responsible for the long-distance spread of the disease; the recent outbreak of black sigatoka in South Florida almost certainly resulted from the importation of infected germplasm by local growers; chemical control of first yellow, and then black sigatoka has evolved considerably over the last 65 years; >>> Bordeaux mixture, first used in the mid-1930s, has been replaced by several succeeding generations of protectant and, later, systemic fungicides; presently, a sterol biosynthesis inhibitor, tridemorph, several different sterol demethylation inhibitors, most importantly propiconazole, and the methoxyacrylate, azoxystrobin, are the most commonly used systemics; since there is a tendency for resistance or tolerance to develop in *M. fijiensis* towards the systemic fungicides, they are usually applied in combination or alternation with broad-spectrum, protectant fungicides, such as the dithiocarbamates and chlorothalonil; with the exception of chlorothalonil, these fungicides are usually mixed with petroleum-based spray oils; the oils themselves are fungistatic and retard the development of the pathogen in the infected leaf; when they are mixed in water emulsions with fungicides, the resulting "cocktails" provide superior disease control; in export plantations, black sigatoka is controlled with frequent applications of fungicides and cultural practices, such as the removal of affected leaves, and adequate spacing of plants and efficient drainage within plantations; in total, these are very expensive fungicides, e.g., application includes the use of airplanes or helicopters, permanent landing strips and facilities for mixing and loading the fungicides, and the high recurring expense of the spray materials themselves; in total, it has been estimated that the costs of control are ultimately responsible for 15–20% of the final retail price of these fruit in the importing countries; their great expense makes them essentially unavailable to small-holder farmers who grow this crop; it is these producers who are affected most by this important disease *phyt*

black vine weevils: Insects known to feed on some ornamental plants; the adult vine weevil is a dark brown beetle which measures about 1.5 cm; both the vine weevil and its larvae can cause extensive damage *phyt hort*

blade: The expanded portion of a leaf, petal, or sepal *bot* >>> Table 30

blade joint: The flexible union between the leaf blade and the leaf sheath *bot*

blanch: A method to whiten or prevent from becoming green by excluding light; blanching is applied to the stems or leaves of plants (e.g., celery, lettuce, and endive); it is done either by banking up the soil around the stems, tying the leaves together to keep the inner ones from light, or covering with pots, boxes, etc. *meth hort*

blanking: >>> beat(ting) up

BLAST: >>> basic local alignment search tool

blasting: A plant symptom characterized by shedding of unopened buds; leads to a failure to produce fruits or seeds *phyt agr*

blaze: To mark a tree, usually by painting and/or cutting the bark; boundaries of forest properties frequently are delineated by blazing trees along the boundary line *fore*

bleached cotton linters: Linters that have been bleached, ready for further processing *agr*

bleeding: Exudation of the contents of the xylem stream at a cut surface due to root pressure *bot*

blemishes: Often on fruit and vegetables, caused by crop parasites; since the development of synthetic crop protection chemicals, it has become fashionable to see only blemish-free produce on sale; however, blemishes are an indication of freedom from pesticides and are more accepted for this reason by lovers of organic food *hort phyt*

blend: A term applied to mechanical seed mixtures of different crop varieties or species that have been mixed together to fulfill a specific agronomic purpose *seed*

blended variety: >>> multiline variety >>> blend

blending: >>> blend

blending inheritance: Inheritance in which the characters of the parents appear to blend into an intermediate level in the offspring with no apparent segregation in later generations *gene*

blending (theory of) inheritance: Inheritance in which the characters of the parents appear to blend into an intermediate level in the offspring with no apparent segregation in later generations; an early disproven theory that the genetic elements of the parents fuse during the fertilization and lose their purity *gene*

blight: A disease characterized by rapid and extensive death of plant foliage, and applied to a wide range of unrelated plant diseases caused by fungi, when leaf damage is sudden and serious (e.g., fire blight of fruit trees, halo blight of beans, potato blight, etc.) *phyt*

blind: Without flowers; sterile *bot*

blind cultivation: Cultivation that is undertaken before a crop emerges *agr*

blind floret: >>> blind

blindfold (trial): A trial to study soil heterogeneity (i.e., variation in the soil fertility); all plots contain the same genetically uniform plant material; the study may show that the growing conditions provided by a particular field may appear homogeneous when observed in some season and for some trait of a crop, but they may appear heterogeneous when observed in a different season or for some trait of a different crop; for a given crop, different traits may differ with regard to their capacity to show soil heterogeneity *stat meth*

B line: The fertile counterpart or maintainer line of an A line; does not have fertility restorer genes; used as the pollen parent to maintain the A line; used in hybrid seed production *seed* >>> A line >>> Figures 2, 55

blister stage: This stage is initiated when significant starch accumulation begins, approximately 12–17 days after pollination; kernels appear white and translucent, and endosperm and its inner fluid are clear; endoreduplication, an increase in DNA content in endosperm cells that begins in the transition stage, peaks about 16–18 days after pollination *phys meth* >>> Table 13

block: A number of plots that offer the chance of equal growing conditions; comparisons among the entries that are tested in the same block offer unbiased estimates of genetic differences *stat meth* >>> bias >>> design of experiment >>> Tables 25, 26

block designs: This theory of design of experiments came into being largely through the work of R. A. FISHER and F. YATES in the early 1930s; they were motivated by questions of design of careful field experiments in agriculture; when it is desired to compare the yield of different varieties of grain, then it is quite possible that there is an interaction between the environment (type of soil, rainfall, drainage, etc.) and the variety that would alter the yields, so blocks (sets of experimental plots) are chosen in which the environment is fairly consistent throughout the block; in other types of experiments in which the environment might not be a factor, blocks could be distinguished as plots that receive a particular treatment (e.g., a particular type of fertilizer); in this way, the classification of the experimental plots into blocks and varieties can be used whenever there are two factors that may influence yield; the obvious technique of growing every variety in a plot in every block may, for large experiments, be too costly or impractical; to deal with this, smaller blocks are used that do not contain all of the varieties; the problem is to minimize the effects of chance due to incomplete blocks and to design the blocks so that the probability of two varieties can be compared; this property is called "balance in the design"; statistical techniques, in particular analysis of variance, can then be used to reach conclusions about the experiment *stat meth* >>> design of experiment

blocking: The procedure by which experimental units are grouped into homogeneous clusters in an attempt to improve the comparison of treatments by randomly allocating the treatments within each cluster or block; one measure of effectiveness of blocking is the F ratio for blocks; a more precise measure of the efficiency of >>> randomized block design relative to the completely randomized design for a given situation is obtained by computing the relative efficiency *stat* >>> block >>> design of experiment >>> thinning >>> Table 25

block mutation: A term used to denote a change in, or the omission of, a group of adjacent genes *gene*

bloom: The white powdery deposit often present on the surface of the stem, leaves, and ears of cereals or sorghum, often of a waxy nature; in general, the flower of a plant or the state of blossoming *bot* >>> waxiness

blooming (bloom period): In grasses, the period during which florets are open and anthers are extended *bot agr*

blossom: The flower of a plant, especially of one producing an edible fruit; the state of flowering *bot*

blot: The transfer of DNA, RNA, or proteins to an immobilizing binding matrix, such as nitrocellulose, or the autoradiograph produced during certain blotting procedures (>>> Southern blot, >>> Northern blot, >>> Western blot, etc.) *meth gene biot* >>> Figure 48

blotch: A disease characterized by large and irregularly shaped spots or blots on leaves, shoots, and stem *phyt agr*

blue-number method: A method of indirect determination of alpha amino nitrogen of proteins using copper reagent, e.g., in sugarbeet; it was developed during the 1930s and used until now; there is a good correlation between the amount of nitrogen in proteins and the so-called blue-number nitrogen; the method was used for the selection of high-protein fodder beets; the method is fast and accurate, and the reagent can be prepared and stored for the whole campaign *meth chem*

blue-stain fungus: Most common form of fungal stain occurring in sapwood; conifers are most susceptible but it may also occur in light-colored heartwood of perishable timbers; commonly develops in dead trees, logs, lumber, and other wood products until the wood is dry; it reduces the grade of wood, but does not significantly reduce the strength; some blue-stain lumber is even highly valued for specialty products *phyt*

blunt-end ligation: Ligation of DNA with blunt ends requires higher concentration of DNA ligase than sticky-end ligation; it is inhibited by ATP concentrations >1 mm *biot*

blunt ends: DNA fragments that are double-stranded paired over the whole length, usually produced by certain types of restriction enzymes *gene* >>> Table 39

BLUP: >>> best linear unbiased prediction

Boerner divider: >>> conical divider

bog: A poorly drained, acidic, freshwater wetland that depends primarily on precipitation, snowmelt, and fog for water and is characterized by a buildup of peat, usually from *Sphagnum* mosses *eco* >>> bog garden

bog garden: Waterlogged area used to grow plants found in bogs, marshes, and/or wet pasture *hort meth* >>> bog

boleless: Without a trunk *bot*

boll: The fruiting structure of a cotton plant; it is made up of separate compartments called locks, in which cotton seeds and lint grow *bot*

Bollgard® insect-protected cotton: A genetically improved cotton that offers protection against the cotton bollworm; the bollworm can cause significant damage to a cotton crop and require repeated applications of insecticides; Bollgard® cotton contains an insecticidal protein from a naturally occurring soil microorganism, *Bacillus thuringiensis*, that gives cotton protection from bollworms *phyt biot*

boll size: Weight in grams of seed cotton from one boll *agr meth*

bolt: Formation of an elongated stem or seedstalk; in the case of biennial plants, this generally occurs during the second season of growth *bot* >>> bolter

bolter: Develops in long cold springs with morning frosts and low temperatures, not exceeding +5°C, causing vernalization of the plants (e.g., in sugarbeet) *phys* >>> bolting

bolting: Production of seed stalks the first season in a biennial crop (e.g., in beets); shoot elongation (bolting) starts after a period of low temperatures; the dominant allele of locus B causes early bolting without cold treatment; this allele is abundant in wild beets whereas cultivated beets carry the recessive allele *phys agr* >>> bolter

bolting resistance: >>> bolter >>> bolting

bonsai: A tree or shrub grown in a container or special pot and dwarfed by pruning, pinching, and wiring to produce a desired shape *meth hort*

boot: The lower part of a cereal plant *bot*

boot(ing) stage: Refers to the growth stage of grasses at the time the head is enclosed by the sheath of the uppermost leaf; in cereals, the development of the flag leaf from whose sheath the inflorescence eventually emerges *agr* >>> Table 13

bootstrap analysis: A method in cladistic analysis to infer the "strength" or "confidence" of a branch on a phylogenetic tree, obtained by generating trees many times from a sample distribution of characters; bootstrap values theoretically can vary from 0% (poor support) to 100% (excellent support) *stat meth biot*

Bordeaux mixture: The first, and also the most spectacularly successful, of all man-made fungicides, discovered in Bordeaux, France, by the botanist P.-M.-A. MILLARDET in 1882; the mixture is prepared by mixing a solution of copper sulphate with freshly slaked lime; this fungicide saved the French wine industry from ruin by the newly introduced downy mildew (*Peronospora viticola*), and it also controlled potato blight, caused by *Phytophthora infestans phyt hort agr* >>> Table 51

border effect: The environmental effect on plots that are on the edge of an experimental area *stat meth*

border strip: A demarcation surrounding a plot, usually given the same treatment as the plot; it is arranged in order to minimize border effects *meth agr hort*

boring platform: Sterile bottom half of a petri dish used for preparing explants with a cork borer *meth biot*

boron (B): A non-metallic element occurring naturally only in combination, as in borax or boric acid; boron can cause toxicity in several crop plants; as a micronutrient, deficiency of boron can be as severe *chem agr*

boss: A dense group of stamens *bot*

botanical pesticides: Pesticides whose active ingredients are plant-produced chemicals such as nicotine, rotenone, or strychnine *phyt* >>> biological control

botany: The science of plants; the branch of biology that deals with plant life; the plant life of a region; the biological characteristics of a plant or group of plants *bot*

bottleneck (effect): A period when a population becomes reduced to only a few individuals, in other words, the reduction of a population's gene pool and the accompanying changes in gene frequency produced when a few members survive the widespread elimination of a species; it is a form of >>> genetic drift that occurs when a population is drastically reduced in size; some genes may be lost from the gene pool as a result of chance *gene eco* >>> Figure 68

bottom recessive: An individual homozygous for the recessive alleles of all genes under investigation *gene* >>> backcross ratio

botuliform: Cylindrical with rounded ends, sausage-like in form *bot*

bough: The main arm or branch of a (fruit) tree *bot hort*

Bouillie bordelaise: >>> Bordeaux mixture

boundary mark: >>> landmark

bouquet stage: A meiotic prophase stage of some organisms during which the chromosome orients one or both ends toward one point in the nuclear envelope *cyto*

bowl-shaped: Describes a flower that is hemispherical with the sides straight or very slightly spreading on the tips *bot*

Boyage system: >>> chopping

bp: >>> base pair

brace root: Aerial roots that extend downward from lower above-ground nodes; when coming into contact with the soil, true roots then develop and function as support and feeder roots *bot*

brachyomeiosis: An abnormal meiosis characterized by omission of the second meiotic division *cyto*

brackling: A term used to describe bending over or breaking at the top node; it appears to be a term particular to the U.K. where cereal varieties are rated routinely for normal lodging and brackling; brackling is more common with barley than wheat; it is genotype-specific, and it is influenced by the environment, frequently associated with wet weather near maturity;

the phenomenon of bending at the top node appears to be more widely known than the term "brackling" *agr* >>> lodging >>> Figure 43

bract: A modified leaflike structure occurring in the inflorescence *bot*

bracteole: A little bract borne on the flowerstalk above the bract and below the calyx; they are small, united bracts, which form a cup-like involucre *bot*

bramble: Any shrub with thorns in the rose family; usually refers to blackberries and raspberries *hort*

bran: Compromises aleurone and pericarp cell layers; the bran and germ are separated during milling *agr meth*

branch: An axillary (lateral) shoot or root *bot*

branch: >>> ramify

branch crown: Plant tissue that is the junction of the roots and stem that forms on the side of a strawberry plant *hort*

branching: >>> ramification

branching agent: A substance inducing and/or increasing branching *hort*

brand: A legal trademark registered by a particular company or distributor for its exclusive use in marketing; a product such as seeds or plants seed *agr*; in plant pathology, a leaf disease caused by a microscopic fungus (e.g., a rust or smut); sometimes names the fungi *phyt* >>> Table 52

***Brassica*:** A genus within the Brassicaceae (Cruciferae), commonly known as the mustard family; the family of about 375 genera and 3,200 species including crops, ornamentals, and many weeds; Brassica contains about 100 species, including rapeseed, cabbage, cauliflower, broccoli, Brussels sprouts, turnip, various mustards, and weeds *agr hort*

brassinosteroids: Endogenous, plant growth-promoting natural products with structural similarities to animal steroid hormones; they affect cell elongation and proliferation, distinct from that of auxins, cytokinins, and gibberellic acids, although they interact with them *phys* >>> biological control >>> hypocotyl

breakage-reunion hypothesis: The classical and generally accepted model of crossing-over by physical breakage and crossways reunion of broken chromatids during meiosis *gene*

breakdown of resistance: The inability of a plant to maintain resistance when attacked by a pest biotype that has a gene for virulence at every locus corresponding to a gene for resistance in the host *phyt gene*

breakpoint: When chromosome mutations occur, the site at which the single or double strand of DNA breaks along a chromosome and/or chromatid *cyto gene*

breakpoint mapping: The localization of breakpoints along a chromosome; the method may involve the sorting of a great amount of somatic chromosomes (~150,000) and their micro-array hybridization to well-characterized genomic clones; sites of breaks are identified that are the prerequisite for deletions, translocations, or inversions *biot gene*

breathing root: >>> pneumatophore

breed: An artificial mating group derived from a common ancestor or for genetic analysis; in breeding, a line having the character type and qualities of its origin; in general, a group of plants, developed by humans, that will not keep their characteristics in the wild

breeder: Someone who raises plants primarily for breeding purposes

breeder's exemption: As breeding is generally considered as incremental, breeders have to build on existing varieties to develop improved ones; the >>> UPOV convention contains an exception to breeders' rights, i.e., it allows the utilization of another protected or new cultivar as an initial source of variation for the purpose of creating other new varieties including its marketing *seed agr*

breeder's seed: Seed or vegetative propagating material increased by the originating or sponsoring plant breeder or institution; it represents the true pedigree of the variety; it is used for the production of genetically pure, foundation, registered, and certified seeds *seed* >>> Table 28, 57

breeder's collection: >>> working collection >>> stock

breeder's preference: A general impression of the breeder concerning a material that is under selection *meth* >>> Tables 33, 57

breeder's rights: Varietal protection; the legal rights of a breeder, owner, or developer in controlling seed production and marketing of crop varieties >>> PPA >>> Table 57

breeding: The propagation and genetic modification of organisms for the purpose of selecting improved offspring; several techniques of hybridization and selection are applied *meth* >>> cooperative breeding >>> Figure 60 >>> Tables 35, 57

breeding cycle: The shortest period between successive generations from germination of a seed to reproduction of the progeny (i.e., the seed-to-seed cycle) *meth* >>> Table 57

breeding line: A group of plants with similar traits that have been selected for their special combination of traits from hybrid or other populations; may be released or used for further breeding approaches *meth* >>> Table 57

breeding method: >>> breeding system >>> Table 57

breeding orchard: A planting of selected trees, usually propagated by cloning or grafting; designed to ease breeding work *meth*

breeding population: A group of individuals selected from a wild, experimental, or crossing population for use in a breeding program; usually phenotypically selected for desirable traits *meth* >>> Table 35 >>> Figures 43, 51

breeding rotation: >>> breeding cycle

breeding size: The number of individuals in a population involved in reproduction during particular generations and breeding procedures *meth* >>> Figure 51

breeding strategy: Prescription for breeding; a sound breeding strategy searches for an optimal compromise between genetic gain, gene diversity, cost, time, and other factors *meth* >>> Figure 43

breeding system: The system by which a species reproduces; more specifically, the organization of mating that determines the degree of similarity and/or difference between gametes effective in fertilization *meth* >>> Figures 31, 51 >>> Table 35

breeding triangle: Plant breeding can be seen as a virtual triangle, inside which the breeder constantly manages resources and germplasm in order to meet breeding goals and objectives; breeding programs are balanced when breeders have free and equal access to both germplasm and the resources needed for success; each breeding triangle exists inside specific and dynamic conditions made of a combination of interacting factors, e.g., natural, geographic, cultural, social, legal, economic, and political ones *meth bree*

breeding true: Producing offspring with phenotypes for particular characters that are identical to those of the parents; homozygous individuals necessarily breed true, whereas heterozygotes rarely do so *meth gene*

breeding value: The value of an individual as defined by the mean value of its progeny, either on the basis of individual traits or a selection index *meth*

breeding zone: An area within which a single population of improved trees can be planted without fear of misadaptation *fore*

breed not at risk: Breed where the total number of breeding females and males is greater than 1,000 and 20, respectively, or the population size approaches 1,000 and the percentage of pure-bred females is close to 100%, and the overall population size is increasing *meth gene*

brevicollate: Short-necked *bot*

brewing: The process by which beer is made; in the first stage the barley (or other) grain is soaked in water and allowed to germinate (malting), during which the natural enzymes of the grain convert the seed starch to maltose, and then to glucose; grain is then dried, crushed, and added to water at a specific temperature (steeping) and any remaining starch is converted to sugar; the resulting liquid (wort) is the raw material to which yeast is added to convert sugar to alcohol; hops (female flowers of *Humulus lupulus*) are added during this process to give a characteristic flavor *meth*

brick grit test: A type of seedling emergence (vigor) test utilizing uniformly crushed brick gravel through which seedlings must emerge to be considered vigorous; it was originally developed by HILTNER and IHSSEN (1911) for detecting seed-borne Fusarium infection in cereals; with modifications, the seeds are planted on damp brick grit or in a container of sand covered with 3 cm of damp brick grit, then germinated in darkness at room temperature for a specific time *seed*

bridge-breakage-fusion-bridge cycle: A process that can arise from the formation of dicentric chromosomes; daughter cells are formed that differ in their content of genetic material due to duplications and/or deletions in the chromosomes *cyto gene*

bridge cross(ing): A method of bypassing an incompatibility barrier between two genotypes or species by using a third genotype or species, which is partly compatible with each of them, in an intermediate cross; for example, a wild plant is first crossed with another (wild) species, its progeny is selected for desired features, and these are then crossed with the target variety; the approach is mostly used where the desired trait is easy to select; it is usually a time-consuming process; once the critical characteristic has been incorporated in the target variety, a number of backcrosses are needed to eliminate the undesirable wild characters *meth*

bridge grafting: When the trunk of fruit trees is damaged by various means and the bark is removed, the bridge graft is a method of repairing a girdled trunk; although it can be done by inserting the bridges into cuts made in the wood of the trunk, the common method is to lift the bark and place the exposed cambiums of >>> scion and >>> stock together; it is thus usually a bark graft and is not done until May; in this case it is necessary to collect the dormant scions earlier and store them until the repairs are to be made; healthy, matured suckers of hardy varieties are suitable; where one-quarter or less of the trunk circumference has been girdled, the necessity of bridge grafting may be doubtful, but a wound dressing applied in early spring is always helpful, even to these less extensive injuries; trees which have been in the orchard less than four years are usually too small for successful bridge grafting; if they are completely girdled, or nearly so, and there is still a collar of live bark above the graft union, saw the top off at a point immediately below the injury; apple, pear, and plum will develop a new top without grafting; cherry and peach are unlikely to do so; they require grafting of scions on the remaining trunks, or preferably the planting of a new tree; apple, pear, and plum, when cut back, will produce many shoots; these shoots should not be thinned until they are a year old, then only the most suitable ones should be retained; of the numerous methods of bridge grafting, the channel or inlay method is generally preferred; the two possible locations of the channel depend on whether or not the wound has been treated previously with a protective covering; where the wound has been covered some time before the grafting operation, the channels must be well above and below the treated area; prepare the scions with the bevel on the side opposite the natural bow of the wood; when a scion has been cut to the proper length and beveled for 5 to 8 cm at each end, it is laid over the wound in the position it is to occupy; by outlining the scions on the bark of the stock with the knife point, an almost perfect fit of the scions in the channels is possible; the distance between the extremities of the channels should be slightly less than the length of the scion, allowing for a slight bow of the scion when the job is completed; this slight bend allows better contact of scion and stock and reduces the danger of breaking connections if the tree sways with the wind; with a screwdriver or similar tool, lift and remove the strips of bark from the channels; place the lower (thicker) end of the scion in the bottom channel and nail it there with two 2.5 cm box or basket nails;

then spring the upper end into position and nail it similarly; place scions about 5 cm apart over the injured area and then cover all wounded surfaces thoroughly with a good quality grafting compound that will not shrink or crack *meth hort* >>> inarching

bridge parent: A parent that is sexually compatible with two reproductively isolated species and can be used to transfer genes between them *meth*

bridging species: A species used in a bridging cross in order to bring together the two incompatible species *eco*

bristle: A stiff, slender hair or appendage likened to a hog's bristle; in Setaria, Pennisetum, and a few other grasses, it is a highly reduced branch without a spikelet at the apex *bot*

broad-base terrace: A low embankment that is constructed across a slope to reduce runoff and/or erosion (e.g., in rice or grape cultivation) *agr hort*

broadcast: Scattered upon the ground with the hand (e.g., in sowing seed, instead of sowing in drills or rows) *meth agr hort*

broadcasting: A method of sowing seeds in which the seeds are scattered randomly rather than planting in rows *seed meth*

broadcast seeding: In contrast to space planting or drilling, a method of seeding in agriculture, gardening, and forestry that involves scattering seed, by hand or mechanically, over a relatively large area. This is in contrast to precision seeding, where seed is placed at a precise spacing and depth or hydroseeding, where a slurry of seed, mulch, and water is sprayed over prepared ground in a uniform layer; broadcast seeding is of particular use in establishing dense plant spacing, as for cover crops and lawns; in comparison to traditional drill planting, broadcast seeding will require 10–20% more seed; it is simpler, faster, and easier than traditional row sowing *meth seed agr* >>> space planting

broadcast sprigging: Vegetative turf establishment by broadcasting and covering of stolons, rhizomes, or tillers with soil *agr meth*

broadleaf: Sometimes used to designate a broad group of non-grasslike (weedy) plants *agr bot*

broad-sense heritability: The ratio of total genetic variance to phenotypic variance; it is used to estimate the degree of genetic control of a trait in a population, and it is useful for predicting response to clonal selection *stat meth*

broad wing: The larger of the two parts of the glume of, for example, wheat, which are separated by the keel *bot*

brokens: Pieces of the rice kernel that are less than three-quarters the size of the full kernel *agr*

bromeliads: A family of plants mostly from Central and South America that are often of rosette formation; leaves are frequently spiky and stiff, with colors ranging from green, red, pink, silver, to variegated; a large percentage of this family are epiphytes, air plants that attach themselves to the branches of other plants rather than soil *bot hort*

broom: A symptom in which lateral branches proliferate in a dense cluster on the main branch (e.g., witch's broom) *phyt agr hort*

broomrape: Parasitic plant, having purplish or yellowish flowers and small scale-like leaves that lack chlorophyll, and which grows on the root of other plants (legumes, tobacco, sunflower, etc.); branched broomrape (*Orobanche ramosa*) is a parasitic weed recently spreading in Central Europe and threatening the production of several crops including tobacco, rapeseed, potato, carrot, and tomato; in contrast to other weeds that compete with the crop for resources, *O. ramosa* is directly attached to the host root and takes up all necessary water, nutrients, and assimilates directly from its host; this leads to significant yield and quality losses; because the parasitic weed is attached to the crop and because it spends about 90% of its life underground, it is very difficult to control >>> *Orobanche* ssp. (Orobanchaceae) *phyt* >>> Figure 54

browning: Discoloration due to phenolic oxidation of freshly cut surfaces of explant tissue; in later culture this phenomenon may indicate a nutritional or pathogenic problem, generally leading to necrosis *biot*

brown plant hopper: Now a menace to rice crops; the pest (*Nilaparvata lugens*) is continuously occurring in epidemics in many regions of India and other countries, leading to heavy damage *phyt*

brown stem rot: A plant disease that can be caused by the soilborne fungus *Phialaphora gregata* in the soybean plant; some soybean varieties are genetically resistant *phyt*

brown rice: Dehusked paddy, often referred to as unpolished rice *agr* >>> rice

bruising (in potato): A gray or blue-black localized discoloration that develops in the tuber flesh as a result of physical impact *agr*

brush: The tuft of hair at the top of, for example, wheat grain; or a collective term that refers to stands of vegetation dominated by shrubby, woody plants or low-growing trees *bot* >>> coma

brushing: Spreading spores of fungi by brush on leaves or flowers for infection experiments phyt; in crossing experiments, spreading pollen on stigmata of (emasculated) flowers in order to initiate fertilization *meth*

bryophyte: Member of the subphylum Bryophyta, a moss or liverwort *bot tax*

BSA: >>> bulked segregant analysis

Bt: An abbreviation for *Bacillus thuringiensis*; it is a naturally occurring soil bacterium used as a biological pesticide (biopesticide); engineered plants have a gene from Bt inserted into their own genetic material; this new gene produces a natural protein that kills insects after the protein is ingested; the toxins are specific to a small subset of insects; for example, cotton has been genetically altered to control the tobacco budworm, bollworm, and pink bollworm; potatoes have been altered to control the Colorado potato beetle; hybrids of "Bt-maize" are altered to be resistant to the European corn borer; the Bt toxin degrades rapidly to non-toxic compounds *phyt* >>> biological control

***Bt* cotton:** A genetically engineered cotton carrying the *Bacillus thuringiensis* (Bt) gene that produces, in every cotton plant cell, protein crystals toxic to some insect pests *biot phyt* >>> *Bt* >>> cotton

Bt gene: A gene from the bacteria *Bacillus thuringiensis* that gives resistance to lepidopterous insects; by biotechnological means it was successfully transferred to cotton, tobacco, etc. *biot*

BTT: >>> balanced tertiary trisomic

bubbles: The nucleic acid configuration during replication in eukaryotic chromosomes, or the shape of heteroduplex DNA at the site of deletion or insertion *gene biot*

bud: An immature shoot, protected by tough scale leaves, from which the stem and leaves or flowers may develop *bot*

bud-bearing: >>> gemmiferous

bud blast: The most common specific fungal diseases that affect rhododendrons; it causes a lack of flowers; first, flower buds go brown and die but remain attached; later, the buds may turn silvery grey before becoming covered in small black bristles; *P. azaleae* is a cosmopolitan fungus causing bud blast and twig blight of azaleas and rhododendrons; it is a member of Micromycetes (*Pycnostysanus azaleae*) *phyt hort*

budded: Grown from a bud grafted onto a desirable understock *hort meth* >>> budding

budding: A method of asexual reproduction in which a new individual is derived from an outgrowth (bud) that becomes detached from the body of the parent; in horticulture, a form of grafting in which a single vegetative bud is taken from one plant and inserted into stem tissue of another plant so that the two will grow together; the inserted bud develops into a new shoot; in microbiology, among fungi, budding is characteristic of the yeast Saccharomyces cerevisiae *meth hort bio*

budding strip: A strip of rubber or other material used to hold grafts *meth hort*

budding union: >>> bud union

bud dormancy: >>> dormancy

bud eye: A dormant bud in the axil of a leaf; used to propagate through bud-grafting *meth hort*

bud graft: Used to asexually propagate stone fruit (peaches, plums, etc.) and many other plants; this method has the advantage of using very little scion material (one bud), and survival is usually higher than with other grafting methods; a disadvantage is that it must be done during a brief period in the spring when the bark slips; to make this graft, a chip of wood containing a leaf with its axillary bud is removed; then a "T" cut is made in the stock and the leaf blade is clipped off; just the petiole (leaf stalk) is left; it is stubbed to serve as a convenient handle for manipulating the chip; then the wood adhering to the bark of the chip is removed; the cut areas should not be touched with the fingers; a knife point is used for that; the bark with its bud and petiole is inserted into the "T" cut; the whole is wrapped with rubber budding tape or a suitable substitute; it is not necessary to use grafting wax, nor is it necessary to remove the wood chip meth hort; the type of graft in which a vegetative bud is removed from its parent plant and used as a scion to be grafted onto a stock; the bud is normally removed with a portion of green bark, which is then inserted under the green bark of the stock; this technique is widely used with fruit trees, such as stone and pome fruits, and >>> citrus, as well as other trees such as rubber, in order to grow a susceptible scion on a resistant rootstock; interspecific and intergeneric grafts are often possible *meth hort*

bud imprint: The outline of the margins, teeth, or other features of one leaf impressed on another leaf while the two leaves are pressed together in the bud, remaining as a permanent marking after the leaves become separated; this effect is particularly marked in certain species of Agave but also in cabbages or even cereals *bot*

bud mutation: >>> bud sport

bud pollination: A procedure utilized in maintaining self-incompatible parent lines by self-pollination; hybrid seed production in plant species with a sporophytic self-incompatibility system is dependent upon the production of inbred lines homozygous for a self-incompatibility allele S; in these species, a protein secretion covers the stigmatic surface just prior to anthesis and acts as a barrier to penetration of the stigma by germinating pollen grains; when buds are opened and the pollen applied before the protein barrier is formed, seed set can be obtained (e.g., in rapeseed) *meth* >>> Figure 55

bud pruning: Removal of lateral buds from a stem to prevent them from developing into branches *hort fore*

bud scale: A modified leaf, without lamina, protecting a bud *bot*

bud scar: A scar left on a shoot when the bud or bud scales drop *meth hort*

bud sport: A somatic mutation occurring in a bud of a plant; it results from local genetic alteration and produces a permanent modification; it is usually retained by grafting; this sort of mutation is often used in fruit tree breeding *gene meth hort*

bud union: The site of junction on a stem, usually swollen, where a graft bud has joined the stock following the process of budding; frequently found at or near soil level *bot hort*

buffer: A solution mixed by a weak acid and a weak alkali; it prevents changes of the pH value; therefore, it is a suitable medium for enzyme reactions *chem prep*

buffering gene(s): A complex of polygenes controlling the expression of a major gene and reducing the variability of its phenotype by toning down its reaction to environmental differences *gene*

bulb: An underground storage organ, comprising a short, flattened stem with roots on its lower surface, and above its fleshy leaves or leaf base, surrounded by protective scale leaves; estimated world production of flower bulbs is about 33,000 ha, mostly in the Netherlands, U.K., and France, and the genera *Tulipa, Lilium, Narcissus, Gladiolus, Hyacinthus, Crocus*, and *Iris* (bulbous) *bot*

bulbil: A small, bulblike structure, usually formed in a leaf axil, that separates from the parent and functions in vegetative reproduction *bot* >>> Figure 28

bulbosum technique: >>> *Hordeum bulbosum* procedure

bulbous: Tuber-shaped; forming tubers *bot*

bulbous plant: >>> bulbous

bulb planter: A sharp-edged, tapered cylinder used to remove a plug of soil or sod in which a bulb is placed; the plug is then returned to the hole in order to cover the bulb *meth hort*

bulgur (*syn* **bulghur** *syn* **burghul** *syn* **bulgar**): A cereal food made from several different >>> wheat species, most often from >>> durum wheat; in the U.S. it is most often made from white wheat; its use is most common in Middle Eastern cuisine, Iran, Turkey, Greece, Armenia, and Bulgaria; in India it is called "lapsi" *eco*

bulk breeding: The growing of genetically diverse populations of self-pollinated crops in a bulk plot with or without mass selection, generally followed by a single-plant selection; it is a procedure for inbreeding a segregating population until the desired level of homozygosity is achieved; the seeds to grow each generation are samples of those harvested from plants of the previous generation; this method is usually used for the development of self-pollinated crops; it is an easy way to maintain populations during inbreeding; natural selection is permitted to occur, which can increase the frequency of desired genotypes compared with an unselected population; it can be used in association with mass selection with self-pollination; disadvantages are (1) plants of one generation are not all represented by progeny in the next generation, (2) genotypic frequencies and genetic variability cannot be clearly defined, and (3) natural selection may favor undesirable genotypes *meth* >>> Figures 16, 51 >>> Table 24

bulk generation: In the breeding process, one or more generations in which plants of a genetically diverse population of autogamous crops are grown in a bulk plot with or without mass selection *meth* >>> bulk breeding >>> Figure 16 >>> Table 24

bulk population selection: Selection procedure in >>> self-pollinating crops; segregating populations are propagated as bulks until segregation has virtually ceased, at which time selection is initiated *meth* >>> Figure 16 >>> Table 24 >>> bulk breeding

bulk screening: A technique for obtaining a fair degree of homozygosity for the purposes of late selection; a heterozygous population of an inbreeding species is multiplied for several generations in the field with minimal or zero selection in the early stages; such early selection as does occur involves only single gene characters such as marker genes; however, >>> single seed descent in a greenhouse is usually preferable, because it is faster *meth*

bulked segregant analysis (BSA): A rapid mapping strategy suitable for monogenic qualitative traits, i.e., a technique used to identify genetic markers associated with a mutant phenotype; when DNA of a certain number of plants is bulked into one pool, all alleles must be present; two bulked pools of segregants, differing for one trait, will differ only at the locus harboring that trait; these two bulked samples can then be analyzed using techniques such as >>> RAPD to detect similarities and differences in the various loci of the genome; the two groups will have a random distribution of alleles in all loci of the genome except for loci that are associated with the mutation; a consistent difference on a locus between the two bulked samples likely means that the locus is associated with the mutation of interest; this allows geneticists to discover genes conferring disease resistance or susceptibility *biot*

bullet planting: Setting out young trees grown in bullet-shaped rigid plastic tubes, which are injected into the ground by a spring-loaded gun, sometimes into prepared holes *fore hort*

bumblebee: Any of several large, hairy social bees of the family Apidae, sometimes utilized for pollination of special crop plants; frequently, bumblebees are used for successful seed multiplication of legume and Brassica crops in greenhouses, especially of genebanks *zoo seed*

bumper mill: A machine designed to clean timothy seed by a continuous bumping action on an inclined plane; the uncleaned seed is metered onto the plane, which is continuously bumped by sets of knockers; the cylindrical timothy seeds are rolled into separate grooves while noncylindrical contaminants are jarred off the end of the inclined plane and separated *seed*

bundle sheath: A layer of cells enclosing a vascular bundle in a leaf *bot*

bunker storing: >>> clamping

Bunsen burner: A hot-flame burner using a mixture of gas and air ignited at the top of a metal tube; this device is used for sterilizing tools and container openings during aseptic transfer *in vitro* experiments; after R. W. BUNSEN (1811–1899) *prep*

bund: A low earth bank to retain irrigation water on a field, i.e., a field or section of a field is bunded *agr meth*

bundle sheath cell(s): Cells forming a sheath around a vascular bundle (>>> vein) of >>> vascular plants; in >>> C4 plants these cells become highly specialized for part of C4 photosynthesis *bot phys* >>> photosynthesis

bunt: Stinking smut; a seed-borne disease of grasses caused by Tilletia spp.; the grain is replaced by masses of fungal spores that have a characteristic fishy smell *phyt*

bur: The rough, prickly covering of the seeds of certain plants (e.g., chestnut) *bot*

burl: A woodknob (bulge, >>> bulb) on a stem, branches, or roots of broad-leaved and, less frequently, conifer tree species; it irregularly appears in places of abundant development of shoots and growth of tightly located dormant and accessory buds; it results from the entwined growth of a cluster of adventitious buds and has contorted grain; the growth rate of the burls' timber is 1.5–3 times higher than that of the normal timber; a burl is often formed on walnut, birch, black poplar, and white maple *fore*

burlap: A loosely woven fabric made of jute or hemp; used to protect newly seeded lawns from wind, water, and birds *meth agr hort*

burr: Prickly, spiny, or hooked fruit; sometimes, woody outgrowth on the trunk of the trees *bot fore*

bursiculate: Baglike *bot*

bushel: An English imperial measure of dry volume of a container or basket used to measure such a volume, typically used to measure dry goods such as grain or fruit; such a cylindrical container would be 45 cm in diameter and 20 cm deep; a bushel is equal to 4 pecks or 8 gallons (2,150.42 cubic inches or 36.368 liters); a bushel of wheat and soybeans each weighs ~60 pounds; a bushel of maize, rye, grain sorghum, and linseed each weighs ~56 pounds; a bushel of barley, buckwheat, and apples each weighs ~48 pounds *agr*

bushy grasses: Grasses forming tufts *bot*

button: The small heads of broccoli or cabbage that form as a result of seedlings being exposed to freezing temperatures *hort*

buttress root: Swollen or fluted tree trunk that aids stability in shallow rooting conditions *fore*

butyrous: Butter-like *bot*

C

C: >>> cytosine

C1, C2, C3…: The first, second, and third "generations" and/or cycles of vegetative propagation *meth*

CA: >>> combining ability

CAAS: >>> Chinese Academy of Agricultural Science

CAAT box: >>> CAT box

"Cabernet Sauvignon": The principal grape cultivar of Bordeaux, France, producing the red wine known as "claret" in England *hort*

caco-2 test: Caco-2 cell line is an immortalized line of heterogeneous human epithelial colorectal adenocarcinoma cells, developed by the Sloan-Kettering Institute for Cancer Research, U.S., through research conducted by Jorgen FOGH; these cells are very similar to noncancerous epithelial cells between the passage of 6-35 before losing the ability to differentiate;

it became widely used with *in vitro* assays to predict the absorption rate of candidate drug compounds across the intestinal epithelial cell barrier; the assay requires that drug absorption rates be determined 21 days after caco-2 cell seeding to allow for monolayer formation and cell differentiation *biot*

cactus Greek: kaktos; a prickly plant, e.g., Spanish artichoke from Sicily; in general, a spiny, succulent, dicotyledonous plant of the family Cactaceae native mainly to arid regions of North and South America, although some species are rainforest epiphytes; the fleshy stems and branches are characteristically furnished with tufts of hairs or spines coming from a common areole, a structure unique to this family; a few species of cacti have leaves; in the majority of species, >>> photosynthesis is carried out on the green surface of enlarged stems that also serve as water storage organs *bot tax*

cadang-cadang disease (of coconuts): A lethal disease in the Philippines, caused by a viroid; this disease should be considered a grave phytosanitary risk in all other coconut areas *phyt*

cadastral gene: A plant gene that controls the expression of floral homeotic genes *biot*

cadastre: An official register of the ownership, extent, and value of real property in a given area; used as a basis of land taxation *agr*

cadmium (Cd): A toxic heavy metal that occurs naturally in soils; durum wheat is known to accumulate generally more Cd than other cereal crops; the uptake of Cd in durum wheat is governed by the gene *Cdu1*, which co-segregates with several DNA markers, such as the co-dominant marker *usw47* and the dominant marker *ScOPC20 phys*

caduceus: Plant parts that fall off early or prematurely *bot*

caespitose (cespitose): Tufted; several or many stems in a close tuft *bot*

CAF1: >>> chromatin assembly factor 1

caffeine: A white, crystalline, bitter alkaloid, $C_8H_{10}N_4O_2$, usually derived from coffee or tea; it can be an efficient poison against garden snails; as spray applied to the leaves with a concentration of 0.01% it prevents snail attacks; concentrations between 1 and 2% are even lethal *chem phys phyt*

caging: If safe isolation distances are not available and time isolation is not suited to the plant in order to prevent cross pollination, caging or bagging techniques are used to insure against crossing; in caging, plants are protected by cages covered with mesh or fabric; the weave of the covering must be small enough to prevent passage of insects or pollen (depending on whether the plant is insect- or wind-pollinated); bags over individual flowers or flower heads can be used for self- or wind-pollinating plants; in this case, the bags simply act as tiny "cages" *meth*

Cajal body (CB): A nuclear structure that is found in both plants and animals; it contains components of at least three RNA-processing pathways *biot*

calceolate: Shoe-like in form *bot*

calcicole: A plant that grows best in calcium-rich soils *bot agr*

calcicolous plant: >>> calcicole

calciphobe: Reduced growth on calcium-rich soils *bot*

calciphyte: >>> calcicole

calcium (Ca): A silver-white bivalent metal, combined in limestone or chalk *chem*

calcium deficiency: Condition which describes a plant that is not getting enough calcium; among other things, a deficiency of calcium can cause the leaves to pale and become deformed or twisted *phys agr*

caliche: A zone near the soil surface that is more or less cemented by secondary carbonates of calcium or magnesium precipitated from the soil solution; it may occur as a soft, thin soil horizon, a hard, thick bed, or a layer exposed by erosion *agr*

calicle: >>> callycle

caliper: An instrument to measure diameters of trees or logs *fore* >>> hypsometer

callogenesis: >>> callus

callose: Hard or thick and sometimes rough organic matter *bot*

callus (calli pl): Tissue that forms over a wound or that develops from actively dividing plant tissue in a tissue culture; usually a disorganized mass of undifferentiated cells biot; in botany, the thickened part of the base of, e.g., oat grain *bot* >>> Figure 58

callus culture: The *in vitro* culture of callus, often as the first stage in the regeneration of whole plants in culture *biot* >>> tissue culture >>> Figures 57, 58

callus induction: Undifferentiated plant tissue is produced at wound edges; callus can also be induced and grown *in vitro* by varying the ratio of hormones (e.g., auxin and cytokinin) in the growth medium *biot* >>> Figures 57, 58

callycle: A protective structure around a flower formed collectively by the sepals *bot*

Calvin cycle: The second stage in the process of >>> photosynthesis (it is also called the CALVIN–BENSON cycle or the carbon fixation cycle); in the CALVIN cycle, carbon molecules from carbon dioxide, CO_2, are fixed into the sugar glucose, ($C_6H_{12}O_2$) (in six repeats of the cycle); it takes place in the stroma of eukaryotic chloroplasts; the major enzyme that mediates the CALVIN cycle is rubisco (ribulose-1-5-biphosphate carboxylase); the CALVIN cycle was first investigated in the late 1940s and early 1950s by the Nobel Prize-winning chemist Melvin CALVIN (1911–1997) *chem phys*

calycular: Cuplike *bot*

calypter: >>> calyptra

calyptra: A cap or hood covering a flower or fruit; in mosses and liverworts, a thin hood fitting over the top of the spore >>> capsule *bot* >>> root cap

calyx (calyces pl): The outer part of a flower; all the sepals of a flower *bot*

CAM: >>> crassulacean acid metabolism

cambium: In the stem and roots of vascular plants, a single layer of cells lying between the xylem and phloem *bot*

camerate: Chamberlike *bot*

campanulate: Bell-like in form *bot*

campylotropous: Form of ovule in which the micropile and chalaza are laterally placed to the placenta *bot*

Canada balsam: Resin distilled from the bark of *Abies balsamea* (balsam fir) and other similar species; used in cytology for mounting (e.g., chromosome spreads) *cyto micr*

candidate gene: A gene whose function suggests that it may be involved in the genetic variation observed for a particular trait (phenotype, disease, or condition), e.g., the gene for growth hormone is a candidate gene for straw length in cereals or dry matter production; candidate genes can be divided into two categories: positional and functional; a positional candidate gene is one that might be associated with a trait, based on the location of a gene on a chromosome; a functional gene is one whose function has something in common with the trait under investigation; positional candidate genes are identified through QTL- and map-based cloning approaches, whereas functional genomics approaches, such as transcriptomics and expression genetics, provide the set of functional candidate genes *gene meth biot* >>> QTL mapping >>> Figure 43 >>> Table 58

candidate-gene strategy: An experimental approach in which knowledge of the biochemistry and/or physiology of a trait is used to draw up a list of genes whose protein products could be involved in the trait *biot* >>> Figure 43

candidate population: In forestry, trees, which are planted to serve as a base for forward selection or in some way selected from that for further studies; for example, selected phenotypic selections may serve as a candidate population, which is subject to further progeny testing before reselection to the breeding population; the genotypes taken into consideration for the breeding population *meth fore* >>> Figure 51

candidate single nucleotide polymorphism: Any single nucleotide polymorphism in an exon of a gene that can be expected to have an impact on the function of the encoded protein *biot*

>>> causative single nucleotide polymorphism >>> clone overlap single nucleotide polymorphism >>> coding single nucleotide polymorphism >>> SNP

candidate SNP: >>> candidate single nucleotide polymorphism

candidate tree: A tree that has been tentatively selected for inclusion in a breeding program, but has not yet been measured or compared with surrounding trees *meth fore*

candidate variety: Breeding strains, lines, or hybrids of high grade that by a breeder or institution are announced for official national (and international) performance testing in order to be released as a certified variety *seed*

candle: The new shoot growth on needled evergreens before the needles expand *hort fore*

cane: A long and slender, jointed, rigid, woody stem that is hollow or pithy (e.g., in grasses, palms, rattan, bamboo, or sugarcane) bot; in viticulture, a mature, woody, brown shoot as it develops after leaf fall; canes were last year's fruiting or renewal shoots; the buds on the canes will produce this season's fruiting shoots *hort*

canescent: Densely covered with grayish or whitish, short, soft hairs *bot*

cane sugar: >>> saccharose

canker: A plant disease in which there is sharply limited necrosis of the cortical tissue (e.g., in apple); in rape, it causes leaf spotting over winter and cankers on the stem later; the latter are the more serious and appear after flowering caused by Leptosphaeria maculans, asexual stage Phoma lingam phyt; sharply defined dead area of tissue on stem *phyt* >>> Figure 62

cannabinol (THC): >>> hemp

Canola™: A type of rapeseed variety (Brassica napus, B. rapa, and B. juncea) that has been developed and grown in Canada; name after the "Canadian-Oil-Low-Acid" breeding program; Canola™ is a registered trademark, corresponding to specified low contents in erucic acid in oil and in glucosinolates in meals equivalent to double "0" in the European Union standard; it was initially obtained by conventional breeding, but in recent years herbicide-tolerant varieties have been developed *agr*

canonical self-incompatibility: A form of self-incompatibility characterized by full expression in very reproductive phase *gene bot*

canopy: The vertical projection downward of the aerial portion of plants, usually expressed as percent of ground so occupied *bot meth*

canopy temperature depression (CTD): The cooling effect exhibited by a leaf as transpiration occurs; it gives an indirect estimate of stomatal conductance, and is a highly integrative trait being affected by several major physiological processes including photosynthetic metabolism, evapo-transpiration, and plant nutrition; it has potential for complementing early generation phenotypic selection in plants *phys*

cantharophily: Pollination by beetles *bot*

cap: A chemical modification that is added to the 5′ end of a eukaryotic mRNA molecule during post-transcriptional processing of the primary transcript; it is introduced by linking the terminal phosphate of 5′ GTP to the terminal base of the mRNA; the added G is methylated, giving a structure of the form 7MeG5′ppp5′Np *biot* >>> cap site

Cape St. Paul wilt disease: The Ghanaian form of lethal yellowing disease (LYD) of coconut, caused by a phytoplasma and has been active in Ghana since 1932 *phyt* >>> Figure 62

capillarity: The process by which moisture moves in any direction through the fine pores and as films around particles *agr*

capillarity moisture: The amount of water that is capable of movement after the soil has drained; it is held by adhesion and surface tension as films around particles and in the finer pore spaces *agr*

capillary: Very slender or hair-like *bot*

capillary action: It is the movement of water as it is pulled upwards through tubes (xylem) within a plant's roots, stems, and leaves; the water (containing minerals and dissolved nutrients) is driven against gravity by adhesion of the water molecules (they stick to the sides of the tubes), cohesion of those molecules (the water molecules sticking together), and surface tension (the forces of the molecules on surface of the upward-moving water) *phys*

capillary electrophoresis: Biochemical technique to separate DNA fragments in an electric field, carried out within narrow tubes (capillaries) *meth chem*

capillary matting: A material used to promote the process of capillary action for the purpose of providing plants with the correct amount of water; self-watering devices which use capillary matting, for example, include the Watermaid *meth phys hort*

capitate: In a globular cluster or head *bot*

capitellate: Possessing a minute swelling at the apex *bot*

capitulum: Flower head; an aggregation of flowers on a flat platform and edged by bracts *bot*

cappiliform: Hair-like *bot*

capping: Modification of the 5′ end of RNA, which has consequences for translation efficiency *biot*; in plant propagation, a hardened layer of surface soil sometimes caused by top-watering; capping can impede the proper distribution of water through the soil *agr hort phys*

CAPS: >>> cleaved amplified polymorphic sequences

capsid: Protein coat that encloses DNA or RNA molecules of bacteriophage or virus *biot*

cap site: The probable transcription initiation site of a eukaryotic gene; the primary transcripts of most eukaryotic mRNAs have an adenine (A) in the first position and the cap is added 5′ to it *gene biot* >>> cap >>> Table 38

capsular fruit: >>> capsule

capsule: A dehiscent fruit with a dry >>> pericarp usually containing many seeds; it is composed of more than one >>> carpel (e.g., Brazil nut); in moss or liverwort, the spore-bearing structure *bot*

capture probe: Phage or antibody probes that bind proteins in a sample such that their relative expression levels can be detected *biot*

carbohydrate: An organic compound based on the general formula $C_x(H_2O)_y$; the dominating substances of the cell sap are soluble carbohydrates; they occur as disaccharides, such as saccharose and maltose, or as monosaccharides, such as glucose and fructose; the simplest carbohydrates are the sugars (saccharides), including glucose and sucrose; polysaccharides are carbohydrates of much greater m.w. and complexity; examples are starch, which serves as an energy store in plant seeds and tubers; cellulose and lignin that form the cell walls and woody tissue of plants; glycogen, etc. *chem phys* >>> Table 16

carbonate: A salt or ester of carbonic acid *chem*

carbon cycle: The biological cycle by which atmospheric carbon dioxide is converted to carbohydrates by plants and other >>> photosynthesizers, consumed and metabolized by organisms, and returned to the atmosphere through respiration or decomposition *phys*

carbon dioxide: A gaseous compound that is formed when carbon combines with oxygen *chem phys*

carbon intensity: The quantity of greenhouse gas emission, assessed as CO_2 equivalents, attributable to one ton of crop produce *phys*

carboxyl group: The univalent group COOH, characteristic of organic acids *chem phys*

carboxylase: Any of the class of enzymes that catalyze the release of carbon dioxide from the carboxyl group of certain organic acids *phys*

carboxylation: Introduction of a carboxyl group into an organic compound *phys*

carcinogen: An agent or substance that causes cancer *bio*

carded: A yarn preparation; during the carding process raw cotton is separated, opened, cleaned, and made into sliver *agr*

carding: The process of untangling and partially straightening fibers by passing them between two closely spaced surfaces *meth agr*

carinate: Keeled, boat-like *bot*

carmine: Used for preparation of carminic acid; this is a red dye used for coloring or staining chromosomes and other cytological material; carmine is prepared from cochineal *meth cyto* >>> cochineal

carmine staining: >>> aceto-carmine staining

Carnoy's fixative A fixator solution, which consists of six parts ethanol, three parts chloroform: one part acetic acid; used in chromosome analysis *micr cyto*

carotene: A compound of carbon and hydrogen that occurs in plants; it is a precursor of vitamin A; its reddish-orange plastid pigment is involved in light reactions in >>> photosynthesis *chem phys*

carotenoid: ·Red to yellow pigments responsible for the characteristic color of many plant organs or fruits, such as tomatoes, carrots, etc.; oxidation products of carotene are called xanthophylls; carotenoids serve as light-harvesting molecules in photosynthetic assemblies and also play a role in protecting prokaryotes from the deleterious effects of light *phys chem*

carotin: >>> carotene

carpel: One of the female reproductive organs of the flower, comprising an ovary and usually with a terminal style tipped by the stigma *bot* >>> Figure 35

carrier: Typically, an individual that has one recessive mutant allele for some defective condition that is "masked" by a dominant normal allele at the same locus, i.e., an individual that is heterozygous for a recessive harmful allele and a dominant normal allele; the phenotype is normal, but the individual passes the defective (recessive) allele to half of its offspring *gene*

carrier DNA: DNA of undefined sequence content, which is added to the transforming (plasmid) DNA used in physical DNA-transfer procedures; this additional DNA increases the efficiency of transformation in electroporation and chemically mediated DNA-delivery systems *biot*

carrying capacity: The density of a density-regulated population at equilibrium *gene*

Cartagena Protocol on Biosafety: The first international treaty dealing with the movement of genetically modified organisms (GMOs) across country borders; the protocol was drawn up under the Convention of Biological Diversity and came into force in September 2003; more than 100 countries have ratified the agreement; although the biosafety protocol was pushed for by the South and drafted as a promise of legal protection against the introduction of GMOs, the weakness of its provisions means that the protocol and the national biosafety laws following its introduction are being steadily turned into tools to facilitate the introduction of GMOs *biot*

cartenoid: A yellow, orange, red, or brown pigment that is located in the chloroplast and chromoplast of plants; it acts as a >>> photosynthetic accessory pigment *bot*

cartilaginous: Firm and tough, but flexible; like cartilage *bot*

caruncle: A reduced aril in the form of a fleshy, often waxy or oily outgrowth near the hilum of some seeds *bot*

caryophyllaceous: Refers to petals that have a long claw at the base *bot*

caryopsis (caryopses *pl*): The single-seeded, dry, and nutlike fruit of the grasses in which the mature ovary wall (pericarp) and the seed coat (testa) are fused (e.g., in cereals as wheat, rye, or barley) *bot*

casein: A protein precipitated from milk; sometimes used for *in vitro* techniques *chem biot*

cash crops: A readily salable crop that is grown and gathered for the market (as vegetables or cotton, tobacco, coffee, sisal, etc.) *agr* >>> truck crop

Casparian strip: A band-like wall formation within primary walls that contains suberin and lignin; typical of endodermal cells, in which it occurs in radial and transverse anticlinal walls *bot*

cassava root rot disease: An increasing problem in Africa, particularly in the sub-Saharan region where yield losses of about 80% have been recorded and where cassava accounts for approximately one-third of the total staple food production; this disease is caused by different root rot fungi (*Botryodiplodia theobromae, Nattrassia mangiferae,* and *Fusarium* ssp.) *phyt* >>> Figure 62

cassette mutagenesis: A procedure in which >>> mutations or alterations of individual >>> DNA >>> codons are systematically induced in order to determine the effect of those alterations on protein folding or function; a cloned gene is mutated by synthesis of short segments of the gene with random base substitutions, insertion of these altered polynucleotide sequences into the gene and then transformation of cells with the mutated gene; cells that contain a functional protein are selected and the protein (or gene) is examined to identify the successful mutation; testing of many such short segments until the entire structural gene is examined maps the areas of tolerance and of sensitivity of the gene product to substitutions *biot meth* >>> TAL effector nuclease

castrate: >>> emasculate

CAT: >>> chloramphenicol acetyl transferase

catabolic: Pertaining to an enzymatic reaction leading to the breakdown of a complex biological molecule into less complex components, which may either yield energy in the form of >>> ATP or be used in subsequent anabolic reactions *chem phys*

catalase: An enzyme that catalyzes the degradation of hydrogen peroxide to water and the oxidation by hydrogen peroxide of alcohols to aldehydes during seed germination *phys*

catalyst: A substance that initiates or accelerates a chemical reaction without apparent change in its own physical or chemical properties *chem*

catalyze: To induce or accelerate a chemical reaction by a substance that remains unchanged in the process *chem phys*

cataphyll: In cycads, a scale-like modified leaf that protects the developing true leaves *bot*

cataphyllary leaf: >>> cataphyll

catapult fruits: Those fruits that discharge their seeds forcefully *bot* >>> ballistic fruits

CAT box: A conserved nucleotide sequence within the promoter region of numerous eukaryotic structural genes *gene biot*

catch crop: A method of increasing agricultural or horticultural productivity by filling in the empty spaces; for example, it is created when slower-growing vegetables are harvested with fast-growing crops; in general, a short-duration crop grown in between two main crops in a rotation to maximize cropping intensity, e.g., summer >>> greengram grown in between two main cereal crops (wheat–greengram–maize); usually grown without any extra nutrient application and expected to feed on the residues of nutrients applied to the main crops *agr* >>> stubble crop >>> underplant crop

catechin: A polyphenolic antioxidant plant metabolite; sometimes refers to the related family of flavonoids and the subgroup flavan-3-ols (flavanols); the term "bioflavonoid" was first used to describe the flavonols, but as an imprecise term has been loosely applied to the larger family of flavonoids, including also the polymeric hydroxyl – only containing flavan-3-ols; catechins are abundant in >>> teas (*Camellia sinensis*) as well as in some >>> cocoas (*Theobroma cacao*); the name of the catechin chemical family derives from "catechu" which is the juice or boiled extract of *Mimosa catechu* (*Acacia catechu*) *phys chem*

catenation: The formation of rings or chains by chromosomes at diakinesis *cyto*

caterpillar: Larva of a moth, butterfly, or sawfly *phyt*

cat gene (CAT): Chloramphenicol acetyl-transferase gene (and protein); it is used as a selective marker for cloning vectors and as a reporter gene *biot* >>> chloramphenicol acetyl-transferase

cation: A positively charged ion in solution *phy chem*

cation exchange: The exchange between cations in solution and another cation held on the exchange sites of minerals and organic matter *chem agr*

catkin: A pendulous spike, usually of simple, unisexual flowers *bot*

caudate: Having a tail *bot*

caudex: A persistently enlarged, woody base of the stem or trunk (located just below the ground) on some plants used for water storage; many desert plants have a caudex, an adaptation to dry conditions; some palms, cycads, and succulents have a caudex *bot*

caudiciform: Having a caudex >>> caudex

caudicle: An extension of tissue derived from the anther and connected to pollinia (e.g., in orchids) *bot hort*

caulescent: Becoming stalked, having a stem *bot*

cauliflorous: Borne on the trunk *bot*

cauliflory: >>> cauliflorous

cauliflower mosaic virus (CaMV): A virus that infects cauliflower and other Cruciferae; it is transmitted by insects; the genome size is about 8 kb; it consists of double-stranded DNA with some single-stranded segments; in molecular genetics, it is used as a vector for transformation experiments *phyt gene biot*

cauliflower mosaic virus 35S promoter (CaMV 35S): A promoter (specific sequence of DNA) that is often utilized in genetic engineering to control expression of inserted gene; in other words, synthesis of desired protein in a plant *biot*

cauline leaf: A leaf formed on the florescence stem of a rosette plant (e.g., in >>> thale cress) *bot*

caulocarpous: Bearing fruits on the stalk or trunk *bot*

caulogenesis: Shoot formation or stem organogenesis bot; as de novo shoot development from callus *biot* >>> Figure 58

causal agent: An organism or agent that incites and governs disease or injury *phyt*

causative single nucleotide polymorphism: Any single nucleotide polymorphism that is in linkage disequilibrium to a disease phenotype and therefore a responsible candidate for the disease *biot* >>> candidate single nucleotide polymorphism clone overlap single nucleotide polymorphism >>> coding single nucleotide polymorphism >>> SNP

causative SNP: >>> causative single nucleotide polymorphism

cavitation: Spontaneous occurrence of a vapor phase in a liquid under tension; it occurs in the xylem of plants that are under moderate to severe water-deficit stress *phys*

cavity: A tunnel left inside the maize stalk, for example, from a European corn borer feeding, or a hole or hollow area, especially inside a tree *phyt*

C banding: A cytological staining technique for chromosomes that labels regions around the centromere with Giemsa stain; usually a bandlike and darkly stained structure appears, which consists of heterochromatin; the technique is intensively used in chromosome identification and genome characterization, including structural changes and polymorphisms *cyto meth* >>> banding >>> Giemsa staining >>> Table 43

CBF/DREB transcription factor(s): CCAAT-binding factor (CBF) and >>> dehydration responsive element binding (DREB) are a group of plant >>> transcription factors that activate expression of genes in response to >>> abiotic stress; the products of the expressed genes act to alleviate the effects of the stress *phys*

CBP: >>> CHORLEYWOOD baking process

CCC: >>> chlorocholine chloride

cccDNA: Covalently closed circles of DNA; it does not show nicks; only cccDNA can be supercoiled *biot*

^{14}C dating: >>> radiocarbon dating

cDNA: >>> complementary DNA

cDNA cloning: A method of cloning the coding sequence of a gene, starting with the >>> mRNA transcript; it is normally used to clone a DNA copy of a eukaryotic mRNA *biot*

cDNA library: A collection of >>> cDNA clones that were generated *in vitro* from the mRNA sequences isolated from an organism, a specific tissue, cell type, or population of an organism *biot meth*

cecidium (cecidia pl) A plant gall generally caused by an insect but sometimes by a fungus *phyt* >>> www.pflanzengallen.de

cefotaxime (sodium, $C_{16}H_{17}N_5O_7S_2$): A third-generation cephalosporin antibiotic and selective antibiotic used in DNA transformation studies; like other third-generation cephalosporins, it has broad-spectrum activity against Gram-positive and Gram-negative bacteria; it inhibits bacterial cell wall synthesis by binding to one or more of the penicillin-binding proteins which in turn inhibits the final transpeptidation step of peptidoglycan synthesis in bacterial cell walls, thus inhibiting cell wall biosynthesis; bacteria eventually lyse due to ongoing activity of cell wall autolytic enzymes (autolysins and murein hydrolases) while cell wall assembly is arrested; in many life science laboratories, the *in vitro* culturing of bacterial, plant, and animal cells is a routine task; antibiotics can be used to ensure successful growth of cells by eliminating unwanted bacterial strains and fungi, while maintaining the health and vitality of the desired cells *chem biot*

c-effects: Non-genetic causes of variation, for example, maternal or cloning effects; initially "c" meant "common" effects *stat*

cell: The basic structural and functional unit of a plant; it is a system surrounded by membranes and is compartmentalized into specific functional areas and/or organelles with special tasks *bot*

cell adhesion: The contact between cells that is involved in cell aggregation and intercellular communication *cyto*

cell culture: The growing of dispersed cells *in vitro* derived from multicellular tissue, organs, or organisms *biot* >>> biotechnology

cell cycle: The sequence of events that occurs between the formation of a cell and its division into daughter cells; it is conventionally divided into G0, G1, (G standing for gap), S (synthesis phase during which the DNA is replicated), G2, and M (mitosis) *cyto*

cell division: The reproduction of a cell by karyogenesis and cytogenesis *bot*

cell envelope: The different surface components of the cell that are present outside the cytoplasmic membrane *bot*

cell furrow: The wall formed between the two daughter nuclei at the end of telophase *cyto*

cell fusion: Fusion of two previously separate cells, occurs naturally in fertilization; it can be induced artificially by the use of fusogens such as polyethylene glycol; fusion may be restricted to cytoplasm, nuclei may fuse as well; a cell formed by the fusion of dissimilar cells is referred to as a heterokaryon *bio biot* >>> biotechnology

cell generation time: The time span between consecutive divisions of a cell *cyto* >>> cell cycle

cell heredity: Inheritance of a cellular level *gene*

cell hybridization: The fusion of a somatic cell *in vitro* and formation of viable cell hybrids *biot* >>> cell fusion

cell line: A population of cells that derives from a primary cell culture *cyto biot*

cell manipulation: >>> biotechnology

cell membrane: A component of the cell surface with a discrete structure and function *bot*

cell nucleus: >>> nucleus

cellome: The entire complement of molecules and their interactions within a cell; it is the information held within the cellome that defines the temporal and spatial interactions of cellular components, and thus normal and abnormal functions; the knowledge base of the cellome is built by connecting layers of these interactions into the pathways and networks that govern all aspects of cellular life *cyto biot*

cell plate: The structure formed between daughter nuclei after karyokinesis *cyto*

cell population: A group of cells that is static (without mitotic activity), expanding (showing scattered mitosis), or renewing (in which mitosis is abundant) *bot*

cell proliferation: The increase in cell number as a result of cell division; the accurate assessment of cell proliferation is useful in many biological assays and is a key readout in a wide range of pharmacological and regulatory studies; observations by HOWARD and PELC led to the introduction of the concept of the cell cycle and its subdivision into several phases; DNA synthesis (>>> replication) and doubling of the genome take place during the >>> S phase; this is preceded by a period of variable duration known as the rst >>> gap phase (G1), which separates the S phase from the previous mitosis (>>> M phase); the S phase is followed by a period of apparent inactivity known as the second gap phase (G2), which comes before the next mitosis; >>> interphase comprises successive G1, S, and G2 phases, and forms the largest part of the cell cycle; for a typical DNA histogram one peak represents the G1 and another (with twice the value) represents the G2/M phase of the cell cycle; S phase cells are spread between the two peaks *cyto*

cell recognition: The mutual recognition of cells due to antigen–antibody or enzyme–substrate reactions *phys*

cell sap: The interorganelle fluid of the cell *bot*

cell selection: Selection within a population of genetically different cells *in vitro* by different means and different approaches *biot*

cell sorting: A procedure that uses a mechanical device in order to separate mixtures of cells by their size, DNA content, etc. *cyto meth* >>> sorting

cell strain: A population of cells derived either from a primary culture or from a single cell *biot*

cell surface: The multicomponent structure surrounding a cell *bot*

cell suspension: Cells and small aggregates of cells suspended in a liquid medium, often used to describe suspension cultures, of single cells or cell aggregates *biot*

cell synchrony: When a population of cells proceeds through the stages of cell cycle with synchrony (i.e., it divides at one time) *biot phys cyto*

cell tetrad: >>> tetrad

cell transformation: A stable heritable alteration in the phenotype of a cell, usually brought about by viral or bacterial infection, but also by experimental means *biot*

cell tray: >>> growing tray

cellular differentiation: The transformation of apparently identical cells, arising from a common progenitor cell, into diverse cell types with different biochemical, physiological, and structural specializations; in plants, cell differentiation is frequently reversible, particularly when plant tissue is excised and maintained in culture; such differentiated cells can reinitiate cell division and, given appropriate nutrients and hormones, even regenerate whole plants; differentiated cells, with some exceptions, retain all the genetic information (encoded in DNA) required for the development of a complete plant, a property called >>> totipotency; this is because most cells differentiate by regulating gene expression, not by altering their genome; exceptions to this include cells that lose their nuclei (phloem sieve tube cells or xylem tracheids) *phys*

cellular endosperm: A type of endosperm in which the early development is characterized by cell wall formation accompanying each nuclear division *bot*

cellular pathways: About 20 μm between the cell membrane and the genetic material in the cell nucleus is the space of the molecules of signal transduction, of the intricate and multifaceted redundancy of the pathways that take signals from the membrane and convert them into the exquisitely selective control of our genes; within these pathways, the regulation of gene transcription is carried out by a multitude of hormones and growth factors *phys*

cellular respiration: A process in which energy is produced from various molecules (like glucose), producing ATP (>>> adenosine triphosphate); during cellular respiration, oxygen is used and carbon dioxide is produced; cellular respiration occurs in the mitochondria of eukaryotes, and in the cytoplasm of prokaryotes *phys*

cellulase: An enzyme that digests cellulose; sometimes used for maceration of plant tissue in order to improve spreading of chromosomes *phys micr cyto*

cellulose: A long-chain complex carbohydrate compound (polysaccharide); it is the chief substance forming cell walls and the woody parts of a plant *bot*

cell wall: The steadfast external coat that surrounds the cell *bot*

cenospecies: A closely related independent species, capable of interbreeding and thereby gene exchange *bot*

census number: The actual counted number *meth*

center of diversity: A geographical location or local region where a particular taxon exhibits greater genetic diversity than it does anywhere else; N. I. VAVILOV developed this concept; he considered that the centers of diversity are also the centers of origin of a crop species; but the centers of diversity and the center of origin have subsequently been found to be distinct phenomena; the global centers of origin of crop plants and/or centers of diversity are summarized as follows (1) China (mountains of Central and Western China and adjoining areas): soybean, Brassica spp., radish, poppy, millets, buckwheat, fruit trees, mulberry, naked oat, naked barley; (2) India and Indo-Malaya (India [without the Northwest], Burma, Indochina, Malaysia): rice, sugarcane, banana, cocos palm, pepper, jute; (3) Central Asia (northwest India, Pakistan, Afghanistan, Tadzhikistan, Tienshan, Uzbekistan): bread wheat, broad bean, pea, lentil, carrot, onion, grape, spinach, apricot; (4) West Asia (Transcaucasia region, Iran, Turkmenistan, Asia Minor): emmer wheat, einkorn wheat, rye oats, barley, vetches, alfalfa, clovers, plums, pea, lentil, fig; (5) Mediterranean coastal and adjacent regions (regions surrounding Mediterranean Sea): vegetables, rape, lupines, beets, clovers, pea, lentil, flax, olive, broad bean, seradella; (6) South Mexico–Central America (South Mexico, Central America, Antilles): maize, Phaseolus beans, sweet pepper, sweet potato, cotton, sisal, cacao, tomato, cucumber, pumpkins; (7) South America (Peru, Chile, Bolivia, parts of Brazil): maize, potato, tomato, cotton, peanuts, bananas, tobacco, rubber tree; (8) North America: lupines, grape, strawberry, sunflower; (9) East Africa: coffee, ricinus, sorghum millet, emmer wheat, barley, linseed *gene tax evol* >>> center of origin >>> center of domestication

center of domestication: The area believed to be that in which a particular crop species was first cultivated *gene* >>> center of diversity

center of origin: An area from which a given taxonomic group of plants has originated and spread and/or where wild-type species are found in greatest genetic variation; the theory was first published by V. I. VAVILOV in 1922 *gene* >>> center of diversity >>> Table 17

center pivot irrigation: A method of agricultural irrigation using a long, wheeled arm with many nozzles that pivots about the center of a circle; used primarily in arid regions *agr meth*

centgener method: One of the earliest established pure-line systems of plant breeding based on 100 selected plants *meth*

centimorgan: Equals 1% crossing-over gene >>> MORGAN unit

central axis: The main axis of the inflorescence *bot*

central cell: The largest cell in the center of the embryo sac surrounding the egg apparatus at the micropylar end *bot*

central dogma: Refers to F. CRICK's seminal concept that in nature genetic information goes from DNA to RNA to protein, i.e., the basic concept that in the nature genetic information generally only can flow from DNA to RNA; however, it is now known that information contained in RNA molecules of certain viruses (retroviruses) can also flow back to DNA *gene*

central groove: A longitudinal depression in the sides of the pedicels of some species *bot*

central mother cell: A large, vacuolated subsurface cell in a shoot apical meristem *bot biot*

central nervure: >>> mid rip

central spine(s): >>> spine

centric: Chromosomes having a centromere, as opposed to acentric (having no centromere) *cyto* >>> Figure 11

centric constriction: The visible bight along a chromosome that bears the centromere *cyto* >>> Figure 11

centric fission: A chromosomal structural change that results in two acrocentric or telocentric chromosomes from one metacentric chromosome; as opposed to centric fusion *cyto* >>> Robertsonian translocation >>> Figure 37

centric fusion: The whole-arm fusion of chromosomes by the joining together of two telocentric chromosomes to form one chromosome *cyto* >>> Robertsonian translocation >>> Figure 37

centric region: The region where the centromere is placed *cyto* >>> Figure 11

centrifugal divider: A seed separator whose mode of operation is based on centrifugal forces *seed*

centrifugal spreader: An applicator from which dry particulate material is broadcast as it drops onto a spinning disk or blade beneath the hopper *agr*

centrifuge: An apparatus that is used to spin liquids in a circular motion at high rates of speed; particles that are suspended in a liquid medium can be separated according to their density, the heavier particles collecting at the outer rim of the circle and the less dense ones collecting in layers toward the center *meth prep*

centriole: In mitosis, this small spherical body forms the center of the astral rays *cyto*

Centro Internacional de Agricultura Tropical (CIAT, Cali, Colombia): Responsible for research on dwarf beans, cassava, and forage crops breeding and research *org* >>> https://ciat.cgiar.org/

Centro Internacional de Mejoramiento de Maiz y Trigo (CIMMYT, Mexico DF, Mexico): Responsible for wheat, maize, barley, and triticale breeding and research *org* >>> www.cimmyt.org/

Centro Internacional de al Papa (CIP, Lima, Peru): Seeks to reduce poverty and achieve food security on a sustained basis in developing countries through scientific research and related activities on potato, sweet potato, other root and tuber crops, and on the improved management of natural resources in the Andes and other mountain areas *org* >>> https://cipotato.org/

centromere: The structure to which the two halves of a chromosome, the chromatids, are joined; the centromere is generally flanked by repetitive DNA sequences and it is late to replicate; the centromere is an A–T region of about 130 bp; it binds several proteins with high affinity to form the >>> kinetochore; it contains the kinetochore that attaches to the spindle during nuclear division; thus, the centromere is the DNA region of the eukaryotic chromosome that determines kinetochore formation and sister chromatid cohesion; centromeres interact with spindle microtubules to ensure the segregation of chromatids during mitosis and of homologous chromosomes in meiosis; the origin of centromeres, therefore, is inseparable from the evolution of cytoskeletal components that distribute chromosomes to offspring cells; the centromeres originated from telomeres; the breakage of the ancestral circular genophore activated the transposition of retroelements at DNA ends that allowed the formation of telomeres by a recombination-dependent replication mechanism; afterward, the modification of the tubulin-based cytoskeleton that allowed specific subtelomeric repeats to be recognized as new cargo gave rise to the first centromere; this switch from actin-based genophore partition to a tubulin-based mechanism generated a transition period during which both types of cytoskeleton contributed to fidelity of chromosome segregation; during the transition, pseudodicentric chromosomes increased the tendency toward chromosomal breakage and instability; this instability generated multiple telocentric chromosomes that eventually evolved into metacentric or holocentric chromosomes *cyto* >>> Figures 11, 70

centromere (centromeric) index (CI): Chromosome length of the short arm divided by the total length of the chromosome × 100 *cyto*

centromere interference: An inhibitory influence by the centromere on crossing-over and the distribution of chiasmata in its vicinity *cyto*

centromere mapping: The localization of the centromere on individual chromosomes by, e.g., fluorescent *in situ* hybridization *cyto meth biot* >>> FISH

centromere misdivision: A transverse instead of lengthwise division of the centromere resulting in telocentric chromosomes *cyto* >>> Figure 37

centromere orientation: The process of orientation of centromeres during prometaphase of mitosis and meiosis that contributes to a proper segregation of chromatids or chromosomes during anaphase *cyto*

centromere repulsion: The mutual repulsion of the centromeres of paired chromosomes toward the end of the meiotic prophase *cyto*

centromere shift: The displacement of centromeres by structural changes of the chromosomes (e.g., translocations, inversions, etc.) *cyto*

centrosome: The site of spindle fiber organization, which provides a polarity to the dividing cell; centrosomes are also referred to as "microtubule organizing centers"; centrosome replication precedes the onset of DNA synthesis; centrosomes roughly double in size; a substantial amount of the proteins in centrosomes appear to be >>> tubulins; then, DNA synthesis occurs; this is the S phase of the cell cycle; finally, centrosomes divide and begin migration to the poles; as the centrosomes migrate to opposite poles, they play out polar spindle fibers, such that fibers from one pole are interdigitated with fibers from the opposite pole; these polar fibers will be used in anaphase to push the poles of daughter cells in opposite directions *cyto*

cephalobrachial: >>> acrocentric

ceraceous: Waxy or wax-like *bot*

cereals: Members of the grass family in which the seed is the most important part, used for food and feed *agr bot* >>> caryopsis >>> Tables 15, 48

cereal leaf beetle (*Oulema melanopus*): A pest of cereals, grains, and various grasses; it has the potential to cause significant economic losses as was demonstrated in Hungary (1891), Romania (1931), and Spain (1938–1939); all cereals, such as barley, wheat, oats, rye, maize, and wild grasses, may serve as hosts; its distribution ranges from Asia (Afghanistan, Azerbaijan, China, Cyprus, Georgia, Iran, Israel, Kazakhstan, Mongolia, Siberia, Syria, Turkey, Turkmenistan, Uzbekistan) to Europe, North Africa (Algeria, Morocco, Tunisia), and North America (Canada, United States); adult beetles overwinter in clusters in protected places, such as in the crevices of tree bark, under field trash, inside rolled leaves, etc.; exposure to –15°C for about a week kills 90% of the overwintering population; adults become active in the spring when the temperature reaches +20°C and feed initially on wild grasses; later, they move onto young cereal plants; egg laying begins about 14 days after the emergence of the adults; each female may lay 100 to 400 eggs over a 50-day period; the larvae hatch in about 5 days and begin feeding; after feeding is completed, the full-grown larva enters the soil and pupates in earthen cells about 2.5 cm beneath the surface; the pupal stage lasts 2–3 weeks, at which time the new adults emerge; the total time required to complete the life cycle is 46 days; adults and larvae severely damage plants by chewing out long strips of tissue between the veins of leaves, leaving only a thin membrane; when damage is extensive the leaves turn whitish and the plant takes on the appearance of frost damage; the plant may be killed or the crop may be seriously reduced *phyt zoo*

Cercospora leaf spot disease: A disease found in sugarbeet caused by the fungus *Cercospora beticola phyt* >>> Figure 62

cereous: >>> ceraceous

certation: The competition in growth rate between pollen tubes of different genotypes resulting in unequal chances of accomplishing fertilization *bot*

certification: >>> approbation statement

certified seed: Seed produced under an officially designated system of maintaining the genetic identity of, and provisions for, seed multiplication and distribution of crop varieties; is the progeny of breeder, select, foundation, or registered seed; it is grown in compliance with regulations determining standards of germination, freedom from diseases and weeds, and trueness to type *seed* >>> Table 28

certify: >>> certification >>> certified *seed*

cesium (Cs): A rare, highly reactive, soft metallic element of the alkali metal group *chem* >>> cesium-chloride-density-gradient centrifugation

cesium-chloride-density-gradient centrifugation: A method for purification of DNA by means of centrifugation in a cesium chloride solution, developed by MESELSON and STAHL *meth*

cespitose: >>> caespitose

CFU: >>> colony-forming units

CGIAR: >>> Consultative Group of International Agricultural Research

CHA: >>> chemical hybridizing agent

chaff: The glumes, husks, scales, or bracts found with mature inflorescences and that separate from seeds during threshing, winnowing, or processing; in general, straw or hay that has been finely cut for animal feed *agr*

chaffy: >>> chaff

chaffy grass divider: A subsampling device used to divide a sample of chaffy grass seed into a working sample *seed*

chalaza: The base of an ovule bearing an embryo sac surrounded by integuments *bot*

chalky: A color descriptor characterizing kernel endosperm of cereal grains (e.g., a rice grain with a high level of chalk is generally undesirable); the chalky appearance arises from the structure of the endosperm; voids cause light to be refracted, and hence the endosperm appears white to reflected light and opaque to transmitted light *meth agr*

chamaephyte: Low woody or herbaceous plant with perennating tissue within ca. 25 cm of soil surface *bot eco*

chambered: >>> camerate

change of environment (during selection process): >>> shuttle breeding

change of sowing time (during selection process): A change of sowing time (early and late spring; early and late autumn) is applied to select for >>> daylength insensitivity, reduced >>> vernalization requirement, and yield/quality stability in different length of growth period; the method is used both in >>> autogamous and >>> allogamous crops *meth* >>> selection method

chapati: A flat pancake-like bread of India, usually of whole-wheat flour, baked on a griddle *meth*

chaperone: Molecules that associate with an immature protein and cause it to fold into its final and active structure *phys*

character: An attribute of a plant resulting from the interaction of a gene or genes with the environment *gene*

characteristic: >>> character

CHARGAFF's rules Discovered by Austrian chemist Erwin CHARGAFF (1905–2002); he stated that >>> DNA from any cell of all organisms should have a 1:1 ratio (base pair rule) of pyrimidine and purine bases (A – T = C – G) and, more specifically, that the amount of >>> guanine (G) is equal to >>> cytosine (C) and the amount of >>> adenine (A) is equal to >>> thymine (T); this pattern is found in both strands of the DNA; the first rule holds that a double-stranded DNA molecule has percentage base pair equality: %A = %T and %G = %C; the rigorous validation of the rule constitutes the basis of >>> WATSON–CRICK pairs in the DNA double helix; the second rule (parity rule) is that the composition of DNA varies from one species to another; in particular in the relative amounts of A, G, T, and C bases (A – T/A + T … C – G/C + G); such evidence of molecular diversity, which had been presumed absent from DNA, made DNA a more credible candidate for the genetic material than protein *biot*

charate: Charred wood containing leachable chemicals that stimulate seed germination in some plant species *bot*

charged-coupled device (CCD) camera: A camera used for digital imaging; it contains a light-sensitive silicon chip; when light falls on that chip, it creates an electrical charge at a specific location *micr*

chartaceous: Paper-like *bot*

chasmogamy: Fertilization after opening of flower, as opposed to cleistogamy, i.e., maturation of flowers before the anthers burst, promoting cross fertilization *bot*

check cross: The crossing of an unidentified genotype with a phenotypically similar individual of defined genotype; F2 segregation analysis serves to establish whether the phenotype resulted from the action of identical or non-identical alleles of the same gene locus or from the action of non-allelic genes *meth*

check cultivar: A commercial cultivar or experimental strain with well-known characteristics and performance that normally is included for comparison purposes with other selections in all testing procedures; the standard check is an appropriate and generally satisfactory base for comparisons; however, check varieties have a variable but nevertheless short life span; with each change of check variety there is a break in the continuity of comparisons over time; a standard check variety is a also a single genotype with homeostatic properties that may give a response at variance with other genotypes under test; sometimes there are even clearly situations where the standard check varieties are not relevant to the breeder's needs (e.g., regional or international nurseries); alternatives can be (1) the use of existing entries or addition of new entries in a nursery or (2) the use of derived statistics generated by the nursery entries; the latter can be realized by so-called floating checks; a floating check is a hypothetical nursery entry (a statistic) not tied to a genetic entity *meth* >>> design of experiment

check plot: In field testing of breeding material, an experimental plot system is usually applied; since variability of the land was recognized as a serious problem, first a duplicate plot system was used and later the check plot system; in the latter, standard or check varieties are sown in every 5th, 10th, 20th, or more plot, depending upon circumstances; a map of the field is prepared, and plots are grouped around a check plot according to their nearness or, where the soil is highly variable, according to the character of the soil; the average yield of all check plots is determined; then, additions are made to below-average check plot yields to bring them up to the average, and subtractions are made from above-average yields *meth agr* >>> design of experiment

check strip: >>> check plot

chelate: A complex organic molecule that can combine with cations and does not ionize; a claw-structure formed as a result of the reaction of a metal ion with two or more groups on a ligand; mugeinic acid is one of the many natural chelates; it can play an important role in uptake of metal ions from the soil; some plants (e.g., rye) may exude chelates into the rhizosphere in order to supply the plant with micronutrients at slow steady rates *chem phys agr*

chelating agent: Organic chemical (e.g., ethylene-diaminetetraacetic acid) that combines with metal to form soluble chelates and prevent conversion to insoluble compounds, or, micronutrients which have been treated to keep them readily available for absorption once they are introduced into the soil; if not chelated, many micronutrients would react with other elements in the soil in ways that would soon make them unavailable to the plant; some commonly chelated elements are copper, iron, magnesium, manganese, and zinc *phys* >>> zinc >>> www.zinc-crops.org/index.html

chelation: The trapping of a multivalent ionic species by ionic bonding to a larger water-soluble molecule so as to render the ion inactive in the biological matrix and to aid excretion *phys*

chelator: A compound that combines with a metal and keeps it in solution, i.e., it binds a ligand, especially a metal ion, by several functional groups whose combined effect results in a high-affinity interaction; e.g., ethylenediaminetetraacetate (>>> EDTA) *phys*

chemical desiccation: A method sometimes used to screen for postanthesis stress tolerance as destruction of the plant's photosynthetic system; the chemical desiccants most commonly used for cereals are magnesium chlorate, potassium iodide, and sodium chlorate *meth >>>* www.plantstress.com

chemical-hybridizing agents (CHA): Compounds applied to plants prior to anthesis to selectively induce male sterility, e.g., tribenuron-methyl (TBM), represent an important approach for practical utilization of >>> heterosis in rapeseed; however, when spraying the female parents with TBM to induce male sterility the male parents must be protected with a shield to avoid injury to the stamens, which otherwise complicates the seed production protocol and increases the cost of hybrid seed production; an application of a herbicide-resistant cultivar in hybrid production by using TBM-resistant mutants of rapeseed may improve the method *meth >>>* gametocide

chemical mutagen: A chemical capable of causing genetic mutation in DNA above the spontaneous background level, e.g., >>> alkylating mutagens such as methyl or ethylmethane sulfonate alkylate guanine compounds *gene >>>* mutagen

chemigation: Application of various pesticides through an irrigation system *agr hort*

chemiluminiscence: The emission of light from chemical reaction *chem phy*

chemostat: An open continuous culture in which cell growth rate and cell density are held constant by a fixed rate of input of a growth-limiting nutrient *phys biot*

chemotaxis: Oriented movement toward or away from a chemical stimulus *bot*

chemotherapy: Control of a plant disease with chemicals (chemotherapeutants) that are absorbed and translocated internally *phyt*

chemotrophic: Any organism that oxidizes inorganic or organic compounds as its principal energy source *bot*

chiasma (Xta, chiasmata pl): A cross-shaped structure forming the points of contact between non-sister chromatids of homologous chromosomes, first seen in the diplotene stage of meiotic prophase I *cyto*

chiasma interference: The occurrence, less frequent or more frequent than expected by chance, of two or more crossing-over and chiasmata in a given segment of a chromosomal pairing configuration and/or chromosome *cyto*

chiasma localization: The physical position of a chiasma in a pairing configuration and/or chromosome *cyto*

chiasmate: Meiosis with normal chiasma formation *cyto*

chiasma terminalization: The progressive shift of chiasmata along the arms of paired chromosomes from their points of origin toward terminal positions *cyto*

chilling damage: Damage to plants at low temperatures in the absence of freezing; common in plants of tropical or subtropical origin at temperatures < +10°C; a change in viscosity of lipids in membranes might be the reason *phys*

chilling injury: >>> chilling damage

chimera: A tissue containing two or more genetically distinct cell types, or an individual composed of such tissue; chimeral plants may originate by grafting, spontaneous mutation, induced mutation, sorting-out from variegated seedlings, mixed callus cultures, or protoplast fusion; one of the earliest described cases of a graft chimera was the "Bizzaria" orange, which arose after a scion of sour orange had been grafted onto a seedling of citron late in the 17th century; the vast majority of variegated-leaf chimeras have arisen by spontaneous nuclear or plastid mutation; colchicine has been widely used to induce cytochimeras of fruiting plants; structural classification of chimeras includes periclinal, mericlinal, and sectorial chimeras; periclinal describes the stable, "hand-in-glove" arrangement of the tunica-corpus region; mericlinal describes a type of periclinal where only part of a layer is mutant; and sectorial describes a form where a solid sector through all apical layers is mutant; the conventional method of describing the genotypes of the tunica and corpus regions is the use of the abbreviations L.I, L.II, and L.III, which represent the outermost layer, the next tunica layer in, and the corpus, respectively; a plant chimeral for ploidy level, or a cytochimera, with a diploid L.I, tetraploid L.II, and tetraploid L.III, would be 2-4-4; a variegated chimeral plant possessing a mutant chlorophyll deficient (albino) outer tunica layer overlying normal inner tissue would be labeled a WGG chimera (W indicating white, or albino, tissue; G indicating green tissue); while a plant with the outer layer normal, the next layer in mutant, and the inner corpus normal, would be designated GWG, and so on; such designations are, in the case of chlorophyll chimeras, generally based on the appearance of leaves and other organs produced by derivatives of the apical meristem, and thus may not refer to precise meristem cell layers, since chlorophyll is not synthesized and therefore is not detectable in the tunica and corpus cells of the meristem itself *bot* >>> ectochimera >>> endochimera >>> mesochimera >>> mutation >>> xenia >>> valence cross >>> Figure 56

chimeric: >>> chimera

chimeric DNA: Recombinant DNA molecules containing unrelated genes *biot*

chimeric gene: A semisynthetic gene, consisting of the coding sequence from one organism, fused to promoter and other sequences derived from a different gene; most genes used in transformation are chimeric *biot*

chimeric selectable marker gene: A gene that is constructed from parts of two or more different genes and allows the host cell to survive under conditions where it would otherwise die *biot*

Chinese Academy of Agricultural Science (CAAS) Established in 1957; different from provincial agricultural research institutes; China's national agricultural research organization, directly affiliated with the Ministry of Agriculture *org* >>> www.caas.cn/en/

Chinese Spring (wheat): A famous wheat variety worldwide; it is generally accepted as the standard variety for cytogenetic and molecular research in wheat; from this variety more than 300 aneuploids were developed; agronomically, however, Chinese Spring has some serious faults, such as shattering, susceptibility to almost all wheat diseases and insects, and poor adaptation to the world's major wheat-growing regions; the answer as to why the variety was chosen in which to produce monosomes, trisomes, tetrasomes, and almost all of the other aneuploids is simple – it came into cytogenetic use by accident; E. R. SEARS made hybrids between wheat and rye in 1936 in attempts to induce chromosome doubling by heat shocks; Chinese Spring was used since it was known to cross readily with rye; among the wheat–rye hybrids a few wheat haploids were obtained; one of these haploids was pollinated by euploid Chinese Spring; 13 viable seeds derived from this backcrossing, showing chromosome numbers of 2n = 41,42,43 and reciprocal translocations; nullisomes were eventually obtained from the monosomes and one nullisome proved to be 3B, which is partially asynaptic and was therefore a good source of additional monosomes and trisomes; it seems that Chinese Spring came to the Western world from the Sichuan province of China; British representatives in foreign countries used to be encouraged to collect plants that would be of possible value in their homeland; it was R. BIFFEN, then director of the Plant Breeding Institute at the University of Cambridge, who received from Sichuan at about the beginning of the last century a wheat that he called "Chinese White"; this type of wheat was at that time of interest because it was early-maturing, set a high number of seeds per spikelet, and was tolerant to drought (BIFFEN et al.); in North America, "Chinese White" first appeared in North Dakota, where the pioneering wheat breeder L. R. WALDRON obtained it from R. BIFFEN in 1924; WALDRON shared it with other breeders, one of whom passed it on to J. B. HARRINGTON at Saskatoon (Canada); a sample of wheat called Chinese Spring came to University of Missouri in 1932 from Saskatoon; it was known at Missouri that this wheat is highly crossable with rye and therefore acquired by L. J. STADLER; the latter was strongly interested in research on polyploidy and amphiploid hybrid production; it seems clear that the Chinese Spring, which E. R. SEARS used in his crosses with rye, was the same as BIFFEN's "Chinese White"; BIFFEN's "Chinese White" is still maintained at the John Innes Centre Collection, Norwich (United Kingdom), and is indistinguishable from Chinese Spring; Y. ZOU (Oregon State University) noticed a strong resemblance of Chinese Spring to certain Chinese Sichuan varieties *agr* >>> wheat

chip budding: Grafting a small section of scion wood that contains a bud to a rootstock *hort meth*

chiral product: Many chemical substances for crop protection products occur in two forms that are mirror images of each other, as the right hand is of the left hand; the "image" and "mirror image" can have completely different effects – e.g., one form of the amino acid asparagine is used as a sweetness enhancer, while the other is perceived as bitter *chem biot*

chirepterophily: Pollination by bats *bot*

chisel: A type of minimum tillage that breaks up the aggregates of the soil more than discing *agr meth* >>> chisel plow

chisel plough (plow): An implement, with points about 30 cm apart, used to till the soil some 30–45 cm deep *agr* >>> chisel

chi sequence: An octane, non-palindromic that provides a hotspot for recBCD-mediated genetic recombination in Escherichia coli; wild-type lamda phage lacks a chi sequence; a chi sequence has been added artificially to some lamda cloning vectors, which cannot make concatemers by sigma-type replication; the chi sequence stimulates the formation of lamda dimers by the host recombination functions *biot* >>> lamda phage

chi square (chi² = X²): A statistical procedure that enables researchers to determine how closely an experimentally obtained set of values fits a given theoretical expectation *stat*

chi-squared test (chi² = X² test): A significance test used to statistically assess the goodness of fit of observed data to a prediction *stat*

chloramine (NH$_2$Cl): A compound which combines chlorine with ammonia; it is used in about 20% of treated municipal water; while the chlorine acts as a disinfectant, the ammonia serves to stabilize the chlorine; as a result, the chlorine cannot readily escape into the air; this makes water treated with chloramine potentially more harmful than water that is simply chlorinated; symptoms caused by chloramine are the same as those caused by excessive chlorine, that is, leaf tip burn and decreased flowering *phys meth hort*

chloramphenicol (chloromycetin): An antibiotic produced by Streptomyces venezuelae; it is a potent inhibitor of protein synthesis *phys gene biot*

chloramphenicol acetyl-transferase (CAT): Coded by a particular bacterial gene cat; it derives from a certain transposon of a plasmid; the resistance to chloramphenicol is commonly used as a reporter gene in genetic experiments for investigating physiological gene regulation and was incorporated in some cloning vectors; beta-galactosidase and luciferase genes can also be used for the same purpose *phys biot* >>> cat gene >>> Figure 46

chlorenchyma: The general term for chloroplast-containing parenchyma cells as leaf mesophyll tissues *bot biot*

chlorocholine chloride (CCC): A growth regulator used for inhibition of internode growth in cereals in order to reduce straw length and thus to increase the lodging resistance *agr*

chloromycetin: >>> chloramphenicol

chlorophyll: The green photosynthetic pigment generally localized in intracellular organelles (chloroplasts) *bot*

chloroplast: A membrane-enclosed, semiautonomous, subcellular organelle containing chlorophyll; it is a site where photosynthesis takes place; it contains DNA and polysomes, and it is capable of replication; chloroplasts of vascular plants contain about 100 genes, most of which encode components of the photosynthetic electron transport machinery and elements of the transcriptional and translational apparatus; although the progenitor of the chloroplast was a free-living prokaryote, the loss of genetic information to the nucleus to control plastid gene expression has largely placed the chloroplast in a "receptor" role, where it responds to nuclear signals; nonetheless, reverse signaling also occurs, demonstrating the interdependence and need for coordination between the cellular compartments; the chloroplast genome is a circular double-stranded DNA molecule located in the stroma of chloroplast; they are highly conserved among plant species; there is more than one copy of the genome in each chloroplast; most chloroplast genomes are between 120 to 160 kb in size and contain about 120–140 genes; chloroplast >>> transformation has several unique advantages; the highest levels of expression in the published literature for engineering agronomic traits or human therapeutic proteins were achieved using this concept; in addition, integration of foreign genes into the chloroplast genome offers transgene containment from pollen transmission because of maternal inheritance of the chloroplast genome in most crops; chloroplast >>> genetic engineering offers a number of other unique advantages including multi-gene engineering in a single transformation event, lack of gene silencing or position effects due to site-specific transgene integration and minimal or lack of pleiotropic effects due to subcellular compartmentalization of toxic transgene products; often tobacco served as the model system to confer herbicide, insect, or disease resistance, drought or salt tolerance, or phytoremediation; more recently, chloroplast genomes of major crops including cotton and soybean, vegetables (carrot, cauliflower, cabbage, eggplant, lettuce, sugarbeet), tubers (potato), fruits (tomato), and trees (poplar) have been transformed; significant progress has also been made in expressing vaccine antigens against human bacterial, viral, and protozoan pathogens in chloroplasts, and animal studies demonstrated their efficacy against pathogen or toxin challenge; most importantly, oral delivery of vaccine antigens bioencapsulated in plant cells was shown to be more efficacious than injectable vaccines or in developing oral tolerance against autoimmune disorders *bot gene*

chloroplast DNA (ctDNA): >>> chloroplast

chlorosis: Yellowing or whitening of normally green plant tissue; the loss of chlorophyll and associated pigments from small lesions or from whole leaves *bot*

chlorox soak test: Used for evaluation of seed damage during harvest or seed conditioning, often in soybean and dry edible beans; materials needed are chlorox bleach (5.25% sodium hypochlorite) and PETRI dishes; first a 5% chlorox solution is made up; a certain number of seeds is placed in the PETRI dishes; obvious damaged seeds are excluded; the seeds are covered with chlorox solution for 12 min; afterward, the chlorox solution is poured off; seeds are spread on toweling for evaluation; the number of swollen seeds is counted in each replicate; the numbers are averaged over replicates; if swollen seeds exceed 10% harvest machines have to be adjusted *seed meth*

chondriome: >>> mitochondrial genome >>> genome >>> plasmon

CHOPIN alveograph: A set of equipment to predict the baking quality of certain cereals (e.g., wheat) in the absence of baking tests; the test consists of inflating a disc of dough with air until it bursts; the maximum pressure required and the time taken are measured; from this the strength and extensibility of the dough is determined *meth*

chopping: A mechanical cut-back of young shoots of sugarbeets for seed production in order to increase the seed quantity; also called "Boyage system" *agr seed*

Chorleywood baking process (CBP): A system of monitoring bread making developed in the United Kingdom in the early 1960s; the system uses high-speed mixing and the use of improvers, special fats, and yeasts to reduce fermentation time; this makes it possible to use lower protein and lower quality wheat varieties in the grist, yet produce good bread quality *meth*

chromatic aberration: Inaccurate focusing of red, blue, and green light either along or at right angles to the optical axis; axial chromatic aberration results in the red image being focused farther along the optical axis than the green image; lateral chromatic aberration results in a slightly bluer image *micr*

chromatid: One of the two daughter strands of a chromosome that has undergone division during >>> interphase; they are joined together by a single centromere *cyto*

chorology: The study of the geographical distribution of organisms *eco*

chromatid aberration: Chromosomal changes produced in one chromatid as a consequence of spontaneous or induced mutations *cyto*

chromatid break: A discontinuity in only one chromatid of a chromosome *cyto*

chromatid bridge: A bridge-like structure caused by a dicentric chromatid with the two centromeres passing to opposite poles during anaphase; the frequency of chromatid bridges in AII of meiosis is sometimes used as a measure for the level of cytological disturbances (e.g., in induced autopolyploids and allopolyploids) *cyto*

chromatid exchange: >>> sister chromatid exchange

chromatid segregation: Segregation of two sister-chromatid segments of a chromosome *cyto gene*

chromatin: From the Greek word chroma = color; named by W. FLEMMING in 1879; the deoxyribonuclein-histone complex of chromosomes; it is readily stained by basic dyes and is therefore easily identified and studied under the microscope; chromatin in the >>> interphase nucleus is organized into discrete domains defined by sites of attachment to the nuclear matrix; chromatin DNA gradually coils itself around flexible rods of histone protein during the prophase, forming two parallel compact cylinders (>>> chromatids) connected by a knot-like structure (>>> centromere) at their middles; in appearance they are rather similar to two rolls of carpeting standing side-by-side that are tied together with rope at their middles; these cylinders are homologous chromosomes (i.e., the genes of the two chromosomes are linked in the same linear order within the DNA strands of both chromosomes); while they are joined at their middles, these paired chromosomes appear X-shaped; chromatin is usually not visible during the interphase of a cell, but can be made more visible during all phases by reaction with basic stains (dyes) specific for DNA *cyto* >>> chromatin domain

chromatin assembly factor 1 (CAF1): A three-subunit H3/H4 histone chaperone responsible for replication-dependent nucleosome assembly *cyto*

chromatin domain: A region of chromatin, the exact character and size of which depends on experimental context; it can be a single nucleosome or can extend to an array of more than 100 nucleosomes; the DNA is organized into loop domains by stable attachment to the nuclear matrix at approximately 50,000 bp intervals; most domains are condensed into higher order chromatin structures; the DNA of active domains is extended by multiple sequence-specific dynamic associations with the nuclear matrix; the chromatin is anchored during >>> interphase to the periphery of the nucleus; the protein matrix to which chromatin is anchored is referred to as the nuclear matrix; on the average, attachment to the matrix occurs every 30 to 100 kb; thus, chromatin is organized into discrete loops, each of which may contain one or a few genes; there are two kinds of matrix: a peripheral matrix, which is primarily on the periphery of the nucleus, and the fibrillar, or internal matrix, which is primarily in the interior; there is good evidence that DNA replication and transcription of genes takes place primarily in regions in contact with the internal matrix; DNA is threaded through the matrix attachment sites, until the appropriate gene or origin of replication is found; then replication complexes or transcription complexes open up the chromatin further, and carry out their functions; each domain can be independently regulated; to be transcriptionally active, a domain must be extended into the fibrillar nuclear matrix; domains that remain coiled are clustered at the periphery of the nucleus; these domains remain transcriptionally inactive; extended domains are potentially active, but require further developmental or environmental signals to turn on transcription *biot cyto*

chromatin insulator: A DNA element that protects a gene from position effects *gene*

chromatin reconstitution: The reconstitution of chromatin with chromosomal constituents previously removed by chromatin dissociation *cyto*

chromatin remodeling: Refers to a reshaping (at molecular scale) of chromatin, that alters specific genes so that DNA subsequently gets expressed; it can be caused by >>> short interfering RNA (>>> siRNA) or certain transcription activators *biot*

chromatography: A technique used for separating and identifying the components from mixtures of molecules having similar chemical and physical properties; molecules are dissolved in an organic solvent miscible in water, and the solution is allowed to migrate through a stationary phase; since the molecules migrate at slightly different rates they are eventually separated *meth*

chromocenter: A central aggregation of heterochromatic chromosomal elements of the cell nucleus; the euchromatic chromosome arms extend from the chromocenter *cyto*

chromogene: A stain-producing material *bot*

chromomere: A small beadlike structure visible in a chromosome during prophase of meiosis and mitosis, when it is relatively uncoiled *cyto* >>> knob

chromomere pattern: The linear order and distribution of chromomeres along a chromosome; this has been extensively studied in crops, such as tomato, maize, rye, etc. *cyto*

chromonema: The smallest light-microscopically observed strand in chromosomes or chromatids *cyto*

chromoplast: A carotinoid-containing plastid that colors ripe fruits and flowers *bot*

chromosomal: Referring to the structure, constituents, and function of chromosomes *cyto*

chromosomal aberration: An abnormal chromosomal complement resulting from the loss, duplication, or rearrangement of genetic material *cyto* >>> chromosome mutation

chromosomal domain: A region of a chromosome, the exact nature and size of which depend on experimental context; a domain can be a region of chromosomal packaging, such as a loop extending from two adjacent attachments to a chromosomal axis, and can vary in size from an array of less than 100 nucleosomes (~30 kb) to potentially more than 500 nucleosomes; a chromosomal domain might also represent a functional unit of chromosomal structure defined by boundary elements or insulators at the edges of the domain *cyto*

chromosomal numerical aberration: The occurrence of an abnormal number of chromosomes in a >>> karyotype, e.g., the inheritance of three instead of two (common) homologous chromosomes, i.e., trisomy *cyto*

chromosomal structural change: A change in chromosome structure spontaneously or experimentally induced *cyto* >>> translocation

chromosome: A DNA-histone protein thread, usually associated with RNA, occurring in the nucleus of a cell; it bears the genes that constitute hereditary material; each species has a constant number of chromosomes; in 1999, a first plant chromosome of the weed >>> thale cress (*Arabidopsis thaliana*) was genetically decoded; the eukaryotic chromosome is a single DNA molecule complexed with chromatin proteins; it is organized to allow for a hierarchical packing scheme, i.e., (1) DNA helix is wound twice around a core particle of histone proteins, (2) 30 nm fiber, six histone core particles per turn, and (3) loops of 30 nm fibers are formed by attachment of chromatin to the nuclear matrix roughly every 30 to 100 kb; it is important to make the distinction between decondensed interphase chromosomes and condensed mitotic chromosomes; during interphase, most of the chromosomal material needs to be in an open configuration to allow for gene expression to occur; during mitosis, the chromatin needs to be condensed; the term was proposed by WALDEYER (1888) for the individual threads within a cell nucleus cytoplasm; a first synthetic chromosome was produced in 2014 for the eukaryotic baker's yeast (*Saccharomyces cerevisiae*), derived from synthesis of a functional 272,871-base pair designer eukaryotic chromosome, synIII, which is based on the 316,617-base pair native *S. cerevisiae* chromosome III; changes to synIII include TAG/TAA stop-codon replacements, deletion of subtelomeric regions, introns, transfer RNAs, transposons, and silent mating loci as well as insertion of *loxPsym* sites to enable genome scrambling; chromosome "synIII" is functional in *S. cerevisiae*; scrambling of the chromosome in a heterozygous diploid reveals a large increase in a-mater derivatives resulting from loss of the *MATα* allele on "synIII" *gene*

chromosome arm: One part of a chromosome apart from the centromere *cyto* >>> Figure 11

chromosome banding: The experimental production of differentially stained regions because of the distribution of different chromatin constituents along a chromosome *cyto* >>> C banding

chromosome breakage: Induced or spontaneous breaks across the entire cross-section of the chromosome *cyto*

chromosome bridge: A dicentric chromosome that forms a bridge between the separating groups of anaphase chromosomes because its two centromeres are being drawn toward opposite poles *cyto* >>> chromatid bridge

chromosome coiling: The spiral or helical coiling of the chromonemata of the chromosomes during some phases of mitosis and meiosis *cyto*

chromosome complement: The group of chromosomes derived from a particular gametic or zygotic nucleus *cyto* >>> genome >>> cf List of Important Crop Plants

chromosome configuration: Any association by chromosome pairing of chromosomes at meiosis *cyto* >>> Figure 15

chromosome conjugation: Joining of homologous chromosomes during meiotic prophase *cyto* >>> Figure 15

chromosome contraction: The coiling and shortening of chromosomes during mitosis and meiosis either in a natural way or experimentally by using specific chemical or cold treatment *cyto* >>> chromosome coiling

chromosome-counting method: The way to determine the number of chromosomes per cell *meth*

chromosome counts: >>> chromosome counting method

chromosome doubling: Induced or spontaneous doubling of chromosome sets leading to rediploidization or to polyploids *cyto meth* >>> polyploidization >>> doubled haploid >>> speargrass

chromosome elimination: The loss of chromosomes from nuclei during certain mitotic or meiotic stages; it is common in several artificial autopolyploids and allopolyploids *cyto* >>> *Hordeum bulbosum* procedure >>> speargrass

chromosome engineering: Manipulation of whole chromosome sets, individual chromosomes, or even chromosome segments, by different means, for scientific analysis or improvement of performance of crop plants *meth* >>> chromosome-mediated gene transfer >>> biotechnology

chromosome jumping: A technique that allows two segments of duplex DNA that are separated by thousands of base pairs (~200 kb) to be cloned together; after subcloning, each segment can be used as a probe to identify cloned DNA sequences that, at the chromosome level, are roughly 200 kb apart *biot*

chromosome length polymorphism (CLP): The phenomenon that chromosomes can substantially vary in length among individuals of a population within a species, usually due to spontaneous structural changes of chromosomes (e.g., in allogamous rye) *gene meth*

chromosome map: A map showing the location of genes on a chromosome, deduced from genetic recombination and cytological experiments *meth* >>> mapping

chromosome-mediated gene transfer: The transfer of genes within and between varieties, species, or genera by means of chromosome manipulations, such as additions, substitutions, translocations, or directed recombinations utilizing specific crossing techniques, cell manipulations, or micromanipulation of chromosomes; more specifically, the use of isolated metaphase chromosomes as a vehicle for the transfer of genes between cultured cells *meth cyto* >>> biotechnology

chromosome microdissection: The removal of relatively large sections (chromosome arm, satellite) from an isolated chromosome using LASER microdissection or atomic force microscopy;

the dissected pieces may still harbor >10^6 bp and are often used for the establishment of subgenomic libraries in bacterial artificial chromosomes or map-based cloning approaches *cyto biot meth*

chromosome mosaicism: The presence of cell populations of various karyotypes in the same individual *cyto*

chromosome movement: The movement of chromosomes during mitosis and meiosis as a prerequisite for the anaphase separation of chromatids and/or chromosomes *cyto*

chromosome mutation: Any structural change involving the gain, loss, or translocation of chromosome parts; it can arise spontaneously or be induced experimentally by physical or chemical mutagens; the basic types of chromosome mutations are deletions (deficiencies), duplications, inversions, and translocations *cyto*

chromosome number: The specific somatic chromosome number (2n) of a given species or a crop derivative of it; the Adder's Tongue Fern, *Ophioglossum reticulatum* (Ophioglossaceae) has the highest known diploid chromosome number 2n = ~1,400, while *Haplopappus gracilis* (Asteraceae) shows the smallest number, 2n = 2x = 4 *cyto* >>> genome >>> Table 1, 14 >>> *cf* List of Crop Plants

chromosome orientation: >>> centromere orientation

chromosome painting: Fluorescent *in situ* hybridization of a specifically labeled >>> DNA probe or probes that hybridizes to the entire chromosome of the probe's origin; using different fluorescent dyes and probes, a pattern of multicolored chromosomes or chromosome segments appears *cyto meth* >>> Figures 52, 53, 54

chromosome pairing: The highly specific side-by-side association of homologous chromosomes during meiotic prophase *cyto* >>> synaptonemal complex >>> Figure 15

chromosome polymorphism: The presence of one or more chromosomes in two or more alternative structural forms within the same population *cyto*

chromosome puffing: Despiralization of the deoxyribonucleoprotein of discrete regions of a chromosome during particular cell stages *cyto* >>> polyteny >>> polytene chromosome >>> puff >>> Figure 66

chromosome pulverization: The destruction of chromosome structure, varying from an apparently total fragmentation of the chromatin to various degrees of defective condensation and erosion *cyto*

chromosome rearrangement: The structural change of the chromosome complement by chromosome mutations *cyto* >>> translocation

chromosome reduplication: The synthesis of all compounds that result in an identical copy of the original chromosome *gene*

chromosome segment substitution lines (CSSLs): A method that combines QTL mapping with the production of novel germplasm for cultivar development, e.g., in the *Festuca/Lolium* system, peanut, or rice *meth*

chromosome segregation: The separation of the members of a pair of homologous chromosomes in a manner such that only one member is present in any postmeiotic nucleus *gene cyto*

chromosome set: The minimum viable complement of indispensable chromosomes (each is represented once) of an individual *cyto* >>> genome

chromosome size: The physical dimensions of a chromosome *cyto*

chromosome sorting: >>> sorting

chromosome staining: The pretreatment and treatment of chromosomes with different dyes in order to make them more suitable for chromosome counting or specific analyses *cyto* >>> chromosome banding >>> opuntia >>> orcein >>> aceto-carmine staining >>> FEULGEN stain >>> FISH >>> GISH

chromosome stickiness: Chromosome agglutination that results in a pycnotic or sticky appearance of chromosomes; sometimes caused by gene mutations or by treatment with chemical or physical agents *cyto*

chromosome substitution: The replacement of one or more chromosomes by others from another source by spontaneous events or a crossing scheme *cyto*

chromosome theory of inheritance: States that the chromosomes, as the carriers of genetic information, represent the material basis of nuclear inheritance *bio gene*

chromosome-walking technique: A procedure that is used for the determination of a gene on a particular DNA clone of a DNA library; the total DNA of a chromosome has to be available as a series of overlapping DNA fragments; such fragments are produced either by DNA shearing or by cleavage using >>> restriction enzymes; the fragments are used for series of hybridizations; it starts with a cloned gene, which is already identified on the same chromosome; this known gene serves as a probe for the detection of clones (fragments) that contain neighboring DNA sequences; during the following hybridization, that DNA sequence is used as a probe for detecting the next neighboring sequences, and so on; through each hybridization there is progress away from the known gene toward the unknown chromosomal site; it seems that the method is more practical for plants with small genomes than for crop plants such as wheat, because the ratio of kilobases of DNA to the genetic map units is roughly proportional to the size of the genome; the cloning of genes by this method becomes much easier if the entire genome is already represented by a contiguous array of ordered DNA clones *meth gene* >>> Table 39 >>> Figures 45, 46

chromotype: The chromosome set *cyto* >>> genome

chronic symptoms: Symptoms that appear over a long period of time *phyt*

CIAT: >>> Centro Internacional de Agricultura Tropical

Cibacron blue: Affinity matrix used for the purification of >>> restriction endonucleases *biot meth* >>> Table 39

cilia: Hairs growing along margins (e.g., in sunflower) *bot*

ciliate: Fringed with hairs on the margin *bot*

ciliolate: Minutely ciliate bot

CIMMYT: >>> Centro Internacional de Mejoramiento de Maiz y Trigo

cincinnus: >>> scorpioid cyme

CIP: >>> Centro Internacional de al Papa

circadian rhythm: A type of rhythmic plant growth response that appears to be independent of external stimuli *phys*

circinnate vernation: The coiled arrangement of young leaves, for example, in ferns or cycads *bot*

circularization: A DNA fragment generated by digestion with a single restriction endonuclease will have a complementary 5′ and 3′ extensions, i.e., sticky ends; if these ends are annealed and ligated, the DNA fragment will have converted to a covalently closed circle or circularized *chem biot*

cirrate: Rolled round, curled, or becoming so *bot*

cirrose: >>> cirrate

cis-arrangement: >>> coupling of factor

cisgene(s): Are natural genes and intragenes that are composed of functional parts of natural genes from the crop plant itself or from crossable species *biot*

cisgenesis: The recombination of species-specific genes with marker-free transformation, mimicking linkage drag free introgression breeding in one step; cisgenesis is a new sub-invention in the traditional breeding field and indicates the need for reconsideration of genetically modified directives; inventions frequently contain not only hardware elements, but also software and orgware elements; for cisgenesis it is foreseen that the technical (hardware) and bioinformatic (software) elements will develop smoothly, but that implementation in society is highly dependent on acceptance and regulations (orgware); cisgenesis has been exemplified for resistance breeding of potato to *Phytophthtora infestans*; by 2016, using cisgenesis/intragensis approaches, plants such as alfalfa, apple, barley, durum wheat, ryegrass, poplar, potato, and strawberry were genetically modified *biot* >>> intragenesis

cisgenic plant(s) Plants created by genetic manipulation using >>> RNAi (sometimes called: "GM-lite'") that pose less risks than the addition of genes from other species (transgenics) as new proteins are unlikely to be produced *biot* >>> transgenic plant

cistron: A section of the DNA or RNA molecule that specifies the formation of one polypeptide chain; the functional unit of the hereditary materials; it codes for a specific gene product, either a protein or an RNA *gene* >>> structural gene >>> gene

CITES: >>> Convention on International Trade of Endangered Species

citrate: A salt or ester of citric acid *chem*

citric acid cycle: The metabolic sequence of enzyme-driven reactions by which carbohydrates, proteins, and fatty acids produce carbon dioxide, water, and >>> ATP *phys*

citrus: A term applied to a variety of popular and nutritious fruits, including oranges, grapefruit, lemons, and limes; breeding is tasked with improving the varieties, which takes usually decades; great progress has been made in fruit improvement; in the last 20 years a disease called Huanglongbing, or "HLB" (also known as "citrus greening") has emerged in the U.S. and other countries and threatens the citrus industry; recently the focus is on how to grow improved citrus varieties that can survive this insidious disease *phyt hort*

citrus greening: >>> citrus

citrus leafminer: Damage is highly dependent on the citrus flushing pattern; chemical control is only required in young trees, both in nurseries and in newly established orchards; however, this situation is completely different in countries where the causal agent of citrus canker, the bacterium *Xanthomonas axonopodis pv. citri*, exists; citrus leafminer infestation results in a higher incidence of citrus canker infection, *Phyllocnistis citrella zoo phyt* >>> Figure 62

citrus tristeza virus (CTV): One of the most important citrus pathogens; it causes the death of millions of trees grafted on sour orange (*Citrus aurantium*); however, this rootstock is very well adapted to the Mediterranean, semiarid conditions; some species confer resistance, for example, *Poncirus trifoliata phyt*

clade: A group of all organisms sharing a particular common ancestor (and therefore having similar features); the members of a clade are related to each other; a clade is monophyletic *tax*

cladistics: A method of classifying organisms based on common ancestry and the branching of the evolutionary family tree; organisms that share common ancestors (and therefore have similar features) are grouped into taxonomic groups called clades; cladistics can also be used to predict properties of yet-to-be discovered organisms *tax* >>> Figures 68, 69 >>> phenetics

cladode: >>> cladophyll

cladogenesis: A mode of evolution (i.e., the splitting of an evolutionary line) such as a species; i.e., evolutionary change characterized by treelike or dendritic branching, illustrating phylogenetic relationships *tax evol* >>> Figures 68, 69

cladogram: A tree-like diagram showing evolutionary relationships; any two branch tips sharing the same immediate node are most closely related; all taxa that can be traced directly to one node are said to be members of a monophyletic group; former cladograms based on physical, easily observed characteristics; today, more reliable information like genetic and biochemical analysis to determine the relationships between different species is available *meth tax* >>> Figures 68, 69

cladophyll: A leaflike flattened branch that resembles and functions as a leaf *bot*

claim: In patenting biological material, a comprehensive and precise description that defines the scope of an invention *adm* >>> Intellectual Property Rights

clamp connection: Bridge or buckle protrusion found at the septa of hyphae In basidiomycetous fungi; associated with cell division *phyt*

clasping: To hold parts together; holding *bot*

class: A group of related or similar organisms; it contains one or more orders; a group of similar classes forms a phylum *tax*

class (of seed and seed crop): Refers to the generations of pedigreed seed and seed crops, such as breeder, select, foundation, registered, and certified, which have met the standards prescribed by recognized seed and seed crop certification agencies *seed*

classing: The process of describing cotton quality so that its value can be determined, e.g., for staple length, grade, color *agr*

class interval(s): Collection of observations into groups; depending upon the need of investigator, they may be large or small *stat*

classification: Helps in the study of organisms; cladistics is a method based on common ancestry; the Linnaean system is based on a simple hierarchical structure *tax*

class-uniform (resolvable design): Resolvable design in which the multiset of block sizes in each parallel class of the resolution is the same *stat* >>> design of experiment

claw: >>> auricle

clay: Mineral material < 2 μm texture, a class of texture, or silicate clay minerals *agr* >>> Table 44

clean cultivation: The practice of periodic soil tillage to eliminate all vegetation other than the crop being grown; all plant residues are taken off the field and are not recycled *in situ agr meth*

cleaning: The process of removing foreign material or impurities from the crop product, e.g., cereals *seed*

cleaning crop: A crop, such as potatoes or sugarbeet, that is used in the crop rotation to help suppress weeds; it does this by shading out the young weeds, which can be finally destroyed by cultivation *agr phyt*

clean seed: Sometimes refers to endophyte-free seed (e.g., in grasses) *seed phyt*

cleared lysate: Cell extract after removal of debris by centrifugation *biot*

cleavage: The processes by which a dividing egg cell gives rise to all the cells of the organism *cyto*

cleave: To make a double-stranded cut in DNA with a >>> restriction endonuclease *gene biot* >>> Table 39

cleaved amplified polymorphic sequences (CAPS): PCR-amplified DNA (>>> STS, >>> EST, or >>> SCAR products) that is digested with >>> restriction endonucleases to reveal polymorphisms in restriction sites *biot* >>> Table 39

cleft grafting: >>> split grafting

cleistogamous: Designating a self-pollinated plant that produces inconspicuous flowers that never open (e.g., wheat or tomato) *bot* >>> Table 18

cleistogamy: Flowers that remain closed until after anthers burst so that fertilization by selfing is likely >>> cleistogamous

climate: The variations of cold and heat, dryness and moisture, calm and wind in a given region or country; in general, the combined result of all the meteorological phenomena of any region, as affecting its crop or vegetable production *eco*

climate chamber: >>> phytotron

climax: A successional community of plants capable of optimal development under the prevailing environment and in dynamic equilibrium with its environment; it is in the terminal stage of an ecological succession *eco*

climber: A plant that clambers upward by attaching itself to other plants or objects; climbers can be distinguished as (1) stem climbers, which wind upward around an erect support, and (2) as tendril climbers, which cling to nearby objects by slender, coiling tendrils (e.g., grape) *agr hort bot*

climbing plants: A vine or other plant that readily grows up a support (climbing rose), twines up a slender support (hop, honeysuckle), or grasps the support by special organs such as adventitious aerial roots (English ivy, poison ivy, trumpet creeper), tendrils, hook-tipped leaves (gloriosa lily, rattan), or stipular thorns (catabrier); some climbing plants when not supported become trailing plants (English ivy); climbing types are to be found in nearly every group of plants, e.g., the ferns (climbing fern), palms (rattan), grasses (some bamboos), lilies (gloriosa lily), and cacti (night-blooming cereus); woody-stemmed tropical kinds – usually called lianas – are particularly abundant; a sturdy vine may strangle a supporting tree and then, as with the strangler fig, become a tree itself *bot hort*

cline: An environmental gradient and a corresponding phenotypic gradient in a population of plants; when clines are evaluated by provenance tests, they are often found to have a genetic basis *fore hort*

clinometer: >>> hypsometer

clipping: Breaking or cutting off the shoot tips in sugarbeet seed production; it stimulates the formation of side branches and thus increases the seed quantity *meth seed* >>> chopping

clonal: Genetically identical *gene* >>> Table 35

clonal expansion: The population of cells produced from a single cell; it is synonymous with clone, but is used in particular context (e.g., a cell with a particular chromosomal abnormality can, by clonal expansion, produce a population of the same type of cell within the organism) *biot*

clonal propagation: Vegetative (asexual) propagation from a single cell or plant *biot* >>> Table 35 >>> Figure 50

clonal seed orchard: Established by setting out clones as grafts or cuttings for seed production *hort fore* >>> seed orchard >>> Figure 50

clonal selection: Choosing the best clones from a clonal testing (e.g., in potato or forest trees) *meth hort fore* >>> Table 35 >>> Figure 50

clonal test: Evaluation of genotypes by comparing clones in a plantation *hort fore* >>> Figures 50, 60

clone: A group of genetically identical cells or individuals, derived from a common ancestor by asexual mitotic division *gene*; in molecular biology, a population of genetically identical organisms or cells; sometimes refers to cells containing a recombinant DNA molecule or to the recombinant DNA molecules themselves *biot*; in horticulture or agriculture, a group of individuals originally taken from a single specimen and maintained in cultivation by vegetative propagation; all clone specimens are exactly alike and identical to the original *hort agr* >>> Table 35 >>> Figure 50

clone-based map: Any physical map of a genome that is based on the alignment and sequencing of overlapping *biot meth* >>> BAC clones

clone coverage: The extent to which a genome is represented in clones of, e.g., a >>> BAC or plasmid library; clone coverage is measure of the amount of physical DNA coverage of the target genome *biot*

clone library: >>> genomic library

clone overlap single nucleotide polymorphism: Any >>> SNP detected by sequencing overlaps of two or more >>> BACs and comparison of resulting sequences *biot meth* >>> candidate single nucleotide polymorphism >>> causative single nucleotide polymorphism >>> SNP

clone overlap SNP: >>> clone overlap single nucleotide polymorphism

clone variety: Refers to a crop variety that consists of individuals deriving from a single clonal genotype (monogenotypic), e.g., in potato, cassava, sweet potato, rubber, mango, avocado, apple, pear, banana, pineapple, strawberry, brambles, grape, peach, cherry, almond, citrus, date, Jerusalem, artichoke, >>> yams, black pepper, olive, fig, pistachio, or edible aroids *meth* >>> Figure 50 >>> Table 35

clonic: >>> clonal

cloning: The process to vegetatively propagate a certain crop and/or plant *bot hort*; in molecular genetics, the cloning of DNA molecules from prokaryotic or eukaryotic sources as part of a bacterial plasmid or phage replicon *biomes* of spent medium; usually cells are separated mechanically from outflowing medium and added back to the culture *biot* >>> Figure 46, 50

cloning site: Restriction site, usually unique in a vector, where DNA can be inserted (cloned) *biot* >>> cloning vector >>> Table 39 >>> Figure 46

cloning vector: A plasmid or phage suitable for insertion and propagation of DNA; many cloning vectors have special properties (e.g., for expression of cloned genes or for the detection of cloned promoters) *biot*

close breeding: >>> inbreeding

closed continuous culture: *In vitro* culture or a bioreactor processing in which inflow of fresh medium is balanced by outflow of corresponding volume; a group of genetically identical cells or individuals, derived from a common ancestor by asexual mitotic division gene; in molecular biology, a population of genetically identical organisms or cells; sometimes it refers to cells containing a recombinant DNA molecule or to the recombinant DNA molecules themselves *biot*; in horticulture or agriculture, a group of individuals originally taken from a single specimen and maintained in cultivation by vegetative propagation; all clone specimens are exactly alike and identical to the original *hort agr* >>> Figure 50 >>> Table 35

closed-ended questions: Questions that provide respondents with a fixed set of alternatives from which they are to choose *stat*

closed population: A group of interbreeding plants (occurring in a certain area or in an experimental design) or a group of plants originating from one or more common ancestors, where there is no immigration of plants or pollen, that is, no genetic input other than by mutation *gene eco stat* >>> population >>> Figure 51 >>> Table 37

clove: One of the small >>> bulbs formed in the axils of the scales of a mother bulb, as in, for example, garlic *bot*

clover yellow vein virus (ClYVV): The virus is capable of causing severe damage to common bean (*Phaseolus vulgaris*) production worldwide; the snap bean market class is particularly vulnerable because infection may lead to distortion and necrosis of the fresh green pods and rejection of the harvest; three putatively independent recessive genes (*cyv, desc, bc-3*) have been reported to condition resistance to ClYVV; their allelic relationships have not been resolved; two novel *PveIF4E* alleles associated with resistance to ClYVV, *PveIF4E³* and *PveIF4E⁴*, are also revealed *phyt*

ClYVV: >>> Clover yellow vein virus

clubroot (disease of crucifers): This soilborne disease, present in many Brassica species, causes swollen and distorted roots by the fungus *Plasmodiophora brassicae*; it is one of the most important diseases of crucifers worldwide, reducing both crop yields and quality; in the past decade, clubroot has emerged as a major issue in the production of rapeseed (*Brassica napus*), a multi-billion dollar industry; this has fostered renewed interest and much research on this disease *phyt*

clump: A single plant with two or more stems coming from a root or rhizome (e.g., in sunflower) *bot*; in horticulture and forestry, the aggregate of stems issuing from the same root, rhizome system, or stool *hort fore*

cluster: A number of similar things, for example, bracts or spikelets, grouped together *bot*

cluster analysis: A technique of statistical analysis in which similar variances are grouped or clustered; the results of statistical calculations are often shown as dendrograms; particularly, in cross breeding the cluster analysis is used in order to select most diverse parents for crossing *stat*

clustered regularly interspaced short palindromic repeats (>>> CRISPR): DNA loci containing short repetitions of base sequences; each repetition is followed by short segments of >>> spacer DNA from previous exposures to a virus; >>> CRISPRs are found in approximately 40% of sequenced bacteria genomes; CRISPRs are often associated with *cas* genes that code for proteins related to CRISPRs; since 2013, the CRISPR/Cas system has been used for gene editing (adding, disrupting, or changing the sequence of specific genes) and gene regulation in species; by delivering the Cas9 protein and appropriate guide RNAs into a cell, the organism's genome can be cut at any desired location; it may be possible to use CRISPR to build RNA-guided gene drives capable of altering the genomes of entire populations *biot*

cluster sampling: Sampling by dividing the population into groups or clusters and drawing samples from only some groups *stat meth* >>> area sampling

clypeate: Pollen wall, in which the exine is subdivided into shields *bot* >>> pollen

C mitosis: Mitosis in single-celled organisms is responsible for the production of new individuals (asexual reproduction); mitosis in multicellular organisms, such as plants, is responsible for the growth of the organism and the repair of damaged tissues; first, DNA must be replicated so that there is a duplicate set of genetic information to be given to each daughter cell; second, the DNA must be divided so that each daughter cell gets the same set of information; mitosis is a three-step process: (1) replication of genetic material in the mother cell, (2) separation of the replicated genetic material, and (3) formation of the two daughter cells; when polyploidy is induced with colchicine, an alkaloid of the meadow saffron (Colchicum autumnale) that inhibits mitosis, the development of the nuclear spindle is hampered; the mitosis that takes place after treatment with colchicine is called a C mitosis; it also enables an easier detection and identification of chromosomes than a normal mitosis does; during the prolonged metaphase of a C mitosis, the chromosomes form an X-shaped structure since the chromatids are still connected at the centromere though they may repel each other; after some time, the chromatids finally part, but they do not segregate; they become enclosed by a new nuclear membrane and proceed to their interphase state; the number of chromosomes has now doubled – diploid nucleus has developed into a tetraploid one *cyto* >>> Figure 58 >>> Table 46

CMS: >>> cytoplasmic male sterility

coadaptation: The process of selection by which harmoniously interacting genes become accumulated in the gene pool of a population, i.e., also the selection process that tends to favor individuals that have mutually beneficial phenotypic associations with each of the other populations *gene*

coadaptation: The process of selection by which harmoniously interacting genes become accumulated in the gene pool of a population *gene*

coancestry: A quantification of relatedness between two individuals; the probability that genes taken from two individuals are identical by descent; the coancestry of mates becomes the >>> inbreeding of the progeny; coancestry of full sibs (from unrelated non-inbred parents) is 0.25, of halfsibs 0.125, and of first cousins 0.0625; it is possible to talk about self-coancestry, thus an individual's relatedness with itself; that is, for non-inbred individuals = 0.5; coancestry and inbreeding are relative concepts; their values depend on how far back ancestry is traced; thus, it is convenient to use them against some reference *stat meth*

coancestry matrix: Square matrix with coancestry among individuals in the cells; on the diagonal appears self-coancestry, in the other cells pairwise coancestry; the matrix is symmetric *stat meth*

coarse shaker (at a harvester): The straw walkers convey the straw to the rear of the straw chamber where it either falls to the ground or is fed into a straw chopper *seed*

coat protein: Protective layer of protein surrounding the nucleic acid core of a virus *phyt*

cob: The rachis of a female maize inflorescence or a heteromorphic condition in the family of Plumaginaceae characterized by short stigmatic papillae *agr bot*

cocaine: A bitter, white, crystalline alkaloid, $C_{17}H_{21}NO_4$, obtained from coca leaves, used as a local anesthetic *chem phys*

coccus: One of the separate divisions of a divided seedpod; it splits up into one-seeded cells *bot*

cochineal: An insect or a red dye prepared from the dried bodies of the females of the cochineal insects, Coccus cacti or Dactylopius coccus, which lives on cactuses *zoo* >>> opuntia >>> carmin

cocoa bean: The seed of the cacao tree, which is only called a cocoa bean once it is removed from the >>> pod in which it grows *bot* >>> cocoa

cocoa pod: The leathery oval pod that contains >>> cocoa beans *bot*

coconut milk: >>> coconut water

coconut water: Liquid endosperm from the center of the coconut seed; a complex, undefined addendum of variable quality and effects in some nutrient solutions (2–15% v/v) for plant tissue culture; it shows growth-promoting effects and cell division factors; it is replaceable in some cases by cytokinins and/or sugar *bot biot*

cocultivation: A technique for transforming protoplasts and other explants, or for *in vitro* selection by incubating them with a low density of transformed bacteria or a certain concentration of selective substances *biot*

coculture: >>> cocultivation

coding sequence: The part of a gene that determines the sequence of amino acids of a protein, as opposed to non-coding sequences, such as promoter, operator, intron, or terminator regions *gene*

coding single nucleotide polymorphism: Synonymous with a single nucleotide polymorphism that is located within an exon of a gene *biot* >>> SNP

coding SNP: >>> coding single nucleotide polymorphism

coding strand: The strand of duplex DNA that is transcribed into a complementary mRNA molecule *gene*

codogenic: That strand of double-stranded DNA (sense strand) used for genetic transcription *gene*

codominance: The expression of both alleles in the heterozygote with equal effect on the phenotype, as opposed to recessive and dominant *gene*

codominant: A heterozygote that shows fully the phenotypic effects of both alleles at a gene locus *gene*; in forestry, a tree receiving full light from above, but comparatively little from the sides; such trees usually have medium-sized crowns; in ecology, a species that shares equal dominance with another species in a plant community *gene fore eco*

codominant marker: Any molecular marker that detects both alleles of a particular genomic locus, e.g., >>> RFLPs or sequence-tagged microsatellite site markers belong to this category *biot gene*

codon: The triplet sequence of nucleotides in mRNA that acts as a coding unit for an amino acid during protein synthesis; it binds by base pairing to a complementary sequence, the anticodon, in tRNA *gene*

codon bias: Although several codons code for a single amino acid, a plant may have a preferred codon for each amino acid; this is called codon bias *gene*

coefficient of coancestry: >>> coancestry

coefficient of correlation: A number that measures the linear dependence between two random variables; limiting values of −1 and +1 indicate perfect negative and perfect positive correlation, respectively; a correlation of zero suggests a complete lack of association between the two variables *stat*

coefficient of inbreeding: >>> inbreeding coefficient

coefficient of parentage (COP): The probability that a random gene from one individual is identical by descent with the a random gene from another individual (e.g., two varieties having one parent in common statistically showing on 50% of loci the same alleles [i.e., cop = 0.5]), or having one grandparent in common showing 25% of loci the same alleles (i.e., cop = 0.25), or no parent in common cop = 0, respectively *stat meth*

coefficient of relatedness: $r = n \times (0.5)^L$ where n is the alternative routes between the related individuals along which a particular allele can be inherited; L is the number of meiosis or generation links *stat*

coefficient of relationship: The probability that two individuals have inherited a certain gene from a common ancestor *stat* >>> coefficient of parentage

coefficient of selection: A measure of the relative change in gene frequency between generations as a result of differential selection *stat gene* >>> Figure 60

coefficient of variation: The standard variation expressed as percentage of the mean *stat*

coenobium: Refers to a colony of a fixed number of cells *bot*

coenocytes: An organism or a portion thereof that is multinucleate; the nuclei are not each separate in one cell, such as in some protoplast or cell fusion products *bot biot*

coenocytic: Multinucleate or formed by nuclear division without cross wall formation *bot*

coenocytic endosperm: Multinucleate condition in which nuclei divide within a common cytoplasm without being separated from each other by a plasmalemma or cell wall *bot*

coenospecies: A group of individuals of common evolutionary origin comprising more than one taxonomic species *bot evol* >>> Figures 68, 69

coenzyme: Non-protein, organic substance that acts as cofactor for an enzyme *phys*

coevolution: Joint evolution of two unrelated species that have a close ecological relationship resulting in reciprocal adaptations as happens between host and parasite, and plant and insect *evol* >>> Figures 68, 69

cofactor: Non-protein component that is required by an enzyme in order for it to function, and to which it may be either tightly or loosely bound *phys*

coffee berry disease (CBD): The most serious disease of coffee, caused by the fungus Colletotrichum coffeanum; at present it is confined to eastern Africa; the fungus resides in the bark and parasitizes the berries only; in a susceptible tree, there is a total loss of all berries several months before harvest, and this represents the minimum level of horizontal resistance; in resistant trees selected recently in Ethiopia, there is no loss of berries, and this represents the maximum level of horizontal resistance; in other countries where the disease occurs, it is controlled with fungicides *phyt*

coffee leaf miner: A serious pest in coffee caused by Perileucoptera coffeella *phyt*

coffee leaf rust: This disease (*Hemileia vastatrix*) is of interest because, when arabica coffee was taken as one pure line to the New World, all of its pests and diseases were left behind in the Old World; this gave Latin America a commercial advantage, and it now produces about 80% of the world's coffee; when the rust was accidentally introduced into Brazil in 1970, there were fears of a major disruption of the world supply; this disease is an apparent exception to the rule that vertical resistance will evolve only in the seasonal tissue of a discontinuous pathosystem; because coffee is an evergreen perennial *phyt*

coffein: >>> caffeine

cohesin: During mitosis and meiosis >>> sister chromatids are held together by protein complexes; this cohesion is important not only for pairwise alignment of chromosomes on the mitotic spindle but also for the generation of tension across centromeres – it counteracts the pulling force of spindle microtubules, which ensures the bipolar attachment of chromosomes; chromosome cohesion thus enables accurate chromosome segregation in both mitosis and meiosis; cohesion is mediated by the cohesin complex, which contains four core subunits: two subunits of the structural maintenance of chromosomes protein family, Smc1 and Smc3; the kleisin family protein Scc1/Rad21; and an accessory subunit, Scc3/Psc3; in vertebrates, Scc3 has two isoforms: SA1 and SA2; another protein, Pds5, is weakly associated with the cohesin complex and may regulate the dynamic interaction of cohesin with chromatin; the cohesin complex has been proposed to form a ring structure that encircles sister chromatids *cyto biot* >>> kleisin

cohesive ends: DNA with single-stranded ends, which are complementary to each other; enabling the different molecules to join each other, also known as >>> sticky ends, overhang, or protruding ends *biot* >>> circularization

cohort: Refers to an inclusive group of all genotype selections entering the trials at one time; some kind of peer grouping is necessary for valid comparisons in breeding programs because only selected material enters (e.g., yield nurseries in seasonal or yearly groups); with the passage of time each cohort is unique in its number of possible data years (e.g., cohorts may enter the system on an annual basis and may be melded into a conglomerate cohort, which each year may contain one plot of every selection undergoing evaluation); the cohort (master nursery) is reasonably stable in size; the addition of the annual new cohort approximately balances the attrition by discard from all older cohorts; this basic nursery serves as a resource pool from which outstanding selections periodically can be chosen to become entries in replicated advanced trials *meth*

coiling: When chromosome cores first become visible in late prophase of mitosis, sister cores are adjacent to one another and run along the inner sides of sister chromatids; as prophase proceeds, the proteinaceous cores separate, and sister chromatids coil; the chromatids coil because cores actively shorten; probably the contraction of a core located to one side of a chromatid causes the chromatid to coil with the core to the inside and chromatin to the outside *cyto*

co-integrate vector: A two-plasmid system for transferring cloned genes to plant cells; the cloning vector shows a T-DNA segment containing cloned genes; after introduction into an *Agrobacterium tumefaciens*, the cloning vector DNA undergoes homologous recombination with a resident disarmed Ti plasmid to form a single plasmid carrying the genetic information for transferring the genetically engineered T-DNA region to plant cells *biot*

coir: Course fiber extracted from the fibrous outer shell of the coconut *agr*

colcemid: A synthetic equivalent of >>> colchicine *cyto meth*

colchicine: A poisonous alkaloid drug ($C_{22}H_{25}NO_6$) that is obtained from meadow saffron (*Colchicum autumnale*); it has a disruptive effect on microtubular activity; thus, it affects tissue metabolism generally, and mitosis and meiosis in particular; it is used for induction of polyploidy *chem phys cyto* >>> C mitosis >>> polyploidization >>> doubled haploid >>> Figure 58 >>> Table 46

colchiploidy: Polyploidy induced by application of colchicine *cyto*

cold frame: A bottomless box consisting of a wooden, concrete, stone, or metal frame with a glass or polyethylene top; it is placed on the ground over plants to protect them from cold or frost in order to speed up germination or to get an earlier harvest *hort meth*

cold spot: Any sequence within a gene or chromosome at which mutations occur at a significantly lower frequency than normal *gene*

cold stimulus: >>> stratification

cold test: A type of stress test that shows the performance of seeds in cool, moist soil in the presence of various soil microorganisms; the test is conducted by planting the seeds in moist, unsterilized field soil, exposing them to cool (+5 to −10°C) temperatures for about one week, then allowing them to germinate in the same soil at warmer conditions *seed* >>> www.plantstress.com

cole (crops): Vegetables of the genus Brassica, including cauliflower, broccoli, cabbage, and turnips *hort* >>> Figure 8

Coleoptera: Beetles, the largest subgroup of insects; they are characterized by their rigid outer wings, which cover a more delicate set of inner wings; their length can vary from less than a millimeter to up to 17 cm; worldwide, there are about 350,000 species of beetles; they pass through several stages of development (their larvae hatch from an egg and pupate; after pupation, adult beetles hatch from their cocoon and live for only a few weeks); they may feed on living and dead plant material, other insects, carrion, and dung; some species are serious agricultural pests, e.g., the western corn rootworm *zoo phyt*

coleoptilar stage: The embryo develops a prominent shoot apical meristem surrounded by the coleoptilar ring, and backed by the flattened, spade-shaped scutellum; it occurs about 12–14 days after pollination *phys meth* >>> Table 13

coleoptilar tiller: >>> axillary tiller

coleoptile: The first leaf in grasses, which ensheathes the >>> plumula *bot*

coleorhiza: A transitory membrane covering the emerging radicle (root apex) in some species; it serves the same function for the root as the coleoptile does for the plumule *bot*

colinearity: The correspondence between the order of nucleotides in a section of DNA (cistron) and the order of amino acids in the polypeptide that the >>> cistron specifies *gene*

collar: The structure at the top of the >>> culm above which lies the ear in cereals; in barley, the type of >>> lodicules that appear to enwrap the base of the >>> caryopsis when seen in position by removal of the lemma *bot* >>> Table 30

collateral mutation: Any random second-site mutation that is introduced into the target DNA during site-directed mutagenesis *gene meth* >>> TAL effector nuclease

collective fruit: >>> multiple fruit

collenchyma: A supporting tissue composed of more or less elongated living cells with non-lignified primary walls *bot*

colloid: A substance that is composed of two homogenous phases, one of which is dispersed in the other *chem*

colony: In tissue culture, a visible mass of cells biot; sometimes, a group of plants where all plants arise from one root system (e.g., in sunflower) *bot*

colony-forming units (CFU): A measure for number of viable bacteria *biot*

colony hybridization: A technique for using *in situ* hybridization to identify bacteria carrying a specific clone; it is only suitable for DNA fragments cloned onto multicopy vectors *biot* >>> Figure 46

Colorado potato beetle: A parasite (*Leptinotarsa decemlineata*) of the wild *Solanum rostratum* (buffalo burr, or prickly potato) in Colorado; this beetle moved on to cultivated potatoes as a new encounter parasite, and became one of the worst insect pests in the whole of agriculture; it is a yellow and black striped beetle, the same shape as a ladybird, but much larger, being 1.5 cm long; the larvae and beetles are voracious eaters of potato leaves and, if not controlled, they can destroy a potato crop; originally controlled with compounds of lead and arsenic, the beetles are now controlled with synthetic insecticides *phyt*

colorimeter: >>> colorimetry

colorimetry: The methods used to measure color and to define the result of the measurement *micr*

color separator: A machine that separates seed on the basis of their surface color; it is used for seed cleaning or type separation *seed*

colter: >>> coulter

columella: An elongated floral axis that supports the carpels in certain plants *bot*

column: Part of the gynoecium of orchids, homologous with the style, which is a gynostegium, carrying the >>> pollinia as well as the stigmatic cavity; sometimes the lower, undivided part of the awns of certain *Aristida* species *bot*

column chromatography: The separation of organic compounds by percolating a liquid containing the compounds through a porous material (e.g., ion exchange resin) in a cylinder *meth*

column-complete Latin square: A >>> Latin square is column-complete if each ordered pair of distinct symbols occurs precisely once in consecutive positions in a column of the square *stat* >>> design of experiment

column diagram: >>> histogram

column-quasi-complete Latin square: A >>> Latin square is column-quasi-complete if each unordered pair of distinct symbols occurs precisely twice in consecutive positions in a column of the square *stat* >>> design of experiment

coma: A tuft of hairs attached to a seed like the brush on wheat grains, or such as milkweed (Asclepias ssp.) and cotton (Gossyplium ssp.) seeds *bot*

combed: An industrial cotton yarn preparation; during the combing process, fibers are combed to make them parallel in the sliver and short fibers are removed *agr*

combination ability: >>> combining ability

combination breeding: A breeding method that utilizes the genetic diversity of individuals or varieties in order to create and to select new phenotypes on the basis of genetic recombination of useful characters of parental material *meth* >>> Figures 5, 7, 16 >>> Tables 5, 35

combine (harvester): A self-propelled grain harvester; in one operation it combines cutting, threshing, separation, cleaning, and straw dispersal *agr*

combine harvester: A machine that harvests and threshes a crop in one operation *agr*

combing: Using a comb to lift >>> stolons and >>> procumbent shoots so that they can be cut by a mower; the comb typically has metal teeth or flexible tines and is fastened immediately in front of a reel mower *agr*

combining ability (CA): The average performance of a strain in a series of crosses (general CA); deviation in a particular cross from performance predicted on the basis of general combining ability (specific CA) *gene* >>> Figures 19, 51 >>> Table 35

common haploid: Any haploid that is characteristic for a majority of individuals of a distinct population *gene*

common single nucleotide polymorphism: Any >>> SNP whose minor allele occurs in a more than 10% of the genomes of a population *biot* >>> causal single nucleotide polymorphism >>> candidate single nucleotide polymorphism >>> copy single nucleotide polymorphism >>> SNP

common scab (of potato): A fungal disease (Streptomyces scabies) affecting the tuber skin; diseased tubers are unattractive, but yield and cooking quality remain unaffected *phyt*

common SNP: >>> common single nucleotide polymorphism

Commonwealth Scientific and Industrial Research Organization (CSIRO, Australia) This organization carries out research in science, industry, and agriculture and is partly funded by industry *org* >>> www.csiro.au

community: A naturally occurring group of various organisms that inhabit a common environment, interact with each other, and generally are independent of other groups *eco*

compact: Short and dense *bot*

compactation: Increase in bulk density of soil due to mechanical forces, such as tractor wheels or combines *agr*

companion cell: A type of cell that pumps nutrients (sugars) into phloem cells *phys*

companion crop: Two crops grown together for the benefit of one or both; it is particularly used of the small grains with which forage crops are sown *agr* >>> catch crop

companion planting: >>> companion crop

companion species: Usually a weed species that grows in close proximity to a crop species and which may be ancestral to the crop; the two species may exchange genes by viable spontaneous hybrids (e.g., Aegilops spp. near a wheat field, as it is seen sometimes in Turkey or other countries of the Middle East or the American Midwest) *eco*

comparative genomic hybridization (CGH): A molecular cytogenetic technique that allows detection of DNA sequence copy number changes throughout the genome in a single hybridization *meth biot cyto* >>> Table 37

comparative (gene) mapping: Localization (mapping) of a common set of >>> DNA probes onto linkage maps of different species; the results of these comparisons indicate substantial conservation of blocks of genes and even large segments of chromosomes between species; this approach shows synteny of markers among related species or genera; such maps have been established for cereals (wheat, maize, oats, rye, rice, sorghum, millet) and Solanaceae, such as potato, tomato, and paprika *gene meth*

comparative positional candidate gene: A gene that is likely to be located in the same region as a DNA marker that has been shown to be linked to a single-locus trait or to a quantitative trait locus (QTL), where the gene's likely location in the genome of the species in question is based on its known location in the map of another species (i.e., is based on the comparative map between the two species) *biot* >>> QTL mapping >>> Table 58

compartmentalization: Isolation of a specific tissue area by host barrier tissues *phyt*

compatibility: In molecular biology, compatible plasmids have different replication functions and can coexist in the same cell; incompatible plasmids are very similar or identical in their replication functions and, in the absence of selection, one of the two plasmids will survive in the cells, while the other is lost spontaneously from a culture; the smaller, faster-replicating plasmid is generally favored in this process *biot*

compensating diploid(s): One normal chromosome that is replaced by two telocentric chromosomes representing both arms or four telocentric chromosomes that compensate for a pair of normal (homologous) chromosomes *cyto*

compensating trisomic(s): An aneuploid individual with an extra chromosome in which a missing standard chromosome is compensated for by two novel interchange chromosomes; the two novel chromosomes carry two different arms of the missing chromosome *cyto* >>> Figures 14, 17, 37

compensation point: The concentration of CO_2 at which photosynthesis balances respiration by a leaf, in defined conditions *phys*

compensatory mutation: Any mutation in an exon of a gene that neutralizes the effect of another mutation in the same exon, or another exon of the same gene *gene*

competence: Ability of a bacterial cell to take up DNA molecules and become genetically transformed *biot*

competency: An ephemeral state, induced by treatment with cold cations, during which bacterial cells are capable of uptaking foreign DNA; in plants, the endogenous potential of a given cell or tissue to develop in a particular way, e.g., embryogenically competent cells are capable of developing into fully functional embryos; the opposite is non-competent or morphologically incapable *biot* >>> dedifferentiation >>> totipotency

competent: In tissue culture, able to function or to develop; in molecular biology, competent (cells) have the ability to take up DNA; *Escherichia coli* is artificially made competent by washing in the cold with $CaCl_2$; some bacteria become naturally competent during certain growth phases *biot*

competition: Interaction of two or more organisms restricting each other's survival when at least one resource (e.g., water, nutrients, light, space) is limiting *eco*

competition for food: >>> survival of fittest

competitive allele specific PCR (KASP™): A homogenous, fluorescence-based genotyping variant of polymerase chain reaction; it is based on allele-specific oligo extension and fluorescence resonance energy transfer for signal generation; a single-nucleotide polymorphism (>>> SNP) occurs when a single nucleotide in a DNA sequence differs between members of the same species or a paired chromosome; SNPs work as >>> molecular markers that help locate >>> genes associated with disease and are used for genotype sequencing; genotyping by >>> next-generation sequencing using SNPs is expensive, time-consuming, and has some missing data; there are many other SNP techniques that can be used depending on the purpose of the research considering throughput, data turnaround time, ease of use, performance (sensitivity, reliability, reproducibility, accuracy) flexibility, requirements, and cost; for highest throughput for large-scale studies it is best to choose multiplexed chip-based technology; multiplex technologies generate anywhere from 100 to over a million SNPs per run but are not economical to use for small to moderate numbers of SNPs; for a smaller number of SNPs, a uniplex assay like KASP can be used *meth gene*

competitive interaction: A condition that may arise in heteroallelic pollen expressing two different pollen S-alleles; they effectively compete with one another such that pollen is not inhibited in pistils expressing either of the matching S-haploids *gene*

compilospecies: A genetically aggressive species that assimilates genomes from related species; it may completely assimilate a species, causing extinction *bot*

compilospecies: A genetically aggressive species that assimilates genomes from related species *bot*

complementary DNA (cDNA): DNA complementary to a purified mRNA and produced by RNA-dependent DNA polymerase; it lacks the introns present in corresponding genomic DNA; it is most commonly made to use in PCR to amplify RNA (RT-PCR); e.g., large-scale DNA sequencing can be carried out on genomic DNA or cDNA; the use of cDNS is more cost-efficient, interpretation of results is more precise, cDNA helps in annotation and identification of exons and introns, and cDNA helps to gain information about transcriptomes *gene biot*

complementary effect: Genes that by interaction produce a phenotype qualitatively distinct from the phenotype of any of them separately *gene*

complementary genes: Mutant alleles at different loci, which complement one another to give a wild-type phenotype; dominant complementarity occurs where the dominant alleles of two or more genes are required for the expression of a particular trait; recessive complementarity is the case of suppression of a particular trait by the dominant allele of either gene, so that only the homozygous double recessive displays the trait *gene* >>> Table 6

complementary herbicide: Herbicides used in conjunction with a specially designed, herbicide-tolerant crop; if a soybean cultivar is genetically modified to have tolerance to an herbicide, the herbicide is considered that soybean cultivar's complementary herbicide; complementary herbicides for herbicide-tolerant plants developed with genetic engineering are generally "non-selective" or "broad-spectrum" herbicides; these affect central sites of plant metabolism and are thus effective against a wide range of plants; an herbicide-tolerant crop and its respective complementary herbicide constitute an herbicide tolerance system; in this combination, wide-spectrum herbicides like Roundup™ can be applied to kill nearly all weeds without harming the crop *biot agr*

complementary homopolymeric tailing: The process of adding complementary nucleotide extensions to different DNA molecules, for example, deoxyguanosine (dG) to the 3'-hydroxyl ends of one DNA molecule and deoxycytidine (dC) to the 3'-hydroxyl ends of another DNA molecule to facilitate, after mixing, the joining of the two DNA molecules by base pairing between the complementary extensions *biot*

complementary interaction: The different components of the system must work together to exert an effect *gene*

complementary nucleotides: Members of the pairs adenine–thymine, adenine–uracil, and guanine–cytosine that have the ability to hydrogen bond to one another *chem biot* >>> Table 38

complementation: Interaction of two or more genetic loci in the >>> expression of a single phenotypic character, or the complementary action of gene products in a common >>> cytoplasm; the support of a function by two homologous pieces of genetic material present in the same cytoplasm, each carrying a recessive mutation and unable by itself to support that function *gene*; in molecular biology, it refers to the ability of a cloned gene to overcome a host mutation *biot*

complementation test: A genetic test to determine if two mutations with similar phenotypes are allelic (occur in the same gene) or are non-allelic (occur in separate genes); individuals homozygous for each mutation are crossed; if the mutations are allelic, the offspring will have the mutant phenotype; if they are non-allelic the offspring will be heterozygous for both alleles and will have the wildtype phenotype *gene meth*

complete block: A simple experimental design in which all testers are included in each replication of the experiment and are arranged in a random order within replication; for accuracy, it should be used with small numbers of entries only; replications may be placed end to end or opposite each other, so that the total area covered by the experiment will be as nearly square in shape as possible; entries with apparent weakness may be discarded before harvest and data may still be analyzed by an analysis of variance *meth stat* >>> design of experiment >>> Table 25

complete cross: >>> diallel cross

complete diallel: A mating design and subsequent progeny test resulting from the crossing of a certain number of parents in all possible combinations including selfs and reciprocals; because of severe inbreeding depression in the selfs, these are often skipped, nevertheless the test is still called a full diallel *meth* >>> design of experiment

complete digest: The treatment of a DNA preparation with an >>> endonuclease for sufficient time for the entire potential target sites within that DNA to have been cleaved *biot*

complete dominance: >>> dominance

complete flower: A flower that has pistils, stamens, petals, and sepals *bot*

complete Latin square: A >>> Latin square is complete if it is both row-complete and column-complete *stat* >>> design of experiment

completely randomized design: The structure of the experiment in a completely randomized design is assumed to be such that the treatments are allocated to the experimental units completely at random *stat* >>> design of experiment

complete penetrance: The situation in which a dominant gene always produces a phenotypic effect or a recessive gene in the homozygous state always produces a detectable effect *gene*

complex heterozygous: Special type of genetic system based on the heterozygosity for multiple reciprocal translocations *cyto*

complex locus: A cluster of two or more closely linked and functionally related genes constituting a pseudoallelic series *gene*

component of variance: >>> variance

component weight: For each stage of sampling, the component weight is equal to the reciprocal of the probability of selecting the unit at that stage *stat*

composite: A plant of the immense family Compositae, regarded as comprising the most highly developed flowering plants *bot*; a mixture of genotypes from several sources, maintained by normal pollination *seed*

composite cross: A population derived from the hybridization of several parents, either by hand-pollination or by the use of male sterility *meth* >>> design of experiment

composite-cross population: A population generated by hybridizing more than two varieties and/ or lines of normally self-fertilizing plants and propagating successive generations of the segregating population in bulk in specific environments so that natural selection is the principal force acting to produce genetic change; artificial selection may also be imposed; the resulting population is expected to have a continuously changing genetic makeup *meth*

composite fruit: A seed distribution unit that includes many ovaries connected by fruit walls or other suitable tissue; if the flower basis (receptacle) or other flower components are thick and fleshy (e.g., in strawberry, apple, or fig), it is called a false fruit or pseudocarp *bot*

composite interval mapping: In this method, an >>> interval mapping is performed using a subset of marker loci as covariates; these markers serve as proxies for other >>> QTLs to increase the resolution of interval mapping, by accounting for linked QTLs and reducing the residual variation; the problem concerns the choice of suitable marker loci to serve as covariates; once these have been chosen, mapping turns the model selection problem into a single-dimensional scan *biot meth* >>> QTL mapping >>> Table 58

composite mixture: Breeder seed obtained by mechanically combining seed from two or more strains; the mixture is increased through successive steps in a certified seed program and distributed as a synthetic variety *seed* >>> Figure 51

composite transposon: A transposable element formed when two identical or nearly identical transposons insert on either side of a non-transposable segment of DNA, such as the bacterial transposon Tn5 *biot*

composite variety: A plant population in which at least 70% of progeny result from cross of the parent lines *seed*

compost: Plant and animal residues that are arranged into piles and allowed to decompose *agr hort*

compound chromosome: Formed by the union of two separate chromosomes *cyto*

compound cross: A combination of desirable genes from more than two inbred lines, breeding strains, or varieties *meth*

compound cyme: A determinate inflorescence where there is secondary branching, and each ultimate unit becomes a simple cyme *bot*

compound leaf: A leaf that is divided into many separate parts (leaflets) along a midrib (the rachis); all the leaflets of a compound leaf are oriented in the same plane; when a compound leaf falls from the tree, it falls as a unit; a double compound leaf is one in which each leaflet of a compound leaf is also made up of secondary leaflets *bot*

computational biology: The use of techniques from applied mathematics, informatics, statistics, and computer science to solve biological problems *stat biot* >>> bioinformatics

compressed: Flattened strongly, typically laterally; keeled *bot*

co-mutagenesis: The occurrence of two or more mutations at closely linked loci with the genome *gene* >>> TAL effector nuclease

concatemer: Tandem repeats of identical DNA molecules; lamda phage DNA must be concatemers in order to be packaged *biot* >>> lamda phage >>> http://wheat.pw.usda.gov/ITMI/Repeats/index.shtml

concatenate: Interlocked circles (e.g., plasmids) *biot*

concave: Shaped like the inside of an egg *bot*

concatemer: DNA segment made up of repeated sequences linked end to end *biot*

concerted evolution: The ability of two related genes to evolve together as though constituting a single locus *evol gene* >>> Figures 68, 69

condensation of chromosomes: >>> chromosome contraction >>> coiling

condenser: A lens or combination of lenses that gathers and concentrates light in a specified direction, often used to direct light onto the projection lens in a projection system *micr*

condenser iris diaphragm: The substage iris diaphragm located at the front focal plane of the condenser lens of a microscope; with >>> KOEHLER illumination, the iris lies in a plane conjugate with the rear focal plane of the objective lens *micr*

conditional mutation: Any mutation that is only expressed under specific conditions, e.g., temperature effects *gene*

conditional lethal genes: Genes that are expressed in response to a specific environmental stimulus such as a specific chemical; when expressed, they are lethal to the plant *gene*

conditional mutation: A mutation that has the wild-type phenotype under certain environmental conditions (temperature, age, nutrition) and a mutant phenotype under other conditions *gene*

conditioned dominance: Dominance affected by the presence of other genes or by environmental influence *gene*

conditioned storage: Storage of seed under controlled conditions of temperature and relative humidity *seed meth*

conditioner: A material or substance added to a fertilizer that keeps it flowing free *meth agr*

conditioning: The term used to describe the process of cleaning seed and preparing it for market; sometimes called processing *seed* >>> Table 11

conduction: Plasmid mobilization involving cointegrate formation *biot*

conductivity test: An electrical conductivity test that associates the concentration of leachates from seeds, after soaking in water, to their quality *seed*

cone: A fruit with overlapping scales in which seeds are formed *fore*

cone collection: Harvesting of cones after seed maturation but before their dispersal *fore hort*

confidence belt: >>> confidence limit

confidence limit: A term for a pair of numbers that predict the range of values (confidence interval) within a particular parameter *stat*

confined research field trial: The release of a plant with a novel trait, for research purposes, under terms and conditions of confinement designed to minimize any impact the plant with a novel trait may have on the environment; these terms and conditions include, but are not limited to, reproductive isolation, site-monitoring, and post-harvest land use restriction *biot eco meth*

confocal optics: A microscope optical system in which the condenser and objective lenses both focus onto one single point in the specimen *micr*

congeneric species: Those plant species that belong to the same genus, but form two or more different species *bot*

congenic strain: A variant plant strain that is obtained by backcrossing a donor plant strain to an inbred parental strain for at least eight generations while maintaining by appropriate selection the presence of a small genetic region derived from the donor strain *gene*

congruent crossing: In cross breeding, when parents match exactly in morphology, physiological, and/or cytogenetic behavior *meth*

conical divider: An inverted metal cone below a spout from a hopper; the seeds fall over the cone to be evenly dispersed; a series of bugle or riffle dividers separate the seeds into channels *seed*

conidiophore: A threadlike stalk upon which conidia (spores) are produced; a specialized hypha upon which one or more conidia may bear *bot*

conidium (conidia pl): Any asexual spore formed on a conidiophore *bot*

conifer: A species of plant that bears its naked seeds in cones (a woody strobilus); their flowers are in cones, and male and flower cones are separate; the oldest (bristlecone pine) and the largest (sequoia) extant organisms belong to this class; it belongs to the Gymnospermae which includes needle-leafed trees such as pines and cypresses; most conifers are evergreen trees and shrubs, e.g., pine, fir, larch, and spruce trees; Mesozoic-era conifers included redwoods, yews, pines, the monkey puzzle tree (*Araucaria* ssp.), cypress, and *Pseudofrenelopsis* ssp. (Cheirolepidiaceae); towards the end of the Mesozoic, flowering plants flourished and began to overtake conifers as the dominant flora; their unique feature is the inheritance of cytoplasmic DNA (chloroplasts) via pollens *bot*

coniferous tree: >>> conifer

conjugate (of a Latin square): A >>> Latin square obtained by permuting the roles of "rows", "columns", and "symbols" *stat* >>> design of experiment

conjugation: A process whereby organisms of identical species, but opposite mating types, pair and exchange genetic material (DNA) gene; in molecular biology, natural process of DNA transfer between bacteria in which the DNA is never exposed; it is insensitive to externally added >>> DNase *biot*

connate: Fusion of like parts, such as sheath margins to form a tube *bot*

connective: The tissue joining the two cells of an anther *bot*

conoidal: Nearly conical *bot*

consanguineous matings: Matings between two individuals who share a common ancestor in the preceding two or three generations *gene*

consanguinity: >>> coancestry

consensus chloroplast simple (single) sequence repeat (ccSSR): Average or most typical form of a sequence that is reproduced with minor variations in a group of chloroplast DNA sequences; the consensus sequence of chloroplast DNA shows the nucleotide or amino acid most often found at each position; the preservation of a consensus implies that the sequence is functionally important; the simple or single sequence repeat, that is, micro-satellite sequences repeat, appears repeatedly in sequence within the DNA molecule in a manner enabling them to be used as markers *biot meth* >>> Figure 63

consensus sequence: If a particular nucleotide sequence is always found with only minor varia-tions, then the usual form of that sequence is called consensus sequence; the term is also used for genes that encode the same protein in different organisms *gene*

conservation: Maintenance of environmental quality and resources *seed*

conservation tillage: Seed bed preparation systems that have about 30% or more of the residue cover on the surface after planting; two key factors influencing crop residue are (1) the type of crop, which establishes the initial residue amount and its fragility, and (2) the type of till-age operations before and during planting; there are three tillage systems: (1) no-till – soil left undisturbed from harvest to planting except for nutrient injection; planting or drilling is accomplished in a narrow seedbed or slot created by coulters, row cleaners, disk open-ers, in-row chisels, or roto-tillers; weed control is accomplished primarily with herbicides; cultivation may be used for emergency weed control, (2) ridge-till – soil left undisturbed from harvest to planting except for nutrient injection; planting, however, is completed in a seedbed prepared on ridges with sweeps, disk openers, coulters, or row cleaners; residue is left on the surface between ridges; weed control is accomplished with herbicides and/or cultivation; ridges are built during cultivation, (3) mulch-till – soil is disturbed prior to planting; tillage tools such as chisels, field cultivators, disks, sweeps, or blades are used; weed control is accomplished with herbicides and cultivation *agr*

conserved orthologous set (of markers) (COS): Used for comparative mapping between closely related plant species; for a given group of species, a COS is formed by identifying genes from each species that are orthologous to genes of other species in the set *meth biot*

consocies: Part of a plant association lacking one or more of its dominant species *eco*

conspecies: Individuals of the same species, or those distinct sympatric species that are distributed or inhabit the same geographic area *eco*

constitutive: A plant is said to be constitutive for the production of an enzyme or other protein if that protein is always produced by the cells under all physiological conditions *biot*

constitutive enzyme: An enzyme that is synthesized continually regardless of growth conditions *chem phys*

constitutive gene: A gene that is continually expressed in all cells of a plant *gene*

constitutive heterochromatin: The material basis of chromosomes or segments that exhibit het-erochromatic properties under most conditions (e.g., centromeric or telomeric heterochro-matin) *cyto* >>> Table 43

constitutive mutation: Causes genes that usually are regulated to be expressed without regulation *gene*

constitutive promoter: An unregulated promoter that allows for continual transcription of its associated *gene biot*

constitutive resistance: Genetically controlled, inherited resistance *phyt gene*

constitutive synthesis: Continual production of RNA or protein by an organism *bio*

constriction: An unspiralized segment of fixed position in the metaphase chromosomes (nucleolar ~, primary or centric ~, secondary ~) *cyto*

construct: An engineered DNA fragment (e.g., plasmid) which contains, but is not limited to, the DNA sequences to be integrated into a target plant's genome *biot*

Consultative: Group on International Agricultural Research (CGIAR) An informal association of 58 public and private sector members supporting 16 international agricultural research centers; the centers develop advanced breeding material for adoption and use by national agricultural research systems in developing countries; under its auspices are the institutes as follows: CIAT, CIMMYT, CIP, ICARDA, ICRISAT, IITA, IRRI; a group of donors established the CGIAR in the early 1970s to fund agricultural research around the world;

it does this via 16 International Agricultural Research Centers, which now call themselves Future Harvest centers comprising more than 8,500 scientists and support staff working in more than 100 countries; the CGIAR is the biggest institutional force guiding research and development for the crops that feed people in the South; as government funding is drying up, the CGIAR is increasingly looking to partnerships with industry to keep itself alive *org* >>> www.cgiar.org

contact: Getting an herbicide to touch the surface of a plant, either foliage or roots *phyt*

contact herbicide/fungicide: Herbicide/fungicide that does not move from point of initial uptake; it remains on the surface where it is applied *phyt*

containment: Measures taken to prevent release of recombinant DNA molecules into the natural environment; biological and physical methods are applied *biot*

contaminant: Growth of microorganisms along the side of the plant tissue *in vitro biot*

contig: A set of overlapping clones that provide a physical map of a portion of a chromosome *biot* >>> contiguous (contig) map

contiguous gene syndromes: Disorders caused by microdeletions or microduplications in neighboring functional genes; inheritance is usually sporadic but recurrences are possible *gene*

contiguous (contig) map: The alignment of sequence data from large, adjacent regions of the genome to produce a continuous nucleotide sequence across a chromosomal region *biot*

contingency tables (cross-tabulations): For cross-classified data the frequency for the values that fall into each possible combination of levels from two different factors is recorded; one of the factors is associated with the columns of the contingency table; the other factor is associated with the rows of the contingency table; individual qualitative data values are simply grouped by the values of the classification variable; for individual quantitative data values, a classification may be created by defining intervals or cutpoints to bin the data into separate groups *stat*

continuation: A filing, while a patent is active, which contains additions or changes to the previous claims *adm* >>> Intellectual Property Rights

continuous: A rachis or other organ that does not disarticulate *bot*

continuous culture: An *in vitro* suspension culture continuously supplied with nutrients by the inflow of fresh medium; the culture volume is normally constant *biot* >>> closed continuous culture

continuous scale: A scale for scoring quantitative data for which the number of potential values is not predefined and is potentially limitless (e.g., seed weight in grams) *stat*

continuous spindle fiber(s): Connect the two polar regions with each other and remain from early prometaphase to early >>> prophase *cyto*

continuous variables: Variables that theoretically have an infinite number of values *stat*

continuous variation: Variation in the expression of inherited traits in which a series of non-discrete, intermediate types, which cannot be divided into separate categories, connect the extremes with no obvious breaks between them *gene* >>> quantitative character

contour: An imaginary line on the land connecting points of the same elevation; a line drawn on a map to show the location of points at the same elevation; a series of such contours serving to delineate the topography of the land *agr*

contour plowing: A system of plowing in which the furrows follow the land contours in order to minimize soil erosion *meth agr*

contracted: Narrow or dense inflorescences, the branches being appressed or short *bot*

contrasting genetic character: A character with marked phenotypic differences *gene*

control: An economic reduction of crop losses caused by plant diseases *phyt agr*

control group: The subjects in an experiment who are not exposed to the experimental stimulus *stat*

controlled breeding: The reproduction of desired characteristics >>> breeding

controlling element: A mobile (autonomous or non-autonomous) genetic component capable of producing an unstable, mutant target gene; originally found in maize by its genetic property; it is a transposable element; it may be autonomous (transpose independently) or non-autonomous (transpose only in the presence of an autonomous element) *gene cyto*

controlling gene: A gene that is involved in turning on or off the transcription of structural genes; two types of genetic elements exist in this process: a regulator and a receptor element; the receptor element is one that can be inserted into a gene, making it a mutant, and can also exit from the gene; both of these functions are under control of the regulator element *gene*

control pollination: In horticulture and forestry, to purposely pollinate the female flowers of a tree with pollen from a known source; usually the flowers are isolated from undesirable pollen by covering them with a pollen-tight cloth or paper bag before they are receptive; it is a way to produce full-sib families *meth hort fore* >>> pollination bag

control treatment: >>> treatment

control variables: Variables whose value is held constant in all conditions of an experiment *stat*

convariety (convar.): A group of similar cultivars within a variable species or hybrids between two species; the term has now been replaced in most cases by the word "group" *tax*

convenience sampling: Used in exploratory research where the researcher is interested in getting an inexpensive approximation of the truth; the samples are selected because they are convenient; this non-probability method is often used during preliminary research efforts to get a gross estimate of the results, without incurring the cost or time required to select a random sample *stat*

Convention on International Trade of Endangered Species (CITES): Provides regulations for the international trade in listed (endangered) species of plants and animals *org* >>> www .cites.org

conventional breeding: The selection process used to try and improve the characteristics of a plant; genetically superior parents are selected for breeding in a cyclic process until the desired trait is received; the goal of breeding is to produce genetically improved populations *meth* >>> Figure 60

conventional tillage: Primary and secondary tillage operations normally performed in preparing a seedbed and/or cultivating for a given crop grown in a given geographical area, usually resulting in <30% cover of crop residues remaining on the surface after completion of the tillage sequence *agr*

convergence: The evolution of unrelated species occupying similar adaptive areals, resulting in structures bearing a superficial resemblance *evol* >>> Figures 68, 69

convergence breeding (convergent improvement): A breeding method involving the reciprocal addition to each of two inbred lines of the dominant favorable genes lacking in one line and present in the other; backcrossing and selection are performed in parallel, each of the original lines serving as the recurrent parent in one series *meth* >>> Figure 31

convergence-divergence selection: A breeding scheme in which selection of promising genotypes is made in a bulk population at different locations followed by massing of selection and allowing mating among them in a pollination field; the harvested bulk seeds constitute the basis for the next propagation cycle *meth* >>> Figure 51

convergent coorientation: In a trivalent, two chromosomes face to one pole, while a third orients to the other pole *cyto*

convergent crossing: >>> convergence breeding

convergent evolution: Evolution of two or more different lineages towards similar morphology due to similar adaptive pressures *evol* >>> Figures 68, 69

convex: Shaped like the outside of an egg *bot*

cool morning method: >>> DIF method

cool-season crop (or plant): A crop that grows best during the cool temperatures of spring and fall or a plant that makes most of its growth during winter and spring and sets seed in late spring or early summer *hort agr*

cooperative breeding: A breeding system where older siblings or adults other than the parents help rear the current-year's brood *eco* >>> breeding

coorientation: >>> centromere orientation

COP: >>> coefficient of parentage

copal: A resinous substance exuded from some species of tropical trees and hardening in air into a glassy solid ranging in color from red to yellow or brown *bot*

copper (Cu): A malleable ductile metallic element having a characteristic reddish-brown color; as a trace element it is needed by plants; deficiency can cause severe problems of growth; as with iron efficiency, copper efficiency is genetically controlled (e.g., on rye chromosome arm 5RL a dominant gene and/or gene complex is located, increasing Cu efficiency not only in rye but also in wheat when the gene is transferred into the recipient) *chem phys* >>> white leaf disease >>> mugeinic acid >>> chelate

coppice: Natural regeneration originating from stump sprouts, stool shoots, or root suckers, or to prune tree and shrubs close to the ground level periodically to promote strong growth *fore hort*

coppice-of-two-rotations method: A coppice method in which some of the coppice shoots are reserved for the whole of the next rotation; the rest being cut *fore*

coppice method: A method of regenerating a forest stand in which the cut trees produce sprouts, suckers, or shoots *fore*

coppice selection method: A method in which only selected shoots of usable size are cut at each felling, leading to uneven-aged stands *fore*

coppice shoot: Any shoot arising from an adventitious or dormant bud near the base of a woody plant that has been cut back *fore hort*

coppice-with-standards method: Regenerating a forest stand by coppicing; selected trees grown from seed are left to grow to a larger size than the coppice beneath them; the method is used to provide seeds for natural regeneration of standards in subsequent rotations *fore*

copulation: The fusion of sexual elements *gene*

copy-choice hypothesis: The interpretation of intrachromosomal genetic recombination that is not regarded as a physical exchange of preformed genetic strands *gene*

copy error: An error in the DNA replication process giving rise to a gene mutation *gene*

copy gene: Genetic material incorporating the genetic code for a desirable trait, which has been copied from DNA of the donor to the host organism *biot*

copy number: The number of molecules per genome, of a plasmid or a gene, that a cell contains; copy number variants are genomic rearrangements resulting from gains or losses of DNA segments; typically, the term refers to rearrangements of sequences larger than 1 kb; this type of polymorphism has been shown to be a key contributor to intra-species genetic variation, along with single-nucleotide polymorphisms (SNP) and short insertion-deletion polymorphisms; over the last decade, a growing number of studies have highlighted the importance of copy number variation; in plants, initial genomic analyses indicate that copy number variations are prevalent in plants and have greatly affected plant genome evolution; many copy number variation events have been observed in outcrossing and autogamous species; they are usually found on all chromosomes, with hotspots interspersed with regions of very low genetic variation; although copy number variation is mainly associated with intergenic regions, many copy number variations encompass protein-coding genes; data suggest that copy number variation mainly affects the members of large families of functionally redundant genes; there are many cases in which copy number variations for specific genes have been linked to important traits such as flowering time, plant height, and resistance to biotic and abiotic stress; reports suggest that copy number variations may form rapidly in response to stress *gene biot* >>> Figures 68, 69

copy SNP: >>> coding single nucleotide polymorphism

cordage: Ropes *agr*

cordate: The heart-shaped base and pointed apex of a structure, e.g., a leaf blade *bot*

cordon: An extension of the grapevine trunk, usually horizontally oriented and trained along the trellis wires; it is considered permanent (or perennial) wood *hort*

core collection: The basic sample of a germplasm collection; it is designed to represent the wide range of diversity in terms of morphology, geographic range, or genes; it contains, with a minimum of repetitiveness, the genetic diversity of a crop species and its wild relatives; it is not intended to replace existing genebank collections but to include the total range of genetic variation of a crop in a relatively small and manageable set of germplasm accessions; core collections are an integral part of biotechnology-aided modern-day crop improvement programs and are utilized for a variety of applications including conventional plant breeding, association mapping, resequencing, among others; since their advent, determination of core collection size has been based on the size of the whole collection; results support the hypothesis that the less frequent alleles seldom contribute to the genetic distance when compared with common alleles; presize can be efficiently utilized in any crop for the precise estimation of core collection size *meth seed*

core eudicot: Members of a large monophyletic eudicot group that includes most of the angiosperms *evol*

core genome: Portion of the >>> pan genome common to all individuals in the species; it represents the genes present in all strains of a species; it typically includes housekeeping genes for cell envelope or regulatory functions *biot* >>> pan genome >>> dispensable genome >>> Figure 71

corepressor: A metabolite that in conjugation with a repressor molecule binds to the operator gene present in an operon and prevents the synthesis of a repressible enzyme *gene*

coriaceous: Leathery *bot*

cork: In woody plants, a layer of protective tissue that forms below the epidermis *bot*

cork cambium: >>> phellogen

cork layer: Layer of dead protective tissue between the bark and cambium in woody plants *bot*

corky root rot (of tomato): A disease caused by Pyrenochaeta lycopersici *phyt* >>> Figure 62

corm(us): An underground storage organ formed from a swollen stem base, bearing adventitious roots, and scale leaves; it may function as an organ of vegetative reproduction or in perennation *bot*

corn U.S.: The edible seed of cereal plants other than maize *bot* >>> caryopsis >>> maize

Corn Belt: An agricultural region of central U.S. primarily in Iowa and Illinois, but also parts of Indiana, Minnesota, South Dakota, Nebraska, Kansas, Missouri, and Ohio, growing maize at a large scale *agr*

corn bran: The fibrous outer coating of the maize kernel, regarded as a low-grade food for cattle or a high-grade food for humans *bot*

corneous: It refers to hard, vitreous, or horny endosperm in cereal grains *bot agr*

corn-loft: >>> granary

corolla: A collective term for all the petals of a flower; a non-reproductive structure; often arranged in a whorl; encloses the reproductive organs *bot*

corpusculum: The gland connecting the two waxy pollen grain masses, e.g., in milkweeds (*Asclepiadaceae* ssp.) *bot*

correlation: The degree to which statistical variables vary together; measured by the correlation coefficient, which has a value from zero (no correlation) to −1 or +1 (perfect negative or positive correlation) *stat*

correlation breaker: >>> outlier

correlation coefficient: A measure for the degree of association between two or more variables in an experiment; it may range in value from −1 to +1 *stat* >>> correlation

correspondence analysis: A complementary analysis to genetic distances and dendrograms; it displays a global view of the relationships among populations; this type of analysis is more informative and accurate than dendrograms, especially when the number of loci is fewer than 30; using the same allele frequencies that are used in phylogenetic tree construction, correspondence analysis can be performed on the VISTA or MVSP computer programs *biot gene*

corresponding gene pair: A pair of genes in a parasite that corresponds with a pair of genes in a host, which function together to bring about a specific outcome *phyt*

cortex: >>> rind

corymb: A racemose inflorescence in which the lower pedicels are longer than the upper so that the flower lies as a dome or dish, and the outline is roundish or flattish *bot*

COS: >>> conserved orthologous set (of markers)

cos end: The 12-base, single-strand, complementary extension of bacteriophage lambda DNA; also known as cos site *biot* >>> cosmid >>> cos site

co-sexual: Having male organs (stamen) and female organs (carpel, pistil) present in the same flower *bot* >>> bisexual

cosmid: A synthetic word derived from the designations cos and plasmid; a cosmid is a plasmid (e.g., pBR322) with so-called cos-sites of the DNA; they offer the chance of incorporation of alien DNA fragments of sizes between 32 and 45 kb *gene* >>> cos end >>> cos site

cosmopolite: Plant of worldwide distribution *eco*

cosmopolitan cultivars: Cultivars that have a wide geographical and environmental range *agr eco*

cos site: The site of the circular form of lamda phage or others that is cleaved by the terminase to generate the cohesive 12 bp 5′ overhang ends of the linear phage as it is packaged into the capsid *biot* >>> cosmid >>> lamda phage

costa (costae pl): A ridge or midrib of a frond or leaf, also, the marginal longitudinal vein of insect wings *bot zoo*

cosuppression: Silencing of a gene by addition of transgenic DNA copies or infection by a virus; this term, which can refer to silencing at the posttranscriptional (PTGS) or transcriptional (TGS) level, has been primarily adopted by researchers working with plants *biot* >>> post-transcriptional silencing

cotransformation: An event of two plasmids entering the same cell by transformation *biot*

cottage garden: A (usually) small, informal garden making optimal use of space *hort*

C_0t curve: Graphic representation of the progress of a (liquid) hybridization experiment; used to determine the complexity of DNA mixtures (e.g., the size of the genome) *gene biot* >>> C_0t value

cotton gin: A machine that separates seeds, hulls, and other unwanted materials from cotton after it has been picked *agr*

cotton square: Fruiting bud of the cotton plant *bot agr*

C_0t value: An expression for the rate of DNA renaturation (annealing-reannealing); DNA renaturing at low C_0t is composed of highly repetitive sequences and DNA renaturing at high C_0t values is minimal or non-repetitive; it is used to study genome structure and organization and has also been used to simplify the sequencing of genomes that contain large amounts of repetitive sequence; it was first developed and utilized by Roy BRITTEN and his colleagues at the Carnegie Institution of Washington, U.S., in the 1960s *gene biot* >>> C_0t curve

cotyledon: The leaf-forming part of the embryo in a seed; it may function as a storage organ from which the seedling draws food, or it may absorb and pass on to the seedling nutrients stored in the endosperm; once it is exposed to light it develops chlorophyll and functions photosynthetically as the first leaf *bot* >>> Table 16

cotyledonary node: The point of attachment of the cotyledons to the embryonic axis *bot*

coulter: A sharp blade or wheel attached to the beam of a plow, used to cut the ground in advance of the ploughshare *agr*

coumarin: A white crystalline compound ($C_9H_6O_2$) with a vanilla-like odor; it gives sweetclover its distinctive odor; it is also known as a chemical growth inhibitor that has germination-inhibiting capability *phys*

coumestrol: Estrogenic compound occurring naturally in forage crops, e.g., in Latino clover, strawberry clover, and alfalfa *chem* >>> coumarin

couple method (of breeding): A breeding method exclusively used in breeding of allogamous plants; from an original population (e.g., of sugarbeet), single plants are selected and, subsequently, pairwise crossed, preventing unwished-for pollination; the crossing partners should be as similar as possible in spite of color, growth habit, etc.; the offspring is grown in separate plots during the following year; the selection of individuals from the plots and a repeated pairwise crossing can be realized during the fourth year; during the fifth year offspring is grown in plots and again selected for progeny testing *meth* >>> Figures 41, 51 >>> Table 35

coupling: The phase state in which either two dominant or two recessive alleles of two different genes occur on the same chromosome, i.e., linked recessive alleles occur in one homologous chromosome and their dominant alternatives occur in the other chromosome; it is also called cis configuration; opposed to >>> repulsion in which one dominant and one recessive occur in each member of the pair of homologous chromosomes *gene*

coupling of factors (cis-arrangement): Linkage in which both dominant alleles are in the one parent, or the condition in which a double heterozygote has received two linked mutations from one parent and their wild-type alleles from the other parent, for example, a b / + + (as opposed to a + / + b) *gene* >>> linkage

covariance analysis: An analysis of the mean of the product of the deviation of two variates from their individual means; it measures the interrelationship between variables *stat*

coverage error: In an estimate, results from the omission of part of the target population *stat*

cover crop: A crop grown between orchard trees or on fields between the cropping seasons of a main crop, to protect the soil against erosion and leaching and for improvement of soil *agr*

cowpea mottle carmovirus (CPMoV): It causes grain yield losses of up to 75% in >>> cowpea (*Vigna unguiculata*); there is no resistance to this virus among cultivated cowpea lines, but a high level of resistance exists in *Vigna vexillata*, a wild *Vigna* species; it is controlled by a single dominant gene, and the level of resistance conferred by this gene in *V. vexillata* is very high *phyt*

C3 pathway: Most common pathway of carbon fixation in plants; this >>> photosynthesis produces at first a 3-carbon (C3) compound (phosphoglyceric acid); in C3 plants, about 25% of the net carbon uptake is revolved immediately in photorespiration *phys* >>> C4 pathway >>> Figure 67

C4 pathway: A carbon fixation found in some plants that have high rates of growth and >>> photosynthesis and that are adapted to high temperatures, strong light, low carbon dioxide levels, and low water supply; this photosynthesis produces at first a 4-carbon (C4) compound (phosphoenolpyruvate, PEP); in C4 plants, photorespiration is suppressed to a very large extent due to the presence of a very efficient C2-concentrating mechanism; in 2015, a breakthrough came by the introduction of a C4 pathway into a >>> rice plant to extract energy from sunlight far more efficiently than it does now; key C4 photosynthesis genes carried out a rudimentary version of the supercharged photosynthesis process in the host plant; despite the genetic changes, the altered rice plants still rely primarily on their usual form of photosynthesis; to get them to switch over completely, it is necessary to engineer the plants to produce specialized cells in a precise arrangement: one set of cells to capture the carbon dioxide, surrounding another set of cells that concentrate it, it is the distinctive wreath anatomy found in the leaves of C4 plants; however, not all the genes that are involved in producing these cells are known; it could be number in the dozens; >>> genome editing methods that allow the precise modification of parts of plant genomes could help solve the problem; once it has solved the C4 puzzle in a plant such as rice, the method can be extended to dramatically increase production of many other crops, including wheat, potatoes, tomatoes, apples, and soybeans *phys* >>> C3 pathway >>> Figure 67

CpG island: Repetitive CpG doublets creating a region of DNA greater than 200 bp in length with a G–C content of more than 0.5 and an observed/expected presence of CpG more than 0.6; usually associated with transcription-initiation regions of (housekeeping) genes transcribed at low rates that do not contain a TATA box; the CpG-rich stretch of 20–50 nucleotides occurs within the first 100–200 bases upstream of the start site region; a transacting transcription factor called SP1 recognizes the CpG islands *biot*

C3 plants: A class of plants in which the first product of CO_2 fixation is the 3-carbon compound, phosphoglyceric acid; these are usually temperate plants and are characterized by lower dry matter per unit water used, occurrence of photorespiration, and need for greater CO_2 concentration as compared to C4 plants; photosynthetically these plants are less efficient than C4 plants; among C3 plants are wheat, rice, and barley; biofertilizers for C3 plants include Rhizobium and Azotobacter; in 2012, wheat plants were genetically transformed by C4 genes of maize; a full-length cDNA for phosphoenolpyruvate carboxylase gene (*pepc*) was isolated from *Zea mays* by PCR; the *pepc* gene was introduced into wheat plants (*Triticum aestivum*) by particle bombardment transformation; physiological studies showed that phosphoenolpyruvate carboxylase activity in the transgenic plants was 140% higher than that of the untransformed plants; the highest rate of photosynthesis in the transgenic plants was 31.95 µmol $CO_2/m^2/s$, which was 26% greater than the rate in untransformed plants; the light saturation point and carboxylation efficiency of *pepc* transgenic wheat were 20% and 22.57% higher, respectively, than those of untransformed wheat; as a result, the weight of seed per spike and thousand-grain weight were 0.23 g and 1.21 g higher than those of untransformed plants *phys agr* >>> C4 plants

C4 plants: Found principally in hot climates whose initial fixation of carbon dioxide in >>> photosynthesis is by the HATCH–SLACK–KORTSHAK (HSK) pathway; the enzyme responsible is PEP carboxylase (carboxylates phosphoenolpyruvate, PEP, to give oxaloacetate), whose products contain four carbon atoms; subsequently, the carbon dioxide is released and refixed by the CALVIN–BENSON cycle; the presence of the HSK pathway permits efficient photosynthesis at high light intensities and low carbon dioxide concentrations; most species of this type have little or no photorespiration; among C4 plants are sugarcane, maize, and tropical grasses; Azospirillum is the main biofertilizer for C4 plants *phys* >>> C3 plants

crassulacean acid metabolism (CAM): A metabolic pathway adaptation of certain plants that allows them to take up CO_2 at night instead of during the day, greatly reducing transpirational water loss during >>> photosynthesis; this type of metabolism is common in xerophytes *phys*

crease: The fold (the longitudinal furrow on the ventral side) on a cereal grain *bot*

creeping growth habit (spreading growth habit): Plant development by extravaginal stem growth at or near the soil surface, with lateral spreading by rhizomes and/or stolons *bot*

Criollo cocoa beans: A premium variety of cocoa beans grown primarily in South and Central America >>> cocoa bean

crisped: Leaves have a tightly curled margin (e.g., parsley and kale) *bot*

CRISPR: A segment of genetic material found in the genomes of prokaryotes that consists of repeated short sequences of nucleotides interspersed at regular intervals between unique sequences of nucleotides derived from the DNA of pathogens which had previously infected the bacteria and that functions to protect the bacteria against future infection by the same pathogens; the CRISPR segment encodes, via transcription, short RNA sequences that pair with complementary sequences of viral DNA, the pairing is used to guide an enzyme to cleave the viral DNA and prevent further infection; in practice, a gene-editing technique in which CRISPR and the RNA segments and enzymes it produces are used to identify and modify specific DNA sequences in the genome of other organisms *biot* >>> clustered regularly interspaced short palindromic repeats

criss-crossing: A continuous, rotational crossbreeding system alternately using males or pollinators of two different breeds; this system is simple to manage and breeds its own replacements; it utilizes the benefit of hybrid vigor; compared to the common F1, some hybrid vigor can be lost, but that loss is more than compensated for by reduced management effort and cost *meth*

criss-cross inheritance: The transmission of a gene from mother to son or father to daughter *gene* >>> criss-crossing

cristae: The multiply-folded inner membrane of a cell's mitochondrion that are finger-like projections; the walls of the cristae are the site of the cell's energy production (where ATP is generated) *phys* >>> mitochondrion

critical breed: A breed where the total number of breeding females is less than 100 or the total number of breeding males is less than or equal to five; or the overall population size is close to, but slightly above, 100 and decreasing, and the percentage of >>> purebred females is below 80% *gene meth*

critical difference: A value indicating least significant difference at values greater than which all the differences are significant *stat*

critical period: The time span during which crops must be kept weed-free to maximize yield *phyt agr*

critical set (in a Latin square): A set of entries in a square grid which can be embedded in precisely one >>> Latin square, with the property that if any entry of the critical set is deleted, the remaining set can be embedded in more than one Latin square *stat* >>> design of experiment

crock(s): Broken pieces from a broken clay pot, used to cover drainage holes in containers in order to modify drainage or improve air circulation to the roots *hort meth*

crook stage: The stage after some legume seedlings have broken through the soil but before the stem has become erect *agr*

crop: A plant species expressly cultivated for use *agr* >>> Table 35 >>> crop plant

crop calendar: A list of the standard crops of a region in the form of a calendar giving the dates of sowing and the agricultural operations and various stages of their growth in years of normal weather *agr meth*

crop diversification: Cropping system where a number of different crops are planted in the same general area and may be rotated from field to field, year after year *agr* >>> crop rotation

crop divider (at the harvester): Separates the standing crop from the material being cut *agr*

crop evolution: The adaptation of a crop over generations of association with humans *evol* >>> Figures 68, 69

crop ferality: Feral plants are often semidomesticated, escaped from the field a long time ago, having been domesticated and now growing wild (*Cyperus rotundus, Cynodon dactylon, Echinochloa crus-galli, Echinochloa colonum, Eleusine indica, Sorghum halepense, Imperata cylindrica, Portulaca oleracea, Chenopodium album, Digitaria sanguinalis, Convolvulus arvensis, Avena fatua, Amaranthus hybridus, Amaranthus spinosus, Cyperus esculentus, Paspalum conjugatum, Rottboellia exaltata*); growing wild has to be circumscribed properly: only in exceptional cases does a feral crop really grow in the wild; usually such populations stick to ruderal places, to disturbed habitats; in the dense competitive environment of natural habitats such as dry meadows it is very difficult for weeds and feral plants to establish permanently; weeds and feral crops are perfectly adapted to life conditions in anthropogenically disturbed areas; thus, surviving strategies of weeds and feral crops are very diverse (germination requirements, discontinuous germination, rapid growth through vegetative phase to flowering, continuous seed production for as long as growing conditions permit; self-compatible but not completely autogamous or apomictic; when cross-pollinated, unspecialized visitors or wind-pollinated; very high seed output under favorable environmental circumstances, adaptations for short- and long-distance dispersal, if a perennial, vigorous vegetative reproduction or regeneration from fragments, etc.) *agr biot*

crop hygiene: The removal and destruction of heavily infested or diseased plants from a crop so that they do not form sources of reinfestation *phyt*

crop mimicry: >>> Vavilovian mimicry

cropping intensity: The number of years of cropping multiplied by 100, and divided by the number of years of the rotation; it is expressed as "R" (e.g., 4 years crop, 6 years fallow = 10-year rotation; thus R = (4 × 100)/10 = 40) *agr*

cropping pattern: The yearly sequence and spatial arrangement of the crops or of crops and fallow on a given area; it includes sequential or multiple cropping, intercropping, mixed cropping, relay cropping, etc. (e.g., rice followed by wheat, maize followed by wheat followed by greengram) *agr meth* >>> crop rotation

crop plant: A plant expressly cultivated for use; the majority of crops can be classified as (1) root and tuber crops (potato, yams), (2) cereals (e.g., wheat, oats, barley, rye, rice, maize), (3) oil and protein crops (rapeseed, pulses), (4) sugar crops (sugarbeet, sugarcane), (5) fiber crops (cotton, jute), or (6) forage crops (grasses, legumes); agronomic crops can be classified as (1) green manure crops, (2) cover crops, (3) silage crops, or (4) companion crops; about 2% of the 250,000 higher plant species are used in agriculture, horticulture, etc. (about 1,700–2,000); economically, the most important families are the legumes and the grasses, which account for more than a quarter of the total species; they are followed by Rosaceae, Compositae, Euphorbiaceae, Labiatae, and Solanaceae, all with more than 100 taxa; among the families with 50 to 100 crop species, Liliaceae, Agavaceae, and Palmae are worth mentioning, whereas more than 50% of the families have fewer than ten crop species *agr* >>> Tables 1, 35, 56

crop residue: That portion of a plant left in the field after harvest (maize stalk or stover, stubble) *agr*

crop rotation: The alternation of the crop species grown on a field; usually, this is done to reduce the pest and pathogen population or to prevent one-track exhaustion *agr*

crop tree: A tree identified to be grown to maturity and which is not removed from the forest before the final harvest cut; it is usually selected on the basis of its location with respect to other trees and its quality *fore meth*

Crop Science Society of America (CSSA): A prominent international scientific society headquartered in Madison, Wisconsin; society members are dedicated to the conservation and wise use of natural resources to produce food, feed, and fiber crops while maintaining and improving the environment *org* >>> www.crops.org

crop wild relative: A wild plant closely related to a domesticated plant; it may be a wild ancestor of the domesticated plant, or another closely related taxon; the wild relatives of crop plants constitute an increasingly important resource for improving agricultural production and for maintaining sustainable agro-ecosystems; genetic material from crop wild relatives has been utilized by humans for thousands of years to improve the quality and yield of crops; farmers have used traditional breeding methods for millennia; wild maize (*Zea mexicana*) is routinely grown alongside maize to promote natural crossing and improve yields; recently, plant breeders have utilized wild relative genes to improve a wide range of crops; they have contributed many useful genes to crop plants, and modern varieties of most major crops now contain genes from their wild relatives *bree*

cross: Bringing together of genetic material from different individuals in order to achieve genetic recombination; in breeding, there are applied simple crosses (A × B), triple crosses

[(A × B) × C], or double crosses [(A × B) × (C × D)]; moreover related (between parents closely related) or wide crosses (between parents differentiated by a big number of traits/ gens) are used *meth* >>> Table 49

cross back: >>> backcrossing

crossability: The ability of two individuals, species, or populations to cross or hybridize *bot eco* >>> cut style >>> mentor pollen method

crossbred: >>> self

crossbreeding: Outbreeding or the breeding of genetically unrelated individuals; this may entail the transfer of pollen from one individual to the stigma of another of a different genotype *meth* >>> Figure 7 >>> Tables 35, 49

crossbreeding barrier: A pre- and/or postfertilization condition (i.e., progamous or postgamous incompatibility) that prevents or reduces crossbreeding or any form of gene transfer; it is caused by genetic, environmental, physical, or chemical influences >>> Table 35

cross cells: These cells are only found in grasses; they are formed from cells of the pericarp that elongate transversely, lose their chlorophyll, and become lignified; they lie between the tube cells and the parenchyma of the pericarp *bot*

cross classification: Classification according to more than one attribute at the same time *stat*

cross coancestry: Refers to the average of the elements in a coancestry matrix excluding the self-coancestry on the diagonal; thus, the expected inbreeding following random mating in a population without inbreeding *stat meth* >>> pairwise coancestry >>> group coancestry >>> coancestry

cross-fertilization: The fusion of male and female gametes from different genotypes or individuals of the same species, as base of genetic recombination *bot* >>> allogamy >>> cross-pollination >>> Table 35

cross-hybridization: In biotechnology, the hydrogen bonding of a single-stranded DNA sequence that is partially but not entirely complementary to a single-stranded substrate; often, this involves hybridizing a DNA probe for a specific DNA sequence to the homologous sequences of different species *biot*

crossing barrier: Any of the genetically controlled mechanisms that either entirely prevent or at least significantly reduce the ability of individuals of a population to hybridize with individuals of other populations *gene* >>> crossbreeding barrier >>> incompatibility

crossing block: A crop plant nursery containing the parental stocks for a breeder's crossing program *meth* >>> design of experiment

crossing groups Any group of individuals that comprises a unique set of parents: (1) diallel crossing group – controlled crosses are made between each pair of parents in the group but crosses with parents outside the group are excluded; (2) factorial crossing group – a limited number of parents are used as male testers in controlled crosses with an unlimited number of female parents; (3) open-pollinated crossing group – all parents in a breeding population are included in a progeny test or series of tests *meth fore hort* >>> Table 35

crossing-over: The exchange of genetic material between homologous chromosomes by breakage and reunion; it occurs during pairing of chromosomes at prophase I of meiosis; the temporary and visible joins between chromosomes during crossing-over are called chiasmata *gene cyto* >>> Figure 24

crossing-over map: A genetic map made by utilizing crossing-over frequencies as a measure of the relative distances between genes in one linkage group (chromosome) *gene*

cross(ing)-over unit: A 1% crossing-over value between a pair of linked genes *gene* >>> Morgan unit

cross-inoculation: Inoculation of one legume species by the symbiotic bacteria from another *agr*

cross-inoculation groups: Groups of legumes, any one of which will form nodules when inoculated with rhizobia isolated from another legume in the group *agr*

cross-licensing: In patenting biological material, agreement in which two or more firms with competing and similar technologies strike a deal to reduce need for legal actions to clarify who is to profit from applications of the technology *adm* >>> Intellectual Property Rights

cross-pollinating crop: >>> crossbreeding >>> cross-pollination >>> xenogamy >>> Table 35

cross-pollination: The transfer of pollen from the stamen of a flower to the stigma of a flower of a different genotype but usually of the same species, with subsequent growth of the pollen tube *bot* >>> allogamy >>> Table 35

cross-protection: Plant protection conferred on a host by infection with one strain of, for example, a virus that prevents infection by a closely related strain *phyt*

cross-resistance: Resistance associated with a change in one genetic factor that results in resistance to different chemical pesticides that were never applied *phyt*

cross-sterility: The failure of fertilization because of genetic or cytological conditions (incompatibility) in crosses between individuals >>> crossing barrier >>> Table 35

cross-tabulations: >>> contingency tables

cross validation: A useful approach for obtaining unbiased estimates of QTL effects and determining the magnitude of bias of predictive power in genome-wide association mapping and genome selection *stat*

crotch: The angle measured from the trunk of a tree to the upper surface of a branch *hort*

crown: The stem–root junction of a plant (e.g., the overwintering base of an herbaceous plant) *hort*; the term is also used for the treetop *bot fore*

crown class: Measure of stand structure classifying trees within a stand as dominant (crowns rise through or above general canopy and receive full light from above and partial light from the sides), codominant (crowns in upper canopy but are blocked from receiving light from the sides by neighboring crowns), emergent (crowns completely above main canopy), intermediate (crowns receive little light from above and none from the side), overtopped or suppressed (one or more neighboring trees completely overtop crowns), and seedlings *fore meth*

crown cover: The canopy of green leaves and branches formed by the crowns of all trees in a forest *fore*

crown gall: A common and widespread plant disease, which can affect a very wide range of woody and herbaceous plants (fruit trees, roses, etc.); it is caused by the bacterium *Agrobacterium tumefaciens*; galls are formed at the crown (stem–root junction) or, less frequently, on roots, stems, or branches *phyt* >>> Figure 62 >>> www.pflanzengallen.de

crown roots: Roots arising directly from the crown *bot* >>> Figure 62

crozier: Coiled juvenile frond of a fern, similar in a form of bishop's staff *bot*

crucifer: A plant belonging to the Brassicaceae or mustard family, a large dicotyledonous family of important crop and ornamental plants (turnip, cabbage, etc.) *bot* >>> Figure 8

crust: A surface layer of soil that becomes harder than the underlying horizon when dry *agr*

Cry1A protein: Derived from the bacterium *Bacillus thuringiensis* that is toxic to some insects when ingested; this bacterium occurs widely in nature and has been used for decades as an insecticide although it constitutes less than 2% of the overall insecticides used *phys phyt biot*

cryoability: The ability of plant material (seeds, tissue, organs) to be preserved or stored under very low temperatures, usually in liquid nitrogen (–196°C) *phys*

cryobank: Place for the preservation or storage of plant material under very low temperatures, usually in liquid nitrogen (–196°C) *meth seed*

cryodamage: Damage caused by exposure to cold conditions *agr*

cryopreservation: The preservation or storage of plant material in very low temperatures, usually in liquid nitrogen (–196°C), e.g., in strawberry, cryopreservation using an aluminum cryoplate is successfully applied to *in vitro*-grown shoot tips; the shoots are cold-hardened at

+5°C for 3 weeks; the shoot tips (1.5–2.0 mm × 0.5–1.0 mm) are dissected from the shoot and pre-cultured at +5°C for 2 d on >>> MURASHIGE and SKOOG medium containing 2 M glycerol and 0.3 M sucrose; the pre-cultured shoot tips can be placed on aluminum cryo-plates containing ten wells embedded in alginate gel; osmoprotection can be performed by immersing the cryo-plates in a loading solution (2 M glycerol and 0.8 M sucrose) for 30 min at +25°C; dehydration is performed by immersing the cryo-plates in plant >>> vitri-fication solution 2 for 50 min at +25°C; then, the cryo-plate with shoot tips is transferred into an uncapped cryotube that was held on a cryo-cane and directly immersed into liquid nitrogen; using this procedure, the average regrowth level of vitrified shoot tips reaches about 80% *meth seed*

cryoprotectant: A chemical which is used to protect seeds, cultured material, tissue, organs, or cells from the low temperatures in cryopreservation (e.g., glycerol) *prep*

cryoscopy: A technique for determining the m.w. of a substance by dissolving it and measuring the freezing point of the solution *meth*

cryostat: A device designed to provide low-temperature environments in which experiments may be carried out under controlled conditions *prep*

cry protein(s): Protein crystals produced naturally by >>> bacteria (e.g., *Bacillus thuringiensis*) that are toxic to some insects; the gene that produces this protein can be inserted into the maize or other genomes giving it >>> European corn borer resistance *phyt phys biot*

crypsis: Where an organism or organ is difficult to perceive or see *bot*

cryptical structural hybridity: Hybrids with chromosomes structurally different but not visible by metaphase-I chromosome pairing *cyto*

cryptic error: The biological error arising when plots in an experiment differ from the fields they are meant to represent *stat*

cryptic self-incompatibility: A form of physiological self-incompatibility found in self-compatible plants in which outcrossing is promoted through differential pollen-tube growth between outcross and self-pollen tubes *gene*

cryptochrome: >>> photomorphogenesis

cryptogam: A plant (e.g., fern) that reproduces by means of spores rather than seeds *bot*

cryptogam(ous): Reproduction by spores or gametes rather than seeds *bot*

cryptogams: >>> cryptogam(ous)

cryptomeric gene: >>> cryptomerism

cryptomerism: The phenomenon that a gene or an allele does not show a phenotypic effect unless it is activated by another genetic factor, which leads to a sudden change of qualities in the progeny not recognized among the ancestors *gene*

crystalline wax: Crystalline-looking wax occurring on leaf surfaces that is difficult to wet *bot* >>> waxy

crystallography: The study of, for example, a protein structure by crystallizing the protein and examining the crystals using X-rays; the diffraction angles of the X-rays are used to compute the relative positions of components of the protein, and thus its structure *phy*

CsCl$_2$ gradient: A method used to separate DNA or phages according to buoyant density *biot*

CSIRO: >>> Commonwealth Scientific and Industrial Research Organization

cSNP: >>> coding single nucleotide polymorphism

CSSA: >>> Crop Science Society of America

CTD: >>> canopy temperature depression

C terminus: Carboxyl terminus, defined by the –COOH group of an amino acid or protein *chem biot*

CTV: >>> citrus tristeza virus

cubing: The process of forming hay or straw into high-density cubes to facilitate transportation, storage, and feeding *agr*

cuckoo chromosome: An alien chromosome that shows a preferential transmission during generative reproduction; found in certain wheat-*Aegilops* crossing progenies; T. E. MILLER (1983) named it *cyto*

cuckoo gene: Refers to a gene conferring preferential transmission from the maternal parent (e.g., in *Lablab purpureus*; KONDURI et al. 2000) *gene*

cucullate: Hood- or cowl-like in form *bot*

cucumber mosaic virus (CMV): One of the most widely occurring plant viruses; infects more than 800 plant species and has a considerable negative impact on agriculture worldwide; in addition to an RNA genome, certain strains of CMV naturally harbor an RNA of 330–400 nucleotides, called sat-RNA, that depends on the virus for its entire cycle, has no significant similarity with the viral genome, and apparently does not encode protein; sat-RNAs are capable of attenuating CMV strains, resulting in a sharp decrease in virus titer in infected plants, and an almost complete lack of symptoms (e.g., lethal tomato necrosis) *phyt*

culled: >>> off-grade

cull(ing): The postharvest removal of pathogen-infected or damaged fruit, seeds, or plants by screening procedures; the culled or off-graded material can later be individually analyzed or discarded *meth*

culm: The jointed stem in cereals, grasses, or sedges; filled with pith or solid *bot*

cultigen: A cultivated plant or group of plants for which there is no known wild ancestor (e.g., maize, *Zea mays*), i.e., a plant that has been deliberately altered or selected by humans; it is the result of artificial selection; the anthropogenic plants are plants of commerce that are used in agriculture, horticulture, and forestry *bot tax*

cultivar: A contraction of "cultivated variety" (abbreviated cv.); refers to a crop variety produced by scientific breeding or a farmer's selection methods; after International Code of Nomenclature for Cultivated Plants (ICNCP-1995); "cultivar" is synonymous with "Sorte" (German), "variety" (English), or "variété" (French); in Chapter 3: "A cultivar is a taxon that has been selected for a particular attribute or combination of attributes, and that is clearly distinct, uniform and stable in its characteristics and that, when propagated by appropriate means, retains those characteristics" *bot* >>> variety

cultivar development: Any activity of crossing, transformation, and/or selection (including marker-assisted selection) among plants, which has the direct purpose of releasing a crop variety *meth*

cultivar identification system: A classification system based on sequence-tagged microsatellite loci analysis with fluorescent primers and suitable computer software; allows unequivocal identification of varieties, paternity testing, and duplicate identification *meth biot*

cultivar mixture: A mixture of different varieties in order to improve the environmental adaptability or the resistance to pathogens *seed*

cultivation: The art or process of agriculture *agr*

cultivator: An implement drawn between rows of growing plants to loosen the earth and destroy weeds *agr*

culton A systematic group of >>> cultivated plants; there are two types of culta: the >>> cultivar and the cultivar group *tax*

cultural requirements: The conditions which provide optimal growth of a plant in cultivation *meth hort*

culture: A growth of one organism or of a group of organisms for the purpose of production, trade, and utilization, or for experiments *agr*

culture collection: A collection of cultures of more or less defined or characterized viruses, bacteria, and other organisms; usually used for reference and comparison with new isolates *phyt*

culture medium: Medium on or in which tissues, organs, or cells are cultured; supplies the mineral and hormonal requirements for the growth *meth biot*

culture room: Room for incubating cultures with controlled light, temperature, and humidity *biot*

culture tube: A tube in which tissue, organs, cells, or organisms are cultured *prep*

cumulative genes: Polymeric non-allelic genes *gene*

cuneate: Wedge-shaped *bot*

cupule: A cup-like structure at the base of some fruits *bot*

curative pesticide: A pesticide that can inhibit or eradicate a disease-causing organism after it has become established in the plant *phyt*

curd grafting: Breeding work on cauliflower is hampered in northwest Europe by difficulties in obtaining seed from field-grown plants; this particularly applies to material which produces mature curds in autumn; selected plants will not survive the winter in the field, and if lifted and transferred to the glasshouse will invariably rot and die; WATTS and GEORGE (1963) described how curd portions grafted on to young, pot-grown cauliflowers continued their floral development, and seed can be produced from the grafted curd in time for sowing in the next season *meth*

curled cotyledon stage: The stage at which the cotyledons curve down towards the suspensor; at this time protein deposition in the cotyledons begins *phys*

current agriculture: Science-based area-specific agriculture that makes use of mineral fertilizers, organic manures, bioinoculants, and plant protection chemicals, adopting best management practices; it is also termed as "conventional agriculture" *agr* >>> agriculture

curry powder: In India, many good cooks make their own curry powders, and there are as many recipes as there are good cooks; most curry powders contain about 25% >>> turmeric (Cucurma domestica), 25% >>> coriander (*Coriandrum sativum*) seeds, and various amounts of >>> cumin (*Cuminum cyminum*) seeds, >>> cardamoms (*Elettaria cardomomum*), >>> fenugreek (*Trigonella foenum-graecum*) seeds, >>> chilies (*Capsicum annum*), >>> ginger (*Zingerber officinale*), >>> black pepper (*Piper nigrum*), and dill (*Anethum graveolens*) seeds *hort*

cushion plants: Have small, hairy, or thick leaves borne on short stems and forming a tight hummock *bot*

cut-and-come-again: Applied to any plant that is cut or sheared after flowering and blooms again (e.g., *Petunia* spp., pansy) *hort*

cut flowers: Flowers that are cut off the plant and used as decoration *hort*

cuticle: A thin, waxy, protective layer covering the surface of the leaves and stems *bot*

cutin: The complex mixture of fatty-acid derivatives with waterproofing qualities of which the cuticle is composed *bot*

cutinase: >>> esterase >>> enzymes that break down the way >>> cuticle, especially, of stigmatic papillae *bot phys*

cutinize: To impregnate a cell or a cell wall with cutin, a complex fatty or waxy substance, which makes the cell more or less impervious to air and moisture *bot*

cutout: The occurrence of physiologically indeterminate growth (e.g., in cotton) *agr*

cut style: A crossing method when pollen grains germinate on the flower stigma, but are not able to grow far enough into the style that the ovary can be fertilized; in some crop species, fertilization can be achieved by cutting part of the style of the female parent and applying pollen, mixed with stigma juice, to the surface of the cut; the pollen tube now has to grow a shorter distance (i.e., increasing the chance of fertilization); this approach is mostly applied in ornamental plants with long styles *meth* >>> crossing method >>> incompatibility

cut surface: >>> cutin >>> cutinize

cutting: A section of a plant that is removed and used for propagation; cuttings may consist of a whole or part of a stem (leafy or non-leafy), leaf, >>> bulb, or root; a root cutting consists of root only; other cuttings have no roots at the time they are made and inserted; as opposed to division, a kind of propagation that consists of part of the crown of a plant or of its above-ground portion and roots; several types of cuttings can be taken from the parent stock, which depends on the point on the parent stock where the cutting is taken; four major categories of cuttings are: (1) stem cuttings, (2) leaf cuttings, (3) leaf-bud cuttings, and (4) root cuttings; stem cuttings are severed twigs that have been placed into a growing medium and encouraged to develop roots; stem cuttings are broken down into subclasses consisting of hardwood, semihardwood, softwood, and herbaceous cuttings; hardwood cuttings are taken from the current season's growth; this portion of the tree offers young tissue; cuttings should be taken during the winter season; semihardwood cuttings are produced from woody, broadleaf evergreens, and leafy summer cuttings; they are taken from a partially matured portion of the plant, usually during the summer growing months just after new shoot development; softwood cuttings are taken from new, soft, succulent spring growth from either deciduous or evergreen species; although softwood cuttings usually root easier and quicker than other cuttings, they also require more labor and equipment; this is because the cuttings are made with their leaves still attached; herbaceous cuttings are taken from succulent, herbaceous plants; this type of cutting roots fast, but is not used in forestry practice; leaf cuttings are not used extensively in forestry applications; this form of propagation utilizes the leaf to promote new plant growth; a root and shoot will form and develop from the leaf cutting into a new plant; the original leaf cutting does not remain as part of the new formed plant; leaf-bud cuttings are not used extensively in forestry applications; the leaf-bud cutting includes the leaf itself, petiole, and a small piece of stem with the axial bud; this form of cutting propagation is useful when material is scarce, because the same amount of stock will produce twice as many new plants as that of stem cuttings; root cuttings, which are used in forestry propagation, should be taken from the young plant stock during the winter and spring months to ensure that they are saturated with stored foods; this time frame also prevents cutting during the time the parent plant is rapidly expanding shoot growth; cutting during active expansion will take food stores away from the root system *hort fore meth*

cutting cycle: The planned time interval between major harvesting operations in the same tree stand; the term is usually applied to uneven-aged stands; for example, a cutting cycle of ten years means that every ten years a harvest would be carried out in the stand *fore*

cutworm: Larva of certain *Noctuidae* sp. (Lepidoptera) that live in the soil, emerging at night to eat foliage and stems; serious pests of root crops and of many other crops as seedlings *phyt*

cv.: >>> cultivar

C value: The DNA quantity per genome (i.e., per chromosome set); the content of diploids is referred to as the 2C; haploid cells contain the 1C amount of DNA *cyto* >>> endopolyploid

C value paradox: C-values, i.e., genome sizes, vary enormously among species; however, there is no relationship to the presumed number of genes as reflected by the complexity of the organism *gene*

cyanobacteria: Prokaryotic oxygenic phototrophs containing chlorophyll a and phycobilins *evol*

cyanogenesis: The release of cyanide following tissue damage; it is common in plant species, e.g., in >>> cassava and >>> white clover; the polymorphism is controlled by independently segregating gens *gene*

cyathium (*pl* **cyathia):** Very special pseudanthia in the genus *Euphorbia*; it consists of five (rarely four) >>> bracteoles; their upper tips are free and in the beginning cover the opening of the involucre (like the shutter of a camera); with these alternate five (one to ten) nectar glands, which are sometimes fused; one extremely reduced female flower standing in the center at the base of the involucre, consisting of an >>> ovary on a short stem with >>> pistil, and surrounded by five groups (one group at the base of each bracteole) of extremely reduced male flowers, which each consist of a single >>> anther on a stem; the flower-like characteristics of the cyathia are underlined by brightly colored nectar glands and often by petal-like appendages to the nectar glands, or brightly colored, petal-like bracts positioned under the cyathia *bot*

cybrid: The hybrid formed from the fusion of a cytoplast and a whole cell; the cytoplast may transmit cytoplasmic components independently of the cell genome *bot*

cycle sequencing: A method that recruits PCR technology to assist sequencing of small amounts of a single-stranded DNA template; a heat-stable polymerase, a single primer, deoxynucleotides, and small amounts of one of the four labeled dideoxynucleotides are incubated with the template in each of four incubations; in each of a series of cycles of annealing, polymerization, and denaturation, a spectrum of single-stranded DNA fragments is generated that can be electrophoretically separated to produce a ladder from which can be assigned nucleotide positions relative to the primer *biot meth*

cyclical parthenogenesis: A life history in which a sequence of apomictic generations is followed by amphimictic generations *bot*

cyclic design: An incomplete-block design is a cyclic design if its blocks can be partitioned into sets of blocks such that each set is a thin cyclic design *stat* >>> incomplete-block design

cycling: A round of recombination, testing, and selection; it may often refer to mating *bree*

cycling strategy: Choice of methods of recombination, testing, and selection in repeated cycling *meth bree* >>> breeding strategy

cyclohexanediones (CHDs): As aryloxyphenoxypropionates (APPs), it is a specific graminicide herbicide; chloroplastic acetyl CoA-carboxylase (ACCase) is the target of the chemical reaction; the wide application resistance to these compounds became a worldwide, increasing problem *phyt phys*

cycloheximide (actidione): An antibiotic from Streptomyces griseus; antibacterial and antifungal *biot*

cylindrical: Shaped like a tube, round in cross section with parallel margins *bot*

cyme: An inflorescence in which each axis ends in a flower; it is usually a flat-topped or round-topped, determinate inflorescence in which the terminal flower blooms first *bot*

cypsela: A single-seeded fruit that develops from an inferior ovary thus surrounded by other floral tissues in addition to the ovary wall, e.g., in Asteraceae, which commonly bears a feathery pappus which is derived from the >>> calyx and promotes dispersal by wind *bot*

cysteine (Cys): An aliphatic, polar alpha-amino acid that contains a sulfydryl group *chem phys* >>> Table 38

cytochimera: Different tissues or parts of them differ in chromosome number *cyto* >>> chimera >>> Figure 56

cytochrome (Cyt): An iron-containing pigment that plays a major role in respiration; more detailed, one of a group of hemoproteins, which are classified into four groups designated a, b, c, and d; they function as electron carriers in a variety of redox reactions in virtually all aerobic organisms *phys*

cytochrome P450 monooxygenase: A heme-containing oxidoreductase that reduces one atom of molecular oxygen to water and, in its most common function, introduces the second atom of oxygen into a substrate; other oxidative reactions, such as introduction of a double bond, may also occur *phys*

cytodifferentiation: The sum of processes by which, during the development of the individual, the zygote, specialized cells, tissue, and organs are formed *cyto*

cytogamy: The fusion or conjugation of cells *cyto*

cytogenetic breeding: Breeding techniques that involve the manipulation or alteration of genetic material in cells (e.g., exposure to radiation or application of chemicals such as colchicine) *meth*

cytogenetic map: A map showing the locations of genes on a chromosome (i.e., the visual appearance of a chromosome when stained and examined under a microscope); particularly important are visually distinct regions, called light and dark bands (or colored bands), which give each of the chromosomes a unique appearance *cyto gene*

cytogenetics: Scientific discipline that combines cytology with genetics; it includes (1) the organization of chromosomes in the nucleus with the requirements of gene expression in different developmental and environmental contexts; (2) the behavior of chromosomes with transmission of phenotypes to progeny; (3) changes in chromosomal structure and number with speciation; and (4) the evolution of the genome with the evolution of the species *gene cyto* >>> Figures 68, 69

cytogenic male sterility: >>> cytoplasmic male sterility

cytogony: The reproduction by single cells *cyto*

cytohet: A cell containing two different cytoplasmic genomes (e.g., mitochondria) that differ in one or more genes contributed by two parents; thus, the individual is cytoplasmatically heterozygous *gene*

cytokinesis: During the division of a cell, the division of the constituents of the cytoplasm; it usually begins in early telophase with the formation of a cell plate, which is assembled within the phragmoplast across the equatorial plane; the phragmoplast is a complex array of Golgi-derived vesicles, microtubule, microfilaments, and endoplasmatic reticulum that assembles during the late anaphase and is dismantled upon completion of the new cell wall *cyto*

cytokinin: One of a group of hormones, including kinetin, that act synergistically with auxins to promote cell division but, unlike auxins, promote lateral growth *phys*

cytology: The branch of biology dealing with the structure, function, and life history of the cell *cyto*

cytological hybridization: >>> *in situ* hybridization

cytolysis: Breaking up or solution of the cell wall *cyto*

cytomixis: The extrusion or passage of chromatin from one cell into the cytoplasm of an adjoining cell *cyto*

cytoplasm: The part of a cell that is enclosed by the plasma membrane, but excluding the nucleus *cyto*

cytoplasmic inheritance: A non-Mendelian (extra-chromosomal) inheritance via genes in cytoplasmic organelles (mitochondria, plastids) *gene* >>> conifer

cytoplasmic male sterility (CMS): Pollen abortion due to cytoplasmic factors, which are maternally transmitted, but which act only in the absence of pollen-restoring genes; this type of sterility can also be transmitted by grafting; in 2008, Australian scientists proposed a *Hordeum chilense*-based CMS–fertility restorer system to wheat, designated as msH1; it has been identified during the process of obtaining alloplasmic bread wheat in different >>> *Hordeum chilense*; when using the *H. chilense* H1 accession, the corresponding alloplasmic line was male sterile; this alloplasmic wheat is stable under different environmental conditions, and it does not exhibit developmental or floral abnormalities, showing only slightly reduced height and some delay in heading; on examining microsporogenesis in the alloplasmic line, it was found that different stages of meiosis were completed normally, but abnormal development occurred at the uninucleate-pollen stage at the first mitosis, resulting in failure of anther exertion and pollen abortion; fertility restoration of the CMS phenotype caused by the *H. chilense* cytoplasm was associated with the addition of chromosome 6H^{ch}S from *H. chilense* accession H1; thus, some fertility restoration genes appear to be located in this chromosome arm *gene* >>> Figure 2 >>> genic male sterility

cytoplasmon: All cytoplasmic hereditary constituents of a cell excepting those localized in the plastids and mitochondria *cyto* >>> Table 53

cytoplasm-restorer: >>> cytoplasmic male sterility

cytoplast: The cytoplasm as a unit, as opposed to the nucleus *bot*

cytosine (C): A pyrimidine base that occurs in both DNA and RNA *chem gene* >>> Table 38

cytosol: The water-soluble components of cell cytoplasm, constituting the fluid portion that remains after removal of the organelles and other intracellular structures *bot*

cytostatic: Any physical or chemical agent capable of inhibiting cell growth and cell division *phys*

cytosterility: A genetically controlled condition of male sterility in breeding lines of a self-pollinating crop variety; a breeding line with cytosterility is used as the seed-producing parental line in the production of hybrid seed *meth* >>> Figure 55

cytotaxonomy: The study of natural relationships of organisms by a combination of cytology and taxonomy *tax* >>> Table 17

cytotype: Any variety of a species whose chromosome complement differs quantitatively or qualitatively from the standard complement of that species *cyto*

D

dalton (da): A unit equal to the mass of the hydrogen atom (1.67×10^{-24} g); the unit was named after J. DALTON, a chemist of the 19th century *chem*

damage: The adverse effect on plants or crops due to biotic or abiotic agents, resulting in a reduction of yield and/or quality *agr*

dam gene: DNA adenine methylation gene of Escherichia coli; it methylates the sequence GATC (Sau3A cleaves methylated and unmethylated DNA, MboI cleaves only unmethylated DNA, DpnI cleaves only the methylated sequence) *biot* >>> Table 38

dammar resin: A hard, lustrous resin derived from Asian trees of the monkey-puzzle family *prep*

damping-off: A disease of young seedlings in which the stems decay at ground level and the seedlings collapse; it can be prevented by planting in sterile soil, treating seeds with a fungicide, or soaking the soil with a fungicide, also avoiding overwatering and planting warm-season plants in cold soil *phyt*

dandelion weeder: >>> asparagus knife

dark-field: >>> dark field microscopy

dark field microscopy: A microscope designed so that the entering center light rays are blacked out and the peripherical rays are directed against the object from the side; as the result the object being viewed appears bright upon a dark background *micr*

dark reaction: The phase of >>> photosynthesis, not requiring light, in which carbohydrates are synthesized from carbon dioxide *phys*

dark respiration: >>> dark reaction

dark room: A room in which film/photographic material is handled or developed and from which the actinic rays of light are excluded prep; a dark room is also used in order to simulate short-day conditions; it is applied either to prevent flowering of >>> long-day plants or to induce flowering of >>> short-day plants under long-day conditions *meth*

DArT marker: >>> diversity arrays technology (DArT) marker

Darwinism: The theory that the mechanism of biological evolution involves natural selection of adaptive variations *bio* >>> Figures 68, 69

database: A store of a large amount of information (e.g., in a form that can be handled by a computer); recently, in breeding, numerous databanks are in use, such as field registration, nursery plans, selection data, data of international observation trials, statistical values, data recording, etc. *stat*

data management: The organization of the recording of data, its preparation for analysis, and its interpretation *prep*

data sheet: A specially prepared form on paper or computer for recording data *meth*

dauermodification: >>> persistent modification

daughter cell: The cells resulting from the division of a single cell *cyto*

daughter chromosome: Any of the two chromatids of which the replicated chromosome consists after mitotic metaphase or anaphase II of meiosis *cyto*

daughter nucleus: The nuclei that result from the division of a single nucleus *cyto*

daylength: The number of hours of light in each 24-hour cycle *phys*

daylength insensitivity: These plants will flower independent of daylength (e.g., tomato or cotton) *phys*

daylength response: >>> daylength sensitivity

daylength sensitivity: Plants that will flower only when the daily photoperiod is of a critical length *phys* >>> short-day plants >>> long-day plants >>> day-neutral plants

day-neutral plants: No daylength requirement for floral initiation *phys*

days post anthesis (dpa): The number of days after flowering and fertilization *phys bot*

days to flower: Number of days required for 50% blooming or flowering in a field *phys meth*

days to harvest: The least number of days, established by law, between the last pesticide application and the harvest date, as set by law to prevent exposure to a hazardous amount of pesticide *phyt agr*

days to heading: Number of days required to 50% panicles/tassel exertion from the flag leaf of grasses or cereals in a field or study *phys meth* >>> days to maturity

days to maturity: Number of days required from seeding to seed/grain ripening phys *meth* >>> days to heading

dead cotton: An extreme form of immature cotton, having a thin fiber wall; it can result from disease, pest attack, or a foreshortened ripening period *agr*

deadhead(ing): The removal of dead flowers from a plant by pinching the flower (including the ovary) from the stem; this is done to encourage further blooming by preventing the plant from wasting energy by producing seeds *hort meth* >>> days to maturity

debearder: In seed precleaning procedures, it has a hammering or flailing action that removes awns, beards, or lint from seed and tends to break up seed clusters of the chaffy grasses, as well as multiple seed units of non-chaffy forms *seed* >>> Table 11

decarboxylase: An enzyme that removes the carboxyl group from an organic compound *chem phys*

decay: The destruction of plant material by fungi and bacteria *agr*

decentralized plant breeding: A defined set of breeding experiments carried out in a variety of local sites (communities, farmer fields) that represent real farming conditions, as opposed to a single, central, research station site that does not represent a real farming context *agr meth*

deciduous: A plant whose leaves are shed at a season or growth stage, by abscission; the function of this habit is usually to escape an adverse season, such as a winter, or a tropical dry season; the deciduous habit also has advantages in the control of leaf parasites by providing a discontinuous pathosystem in which a gene-for-gene relationship can operate as a system of biochemical locking *bot*

decimal code for the growth of cereal plants: A decimal code used for describing different growth stages of cereal plants; it is applied for comparison of several morphological stages; there are several schemes of description *phys meth* >>> Table 13

declining vitality: Seeds that are aged or have been subjected to unfavorable storage conditions; they usually show a slow germination; some of the essential plant parts are frequently stunted or lacking; saprophytic fungi may also interfere with the growth of the seedlings *seed*

decompose: To rot or putrefy *agr*

decomposer: An organism that breaks down dead tissues into simple chemical components, thereby returning nutrients to the physical environment of plants *bio*

decondensation stage: A stage between >>> interphase and prophase of mitosis in which heterochromatin is decondensed for a short period *cyto* >>> Table 43

decontaminate: To free from contamination; purify *meth*

decumbent: Lying on the ground with the end ascending *bot*

decurrent: Describes the open type of collar in barley where the margin of the platform is incomplete, merging with the neck; in general, extending downward from the point of insertion *bot*

dedifferentiation: A loss of specialization of a cell; it can be observed when differentiated cells are placed *in vitro* culture, followed by redifferentiation with the ability to reorganize into new organs *biot* >>> competency >>> totipotency

deep intronic mutation: Mutation that is located within an intron but not adjacent to an intron-exon border *gene*

deep sequencing: Refers to **sequencing** a genomic region multiple times, sometimes hundreds or even thousands of times; this >>> next-generation sequencing approach allows the detection of rare clonal types, cells, or microbes comprising as little as 1% of the original sample *biot*

de-etiolation: Anatomical and physiological changes in dark-grown seedlings after exposure to light; this typically occurs in seedlings after emergence from soil and includes greening, due to chloroplast development and chlorophyll synthesis, and a reduction in elongation growth *phys*

defense response: The active response of a plant to pathogen attack; it includes the elements that inhibit pathogen development; a defense response is activated in compatible and incompatible interactions *phyt*

deficiency: The absence or deletion of a segment of genetic material *cyto* >>> chromosome mutation

deficiency disease(s): Disease caused by the lack or insufficiency of some nutrient, element, or compound (e.g., copper, zinc, or iron deficiency in cereals); deficiency diseases are among the non-parasitic physiological disorders that are due mainly to nutritional deficiencies or toxicities; each nutritional element produces its own deficiency symptoms; within one plant, mobile elements can be taken from old tissues to feed the young tissues, and the symptoms then appear mainly in the older tissues; conversely, immobile elements cannot be reallocated in this way, and the main deficiency symptoms then appear in the youngest tissues; deficiency symptoms are easily confused with herbicide injury *phyt phys* >>> zinc >>> www.zinc.org/

definite host: The host in which the parasite attains sexual maturity *phyt* >>> host

deflexed: Bent sharply downward *bot*

defoil: To strip a plant of leaves *meth*

defoliant: A chemical or method of treatment that causes only the leaves of a plant to fall off or abscise (e.g., it is applied before harvest of potato) *agr*

defoliated: >>> defoliation

defoliation: The process of leaves being removed from a plant (e.g., to make harvest easier) *bot*

degeneracy of genetic code: >>> degenerated code

degenerated code: A term applied to the >>> genetic code because a given amino acid may be encoded by more than one codon *gene* >>> Table 38

degenerate oligonucleotide primed-PCR: A molecular marker technique that uses partially degenerated primers for polymorphism detection in comparative genomics studies *biot meth*

degermed: Grains from which the embryo (germ) has been removed *agr meth seed*

degermination: The process of removing the grain's embryo (or germ) by mechanical devices, a term restricted almost exclusively to maize and rice milling *meth*

degradation: The progressive decrease in vigor of successive generations of plants, usually caused by unfavorable growing conditions or diseases; viruses may cause great loss of vigor; in agriculture, the change of one kind of soil to a more highly leached soil *agr* >>> Table 44

degree of dominance: >>> dominance

degree of freedom (DF): The number of items of data that are free to vary independently; in a set of quantitative data, for a specified value of the mean, only $(n - 1)$ items are free to vary, since the value of the nth item is then determined by the values assumed by the others and by the mean *stat*

degree of genetic determination: The portion of total variance that is genetically determined *gene*

degree of milling: Expression used to indicate the amount of bran removed in the milling process *meth*

dehiscence: The bursting open at maturity of a pod or capsule along a definite line or lines *bot*

dehiscent fruit: >>> dehiscence

dehull: Removal of outer seed coat (hull) or to remove the glumes of cereal caryopses *seed*

100-dehulled grain weight: Weight of the 100-dehulled grain having pericarp (seed coat) *meth*

dehusk: To remove the husk (e.g., the leaf sheath of a maize cob or the outer layer of a coconut) *seed*

dehydration: The elimination of water from any substance *phys*

dehydrins: Proteins produced during the late stages of plant embryo development and following any environmental stimulus involving >>> dehydration *phys*

dehydrogenase: An enzyme that catalyzes the removal of hydrogen from a substrate *phys*

delayed maturity: A breeding goal for fruits and vegetables; in recent horticulture, many fruits like tomatoes, melons, and bananas are usually harvested before ripening; the firmness of unripe fruit facilitates mechanical harvesting and extends the period of time allowed for transport and storage; for several fruits and vegetables, transgenic procedures have been used to develop cultivars with delayed maturity; it is achieved by suppressing of the synthesis of polygalacturonase, an enzyme responsible for fruit softening, either by inhibiting the production of >>> ethylene (a plant hormone that triggers the ripening process) by introducing genes for enzymes that degrade ethylene precursors, or by blocking the polygalacturonase gene using antisense technology; at first transgenic tomatoes with extended shelf life were introduced to the U.S. market *biot meth* >>> genetically modified organisms

deletion: The loss of a chromosomal segment from a chromosome set; the size may vary from a single nucleotide to sections containing several genes *cyto* >>> deficiency >>> chromosome mutation

deletion hotspot: Region of a genome in which deletions occur more frequently than in the rest of the genome *gene*

deletion mapping: The use of overlapping deletions to localize the position of an unknown gene on a chromosome or linkage map, i.e., the localization of the position of deletions in the DNA of an organism or the localization of a specific, yet unidentified gene on a chromosome by using overlapping deletions *gene*

deletion mutation: A mutation in which one or more bases are removed from the DNA sequence of a gene *gene*

deletion TILLING: A variant of the conventional >>> TILLING technique that uses γ- and X-rays, and fast neutron bombardment instead of chemicals as inducing mutagens to produce mutant populations *gene biot meth*

DELLA proteins: Produced by a family of genes (of which Rht = "reduced plant height" in cereals is one) that repress the activity of key genes involved in plant growth; they are known from a number of plants; >>> gibberellin destabilizes DELLA proteins, reducing the amounts of the proteins in cells and their repressive effects on cell growth, e.g., in dwarf wheat a mutation makes the DELLA proteins produced by the *Rht* gene insensitive to gibberellin, and so they are not destabilized by gibberellin and cell growth is inhibited *phys gene* >>> *Rht* gene

deltoid: Shaped like the Greek letter "delta" *bot*

deme: Interbreeding group in a population, i.e., a population of potentially interbreeding individuals *eco*

demic selection: Special type of intergroup selection that does not necessarily involve direct competition; it has an effect on the general genetic composition of a population if subsets of a population have different gene frequencies *meth*

democratic plant breeding: The converse of autocratic plant breeding; with democratic plant breeding, as many breeders as possible are producing as many cultivars as possible so that the farmer has a wide choice of cultivars; this approach is possible with the use of horizontal resistance because breeding for this kind of resistance is so easy; in many cases, farmers can do their own plant breeding

demonstration strip design: The simplest design type for a field trial; varieties or breeding lines, herbicide or fertilizer applications are compared using demonstration strips on a field; the emphasis is on visual impact, not on measured results that are critically compared; usually the comparison is between two to four treatments, up to a maximum of ten; although this design is aimed at visual comparison, some measurements may be gathered for general comparison; the simplest measurement method is to take one randomly located sample for a variable such as yield, from near the center of each strip; the measured data from individual strips are compared to one another *agr meth stat* >>> treatment >>> design of experiments

denaturation: Reversible or irreversible alterations in the biological activity of proteins or nucleic acids that are brought about by changes in structure other than the breaking of the primary bonds between amino acids or nucleotides in the chain *chem gene*

denatured protein: A protein whose properties have been altered by treatment with physical or chemical agents *chem*

dendrogram: A genealogical diagram that resembles a tree; an evolutionary tree diagram may order objects, individual genes, etc., on the basis of similarity *evol meth* >>> Figures 68, 69 >>> cluster analysis

dendrology: The branch of botany dealing with trees and shrubs *bot*

denitrification: The conversion of nitrate or nitrite to gaseous products, chiefly nitrogen and/or nitrous oxide, resulting in the loss of nitrogen into the atmosphere and therefore undesirable in agriculture *chem agr eco*

de novo **sequencing:** The sequencing of a complete genome base by base for which neither sequence data nor any fragment library (>>> BAC library) are yet available *biot meth*

dense: Inflorescences having crowded spikelets *bot*

density-dependent: Of or referring to a factor, such as nutrient shortage, that limits the growth of a population more strongly as the density of the population increases *eco* >>> density-dependent selection

density-dependent selection: The limiting of the size of a population (e.g., a vertical pathotype) by mechanisms that are also controlled by the size of population; this is a probable genetic mechanism for controlling the system of locking of the n/2 model, ensuring that all the n/2 biochemical locks and keys occur with an equal frequency; the rarity of a vertical pathotype or pathodeme is a reproductive advantage that leads to commonness; commonness is a reproductive disadvantage that leads to rarity *eco*

density gradient centrifugation: The separation of macromolecules or subcellular particles by sedimentation through a gradient of increasing density under the influence of a centrifugal force; the density gradient may either be formed before the centrifugation run by mixing two solutions of different density (e.g., in sucrose density gradients), or it can be formed by the process of centrifugation itself (e.g., in $CsCl_2$ and Cs_2SO_4 density gradients) *meth*

dent maize (corn): A variety of maize, Zea mays ssp. indentata, having yellow or white kernels that become indented as they ripen *agr*

deoxyribonucleic acid (DNA): A nucleic acid, characterized by the presence of a sugar deoxyribose, the pyrimidine bases cytosine and thymine, and the purine bases adenine and guanine; its sequence of paired bases constitutes the genetic code; e.g., roughly 3 m of DNA is tightly folded into the human nucleus of every cell of the body; this folding allows some genes to be expressed, or activated, while excluding others; genes are made up of >>> exons and >>> introns – the former being the sequences that code for protein and are expressed, and the latter being stretches of noncoding DNA in-between; as the genes are copied, or transcribed, from DNA into RNA, the intron sequences are cut or spliced out and the remaining exons are strung together to form a sequence that encodes a protein; depending on which exons are strung together, the same gene can generate different proteins; specific genes and even specific exons, are placed within easy reach by folding; the three-dimensional structure of the genome can influence the splicing of genes; the genome is folded in such a way that the promoter region – the sequence that initiates transcription of a gene – is located alongside exons, and they are all presented to transcription machinery; this supports a new way of looking at genetics, one in which the genome is folded around transcription machinery, rather than the other way around; those genes that come in contact with the transcription machinery get transcribed, while those parts that loop away are ignored *chem gene* >>> Table 38

deoxyribonuclease: Any of several enzymes that break down the DNA molecule into its component nucleotides *phys*

depauperate: Stunted *bot*

dependent variable: The predictand in a regression equation (plotted on the x axis); the variable of interest in a regression equation, which is said to be functionally related to one or more independent or predictor variables *stat*

dephosphorylation: The removal of a phosphate group from an organic compound, as in the changing of >>> ATP to >>> ADP *chem phys*

deployment: The physical movement of clones (ramets) or other genetic units from one site (usually a nursery) to another (usually plantations), often including their spatial configuration on the recipient site *meth bree hort* >>> Figure 50

derivative hybrid: A hybrid arising from a certain cross between two hybrids

dermal: Having to do with the epidermis *bot*

dermal tissue system: The tissue system in plants that forms their outer covering *bot*

dermatogen: A specialized meristem in flowering plants in which floral induction begins; gives rise also to the epidermis *bot*

dermatogen stage: The stage at which the embryo has undergone three mitotic divisions and contains 16 cells; the outer protodermal cells will give rise to the epidermis through a series of anticlinal divisions; development of the >>> suspensor is complete at this stage *phys*

descendant: An individual resulting from the sexual reproduction of one parental pair of individuals *gene*

descent: The act, process, or fact of descending *gene*

descriptive statistics: Procedures that assist in organizing, summarizing, and interpreting the sample data we have at hand *stat meth*

descriptor: An identifiable and measurable characteristic used to facilitate data classification, storage, retrieval, and use *stat*

desert: A very dry area that receives less than 250 mm of precipitation each year *eco*

desiccant: A chemical applied to crops that prematurely kills their vegetative growth; often used for legume seed crops so the seed can be harvested prior to normal plant >>> senescence *phys meth* >>> defoliation >>> chemical desiccation

desiccate: >>> desiccation

desiccation: The process of drying out *phys* >>> chemical desiccation

desiccator: A glass jar with an air-tight lid that is used for drying out small quantities of plant tissue, such as seeds or root nodules, with a desiccating chemical; dry calcium chloride is a powerful desiccating chemical, but it is toxic and must be kept well-separated from living tissues; alternatively, silica gel is harmless, but it is less powerful in its drying action meth *cyto seed*

desinfectant: A physical or chemical substance used for the destruction of pathogenic microorganisms *phyt meth*

desinfection: The destruction of germs of infectious diseases *phyt*

desiccation-intolerant seed: >>> desiccation-tolerant *seed*

desiccation-tolerant seed: There are basically two types of seed, (1) desiccation-tolerant and (2) desiccation-intolerant; most of the plants produce desiccation-tolerant seeds, which means they can be safely dried for long-term storage; exceptions include many aquatic plants, large-seeded plants, and some trees (oaks, buckeyes, etc.), any of which produce desiccation-intolerant seeds which will die if allowed to dry; they do not enter dormancy after maturing; instead, respiration and other physiological processes continue *phys*

designated host: A genetically stable host (i.e., a clone or pure line) that has been chosen for use in the one-pathotype technique in a >>> horizontal resistance breeding program; it has a resistance that is matched by the designated pathotype, which is cultured on that host for the entire duration of the breeding program; all the original parents of the breeding population are chosen on the basis of their susceptibility to the designated pathotype, which is used to inoculate every screening population; it ensures that all vertical resistances are matched during the screening for horizontal resistance, regardless of how the vertical resistance genes may have recombined during the crossing process *phyt* >>> Figures 50, 51

designer chromosome: Gene cloning and transformation can be used to circumvent linkage drag effects that can plague conventional interspecific gene transfers; these techniques can also be used to create desirable genetic linkages (e.g., the use of *Nicotiana glutinosa* N-gene mediated tobacco mosaic virus [TMV] resistance in flue-cured tobacco, *N. tabacum*, has been limited due to linkage drag effects); transformation was used to introduce the cloned N-gene into "NC152", a chromosome addition line possessing a chromosome pair from *N. africana*; this chromosome has been proposed to be used as a "designer chromosome" into which numerous transgenes could be inserted to form a desirable linkage package; the system was proposed to shuttle a large number of transgenes from genotype to genotype *cyto biot* >>> artificial chromosome

design of experiments A procedure which can be used interactively to form experimental designs of various types; there several types of designs: (1) orthogonal hierarchical designs, such as >>> randomized blocks, split-plots, or split-split-plots; (2) factorial designs (with blocking) – these have several treatment factors and a single blocking factor (giving strata for blocks and plots within blocks); the blocks are too small to contain a complete replicate of the treatment combinations and so various interaction are confounded with blocks; (3) fractional factorial designs (with blocking) – several treatment factors but the design does not contain every treatment combination and so some interactions are aliased; there can also be a blocking factor and some interactions will then be confounded with blocks; (4) lattice designs – designs for a single treatment factor with number of levels that is the square of some integer k; the design has replicates, each containing k blocks of k plots, and different treatment contrasts can be confounded with blocks in each replicate; (5) lattice squares – similar to lattices except that the blocking structure with the replicates has rows crossed with columns; different treatment contrasts can be confounded with the rows and columns in each replicate; (6) Latin squares – available for any number of treatments; where feasible, more than one orthogonal treatment factor can be generated to form Graeco-Latin squares; (7) Latin squares balanced for carry-over effects – relevant when the same plots or subjects are treated during several successive time periods, and there is interest both in the direct effect of a treatment during the period in which it is applied and its carry-over (or "residual") effect during later periods; (8) semi-Latin squares (i.e., n × n Latin squares whose individual plots are split into k sub-plots to cater for a treatment factor with n × k levels); three types are available: Trojan squares, interleaving Latin squares, and inflated Latin squares; (9) alpha designs – these have a single treatment factor but there is no constraint on the number of levels; the blocking structure has replicates and blocks within replicates; (10) cyclic designs – these are designs with a single blocking factor that defines blocks that are too small to contain every treatment; usually there is a single treatment factor; however, in a cyclic superimposed design, two treatment factors can be considered; (11) balanced-incomplete-block designs, where the experimental units are grouped into blocks such that every pair of treatments occurs in an equal number of blocks; all comparisons between treatments are made with equal accuracy, so the design is balanced; and (12) neighbor-balanced designs that allow an adjustment to be made for the effect that a treatment may have on adjacent plots; lastly, there are (13) central composite designs used to study multidimensional response surfaces *stat*

desmosome: A plaque-like site on cell surfaces that functions in maintaining cohesion with an adjacent cell *bot biot*

desynapsis: The premature separation of paired chromosomes during >>> diplotene or >>> diakinesis of meiotic prophase; it is often genetically controlled but can also be induced by special environmental conditions (e.g., heat) *cyto*

desyndesis: >>> desynapsis

detasseling: Artificially removing (cutting or pulling) the tassel of the female parent to prevent selfing during hybrid seed maize production *meth seed*

detergent: Any synthetic organic cleaning agent that is liquid- or water-soluble and has wetting-agent and emulsifying properties *chem*

determinant flowering: Inflorescence in which the terminal flower blooms first, halting further elongation of the main axis *bot*

determinate: Descriptive of an inflorescence in which the terminal flower opens first, thus arresting the prolongation of the floral axis *bot*

determinate growth: The type of development in which a plant ripens all of its seeds at approximately the same time *phys*

deterministic equilibrium frequency: Equilibrium frequency of alleles expected in the absence of stochasticity *gene eco*

deterministic process: >>> stochastic

detoxicate: >>> detoxication

detoxication: The metabolic process by which toxins are changed into less toxic or more readily excreted substances *phys*

detrimental: To describe a gene, allele, or mutation that causes an impairment or disadvantage to the individual that carries it; it is often used as a noun, e.g., recessive lethals and detrimentals are common in populations of plants, especially outbreeders *gene*

detritivore: A consumer that relies on dead tissues for nutrients *eco*

developmental cycle: The gradual progression of phenotypic modifications of an organism during development *phys* ·

developmental genetics: The study of mutations that produce developmental abnormalities in order to gain understanding of how normal genes control growth, form, behavior, etc. *gene*

developmental stage: >>> growth stages

devernalization: The reversion of vernalization by non-vernalizing temperatures or other means *phys* >>> vernalization

deviation: The departure of a quantity from its expected value *stat*

dew point: The temperature at which relative humidity reaches 100% and water vapor is able to condense into water droplets; the dew point varies depending on the absolute water vapor content of the air *phy*

dextrin(e): A soluble gummy substance, formed from >>> starch by the action of heat, acids, or enzymes *chem phys*

dextrose: An aldohexose monosaccharide ($C_6H_{12}O_6$) that is a major intermediate compound in cellular metabolism; the dextrorotatory form of glucose, occurring in fruits and commercially obtainable from starch by acid hydrolysis *chem phys*

DF: >>> degree of freedom

DFP: >>> DNA fingerprint(ing)

DFR: >>> dihydroflavonol 4-reductase

DH lines: >>> doubled-haploid lines >>> *Hodreum bulbosum* procedure >>> *Imperata cylindrica* procedure >>> sorghum >>> speargrass

diadelphous: Showing stamens united in two sets by their filaments (e.g., in pea and bean flowers); nine out of ten stamens are usually united while one is by itself *bot*

diakinesis: The last stage in the prophase of meiosis I, when the paired homologous chromosomes are highly contracted but before they have moved onto the metaphase plate *cyto* >>> Figure 15

diallel: In either the complete or incomplete diallel, identities of both seed and pollen parents are maintained for each family *meth* >>> diallel cross >>> complete diallel >>> incomplete diallel >>> design of experiment

diallel cross: The crossing in all possible combinations of a series of genotypes *meth* >>> design of experiment >>> Table 24

diallel crossing group: >>> crossing groups

diallelic: >>> diallel cross

diallelic map: A genetic map that is based on molecular markers, e.g., >>> SNPs, of which both parental alleles are known *biot meth*

diallel mating: >>> diallel cross

dialysis: Technique for separating molecules of different sizes through a membrane of known pore diameter, usually across an electropotential gradient *meth*

diapause: A period of dormancy phys; in insects, a state during which growth and development is temporarily arrested *zoo*

diaschistic: >>> anaschistic

diatomaceous earth: A fully inert, non-volatile substance sometimes recommended as an alternative to traditional chemical treatments for controlling a number of insects and other pests, including soil bugs; it is made from the skeletal remains of diatoms, a microscopic form of algae; when processed these skeletal remains form razor-sharp particles which cut into the soft bodies of small insects *hort meth*

diazinon: An insecticide sometimes recommended for controlling >>> black vine weevils *phyt hort meth*

dibber: >>> dibble

dibble: A pointed tool used to make holes in the ground for seeds and seedlings *prep hort* >>> pricking-out peg

dibble planting: >>> dibbling

dibbling (seed): Sowing in holes made by a pointed tool *meth hort fore*

dicentric: A chromosome or chromatid with two centromeres *cyto* >>> Figure 11

2,4-dichlorophenoxy acetic acid (2,4-D): A crystalline powder, $C_8H_6O_3Cl_2$, used for killing weeds and as a growth hormone *phys biot*

dichogamous: Having pistils and stamens that mature at different times to prevent self-fertilization *bot* >>> dichogamy

dichogamy: The condition in which male and female parts of a flower mature at different times *bot*

dichophase: The phase of the mitotic cycle in which a cell is determined for further mitotic differentiation or for special cell functions *cyto*

dichotoma gene: A gene involved in promoting asymmetric development of petals; in snapdragon plants where a mutation has inactivated the dichotoma gene, the flower petals develop more symmetrically than normal; the active form of the gene normally causes the development of asymmetry in the petals; the gene is active at a very early stage in flower development, when the flower bud is only about half a millimeter across; at this time it is switched on in only one-half of the developing petal; this causes the petal to grow in a more asymmetric way; asymmetry is common in nature; there are cases where asymmetry is very consistent (e.g., while most leaves are symmetrical about their midline, leaves of Begonia plants are consistently asymmetrical, one half of the leaf being bigger than the other half); Begonia leaves come through in two mirror-image forms, left or right, depending on which side of the leaf is bigger *phys gene biot* >>> dichotomous

dichotomous: Branched or forked into two more or less equal divisions *bot* >>> dichotomous key

dichotomous key: A tool used in the identification of unknown plants by stating the conspicuous features by which the various taxa can be recognized; two contrasting choices are presented at each step; one choice in the couplet is accepted and the other rejected, weaving a path to the identity of the plant *bot tax*

dichotomous ramification: Branching, frequently successively, into two more or less equal arms *bot*

diclinous species: Species having pistils and stamens on different flowers *bot* >>> Table 32

dicliny: Where not all >>> genets in a population are regularly >>> hermaphrodite, i.e., males, or females, or both occur *eco*

dicot: An abbreviated name for dicotyledon, which refers to plants having two seed leaves *bot* >>> dicotyledons

dicotyledonous: >>> dicotyledons

dicotyledons: Plant species having two cotyledons; flower parts arranged in fours or fives or multiples thereof; net-veined leaves and vascular bundles in the stem arranged in a ring *bot* >>> Table 32

didiploid (*syn* **allotetraploid**): Two different diploid chromosome sets present in one cell or organism *cyto*

didynamous: Having stamens arranged in two pairs of unequal length *bot*

dieback: The death of tips or shoots due to damage or disease *phys*

dietary fiber: The plant polysaccharides and lignin, which are resistant to hydrolysis by the digestive enzymes of man *phys* >>> dietary fiber complex

dietary fiber complex: In addition to indigestible polysaccharides and lignin, minor components, such as phenolic compounds (e.g., lignans) or strongly associated cell wall proteins, also may be included in the dietary fiber concept; the presence of such minor constituents, as well as structural variations in the dietary fiber components, can be of great importance for the functional and/or physiological properties of dietary fiber *phys*

differential AFLP: DNA fragments that were amplified by using >>> PCR method and that differ in length between organisms of the same species due to allelic variation *meth biot* >>> AFLP

differential centrifugation: A method of separating subcellular particles by centrifugation of cell extracts at successively higher speeds; it is based on differences in sedimentation coefficients that are roughly proportional to particle size; in other words, large particles (>>> nuclei, >>> chloroplasts, or >>> mitochondria) are sedimented at lower speeds than small particles (>>> ribosomes) *prep meth biot*

differential host: A plant that gives distinctive symptoms when infected with a specific virus, allowing the virus to be distinguished from others *phyt*

differential medium: An *in vitro* cultural medium with an indicator (e.g., a special dye) that allows various chemical reactions to be distinguished during plant growth *meth*

differential selection: The difference between a selected plant, family, or clone, and the average of the population from which it is taken *meth* >>> Figure 50

differential staining: In microbiology, staining procedures that divide bacteria into separate groups based on staining properties; in cytology, staining procedures that divide >>>

chromosomes or >>> genomes into separate segments based on structural and biochemical properties *meth cyto* >>> Table 53

differential variety: A host variety, part of a set differing in disease reaction, used to identify physiologically specialized forms of pathogens *phyt* >>> differential host

differentiation of cells: The development of specialized kinds of cells from non-specialized cells in a growing tissue *bot phys*

diffuse: Open and much-branched, widely spread *bot*

diffuse stage: A meiotic >>> prophase stage in which the chromosomes become reorganized; they even may disappear *cyto*

diffusion: Passive movement of molecules from areas of high concentration of the molecule to areas of low concentration; one of three processes by which soil nutrients become available to plants; as nutrients are absorbed, a lower concentration is formed at the root interface; to even out the concentration throughout the soil, nutrients move from areas of higher concentration to areas of lower concentration around the roots *phy hort phys* >>> Table 44

diffusion approximation: In population genetics, an approximation of a stochastic process that is discrete in state, i.e., number of genes, and time, i.e., generation by a stochastic process that is continuous in state and time *stat meth*

DIF method: Controlling plant height in greenhouses by temperature, using a method known in the industry as DIF; it was developed by researchers from Michigan State University back in the 1980s; their research showed that the average temperature (the average day plus night temperature) affects a plant's growth rate, with higher averages resulting in more rapid growth and development; the difference between day temperatures and night temperatures affects stem elongation and height; DIF is calculated as the day temperature minus the night temperature and can be either positive DIF (day temperature is higher than night), zero DIF (day temperature = night), or negative DIF (day temperature is less than night); trying to keep a greenhouse in negative DIF is a difficult task for most; another way was found that accomplished reducing a plant's height; it is the easiest DIF treatment to use; it is called "cool morning pulse"; reducing the greenhouse temperature −15 to −12°C lower than the night temperature for 2 to 3 h, starting 30 min before dawn, reduced plant height as effectively as negative DIF and was easier to do; for example, night temperature of 20°C, 2-h drop to 15°C (30 min before dawn), and then 18°C maintained during the day, or for warmer plants 22°C at night; 17°C pre-dawn for 2 to 3 h; 18°C d; there are thermostatically controlled devices that can be set up to automatically change these temperatures at the correct times; salvia, rose, snapdragon, and fuchsia had very good responses to this DIF control compared to aster, French marigold, tulip, and squash, which show little or no response *meth hort*

dig: To break up and turn over piecemeal the soil or ground *meth agr* >>> Table 44

digametic: >>> heterogam(et)ic

digenic: It refers to an inheritance that is determined by two genes *gene*

digestibility: The attribute of forage biomass to be digested by grazing animals – an important selection criterion in forage crop breeding *agr* >>> Table 33

digital karyotyping: A technique of the quantitative and high-resolution detection of copy number changes, i.e., amplified and deleted chromosomal regions, on a genome-wide scale; it uses short sequence tags derived from specific genomic loci at ~4 kb intervals along the entire genome by enzymatic digestion *biot meth*

digital plant phenotyping: Accurate characterization of complex plant phenotypes is critical to assigning biological functions to genes through forward or reverse genetics; it can also be vital in determining the effect of a treatment, genotype, or environmental condition on plant growth or susceptibility to insects or pathogens; although techniques for characterizing complex phenotypes have been developed, most are not cost-effective or are too imprecise or subjective to reliably differentiate subtler differences in complex traits like growth, color change, or disease resistance; therefore, low-cost genotyping and phenotyping address some of the environmental, agricultural, and industrial sustainability challenges of the future; since the human eye has restricted sensitivity, camera systems can expand this sensitivity dramatically; usually, it is an automated image-based technology, including specific computer, camera, and facility hardware; plant shoots and roots are fully automated and comprehensive imaged; scanning can be arranged in different wavelength and modes (infrared light – shoot, visible light – shoot, near infrared – root, near infrared light – shoot, fluorescent light – shoot); highly sensitive measurements lead to reproducible imaging protocol considering the dynamics of leaf temperature; high-resolution color images result in comprehensive morphological and growth phenotyping; color classification allow measurements of plant health, stress, nutrients, and senescence; calculation of spatial distribution of water content in soil is feasible, including dynamic root growth patterns *biot meth*

digitate: Fingerlike, or a compound, with the members arising together at the apex of the support *bot*

digoxigenin (DIG): Antigenic alkaloid from Digitalis spp. (foxglove), which is used to label DNA *biot* >>> *in situ* hybridization

digynoid: In parthenogenesis, the progeny derives from an unreduced egg cell *bot*

dihaploid: A haploid cell or individual containing two haploid chromosome sets – not to be confused with doubled-haploid *cyto*

dihybrid: A cross between individuals that differ with respect to two specified gene pairs *gene*

dihydroflavonol 4-reductase (DFR): Bulb color in >>> onions (*Allium cepa*) is an important trait, but the mechanism of color inheritance is poorly understood; a study showed that inactivation of the DFR gene at the transcriptional level resulted in a lack of >>> anthocyanin production in yellow onions; a deletion mutation in the yellow DFR-A gene also results in the lack of anthocyanin production in yellow onions *phys biot*

diisosomic: A cell or an individual which has a pair of homologous isochromosomes for one arm of a particular chromosome *cyto* >>> aneuploid >>> Figure 37

diisotrisomic: A cell or an individual that lacks one chromosome but carries two homologous isochromosomes of one arm of a particular chromosome *cyto* >>> aneuploids >>> Figure 37

dikaryon: A dinucleate cell *cyto*

dikaryotic: A cell showing two nuclei *bot*

dilated: Expanded, enlarged, or wider *bot*

dilatory stage: The initial few days following fertilization, during which time very slow dry matter accumulation occurs in the kernel; in the endosperm, the primary endosperm nucleus divides 2–4 h after fertilization, followed by rapid synchronous divisions to produce a free nuclear tissue; the fertilized egg undergoes its first division about 10–12 h after fertilization, and by the end of this stage will have developed into the proembryo with about 12–24 cells *phys meth*

DIMBOA: Acronym for 2,4-dihydroxy-7-methoxyl-1,4-benzoxazin-3-one, a naturally occurring compound that confers resistance to certain insect pests in maize *phyt*

dimensional sampling: A sampling technique designed to enhance the representativeness of small samples by specifying all important variables and choosing a sample that contains at least one case to represent all possible combinations of variables *stat meth* >>> cluster sampling >>> area sampling

dimer: A protein that is made up of two polypeptide chains or subunits paired together *chem*

dimerization: >>> dimer

dimethyl sulfoxide (DMSO): A liquid solvent, C_2H_6OS, approved for better penetration of specific substances through the cell wall *chem meth*

dimorph: >>> dimorphism

dimorphism: The occurrence of two forms of individuals within one population or other taxa (e.g., sexual dimorphism or the presence of one or more morphological differences that divide a species into two groups) *bot*

dimorphous branching: Some crop species (e.g., arabica coffee, cotton, black pepper) have two kinds of branches; the orthotropic branch is the branch that grows vertically, and it produces side branches, called plagiotropic branches, that tend to grow horizontally; it is usually the plagiotropic branches that bear the flowers and seed; cuttings must be taken from the orthotropic branch, and this severely limits vegetative propagation *bot meth hort*

dinitroaniline(s): A family of herbicides that inhibit microtubule assembly and belong to the seedling growth inhibitor mode of DNA action *phyt phys*

dioecious: Possessing male and female flowers or other reproductive organs on separate, unisexual, individual plants (e.g., in hemp, hops, asparagus, or spinach), e.g., 6% (= 990 genera) of angiosperms are dioecious *bot* >>> sex chromosome >>> Figure 65

dioecism: The phenomenon of plants showing either male or female sex organs *bot* >>> dioecious >>> sex chromosome >>> Figure 65

diphasic: Chromosomes that show both euchromatic and heterochromatic segments *bot*

diphyletic: A group of species that share two ancestries *evol* >>> monophyletic >>> polyphyletic

diplandroid: In parthenogenesis, the progeny derives from an unreduced sperm cell *bot*

diploid: A cell with two chromosome sets or an individual with two chromosome sets in each cell; a diploid state is written as "2n" to distinguish it from the haploid state "n" *gene* >>> Table 1 >>> Figure 8

diploid parthenogenesis: A type of gametophyte apomixis by which a diploid embryo sac cell results in a diploid embryo; the diploid condition is the result of cytogenetic mechanisms occurring in the egg stage *bot* >>> Figure 28

diploidization: In polyploids, a natural or induced mechanism in which the >>> chromosomes pair completely or partially as bivalents, although >>> polyploid sets of chromosomes are present; it may be caused by a structural differentiation of homologous chromosome sets or by genetic control; for example, in bread wheat three homoeologous genomes are available (AABBDD); they do not pair as hexavalents but exclusively as bivalents *cyto* >>> Table 53

diploidizing mechanism: A mechanism whereby the chromosomes of a >>> polyploid sometimes form exclusively or partly bivalents instead of multivalents during meiosis; usually it is under strict genetic control [e.g., the pairing homologous (*Ph*) locus on chromosome 5B of hexaploid wheat] *cyto*

diploidy: The presence of two homologous sets of chromosomes in somatic cells *cyto* >>> Table 1 >>> Figure 8

diplonema: >>> diplotene

diplontic: >>> diploid

diplophase: The diploid generation phase after fertilization to meiosis *cyto*

diplospory: A type of agamospermy (apomixis) in which a diploid embryo sac is formed from archesporial origin *bot* >>> Figure 28

diplotene: The stage in the prophase of first meiosis, when the paired homologous chromosomes separate, except where they are held together by chiasmata *cyto*

dipping: The immersion of seedling roots in a solution or water prior to planting *fore hort meth*

dip treatment: The application of a liquid chemical to a plant by momentarily immersing it, wholly or partially, under the surface of the liquid to coat the plant with the chemical *phyt*

***Dir1* gene:** The abbreviation of "defective induced resistance"; the name given to the single gene that has been identified as the basis of the loss of systemic acquired resistance response in a >>> thale cress (*Arabidopsis thaliana*) mutant line; *Dir1* is a 306 nucleotide open reading frame that encodes a 102 amino acid peptide of 10.6 kDa; *Dir1* shares a region of 54 nucleotides with a *Phaseolus vulgaris* nonspecific lipid transfer-like protein and has 53% sequence similarity throughout the protein; the *Dir1* protein contains a hydrophobic

N-terminal signal sequence and the eight >>> cysteine residues conserved in all lipid transfer proteins; this suggests that *Dir1* encodes an apoplastic lipid transfer protein that is involved in export and/or transport of an apoplastic signal molecule; this gene is central to the plant's ability to develop long-lasting immunity that is effective against a wide range of different diseases; *Dir1* gene is involved in production of the signal itself or affects the export of the signal molecule from infected cells and its transport around the plant *phyt biot* >>> resistance >>> systemic acquired resistance >>> Table 38

direct drilling: Drilling seed directly into soil without any mechanical seedbed preparation, for example, tilling since the previous crop *agr*

directed selection: The process of controlling genetic change in domesticated plants through manipulation of the plants' environment and their breeding process *meth* >>> Figure 60

direct embryogenesis: Embryoid formation directly on the surface of zygotic or somatic embryos or on seedling plant tissues in culture without an intervening callus phase *biot* >>> Figure 58

direct organogenesis: Organ formation directly on the surface of relatively large intact explants without an intervening callus phase *biot* >>> Figure 58

directed dominance: >>> directional dominance >>> dominance

directional cloning: In biotechnology, DNA inserts and vector molecules are digested with two different >>> restriction enzymes to create non-complementary sticky ends at either end of each restriction fragment; it allows the insert to be ligated to the vector in a specific orientation and prevents the vector from recircularizing *biot* >>> Table 39

directional dominance: A type of dominance in which the majority of dominant alleles have positive effects in one direction *gene*

directional mutation: A genetic change that favors a certain genotype or population *gene*

directional selection: Selection resulting in a shift in the population mean in the direction desired by the breeder *meth* >>> Figure 60

directochimera: >>> chimera >>> Figure 56

direct repeat: Two or more stretches of DNA within a single molecule which have the same nucleotide sequence in the same orientation, e.g., TATTA ... TATTA/ATAAT ... ATAAT *biot*

direct seeding: Offers several benefits over conventional tillage, including improved yield potential and time savings for field operations; however, the change to direct seeding is a change to a different cropping system; it requires a new approach to crop residue, and weed and fertility management; it may also require crop rotation changes to prevent specific pest problems that were kept in check by tillage *agr*

direct (seed) treater: The most recent development in seed treatment, includes the "panogen" and "mist-o-matic" treaters; these two were initially designed to apply undiluted liquid treatment; instead of applying 23 cm^3 of material per 3.5 kg of wheat, as in slurry treaters, they apply 14 to 21 cm^3 (15 to 20 g) per 35 l of wheat; this small quantity of material is suitable only with liquid materials that are somewhat volatile and do not require complete, uniform coverage for effective action; other modifications for treaters include dual tanks that permit simultaneous addition of a fungicide and an insecticide, and adaptations for the application of slurries; the metering device in both treaters is similar to that of the slurry treater, since it is attained through synchronization of a treatment cup and seed dump; otherwise, they differ decidedly from the slurry treater and from each other; both of these direct treaters have an adjustable dump pan counterweight to adjust the weight of the seed dump *seed* >>> seed treater

dirty crop approach: Partially resistant multilines where each component of the multiline carries a single gene for resistance but none of the resistances are effective against all known races of the pathogen (MARSHALL 1977); it has been advocated as a dynamic, natural biological system for the control of crowd diseases in crop plants; multilines could stabilize the race structure of the pathogen population *meth* >>> multiline

dirty seed: Endophyte-infected seeds (e.g., in grass) *seed* >>> clean seed

disaccharide: Any of a group of carbohydrates, such as sucrose, that yield monosaccharides on hydrolysis *chem phys*

disarm: To delete from a plasmid or virus genes that are cytotoxic or tumor-inducing *biot*

disarticulate (disarticulation): Separating at the nodes (joints) naturally at maturity *bot*

disassortative mating: Occurs if the plants mating resemble each other less than plants belonging to pairs of random plants, with regard to some trait *gene* >>> assortative mating

disbud(ding): Removing buds, shoots, or growing tips (with finger and thumb) of a tree, vine, or flowering plant to encourage production of side shoots or high-quality flowers and fruits; >>> pinching out is also used when small side shoots are completely removed; it is done when single stems are desired, especially when training to form the "trunks" of standard (tree-form) specimens *hort*

disc floret: A small flower, usually one of a dense cluster (e.g., sunflower) *bot*

disc flower: The tubular flowers of the head, as distinct from the ray (e.g., in Asteraceae or sunflower) *bot*

disc grain-grader: Discs revolve through a seed mass and a certain size of seeds are lifted and discharged, while the other size (e.g., the longer ones) is rejected by the disc indents *seed*

disc harrow: A tractor-drawn implement composed of circular plates arranged on an axle at an angle to the direction of travel; most usually, there will be two pairs of disc axles arranged symmetrically about the longitudinal axis of travel direction and with the tandem axles at opposite angles to work the soil in two directions with each pass; a drag is commonly pulled behind to smooth the soil; it is used to prepare soil beds for seeding and to control weeds *tech meth*

disc plow: >>> disk plow

discoid: Resembling a disk or platter *bot*

discolored grains: Grains, which have changed to a yellowish or brownish or black color because of heat damage during storage *agr*

discontinuous character: Variation in which discrete classes can easily be recognized (e.g., flower color, straw length) *gene* >>> qualitative character

discovery rights: In patenting biological material, selling only research findings while keeping rights to all the knowledge that is uncovered along the way *adm* >>> Intellectual Property Rights

discrete distribution: A probability distribution that only has a finite or countably infinite number of possible values *stat*

discrete variables: Variables with a finite number of distinct and separate values *stat*

disease: A condition in which the use or structure of any part of a living organism is not normal; harmful deviation from normal functioning of physiological processes; six types of causal agents can be considered: (1) fungi, (2) bacteria, (3) viruses, (4) nematodes, (5) insects, and (6) plant parasites *phyt* >>> seed-borne pathogens

disease avoidance: Avoiding the disease by, for example, growing the crop sufficiently early so that the vulnerable part of the plant's growing cycle is over before the disease-causing organism arrives in the area; this stops the disease from starting; it is sometimes called "passive resistance" *phyt*

disease control Several types of disease control can be classified: (1) disease resistance, (2) protection, (3) avoidance, (4) exclusion, (5) eradication, (6) therapy *phyt*

disease cycle: A cyclical sequence of host and parasite development and interaction that results in disease and in reproduction of the >>> pathogen *phyt*

disease eradication: This control measure is applied to a situation in which the disease is present in the area; it involves removing the disease by, for example, burning all stubble of the diseased crop, in order to prevent transfer of the disease from the previous crop to the next season *phyt*

disease escape: For a variety of reasons, some individuals in a screening population may remain free of pests or disease; it also known as chance escape; this phenomenon can be very misleading because it is so easily confused with resistance *phyt*

disease protection: Involves protecting the plant with a chemical; this is applied before the disease starts and prevents the beginning of problems; systemic fungicides can penetrate and move inside the plant; they therefore have a greater exposure to the pathogenic organism; some are taken up by roots, others by leaves *phyt* >>> systemic pesticide

disease resistance: The ability to resist disease or the agent of disease and to remain healthy *phyt*

disease therapy: This applies to removing the particular part of the plant that is diseased; often applied to large and valuable plants, such as fruit trees, where a diseased branch may be removed by a tree surgeon *phyt*

disease triangle The three conditions required for a disease to occur, namely: a susceptible host, a suitable pathogen, and an appropriate environment *phyt*

disequilibrium mapping: >>> association mapping

disinfectant: A chemical treatment used to disinfect seed for planting; it is commonly useful for surface-borne pathogens *seed meth*

disjunction (of daughter chromosomes): The separation of homologous chromosomes at the anaphase stage of mitosis and meiosis, and movement toward the poles of the nuclear spindle *cyto*

disjunctional separation: >>> alternative disjunction

disk: Enlarged growth of the head made up of a circular arrangement of fused petals (e.g., in sunflower) *bot* >>> disc flower

disk flower: >>> disc flower

disk plow: A plow with saucer-shaped units for breaking the soil *agr*

dislocation: The displacement of a chromosome segment away from its original position in the chromosome *cyto*

disome: A cell or an individual showing two homologous sets of chromosomes *cyto* >>> aneuploid >>> Figure 37

disomic: >>> disome

disomo-autotetraploid: >>> aneuploid

disomy: >>> disome

disorder: Any harmful deviation from normal plant physiological processes due to abiotic factors *phyt*

dispensable genome: Portion of the pan genome that is only in a subset or unique to individuals; it refers to genes not present in all strains of a species; these include genes present in two or more strains or even genes unique to a single strain only, for example, genes for strain-specific adaptation such as antibiotic resistance *biot* >>> pan genome >>> core genome >>> Figure 71

dispermy: The entering of two sperm cells into one egg cell *bot*

dispersal: The spread of a pathogen within an area of its geographical range *phyt*

dispersing agent: A chemical added to a pesticide formulation to aid the efficient distribution of particles of the active ingredient *phyt prep*

disporic: Having two spores *bot*

disruptive selection: A selection that changes the frequency of alleles in a divergent manner, leading to the fixation of alternative alleles in members of the population; the result after several generations of selection should be two divergent phenotypic extremes within the population *meth*

dissecting microscope: Usually, a low-power microscope (50× magnification) used to facilitate dissection, examination, or excision of small plant parts; however, in recent biotechnology high-power microscopes also are applied for dissections *prep*

dissepiment: A partition within an organ of a plant (e.g., the membrane that separates sections of the orange and other citrus fruits) *bot*

dissemination: The spread of seeds or spores; in plant pathology, the transport of inoculum or pest from a diseased to a healthy plant *eco phyt* >>> dispersal

dissimilation: >>> assimilation

disseminule: An organ of reproductive dispersal; >>> propagule *bot*

distal: Farthest from the point by which it is attached to the starting point gene *cyto*

distance isolation: Several plants can be protected from cross-pollination by separation by a certain distance, such as lettuces (7.5 m) or eggplants (150 m); insect-pollinated plants, such as the cabbage family (collards, broccoli, etc.), squashes, and okra, require from 0.4 to 1.5 km for complete safety; maize, a wind-pollinated plant, can require 1.5 km or more for safe distance isolation, and members of the beet family may need as many as 8 km; the exact distance for safely isolating a particular crop depends on a number of factors (e.g., the type of plant and how it is pollinated, i.e., wind, insects, selfing, or a combination of these), the location, climate, prevailing wind patterns and surrounding terrain and vegetation features, the relative size of plantings, etc. *meth*

distant hybridization: The crossing and/or hybridization of members of different genera *meth*

distichous: Obviously 2-ranked and appearing as the rungs of a ladder *bot*

distinguishable hybrid: A type of hybrid in which intermediate inheritance is phenotypically expressed (i.e., the heterozygous gene constitution is visible by the phenotype) *gene*

distyly: The presence of either pin or thrum flowers; pin × pin and thrum × thrum crosses are incompatible due to alleles at a single locus; the thrum morphology is controlled by a dominant allele S and the pin morphology by the recessive alleles s *bot* >>> self-incompatibility

disulfide bridge: A covalent bond formed between two sulfur atoms; it is a particular feature of peptides and proteins, where it is formed between the sulfhydryl groups of two cysteine residues, helping to stabilize the tertiary structure of these compounds *chem* >>> Table 38

ditelocentric: >>> ditelosomic

ditelomonotelosomic: A cell or an individual that has a pair of telocentric chromosomes for one arm and a single telocentric chromosome for the other arm *cyto* >>> aneuploid >>> Figure 37

ditelosomic: A cell or an individual that has two telocentrics of one chromosome arm *cyto* >>> aneuploid >>> Figure 37

ditelotrisomic: A cell or an individual that has two telocentric chromosomes of one arm plus one complete chromosome of the homologue *cyto* >>> aneuploid >>> Figure 37

ditertiary compensating trisomic: A cell or an individual with a compensating trisomic chromosome in which a missing chromosome is compensated by two tertiary chromosomes *cyto* >>> Figure 14

diurnal: With activity taking place only in daylight, i.e., a flower that opens only during the day *bot*

diurnal rhythm: >>> circadian rhythm

divalent: >>> bivalent

divergence time: Time since the most recent common ancestor *evol eco*

divergent: Set at an angle to one another herring-bone fashion, as in the lateral spikelets of the two-row barley spike *bot*

divergent evolution: A kind of evolutionary change that results in increasing morphological difference between initially more similar lineages *evol* >>> Figures 68, 69 >>> crop evolution

diversification rate: The rate of reproduction of new forms such as species or alleles *eco evol*

diversifying selection: Selection in which two or more genotypes show optimal adaptation under different environments *meth* >>> Figure 60

diversity arrays technology (DArT) marker: A DArT marker is a segment of genomic DNA, the presence of which is polymorphic in a defined genomic representation; DArT markers are >>> diallelic and behave in a dominant (present versus absent) or codominant (two doses versus one dose versus absent) manner; to identify the polymorphic markers, a complexity reduction method is applied on the metagenome, a pool of genomes representing the germplasm of interest; the genomic representation obtained from this pool is then cloned and individual inserts are arrayed on a >>> microarray resulting in a "discovery array"; labeled genomic representations prepared from the individual genomes included in the pool are hybridized to the discovery array; polymorphic clones (DArT markers) show variable hybridization signal intensities for different individuals; these clones are subsequently assembled into a "genotyping array" for routine genotyping; the method was developed to provide a practical and cost-effective whole-genome fingerprinting tool; DArT has three key attributes of interest to plant breeders and scientists studying and managing genetic diversity: (1) it is independent from DNA sequence, (2) the genetic scope of analysis is defined by the user and easily expandable, and (3) the method provides for high-throughput and low-cost data production; the discovery of polymorphic DArT markers and their scoring in subsequent analysis does not require any DNA sequence data; it makes the method applicable to all species, regardless of how much DNA sequence information is available for that species; however DArT markers are sequence-ready clones of genomic DNA; for each species, the method is developed on the "metagenome", the pooled genomes from the germplasm of interest to the user, e.g., the metagenome may include DNA from the cultivated varieties of a particular region or the lines used in a breeding program; alternatively, the metagenome may cover the genetic diversity within the entire species and even extend to its wild relatives; importantly, the diversity surveyed by DArT can be expanded if new individuals with marked genetic differences are incorporated into the analysis at a later stage; in DArT, several hundred polymorphic markers are identified in parallel; the efficiency of this marker discovery effort is only dependent on the level of genetic diversity within the species (5–10% of wheat and barley DArT clones and 25–30% of cassava DArT clones were polymorphic); the microarray platform enables a high level of multiplexing: ~5,000–8,000 genomic loci are typically surveyed in parallel in single-reaction assays to discover polymorphic markers; DArT markers can be used as any other genetic marker; with DArT, comprehensive genome profiles are becoming affordable for virtually any crop, regardless of the molecular information available for the crop; DArT genome profiles are very useful for the recognition and management of biodiversity; DArT genome profiles enable breeders to map QTL in one week, thereby allowing them to focus on the most crucial factor in plant breeding – reliable and precise phenotyping; once many genomic regions of interest are identified in many different lines, DArT profiles accelerate the introgression of a selected genomic region into an elite genetic background; furthermore DArT profiles can be used to guide the assembly of many different regions into improved varieties; for that purpose, dense genome cover is essential in order to follow many regions simultaneously *meth biot* >>> QTL mapping >>> Figure 71 >>> Table 58 >>> www .diversityarrays.com/

division: A method of propagation by which a plant clump is lifted and divided into separate pieces; it includes roots and a growing point *meth hort*

divisional patent: A patent that covers the same specification as a previous (parent) patent but claims a different invention *adm* >>> Intellectual Property Rights

dizygous: Also called fraternal twins; when two different eggs, each fertilized by a different gamete, have different genotypes *gene cyto*

d-limonene: One of the active compounds in citrus oils, extracts of citrus peels known to have natural insecticidal properties *phyt chem*

DMSO: >>> dimethyl sulfoxide

DNA: >>> deoxyribonucleic acid

DNA-amplified fingerprinting: A technology based on amplification of random genomic DNA sequences achieved by a single short (5–8 bases) oligonucleotide primer of arbitrary sequence; it produces a characteristic spectrum of short DNA pieces of varying complexity that are resolved on polyacrylamide gel (PAGE) following silver staining; it is used to detect genetic differences between genotypes as well as for detecting polymorphism even between organisms that are closely related, such as near isogenic lines *biot meth* >>> DNA fingerprint(ing) >>> Figure 48

DNA banking: In addition to the *in situ* and *ex situ* conservation of plant genetic resources, this is a complementary conservation strategy and an effective tool for more efficient management of genebanks and conservation research; however, it is still not a common practice since technical capacity and financial resources are currently limited, particularly in developing countries; in future DNA banks can be the libraries for reference sequences that can be applied to estimate genetic diversity and relative change within populations or to identify samples when genotypes confound morphologically based identification; DNA is an unusually stable biomolecule that often outlasts the organism it encodes; DNA banks provide an accessible means for genetic characterization and electronic integration of many existing biomaterial collections *seed biot*

DNA barcoding: The use of a short DNA sequence or sequences from a standardized locus (or loci) as a species identification tool of plants; it is useful for identification of different life stages, e.g., seeds and seedlings; for identification of fragments of plant material, for biosecurity and trade in controlled species, or for plant inventory and ecological surveys; barcoding of animals is already in progress using the cytochrome C oxidase 1 (*cox1*) gene; in plants, it is more difficult since there is often no sequence variation among species within a genus, and therefore the gene is not suitable as a plant barcode; barcoding is a valuable taxonomic classification tool *meth biot* >>> bar code

DNA binding motif: Common sites on different proteins which facilitate their binding to DNA, e.g., leucine zipper and zinc finger proteins *biot* >>> Table 38

DNA chip (technology): High-throughput screening technique based on the hybridization between oligonucleotide probes and complementary DNA sequences in microarrays; chips may consist of arrays of amplified DNA immobilized on miniature glass or nylon substrates that are then tested for hybridization to series of fluorescently labeled oligonucleotide probes; alternatively, arrays of oligonucleotide probes may be synthesized, followed by the exposure to fluorescently labeled PCR samples; hybridization signals are determined by LASER scanning from which data on sequence variation are obtained; microarrays may comprise up to 250,000 features/cm²; applications of the DNA chip technology include diagnostics (mutation and polymorphism detection), gene discovery, gene expression, and gene mapping *biot meth*

DNA clone: A section of DNA that has been inserted into a bacterium, phage, or plasmid vector and has been replicated many times *gene*

DNA content: The capacity of some dyes to bind to DNA in a stoichiometric manner means that the amount of DNA present in a nucleus can be determined, at least in comparison with some reference standards; the reliability of the >>> FEULGEN reaction for localizing DNA within the cell was universally accepted many years ago; today, the FEULGEN reaction, when properly controlled, is beyond any doubt specific and stoichiometric for DNA; usually, the total DNA amount per nucleus, given as picograms; the nuclear DNA content (>>> C-value) is a fundamental parameter of eukaryotic cells; usually, flow cytometry provides the most convenient and accurate way to determine C-values of plants, but sampling of the ca. 480,000 species of >>> angiosperms (flowering plants) remains very limited, only about 1% being reported; for plant species for which C-values are available, these values span an extraordinary range, appr.1,250-fold (0.2 pg to 254.8 pg); for plants whose C-values are unknown, this complicates analysis using flow cytometry, when the range of possible C-values exceeds the dynamic range of the >>> flow cytometer, since it requires adjustment of various instrument settings (amplification values, PMT high voltages, and LASER outputs) to optimally position the 2C peak values for the unknowns and the appropriate controls within the fluorescence frequency distributions; although this can be done empirically, it considerably slows down the sample throughput; the Accuri C6 Flow Cytometer® is ideally suited for determining C-values for new plant species; the totally digital C6 has state-of-the-art electronics that make it possible to simultaneously collect over 16 million channels of digital data; this wide dynamic range allows simultaneous analysis of species whose C-values span almost the entire range of the described angiosperms; DNA content can range between 145 Mb in *Arabidopsis thaliana*, 389 Mb in *Oryza sativa*, 673 Mb in *Phaseolus vulgaris*, 760 Mb in *Sorghum bicolor*, 870 Mb in *Musa* sp., 953 Mb in *Lycopersicon esuclentum*, 1,115 Mb in *Clycine max*, 2,504 Mb in *Zea mays*, 2,702 in *Capsicum annuum*, 4,873 Mb in *Hordeum vulgare*, 11,315 *Avena sativa*, 15,290 Mb in *Allium cepa*, and 159,066 Mb in *Triticum aestivum*, >>> List of Crop Plants, attached *cyto*

DNA deletion: The removal of DNA sequences of various length, parts of chromosomes, or whole chromosomes from a genome in evolutionary times or during developmental pathways *gene cyto evol* >>> Figures 68, 69, 71

DNA-DNA hybridization: The annealing of two complementary DNA strands to produce hybrid nucleic acid molecules; it is used to identify the base sequences in two polynucleotide chains from different sources *gene meth*

DNA fingerprint(ing) (DFP): The unique pattern of DNA fragments identified by SOUTHERN hybridization (using a probe that binds to a polymorphic region of DNA) or by polymerase chain reaction (PCR) using primers flanking the polymorphic region *meth gene* >>> DNA-amplified fingerprinting >>> Figure 48

DNA hybridization: Base pairing of DNA from two different sources; in biotechnology, a technique for selectively binding specific segments of single-stranded (ss) DNA or RNA by base pairing to complementary sequences of ssDNA molecules that are trapped on a nitrocellulose membrane *gene biot*

DNA library: >>> clone library

DNA ligase (polynucleotide ligase): An enzyme that creates a phosphodiester bond between the 5′-PO4 end of one polynucleotide and the 3′-OH end of another, thereby producing a single, larger polynucleotide *biot*

DNA methylation: The methylation of DNA bases by endogenic methylases; a biochemical process that is important for normal development in higher organisms; it involves the addition of a methyl group to the 5 position of the cytosine pyrimidine ring or the number 6 nitrogen of the adenine purine ring (cytosine and adenine are two of the four bases of DNA); this modification can be inherited through cell division; DNA methylation is a crucial part of normal development and cellular differentiation in higher organisms; DNA methylation stably alters the gene expression pattern in cells; it suppresses the expression of genes and other deleterious elements that have been incorporated into the genome of the host over time; DNA methylation also forms the basis of chromatin structure, which enables cells to form the myriad characteristics necessary for multicellular life from a single immutable sequence of DNA; DNA methylation includes >>> gene silencing, >>> gene imprinting, >>> paramutation, and biotic or abiotic stress factors *gene chem* >>> Figure 71 >>> Table 43

DNA micro arrays: A powerful, versatile, and economical molecular technique for screening of genetic aberrations; high-density gene sequences are printed onto glass slides; fluorephore-labeled genomic or complementary DNA (cDNA) is hybridized to slides with fixed signature patterns and resolved using computer-driven fluorescent images *biot*

DNA moving pictures: Spectacular images of intact chromosomal DNA molecules as long as 100 megabases *biot*

DNA packing: The highly organized way in which large amounts of DNA are packed into the cells of eukaryotic organisms *gene*

DNA polymerase: A group of enzymes mainly involved in copying a single-stranded DNA molecule to make its complementary strand; eukaryotic DNA polymerases participate in chromosomal replication, repair, crossing-over, and mitochondrial replication; to initiate replication, DNA polymerases require a priming RNA molecule; they extend the DNA using deoxyribonucleotide triphosphates (dNTP) as substrates and releasing pyrophosphates; the dNMPs are added to the 3′ OH end of the growing strand (i.e., DNA replication proceeds from 5′ to 3′ end) *gene phys*

DNA polymerase I (Kornberg enzyme): Used for nick translation and for the production of the >>> KLENOW fragment *biot* >>> Figures 52, 53, 54

DNA polymorphism: One of two or more alternate forms (alleles) of a chromosomal locus that differ in nucleotide sequence or have variable numbers of repeated nucleotide units *biot* >>> http://wheat.pw.usda.gov/ITMI/Repeats/index.shtml

DNA profile: The distinctive pattern of DNA restriction fragments or >>> PCR products that can be used to identify with great certainty an organism from the environment *biot*

DNA probe: A more or less defined piece of a DNA that is used for DNA-DNA or DNA-RNA hybridization experiments *gene meth*

DNA radical(s): Any one of a series of highly reactive radicals of purines, pyrimidines, or sugars in DNA, that are generated by UV irradiation or copper ions; DNA radicals can be detected by electron spin resonance since they are paramagnetic or immuno-spin trapping *chem*

DNA repair: The removal of damaged segments, e.g., pyrimidine dimers, from one strand of double-stranded DNA and its correct resynthesis; in general, the reconstruction of DNA molecule after different sorts of DNA-strand damages by endogenic enzymes; both normal metabolic activities and environmental factors such as UV light can cause DNA damage, resulting in as many as 1 million individual molecular lesions per cell per day, for example, in humans; many of these lesions cause structural damage to the DNA molecule and can alter or eliminate the cell's ability to transcribe the gene that the affected DNA encodes; other lesions induce potentially harmful mutations in the cell's genome, which affect the survival of its daughter cells after it undergoes mitosis; it is involved in the recombination process and promotes the survival of an organism after partial DNA damage; the DNA repair process must be constantly active so it can respond rapidly to any damage in the DNA structure; the rate of DNA repair is dependent on many factors, including the cell type, the age of the cell, and the extracellular environment; a cell that has accumulated a large amount of DNA damage, or one that no longer effectively repairs damage incurred by its DNA, can enter one of three possible states: an irreversible state of dormancy, known as >>> senescence, cell suicide, also known as apoptosis or programmed cell death, or unregulated cell division, which can lead to the formation of a tumor that is cancerous or deleterious *gene*

DNA replication: The process whereby a copy of a DNA molecule is made, and thus the genetic information it contains is duplicated; the parental double-stranded DNA molecule is replicated semi-conservatively (i.e., each copy contains one of the original strands paired with a newly synthesized strand that is complementary in terms of AT and GC base pairing) *gene*

DNase: Nuclease specific for DNA *biot* >>> deoxyribonuclease

DNase I: An >>> endonuclease that makes random single-stranded nicks in DNA; used for nick translation *biot* >>> Figures 52, 53, 54

DNA sequencing: Methods and procedures for determining the nucleotide sequence of a DNA fragment and/or chromosome *biot gene*

dn/ds ratio: In molecular phylogenetic studies, the ratio of the number of non-synonymous nucleotide substitutions to the number of synonymous nucleotide substitutions; in the case of functionally important (or otherwise constrained) genes, ds is expected to exceed dn (*dn/ds* < 1) because most amino acid changes will disrupt protein structure and those non-synonymous substitutions (*dn*) causing them will not be maintained; in a non-functional >>>

pseudogene, there will be no discrimination between them and equal numbers of dn and ds are expected ($dn/ds = 1$); when natural selection is acting to favor changes at the amino acid level, it is predicted that dn will exceed ds, hence a high dn/ds ratio; in classical loci, in the peptide-binding regions (allele-specific sequences) because of heterozygote advantage/frequency-dependent selection, there is always a high dn/ds ratio (>1), whereas in the remainder of the gene $dn/ds < 1$ (due to functional constraints); this suggests balancing selection is acting on peptide-binding regions *biot evol*

dodder: A parasitic seed plant (*Cuscuta* ssp.) without leaves, however a yellow filamentous vine *phyt*

9-dodecenyl acetate: A pheromone containing the (E) and (Z) isomers; it is used in manufacturing pheromone-based products to control moths; the pheromone is used in traps, dispensers, and sprays to help control destructive moths in forests and agricultural applications *fore agr phyt*

domain: Region of a protein with a distinct tertiary structure and characteristic activity *chem biot*

domain fusion: The combination of two or more naturally unrelated or synthetically produced sequences in a single DNA molecule that encode specific protein domains; the shuffling of domains generates new proteins, of which a minor fraction also has novel and/or improved functions *chem gene*

domain walking: An *in silico* approach to detect DNA sequences encoding flanking regions adjacent to a specific domain; the sequence of the latter represents the beginning of the search for adjacent sequences in the database *meth biot* >>> chromosome walking >>> *in silico* mapping

dome: In cereals, the zone of cells at the tip of the apical meristem (shoot apex), which, by cell division, forms the site for production of leaf and spikelet primordia *bot* >>> meristematic tip

domesticated populations: >>> domestication

domestication: The selective breeding by humans of species in order to accommodate human needs; there have been significant changes in plant characters, such as synchrony, free threshing, gigantism, increased harvest index, and altered genetic population *evol*

dominance: The quality of one of a pair of >>> alleles that completely suppresses the expression of the other member of the pair when both are present; the degree of dominance is expressed by the ratio of additive genetic variance to total phenotypic variance; in case the ratio equals 1, the trait shows complete dominance, if the ratio is greater than 1, the trait shows >>> overdominance, if the ratio is less than 1, the trait shows >>> incomplete dominance *gene* >>> Tables 2, 6, 9, 20, 21

dominance hypothesis: In hybrid breeding, dominant alleles of genes that should have stimulating effects on >>> heterosis, while recessive ones should show inhibitory effects *gene* >>> dominance >>> Figure 18

dominance–recessiveness relation: >>> dominant

dominance variance: That portion of the genetic variance attributable to dominant gene effects *stat* >>> dominance

dominant: In diploid organisms, a gene that produces the same phenotypic character when its alleles are present in a single dose (heterozygous) per nucleus, as it does in a double dose (homozygous); a gene that is masked in the presence of its dominant allele in the heterozygous state is said to be recessive to that dominant gene *gene* >>> dominance >>> Figures 6, 18 >>> Tables 6, 20, 21

dominant epistasis: One dominant factor A is epistatic of another factor B, or B is hypostatic to A *gene* >>> Table 6

dominant marker: Any molecular marker that does not discriminate between the alleles of a gene locus *gene biot*

dominant-negative mutation: A (heterozygous) dominant mutation on one allele blocking the activity of wild-type protein still encoded by the normal allele causing a loss-of-function phenotype; the phenotype is indistinguishable from that of homozygous dominant mutation *gene*

dominant species: The species with the greatest impact on both the biotic and abiotic components of its community *eco*

donator: >>> donor

donor: An individual, line, population, or variety from whom pollen or genetic material is used for transfer to another *meth* >>> Figure 51

donor parent: The parent from which one or a few genes are transferred to the recurrent parent in backcross breeding *meth*

donor plant: The source plant used for propagation, crossing, etc., whether a simple individual, an explant, graft, or cutting *meth*

dormancy: A resting condition with reduced metabolic rate found in non-germinating seeds and non-growing buds; in cereals, it can be artificially broken (1) by pre-drying seeds at +35 to +40°C for 5–7 d (with provision for circulation of the air); (2) by prechilling seeds at +10°C for 5 days; after the pretreatment seeds are germinated on blotters moistened with 500 ppm >>> gibberellic acid (GA3) for 7 d at +20°C (do not overwater the seeds), or (3) by heat treatment (the grains should be at low moisture content of about 15% or lower) at +40°C for 2–3 d; wheat dormancy is usually not expressed at low temperatures (<15°C) and does not cause problems for October–November emergence; >>> abscisic acid (ABA) is a plant hormone that also regulates seed dormancy and germination; seeds undergo changes in both ABA content and sensitivity during seed development and germination in response to internal and external cues; recent advances in functional genomics have revealed the integral components involved in ABA metabolism (biosynthesis and catabolism) and perception, the core signaling pathway, as well as the factors that trigger ABA-mediated transcription; these allow for comparative studies to be conducted on seeds under different environmental conditions and from different genetic backgrounds *phys*

dormant: >>> dormancy

dormant bud: >>> dormancy

dormant seeding: Sowing during late autumn or early winter after temperatures become too low for seed germination to occur until the following spring *meth agr*

dormant spray: A pesticide applied to dormant, leafless plants to control insects and diseases phyt

dorsal: In general, upon or relating to the back or outer surface of an organ (abaxial); the side of the caryopsis on which the embryo is situated *bot* >>> adaxial

dosage compensation: A genetic process that compensates for genes that exist in two doses in the homozygous dominants, so that the heterozygotes produce the same amount of gene product as the homozygotes *gene* >>> Table 36

dosage effect: The influence upon a phenotype of the number of times a genetic element is present *gene*

dose effect: >>> dosage effect

dot-blot analysis: A variant of the >>> SOUTHERN transfer; different concentrations of >>> RNA or >>> DNA may be determined; non-radioactive DNA will be denatured and with different concentrations transferred to nitrocellulose filters; only small dots are transferred; that DNA may hybridize with radioactively labeled probe; after autoradiography, the intensity of blackness is used as a simple measure of DNA concentration or DNA homology *meth biot* >>> Figure 48

double-blind (experiment): An experimental protocol whereby neither the experimental subjects nor the administrators know whether a substance is being administered; double-blind protocols are used to eliminate bias *meth stat*

double cropping: The more or less contemporary growing of two crops on the same field; for example, it might be to harvest a wheat crop by early summer and then plant maize or soybeans on that acreage for harvest in autumn; this practice is only possible in regions with long growing seasons *agr*

double cross: A cross between two F1 hybrids; the method used for producing hybrid seed; four different lines (A, B, C, D) are used; A × B = AB hybrid and C × D = CD hybrid; the single-cross hybrids (AB and CD) are then crossed and the double-cross hybrid (ABCD) seed is used for the commercial crop *meth seed* >>> Figures 31, 55

double-cross hybrids: Hybrids resulting from crossing two single cross hybrids *seed* >>> double cross >>> Figures 22, 31, 55

double crossing-over: The situation in which two crossing-overs take place within a tetrad *cyto gene*

double digestion: Cleavage of a DNA molecule with two different restriction enzymes *biot* >>> Table 39

double dig: A method of digging a garden bed which involves removing the soil to the depth of one spade blade and then digging down an equal distance, breaking up and mixing the soil *hort*

double ditelocentric: The phenomenon in which both arms of a certain chromosome are present as telocentrics, and each telocentric with its homologue *cyto* >>> Figure 37

double fertilization: The union of one sperm nucleus with the egg nucleus to form the diploid >>> zygote and of the other sperm nucleus with the two polar nuclei to form a >>> triploid endosperm nucleus; the male gametophyte, >>> pollen grain plus pollen tube, actually contains three sperm nuclei, but one, the vegetative nucleus, degenerates once double fertilization has been accomplished *cyto* >>> Figure 35

double flower: Flowers that have more than one row of petals; stamens and sometimes the pistils are transformed into petals or sometimes the petals split to form several more; completely double flowers have usually lost their reproductive organs and are therefore unable to produce seeds; these varieties of plants are bred to stay fresh longer than single flowers when cut; commonly, plants do not show double flowers; they can only be formed when three specific genes are simultaneously mutating; these genes are the main regulators for flower formation; their DNA sequences are almost identical; by induced mutations and subsequent combination of those genes double flowers can be induced (e.g., in *Arabidopsis*) *bot gene hort*

doubled haploid: A diploid plant, which results from spontaneous or induced chromosome doubling of a haploid cell or plant, usually after anther or microspore culture by using different means *biot* >>> *Hordeum bulbosum* procedure >>> *Imperata cylindrica* procedure >>> Figures 17, 26, 58 >>> Table 7

doubled-haploid lines (DH lines): Homozygous lines derived from haploidization and doubling again the chromosome number *biot meth* >>> doubled haploid (DH) method >>> *Hordeum bulbosum* procedure >>> *Imperata cylindrica* procedure >>> Figures 17, 26, 58 >>> Table 7

doubled haploid (DH) method: A method used to speed up the production of homozygotes and to decrease the population size for selection; in other words, generating haploid plants by parthenogenesis or by anther culture followed by doubling of the number of their chromosomes (spontaneously or induced) *biot meth* >>> doubled-haploid lines (DH lines) >>> *Hordeum bulbosum* procedure >>> *Imperata cylindrica* procedure >>> Figures 17, 26, 58 >>> Tables 7, 36

double helix: A structure of DNA consisting of two helices around a common axis *chem gene*

double heterozygote: An individual that is heterozygous at two loci under investigation *gene*

double hybrid: >>> double cross

double monoisosomic: The presence of two isochromosomes, one for each arm; it can derive from a double monotelosomic individual *cyto* >>> Figure 37

double monotelosomic: The presence of two telocentrics, one for each arm *cyto*

double mutant: An individual whose DNA has suffered from two independent mutations *gene*

double potting: A method of providing leaf support to ornamental plants; the method involves placing the pot inside a larger pot and filling around the inner pot with soil; done in this manner, the outer pot will provide support to the leaves *meth hort*

double recessive: An individual that is homozygous for a recessive allele (e.g., *aa*), as opposed to homozygous for a dominant allele (e.g., *AA*) *gene*

double reduction: The genetic outcome of chromatid segregation, as opposed to chromosome segregation, whereby two sister-chromatid segments are included in the same meiotic product *cyto*

double ridge: Describes the transition from the "vegetative" to the "floral" state in the shoot apex of cereals, such as wheat or barley; growth of the leaf primordium (ridge) is suppressed and the floral bud primordium (ridge) begins to grow out, giving a "double ridge" appearance to the dissected shoot apex when viewed under the microscope; at this time the plant apex no longer produces leaves but is "florally initiated" and only flower parts will subsequently develop *phys bot*

double telotrisomic: The presence of two telocentrics, one for each arm of a missing whole chromosome, but together with the complete homologue *cyto* >>> aneuploid >>> Figure 37

doubling time: The average time required to double the number of individuals of a population *gene*

dough quality: >>> "strong" flour

dough stage: One of the ripening stages when endosperm starts hardening, i.e., developmental stage of plant in which seeds are nearly mature and of dough-like consistency; it is followed by the >>> milk stage *phys meth* >>> Table 13

dough strength: >>> "strong" flour

down promoter mutations: Mutations that decrease the frequency of initiation of >>> transcription; they lead to the production of less >>> mRNA than is the case in the non-mutated state *biot*

downstream: A term used for description of the position of a DNA sequence within a DNA or protein molecule; it means that the position of the sequence lies toward the direction of the synthesis of a DNA or protein molecule *gene*

downy: Covered with soft hairs or down *bot*

downy mildew (of rape): The most common rapeseed disease (*Peronospora parasitica*); it causes yellow discoloration of the upper leaf surface and white fungal growth on the lower surface; though most severe in the autumn, the disease can reach high levels in cool wet springs; severe infections limit the ability of small plants to respond to good growing conditions *phyt*

dpa: >>> days post-anthesis

drain: Removal of water from the soil by artificial means (e.g., drainage ditches, buried perforated plastic pipes, or a gravel sump) *meth agr* >>> Table 44

drainage: Excess water can be harmful to crop production; wet soils are usually low in temperature and low in oxygen content; drainage can be facilitated by open ditches or different subsoil drains *agr* >>> Table 44

dressing: Manure, compost, or other fertilizers *agr*

drift: Changes in gene and genotypic frequencies in small populations due to random processes *gene*

drill: A machine for seeding crops by dropping them in rows and covering them with earth; the term is also used for a row of seeds deposited in the earth (i.e., the trench or channel in which the seeds are deposited); in general, to sow seeds in rows (i.e., the field was drilled, not sown broadcast) *agr*

drill-row: >>> drill

drip irrigation: A system of watering by which moisture runs through a porous hose; the water is slowly released through tiny holes or emitters to the plant roots; it is one of the most efficient of irrigation technologies *agr*

drip line: A line encircling a tree corresponding to the furthest extension of the branches of a tree *hort*

drip tip: A sharp elongated point on which atmospheric water accumulates from dews, mists, and rain until the water drop is large enough to fall off the tip to the ground around the roots of the plant; drip tips are common on the leaves of tropical trees in areas of high rainfall; the spines of some Cactaceae also function in this way *bot*

drive: Occurs when chromosome number in the gametes is greater than Mendelian expectations; in some cases, this can occur at meiosis, e.g., when B chromosomes can migrate to one pole of the spindle during the first anaphase and then pass preferentially into the nucleus that is destined to form the egg cell; alternatively, drive can occur in the first pollen grain mitosis, e.g., when the B-chromatids fail to separate and both pass into the generative nucleus *cyto* >>> B chromosomes >>> pollen grain mitosis

***Drosera*-scheme pairing:** A triploid hybrid showing 2× bivalent + 1× univalent configurations at diakinesis or metaphase I *cyto*

drought hardening: Adapting plants to survive periods of time with little or no water by stepwise reducing water supply or germinating and/or growing under insufficient moisture conditions *meth* >>> Table 33

drought resistance: The ability of a plant to withstand drought; this property can be very valuable in areas of uncertain rainfall, e.g., sorghum has greater drought resistance than maize, and is grown in many semi-arid areas for this reason *agr*

drought stress: >>> stress proteins

drought-tolerant: Plants that can survive periods of time with little or no water; it appears that seed and cereal crops are typically most drought-sensitive around the flowering stage; in peas, for example, the reduction in yield can be as much as 10% for a single so-called "stress day" during the flowering phase; a "stress day" is an expression of how much the actual transpiration has been reduced compared with the potential transpiration; it was found that the flowering, but thirsty plants had fewer grains, because the young seed ovules were aborted; this is due to the specific genes controlling the plant metabolism being down-regulated *agr* >>> Table 33

drupe: A fleshy fruit, such as a plum, cherry, coconut, walnut, peach, or olive, containing one or more seeds, each enclosed in a stony layer that is part of the fruit wall (hard endocarp) *bot*

dry deposition: Gases or particles in the atmosphere arriving at a surface without involvement of water, as opposed to wet deposition where rain or snow is involved *agr*

dry farming: A method of farming in arid and semiarid areas receiving less than 500 mm rainfall per year without using irrigation, the land being treated so as to conserve moisture; the technique consists of cultivating a given area in alternate years, allowing moisture to be stored in the fallow year; moisture losses are reduced by producing a mulch and removal of weeds *agr*

dry matter (DM): The substance in a plant or plant material remaining after oven drying to a constant weight at a temperature slightly above the boiling point of water *meth*

dry rot: A disease characterized by the formation of dry, shriveled lesions, often caused by the fungus *Phoma lingam phyt*

dry season: A period each year during which there is little precipitation *eco*

dry stigma: In dry stigmas, stigma exudates are absent on the stigmatic papillae, i.e., they lack free-flowing secretion; often they have intact papillar cells covered by primary cell wall, a waxy cuticle, and an extracellular protein layer *bot* >>> wet stigma

dry weight: Moisture-free weight *meth agr*

dsDNA: Double-stranded DNA *biot*

dsRNA: Double-stranded RNA; RNA with two strands instead of the typical one; dsRNAs longer than 30 nucleotides are normally referred as long dsRNAs; long dsRNA is the precursor of the >>> siRNA that can trigger >>> RNAi *biot chem*

3D tomography: A >>> LASER slicing technique was developed in order to explore the three-dimensional structure of small objects, such as roots, pollen grains, etc.; a nanosecond-pulse LASER slices many identically spaced, e.g., root samples, per second; it is done by placing the root on a moveable platform beneath the LASER, leaving a series of clear surface images, which can be combined with software to make a 3D rendering of the interior and exterior of the sample, this LASER tomography method is novel in that it provides high-contrast, full-color images without the use of contrast-enhancing agents; it allows nuanced compositional differences to be seen in the samples; additional benefits of the LASER tomography method are its speed, in the order of minutes, and that in most cases no preparation is required for the small biological specimens studied; the technology is being applied to biological specimens, as it stands out as a promising solution for obtaining information about anatomical features rapidly for phenotyping, understanding physiology, or gene expression *micr meth*

duff: Partially decomposed organic matter lying beneath the litter layer and above the mineral soil; it includes the fermentation and humus layers of the forest floor *agr fore*

dual-specificity chimeric: An artificial gene made by combining parts of two genes to achieve two specificities within one gene *biot*

duftmale: An area of a petal that secretes scent *bot*

Dutch elm disease: A fungus spread by bark beetles that causes wilting and dieback on elms; the causative agents are ascomycete microfungi; three species are recognized (*Ophiostoma ulmi*, which afflicted Europe in 1910, reaching North America on imported timber in 1928, *Ophiostoma himal-ulmi*, a species endemic to the western Himalayas, and the extremely virulent species, *Ophiostoma novo-ulmi*, which was first described in Europe and North America in the 1940s and has devastated elms in both areas since the late 1960s); the disease is spread in North America by two species of bark beetles (Curculionidae) *phyt* >>> Figure 62

dummy trial: >>> blindfold (trial)

dummy variable: An artificial variable used to represent the level of a qualitative variable in a multiple regression analysis *stat*

duplex type: A polyploid plant that shows two dominant alleles for a given locus *gene* >>> nulliplex type >>> autotetraploid >>> Table 3

duplicate gene(s): Two or more pairs of genes in one diploid individual, which alone or together produce identical effects *gene* >>> Figure 36

duplication: A chromosomal aberration in which more than one copy of a particular chromosomal segment is produced within a chromosome set *gene* >>> Figure 36

duplication of germplasm: A duplicated seed sample that is prepared for safety reasons; usually the two samples are kept at different locations *seed*

durable resistance: Resistance that remains effective and stable during the agronomic life of a crop; this type of resistance is usually determined by several genes and is a main target of resistance breeding *phyt*

DUS testing: The methods and standards of the identification and description of varieties are elaborated by different teams of an international organization, the >>> Union for Protection of New Varieties of Plants (UPOV) established for the protection of plant varieties; the so-called TG/01/2/1/ contains the most important prescriptions; these procedures, prescriptions, and methods are called DUS testing (D = distinctness, U = uniformity, S = stability); for the legal protection of varieties, the candidates (1) have to have names that have not yet been used for the registration of recognized varieties, (2) must have been discovered distinct of the other "known" varieties (D), (3) have to be uniform and homogeneous (U), and (4) have to be stable (S); the characteristics involved in DUS testing can be divided into two main types: (1) measured characteristics and (2) visually observed (bonitated, qualitative) characteristics; the measured characteristics have values on the continuous scale; visually observed data have values in the interval 1, 2 to 9; in other words, visually observed traits are of the so-called ordinal type, while measured data are on the so-called interval or ratio scale; phenological data must also be mentioned as a special type of measured characteristic (e.g., number of days until flowering, etc.) *seed*

dust: Pesticide formulation in dry, finely divided form (with particle size less than 30 μm) designed for application as a dry dressing without further preparation or dilution; a method of pesticide control in which a dry substance is applied by spraying seeds, tubers, or >>> bulbs *phyt*

dust mulch: A loose, dry surface layer of a soil under cultivation *agr*

duster: A device consisting of a bin, a wand with a nozzle, and a crank mechanism; it is used for applying dust or powder to plants *phyt meth*

dusting: >>> dust

dust treater: The amount of seed that flows into the weigh pan, which is just beneath the feed hopper on top of the treater, is controlled by opening or closing the gates of the seed hopper by means of the hand wheel on the side of the hopper; the scale on the hopper shows how far gates are open in inches; gates should be open to whatever number of inches it takes to keep the weight pan filled to the required number of pounds per dump as it tilts in either direction; the number of pounds per dump is adjusted by correctly setting the counterweight up or down on the counterweight arm; to ascertain that the correct amount of power is being applied to the seed flow, a preliminary test must be carried out in which a given number of pounds of seed is run through the feeder; during this run, the measuring cup provided with the feeder should be used to catch the powder as it comes off the vibrator; after the given amount of seed has run through, the powder should be weighed so as to determine how much is being applied to that amount of seed; the vibrator speed can then be adjusted accordingly *seed* >>> seed treater

dwarf fruit trees: A small fruit tree reaching a height at maturity of 150–200 cm; bred for convenient harvest technology as well as bearing early and normal fruits; dwarf varieties of apple trees have become popular among growers during the past three decades; while normal-sized trees commonly grow to a height of 5–6 m, dwarf varieties grow only 3–4 m tall; through genetic mapping, molecular markers were found for the dwarfing characteristic in apple rootstocks *meth hort* >>> Table 33

dwarfing gene: One category of genes that control the height of the plant; according to MULTANI et al. (Science 302: 81–84, 2003), compact stalks of maize and dwarf sorghum mutants are caused by a loss of an auxin transporter class of P-glycoproteins (multidrug resistant); while not currently used as a source of height reduction in maize, *br2* may provide an agronomic advantage to dwarf maize; interestingly, *dw3* of sorghum was found to have a direct repeat in one of its exons (exon V); occasional unequal crossing-over in this exon produces mutant reversions (tall plants); the *dw3* locus has often been used in elite sorghum lines and the occasional tall plant has, until now, been something of a mystery *gene* >>> *Rht* gene >>> Table 33

dyad: A pair of cells (i.e., one of the products of the disjunction of the tetrads at the first meiotic division, contained in the nuclei of secondary gametocytes) *cyto*

dyeing flowers: Blossoms capable of producing dyes *hort*

dysgenic: >>> eugenic

dysploid: A plant or species in which the chromosome number is more or less than the expected normal euploid number *cyto*

dysploidy: Abnormal ploidy (e.g., the appearance of >>> diploid or >>> triploid individuals in a normally tetraploid population or of triploid and tetraploid ones in a normally diploid population) *cyto* >>> anisoploid

E

E1, E2, E3 ...: First, second and third generation following irradiation with X-rays *meth*

ear: >>> spike

ear-bed method: >>> spike-bed method

ear lifter (on a harvester): Improves cutting efficiency with laid crops or overhanging ears *agr*

early-generation selection or test(ing) (EGT): Selection schemes in which poor recombinants are discarded already in F2 and F3 generations; despite widespread use of early-generation testing and selection for self-pollinated crops, its effectiveness remains largely an unresolved issue; applying selection response at one or more early generations relative to the response to direct selection at homozygosity and the probability of retaining superior lines selected during EGT, the selection response to EGT was analyzed for a model with quadratic genetic components due to additive, dominance, additive × additive, and linkage effects in selfed populations derived from a cross between two inbreds; the response to one cycle of EGT is less than the response to direct selection and decreases with non-additive effects, repulsion linkage and reduced heritabilities; the cumulative response to two or more cycles of EGT is greater than the response to direct selection unless there are strong non-additive effects, strong repulsion linkage, and low heritabilities; the probability of retaining a superior line at EGT decreases with increased non-additive effects and low heritabilities; the proportion of lines needed to minimize the risk of erroneously culling superior lines increases with increased non-additive effects and low heritabilities; thus, EGT should be used for populations or traits with little non-additive effect, coupling linkage, and high heritability *meth stat* >>> Table 57

early milk (stadium): Description of the young cereal grain; when punctured the grain contents start to look milky *phys bot*

early selection: It is conducted on highly heterozygous individuals which become the parents of the next screening generation *meth* >>> late selection >>> Table 57

early wood: An annual ring of secondary xylem formed early in the growing season, with relatively large, thin-walled cells compared to cells formed late in the growing season *fore bot*

ear rot (in maize): Caused by different *Fusarium* spp., ear rots are one of the most dangerous food and feed safety challenges in >>> maize production; at present, the majority of the >>> inbreds and hybrids are susceptible; *Gibberella* and *Fusarium* ear rots (caused by *F. graminearum* and *F. verticillioides*, respectively) are the two main diseases, but more than ten further *Fusarium* spp. cause ear rots; natural infection is initiated by a mixture of the local *Fusarium* spp., but usually one species predominates; breeders rely on natural infection to create sufficient levels of disease severity for selection-resistant genotypes; significant differences in genotypic resistance after >>> inoculation exist; resistance to the two major modes of fungal entry into the ear, via the silk or through kernel wounds, is not correlated in all genotypes; both native and exotic sources of resistance are under investigation; inbreds differ in general and specific combining ability for ear rot resistance; the >>> expression of resistance to disease severity and resistance to toxins is often used as synonyms; higher resistance to visual disease severities mostly results in lower toxin contamination; the mode of inheritance of resistance appears to differ: additive, possibly non-additive effects, digenic (dominant), and polygenic patterns have been identified *phyt*

earsh: >>> stubble

ear shank: The part of the ear (>>> cob) that is attached to the stalk of the maize plant; it is a transport tissue for water and nutrients from the plant to the ear *bot*

ear-to-row planting: >>> ear-to-row selection

ear-to-row selection: A separate growing of progenies (i.e., the separate sowing of lines or families); a procedure developed by the German breeder T. ROEMER (*syn* Ohio method) *meth* >>> Illinois method

EC50: The median effective concentration (ppm or ppb) of the toxicant in the environment (usually water) that produces a designated effect in 50% of the test organisms exposed *phyt*

ecad: A plant type which has resulted from adaptation to the selective effect of environment *phys*

eccDNA: >>> extrachromosomal circular DNA

echinate: Having sharply pointed spines *bot* >>> prickly

eclectic species concept: A philosophy that species are defined and formed and maintained by a variety of morphological, interbreeding, ecological, and phylogenetic factors *eco* >>> ecological species concept >>> nominalistic species concept

eclipsed antigen: Antigen borne by the parasite that is common to both the host and the parasite but which genetically is of parasitic origin *phyt*

eclipse period: A phase or period in the developmental cycle of viruses, occurring immediately after infection, in which infective particles cannot be detected *phyt*

ecocline: A series of biotypes within a species that shows a genetic gradient correlated with a gradual environmental gradient *eco gene*

ecodeme: A population or gene pool adapted to a particular environment *eco*

ecological dominance: The state in plant communities in which one or more species, by their size, number, or coverage, exert considerable influence or control over the other species *eco*

ecological farming: A farming system that aims to develop an integrated, humane, and environmentally and economically sustainable agricultural production system *agr eco*

ecological genetics: The analysis of genetics of natural populations and of the adaptations of them to the environment *eco*

ecological longevity: Average length of life of individuals of a population under stated conditions *eco*

ecological niche: The position occupied by a plant in its community with reference both to its utilization of its environment and its required associations with other organisms *eco*

ecological race: A group of local populations within a species in which individuals have similar environmental tolerances; wide-ranging species may consist of many ecological races *eco gene*

ecological species concept: A philosophy that ecological constraints are the primary factor in forming and maintaining a species *eco* >>> eclectic species concept >>> nominalistic species concept

ecology: The study of the interrelationships between individual organisms and between organisms and their environment *eco*

economic trait loci (ECL): Sites in the genome that determine characteristics of economic importance *gene*

ecophene: The range of different phenotypes produced by one genotype within a certain environment *eco*

ecospecies: A locally adapted species; it shows minor changes of morphology and physiology compared to another species, which are related to a habitat and are genetically determined *eco*

ecosystem: The complex of an ecological community together with a biological component of the environment, which function together as a stable system *eco*

ecosystem service: Environmental benefits which sustain and enhance human activities and the general environment *eco*

EcoTILLING: A high-throughput, low-cost technique for rapid discovery of polymorphisms in natural populations of plants; it is similar to TILLING but it differs from TILLING in that natural polymorphisms are detected rather than polymorphisms induced through chemical mutagenesis; on the other hand, single nucleotide polymorphisms (>>> SNPs), small insertions and deletions, and variations in microsatellite repeat number can be efficiently detected using the EcoTILLING technique *meth biot* >>> targeting-induced local lesions >>> TAL effector nuclease

ecotype: A locally adapted population of a widespread species; it shows minor changes of morphology and physiology that are related to a habitat and are genetically determined; the individuals of an ecotype are only uniform in regard to the traits that provide them with special adaptation to specific environments; all the other characters may vary; ecotypes may be found in perennial clover, alfalfa, grasses, or other forage crops *eco*

ecovalence: This parameter is a quantitative measure for the evaluation of ecological adaptability *eco*

Ecovar™ (ecological variety): A seed source of a plant species that can be licensed and that is the result of merging plant collections from a diversity of populations and environments within an ecozone with the objective of providing a diverse commercial seed source *agr*

ectocarp: An outermost layer of pericarp *bot*

ectochimera: >>> chimera >>> Figure 56

ectoparasite: An external parasite *phyt* >>> opuntia

ectopic expression: Expression of a gene out of its expected time or place *gene biot*

ectopic pairing: Chromosome pairing between an allele and non-allelic (ectopic) homologous sequences due to some homology searching mechanism *cyto*

ectosite: >>> ectoparasite

ectozoon: >>> ectoparasite

ED50: The median effective dose of a pesticide, expressed as mg/kg of body weight, which produces a designated effect in 50% of the test organism exposed *phyt*

edaphon: All (micro)organisms living in the soil close to the plant *agr* >>> Table 44

eddish: >>> stubble

edema: Swelling or blistering on leaves or other plant organs under conditions of high moisture and restricted transpiration *phys*

edge effect: The phenomenon of an edge community, or ecotone, having greater ecological diversity than the neighboring communities *eco*

edging: A row of plants set along the border of a plot or flower bed *agr meth*

EDTA: >>> ethylenediamine tetra acetic acid

eelworm: In cereals, there are two species of eelworm that attack plants, the cereal cyst nematode (*Heterodera avenae*), which can infest cereal crops, and the oat stem eelworm (*Ditylenchus dipsaci*), which attacks oats and rye, as well as several other crops and weeds *phyt* >>> nematode

effective breeding population (size): In general, the size of a population, which is adjusted mathematically to permit comparisons with others; more specifically, it is the size of an equivalent ideal population, which is expected to experience the same increase in homozygosity over time (i.e., drift) as the population in question; the ideal population is one in which mating is at random in the absence of selection and in which all individuals have the same expected contribution to the next generation *meth* >>> Figure 51

effective population size (N or N_e): Number of individuals contributing "unique" chromosomes to the next generation (Nf = number of mothers in a population; relevant in the calculation of number of generations for the fixation of a mitochondrial allele); it is always less than or equal to the actual population size; inbreeding effectively reduces Ne because of the identity (not unique) of most chromosomes in the population *stat meth*

effective seedling: Any seedling that has survived in reasonable vigor for some arbitrary time and is so sited that it should make an effective contribution to the crop *agr fore*

efficiency of plating: Number of plaques formed by a phage lysate; it can be different for different hosts *biot*

efflorescence: The time or state of flowering; in general, blooming or flowering *bot* >>> anthesis

EFSA: >>> European Food Safety Authority

egg: A female gamete or germ (egg) cell *bot* >>> Figure 25

egg apparatus: The structure containing the embryo sac, within which is the egg cell *bot* >>> Figure 25

egg cell: >>> egg

egg mother cell: A megasporocyte from which an egg cell is derived during megasporogenesis *bot*

EGT: >>> early generation test(ing)

ektexine: Outer exine wall of the pollen *bot*

elaiophore: An oil-secreting gland on a flower *bot*

elaiosome: A seed appendage (gelatinous projection) on some plants (e.g., *Viola* ssp., *Helleborus* spp.) that contains oily substances attractive to ants; ants often aid in seed dispersal when these appendages are present *bot*

electric-aided pollination: In Brussels sprouts and Savoy cabbage the sporophytic incompatibility barrier can be broken by applying a direct electric potential difference of 100 V between pollen and stigma during >>> pollination; the effect, measured as the number of seeds per flower, depends on the kind of plant and the degree of incompatibility; the mechanism underlying this phenomenon remains unclear *meth*

electrical conductivity test: Tests are used specifically for certain crops (e.g., pea); this test measures the soluble salts, sugars, and amino acids that leach into the soak water; the electrical properties of the soak water are measured by a conductivity meter (expressed as micro Siemens, μS); the higher the leachate, the lower the vigor; this test is particularly useful in detecting mechanical and frost damages of cells *seed*

electrical field fusion: >>> cell fusion

electrofusion: >>> cell fusion

electromorphs: Allozymes that can be distinguished by electrophoresis *phys chem*

electron acceptor: A substance that accepts electrons during an oxidation-reduction reaction, i.e., an oxidant *chem*

electron donor: A compound that donates electrons in an oxidation-reduction reaction; i.e., a reductant *chem*

electronic single nucleotide polymorphism: Any >>> SNP derived from >>> expressed sequence tag (EST) database by data mining; the EST sequences from many different individuals of plant or animal species aligned and the sequence is screened for SNPs; the SNPs are therefore discovered solely *in silico meth biot* >>> SNP

electron microscope: A microscope that permits magnification of particles up to 200,000 diameters; instead of having the specimen exposed to a light source, a stream of electrons is directed on the object; the higher resolving power of the electron microscope is largely the result of the shorter wavelength associated with electrons; the electrons are accelerated in a high vacuum through electromagnetic lenses and focused on the specimen; they are projected on a fluorescent screen where the image of the particle may be viewed, or onto a photographic plate and/or film *micr*

electropermeabilization: >>> electroporation

electrophoresis: The migration of charged particles under the influence of an electric field within a stationary liquid; the latter may be a normal solution or held upon a porous medium such as starch acrylamide gel or cellulose acetate *meth*

electrophoretic karyotypes: Refers to descriptions of chromosome number and size produced by pulse-field gel electrophoresis in species with chromosomes that are too small to be seen by light microscopy (e.g., fungi) *cyto meth*

electroporation: The application of a short sharp electrical shock to protoplasts in order to force the incorporation or uptake of DNA by producing transient holes in the cellulose membrane *biot*

electrostatic separator: A machine that separates seeds on the basis of their ability to accept and retain an electrical charge *seed*

elemental mobility: Relative measure of how efficiently elements move through plant tissue; mobility of elements is classified from very mobile to very immobile; the mobility of specific elements will affect which portions show symptoms of a deficiency for that element; thus, a deficiency of a very mobile element will, first, exhibit symptoms in older growth, while a deficiency of very immobile elements will, first, exhibit symptoms in the newer growth *phys*

elemental pesticide: Any pesticide in which the active ingredient is a natural element, e.g., sulfur which is used to treat various fungi; it is in contrast with organic and synthetic pesticides *phyt*

elicitor: A molecule produced by the pathogen host (or pathogen) that induces a response by the pathogen (or host) *phyt*

elimination: >>> culling

ELISA: >>> enzyme-linked immunosorbent assay

elite: High-grade seed used for both seed and ware production; or, an agronomically superior and high-performing local cultivar *seed* >>> base seed >>> Figure 59

elite experimental line: A crop line that is very close to release as a cultivar *meth*

elite germplasm: Germplasm that is adapted (i.e., selectively bred and optimized to new environment); for example, maize, which is native to Mexico, has been adapted to many locations in the world *gene seed*

elite hybrid variety: >>> elite

elite plants: >>> elite

elite population: Genetically advanced intensively managed population in a short-term breeding program; sometimes used synonymously with nucleus population *meth* >>> Figures 51, 59

elite seed: The class of pedigreed seed corresponding to the foundation seed class in some countries *seed* >>> base seed >>> Figure 59

elite strains: >>> elite

elite tree: A tree that has been shown by progeny testing to produce superior offspring *meth hort fore*

elliptic: Arching margins of leaf, pointed at both ends, about two times longer than wide *bot*

elongate: Narrow, the length several times the width or thickness *bot*

elongation step: The last temperature step in >>> PCR procedure; during this step a >>> DNA polymerase adds nucleotides to the growing DNA strand *biot*

Elston index: >>> index selection

eluate: A liquid solution resulting from eluting *prep*

emarginate: Slightly notched apex *bot*

emasculate: To remove the anthers from a bud or flower before pollen is shed; a normal preliminary step in crossing to prevent self-pollination; there are basically two methods used for emasculation: individual emasculation and mass emasculation; mass emasculation is applied, for example, in monoecious maize by mechanical detasseling, in hemp by removing male plants from the dioecious population, in rice by spike treatment with high temperatures (warm water of about +42°C), in wheat by using gametocides, or in tobacco by dark-phase treatment during anthesis *meth*

emasculation: >>> emasculate

embedded: Appearing or growing as part of another structure *bot*

EMBL3 vector: A replacement vector for the cloning of large (~20 kb) DNA fragments; it derives from phage lamda *biot* >>> lamda phage

embryo: The rudimentary plant within a seed that arises from the >>> zygote, sometimes from an unfertilized >>> egg cell, or from progressive differentiation in cell culture *bot* >>> Table 8

embryo culture: A method of inducing the artificial growth of embryos by excising the young embryos under septic conditions and placing them on suitable nutrient media; the method is often applied when postgamous incompatibility exists (e.g., in wide crosses) *biot* >>> Table 47

embryogenesis: The formation of an embryo; after double fertilization, seed development begins with endosperm, embryo growth, and differentiation; during subsequent seed maturation, the seed storage proteins, starch, and lipids accumulate before the seed undergoes desiccation and dormancy; during late embryogenesis endosperm and mature embryo express seed storage reserve genes; many other hydrophilic proteins with certain functions that share similar sequence repeats also are accumulated; they are called "late embryogenesis-abundant proteins", protecting cells during seed desiccation (e.g., abscisic acid) *bot* >>> Table 8

embryogenic: Related to or like an embryo *bot*

embryoid: An embryolike structure *bot biot*

embryonic axis: The major part of the embryo, consisting of the shoot, >>> mesocotyl, and the root *bot*

embryonic callus: Callus cultures that under suitable conditions are capable of producing embryos; conversely callus cultures that have lost this capability are termed "non-embryonic" *biot* >>> Table 47

embryonic leaves: Initial leaves, which are easily discernible in the germ of a mature grain *bot*

embryo percent: The amount of embryo compared with endosperm and other seed parts, and/or the percent of embryo in the whole seed *agr meth*

embryo rescue: When cross-pollination is made or occurs between genetically widely different plants, the resulting embryo may be aborted because of parental mutual incompatibility; such embryos may be excised during an early stage and grown on a congenial medium such as nutrient agar *biot* >>> embryo culture >>> Table 47

embryo sac: The female gametophyte is formed by the division of the haploid megaspore nucleus – the site of fertilization of the egg and development of the embryo *bot* >>> Figures 25, 28, 35

embryo sac development: >>> embryo sac

embryo sac mother cell: >>> embryo sac

EMC: >>> embryo sac mother cell >>> embryo sac

emergence: Germination of a seed and the appearance of the shoot, or the time when the first leaves of the crop plant come through the ground *bot*

emergence date: The time or date when plants start to emerge spikes, buds, flowers, or showing coleoptiles, first leaves, etc. *agr meth*

empirical distribution: The cumulative distribution function of a sample of size n, denoted Fn(x); it is 1/n × x (the number of values in the sample less than or equal to x) *stat*

empty fruits: In seed production, a fruit without a seed; for example, it happens in the composite family (sunflower, chicory, lettuce, etc.); the so-called seed is a fruit (achene) of which the outer structures are comparable to the pod in the legumes; the true seed may or may not form inside the fruit, which is not (clearly) visible from outside *seed*

empty glume: >>> spikelet glume

enabling technologies: Technical processes that allow genetic modifications in organisms; this includes the ability to activate genes specifically or to influence them in their activity or the use of marker genes *biot*

enantiomer: Mirror image form of a molecule; there may be more than two, as several parts of the molecule may have a mirror image form *chem phys*

enation: An outgrowth from the surface of a leaf or other plant part *bot*

encapsidation (of viruses): The process when a virus forms the protein shell that surrounds the virus nucleic acid *phyt*

encapsulation: The process of enclosing fragile organic material in a protective casing, sometimes of a semisolid nature; it is used for planting or moving somatic embryos *seed meth biot*

3′ end: Describes the different, complementary ends of a DNA single strand; ends with an OH-group *biot gene*

5′ end: Describes the different, complementary ends of a DNA single strand; ends with a phosphate group *biot gene*

endangered species: A species on the verge of extinction *eco*

endarch: Primary xylem in which the development is from the center towards the outside; typical of stems *bot* >>> exarch

endemic: Confined to a given geographic region; diseases continuously occurring in a particular area *eco phyt*

endemic species: >>> endemic

endexine: Inner exine wall of pollen *bot*

end-filling: Conversion of a sticky end (5′ overhang only) to a blunt end by enzymatic synthesis of the complement to the single-stranded overhang *biot*

end-labeling: Attaching a radioactive or non-radioactive label specifically to the ends of a DNA molecule *biot*

endocarp: Inner layer of the fruit wall (pericarp); it may be hard and stony (peach pit), membranous (apple core), or fleshy (orange pulp) *bot*

endochimera: >>> chimera >>> Figure 56

endocytosis: The uptake of cellular material through the cell membrane by formation of a vacuole *bot*

endodermis: A specialized tissue in the roots and stems of vascular plants, composed of a single layer of modified parenchyma cells forming the inner boundary of the cortex *bot*

endoduplication: The doubling of the haploid chromosome complement owing to the failure of cell-wall formation *cyto*

endogamy: Sexual reproduction in which the mating partners are more or less closely related *bot* >>> inbreeding >>> intrabreeding

endogenous rhythm: A type of rhythmic plant response or growth capacity that is not affected by external stimuli *phys*

endomitosis: A doubling of the chromosomes within a nucleus that does not divide, thus producing a polyploid; the doubling may be repeated a number of times, which leads to endopolyploidy *cyto* >>> endopolyploid >>> C value

endonuclease: Any >>> enzyme that cuts DNA at specific sites corresponding to specific base sequences within a polynucleotide chain; each endonuclease cuts at its own specific site; it is used to identify genomes, genotypes, and chromosomes, and it acts as a genetic marker *chem phys* >>> exonuclease >>> Table 39

endophyte: (Micro)organisms living inside host plants; seedborne, non-pathogenic, fungal endophytes are commonly found in symbiotic relationships with many members of the cool-season grass subfamily Poideae; there are beneficial effects on plants possessing fungal endophytes, and detrimental effects on consumers of fungal endophyte-infected plants; for example, fungi of the genus Neotyphodium were found in the diploid *Triticum* species (*T. dichasians*), a second endophyte, an *Acremonium* species, can be found in *T. columnare, T. cylindricum, T. monococcum*, etc. *phyt*

endoplasm: The granular central material of the cytoplasm *bot*

endoplasmic reticulum (ER) A system of minute tubules within the cytoplasm; two types are recognized: rough and smooth; these are particularly concerned with pathways of protein and steroid synthesis *bot*

endopolyploid: Diploid individuals whose cells contain 4C, 8C, 16C, 32C, etc., amounts of DNA in their nuclei *gene cyto* >>> endomitosis >>> endoduplication >>> DNA content

endoreduplication: Chromosome reduplication in the >>> interphase of the mitotic cell cycle *gene cyto* >>> C value

endosome: A nucleolus-like organelle that does not disappear during mitosis *cyto*

endosperm: In grasses, the reserve food material in the caryopsis lying outside the embryo; it is a starchy tissue that is formed during grain development; it provides nourishment for the developing embryo and for the seedling after germination until it can establish itself; usually the endosperm is triploid and originates during the double fertilization when one of the two sperm nuclei is fused with the two >>> polar nuclei *bot* >>> Tables 8, 16

endosperm balance number hypothesis Normal endosperm and seed development occurs if the endosperm tissue results from a fusion of maternal and paternal nuclei in a ratio of 2 maternal: 1 paternal *cyto*

endosperm texture: The tendency of, for example, wheat endosperm to fracture either along the outlines of the cells (hard) or across the cells in a more irregular way (soft); it can be assessed by "grinding time" or by inspection of the grain cross-section *bot meth*

endospory: A condition in which the gametophyte develops within the spore wall, rather than externally *bot*

endotoxin: A poison produced within a cell and released only when the cell disintegrates *phys*

endozoochorous: Dispersed within animals, i.e., eaten of a seed or fruit *bot*

engraft: >>> ingraft

enhancer: A modifier gene that may enhance the action of another gene *gene*

enhancer mutation: Any mutation occurring in an enhancer sequence; enhancer mutations (deletions or point mutations) can be detected by their effect(s) on the transcription of associated genes *gene*

enhancer single nucleotide polymorphism: Any polymorphism between two genomes that is based on a single nucleotide exchange, small deletions, or insertions within the enhancer sequence; it can be neutral but may also effectively prevent the binding of activator proteins; in consequence, the corresponding gene is less efficiently transcribed *biot meth* >>> SNP

enhancer SNP: >>> enhancer single nucleotide polymorphism

enneaploid: A polyploid plant with nine chromosome sets *cyto*

entire leaf: A leaf margin that has no teeth (i.e., smooth) *bot*

entity: Something that exists as a particular and discrete unit, e.g., persons and corporations are equivalent entities under the law; in general, the fact of existence (being) or the existence of something considered apart from its properties *gene*

entomologist: An insect specialist *phyt*

entomology: The branch of zoology dealing with insects *phyt*

entomophilous: Insect-borne pollen; entomophilous pollen is usually characterized by a sticky, mainly liquid layer that fills in the interstices of the sculpted pollen wall, or exine, and is responsible for its adhesion to insect and other vectors; this liquid pollen coat, the tryphine, is also responsible for pollen adhesion to the stigmatic surface in many species *bot* >>> Table 35

enucleate (of cells): Removing the nucleus from a cell *cyto biot*

environment: The sum of biotic (diseases, pests, weeds, animals) and abiotic (rainfall, temperature, nutrients, water, light, wind, growing space, soil characters) factors that surround and influence an organism *eco*

environmental mutagen: A substance that may act as a mutagen in the environment of an organism *gene eco*

environmental resistance: All genetically based characteristics that protect the reproduction of plants against negative influences of the environment *eco*

environmental variance (EV): That portion of the phenotypic variance caused by differences in the environments to which the individuals in a population have been exposed *stat*

enzyme: A molecule, wholly or largely protein, produced by a living cell that acts as a biological catalyst of chemical reactions; in agriculture, enzyme preparations have been reported to improve the utilization of feeds for ruminant animals; new enzyme preparations are constantly being developed; biological silage inoculants frequently contain enzymes in addition to lactic acid bacteria; the enzymes in such products partially breakdown some of the cell wall components of the plant material to be ensiled into soluble sugars; these liberated sugars are then metabolized by the natural or applied lactic acid bacteria such as *Lactobacilli* or *Pediococci* into lactic acid which reduces the pH and so ensiles the crop; the use of enzymes in arable agriculture especially in the processing of some major crops and in waste disposal systems are areas which have not been fully investigated *phys agr*

enzyme-linked immunosorbent assay (ELISA): A highly sensitive immuno assay based on enzyme reactions; usually two antibodies are used; the first antibody binds to an antigen and mutates itself toward the second antibody, the antiglobin; the antiglobin is linked to an enzyme (e.g., horseradish peroxidase); the enzyme activity has to be easily detected (e.g., by a staining reaction); the degree of enzyme reaction quantitatively shows how many anti-bodies of the first type (i.e., of the original antigen) are present *meth*

e.o.p.: >>> efficiency of plating

ephemeral: Short-lived, temporary *bio*

epialleles: Alternative states of a gene that have an identical DNA sequence but differ in methyla-tion or chromatin structure and, hence, level of expression *gene* >>> Table 43

epibiotic plant: >>> allogenous flora

epiblast: The primordial outer layer of a young embryo; that is, a small flap of tissue on the side of the grass embryo axis opposite the scutellum *bot*

epicarp: The outermost layer of a pericarp, as the rind or peel of certain fruits *bot*

epicormic branching (or sprouting): Branches which grow out of the main stem of a tree, arising from buds under the bark; epicormic branching may increase knottiness *fore*

epicotyl: The portion of the embryo or seedling above the cotyledons *bot*

epidemic: A rapid increase in disease over time and in a defined space *phyt*

epidemic potential: The biological capacity of a pathogen to cause disease in a particular environ-ment *phyt*

epidemic rate: The amount of increase of disease in a plant population per unit of time *phyt*

epidemiology: The study of disease epidemics, with an effort to trace down the cause *phyt*

epidermal-strip test: An electron microscope technique by which a virus can be quickly examined in crude-sap extracts *cyto meth phyt*

epidermis: The outermost layer of cells; it protects them against drying, mechanical injury, or pathogens *bot*

epigeal: Seed germination in which the cotyledons are raised above the ground by elongation of the hypocotyl (e.g., bean) *bot*

epigeal germination: Characterizes a type of germination in which the >>> cotyledons are raised above the ground by elongation of the >>> hypocotyl *bot*

epigenesis: The concept that an organism develops by the new appearance of structure and func-tions, as opposed to the hypothesis that an organism develops by the unfolding and growth of entities already present in the egg at the beginning of development *bio*

epigenetic: Reversible, non-hereditary variation that may be the result of changes in gene expres-sion *gene*

epigenetic code: The specific distribution of methylated cytosines along the DNA of a chro-mosome and/or the specific side-chain modification of histones in the chromatin of the chromosome; since both the cytosine methylation patterns as well as histone side-chain modifications in a specific region of the genome vary with time, so does the epigenetic code *cyto biot*

epigenetic gene silencing: The switch-off of a gene by a mechanism that does not Involve changes in the underlying DNA sequence of the plant *biot meth*

epigenetic regulation: Regulation of gene activity mediated by a reversible change in DNA modi-fication or chromatin structure *gene*

epigenetic trait: A distinguishable feature, characteristic, quality of character, or phenotypic feature of a developing or developed individual that has arisen as a result of mechanisms regulating the expression of the genes rather than differences in the gene sequence *gene*

epigenetics: The study of the mechanisms by which genes bring about their phenotypic effects, i.e., the study of the heritable or also acquired, non-hereditable differences in gene expression patterns not caused by changes in the primary sequence of the DNA, but rather on changes in DNA methylation and chromatin modifications; the fact that the functioning of genes is also affected by epigenetic marks has been known for decades, e.g., the nucleotide cytosine can be changed into a methylcytosine; this cytosine methylation, which is one type of epigenetic mark, is typically associated with repression of gene activity; epigenic marks can pass on for many generations in a stable manner *gene*

epigenome: The complete set of genes involved in genetic imprinting, or the pattern of methylated cytosines and of modified histones in a genome; in general, a network of chemical compounds surrounding DNA that modify the genome without altering the DNA sequences; the compounds play a role in determining which genes are active in a particular cell *gene* >>> Figure 71

epigenome profiling: The genome-wide analysis of cytosine methylation, DNA replication, distribution of DNA-binding proteins, and histone modification patterns *gene biot meth*

epigenomics: All techniques that allow the analysis of epigenetic parameters, such as the methylation pattern of cytosine residues in genes or >>> promotors during the activation or silencing of these genes, in different developmental stages, or after treatment of cells, i.e., the methods for the identification of genes involved in genetic >>> imprinting *gene biot meth*

epigenotype: >>> epimutation

epimutation: A heritable change in gene activity that is not due to a change in base sequence; generally applied to an abnormal change in gene activity (e.g., in a tumor or a cell culture); in development, normal changes in gene activity occur, and these are often somatically inherited; such changes are not epimutations; the pattern of epimutations in an individual or in a cell is referred to as the epigenotype *gene*

epinasty: A distortion of a leaf blade such that its edges curl *bot*

epiphyll: A plant that uses the leaf of another plant for support, but that draws no nutrients from the host plant *bot*

epiphytic: Growing on other plants for physical support but not drawing nourishment from them *bot eco*

epiphyty: Of plants which live rooted to the aerial parts of other plants, e.g., on tree trunks *bot*

epiphytotic: An unarrested spread of a plant disease, or sudden and unusually widespread development of a destructive disease in plants *phyt*

episome: A bacterial plasmid that can integrate reversibly with the bacterial chromosome and replicate it *bot biot*

episperm: >>> seed coat

epistasis: The non-reciprocal interaction of non-allelic genes; the situation in which one gene can interfere with the expression of another; more specifically, gene interaction in which one gene interferes with the phenotypic expression of another non-allelic gene so that the phenotype is determined effectively by the former; the latter is described as hypostatic; inter-locus interaction of neighboring loci are implicated in natural polymorphisms; epistatic effects in qualitative traits (disease resistance) are often described as the masking of the expression of a gene by a gene at another locus (non-allelic gene interaction) *gene* >>> Tables 6, 20, 21

epistasy: >>> epistasis

epistatic: >>> epistasis

epistatic variance: >>> variance >>> epistasis

epithet: The second (uncapitalized) word in the scientific name of a species, following the name of the genus; a complete species name consists of the name of the genus to which the species belongs, plus the specific epithet, plus the author of the species *tax*

epitope: A site on an antigen at which an antibody can bind; the molecular arrangement of the site determining the specific combining antibody *chem sero*

epitope mapping: Methods used for studying the interactions of antibodies with specific regions of protein antigens; important applications of epitope mapping are found within the area of immunochemistry *meth chem phys*

EPSP synthase: An enzyme in plants' metabolic pathways that leads to aromatic amino acid production necessary for the development of proteins essential to plants' growth *phys biot*

equational division: The division of each chromosome during the metaphase of mitosis or meiosis into two equal longitudinal halves (i.e., sister chromatids), which are distributed to the two cell poles, and then incorporated into separate daughter nuclei *cyto*

equational separation: >>> equational division

equatorial plane: The plane between the two daughter nuclei of a dividing cell *cyto*

equatorial plate: An arrangement of the chromosomes in which they lie approximately in one plane, at the equator of the spindle; it is seen during metaphase of mitosis and meiosis *cyto*

equilateral: The type of panicle (e.g., in oats) in which the branches appear to spread equally on all sides *bot*

equilibrium moisture content: The moisture content at which a seed is in equilibrium with the relative humidity of the surrounding air *seed phys*

equireplicate (block design): A >>> block design is called equireplicate if each treatment occurs in the same number of plots; that is, if the number of occurrences of a given treatment in each block is summed, the result is independent of the treatment chosen; if the block design is binary, then this condition is equivalent to regularity of the corresponding hypergraph, i.e., each point is in a constant number of blocks *stat* >>> design of experiment

ER: >>> endoplasmic reticulum

eradicant: A chemical substance that destroys a pathogen at its source *phyt*

eradicate: To remove entirely; to pull up by the roots *meth hort agr* >>> disease eradication

eradication: Control of plant disease by eliminating the pathogen after it is established or by eliminating the plants that carry the pathogen *phyt agr* >>> disease eradication

erect: Upright *bot*

erectoides mutant: In cereals, a mutant form showing upright tillers and/or leaves *gene*

ergot: A disease that affects many grasses, including cereals; the conspicuous, hard, black sclerotia of the fungus (*Claviceps purpurea*) replace the ovaries (i.e., the grain in the spikelet of an infected plant); the germination of the sclerotia takes place after overwintering, the asco-spores, from perithecia in stalked capitula (stromata), causing infection of the stigmas of susceptible plants; in the sphacelia stage, which comes after infection, there is a sweet liquid or nectar with an attraction for the insects, which take the conidia from plant to plant; later in the year, the life cycle is completed by the development of sclerotia; the sclerotium contain alkaloids, which can cause severe poisoning or even death if ingested by animals or humans; most ergot for medicine comes from naturally or artificially diseased rye; it is chiefly of use in gynecology and obstetrics; its properties are dependent on a number of alkaloids of which ergotamine and ergotoline are the most important physiologically *phyt* >>> rye

erose: Irregularly toothed on the margins or at the apex; appearing to be chewed off *bot*

error of estimation: The difference between an estimated value and the true value *stat*

error variance: Variance arising from unrecognized or uncontrolled factors in an experiment with which the variance of recognized factors is compared in tests of significance *stat*

ersh: >>> stubble

erucic acid: The major fatty acid found in rapeseed oil used for industrial purposes; varieties of oil with low erucic acid content such as >>> Canola™ oil have nutritional value *agr phys chem*

escape: Applied to a plant that has escaped from cultivation and naturalized more or less permanently *eco*; in plant pathology, the failure of inherently susceptible plants to become diseased, even though disease is prevalent *phyt*

Escherichia coli (E. coli): An aerobic bacterium; because of its rapid multiplication, it is extensively used in molecular genetics and genetic engineering *biot*

eSNP: >>> electronic single nucleotide polymorphism

espalier: A trelliswork of various forms on which the branches of fruit trees, grapevine, etc., are extended horizontally, in a fan or other shape, in a single plane in order to provide better air circulation and sun exposure for the plants *hort meth*

ESS: >>> evolutionarily stable strategy

essential element Any chemical element which is essential to the normal growth and reproduction of a plant; there are 16 essential elements: boron (B), calcium (Ca), carbon (C), chlorine (Cl), copper (Cu), hydrogen (H), iron (Fe), magnesium (Mg), manganese (Mn), molybdenum (Mo), nitrogen (N), oxygen (O), phosphorus (P), potassium (K), sulfur (S), and zinc (Zn) *phys* >>> micronutrient >>> trace element >>> zinc >>> www.zinc.org/

essentiality: Determination of whether an element is essential to the normal growth and reproduction of plants; it is determined by three criteria: (1) absence of a specific element will result in abnormal growth, incomplete life cycle or premature death; (2) the element cannot be replaceable by another; and (3) the element must have a direct effect on the growth and metabolism of plants *phys*

essential oils: The volatile, aromatic oils obtained by steam or hydrodistillation of botanicals; most of them are primarily composed of terpenes and their oxygenated derivatives; different parts of the plants can be used to obtain essential oils, including the flowers, leaves, seeds, roots, stems, bark, and wood; certain cold-pressed oils, such as the oils from various citrus peels, are also considered to be essential oils, but these are not to be confused with cold-pressed fixed or carrier oils, such as olive, grapeseed, and apricot kernel, which are non-volatile oils composed mainly of fatty acid triglycerides *chem phys*

ESTs: >>> expressed sequence tags

establishment: The process of developing a crop to the stage at which the young plant may be considered established (i.e., safe from juvenile mortality and no longer in need of special protection) *agr fore*

establishment period: The time elapsing between the initiation of a new crop and its establishment *agr*

ester: A compound that is formed as the condensation product; water is removed from an acid and an alcohol, while water is formed from the OH of the acid and H of the alcohol *chem phys*

esterase: A member of hydrolases that may catalyze esters; esterases were frequently used as biochemical markers in breeding experiments *chem phys*

estimate: A numerical value obtained from a statistical sample and assigned to a population parameter; the particular value yielded by an estimator in a given set of circumstances or the rule by which such particular values are calculated *stat*

estimation: The inference about the numerical value of unknown population values from incomplete data, such as a sample; if a single figure is calculated for each unknown parameter, the process is called point estimation; if an interval is calculated within which the parameter is likely, in some sense, to lie, the process is called interval estimation *stat*

estimation error: The amount by which an estimate differs from a true value; this error includes the error from all sources (e.g., sampling error and measurement error) *stat*

estivation: Stagnating or otherwise non-functional during the summer period *bot agr hort*

étagère: A series of open shelves for growing and displaying plants or *in vitro* cultures *hort biot meth*

ethanoic acid: >>> acetic acid

ethanol: A liquid solvent *chem*

ethene: >>> ethylene

ether extract: Fats, waxes, oils, etc., that are extracted with warm ether in chemical analysis *prep meth*

ethereal oil: >>> essential oil

ethidium bromide (EB): Fluorescent molecule that intercalates between base pairs of DNA and RNA *chem biot*

Ethrel: The trade name for ethephon, which is 2-(chloro-ethyl)-phosphonic acid; it is an ethylene generator when applied to plant surfaces; ethylene has numerous physiological effects, such as inducing synchronous flowering and fruit ripening, which assists mechanical harvesting, etc.; Ethrel is also used as a male gametocide to induce random cross-pollination for recurrent mass selection in autogamous cereals, such as wheat or barley *phys*

ethyl alcohol: >>> ethanol

ethyl methane sulfonate (EMS): A chemical compound that acts as a mutagen in plants *chem meth* >>> mutagen

ethyl alcohol: >>> ethanol

ethylene: A volatile plant hormone; synthesis is promoted by auxin or damage in seedlings, in shoot apex, and various organs; it is known as a ripening hormone; it stimulates flowering and fruit ripening; it inhibits elongation of stems, roots, and leaves *chem phys*

ethylenediamine tetra acetic acid: Known as edetic acid or tetra acetic acid, this chelator is a synthetic amino acid having an m.w. of 292.25 and a molecular formula of $C_{10}H_{16}N_2O_8$; it is used in a wide range of biochemical and chemical procedures *chem biot*

ethyloxanilate: A new class of male >>> gametocide for wheat (*Triticum aestivum*), the most active example of this class being ethyl 4-fluorooxanilate (E4FO); the latter induces male sterility, specifically, without detectable effects on various agronomic features and female fertility; plants sprayed once with 0.15% exhibited 100% pollen and floret sterility without causing a significant reduction in total yield; the induction of male sterility by deployment of male gametocides holds immense potential in >>> heterosis breeding of wheat *meth chem*

etiolation: A plant syndrome caused by suboptimal light, consisting of small, yellow leaves and abnormally long internodes *phys meth*

etiology: The study of causes of diseases *phyt*

E-type: In sugarbeet breeding the high-yielding (E = Ertrag = yield) varieties with average sugar content *seed*

euapogamy: A form of apomixis in which the sporophyte develops from a gametophyte without fertilization and formation of zygotes *bot*

EUCARPIA: >>> European Association for Research on Plant Breeding

eucell: A eukaryotic cell showing a nucleus, nuclear envelopes, chromosomes, and nuclear divisions *bot*

eucentric: A chromosomal interchange by which the translocated segment does not change the relative position to the centromere *cyto*

euchromatin: Chromatin that shows the staining behavior of the majority of the chromosome complement; it is uncoiled during >>> interphase and condenses during mitosis, reaching a maximum density at metaphase *cyto* >>> Table 43

euchromatization: The induced or spontaneous change of heterochromatin into euchromatin *cyto* >>> Table 43

euchromosome: A chromosome showing the typical features of the standard complement of a given species *cyto* >>> autosome

eugenic: Favorable to the genetic quality of a population, as opposed to dysgenic *gene*

euhaploid: A haploid genome showing no deviating number of chromosomes compared with the standard genome of the species *cyto*

eukaryon: The highly organized nucleus of a eukaryote *bot*

eukaryote: An organism whose cells have a distinct nucleus *bot*

eukaryotic: >>> eukaryote

eupatorium yellow vein virus (EpYVV): A relative of viruses found in honeysuckle, *Ageratum conyzoides* (a common weed), and cotton; it shows a DNA genome and belongs to the geminiviruses (so-called because the virus particles occur as pairs); EpYVV causes the spectacular and beautiful symptoms first described in a Japanese poem written by E. KOKEN, in 752 AD, and it is thought to be the world's first record of a plant virus; in Japan, eupatorium plants frequently show striking yellow net-like patterns in their foliage that resemble autumn coloring, but in mid-summer; the ancient Japanese valued the beautiful effects caused by the virus, but the virus itself has proved to be of scientific interest because it includes both the main virus component and a satellite component – and both are involved in infection and symptom development *phyt biot* >>> geminivirus

euploid: A cell and/or plant having any number of complete chromosome sets *cyto*

euploidy: >>> euploid

eupycnotic: Normally coiled and normally stainable chromosomes *cyto*

European Association for Research on Plant Breeding (EUCARPIA): Promotes scientific and technical cooperation in the field of plant breeding in order to foster its further development; to achieve this purpose, it arranges and sponsors meetings of members, or groups of members, to discuss general or specific problems from all fields of plant breeding and genetic research; the association has ten sections and a number of working groups focusing on particular crop species; it was established in 1956 in Wageningen (Netherlands) *org* >>> www.eucarpia.org

European Food Safety Authority (EFSA): Established in 2002 as the central authority for the scientific evaluation of food and feed safety in the European Union (EU); as of 2005, EFSA is permanently based in Parma, Italy; it was established based on the legal mandate of regulation (EG) 178/2002 of the European Parliament and European Council addressing basic principles of food law; new food law legislation and hence EFSA were created in response to a number of food scandals that shook consumer confidence; it addresses two main areas: (1) scientific risk evaluation for all questions related to food and feed safety and (2) informing the public about potential risks; to assist with safety evaluations, EFSA is supported by eight scientific panels composed of independent researchers from various EU member states; the GMO panel is responsible for GMOs and genetically modified food and feed; it offers a solid scientific basis for making informed political decisions *org*

European maize (corn) borer (*Ostrinia nubilalis*): A major insect pest in maize that costs millions of dollars annually in control expenditures and yield loss; the European maize borer larva feed on the leaves and "bore" holes into the stalks where they tunnel *phyt*

eusexual: Showing regular alternation of >>> karyogamy *bot*

eusom: A plant showing each member of the chromosome complement with the same copy number *cyto*

eutriploid: With exactly three sets of >>> homologous or >>>homoeologous >>> chromosome, i.e., >>> genomes *cyto*

eutrophication: Degradation of water quality owing to enrichment by nutrients, primarily nitrogen and phosphorus, which result in excessive plant growth (e.g., algae) and decay *eco*

evaluation: The recording of those characters whose expression is often influenced by environmental factors *meth*

evapotranspiration: Loss of water to the atmosphere from vegetated ground from a combination of >>> evaporation from the soil and >>> transpiration through the plants *phy phys*

even-aged forest: A forest in which all of the trees present are essentially the same age (e.g., within 10 to 20 years) *fore* >>> uneven-aged forest

event: A collection of sample points *stat*

evaluation: The recording of those characters whose expression is often influenced by environmental factors *meth*

everbloomer: A plant that blooms continuously throughout the growing season, for example, Busy Lizzie (*Impatiens walleriana*) *hort*

evolution: The process by which new species are formed from preexisting species over a period of time; recently, C. WOESE (2002) suggested, contrary to the Darwinian hypothesis of evolution, a polyphyletic origin of evolution – that is, the origin of the biological evolution based on several cell lines and types that exchanged genes and proteins by >>> horizontal gene transfer *evol* >>> Figures 68, 69 >>> Tables 12, 32, 45 >>> plant evolution

evolution pressure: The joint action of mutation, immigration, hybridization, and selection pressure *evol* >>> Figures 68, 69

evolutionarily stable strategy (ESS): Where an equilibrium results from divergent but balanced selection pressures *stat eco*

evolutionary: >>> evolution

evolutionary breeding: Breeding procedure in which the variety is developed from an unselected progeny of a cross or multiple crosses that have undergone evolutionary changes *meth*

evolutionary divergence: The mode of evolutionary change whereby an ancestral population is split into different genotypic populations or phyletic lines *evol* >>> Figures 68, 69

evolutionary genetics: Provides useful information for guiding gene mapping, e.g., understanding the phylogeny helps to manipulate horizontal gene transfer between microorganisms *meth* >>> Figures 68, 69

evolutionary plasticity: The degree of genetic adaptability of a species or other taxa *evol* >>> Figures 68, 69

evolve: To evolve is to develop by the process of evolution, changing in some way as an adaptation to the environment *evol*

exarch: Primary xylem in which the development is from the outside, toward the center; typical of roots *bot* >>> endarch

exchange pairing: The type of pairing of homologous chromosomes that allows genetic crossing-over to take place *cyto gene*

excised embryo test: A quick method for evaluating the growth potential of a root–shoot axis that has been detached from the remainder of the seed *seed meth*

excision repair: The repair of DNA defects by excision of defective oligonucleotides in one of the two DNA strands and the subsequent resynthesis of the excised nucleotide sequence, utilizing the complementary base pair code in the intact strand *gene*

exclusion net(s): A system successfully used in France against codling moth *Cydia pomonella* since the early 2000s, it has also been adapted for North American conditions and was tested in an experimental apple orchard in southern Quebec, Canada, from 2012 to 2016; evaluation of insect and disease damage, as well as physical and physiological damage, was made in complete exclusion plots and in unnetted control plots; the exclusion system proved to be an effective protection device for the vast majority of key pests of apple fruit in most years; damage from key insect pests such as the apple maggot *Rhagoletis pomonella*, the tarnished plant bug *Lygus lineolaris*, and the codling moth was significantly lower in netted plots than in unnetted plots *phyt meth*

exerted: Protruding (e.g., the inflorescence from the sheath) *bot*

exhauster: >>> aspirator

exhaustion test: A type of vigor test that measures the ability of seeds to grow rapidly under rigidly controlled conditions of high temperatures, relative humidity, and moisture content in continuous darkness *seed meth*

exine: The outer, decay-resistant coat of a pollen grain or spore; it shows different characteristics on the surface, which allows taxonomic differentiation between genera and even species; there are several scanning electron microscopy studies made on crop plants for comparisons in crop plant evolution *bot*

exocarp: The pericarp or ovary wall of angiosperm fruits is composed of three different layers; the outer layer is the exocarp *bot*

exocytosis: The extrusion of cellular material from a cell, as opposed to endocytosis *bot*

exogamy: Outbreeding; all forms of sexual reproduction in which the mating of unrelated and/or more distantly related partners dominates *bot*

exon: The portion of a gene that is transcribed into >>> mRNA and is translated into protein (i.e., a DNA sequence in the encoding part of a gene) *gene*

exonic single nucleotide polymorphism: Any >>> SNP that is present in an exon of a gene *gene biot* >>> SNP

exon shuffling: The generation of new genes through intron-mediated recombination of coding sequences (exons) that previously specified different proteins or different parts of one and the same protein *biot meth*

exon SNP: >>> exonic single nucleotide polymorphism

exonuclease: A DNase enzyme that digests DNA beginning at the ends of the strands (e.g., exonuclease III degrades dsDNA starting from the 3' end, lamda exonuclease degrades ssDNA and dsDNA) *gene phys biot* >>> endonuclease >>> Table 39

exotic species: A species that is not native to a region *agr hort*

exotoxin: A poison excreted by a plant into the surrounding medium *phys*

expanded genetic code: An artificially modified genetic code in which one or more specific codons have been re-allocated to encode an amino acid that is not among the 20 common naturally encoded proteinogenic amino acids; prerequisites to expand the genetic code are the non-standard amino acid to encode, an unused codon to adopt, a tRNA that recognizes this codon, and a tRNA synthetase that recognizes only that tRNA and only the non-standard amino acid *biot*

ex parte: A legal proceeding where only one party is represented; patent prosecution is an *ex parte* procedure *adm* >>> Intellectual Property Rights

experimental alteration of germplasm: >>> mutation

experimental design: The planning of a process of data collection; also used to refer to the information necessary to describe the interrelationships within a set of data; it involves considerations, such as number of cases, sampling methods, identification of variables and their scale-types, identification of repeated measures, and replications *stat* >>> design of experiment >>> pasture experiments >>> variety trial

experimental error: There are a number of sources of variation in field experiments contribution to experimental error; some of the variations can be reduced by the use of appropriate experimental design, some cannot; plot size and shape, >>> experimental design, layout of the experiment, and other factors have an effect on experimental error; main sources are plant variability, seasonal variability, and soil variability; plant variations include type of plant, competition among plants within plots, and variability between plots; ordinarily, there is a larger variation among larger plants (e.g., maize) than among smaller plants (e.g., cereals); similarly, the variation among closely spaced plants is smaller than it is among widely spaced ones; and variation is lower for plots with a uniform stand than it is for plots with a patchy stand; plot-to-plot variation is another important factor contributing to experimental error; plot that are surrounded by similar plots are usually less variable than plots next to alleys, borders, plots with plants of different height, etc.; this variability is due to differential competition between plants at the edges of the plots and sometimes called >>> "border effect"; the seasonal variability is caused by the climatic factors varying from year to year; the soil variability is present almost everywhere, even over a small area in a single field; this is a result of different texture, depth, moisture, drainage, or nutrients of the soil *stat meth*

experimental lines: A group of individuals of a common ancestry and more narrowly defined than a strain or variety; a pure line is a >>> clone; in plant breeding, >>> "line" refers to any group of genetically uniform individuals formed from a common parent *meth* >>> Figure 50

experimental stimulus: >>> treatment

experimental use: In patenting biological material, the practice of a patented invention solely with the intention of experimentation or perfection of the invention *adm* >>> Intellectual Property Rights

experimental variability: Variation in a dependent variable produced by an independent variable *stat*

explant: An excised fragment of a tissue or an organ used to initiate an *in vitro* culture *biot*

explantation: >>> explant

explosion method: A technique for gene-into-cell introduction in which the gene or >>> foreign DNA is driven into plant cells by the force of an explosion (vaporization) of a drop of water to which the gene and gold particles have been added; the explosion is caused by application of high-voltage electricity to the drop of gene-laden water; the water is then vaporized explosively, driving the shot (gold particles) and genetic material through the cell membrane; the plant cell then heals itself, incorporating the new gene into its genetic complement *biot* >>> gene transfer >>> particle gun

exponential stage: The phase with increased metabolic activity and rapid kernel development that links the dilatory phase with the linear grain-filling period; it occurs approximately 5–12 days after pollination; in the endosperm, cell walls are laid down beginning about 5 d after pollination, changing the free nuclear tissue into a cellular one; the embryo continues in the proembryo stage through about 10 d after pollination *phys meth* >>> Table 13

expressed single nucleotide polymorphism: >>> exonic single nucleotide polymorphism

expressed sequence tags (ESTs): ESTs correspond to regions of expressed genes; they are used as a gene discovery tool and also to demonstrate which genes are expressed in a particular tissue at a specific time; sequencing ESTs is a powerful method of gene discovery and gene sequence determination, and they are also a valuable resource >>> of SNP discovery; DNA sequences derived by sequencing an end of a random cDNA clone from a library of interest; usually, tens of thousands of such ESTs are generated as part of a given genome project; these ESTs provide a rapid way of identifying cDNAs of interest, based on their sequence "tag" *biot* >>> Figure 43

expression imbalance map (EIM): A visualization method for detecting >>> mRNA expression imbalance regions, reflecting genomic losses and gains at a much higher resolution than conventional technologies such as comparative genomic hybridization; simple spatial mapping of the microarray expression profiles on chromosomal location provides little information about genomic structure, because mRNA expression levels do not completely reflect genomic copy number and some microarray probes would be of low quality *meth biot cyto*

expression mapping: Creation of quantitative maps of protein expression from cell or tissue extracts, e.g., >>> expressed sequence tag maps; the approach relies on 2D gel maps and image analysis, and opens up the possibility of studying cellular pathways and their perturbation by disease or other biological stimuli at the whole-proteome level *meth biot*

expression marker: Any expressed sequence (cDNA), i.e., a tag derived from serial analysis of gene expression (>>> SAGE), or an >>> EST that has been identified by high-throughput expression profiling; it serves as a diagnostic marker for disease *biot meth*

expression variegation: A type of variation in gene expression during development that causes streaks or patches of cells to have different phenotypes *gene*

expression vector: Vector constructs with a promoter sequence showing a highly efficient transcription of an inserted gene; the cell shows a high concentration of the gene product (i.e., protein) *biot*

expressivity (of a gene): The degree to which a particular genotype is expressed in the phenotype *gene* >>> Figure 60

ex situ: Out of place or not in the original environment (e.g., seeds stored in a genebank) *seed meth*

ex situ **conservation:** A conservation method that entails the removal of seed, pollen, sperm, or individual organisms from their original habitat, keeping these resources of biodiversity alive outside of their natural environment; the storage of plant genetic resources as seeds in genebanks at subzero temperatures is the most widely applied method; approximately 90% of more than 6 million accessions stored *ex situ* worldwide are maintained as seeds; other methods include maintenance of whole plants in field genebanks, as tissue culture in *in vitro* genebanks, and immersion of tissue, embryo, or seeds in liquid nitrogen – that is, cryopreservation; the latter two methods are particularly used for conservation of vegetatively propagated species and for species with recalcitrant seeds *meth seed*

extension: The 5′ and 3′ extensions are ssDNA regions at the ends of dsDNA; sometimes they are called "overhangs" or "sticky ends" *biot*

extension growth: New growth made during one season *hort meth*

extensive crop: A crop that has low production costs and profit margins; for example, soybeans, maize, and wheat are typical extensive crops in North America *agr*

extensograph: Standard equipment for evaluating dough extensibility of cereal flour; used for measuring the extensibility and resistance-to-extension of a wheat-flour dough; the machine records extension force against time until the point of dough rupture; it is an important quality parameter for biscuit making and thus a breeding target *meth*

exterminate: >>> eradication

extine: >>> exine

extrachromosomal: Structures or processes outside the chromosomes *gene*

extrachromosomal circular DNA (eccDNA): It is one characteristic of the plasticity of the eukaryotic genome; it was found in various non-plant organisms from yeast to humans but also in *Arabidopsis thaliana* and *Brachycome dichromosomatica*, reflecting a normal phenomenon that occurs in wild-type plants as well; eccDNA is heterogeneous in size and contains sequences derived primarily from repetitive chromosomal DNA; the size of plant eccDNA ranges from >2 kb to <20 kb, which is similar to the sizes found in other organisms; these DNA molecules correspond to 5S ribosomal DNA (rDNA), non-coding chromosomal high-copy tandem repeats, and telomeric DNA; circular multimers of the repeating unit of 5S rDNA were identified; such circular multimers of tandem repeats were found in animal models, suggesting a common mechanism for eccDNA formation among eukaryotes; this mechanism may involve looping-out via intrachromosomal homologous recombination *cyto biot*

extrachromosomal element: All genetic elements that are not at all times part of a chromosome, for example, plasmids, phages, transposons, insertion sequences, plastid (mitochondrial and chloroplast) genomes *gene* >>> Figure 46

extrachromosomal inheritance: Inheritance that is not controlled by chromosomal determinants but by cytoplasmic components *gene*

extra chromosome: >>> B chromosome

extraction rate: In milling technology, flours are characterized by the rate of extraction; extraction rate is the proportion of flour, derived by milling, from a known quantity of grain; it is used to define various types of flours, e.g., white wheat flour has an extraction rate of 75–78% or less, whole grain flour has an extraction rate of 100% *meth*

extranuclear: Structures and processes outside the nucleus *gene*

extrorse: Of an anther which dehisces away from the center of the flower *bot*

exudate: A substance issuing from a tissue either as a normal process or because of disease in the tissue *phys* >>> NUSSINOVITCH (2009)

ex vito: Describes plants or organs that are transplanted from culture to soil or to pots *meth*

eye: The center of a flower when it is a different color from the petals; the term is also used for an undeveloped bud on a tuber (e.g., potato or dahlia), or for a cutting with a single bud *bot hort*

eye depth (in potato): Common scaling 1 = very deep; 3 = deep; 5 = medium; 7 = shallow; 9 = very shallow *meth prep*

eyepiece: The lens or combination of lenses in an optical instrument (e.g., microscope) through which the eye views the image formed by the objective lens or lenses *micr*

eyespot disease (of wheat, barley, rye, *Pseudocercosporella* syn *Cercosporella herptotrichoides*, of maize, *Kabatiella zeae*): Common in autumn-sown wheat and barley; early symptoms appear as brown smudges on the leaf sheath below the first node; the lesions develop to become eye-shaped with diffuse brown margins and paler centers, which may carry black dots in their center; severe infections girdling the stem may result in premature ripening and white heads and weaken the stem to cause straggling and lodging; *Oculimacula acuformis* is one of two species of soilborne fungi that cause eyespot of wheat, the other being *Oculimacula yallundae* syn *Tapesia yallundae*; both pathogens can coexist in the same field and produce elliptical lesions on stem bases of wheat that are indistinguishable; *Pch1* and *Pch2* are the only two eyespot resistance genes readily available to wheat breeders, but neither provides complete control; a new source of eyespot resistance was identified from *Aegilops longissima* ($2n = 2x = 14$, S^lS^l), a wild relative of wheat; three QTL for resistance to *O. acuformis* were mapped in chromosomes $1S^l$, $3S^l$, and $5S^l$ using a recombinant inbred line population developed from the cross *Ae. longissima* accessions; the three QTL explained 66% of phenotypic variation by β-glucuronidase score (GUS) and 84% by visual rating; QTL associated with resistance to *O. acuformis* have similar chromosomal locations as some for resistance to *O. yallundae*, except that a QTL for resistance to *O. yallundae* was found in chromosome $7S^l$ but not for *O. acuformis*; thus, it appears that some genes at the same locus in *Ae. longissima* may control resistance to both eyespot pathogens; QTL effective against both pathogens will be most useful for breeding programs and have potential to improve the effectiveness and genetic diversity of eyespot resistance *phyt* >>> *Tapesia yallundae*

F

F1: >>> filial generation

F2: The progeny produced by intercrossing or self-fertilization of F1 individuals *meth* >>> Tables 27, 35, 53

F3: Progeny obtained by self-fertilizing F2 individuals *meth*

F1 synthetic variety (Syn-1): A variety derived by intercrossing a specific set of clones or seed-propagated lines; they may include varieties of normally cross-fertilizing or self-fertilizing crops into which mechanisms have been introduced to maximize cross-fertilization, for example, "Vitagraze" rye *seed* >>> Figures 50, 51

F2 variety: Refers to the next generation seed derived from the hybrid (F1) generations; the variety cannot be perpetuated by growing additional generations, for example, the tomato variety "Foremost F2" *seed meth*

faciation: A subdivision of a plant >>> association that lacks some of the typically dominant species due to local differences in climate *eco*

facilitated recurrent selection: A type of recurrent selection in which genetic male sterility is maintained in the population to maintain heterozygosity and genetic diversity and to permit the recombination and shifting of gene frequencies *meth*

facilitation: A positive effect of one plant or plant species upon another *eco* >>> interference >>> competition

factor: Synonymous with gene gene; an independent variable under examination in an experiment as a possible cause of variation *stat* >>> Table 40

factor(ial) analysis: A multivariate statistical analysis in which the independent variables are grouped into factors describing the variance of the dependent variables; it is used in cluster analysis *stat*

factorial crossing group: >>> crossing groups

factorial design: In a factorial design, the effect of two or more factors can be simultaneously observed (e.g., planting time, fertilizer, watering regime), i.e., the design provides information on the average effect of the individual factors as well as the interaction between these factors; this design type also allows a wider application of the conclusions reached on the effect of each factor because each factor is tested over a wide range of conditions of other pertinent factors; the statistical analysis used is >>> ANOVA followed by a DUNCAN's multiple range test; because the total number of treatment combinations (the product of the numbers of levels of the factors) may be very large, it is often necessary to test only some of these combinations, giving a "fractional factorial design"; a suitable fraction may be an orthogonal array, or a subgroup of the direct product of abelian groups of orders equal to the number of levels of each factor *agr meth stat* >>> design of experiment >>> factorial trial

factorial trial: Experimental design in which the effects of a number of different factors are investigated simultaneously; the factorial set of treatments consists of all combinations that can be formed from the different factors; the trial can include randomized complete blocks and >>> Latin squares (e.g., treatments A and B at levels x and y); the sources of variance are replicates A, B, A × B, and error *stat* >>> design of experiment >>> factorial design

factor-pair: >>> allelomorph

facultative apomict: Apomicts that retain sexuality so that aberrant offspring may occur *bot*

facultative growth habit: >>> winter-and-spring wheat

facultative heterochromatin: Heterochromatin that is present in only one of a pair of homologues or not permanently present *cyto* >>> Table 43

facultative parasite: A mainly saprophytic organism with weakly pathogenic properties *phyt*

facultative-type (of growth habit): Can be winter-and-spring type *phys*

facultative weed: A weed found growing both wild and in association with man *phyt*

fading: >>> photobleaching

falcate: >>> falciform

falciform: Curved or sickle-shaped (e.g., leaves or leaf hairs) *bot*

falling number (after HAGBERG): The test provides an indication of alpha amylase activity and depends on the action of this enzyme in reducing the viscosity of a heated flour-and-water slurry; alpha amylase is an enzyme involved in the degradation of starch to sugars and is usually associated with germination; sprouted grains give low falling numbers; a high number is required for better bread-making quality (<200 low; 230 medium; >250 high; about 290 very high; >310 extremely high) *meth*

fallow (ground or cropland): Leaving the land uncropped for a period of time; it may contribute to moisture accumulation, to improvement of soil structure, or to mineralization of nutrients; it may be tilled or sprayed to control weeds and conserve moisture in the soil *agr* >>> Table 44

falls: The drooping or horizontal petals of irises *bot hort*

false color: Representation in colors differing from the original scene *micr*

false fruit: >>> pseudocarp

false negative: An experimental outcome that incorrectly yields a negative result; false negatives can complicate diagnosis *meth* >>> false positive

false node: An abnormal node that occurs in some varieties of, for example, oats; a true node, appearing devoid of branches, occurs a little distance below it, but the branches remain fused with the rachis and appear at the false node *bot*

false positive: An experimental outcome that incorrectly yields a positive result; false positives can frustrate the assessment of the performance of a lead substance *meth* >>> false negative

family: A group of individuals directly related by descent from a common ancestor, with at least one parent in common (one parent in common = half sibs; both parents in common = full sibs, selfing family = family obtained by self-pollinating a genotype) *tax* >>> sib >>> polycross >>> half-sib >>> full-sib >>> open pollination >>> wind pollination >>> selfing >>> List of Crop Plants

family forestry: Tested open-pollinated, polycross, or full-sib families are deployed as single families to commercial plantations *meth fore* >>> Figure 51

family selection: The selection of progeny families on their mean performance; in addition, the best individuals are usually selected in the best families *meth*

fangy root (of beets): Forked roots (in beets) caused by diseases, soil, or weather conditions *agr phyt*

fanning machine: >>> winnower

fanning mill: The air-screen machine that utilizes airflow and sieving action in separating and cleaning seeds *seed* >>> air screen cleaner >>> winnower

fan-shaped: >>> flabellate

fan-training: >>> espalier

FAO: >>> United Nations Food and Agriculture Organization

farina: White-mealy *bot*

farinogram: A curve on a kymograph chart; it provides accessory information on dough properties, such as absorption, optimum mixing time, and mixing tolerance; it may be used to estimate absorption and mixing time of a flour to be backed; specific correlations between those traits have been established *meth*

farinograph: An instrument used for measuring the mixing properties of a wheat-flour dough; the machine records mixing torque as a function of time *meth* >>> mixograph

farmer selection (*syn* **participatory plant breeding**): An aspect of some plant breeding programs in which the farmers make the final selection of cultivar; each farmer is given a different group of new clones or pure lines of a crop, emerging from a breeding program; they then may grow them and choose those they like best; their favorites become their own property, with the sole provision that the breeder may have some of them for the purpose of further breeding; the farmers can then grow that material for their own use, and give or sell propagating material to their friends and neighbors *agr* >>> Figure 50

Farmer's fixative 3 parts of anhydrous ethanol: 1 part of glacial acetic acid; a fixing and dehydrating agent used in histology; also used in conjunction with FEULGEN stain and carbol-fuchsin stain for chromosome analysis *prep cyto*

farmer's privilege: A clause in the plant breeder's rights legislation of most countries that permits a farmer to use some of their own crop of a registered cultivar for seed on their own farm only; a farmer may not sell any of that crop for seed unless licensed to do so; some seed companies deny this right, particularly with respect to >>> genetically modified organisms, by a special clause in the sale contract *agr*

farmer's rights: The recognition of farmers (past, present, and future) as *in situ* agricultural innovators who collectively conserve and develop agricultural genetic resources around the world; as such, farmers are recognized as innovators entitled to intellectual integrity and to compensation whenever their innovations are commercialized *agr*

farm-saved seed: For certain crop species – particularly small-grain cereals – growers can opt to save their own seed for sowing the following year provided care is taken to ensure that the crop remains healthy and free from impurities, and that the resulting seed is carefully conditioned and cleaned; without independent testing for germination and freedom from seedborne diseases, however, farm-saved seed can harbor risks to growers which may not become apparent until well into the growing season; many growers choose >>> certified seed for the peace of mind that quality and performance is independently assured *seed agr meth*

farm-scale evaluation: A recent research exercise in the U.K. to determine the on-farm effects to fauna and flora of growing and managing herbicide-tolerant crops to non-tolerant varieties of the same crop *eco biot*

farro: Known as hulled wheat; this means that the caryopsis retains its hull or husk during harvest and must be dehulled prior to further processing; its nutty flavor has long been popular in Europe, where it is also known as farro (Italy) and Dinkel (Germany); in Roman times it was called "farrum", an ancient grain believed to have sustained the Roman legions; origins can be traced back to early Mesopotamia; spelt (*Triticum spelta*) is ancient and related to modern wheat (*Triticum aestivum*); it is one of the oldest of cultivated grains, preceded only by "emmer" and "einkorn"; in North America, this fine grain is commonly known as spelt; while there are occasional descriptions of spelt as not "true" farro, the International Plant Genetic Resources Institute, via its report on Underutilized Mediterranean Species, states that the only registered varieties of farro belong to spelt; it is an old Italian favorite (a famous wedding soup of these regions is called "Confarrotio"); for centuries, farro has been a mainstay of Tuscany, Lazio, Umbria, and Abruzzi in the northern part of Italy; these are relatively poor areas; pushed aside in recent decades by easier to grow and harvest varieties of common wheat, farro is making a comeback among health-conscious cooks and consumers; from a cross-country reading of the culinary winds, it appears that farro has finally made it to the New World; used in soups, salads, and desserts, the little light brown grain is an intriguing alternative to pasta and rice; now farro (pronounced FAHR-oh) appears to be moving from rustic tables into fashionable restaurants not only in Tuscany and northern Italy (where it suddenly seems ubiquitous on menus), but also in the United States, particularly on the West and East Coasts *agr* >>> wheat

fasciata-type of pea: A leaf mutant in peas; the leaves mutated into tendrils; it became a breeding target; several varieties are commercially used *gene meth agr*

fascicle: A bundle of needles on a pine tree *bot*

fasciculated root: Fibrous roots, in which some of the branches are thickened *bot agr*

fast green test: Used for evaluation of harvest damage to seed coat, e.g., in maize, alfalfa, sweet clover, and other legumes; materials needed are fast green (FCF) and water; a 0.1% solution is prepared by adding 1 g fast green to 1,000 ml water; a certain number of seeds is counted and placed in a 250 ml beaker; seeds are covered with fast green solution and stirred for 30 seconds, and for an additional 2 minutes; the fast green solution is poured off and seeds are rinsed thoroughly under running water; later the seeds are spread on toweling to dry; the damage is determined by counting seeds that show stained pericarp breaks *seed meth*

fat: Used by most plants as an energy-rich storage substance, usually present in the seeds; in some seeds it may amount to about 70% of the dry matter; plant fats are usually liquid at room temperature, and they are mixtures of glycerine ester of many fatty acids *bot phys* >>> Table 16

fat body: >>> fat

father plant: The individual or species from which pollen was obtained to create a hybrid *meth*

fatuoid: A mutation that arises spontaneously in cultivated oats; the plant and grain closely resemble the variety in which the mutation occurs but the grain shows, to a varying degree, certain characteristics of wild oats (Avena fatua), for example, a strong geniculate awn, a "horseshoe" base, and dense hairs on the callus and the rachilla *bot gene*

fatty acid: A long-chained, predominantly unbranched, carboxylic acid, in which a side-chain of carbon atoms is attached to the carboxyl group, and hydrogen atoms to some or all of the carbon atoms in the side chain *chem phys*

FCF: >>> fast green test

fecundity: The potential number of offspring produced during a unit of time, or the ability to produce, particularly as the production of seeds per >>> *gene bot* >>> resilience

FDR: >>> first division restitution

fecundity: The potential number of offspring produced during a unit of time *bot* >>> resilience

fecundity selection: The forces acting to cause one genotype to be more fertile than another genotype *evol*

Federal Seed Act: A regulation of 1938 and improved in 1976; the U.S. government statute governing aspects of seed production, handling, and sales *seed*

feedback mechanism: A control device in a system; homoeostatic systems have numerous negative-feedback mechanisms, which tend to counterbalance positive changes and so maintain stability *phys*

feeder root: One of the numerous small roots of a plant, through which moisture and nutrients are absorbed from the soil *bot hort agr*

feed grain: Any of several grains most commonly used for livestock feed, including maize, grain sorghum, soya, lupines, oats, rye, barley, etc. *agr*

Feekes scale: Scale expressing stages in cereal seedling development *meth phys seed* >>> Table 13 >>> physiological maturity >>> decimal code for the growth of cereal plants

female sterility: Some crops (e.g., banana) do not produce true seed because of a female sterility; however, >>> male sterility is much more common, and is more useful in plant breeding as a technique for achieving cross-pollination *bot*

fen: Low land covered wholly or partially with water *agr*

feral: Wild, i.e., not cultivated *eco*

ferment: >>> enzyme

fermentation: Anaerobic respiration; usually applied to the formation of ethanol or lactate from carbohydrate *phys* >>> biotechnology

ferredoxin: A non-heme iron protein with a low redox potential that functions as an electron carrier in both >>> photosynthesis and nitrogen fixation *phys*

ferric chloride test: A fast and useful test to estimate the percentage of abnormal seedlings; materials needed are: $FeCl_3$, mortar and pestle or some other means of grinding, a liquid detergent, and PETRI dishes; first a 20% $FeCl_3$ solution is prepared by adding 4 parts water to 1 part $FeCl_3$ by weight; a certain number of seeds is placed in PETRI dishes; seeds are covered with $FeCl_3$ solution; black staining seed are removed within 5 min of adding the solution; separating black seeds is continued for 15 min after applying solution; finally, the number of black seeds in each replicate is counted; the average number of black seeds over replicates is estimated; generally all stained seeds are dead or develop abnormally *seed meth*

fertigation: Application of soluble fertilizers through a drip irrigation system *hort agr fore*

fertile: A plant which produces seed capable of germination or which produces viable gametes *cyto gene*

fertile crescent: An archaeological term used to describe the fertile area of ancient agriculture that extends from modern Israel in a wide arc to the valley of the Tigris and Euphrates rivers *agr* >>> SCHLEGEL (2007)

fertility: Fruitfulness, ability to produce viable offspring *bot gene*

fertilization: The union of two gametes to produce a zygote that occurs during sexual reproduction *bot*

fertilize: Bringing together two gametes to produce a zygote *meth*

fertilizer: A material that is added to the soil to supply one or more plant nutrients in a readily available compound *agr*

fertilizer grade: An expression that indicates the percentage of plant nutrients in a fertilizer *agr phys*

festulolium: An artificial grass hybrid of *Festuca pratensis* (2n = 2x = 14) × *Lolium perenne* (2n = 2x = 14), which is used in agriculture as a forage crop *bot agr*

FEULGEN method: >>> FEULGEN reaction

Feulgen reaction: A cytochemical test that utilizes SCHIFF's reagent as a stain and DNA hydrolysis; it is highly specific for DNA detection; the method allows a wide range of chromosome studies and quantitative determination of DNA contents of nuclei applying the so-called cytophotometry; it was discovered in 1912 by the German Robert FEULGEN (1884–1955); the methods of fluorescence image cytometry are being developed as an alternative for the determination of DNA content on a single-cell basis when using DNA-specific fluorochromes *micr*

Feulgen staining A histochemical and/or cytochemical reagent for quantifying nuclear DNA content or for staining chromosomes in cells; the FEULGEN reaction is done in two separate steps: (1) acid hydrolysis to remove the purine bases from DNA molecules (depurination) and unmask the aldehyde groups of deoxyribose; (2) staining the aldehyde groups by a chemical reaction between the exposed aldehyde groups and the SCHIFF's reagent leading to a color complex; two dyes can be used as SCHIFF's reagent: pararosalinine or thionine, leading both to colored compounds, respectively, magenta and blue; it is prepared as follows: 900 ml distilled water is boiled in a 2-l flask, then 5.0 g basic fuchsine is slowly added; it is swirled for 1 min and then vacuum filtered through two layers of filter paper in a Buchner funnel into a 1-l flask; it is then cooled to 50°C; while swirling 100 ml hydrochloric acid (1.0 N), 10.0 g potassium metabisulfite ($K_2S_2O_5$) is added and swirled for 2 min; the cloudy red solution will become clear blood red; it is stored in the refrigerator for 24 h, then it is removed, warmed to room temperature, and supplemented by 3.75 g activated charcoal; it is shaken vigorously for one minute and then vacuum filtered; the reagent should be clear and colorless; if not, the charcoal step must be repeated; the reagent has to be stored in the refrigerator and used within 48 h; it remains stable for months *meth cyto*

few-seeded: >>> oligospermous

fiber: An elongated, thick-walled, often lignified cell (sclerenchyma) present in various plant tissues, usually providing mechanical support *bot*

fiber crop: A crop plant mainly used for production of fiber (e.g., flax, abaca, hemp, etc.) *agr*

fiber FISH: >>> fiber fluorescence *in situ* hybridization

fiber fluorescence *in situ* hybridization: A variant of conventional >>> FISH technique for the detection and quantification of target sequences in which different probes labeled with different fluorochromes are hybridized to chromosome fibers; the latter are generated by molecular combing *biot cyto meth*

fibril: A small thread or very fine fiber; normally a fiber is constituted of a bundle of fibrils *bot*

fibrillar: >>> fibril

fibrillarin: A nucleolar protein and a component of a nucleolar small nuclear ribonucleoprotein particle (snRNP) thought to participate in the first step in processing pre-rRNA, and associated with U3, U8, and U13 snRNP >>> RNAs; it is the nucleolar scleroderma antigen, rich in N^G-dimethylarginine, and common to the major family of nucleolar snRNPs *bot cyto biot*

fibrograph test beard: The portion of a test specimen (e.g., of cotton) that has been combed into a beardlike shape and that protrudes from the outside of a pair of fibrograph combs *meth*

fibrous: Resembling or having fibers *bot*

fibrous root: A fine, densely branching root that absorbs moisture and nutrients from the soil, for example, in grasses *bot agr*

fibrous root system: Composed of profusely branched roots with many lateral rootlets but with no main or taproot development *bot agr*

Ficoll®: The brand name for an inert, synthetic, highly soluble polymer used as an osmotic agent; sometimes used for suspending protoplasts *prep biot*

field book: It contains the records of single plants or plots made by hand on the experimental field; usually, the running plant/plot numbers are given, and the description of material and the scores are given to the entries by measuring or visual estimation; a free and flexible field note-taking software (Field Book v2.0) has been developed and is permanently updated; it shows high app stability, an in-app tutorial, a new user interface, in-app trait definition, a virtual field map with analysis, and multiple language options; the electronic Field Book is a free, open-source application released under GNU/GP and intended for research purposes only and comes with no warranty; additional Android tablet tools for seed inventory and weighing as well as confirming ordering of planting seed packets are also available at the wheat genetics site *stat meth*

field border planting: Vegetation established on the margins of fields to conserve soil and to provide food and shelter for wildlife, like native shrubs *agr*

field burning: Burning plant residue after harvest (1) to aid in insect, disease, and weed control, (2) to reduce cultivation problems, and (3) to stimulate subsequent regrowth and tillering of perennial crops *agr meth phyt*

field capacity: Water that remains in soil after excess moisture has drained freely from the soil; usually expressed as a percentage of oven-dry weight of soil *agr* >>> Table 44

field collection: A collection of germplasm maintained as living plants that would otherwise be difficult to maintain as seed is commonly maintained in field collections *seed meth*

field crops: >>> cash crops

field diaphragm: A variable diaphragm located in the illumination pathway *micr*

field experiment(ation): An evaluative test whereby the field performance of experimental plants is assessed in comparison to controls *meth* >>> design of experiments

field genebank: A collection of accessions kept as plants in the field (e.g., perennial entries) *meth*

field germination: A measure of the percentage of seeds in a given sample that germinate and produce a seedling under field conditions *meth agr* >>> ground germination rate

field grafting: Grafting a new variety on to an established rootstock already growing in the orchard *hort*

field laboratory: >>> field test

field moisture capacity: The water that soil contains under field conditions *agr* >>> field capacity >>> Table 44

field plane: The set of planes that are conjugate with the focused specimen; in a microscope adjusted for >>> August KOEHLER (1866–1948) illumination it includes the planes of the specimen, the field diaphragm, the intermediate image plane, and the image on the retina *micr*

field plot size: >>> plot size

field recording form: A form used in collecting field information, usually in a format compatible with database management systems, statistical calculations, or computing *stat*

field resistance: Synonymous with general resistance; it is under polygenic control (i.e., controlled by many genes with minor individual effects); in general, field resistance is longer lasting than race-specific resistance; field resistance slows down the rate at which disease increases in the field *phyt*

field test: An experiment conducted under regular field conditions (i.e., less subject to control than a precise contained experiment); more general, the study of a data collection activity in the setting where it is to be conducted *stat meth agr* >>> field experiment(ation)

field trial: Experiments carried out in the field *meth* >>> field test

filament: The stalk of a stamen, which bears the anther *bot* >>> Figure 35

filamentous: Threadlike *bot* >>> filiform

filial generation (F1): The offspring resulting from first experimental crossing of the plants; the parental generation with which the genetic experiment starts is referred to as P1 *gene* >>> F1

filiation: >>> descent

filiform: Threadlike, long, and slender *bot*

filiform apparatus: Outgrowth on one or both synergid nuclei, which engulf the pollen tube apex just prior to double fertilization in the >>> angiosperms *bot*

filing date: In patenting biological material, the date on which a complete patent application is received by the U.S. Patent and Trademark Office *adm* >>> Intellectual Property Rights

filling: >>> beat(ting) up >>> fill planting

fill planting: The planting of plants in areas of inadequate stocking to achieve the desired level of stocking (density), either in plantations, areas of natural regeneration, or other trials *meth fore*

filter hybridization: Hybridization of nucleic acid fragments on a filter (e.g., nitrocellulose) as a carrier *meth gene* >>> Southern transfer >>> Figure 48

final host: >>> definite host

fine(-scale) mapping: Construction of high-resolution genetic linkage maps of numerous DNA markers and genes at a level of less than 0.1 cM using thousands of progeny from a single cross to enable detection of infrequent recombination events *biot meth*

fingerling potato: Any of many specialty potato varieties named for their finger-like shape; fingerlings can be prepared like other potatoes: mashed, baked, roasted, grilled, or boiled; depending on the variety, they can be dry and mealy, or moist and waxy; they come in all the standard colors: red, white, yellow, blue; size varies as well *agr*

fingerprint: The characteristic spot pattern produced by electrophoresis of the polypeptide or DNA fragments obtained through denaturation of a particular protein or DNA with a proteolytic enzyme or other means *biot* >>> fingerprinting >>> Figure 48

fingerprinting: The method for combining electrophoresis and chromatography to separate the components of a protein or DNA; the protein is denatured by means of a proteolytic enzyme and the resulting polypeptide fragments produce a characteristic spot pattern, referred to as a fingerprint, after electrophoresis; it is also used with hydrolyzed fragments of nucleic acids; it became a common and very sensitive method for identification of different plant genotypes *gene* >>> Figure 48

fire blight (*Erwinia amylovora*): A disease of fruit trees, especially of pears and apples that blackens the foliage and is caused by bacterium; it is one of the most destructive diseases of apple (*Malus* ssp.) worldwide; no major, qualitative gene for resistance to this disease has been identified to date in apple *phyt*

firm seeds: Sometimes applied to grass caryopses that are dormant due to seed coats that are impervious to water or gases; in potato breeding, a descriptive term for tubers that are not shriveled or soft (flabby) *bot*

first division restitution (FDR): Results from an abnormal orientation of the spindles right before meiotic anaphase II; non-sister chromatids end up in the same nucleolus; it was found in several crop species, such as potato, rye, etc. *cyto*

first-generation synthetic varieties (Syn-1): Progenies derived by intercrossing a specific set of clones or seed-propagated lines; these may include varieties of normally cross-fertilizing or self-fertilizing crops into which mechanisms have been introduced to maximize cross-fertilization such as male sterility or self-incompatibility; these varieties usually contain mixtures of seed that result from cross-, self-, and sib-fertilization; the variety consists of only the first-generation progenies after intercrossing and cannot be reproduced from seed of the first generation, e.g., "Gahi" pearl millet, "Vitagraze" rye, or "Tempo" alfalfa *bree meth*

first glume: The lowermost of the two typical glumes of a spikelet; attached to the rachilla; empty without flower parts in the axil; odd veined *bot* >>> glume

FIS: International Federation of Seed Trade

FISH: >>> fluorescence *in situ* hybridization

fishtail weeder: >>> asparagus knife

fissable: >>> fissile

fissile: >>> fission

fission: In genetics, the division of one cell by cleavage into two daughter cells, or a chromosome into two arms *cyto*

fitness: The relative ability of a plant to survive and transmit its genes to the next generation *eco* >>> resilience

fixation: The first step in making permanent preparations of tissue, etc., for microscopic study; the procedure aims at killing cells and preventing subsequent decay with the least distortion of structure *prep cyto*

fixation agent: A solution used for the preparation of tissue for cytological or histological studies; it precipitates the proteinaceous enzymes and prevents autolysis, destroys bacteria, etc. (e.g., acetic acid, formalin, Farmer's fixative, etc.) *prep*

fixation index (F_0): A measure of the genetic effects of >>> inbreeding, based on the deficiency of the frequency of heterozygotes compared with those predicted by the >>> HARDY–WEINBERG law *stat*

fixative: >>> fixation agent

fixed effects model: An effect of a treatment, in any experiment, which is concerned only with a certain, particular set of treatments rather than with the whole range of possible treatment effects *stat*

fixed polyploidy: Heterozygosity within the genome is fixed as a result of pairing only between homologous chromosomes derived from the same diploid parent; the largest group of polyploids; the allopolyploids (disomic polyploids) have fixed heterozygosity in the two or more divergent genomes they possess *cyto*

flabellate: fan-shaped bot

flabelliform: >>> flabellate

flaccid: A tissue or body becomes soft and weak, lacking firmness *bot*

flag: An isolated, wilted, or necrotic branch with dead leaves attached *phyt*

flagging: The loss of rigidity and drooping of leaves and tender shoots preceding the wilting of a plant *phys phyt* >>> Figure 62

flag leaf: The uppermost leaf on the grass stem and the last to emerge before the spike *bot*

flag stage: Stage of development in cereals and other grasses at which the sheath and leaf have been produced; in horticulture, the early post-emergence stage of onion seedlings between the crook stage and the emergence of the first true leaf; the bent tip of the seed leaf resembles a flag attached to a staff; sometimes used to characterize the knee stage in onions *bot hort*

flaming: Sterilization of instruments (forceps, scalpel) by an open flame after dipping in alcohol *biot*

flanking region: The DNA sequences extending on either side of a specific locus or gene *biot gene*

flash tape: A metalized plastic tape that produces bursts of light in response to breezes; it is suspended over crops to scare away birds *agr*

flat: A container for holding packs of plant starter cells *hort*

Flc: >>> flowering locus C

fleck: A minute spot *phyt*

flexible genome: >>> dispensable genome

flexuous (spike neck): Wavy or in a more or less zigzag line *bot*

flint maize: A variety of maize, Zea mays ssp. indurata, with very hard-skinned kernels *agr*

floating check: >>> check variety

floating leaf: A leaf swimming on the surface of water (e.g., in lotus) *bot*

floating row cover: A fiber sheet, water- and air-penetrable; it is placed over a row or bed of plants for protection from heat, cold, or insects *hort meth*

flora: The plants of a particular region or period, listed by species and considered as a whole; in general, plants, as distinguished from fauna *bot*

floral axis: The structure through which the palea and flower are attached to the rachilla *bot*

floral induction: The morphological changes in the development of a reproductive meristem from a vegetative meristem; it is the morphological expression of the induced state and usually occurs inside the meristem *phys* >>> Table 42

floral initiation: >>> floral induction

floral meristem: A meristem that gives rise to a flower *bot* >>> Table 42

floral meristem identity gene (Lfy): A gene that acts within the floral primordium to promote the expression of genes that specify floral organ identity *biot gene*

floral primordium: >>> floral meristem

florescence: Anthesis or flowering time or the state of being in bloom *bot*

floret: A single flower consisting of the ovary, stamens, and lodicules together with its enveloping lemma and palea in grasses *bot*

floret initial: >>> floral primordium

floricanes: On raspberries and blackberries, two-year-old canes which bear fruit and then die *hort*

floriculture: The cultivation of flowers or flowering plants *hort*

Florida test: In potato breeding, an integral step in the seed potato >>> certification process of the U.S.; seed growers seeking certification must submit one or more representative samples from each seed lot or from each field for planting in a special (Florida) test plot to test for accumulation of >> virus and viroid diseases; the test results shall show a total less than 5% virus (mosaics, leaf roll, other virus) and spindle tuber viroid as well as 0.25% varietal mixture and 0% >>> bacterial ring rot for certified seed; for >>> foundation seed, the results shall be less than 0.5% of the above viruses and viroids and the same tolerances for bacterial ring rot and varietal mixture; tags for foundation seed cannot be issued until winter test results are available at the end of January each year *seed*

floridean starch: An algal reserve resembling glycogen or amylopectin *bot*

floriferous: Bearing flowers *bot*

florigen: The universal hormone that supposedly causes plants to change from the vegetative to the reproductive state; in 2005, after 70 years of hypothesis O. NIELSON et al. (Science 2005) identified >>> mRNA of the florigen gene *Ft* of >>> thale cress (*Arabidopsis thaliana*) that is produced in leaves and induces flowering when transported to apex tissue; the florigen is triggered by a gene *Constans* (a transcription factor) that activates the formation of the flowering protein; on apex there is another protein *Fd* that responds to the arrival of *Ft* protein; together they initiate flowering of the plant *phys*

flour: The finely ground meal of grain (e.g., from cereals) separated by bolting *agr prep*

floury: >>> chalky

flow cytometry: >>> sorting

flow densitometry: >>> sorting

flower: A typical flower of angiosperms or plants whose seeds are enclosed in an ovary; it is composed of petals, sepals, stamens, and a pistil; the flower morphology contributes to the relative importance of self- and cross-pollination; the two structures directly involved in sexual reproduction are the male stamina and the female pistil; a stamen consists of an anther, which contains the pollen grains, and a filament on which the anther is borne; some flowers (called >>> perfect flowers) have both male and female reproductive organs; some flowers (called >>> imperfect flowers) have only male reproductive organs (stamens) or only female reproductive organs (ovary, style, and stigma); some plants have both male and female flowers, while other have males on one plant and females on another; complete flowers have a stamen, a pistil, petals, and sepals; incomplete flowers lack one of these parts *bot* >>> self-incompatibility >>> Tables 18, 42

flower bud initiation: >>> floral primordium

flower debris: The remains of spent flowers which drop and often accumulate on the surface of the soil; the presence of flower and other plant debris may attract certain insects which feed on or take shelter beneath decaying plant matter *hort*

floweret: One of the segments of a cauliflower head *hort*

flowering hormone: >>> florigen

flowering locus C (*Flc*): A gene that encodes a MADS-box protein that may act as a transcriptional repressor of flowering and plays a central role in the vernalization response *biot phys*

flowering plants: Angiosperms; they produce seeds enclosed in fruit (an ovary); their flowers are used in reproduction; they are the dominant type of plant today; there are over 250,000 species; angiosperms evolved about 140 mya, during the late Jurassic period; they became the dominant land plants about 100 mya (edging out conifers, a type of gymnosperm) *bot*

flow sorting: >>> sorting

fluid drilling: A mechanical procedure for planting seed; pregerminated seeds are suspended in a gel and sowed through a fluid drill seeder; this technology is potentially adaptive for sowing artificial seed, such as somatic embryos or embryoids *agr biot*

fluorescence: Property of certain molecules to absorb energy in the form of light and then release this energy at a longer wavelength than the wavelength of absorption (i.e., at a lower energy level) *micr phy*

fluorescence activated cell sorter: A machine that is used to sort specific cells from a mixed group of cells; the desired cells are first labeled with a specific fluorescent dye, or a gene for a fluorophore is inserted, then passed through a flow chamber that is illuminated by a LASER beam, which causes the labeled cells to fluoresce; the molecules of the fluorescent dye, which are associated with only one type of cell in the mixture, contain chromophores that can be elevated to an excited, unstable state via irradiation with specific wavelengths of light; those chromophores remain in that excited state for a maximum of 10^{-9} sec before releasing their energy by emitting light, and returning to their unexcited state; this fluorescence is a measurable property and the machine utilizes it to separate the desired cells from the rest of the mixture *cyto biot*

fluorescence dye: Several dyes are critical for specific light specters, for example, Hoechst, DAPI, AMCA, Cascade Blue, Fura, Dansylchloride, Fluorescine FITC for green light, Lucifer Yellow, Quinacrin, Chromomycin A3, FITC, NBD Chloride for yellow light, Phycoerythrin, Propdium Iodide, FEULGEN, Auramine, DiOC, Ethidium Bromide for orange light, or Rhodamine or Texas Red for red light *micr* >>> fluorescence *in situ* hybridization >>> fluorescence staining

fluorescence *in situ* hybridization (FISH): A technique for visual detection in the microscope of specific DNA sequences on cytological fixed chromosomes, after hybridization with DNA probes labeled with a fluorochrome; it can be done on both >>> interphase and metaphase chromosomes *cyto meth* >>> chromosome painting >>> BAC clone >>> Figures 52, 53, 54

fluorescence microscopy: The common method of microscopic examination based on observing the specimen in the light transmitted or reflected by it; fluorescence preparations are self-luminous; the tissue is stained by fluorochromes, dyes that emit light of longer wavelength when exposed to blue or ultraviolet light; the fluorescing parts of the stained object then appear bright against a dark background; the staining technique is extremely sensitive *micr* >>> fluorescence dye >>> chromosome painting

fluorescence staining: Very few biological samples are inherently fluorescent such that they can be imaged directly; instead, fluorescence microscopy essentially always uses a fluorochrome that is introduced through some form of staining procedure; fluorescence staining for flow cytometric, fluorescence, or light microscopic analysis; the fluorescence can be directly applied, such as 4'-6-diamidino-2-phenylindole (DAPI); it is known to form fluorescent complexes with natural double-stranded DNA, showing a fluorescence specificity for AT, AU, and IC clusters; because of this property, DAPI is a useful tool in various cytochemical investigations; when DAPI binds to DNA, its fluorescence is strongly enhanced, which has been interpreted in terms of a highly energetic and intercalative type of interaction, but there is also evidence that DAPI binds to the minor groove, stabilized by hydrogen bonds between DAPI and acceptor groups of AT, AU, and IC base pairs; other applications use single-stranded DNA sequences with a fluorescent label to hybridize with its complementary target sequence in the chromosomes, allowing it to be visualized under ultraviolet light; fluorescence staining methods offer several advantages, such as high resolution, live cell staining, and the possibility of dual or multiple labeling; long shelf-life for probes and nucleotides, since there is no radioactive decay, eliminates time-consuming handling, monitoring, and clean up; does not need designated special "hot" areas in lab, and is often quicker than doing autoradiography *cyto meth* >>> fluorescence microscopy >>> fluorescence dye

fluorescent: The color exhibited when the grain or glumes of certain oat varieties are viewed under ultraviolet light or other radiation *phys*

fluorescin: A red crystalline compound, $C_{20}H_{12}O_5$, that in alkaline solutions produces an intense green *chem*

fluorescin diacetate (FDA) staining: Living cells stained with FDA fluorescence in the presence of UV light; the stain is used to assess cell viability of cell cultures, etc. *meth*

fluorochrome: >>> fluorescence dye

flush cut: A pruning cut to remove a tree limb in which the cut is completely flush with the tree; the resulting scar is usually too large to heal efficiently *meth hort*

flush ends: >>> blunt ends

flush season: The plant growth that is produced during a short period *agr*

flux: A flow of matter or energy of which direction, rate, and density can be determined *phys*

focal plane: A plane through a focal point and perpendicular to the axis of a lens, mirror, or other optical system *micr*

focus: The ability of a lens to converge light rays to a single point *micr*; in plant pathology, the site of local concentration of infection of infestation from which secondary spread may occur *phyt*

fodder: Harvested grass or other crop parts for animal feed *agr*

fodder crop: Any crop that is grown for feeding farm animals, such as hay, turnips, mangolds, fodder beet, fodder legumes, and fodder grasses, etc. *agr*

fog treatment: The application of a pesticide as a fine mist for the control of pests *phyt*

foliaceous: Leaflike shape *bot*

foliage: The leaves of plants *bot*

foliage blight: >>> late blight

foliar feeding: Feeding plants by spraying liquid fertilizer on the leaves *agr hort*

foliar nutrient: Any liquid substance applied directly to the foliage of a growing plant for the purpose of delivering an essential nutrient in an immediately available compound *agr*

foliar treatment: Treating plants by spraying liquid or dry insecticides, pesticides, or herbicides on their leaves *agr hort phyt*

foliation: Leafing *bot*

foliole: Leaflet *bot*

foliose: Leafy *bot*

follicetum: A fruiting structure made up of an aggregate of follicles *bot*

follicle: A fruit with a simple pistil that at maturity splits open along one suture, e.g., in milkweed *bot*

follicular fruit: >>> follicle

food crop: >>> food plant

food grain: Cereal seeds most commonly used for human food, chiefly wheat and rice *agr*

food legume: Legume plants with nutritive value for humans, directly and indirectly consumed *agr*

food plant: Plants with nutritive value for humans, directly and indirectly consumed *agr* >>> Table 55

food-processing enzyme: An enzyme used to control food texture, flavor, appearance, or nutritional value, e.g., of such proteins including amylases that break complex polysaccharides down to simpler sugars and proteases that tenderize meat proteins *biot meth*

food species: >>> food plant

foot (ft): Equals 30.48 cm

foot rot: Rotting involving the lower part of the stem–root axis, but not the distal parts of the roots *phyt* >>> eyespot disease

footprinting: A method used to determine the length of nucleotide chains that are close to a protein (which bind to DNA); for example, certain types of drugs act by binding tightly to certain DNA molecules in specific locations *gene meth* >>> genetic footprinting >>> fingerprinting

forage: Feed from plants for livestock such as hay, pasturage, straw, silage, or browse *agr*

forage crop: Crop plants for feeding of livestock (e.g., alfalfa, clover, maize) *agr* >>> forage

forage shrub: >>> forage

forb: A non-grass-like herb of the range or a small grass-like plant *agr*

forceps: >>> pincers

forcing: The practice of bringing a plant into growth or flower (usually by artificial heat or controlling daylight) at a season earlier than its natural one; it is sometimes applied in order to synchronize flowering dates of parental plants for crosses in the greenhouse *meth*

forecrop: A section of a crop that is harvested first *agr*

fore-ground selection: YOUNG and TANKSLEY (1989), demonstrated that a large amount of DNA from the donor can remain around the target gene even after many generation of backcrossing in backcross breeding approaches; this surrounding material contributes toward "linkage drag" especially if the donor parent is a wild relative; so markers are used to select the same progeny in which recombination near the target gene have as little chromosome segment as possible; this is called "fore-ground selection" *meth*

foreign DNA: DNA that is not found in the normal genome concerned; usually it is directly or indirectly introduced into a recipient cell by several experimental means *biot meth*

foreign matter: Foreign objects, such as stones, sand, chaff, straw, or other seeds mixed with the crop product *seed agr*

forest belt: A protective forest plantation in the form of lines made around plots of arable land, pastures, gardens, along irrigation channels and roads and gullies, on mountain and hill slopes, etc.; the forest belt is intended for improvement of the hydrological situation in a given territory and its microclimate, as well as for prevention of water and wind erosion *agr*

forestero cocoa beans: The most commonly grown and used cocoa beans; these beans make up about 90% of the world's production and are grown primarily in West Africa *agr* >>> cacao

forest tree breeding: The genetic manipulation of trees, usually involving crossing, selection, testing, and controlled mating, to solve some specific problem or to produce a specially desired product *fore gene*

forked: Sometimes a morphological deviation of the common root in beets (e.g., sugarbeet) *agr* >>> fangy root

form: A botanical category ranking below a variety and differing only trivially from other related forms (e.g., in waxiness of leaves) *tax* *forma*: Lowest category of species with sporadic variation on one or two characters *bot* >>> *forma specialis*

forma specialis [(*f.* sp.): *pl formae speciales* (*ff.* sp.)] A >>> taxon characterized from a physiological standpoint (especially host adaptation); a subspecific taxon usually denoting the host genus of which the organism is a parasite, e.g., *tritici* to denote wheat in *Puccinia graminis f.* sp. *tritici*; in general, biotypes of pathogen species that can infect only plants that are within a certain host genus or species *phyt tax* >>> Tables 12, 51

formaldehyde: A colorless gas readily soluble in water *chem phys*

formalin: An aqueous solution of formaldehyde commonly used as a fixative that functions through cross-linking protein molecules *chem prep*

formal seed system: Seed production, control, and distribution activities carried out by the public and commercial sector; it may include breeding *meth*

formazan: Colorless when dissolved in water; the chemical 2,3,5-triphenyl tetrazolium chloride is reduced to the red-colored chemical triphenyl formazan on contact with living, respiring tissue; the amount of formazan formed is used as a measure of seed viability, as it reflects oxidative metabolism *chem seed* >>> tetrazolium test

forward genetics: A genetic method of gene identification, starting with a genetically distinct phenotype and then working towards the identification of the gene underlying the phenotype *gene meth biot*

forward mutation: A mutation that alters (usually inactivates) a wild-type allele of a gene; they occur at a rate of about 10^{-6} per locus and generation *gene* >>> backmutation

forward selection: Choosing good individuals out of a progeny test for possible use in seed orchards and/or subsequent generations of breeding *fore hort* >>> backward selection

fuscous: Dusky grayish brown *bot*

fosmid: A single-copy F factor-based >>> cosmid, which is a cloning vector that allows the packing of cosmid-sized >>> DNA fragments; a fosmid vector permits efficient cloning of DNA fragments of ~40 kb in a single-copy vector based on the F factor of >>> *Eschericha coli biot*

fosmid library: A collection of >>> DNA fragments that is carried out by a >>> fosmid *biot*

fosmid *in situ* hybridization: >>> fosmid >>> *in situ* hybridization

fossil: Markedly outdated; a remnant or an impression on an organism preserved from a past geologic age *evol* >>> Figures 68, 69 >>> crop evolution >>> Table 32

foundation population: >>> founder population

foundation seed: Seed stocks increased from >>> breeder seed, i.e., the progeny of breeder, select, or foundation seed; handled so as to closely maintain the genetic identity and purity of the variety; it is a sort of certified seed, either directly or through >>> registered seed *seed* >>> Table 28 >>> base seed

foundation single cross: It refers to a single cross in the production of a double, three-way, or top cross *meth* >>> Table 37

foundation stock: The original source of seed from which all other grades of seed are produced *meth seed* >>> foundation seed

founder effect: Genetic drift due to the founding of a population by a small number of individuals *gene*

founder mutation: Any mutation that occurs in the genome of one single individual and is subsequently transferred to many other individuals of the same species; usually, it is a single nucleotide exchange and embedded in flanking DNA, which is highly conserved in the different progeny individuals, i.e., it is transmitted to the progeny as a sort of block *gene*

founder population: The first generation of a breeding population, e.g., in forest tree breeding often the initial plus trees; this is usually the starting point of calculations *fore hort meth* >>> base population >>> Figure 51

founders: >>> founder population

four-flap grafting: Using this technique, a stock plant with a primary stem or lateral limb of 1–3 cm diameter is used; a cut is made straight across the trunk or limb with sharp pruning shears at the point the graft is wished; if possible, one or two side branches below the grafting point are left, however, they are cut back to about 15 cm; it keeps the tree vigorous and protects it from sunscald; it also keeps the >>> scion from becoming too tall or whip-like and breaking off; a rubber band is rolled 7 cm down from the top to the stock; on the stock plant where the horizontal cut was made, four vertical, equally spaced cuts 1–3 cm long are prepared; the cuts have to penetrate the bark down to the interior wood *hort meth*

four-way cross: >>> double cross

foxtailing (in pine trees): When pines (e.g., >>> Monterey pine [*Pinus radiata*]) are transferred south to tropical climates, it is fairly common to see a few individuals that fail to branch; there are also examples of extremely narrow-crowned phenotypes even in temperate climates, e.g., Jeffrey pine (*Pinus jeffreyi*) growing in the Sierra Nevada Mountains of California; there are both genetic and environmental components involved in foxtailing; e.g., a selected strain of Caribbean pine that was certified not to foxtail in Australia reportedly exhibited 80% foxtailing when grown in Puerto Rico; foxtailing decreases with altitude, stand density, and soil quality; the cause is thought to be due to hormone imbalances induced by exotic environments *fore phys*

fractional factorial design: An experiment design in which parameters are studied at several levels with only a fraction of the possible parameter-level combinations included in the experiment *stat* >>> design of experiment >>> Table 40

fractional replication: As the name implies, fractional replication of factorial designs involves eliminating some of the treatment combinations from a factorial design instead of simply confounding them with groups; a fractional replication of a factorial design might contain only a small fraction of the number of possible treatment combinations, e.g., in a fractional replication of a $2 \times 2 \times 2 \times 2$ design only a half or only a fourth of the treatment combinations might be included in the experiment; in a fractional replication of a 3×3 design only a third of the treatment combinations might be included; fractional replication designs are similar to confounded factorial designs in that modular arithmetic is used to divide a factorial design into independent sets of treatment combinations; fractional replication designs differ from confounded factorial designs in that in fractional replication some of the sets of treatment combinations are eliminated from the design instead of being confounded with groups *stat* >>> design of experiment >>> fractional factorial design

fragment banding pattern: Bands made when many copies of the same DNA fragment move together through an electrophoresis gel; the pattern of bands reveals the number of different size fragments in a sample *chem biot*

fragrant: The odor or smell that a plant exudes from either its flowers or its foliage *hort*

frameshift mutation: A mutation that is caused by a shift of the reading frame of the mRNA (usually by the insertion of a nucleotide) synthesized from the altered DNA template *gene*

frankenfood: A derogatory reference to food derived from genetically engineered plants

fraternal twin(s): >>> dizygous

freco bract: A mutant bract type in cotton in which the bracts curl outward, exposing flower buds and bolls *bot*

free nuclear division: Refers to mitotic division of nuclei without accompanying cytokinesis, i.e., nuclei divide in a common cytoplasm with the cells walls only forming around each later *bot*

free radicals: Reactive oxygen species (ROS) and highly toxic compounds produced by many living systems as a by-product of normal metabolism; they have a two-edged role in biology; in animals, they are known to cause cancer by damaging DNA but they also play an important role in the immune system where they are used to kill viruses and bacteria invading the body; they have a similar two-edged role in plants; these powerful toxins play an essential role in plant growth and all cell growth is controlled by the production of these highly reactive and therefore very toxic free radicals; a gene (*Rhd2*) has been identified that makes a protein, which produces free radicals; it was demonstrated that controlled production of free radicals by *Rhd2* stimulates calcium channels in the membranes of cells resulting in calcium being taken up by the cells; the accumulation of calcium in turn activates cell expansion; in plants where the *Rhd2* gene was inactivated by a mutation, the roots and root hairs were stunted *phys biot chem*

free-threshing: Spikes with brittle (fragile) rachis; for example, in wheat, it is controlled by a single recessive allele; the fragility of the spike reveals the main difference between wild and cultivated forms *bot gene agr*

freeze drying: A technique used in the preservation of living material, whereby water is removed under vacuum while the tissue remains frozen *meth seed*

freeze preservation: >>> cryopreservation

freezing injury: A type of winter injury caused by the combined effects of low temperatures, wind, and insufficient soil moisture; the low-temperature injury is associated with ice formation in the extracellular spaces resulting in freeze-induced dehydration and metabolic changes; the plasma membrane remains attached to the cell wall, causing the cell to collapse *phys agr*

frego-bract mutant: A mutant bract type in cotton in which the bracts curl outward, exposing flower buds and bolls *gene*

frequency distribution: A specification of the way in which the frequencies of members of a population are distributed according to the values they exhibit *stat*

frequency histogram: A step-curve in which the frequencies of various arbitrarily bounded classes are presented as a graph *stat*

frequency-dependent selection: Selection where the fitness of a type varies with its frequency (i.e., whereby a genotype is at an advantage when rare and at a disadvantage when common) *gene meth*

frequency table: A way of summarizing a set of data; a record of how often each value (or set of values) of the variable in question occurs; may be enhanced by the addition of percentages that fall into each category; used to summarize categorical, nominal, and ordinal data; may also be used to summarize continuous data once the data set has been divided up into sensible groups *stat*

friabilins: Proteins that determine the adhesion of the starch granules; for example, in wheat they determine the difference between hard and soft wheat; soft wheats show strong friabilins, which bind the granules and hence the endosperm fractionates into large fragments, whilst hard wheats contain weaker friabilins and hence fracture into small fragments; the latest molecular and biochemical evidence associates them with differences in the structure of proteins (puroindolines) *phys meth* >>> http://wwql.wsu.edu/

friction polisher: A type of whitener using the friction between the rice grains to remove the bran layer *seed meth*

Friedman test: A non-parametric test to compare t > 2 treatments when the randomized block design has been used; the data are ranked within each block, and the rank sums for each of the treatments are used in the test statistic "X^2" *stat*

Fri: >>> FRIGIDA

FRIGIDA (Fri): A plant gene that controls whether or not a plant needs a winter period before it will flower; it ensures that flowering is delayed until after winter so that the plant flowers in the favorable conditions of spring; this process of >>> vernalization (the acceleration of flowering by a cold period of 3–8 weeks at +4 to +8°C) is an important characteristic in many plants; *FRIGIDA* is a single copy gene in >>> thale cress and encodes a protein product of 609 amino acids; the protein has no significant match with any protein sequences or sequence fragments of known function; the action of an active *Fri* is dependent on an active *Flc* gene, and *Fri* is known to increase the level of RNA from *Flc*; *FRIGIDA* is dominant; summer annual types are recessive biot *gene* >>> *Flc*

frilling: A method of killing trees by inflicting a series of cuts around the bole or stem and applying an herbicide to the wounds; frilling or >>> girdling of trees may be used to reduce the density of a stand or to kill individual undesirable trees *fore*

frond: The leaves of ferns and other cryptogams; includes both stipe and blade; commonly used to designate any fernlike or featherlike foliage *bot*

frost crack: A vertical split in the wood of a tree, generally near the base of the bole, due to internal stresses and low temperatures *fore hort*

frost damage: >>> killing frost

frost killing: >>> killing frost

frost-lifting of seedlings: >>> heaving

frost mold: Certain strains of at least three species of epiphytic bacteria (*Pseudomonas syringae*, *P. fluorescens*, *Erwinia herbicola*) are present on many plants; they serve as ice-nucleation-active catalysts for ice formation at temperatures as high as –1°C *phyt*

frost resistance: The capacity to survive temperatures below 0°C *phys*

frost tolerance: The ability of plants to survive very harsh winter conditions (i.e., to withstand subzero temperatures) *agr*

fructiferous: Bearing fruits bot

fructification: The process of forming a fruit body or the fruit body itself *bot*

fructose: A sugar ($C_6H_{12}O_6$) that occurs abundantly in nature as the free form, but also with glucose in the form of the disaccharide sucrose *chem phys*

fruit: Strictly, the ripened ovary; more loosely, the term is extended to the ripened ovary and seeds together with the structure with which they are combined *bot*

fruit acidity: An important characteristic to determine the marketability of >>> apple, *Malus domestica*; low acidity in apple is presumably determined by one recessive gene; an allele of a major gene appears to have complete dominance with a function of lowering acidity; sweetness and acidity in apple and >>> pear inherit independently and can be organoleptically evaluated separately, but less accurately in pear than in apple; in apple, the acidity – decreasing with time – of the unripe fruit is strongly indicative of that of the eating-ripe fruit; sugar-increasing with time-not before the fruit is picking ripe; sugar content in apple and pear, and the pH in pear, appear to be normally distributed; the pH in apple shows a >>> segregation into an acid and a low-acid group, which occurred in both the unripe and ripe stage; the segregation ratio between these groups is found to be highly variable; the mean acidity and sugar content of apple and pear progenies is significantly determined by that of the parents *phys*

fruit-bearing branchlet: In some fruit trees (e.g., in apple) the main type of fruit-bearing wood *hort*

fruit cycle: The period or length of time between fruit set and maturity *bot*

fruit drop: The premature abscission of fruit before it is fully ripe; it is a common process; in many fruits there are certain peak periods of fruit drop, e.g., apple fruits are lost immediately following pollination (post-blossom drop), when the embryos are developing rapidly (June drop), and during ripening (preharvest drop); as with leaf fall, fruit drop is associated with low auxin levels; auxin sprays have been used to prevent excessive fruit drop *hort phys*

fruiting body: A specialized organ for producing spores (e.g., in mushrooms) *bot*

fruiting wood: On grapevine, the 1-year-old canes that will produce the current year's fruit *hort*

frutex: >>> shrub

fruticous: >>> shrubby *bot*

f. sp.: >>> forma specialis

fuchsine: A greenish, water-soluble, solid, coal-tar derivative, obtained by the oxidation of a mixture of aniline and the toluidines, that forms deep-red solutions; used chiefly as a dye *micr cyto* >>> FEULGEN staining

full bloom: The developmental stage at which essentially all florets in the inflorescence are in anthesis *bot*

full diallel: >>> complete diallel

full pedigree: All parents, grandparents, etc., of a particular genotype are known back to natural population *gene* >>> www.geneflowinc.com

full sibs: Individuals that have both parents in common; the mating of full sibs is the most extreme form of inbreeding that can occur in bisexual, diploid plants *gene meth*

full-tree harvesting: Cutting and removing an entire upper portion of a tree consisting of trunk, branches, and leaves or needles *fore meth*

fumigant: Vaporized >>> pesticide used to control pests in soil, storage rooms, and greenhouses (e.g., methyl bromide) *phyt seed*

fumigate: >>> fumigation

fumigation: To expose to smoke or fumes, as in disinfecting or in exterminating vermin, fungi, etc. *seed meth*

functional diversity: Genetic diversity as assessed by variation in transcribed regions of the genome that is known to be associated with a biological function *gene biot*

functional food: A functional food is consumed as part of a usual diet that is similar in appearance to, or may be, a conventional food, and is demonstrated to have physiological benefits and/or reduce the risk of chronic disease beyond basic nutritional functions *agr*

functional genomics: The science of how the genes in organisms interact to express complex traits – that is, the field of research that aims to determine the function of newly discovered genes; attempts to convert the molecular information represented by DNA into an understanding of gene functions and effects; functional genomics also entails research on the protein function (proteomics) or, even more broadly, the whole metabolism (metabolics) of an organism *gene biot* >>> Figure 43

functional hemizygosity: An autosomal gene is functionally hemizygous if only one of the two copies in a diploid cell is expressed *gene*

functional map: A physical map of a genome in which the locations of genes with known functions are depicted and/or any graphical depiction of molecules that interact with each other and whose individual functions are experimentally proven *biot meth*

fungal pathogen: >>> pathogen

fungicide: A chemical, physical, or biological agent that inhibits fungal growth *phyt*

fungicide control: >>> fungicide

fungus (fungi *pl***):** Any member of the kingdom Fungi (or division Thallophyta of the kingdom Plantae), comprising single-celled or multinucleate organisms that live by decomposing and absorbing the organic material in which they grow (e.g., mushrooms, molds, mildews, smuts, rusts, and yeasts) *bot phyt* >>> Tables 51, 52 >>> https://english.tau.ac.il/

funicle: The stalk connecting the ovule with the placenta on the ovary wall *bot*

funiculus: >>> funicle

furlong English unit of length equal to 660 feet, 220 yards, 40 poles, 10 chains or one-eighth of an English mile (metric: 201.168 meters); literally a "furrow long" from Old English *furlang*, the length of a single furrow that a horse could plow without stopping and hence the length of a medieval field *agr*

furrow: A trench in the earth, made by a plow *agr*

furrow application: Placement of pesticides with seed in furrow at time of sowing *agr phyt*

furrow irrigation: Small, shallow channels guide water across the surface of a leveled field; crops are typically grown on a ridge or raised bed between the furrows *agr*

furrow weed: A weed that grows on ploughed (plowed) land *phyt agr*

6-fururylaminopurine: >>> kinetin

Fusarium **ear blight:** All cereals may be infected during wet warm weather at flowering; the disease is most common in wheat; it starts with small brown spots on the glumes that develop to leave florets or whole spikelets prematurely bleached; pink or orange colonies of the fungus can often be seen at the base on the spikelet *phyt*

Fusarium wilt (of melon, *Fusarium oxysporum f. sp. melonis*): It causes serious economic losses in melon (*Cucumis melo*); two dominant resistance genes have been identified, *Fom1* and *Fom2*, which provide resistance to races 0 and 2 and races 0 and 1, respectively; however *Fusarium* race 1.2 overcomes these resistance genes; a partial resistance to *Fusarium* race 1.2 that has been found in some Far East accessions is under polygenic control *phyt* >>> Figure 62

fused polar nucleus (primary endosperm nucleus): The dikaryon in the angiosperm embryo-sac which after fusion with a sperm cell forms the endosperm *bot*

fusiform: Shaped like a spindle or cigar, tapering at both ends *bot*

fusion gene: A hybrid gene created by joining portions of two different genes (to produce a new protein) or by joining a gene to a different promoter (to alter or regulate gene transcription) *biot*

fusogen(ic): A fusion-inducing agent used for protoplast agglutination in somatic hybridization studies (e.g., polyethylene glycol [PEG]) *biot*

fuzz: beard of grain *bot* >>> lint (linters)

G

G1, G2, G3 …: First, second, and third self-bred generations >>> I1, I2, I3 …

G1: The period of the eukaryotic cell cycle between the last mitosis and the start of DNA replication (synthesis) *cyto*

G2: The period of the eukaryotic cell cycle between the end of DNA replication (synthesis) and the start of next mitosis *cyto*

GA: >>> gibberellic acid

GA testing: >>> gibberellic acid (response) testing

gap: Single-stranded region in dsDNA *biot*

gapping: >>> beat(ting) up

galactose (Gal): A component ($C_6H_{12}O_6$) of milk sugar (lactose) *chem phys*

gall: A localized proliferation of plant or parasite tissue that produces an abnormal growth or swelling, usually caused by pathogenic organisms, nematodes, or insects *phyt* >>> www.pflanzengallen.de/

gallery: An insect tunnel in bark or wood of plants *phyt*

gall midge Any midge of the family Cecidomyidae, the larvae of which form characteristic galls on plants *zoo phyt* >>> www.pflanzengallen.de/

Galton's regression law: Individuals differing from the average character of the population produce offspring, which, on the average, differ to a lesser degree but in the same direction from the average as their parents *stat gene*

gamete: A specialized haploid cell whose nucleus and often cytoplasm fuses with that of another gamete in the process of fertilization, thus forming a diploid zygote *bot* >>> gametocide >>> Tables 2, 9

gamete and embryo storage: Storage of ovules, pollen, or zygotes outside their original source; almost invariably this means >>> cryopreservation in liquid nitrogen *meth biot*

gametic (phase) disequilibrium: In relation to any two loci, the occurrence of gametes with a frequency greater than or less than the product of the frequency of the two relevant alleles *gene* >>> association mapping >>> gametic (phase) equilibrium

gametic (phase) equilibrium: In relation to any two loci, the occurrence of gametes with a frequency equal to the product of the frequency of the two relevant alleles, e.g., loci A and B are in linkage equilibrium if the frequency of the gamete A_nB_n equals the product of the frequencies of alleles A_n and B_n *gene* >>> association mapping >>> gametic (phase) disequilibrium

gametic selection: The influences acting to cause differential reproductive success of one allele over another in a heterozygote *gene*

gametocidal: >>> gametocide

gametocidal gene: A gene encoding a product that destroys cells that divide to produce the gametes *gene* >>> cuckoo chromosome

gametocide: A chemical agent used to selectively kill either male or female gametes; it is used in hybrid seed production of autogamous crops (e.g., barley or wheat); there are approaches to replace the (non-optimal) gametocyte-sterilization technology as conventional method; to achieve a conditional male sterility of the female line a barnase gene is split into two fragments and both fragments are placed under control of a tapetum-specific promoter; by *in vivo* complementation of two inactive barnase fragments *via* intein-mediated *trans*-splicing, a functional barnase protein is produced in the tapetum and leads to pollen ablation; additionally, by using an *in vivo* recombination system, the complementary gene fragments can be located on the same locus on homologous chromosomes ("linked in repulsion"); in the female crossing line both gene fragments are present, thus causing the plant to be male-sterile; as a result of the subsequent hybridization step, each progeny inherits only one gene fragment, rendering all hybrid progenies fully fertile *meth* >>> Figure 55

gametoclonal: Regenerated from a tissue culture originating from gametic cells or tissue *biot*

gametoclonal variation: Variation among regenerants obtained from pollen and/or anther culture *biot* >>> Figure 58

gametoclone: Plants regenerated from cell culture derived from meiocytes or gametes *biot*

gametocyte: A cell that will undergo meiosis to form gametes *bot*

gametogamy: The fusion of the sexual gametes and the formation of the zygotic nucleus *bot*

gametogenesis: The formation of gametes from gametocytes *bot*

gametophyte: A haploid phase of the life cycle of plants during which gametes are produced by mitosis; it arises from a haploid spore produced by >>> meiosis from a diploid sporophyte *bot* >>> Figure 28 >>> Table 16

gametophytic apomixis: Agamic seed formation in which the >>> embryo sac arises from an unreduced initial; it includes both diplospory and apospory *bot*

gametophytic self-incompatibility: Self-incompatibility is based on the genotypic and phenotypic relationship between the female and male reproductive system; alleles in cells of the pistil determine its receptivity to pollen; the phenotype of the pollen, expressed as its inability to effect fertilization, may be determined by its own alleles, referred to as gametophytic incompatibility *gene* >>> self-incompatibility

gamet precision divider: A type of mechanical halving device for subdividing a large seed sample to obtain a smaller working sample for germination or purity analysis; it has an electrically operated rotating cup into which the seed is funneled to be spun out and into one of two spouts *seed*

gamopetalous: Having the petals of the corolla more or less united *bot*

gangrene of potato: Necrosis or death of soft tissue due to obstructed circulation, usually followed by decomposition and putrefaction (*Phoma exigua* var. *foveata*) *phyt*

gap: Single-stranded region in dsDNA *biot*

gapped DNA: A duplex DNA molecule with one or more internal single-stranded regions *chem*

gas chromatograph: An analytical technique for identifying the molecular composition and concentrations of various chemicals in agricultural products, water, or soil samples *meth prep*

gas transfer: The rate at which gases are transferred from gas into solution; it is an important parameter in fermentation systems because it controls the rate at which the organism can metabolize; gas transfer can be done by several methods, including use of small gas bubbles that diffuse faster than larger ones thanks to their larger surface area per unit of volume, or spreading the liquid out, e.g., in a thin sheet, or in a thin permeable tube, as in a hollow fiber bioreactor *biot*

GAUSS distribution: >>> normal distribution

Gaussian curve: >>> normal distribution

G band: >>> G banding

G banding: A special staining technique for chromosomes that results in a longitudinal differentiation by >>> Giemsa stain, which is a complex of stains specific to the phosphate groups of DNA; the characteristic bands produced are called G bands; these bands are generally produced in AT-rich heterochromatic regions *cyto meth* >>> differential staining

GCA: General combining ability >>> combining ability (CA)

GC box: A component of many eukaryotic promoters, especially those from constitutively expressed genes; the consensus sequence for the GC box is 5′-GGGCGG-3′ *biot*

geitonogamy: When neighboring flowers of the same plant can achieve pollination, as opposed to xenogamy *bot*

gelatin(e): A nearly transparent, glutinous substance obtained by boiling the bones, ligaments, etc., of animals; used in making jellies *chem prep*

GEM: Genetically engineered microorganism *biot*

geminivirus: A single-stranded DNA virus that causes serious diseases in cereals, vegetables, and fiber crops worldwide; like nanoviruses, it is transmitted by either whiteflies or leafhoppers; whitefly-transmitted geminiviruses are all in the genus Begomovirus, which typically have bipartite genomes (DNA A and DNA B) comprising circular DNAs of ~2,700 nucleotides, although some have been shown to be monopartite, having only a DNA A component; in contrast, the nanoviruses are a recently established group of plant viruses that are transmitted by either aphids or planthoppers, and have multipartite genomes comprising circular DNAs of ~1,000 nucleotides *phyt*

gemma (gemmae *pl*): A cell or cluster of often bud-like cells, borne on the >>> gametophyte, that can reproduce the plant vegetatively *bot*

gemmiferous: Bearing buds *bot*

gemmule: Imagined particles of inheritance proposed by C. DARWIN (1809–1882) as part of his >>> "Pangenesis" theory; this appeared in his book *The Variation of Animals and Plants under Domestication*, published in 1868; gemmules, also called plastidules or pangenes, were assumed to be shed by the organs of the body and carried in the bloodstream to the reproductive organs where they accumulated in the germ cells or gametes; they thus provided a possible mechanism for the inheritance of acquired characteristics, as proposed by J.-B. LAMARCK (1744–1829, which DARWIN believed to be a cause of the observed variation in living organisms *bio evol*

gender: Differences between any two complementary organisms of the same species that render them capable of mating *gene*

gene: The hereditary unit that occupies a fixed position on the chromosome, which through transcription has a specific effect upon phenotype gene; it may mutate to various allelic forms; in molecular biology, a segment of DNA including regulatory sequences (promoter, operator, terminator) that encodes an RNA and/or protein molecule *biot* >>> cistron >>> average effect of a gene >>> Table 9

gene action: The expression of the gene based on the transcription into complementary RNA sequences, the subsequent translation of mRNA into polypeptides, which may form a specific protein *gene*

gene activation: Different mechanisms of repression and activation of genes *gene*

genealogical depth: Age of the most recent common ancestor of a sample of alleles *eco evol*

genealogy: Descent from an original form or progenitor and its record >>> ancestry *evol*

gene amplification: The more or less specific production of multiple copies of a gene *gene*

genebank: An establishment in which both somatic and hereditary genetic material are conserved (seeds, pollen, whole plants, extracted DNA); it stores, in a viable form, material from plants that are in danger of extinction in the wild and cultivars that are not currently in popular use; the stored genetic material can be called up when required; the normal method of storage is to reduce the water content of seed material to around 4% and keep it at 0°C or less (–20°C); all stored stocks are periodically checked by germination tests; although 7.4 million plant accessions are stored in 1,750 germplasm banks around the world, only a small portion of the accessions has been used so far to produce commercial varieties *meth* >>> gene library

gene-based single nucleotide polymorphism: Any >>> SNP that is located in an exon, an intron, or a >>> promoter of a gene *biot gene* >>> SNP

gene-based SNP: >>> gene-based single nucleotide polymorphism

gene center: Refers to the center of origin of a given crop plant *evol gene* >>> center of diversity >>> center of origin

gene chip: A device in which a large number of different DNA probes are carefully placed at specific locations on a glass slide (e.g., spotted arrays) or by putting probes at specific positions on some surface; it involves labeling the sample instead of the probe, propagating thousands of copies of the labeled sample across the chip and then washing away any copies of the sample that do not remain attached to some probe; because the probes are attached to specified positions on the chip, if a labeled sample is detected at any position on the chip, it can easily be known which probe was able to hybridize its complement; is most commonly used to measure the expression level of various genes in an organism; each expression level gives a picture of the rate at which a specific protein is being produced in an organism's cells at any given time *biot meth*

gene cloning: Insertion of a DNA fragment carrying a gene into a cloning vector; subsequent propagation of the recombinant DNA molecule in a host organism results in many identical copies of the gene (clones) in a form that is more easily accessible than the original chromosomal copy *gene biot*

gene conservation population: A population used to maintain original genetic variation in species *meth*

gene construction: An experimentally engineered gene with functional and non-functional properties *gene biot*

gene containment: >>> biological containment

gene content: The absolute number of genes per genome or chromosome *gene* >>> Figure 71

gene conversion: A process whereby one member of a gene family acts as a blueprint for the correction of the other; this can result in either the suppression of a new mutation or its lateral spread in the genome *gene*

gene dosage (dose): The number of times a given gene is present in the nucleus of a cell and/or individual *gene*

gene duplication: A process in evolution in which a gene is copied twice; the two copies lie side by side along the same chromosome *gene*

gene editing: The process of inserting, deleting and/or silencing, or replacing DNA sequences within a genome; such techniques utilize methods that frequently can be observed in nature; for example, gene silencing techniques make use of a mechanism found in the nematode worm *Caenorhabditis elegans*; the purpose of gene modification, besides scientific research, is for gene therapy; examples of gene editing methods are >>>

CRISPR, >>> gene silencing, >>> gene transfer, gene modification, and *in vitro* transcription *biot*

gene expression: The phenotypic manifestation of a gene depending on the different levels of gene activation, or the process by which the information in a gene is used to produce a protein; in molecular genetics, the full use of the information in a gene via transcription and translation leading to production of a protein and hence the appearance of the phenotype determined by that gene; gene expression is assumed to be controlled at various points in the sequence leading to protein synthesis; this control is thought to be the major determinant of cellular differentiation; serine/arginine-rich proteins are major modulators of alternative splicing, a key generator of proteomic diversity and flexible means of regulating gene expression likely to be crucial in plant environmental responses; in 2011, data on >>> *Populus* ssp. demonstrate that plants may influence and regulate their gene expression depending stress situations, such as drought, infections, or malnutrition *gene* >>> nursery effect >>> Table 9

gene expression profiling: The analysis of gene expression level of many genes at the same time; technologies that can be used to obtain this information include Northern blot, >>> DNA chips, high-density array spotted on glass or membranes, and quantitative techniques >>> PCR *biot*

gene family: A group of similar or identical genes, usually along the same chromosome, that originate by gene duplication of a single original gene; some members of the family may work in concert, others may be silenced and become >>> pseudogenes *gene*

gene flow: The spread of new genes, which takes place within an interbreeding group as a result of crossing with immigrants, or the transfer of genetic material by interbreeding from one population of a species to another population (same or related species), thereby changing the composition of the gene pool of the receiving population *evol*

gene-for-gene theory: In certain plant–pathogen interactions, a gene for resistance in the host corresponds to and is directed against a gene for >>> virulence in the >>> pathogen; in flowering plants, four families of disease resistance genes confer gene-for-gene resistance to a wide array of pathogens by recognizing the products of the corresponding pathogen avirulence genes *phyt biot*

gene frequency: The number of loci at which a particular allele is found divided by the total number of loci at which it could occur for a given population, expressed as a proportion or percentage *gene*

gene genealogy: Relationship among genes through direct descent *eco*

gene gun: >>> particle gun >>> gene transfer

gene insertion: The addition of one or more copies of a normal gene into a defective chromosome *biot*

gene interaction: Modification of gene action by a non-allelic gene or genes, generally the interaction between products of non-allelic genes *gene*

gene island(s): Cluster of genes that is separated from neighboring clusters by regions of repetitive DNA *gene*

gene knockdown: The reduction of a gene's activity to very low levels through various mechanisms, e.g., RNA interference; the method is a choice if >>> gene knockout is lethal for the organism *biot meth*

gene knockout: The elimination of a functioning gene from an organism; the resulting organism has all but one of its normal complement of genes, allowing determination of the function of the missing genes by observing the biochemistry, gene expression, or phenotype; the two most popular techniques for knocking-out genes are insertional inactivation (of which homologous recombination is a specific technique) and antisense inactivation *biot*

gene library: In molecular genetics, a random collection of cloned DNA fragments in a number of vectors that ideally includes all genetic information of that species *biot*

gene linkage: >>> linkage

gene location: Determination of physical or relative distances of a gene on a particular chromosome *gene*

gene locus: The fixed position that a gene occupies on a chromosome *gene*

gene machine: In common literature, an idiomatic description of an automated oligonucleotide synthesizer *biot*

gene map: A graphic presentation of the linear arrangement of a chromosome or segment; it shows the relative distance between loci gained in linkage experiments *gene* >>> Figure 43

gene mapping: Determination of the position of genes on a DNA molecule *biot* >>> Figure 43 >>> genetic mapping

gene mining: >>> allele mining

gene mutation: A heritable change of gene revealed by phenotypic modifications; it occurs in the coding region of a gene leading to the synthesis of a defective polypeptide, or the promoter is leading to an aberrant regulation of the adjacent gene *gene*

gene number paradox: >>> C value paradox

gene pairs: The two copies of a gene present in a diploid, one on each homologous chromosome *gene*

gene patenting: Protection provided by governmental or non-governmental institutions to the discoverer of new genes, genotypes, strains, or testing procedures so that the detailed information can be declared publicly; currently, a synthetic gene but not a natural gene itself can be patented, however its sequenced functional unit or specific utilization can be a matter of patent *biot adm* >>> Intellectual Property Rights

gene pool: The reservoir of different genes of a certain plant species or lower and higher taxa available for crossing and selection; it is possible to differentiate between (1) primary gene pools (consists of those species that readily hybridize, produce viable hybrids, and have chromosomes that may freely recombine), (2) secondary gene pools (consists of those species with a certain degree of hybridization barrier due to ploidy differences, chromosome alterations, or incompatible genes), and (3) tertiary gene pools (consists of distinct species or higher taxa with strong crossing barriers); in general, the total number of genes or the amount of genetic information that is possessed by all the reproductive members of a population of sexually reproducing organisms *gene* >>> primary gene pool >>> secondary gene pool

gene pool system: It consists of three informal categories in order to provide a genetic perspective and focus for cultivated plants *meth gene*

general combining ability: >>> combining ability (CA)

generalist: A species that tolerates a broad range of environmental conditions; a generalist has a broad ecological niche *eco*

generalized incomplete Trojan squares: Require the omission of two or more complete rows from a complete >>> Trojan square; unlike simple incomplete Trojan squares, not all generalized incomplete Trojan squares of given size are equally efficient, and the arbitrary omission of two or more rows from a complete Trojan square will not necessarily give the most efficient possible design; for designs where YOUDEN rectangles exist, an efficient (m × n)/k incomplete Trojan square can be obtained by superimposing k suitable (m × n) YOUDEN rectangles and regarding each row-by-column intersection of the superimposed design as a block of size k; incomplete Trojan designs with m < (n − 1) normally lack factorial balance in the plots-within-blocks stratum and do not have a simple factorial analysis of variance *stat* >>> design of experiment

generalized lattice design: Each replicate is subdivided into incomplete blocks typically of five or six plots; sometimes referred to as "alpha design" (PATTERSON and WILLIAMS 1976) *meth stat* >>> design of experiment

generalized resistance: >>> general resistance

general resistance: Resistance against all biotypes of a pathogen; non-specific host-plant resistance *phyt*

general seed blower: A precision seed blower used to aid in separating light seed and inert matter from heavy seed *seed*

generation time: The time required for a culture to double its cell number *biot*

generative: Sexual processes *bot*

generative meristem: Gives rise to parts, such as floral organs, that ultimately produce fruits and seeds *bot*

generative nucleus: A haploid nucleus of a pollen grain that produces two sperm nuclei by mitosis (pollen grain mitosis) *cyto* >>> sperm nucleus >>> Figures 25, 35

gene recombination: >>> recombination

gene redundancy: The presence of a gene in multiple copies due to polyploidy, polytenic chromosomes, gene amplification, or chromosomal duplications *gene*

gene repair: The correction of mutations in a gene within living cells, using bifunctional oligonucleotides or chimeric oligonucleotide-directed >>> gene targeting *biot meth*

gene retention: Retention of gene duplicates is facilitated by coordinated reduction of expression levels, rendering both copies necessary to fulfill the biological role of the ancestral gene, while creating redundancy at the level of their molecular function; however, these differences in gene duplicability can reflect relative strengths of purifying selection in the three lineages imposed by factors, such as effective population sizes; the greater permissiveness of slightly deleterious events in organisms would allow the accumulation of redundant gene copies that would be purged under more intensive selection pressures; selection intensity on gene retention can be deduced from phyletic distributions of member genes; their presence or absence in the considered species; constraints on gene duplicability may be manifested in gene copy-number variations in independently evolving lineages; detailing ancestral gene relations therefore enables quantification of properties of ortholog sequence divergence, phyletic retention, and copy-number variation; although orthologous relations are not defined by gene function, identifying "equivalent" genes in modern species nevertheless provides a hypothesis of similar functionality, especially for single-copy orthologs *biot*

generic: Referring to the genus *tax*

generic diversity: The differences between individuals of different genera *evol eco tax*

gene sanctuaries: *In situ* conservation of germplasm under natural conditions, an area protected from human interference *bot eco seed*

gene silencing: Any mechanism that silences a gene, such as various sequence homology-dependent silencing mechanisms (>>> RNAi, >>> quelling, >>>PTGS, >>> TGS, >>> VIGS), gene knockout by homologous recombination, imprinting, DNA methylation, and transvection; it is the suppression of >>> gene expression, e.g., of the gene for polygalacturonase which causes fruit to ripen, of the gene for P34 protein in soybeans, *via* a variety of methods, e.g., *via* RNA interference (RNAi), chemical genetics, effect of certain viruses, zinc finger proteins, or sense or antisense genes; the term was introduced by the American A. FIRE in 1998 *biot meth* >>> bean golden mosaic virus >>> chloroplast >>> target validation

gene space: Long gene-rich regions on a chromosome that contain the vast majority of genes, separated by long gene-poor regions in a >>> genome of given species; it is a common feature of plant species, which have a large genome size owing to the abundance of repetitive DNA (transposons and retrotransposons) in their genome *cyto gene*

gene splicing: The enzymatic attachment (joining) of one gene or part of a gene to another; also removal of introns and splicing of exons during mRNA synthesis *biot*

gene stacking: In general, the process of introducing several traits into one plant or plant type by either selective (human) or natural breeding methods; in genetic engineering, the insertion of two or more (possibly synthetic) genes into the genome (e.g., the bat gene from *Bacillus thuringiensis* and a gene for resistance to a specific herbicide) *biot*

gene string: Formerly, the central chromosome thread on which the genes are, assumedly, arranged in sequence *gene*

gene substitution: The replacement of one allele by another mutant allele in a population by natural or directed selection *gene*

gene surfing: The identification of genic DNA sequences in an anonymous DNA by comparing it against sequences in genome and/or protein databases that already have assigned functions *meth biot*

gene symbol: Designating a gene, usually by an abbreviation of the name or description of a given gene; in the past, genes have been described by Latin names, and subsequently the one- to three-letter abbreviations; currently, several systems of naming and symbolization are in use, although a comparable, uniform symbolization is sought *gene* >>> genetic nomenclature

genet: A genetic individual, resulting from a single sexual fusion (zygote), consisting of one to many >>> ramets, and usually genetically distinct from all other genets *gene* >>> clone

gene tagging: The labeling of a gene by a marker gene or specific DNA sequence closely linked with the gene in question *gene biot* >>> Figure 43

gene targeting: The insertion of antisense DNA molecules *in vivo* into selected cells of the body in order to block the activity of undesirable genes; these genes might include oncogenes or genes crucial to the life cycle of parasites *biot* >>> gene repair >>> Figure 43

gene technology: In a broad sense, it is the artificial transfer of genes between cells or individuals by means of molecular and *in vitro* techniques; prerequisites are (1) the presence of a gene, which is available as a DNA fragment, (2) a cloneable DNA, (3) the fragment has to be transferable by different systems, (4) the incorporation of the DNA fragment into a recipient cellular genome has to be feasible, (5) the transformed cells have to be regenerable into a normal plant, and (6) the gene that was transferred must be expressed in the alien genetic background *biot*

genetic: Pertaining to the origin or common ancestor or ancestral type *gene*

genetic advance: The expected gain in the mean of a population for a particular quantitative character by one generation of selection of a specified percent of the highest-ranking plants *gene*

genetically enhanced: An organism that has undergone the process of inserting new genetic information into existing cells in order to modify the organism for the purpose of improving one or more of its characteristics *biot*

genetically modified organisms (GMO): A term, currently used most often in official discussions, that designates crops, which carry new traits that have been inserted through advanced genetic engineering methods, e.g., flavor-saver ("FlavrSavr") tomato, Roundup™-ready soybeans, *Bt* cotton, or *Bt* maize; since their introduction during the 1990s, worldwide there have been no verifiable ill effects reported from the extensive consumption of products from GM crops over seven years by humans and livestock; it is concluded that the risks to human health from GM crops currently on the market are very low; for the current generation of GM crops, the most important issue was their potential effect on farmland and wildlife; detailed field experiments on current GM crops show that in a range of environments they are very unlikely to invade the countryside, nor are they likely to be toxic to wildlife *agr biot* >>> delayed maturity >>> (SCHLEGEL 2007)

genetic architecture: The distribution of genetic variation in a species, usually described hierarchically as variation at the regional, local, family, and individual levels, and also relating to proportions of additive and non-additive inheritance *meth*

genetic assimilation: Eventual extinction of a natural species as massive pollen flow occurs from another related species and the natural species becomes more like the related species *tax gene* >>> gene flow

genetic association: Refers to a concurrence greater than predicted by chance between a specific allele and another trait that may or may not have a genetic basis; evaluation of association requires the study of unrelated individuals; association studies may prove useful in identifying a genetic factor in a trait; except when linkage disequilibrium exists, association is not due to genetic linkage and should not be confused with it *gene* >>> association mapping >>> Figure 43

genetic background: The remaining genetic constitution when a particular locus or allele of a given individual or taxon is studied *gene*

genetic balance: The optimal interaction of coadapted genes, alleles, or genetic systems within a given individual *gene*

genetic block: The reduction or termination of an enzyme activity caused by a specific gene mutation *gene*

genetic code: The set of correspondences between base triplets in DNA and amino acids in protein; these base triplets carry the genetic information for protein synthesis *gene* >>> Table 38

genetic complement: >>> genome

genetic complementation: The complementary action of homologous sets of genomes *gene* >>> Figure 71

genetic correlation: The correlation between the genotypic values of two characters with respect to the genetic character *stat gene*

genetic death: Death of an individual without reproducing; caused by mutationally arisen alleles that reduce the fitness of a genotype and/or taxon *gene evol*

genetic determinism: Doctrine that all acts, choices, and events are the inevitable consequence of antecedent sufficient causes, genetic makeup, or the sum of one's gene *gene*

genetic distance: A measure of gene differences between individuals or populations measured by the differences of several characters; such distance may be based on phenotypic traits, allele frequencies, or DNA sequences (e.g., genetic distance between two populations having the same allele frequencies at a particular locus and based solely on that locus is zero); the distance for one locus is maximum when the two populations are fixed for different alleles; when allele frequencies are estimated for many loci, the genetic distance is calculated by averaging over these loci *gene*

genetic distance estimation by PHYLIP: The most popular phylogenetics computer program that can be used to estimate genetic distance between populations; most components of PHYLIP can be run online; one component of the package, GENDST, estimates genetic distance from allele frequencies using one of the three methods; GENDST can be run online using the default options to obtain genetic distance matrix data; the PHYLIP program CONTML estimates phylogenies from gene frequency data by maximum likelihood under a model in which all divergence is due to >>> genetic drift in the absence of new mutations and draws a tree *stat* >>> Figure 43

genetic distancing: The collection of the data on phenotypic traits, marker allele frequencies, or DNA sequences for two or more populations, and estimation of the genetic distances between each pair of populations; from these distances, the best representation of the relationships among all the populations may be obtained *biot gene* >>> Figure 43

genetic drift: The random fluctuations of gene frequencies in a population such that the genes amongst offspring are not a perfectly representative sampling of the parental genes; it is often observed in small isolated populations; small populations are, on a simple basis of probability, more likely to have atypical mixtures of genes than large ones; as a consequence, >>> recessive genes are more likely to come together and be expressed; other gene combinations unlikely in "normal" populations are also likely to occur; the result is that the population will, through successive generations, drift away from the norm *gene*

genetic engineering: The manipulation of DNA using >>> restriction enzymes, which can split the DNA molecule and then rejoin it to form a hybrid molecule – a new combination of non-homologous DNA; the technique allows the bypassing of all the biological restraints to genetic exchange and mixing, and even permits the combination of genes from widely differing species *biot >>>* chloroplast *>>>* Table 39

genetic equilibrium: An equilibrium in which the frequencies of two alleles at a given locus are maintained at the same values generation after generation; a tendency for the population to equilibrate its genetic composition and resist sudden change is called genetic homoeostasis *gene >>>* homoeostasis

genetic erosion: The loss of genetic information that occurs when highly adaptable cultivars are developed and threaten the survival of their more locally adapted ancestors, which form the genetic base of the crop *gene evol*

genetic footprinting: An approach to the identification of the function of a large number of putative genes of a microorganism, such as may have been uncovered by genome sequencing; a transposable element is inserted into a large number of sites in the genome of the microorganism, and the mutagenized population is then grown under a wide variety of conditions that may suggest a gene's function; after many population doublings under each set of conditions, the DNA is extracted and used as a resource for analysis of as many of the putative genes as is desired; PCR primers are constructed to hybridize with the transposable element and with a putative gene; PCR amplification using the primer pair will produce a spectrum of bands on separation by polyacrylamide-gel electrophoresis, one from each mutation that can grow under the set of conditions, i.e., a genetic footprint of the gene *biot meth*

genetic gain: The change achieved by artificial selection in a specific trait; the gain is usually expressed as the change per generation or the change per year; it is influenced by selection intensity, parental variation, and heritability *gene*

genetic homoeostasis: *>>>* genetic equilibrium

genetic homology: The identity or near identity of DNA sequences, genes, or alleles *gene*

genetic information: The information contained in a sequence of nucleotide bases in a nucleic acid molecule *gene >>>* Table 38

genetic instability: Different mechanisms that give rise to phenotypic variation *gene*

genetic interaction: The interaction between genes resulting in different phenotypic expressions *gene*

geneticist: A specialist in genetics *gene*

genetic load: The average number of lethal mutations per individual in a population *gene evol*

genetic (linkage) map: The linear arrangement of gene loci on a chromosome, deduced from genetic recombination experiments; a genetic map unit is defined as the distance between gene pairs for which one product of meiosis out of a hundred is recombinant (i.e., it equals a recombination frequency of 1%) *gene*

genetic mapping: The process of determination of a genetic map; in former times, genetic mapping was usually on a gene-by-gene basis; until molecular markers, there were no specific projects to map entire genomes; therefore, genetics will never be the same because of greater precision (<0.5 cM), fewer crosses, use of F2 data, fewer breeding generations, more reliance on maps, mapping kits, etc. *gene*

genetic marker: Any phenotypic difference, controlled by genes, that can be used for studying recombination processes or selection of a more or less closely associated target gene *gene*

genetic material: All single- or double-stranded DNA carrying genetic information or that is a substantial part of the genetic information *gene >>>* Table 38

genetic nomenclature: The designation of genes by abbreviated gene descriptions or symbols, usually a beginning capital letter of the abbreviation represents a dominant allele, while a small beginning letter refers to a recessive allele *gene >>>* gene symbol

genetic pollution: Uncontrolled spread of genetic information (frequently referring to transgenes) into the genomes of organisms in which such genes are not present in nature *agr eco*

genetic polymorphism: An occurrence in a population of two or more genotypes in frequencies that cannot be accounted for by recurrent mutation *gene*

genetic recombination: A number of interacting processes that lead to new linkage relationships of genes *gene*

genetic relatedness (r): A quantitative measure of genetic relatedness between individuals; in diploid species, r = 1/2 between full >>> siblings, or parent and offspring *gene stat*

genetic resistance: Resistance against pathogens or pests due to specific or general gene action *gene* >>> passive resistance

genetic resources: The gene pool in natural and cultivated stocks of organisms that is available for human exploitation *gene*

genetics: The scientific study of genes and heredity *gene*

genetic segregation: >>> segregation

genetic sterility: A type of male sterility conditioned by nuclear genes, as opposed to cytoplasmic sterility *gene*

genetic stock: A variety or strain known to carry specific genes, alleles, or linkage groups *gene*

genetic system: The organization of genetic material in a given species and its method of transmission from the parental generation to its filial generations *gene*

genetic targeting: >>> gene targeting

genetic thinning: In seed orchards, it refers to the removal of orchard genotypes based on their supposed breeding value *meth hort fore* >>> rouging >>> thinning

genetic trait: A distinguishable feature, characteristic, quality of character, or phenotypic feature of a developing or developed individual that is linked to a physical or genetic marker *gene* >>> epigenetic trait

genetic trespass: The movement of unwanted genes into a crop by crosspollination from a crop in a neighboring field *gene*

genetic variability: Individuals differing in their genotypes due to mutational, recombinational, or selective mechanisms *gene* >>> Figure 60

genetic use restriction technology (GURT): >>> terminator technology

genetic variance (V_{gen}): A portion of phenotypic variance that results from the varying genotypes of the individuals in a population; together with the environmental variance, it adds up to the total phenotypic variance observed amongst individuals in a population; it is divided into additive (resulting from differences between >>> homozygotes, V_{add}) and dominance variance (resulting from specific effects of various alleles in heterozygotes, V_{dom}); $V_{gen} = V_{add} + V_{dom}$; the quotient V_{add}/V_{dom} is termed heritability in a narrow sense *stat* >>> heritability >>> Figure 60 >>> Table 36

genetic vulnerability: The potentially dangerous condition which results from a narrow genetic base, i.e., the susceptibility of genetically uniform crops to damage or destruction caused by outbreaks of a disease or pest or unusually poor weather conditions or climatic change [e.g., the Irish potato famine of the 1840s in which more than 1 million Irish starved to death as a consequence of a massive attack of late blight (*Phytophtora infestans*) that destroyed the Irish potato crop] *agr phyt*

gene tracking: Following the inheritance of a particular gene from generation to generation gene *meth*

gene transfer: The physical transfer of a gene by crossing, chromosomal manipulation, and molecular means; in biotechnology, different methods are described, such as (1) microinjection, (2) insertion via microprojectiles (particle gun, particle bombardment) using silicon fibers as carriers of the DNA, (3) direct transfer, (4) electroporation, (5) liposome fusion, or (6) a vector-mediated transfer *biot* >>> nano-biolistic >>> Figure 27

gene translocation: The transfer or movement of a gene or gene fragment from one chromosomal location to another; often it alters or abolishes expression *gene cyto biot* >>> position (positional) effect

genic male sterility (GMS): Male sterility caused by genetic components, e.g., the chromosomal XYZ-4E-*ms* system; in 2006, Chinese scientists proposed a cytogenetic method for producing hybrid seed; using a nuclear gene for male sterility and an alien chromosome to obtain a pollination control system has been proposed in bread wheat; they transferred the alien chromosome 4E to the monogenic recessive male-sterile mutant "Lanzhou", and then established an efficient cytogenetic system of maintaining the male sterility of that mutant; after crossing the 4E disomic addition line "81529" (2n = 6x = 44) as male to "Lanzhou" (msms), they obtained the 4E monosomic addition line [2n = 43 (*msms*)] in the F3 generation that showed light-blue seed color; the line was homozygous for the male-sterile gene (msms) of the host wheat and also had good self-fertility; the self-fertilized seeds segregated into 64.3% white (non-blue) with genotype 2n = 42 (*msms*), 32.1% light-blue with genotype 2n = 43 (*msms*), and 3.6% deep-blue with genotype 2n = 44 (*msms*) grains; all plants grown from the white grains were completely male sterile; all plants grown from the light-blue grains had good self-fertility whose progeny seeds segregated into white, light-blue, and deep-blue grains again in succeeding generations; all plants grown from the deep-blue grains were self-fertile and retained their deep-blue color in the succeeding generations; when white grain male-sterile lines were crossed using any cultivar as the male parent, the hybrid plants had male fertility restored; this is the basis for the 4E-ms system of producing hybrid wheat *breed meth* >>> cytoplasmic male steriliy

geniculate: Bent abruptly like a knee *bot*

genocopy: A gene or genotype causing the same phenotype as another gene or genotype; genocopies are the basis of genetic heterogeneity and important in genetic diagnosis and counseling *gene*

genome The total genetic information carried by a single set of chromosomes in a haploid nucleus; example genome sizes: lamda phage 48.5 kb, *Escherichia coli* 4,500 kb, yeast 1.6×10^4 kb, *Drosophila* 1.2×10^5 kb; in 2000, the total nucleotide sequence of a plant genome, *Arabidopsis thaliana*, was identified for the first time; the sequenced genome contains about 120 kb; common wheat has one of the largest genomes with over 150,000 genes as compared to humans with 35,000 or rice with about 40,000; in 2006, the smallest genome size was discovered; the complete genome sequence of the psyllid symbiont, *Carsonella ruddii*, consists of a circular chromosome of 159,662 bp, averaging 16.5% GC content; it is by far the smallest and most AT-rich bacterial genome yet characterized; the genome has a high coding density (97%) with many overlapping genes and reduced gene length; genes for translation and amino acid biosynthesis are relatively well-represented, but numerous genes considered essential for life are missing, suggesting that *Carsonella* may have achieved organelle-like status *gene* >>> Figure 71 >>> Tables 8, 14, 53 >>> lamda phage

genome affinity index (GAI): Used for comparing the degree of homology among parental genomes; the mean number of bivalent-equivalents is divided by the basic chromosome number *cyto*

genome allopolyploids: >>> allopolyploid

genome analysis: The study of the genome by combination of cytogenetics, karyotyping, and crossing *cyto gene* >>> Figure 71 >>> Table 8

genome doubling: >>> autopolyploidization

genome editing methods: Techniques such as >>> CRISPR/Cas, >>> TALENs, >>> zinc-finger nucleases, >>> meganucleases, >>> oligonucleotide-directed mutagenesis (ODM), and >>> base editing provide options for simple, time-saving, and cost-effective breeding, enabling a precise modification of DNA sequences; they have already been promoted for a wide range of plant species; although the application of genome-editing induces less unintended modifications (>>> off-targets) in the genome compared to classical mutagenesis techniques, off-target effects are a prominent point of criticism as they are supposed to cause unintended effects, e.g., genomic instability or cell death; the rapid adoption in plant breeding is demonstrated by a considerable number of market-oriented applications *meth biot*

genome equivalent: A statistical measure for the extent of representation of a genome in a genomic library *biot stat*

genome fingerprint map: >>> BAC map

genome formula: >>> genome symbol

genome map: Any graphical depiction of the linear order of all sequence elements (genes, promoters, or repetitive DNA) in a genome as composed by, e.g., genome mapping *meth biot*

genome mutation: Spontaneous or induced changes in the number of complete chromosomes that result either in >>> polyploids or >>> aneuploids *cyto gene* >>> Figure 37

genome scrambling: Independent assortment and recombination of gene, genetic factors, and genomic structures, e.g., in the F2 population, two inbred lines were mated to form a heterozygous F1 generation; these F1 organisms are then interbred to form an F2 generation; independent assortment and recombination will scramble the genomes in these F2 organisms, generating genotypic variation; biotechnical approaches or mutation experiments may also scramble genomes by breaking and joining the genetic material in inappropriate places, i.e., genetic engineering results in widespread mutations – within the inserted gene, near its insertion, and in hundreds or thousands of locations throughout the genome *gene biot*

genome symbol: The description of specific genome by a symbol, usually a capital letter with or without a specification *gene* >>> List of Crop Plants

genome transplantation: A procedure in which DNA from one species is transplanted into a cell of another species, and after repeated division of hybrid cells the original cells are destroyed; only achieved in bacteria so far *biot*

genomic imprinting: The phenomenon whereby genes function differently depending on whether they are inherited from the maternal or paternal parent; this is thought to be caused by information superimposed on DNA sequences, which is different in male and female gametes; such information is transmitted, or inherited, in somatic cells but usually erased and reset in the germ line; it is due to methylation of one of the alleles depending on its origin *gene* >>> Table 43

genomic *in situ* hybridization (GISH): An *in situ* hybridization technique that uses total genomic DNA of a given species as a probe and total genomic DNA of another species as a blocking DNA; it is based on >>> fluorescence *in situ* hybridization; it is a useful method to detect interspecific or intergeneric genome differentiation, chromosome rearrangements (translocations), and substitutions or additions *cyto meth* >>> Figures 52, 53, 54

genomic library: A type of DNA library in which the cloned DNA is from a genomic DNA of the plant and/or organism; since genome sizes are relatively large compared to individual cDNAs, a different set of vectors is usually employed in addition to plasmid and phage *biot* >>> bacterial artificial chromosomes (BAC) >>> yeast artificial chromosomes (YAC) >>> cosmid >>> Figure 46

genomic heteromorphism: Variation in genomic nucleotide sequences in different genotypes due to insertion or deletion of some nucleotide sequences, or accumulation of base-pair changes *gene biot*

genomics: The scientific study of genes and their role in the structure, growth, health, and disease of a plant (e.g., how a certain number of genes contributes to the shape, function, and the development of the organism); by mapping the genetic makeup, or genome, of crop species, one can identify the exact position and function of individual genes; genome mapping has revealed striking similarities in the genomes of different crop species, such as rice, wheat, barley, and rye; this information is already helping to broaden the scope and precision of current breeding programs *biot*

genomic selection (GS): A method in which the number of polymorphic bands resembling the recurrent parent is used as the selection criterion; despite important strides in marker technologies, the use of marker-assisted selection has stagnated for the improvement of quantitative traits; biparental mating designs for the detection of loci affecting these traits (QTL) impede their application, and the statistical methods used are ill-suited to the traits' polygenic nature; GS has been proposed to address these deficiencies; it predicts the breeding values of lines in a population by analyzing their phenotypes and high-density marker scores; a key to the success of GS is that it incorporates all marker information in the prediction model, thereby avoiding biased marker effect estimates and capturing more of the variation due to small-effect QTL; in simulations, the correlation between true breeding value and the genomic estimated breeding value has reached levels of 0.85 even for polygenic low-heritability traits; this level of accuracy is sufficient to consider selecting for agronomic performance using marker information alone; such selection would substantially accelerate the breeding cycle, enhancing gains per unit time; it also would dramatically change the role of phenotyping, which would then serve to update prediction models and no longer to select lines; conventional genomic selection approaches use breeding values to evaluate individual plants or animals and to make selection decisions; multiple variants of breeding values and selection approaches have been proposed, but they suffer two major limitations; first, selection decisions are not responsive to changes in time and resource availability; second, selection decisions are not coordinated with related decisions such as mating and resource allocation; in 2018, three new genomic selection approaches were proposed that attempt to address these two limitations *biot*

genomic technology: Includes DNA synthesis, sequencing, genotyping, and expression profiling, >>> proteomics (peptide synthesis, protein sequencing, and mass spectrometry); more recently, it includes applications of nanobiotechnology, isolation, imaging, and characterization of single molecules *meth biot*

genotoxic: Refers to substances and circumstances inducing mutants and damage of the heritable material *gene*

genotype: The genetic constitution of an organism, as opposed to its physical appearance (phenotype); usually, it refers to the specific allelic composition of a particular gene or set of genes in each cell of an organism, but it may also refer to the entire genome *gene* >>> Figure 60 >>> Tables 2, 9

genotype–environment interaction: Besides allelic and non-allelic interactions, a third type of interaction, namely between genes and their environment; several statistical procedures, such as analysis of variance, the AZZALINI/COX test, the HILDEBRAND procedure, the KUBINGER approach, and the De KROON/Van der LAAN technique, can be applied for the analysis of genotype environment interactions in cross-classified data sets from cultivar performance yield trials with rows = cultivars and columns = environments (locations and/or years); the procedures HILDEBRAND/KUBINGER and De KROON/Van der LAAN are non-parametric methods based on ranks, while analysis of variance and the AZZALINI/COX test proceed from the original absolute yield data; the AZZALINI/COX and De KROON/Van der LAAN methods are based on the crossover concept of interaction (different rank orders), while the other methods are based on the usual concept of interaction (deviations from additivity of main effects); for an analysis of usual interactions the procedures HILDEBRAND/KUBINGER and analysis of variance are approximately equivalent; for the crossover concept of interaction, the AZZALINI/COX approach might be recommended, especially if one is particularly interested in rank changes between environments within genotypes *gene stat* >>> Figure 60

genotypic: Phenomena and processes that are associated with the genotype *gene* >>> Figure 60

genotypic variance: >>> genetic variance (V_{gen})

genotyping: The process of determining the genotype, i.e., the total of all genetic information contained in an individual plant and/or organism *biot meth*

genus (genera pl): A taxonomic grouping of similar species *tax* >>> List of Crop Plants

genus name: Name of a taxonomic group *tax* >>> List of Crop Plants

geologic time: Division and history of life on earth *evol*

geometric mean: The square root of the product of two numbers *stat*

geophyte: A land plant that survives an unfavorable period by means of underground food-storage organs (rhizomes, tubers, >>> bulbs, e.g., onions, tulips, potato, asparagus, Jerusalem artichoke, etc.) *bot*

geosmin: A chemical produced by a common bacterium, *Streptomyces coelicolor*, that is found in most soils; it gives freshly turned soil its distinctive smell ("earthy" smell); geosmin is a Greek word and translates as "smell of the earth" *agr*

geotaxis: Oriented movement of a motile organism toward or away from a gravitational force *bot*

germ: The embryo; or a collective name given to the embryonic roots and shoot, and the scutellum tissue of the grain *bot*

germability: The degree of potential for germination *bot* >>> germination test

germ cell: >>> gamete

germicidal: Refers to any substance or condition that kills the embryo *seed*

germinal: Referring to the germ or germination *bot*

germinal selection: The selection during gametogenesis against induced mutations that retard the spread of mutant cells *gene*

germination: The beginning of growth of a seed, spore, or other structure, usually following a period of dormancy and generally in response to the return of favorable external conditions; when it takes place, a root is produced that grows down into the soil (e.g., in wild wheat the dispersal unit bears two pronounced awns that balance the unit as it falls; the awns are also able to propel the seeds on and into the ground; the arrangement of cellulose fibrils causes bending of the awns with changes in humidity; silicified hairs that cover the awns allow propulsion of the unit only in the direction of the seeds; the dead tissue is analogous to a motor, fueled by the daily humidity cycle, the awns induce the motility required for seed dispersal); at the same time, a stem and leaves are growing upward *bot* >>> awn >>> Table 47

germination test: A procedure to determine the proportion of seeds that are capable of germinating under particular conditions; commonly, a standard germination is conducted on a 400-seed sample at +25°C for 7 days and seedlings are evaluated in accordance with the >>> Association of Official Seed Analysts (AOSA) Rules for Testing Seeds; several analysts evaluate each sample, all fungal species are identified, and when abnormal seedlings exist the primary abnormal-type is noted *seed* >>> germinator

germinative cell: >>> germ cell

germinator: An apparatus with which seed germination is realized under more or less controlled conditions *seed prep*

germ line: The lineage of cells from which the gametes are derived and which therefore bridge the gaps between generations, unlike somatic cells in the body of an organism *bot*

germplasm: The hereditary material transmitted to offspring through the germ cells and giving rise in each individual to the cells *gene* >>> Figure 43

germplasm bank: >>> genebank

germplasm collection: >>> genebank

germplasm enhancement: Any activity that includes gene transfer via sexual and asexual means from germplasm accessions, or increasing the frequencies of desirable genes in crop gene pools that will be used for developing parents or cultivars *meth seed*

germ pore: An area, or hollow, in a spore wall through which a germ tube may come out *bot*

germ tube: The filament that emerges when a spore germinates *bot phyt* >>> pollen tube

GGE biplot analysis: GGE stands for genotype main effect (G) plus genotype by environment interaction (GE), which is the only source of variation that is relevant to cultivar evaluation; mathematically, GGE is the genotype by environment data matrix after the environment means are subtracted; first developed by GABRIEL (1971) it is a scatter plot that graphically displays both the entries (e.g., cultivars) and the testers (e.g., environments) of two-way data: in breeding and genetics data, testers can also be traits, genetic markers, etc.; when the two-way data are subjected to singular value decomposition, they are decomposed into three matrices (singular value matrix; entry eigenvector matrix; tester eigenvector matrix); the singular value matrix is a diagonal matrix, and can be somehow partitioned into the entry and tester eigenvector matrices; after singular values are partitioned, the positions of the entries in the biplot are defined by the entry eigenvector matrix and those of the testers by the tester eigenvector matrix *stat*

giant cell(s): Enlarged, multinucleate cell formed in roots of plants by repeated nuclear division without cell division; it can be a natural phenomenon or can be induced by secretion of certain plant-parasitic nematodes *phyt*

gibberella ear rot (in maize): Caused by the fungal pathogen *Fusarium graminearum*; a serious disease of maize grown in northern climates; the infected maize grain contains toxins that are very harmful to livestock and humans; a maize gene that encodes a putative 267-amino acid guanylyl cyclase-like protein (ZmGC1) was characterized and shown to be associated with resistance to this disease; the putative ZmGC1 amino acid sequence is 53% identical and 65% similar to AtGC1, an Arabidopsis guanylyl cyclase *phyt*

gibberellic acid (GA): A group of growth-promoting substances; they regulate many growth responses and appear to be a universal component of seeds and plants *phys* >>> short-straw mutant >>> aleurone layer

gibberellic acid (response) testing (GA testing): The dwarfing alleles *Rht-B1b* and *Rht-D1b* (previously known as *Rht1* and *Rht2*, respectively) have been used in wheat improvement programs worldwide; dwarf and semidwarf cultivars generally have higher grain yield potential and improved lodging resistance under irrigated and high input conditions; those characteristics enabled the >>> Green Revolution in wheat; the deployment of the *Rht* genes by >>> CIMMYT and by many other public breeding programs has led to the replacement of taller wheats with shorter cultivars; in the late 1990s, over 80% of registered wheat cultivars globally carried at least one dwarfing allele; the *Rht-B1b* and *Rht-D1b* alleles, which occur at homoeologous loci on chromosomes 4B and 4D of wheat, respectively, reduce sensitivity to gibberellic acid, which is necessary for stem elongation; in favorable environments, the reduced demand for assimilates by a shorter stem results in improved assimilate partitioning to the developing head, leading to higher spikelet fertility and more but smaller grain per head; semidwarf wheats have smaller leaves, but compensate with increased photosynthetic rates, resulting in a biomass similar to that of tall lines; the relative yield advantage of dwarf and semidwarf cultivars varies with spring or winter habit, genetic background, and environmental conditions; the benefits of the dwarfing alleles are more pronounced in high-yielding winter wheat environments and in high-yielding spring wheat locations at latitudes less than 40°, however, under heat and drought stress, there may be no benefit of the dwarfing alleles in spring wheat; determination of the *Rht* genotypes is nearly impossible on the basis of plant height alone as an indicator; the major genes determining plant height are normally classified into two groups depending on their reaction to gibberellic acid: *Rht-B1b* and *Rht-D1b* are insensitive to GA, so exogenous application of GA to plants with those alleles does not restore the tall phenotype; the presence of dwarfing alleles can be determined by lack of seedling response to GA; for fast testing seedlings are treated with 10 ppm GA3 and grown at +20°C until the three leaf stage, when the distance between the stem base and the end of the second leaf sheath is measured in millimeters; usually

15 seedlings per genotype are scored for statistical comparison along with 10 control plants that are grown under the same conditions but omitting GA3; gibberellic acid may also be used as a >>> gametocide in common onion *meth phys* >>> gibberellins >>> gibberellic acid >>> short-straw mutant

gibberellin: The generic name of a group of plant hormones that stimulate the growth of leaves and shoots; they tend to affect the whole plant and do not induce localized bending movements; they are thought to act either at a transcriptional level or as inducers of enzymes; first isolated from the fungus *Gibberella fujikuroi*, which causes the Bakanae disease in rice *phys phyt* >>> aleurone layer >>> short-straw mutant >>> www.plant-hormones.info/gibberellins.htm >>> P. HEDDEN, 2012, Gibberellin biosynthesis; DOI: 10.1002/9780470015902. a0023720

gibbous: Swollen on one side (e.g., at the second glume of *Sacciolepis*) *bot*

Giemsa stain: >>> banding

gigantism: Abnormal overdevelopment due to an increase in cell size (hypertrophy); for example, in roots of crucifers infected with club root (*Plasmodiophora brassicae*); abnormal overdevelopment as a result of an increase in the number of cells in response to a disease-production agent (hyperplasia), for example, witches broom, cankers, galls, leaf cure, or scab *phys phyt* >>> Figure 62 >>> www.pflanzengallen.de

ginned lint: Cotton fibers after they have been removed from the seed *agr*

ginning outturn: The ratio of lint to seed cotton produced by the ginning process *agr*

GIPB: >>> Global Partnership Initiative for Plant Breeding Capacity Building

girdle: The act of removing a band of bark from around a tree trunk *hort fore*

GISH: >>> genomic *in situ* hybridization

glabrescent: Becoming glabrous (without hairs) as the plant structure ages or matures *bot*

glabrous: Without hair or smooth *bot*

gland: Organs or swellings that usually secrete a watery or characteristic substance; many oily and aromatic products are glandular in origin *bot*

glandular: Having or showing glands *bot*

glass slide: >>> heating

glasshouse: >>> greenhouse

glassy grain: >>> hyaline grain

glaucous: With waxy bloom present on the surface of the plant structure; a whitish, grayish, or bluish appearance is often imparted *bot*

gliadin: Any >>> prolamin and a simple protein of cereal grains that imparts elastic properties to flour; it is a monomeric molecule between 30,000 and 75,000 kDa; it is divided in alpha-, gamma-, and omega-gliadins; it may form large polymeric structures as a result of inter-molecular disulfide bonds *chem phys* >>> Table 15

Global Partnership Initiative for Plant Breeding Capacity Building (GIPB): A multi-party initiative of knowledge institutions around the world that have a track record in supporting agricultural research and development, working in partnership with country programs committed to developing stronger and effective plant breeding capacity; it was elaborated in Madrid in June 2006; the goal of the initiative is to enhance the capacity of developing countries to improve crop productivity for food security and sustainable development through sustainable use of plant genetic resources using better plant breeding and seed delivery systems; an internationally facilitated partnership forms the basis for achieving the goal of the initiative by catalyzing and supporting national, regional, and global action among relevant international organizations, foundations, universities and research institutes, private sector, civil societies, and national and regional bodies; the office is hosted by >>> FAO Plant Production and Protection Division, Rome, Italy *org*

globulin: One of a group of globular, simple proteins, which are insoluble or only sparingly soluble in water, but soluble in salt solutions; they occur in plant seeds (mainly in dicots), where they have a variety of functions (e.g., legumin, vignin, glycinin, vacilin, or arachin) *chem phys* >>> Table 15

glochid: A tuft of short, barbed spines that are found at the areoles of *Opuntia* cacti *bot*

glomerule: A very compact cyme; a cluster of flowers *bot*

glucide: >>> carbohydrate

glucose: >>> dextrose

glucoside: Glucosides are soluble in water and alcohol; some of them are highly poisonous (e.g., saponin, from tung tree, cf List of Crop Plants); they are found in vegetative organs and in some seeds (e.g., salicin in bark and leaves of willows; amygdalin in seeds of almonds, peaches, or plums; sinigrin in black mustard; aesculin in horse chestnut seeds; quercitron in the bark of oaks) *phys*

glufosinate: >>> pat gene

glumes: The outermost pair of bract-like structures of each spikelet; a chaff-like bract *bot* >>> first glume >> second(ary) glume >>> Table 34

glume surface: The upper external surface of the broad wing of the glume (e.g., in wheat, which is described as being rough and/or smooth when scratched with a needle point) *bot*

glutamate (Glu): A salt or ester of glutamic acid *chem phys* >>> Table 38

glutamic acid (Glu): An amino acid ($HOOC(CH_2)_2CH(NH_2)COOH$) involved in purine biosynthesis, occasionally added to plant tissue culture media; it may replace ammonium ions as the nitrogen source; it is of key importance in pollen growth *in vitro chem phys* >>> Table 38

glutathione: A tripeptide containing glutamic acid, cysteine, and >>> glycine capable of being alternately oxidized and reduced; it plays an important role in cellular oxidation *chem phys* >>> Table 38

gluten: A term that is utilized to refer to a naturally occurring mixture of two different proteins (glutenin and gliadin) in the seeds of, for example, wheat; it is the principal protein in cereal seeds; it consists of a long polypeptide chain; in wheat, it possesses particular elasticity, which allows production of high-quality bread (strength and elasticity of the flour); for example, more of the high-molecular-weight glutenin (which is "stretchy" and imparts physical strength to a dough) results in a flour that is better suited to manufacturing high-quality yeast-raised bread products *chem phys* >>> gliadin >>> glutenin >>> Table 15

glutenin: It is soluble in aqueous or saline solutions or ethyl alcohol; it can also be extracted with strong acid or alkaline solutions; it is found in cereal seeds (e.g., glutenin in wheat or oryzinin in rice); it is divided into a low-molecular-weight glutenin subunit (LMW-GS) and a high-molecular-weight glutenin subunit (HMW-GS) of about 65,000–90,000 kDa; it is a bound structural protein associated with membranes and corresponding to the "matrix" protein of the cell *chem phys* >>> Table 15

glycerin(e): A three-carbon trihydroxy alcohol that combines with fatty acids to produce esters, which are fats and oils; it may serve as a cryoprotectant *chem phys* >>> cryoprotectant >>> refraction index

glycerol: >>> glycerin(e)

glycine (Gly): An amino acetic acid; the simplest alpha amino acid *chem phys* >>> Table 38

glycinebetaine: A small N-trimethylated amino acid, existing in zwitterionic form at neutral pH; this substance is called glycine betaine to distinguish it from other betaines that are widely distributed in microorganisms and plants; the original betaine, *N,N,N*-trimethylglycine, was named after its discovery in sugarbeet (*Beta vulgaris*) in the 19th century; it has been reported to increase in response to >>> abiotic stresses in many species; it is an amphoteric compound that is electrically neutral over a wide range of physiological >>> pH values and is extremely soluble in water despite a non-polar hydrocarbon moiety that consists of three methyl groups; in addition to its role as an osmoregulator, glycinebetaine is thought to interact with both hydrophilic and hydrophobic domains of macromolecules, such as protein and membrane complexes, thus stabilizing their structures and activities against the damaging effects of excessive salt, cold, heat, and freezing *phys*

glycinin: >>> globulin

glycoll: >>> glycine

glycoprotein: A conjugated protein that consists of a carbohydrate covalently lined to a protein *chem phys*

glycoside: >>> glucoside

glycosylation: The attachment of a carbohydrate to another molecule *phys*

glyoxaline: >>> imidazole

glyphosate: A chemical compound used as a herbicide; glyphosate resistance is a subject of biotechnological approaches (e.g., tobacco and tomato transformants may show an overexpression of 5-enolpyruvyl shikimate-e-phosphate synthase [EPSPS], which is usually blocked at normal concentrations by the herbicide); transformed cells, callus, or individuals have proved to be tolerant to high glyphosate concentrations *phyt biot*

GM crop: >>> genetically modified organisms

GM food: >>> genetically modified

GMO: >>> genetically modified organisms

GMO testing: Polymerase chain reaction detects the presence of a certain DNA sequence and is used to test maize, soybean, rice, cotton, and >>> Canola™ tissue resulting in qualitative results; tests for the >>> 35S promoter gene associated with GMO; DNA is extracted from ground material; a certain section of the DNA is replicated numerous times to attain sufficient DNA to identify the promoter through electrophoresis; electrophoresis separates different size particles, allowing the 35S promoter to be detected when compared to a known 35S sample *meth seed*

GoldenGate®™-assay: A proven technology that has been used in key genotyping projects developed by Illumina, Inc. (San Diego, U.S.); it is based on methylation analysis and monitors differences in methylation at specific >>> CpG sites with single-nucleotide resolution; bisulfite-treated genomic DNA is hybridized to primers that target specific CpG sites, either methylated or unmethylated; after primer extension and ligation, each CpG sequence is >>> PCR amplified with fluorescently labeled, common primers; methylated and unmethylated DNA is labeled with Cy5 and Cy3, respectively, to allow quantitative measurement of methylation at each CpG site; labeled DNA is then hybridized to an Illumina Inc. universal bead array; the combination of multiplexing and multisample processing decreases screening cost significantly, while ensuring high call rates; it allows measurement of the DNA methylation status of up to 1,536 targeted >>> CpG sites in 96 samples simultaneously *biot meth*

Golden rice: A biotechnology-derived rice (*Oryza sativa*) created in the 1990s by I. POTRYKUS and P. BEYER, which contains large amounts of beta-carotene, i.e., precursor of vitamin A, in its seeds; the researchers utilized *Agrobacterium tumefaciens* bacteria to genetically engineer the rice plant, i.e., by inserting the following genes from daffodil and from the bacterium *Erwinia uredovora*: (1) phytoene synthase – from daffodil (narcissus), which converts geranylgeranyl-diphosphate into phytoene; (2) CRTL gene – from *Erwinia uredovora*, which codes for phytoene desaturase, which causes the rice plant to convert phytoene (a "light harvesting" carotenoid involved in photosynthesis) into lycopene (a carotenoid which is then utilized by the rice plant in the production of beta-carotene); and (3) lycopene beta-cyclase – from daffodil, which converts lycopene into beta-carotene *biot*

Golgi apparatus: A system of flattened, smooth-surfaced, membranaceous cisternae, arranged in parallel 20–30 nm apart and surrounded by numerous vesicles; a feature of almost all eukaryotic cells; this structure is involved in the packaging of many products of cell metabolism *bot*

goodness of fit: Methods to test the conformity of an observed empirical distribution function of data with a posited theoretical distribution function (e.g., Chi–square test) by comparing observed and expected frequency counts; the KOLMOGOROV–SMIRNOV test calculates the maximum vertical distance between the empirical and posited distribution functions *stat*

gossypol: A dark pigment, $C_{30}H_{30}O_8$, derived from cottonseed oil *chem phys*

gradation: The successive increase of organisms in a more or less cyclic or spontaneous pattern *evol eco*

grader: A truck which smooths the surface of gravel roads and raw soil *agr meth*

gradient: A gradual change in a quantitative property over a specific distance or time *prep*

Graeco-Latin square: A >>> Latin square of order n is a square array of size n that contains symbols from a set of size n; the symbols are arranged so that every row of the array has each symbol of the set occurring exactly once, and so that every column of the array has each symbol of the set also occurring exactly once; two Latin squares of order n are said to be orthogonal if one can be superimposed on the other, and each of the n^2 combinations of the symbols (taking the order of the superimposition into account) occurs exactly once in the n^2 cells of the array; such pairs of orthogonal squares are often called Graeco-Latin squares since it is customary to use Latin letters for the symbols of one square and Greek letters for the symbols of the second square *stat* >>> design of experiment >>> semi-Latin squares

graft: To transfer a part (a small piece of tissue or organs) of an organism from its normal position to another position on the same organism (autograft) or to a different organism or species (heterograft); the stem or shoot that is inserted into a rooted plant is called the scion; the plant or part of a plant into which the scion is inserted is called the stock or understock; there are many different methods of grafting (e.g., wood grafting, cleft grafting, stub grafting, bark grafting, awl grafting, veneer grafting, bud grafting, flat grafting, split grafting, or side grafting); sometimes the term is used in order to describe the point where a scion is inserted in the stock *hort* >>> transplant >>> scion

graft chimera: >>> graft hybrid

graft hybrid: A plant made up of two genetically distinct tissues due to fusion of host and donor tissues after grafting *hort*

grafting: The joining together of parts of plants by holding cut surfaces in position until a union of living cells forms; the united parts will continue their growth as one plant; this technique is used regularly to asexually propagate fruit trees and other woody plants that do not readily root from cuttings; while nearly all of the plants that are commonly grafted can be produced from seed, the variability in the plants produced is so great that it is more practical to clone these plants by grafting; grafting was a horticultural art for several thousand years, and in the past three centuries, the potential was realized for using selected rootstocks to affect growth and performance of scions of plants; grafting is used to create novelty items such as potato plants that produce tomatoes, moon cactus, and fruit trees that bear more than one variety; grafting also is used to incorporate both maleness and femaleness into the same dioecious plant, such as holly, so that it bears fruit; it is used to accelerate a breeding program, to test for compatibility, and to determine if symptoms are virus-caused; four common graft methods are bridge, bud, cleft, and whip; molecular studies from 2009 show evidence for genetic recombination in cells of crafted hybrids, e.g., in tobacco, what was not accepted by classical genetics over 50 years *hort* >>> gemmule >>> horizontal gene transfer >>> rootstock >>> Figure 50

grafting knives: Knives specially designed for grafting purposes; the blade is beveled only in one side, and they are kept razor sharp so they will cut smoothly; they are available in either left or right hand *hort*

grafting strip: A rubber strip used to hold scion in place until knitting has occurred *hort*

grafting tape: Tape backed with biodegradable cloth; used in budding and grafting operations and in banding tree wounds *hort*

grafting thread: A fine-waxed string used in budding and grafting operations *hort prep*

grafting wax: A wax or related substance that is used to cover all injured parts of the rootstock and the scion after grafting and thus prevent infection by fungi or bacteria *prep hort*

grain: A cereal caryopsis that may or may not be enclosed by the lemma and palea *bot* >>> corn

GRAIN: A small international non-profit organization that works to support small farmers and social movements in their struggles for community-controlled and biodiversity-based food systems; GRAIN's work goes back to the early 1980s, when a number of activists around the world started drawing attention to the dramatic loss of genetic diversity on our farms; by the mid-1990s, GRAIN reached an important turning point; around the world, and at local level, many groups had begun rescuing local seeds and traditional knowledge and building and defending sustainable biodiversity-based food systems under the control of local communities, while turning their backs on the laboratory-developed "solutions" that had only got farmers into deeper trouble; by the turn of the century, GRAIN had transformed itself from a mostly Europe-based information and lobbying group into a dynamic and truly international collective – functioning as a coherent organization – that was linking and connecting with local realities in the South as well as developments at the global level *org*

grain filling In cereal crops, a period that lasts for about 20 days in total; it can be divided into three subperiods: 11–16, 17–21, and 21–30 days after flowering; they correspond to the "medium milk", "soft dough", and "hard dough stages" used by farmers to describe the grain in the field; at the beginning of the grain filling, the developing grain is still very soft with the maternal pericarp a mint green color under the outer, clear epidermis; the green color comes from chloroplasts in the cross cell layer; the embryo is still easy to dissect out; the grain is at the "medium milk stage"; the meristematic cells of the endosperm continue to divide and compartments start to form within the endosperm; the first large, "type A" starch grains are seen at about 16 days after flowering; lipid and protein bodies are also seen at this time; the cell layers surrounding the embryo sac continue to change their character, cell walls become thickened, and what will become the aleurone layer is recognizable for the first time; typical aleurone cells are first visible near the nucellar projection at 12 days after flowering; the aleurone cells close to the ventral groove, which acts as transfer cells for the uptake of assimilates into the endosperm, stop growing early and develop special characteristics; in comparison the cells of the dorsal aleurone are still dividing, enabling the grain to continue to expand; the aleurone cells that interface with the embryo near the scutellum are also changing even though the embryo is still physically separate; the embryo is developing rapidly and has an elongated shape; the densely cytoplasmic cellular endosperm, still present at 11 days after flowering, has been completely consumed by the embryo by the 16th day; at 16 days after flowering the scutellum is clearly defined and the embryo now uses the endosperm starch reserves near the scutellum for its own development *phys* >>> Table 13

grain grade: A market standard established to describe the amount of contamination, grain damage, immaturity, test weight, and marketable traits *seed meth*

grain loss: The loss in weight, occurring over a specified period and expressed on a moisture-free basis *phys*

graminicide: Herbicides that target the grass family, i.e., it refers to the lipid synthesis inhibitor that is only effective on grasses *phyt phys*

Gramineae: Any of the monocotyledonous, mostly herbaceous plants, having jointed stems, slender sheathing leaves, and flowers born in spikelets of bracts; the species in which the cereals are included appeared during the Cretaceous period (136–65 mya) bot evol; generally, in grassland agriculture the term does not include cereals when grown for grain but does include forage species of legumes often grown in association with grasses *agr* >>> crop evolution >>> Figure 10 >>> Table 32

gramineous: >>> Gramineae

graminoid: A grass or grass-like monocot; includes the grass (Poaceae), sedge (Cyperaceae), rush (Juncaceae), cattail (Typhaceae), and arrowgrass (Juncaginaceae) families *bot*

Gram's stain: An important bacteriological staining procedure discovered empirically in 1884 by the Danish scientist C. GRAM; a technique used to distinguish between two major bacterial groups based on stain retention by their cell walls; bacteria are heat-fixed, stained with crystal violet, a basic dye, then with iodine solution; this is followed by an alcohol or acetone rinse; GRAM-positive bacteria are stained bright purple; GRAM-negative bacteria are decolorized; safranin is used to stain them *meth*

granary: A storehouse or repository for grain *agr*

granulated fertilizer: Also called pelletized fertilizer; a fertilizer which has been processed into granules, or pellets, which are to be applied directly to the soil; granulated fertilizers are often coated to produce a slow-release formula *phys agr hort meth*

granum (grana *pl*): Stacks of circular thylakoids, composed of lamellae in higher green plant chloroplasts, containing pigments and other essential components of photosynthetic light reactions *bot*

grape sugar: >>> dextrose

grasses: >>> Gramineae

grass flower: The reproductive axis subtended by the palea and lemma and consisting of lodicules, stamens, and carpels, although it may also be unisexual *bot*

gravitational water: The water that flows freely through the soil in response to gravity *agr* >>> Table 44

gravity separator: A machine utilizing a vibrating porous deck and air flow to separate seeds on the basis of their different specific gravities *seed*

gray level: The brightness of pixels in a digitized video and/or computer image; for an 8-bit signal it ranges from 0 (black) to 225 (white) *micr*

gray mold: Fungus disease characterized by the gray hairy appearance of affected parts *phyt*

graze: The eating of crops by animals in the field *agr*

green bridge: Living plant material used by biotrophs to overwinter *eco*

green chop: Green plants cut into small sections for animal feed *agr*

green fluorescent protein (GFP): Protein from the jellyfish, *Aequorea victoria* (Scyphozoa); the gene is used as a reporter gene; it fluoresces in UV light; several variants have been developed, each of which exhibits characteristic spectra; significant advantages are that the protein can be seen in living tissue and is not toxic *biot*

greenhouse: A building, room, or area, usually of glass, in which the temperature is maintained within a desired range; used for cultivating tender plants or growing plants out of season *prep hort*

greenhouse effect: The trapping of heat from sunlight by atmospheric gases, including carbon dioxide, methane, and water, which raises planetary temperatures above those expected from a simple model of heat radiation received from the sun and re-radiated into space *eco*

green manuring: >>> green manure crops

green manure crops: Crops that are grown for the purpose of being ploughed (plowed) into the soil to improve soil fertility and organic content; phacelia may be a green manure crop in some regions, as are various legumes, such as lupines; the crops are ploughed into the soil while they are still green *agr*

Green Revolution: A dramatic increase in agricultural yields that occurred in the 1950s through 1960s; the Green Revolution was based upon many improvements in plant science, including the genetic improvement of many plants (including new, high-yield hybrid varieties), improved irrigation, more efficient machinery, new fertilizers, and pest controls that increased plants' disease-resistance, improved their hardiness, and increased their productivity (especially rice, wheat, and maize) *agr* >>> CIMMYT >>> SCHLEGEL (2007)

green rice leafhopper (*Nephotettix cincticeps*): One of the most serious insect pests affecting cultivated rice (*Oryza sativa*) in temperate regions of East Asia; an accession of the wild rice species, *O. rufipogon*, was found to be highly resistant to the insect by an antibiosis test *phyt*

green seed: A Canola™ production problem in which an early frost inhibits the degradation of the green pigment chlorophyll; the presence of chlorophyll in the oil lowers the yield and grade of oil obtained from the seed *agr*

greenwood cutting: A cutting taken from a stem that is in a quickly grown, vegetative state *hort meth* >>> hardwood cutting >>> semiripe cutting >>> root cutting

grex: A group of hybrids of different origins but similar parentage, or a collective term applied to all the progeny of an artificial cross from known parents of different taxa; mainly used in orchids and rhododendrons *gene*

grid design: For the grid design, plants or variants are divided into blocks and the best ones chosen from each *stat* >>> Figure 32

grist: A mixture of grain (e.g., wheat or barley) utilized by, for example, the miller for grinding or the maltster for producing malt; grist may contain a mixture of several varieties meth *prep agr*

grits: >>> semolina

groat: The caryopsis of oats after the husk has been removed *agr seed*

ground germination rate: >>> field germination

groundkeeper: A self-sown plant, for example, a potato *bot*

group coancestry: The probability that two genes taken from a gene pool of a population are identical by descent, or the average of the cells in a coancestry matrix for the population concerned *stat meth* >>> coancestry

group merit: The genetic merit of a group as a function (weighted average or index) of its breeding value and gene diversity; it is an index to quantify the merit of a group as a weighted average of its advance in breeding value, and its loss in gene diversity relative to some reference population *stat meth*

group merit progress: Group merit changes over generations, mainly as breeding produces a genetic gain but at the same time a loss; group merit progress (per year) takes genetic gain, gene diversity, and time into consideration, and ought thus to be a good measure of the progress in breeding *meth stat*

group merit selection: Maximizing the group merit of selections given the candidates and group merit measure; the selection method was initially called "population merit selection" *stat meth*

group-selection method: A method of regenerating and maintaining uneven-aged stands in which trees are removed in small groups *meth fore*

growing degree days (GDD): Method for calculating the amount of heat units based on daily temperatures that contribute to the physiological growth and development of plants (GDD = $\text{Temp}_{min} + \text{Temp}_{max}/2 - \text{base} + 4°C \text{ or } + 10°C$) *hort*

growing out: >>> sprouting

growing tray: A tray having compartments like an ice-cube tray, used for starting seeds *hort prep*

growth analysis: A mathematical analysis of crop or plant growth using relative growth rate, net assimilation rate, leaf area growth rate, and crop growth rate *agr phys*

growth crack(s): A physiological disorder of the potato tuber in which the tuber splits while growing; the split often heals but leaves a fissure in the tuber; cracks generally start at the bud or apical end of the potato and can extend lengthwise; growth cracks make fresh-market tubers unattractive; growth crack incidence increases when growing conditions are uneven or sudden environmental changes occur *agr phyt*

growth curve: A curve showing the change in the number of cells in a growing culture as a function of time *phys*

growth form: The form of a plant; the habit in which a plant grows (e.g., shrubby plant, climbing plant, leaf plant, etc.) *bot* >>> growth habit >>> habit(us)

growth habit: The mode of growth of a plant; crops may be classified as (1) annuals (e.g., barley), (2) biennials (e.g., sugarbeet), or (3) perennials (e.g., alfalfa) *phys*

growth inhibitor: Any substance that retards the growth of a plant or plant part; almost any substance will inhibit growth when concentrations are high enough; common inhibitors are abscisic acid and ethylene; other inhibitors, such as phenolics, quinones, terpens, fatty acids, and amino acids, affect plants at very low concentrations *phys* >>> growth promoter

growth promoter: A growth substance that stimulates cell division (e.g., cytokinin) or cell elongation (e.g., gibberellin) *phys* >>> growth inhibitor

growth rate (of crop): The crop growth rate is a specific plant growth analysis term denoting the absolute growth rate of mass per unit land *phys agr*

growth regulator: Despite natural growth regulators, a synthetic compound that, when applied to a plant, promotes, inhibits, or otherwise modifies the growth of that plant *phys* >>> brassinosteroid >>> cytokinin >>> gibberellin

growth stages (*syn* developmental stage): The discrete portion of the life cycle of a plant, such as vegetative growth, reproduction, or >>> senescence; there are systems for various crops in order to subdivide the broad physiological and/or morphological stages; for example, in grain crops: (1) tillering stage, when a plant produces additional shoots from a single crown, as in wheat; (2) jointing stage, when the internodes of the stems are elongating; (3) boot stage, when the leaf sheath swells up due to the growth of developing spike or panicle; (4) heading stage, when the seed head of a plant begins to emerge from the sheath *phys*

growth substance: A naturally occurring compound, other than a nutrient, that promotes, inhibits, or otherwise modifies the growth of a plant *phys*

grub hoe: A heavyweight hoe for digging roots *hort agr*

GS: >>> genomic selection

guanidine: A crystalline, alkaline, water-soluble solid, CH_5N_3, used in making resins *chem phys*

guanine (G): A purine base that occurs in both DNA and RNA *chem gene* >>> Table 38

guanylic acid: Guanosine monophosphate; a ribonucleotide constituent of ribonucleic acid that is the phosphoric acid ester of nucleoside guanosine *chem*

guard cell: A specialized type of plant epidermal cells; two of which surround each stoma; changes in their turgidity cause stomatal opening and closing *bot phys* >>> photosynthesis >>> respiration

Guidelines for Classifying Cultivated Plant Populations: A regulation act of the U.S., approved in 1978; it appeared as an appendix to the Federal Seed Program Review (1980), which gives more precise definitions of the various categories of cultivated varieties *seed*

guidepost: >>> landmark

GUS gene: A gene that codes for production of beta-glucuronidase (GUS protein) in *Escherichia coli* bacteria *biot*

guttation: The tear-like extrusion of water and sometimes salts from the aerial parts of plants, particularly at night when transpiration rates are low; it is the process of water being exuded from hydathodes at the enlarged terminations of veins around the margins of the leaves *bot*; exudation of specific proteins made by artificially inserted genes *biot*

Guttman scale: A measurement scale in which the items have a fixed progressive order and that has the characteristic of reproducibility *stat*

gymnosperm(ous): A kind of plant that produces seeds but not fruits; the seeds are not borne within an ovary and are called naked *bot*

gymnosperms: >>> gymnosperm(ous)

gynaecium: >>> gynoecium

gynandromorph: An individual exhibiting both male and female sexual differentiation *gene*

gynic: Female *bot*

gynodioecious: Plant species with female and >>> perfect flowers on separate plants *bot* >>> gynodioecy

gynodioecy: Plant species or population in which female plants as well as hermaphroditic plants occur *bot*

gynoecious: >>> gynoecium

gynoecium: The collective term for the female reproductive organs of a flower, comprising one or more carpels *bot* >>> pistil

gynoecy: Femaleness *bot*

gynomonoecious: Plant species with female and >>> perfect flowers on the same plant *bot* >>> gynomonoecy

gynomonoecy: Where a hermaphrodite bears both female and hermaphrodite flowers *bot*

gynophore: The stalk that pushes pollinated peanut flowers into the soil *bot*

gynostegium: A complex gynoecium on which are borne the anthers, e.g., in Orchidaceae

H

h²: The ratio between genetic variance and phenotypic variance used by plant breeders in selecting for yield, resistance, or other characteristics *stat* >>> heritability

habit(us): The general appearance of a plant *bot*

habitat: The living place of an organism or community characterized by its physical or biotic properties *eco*

habituation: The diminishing requirement of some tissue cultures for growth-regulatory substances, possibly due to endogenous production *biot*

haft: Narrow or constricted base of an organ, particularly on the fall petals of *Iris* flowers *bot*

hair: A slender outgrowth of the epidermis, common on certain leaf or stalk structures *bot*

hairpin loop: Binding of complementary DNA sequences to each other to form a hairpin loop (also called stem loop); if this happens in a PCR primer, it will not function *biot*

hairy root culture: A fairly recent development in plant culture, consisting of highly branched roots of a plant; a plant tissue is treated with a culture of the bacterium *Agrobacterium rhizogenes*, which transfers part of its own plasmid DNA to the cells of an infected plant; this alters the plant's metabolism, including alterations in hormone levels, which in turn cause the explant to grow highly branched roots from the sites of infection; the roots branch much more frequently than the usual root system of that plant, and are also covered with a mass of tiny root hairs; their most significant feature is that they produce secondary metabolites at levels similar to those made in the original plant; thus they can be used as replacement plants for making such compounds as food flavors or fragrances *biot*

hairy root disease: A disease in some dicots; rootlike tissue is proliferated along segments of the stem; it is caused by *Agrobacterium rhizogenes*, if it carries a Ri plasmid *phyt biot*

half-diallel cross: The crossing of a series of genotypes in all combinations except reciprocal combinations *meth*

half-grain method: A method in cereals separating the embryo from the endosperm by transversal dissection; it is applied when the endosperm is used for biochemical and molecular studies while the embryo can be grown for seed production; in this way preselected individuals and/or genotypes can be multiplied; it can reduce the breeding population and experimental costs *meth biot*

half-hardy annuals: Plants that will survive some frost, but not a long freeze *phys*

half-life (time): The time required for one-half the atoms of a given amount of a radioactive substance to decay *phy biot* >>> radio carbon dating >>> isotope

half-meiotic polyploidization: >>> unilateral sexual polyploidization

half-shrub: >>> suffrutex

half sibs: Progeny with only one common parent *gene*

half-sib progeny selection: >>> method of overstored seeds (*syn* remnant seed procedure) >>> Figure 4

haloing: Condition which describes a leaf which has chlorosis along the edge; edges of the leaf appear yellow; the condition is most often associated with a nutrient imbalance, such as a deficiency of boron, molybdenum, nitrogen, or phosphorus *phys*

halomorphic: A soil that has high levels of salt *agr* >>> Table 44

halophilic: >>> halophilous

halophilous: A salt-loving organism, adapted to a high-salt environment (e.g., in a salt marsh) *bot*

halophytes: Plants that grow in saline soil *bot* >>> Table 44

HANAHAN transformation procedure: An optimized procedure for the transformation of *Escherichia coli* with plasmid DNA using $CaCl_2$ *biot*

hand-dibbed: Sowing individual seeds by hand according to a special plot design; mainly used for F1 seeds when they are rare or seeds that need special care *meth*

hand weeding: Manually removing the undesirable species inhibiting the growth of valued species *meth agr*

hanging-drop culture (technique): A method of microscopic examination of organisms or particles suspended in a drop on a special concave microscopic slide *meth micr biot* >>> microdroplet array

hapblock: >>> haplotype block

***hap* inducer:** A mutant in barley and maize that induces haploids by the dominant initiator *hap* gene; in *hap* mutants, induced by ethyl methane sulfonate treatment, up to 40% of the progeny are haploid; the key process of haploid embryo formation is a failed fusion of the egg and sperm cell, while the endosperm develops normally *gene meth* >>> haploid >>> double haploid >>> DH line

haplodiploidy: Sex differentiation in which males are haploids and females diploids *bot* >>> *Hordeum bulbosum* procedure

haploid: Applied to a cell nucleus that contains one of each type of chromosome (i.e., one set of chromosomes); designated "n" *gene* >>> monoploid >>> *Hordeum bulbosum* procedure

haploid breeding: The production of >>> homozygous lines for >>> true-breeding lines; the desired trait will be carried through unchanged to subsequent generations only if the trait is present in homozygous form; to obtain homozygous breeding lines, originally self-pollination was needed over six to eight generations, which is a very time-consuming and costly process; nowadays, homozygous lines of some plant species (e.g., tobacco, barley, potatoes, rape, wheat, etc.) can be produced from >>> gametes, which contain only one (haploid) set of chromosomes; in most cases unripe pollen is placed on a suitable >>> culture medium, where it develops into plants with a single set of chromosomes (haploid >>> androgenesis); ovaries may also be used as the source material, although this is less common (haploid >>> parthenogenesis); following a cultivation period of 3–4 weeks, the haploid plantlets are treated with >>> colchicines in order to induce duplication of the chromosomes; resulting cells produce >>> "double haploids", fully homozygous plants which produce identical offspring *meth* >>> *Hordeum bulbosum* procedure >>> *Imperata cylindrica* procedure >>> Figure 51

haploid-initiator gene: In >>> barley, a partially dominant haploid-initiator gene (>>> *hap* gene) that controls the abortion or the survival of abnormal embryos and endosperms; plants homozygous for the *hap* gene produce progenies that include 10–14% haploids *bio gene*

haploidization: The process whereby diploid somatic cells become haploid during a parasexual cycle or by experimental means, for example, in barley by crossing with wild (>>> bulb) barley (*Hordeum bulbosum*), with *Imperata cylindrical*, or in wheat by crossing with pearl millet (Pennisetum glaucum) cyto meth >>> *Hordeum bulbosum* procedure >>> Imperata cylindrica procedure

haploid parthenogenesis: The situation in which the unfertilized egg hatches and develops normally to produce a viable male adult whose cells contain only the haploid number of chromosomes cyto >>> *Hordeum bulbosum* procedure

haploid production: >>> haploidization

haplo-insufficiency: Situation where one normal copy of a gene alone is not sufficient to maintain normal function; it is observed as a >>> dominant mutation on one allele (or deletion of it) resulting in total loss-of-function in a diploid cell because of the insufficient amount of the wild-type protein encoded by the normal allele on the other >>> haplotype gene >>> hemizygote

haplontic: Organisms in which meiosis occurs in the zygote resulting in four haploid cells *bot*

haplophase: That part of the life cycle in which the gametic chromosome number is found in reproduction cells *bot*

haplosis: The meiotic division resulting in haploid cells and/or gametes *bot*

haplosomic: The situation in which the homologue of a pair of chromosomes is missing in somatic cells *gene* >>> monosomic >>> Figure 37

haplotype: A combination of alleles of closely linked loci found in a single chromosome; sometimes, a combination of particular nucleotide variants within a given DNA sequence, i.e., the linear arrangement of alleles along a region in DNA; in general, the collective genotype of a number of closely linked loci; the constellation of alleles present at a particular region of a chromosome *gene biot*

haplotype block: Any one of the relatively large genomic regions with which defined sequences are associated, e.g., specific genes but also >>> single nucleotide polymorphisms *biot*

haplotype map: A variant of a genomic map in which the haplotype blocks of the genome of an organism are depicted *biot meth*

haplotype signature: Any characteristic configuration of specific alleles *biot*

haplotype single nucleotide polymorphism (htSNP *syn* **haplotype tag SNP):** Any >>> SNP contained with a haploid block *biot meth*

haplotype tag SNP: >>> haplotype single nucleotide polymorphism

haplotype tagged single nucleotide polymorphism: Any >>> SNP that is identified and/or tagged from larger SNP databases; it is located in a specific genomic region, and used for definition of haploids and haploid structures *biot meth*

hard dough: Description of the cereal grain at maximum dry weight; the grain is hard to puncture and has started to dry off *phys bot*

harden(ing off): The gradual process of acclimating plants started indoors to outside conditions (e.g., placing them in a sheltered location outdoors for increasing lengths of time over a period of days) *meth phys hort agr* >>> acclimatization

hardiness The capability of a plant to withstand environmental stress phys agr >>> www.plant-stress.com

hard-leafed: >>> sclerophyllous

hardpan: A hardened or cemented soil horizon or layer; the soil material may be sandy or clayey and may be cemented by iron oxide, silica, calcium carbonate, or other substances *agr* >>> Table 44

hard red spring wheat: Contains the highest percentage of protein, making it an excellent bread wheat, with superior milling and baking characteristics; chiefly grown in Montana, North Dakota, South Dakota, and Minnesota (U.S.); seeded in the spring, and may have a hard or soft endosperm; subclasses are Dark Northern Spring, Northern Spring, and Red Spring wheats *agr seed* >>> wheat

hard red winter wheat: The class of wheat used mostly for bread and all-purpose flour; seeded in the fall; ranges from medium to high in protein; may have either a hard or soft endosperm; accounts for more than 40% of the U.S. wheat crop and more than half of U.S. wheat exports; produced in the Great Plains, the large interior area extending from the Mississippi River west to the Rocky Mountains, and from the Dakotas and Montana down to Texas; it shows a wide range of protein, and good milling and baking characteristics; it is used to produce bread, rolls, and, to a lesser extent, sweet goods and all-purpose flour *agr seed* >>> wheat

hard seed: A seed that is dormant due to the nature of its seedcoat, which is impervious to either water or oxygen *seed*

hardseedness: >>> hard seed

hard-shelled seed: >>> hard seed

hard wheat: Tetraploid wheat, *Triticum durum*, used for high-quality noodles, bread, and pastas *agr* >>> cf List of Important Crop Plants >>> wheat >>> Table 1

hard white wheat: The newest white class of wheat to be grown in the U.S.; it is closely related to red wheats (except for color genes); this wheat has a milder, sweeter flavor, equal fiber, and similar milling and baking properties; used in yeast breads, hard rolls, bulgur, tortillas, and oriental noodles; it is used in domestic markets, and exported in limited quantities *agr seed* >>> wheat

hardwood: A term used to describe broadleaf, usually deciduous, trees such as oaks, maples, ashes, elms, etc.; it does not refer to the hardness of the wood *fore hort*

hardwood cutting: A cutting taken from a stem whose bark is mature and the wood has fully hardened for the winter *hort meth* >>> greenwood cutting >>> semiripe cutting >>> root cutting

Hardy–Weinberg law (equilibrium): The stable frequency distribution of genotypes, AA, Aa, and *aa*, in the proportions p2, 2pq, and q2, respectively (where p and q are the frequencies of the alleles, A and a); that is, a consequence of random mating in the absence of mutation, migration, natural selection, or random drift; it states that in an infinitely large, interbreeding population in which mating is random and in which there is no selection, migration, or mutation, gene and genotype frequencies will remain constant from generation to generation *gene bio*

harlequin chromosome: >>> sister chromatid exchange (SCE)

harrow: An agricultural implement with spike-like teeth or upright disks, for leveling and breaking up clods in plowed land or experimental plots; to draw a harrow over soil *agr*

harvest index (HI): A measurement of crop yield; the weight of a harvested product (e.g., grain) as a percentage of the total plant weight of a crop, or more generally, the proportion of the biological yield to economic yield *phys* >>> Table 33

harvest intervals: >>> days to harvest

HarvestPlus: A global alliance of institutions and scientists seeking to improve human nutrition by breeding new varieties of staple food crops consumed by the poor that have higher levels of micronutrients, through a process called >>> biofortification *org agr* >>> www.harvestplus.org

harvest ripe: The (cereal) grain has dried off and is ready to be harvested *phys agr*

HASEMAN–ELSTON regression: A statistical method developed in the 1972, in common use to detect genetic linkage between a quantitative trait and a genetic marker; it detects quantitative trait loci by linkage to a marker; the squared sib-pair trait difference is regressed on the proportion of marker alleles the pair is estimated to identical by descent: a significantly negative regression coefficient suggests linkage; although the technique has been improved in a number of ways, it predicts a high rate of false positive quantitative trait loci (QTL) because it is based on a single-QTL model; it has been shown that a maximum likelihood method that directly models the sib-pair covariance has more power *stat biot* >>> QTL mapping >>> Table 58

haulm: Stems or stalks collectively, as of peas, beans, potato, or hops, or a single stem or stalk *bot*

haulm killer: Chemical for killing off (e.g., potato plant tops to facilitate harvest) *agr*

Haun scale: Scale expressing stages in cereal seedling development *meth phys seed* >>> Table 13 >>> decimal code for the growth of cereal plants >>> physiological maturity

haustorium (haustoria *pl*): In certain parasitic fungi, an outgrowth from hypha that penetrates a host cell in order to absorb nutrients from it; in some parasitic angiosperms, outgrowth of the roots; in endosperm development, nutrient-gathering outgrowths toward surrounding tissue of the developing endosperm *bot phyt phys*

hay: Herbage, as grass, clover, or alfalfa, cut and dried for use as forage *agr*

head: An inflorescence in which the floral units on the peduncle are tightly clustered, surrounded by a group of flowerlike bracts called an involucre (e.g., in sunflower) *bot*

head back: To cut back the main branches of a tree or shrub by at least one-half of their length *hort meth*

head components: Generally, all components of the inflorescence of grain and grass crops *agr*

head shattering: A preharvest loss of kernels in cereals caused by loose seeds inside the spikelets and mechanical shattering (wind, etc.) *agr*

heading: Emerging spikes (i.e., from initial emergence of the inflorescence from the boot until the inflorescence is fully exerted) *phys*; in viticulture, to shorten or prune the trunk when it reaches the desired height; done in an effort to focus growth on the lower shoots *hort*

headland: Outside area of cropped field, typically between 6 and 24 m wide; it may be sown with the crop, left unsown or sown with an alternative species *agr meth*

heart stage: Embryonic stage at which the cotyledon primordia have enlarged enough to form two mounds on either side of the apical meristem *phys*

heartwood: The inner core of a woody stem, wholly composed of non-living cells and usually differentiated from the outer enveloping layer (sapwood) by its darker color *fore hort*

heat filter: Absorption glass filter that attenuates infrared radiation, but transmits light in the visible wavelength range *micr*

heath: Species within the plant families Ericaceae, Empetraceae, and/or Diapensiaceae; a heathland is dominated by species in those families *bot eco*

heating: Gentle heating of the slide over a spirit or other flame flattens the cells, sticks them to the glass slide and cover slip, and spreads the chromosomes, whether in prophase or metaphase *prep cyto*

heat shock protein (HSP) When certain plants are exposed to high temperature, heat shock proteins are synthesized; they provide thermal protection to subsequent heat stress *phys* >>> stress proteins >>> www.plantstress.com

heat shock response: A ubiquitous and highly conserved defense mechanism for protection of cells from harmful conditions such as heat shock, UV irradiation, toxic chemicals, infection, transformation, and appearance of mutant and misfolded proteins; heat shock proteins (HSPs) also function as accessory molecules in antigen presentation; high levels of HSPs prevent stress-induced apoptosis, and may have a transforming potential *phys biot*

heat stress: A decline in crop production caused by heat stress is one of the biggest concerns resulting from future climate change; cereal spikelets are most susceptible to heat stress at flowering; e.g., in rice the early-morning flowering (EMF) trait mitigates heat-induced spikelet sterility at the flowering stage by escaping heat stress during the daytime; there is an EMF locus in >>> wild rice, *Oryza officinalis* (CC genome); it was detected as a stable quantitative trait locus (QTL) for flower opening time (FOT) on chromosome 3; it was designated as *qEMF3*, that shifts FOT by 1.5–2.0 h earlier for cultivated rice "Nanjing 11" in temperate Japan and "IR64" in the Philippine tropics; introgression lines for EMF mitigate heat-induced spikelet sterility under elevated temperature conditions, completing flower opening before reaching +35°C, a general threshold value leading to spikelet sterility *gene*

heaving: Lifting effect of the soil due to alternate freezing and thawing; it may result in the lifting up of plants and may tear them loose from the soil or may shear off roots *agr* >>> Table 44

heavy feeder: A plant that requires a great amount of nitrogen because of its speedy growth (e.g., squash, potato, tomato, etc.) *agr*

heavy soil: A soil that has a high content of clay and is difficult to cultivate *agr* >>> Table 44

hectare: Equals 10,000 square meters or 2.471 acres agr

hedge: A border, fence, or boundary formed by a row of closely planted trees, shrubs, or bushes often trimmed to a formal shape; the number of species forming a hedge tends to increase in proportion to its age *hort*

hedgerow: >>> hedge

heel: A piece of the old branch or shoot that is detached from the old branch or shoot along with a cutting *hort meth*

heeling in: Temporarily covering the base of a plant with soil for a short time (e.g., stecklings of sugarbeet during the winter) in order to prevent frost damage of cold-sensitive genotypes *agr hort meth*

heirloom plant (heirloom variety *syn* **heritage fruit** *syn* **heirloom vegetable):** An old cultivar that is maintained by gardeners and farmers, particularly in isolated or ethnic minority communities in Western countries; these may have been commonly grown during earlier periods in human history, but are not used in modern large-scale agriculture *agr evol* >>> heirloom variety

heirloom vegetable: >>> heirloom plant

heirloom variety: >>> heritage seeds >>> heirloom plant

height selective application(s): Use of equipment such as rope or roller applicators that utilize the height difference between tall weeds and a crop to apply herbicides only to the parts of weed above the crop *agr meth phyt*

helicase: An enzyme that unwinds the double DNA helix near the replication fork before DNA polymerase acts on it; replication fork moves from 3′ to 5′ of the leading strand; unwinding is also necessary for DNA repair gene

helobial endosperm: A type of endosperm that develops in an intermediate way between that by which all cellular and nuclear endosperm develop *cyto*

helophyte: Freshwater plant with perennating tissue in water or saturated soil beneath water *bot eco*

hematoxylin staining: A method used as an early indicator of aluminum toxicity effects on the apices of young, developing roots of cereals grown in nutrient solution *meth*

hemialloploid: Not a normal full alloploid but a segmental alloploid, in which some parts of the unified genomes show some degree of structural conformity *cyto*

hemiautoploid: Autopolyploids with a certain degree of differentiation between the diploid sets of chromosomes, either by a subsequent differentiation of previously homologous sets of chromosomes or by spontaneous or induced intervarietal or subspecies hybridization *cyto*

hemicellulose: A heterogeneous group of compounds that in plant cell walls form part of the matrix within which cellulose fibers are embedded *chem phys*

hemichromosome: A chromosome split into chromatids without previous reduplication at >>> interphase *cyto*

hemicryptophyte: Herbaceous perennial or biennial plant with perennating tissue at the soil surface *bot*

hemigamy: A form of parthenogenesis where gametes fuse but the nuclei do not, forming a dikaryon, i.e., resulting in a haploid >>> chimera; it is a type of facultative >>> apomixes in which the male sperm nucleus does not fuse with the egg nucleus after penetrating the egg in the embryo sac, e.g., in cotton *bot*

hemihaploid: Individuals with half of the normal haploid chromosome number *cyto*

hemimethylation: The state in which a DNA duplex is methylated in one strand but unmethylated in the other *gene biot* >>> Table 43

hemiploid: Individuals with half of the somatic chromosome number *cyto* >>> haploid

hemizygous: An individual, generally diploid, having a given gene and/or allele present once (e.g., in monosomics or haploids) *gene* >>> Figure 37

herb(s): A small, non-woody, seed-bearing plant in which all the aerial parts die back at the end of each growing season *bot*

herbaceous: Non-woody, as applied to kinds of plant growth *bot* >>> herbs

herbage: Plant material used for animal feed *agr*

herbal: A book, often illustrated, that describes the appearance, medicinal properties, and other characteristics of plants, often used in herbal medicine *bot* >>> herbarium

herbarium: A collection of dried, preserved, and systematically classified plants *bot tax* >>> herbal

herbarium beetle: Cartodere filum; eats the spores of certain fungi (e.g., *Lycoperdon*, smuts, etc.) that are attached to the plant material of an herbarium *phyt*

herbarium glue: An adhesive that minimizes cracking, discoloration, and shattering with age; it is used in fastening plant specimens to the herbarium sheet *meth bot*

herbarium paste: >>> herbarium glue

herbicide: A chemical substance that suppresses or eliminates plant growth; it may be a non-selective or selective weed killer *phyt*

herbicide tolerance: The ability of some plants to tolerate herbicides; it is a task of genetic engineering to modify crop plants for this trait in order to apply herbicides against weeds in the field *phyt biot*

hercogamy: >>> herkogamy

hereditary: Transmissible from parent to offspring or progeny *gene*

hereditary determinant: Any genetically acting unit of an organism that is replicated and conserved, transferred from generation to generation *gene*

hereditary factor: >>> hereditary determinant

heredity: The transmission of genetic characters from one generation to the next generation; it operates primarily by the germ cells in sexually reproducing species *gene*

heritability: A measure of the degree to which a phenotype is genetically influenced and can be modified by selection; it is represented by the symbol h^2; this equals V_{gen}/V_{phe} where V_{gen} is the variance due to genes with additive effects and V_{phe} is the phenotypic variance; there are two types of heritability: (1) broad-sense heritability, $h_b2 = V_{gen}/V_{phe}$, and (2) narrow-sense heritability, $h_n2 = V_{add}/V_{dom}$ *gene stat* >>> genetic variance (V_{gen}) >>> Figures 38, 60 >>> Table 34

heritable: >>> hereditary

heritage fruit: >>> heirloom plant

heritage seeds: Non-hybrid seeds of old varieties that have been passed from generation to generation *agr*

herkogamy: Pollination by the neighbor individual, population, or species *bot*

hermaphrodite: A plant having both female and male reproductive organs in the same flower, e.g., cotton, sugarbeet, alfalfa, rapeseed, rye, or sunflower *bot* >>> Table 18

hermaphroditic: Reproductive organs of both sexes present in the same individual or in the same flower in higher plants *bot* >>> bisexual >>> Table 18

hesperium (hesperia pl): A berrylike fruit with papery internal separations or septa and a leathery, separable rind (e.g., orange, lemon, lime, and grapefruit) *bot*

heteroallele: An allele that differs from other alleles of the same gene by nucleotide differences at different sites within the gene; in contrast with "true" alleles, of which only four are possible at each site within the gene *gene*

heteroallelic: >>> heteroallele

heteroauxin: An obsolete term for the auxin 1H-indole-3-acetic acid (IAA) chem phys >>> indole-3-acetic acid

heterobeltosis: >>> heterosis

heterochromatic: Of chromosome regions or whole chromosomes that have a dense, compact structure in telophase, >>> interphase, and early prophase *cyto*

heterochromatin: The chromosome material that accepts stains in the >>> interphase nucleus (unlike euchromatin); such regions, particularly those containing the centromeric and >>> nucleolus organizers, may adhere to form a chromocenter; some chromosomes are composed primarily of heterochromatin; these are termed heterochromosomes, such as the Y chromosome in some species; heterochromatin is condensed, darkly staining, and transcriptionally inactive with nucleosomes which show histone H1; it has a peripheral nuclear matrix, late-replicating DNA, highly repetitive DNA, and is highly methylated; while >>> euchromatin is extended, lightly staining, transcriptionally active, with nucleosomes which show high-mobility group proteins (HMGs); this class of proteins is typically found in place of histone H1 in transcriptionally active chromatin; euchromatin has a fibrillar internal nuclear matrix, early replicating DNA, single copy or middle repetitive, and is less methylated; heterochromatin may be constitutive or facultative: constitutive heterochromatin is always condensed during interphase, while facultative heterochromatin is composed of sequences which may be euchromatin in some developmental or physiological states, and heterochromatin in others; part of the program of cellular differentiation is carried out by having some genes in euchromatin when they need to be expressed, and in heterochromatin when they need to be off *cyto* >>> euchromatin >>> Table 43

heterochromosome: Any chromosome that differs from the autosomes in size, shape, and behavior *cyto*

heteroduplex analysis: A method of detecting gene mutation by mixing PCR-amplified mutant and wild-type DNA followed by denaturation and reannealing; the resultant products are resolved by gel electrophoresis, with single base substitutions detectable under optimal electrophoretic conditions and gel formulations; large base pair mismatches may also be analyzed by using electron microscopy to visualize heteroduplex regions *meth biot cyto*

heteroduplex DNA: A double-stranded DNA molecule formed by the annealing of strands from two different sources, as opposed to homoduplex, which has homologous strands; as a result, there are regions that are non-complementary and show abnormalities in the form of extra loops *gene*

heteroecious: A species that produces male and female gametes on different individuals *bot* >>> dioecious; the requirement of a pathogen for two host species to complete its life cycle *phyt*

heterofertilization: Fertilizing of the nuclei of endosperm and embryo-forming cells by genetically different gametes *bot*

heterogamete: >>> anisogamete

heterogam(et)ic: A species that sexually reproduces by two types of gametes *bot*

heterogamy: Reproduction involving two types of gametes *bot*

heterogeneity index: A measure for genetic differences within populations *gene*

heterogenetic (chromosome pairing): Chromosome pairing between more or less different genomes in amphiploids *cyto* >>> homogenetic chromosome pairing

heterogenic: Gametes or populations differing in alleles or genes *gene*

heterograft: Heterologous graft; the scion and rootstock derive from different species *hort* >>> xenograft

heterohistont: An individual or cell aggregate that is composed of tissues of genetically different origin *bot*

heterokaryon: Cells with two or more nuclei that are genetically non-identical *bot*

heterokaryotype: A chromosome complement that is heterozygous for any sort of chromosome mutations *cyto*

heterolabeling: A chromosomal labeling pattern due to induced or spontaneous exchange of labeled and non-labeled half-chromatids *cyto*

heterologous encapsidation (transcapsidation): The creation of "new" viruses by surrounding one virus with the envelope protein of another virus; it is a natural process that can occur when plants are co-infected by different strains of viruses; it has also been observed under laboratory conditions for transgenic, virus-resistant plants that produce viral envelope (>>> capsid) proteins; since envelope proteins are responsible for virus transmission (e.g., by insects like aphids), heterologous encapsidation can change a virus' mode of transmission; this can enable animals to transmit plant viruses in ways that were not previously possible; it is only effective for one virus generation; because the virus genome is not modified, the original envelope proteins will be produced by the next generation phyt biot

heterologous gene expression: Expression of a gene in another host *biot*

heterologous probing: Probing at low stringency with a DNA fragment that originates from another organism and thus does not have an identical counterpart in the target DNA; often it gives significant signals because of sequence conservation *biot*

heteromeric: Genes that control the determination of a trait by joint gene action, but each of them shows a definitely different contribution to the final product *gene*

heteromorphic: Chromosomes that differ in size and/or shape; the term is also used for meiotic pairing configuration, which is composed of different chromosomes and/or chromosome segments *gene*

heteromorphic bivalent: A bivalent consisting of non-homologous chromosomes or segments *cyto*

heteromorphous: >>> heteromorphic

heteromorphy: >>> heteromorphic

heterophylly: The production of more than one leaf form on a plant species; in developmental heterophylly, juvenile leaves may differ from adult ones *bot*

heterophyte: A plant that is dependent upon another, obtaining its nourishment from other living or dead organisms, such as parasites or saprophytes *bot*

heteroplasmic: >>> alloplasmic

heteroplasmonic: >>> alloplasmic

heteroplasmy: Heteroplasmy occurs when copies of an organellar genome (plastid or mitochondrial) differ from one another either within a cell or among cells within an individual; this phenomenon was first discovered in plastids over 100 years ago, though "heteroplasmy" was not formally defined until decades later; mitochondrial and plastid heteroplasmy have since been discovered in diverse taxa, including numerous plants, particularly those with the gynodioecious breeding system; though heteroplasmy can arise through mutations in organellar genomes, biparental inheritance of organelles often generates heteroplasmy; in other words, the maternal inheritance of cytoplasmic organelles – chloroplasts and mitochondria – to the sexual progeny has been considered a genetic axiom in eucaryotic organisms; one-third of the higher plant genera inherit plastids biparentally, at least occasionally; some angiosperm species (e.g., *Pelargonium zonale* and *Portulaca grandiflora*) possess both mitochondrial (mt) and chloroplast (ct) DNA in mature pollen grains, whereas other species have organellar DNA of only one type or have none in the male gametes; paternal organellar DNA after pollination can meet one of two fates: it is either selectively destroyed or successfully replicated and maintained in the zygotes; the mechanisms controlling the recognition of paternal and/or maternal organellar DNA and their selective replication or destruction are poorly understood, although it is believed that the inheritance of both mt- and ctDNA are under independent control *gene*

heteroplastidic: Cells whose plastids are different in shape *bot*

heteroploid: Deviating chromosome numbers from the standard chromosome set *cyto*

heteropolar: Pollen grain with different proximal and distal faces, e.g., in *Carex alba bot* >>> isopolar >>> pollen

heteropycnotic: Chromosomes or chromosomal segments that show a different coiling or staining pattern *cyto*

heterosis: The increased vigor of growth, survival, and fertility of hybrids, as composed with the two >>> homozygotes; it usually results from crosses between two genetically different, highly >>> inbred lines; it is always associated with increased heterozygosity; in breeding, three types of heterosis are distinguished: (1) F1 yielding more than the mean of the parents, (2) F1 yielding more than the best-yielding parents, (3) F1 yielding more than the best-yielding variety; for the genetic basis of heterosis, two hypotheses have received the most attention: dominance hypothesis and overdominance hypothesis; in general, hybridization is a prominent process among natural plant populations that can result in phenotypic novelty, heterosis, and changes in gene expression; when extensive >>> transcriptome data sets are investigated (1) potential candidate transcripts for heterosis can be obtained, and (2) changes in the expression of the candidate transcripts can be studied; sometimes the term "negative heterosis" is used; it describes the negative yields of hybrids in comparison to the parental genotypes crossed; SCHNELL (1961) accepted the term only when the performance of the hybrids ranges below that of the homozygous parents; "absolute heterosis" opposed to "relative heterosis" after FISCHER (1978) is related to the absolute increase of yield; "adaptive heterosis" after GUSTAFSON (1951) is observed in plants that are well-adapted to their environments (*syn* adaptive heterosis or >>> euheterosis); "balanced euheterosis" after DOBZHANSKY (1949, 1950, 1952) refers to adaptive heterosis or euheterosis based on heterozygosity (cf. mutational heterosis); hybrid vigor *syn* heterosis; "beneficial heterosis" after POWERS (1945) refers to positive heterosis; "cisheterosis" after HIORTH (1963) refers to normal growth and vitality as opposed to "transhetrosis"; "true heterosis" after MALINOWSKI (1952) points to a type of "relative heterosis" related to the performance of best cross parent (*syn* heterobeltiosis or true heterosis); "one-gene heterosis" refers to a single heterozygous locus-based heterosis (*syn* "monohybrid heterosis", "monofactorial heterosis", "single-gene heterosis", or "single-locus heterosis"); "fixed heterosis" refers to long-lasting hybrid effects, i.e., over several generations; "gigas heterosis" after FISCHER (1978) refers to heterosis associated with strong or gigas growth of the plant; "heterosis effect", "hybrid effect", "hybrid vigor", "effect of heterozygosity", or "superdominance" are synonymous terms for "heterosis"; "homogametal heterosis" after MALINOWSKI (1952) refers to heterosis of female individuals with two X-chromosomes as compared to males with X- and Y-chromosomes; "hypothetical heterosis" after MALINOWSKI (1952) refers to positive yield performance of hybrids as compared to the mean of parents' performance, that does not exceed the performance of the better parent (when it exceeds the better parent then "true heterosis" is given); "combination effect" – sometimes used instead of heterosis; "combination heterosis" after HAGBERG (1952) refers to heterosis based on defined genes or traits; "cumulative heterois" after HALDANE (1957) refers to heterosis caused by accumulation of genes or genetic effects, i.e., transgression or additive polygeny; "luxuriance" is synonymous with "heterosis", "gigas heterosis", or "somatic heterosis" (used by DOBZHANSKY instead of pseudo heterosis); "mitochondrial heterosis" refers to heterosis by mixing of genetically different mitochondria; "mutational heterosis" after DOBZHANSKY (1949, 1950, 1952) refers to a type of adaptive heterosis or euheterosis based on dominant genes dominating lethal or sublethal genes (cf. balanced euheterosis); "nonbeneficial heterosis" after POWERS (1945) refers to >>> "negative heterosis"; "permanent heterosis" *syn* "fixed heterosis"; "plasmatic heterosis" refers to heterosis caused by interaction of cytoplasms and/or cytoplasms and genotypes; "potential heterosis" after FISCHER (1963) refers to the power of a certain amount of heterosis; "pseudogamous heterosis" after HASKELL (1960) refers to growth effects caused by heterozygous endosperms and/or hybrid endosperms; "reproductive heterosis" after GUSTAFSSON (1951) points to heterosis of increase fertility; "somatic heterosis" syn "vegetative heterosis" refers to increase of growth of vegetative organs; *gene* >>> Figure 2 >>> Tables 35, 36

heterosomal: A chromosome mutation involving non-homologous chromosomes *cyto*

heterosome: A chromosome that deviates from the standard chromosomes in size, shape, or behavior *cyto*

heterospory: The production of spores of two different types of the same plant *bot*

heterostyly: A polymorphism among flowers that ensures cross-fertilization through pollination by visiting insects; flowers have anthers and styles of different length *bot*

heterothallic: Describes yeast strains that have a fixed mating type and as a result can only mate when mixed with a strain of the opposite mating type *bot*

heterotic group(s): A group or groups of germplasm that when crossed maximize heterosis; heterosis is a phenomenon in which heterozygotes in a population often have higher fitness than the homozygotes *eco meth*

heterotroph: An organism that is unable to manufacture its own food from simple chemical compounds and, therefore, consumes other organisms *phys*

heterozygosity: The presence of different alleles of a given gene at a particular gene locus *gene* >>> Tables 9, 10

heterozygote: A diploid or polyploid individual that has different alleles on at least one locus *gene* >>> Tables 9, 10

heterozygotic: >>> heterozygous

heterozygous: The condition of having unlike alleles at corresponding loci *gene*

hexaploid: With six sets of chromosomes *cyto*

hexasomic: A cell or individual showing one chromosome six times *cyto*

hexokinase: An enzyme that catalyzes the phosphorylation of hexose sugars *chem phys*

hexose: A monosaccharide sugar that contains six carbon atoms *chem phys*

HI: >>> harvest index

hibernaculum: The winter resting body of some plants, generally a bud-like arrangement of potential leaves *bot*

hibernating organ: >>> hibernaculum

Hidden hunger: The nutritional status of a growing plant when it is experiencing the deficiency of a nutrient but this effect is not severe enough to produce visible deficiency symptoms – that is, one stage below the critical level; it results in loss of yield *phys agr*

Hieracium **type:** Apospory where the embryo sac is derived from a somatic cell, usually from the center of the nucellus; three mitoses lead to a mature eight-nucleate, unreduced embryo sac *bot*

hierarchical classification: The grouping of individuals by a series of subdivisions or agglomerations to form characteristic "family trees" or dendrograms of the group *stat*

high-density map: Any genetic or physical map of a genome that contains a larger number of mapped markers such that markers are spaced at recombination frequencies of 1–5 cM, or about 100–550 physical map units *biot meth stat*

high oleic: In sunflower breeding, seeds that contain a trait for high oleic fatty acid content in their oil; premium oil used in the snack food industry *agr*

high-resolution mapping: The production of a detailed genetic map with recombination frequencies of less than 1 cM either for the full genome or for a defined region of a chromosome *biot meth*

high-throughput DNA sequencing: The industrialized process of determining the exact order of the four letter code (bases, GAT or C) that composes all genetic material; because a typical higher organism contains >20,000 genes, each with several thousand bases of DNA, data storage and analysis is one of the most important parts of this process *biot*

high-throughput screens: An automated process using small microwell plates that allows the testing of many hundreds of thousands of chemicals for potential activity; these tests require very small quantities (a few micrograms) of test substances *biot*

high tunnel: Structures that resemble plastic-covered greenhouses but are considered non-permanent structures that have no automated heating or ventilation and are covered with a single layer of 0.6 mm plastic *hort meth* >>> low tunnel

high-velocity microprojectile transformation: A procedure for the introduction of DNA into plant cells; for example, gold particles or nanoparticles are coated with DNA and propelled at high speed through the target cell walls by means of an electrical or gunpowder discharge *biot* >>> nano-biolistic

high-yielding variety: The phrase refers to "modern" (post-1960) crop varieties having a much higher grain yield potential (5–10 t/ha) than traditional tall varieties; in maize, usually hybrids or composites are meant; in wheat and rice, usually stiff-strawed, short-statured varieties, or developed hybrids; high-yielding varieties are more responsive to fertilizer application than local tall varieties and their large-scale cultivation triggered the >>> Green Revolution *agr*

hiller: A mechanical device or implement used to move loose soil from the space between, e.g., the potato rows, and deposit it beneath and against the potato stems to form a long, continuous "hill" the full length of the row; the soil covers and protects potato tubers growing near the surface from frost damage and greening; in former times, hilling was done by hand with a hoe; modern hillers are tractor-mounted and operated hydraulically; rotary hoes, discs, moldboards, and power hillers equipped with a metal mold are commonly used to shape loose soil into a hill; the >>> hilling implement should be adjusted to produce a wide, flattened hill ideal for protecting the tubers from sunlight, late blight spores, and frost *tech meth*

hilling: The tillage operation necessary in the production of potatoes after planting; the objective of hilling is to control weeds, to maintain coverage of soil over the growing tubers to prevent greening, to minimize infection with late blight, to minimize frost damage, to improve drainage in the area of tuber formation, and to facilitate harvest; to effectively control weeds, hilling must take place before the weeds get past the two true leaf stage; the preparation can be performed pre- or post-emergence *meth tech* >>> hiller

hill planting: Grouping plants in a cluster, not necessarily on an elevated mound *agr hort*

hill plot: >>> Figure 33

Hill–Robertson effect: In a fundamental study, in 1966, the authors used computer simulations to investigate multi-locus interactions under realistic conditions, including selection and recombination in finite populations, where they investigated the probability of fixation of a beneficial mutation in the presence of another segregating beneficial mutation; the study confirmed R. A. FISHER's predictions, and showed that selection at one locus interferes with selection at a second beneficial mutation, reducing its probability of fixation; that is, in natural populations (with finite size), selection at more than one site should cause an overall reduction in the effectiveness of selection; the magnitude of this effect depends on selection intensities and the initial frequencies of the alleles; more important, the degree of interference increases with genetic linkage between the loci under selection *evol stat*

hilum (hila pl): The scar on a seed that marks the point at which it was attached to the plant *bot*

hips: Seed pods (fruits), for example, in roses or apples, that are formed after a flower's petals fall if the bloom was pollinated *hort*

hirsute: Covered with coarse, straight, rather stiff hairs, usually perpendicular to the surface *bot*

hispid: Rough with erect, bristly hairs *bot*

histidine (His): A basic, polar amino acid that contains an imidazole group *chem phys* >>> Table 38

histogram: A bar graph of a frequency distribution in which the bars are displayed proportionate to the corresponding frequencies *stat*

histone: One of a group of globular, simple proteins that have a high content of the amino acids arginine and lysine; it forms part of the chromosomal material of eukaryotic cells and appears to play an important role in gene regulation; highly conserved basic proteins that are involved in the packing of DNA; histone proteins and the nucleosomes form, with DNA, the fundamental building blocks of eukaryotic chromatin; they bind to the phosphate groups of DNA by their amino termini; there are five major types of histone proteins; two copies of H2A, H2B, H3, and H4 bind to about 200 base pairs of DNA to form the repeating structure of chromatin (nucleosome), with H1 binding to the linker sequence; histone genes do not encode for poly-A tail; possible post-translational modifications of histone molecules include deacetylation of >>> lysine, methylation of lysine and arginine, ubiquitination, or phosphorylation; while histone acetylation and possibly phosphorylation correlate with gene activity, histone methylation seems to have diverse functions; in general, histones are subjected to a variety of post-translational modifications, including acetylation, phosphorylation, ADPribosylation, methylation, and ubiquitination, chem *phys cyto* >>> Tables 38, 43

hitch-hike: Where genes may become linked to an advantageous gene or >>> supergene and segregate with it disadvantageously in linkage equilibrium *gene eco*

hitchhiking effect: Genes favored in a population by close linkage with other genes, which are positively selected evol *gene*

HMG-box protein(s): High-mobility group (HMG) proteins of the HMGB family containing a highly conserved HMG box; they are chromatin-associated proteins that interact with DNA and nucleosomes and catalyze changes in DNA topology, thereby facilitating important DNA-dependent processes, e.g., the genome of *Arabidopsis thaliana* encodes 15 different HMG-box proteins that are further subdivided into four groups: HMGB-type proteins, ARID-HMG proteins, 3xHMG proteins that contain three HMG boxes, and the structure-specific recognition protein 1 (SSRP1); typically, HMGB proteins are localized exclusively to the nucleus, like *Arabidopsis* HMGB1 and B5; however, these *Arabidopsis* HMGB proteins showed a very high mobility within the nuclear compartment; studies revealed that *Arabidopsis* HMGB2/3 and B4 proteins are predominantly nuclear but also exist in the cytoplasm, suggesting an as yet unknown cytoplasmic function of these chromosomal HMG proteins *biot cyto*

holandric gene(s): The genes located on Y chromosomes *cyto gene*

holdfast: An adhesive material at a localized position on a cell, enabling the cell to attach to a surface *phyt*

hole seeding: >>> dibbling (seed)

holocentric: Applied to chromosomes with diffuse centromeres such that the properties of the centromere are distributed over the entire chromosome; holocentric chromosomes occur in a number of independent eukaryotic lineages; they form holokinetic kinetochores along the entire poleward chromatid surfaces, and owing to this alternative chromosome structure, species with holocentric chromosomes cannot use the two-step loss of cohesion during meiosis typical for monocentric chromosomes; in the plant *Luzula elegans* (or *Cuscuta* ssp., *Rhynchospora* ssp., *Cyperus* ssp.) a holocentric chromosome architecture and behavior are maintained throughout meiosis, and in contrast to monopolar sister centromere orientation, the unfused holokinetic sister centromeres behave as two distinct functional units during meiosis I, resulting in sister chromatid separation; homologous nonsister chromatids remain terminally linked after metaphase I, by satellite DNA-enriched chromatin threads, until metaphase II; they then separate at anaphase II; thus, an inverted sequence of meiotic sister chromatid segregation occurs; this alternative meiotic process is most likely one possible adaptation to handle a holocentric chromosome architecture and behavior during meiosis *cyto* >>> Figure 70

holokinetic: >>> holocentric

homeobox: A characteristic DNA sequence of 180 bp that codes for the 60-amino-acid DNA-binding domain of some developmentally important regulatory genes; mutations in the homeobox can have homeotic effects *gene biot*

homeostatic mechanism: A physiological process that contributes to the maintenance of a relatively stable internal environment in a multicellular organism *phys*

homeotic genes: A class of genes that determines the identity of an organ, segment, or other structural unit during development; it controls the identity of, for example, floral organs *gene*

homeotic mutation: A mutation that causes one body structure to be replaced by a different body structure during development *gene*

homoallelic: Applied to allelic mutants of a gene that has different mutations at the same site *gene*

homoduplex DNA: >>> heteroduplex DNA

homoeologous: Partially homologous; chromosomes or genomes that are believed to have originated from ancestral homologous chromosomes *cyto*

homoeologous group: Series of two or more chromosomes with similar but not homologous chromosomes (e.g., in hexaploid wheat or oat) *cyto*

homoeostasis: The tendency of a biological system to resist change and to maintain itself in a state of stable equilibrium *bio* >>> genetic equilibrium

homogametic: Producing only male or female gametes *bot*

homogamic: Of matings between individuals from the same population or species *gene*

homogamous: Hermaphroditic flowers showing synchronized function of male and female sex organs, i.e., pistils and stamens that mature at the same time *bot*

homogamy: The preference of individuals to mate with others of a similar genotype or phenotype; in botany, the condition in which male and female parts of the flower mature simultaneously *bot*

homogenetic chromosome pairing: Synapsis of homologous chromosomes of the same genome *cyto*

homogenic: Sometimes refers to chromosome pairing between morphologically identical genomes in amphiploids *cyto*

homogenotization: A genetic technique used to replace one copy of a gene, or other DNA sequence within a genome, with an altered copy of that sequence; the DNA is first cloned and then altered in some way, for example, a transposon is inserted into a gene; the mutated gene copy can be used to replace the original gene by recombination *in vivo*; the incorporation of the mutated gene is usually selected, for example, by virtue of its containing a transposon-encoded antibiotic resistance *biot meth*

homograft: >>> heterograft

homologous: Applied to organs and chromosomes; both showing identical structures *cyto bot*

homologous genes: Genes with a common ancestor, generally used to describe genes from different species but which are similar and have the same function *gene*

homology: Fundamental similarity; in molecular biology, the degree of identity between two sequences from related organisms *biot*

homomeric: Genes that control the determination of a trait by joint gene action, and each of them shows a similar contribution to the final product *gene*

homomorphic bivalent: A bivalent that is composed of two homologous chromosomes (i.e., in size and shape) *cyto*

homoplasmic: Cells or individuals that carry two or more different types of cytoplasmic components *bot*

homoplasmon: Extranuclear gene – alike (analogous to homozygote for nuclear genes) *gene* >>> Table 53

homoplasmonic: >>> homoplasmic

homoplasy: A term in cladistic analysis that refers to the proportion of parallelisms and reversals on a phylogenetic tree; in biotechnology, also used for different DNA fragments of identical size that cannot be distinguished by gel electrophoresis *bot biot meth*

homopolymeric tailing: >>> homopolymer-tailing >>> tailing

homopolymer-tailing: Attachment of identical nucleotides to the 3′ end of a DNA molecule that can be achieved with terminal deoxynucleotide transferase *biot*

homosequential linkage map: A linkage map of a genome "A" that shares a similar or even identical marker order with a linkage map from a genome "B" *biot meth*

homothallic strains: In yeast, strains that are capable of switching mating type and thus of mating with themselves, forming zygotes, asci, and spores under appropriate conditions *bot*

homospory: A condition in which an organism produces only one type and size of spore, e.g., microspores *bot*

homozygosity: The presence of identical alleles at one or more loci in homologous chromosomal segments *gene* >>> Table 9

homozygosity mapping: Recessive characters require two copies of an allele for expression; because of >>> linkage disequilibrium, loci surrounding the trait locus will tend to be homozygous in affected individuals; searching for homozygous segments in mutated individuals helps to locate the trait gene *meth biot* >>> association mapping

homozygote: A cell or organism having the same allele at a given locus on homologous chromosomes *gene* >>> Table 9

homozygous: Having identical rather than different alleles in the corresponding loci of a pair of chromosomes and therefore breeding true *gene* >>> Table 9

honeybee: Any bee that collects and stores honey (e.g., Apis mellifera) and contributes to improved seedsetting in cross-pollinating crops (e.g., rapeseed, fruit trees, etc.); INRA and CNRS French scientists and a UFZ German scientist found that the worldwide economic value of the pollination service provided by insect pollinators, bees mainly, was €153 billion in 2005 for the main crops that feed the world; this figure amounted to 9.5% of the total value of the world agricultural food production; the study also determined that pollinator disappearance would translate into a consumer surplus loss estimated between €190 to €310 billion; three main crop categories (following FAO terminology) were of particular concern: fruits and vegetable were especially affected with a loss estimated at €50 billion each, followed by edible oilseed crops with €39 billion; the impact on stimulants (coffee, cocoa), nuts and spices was less, at least in economic terms *zoo seed*

honeycomb design: In a honeycomb design, the plant at the center of the hexagon is compared with every other plant within the hexagon; a plant is chosen only if it is superior to every other plant in the hexagon; it was developed for selecting individual plants in a population; seeds or plants are usually spaced equidistantly from one another in a hexagon pattern; plants are spaced far enough apart that they cannot compete with adjacent individuals; homogeneous checks can be included; the size of the hexagon determines the selection intensity; it is used to minimize adverse effects of interplant competition *stat* >>> design of experiment >>> Figure 32

honeydew: A sticky exudate (>>> NUSSINOVITCH 2009) containing conidia, which is produced during one stage of the life cycle of the fungus *Claviceps purpurea phyt bot* >>> ergot

hook climber: A plant that climbs by the aid of hooks or prickles (e.g., roses) *hort* >>> climbing plants

hop blower: Rotary-blower duster for hops *agr*

"Hope" (wheat): A variety of common wheat showing resistance to stem rust; it was transferred from *Triticum turgidum* ssp. *dicoccum* (cv. "Yaroslav") into hexaploid wheat by McFADDEN (1930); HARE and McINTOSH (1979) determined that stem rust resistance in "Hope" was largely controlled by a single gene located on the short arm of chromosome 3B; the resistance gene was named *Sr2* and assumed to have come from *T. turgidum*, though the original tetraploid accession was lost and could not be confirmed to carry the gene; the cultivar "Hope" was used in Mexico during the 1940s as the donor for developing the stem rust-resistant wheat cultivar "Yaqui 48" (BORLAUG 1968); since then, the *Sr2* gene has been employed widely by CIMMYT's global wheat improvement program in Mexico, and germplasm exchange from CIMMYT has distributed it to many wheat production regions of the world; the gene has provided durable, broad-spectrum rust resistance effective against all isolates of *Puccinia graminis* worldwide for more than 50 years; on the basis of its past performance, *Sr2* has been described as one of the most important disease-resistance genes deployed in modern plant breeding (McINTOSH et al. 1995, SHARP and LAGUDAH 2003) *agr gene*

hordecale: An amphiploid hybrid between cereal species of the genera Hordeum and Secale, in which Hordeum species served as donors of the cytoplasm *bot*

hordein: >>> prolamin

***Hordeum bulbosum* procedure:** A method for producing zygotic haploids in barley by crossing *Hordeum bulbosum* with *Hordeum vulgare* genotypes; after formation of zygotes, the wild *H. bulbosum* chromosomes are subsequently eliminated during embryogenesis, which results in haploid *H. vulgare* plants; the procedure involves: day 0 – emasculation of florets inside the spike; day 2 – pollination with *H. bulbosum* pollen; day 3 (to 5) – 40% of the embryonic cells are haploid, endosperm abortion occurs, GA3 treatment enhances retention of florets; day 11 – 94% of the embryonic cells are haploid; day 14 (to 16) – embryos are dissected and cultured in the dark at 18 to 22°C, embryos develop *in vitro*; day 22 (to 28) – embryos are transferred to light for seedling development *biot* >>> haploid >>> haploid production >>> Figures 17, 26 >>> Table 7

horizontal gene transfer: Horizontal gene transfer, also lateral gene transfer, is any process in which an organism incorporates genetic material from another organism without being the offspring of that organism; by contrast, vertical transfer occurs when an organism receives genetic material from its ancestor, e.g., its parent or a species from which it has evolved; most thinking in genetics has focused upon vertical transfer, but there is a growing awareness that horizontal gene transfer is a highly significant phenomenon and amongst single-celled organisms perhaps the dominant form of genetic transfer; artificial horizontal gene transfer is a form of genetic engineering *evol biot* >>> evolution >>> grafting

horizontal resistance: Resistance conditioned by polygenes or quantitative genes; it is race non-specific in nature and does not reveal a gene-for-gene hypothesis; this type of resistance is difficult to identify *phyt*

hormone: A regulatory substance, active at low concentrations, that is produced in specialized cells but that exerts its effect either on distant cells or all cells to which it is conveyed via tissue fluids in the organism *phys*

horticulture: The science or art of cultivating flowers, fruits, vegetables, or ornamental plants (e.g., in a garden, orchard, or nursery) *hort*

hortus siccus: A collection of specimens of plants carefully dried, preserved, and described for botanical purposes and comparisons of mutants, etc. *bot meth* >>> herbarium

host A living organism harboring a parasite; its cells and metabolism are used for the growth of pathogens; plant hosts can be classified by: (1) importance (a) primary or principal hosts, (b) secondary hosts, (c) intermediate hosts, (d) accessory hosts, (e) accidental hosts, and (f) definite or final hosts; (2) season (a) winter hosts and (b) spring hosts; (3) other functions (a) alternative, alternate, or differential hosts, (b) transport or transfer hosts *phyt*

host cell: A cell whose metabolism is used for the growth of a pathogen *phyt*

host plant: >>> host

host plant resistance: A method of pest control in which resistant, tolerant, or unattractive host organisms are used; the inherited qualities of resistance influence the extent of pest damage *phyt*

host-parasite specificity: The ability of a pathogen to pathogenize a specific group of plants *phyt*

host range: The spectrum of genotypes that can infect a specific pathogen or pest; in molecular biology, hosts in which a phage or plasmid can replicate; restriction is one factor that can limit the host range of plasmids or phages *phyt biot*

host resistance: The result of genetic manipulation of the host which renders it less susceptible to pathogens that would or do attack the host *phyt*

host species: >>> heteroecious

host-mediated restriction: A mechanism by which bacteria prevent infection by phages originating from other bacteria; restriction also acts against unmodified plasmid DNA in transformation experiments; restriction endonucleases cleave the foreign DNA while the host DNA is protected from cleavage by specific methylation *biot* >>> Table 39, 43

hotbed: A bed of soil enclosed by a structure with a top of glass, heated, often by manure, for forcing or raising seedlings *hort*

hot spot: In genetics, one of the sites tending to mutate frequently *gene*

housekeeping enzymes: Enzymes present in all cells capable of normal metabolism; they are essential for the synthesis or breakdown of proteins, nucleic acids, and lipids for glycolysis and respiration, and for many standard metabolic pathways *phys*

housekeeping genes: Genes whose products are required by all cells at all times *gene* >>> housekeeping enzymes

htSNP: >>> haplotype single nucleotide polymorphism

huanglongbing: >>> citrus

hull: Usually the hard, tightly adhering, outer covering of a seed or caryopsis, which is composed of the pericarp in some species and the lemma and palea in others seed; the persistent calyx at the base of some fruits, such as strawberry *bot*

hull: To remove the hull *seed prep* >>> dehull

huller-scarifier: A seed-conditioning machine; it removes hulls or pods from seeds by an abrading or rubbing action *seed* >>> Table 11

humic acid: A mixture of dark-brown organic substances that can be extracted from soil with dilute alkali *agr* >>> Table 44

humification: The development of humus from dead organic material *agr*

humus: Decomposed organic matter of soils *agr* >>> Table 44

husk: The leaf sheaths of an ear of maize; the lemma and palea in other grass species or the dry outer cover of a coconut; the dry outer covering of some fruits or seeds (e.g., cereal grain); husk leaves surround the >>> cob, as opposed to the foliar leaves born on the stem *bot* >>> dehusk

hyaline: Clear, colorless, transparent bot

hyaline grain: >>> hyaline

hybrid: Any sort of sexual or somatic combination of genetically more or less differentiated parental cells, individuals, or taxa; specifically, an individual plant from a cross between parents of differing genotypes; any heterozygote represents dissimilar alleles at a given locus; a hybrid graft *gene* >>> Figures 2, 18, 31

hybrid arrested translation: A method used to identify the proteins encoded by a cloned DNA sequence; a crude cellular >>> mRNA preparation, composed of many individual types of mRNA, is hybridized with cloned DNA; only mRNA molecules homologous to the cloned DNA will anneal to it; the rest of the mRNA molecules are put into an *in vitro* translation system and the protein products are compared with the proteins obtained by use of the whole mRNA preparation *biot meth* >>> hybrid released translation

hybrid breakdown: >>> hybrid lethality

hybrid breeding (*syn* **heterosis breeding**): The discovery of heterosis has been recognized as one of the major landmarks of plant breeding; in comparison to inbred lines and homozygous material, the phenotypic superiority of heterozygotes is the basis of hybrid breeding; it is exploited for production of hybrid, synthetic, and composite varieties; hybrid breeding can be performed using traditional breeding techniques, or the process can be hastened using gene marker technology to rapidly identify parents with desired genes for certain attributes; numerous commercial crops are hybrids with increasing tendency; seeds from a hybrid variety, if planted, will not deliver the same benefits as the original seeds and after several offspring will have lost the desired qualities from the original hybridization *meth* >>> gametocide >>> Figure 51 >>> Table 37

hybrid chlorosis: Plant and/or leaf chlorosis due to interacting genes and/or cytoplasm of parental lines in a hybrid *gene*

hybrid complex: Masking morphological differences of parental lines in a hybrid plant *gene*

hybrid heterosis: >>> heterosis

hybrid inviability: Reduced vigor of hybrid plants compared to their crossing parents *gene*

hybridization: A method of breeding new varieties that applies crossing to obtain genetic recombination; in genetics, the fusion of unlike genetic material, such as sexual organs or DNA; in molecular biology, pairing of complementary DNA and/or RNA *meth biot* >>> gametocide >>> Figure 31

hybridization probe: Labeled nucleic acid molecule used to detect complementary DNA sequences after hybridization *biot*

hybridization stringency: The percentage of nucleotides, which must match on two unrelated single-stranded nucleic acid molecules before they will base pair with each other to form a duplex, given a certain set of physical and chemical conditions; the hybridization stringency is used to determine when a hybridization probe and a target nucleic acid will come together, and can be set by varying the conditions; in general, if the percentage of matching nucleotides is lower than 70%, the two single-stranded nucleic acid molecules are considered non-homologous and any hybridization is considered non-stringent *prep biot*

hybrid lethality (*syn* **hybrid sterility**): The failure of hybrids to produce viable offspring *gene*

hybrid necrosis: >>> hybrid lethality

hybrid plant: >>> gametocide >>> hybrid

hybrid released translation: A method used to detect the proteins encoded by cloned DNA; the cloned DNA is bound to a >>> nitrocellulose filter and a crude preparation of mRNA is hybridized to the filter-bound DNA; only mRNA sequences homologous to the cloned DNA will be retained on the filter; these mRNA molecules can then be removed by high temperatures or by using formamide; the purified mRNA is then placed in an *in vitro* translation system, and the proteins encoded by the message can be analyzed by electrophoresis through a polyacrylamide gel *biot meth* >>> hybrid arrested translation

hybrid seed production: The production of hybrid seeds by combination of more or less defined parental forms; usually those hybrid seeds are more productive or more suitable than pure lines; they are used for subsequent growing and commercial production of a crop *seed* >>> Figures 2, 18, 22, 23, 29, 51, 55 >>> Tables 5, 37

hybrid selection: The process of choosing plants possessing desired traits among a hybrid population *meth*

hybrid sterility (*syn* **hybrid lethality**): The failure of hybrids to produce viable offspring *gene*

hybrid swarm: A group of morphologically distinctive individuals that results from the creation of hybrids between two parent species, then the backcrossing of the offspring to members of the parent species and the interbreeding among the hybrid individuals *gene*

hybrid variety: A variety produced from the cross-fertilization of inbred lines with favorable combining ability; the progeny is homogeneous and highly heterozygous; it can be produced by (1) two inbred lines, (2) single crosses, (3) a single cross and an open-pollinated or a synthetic variety, or (4) two selected clones, seed lines, varieties, or species *seed* >>> gametocide >>> Figures 2, 29, 50, 51

hybrid vigor: The increase in vigor of hybrids over their parental inbred types *gene* >>> heterosis >>> Figure 18

hybrid weakness: The decrease in vigor of hybrids below their parental inbred types; for example, in rice a hybrid weakness phenomenon is controlled by a set of complementary genes, *Hwc1* (hybrid weakness c) and *Hwc2*; the *Hwc2* gene is prevalent among temperate Japonica rice but not among tropical Japonica or Indica rices; the chromosomal location of the *Hwc2* locus was determined from the segregation in the F1 hybrids made between recombinant inbred lines and the cultivar "Jamaica"; *Hwc2* was located between the two restriction fragment length polymorphism loci, *XNpb264* and *XNpb197* on chromosome 4 (K. ICHITANI et al. 2001) *gene* >>> Table 39

hybrid zone: A geographic area where different populations of species meet and hybridize after a period of geographic isolation *eco*

hydathode: An epidermal structure specialized for the secretion or exudation of water *bot* >>> guttation

hydratation: The status of imbibition of the cytoplasm *phys*

hydrate: Any of a class of compounds containing chemically combined water *chem*

hydration: The process whereby a substance takes up water *agr*

hydraulic seeding: A method of planting grass seed by spraying it in a stream of water, which may contain other materials such as nutrients *meth agr*

hydrogen peroxide test: A quick test to determine seed viability; in response to a hydrogen peroxide soak, viable seeds elongate their roots through a cut in the seedcoat; frequently used in conifer seeds *seed*

hydrolase: An enzyme that catalyzes reactions involving the hydrolysis of a substrate *chem phys*

hydrolysat(e): Any compound formed by hydrolysis *chem*

hydrolysis: In soil science, the process whereby hydrogen ions from water are exchanged for cations such as sodium, potassium, calcium, and magnesium, and the hydroxyl ions combine with the cations to give hydroxides *chem agr*

hydrophytes: Freshwater plant with perennating tissue below the surface of the water; vegetative shoots are submerged and leaves are submerged or floating; only flowers rise above the water surface >>> hygrophytes

hydroponis: The cultivation of plants by placing the roots in liquid nutrient solutions rather than in soil *hort*

hydroseeding: Dissemination of seed hydraulically in a water medium, supplemented by mulch, lime, and fertilizer *meth fore hort*

hydrotaxis: Movement of an organism toward or away from water *bot*

hydroxide: A chemical compound containing the hydroxyl group *chem*

hygrometer: Any instrument for measuring the water-vapor content of the atmosphere *prep*

hygrophytes: Plants that can tolerate an excess of water *bot*

hygroscopic(al): Water attracting; plants or part of them becoming soft in wet air and hard in dry air *bot*

hygroscopic water: Water that is adsorbed onto a surface from the atmosphere *agr*

hypanthium (hypanthia *pl*): A cup-shaped extension of the floral axis, usually formed from the union of the basal parts of the calyx, corolla, and stamens, commonly surrounding or enclosing the pistils *bot*

hypermorph: A mutant gene, which causes an increase in the activity that it influences *gene*

hypermutation: The process of dramatically increasing the mutation rate of a distinct DNA molecule, a genomic region, or a whole chromosome above background *gene*

hyperplasia: The enlargement of tissues by an increase in the number of cells by cell division bot

hyperploidy: Having additional chromosome complements compared to the standard chromosome set *cyto*

hypersensitive resistance: >>> hypersensitivity

hypersensitive site: A region of DNA located in a chromatin structure that makes it more sensitive to attack by >>> endonucleases than DNA sites, located elsewhere in the chromatin; the presence of hypersensitive sites is correlated with transcription of adjacent DNA sequences in eukaryotic cells *gene*

hypersensitivity: The response to attack by a pathogen of certain host plants in which the invaded cells die promptly and prevent further spread of infection *phyt* >>> resistance

hypertonia: >>> hypertonic

hypertonic: A solution whose osmotic potential is less than that of living cells, causing water loss, shrinkage, or plasmolysis of cells *phys*

hypertonicity: >>> hypertonic

hypertrophy: The enlargement of tissues by an increase of the size of the cells *bot* >>> hyperplasia

hypha (hyphae *pl*): A tubular, threadlike filament of fungal mycelium *bot*

hypocotyl: Part of the embryonic shoot or seedling located below the cotyledon and above the radicle; the plant hormones >>> brassinosteroids promote hypocotyl elongation, e.g., of >>> *Arabidopsis thaliana* both under light and dark (etiolated) conditions; there is a common assay to determine if a >>> mutant or transgenic line is affected in brassinosteroid biosynthesis or response is a sensitivity assay to brassinazole, an inhibitor of P450 cytochromes seedlings specific to brassinosteroid biosynthesis *bot*

hypodermis: The cell layer beneath the epidermis of the pericarp *bot*

hypogeal: Living or growing underground *bot*

hypogeal emergence: A type of emergence characterized by the elongation of the epicotyl and the cotyledons remaining below ground that is common in monocots *bot*

hypogean: >>> hypogeal

hypgynous: Flower with a superior ovary, i.e., with ovary borne distal to the receptacle *bot*

hypophysis: The uppermost cell or suspensor from which part of the root and root cap in the embryo of angiosperms are derived *bot*

hypoplasia: Abnormal deficiency of cells or structural elements *bot*

hypoploidy: Missing chromosome complements compared to the standard chromosome set *cyto*

hypostatic epistasis: >>> epistasis

hypotonic: Of or designating a solution of lower osmotic pressure than another, as opposed to hypertonic *phys*

hypoxia: Low oxygen supply in the plant rhizosphere due to waterlogging; it reduces growth and yield *phys*

hypsometer: Any of several tools or instruments designed to measure the height of trees *fore*

I

I1, I2, I3, etc.: The first, second, third, etc., generations obtained by inbreeding *gene meth*

IAA: >>> indole-3-acetic acid

IAEA: >>> International Atomic Energy Agency

IBPGR: >>> International Board for Plant Genetic Resources >>> bioversity

ICARDA: >>> International Center of Agricultural Research in the Dry Areas

ICBN: >>> International Code of Botanical Nomenclature

ice-cold water (treatment): A pretreatment with ice-cold water is effective and widely used for cereal chromosome condensation prior to fixation, similar to treatment with 8-hydroxyquinoline, >>> colchicine, >>> alpha-bromonaphtalen, or paradichlorobenzene *cyto*

ICIA: >>> International Crop Improvement Association

ICNCP: >>> International Code of Nomenclature for Cultivated Plants

ICRA: >>> International Cultivation Registration Authority

ICRISAT: >>> International Crops Research Institute for the Semi-Arid Tropics

identical by descent: Two genes that are identical in nucleotide sequence because they are both derived from a common ancestor *gene*

identical in structure: Two genes that are identical in nucleotide sequence, regardless of whether or not they are both derived from a common ancestor *gene*

identity preservation (IP): A system of crop or raw material management that preserves the identity of the source or nature of the materials *agr*

ideotype: Crop plant with model characteristics known to influence photosynthesis, growth, and grain production; "ideotype" was coined by DONALD (1968); it means literally a form denoting an idea, and in the broadest sense it is a biological model which is expected to perform or behave in a predictable manner within a defined environment; more specifically a crop ideotype is a plant model which is expected to yield a greater quantity of grain, oil, or other useful product when developed as a cultivar *phys gene* >>> idiotype

ideotype breeding: A method of breeding to enhance genetic yield potential based on modifying individual traits where the breeding goal for each trait is specified *meth*

idioblast: A plant cell committed to develop into a cell type that differs from the surrounding tissue *bot*

idiochromosome: A chromosome that contributes to the determination of sex *cyto*

idiogamy: Combination of male and female gametes from the same individual *bot*

idiogram: A diagrammatic representation of the karyotype of a plant *cyto* >>> Figure 12

idiomorph: Used to describe fungal mating types which are extremely dissimilar from each other and do not show homology between strains of the opposite sex (as opposed to the allelic relationship in most polymorphic systems) *gene*

idioplasm: All hereditary determinants of a plant including genotype and plasmotype *gene* >>> germplasm

idiotype: The sum of the hereditary determinants of a cell or plant consisting of the genotype and plasmotype; idiotype is an established term, first used by SIEMENS (1921); "idio" in idiotype is derived from the Greek word "idios", meaning one's own, personal, separate, distinct *gene* >>> ideotype

IGFRI: >>> Indian Grassland and Fodder Research Institute

IITA: >>> International Institute of Tropical Agriculture

I-line: >>> inbred line

illegitimate: Pollination of a flower of the same self-incompatibility genotype as the pollen parent *bot* >>> legitimate

illegitimate crossing-over: >>> unequal crossing-over

illegitimate recombination: Recombination between DNA fragments that do not share extensive DNA sequence homology; >>> transposons and insertion sequences have special functions that catalyze illegitimate recombination *biot* >>> Figure 46

illegitimate seed: Resulting from natural crosspollination between plants or clones where the male parent cannot be ascertained *seed* >>> Figure 50

Illinois method: The separate sowing of lines or families and the selection of the best plants from the best families *meth* >>> ear-to-row selection

Illumina HiSeq 2500™ or MiSeq™: The HiSeq 2500 or similar sequencers deliver the highest daily throughput of any currently available sequencing instrument; sequencing systems enable individual labs to process large numbers of samples and decode complex genomes; the generation and analysis of high-throughput sequencing data are becoming a major component of many studies in molecular biology and medical research; Illumina's Genome Analyzer (GA) and HiSeq instruments are during currently the most widely used sequencing devices *biot meth*

image processing: Various mathematical procedures to improve the signal-to-noise and contrast, and to obtain quantitative intensity data from images *micr*

imazamox (2-[4,5-dihydro-4-methyl-4-(1-methylethyl)-5-oxo-1*H*-imidazol-2-yl]-5-(metho xymethyl)-3-pyridinecarboxylic acid): A herbicide with the genome formula $C_{15}H_{19}N_3O_4$; imazamox is marketed under the brand name BEYOND by BASF; in the U.S., it is used for the control of a variety of grass and broadleaf weeds including jointed goatgrass, morning-glory, Canadian and Russian thistle, cocklebur, and chickweed; investigations indicated that tolerance to imazamox in cotton is controlled by a partially dominant single gene; an allelism test revealed that the tolerance genes in the four mutants are either alleles in the same locus or are very tightly linked *phyt chem*

imbibe: >>> imbibition

imbibition: The adsorption of liquid, usually water, into ultramicroscopic spaces or pores found in material such as cellulose, pectin, and cytoplasmic proteins in seeds *bot*

imbibition damage: Caused by rapid uptake of water in very dry seeds *seed*

imbricate: Overlapping, as do shingles on a roof *bot*

imidazole: A compound whose molecule forms a pentagonal ring of C and H atoms with an N and NH group attached *chem*

imidazolinone: A herbicide which inhibits acetolactate synthase in susceptible plants *phyt phys*

imino acid: An acid derived from an imine in which the nitrogen of the imino group and the car-boxyl group are attached to the same carbon atom *chem*

immature: Not mature or ripe *phys*

immature grains: Grains which are underdeveloped or not fully developed; lacking in size and weight, compared to a fully mature grain *seed*

immaturity: >>> immature

immediate early gene: A viral gene that is expressed promptly after infection *biot phyt*

immersed: Embedded in a substrate *meth*

immersion lens: A special microscopic lens adjusted to place material between the uppermost surface of a microscopic sample (>>> slide or >>> coverslip) and the objective (e.g., >>> immersion oil) *micr* >>> immersion medium

immersion medium: Material placed between the uppermost surface of a microscopic sample (slide) and the objective (e.g., >>> immersion oil) *micr*

immigration: In genetics, the movement or flow of genes into a population, caused by immigrating individuals, which interbreed with the residents *gene eco*

immiscible: Two solvents that are not able to dissolve into each other when mixed *chem*

immobile nutrients: Nutrients that cannot move around in the plant and, consequently, their defi-ciency symptoms appear first in the young leaves; immobile nutrients include calcium, boron, sulfur, iron, and copper *phys* >>> chelating agent >>> chelator

immobilization: The conversion of a chemical compound from an inorganic to an organic form as a result of biological activity *phys agr*

immune: Not affected by pathogens; exempt from infection; the condition of having qualities that do not allow the development of a disease *phyt*

immune reaction: The reaction between a specific antigen and antibody; when plants are inocu-lated (e.g., with the BUK strain of tomato blacking nepovirus) they are subsequently pro-tected against secondary infection with a similar viral strain, but not against a dissimilar strain of virus *phyt*

immune response: >>> immune reaction

immune system: Active defense of plants against infections and other invasive aggressions; it detects antigens on the invading entities and creates new antibodies to destroy them *phys phyt*

immunity: A natural or acquired resistance of a plant to a pathogenic microorganism or its prod-ucts; it means that a host cannot be parasitized by a particular species of parasite; thus, coffee is immune to wheat rust, and wheat is immune to coffee rust; immunity is a non-variable; the maximum level of horizontal resistance may be an apparent immunity, but it is not true immunity because it is variable, and it can be eroded; vertical resistance has often been called immunity, but it too is an apparent immunity because it operates only against non-matching strains of the parasite *phyt* >>> Table 52

immunity breeding: >>> resistance breeding

immunize: To make immune *phyt*

immuno-electrophoresis: A technique for the differentiation of proteins in solution, based on both their electrophoretic and immunological properties; initially the proteins are separated by >>> gel electrophoresis; they are then reacted with specific antibodies by double diffusion through the gel; the pattern of precipitating arcs thus formed can be used to identify the proteins *meth sero*

immunofluorescence: Any of various techniques for detecting an antigen or antibody in a sample by coupling its specifically interactive antibody or antigen to a fluorescent compound, mixing with the sample, and observing the reaction under an ultraviolet-light microscope meth

immunogenetics: Studies using a combination of immunologic and genetic techniques, as in the investigation of genetic characters detectable only by immune reactions *gene*

immunochromatography method (for field-level detection of preharvest sprouting in wheat): The method detects >>> alpha-amylase isozymes, using specific antibodies; alpha-amylase from ground grain is extracted with a salt solution, and two drops of the extract are added to a zone on a disposable card; the result appears in <5 min; if the grain is sprouted, amylases in the samples become sandwiched between gold-labeled and immobilized antibodies, and a maroon band appears in the test window; the color intensity depends on the extent of weather damage, with good (negative) correlations between test color and "falling number"; precision is as good as or better than the "falling number" test *meth*

immunological screening: Use of an antibody to detect a polypeptide synthesized from a clone *biot*

immunoprecipitation: Precipitation of antigens with the help of antibodies *biot*

immunosuppressants: A substance that results in or affects immunosuppression *meth sero*

impeder: An individual of any value actually impeding the development of another individual of a higher grade *gene meth*

***Imperata cylindrica* procedure:** A method of haploid production and chromosome doubling in >> DH breeding of wheat; the approach requires hand emasculation of the maternal spikes followed by >>> pollination with *I. cylindrica* pollen; the pace of this endeavor can be enhanced by utilizing asynchronous flowering by direct pollination without emasculation followed by morphological marker-assisted screening of selfed and crossed seeds; although the proportion of >>> pseudoseed formation is lower in case of direct pollination, yet in some genotypes, it is comparable to that of pollination after emasculation *meth biot* >>> haploid >>> doubled haploids >>> *Imperata cylindrica* procedure

imperfect flowers: Unisexual flowers; flowers lacking either male or female parts *bot*

imperfect state (of fungi): The asexual state of a fungus (i.e., the state in which no sexual reproduction occurs) *phyt bot*

implant: Material artificially placed in an organism *biot*

imprinting: >>> genomic imprinting

improvement planting: Any planting done to improve the value of a stand and/or experiment and not to establish a regular plantation *meth*

inarable: A field or land not arable and/or not capable of being plowed or tilled agr

inarching (*syn* side grafting, *syn* bridge grafting): A method of grafting, usually a new plant growth onto a stronger root system; it is carried out by establishing young plants (sometimes, one that is in a pot) near an existing tree; at the point where they meet, at the matching areas the bark is removed; the two cut surfaces are then fitted together and bound with soft tying material until they grow together; later they can be gradually separated with the new branches attached to the older rootstock *hort* >>> graft

inbred: A plant resulting from successive self-fertilization of parents throughout several generations *gene*

inbred line: A line produced by continued inbreeding; usually a nearly homozygous line originating by continued self-fertilization, accompanied by selection; inbred lines are in general
smaller and show reduced vigor and fertility; some inbred lines are even not viable; this is
caused by the inevitable fixation of inferior recessive genes *gene* >>> Figure 51

inbred pure lines: Involves inbreeding of annual seed-propagated material; homogeneous and
homozygous lines are isolated by selection of desired recombinants or segregates in F2 to
F7 generations of crosses between parental pure lines (generally monogenotypic lines can
be blended to form multilines, e.g., in tomato, lettuce, soybean, pea, cowpea, snapbean,
field bean, Arabian coffee, Capsicum pepper, eggplant, okra, lentil, and papaya) *meth* >>>
Figure 51

inbred-variety cross: The F1 cross of an inbred line with a variety *meth* >>> toppers >>>
Table 37

inbreeding: The crossing of closely related plants; one important purpose of induced inbreeding is
the development of genotypes that can be maintained through multiple generations of seed
production; self-pollinated cultivars are reproduced for many generations by inbreeding;
inbreeding is also used to reduce the frequency of deleterious recessive alleles in genotypes that serve as parents of a synthetic or a vegetatively propagated cultivar; inbreeding
increases the genetic and phenotypic variability among individuals in a population; four
mating systems are used to increase the homozygosity in a breeding population *meth* >>>
Figures 50, 51

inbreeding coefficient: The probability that the two genes at any locus in a diploid individual are
identical by descent (i.e., they originated from the replication of one gene in a previous
generation) *gene meth* >>> Table 10

inbreeding depression: The reduction in vigor often observed in progeny from matings between
close relatives; it is due to the expression of recessive deleterious alleles; it is usually severe
in open-pollinated outcrossing species; an effect opposite to heterosis *gene* >>> heterosis

inbreeding load: The extent to which a population is impaired by inbreeding *bio evol* >>>
Table 10

inbreeding population: >>> inbreeding

incertae sedis: Of uncertain taxonomic position *tax*

inch (in): A measure of length; equals 2.54 cm

incipient species: Populations that are too distinct to be considered as subspecies of the same
species, but not sufficiently differentiated to be regarded as different species; sometimes
called "semispecies" *táx*

inclined draper: A device for separating seeds using an inclined endless belt onto which seeds
are metered; seeds are separated on the basis of their different tendencies to roll down the
plane or to catch and be carried up and into a separate discharge spout *seed*

incompatibility (homomorphic or heteromorphic): A genetically determined inability to obtain
fertilization and seed formation after self-pollination or cross-pollination; there are several
types of progamous or postgamous incompatibility; in contrast to heteromorphic incompatibility (e.g., >>> heterostyly in Primula ssp.), homomorphic incompatibility is not
associated with morphological differences *gene* >>> cross-sterility >>> cross-breeding
barrier >>> style grafting >>> cut style >>> mentor pollen method

incompatibility group: Plasmids that are incompatible with each other belong to the same incompatibility group *biot*

incomplete block design: A >>> block design in which not every treatment occurs in a block; it is
preferable where large numbers of cultivars are compared in a single yield trial; the entries
in each replication are subdivided into smaller blocks, in a manner designed to reduce
the error caused by soil variation; usually, it refers to a lattice design, considering the

restriction that (1) the number of testers must be harvested, (2) inferior strains cannot be discarded prior to harvest to reduce harvest expenses, and (3) researchers still must analyze the experiment as a lattice design *meth stat*

incomplete diallel: A partial sampling; any individual family or type of family may be omitted *meth* >>> complete diallel

incomplete dominance: >>> partial dominance

incomplete Latin squares: Row-and-column designs for n − 1 replicates of n treatments arranged in n − 1 rows and n columns; can be constructed by omitting any row from an n × n Latin square; the n − 1 rows comprise a set of replicate blocks of size n whereas the n columns comprise a set of balanced incomplete blocks of size n − 1; the designs have a balanced partitioning of treatment effects and have an orthogonal factorial structure for factorial treatments designs; if n is too large for a simple incomplete Latin square, a design of size m × n where m < n − 1 can sometimes be useful; if the rows are complete replicate blocks and the columns are balanced incomplete blocks, the designs are called a YOUDEN rectangle and have similar properties to incomplete Latin squares; where available, YOUDEN designs can be very useful for row-and-column type designs, but the availability of balanced YOUDEN designs is extremely limited; also, if m is substantially smaller than n, a substantial amount of treatment information will be confounded between columns and some method for recovery of information from the columns stratum of the analysis will be necessary for an efficient analysis of variance *stat*

incomplete resistance: A type of resistance that is not complete and shows slow susceptibility to the pathogen *phyt*

increase (seed): To multiply a quantity of seed by planting it, thereby producing a larger quantity of seeds *seed*

incubation: The act or process of incubating *meth*

incubation period: The period between infection and the appearance of visible disease symptoms *phyt*

incubator: An apparatus in which media inoculated with microorganisms are cultivated at a constant temperature and/or air humidity *prep*

indehiscent: Applied to fruits that do not open to release their seeds *bot* >>> dehiscent

indehiscent fruit: >>> indehiscent

InDel: A molecular term for the **in**sertion or the **del**etion of bases in the DNA of an organism; it has slightly different definitions between its use in evolutionary studies and its use in germline and somatic mutation studies; in evolutionary studies, InDel is used to mean an insertion or a deletion and InDels simply refers to the mutation class that includes both insertions, deletions, and the combination thereof, including insertion and deletion events that may be separated by many years, and may not be related to each other in any way; in germline and somatic mutation studies, InDel describes a special mutation class, defined as a mutation resulting in both an insertion of nucleotides and a deletion of nucleotides which results in a net change in the total number of nucleotides, where both changes are nearby on the DANN; a micro InDel is defined as an InDel that results in a net change of 1 to 50 nucleotides; in coding regions of the genome, unless the length of an InDel is a multiple of 3, it will produce a frameshift mutation; InDels can be contrasted with a point mutation; an InDel inserts and deletes nucleotides from a sequence, while a point mutation is a form of substitution that replaces one of the nucleotides without changing the overall number in the DANN; InDels can also be contrasted with >>> tandem base mutations (TBM), which may result from fundamentally different mechanisms; InDels, as defined as either an insertion or deletion, can be used as genetic markers in natural populations, especially in phylogenetic studies; it has been shown that genomic regions with multiple InDels can also be used for species-identification procedures *gene biot*

indent cylinder separator: A seed separator utilizing a rotating indented cylinder through which seeds are passed for cleaning; it lifts shorter seeds from longer seeds, thus separating them *seed*

indent disk separator: A seed separator utilizing multiple rotating disks inside a cylinder through which seeds are moved; it lifts seeds from longer seeded types, thus separating them seed

independence: The relationship between variables when the variation of each is uninfluenced by that of others, that is, correlation of zero *stat*

independent assortment (of genes) The random distribution in the gametes of separate genes; if an individual has one pair of alleles A and a, and another pair B and b, then it should produce equal numbers of four types of gametes: AB, Ab, aB, and ab; it is asserted in Mendel's second law – the law of independent assortment *gene* >>> Figure 6

independent variables: If two random variables "a" and "b" are independent, then the probability of any given value of "a" is unchanged by knowledge of the value of "b" *stat*

indeterminant flowering: Inflorescence in which the outer or lower flowers open first *bot*

indeterminate: Descriptive of an inflorescence in which the terminal flower is last to open; the flowers arise from axillary buds and the floral axis may be indefinitely prolonged by a terminal bud; some crops, such as haricot beans, can have either the determinate or the indeterminate habit; with the former, they are self-supporting, bushy plants, and with the latter, they grow as vines; on the other hand, potatoes are determinate plants but, when grafted on to tomatoes, they become indeterminate, and this is a very useful technique when many flowers are needed for the production of true seed for breeding purposes *bot*

indexing: Testing of a plant for infection, often by mechanical transmission or by grafting tissue from it to an indicator plant *phyt meth*

index selection: A form of intentional simultaneous selection; with the index selection some index value is assigned to each candidate; the index value indicates the aggregate value of each candidate across several traits; the index selection consists of truncation selection with regard to the index values; breeding program selection usually has to consider several traits; in population improvement it is important to consider that the breeding material undergoes selection as well as recombination in recurrent cycles; discrete simulation is used to solve the numerical problem of model calculations to take recombination into account; two negatively correlated traits with high, as well as high and low, heritabilities are considered in a population with initially equal gene frequencies for the favorable and unfavorable allele; when the linear SMITH–HAZEL, PESEK–BAKER, and heritability indices, as well as the non-linear ELSTON index, are compared with another type of non-linear selection index over 40 cycles of recurrent selection, the new type of selection index shows two characteristics for population improvement: (1) more genetic variation is maintained and less favorable alleles are lost in a recurrent selection program compared to other indices; (2) it can be shown that situations exist were the classical SMITH–HAZEL index is not optimal in medium- and long-term population improvement; (3) moreover, if two negatively correlated traits have considerably different heritabilities, the selection response is negative for the low-heritable trait for many recurrent selection cycles with all selection indices except for the PESEK–BAKER index and the new type of selection index *meth stat*

indexing (U.S.): The process used to test vegetatively reproduced plants for freedom from virus diseases before multiplying those plants *seed*

index of prevalence: Proportion of plant stands affected by disease or infestation in a defined area *phyt*

Indian bread (*syn* **llao-llao,** *syn* **pan de Indio**)**:** A fungus (*Cyttaria darwiini*) which parasitizes on species of the Antarctic beech (e.g., *Nothofagus betuloides*); the edible ascomycete forms doughy and slightly sweet yellow knobs on the branches and causes heavy malformation or disorder on the wood *phyt*

Indian Grassland and Fodder Research Institute (IGFRI): Established in 1962; a national institute under the administrative control of the Indian Council of Agricultural Research, IGFRI is mandated to conduct basic, strategic, applied, and adaptive research, development, and training in forage production and its utilization; with more than 30 years of experience in forage research and development, IGFRI today stands as the premier R&D institution in South Asia for sustainable agriculture through quality forage production for improved animal productivity *org* >>> https://icar.org.in/igfri/index.html

indicator plant: Plants that are indicative of specific site or soil conditions; react markedly to the deficiency of a specific input/condition such as the deficiency of plant nutrients, soil moisture, stress, etc.; for example, alfalfa, turnip, cauliflower, apple, and pear to boron; lettuce to chlorine; barley, maize, oats, onion, apple, tobacco, lettuce, tomato, and citrus to copper; sorghum, rice, barley, citrus, peach, and cauliflower to iron; oats, maize, bean, pea, radish, apple, apricot, cherry, and citrus to manganese; cabbage, cauliflower, citrus, legumes, oats, and spinach to molybdenum; maize, onion, citrus, and peach to zinc *bot agr eco*

indigenous: An organism existing in, and having originated naturally in, a particular area or environment *bot eco*

indirect embryogenesis: Embryoid formation on >>> callus tissues derived from zygotic or >>> somatic embryos, seedling plants, or other tissues in culture *biot*

indirect fluorescence: Fluorescence emitted by fluorophores that are not an endogenous part of the specimen; usually introduced into a specimen as a stain or probe *micr*

indirect organogenesis: Organ formation on callus tissues derived from explants *biot* >>> Figure 58

indirect selection: The direct selection for specific traits may imply unintentional indirect selection with regard to many other traits *meth*

indole-3-acetic acid (IAA): A substance ($C_{10}H_9NO_2$) that acts as a growth hormone or auxin in plants, where it controls cell enlargement and, through interaction with other plant hormones, also influences cytokinesis *phys*

indoor culture: Growing plants indoors using natural and/or artificial light and additional heating; it is used for subtropical or tropical plants or for plant propagation *hort meth*

indoor plant: >>> indoor culture

indoxyl acetate test: A rapid test for evaluating damage to large-seeded legumes; materials needed are: indoxyl acetate, ethyl alcohol, ammonia solution (20%), cotton balls, and a spray bottle; a certain number of legume seeds are soaked in the solution containing 0.1% indoxyl acetate in ethyl alcohol for 10 s; then seeds are sprayed with ammonia solution; seeds are then allowed to air-dry; damage is assessed by observing purplish-green abrasions on the seed coat *seed meth*

induced mutation: A change in a gene caused by a treatment *gene*

induced resistance: A form of resistance activated in plants by an external stimulus – for example, elicitor treatment or pre-inoculation with a microorganism; it can be a local or systemic acquired resistance *phyt biot*

inducer: An effector molecule responsible for the induction of enzyme synthesis *phys*

inducing medium: >>> induction media

induction: Transcription of genes can be induced by inactivation of a repressor or by the action of an activator *biot*

induction media: Media that can induce organs or other structures to form, or a medium that will cause variation and/or mutation in the tissue exposed to it *biot*

induction of flowering: The initiation of the production of flowers, possibly stimulated by florigen *phys* >>> florigen

induction of mutation: The process of causing a variation or mutation *gene* >>> mutagenesis >>> TAL effector nuclease

indurate: Hard *bot*

industrial crop: Crops that are processed on a large scale by industrial means (e.g., potato, pea, lupine, dwarf bean, or cereals for starch production; sugarbeet, beets, sweet sorghum, chicory, or Jerusalem artichoke for sugar processing; linseed, false flax, common marigold, crambe, caper spurge, *Cuphea* ssp., meadowfoam, or jojoba for oil processing; flax, hemp, or nettles for fiber processing; foxglove, poppy, yellow bark, or cocoa for pharmaceutical utilization); industrial uses account for a relatively small but growing and potentially much larger share of the market for agriculture commodities *agr*

inert: A chromosomal segment that is supposed to be genetically inactive or without coded genetic information *gene*

inert matter: One of the four components of a purity test of seed investigation; it includes non-seed material and seed material that are classified as inert according to the rules for testing seeds, e.g., (1) seed units in which it is readily apparent that no true seed is apparent; (2) pieces of broken or damaged seed units half or less than half the original size; (3) those appendages not classed as being part of pure seed in the pure seed definitions must be removed and included in the inert matter; (4) seeds of Leguminosae, Cruciferae, Cypressaceae, Pinaceae, and Taxodiaceae with the seed coat entirely removed; (5) unattached sterile florets, empty glumes, lemmas, paleas, chaff, stems, leaves, cone scales, wings, bark, flowers, nematode galls, fugues bodies – such as ergot – sclerotia and smut balls, soil, sand stones, and all other non-seed matter; and (6) all material, except other seeds, left in the light fraction when the separation is made by the uniform blowing method *seed* >>> www.pflanzengallen.de

infect: Of a pathogen, to enter and establish a pathogenic relationship with an organism; to enter and persist in a carrier; to make an attack on a plant *phyt*

infection: The invasion of the tissue of a plant by a pathogenic microorganism *phyt*

infection court: The site on a host plant at which infection by a parasitic organism is affected *phyt*

infection peg: A thickening of the host cell wall in the vicinity of the penetrating >>> hypha; lignin, callose, cellulose, or suberin may be deposited at this site *phyt*

infection thread: Specialized hypha of a pathogenic fungus that invades tissue of the susceptible plant *phyt*

inferior: Applied to an ovary when the other organs of the flower are inserted above it *bot*

inferior pelea: >>> lemma

infertile: Not able to reproduce or not able to produce viable gametes *bot*

infertility: The situation in which a plant is unable to produce viable offspring *bot*

infest: Attacked by animals (e.g., insects), or sometimes used of fungi in soil in the sense of contaminated *phyt*

infilling: >>> beat(ing) up

inflated: Puffed up, bladdery *bot*

inflected: When the keel of the, for example, wheat glume is bent inward in the upper third bot

inflorescence: A flower structure that consists of more than a single flower; the flower head terminates the culm in grasses; it may be determinate (solitary flower, simple >>> cyme, compound cyme, scorpioid cyme, >>> glomerule) or indeterminate (>>> raceme, >>> panicle, >>> spike, >>> catkin, spadix, >>> umbel, head); determinate flowers are those in which the axis terminates as a flower; indeterminate flowers terminate in a bud, which continues to grow and produce flowers throughout the growing season; the latter results in flowers of different maturity within the same inflorescence *bot*

inflorescence meristem: The relatively undifferentiated, dividing plant tissue that gives rise to the inflorescence *bot*

infrared light: The part of the invisible spectrum that is contiguous to the red end of the visible spectrum and that comprises electromagnetic radiation of wavelengths from 800 nm to 1 mm *phy*

infructescence: A fruiting structure that consists of more than a single fruit *bot*

infundibular: Funnel-shaped *bot*

infundibuliform: >>> infundibular

ingraft: To insert, as a scion of one tree or plant into another, for propagation *hort meth*

ingress: The act, by a plant pathogen, of gaining entrance into the tissues of a susceptible plant *phyt*

inherit: Receive from one's predecessors, i.e., in organisms, chromosomes and genes are transmitted from one generation to the next *gene*

inheritance: The transmission of genetic information from parents to progeny *gene*

inhibitor: A chemical substance that retards or prevents a growth process such as germination *phys*

injection: The act of injecting *meth*

injector(s) Devices which are able to inject (mix) concentrated fertilizers and chemicals with flowing water in an irrigation system; there are two types of injectors: venturi (vacuum type) and hydraulic type (water drives a piston pump) *hort* >>> venturi injector

inoculant: A preparation containing specific nitrogen-fixing bacteria that is added to legume *seed* prior to planting to assure that the resulting crop will have nitrogen fixation ability *seed*

inoculate: To place inoculum deliberately where it will reproduce *meth*

inoculation: The act or process of inoculating meth; addition of effective *Rhizobia* (bacteria) to legume seed prior to planting for the purpose of promoting >>> nitrogen fixation *agr* >>> inoculate

inoculum: Spores of other diseased material that may cause infection *phyt*

inosine (I): A modified nucleotide that occurs in >>> tRNA (anticodon) and can pair with A, U(T), or C in the codon *gene chem*

inositol: A carbocyclic or sugar alcohol that is widely distributed in plants; important for growth and development; the inositol biosynthesis is regulated by the enzymes myo-inositol phosphate synthase (MIPS) and myo-inositol monophosphatase (IMP); myo-inositol phosphate synthase (mips) genes have been identified in plants *chem phys*

inositol triphosphate (InsP3): A chemical compound ($C_6H_{15}O_{15}P_3$) that contributes to the spatial orientation of a plant; for example, when wheat or maize plants are pressed down to the ground, a change of orientation of starch granules occurs in the cells; after a short time (30–120 min) InsP3 is accumulated on the lower side of the leaves; so-called motor cells are activated; they grow longitudinal and stepwise to upright the plant *phys*

input trait: A genetic character that affects how the crop is grown without changing the nature of the harvested product, e.g., plant resistances to diseases or herbicides are agronomic useful input traits in the context of crop management, but they do not alter seed quality or other so-called output traits *gene* >>> output trait

unprincipled: >>> non-dormant >>> sprouting

INRA: >>> Institut National de Recherche Agronomique

inrolled: Having margin rolled toward the midvein; involute *bot*

insect herbivory: Plant damage by certain insects (e.g., fall >>> armyworm, *Spodoptera frugiperda*); it is responsible for about 15% of the world's crop losses each year *phyt*

insecticide: Pesticide used to control insects which feed on plants; in its exact meaning, an insecticide kills insects; however, the term more generally applies to any substance which controls insects; as such, an insecticide may only deter insects without actually killing them; insecticides react with insects in different ways: some are formulated as poisons which are applied superficially to a plant, where they remain until ingested by feeding insects; other insecticides are formulated to act as contact pesticides or systemics; examples of synthetic insecticides include acephate, diazinon, dimethoate, and malathion; examples of organic insecticides include >>> neem; many general-use or household insecticides are formulated as aerosols or baits; other insecticides, especially those used by commercial growers, are formulated as emulsifiable concentrates, soluble powders, and wettable powders *phyt meth*

insect-pollinated plant: >>> cross-pollination

insert: A piece of foreign DNA introduced into a phage, plasmid, or other vector DNA *biot*

insertion: A genetic mutation in which one or more nucleotides are added to DNA, or the process and the result of transferring a foreign DNA or chromosome fragment into a recipient *biot*

insertion sequence: DNA sequence, which can excise and integrate into DNA without the need for extensive DNA homology *biot*

insertion site: Any unique restriction site of a cloning vector molecule into which foreign DNA can be inserted, or the integration site of transposons or insertion DNA sequences *gene biot*

insertion vector: Cloning vector where the cloned DNA is inserted into a restriction site, as opposed to replacement vectors where a piece of DNA is replaced in the process of cloning *biot* >>> Table 39 >>> Figure 46

insertional duplication: Insertion of extra homologous base pairs into a recipient genome, which results in a mutation (by duplicated segments) *cyto gene*

insertional inactivation (of a gene): Insertion of a DNA fragment into the coding sequence of a gene; usually leads to the inactivation of this gene *biot*

in silico: Modern term used to characterize biological experiments carried out entirely on a computer *stat bio*

in silico **mapping:** As an alternative to designed mapping experiments using F2 or backcross mapping population, an *in silico* mapping was developed to detect genes by simultaneously exploiting existing genotypic, phenotypic, and pedigree data available from breeding programs or genomic databases *gene stat meth*

in situ: In place; where naturally occurring *meth*

in situ **conservation:** Refers to the maintenance of plant populations in the habitats where they naturally occur and have evolved; in crop species, conservation can be achieved in home gardens or on-farm, in wild species through the designation of protected areas, such as natural parks and nature reserves *seed eco*

in situ **hybridization (ISH):** A technique to locate those segments complementary to specific nucleic acid molecules; chromosomes, which are treated to denature the DNA and remove RNA and proteins, are then incubated with radioactive labeled nucleic acids or non-radioactive labeled probes of special properties; the hybridized segments are then visualized by autoradiography or directly through the microscope and photography of fluorescent signals *cyto meth* >>> Figures 52, 53, 54

inspection: The act of inspecting a pedigreed seed crop by an inspector authorized to report to the official organization on the condition of the seed crop offered for pedigreed status; the inspector reports on varietal impurities, difficulty to separate from other crop kinds, isolation, objectionable weed content, previous land use, and the pedigree of the parent seed planted *seed*

instability: Variation that appears to be random and occurs constantly *gene*

instar: The larval or nymph stage of an immature insect between successive molts *phyt*

institutional plant breeding: Plant breeding conducted by large institutes; this kind of breeding is usually expensive and, consequently, it favors cultivars with a wide climatic adaptability *org*

Institut National de Recherche Agronomique (INRA): The National Institute of Agronomic Research was created in 1921 as Institute of Agronomic Research (IRA), on the site of Noisy-le-Roi (France) under the directorship of E. SCHRIBAUX; the function of the station was outlined in a report dated October 25, 1923; the constitution of collections, a storehouse of specialized scientific documentation, experimentation on varieties in collaboration with other regional stations, and evaluation of new varieties ("Noisy was the test field for foreign varieties and the birthplace for hybrids of wheat and oats released by the central station, novel varieties that can be found today in all the regions of France and who start to find their place in our possessions in North Africa"); in 1929, under the directorship of L. ALABOUVETTE the station was transferred to Versailles; from its creation the station attached importance to bringing together as complete as possible collections of wheat, oats, and potato; in 1930, beet was added to these species, as well as two new branches (varietals identification and the study of new varieties with "a register of plants obtained by selection"); a catalogue of wheat varieties was published in 1932, and P. JONARD established in 1936 a classification of tender wheat with the aid of grain and blade characteristics; later, the number of species grew, with tomato, asparagus, apples, and pears added in 1936; those species became objects of important collections, of descriptive studies, of varietal trials, and of selection; recently, maize and wheat are the main subjects of research and breeding; after 1945, the country was not self-sufficient for food; the plants studied principally then became forage, to restart the cultivation of animals and field crop plants (wheat, barley, oats, beetroot, potatoes, maize, Canola™ (rape), turnips); in 1947, A. CAUDERON was involved in an enormous task of collection and experimentation on maize coordinated by L. ALABOUVETTE; he selected F7 lines and F2 offspring of self-fertilizations of plants from a population originating in Lacaune; the lines were used to generate many maize hybrids, such as "INRA258" (258 being the number of days from sowing to harvest); with X. LASCOLS and then M. CAENEN, A. CAUDERON developed a simple hybrid F7 × F2 as crossing partner of the North American lines to create productive, cold-tolerant and relatively homogenous hybrids (>>> double hybrids "INRA 200" or >>> three-way hybrids "INRA 260"); breeding on wheat, undertaken before 1945 by C. CRÉPIN's team in Dijon, then at Versailles, ended in the creation of the variety "Etoile de Choisy" ("Choisy Star"), in which productivity, precocity, and cold resistance traits were associated for the first time; at once disseminated in the southwest of France, the Etoile de Choisy variety made fertilization profitable, and played a big role in the revival of the region; the variety was successful not only in France but also in southern Europe and the USSR; at the end of the 1950s, rapeseed became a target of research at Versailles; the variety "Sarepta" (1956) demonstrated taking "pure lines" for improvement of rapeseed; this permitted J. MOICE to subsequently substitute varieties without erucic acid for normal varieties; the variety "Primor" (1973) became a new basis of this crop development; Versailles with the work of Y. DEMARLY on the genetics of tetraploids, P. DOMMERGUES on mutagenesis and the function of meristems, and N. MAÏA, as well as R. ECOCHARD on cytogenetics led the line "VPM1", the offspring of intergeneric crosses, carrying resistance to root rot in modern varieties of wheat; since the 1960s, the research was decentralized to Clermont-Ferrand, Lusignan, Dijon, Rennes, and Mons-en-Chaussée; the work of J.-P. BOURGIN and J.-P. NITSCH led to new approaches of *in vitro*, another culture; in asparagus, selection and crosses of individuals by L. CORRIOLS-THÉVENIN and C. DORÉ established high-performing and early varieties (>>> double hybrids, e.g., variety "Larac") and then simple hybrids of clones contribute to the optimization of *in vitro* cloning (var. "Aneto", "Desto", "Stéline"); *in vitro* anther cloning permitted haploid, double homozygote, super males (YY) to be obtained and female (XX) plants in which crosses produced only male plants XY; the production of hybrid varieties was rapidly envisaged for endive culture; pollenic competition was used to produce hybrid seeds of different varieties; the work of H. BANNEROT resulted in new hybrid varieties ("Zoom", "Flash", "Béa", "Turbo"); more than 80% of French endive culture today is hydroponic, grown in the dark, in climate-controlled rooms; cytoplasmic male sterility, discovered in Japan by H. OGURA in radish (1968), permitted researchers to envisage the creation of hybrid varieties in Brassica; the transfer of male sterility was necessary for intergeneric crosses between radish and cabbage; G. PELLETIER replaced chloroplasts of radish, the cause of the chlorosis, with those of cabbage; starting in 1968 programs were developed aiming at obtaining peas and beans of quality for the canning and frozen food industry; at the demand of producers, L. BOULIDARD created varieties of sauerkraut cabbage ("Septdor", "Bouledor", "Neuropa"), associating the productivity of Dutch varieties with the quality of Alsatian populations adapted to mechanical harvest and producing a quality sauerkraut; the station also made a large contribution to the improvement of flax, in association with the producers, for oil and fiber; studies on cold resistance ended in the creation of "Oliver", a top variety of seed flax; in 1973, the embargo of the U.S. on the soybean oil cake obliged all the countries of the EEC to seek substitute products rich in proteins; at Versailles, R. COUSIN directed his research towards the improvement of proteinaceous pea; in 1973, the first variety of winter pea ("Frisson") was added to the catalogue of varieties, followed by many other cultivars, such as "Frijaune" or "Frilaine"; for spring pea varieties, the ideotype of the plant and the criteria for selection were defined; the results contributed to the development of a new type of pea "afila" where the leaflets were replaced by tendrils (var. "Rafale") *org* >>> www.inra.fr/

intake auger (at a harvester): The auger tines guide the crop to the chain conveyor, which delivers it to the threshing section; any foreign bodies that may have been ingested fall into the stone trap, which is located between the conveyor and the concave *agr*

integrase: An enzyme involved in the integration of some viruses, such as bacteriophage lambda, and some transposons into host cell chromosomal DNA *biot*

integrated control: >>> integrated plant protection

integrated crop management (ICM): A management system that recognizes and integrates the individual production practices required for success and profitability as dependent variables that can be controlled by the grower in some cases (i.e., drip irrigation to control plant growth or use of row covers to modify/prevent frost damage to crops) *agr hort*

integrated map: The combination of a genetic and a physical map, i.e., an integration of genetic linkage data and the physical distance between markers and/or genes *biot cyto meth*

integrated pest management (IPM): A pest population management system that anticipates and prevents pests from reaching damaging levels by using all suitable tactics including natural enemies, pest-resistant plants, cultural management, and the judicious use of pesticides, leading to economically sound and environmentally safe agriculture *agr phyt*

integrated plant protection: Disease and pest control by combining all available techniques, such as agronomic control, biological control, chemical control, and sanitary procedures *phyt*

integument: The coats of the ovule (mostly two), which develop into the seed coat (>>> testa) after fertilization *bot*

intellectual property: Intangible assets such as patents, trade secrets, trade names, etc. *adm* >>> Intellectual Property Rights

Intellectual Property Rights (IPR): A system of patents that allows ownership over the applications of research like patent, copyright, and trademark; industrial design protection, a special protection for new plant varieties, has been available since 2007; the International Union for the Protection of New Plant Varieties (UPOV) and the TRIPS Agreement provide this protection; the Act of UPOV Convention in 1991 gave this important protection to the plant breeders; the UPOV office is situated in Geneva and it coordinates the protection of new plant species; there is a debate in many countries that the plants developed do not satisfy the non-obvious requirement in a patent application system, as existing techniques are used and the new breed is obvious; therefore, a unique >>> sui generis system is utilized for the protection of new plant varieties; the TRIPS Agreement provides a protection for 25 years in the case of trees and vines, while the protection for other plants is 20 years; in 2014, a new business association, the International Licensing Platform (ILP), was launched to improve global access to and use of plant breeding traits for vegetables; the ILP's licensing system is simple and transparent; if a member wants to take a license to use a fellow member's patented invention, the two parties begin bilateral negotiations; if no agreement is reached within three months, the case is put to arbitration by independent experts; membership is open to all interested parties, regardless of whether they own patents or not *adm*

intense selection: Selection for the traits of interest with little or no retention of lines not having the traits of interest *meth* >>> Figure 60

intensifying screen: A plastic sheet impregnated with a rare-earth compound, such as calcium tungstate, which absorbs beta radiation and emits light; when placed on one side of a piece of X-ray film with a radioactive sample on the other side, the intensifying screen will capture some of the beta emissions which pass through the film, blackening the X-ray film and thus greatly enhancing the sensitivity of the detection; an intensifying screen is used in >>> SOUTHERN and >>> Northern blotting procedures *biot prep*

intensive crop: A crop that has high profit margins and which consequently justifies considerable expense in its production; horticultural crops are commonly intensive crops, while cereals are usually >>> extensive crops *agr hort*

intensive forestry: The practice of forestry with the objective of obtaining the maximum in volume and quality of products per unit of area through the application of the best techniques of silviculture and management *fore*

interagency tags: Labels or tags applied to pedigreed seed moving from one country or state to another *agr seed*

interbreeding: Intercrossing of individuals within a >>> population *meth* >>> Table 37

intercalary: Chromosomal segments located beside terminal regions *cyto*

intercalary meristem: An internodal meristem, situated between differentiated tissues; it produces cells perpendicular to the growth axis, causing internode elongation *bot biot*

intercalary segment: >>> interstitial segment

intercalate (in a >>> Latin square): A subsquare of order 2, that is, a pair of rows and a pair of columns in whose four positions just two different symbols occur, each twice *stat* >>> design of experiments

intercalating agent: A chemical that can insert itself between the stacked bases at the center of the DNA double helix, possibly causing a frameshift mutation *chem meth*

intercalating dye: >>> intercalating agent

interchange: An exchange of segments between non-homologous chromosomes resulting in translocations *cyto* >>> translocation

interchange trisomic: An additional chromosome to the diploid set, which is composed of two different chromosomes via translocation *cyto* >>> balanced tertiary trisomic

interchromosomal: Effects and processes between chromosomes *cyto*

interchromosomal duplication: The addition of one or more segment(s) from one chromosome to another chromosome, mostly by faulty crossing-over; it leads to a functional imbalance of genes in the involved region and usually is the basis for genetic disorder *cyto gene* >>> intrachromosomal duplication

interclass variance: >>> variance

intercropping: Two or more crops produced on the same field at the same time, or planting more than one crop in a field using a regular pattern that interleaves each crop in some pattern; a form of polyculture; the crops need not to be planted nor harvested at exactly the same time, but they are grown together for a significant part of the growing season; intercropping is used extensively in the tropics and subtropics *agr*

intercrossing: Mating of heterozygotes *meth*

interference: The effect of recombination in one interval on the probability of recombination in an adjacent interval gene; in patenting biological material, when two or more patent applications or issued patents claim the same invention; in ecology, the negative effect of one organism upon another, regardless of the presence of a limiting resource; >>> competition is one facet of interference; >>> allelopathy is another; sometimes it refers to both positive and negative interactions between organisms *gene eco adm* >>> Intellectual Property Rights

interference distance: The distance within which further crossing-overs may be formed after the previous has been produced away from the centromere *gene*

interference microscopy: Like the phase microscope, the interference microscope is used for observing transparent structures *micr*

interference range: The distance large enough for forming two crossing-overs without mutual interference *cyto*

intergeneric cross: Spontaneous or experimental crosses of individuals of different genera, for example, wheat (*Triticum aestivum*) and rye (*Secale cereale*); this cross even resulted in a human-made new crop plant "triticale" *meth* >>> Figure 3 >>> cf List of Crop Plants

intergeneric hybrid: >>> intergeneric cross

intergenic: Effects and phenomena between genes *gene*

intergenic DNA (*syn* **intergenic region**): A DNA sequence that is located between two adjacent genes *gene*

intergenic region: >>> intergenic DNA

intergenic suppressor: A mutation that suppresses the phenotype of another mutation in a gene other than that in which the suppressor mutation resides *gene*

intergenotypic competition: >>> allocompetition

interkinesis: A resting stage that may occur between the first and second meiotic division *cyto* >>> interphase

interlocking: During meiotic pairing, the intertwisting of non-homologous chromosomes and/or chromosome configurations *cyto* >>> Figure 13

intermated recombinant inbreds: Another structure of mapping >>> population-alike F2, back-cross, or near-isogenic lines; the intermating of F2 individuals results in new recombination events; therefore, intermated-recombinant-inbred populations have improved genetic resolution *biot* >>> Table 37

intermediary: A plant trait controlled by a heterozygous pair of alleles, which result in an intermediate phenotype as compared to the corresponding homozygous genotypes *gene*

intermediate host: A host essential to the completion of the life cycle of a >>> parasite, but in which it does not become sexually mature *phyt* >>> host

internal hairs (of the glume): The hairs situated across the upper part of the internal surface of the broad wing in the glumes of, for example, wheat *bot*

internal imprint: The mark on the inner surface of, for example, wheat glume caused by the pressure of the enclosed lemma and grain *bot*

internal transcribed spacer (ITS): A spacer DNA situated between the small-subunit ribosomal RNA (rRNA) and large-subunit rRNA genes in the chromosome or the corresponding transcribed region in the polycistronic rRNA precursor transcript; sequence comparison of the ITS region is widely used in taxonomy and molecular phylogeny because of several favorable properties, such as, (1) it is routinely amplified due to its small size associated to the availability of highly conserved flanking sequences, (2) It is easy to detect even from small quantities of DNA due to the high copy number of the rRNA clusters, (3) it undergoes rapid concerted evolution via unequal crossing-over and gene conversion; this promotes intra-genomic homogeneity of the repeat units, although high-throughput sequencing showed the occurrence of frequent variations within plant species, (4) it has a high degree of variation even between closely related species; this can be explained by the relatively low evolutionary pressure acting on such non-coding spacer sequences *biot meth*

International Atomic Energy Agency (IAEA): The world's center of cooperation in the nuclear field; it was set up as the world's Atoms for Peace organization in 1957 within the United Nations family; the agency works with its member states and multiple partners worldwide to promote safe, secure, and peaceful nuclear technologies *org* >>> www.iaea.org/

International Board for Plant Genetic Resources (IBPGR): Rome, Italy; coordinating international plant conservation, recently renamed >>> IPGRI

International Center of Agricultural Research in the Dry Areas (ICARDA) Aleppo, Syria: The institute coordinates research, breeding and production activities for wheat, durum wheat, barley, faba beans, lentil, chickpeas, and alfalfa of dry areas of the world >>> www .icarda.org/

International Code of Botanical Nomenclature: The documentation containing the rules that govern the assignments of names to plants, the ultimate goal being to provide a single correct name for each taxon *bot tax* >>> www.bgbm.org/iapt/nomenclature/code/SaintLouis/0000St.Luistitle.htm

International Code of Nomenclature for Cultivated Plants (ICNCP): It regulates the naming of cultivars, cultivar groups, and graft-chimaeras, e.g., *Clematis alpina* "Ruby", a cultivar within a species; ICNCP operates within the framework of the >>> International Code of Botanical Nomenclature which regulates formal names for plants in general; ICNCP does not regulate trademarks for plants nor the naming of plant varieties; trademarks are regulated by the law of the land involved; W. T. STEARN (1911–2001) was of the first author of Cultivated Plant Code in 1953 *org* >>> http://en.wikipedia.org/wiki/Cultivated_ plant_taxonomy

International Crop Improvement Association (ICIA): Former name of the Association of Official Seed Certifying Agencies (AOSCA) *seed org* >>> AOSCA; it is dedicated to assisting clients in the production, identification, distribution, and promotion of certified classes of seed and other crop propagation materials >>> www.aosca.org/

International Crops Research Institute for the Semi-Arid Tropics (ICRISAT): This institute has a global mandate for the improvement of its mandate crops, such as sorghum, pearl millet, chickpea, pigeonpea, and groundnut; these crops are grown on a large scale worldwide, but generally grown on marginal land by resource-poor farmers *org* >>> www .icrisat.org

International Cultivar Registration Authority (ICRA): An organization responsible for ensuring that each plant cultivar receives a unique, authoritative botanical name; ICRA was established in 1950, and operates under the >>> International Code of Nomenclature for Cultivated Plants (ICNCP); its chief aim is to prevent duplicated uses of cultivar and group epithets within a defined denomination class, and to ensure that names are in accord with the latest edition of the ICNCP; each name designation must be formally established by being published in hard copy, with a description in a dated publication *org*

International Institute of Tropical Agriculture (IITA): Ibadan, Nigeria; responsible for groundnut, soybean, sweet potato, cassava, cowpea, and rice research *org* >>> www.iita.org/

International Rice Research Institute (IRRI): Los Banos, Philippines; IRRI is a non-profit agricultural research and training center established to improve the well-being of present and future generations of rice farmers and consumers, particularly those with low incomes; it is dedicated to helping farmers in developing countries produce more food on limited land using less water, less labor, and fewer chemical inputs, without harming the environment *agr* >>> www.irri.org

International Seed Testing Association (ISTA): Bassersdorf (Switzerland); the primary purpose of ISTA is to develop, adopt, and publish standard procedures for sampling and testing seeds and to promote uniform application of these procedures for evaluation of seeds moving in international trade; the secondary purpose of ISTA is to actively promote research in all areas of seed science and technology (sampling, testing, storing, processing, and distributing seeds), to encourage variety (cultivar) certification, to participate in conferences and training courses aimed at furthering these objectives, and to establish and maintain liaison with other organizations having common or related interests in seed *seed* >>> www.seedtest.org

International Triticeae Mapping Initiative (ITMI): Several research projects were conceived in 1989, originally as a five-year effort to develop RFLP maps for crops of the Triticeae, mainly wheat and barley; the mapping effort was organized around the seven homoeologous chromosome groups of the Triticeae for which co-coordinators were appointed; additional coordinated topics included related diploid genomes, genetics of abiotic stress resistance, and Triticeae informatics (database and RFLP probe repository); some 130 scientists are affiliated with ITMI; the primary activity of ITMI is the holding of two annual workshops; the first of these traditionally takes place at the International Plant and Animal Genome Conference in January, the venue for the second is variable; the meetings are designed around specific topics in mapping and genomics, with invited speakers, and include summary presentations of topic co-coordinators; management and planning sessions also form part of the workshops; the activities of ITMI are organized by its Management Office, originally directed by C. O. QUALSET and P. E. MCGUIRE, University of California, Davis; in January 2001, as part of reorganization plans, the Management Office moved to the Scottish Crop Research Institute; in 2004 the Management Office moved to the Australian Centre for Plant Functional Genomics; the ITMI website is currently operated from the GrainGenes server in Albany, California *org*

internodal cell: >>> internode

internode: The part of a stem between two consecutive nodes *bot*

interphase: A stage in the cell cycle in which there is no visible evidence of nuclear division; therefore it is sometimes called "resting phase" but is a period in which there is intense activity, including replication of chromosomes; during interphase, the chromosomes of eukaryotes decondense, and they occupy distinct regions of the nucleus, called chromosome domains or chromosome territories (CTs); in plants, the >>> RABL's configuration, with telomeres at one pole of nucleus and centromeres at the other, appears to be common, at least in plants with large genomes; it is unclear whether individual chromosomes of plants adopt defined, genetically determined addresses within the nucleus, as is the case in mammals *cyto*

interphase nucleus: A nucleus during the stage of >>> interphase in which there is no visible dividing activity, but in which metabolic and synthetic activities are going on *cyto*

interphase nucleus mapping: A variant of the conventional >>> FISH technique for the visualization of specific genes in interphase nuclei in which the chromosomes are visible; the gene is detected by fluorochrome-labeled gene probe that hybridizes to the target gene; the location of the hybridized probe (gene locus) can be visualized by LASER-induced excitation of the fluorochrome *biot cyto meth*

interplanting: The interplanting of one crop within another for the purpose of trapping pest insects *phyt meth*

interplot competition: It can be avoided and/or decreased by use of plots with multiple rows in which only plants in the center rows are evaluated; in plots with three or more rows, the outermost rows are designated as the border or guard rows; they may prevent plants in adjacent plots from influencing the performance of plants in the center of the plot *stat meth* >>> design of experiments

interplot interference: An interaction between plots in field experiments, arising whenever particular treatments affect plots other than those to which the treatments were applied, for example, increased levels of infestation on resistant test plants due to their proximity to susceptible plants; interference is most likely to occur when plots are small and not bordered *stat* >>> design of experiments

interposon: A recombinant DNA fragment that is used for *in vitro* insertional *mutagenesis gene meth biot*

inter-retrotransposon amplified polymorphism (IRAP): IRAP fragments between two retrotransposons are generated by PCR amplification of sequences, using outward-facing primers annealing to long-terminal-repeats (LTR) target sequences; fragments are separated by high-resolution agarose gel-electrophoresis; this marker system was used for the first time in barley, based on *BARE-1* retrotransposon, followed by *Oryza, Pisum, Musa,* and *Crocus biot meth* >>> retrotransposon amplified polymorphism

interseeding: Seeding between sod plugs, sod strips, rows, or sprigs *agr*

intersex: A class of individuals of a bisexual species that have sexual characteristics intermediate between the male and the female *gene*

inter simple sequence repeat (ISSR): A >>> PCR-based molecular marker assay of genomic sequence lying between adjacent microsatellites; ISSR primers are anchored at their 3′ ends to direct the amplification of the genomic segments between the ISSRs; it is a RAPDs-like approach that accesses variation in the numerous microsatellite regions dispersed throughout the various genomes (particularly the nuclear genome) and circumvents the challenge of characterizing individual loci that other molecular approaches require; microsatellites are very short (usually 10–20 bp) stretches of DNA that are "hypervariable", expressed as different variants within populations and among different species; they are characterized by mono-, di-, or trinucleotide repeats, e.g., AA…, or AG…, CAG…, that have four to ten repeat units side-by-side; in ISSRs, specifically the di- and trinucleotide repeat types of microsatellite are targeted, because these are characteristic of the nuclear genome (mononucleotide types are found in the chloroplast genome) *biot meth* >>> simple sequence repeat (SSR) >>> Figure 63 >>> Table 29

inter small RNA polymorphism (iSNAP): Endogenous non-coding small RNAs usually consist of 20–24 nucleotides; they are found ubiquitously throughout the genome and play important regulatory roles in most eukaryotes; iSNAP primers are designed based on multi-mapped small RNA, including its two end conserved flanking sequences; the technique can be effectively applied in the mapping of rice and other crops, including fingerprinting of species with extremely low genetic diversity, e.g., tobacco *meth biot*

interspecific: Effects and phenomena between species *gene eco*

interspecific cross: A crossing between two species *meth* >>> Figure 3

interspecific hybrid: A hybrid between two or more species *meth* >>> Figures 2, 3

interspecific hybridization: Crossing between species *meth* >>> species hybridization >>> Figures 2, 3

interstitial segment: A chromosome region between the centromere and a site of rearrangement *cyto*

interstock: An intermediate plant part that is compatible with both the scion and the rootstock *hort*

intertilled: A crop planted in rows and cultivated between the rows *agr*

interval mapping: Is currently the most popular approach for QTL mapping in experimental crosses; it was developed in order to overcome the three disadvantages of analysis of variance at marker loci; the method makes use of a genetic map of the typed markers, and, like analysis of variance, assumes the presence of a single >>> QTL; each location in the genome is posited, one at a time, as the location of the putative QTL *biot meth* >>> composite interval mapping >>> QTL mapping

intervarietal: Effects and phenomena between varieties (cultivars) *agr gene*

intervening sequence: A non-coding nucleotide sequence in eukaryotic DNA, separating two portions of nucleotide sequence found to be contiguous in cytoplasmic >>> mRNA *gene*

intine: The outer epidermis of pollen *bot*

intine: The outer epidermis of pollen, i.e., the inner cellulose wall of an angiosperm pollen grain *bot* >>> exine

intrabreeding: A mating type in which only individuals of the same populations are combined *meth* >>> Table 37

intrachromosomal: Within a chromosome *cyto*

intrachromosomal duplication: Any genomic segment that is duplicated within a particular chromosome or chromosome arm; it mediates chromosomal rearrangements *cyto gene* >>> interchromosomal duplication

intragenic: Effects and phenomena within a gene or its physical unit *gene*

intragenic single nucleotide polymorphism: Any sequence polymorphism between two or more genomes that is based on a single nucleotide exchange, small deletion, or insertion within a gene, i.e., coding sequence *gene* >>> SNP

intragenic SNP: >>> intragenic single nucleotide polymorphism

intragenic suppressor: A mutation that suppresses the phenotype of another mutation in the same gene as that in which the suppressor mutation resides *gene*

intragenome duplication: The occurrence of identical sequences, such as genes or gene families, on different chromosomes of the nucleus *cyto gene*

intragenotype competition: >>> isocompetition

intraspecific: Effects and phenomena within a species *eco*

intravarietal: Effects and phenomena within a variety (cultivar) *agr*

introduced species: Species not part of the original flora of a given area, rather, brought by human activity from another geographical region *eco agr*

introgression: The incorporation of genes of one species into the gene pool of another; if the ranges of two species overlap and fertile hybrids are produced, the hybrids tend to backcross with the more abundant species; it results in a population in which most individuals resemble the more abundant parents but also possess some of the characters of the other parent species *meth* >>> Table 37

introgression library: Increasing the genetic diversity of elite breeding materials with exotic germplasm requires techniques that minimize negative side effects attributable to genetic interactions between recipient and donor; this seems achievable by an introgression library approach involving the systematic transfer of a limited and/or restricted number of short donor chromosome segments from an agricultural unadapted source (donor) into an elite line (recipient or recurrent parent); established introgression libraries represent a dynamic resource that can substantially foster breeding programs and provide an opportunity to proceed towards functional genomics *meth biot*

introgressive hybridization: Crossbreeding of plants from different species that results in introgression *meth*

intron: A segment of DNA of unknown function within a gene; it may be transcribed in precursor RNA, but cannot be found in functional mRNA *gene*

intronic single nucleotide polymorphism: Any >>> SNP that occurs in introns of eukaryotic genes; it occurs more frequently than SNPs in coding regions *biot*

intronic SNP: >>> intronic single nucleotide polymorphism

intron SNP: >>> intronic single nucleotide polymorphism

introrse: Of an anther which dehisces towards the center of a flower *bot* >>> extrorse

inulin: A polysaccharide in which about 32 b-fructose units are joined in a chain by glyosidic linkages between the first and second carbon atoms on neighboring sugar units; it is found as a storage carbohydrate in roots, rhizomes, and tubers of many species of >>> Compositae or Asteraceae (e.g., >>> Jerusalem artichoke or >>> dahlia); it can be used as a sweetener in the food industry (with a sweetening power 30% higher than that of sucrose) and is sometimes added to yoghurts as a prebiotic; inulin can be converted to fructose and glucose through hydrolysis *chem phys* >>> chicory

inulin: A polysaccharide in which about 32 b-fructose units are joined in a chain by glyosidic linkages between the first and second carbon atoms on neighboring sugar units; it is found as a storage compound in roots, rhizomes, and tubers of many species of >>> Compositae *chem phys*

invasion: The spreading of a pathogen through tissues of a diseased plant *phyt*

invasiveness: Ability of a plant to spread beyond its introduction site and become established in new locations where it may provide a deleterious effect on organisms already existing there *eco*

inversion: A change in the arrangement of genetic material involving the excision of a chromosomal segment that is then turned 180° and reinserted at the same position in the chromosome *cyto*

inversion polymorphism: The presence of two or more chromosome sequences, differing by inversions, in the homologous chromosomes of a population *gene*

inviability: The inability to survive *bot*

inviable: >>> non-viable

in vitro: Literally, "in glass", but applied more generally to studies and propagation of living plant material that are performed under artificial conditions in tubes, glasses, dishes, etc. *prep*

in vitro **collection:** A collection of germplasm maintained as plant tissue grown in active culture on solid or in liquid medium; it can be maintained as plant tissue ranging from >>> protoplast and cell suspensions to >>> callus cultures, >>> meristems, shoot-tips, and embryos *meth*

in vitro **culture:** The cell, organ, or tissue culture performed under artificial conditions in tubes, glasses, dishes, etc. *biot* >>> Figure 58

in vitro **fertilization:** Pollination performed aseptically *in vitro* by direct application of the pollen to the ovule; it is used to overcome prezygotic incompatibility *meth*

in vitro **marker:** A mutation that allows identification *in vitro* of a cell line possessing the marker *biot*

in vitro **mutagenesis:** Methods for altering DNA outside the host cells; mutagenesis can be random or specific for the site and base change depending on the technique used *biot* >>> TAL effector nuclease

in vitro **pollination:** >>> *in vitro* fertilization

in vitro **propagation:** Propagation of plants under a controlled and artificial environment, usually using plastic or glass vessels, aseptic techniques, and defined growth media *biot* >>> Tabelle 62

in vitro **screening:** Search and selection for particular characters of cells, organs, or tissues performed under artificial conditions in tubes, glasses, dishes, etc., usually in combination with special nutritional media, which allow a differentiated growth of the cells, etc. *biot*

in vitro **selection:** Used to screen large numbers of plants or cells for a certain characteristic before growing them in the field or in glasshouses, e.g., salt tolerance *meth*

in vivo: Literally, "in life"; applied to studies and propagation of whole, living organisms, on intact organ systems therein, or on populations of microorganisms *meth phyt*

involucral bract(s): A series of bracts beneath or around a flower or cluster of flowers *bot*

involucre: A whorl of >>> bracts below an >>> inflorescence *bot*

involute: Having edges that roll under or inwards *bot*

iodine (I): A non-metallic halogen element occurring as a grayish-black crystalline solid that sublimes to a dense violet vapor when heated; used in radiolabeling *chem cyto meth*

iojap: An idiomatic description of a mutant locus in maize that produces variegation *gene*

ion: An atom that has acquired an electric charge by the loss or gain of one or more electrons *chem*

ionomics: It is one of the major pillars for the structural and functional genomic study; a complete set of ions present in an organism is referred to as the ionome of the organism; it is defined as the study of quantitative complement of low-molecular-weight molecules present in cells in a particular physiological and developmental state of the plant; the complete ionomic profiling is done by using a number of analytical tools like ICP-MS, ICP-OES, X-ray crystallography, neutron activation analysis (NAA), and others; all these tools give a complete profile of the ions present in the plants; recently, the data are stored in a database called the Purdue ionomics information management system (PIIMS) *biot meth*

IPGRI: >>> International Plant Genetic Resources Institute, formerly >>> IBPGR >>> Biodiversity International

IPR: >>> Intellectual Property Rights

IRAP: >>> inter-retrotransposon amplified polymorphism

iris diaphragm: A composite diaphragm with a central aperture readily adjustable in size in order to regulate the amount of light admitted to a lens or optical system *micr*

iron (Fe): An element required by plants; it is used in reactions in which rapid reductions occur by the transfer of electrons as in photophosphorylation and oxidative phosphorylation; iron-deficient plants have chlorotic young leaves; at first the veins remain green but later they too become chlorotic; fertilizers containing iron chelates can be added to the soil or sprayed on the leaves to make iron available to the roots and foliage *chem phys*

iron deficiency chlorosis: Iron is one of the more abundant minerals on earth, but may often be present in forms that are not readily available to plants; in alkaline soils, particularly those that are rich in limestone, low-lying plants cannot access the iron that is present, and may become deficient in this element and chlorotic *agr phys*

irradiate: Expose to radiation that may increase the >>> mutation rate of some genes and hence may increase genetic variation; radioactive irradiation can directly and indirectly cause damage to the DNA molecule; not only the energy of the electrons but also the charge of the electrons can cause damage to the molecules; slow and fast electrons may cause >>> mutations *meth*

irradiated callus: Callus that have been exposed to radiation *biot*

IRRI: >>> International Rice Research Institute

irrigation: To supply land with water by artificial means, as by diverting streams, flooding, or spraying; the main types of irrigation are: (1) sprinkler irrigation, (2) surface irrigation, and (3) subsurface irrigation; "alternative drip irrigation" is a special method in order to reduced plant evapotranspiration, i.e., plants are alternatively irrigated from two sides; it temporarily simulates drought stress to the plant on one side while closing stomata and decrease evapotranspiration, however from the other side water is made available again to maintain plant growth; it works in fruit tree and citrus plantage as well as in tomato, wheat, and potato irrigation *agr meth*

iSNAP: >>> inter small RNA polymorphisms

isoallele: An allele whose effect can only be distinguished from that of the normal allele by special tests *gene*

isobrachial: A chromosome with a metacentric centromere position resulting in two chromosome arms with equal length *cyto* >>> metacentric >>> Figure 11

isochromocentric: Nuclei showing as many chromocenters as chromosomes *cyto*

isochromosome: A chromosome with two identical arms; it usually derives from telocentric chromosomes *cyto* >>> Figure 37

isocompetition: Cultivation at high plant density implies the presence of strong interplant competition; when there is no genetic variation, the competition is called isocompetition *stat*

isodicentric chromosome: A structurally abnormal chromosome containing a duplication of part of the chromosome including the centromere; the resulting structure contains two centromeres and a point of symmetry that depends on the position of the breakpoint *cyto* >>> Figure 37

isoelectric focusing (IEF): A technique for the electrophoretic separation of amphoteric molecules in a gradient of pH, usually formed from a combination of buffers held on a >>> polyacrylamide gel support medium; the molecules will move in the gradient, under the influence of an electric field, until they reach their isoelectric pH, where they form a sharp band; separation is achieved because the various molecular species will have different isoelectric values of pH *meth*

isoenzyme: A species of >>> enzyme that exists in two or more structural forms, which are easily identified by electrophoretic methods *phys* >>> Table 29

isogamete: Male and female gametes that are similar to each other *bot*

isogamy: The fusion of gametes that are morphologically alike *bot*

isogeneic: Applied to a graft that involves a scion and stock that are genetically identical *hort*

isogenic: A group of individuals showing the same genotype *gene*

isogenic lines (vs. random lines): Two or more lines differing from each other genetically at one locus only gene

isogeny: The situation when a group of individuals shows the same genotype *gene*

isograft: A graft or transplant among isogenic (i.e., genetically identical) individuals on the same organism *hort*

isolate: In general, to put in isolation; a segment of a population within which assortative mating occurs gene; in plant pathology, to remove an organism (e.g., a fungus) from the plant in pure form *phyt*

isolated microspore culture systems (IMC): Offer an unlimited supply of embryogenic cells and developing embryos, which facilitate studies of key developmental processes relevant to seed formation and metabolism *biot*

isolation: The separation of one group from another so that crossing between groups is prevented *meth* >>> isolation distance >>> isolation requirements >>> Table 35

isolation distance: Distance that separates one field of plants from another; used to maintain seed purity in seed production fields and to minimize cross-pollination between different kinds of plants, for example, transgenic crops and organically grown crops *meth* >>> design of experiments >>> isolation requirements

isolation requirements: The distance required to isolate pedigreed seed crops from other crops which may be a source of pollen or seed contamination; used by most seed certification agencies as one of the requirements to maintain varietal purity of pedigreed seed crops *seed* >>> Tables 29, 35 >>> isolation distance

isoleucine (Ile): A crystalline amino acid, $C_6H_{13}O_2$, present in most proteins *chem phys* >>> Table 38

isolines: >>> isogenic lines

isomer: A chemical compound or nuclide that displays isomerism *chem*

isomerase: An enzyme that catalyzes a reaction involving the interconversion of isomers *chem phys*

isomeric: Genes that can each produce the same or similar phenotype *gene*

isopolar: Pollen grain with identical proximal and distal faces, e.g., in *Bellis perennis bot* >>> heteropolar

isoprene (2-methyl butadiene): A five-carbon compound (C_5H_8) that forms the structural basis of many biologically important compounds, such as terpenes, etc. *chem phys*

isoschizomer: Restriction >>> endonucleases with identical recognition sequences and >>> cleavage sites, isolated from different bacterial species; it can differ in the amino acid sequence and temperature stability, may require different reaction conditions, and may differ in the sensitivity to DNA methylation *biot* >>> Table 39

isosome: A chromosome showing morphologically and genetically identical arms *cyto* >>> Figure 37

isosomic: Cells or individuals showing isosomes (i.e., chromosomes with genetically and morphologically identical chromosome arms), usually derived from telocentric chromosomes *cyto* >>> Figure 37

isotelocompensating trisomic: A compensating trisomic; a missing chromosome is compensated for by one telocentric and one tertiary chromosome *cyto* >>> Figure 14

isotertiary compensating trisomic: A compensating trisomic; a missing chromosome is compensated for by one isochromosome and one tertiary chromosome *cyto* >>> Figure 14

isotope: One of two or more varieties of a chemical element whose atoms have the same numbers of protons and electrons but different numbers of neutrons *phy*

isotopic dating: An approach to determining the age of certain materials by reference to the relative abundances of the parent isotope and the daughter isotope; if the decay constant and the concentration of the daughter isotope are known, it is possible to calculate an age *meth* >>> radiocarbon dating

isotope labeling: >>> isotopic tracer

isotopic tracer: Isotopically labeled precursors of nucleic acids; the labeled compounds are injected or fed to plants; subsequently, the excretions (solid, liquid, or gas) or tissues are analyzed to determine by detection of radioactive tracer how the original compound has been changed; many tracers have been used but most common are ^{3}H, ^{14}C, ^{32}P, and ^{35}S *prep meth*

isotrisomic: When the extra chromosome shows identical arms *cyto* >>> Figure 14

isozyme: >>> isoenzyme

ISSR: >>> inter simple sequence repeat

ISTA: >>> International Seed Testing Association

iteration: A procedure that makes use of repeated trials to find the best fitting value of a parameter from observed data *stat*

ITMI: >>> International Triticeae Mapping Initiative

ITS: >>> internal transcribed spacer

IUCN: >>> International Union for Conservation of Nature and Natural Resources

Ixeris type: Diplospory where a syndetic prophase leads to a restitution nucleus, which divides; it is not followed by a cell division; the resulting megaspore contains two unreduced nuclei; two further mitotic divisions lead to an eight-nucleate embryo sac *bot*

J

jaggedness: Incomplete seed set of cereal spikes caused either by genetic reasons or by unfavorable pollination or fertilization *breed*

jaggery solution: A solution of unrefined country sugar (gur, jaggery), which can be used as a sticker on seeds before treatment with biofertilizer *seed meth*

jarovization: >>> vernalization

jasmonates: A group of cyclopentanone derivatives, which originate biosynthetically from linolenic acid via an inducible octadecanoid pathway consisting of at least seven enzymatic steps; the end product is (+)-7-iso-jasmonic acid, a physiologically active substance that is rapidly converted to its stereoisomer, stable (−)-jasmonic acid *chem phys*

jasmonic acid ($C_{12}H_{18}O_3$): Distributed throughout higher plants, synthesized from linolenic acid via the octadecanoid pathway; an important role seems to be its operation as a "master switch", responsible for the activation of signal transduction pathways in response to predation and pathogen attack; proteins encoded by jasmonate-induced genes include enzymes of alkaloid and phytoalexin synthesis, storage proteins, cell wall constituents, and stress protectants; the wound-induced formation of proteinase inhibitors is an example, in which jasmonic acid combines with abscisic acid and ethylene to protect the plant from predation *chem phys* >>> www.plantstress.com

joining segment: A small DNA segment that links genes to yield a functional gene encoding an immunoglobulin *gene*

joint: The node of a grass culm, spikelet, inflorescence, or any other node *bot* >>> node

jointing stage: In cereals, the growth stage at which the first stem node is visible above ground *bot*

J-root: Seedling roots planted in a manner that forms a J-shaped configuration in the planting slit; such seedlings may grow poorly or die *fore hort*

judgment sampling: The selection of a sample that, according to the judgment and intuition of the sampler, accurately reflects the >>> population *stat meth* >>> Table 37

JUKES–CANTOR model: The simplest MARKOV model of base substitution, assuming that all bases occur with equal probability in the ancestral sequence; it also assumes that the conditional probabilities describing an observable base substitution are all the same *stat*

jumbo pollen: It refers to a pollen trait in diploid alfalfa; homozygous recessive (*jpjp*) plants are characterized by the complete failure of post-meiotic cytokinesis during microsporogenesis resulting in 100% 4n-pollen formation *cyto gene*

jumping gene: >>> transposon

junk DNA: Presently, only the function of a few percent of the DNA is known in plants, the rest has been believed to be "junk"; over 95% of DNA has largely unknown function; the most exhaustive knowledge is about the genes responsible for the morphological structures, the structural genes, which are the simplest part of the system; the knowledge about the most important part of this system, the regulator genes and their incomplete genetic code language is only partially known; it has been reported that the sequences of this unknown DNA are inherited and that some are repetitive; the DNA has been found to mutate; however, the idea that a major part of the DNA in higher organisms is "garbage" ignores the fact that a key feature of biological organisms is optimal energy expenditure; to carry enormous amounts of unnecessary molecules is contrary to this fundamental energy-saving feature of biological organisms; increasing evidence is now indicating many important functions of this DNA, including various regulatory roles; an elaborate nomenclature has been proposed for this DNA: any stretch of DNA is a "nuon"; DNA with an evolutionary potential for development of function is a "potonuon"; DNA that was previously non-functional, or that had a completely different function and has been evolutionarily co-opted for a new function, is a "xaptonuon"; and a non-coopted potonuon is a "naptonuon" *gene biot*

juvenile–mature correlation: Correlation between measurements at different ages *meth*

juvenile resistance: The characteristically vertical resistance of plants in the seedling stage; also known as >>> seedling resistance *phyt gene*

juvenile stage: The immature, reproductively incompetent, and, sometimes, phenotypically distinct phase of plant growth *bot*

juvenillody: A condition in which tissues and organs remain immature *bot*

K

kafirin: A storage protein of sorghum *phys* >>> Table 15

kanamycin: An aminoglycoside antibiotic; kanamycin resistance is used as a selection marker in genetic experiments *phys biot* >>> kanamycin-resistant tissue

kanamycin-resistant tissue: Tissue that is resistant to the lethal effects of the aminoglycoside antibiotic, kanamycin; some cloning vectors have a kanamycin-resistant gene as a selectable marker *biot*

karnal bunt: A fungus disease of, for example, wheat that reduces yields and causes an unpalatable but harmless flavor in flour milled from infected grains *phyt*

karyoevolution: Evolutionary change in the chromosome set, expressed as changes in number and gross structure of the chromosomes; more broadly, evolutionary relationships between taxa as indicated by >>> karyotype differences *evol cyto* >>> Figures 68, 69

karyogamy: The fusion in a cell of haploid (n) nuclei to form a diploid (2n) *cyto*

karyogenesis: Formation of the nucleus (the central structure) of a cell – the smallest, most basic unit of life that is capable of existing by itself; karyogenesis comes from the Greek words karyon meaning "nucleus", and genesis meaning "production"; in a narrow sense, the division of the cell nucleus is distinguished from cytoplasmic division or cytokinesis; it represents a system by which the genetic information contained in the chromosomes of eukaryotes is distributed to the daughter nuclei, which are generally identical to the mother cell nucleus *cyto*

karyogram: >>> idiogram

karyology: The study of the nucleus and its components *cyto*

karyolysis: The disappearance of the >>> interphase nucleus during karyogenesis *cyto*

karyosome: Any of several masses of chromatin in the reticulum of a cell nucleus *cyto*

karyostasis: The stage of the cell cycle in which there is no visible dividing activity of the nucleus, but metabolic and synthetic activity *cyto phys*

karyotype: The entire chromosomal complement of an individual cell or individual, which may be observed during mitotic metaphase *cyto*

keel: The main nerve of, for example, the wheat glume, shaped somewhat like a keel of a boat; in legumes, also a boat-like formation of the flower *bot*

keel flower: Boat-like shape of a flower (e.g., in legumes such as pea) *bot*

keiki: A vegetative offshoot formed at a node (e.g., in some orchids) *bot*

kernel: A whole grain or seed of a cereal plant or the part of the seed inside the pericarp *bot*

kernel weight (1,000 grain weight): A crop yield or crop loss parameter *agr seed*

ketone: Any of a class of organic compounds containing a carbonyl group, CO, attached to two alkyl groups, as CH_3COCH_3 *chem*

key gene: >>> oligogene

key management species: Plant species on which management of a specific unit is based *tax*

kharif: In India, monsoon season crops *agr*

killing frost: A sharp fall in temperature that damages a plant so severely as to cause its death *phys*

kilning: The heating and/or drying process used in the production of malt to stop germination and kill the grain *prep*

kilobases (kb): 1,000 base pairs/bases in a single- or double-stranded nucleic acid, which is used as a common unit of length in molecular genetics *gene*

kilogram (kg): Equals 1,000 grams

kilometer (km): Equals 1,000 meters

kinase: An enzyme that catalyzes reactions involving the transfer of phosphates from a nucleoside triphosphate (e.g., ATP) to another substrate *phys*

kind: One or more related species or subspecies that, singly or collectively, is known by one common name; e.g., soybean, flax, carrot, and radish *tax*

kinetin (6-fururylaminopurine): A degradation product of animal DNA ($C_{10}H_9N_5O$), which does not occur naturally and which has properties similar to those of cytokinins; applied to certain leaves, kinetin delays >>> senescence in its vicinity and attracts nutrients *chem phys*

kinetochore: A dense, plaque-like area of the centromere region of a chromatid, to which the microtubules of the spindle attach during cell division *cyto* >>> centromere >>> Figure 11

kinin: >>> cytokinin

kinship: A potential problem in population analysis, e.g., a collection of inbred lines containing more than one representative of some lines; for two unlinked loci in equilibrium a random sample of lines might have equal numbers of genotypes *AABB*, *AAbb*, *aaBB*, and *aabb*, but if additional representatives of the first and last genotypes were present, the sample might be *AABB*, *AABB*, *AABB*, *AAbb*, *aaBB*, *aabb*, *aabb*, and *aabb*, such kinship could create linkage disequilibrium *gene stat* >>> coancestry

Kjeldahl method: A technique often used for the quantitative estimation of the nitrogen content of plant material (e.g., of cereal grains), after the Danish Johan G. C. T. KJELDAHL (1849–1900) *meth*

kleisin: >>> cohesin

Klenow fragment: Large fragment of >>> DNA polymerase I after proteolytic digestion; in 1970, coined by the Danish Hans KLENOW (1923–2009); it lacks 5′ to 3′ exonuclease activity and can therefore not be used for nick translation but is very useful for filling-in reactions and DNA sequencing by the >>> SANGER method *biot* >>> Figures 52, 53, 54 >>> Table 39

klon: >>> clone

kneading: To work dough into a uniform mixture by pressing, folding, and stretching *meth*

knob: A heavily stainable and quite large chromomere observed along a >>> chromosome of some plants (e.g., in maize it is used as a marker in pachytene analysis) *cyto* >>> chromomere

knockout genes: Transgenes designed to silence an existing *gene biot*

knockouts: A plant in which a gene has been rendered non-functional; by eliminating the gene's function, one can infer its role in the plant's physiology from the phenotype *biot*

knot: A lump or swelling in or on a part of a plant (e.g., the node of grass) *bot*

Koehler illumination: Illumination optics resulting in the image of the light source being out of focus at the specimen plane; it provides homogeneous illumination of the specimen *micr*

Kornberg enzyme: >>> DNA polymerase I

Kosambi formula: Recombination fractions and map distances correspond only over relatively short recombinational distances; as genetic distance increases, the probability of a second (and correcting) recombination also increases, hence the measured recombination for two loci is less than would be apparent if a third intervening locus were present; various mapping functions have been suggested to permit single recombination fractions to be converted to map distances; KOSAMBI presented the simplest and probably most general functions, which is given as $x = 25 \log n [(1 + 2y)/(1 - 2y)]$; x is the map distance (cM) corresponding to the recombination fraction, y; for example, if the recombination value is 0.05 then the distance is 5 cM; if the value is 0.1, then the distance is 10.1 cM, etc. *gene*

Kosena (fertility restoring gene): >>> Ogura

K strategy ramets (or genets): Which assign low-energy allocations to reproduction, having approached the biomass carrying capacity of the environment *eco*

L

label: >>> plant label

labeling: Incorporation of an easily detectable signal into a DNA molecule; radioactive labeling is being increasingly replaced by non-radioactive methods biot; attaching labels to seed lots with information on variety identity, purity, and seed quality *seed*

labellum: In Orchidaceae, the lowest of the three flower petals, which differs from the other two; in lipped flowers, the platform formed by the lowest petal or fused petals *bot*

labiate: A member of the botanical family Labiateae *bot* >>> List of Crop Plants

lacerate: Appearing torn on the margin; irregularly cleft *bot*

lacinate: Deeply cut, into irregular, narrow segments or lobes *bot*

Lac operon: A cluster of structural genes specifying the enzymes acetylase, permease, and beta-galactosidase *gene*

laggard: A >>> chromosome, which is not included in the daughter nuclei after >>> anaphase *cyto* >>> lagging

lagging (of chromosome): Delayed movement from the equator to the poles at anaphase of a chromosome so that it becomes excluded from the daughter nuclei *cyto*

lagging strand: DNA strand growing in the 3′ to 5′ direction, synthesized discontinuously *biot*

lamda phage Lamda temperate bacteriophage, size: 48.5 kb; it infects *Escherichia coli biot*

lamina: A flat, sheet-like structure (e.g., the blade of a leaf) *bot*

laminar flow cabinet: A workspace where sterile air moves over the work area and that allows for the introduction of explants aseptically *biot*

laminarin: A beta-1,3-glucan reserve *phys*

laminated aluminum foil packet(s): Packets constructed of a laminate consisting of an inner layer of polyethylene, a middle layer of aluminum foil, and an outer layer of polyester, often used for seed storage in genebanks *seed meth*

lampbrush chromosome: A particular type of chromosome shape, usually found in the >>> diplotene stage of animals including flies; it is a type of >>> puffed chromosome; the loops of DNA strands form a lampbrush-like shape; puffed chromosome regions may also occur in plants (e.g., in *Phaseolus* beans) *cyto* >>> polyteny >>> polytene chromosome >>> puff >>> Figure 66

lanate: Woolly covering of short dense hair *bot*

lanceolate: Shape or outline like the head of a spear; pointed at both ends and widest below the middle *bot*

land classification: Soils that are grouped into special units, subclasses, and/or classes according to their capability for use and treatments that are required for sustained agriculture, horticulture, and forestry *agr* >>> Table 44

land-grant universities: A network of agricultural focused universities established by MORILL Act in 1862 in the U.S. *adm*

landmark: A labeled stick of different length and manufacturing used for marking fields, experimental plots, paths, or margins, usually after seed bed preparation *prep meth*

landrace: A set of >>> populations or clones of a crop species produced and maintained by farmers; in breeding, a mixture of a great number of different genotypes that are well-adapted to the environmental conditions of its habitat; shows only average, but reliable yield; in countries of highly developed agriculture, landraces have been superseded by highly advanced varieties; however, for selection, landraces are a suitable material in which a great diversity of useful genotypes may be found *tax* >>> Figures 5, 50

landscape of recombination: >>> recombination

land use inspection: An official inspection of a non-pedigreed crop to determine the degree of contamination in the crop which may pose a varietal purity problem in a pedigreed seed crop planned to be grown on the same land the following crop season *seed*

Langley: A unit of solar radiation equivalent to one gram calorie per square centimeter of irradiated surface *hort*

lanuginose: Covered with down or fine short hair, woolly, cottony *bot*

lapsi: >>> bulgur

larva: The wormlike immature form of certain insects; some are called caterpillars, grubs, or maggots *zoo phyt*

LASER: >>> light amplification by stimulated emission of radiation

late blight (*syn* **foliage blight,** *syn* **tuber blight of potato**): A widespread and serious >>> disease (*Phytophthora infestans*) affecting the potato and related species; symptoms include the appearance of brown patches on the leaves, often with white mold on the underside; under damp conditions the entire foliage may collapse; brown lesions also develop on tubers, spreading to involve the entire tuber in a dry brown rot; it reduces yields and marketability and may cause losses in store by encouraging soft rotting; the Mexican potato species *Solanum demissum* is a good source of resistance genes against late blight; the resistance genes *R8* and *R9* have shown broad-spectrum resistance in both laboratory and field conditions *phyt*

late crop: A crop or plant showing late maturation within a given season *agr hort*

latency: The state of being latent; the interval between exposure to a toxin or disease-causing organism and development of a consequent disease *phyt*

latency period: >>> latency

latent: >>> latency

latent infection: A chronic infection in which host–pathogen equilibrium is established without any visible symptoms of disease *phyt*

latent period: The period between infection and the sporulation of the pathogen on the host *phyt* >>> latency

latent virus: A virus that does not induce symptom development in its host *phyt* >>> latent infection

lateral: Belonging to or borne on the sides *bot*

lateral flow strip: A variation on the >>> ELISA tests, using strips as opposed to microtiter wells to detect the presence of a protein produced *chem meth*

lateral gene transfer: >>> horizontal gene transfer

lateral growth: >>> tillering

lateral meristem: A meristem giving rise to secondary plant tissues, such as the vascular and cork cambia *bot biot*

lateral nerves: For example, in wheat, the nerves which run along the length of the broad and narrow wings of the >>> glume; in barley, the two pairs of nerves (inner and outer) lying toward the margins of the lemma and on either side of the median nerve *bot*

lateral root: Roots arising from the main root axis *bot*

lateral shoot: Shoots originating from vegetative buds in the axils of leaves or from the nodes of stems, rhizomes, or stolons *bot*

late replicating: Often the heterochromatic regions of chromosomes, which show a later replication than the euchromatic ones *biot cyto*

late replication: The rolling circle of replication of phage lamda, producing concatemers suitable for packaging in lamda heads *bio*; heterochromatic regions of chromosomes, which show a later replication of DNA than the euchromatic ones *cyto* >>> lamda phage >>> heterochromatin

laterite: A weathering product of rock, composed mainly of hydrated iron and aluminum oxides, hydroxides, and clay minerals but also containing some silica *agr* >>> Table 44

late-sown: Sowing date later than the optimal time for a given crop or variety *agr*

latest safe sowing date: The date until which seeds can be sown without severe yield loss during the following year, usually in cereal crops *agr*

late selection: Applied in self-pollinating crops, that is, in the variable progeny of a cross for three or four generations, using either the >>> bulk breeding method or >>> single-seed descent, and producing a mixed population of relatively homozygous individuals; the selection is made among these homozygous individuals; it is efficient because it produces plants with a reduced hybrid vigor, which can be misleading during the screening process, and it also produces a greater expression of recessive alleles that are exhibited only in the homozygous state; late-selected plants have a higher heritability than those of >>> early-selected plants; the advantage of late selection must be equated with the longer breeding cycle required *meth* >>> early selection >>> Figure 60 >>> Table 37

late wood: A result of secondary plant growth; very young branches have hardly any fascicular cambium; interfascicular cambium develops very early during the year, even before the beginning of secondary growth; the activity of the cambium increases branch diameter, and the vascular bundles become elongated in cross section; far more xylem than phloem elements are produced; annual rings become clearly visible because at the beginning of each vegetation period (in spring) vessels (conducting function) and fibers (supporting function) with a wide lumen are assembled first – the so-called early wood; in the following season, elements with steadily narrowing volumes are produced; in autumn, only a few vascular elements with narrow lumina (late wood) form *bot hort fore*

latex: A white, commonly sticky substance produced in specialized tissues within a plant *bot*

latifoliate: Broad-leafed *bot*

Latin rectangle: A field design that is similar to the >>> Latin square, just differentiated by the number of replications, which is not equal to the number of variants; the number of replications may be a third, a quarter, or a fifth of the number of variants; thus, the number of replications is reduced *stat* >>> design of experiment >>> Figure 9

Latin square: In general, a set of symbols arranged in a checkerboard in such a fashion that no symbol appears twice in any row or column; it is used for subdividing plots of land for agricultural and breeding experiments, so that treatments can be tested even though the soil conditions of the field might vary in an unknown fashion in different areas; it requires that the field is subdivided by a grid into subplots and the differing treatments be performed at consecutive intervals on plants from different subplots *stat* >>> design of experiment >>> Figure 9 >>> Table 26

lattice: Either a partially ordered set in which any two elements have a greatest lower bound and a least upper bound, or a discrete subgroup of the additive group of a Euclidean space; that is, the lattice is closed under addition and negation, and there is a non-zero minimum distance between two points of the lattice *stat* >>> design of experiment >>> lattice design

lattice design: An >>> incomplete block design in which the number of treatments must be an exact square; there are several types of lattice designs, for example, alpha lattice, partially balanced square lattice; the lattice design is "simple" if r = 2; square lattice designs are also called "nets"; they are partially balanced with respect to an association scheme of >>> Latin square type; first associates concur once in blocks, second associates, never; for a "rectangular" lattice design, there are n(n − 1) treatments in r_n blocks of size n − 1; the construction is similar to that for a square lattice design, except that n cells are removed from the square array; for a "cubic" lattice design, there are n^3 treatments in $3n^2$ blocks of size n; the treatments form a cube of order n; the blocks are the one-dimensional slices *stat* >>> alpha lattice design >>> design of experiment >>> lattice square

lattice square: An experimental design for n^2 treatments in r blocks, each of which is an n × n square array; the construction uses a square lattice design in 2r replicates; pairs of replicates in the old design are combined so that their blocks form the rows and columns, respectively, of squares in the new design *stat meth* >>> design of experiment >>> lattice design

lowland rice: >>> paddy rice

lawn: A stretch of open, grass-covered land (i.e., one closely mowed, as near a house, on an estate, or in a park) *agr*

law of diminishing returns: If one factor of production is increased while the others remain constant, the overall returns will relatively decrease after a certain point; for example, if more and more laborers are added to harvest a wheat field, at some point each additional laborer will add relatively less output than his predecessor did, simply because he has less and less of the fixed amount of land to work with *agr*

laws of inheritance: >>> MENDEL's laws of inheritance

layering: Covering stems, runners, or stolons with soil, causing adventitious roots to form at the nodes, enabling propagation by rooted cuttings; this procedure is used commercially to propagate many plants hort; *in vitro* layering is the horizontal placement of cultured shoots or nodal segments on agar growth medium in order to produce axillary bud formation *biot*

LD mapping: >>> LD map >>> association mapping >>> linkage disequilibrium map

leaching: The washing out of material from the soil, either in solution or suspension *agr* >>> Table 44

leader: A terminal leader is the uppermost branch or vertical tip of the tree; it eventually becomes the tree stem or trunk *fore hort*

leader sequence: A nucleotide sequence of the >>> mRNA on which the ribosomes bind; in other words, non-translated sequence at 5′ end of mRNA, or N-terminal sequence of a protein constituting a signal for transport through a membrane, which is later removed *biot*

leading strand: DNA strand synthesized in the 5′ to 3′ direction *biot*

leaf: A thin, usually green, expanded organ born at a node on the stem of a plant, typically comprising a petiole (stalk) and blade (lamina), and subtending a bud in the axil of the petiole; it is the main site of photosynthesis; the leaf type, also called leaf form, is a description of the physical characteristics of a leaf; a leaf type denotes some combination of a leaf's color, shape, size, texture, and amount of hair; specific leaf types include bustleback, crenate, holly, longifolia, oak, ovate, plain, pointed, quilted, red reverse, round, ruffled, scalloped, serrated, spooned, strawberry, supreme, tailored, trumpet, truncate, or variegate *bot*

leafage: >>> foliage

leaf analysis: Chemical analysis, usually of growing plants for their nutrient status and other characteristics; a diagnostic tool for taking corrective measures while the crop is still in the field *agr phys meth*

leaf area index (LAI): The total leaf surface area exposed to incoming light energy, expressed in relation to the ground surface area beneath the plant (e.g., LAI = 3, the leaf area exposed to light is three times that of the ground surface area) *phys*

leaf axil: The angle between a petiole and the stem *bot*

leaf blight: Various diseases that lead to the browning and dropping of leaves *phyt* >>> late blight >>> Figure 62

leaf blight (of maize) One of the most important foliar diseases of maize caused by the fungus *Helminthosporium maydis*; it is widespread in tropical and subtropical parts of the world; there are two races: the "O" race is more prevalent; the "T" race caused heavy damage in the U.S. during the 1970s; the inheritance of resistance to race "O" follows two patterns; when resistance is expressed in number of lesions or percentage of leaf area infested, the expression is quantitative and gene action is additive with quite a high heritability; however, when resistance is in the form of lesion type, then there is a qualitative expression; but polygenic resistance is more common *phyt* >>> Figure 62

leaf bud: A bud producing a stem and leaf, unlike a flower bud, which contains a blossom *bot*

leaf color chart: A tool to optimize the use of nitrogen in cereal (mainly rice) cropping; farmers generally use leaf color as a visual and subjective indicator of the crop's nitrogen status and need for nitrogen fertilizer; the chart provides a simple, easy-to-use, and inexpensive tool for efficient nitrogen management; often, it contains six gradients of green color from yellowish green (1) to dark green (6) and can guide nitrogen top-dressing *agr meth*

leaf curling: A trait that is expressed in several plants caused either by abiotic stress, diseases, or mutant genes, for example, the soft red winter wheat varieties "Kaskaskia" or "Penjamo 62" (registered in the U.S.) exhibit leaf rolling; this trait is pronounced under some heat or drought stress; the expression is most visible just prior to heading; the flag leaves curl up lengthwise; some triticales tightly roll the leaves to give the appearance of an onion leaf *agr*

leaf cutting: A cutting made from a single leaf; a method for propagation (e.g., of succulents); a leaf can be knocked off or cut off the plant; either it spontaneously roots on the ground or it is placed in certain media for rooting *hort meth*

leaf-disk test: A test in which resistance is measured by allowing pests to feed on disks cut from leaves of test plants; resistance is based on the amount of leaf uneaten after a given test period *phyt meth*

leaf fall: Leaf abscission

leaf grade: Describes the leaf or "trash" content of cotton lint; there are eight grades, of which seven are physical and the remaining is a descriptive standard based on the other seven *agr*

leaf index: The ratio of leaf area produced by plants to the area of the ground on which they are growing *phys*

leaf miner: Various insects, which, in the larval stage, produce a tunnel through leaves, feeding on the tissue, and leaving conspicuous traces of their paths; these being larvae of Agromyzidae (Diptera), Lyonetiidae, and Gracillariidae (Lepidoptera), Hispidae (Coleoptera), etc. *phyt*

leaf posture: The characteristic position of the foliage leaves on the stem axis, which imprints the plant habit of the species; it may contribute to optimal utilization of light and thus photosynthesis; there are several approaches to breeding for specific leaf posture in order to improve photosynthetic capacity of, for example, cereals *bot*

leaf primordium: A lateral outgrowth from the apical meristem that develops into a leaf *bot biot*

leaf propagation: Method of reproducing some ornamental plants; it involves removing a leaf from a parent plant, along with 3–4 cm of the petiole; this is called a leaf cutting; this leaf cutting is placed stem-first (or heel-first) into potting soil, water, or some other rooting medium; within 40 to 50 days, a plantlet will begin to emerge, complete with its own root system; this method produces a variety that is the same as the parent plant from which the leaf cutting came (>>> clone or cloning) *hort meth* >>> Figure 50

leaf rolling: A typical response of a plant during water deficit that is observed in various field crops such as rice, maize, wheat, and sorghum; it decreases transpiration by decreasing the effective leaf area, and thus is a potentially useful drought tolerance mechanism in dry areas; some Mediterranean grasses decrease transpiration as much as 46 to 63% by leaf rolling; in rice, it is classified as abaxial leaf roll (both sides of the leaf roll inward along the vein) and adaxial leaf roll (both sides of the leaf roll outward along the vein) according to the direction of rolling; bulliform cells, which are located in the upper epidermis of the leaf near the midrib or vascular bundles of leaves, cause rolling in some Gramineae *phys*

leaf rot (*syn* petiole rot): A condition caused by the application of too much fertilizer; because normal watering is unable to leach them out, fertilizer salts begin to accumulate in the soil and around the rim of the pot; where leaves and stems come in contact with the pot, they begin to develop lesions; eventually, these leaves and stems will wilt and turn mushy, that is, they rot *hort phyt* >>> Figure 62

leaf scorch: Leaf necrosis, usually marginal, due to phytotoxicity or nutrient and water deficiency *phys*

leaf senescence: A type of programmed >>> cell death, during which leaf cells undergo coordinated changes in cell structure, metabolism, and gene expression, resulting in a sharp decline in photosynthetic capacity; a >>> cytokinin class of plant hormones plays a role in controlling leaf senescence because a decline in the cytokinin level occurs in senescing leaves; external application of cytokinin often delays senescence *phys*

leaf sheath: A tubular envelope, as the lower part of the leaf in grasses *bot*

leaf spot: It refers to various plant diseases that cause well-defined areas of tissue to die, creating noticeable spots *phyt* >>> Figure 62

leafstalk: The footstalk or supporting stalk of a leaf *bot* >>> petiole

leaf vein: Vascular bundles in the leaves; in the petiole and the midvein of the leaf the veins are very large; farther out into the mesophyll the veins may consist of only one xylem or phloem element; in these regions, these very small veins are called veinlets *bot*

leaky mutation: Mutation that is very prone to reversion, i.e., it does not completely abolish gene function and allows the synthesis of a protein, which still partly functions *gene*

least squares method: A method of estimation based on the minimization of sums of squares *stat*

lectin: A generic term for proteins extracted from plants (e.g., legumes) that exhibit antibody activity in animals *chem phys*

leghemoglobin: An iron-containing, red pigment produced in root nodules during the symbiotic association between rhizobia and leguminous plants *phys phyt agr* >>> legumes

legitimate: Pollination of a flower of a different self-incompatibility genotype to the pollen parent, likely to result in fertile seed set *bot* >>> illegitimate

legumes: Plants showing a simple or single pistil and characterized by a dry fruit pod that splits open by two longitudinal sutures and has a row of seeds on the inner side of the ventral suture (e.g., bean, pea, soybean, locust); there are many valuable food, forage, and cover species, such as peas, beans, soybeans, peanuts, clovers, alfalfas, sweet clovers, lespedezas, vetches, and kudzu; sometimes referred to as "nitrogen-fixing" plants; legumes are an important rotation crop because of their nitrogen-fixing property; the nuclear genomes of legumes vary greatly in size, from 370 million base pairs (Mbp) in *Lablab niger* to the enormous genome of *Vicia faba* at more than 13,000 Mbp; legume genomes tend to be intermediate in size compared with those of other higher plants; more than 50% of the known legume genome sizes listed on the comprehensive plant C-value data banks are smaller than 1,300 Mbp; most of the cultivated species are modest in genome size; mung bean, cowpea, common bean, chick pea, and clover all have haploid genomes smaller than 1,000 Mbp; the model legumes, *Lotus japonicus* and *Medicago truncatula*, both have compact genomes of approximately 470 Mbp; despite the modest genome size of most legumes, it is striking that some genera have genome sizes that vary by a factor of 10 or more; this is associated with extensive differences in the abundance of retroelements, which account for substantial proportions of these genomes; much of the genome content in these taxa appears to be of recent origin, and this is consistent with the observation that genomes that differ greatly in size (e.g., pea and alfalfa) still have essentially collinear genomes; from a practical perspective, the recent amplification of retroelements in these genomes is consistent with the utility of retroelements as polymorphic markers *bot agr* >>> www.kew.org/

legumin: >>> globulin

leguminous plant: >>> legumes

lemma: Flowering glume; the lower or outer of the two bracts of the floret *bot*

lenticular: Shaped like a biconvex lens, lentil-shaped *bot*

lentiform: >>> lenticular

Lepidoptera: A group of insects known as butterflies and one of the largest orders within the class of insects; ca. 160,000 species of butterflies are known around the world; butterflies go through four stages of development (caterpillars hatch from eggs and feed on plants; caterpillars then form a cocoon, a process called pupation; the adult butterfly emerges, which then feeds by taking up liquid food through its proboscis, usually nectar); some caterpillars are specialists, eating only certain plant species; specialist caterpillars may starve if they cannot find their target plant; some butterflies can be beneficial insects, while others are considered pests; they are responsible for fertilizing many flowering plants; the silk worm is another beneficial lepidopteron, used for the production of silk; conversely, caterpillars are pests for most food crops, forestry products, and fiber crops *zoo phyt*

leptodermous: Thin-walled or thin-skinned *bot*

leptokurtic (distribution): A flat-topped, bell-shaped curve of frequency distribution of a given character in a population *stat*

leptokurtosis: Shape of a graphical curve such that the greater the value on the x-axis, the lesser the reduction of the value of the y-axis becomes per x-axis unit *stat*

leptonema: >>> leptotene

leptotene: During the first meiotic division, the first stage, in which the >>> chromosomes appear as long, widely uncoiled, and single strands; the DNA of each of the chromosomes has replicated; each chromosome consists of two identical members (chromatids) *cyto*

lesion: A visible area of diseased tissue on an infected plant *phyt*

lethal: A gene or genotype that is fatal for the individual *gene*

lethal doses (LD): The concentration of a poison that kills a certain number of cells, individuals, etc.; for example, LD50 = 50% of the cells or individuals are killed *meth*

lethal gene: A gene whose expression results in the premature death of the organism carrying it; dominant alleles kill heterozygotes, whereas recessive alleles kill homozygotes only *gene* >>> Table 36

lethal load: It is usual in many organisms that normal individuals are carriers of genes, which have little effect in heterozygotes, but are recessive and detrimental or harmful in homozygous form; for example, conifers are often carriers of a genetic load which is released by inbreeding (in particular selfing), and actually reduces the amount of selfing (inbreeding) in vital seeds compared to the frequency of selfing pollen *gene fore hort* >>> lethal gene

lethal mutation: A gene mutation whose expression results in the premature death of the organism carrying it *gene* >>> lethal gene

leucine (Leu): An aliphatic, non-polar, neutral amino acid ($HO_2CCH(NH_2)CH_2CH(CH_3)_2$ that, unlike most amino acids, is sparingly soluble in water *chem phys* >>> Table 38

leucine-rich repeat: Conserved domain of disease-resistance genes *biot gene*

leucoplast: A colorless plastid that is involved in the metabolism and storage of starches and oils *bot*

leveling off of grain-filling (in maize): Rate of dry matter accumulation in kernel diminishes as endosperm transfer cells are crushed; all or nearly all kernels are dented (in a dent geno-type), kernels are about 50–55% moisture, husk has begun to dry rapidly; occurs about 35–42 days after pollination; embryo is at Stage 5, fifth leaf primordium has formed, and lateral seminal root primordia can be observed *phys meth* >>> Table 13

levulose: >>> fructose

Lfy: >>> floral meristem identity *gene*

liana: A vigorous woody vine, usually referring to tropical vines *bot*

liber: >>> bast

library: In biotechnology and molecular genetics, a collection of cells, usually bacteria or yeast, that have been transformed with recombinant vectors carrying DNA inserts from a single species (e.g., cDNA), expression, or genomic library *biot* >>> Table 38 >>> Figure 46

license: An agreement to grant rights to a patent or tangible subject *adm* >>> Intellectual Property Rights

lichen: A complex differentiated organism formed by a symbiotic relationship between an alga and a fungus *bot*

lid: The cap of a boxlike seed capsule *bot*

life cycle (*syn*** life history):** In fungi, the stage or series of stages between one spore form and the development of the same spore again; there are commonly two stages in the life cycle (the imperfect, which may have more than one kind of spore, and the perfect), but there may be no development of one or the other *phyt*

life span: The longest period over which the life of any organism or species may extend; in 2008, the oldest plant, a >>> spruce tree (*Picea* ssp.) of Fuluberg (Sweden) was determined as 9,550 years old by a special ^{14}C-dating procedure, i.e., older than the 5,000-year-old pines of North America *phys*

ligand: An atom, ion, or molecule that acts as the electron donor partner in one or more coordina-tion bonds or a molecule (e.g., antibody), which can bind to specific sites on cell mem-branes; or an organic molecule which can form a chelated complex with a metal cation; also called chelating agent, e.g., EDTA, DTPA, or EDDHA *chem phys*

ligase: An enzyme that catalyzes a reaction that joins two substrates using energy derived from the simultaneous hydrolysis of a nucleotide triphosphate; in general, a joining enzyme, which closes single-strand breaks in DNA *phys gene*

light amplification by stimulated emission of radiation (LASER): A device that produces a nearly parallel, nearly monochromatic, and coherent beam of light by exciting atoms and causing them to radiate their energy in phase; it can be used for the elimination or manipu-lation of cell particles via a special microscope device *micr*

light leaf spot (of rape, *Pyrenopeziza brassica,* **asexual stage** *Cylindrosporium concentricum*)**:** Appears as light green or bleached areas on the leaves; small white spore masses bordering the lesions *phyt* >>> Figure 62

light reaction: During photosynthesis, those reactions that require the presence of light *phys*

light soil: A soil that has a coarse texture and is easily cultivated *agr* >>> Table 44

ligneous: Woody *bot*

lignification: Converting into wood; cause to become woody *bot*

lignin: A complex, cross-linked polymer, comprising phenyl propene units, that is found in many cell walls; its function is to cement together and anchor cellulose fibers and to stiffen the cell wall; it reduces infection, rot, and decay *chem bot*

lignotuber: A woody storage structure forming a swelling, more or less at ground level, from which dormant buds can develop; functionally and ecologically similar to >>> burls that occur at ground level of some woody species *bot eco*

ligula: A scale-like membrane that covers the surface of a leaf; in some Compositae, a strap-shaped >>> corolla; sometimes, a fringe of epidermal tissue found at the boundary between the sheath and the blade of a maize leaf *bot* >>> Table 30

ligulate: Strap-like, tongue-shaped *bot*

ligulate flower: >>> ligulate >>> ligula

ligule: >>> ligula

likelihood: The state of being likely or probable; probability *stat*

lime: Compounds of calcium used to correct the acidity in soils *agr* >>> Table 44

liming: >>> lime

limited backcrossing: As opposed to complete backcrossing, which requires at least six cycles, only two or three cycles are coupled with rigorous selection to gain the advantage of >>> transgressive segregation *meth*

limiting nutrient: A nutrient not present in the soil in sufficient quantity to support optimal plant growth *phys agr*

line: A group of individuals of a common ancestry and more narrowly defined than a strain or variety; in breeding, it refers to any group of genetically uniform individuals formed from the selfing of a common homozygous parent *gene* >>> Figure 51 >>> Table 57

0-line: A structure in young embryo which is the lower boundary of shoot apical tissues *bot biot*

lineage: A chart that traces the flow of genetic information from generation to generation *meth gene* >>> Figure 51 >>> Table 57

linear: Long and narrow, with parallel margins *bot*

linear correlation: A measure of the strength of the linear relationship between two variables y and x that is independent of their respective scales of measurement; linear correlation is commonly measured by the coefficient of correlation *stat*

linear variance: >>> variance

line breeding: A system of breeding in which a number of genotypes, which have been progeny tested in respect to some characteristics, are composited to form a variety; examples of line varieties are from normally self-fertilized crops, for example, "Gaines" wheat *meth* >>> line >>> Figure 51 >>> Table 57

line of breeding: >>> line

line of descent: >>> line

line of inbreeding: >>> line

line out: To insert cuttings or to transplant seedlings or new plants in rows in a nursery bed *hort meth*

LINE(s): >>> long interspersed nuclear element(s)

line selection: One of the common techniques of crop improvement along with traditional breeding (hybridization) and genetic engineering; it has been used for many decades, perhaps centuries, to improve appearance and production characteristics; *syn* strain *syn* selection *syn* sub-clonal selection *syn* intraclonal selection *meth* >>> line breeding

line variety: >>> line breeding

lining out: Transplanting seedlings or rooted cuttings in rows in a nursery bed *hort fore meth*

linkage (of genes): The association of genes that results from their being on the same chromosome; linkage is detected by the greater association in inheritance of two or more non-allelic genes than would be expected from independent assortment; the nearer such genes are to each other on a chromosome, the more closely linked they are, and the less often they are likely to be separated in future generations by crossing over; all genes in one chromosome form one linkage group *gene*

linkage disequilibrium: The non-random association of alleles at different gene loci in a population (e.g., when two loci occur close together on the same chromosome and selection operates to keep the allele combinations together); it can provide a route to the identification and isolation of genes controlling quantitative sources of phenotypic variation *gene >>> association mapping >>> Table 37*

linkage disequilibrium mapping: >>> association mapping

linkage drag: The inheritance of undesirable genes along with a beneficial gene due to their close linkage; when transferring a single gene from a donor into the genetic background of a recurrent parent by repeated backcrossing, genetic linkage will cause fragments of the donor genome surrounding the target gene to be dragged along; it is a persistent problem in plant breeding by alien introgression *gene*

linkage group: All genes in one chromosome form one linkage group *gene >>> linkage*

linkage map: An abstract map of chromosomal loci, based on experimentally determined recombinant frequencies, that shows the relative positions of the known genes or other (DNA) markers on the chromosomes of a particular species; the more frequently two given characters recombine, the further apart are the genes that determine them *gene >>> linkage*

linkage value: Recombination fraction expressing the proportion of non-parental or recombinant versus parental types in a progeny; in diploids, the recombination fraction can vary between zero to one-half *gene >>> linkage*

linked (genes): Genes or alleles showing less than 50% recombination, which is typical for unlinked (independent) genes; depending on the strength of linkage, the linked genes tend to be transmitted together *gene >>> linkage >>> linkage value*

Linola: A new form of linseed known by the generic crop name "solin", which produces high-quality edible polyunsaturated oil similar in composition to sunflower oil *agr*

lint (linters): The long fibers of cotton seed; the short fibers generally remain attached to the seed in ginning; sometimes called "fuzz"; they are used mainly for batting, mattress stuffing, and as a source of cellulose; presence or absence of lint and fuzz are controlled by the interaction of four gene loci on non-homologous chromosomes; these loci were designated as *N1*, *N2*, *Li3*, and *Li4*, where *N1N1* confers the presence of fuzzy, *N2N2* confers inhibition of fuzzy initiation and development, and duplicate gene pairs, *Li3Li3* and *Li4Li4*, determine the presence of lint; homozygosity for *li3li3* and *li4li4* might also inhibit fuzz from development; in other words, they are recessive epistatic to fuzz genes (DU, X. M. et al. 2001) *agr*

lipase: An enzyme that degrades fats to glycerol and fatty acids *chem phys*

lipid(e): A member of a heterogeneous group of small organic molecules that are sparingly soluble in water but soluble in organic solvents; included in this classification are fats, oils, waxes, terpenes, and steroids; the functions are equally diverse and include roles as energy-storage compounds, hormones, vitamins, and structural components of cells, such as membranes *chem phys >>> Table 48*

lipid body: >>> lipid(e)

lipid synthesis inhibitors: Mode of action of herbicides that target the production of lipids which are essential in plant growth and function *phyt phys*

lipophilic: Having chemical properties relating to lipids (lipid-like); non-polar compounds that are highly soluble in organic solvents, but not water; hydrophobic *chem*

lipoprotein: An intimate association between a >>> lipid and a >>> protein *phys chem*

liposome(s): Membrane-bound vesicles experimentally constructed to transport biological molecules *biot*

liquid culture: The culturing of cells on or in a liquid medium on supports or in suspension; the culture can be stationary or agitated *prep*

liquid nitrogen: Nitrogen gas that has been condensed to a liquid and has a boiling point of −195.79°C; it is used for storage of tissue, organs, cells, or suspensions, and for several cytological and molecular preparations *chem* >>> cryopreservation

lithophytic: Growing on rocks for support *bot eco*

litter: Recently fallen plant material that is only partially decomposed and is still discernible *eco*

living mulch: A cover crop that is interplanted with the primary crops during the growing season *agr*

L-notch planting: A form of slit planting involving two slits at right angles with the seedling placed at the apex of the "L" *meth fore hort*

loam soil: A soil containing sand, silt, and clay *agr* >>> Table 44

local: >>> indigenous

local infection: An infection just affecting a limited number of a plants *phyt*

localized kinetochore: A chromosome carrying normal permanently localized centromere to which spindle fiber is attached during chromosome separation *cyto* >>> kinetochore

local population: A group of individuals of the same species growing near enough to each other to interbreed and exchange genes *tax* >>> deme >>> Table 37

locule: A cavity of the ovary *bot*

locus (loci *pl*): A specific place on a >>> chromosome where a gene is located; in diploids, loci pair during meiosis and, unless there have been >>> translocations, >>> inversions, etc., the homologous chromosomes contain identical sets of loci in the same linear order; at each locus is one gene; if that gene can take several forms (alleles), only one of these will be present at a given locus *gene*

locus-specific: Plant characters or genetic activity that is exclusively correlated with a particular chromosomal locus *gene*

lodging: A state of permanent displacement of a stem crop from its upright position; it can cause considerable reduction in yield by storm damage, rots, insects, or excess of nitrogen *agr* >>> Table 34, 57

lodging resistance: Plants that can resist lodging by an optimal root system, stiffer straw, or other characteristics (e.g., in cereal breeding, the introduction of "semi-dwarf genes" contributed to shorter plants and thus higher lodging resistance even when nitrogen fertilization is increased) >>> lodging >>> near-isogenic lines >>> semidwarf >>> *Rht* gene

lodiculae: Two small, translucent, scale-like structures situated at the base of the floret *bot*

lodicule: >>> lodiculae

LOD: >>> logarithm of odds

logarithm of odds (LOD): A significance measure used in maximum likelihood statistics; in interval gene/marker mapping, the LOD score is log10 of the ratio of the probability that a QTL is present to the probability that a QTL is absent, for example, if the presence of a QTL is 1,000 times more probable than its absence, then the LOD score = 3, because log10 of 1,000 = 3 *stat biot* >>> QTL mapping >>> Table 58

loess (soil): Unconsolidated, wind-deposited sediment composed largely of silt-sized quartz particles (0.015–0.05 mm diameter) and showing little or no stratification *agr* >>> Table 44

logistic growth: A pattern of growth which is slow at first, then rapidly increases and finally levels off to a plateau; it can be described by a logistic equation *phys*

loment(um): A dry schizocarpic fruit in the form of a >>> legume or >>> siliqua with constrictions formed between the seeds as it matures, so that the final fruit is composed of one-seeded, indehiscent loment segments *bot*

long-branch attraction: A phenomenon in >>> cladistic analyses where strongly unequal rates of evolutionary change in different members of a group cause cladistics to produce incorrect trees *eco biot meth* >>> Figures 68, 69

long-day plant A plant in which flowering is favored by long days (>14 h daylight) and corresponding short dark periods; there are two types: species in which there is an absolute requirement for these conditions, and others in which flowering is merely hastened by them *bot*

longevity: The persistence of an individual for longer than most members of its species, or of a genus and/or species over a prolonged period of geological time *bot phys evol*

long interspersed nuclear elements (LINEs): These interspersed nuclear elements are ancient non-LTR retrotransposons which are highly abundant, heterogeneous, and in contrast to more conserved LTR-retrotransposons, only poorly annotated in plant genomes; they can be used as molecular markers *biot* >>> short interspersed nuclear elements

long-plot design: A specific type of field experiment using preferentially long plots; a plot, the area to which an individual treatment is applied, can be any size, including a single plant growing in a pot, a 5-acre field, or larger; however, there are some considerations, including the equipment to be used in planting, harvesting, and treatment application, that determine the size and shape of plots (e.g., space for the experiment, number of treatments, or specificity of character to be tested); if there is equipment to plant, harvest, and apply treatments to four rows at a time, then the logical plot width would be some multiple of four rows; the lengths of plots are more flexible than their widths (e.g., if the harvest from each plot has to be weighed, the scales may influence the length of plots; if the scales are designed to weigh hundreds of pounds, the plots must be large enough to provide a harvest weight that can be accurately determined by the equipment; increasing the length of plots is an easy way to do that); in general, once the plots are large enough to be representative of a much larger area, further increasing plot size will not significantly improve the accuracy of the results; plots that are larger than necessary take more field space and may increase the amount of work required for an experiment, but they usually will not adversely affect the test results unless the plots get so large that the plots within a block are no longer uniform; plots that are too small may prevent the accurate assessment of treatment effects; if the space available for an experiment is limited, more replications are usually more beneficial than having larger plots as long as plot size allows accurate assessment of treatment effects *meth stat agr* >>> design of experiment

long-term gene pool: A population with wide genetic variability established for long-term breeding objectives; lower selection pressure is applied when it is improved through recurrent selection; it can also provide genetic variability to other gene pools *meth* >>> Table 37

long-term storage: Storage of seeds in a >>> genebank longer than ten years *seed*

loose smut: A disease of plants caused by a >>> fungus of the Ustilaginales in which the masses of spores are exposed at maturity and can be dispersed freely by wind *phyt*

lophate: Pollen wall with lophae, a network-like pattern of ridges formed by the outer exine surrounding window-like spaces or depressions (lacunae), e.g., in >>> *Scorzonera hispanica* *bot* >>> pollen

lopping: A procedure by which all the branches of a tree are cut off, except the leading shoot, as opposed to pruning, in which only some of the branches are cut *hort meth*

lopping shears: >>> pruning shears

lower glume: The glume bract placed below the upper glume bract in a glume *bot*

lower palea: >>> lemma

low-input variety: A crop variety with low claims at macro- and micronutrient fertilizers, pest control, and agronomic measures *agr*

low-temperature acclimatization: Ability of plants to cold acclimate when exposed to gradually decreasing temperatures below a specific threshold; this is the most common mechanism that plants have evolved for adapting to low-temperature stress; it is a gradual process during which there are changes in just about every measurable morphological, physiological, and biochemical character; these changes are determined by genotype–environment interactions; plants can be grouped into three different classes according to their low-temperature tolerance: (1) frost tender plants which are sensitive to chilling injury; they can be killed by short periods of exposure to temperatures just below freezing point; they cannot tolerate ice in their tissues and readily exhibit frost injury symptoms that include a water-soaked flaccid appearance with loss of turgor followed by rapid drying upon exposure to warm temperatures (e.g., beans, maize, rice, tomato); (2) plants that tolerate the presence of extracellular ice in their tissues; their frost resistance ranges from the broad-leafed summer annuals, which are killed at temperatures slightly below freezing point, to perennial grasses that can survive exposure to –40°C; as temperature decreases, the outward migration of intracellular water to the growing extracellular ice crystal causes dehydration stress that will eventually result in irreversible damage to the plasma membranes, which is the primary site of low-temperature injury; (3) very cold hardy plants, which are predominantly temperate woody species; their lower limits of cold tolerance are dependent on the stage of acclimatization, the rate and degree of temperature decline, and the genetic capability of tissues to accommodate extracellular freezing and the accompanying dehydration stress *phys*

luciferin: A pigment of bioluminescent organisms that emits light while being oxidized *chem phys*

lumen: The central cavity of a cell or other structure *bot*

luminescence: The emission of light without accompanying heat *phy*

lumper: An historic heirloom and one of the oldest potato variety introduced in the U.K. around 1806; it is an inconsistently round-oval, sometimes knobby, white-skinned variety with deep eyes; flesh is white and waxy, but taste is poor and culinary use is limited; it was grown widely in Ireland before the 1845–1850 famine because it was prolific and reliable; it was nearly wiped out by *Phytophthora infestans* disease *agr*

lunate: Shaped like a half-moon *bot*

lutein: >>> xanthophyll

Lux: A unit of light measurement (= 0.0929-foot candles) once widely employed but now largely supplanted by photosynthetically active radiation units, such as $\mu mol \times m^{-2s-1}$ ($\mu mol \times E \times m^{-2s-1}$) and $W \times m^{-2}$ *phy*

luxuriance: Hybrids that are larger, faster-growing, or otherwise exceed the parental forms in some traits; usually occurs due to complementary gene action present in the parents and combined in the hybrid *gene* >>> heterosis

luxury consumption: Nutrient absorption by an organism in excess of that required for optimum growth and productivity *phys*

lyase: An enzyme that catalyzes non-hydrolytic reactions in which groups are either removed or added to a substrate, thereby creating or eliminating a double bond, especially between carbon atoms or between carbon and oxygen *phys*

lyse: To destroy or disorganize cells by enzymes, viruses, or other means *meth*

lysimeter: An apparatus for electronically measuring water balance *meth*

lysine (Lys): Non-essential amino acid ($HO_2CCH(NH_2)(CH_2)_4NH_2$ found in legumes, whole grains, and nuts *chem* >>> amino acid >>> Table 38

lysis: Cell rupture and death; it is applied if a bacterial cell is killed upon the release of phage progeny phys

lysosome: A membrane-bound vesicle in a cell that contains numerous acid hydrolases capable of digesting a wide variety of extra- and intracellular materials phys *bot*

lysozyme: An enzyme that is destructive of bacteria and functions as an antiseptic, found in certain plants *phys*

M

M1, M2, M3, etc.: Symbols used to designate first, second, third, etc. generations after treatment with mutagenic agents *meth* >>> X1, X2, X3, etc. >>> Figure 1

M2 population: The progeny derived from selfing M1 plants, which themselves are progeny that arise by selfing plants grown from mutagenized seed; recessive mutations, resulting from the seed mutagenesis, are detected in M2 plants, which are homozygous for the mutation *gene* >>> Figure 1 >>> Table 37

macerate: >>> maceration

maceration: Softening of plant tissue by use of >>> enzymes, hydrolic acid, or other means; usually, the middle lamella of the cell walls is degraded without modification of the cell content meth *cyto*

macerozyme: An enzyme or a mixture of enzymes able to soften plant tissue *phys* >>> maceration

machinability: In quality testing of cereal flour, a test that measures the stickiness of the dough *meth*

macrocarpous: Carrying or forming big fruits *bot*

macroclimate: The general climate of a large area, as that of a continent or country, as opposed to microclimate *eco*

macroelement: Chemical elements, such as >>> nitrogen or >>> phosphorus, which are needed in large amounts as nutrients for plant growth *phys agr* >>> major element

macroevolution: Evolution above the species level (i.e., the development of new species, genera, families, orders, etc.) *evol* >>> Figures 68, 69

macromolecule: A molecule that has a high molecular weight, often a polymer *chem*

macromutant: >>> macromutation

macromutation: A mutation that results in a profound change in an organism, such as a change in a regulatory gene that controls the >>> expression of many structural genes, as opposed to micromutation *gene*

macronutrient: An inorganic element or compound that is needed in relatively large amounts by plants *phys* >>> macroelement >>> major element

macroscopic: Visible to the naked eye *micr*

macrospore: >>> megaspore

macrostylous: Showing long stamen *bot*

macrosynteny: The conserved order of large genomic blocks in megabase range in the genomes of related but also unrelated species; usually it is identified by chromosomal *in situ* suppression hybridization or >>> FISH *biot cyto meth*

maculate: Spotted or blotched *bot*

MAGIC population: >>> multi-parent advanced generation intercrosses

magnesium (Mg): An element that is found in high concentrations in plants; it plays an important role in the chemical structure of >>> chlorophyll and of membranes, and is involved in many enzyme reactions, especially those catalyzing the transfer of phosphate compounds; deficiency can produce various symptoms, including >>> chlorosis and the development of other pigments in leaves *chem phys*

magnesium chlorate: >>> chemical desiccation

magnification: The ratio of the distance between two points in the image to the distance between the two corresponding points in the specimen; the apparent size of the specimen at 25 cm from the eye is considered to be at 1× *micr*

maiden: A tree in its first year *fore hort*

maintainer: Used for maintaining and multiplication of a cytoplasmic male sterile line; usually genotypes containing the normal cytoplasm and recessive at the restorer locus *seed* >>> Figure 23 >>> hybrid breeding >>> heterosis >>> cytoplasmic male sterility (CMS)

maintaining genetic diversity: Central to the process of crop improvement; it is in every breeder's interest to ensure that the gene pool from which new traits are selected remains as extensive as possible; plant breeders created the first >>> genebanks in the 1930s to conserve the valuable genetic diversity within past and present varieties, as well as landraces and wild relatives of cultivated crop species; plant breeding is integral to ongoing initiatives to identify, classify, and conserve existing biodiversity *seed gene eco*

maize gluten: A byproduct of wet milling; it is used as a medium-protein (20–24%) and medium-fiber (10%) foodstuff *agr* >>> Table 15

maize picker: A device to harvest maize *agr*

maize streak virus (MSV): A geminivirus, which has a single-stranded DNA genome that replicates in the nucleus of the host cells to provide a double-stranded replication intermediate and transcription template *phyt*

major element: An essential element required by plants in relatively substantial concentrations compared to those essential elements which are called micronutrients; the nine major elements are calcium (Ca), carbon (C), hydrogen (H), magnesium (Mg), nitrogen (N), oxygen (O), phosphorus (P), potassium (K), and sulfur (S); major elements are sometimes called macronutrients, presumably in contrast to micronutrients; major elements, however, is the preferred term *phys agr* >>> macroelement >>> macronutrient

major gene: A gene with pronounced phenotypic effects; in contrast to modifier genes, which modify the phenotypic expression of another gene *gene* >>> oligogene

majoring: The learning of a rewarding floral patch by a bee; to "major" is the U.S. word for to "specialize in an academic subject" *eco*

major staple: A crop that has a high yield per person-hour, and per unit area, and that is reliable from season to season; produces a food that can be stored, and a food that is easily cooked; there are only three major staples in the world, that is >>> wheat, >>> rice, and >>> maize; every ancient and modern civilization was based on one of those crops *agr*

male gametocide: Any substance that kills the male reproductive cells of a plant (pollen or pollen mother cells), rendering it male-sterile; male >>> gametocides can be used to convert an inbreeder (e.g., barley) into an outbreeder, for purposes of >>> recurrent mass selection; treated plants become the female parent, and untreated plants become the male parents; male gametocides can be used for the commercial production of >>> hybrid seed *meth*

male parent: >>> father plant

male sterility: Producing no functional pollen *bot* >>> Figure 23

malformation: Faulty or anomalous formation or structure *bot gene*

Malpighian layer: A protective layer or layers of cells present in the coats of many seeds; it is characteristically made up of close-packed, radially placed, heavy-walled columnar cells without intercellular spaces; the cells often are heavily cutinized or lignified and are relatively impervious to moisture and gases *bot*

malt: Germinated grains used in brewing and distilling, such as barley and wheat, that have been made to germinate by soaking in water and then have their >>> germination halted by drying in hot air; the process of malting grains stimulates the grain to develop the enzymes required to modify starches to sugars (e.g., >>> monosaccharides, such as glucose or fructose, and disaccharides, such as sucrose or maltose); it is used to make beer, whisky, malted shakes, malt vinegar, confections, such as Maltesers, flavored drinks, such as Horlicks and Ovaltine, and some baked goods, such as malt loaf *meth*

maltose: A disaccharide ($C_{12}H_{22}O_{11}$) that consists of two alpha-glucose units linked by an alpha-1, 4-glycosidic bond *chem phys*

MALTHUS theory: Related to T. R. MALTHUS (1766–1834) and his theory that population numbers tend to increase at a faster rate than means of subsistence, leading to competition for environmental resources in short supply *stat evol*

manganese (Mn): An element that is required in small amounts by plants; it is involved in the light reaction of photosynthesis and also binds proteins; deficiency causes interveinal chlorosis and malformation *chem phys*

mannitol: A polyhydroxy alcohol ($C_6H_{14}O_6$; f.w. 182.17) that can be synthesized chemically by the reduction of mannose and is present in many plants; it is commonly used as a nutrient and osmoticum in suspension medium for plant protoplasts *chem phys biot*

mannose: A hexose, $C_6H_{12}O_6$; f.w. 180.16, obtained from the hydrolysis of the ivory nut and yielding mannitol upon reduction; is occasionally used as a carbohydrate source in plant tissue culture media *chem phys biot*

MANN–WHITNEY test: A statistical test of differences in location for an experimental design involving two samples with data measured on an ordinal scale or better *stat*

mantel test: A test that computes the linear correlation between two proximity matrices (dissimilarity or similarity), used in >>> phenetics to test whether results from different analyses of the same taxa are similar or different *stat bot meth*

manure: Animal excreta with or without a mixture of bedding or litter *agr*

map (chromosomes, genes): As a verb, to determine the relative or physical position of a gene, DNA molecule, or chromosome segment *gene*

map-based cloning: The isolation of important genes by cloning the gene in question on the basis of molecular maps, where the biochemical function is unknown; it involves the identification of a mutant phenotype for the trait of interest (obtained by mutagenesis or from natural variation) and genetic fine mapping using many progeny plants; this map is then used for chromosome walking or landing, with the help of large-insert DNA libraries or physical maps to isolate the gene *biot gene meth*

map distance: The distance between any two markers on a genetic map, based on the percentage of crossing-over; the minimum distance between linked genes is 1% and maximum 50% *gene*

map length: >>> map distance >>> Morgan unit

mapmaker: An idiomatic description of computer software developed for detection and estimation of linkages; this calculation is based on the maximum likelihood method and often applied for construction of molecular marker maps *meth stat biot* >>> KOSAMBI formula

mapping: The process and the result of determination of map distances within or between linkage groups; sometimes it refers to the localization of genes or chromosome segments *gene* >>> physical map >>> optical mapping

mapping kit: A set of molecular markers that are evenly spaced on the chromosomes; if a minimal set of markers can be defined, one can detect linkage by testing a minimal number of probes, for example, if the genome is 1,468 cM in length, and markers evenly spaced at 15 cM distance, then at least 98 markers are needed to detect any gene ($N = \ln(1 - P)/\ln(1 - f)$); N = the number of markers necessary to give a probability P of finding at least one marker linked within d map units of the gene, f = the fraction of the genome represented by a region of d map units on either side of the gene, i.e., 2d/genome size (e.g., a marker is needed in the genome of 1,468 cM and linked within 10 cM of any gene, then $N = \ln(1 - 0.99)/\ln [1 - (2 \times 10 \text{ cM})/1,468 \text{ cM}] = 337$ markers; when a marker should be linked to a gene of interest, then screening of randomly chosen markers requires that a large number of markers be screened to be sure that at least one is linked to the critical gene; however, once a genome has been saturated with markers, it is only necessary to search a small set of evenly spaced markers that together cover the entire genome, that way, no matter where the gene is located, at least one of the markers screened must be closely linked with the gene in question *biot*

mapping population: A set of phenotypes based on inherent genotypes used for genetic and/or segregation studies; if pure lines are available or can be generated with only a slight change of plant vigor in self-fertilizing plants, the mapping populations consist of F2 plants, >>> recombinant inbred lines (RIL), >>> backcross (BC) populations, >>> introgression lines assembled in exotic libraries, and >>> doubled haploid lines (DH); in cross-pollination plants, heterozygous parental plants are used to derive mapping populations such as F1 and backcross lines (BC); this is the case for several tree species, such as apple, pear, and grape and for potato; for the tree species crosses between different cultivars are used to produce F1 progenies to be genotyped; in potato, the heterozygosity of parental lines used for one cross was evaluated to correspond to 57–59%; in the foundation cross population, different alleles are contributed from either parent to individual F1 plants; the >>> linkage among markers is assessed by the production of a genetic map for either parent; in natural populations or breeding pools, the natural variation between individuals of one species can be exploited for mapping; in the case of crop plants, sets of different breeding lines can fulfill this purpose; this approach is suited to mapping complex traits that are influenced by the action of many genes in a quantitative manner; such loci are defined as >>> quantitative trait loci (QTL); it is important that such a collection of different accessions contains a whole spectrum of phenotypes for a given trait; in particular, the availability of extreme phenotypes is advantageous; the underlying idea is that genomic fragments naturally present in a particular genotype are transmitted as non-recombining blocks and that markers, like >>> single-nucleotide polymorphisms (SNPs) and insertions/deletions, can easily follow the inheritance of such blocks; these units are also called >>> haplotypes, and their existence reveals a state of >>> linkage disequilibrium among allelic variants of tightly linked genes; the existence of haplotypes has clearly been shown for >>> maize and >>> sugarbeet; in mapping genes and mutants to physically aligned DNA, recombination frequencies are not equally distributed over the genome; in heterochromatic regions, such as the >>> centromeres, >>> NORs, >>> knobs, or >>> telomeres, recombination frequency is quite reduced; in these cases, cytogenetic maps can provide complementary information because they are based on the fine physical structure of chromosomes; chromosomes can be visualized under the microscope and be characterized by specific staining patterns, e.g., with >>> Giemsa C-banding, >>> fluorescence staining; cytogenetic maps allow association of >>> linkage groups with chromosomes and determination of the orientation of the linkage groups with respect to chromosome morphology; in species such as maize, wheat, barley, and *Arabidopsis thaliana*, lines carrying chromosome deletions, translocation breakpoints, or trisomics can be generated as valuable tools for the cytogenetic approach; numerical chromosome aberrations, together with marker data generated, e.g., by RFLP analysis, can clearly identify chromosomes; defined translocation breakpoints can also localize probes to specific regions on the arms of chromosomes; more recently, techniques have been developed to localize nucleic acids *in situ* on the chromosomes; during the >>> pachytene, a stage during the meiotic prophase, the chromosomes are 20 times longer than at the mitotic metaphase; they display a differentiated pattern of brightly fluorescing heterochromatin segments; it is possible to identify all chromosomes based on chromosome length, centromere position, heterochromatin patterns, and the positions of repetitive sequences such as >>> 5S rDNA and >>> 45S rDNA visualized by >>> fluorescence *in situ* hybridization; this approach has been successful in mapping two genes near the junction of euchromatin and pericentromeric heterochromatin; refined multicolor FISH even allows the mapping of single-copy sequences *gene meth* >>> Table 58

marginal farmland: Land repeatedly farmed without benefit of humus or chemical replacements *agr*

marginally hardy: Close to the limit of hardiness that a plant can withstand; planting plants that are marginally hardy is risky, because under the most severe conditions for that zone, the plant may not survive without extra protection *phys agr*

marker: A gene of known function and location, or a mutation within a gene that allows the inheritance of that gene to be studied; phenotype-based, metabolic-based (polyphenol profiles, flavonoids, carbohydrates, oils, secondary products), protein-based (isozymes, seed storage proteins, total soluble proteins), and DNA-based (>>> RFLP, by hybridization, and >>> RAPD, >>> SSR, STR, STMS, >>> AFLP, CAPS, >>> EST, Inter-SSR, >>>

SNPs, SCAR, PCR-sequencing, by PCR) markers can be distinguished *gene meth* >>> negative marker >>> SMART breeding >>> Figures 43, 63 >>> Table 29

marker gene: >>> marker >>> SMART breeding

marker-aided selection *syn* **marker-assisted selection (MAS):** Indirect selection exploiting the association between the qualitative variation in a trait (isoenzymes, DNA marker) and the quantitative variation in another trait; it is a strategy permitting plant selection at the juvenile stage from early generations; the essential requirements for MAS in plant breeding are: (1) markers should cosegregate or be closely linked (1 cM or less) with the desired trait; (2) an efficient means of screening large populations for molecular markers should be available; (3) the screening technique should have high reproducibility across laboratories, be economical to use, and should be user-friendly; it allows breeders to determine whether desired traits are present in a new variety at an early stage in the breeding program; schemes of MAS are (1) selection without testcrossing of progeny test, (2) selection independent of environments, (3) selection without field or laboratory work, (4) early-stage selection, (5) selection for multiple genes and/or multiple traits, or (6) whole-genome selection *meth* >>> SMART breeding >>> Figures 43, 59 >>> Tables 29, 57

marker-assisted breeding: >>> marker-aided selection

marker-assisted introgression: The use of DNA markers to increase the speed and efficiency of introgression of a new gene or genes into a population; a method to facilitate introgression of desirable genes into recipient plants; first, molecular markers are identified, that are closely linked to the target gene; there the fate of the linked markers in the cross progeny are monitored; the presence of the marker in a progenial individual is then taken as an indicator of the presence of the linked gene *meth bree biot* >>> SMART breeding >>> Figures 43, 59

marker-assisted selection: >>> marker-aided selection

marker chromosome(s): Structurally abnormal chromosomes in which no part can be identified; the significance of a marker is very variable as it depends on what material is contained within the marker; small supernumerary marker chromosomes (sSMC) may originate from each of the chromosomes and have different shapes – minute-like, ring, and inverted duplicated; there are numerous synonyms, such as, small supernumerary marker chromosome (s=SMC), supernumerary marker chromosome (SMC), accessory chromosome (AC/ACH), small accessory chromosome (SAC), extra structurally abnormal chromosome (ESAC), extra marker chromosome, additional marker chromosome, supernumerary micro chromosome, accessory marker chromosome, extra micro chromosome, additional chromosome fragments, minute (centric) fragment, bisatellited marker chromosomes, metacentric chromosome fragment, supernumerary ring chromosome (SRC), small bisatellited additional chromosomes (SBAC), neocentric marker chromosome (NMC), supernumerary minute ring chromosome (SMRC), marker chromosome (MC), cancer-associated neochromosomes (CaNC), or exceptionally sSMC can be acquired in neoplasia *cyto*

marker index: A way to assess the comparative degree of information provided by different molecular marker systems, as assessed by the product of heterozygosity and multiplex ratio *biot meth*

marker rescue: A technique to localize the site of a mutation within a gene; a polynucleotide, the "marker", from the known wild-type sequence is allowed to recombine with the mutant genome such that, if it includes the site of the mutation, it "rescues" the mutant by causing it to revert *biot meth*

MARKOV network(s): Concisely represent complex multivariate statistical distributions by exploiting the sparsity of direct dependencies: each node is directly connected to only a small subset of other nodes, e.g., in a MARKOV network for modeling articulated objects, only neighboring parts are assumed to have direct interaction; in a social network, each individual directly interacts with only a small number of people in the network; in a gene network, a regulatory pathway involves only a small number of genes; such sparse structures are not only essential for enabling effective learning and inference, but also may be the primary goal in knowledge discovery (e.g., discovering the gene regulatory pathway) *stat meth*

marsh: >>> fen

MAS: >>> marker-aided selection

masked symptoms: Plant symptoms (e.g., caused by a virus) that are absent under some environmental conditions but appear when the host is exposed to certain conditions of light and temperature *phyt*

mass emasculation: In hybrid breeding (e.g., in maize), the emasculation of the male flowers by mechanical means *meth* >>> emasculate >>> detasseling >>> Table 35

mass maturity: The stage in development at which seeds attain maximum dry weight *agr phys*

mass pedigree (method) selection: A system of breeding in which a population is propagated in bulk until conditions favorable for selection occur; usually, after mass pedigree selection, pedigree selection follows *meth* >>> Tables 5, 35, 37 >>> www.geneflowinc.com

mass selection (positive or negative): A form of breeding in which individual plants are selected on their individual advantages and the next generation propagated from the aggregate of their seeds; the easiest method is to select and multiply together those individuals from a mixture of phenotypes, which correspond to the breeding aim (positive mass selection); it is still applied in cross-pollinating of vegetable species, such as carrots, radishes, or beetroots, in order to improve the uniformity; when all undesired off-types are rogued in grown crop population, and the remaining individuals are propagated further, the method is termed negative mass selection; negative mass selection is no longer an adequate breeding method for highly advanced varieties; it is usually applied in multiplication of established varieties (i.e., for seed production in order to remove diseased plants, casual hybrids, or other defects) *meth* >>> Figures 39, 40 >>> Tables 5, 35

mass spectrometry: A technique that allows the measurement of atomic and molecular masses; material is vaporized in a vacuum (ionized) and then passed first through a strongly accelerating electric potential, and then through a powerful magnetic field; it serves to separate the ions in order of their charge (i.e., mass ratio); detection is made using an electrometer, which measures the force between charges and hence the electrical potential *meth phy* >>> gas chromatograph

master nursery: >>> cohort >>> check variety

matching: A process of assigning subjects to experimental and control groups in which each subject is paired with a similar subject in the other group *stat*

mate: A unisexual individual that is involved in sexual reproduction *gene* >>> Table 35

maternal effect: Any non-lasting environmental effect or influence of the maternal genotype or phenotype on the immediate offspring *gene*

maternal inheritance: Phenotypic differences found between individuals of identical genotype due to an effect of maternal inheritance *gene* >>> uniparental inheritance >>> biparental inheritance

mat-forming: A low-growth form appearing like a pad *bot*

mating The combination of unisexual individuals with the aim of sexual reproduction; in eukaryotes, the pairwise union of unisexual individuals for the purpose of producing zygotes; four types of mating can be distinguished: (1) genetic assertive mating, (2) phenotypic assertive mating, (3) genetic disassortive mating, (4) phenotypic disassortive mating *gene* >>> Table 35

mating design: The pattern of pollination set up between individuals for an artificial crossing program, e.g., single pair mating, double pair mating (where each parent participates in one or two crosses) *meth*

mating group: A group of individuals that allows for mating among one another on the basis of genetic prerequisites *gene*

mating system The pattern of mating in sexually reproducing organisms; two types of mating systems are: (1) random mating and (2) assortative mating (genetic assortative mating, genetic disassortative mating, phenotypic assortative mating, phenotypic disassortative mating) *gene* >>> Table 35

mating type: The genetic properties of an individual for a particular type of mating *gene* >>> Table 35

matroclinal: With hereditary characteristics more maternal than paternal (e.g., in certain banana hybrids) *bot*

matroclinous: >>> matroclinal

matromorphy (*syn* **diploid parthenogenesis**): Resembling the female parent in morphology; the plants are called matromorphs, matromorphics, or matromorphic plants; matromorphy should not be confused with >>> matrocliny; matroclinous plants are hybrids (formed after fertilization of the egg-cell), which resemble the mother much more closely than the father; this may be caused by a dominant gene action, or dose effects caused by different numbers of genomes from the mother and the father in the hybrid individual; besides matromorphy and matrocliny, patromorphy (diploid androgenesis) and patrocliny may occur; patroclinous plants are hybrids which resemble the father much more closely than the mother *bot cyto*

matted-row: A system of planting where plants are placed off-center or are centered on a diagonal *hort meth*

maturation: The completion of development and the process of ripening; seed maturation includes >>> dormancy, storage reserves accumulation, increase of dry weight, and moisture decrease; ABI3 (abscisic acid insensitivity) genes are central regulators of seed maturation *phys*

maturation division: >>> meiosis

mature: Fully differentiated and functionally competent cells, tissues, or organisms *phys* >>> delayed maturity

mature resistance: >>> adult resistance

maturity: >>> mature

maturity group: Division of soybean varieties in North America into 12 groups based on how flowering responds to dark periods; they are selected for the latitude, so that flowering is initiated for maximum yield *agr*

maximum likelihood method: A statistic method of linkage estimation depending on maximizing of the log likelihood; the method leads to an efficient and sufficient statistic if one exists *stat*

maysin: A naturally occurring C-glycosyl flavone glycoside present in silks of maize; it confers resistance to corn earworm (*Helicoverpa zea*); the adoption of high-pressure liquid chromatography procedures allowed the identification of analogues, such as [apimaysin (AP), 3-methoxymaysin (ME), isoorientin, and other luteolin derivatives] as well as other compounds, such as chlorogenic acid (CHA), which demonstrated antibiotic activity to the maize earworm; the generally accepted level of maysin in silks required to reduce maize earworm larval weights by 50% is approximately 0.2% of silk fresh weight; recently, two new maize populations were derived for high silk maysin; the two populations were named the exotic populations of maize (EPM) and the SOUTHERN inbreds of maize (SIM); QTL analysis was employed to determine which loci were responsible for elevated maysin levels in inbred lines derived from the EPM and SIM populations; the candidate genes consistent with QTL position included the *p* (pericarp color), *c2* (colorless2), *whp1* (white pollen1), and *in1* (intensifier1) loci; the role of these loci in controlling high maysin levels in silks was tested by expression analysis and use of the loci as genetic markers onto the QTL populations; the studies support *p*, *c2*, and *whp1*, but not *in1*, as loci controlling maysin; the *p* locus regulates *whp1* transcription; increased maysin in these inbred lines is primarily due to alleles at both structural and regulatory loci promoting increased flux through the flavone pathway by increasing chalcone synthase activity *phyt phys* >>> QTL mapping >>> Table 58

mDNA: >>> messenger DNA

ME: >>> mega-environment

meal: The byproduct of oilseeds; used as a high-protein animal feed *agr*

mean: The sum of an array of quantities divided by the number of quantities in the group *stat*

mean square: The square of the mean variation of a set of observations around the sample mean *stat* >>> variance

measurement error: Refers to errors in estimates resulting from incorrect responses gathered during the data collection phase of a survey *stat*

mechanical inoculation: A method of transmitting the pathogen from plant to plant; for example, sap from diseased plants or a defined inoculum is rubbed on test-plant leaves that usually have been dusted with carborundum or other abrasive materials; it is applied in experimental testing of plant resistance *phyt*

mechanical purity: Refers to the degree of freedom of a seed lot from seeds of other crop kinds, weed seeds, and inert matter *seed*

media composition: Different supplements to the nutritive substance provided for the growth of a given plant in the laboratory *biot*

median centromere: A centromere that is located midway on chromosomes resulting in two equal-length arms *cyto*

medical plants: Plants that are or have been used medicinally (e.g., chamomile) *hort* >>> chamomile

medium (media *pl*): Any material in or on which cultures are grown prep

medium formulation: In tissue culture, the particular formula for the culture medium; it commonly contains >>> macroelements and >>> microelements, some >>> vitamins, plant growth regulators (>>> auxin, >>> cytokinin, and sometimes >>> gibberellin), a carbohydrate source (usually sucrose or glucose), and often other substances, such as amino acids or complex growth factors; media may be liquid or solidified with agar; the pH is adjusted (5–6), and the solution is sterilized (usually by filtration or autoclaving); some formulations are very specific in the kind of explant or plant species that can be maintained; some are very general *biot prep*

medium milk (stadium): Description of the half-grown grain; when punctured, the grain contents are milky *phys bot*

medium-term seed storage: With a storage time of about ten years; usually, it is the storage of germplasm in active and working collections; it is generally assumed that little loss of viability will occur for approximately ten years; medium-term conservation takes place at temperatures between 0° and +10°C *meth* >>> long-term storage

medoid(s): Is or are representative objects of a data set or a cluster with a data set whose average dissimilarity to all the objects in the cluster is minimal; medoids are similar in concept to >>> means or centroids, but medoids are always members of the data set; medoids are most commonly used on data when a mean or centroid cannot be defined such as 3-D trajectories or in the gene expression context; the term is used in computer science in data-clustering algorithms *stat*

megabase cloning: The molecular cloning of very large >>> DNA fragments (~500–1,000 bp) *meth biot*

megabase mapping: The construction of a linear genetic or physical map using markers that are within 1 Mb of each other *meth biot*

mega-environment (ME): A broad, not necessarily contiguous area, occurring in more than one country and frequently transcontinental, defined by similar biotic and abiotic stresses, cropping system requirements, consumer preferences, and by a volume of production; the concept was introduced by >>> CIMMYT in 1988 to address the needs of diverse wheat-growing areas of the world *eco agr* >>> www.plantstress.com

megagametogenesis: The development of the female gametophyte from a functional megaspore *bot*

megagametophyte: >>> embryo sac

megaspore: One of the four cells formed in the ovule of higher plants as a result of >>> meiosis or sexual cell reduction division; one of the latter undergoes mitosis to give rise to the female gamete *bot*

megaspore mother cell: A diploid cell in the ovary that gives rise, through >>> meiosis, to four haploid megaspores *bot*

megasporocyte: >>> embryo sac

megasporogenesis: The development of the megaspore from the archesporial cell *bot*

mega-yeast artificial chromosomes (mega YAC): A large (>500 bp) piece of DNA that has been cloned inside a living yeast cell; while most bacterial vectors cannot carry DNA inserts that are larger than 50 bp, and standard YACs typically cannot carry DNA pieces that are larger than 500 bp, mega YACs can carry DNA pieces (chromosomes) as large as 1 million bp *biot*

meiocyte: The sporocyte giving rise to the embryo sac and to pollen grains *bot*

meiosis: A type of nuclear division that occurs at some stage in the life cycle of sexually reproducing organisms; by a specific mechanism, the number of chromosomes is halved to prevent doubling in each generation; genetic material can be exchanged between homologous chromosomes; there are different stages: leptotene, zygotene, pachytene, diplotene, diakinesis, metaphase I, anaphase I, telophase I, >>> interphase, metaphase II, anaphase II, telophase II, microspore formation; the duration of meiosis is positively correlated with DNA content per nucleus in diploid species and with mitotic cycle time; meiotic duration is also influenced by chromosomal organization, DNA structure, and the developmental pattern of the species *bot cyto* >>> recombination >>> Figures 15, 28

meiotic crossing-over: >>> chiasma >>> recombination

meiotic cycle: >>> meiosis

meiotic duration: The time needed for completing the cell cycle from prophase to telophase under certain conditions *cyto*

meiotic mutation: Used to elucidate several areas of genetic research, e.g., it involves gene-centromere mapping by half-tetrad analysis with 4x–2x crosses (potato), where the 2x parent forms 2n pollen by either first-division restitution (FDR, ps/ps) or second-division restitution (SDR, pc/pc); since two of four strands of a bivalent are recovered together in 4x progeny from crossing a 4x nulliplex (*aaaa*) with a 2x heterozygous (*Aa*), the frequency of 4x nulliplex progeny provides an estimate of gene-centromere map distance through half-tetrad analysis; the second application involves pyramiding of distinct meiotic mutations in the same genetic background, e.g., clones with the doubly homozygous genotype ps/ps, sy-3/ y-3 are able to produce 2n gametes by a mechanism equivalent to FDR without crossover; these gametes transmit the parental genotype virtually intact to their progenies, the FDR non-crossing-over (NCO) 2x clones provide a homogeneous sample of heterozygous gametes for testing the parental value of 4x clones; factorial 4x–2x crosses using 2x (FDR-NCO) male parents allow estimation of the relative contribution of the random meiotic products (from the 4x parents) and the "somatic" (non-recombinogenic) male genome to the phenotypic expression of >>> quantitative traits; another application of meiotic mutants has been to permit genetic inference about the chromosomal (physical) location of >>> quantitative trait loci (QTL) controlling important traits (e.g., in potato); a large range of 4x cultivars can be crossed with a collection of full-sib 2x clones able to transmit different fractions of their heterozygosity via 2n gametes; tuber yield of the progenies then can be determined for different environments; one group of progenies is derived from FDR with crossing-over (FDR-CO) clones, where the 2x parent transmits about 80% of the heterozygosity to the 4x progeny; the other group is derived from FDR-NCO clones in which the 2x parent transmits almost 100% of its >>> heterozygosity and epistasis to 4x progeny; therefore, one can expect higher yields with 100% transmission of heterozygosity vs. 80% heterozygosity; however, no significant difference in total tuber yield between the two groups is found; these results can be interpreted as meaning that loci with a major effect on yield are located between >>> centromeres and proximal >>> crossovers, since these regions are in common between the two groups of 2x clones; all the genetic analyses using meiotic mutants thus far are converging to one interesting notion: QTL with a major effect on yield are predominantly located in genomic regions with reduced levels of recombination; in this context, breeding programs should develop strategies to maximize the transfer of heterozygosity to proximal loci, since theoretical models also indicated that >>> deleterious mutations would preferentially accumulate in these regions of the 4x potato chromosomes *gene cyto meth* >>> QTL mapping >>> Table 58

meiotic sieve: The tendency of meiosis to only function regularly in tissues without major chromosomal abnormalities; hence meiotic products tend to be chromosomally normal *cyto*

melittophily: Pollination by bees *bot*

melting temperature I: The temperature at which the two strands of a double-stranded DNA molecule come apart; a short (<18 nucleotides) oligonucleotides T_m value (°C) is estimated by the formula: $T_m = $ (number of A + T) $\times$ 2(number of G + C) $\times$ 4 *chem meth*

membrane: A sheet-like structure, 7–10 nm wide, that forms the boundary between a cell and its environment and also between various compartments within the cell; it is composed of lipids, proteins, and some carbohydrates; it functions as a selective barrier and also as a structural base for enzymes *bot*

membrane bioreactors: Bioreactors where cells grow on or behind a permeable membrane, which lets the nutrients for the cell through but retains the cells themselves *biot prep*

Mendelian character: A character that follows the laws of inheritance formulated by >>> G. MENDEL *gene*

Mendelian inheritance: >>> MENDEL's laws of inheritance

Mendelian population: A group of (potentially) interbreeding plants (cross-fertilizing crops), which may occur in a certain area or in an experimental design *gene stat* >>> Table 37

Mendelian ratio: The segregation ratios according to MENDEL's laws of inheritance *gene* >>> Figure 6 >>> Table 2

Mendelism: >>> MENDEL's laws of inheritance

mendelize: To segregate according to MENDEL's laws of inheritance *gene*

Mendel's laws of inheritance (*syn* **Mendelian laws of inheritance**) The inheritance of chromosomal genes on the basis of the chromosome theory of heredity; three laws are considered: (1) law of dominance or of uniformity of hybrids, (2) law of segregation, (3) law of independent assortment *gene* >>> Figure 6 >>> Table 2

mentor method: Used after grafting to direct the training of a young hybrid plant by subjecting it to the influence of another plant, the mentor; it was postulated as to eliminate defects in hybrid plants or to strengthen desirable properties in new varieties; it is based on the profound formative influence of various substances of one of the grafted components (the already formed variety) on the other (the developing hybrid); the Russian I. V. MITSCHURIN (1855–1935) invented the method; he used it to develop varieties of apple ("Bellflower – Chinese apple", "Candella – Chinese apple"), pear ("Bergamot Novik"), cherry ("Krasa Severa"), etc.; the frost resistance of the grafted hybrid seedling seemed to be increased, and the quality of its fruit as well; the changes obtained are fixed only after the vegetative reproduction of the new variety; although graft hybridization has become inextricably linked to the name of T. D. LYSENKO (1898–1976), C. DARWIN (1809–1882) was the first to put forward the concept of graft hybridization, and formulated a >>> pangenetic hypothesis to account for it; in his *The Variation of Animals and Plants under Domestication* he recorded various cases in which shoots that developed from grafted trees exhibited the characters of both stock and scion; to explain the formation of graft hybrids and many other facts, he proposed that cells are not only able to grow and divide but are also capable of "throwing off" >>> gemmules (molecular carriers of hereditary characters and the embryonic form of our modern genes); these gemmules were considered capable of self-replication and were able to freely "diffuse" from cell to cell; in the cases of grafts, such diffusion was thought to occur between the cells of the stock and scion; these ideas of DARWIN have generally been considered to be wrong because the stock and scion have been thought to not exchange their genetic materials; for more than a century there has been a refusal to accept the existence of graft hybrids, which have been regarded only as a type of chimera – an individual composed of genetically different tissues and cells, representing a mixing of stock and scion without any true genetic exchange; STEGEMANN and BOCK (2009) stated that gene transfer occurs but is restricted to the contact zone between scion and stock, indicating that the changes can become heritable only through a lateral shoot formation from the graft site; RUSK (2009) has interpreted this as "molecular evidence that the concept of graft hybridization is untenable"; over the past several decades, the concept of graft hybridization has been tested and the existence of graft hybrids has been confirmed by several independent groups of scientists, e.g., OHTA and CHUONG (1975), working on *Capsicum annuum*, repeatedly showed that genes for fruit color and fruit position could be transferred by grafting; TALLER et al. (1998) showed that some of the characteristics of stock were introduced into the progeny obtained from selfed seeds of the scion and that novel characteristics appeared as a result of graft induction; in the history of genetics, neglecting certain findings and phenomena is not uncommon; for example, both G. MENDEL's laws of heredity and B. McCLINTOCK's study on transposable elements were ignored for decades; graft hybridization is potentially a simple and powerful means of plant breeding; it is to be hoped that the new results reported by STEGEMANN and BOCK (2009) will once more direct the attention of geneticists to a challenging, yet unaccepted and unexplained, phenomenon; STEGEMANN and BOCK have confirmed the exchange of large pieces of plastid DNA at the immediate graft site by ordinary grafting, e.g., in tobacco *breed* >>> mentor pollen method

mentor pollen method: The technique used in order to solve recognition and growth problems of pollen; pollen from the desired male parent is mixed with pollen from the same genotype and/or species as the maternal parent; the latter pollen is partially inactivated by irradiation; this type of pollen still germinates but does not fertilize; the "mentor" pollen germ tube "guides" the pollen tube of the male parent toward the ovary for fertilization; mentor pollen which has not been irradiated, and which has thus retained its vigor, can also be used in such crosses, however, it competes with pollen of the male parent; this approach is used in crossings of ornamental plants or cereals, such as barley *meth* >>> crossing method >>> incompatibility

mercaptan: >>> thiol

mericlinal: Refers to a chimera in which the inner tissue has a different genetic constitution than the surrounding outer tissue *bot* >>> chimera >>> Figure 56

mericloning: A propagation method using shoot tips in culture to proliferate multiple buds, which can then be separated, rooted, and planted out *biot meth*

meristem: A group of plant cells that is capable of dividing indefinitely and whose main function is the production of new growth; meristematic cells are found at the growing tip of a root

or a stem (apical meristem), in cambium (lateral meristem), and also within the stem and
leaf sheaths (intercalary meristem of grasses) *bot phys*

meristematic: Pertaining to the meristem *bot*

meristematic tip: The meristematic dome and one pair of leaf primordia; commonly used as
explants, particularly to produce virus-free plant material *bot hort* >>> meristem

meristem cloning: Artificial propagation of a plant using cells taken from the meristem of a parent
plant and yielding genetically identical offspring *meth biot*

meristem (tip) culture: The culture of an explant consisting only of a meristematic part *biot*

meristemoid: A localized group of cells in callus tissue, characterized by an accumulation of
starch, RNA, and protein, and giving rise to adventitious shoots or roots *biot*

merogony: An individual with the egg cytoplasm from one parent and the egg nucleus from the
other parent *gene*

mescal: Strong alcoholic liquor, similar to tequila, made by distillation of a fermented preparation
of the starchy core of *Agave angustifolia* and other species *prep hort*

mescaline: A hallucinogenic alkaloid found in *Lophophora williamsii* and other species of cactus traditionally used in native American Indian religious ceremonies and medicine *chem*
>>> Table 46

mesenchyme: An embryonic type of connective tissue *bot*

mesocarp: Middle layer of the fruit wall *bot* >>> pericarp

mesochimera: >>> chimera >>> Figure 56

mesocotyl: An elongated portion of the seedling axis between the point of attachment of the scutellum and the shoot apex of, for example, a grass seedling; it is recognized as a compound
structure that is formed by the growing together of the cotyledon and the hypocotyl *bot*

mesoderm: The middle layer of embryonic cells between the ectoderm and the endoderm *bot*

mesophyll: Internal parenchyma tissue of a plant leaf that lies between epidermal layers; it functions in >>> photosynthesis and in storage of starch *bot*

mesophyll explant: An explant prepared from internal parenchyma tissue *biot* >>> mesophyll

mesophyte: A plant with an intermediate water requirement *bot*

mesopolyploid: Individuals that have undergone moderate aneuploid-induced base number shift;
apomicts may contain complete or nearly complete duplicate sets of genes *cyto*

messenger DNA (mDNA): A single-stranded DNA that acts as a messenger of the protein biosynthesis *gene*

messenger RNA (mRNA): A single-stranded RNA molecule responsible for the transmission to
the ribosomes of the genetic information contained in the nuclear DNA; it is synthesized
during transcription, and its base sequencing exactly matches that of one of the strands of
the double-stranded DNA molecule *gene*

meta analysis: In genetics, the judge amount of mapping studies based on its own experimental
populations; each experiment is limited in size and restricted to a single population or
a cross, planted in a certain environment; then QTL effects are limited; therefore, one
approach for QTL analysis is the combination of several studies by meta analysis *stat meth*

metabolic engineering: Aims to optimize substance production in organisms in such a way that
the substances are developed in a greater quantity, a purer form, or in a more favorable
composition; this is achieved through the specific recombination of the genetic material for
metabolic and regulator proteins in the organisms *biot*

metabolic pathway: A sequential series of enzymatic reactions involving the synthesis, degradation, or transformation of a >>> metabolite; the pathway can be linear, branched, or cyclic
and directly or indirectly reversible *gene*

metabolism: The chemical changes within the living cell; it is sum of all the physical and chemical processes by which the living protoplasm is produced and maintained, and by which
energy is made available for the use of the organism *phys*

metabolite: A substance taking part in metabolism *phys*

metabolomics: An extended discipline of biochemistry that involves the analysis (usually high-throughput or broad-scale) of small-molecule metabolites and polymers, such as starch; it involves descriptions of biological pathways and current metabolomic databases, such as the Kyoto Encyclopedia of Genes and Genomes (KEGG) *biot meth* >>> www.genome.jp /kegg

metacentric: Applied to a chromosome that has its centromere in the middle *cyto* >>> Figure 11

metakinesis: Prometaphase *cyto*

metaphase: A stage of mitosis or meiosis at which the chromosomes move about within the spindle until they eventually arrange themselves in its equatorial region; in metaphase I (MI) of meiosis, the chromosomes of a genome line up within the cell at a position referred to as the equatorial plate; spindle fibers form, which link each chromosome of a homologous pair to a different pole of the cell; the orientation of the chromosomes relative to the two poles seems to be random; the number of different combinations of chromosomes that can occur due to their orientation at MI is defined by the formula $2n - 1$, where n is the number of chromosomes in the genome; for example, there are two combinations possible with two chromosome pairs, four combinations with three chromosome pairs, and eight combinations with four chromosome pairs *cyto* >>> Figure 15

metaphase arrest: The stopping of cell division at mitotic or meiotic metaphases, usually by application of specific agents *cyto meth*

metaphase plate: The grouping of the chromosomes in a plane at the equator of the spindle during the metaphase stage of mitosis and meiosis *cyto* >>> Figure 15

metapopulation: The sum of multiple sublined breeding populations may be referred to as a metapopulation *meth* >>> Table 37

metaxenia: The influence of pollen on maternal tissue of a fruit outside the embryo sac (i.e., edible fruit tissues); the phenomenon is seen in >>> dates and sometimes called xenia *bot*

methionine (Met): Sulfur containing non-polar amino acid ($HO_2CCH(NH_2)CH_2CH_2SCH_3$); an intermediate in the biosynthesis of cysteine, carnitine, taurine, lecithin, phosphatidylcholine, and other phospholipids; the only one of two amino acids encoded by a single codon (AUG) in the standard genetic code (tryptophan, encoded by UGG, is the other); the >>> codon AUG is also significant, in that it carries the "start" message for a ribosome that signals the initiation of protein translation from mRNA *chem phys* >>> Table 38

method of overstored seeds: In pedigree breeding of allogamous crop plants (e.g., in rye), an effective method of regulating cross-fertilization; usually, a greater number of individual plants is harvested from a genotypic mixture of a certain population; their progenies are sown as A families in smaller plots, while half of the seed of all elite plants is retained in reserve; those A families meeting all the requirements are not directly multiplied but the remaining seed of the corresponding elite plants is sown in the following year; it enters a so-called A family trial; in this way the economically valuable traits of the A family can be definitely evaluated after maturity; the best A families determined for further breeding have already been pollinated by a pollen mixture that also contains pollen of less valuable plants *meth* >>> Figure 4 >>> Tables 35, 37

methotrexate: A toxic folic acid analogue, $C_{20}H_{22}N_8O_5$, that inhibits cellular reproduction *chem phys biot*

methylation: The introduction of a methyl group into an organic compound; methylation of specific nucleotides within a target site of a restriction enzyme can protect the DNA against attack by that enzyme *chem phys* >>> Tables 39, 43 >>> genomic imprinting

2-methyl-butadiene: >>> isoprene

3-methyl-2-cyclohexene-1-one (MCH): Used in forests to protect live trees from spruce beetles and >>> Douglas fir beetles; the volatile, naturally occurring chemical acts as a beetle repellent; when small amounts of MCH are attached to dead trees, beetles are prevented from aggregating on the dead trees and from large-scale reproduction *fore phyt*

1-methylcyclopropene (MCP): Used for extending the lifetime of cut flowers and potted >>> ornamental plants; it is approved for use only in enclosed spaces such as greenhouses and shipping containers *hort meth*

methylene blue: A vital staining agent for chromosomes *micr*

methyl-methanesulfonate: A frequently used, very efficient chemical mutagen ($C_2H_6O_3S$); it acts by adding methyl to guanine; thus, it causes base pairing errors as it binds to >>> adenine *prep gene* >>> Table 38

metric character: A trait that varies more or less continuously among individuals, which are therefore placed into classes according to measured values of the trait; also called "quantitative character" *gene*

microarray(s): In molecular genetics and biotechnique, a tool to examine the expression levels and intertwined interactions among genes and their products; >>> complementary DNA of interest is affixed to a glass slide in an ordered array; the expression level is determined by the binding to cDNA; a small glass, or filter, square may contain probes for thousands of gene products; in statistics, a microarray is a >>> block design with blocks of size 2; the concept arose in DNA testing, where each array is probed with two DNA samples *biot meth stat* >>> design of experiment >>> gene chip

microbe: A microorganism, especially a disease-causing bacterium *bot phyt* >>> microorganism

microbial: >>> microbe

microbiome: The ecological community of commensal, symbiotic, and pathogenic microorganisms that literally share an organism; Joshua LEDERBERG (1925–2008) coined the term, emphasizing the importance of microorganisms inhabiting an organism in health and disease; several scientific articles distinguish microbiome and microbiota to describe either the collective genomes of the microorganisms that reside in an environmental niche or the microorganisms themselves, respectively *biot*

microbiome, cold-free sample collection and transportation: Shipping microbial samples on dry ice is expensive and, now, unnecessary; instead, microbiome samples can be safely transported at room temperature with DNA/RNA Shield™; this liquid solution inhibits microbial growth and protects both DNA and RNA at ambient temperatures *biot*

microchromosomes: Tiny dot-like chromosomes that are too small for their centromeres and individual arms to be resolved under the light microscope *cyto*

microclimate: The atmospheric characteristics prevailing within a small space *eco* >>> macroclimate

microdissection: >>> micromanipulation

microdroplet array: A technique introduced by KAO and KONSTABEL (1970); it is used to evaluate large numbers of media modifications, employing small quantities of a medium into which are placed small numbers of cells; droplets of liquid are arranged on the lid of a PETRI dish, inverted over the bottom half of the dish containing a solution with a lower osmotic pressure, and the dish is sealed; the cells or protoplasts form a monolayer at the droplet meniscus and can easily be examined *biot prep*

microelement: >>> trace element llelementment >>> micronutrient

microevolution: Evolutionary change within species that results from the differential survival of the constituent individuals in response to natural selection; the genetic variability on which the selection operates arises from mutation and sexual recombination in each generation *evol*

microfibers: A method used to transform tissue culture cells using tiny fibers coated with DNA; these fibers are combined in solution with the cells and shaken vigorously, causing the fibers to stab the plant cells delivering the DNA to them *biot meth*

microgametogenesis: The development of the microgametophyte (pollen grain) from a microspore *bot*

micrografting: A method by which small embryos, either from artificial cultures or from mature seeds, are grafted onto normal stocks with protection against evaporation *meth hort*

microinjection: Injection performed under a microscope using a fine microcapillary pipette (e.g., into a single cell or cell part) *biot*

micromanipulation: Manipulation or surgery done while viewing the object through a microscope and often carried out with the aid of injection or dissection of substances or particles *biot* >>> microinjection

micromanipulator: The facility usually attached to a microscope in order to carry out micromanipulation *micr* >>> micromanipulation

micrometer: A unit of measurement frequently used in microscopy (1 μm = 10^{-6} meter); in older usage also known as micron (1 μm) *meth*

micromutant: >>> macromutation

micron: >>> micrometer

micronaire: The size of an individual cotton fiber taken in cross-section *agr meth*

micronucleolus: A small nucleolus that may arise by nucleolar budding in the course of nucleolar degeneration *cyto*

micronucleus: The smaller nucleus as distinguished from the larger nucleus, produced during telophase of mitosis or meiosis by lagging chromosomes or chromosome fragments derived from spontaneous or induced chromosome aberrations *cyto*

micronutrient: An inorganic element or compound that is needed in relatively small amounts by plants *phys* >>> trace element

microorganism: Literally, a "microscopic organism"; the term is usually taken to include only those organisms studied in microbiology (bacteria, fungi, microscopic algae, protozoa, viruses) *bio*

mycoparasite: Organism (often a fungus) that parasitizes other fungi *phyt*

microphotography: Photography requiring optical enlargement *meth micr*

micropile: The pore by which the pollen tubes gain access to the interior of the ovule *bot* >>> micropyle

microplot: Microplots are used to minimize the amount of seed or space required to evaluate a group of individuals; the number of plants in a microplot differs among crops; when short rows are used as microplots, the plant density is comparable to that of larger row plots; in unbordered microplots, the effect of interplot competition has to be considered when determining an appropriate distance among the plots *meth* >>> design of experiment

microprojectile: >>> high-velocity microprojectile transformation

microprojectile bombardment: >>> high-velocity microprojectile transformation >>> nano-biolistic

micropropagation Vegetative propagation by application of tissue culture; usually conducted in growth chambers; several stages are distinguished: stage I – establishment of small fragments of stock plants in tissue culture; stage II – multiplication of >>> propagules; the most common method is through stimulation of branching and subsequent division of shoot clumps on smaller explants, which are then placed on a fresh medium; through repetition of the process, the number of initial propagules can increase a million times in one year; stage III – preparation of propagules for transfer to normal growing conditions through rooting or elongation of shoots; stage IV – establishment of stage II or III propagules in normal growing conditions – usually in soil or potting mix in a greenhouse *seed biot* >>> *in vitro* propagation >>> Table 62

micropropagule(s): >>> synthetic seeds

micropylar: >>> micropyle

micropyle: A canal in the coverings of the nucellus through which the pollen tube usually passes during fertilization; later, when the seed matures and starts to germinate, the micropyle serves as a minute pore through which water enters *bot*

micro-RNA (miRNA): A tiny piece of >>> RNA, about 21 to 24 bases in length in the genome of plants, that binds to matching pieces of messenger RNA to make it double-stranded and decrease the production of the corresponding protein; a class of non-coding, short RNAs having important roles in regulation of gene expression; micro-RNAs were first discovered in the roundworm Caenorhabditis elegans in the early 1990s and are now known in many species, including plants *biot meth*

micro-RNA based marker: DNA-based molecular markers play a significant role in gene mapping, genetic diversity analysis, germplasm evaluation, and molecular marker-assisted selection; a combination of desirable marker characteristics such as abundant polymorphism, good stability, ease of production, and high efficiency is difficult to achieve when utilizing traditional molecular marker systems; microRNAs are a type of endogenous non-coding RNAs prevalent in the genomes of many organisms; the high conservation of miRNA and pre-miRNA sequences provides an opportunity to develop a novel molecular marker type; miRNA sequences can be downloaded flleleBase databases; from them miRNA-based primers can be designed; a proportion of these primer pairs can be validated for transferability and polymorphism using DNA from tester varieties; there is a good transferability across species; miRNA-based markers provide a novel genotyping technique with low costs, high efficiency, stability, and good transferability *biot meth*

microsatellite DNA: Pieces of 2 to 5 bp DNA sequences that are repeated up to 50 times (appear repeatedly in sequence within the DNA molecule) adjacent to a specific gene within the DNA molecule; thus, microsatellites are linked to that specific gene *biot* >>> microsatellite marker >>> http://wheat.pw.usda.gov/ITMI/Repeats/index.shtml

microsatellite marker: Microsatellites or simple sequence repeats are a type of molecular marker; microsatellites consist of tandem repeats of one to six nucleotide motifs; the repeats usually are in units of ten or more, although repeats as small as six units have been found; the repeats can be (1) perfect tandem repeats, (2) imperfect (interrupted by several non-repeat nucleotides), or (3) compound repeats; they are well-distributed throughout a genome; microsatellites can be amplified by the >>> polymerase chain reaction (PCR) using a pair of primers flanking the repeat sequence; the polymorphism between different individuals is due to the variation in the number of repeat units; each locus can have many alleles; one advantage of microsatellites is that they are mostly codominant, which makes them easily transferable between genetic maps of different crosses in the same or closely related species, in contrast with >>> RAPDs, which are dominant, and therefore new maps have to be generated for every cross; several microsatellite–primer pairs may be used simultaneously, thus reducing time and costs; the relatively simple interpretation and genetic analysis of single-locus markers make them superior to multi-locus DNA marker types such as RAPDs; microsatellites are also called simple sequence repeats (>>> SSRs), simple tandem repeats (STRs) or simple sequences (SSs) *gene* >>> Figure 63 >>> Table 29

microsatellite-primed PCR (MP-PCR): A technique resulting in RAPD-like patterns after agarose gel electrophoresis and ethidium bromide staining; the MP-PCR technique is more reproducible than RAPD analysis because of higher stringency *biot* >>> random amplified polymorphic (RAPD) technique >>> Figure 48

microsatellite repeat sequences: Sequences of 2 to 5 bp repeated up to 50 times such as a TA dinucleotide repeat polymorphism; the variable number of repeats creates the polymorphism; they may occur at 50,000–100,000 locations in a certain genome; microsatellites mutate faster than non-repeat polymorphisms and can be used to estimate evolutionary relationships over shorter time scales *gene meth biot*

microscope: An optical instrument having a magnifying lens or a combination of lenses for inspecting objects too small to be seen distinctly by the unaided eye *micr*

microscopic slide: >>> glass slide

microsomal: The fraction of a cell extract that comprises the organelles *biot*

microsome: >>> minichromosome

microsporangium: A sporangium (e.g., pollen sac) that produces the microspores (pollen) *bot*

microspore: The first cell of the male gametophyte generation of Angiospermae and Gymnospermae, later to form the pollen grain *bot*

microspore culture: Refers to the *in vitro* culture of pollen grains to obtain haploid callus or haploid plantlets directly from the pollen grains; microspore cultures differ from pollen cultures by the stage of development in gametogenesis *biot* >>> Figures 17, 26, 58 >>> Table 7

microspore mother cell: One of the many cells in the microsporangium (anther) that undergo microsporogenesis to yield four microspores; as opposed to megaspore mother cell *bot* >>> Figure 58

microsporocyte: >>> pollen mother cell (PMC)

microsporogenesis: The development of microspores from the microspore mother cell *bot* >>> meiosis

microstylous: Showing a short stamen *bot*

microsurgery: >>> micromanipulation

microsynteny: Genomic relationships at gene level >>> synteny

microtome: A machine for cutting thin slices of embedded tissue; these sections may be stained and examined with the light or electron microscope *prep*

microtuber: Cultured tissue capable of growing into tuberous plant; *in vitro* microtubers are used for a number of experimental procedures such as mass propagation of seed potato stocks, germplasm conservation and exchange, and physiological and molecular studies related to >>> tuberization *biot*

microtubule: A tubular structure, 15–25 nm in diameter, of indefinite length and composed of subunits of the protein tubulin; occurs in large numbers in all eukaryotic cells, either freely in the >>> cytoplasm or as a structural component of organelles; they form part of the structure of the mitotic spindle, which is responsible for the movement of chromosomes during cell division *cyto*

micrurgy: >>> micromanipulation

mictic: >>> amphimictic

micton: Refers to a species of wide distribution, which is the result of hybridization of two or more species; all individuals are cross-fertile and have ancestral genotypes; apomixis is not present *gene*

middlings: A by-product of flour milling that contains varying proportions of endosperm, bran, and germ *agr*

mid-parent value: In quantitative genetics, the average of the phenotypes of two mates *stat gene*

mid rip: The central, thick, linear structure that runs along the length of a plant lamina; it occurs in true leaves as a vein running from the leaf base to the apex; it provides support and is a translocative vessel *bot*

mid-season drainage: The practice in managing irrigated >>> rice of draining the field completely in the middle of the growth of the >>> crop and then reflooding *agr meth*

migration: The movement of individuals or their propagules from one area to another *eco*

mildew: A plant disease in which the pathogen is seen as a growth on the surface of the host; a powdery (true ~) mildew is caused by one of the Erysiphaceae, a downy mildew (false ~) by one of the Peronosphaceae; the first may be controlled by sulfur, the second by copper; latest research shows that even milk of concentration 1:9 can have similar effects, at least in cucumber and grape vine *phyt* >>> Table 51

milk(y) stage: In cereals, the stage of "milk" ripening of caryopses at which the endosperm shows a milky consistency *phys* >>> Table 13

milling: The processes in which cereal grains are subjected to grinding followed by sifting, sizing, or other separation techniques; for example, in wheat, the grain is tempered to approximately 15–17% moisture, which facilitates the separation of the endosperm, the pericarp, and embryo *meth agr*

Millipore™ filter: A brand of disc-shaped synthetic filter having holes of a specified diameter (0.005–8 μm) through its surface *prep*
>>> www.millipore.com

mimicry: >>> Batesian mimicry >>> Vavilovian mimicry

mineralization: The conversion of organic tissue to an inorganic state as a result of decomposition by the organic content *agr*

mineral soil: A soil containing <20% organic matter or having a surface organic layer <30 cm thick *agr* >>> Table 44

mines: Deep holes or tunnels in plant parts caused by burrowing insects or their larvae *phyt*

miniature inverted-repeats transposable elements (MITEs): A class III >>> transposon; recent completion of the genome sequence of rice (*Oryza sativa*) and a microscopic nematode, *Caenorhabditis elegans*, has revealed that their genomes contain thousands of copies of a recurring motif consisting of (1) almost identical sequences of about 400 bp flanked by (2) characteristic inverted repeats of about 15 bp such as 5′ GGCCAGTCACAATGG. ~400 nt. CCATTGTGACTGGCC 3′ 3′ CCGGTCAGTGTTACC. ~400 nt. GGTAACACTGACCGG 5′; MITEs are too small to encode any protein; it is still uncertain how they are copied and moved to new locations; probably larger transposons that do encode the necessary enzyme and recognize the same inverted repeats are responsible; there are over 100,000 MITEs in the rice genome (representing some 6% of the total genome); some of the mutations found in certain strains of rice are caused by insertion of an MITE in the gene *biot*

minichromosome: A very small chromosome, usually as a result of chromosome aberrations; engineered minichromosomes offer an opportunity to improve crop performance; unlike conventional gene transformation technologies, minichromosomes can be used simultaneously to transfer and to express stably (multiple) sets of genes; because they segregate independently of host chromosomes, they provide a platform for accelerating plant breeding; strategies for producing artificial chromosomes may consider engineering of endogenous chromosomes or *de novo* assembly from chromosomal constituents; artificial minichromosomes offer an enormous opportunity to improve crop performance; unlike conventional gene transformation technologies, minichromosomes can be used simultaneously to transfer and to express stably (multiple) sets of genes; because minichromosomes segregate independently of host chromosomes, they provide a platform for accelerating plant breeding and for studying the specific chromatin domains (e.g., centromeric regions) inserted into them; strategies for producing artificial chromosomes follow a top-down (engineering of endogenous chromosomes) or bottom-up (*de novo* assembly from chromosomal constituents) approach; one example of a top-down approach involves telomere-mediated truncation of chromosomes; it was shown that cloned telomeric repeats introduced into cells may truncate the distal portion of a chromosome by the formation of a new telomere at the integration site; using this method, a vector for gene transfer can be prepared by trimming off the arms of a natural plant chromosome and adding an insertion site for large inserts; in maize, an engineered minichromosome was already patented by Chromatin Inc. (Patent No. 7,119,250); the bottom-up method was demonstrated in maize; maize cells were transformed with centromeric sequences and screened for plants that assembled autonomous chromosomes *de novo*; the constructs combined a selectable marker gene with between 7 and 190 kb of genomic maize DNA fragments containing centromeric satellite and retroelement sequences and/or other repeats; after particle bombardment into embryonic maize tissue and subsequent transgene selection, microscopic analysis revealed some cases of fluorescent signals at chromatin fragments independent of the regular maize chromosomes; such fragments were reported to be transmitted mitotically and meiotically; as in non-plant organisms, telomeres were claimed to be unnecessary for minichromosome formation because the artificial chromosomes were presumed to be circular *cyto biot* >>> minute chromosome

minimal tiling path: Any map or table showing the placement and order of a set of clones (e.g., BAC clones) that are completely and contiguously over a specific DNA segment *biot meth*

minimum effective cell density: The inoculum density below which the culture fails to give reproducible cell growth; the minimum density is a function of the tissue (species, explant, cell line) and the culture phase of the inoculum suspension; minimum density decreases inversely to the aggregate size and division rate of the stock culture *biot*

minimum inoculum size: The critical volume of inoculum required for initiating culture growth, due to the diffusive loss of cell materials into the medium; the subsequent culture growth cycle is dependent on the inoculum size, which is determined by the volume of medium and size of the culture vessel *biot meth*

minimum tillage: A production system in which soil cultivation is kept to the minimum necessary for crop establishment and growth, thereby reducing labor and fuel costs, as well as damage to soil structure *agr*

miniprep: An abbreviation for minipreparation; it refers to a small-scale preparation of plasmid or phage DNA commonly used after cloning to analyze the DNA sequence inserted into a cloning vector *biot*

minisatellite: Highly polymorphic DNA markers comprised of a variable number of tandem repeats that tend to cluster near the telomeric ends of chromosomes; the repeats often contain a repeat of ten nucleotides; they are useful tools for genetic mapping *meth biot* >>> microsatellite

minitubers: Small tubers (5–15 mm in diameter) formed on shoot cultures or cuttings of tuber-forming crops, such as potato *biot*

minor crops: Crops that may be high in value but that are not widely grown (e.g., many fruits, vegetables, and trees) *agr*

minor element: >>> trace element

minor gene: A gene that individually exerts a slight effect on the phenotype *gene* >>> modifying (modifier) gene

minor oilseeds: Oilseed crops other than soybeans and peanuts (e.g., in some countries, sunflower seed, rapeseed, safflower, mustard seed, and flax seed) *agr*

minor resistance genes: Confer a low level of resistance, but it is thought that they may, when aggregated in a single genotype, give a high level of resistance that is durable *phyt*

minute fragment (of a chromosome or minute chromosome): Usually a very tiny chromosome segments as a result of chromosome aberrations; it is smaller than the diameter of a single chromatide *cyto* >>> minichromosome

miRNA: >>> micro-RNA

misdivision: Aberrant chromosome division in which transversal but no longitudinal separation of the centromere occurs; the consequence may be telocentric chromosomes *cyto* >>> Figure 37

mismatch repair: Any of several cellular mechanisms for correction of mispaired nucleotides in double-stranded DNA *gene* >>> mismatching

mismatching: A region of DNA in a heteroduplex where bases cannot pair *gene* >>> mismatch repair

mispairing: The presence in one chain of a DNA double helix of a nucleotide that is not complementary to the nucleotide occupying the corresponding position in the other chain *gene*

missense mutation: A mutant in which a codon has been altered by mutation so that it encodes a different amino acid; the result is almost always the production of an inactive or possibly unstable protein and/or enzyme *gene*

missense single nucleotide polymorphism: Any >>> SNP that occurs in the coding region of a gene and changes the amino acid sequence of the encoding protein; such missense SNPs may cause diseases if responsible for a functional unit *biot*

missense SNP: >>> missense single nucleotide polymorphism

missing data: Incomplete data found in available data sets *stat*

missing link: An absent member needed to complete an evolutionary lineage *evol*

missing plots: >>> analysis of variance

mist-o-matic (seed) treater: Applies treatment as a mist directly to the seed; the metering operation of the treatment cups and seed dump is similar to that of the >>> panogen treater; cup sizes are designated by the number of cubic centimeters they actually deliver, for example, 2.5, 5, 10, 20, and 40; the treater is equipped with a large treatment tank, a pump, and a return that maintains the level in the small reservoir from which the treatment cups are fed; after metering, the treatment material flows to a rapidly revolving fluted disc mounted under a seed-spreading cone; the disc breaks the droplets of the treatment into a fine mist and sprays it outward to coat seed falling over the cone through the treating chamber; below the seed dump are two adjustable retarders designed to give a continuous flow of seed over the cone between seed dumps; this is important since there is a continuous misting of material from the revolving disc; the desired treating rate is obtained through selection of treatment cup size and proper adjustment of the seed dump weight *seed* >>> seed treater

MITE: >>> miniature inverted-repeats transposable elements

mites: Very small insects of the genus Arachnida, which includes spiders; they occur in large numbers in many organic surface soils or as pest on plants; they are not an important problem, possibly with the exception of the wheat curl mite, which is a vector of wheat streak mosaic virus (WSMV) *zoo phyt* >>> acarides

mitigation: The act of decreasing or reducing something or the action of lessening in severity or intensity *bio*

mitochondrial DNA (mtDNA): The maternally inherited nucleic acid found in >>> cytoplasm whose homologue in plants is chloroplastic DNA; this small circular DNA codes for tRNAs, rRNAs, and some mitochondrial DNAs; it is more closely related to bacterial DNA than to eukaryotic nuclear DNA; mitochondrial DNA mutates 10–20 times faster than nuclear DNA; it is much more abundant than nuclear DNA, and this is why most ancient DNA studies use mtDNA; mitochondrial genomes may range in size between about 200 kb in *Brassica* species to over 2,000 kb in muskmelon (*Cucumis melo*); this variability stems, in part, from the accumulation of large regions of non-coding DNA; e.g., the maize mitochondrial genome is much larger than the chloroplast genome but contains far fewer genes, most of the genes being involved in the protein synthesis or respiratory electron transport; plant mitochondrial DNA also contains chloroplast DNA, called promiscuous DNA *gene meth*

mitochondrion (mitochondria *pl*): An oval, round, or thread-shaped organelle, whose length averages 2 μm and that occurs in large numbers in the >>> cytoplasm of eukaryotic cells; it is a double-membrane-bound structure in which the inner membrane is thrown into folds (cristae) that penetrate the inner matrix to varying depths; it is a semi-autonomous organelle containing its own DNA and ribosomes and reproducing by binary fission; it is the major site of ATP production and thus of oxygen consumption in cells *bot*

mitomycin C: A form of a family of antibiotics ($C_{15}H_{18}N_4O_5$) produced by *Streptomyces caespitosus*; it prevents DNA replication by crosslinking the complementary strands of the DNA double helix *chem phys*

mitosis: The process of nuclear division by which two daughter nuclei are produced, each identical to the parent nucleus; before mitosis begins each chromosome replicates to two sister chromatids; these then separate during mitosis so that one duplicate goes into each daughter nucleus; in contrast to the prophase of meiosis I, the prophase of mitosis does not involve pairing of chromosomes or crossing-over between the homologous chromosomes; during metaphase, the individual chromosomes line up at the equatorial plate of the cell and a spindle fiber develops that links each of their chromatids to one of the two poles in the cell; the chromatids of each chromosome separate and move to opposite poles at anaphase; at telophase, a nuclear membrane develops around the chromosomes to define the nucleus of the cell; a cell wall is formed *cyto*

mitotic apparatus An organelle consisting of three components: (1) the asters, which form the centrosome; (2) the gelatinous spindle; and (3) the traction fibers, which connect the centromeres of the various chromosomes to either centrosome *bot*

mitotic crossing-over: >>> somatic crossing-over

mitotic cycle: The sequence of steps by which the genetic material is equally divided before the division of a cell into two daughter cells happens *cyto*

mitotic index (MI): The fraction of cells undergoing mitosis in a given sample; usually the fraction of a total of 1,000 cells that are undergoing division at one time *cyto meth*

mitotic inhibition: Induced or spontaneous inhibition of mitotic division *phys cyto*

mitotic poison: Any substance that hampers proper mitosis *cyto phys*

mitotic recombination: The recombination of genetic material during mitosis and the process of asexual reproduction; the mechanism for the production of variation in heterokaryons *cyto*

mitotic spindle: The spindle-shaped system of microtubules that, during cell division, traverses the nuclear region of eukaryotic cells; the chromosomes become attached to the spindle, which separates them into two sets, each of which can be enclosed in the envelope of a separate daughter nucleus *cyto*

mixograph: An instrument used to record the mixing behavior of wheat flour doughs; mixing torque is plotted as a function of time *meth*

mixoploid: Cell populations in which different cells show different chromosome numbers *cyto* >>> mosaicism

mixture: Seeds of more than one kind or variety, each present in excess of at least 5% of the whole *seed* >>> blend

mobile element: >>> transposable element

mock dominance: >>>pseudo-dominance

modal number: The chromosome count which is characteristic of cells from a particular individual or cell line *cyto*

modal value: >>> mode

mode: The value of the variate at which a relative or absolute maximum occurs in the frequency distribution of the variate *stat*

mode of reproduction: Two general modes of reproduction are distinguished: (1) sexual and (2) asexual; sexual reproduction involves the union of male and female gametes derived from the same or different parents; asexual reproduction occurs by multiplication of plant parts or by seed production that does not involve the union of sexual gametes; the breeding procedures are dependent on the mode of reproduction *bio gene*

modern varieties: Varieties developed by breeders in the formal system *seed*

modificability: The ability of phenotypic variation of a particular genotype in response to varying environmental conditions *gene*

modification: Non-heritable morphological or physiological changes induced by varying abiotic or biotic influences *gene*; in molecular biology, modification of DNA by DNA methylases occurs after replication; site-specific methylation protects the DNA (e.g., of bacteria), which synthesize restriction endonucleases *biot* >>> Tables 39, 43

modified single cross: The progeny of the cross between a single cross, derived from two related inbred lines, and an unrelated line *seed meth*

modifying (mollele) alelle: Alleles that change the conditions under which other genes evolve *gene*

modifying (modifier) gene: A gene that modifies the phenotypic expression of another gene *gene*

moisture meter: >>> hygrometer

moisture tension: The force at which water is held by soil: it is expressed as the equivalent of a unit column of water in centimeters *agr*

mol(e) (Mmol): Gram molecular weight *chem* >>> molecular weight (m.w.)

molasses: Residue of beet sugar or sugarcane production *agr*

mold: Any profuse or woolly fungus growth on damp or decaying matter or on surfaces of plant tissue; any fungus of "moldy" appearance (i.e., with abundant, visible, woolly mycelium upon which dusty or powdery conidia can be seen); for example, anther mold of clover (*Botrytis anthophila*); black mold (*Aspergillus niger*); blue mold (*Penicillium* spp.) of apple (*P. expansum*), citrus (*P. italicum*), or tobacco (*Peronospora tabacina*); bread mold (*Monilia sitophilo*); green mold of citrus (*Penicillium digitatum*); gray mold (*Botrytis cinera*); snow mold (*Calonectria nivalis*); tomato leaf mold (*Cladosporium fulvum*); or white mold of sweet pea (*Hyalodendron album*) *phyt*

moldboard: The curved metal plate in a plough (plow) that turns over the earth from the furrow *agr*

moldboard plow: A plow with a point and a heavy curved blade for breaking the soil *agr*

molecular cloning: DNA segments of different sizes of prokaryotic or eukaryotic origin are identically multiplied as a part of bacterial plasmids or phages; the alien DNA is incorporated into the host DNA ring molecule; by the rapid division of the host cells, the alien DNA segments are simultaneously cloned; subsequently, those cloned DNA segments can be excised, separated, and purified for further utilization *biot*

molecular farming: The application of biotechnology for the selected production of pharmaceutical compounds or other health or industrial compounds within a living organism (e.g., microbe or agricultural crop) *biot*

molecular genetics: A branch of genetics that deals with molecular aspects of genetic mechanisms; nowadays those techniques are summarized as "omics", including proteomics, metabolomics, transcriptomics, and phenomics; these underlying developments include advanced gel, cell imaging by light and electron microscopy, high-density microarrays, and genetic readout approaches *gene* >>> Table 38

molecular hybridization: The annealing of previously purified and denatured DNA strands by different means *gene meth* >>> DNA hybridization

molecular marker: Particular DNA sequences and/or segments that are closely linked to a gene locus and/or morphological or other characters of a plant; those segments can be detected and visualized by molecular techniques; roughly, three groups of markers can be classified: (1) hybridization-based DNA markers such as restriction fragment length polymorphisms (RFLPs) and oligonucleotide fingerprinting; (2) PCR-based DNA markers such as random amplified polymorphic DNAs (RAPDs), which can also be converted into sequence characterized amplified regions (SCARs), simple sequence repeats (>>> SSRs) or microsatellites, sequence-tagged sites (STS), amplified fragment length polymorphisms (AFLPs), inter-simple sequence repeat amplification (ISA), cleaved amplified polymorphic sequences (CAPs), and amplicon length polymorphisms (ALPs); and (3) DNA chip and sequencing-based DNA markers such as single nucleotide polymorphisms (>>> SNPs); advantages of molecular markers are: (1) the precision (<1cM) with which molecular markers can be used to map a site on a genome is often better than phenotypic markers because one can use more of them, and there are fewer errors in scoring; (2) linkage can be easily established; when a set of markers is carefully chosen from different parts of the genome, it is possible to map any locus to a marker; (3) the markers do not even have to be genes; they can be any piece of DNA that detects a small number of bands, and for which there is polymorphism in the population; it makes them selectively neutral; and (4) for RFLPs, polymorphism detected does not have to fall inside the region of the probe itself; it only has to be between the probe and the next site for whichever restriction enzyme is used *gene biot* >>> Figure 48 >>> Table 39 >>> http://wheat.pw.usda.gov/ITMI/Repeats/index.s html >>> http://maswheat.ucdavis.edu

molecular weight (m.w.): The sum of the atomic weights of all of the atoms in a given molecule *chem*

molecule: That ultimate unit quantity of a compound that exists by itself and retains all the chemical properties of the compound *chem*

molybdenum (Mo): An element that is required in small amounts by plants and is found largely in the enzyme nitrate reductase; deficiency leads to interveinal chlorosis *chem phys*

monad: A meiocyte-derived individual cell instead of a tetrad as a result of meiotic disturbances *cyto*

monoallelic: Applied to a >>> polyploid in which all alleles at a particular locus are the same (in a tetraploid – *A1A1A1A1*) as opposed to diallelic (in a tetraploid – *A1A1A1A2*), triallelic (in a tetraploid – *A1A1A2A3*), tetraallelic (in a tetraploid – *A1A2A3A4*), etc. *gene*

monobrachial: A chromosome with a terminal centromere *cyto* >>> telocentric >>> Figures 11, 37

monocarpic: Bearing one fruit, i.e., flowering and bearing fruit only once and then dying; term can apply to annuals, biennials, or perennials *bot* >>> polycarpic

monocentric chromosome: A chromosome with only one centromere *cyto* >>> Figure 11 >>> neocentric >>> polycentric

monocentric crop plant: A crop plant species that has only one center of origin *evol* >>> center of diversity

monoclonal: Describing genetically identical cells produced from one clone *gene*

monoclonal antibody: An antibody preparation that contains only a single type of antibody molecule; monoclonal antibodies are produced naturally by myeloma cells; a myeloma is a tumor of the immune system; a clone of cells producing any single antibody type may be prepared by fusing normal lymphocyte cells with myeloma cells to produce a hybridoma *meth sero*

monoclonal blocks: A deployment option for clones or for families in monofamily blocks in which each clone (family) is established in a pure block; diversity may be maintained by a mosaic of blocks of different genetic entries *meth*

monocot: An abbreviated name for monocotyledon, referring to plants having single-seed leaves; flower parts arranged in threes or multiples thereof, parallel-veined leaves, closed vascular bundles arranged randomly in the stem tissue *bot* >>> Table 32

monocotyledonous: Having one cotyledon *bot* >>> monocot

monocropping: >>> monoculture

monoculm mutant: A mutant form that shows only one culm instead of normally more (e.g., in wheat) *gene*

monoculture: The result of large-scale industrial farming systems that focus on one crop; such systems tend to eliminate biodiversity of plant life in an area and consequently lead to a loss of biological balance between pests and predators; this can lead to intensified pest pressure, increased use of pesticides, and therefore higher crop protection costs *agr phyt*

monoecious: Male and female flowers borne on the same plant *bot* >>> monoecism

monoecism: The condition in plants that have male and female flowers separated on the same plant (e.g., in maize, cucumber, melon, or pumpkin) *bot* >>> Table 18

monogenic: A trait controlled by the alleles of one particular locus, as opposed to digenic, trigenic oligogenic, or polygenic *gene* >>> Table 33

monogenic resistance: Resistance determined by a single gene *phyt*

monogenomatic: >>> monohaploid

monogenotypic: >>> clone variety

monogerm(ous): A fruit of, for example, sugarbeet containing only one ovule; in contrast to a multigerm fruit, which represents an aggregate fruit containing several ovule units *bot seed* >>> Table 33

monohaploid: A haploid cell or individual possessing only one chromosome set in the nucleus *gene cyto*

monohybrid: A cross between two individuals that are identically heterozygous for the alleles of one particular gene (i.e., *Aa* × *Aa*) *gene*

monohybrid heterosis: >>> superdominance

monohybrid segregation: A segregation pattern according to a monogenic inheritance *gene* >>> Table 2

monoisodisomic: A cell or individual showing monosomy for one chromosome but an isochromosome for one of the arms of the missing chromosome *cyto* >>> Figure 37

monoisosomic: A cell or individual showing nullisomy of one chromosome but having an isochromosome for one arm of the missing chromosome pair *cyto* >>> Figure 37

monomorphic: Of only one form or genotype, as in a population that is not polymorphic for a gene *gene eco*

monophylesis: Having a single origin in evolutionary history *evol* >>> Figures 68, 69

monophyletic: A group of species that share a common ancestry, being derived from a single interbreeding population *bot evol*

monophyllous: Showing one leaf *bot*

monoploid: Having the basic chromosome number in a polyploid series *cyto* >>> aneuploid

monoplontic: Refers to a haploid individual or monoploid phase of the life cycle *bot*

monopodial: Having branches arise from a single main axis *bot*

monosome: A chromosome that lacks a homologue in a diploid organism *cyto* >>> aneuploid >>> Table 50

monosomic: A genome that is basically diploid but that has only one copy of one particular chromosome type, so that its chromosome number is 2n − 1; monosomic series were developed mainly in wheat and other crops *cyto gene* >>> aneuploid >>> Figure 37 >>> Table 50

monosomic analysis: A common method for >>> gene mapping (e.g., in hexaploid wheat); when genes determining phenotypes of interest for which an aneuploid series is not available, crosses can be made to a monosomic series in a variety with a contrasting phenotypic pattern; the monosomics are used as female parents in order to ensure the majority of progeny (~72%; the transmission of n − 1 gametes is about 75% by the egg cell, but only about 4% by the pollen) will be become monosomic; if the gene involved is both recessive and hemizygous-effective (an uncommon situation), direct phenotypic observations on F1 monosomic progenies enable the researcher to locate the gene on a particular chromosome; only monosomic individuals in the critical cross will exhibit the phenotype, whereas disomic sibs and monosomic (or disomic) progenies from all other crosses will display the dominant phenotype; if the plants of the monosomic series carry the dominant allele of a gene of interest, monosomic individuals in the critical cross exhibit the recessive phenotype, whereas disomic sibs display the dominant phenotype *gene* >>> aneuploid >>> Figure 37 >>> Tables 23, 50

monosomo-autotetraploid: >>> aneuploidy

monosomy: >>> aneuploid >>> monosomic

monospermic: >>> monospermous

monospermous: Bearing one seed *bot*

monostand: Sometimes it refers to a grass community composed of only one cultivar *agr*

monotelic: A mitotic chromosome with one oriented and one unoriented centromere *cyto* >>> centromere

monotelocentric: A cell or individual lacking one chromosome pair but showing one telocentric chromosome for one arm of the two missing homologues *cyto* >>> aneuploid >>> Figure 37

monotelodisomic: A cell or individual lacking one chromosome pair but showing two homologous telocentric chromosomes for one arm of the two missing homologues *cyto* >>> aneuploid >>> Figure 37

monotelomonoisosomic: A cell or individual lacking one chromosome pair but showing a telocentric chromosome for one arm of the missing homologous pair and an isochromosome for the other arm *cyto* >>> aneuploid >>> Figure 37

monotelotrisomic: A cell or individual showing an additional telocentric chromosome to a certain pair of chromosomes *cyto* >>> aneuploid >>> Figure 37

mordant: >>> mordanting

mordanting: In a broad sense, to produce surface conditions by metal ions in the fixed structures that will enable them to hold the particular stains intended for making them visible *cyto*

Morgan unit (morgan, M): A unit of relative distance between genes on a chromosome; one morgan (1 M) represents a crossing-over value of 100%; a crossing-over value of 10% is a decimorgan (dM); 1% is a centimorgan (cM) *gene*

morphogenesis: The developmental processes leading to the characteristic mature form of a plant or parts of it *phys*

morphosis: A modification of the morphogenesis of an individual caused by environmental changes; it encompasses growth and cellular differentiation, but it is a process of a higher order of magnitude which supersedes events occurring in single cells; interactions between cells have a great effect on the collective fates of individual cells in the developmental plant body *phys*

morphotype: >>> habit(us)

mortality: The relative frequency of deaths in a specific population (i.e., death rate) *bio*

mosaic disease: A pattern of disease symptoms displaying mixed green and lighter colored patches caused by certain viruses *phyt*

mosaicism: Intraindividual variation of chromosome numbers or chromosome structure, usually in different tissues *cyto*

motes: In cotton production, it can be either fuzzy (immature seeds on which fiber development ceased at an early stage) or bearded (a piece of seed coat with long fibers attached); both are often termed seed coat neps *agr meth*

mother cell: Special cells in the anther and ovule that give rise to pollen or egg cells *bot*

mother plant: The female ancestor of a hybrid and/or hybrid progeny; in horticulture, a mature plant from which cuttings are taken *meth agr gene hort* >>> donor plant

motte: Small grove or clump of trees *bot*

mottled leaf: Disease caused by a zinc deficiency that reduces the size of leaves and fruits (e.g., in citrus plants); often also caused by viruses *phyt phys*

motif: A characteristic stretch of amino acids in a protein sequence, which is responsible for a biochemical function, e.g., the active site of an enzyme *biot*

mould: >>> mold

mound layering: A method of propagation whereby a branch or stem is scored and then brought into contact with the soil to spur rooting *meth hort*

MP-PCR: >>> microsatellite-primed PCR

mRNA: >>> messenger RNA

MS medium: *In vitro* culture medium named after the description by T. MURASHIGE and F. K. SKOOG *biot*
>>> https: //en.wikipedia.org/wiki/Murashige_and_Skoog_medium

mtDNA: >>> mitochondrial DNA

mucilage: The gummy, sticky complex carbohydrate substances (consisting principally of poly-uronides and galacturonides that chemically resemble the pectic compounds and hemicel-lulose) that cover the seeds, the root tip, bark, or stems of some plants *bot*

mucro: A minute awn or excurrent midvein of an organ (e.g., on a lemma) *bot*

mucronate: Tipped with a short tip or point like an awn, usually the short extension of a vein beyond the leafy tissue *bot*

mugeinic acid: A chelating agent; it plays an important role in the uptake of heavy metal ions from the soil when it is exuded by roots of some gramineous plants (e.g., rye) *chem phys* >>> chelate

mulch: A crumbly intimate mixture of organic and mineral material formed mainly by worms *agr*

mule: A plant hybrid that is self-sterile and usually cross-sterile due to infertile pollen or undevel-oped pistils *gene*

MÜLLER mimicry (Müllerian ~): The sharing of a feature or signal by a number of different species to the mutual benefit of all, as compared to BATES mimicry *eco*

multiallelic: A gene with more than two alleles at a locus *gene*

multi-environment trials: A set of genotypes is evaluated across a number of environments that may represent a target environment in order to select widely or specifically adapted geno-types; a major objective in plant breeding is to assess the suitability of individual crop genotypes for agricultural purposes across a wide range of agro-ecological conditions; meanwhile worldwide networks of testing sites are established for many crop species *agr meth breed eco*

multidimensional scaling (MDS): A multivariate method of showing similarities and/or dissimi-larities of empirical data (e.g., genotypes) within an n-dimensional Euclidean space (e.g., n = 2 or 3) in which the distances between the objects are as best as possible arranged according to the distances of the data matrix *meth stat*

multifactorial: >>> polygenic

multigene family: >>> multigene variety >>> chloroplast

multigene variety: A variety that carries a number of specific genes governing resistance to a particular pathogen *phyt*

multigenic: A trait that is controlled by many genes; as opposed to monogenic *gene* >>> polygenic

multigerm: An aggregate fruit containing several ovules *bot* >>> monogerm(ous)

multihybrid: An individual that is heterozygous for more than one gene *gene*

multiline: A cultivar or variety that is composed by many more or less defined lines *seed* >>> multiline variety

multiline variety: A composite (blended) population of several genetically related lines of a self-pollinated crop, but bearing different genes (e.g., for resistance to pathogens) *seed* >>> Table 37

multilineal variety: >>> multiline variety

multilocation testing: A testing of breeder's strains and varieties on several geographically dif-ferent sites in order to estimate the adaptive environmental response and/or performance stability *meth*

multilocus association model: The >>> single-locus model tests a single >>> SNP at a time and is known to have some drawbacks, e.g., it is hard to locate the region containing the >>> QTL into a small interval because a number of SNPs can interfere with the QTL, etc.; among other, the multilocus association model has the capability of automatically correct-ing and/or controlling the confounding due to population structure and relatedness without having polygenic terms in the model *meth stat* >>> nested association mapping

multilocular: Of a >>> carpel with more than one ovule *bot*

multi-locus probe: A DNA probe that hybridizes to a number of different sites in the genome of an organism *biot* >>> Figure 45

multinucleate: Describes cells that have more than one nucleus *bot*

multi-parent advanced generation intercrosses (MAGIC): The past years have seen the rise of multi-parental populations as a study design, offering great advantages for genetic studies in crop plants; the genetic diversity of multiple parents, recombined over several generations, generates a genetic resource population with large phenotypic diversity suitable for high-resolution trait mapping; while there are many variations on the general design, also populations should be considered where the parents have all been inter-mated, typically termed as multi-parent advanced generation intercrosses (MAGIC); such populations have already been created in model plants, and are emerging in many crop species; however, there has been little consideration of the full range of factors which create novel challenges for design and analysis in these populations *stat meth*

multiple alleles: The existence of several known allelic forms of a gene *gene* >>> allelism

multiple cropping: The growing of more than one crop on the same field in one year *agr*

multiple drop array: >>> microdroplet array

multiple fruit: Developed from a cluster of flowers on a common base, called an >>> inflorescence growing on a >>> catkin; each flower on the catkin produces a fruit (>>> drupelet), but these mature into a single mass; examples are the pineapple, edible fig, mulberry, osage-orange, and breadfruit *bot*

multiple genes: Two or more genes at different loci that produce complementary or cumulative effects on a single, quantitative genetic trait *gene* >>> polygenes

multiple lattice designs: Often applied to test a large number of entries that are to be directly compared for selection; comparison among entries from different lattices is accommodated by the use of checks; while this design has some practical merits, it is statistically expected to be less efficient than alternative designs for a large number of entries; the comparison for a number of settings by using a single large alpha-design shows that the efficiency gain in terms of >>> sample size or variance may be appreciable; therefore, plant breeders should seriously consider using alpha-designs in place of multiple lattices *stat meth* >>> design of experiment

multiple population breeding system: The breeding population is subdivided in several smaller subpopulations that are bred for different objectives, for example, different areas *meth* >>> Table 37

multiplication: The increase in number of individuals produced from seed or by vegetative means *meth* >>> Table 49

multiplication population: Subpopulations of the general breeding population purposefully selected for different sets of traits or deployment destinations *meth* >>> Table 37

multiply: >>> multiplication

multisite mutation: Any mutation that either involves alteration of two or more contiguous nucleotides or occurs repeatedly at many loci in a genome *gene*

multistage sampling: A method of sampling from repeatedly smaller, internally subdivided divisions of a population to make survey activities more efficient; it is synonymous with cluster or nested sampling *stat* >>> Table 37

multitude of genes: >>> polygenes

multivalent: Designating an association of more than two chromosomes whose homologous regions are synapsed by pairs (e.g., in autopolyploids or in translocation heterozygotes) *cyto* >>> Figure 15

multivar: >>> cultivar mixture

mummy: A dried, shriveled fruit or seed colonized by a fungus or parasite *phyt*

mutability: The ability of a gene to undergo mutation *gene*

mutable genes: A class of genes that frequently spontaneously mutate *gene*

mutable site: >>> mutational site

mutagen: An agent that increases the mutation rate within an organism or cell; for example, X-rays, gamma-rays, neutrons, or chemicals [base analogues, such as 5-bromo uracil, 5-bromo deoxyuridine, 2-amino purine, 8-ethoxy caffeine, 1.3.7.9.-tetramethyl-uric acid, maleic hydrazide; antibiotics, such as azaserine, >>> mitomycin C, streptomycin, streptonigrin, actinomycin D; alkylating agents, such as sulfur mustards (ethyl-2-chloroethyl sulfide), nitrogen mustards (2-chloroethyl-dimethyl amine), epoxides (ethylene oxide), ethyle-neimines, sulfates, sulfonates, diazoalkanes, nitroso compounds (N-ethyl-N-nitroso urea); azide (sodium azide); hydroxylamine; nitrous acid; acridines (hydrocyclic dyes), such as acridine orange] *gene* >>> Table 38

mutagenesis: The process leading to a mutant genotype *gene* >>> TAL effector nuclease >>> Table 49

mutagenic: Substances and circumstances inducing mutants *gene*

mutagenic agent: >>> mutagen

mutagenicity: The potential of agents and circumstances to induce mutations *gene*

mutagenicity testing: The assessment of chemical or physical agents for mutagenicity *gene meth*

mutagenize: Treatments that result in mutations *meth*

mutagenized: Cells or individuals that were treated with mutagens *gene*

mutant: A plant bearing a mutant gene that expresses itself in the phenotype *gene* >>> TAL effector nuclease >>> Figure 56 >>> Table 35

mutant site: A site on a chromosome at which a mutation can occur or has occurred *gene*

mutant strain: A strain of cells or individuals that, by one or more mutations, is differentiated from the original strain *gene*

mutation: A change in the structure or amount of the genetic material of an organism; in cyto-genetics, a gene or a chromosome set that has undergone a structural change *gene* >>> unique event polymorphisms >>> chimera >>> directochimera >>> secondary periclinal chimera >>> endochimera >>> ectochimera >>> mesochimera >>> primary periclinal chimera >>> TAL effector nuclease >>> Table 35

mutational hot spot: A site within a gene or genome that frequently mutates *gene*

mutational site: The more or less defined position along a gene at which mutations occur *gene*

mutation breeding: To experimentally introduce or remove a character from a cell or organism by exposure to mutagenic agents followed by screening for the desired attribute; it also refers to several techniques, involving induced mutations, that were utilized (mainly in the 1960s and 1970s) to introduce desirable genes into the plants (e.g., resistance to plant diseases, increased yield, improvements in composition); usually seeds or pollen were soaked in mutation-causing chemicals (mutagens) or treated via bombardment with ionizing radia-tion followed by screening of the resultant plants and selection of the particular mutation (beneficial trait) *meth* >>> nano-biolistic >>> TAL effector nuclease >>> Figure 1 >>> Tables 35, 49

mutation delay: The time lag between the mutation event and its phenotypic expression, e.g., recessive mutations, become obvious if they become homozygous *gene*

mutation map: The frequency of mutations recorded along chromosomes, represented as a dia-grammatic drawing *gene*

mutation pressure: The continued production of an allele by mutations *evol*

mutation rate (μ): The number of mutation events per gene and per unit of time (e.g., per cell generation); this is the only source of variation in asexual organisms; the mutation rate is the likelihood of parentage when findings suggest otherwise; beware of the different units in different mutation rates; in humans, the mutation rate is 1 bp per 10^9 bp per cell divi-sion; this corresponds to 10^{-6} mutations per gene per cell division and because there are 10^{16} divisions in a lifetime, 10^{10} mutations per gene per lifetime *gene* >>> TAL effector nuclease

mutator gene: A gene that may increase the spontaneous mutation rate *gene*

muticous: Blunt, or without a definite point *bot*

muton: The smallest unit of DNA in which a change can result in a mutation (i.e., the single nucleotide) *gene*

mutually orthogonal Latin squares (*syn* **pairwise orthogonal Latin squares**): A set of >>> Latin squares, any two of which are orthogonal *stat*

mycelium (mycelia pl): A mass of hyphae that form the body of a fungus *bot*

mycobacterium (mycobacteria *pl*): Any of several rod-shaped aerobic bacteria of the genus *Mycobacterium bot*

mycology: The science of fungi; the study of mushrooms *bot*

mycoplasma: The smallest free-living microorganism; it lacks a rigid cell wall and is therefore pleomorphic (polymorphic); mycoplasmas cause many diseases in plants; many formerly attributed to viruses are now known to be caused by mycoplasmas *phyt*

mycoprotein: A growth of fungi which is rich in protein and suitable for human consumption *bio*

mycorrhiza: A close physical association between a fungus and the roots (or seedlings) of a plant from which both fungus and plant appear to benefit *bot*

mycotoxins: Toxic substances produced by fungi or molds on agricultural crops *phys chem phyt*

myophily: Pollination by flies *bot*

myrosin cell: Cell containing glucosinolates (mustard oil glucosides) and myrosinases, enzymes which hydrolyze the glucosinolates; they occur in 11 dicotyledon families, the two largest of which are the Brassicaceae and Euphorbiaceae *phys*

N

n: The symbol for the haploid chromosome number

NAC: >>> N-acetylcysteine

N-acetylcysteine (NAC): An acetylated form of the amino acid L-cysteine ($C_5H_9NO_3S$); it is a precursor to glutathione, a powerful antioxidant present in all cells; N-acetylcysteine is well-absorbed and is incorporated into glutathione once in the cells; it is able to raise glutathione levels better than glutathione supplementation by itself; glutathione is a tripeptide made up of three amino acids, >>> glycine, L-cysteine, and L-glutamate; it functions as a coenzyme in many biochemical reactions *chem biot* >>> Table 38

naked (barley or oats): Mutant varieties that thresh free from their husk in contrast to conventional varieties where the husk is held firmly to the grain; molecular studies in barley revealed that naked barley has a monophyletic origin, probably in southwestern Iran; in barley, a single recessive gene, nud, controls the naked caryopsis character, and is located on the long arm of chromosome 7H *bot evol*

naked bud: A bud not protected by bud scales *bot biot*

NAM: >>> nested association mapping

nano-biolistic: Microparticle bombardment is frequently used as an efficient alternative to >>> Agrobacterium >>> co-culture for >>> genetic transformation of plants; the microparticles used for biolistic transformation cannot, however, efficiently transform small cells and may cause significant cell damage to plant cells and tissue; nanoparticles have been reported to inhibit growth and microRNA expression in plants; they are a thousand times smaller than a microparticle, such as a 1 μm gold particle; bombardment with microparticles results in cell damage to about 20% whereas nanoparticles damaged only ~10 % *biot*

nanometer: Equals 10^{-9} meter; equals 1 Ångström *phy*

nanopore DNA sequencing: A unique, scalable technology that enables direct, real-time analysis of long DNA or RNA fragments; it works by monitoring changes to an electrical current as nucleic acids are passed through a protein nanopore; the resulting signal is decoded to provide the specific DNA or RNA sequence; it helps to resolve complex structural variants

and repetitive regions, simplifies *de novo* genome assembly, and improves existing reference genomes *biot*

nanovirus: >>> geminivirus

nap: Large masses of curled and loosely matted fibers found in raw cotton *agr*

naphthalene acetic acid (NAA): A synthetic auxin *chem phys*

napiform: Turniplike in form *bot*

narrow-host-range plasmid: A plasmid that can replicate in one, or at most a few, different bacterial species *biot*

narrow-leafed: >>> angustifoliate

narrow-sense heritability: In quantitative genetics, the proportion of the phenotypic variance that
is due to variation in breeding values *stat gene*

narrow wing: The smaller of the two parts of, for example, the wheat glume, which are separated
by the keel *bot*

nastic movement: >>> nasty

nasty: The response of a plant organ to a non-directional stimulus (e.g., light); it is facilitated by
changes in cell growth or changes in turgor *bot phys*

national listing: >>> testing plant varieties >>> variety listing

native: >>> indigenous

native breed: >>> landrace

native species: A species that is part of the original flora of the area *bot*

native DNA: Double-stranded DNA isolated from a cell with its hydrogen bonds between strands
intact, as opposed to denatured DNA *gene biot*

native species: A species indigenous to an area, that is, not introduced from another environment
or area *eco*

native trait recovery: NT RECOVERY™ is a technology for recovering genes of agronomic
importance from ancestral strains of maize (corn); in the future, the scope of the technology will cover traits that improve high-end yield potential as well as the ability to
tolerate adverse environments; two native traits, *EaSY™* and *CT Start™*, illustrate the
successful application of NT RECOVERY™ technology; *CT Start™* confers tolerance to
the cold temperatures that occur throughout the Midwest of the U.S. during early spring;
it also provides resistance to drought and the high temperatures that may occur later in
the growing season; the first application of NT RECOVERY™ technology was for the
development of *EaSY™* and that initial success was to set the stage for the development
of *CT Start™ biot*

nattō: A traditional Japanese food made from fermented soybeans, popular especially for breakfast; as a rich source of protein, nattō and the soybean paste miso became a vital source of
nutrition in feudal Japan; it can be an acquired taste due to its powerful smell, strong flavor,
and sticky consistency; steamed, fermented, and mashed soybeans have a glutinous texture
and strong cheese-like flavor; breeding of >>> soybean cultivars suitable for nattō involves
selection for seed traits that influence the quality of the product; more than ten nattō quality traits have to be considered to assess the influence of genotypes and environment on
the traits; all of the seed traits, except starch, oil, and seed size, are correlated with one or
more of the other nattō quality traits *foo*

natural cross-pollination: Cross-pollination that occurs naturally, as opposed to artificial or hand-
pollination *bot*

natural host (*syn* **typical host**)**:** A host in which the pathogenic microorganism (or parasite) is
commonly found and in which the pathogen can complete its development *phyt*

naturalize (naturalizing): The technique or practice of planting >>> bulbs that are not native to
an area in places that will allow them to thrive and multiply on their own with little care
hort meth

naturalized plants: Introduced species that have become established in a region *eco*

natural selection: A complex process in which the total environment determines which members of a species survive to reproduce and so pass on their genes to the next generation *gene*

nature reserve: An area of land set aside for nature conservation and associated scientific research, usually with strong legal protection against other uses *eco*

N banding: A special chromosome staining method related to C banding, which reveals specific types of >>> heterochromatin; the pattern of bands and interbands along a chromosome serves as a tool for chromosome or chromosome segment identification; the method has been successfully applied in wheat *cyto* >>> C banding >>> Table 43

near-isogenic lines (NIL): Not fully isogenic; for example, in maize, two distinct composites of F3 lines from a single cross, one consisting of lines homozygous-recessive and the other consisting of lines homozygous-dominant for a certain gene (i.e., there is same genetic background), however, differing only in being homozygous-dominant versus -recessive for the genes; in wheat, near-isogenic lines have been produced for different reduced-height (*Rht*) genes causing different straw length *gene*

neck: The uppermost part of the culm between the flagleaf sheath and the collar *bot*

necrosis: Death of plant tissue, which is usually accompanied by discoloration or becoming dark in color; commonly a symptom of fungus infection *phys*

necrotrophic pathogen(e): A fungal pathogen that causes the immediate death of the host cells as it passes through them; a colonizer of dead tissue *phyt*

nectar: A sticky, sometimes sweet, secretion of flowers that has an attraction for insects *bot*

nectar gland: >>> nectarium >>> nectary

nectariless: In cotton, devoid of leaf and extrafloral necataries *bot*

nectarium: >>> nectary

nectary: Refers to a sugar-secreting gland; nectaries are usually situated at the base of a flower, sometimes in a spur, in order to attract pollinators; nectaries can also be extrafloral (e.g., the gland spin of certain cacti where they attract seed dispersal insects, such as ants) *bot*

needle: A linear, commonly pungent leaf *bot*

needlecast: In conferes, a disease symptom caused by fungi resulting in premature drop of needles *phyt fore*

neem (*syn* azadirachtin): An organic pesticide used to control insects and related pests; it is extracted from the seeds of the neem tree; it discourages feeding by making plants unpalatable to such pests as aphids, beetles, caterpillars, mealy bugs, mites, thrips, or whiteflies; neem is considered virtually non-toxic to humans *phyt meth*

negative assortative mating: A type of non-random mating in which individuals of unlike phenotype mate more often than predicted under random mating conditions *gene*

negative cross-resistance: Occurs when a resistant biotype is more susceptible to other classes of herbicides than the susceptible biotype *phyt biot*

negative-frequency-dependent selection: Natural selection in which the fitness of individual genotypes or phenotypes depend on their frequency in a population; rare genotypes or phenotypes have higher fitness than those that are common, i.e., selection that always favors rare forms; it may result in polymorphic equilibria *eco evol*

negative heterosis: >>> heterosis

negative (selection) markers: With negative selection markers it is possible to achieve genetically modified plants that no longer contain the marker genes necessary for the transformation procedure; once the transformed cells have been identified with the help of the marker genes, these are no longer required in the genetically modified plant; by use of a "chemical signal", plants can be separated out that carry the marker gene; for example, the enzyme N-acetylphosphinothricin deacetylase (DEA) serves as a negative selection marker; it converts a substance that is non-toxic to the plant into one which is toxic; in the case of DEA, an inactive form of a herbicide (N-acetylphosphinothricin) is converted into an active herbicide (phosphinothricin), which attacks the plant's central metabolism and kills it; if the transgenic progeny are treated with the inactive herbicide, those that carry the negative DEA selection marker will die; those plants that are not sensitive to the inactive herbicide remain unaffected; they are marker-gene-free *meth biot* >>> marker >>> www .gmo-safety.eu/

negative screening: A screening technique designed to identify and eliminate the least desirable plants; as opposed to positive screening, which involves identifying and preserving the most desirable plants; the technique is often used with >>> recurrent mass selection, in which the undesirables are weeded out, and the best plants are left to cross-pollinate *meth*

neighbor design: A design in which there is a concept of elements of a block being "neighbors", so that every pair of elements occurs as neighbors in a block the same number of times *stat* >>> design of experiment

neighborhood: A partially isolated subpopulation with a certain degree of inbreeding; it may arise when a large population splits into subpopulations by inbreeding effects; or the theoretical concept of an area containing a number of genets between which there is panmixis *eco stat*

nema: >>> eelworm

nematicid: A substance or preparation used for killing nematodes parasitic to plants *phyt*

nematode: A microscopic soil worm that may attack roots or other structures of cereal, sugarbeet, potato, and other plants and cause extensive damage *phyt* >>> eelworm

neocentric: Secondary centromeres that, under certain conditions, show movement as the primary centromere; sometimes they are observed on chromosome ends when they move toward the poles during anaphase of meiosis I *cyto*

neocentric activity: >>> neocentric

neo-Darwinian evolution: Evolutionary theory incorporating Darwinism and Mendelian genetics; a modified Darwinian paradigm developed during 1930–1940 that incorporated concepts from modern population, ecological, and evolutionary genetics into descent with modification; sometimes referred to as the "modern evolutionary synthesis" *bio evol*

neoendemic: A species whose restricted geographical distribution is a function of its recent origin *eco*

neomorph: Alleles that show qualitatively different effects compared to the wild-type allele gene

neomycin: An aminoglycoside antibiotic ($C_{23}H_{46}N_6O_{13}$) produced by Streptomyces fradiae that functions by interfering with ribosomal activity, thus causing errors in the reading of the >>> mRNA *chem phys gene*

neomycin phosphotransferase (NPT): An enzyme used in gene transfer experiments as a reporter for transient gene expression *chem phys biot*

neonicotinoid(s): A class of neuro-active insecticides chemically similar to nicotine; the development of this class of insecticides began with work in the 1980s by SHELL Inc. and the 1990s by BAYER Inc.; the neonicotinoids show reduced toxicity compared to previously used organophosphate and carbamate insecticides; most neonicotinoids show much lower toxicity in birds and mammals than insects, but some breakdown products are toxic; the neonicotinoid imidacloprid is currently the most widely used insecticide in the world; the neonicotinoid family includes acetamiprid, clothianidin, imidacloprid, nitenpyram, nithiazine, thiacloprid, and thiamethoxam *phyt*

neopolyploid: Polyploid of recent origin *cyto*

neoteny: The persistence of a juvenile condition into an adult stage *phys*

neotuberosum: Potatoes that resulted from an experimental breeding of Solanum andigena in order to confirm that it is the original parent of S. tuberosum; the change is complete after a mere five generations of >>> recurrent mass selection; thus the neotuberosum has provided a considerably widened genetic base for breeding approaches *meth*

nep: In cotton production, a small knot of tangled fiber *agr*

"Nerica" rice: A new rice variety is transforming agriculture in a large portion of West Africa, benefiting about 20 million farmers; it helps to reduce the rice import; the new variety combines the ruggedness of local >>> African rice species with the high productivity traits of the Asian rice that was the mainstay of the >>> Green Revolution; Asian rice species which entered Africa 450 years ago cannot compete with the weeds, so after a crop or two it's time to clear more land; planting the traditional >>> African rice species is not worthwhile for farmers as it simply does not produce enough rice; in 1991, a biotechnology-based program was initiated to combine the best traits of the Asian and African rices; vital to the effort were genebanks that contain seeds of 1,500 African rices – which had faced extinction as farmers abandoned them for higher yielding Asian varieties; research at the West Africa Rice Development Association (>>> WARDA) overcame a series of disappointing failures when they succeeded in crossing two species using embryo rescue techniques; genetic differences in the two species made breeding difficult but also gave the new rices high levels of >>> heterosis; the new rice smothers grain-robbing weeds like its African parents, resists droughts and pests, and is able to thrive in poor soils; the panicles of "Nerica" variety can hold 400 grains compared to the 75–100 grains of its African parents; further improvements in the plant's architecture such as longer panicles with forked branches, strong stems, and panicles that hold grain tightly and prevent shattering allow to out-yield others and produce bountiful harvests with modest fertilization; they mature 30–50 days earlier than traditional varieties, allowing farmers to grow extra crops of vegetables or legumes; they are taller, thus making harvesting easier, and they grow better on the fertile, acid soils that comprise 70% of the upland rice area in the region; in addition, there is 2% more body-building protein in this new variety than in either their African or Asian parents *agr* >>> rice

nervation (of leaf): The arrangement of veins in a leaf *bot*

nervature: >>> nervation

nerve: The line, usually raised, on the surface of a lemma or glume marking the presence of conducting tissue below the surface *bot*

nervure: >>> leaf vein >>> nervation

nested association mapping (NAM): An integrated mapping strategy that allows genome-wide high-resolution mapping in a cost-effective way; it greatly facilitates complex trait dissection in many species; it is used as a large-scale maize mapping resource and as a genomics and bioinformatics tool; it requires large amounts of >>> SNPs and/or individuals, and it is sensitive to genetic heterogeneity; it was originally developed in maize; instead of interrogating a single accession to possess useful genetic diversity, tens of diverse donor lines were explored in parallel, in order to increase the extent of diversity interrogated *stat meth biot*

nested block design: An experimental design in which the set of plots is partitioned into blocks, each of which is partitioned into sub-blocks *stat meth* >>> design of experiment >>> split-plot design

nest planting: Setting out a number of seedlings or seeds close together in a prepared hole, pit, or spot *meth* >>> design of experiment

net blotch disease (of barley): An important disease of winter barley (*Pyrenophora teres*, syn *Helminthosporium teres*, syn *Drechslera teres*); it is favored by cool, damp conditions; diseased leaves bear short, dark brown stripes or blotches consisting of a fine network of lines, running at right angles to, as well as parallel to, the veins; the blotches are usually surrounded by a narrow yellow zone *phyt*

netting: Covering plots or individual plants with nets in order to prevent bird damage *meth*

neuter: Without sexual structures; not having stamens or pistils *bot*

neutral marker diversity: Genetic diversity as assessed by variation in non-transcribed regions of the genome that are not known to be associated with a biological function *gene biot*

neutral substitution: Any exchange of one or more amino acid(s) in a protein without any change of its function *gene*

neutral theory: Originally proposed by KIMURA (1969), neutral theory suggests that much of the variation at the molecular level is due to the interaction between drift and mutation rather than being actively maintained by selection *evol*

nexine: The inner layer of the exine *bot*

next-generation screening(s): Variants of classical screening for >>> SNPs or other mutations that uses high-density glass >>> microarrays onto which thousands of PCR-amplified single loci or gene fragments of individuals are immobilized; each spot on such microarrays corresponds to a single locus of a particular individual and contains the specific genes of it; this microarrays are hybridized with synthetic, fluorescently labeled, allele-specific oligonucleotides complementary to, e.g., disease alleles; the hybridization event is detected by LASER excitation of the corresponding fluorochrome *biot meth*

next-generation sequencing (NGS): Recent scientific discoveries that resulted from the application of next-generation DNA sequencing technologies highlight the striking impact of these massively parallel platforms on genetics; these new methods have expanded previously focused readouts from a variety of DNA preparation protocols to a genome-wide scale and have fine-tuned their resolution to single base precision; the sequencing of RNA also has transitioned and now includes full-length cDNA analyses, serial analysis of gene expression (SAGE)-based methods, and noncoding RNA discovery; NGS has enabled novel applications such as the sequencing of ancient DNA samples, and has substantially widened the scope of metagenomic analysis of environmentally derived samples; taken together, an astounding potential exists for these technologies to bring enormous change in genetic and biological research; DNA nanoball sequencing is a type of high-throughput sequencing technology used to determine the entire genomic sequence of an organism; the method uses rolling circle replication to amplify small fragments of genomic DNA into DNA nanoballs; unchained sequencing by ligation is then used to determine the nucleotide sequence; this method of DNA sequencing allows large numbers of DNA nanoballs to be sequenced per run and at low reagent costs compared to other next-generation sequencing platforms >>> sequencing

niacin: >>> nicotinic acid

niche: The functional position of an organism in its environment, comprising the habitat in which the organism lives or is adapted, the period of time during which it occurs, and the resources it obtains there *eco*

nick: The two parents for producing hybrid seed when they produce high yields of seed of a highly productive and desirable hybrid; in breeding, synchronization of the receptivity of the female organ to the maximum pollen load of the pollinator for cross-fertilization *meth*; in molecular genetics, a single-strand break of DNA *gene biot* >>> nick translation

nicking: Synchronization of the receptivity of the male sterile plant to the maximum pollen load of the pollinator for cross-pollination in hybrid seed production *seed meth*

nick translation: A technique by which a DNA molecule is radioactively labeled with high specificity; such labeled DNA is used by different >>> DNA hybridization methods as a probe (e.g., >>> SOUTHERN transfer); within a double-stranded DNA, several single-strand breaks (nicks) are produced by DNase hydrolysis; each break or gap is extended by $5' > 3'$ exonuclease activity of >>> DNA polymerase I; the removed $5'$ nucleotides are immediately substituted by the polymerase activity of the same exonuclease; however, now ^{32}P-deoxynucleotides are used, which label the newly synthesized DNA *gene biot* >>> Figures 48, 52, 53, 54 >>> Table 39

nick-translated: >>> nick translation

nicotianamide: A soluble crystal amide of nicotinic acid that is a component of the vitamin B complex *chem phys*

nicotine: A colorless, oily, water-soluble, highly toxic liquid alkaloid, $C_{10}H_{14}N_2$, found in tobacco and valued as an insecticide *chem phys* >>> biological control

nicotinic acid (niacin, $C_6H_5NO_2$): An organic compound and a form of vitamin B3, an essential human nutrient; it belongs to the group of pyridinecarboxylic acid and exists in fruits, peanuts, or vegetables; a genetically induced increase in the niacin concentration of maize kernels is important for the improvement of kernel nutrition *chem*

nidus: A center in which infection settles and from which infection spreads; specific locality of a given disease; result of a unique combination of ecological factors that favors the maintenance and transmission of the disease organism *phyt*

nif genes: The genetic designation of genes participating in the process of nitrogen fixation; about 17 genes are organized in the nif operon; by the nif genes-produced proteins the atmospheric nitrogen (N_2) will be fixed as NH_4^+ and NO_3^- ions; many soil bacteria may fix atmospheric nitrogen; there are many research activities dealing with the transfer of the bacterial system of nitrogen fixation into crop plants other than legumes *gene biot*

NIL: >>> near-isogenic lines

NimbleGen capture array technology: In 2008, ROCHE (Basel, Switzerland) NimbleGen Ltd. announced the launch of their Sequence Capture Technology, a groundbreaking technique for enriching targeted genomic regions for high-throughput sequencing, enabling researchers to perform targeted sequencing from any region in the genome; initially offered as a service, the Sequence Capture Technology is now available for delivery, or use in research laboratories; researchers perform their own capture experiment following the instructions of a user's guide, and perform quick sequencing of enriched regions using next-generation technology such as the 454 Genome Sequencer FLX; the entire workflow, from starting genomic DNA, to enriched DNA of target regions, and sequencing results, takes about two weeks, compared to months or years for traditional PCR-based methods coupled with capillary sequencing; hence, the selective enrichment of target DNA using custom-designed high-density oligonucleotide microarrays, consisting of 385,000 probes, is a very efficient alternative to PCR-based approaches, especially in terms of time, cost, and scalability; up to 5 Mb of contiguous or non-contiguous genomic sequence can be captured on a single array; currently, NimbleGen Sequence Capture Arrays and Service are the only commercially available technology for researchers to perform large-scale targeted sequencing with quick turnaround time *biot meth* >>> www.roche.com

nitrapyrin: >>> N-serve

nitrate: The form of nitrogen (N) affects plants at different stages of their development; in particular, nitrate (NO_3^-) can affect their time and rate of seed germination, leaf expansion and function, dry matter partitioning between shoot and root, and root architecture; the magnitude of these effects is dependent on environmental factors outside the supply of N; the mechanism of these effects is variable, mainly influenced by root and shoot NO_3^- assimilation under different environmental conditions *phys* >>> nitrification >>> nitrogen

nitrification: The oxidation of ammonia to nitrite and nitrite to nitrate by chemolithotrophic bacteria *chem phys*

nitrocellulose: A nitrated derivative of cellulose; it is used in the form of a membrane as a filter for macromolecules in blotting techniques *prep meth*

nitrocellulose filter: A filter made up of nitrocellulose membrane prep >>> nitrocellulose

nitrogen (N): An element that is essential to all plants; it is found reduced and covalently bound in many organic compounds, and its chemical properties are especially important in the structure of proteins and nucleic acids; deficiency causes >>> chlorosis and >>> etiolation *chem phys* >>> nitrate

nitrogen use efficiency: A measure of the proportion of nitrogen in a crop relative to what was supplied as fertilizer or what was derived from the soil or both *phys* >>> phosphorus

nitrogenase: An enzyme complex that catalyzes the reduction of molecular nitrogen in the nitrogen-fixation process in which dinitrogen is reduced to ammonia *chem phys*

nitrogen (N) consumers: A crop plant in the crop rotation that takes nitrogen up from soil, as opposed to plants (e.g., legumes) that provide nitrogen to the soil by their nitrogen fixation activity *agr* >>> nitrogen fixation

nitrogen (N) fixation: The reduction of gaseous molecular nitrogen and its incorporation into nitrogenous compounds; it is facilitated by lighting, photochemical fixation in the atmosphere, and by the action of nitrogen-fixing microorganisms (bacteria) *chem phys agr*

nitrogenomics: The study of genomics pertaining to nitrogen utilization and assimilation in plants and other microorganisms; genomics of nitrogen assimilation happens at different levels, from microbes to higher organisms, where different genetic controls regulate the actual assimilation; the term was introduced by K. K. VINOD in 2004 *biot*

nitrous oxide gas (N_2O): Used as a polyploidizing agent in order to induce 2n pollen, i.e., unreduced gametes, particularly in tulip bulbs *meth* >>> mutagen

NMS: >>> nuclear male sterility

nobilization: A term used in the breeding of sugarcane to indicate repeated matings (backcrossing) to the "noble" canes (i.e., restoring intergeneric *Saccharum* hybrids to the phenotype of *Saccharum officinarum*) *agr*

nocturnal: During the hours of darkness; as opposed to diurnal *phys*

nodal bud: The lateral shoot bud located within the root ring at the node *bot*

nodal culture: The culture of a lateral bud and a section of adjacent stem tissue *biot meth*

node: A slightly enlarged portion of a stem where leaves and buds arise, and where branches originate *bot*

nod genes: A category of genes present within the DNA of certain soil-dwelling *Rhizobium* bacteria; when those bacteria are in the presence of specific "signaling molecules", e.g., isoflavones or luteolin produced by roots of legume plants, nod genes code for the bacterial production of specific chemical compounds which then trigger relevant plant genes (e.g., NARK gene in soybean plant) to cause the plant roots to create and/or grow nodules which the bacteria subsequently move into and begin nitrogen fixation *biot* >>> nodulation

nodulation: In legumes, species of *Rhizobium* bacteria fixing nitrogen of the air in association with the roots on which they are the cause of swellings (nodules) *bot agr* >>> nod genes

nodule: A small, hard lump or swelling; root nodules are characteristic of Rhizobium infection and nitrogen fixation in legumes *bot*

nodule bacteria: Refers to several species of nitrogen-fixing *Rhizobium* bacteria, which form ball-like nodules along legume roots *bot agr*

nodus: >>> node

nominalistic species concept: A philosophy that questions the very existence of species, and believes that individuals or interbreeding populations are the only population system with any objective reality *eco meth* >>> eclectic species concept >>> ecological species concept

nominal scale: A scale for scoring quantitative data using a series of predefined values (e.g., flower color) *stat*

non-coding DNA: A certain portion of DNA that obviously does not determine a gene product, such as a protein and/or character *gene*

non-coding region: Parts of a gene that include sequences which are not translated; both 5′ and 3′ untranslated regions (UTRs), upstream promoter region, and introns are classified as non-coding regions *biot*

non-coding single nucleotide polymorphism: SNPs in a con-coding region of the genome (e.g., intron); they are the most frequent types of SNPs and somewhat misleading *biot meth*

non-coding SNAP: >>> non-coding single nucleotide polymorphism

non-conjunction: The failure of metaphase chromosome pairing during meiosis *cyto*

non-demanding variety: >>> low-input variety

non-disclosure agreement: In patenting biological material, an agreement, common between entrepreneurs and potential investors, that allows a company to share protected information while preventing its release *adm* >>> Intellectual Property Rights

non-disjunction: The failure of separation of paired chromosomes at metaphase, resulting in one daughter cell receiving both and the other daughter cell none of the chromosomes in question; it can occur both in meiosis and mitosis *cyto*

non-dropping flowers: Flowers which do not drop; instead, at the conclusion of their cycle, non-dropping flowers will quickly shrivel up until the bloom and its pedicel virtually disappear; prior to the introduction of the non-dropping characteristic, the slightest disturbance would cause flowers to drop *hort*

non-homologous association: Pairing of chromosomes, which obviously are not homologous; however, it is presumed that cryptically homologous segments allow the chromosome association *cyto*

non-homologous chromosomes: The different chromosomes of a haploid chromosome set, which usually cannot pair with one another *cyto*

non-host resistance: Inability of a pathogen to infect a plant because the plant is not a host of the pathogen due to lack of something in the plant that the pathogen needs, or to the presence of substances incompatible with the pathogen *phyt*

non-immune: >>> susceptible

non-infectious disease: A disease that is caused by an abiotic agent (i.e., by an environmental factor, not by a pathogen) *phyt*

non-Mendelian inheritance An unusual ratio of progeny phenotypes that does not reflect the simple operation of Mendel's law, for example, mutant: wild-type ratios of 3:5, 5:3, 6:2, or 2:6 indicate that gene conversion has occurred; in general, it refers to extrachromosomal and/or non-chromosomal inheritance *gene*

non-parametric tests: These are tests that do not make distributional assumptions, particularly the usual distributional assumptions of the normal-theory-based tests; non-parametric tests usually drop the assumption that the data come from normally distributed populations *stat*

non-pedigreed crop: A crop for which a crop certificate has not been issued or recognized by the official organization *seed*

non-preference: A term used to describe a resistance mechanism where parasites prefer to be on some host genotypes more than others; the less preferred genotypes are resistant *phyt* >>> antixenosis

non-probability samples: Samples in which the probability of each population element being included in the sample is unknown *stat*

non-random mating: A mating system in which the frequencies of the various kinds of matings with respect to some trait or traits are different from those expected according to chance *gene*

non-recurrent apomixis: Refers to occasional apomixis, usually caused by haploid parthenogenesis *bot* >>> Imperata cylindrica procedure

non-recurrent parent: A parent that is not involved in a backcross *meth*

non-sampling error: Used to describe variations in the estimates that may be caused by population coverage limitations, as well as data collection, processing, and reporting procedures *stat*

non-selective herbicide: A chemical that kills all or nearly all plants; the contrasting term is "selective herbicide", a chemical that kills some plants but not others *phyt*

non-sense codon: A codon that does not determine an amino acid; it may terminate the translation *gene* >>> stop codon

non-sense mutation: A mutation that alters a gene so that a non-sense codon is inserted; such a codon is one for which no normal >>> tRNA molecule exists, therefore it does not code for an amino acid; usually non-sense codons cause the termination of translation; several non-sense codons are recognized (e.g., amber, ochre, opal) *gene*

non-sense suppressor: A mutation in >>> tRNA that leads to the insertion of an amino acid at the position of a stop codon and often restores enzyme activity in non-sense mutations *biot*

non-sibling chromatid: A chromatid that derives from the other homologue of the two homologous chromosomes or from non-homologous chromosomes of the complement *cyto*

non-sister chromatid: >>> non-sibling chromatid

non-specific resistance: >>> horizontal resistance

non-synonymous single nucleotide polymorphism: Any >>> SNP that occurs in a non-coding region of a gene and changes the encoded amino acid; it may cause the synthesis of a non-functional protein *biot meth*

non-target genes: Genes in the recurrent parent that are not at the locus of the gene of interest; basically, non-target genes are all those genes in the recurrent parent that the breeder would like to retain in the backcrossing program *meth*

non-till rotation: A method of planting crops that involves no seedbed preparation in the rotation other than opening small areas in the soil for placing seed at the intended depth; moreover, there is no cultivation during crop production, despite chemicals that are used for vegetation control *agr*

non-uniform resistance: >>> resistance

non-virulent: >>> avirulent

nopaline: A rare derivate of an amino acid [(1,3-dicarboxypropyl)-L-arginine]; it is produced in some crown galls of plants; the controlling genes are part of the T-DNA of Ti plasmids *gene phys* >>> Figure 62 >>> www.pflanzengallen.de

norm: The description of the characteristics of a variety as supplied by the breeder; also known as a variety description *seed*

norm of reaction: >>> range of reaction

normal curve: >>> normal distribution

normal distribution: The most commonly used probability distribution in statistics; in nature, a vast number of continuous distributions are normally distributed; a continuous symmetrical bell-shaped frequency distribution *stat*

normalized Latin square: A >>> Latin square with symbol set 1, …, n, in which the first row and column contain the symbols in their usual order; any Latin square can be normalized by permuting rows and columns *stat meth* >>> design of experiment

normalizing selection: The removal of genes and/or alleles that produce deviations from the normal phenotype of a population *meth* >>> Table 37

North Carolina design: There are three designs denoted by NCI, NCII, and NCII; they are often used in cross-pollinated crops and to study broad-based populations; in self-pollinated crops many inbred lines are considered to represent a large population; by using NCI, two inbred lines are crossed to produce F2, and then some individuals are randomly selected from the F2 population as males to intermate with other randomly selected females; the offspring derived will be used in genetic studies; by using NCII, a certain number of parental lines are divided into two groups, one as males and the other as females, to produce crosses for all possible combinations; by using NCIII, a certain number of individuals are selected from an F2 population to backcross with two parents, P1 and P2 *bree meth stat*

Northern blot: >>> Northern blotting

Northern blotting: A procedure analogous to >>> SOUTHERN blotting, but transferring RNA, instead of DNA, from a gel to a carrier (like nitrocellulose) and then hybridizing to a radioactively labeled DNA probe complementary to the desired sequence, and visualized by autoradiography; the technique can therefore be used to locate and identify an RNA fragment containing a specific sequence *gene meth* >>> Figure 48

novel foods: Products that have never been used as a food; foods which result from a process that has not previously been used for food, or foods that have been modified by genetic manipulation *biot*

novel plant: >>> intergeneric cross

novel variety: Refers to seed, transplants, or plants showing sufficient distinctness in the sense that the variety clearly differs by one or more identifiable morphological, physiological, or other characteristics *seed*

noxious: Injurious (e.g., a noxious weed is one that crowds out desirable crops, robs them of plant food and moisture, and causes extra labor in cultivation) *agr*

noxious weeds: Undesirable plants that infest either land or water resources and cause physical and economic damage *agr* >>> noxious

NPK analysis: The analysis of the ratio of nitrogen (N), phosphorus (P), and potassium (K) in an organic soil amendment *agr phys*

N-serve (nitrapyrin): A proven substance that selectively inhibits one of the bacteria responsible for nitrification; when added to an NH_4- nitrogen material, it delays its conversion to NO_3- for several weeks; it is most effective when mixed with anhydrous ammonia; this delayed nitrification protects the fertilizer from losses due to denitrification and leaching in seasons when excessive rainfall occurs during the period of inhibition *agr phys*

nsSNP: >>> non-synonymous single nucleotide polymorphism

N-type: In sugarbeet breeding, varieties with normal sugar content and normal yielding capacity (N = normal) *seed*

nucellar embryony: Parthenogenesis in which the embryo arises directly from the nucellus *bot*

nucellar pillar (~ projection): A shaft of tissue that runs along the top of the crease on the inner side of the pigment strand *bot*

nucellus: The mass of tissue in the ovule of a plant that contains the embryo sac; size and shape can be diagnostic for species *bot*

nuclear division: The division of the cell nucleus by mitosis, meiosis, or amitosis *cyto*

nuclear division cycle: The sequence of stages of the division of the nucleus *cyto*

nuclear dye: >>> nuclear stain

nuclear envelope: >>> nuclear membrane

nuclear fragmentation: The degeneration of the nucleus by partition of the nucleus into more or less different parts *cyto*

nuclear gene: A gene that is located on the chromosome of the nucleus *gene*

nuclear male sterility (NMS): Refers to male sterility that is determined by nuclear genes; as opposed to cytoplasmic male sterility *gene seed*

nuclear membrane: The structure that separates the nucleus of eukaryotic cells from the cytoplasm; it comprises two unit membranes, each 10 nm thick, separated by a perinuclear space of 10–40 nm; at intervals, the two membranes are fused around the edges of circular pores, which allows for the selective passage of materials into and out of the nucleus *bot*

nuclear pore: Allows the selective passage of materials into and out of the nucleus *bot* >>> nuclear membrane

nuclear sap: A non-staining or slightly stainable liquid or semiliquid substance of the interphase nucleus *cyto*

nuclear stain: Usually, basic dyes such as >>> methylene blue, methylgreen, crystal violet, green-puron, or azure B bromide that bind preferentially to the cell nuclei (i.e., chromosomes and nucleolus) with slight cytoplasmic effect *micr meth cyto*

nuclear staining: >>> nuclear stain

nuclear transplantation: The transfer of a nucleus into the cytoplasm of another cell *biot*

nuclease: Any enzyme that degrades DNA and RNA *gene* >>> endonuclease >>> exonuclease >>> Table 39

nucleic acid: Nucleotide polymers with high relative molecular mass, produced by living cells and found in both the nucleus and cytoplasm; they occur in two forms (DNA and RNA) and may be double- or single-stranded; DNA embodies the genetic code of a cell or organelle, while various forms of RNA function in the transcriptional and translational aspects of protein synthesis *gene*

nucleolar chromosome: The chromosome that carries the nucleolus organizer (region); it may also be called a satellite chromosome *cyto*

nucleolar constriction: That region of a chromosome that carries the >>> nucleolus organizer; besides the centromeric region, it is observed as a secondary constriction along particular chromosomes; it is not stained by the standard chromosome techniques; by the nucleolar constriction, the chromosome arm appears divided into two parts; the terminal part is called "satellite", the whole chromosome is called the "satellite chromosome" *cyto*

nucleolar dominance: >>> amphiplasty

nucleolar organizer: A region on a loop of DNA emanating from a chromosome in the nucleolus and around which rRNA genes are clustered; it is involved in the regulation of chromosome behavior *cyto*

nucleolar zone (*syn* **nucleolus organizer region, NOR**): A chromosome region that is associated with the formation of the nucleolus during telophase *cyto* >>> nucleolar constriction

nucleolin: One of the non-ribosomal proteins; it is considered to play a key role in the regulation of rDNA transcription, perisomal synthesis, ribosomal assembly, and maturation; it influences the nucleolar chromatin structure through its interaction with DNA and histones; it is involved in cytoplasmic-nucleolar transport of preribosomal particles *phys* >>> ribonucleoprotein

nucleolus: A clearly defined, often spherical area of the eukaryotic nucleus, composed of densely packed fibrils and granules; its composition is similar to that of chromatin, except that it is very rich in RNA and protein; it is the site of the synthesis of ribosomal RNA; the assembly of ribosomes starts in the nucleolus but is completed in the cytoplasm *cyto*

nucleolus organizer: >>> nucleolar organizer

nucleolus organizer region (NOR): >>> nucleolar zone

nucleoprotein: A conjugated protein, composed of a histone or protamine bound to a nucleic acid as the non-protein portion *cyto*

nucleosid(e): A glycoside that is composed of ribose or deoxyribose sugar bound to a purine or pyrimidine base *chem gene*

nucleosome: >>> karyosome

nucleosome mapping: A technique quite similar to DNA footprinting that employs micrococcal nuclease digestion to localize >>> nucleosomes on a stretch of >>> DNA; micrococcal nuclease (mNase) is an enzyme that makes double strand cuts in DNA between nucleosomes when incubated with cell nuclei or permeabilized cells; full digestion with mNase gives mostly mononucleosome-size DNA fragments, while partial digests result in single cuts between nucleosomes and fragments of varying size; generally a range of mNase concentrations are used to produce a collection of cleavage products which can then be run on a gel and probed via >>> SOUTHERN blot to determine nucleosome position; the technique does not give precise position information about nucleosomes but instead indicates the area of the DNA where a nucleosome probably binds *biot meth*

nucleotide: A nucleoside that is bound to a phosphate group through one of the hydroxyl groups of the sugar; it is the unit structure of nucleic acids *chem gene*

nucleotide binding site: A conserved domain of disease-resistance genes *biot*

nucleotide binding site- directed profiling: A molecular marker technique that targets variation at disease-resistance genes and analogues *biot meth*

nucleotide pair: A pair of nucleotides joined by a hydrogen bond that is present on complementary strands of DNA *gene*

nucleotide replacement site: Any position in a >>> codon where a >>> point mutation has occurred *gene*

nucleotide sequence: The order of nucleotides along a DNA or RNA strand *gene*

nucleotide substitution: The exchange of one nucleotide in a DNA molecule for another one; it is neutral if the genetic code is not changed but has massive consequences if the genetic code is altered; the latter may result in the synthesis of a non-functional protein *gene*

nucleotide synthesis: >>> biotechnology

nucleus: The double-membrane-bound organelle containing the chromosomes that is found in most non-dividing eukaryotic cells; it disappears temporarily during cell division; within the nucleus several independent approaches point to the compartmentalization of particular activities such as transcription, RNA processing, and replication; chromosomes are revealed to occupy defined domains and to represent highly differentiated structures; the numerous activities that use DNA and RNA as a template occur with a defined spatial and temporal relationship (e.g., compartmentalization of nuclear functions is particularly seen with replication); DNA moves through a fixed architecture containing the molecular machines directing replication *cyto* >>> Figure 25

nucleus breeding: Breeding scheme where populations in the breeding cycle are divided into intensively managed nuclei with top-ranking genotypes, and the less intensively managed, genetically less advanced, main population *meth fore* >>> Figure 51 >>> Table 37

null allele: A "silent" allele without an obvious expression *gene*

null hypothesis: A hypothesis that there is no discrepancy between observation and expectation based on some sets of postulates *stat*

null hypothesis test: The standard hypothesis used in testing the statistical significance of the difference between the means of samples drawn from two populations; the null hypothesis states that there is no difference between the populations from which the samples are drawn; one then determines the probability that one will find a difference equal to or greater than the one actually observed; if this probability is 0.05 or less, the null hypothesis is rejected and the difference is said to be statistically significant *stat*

nullihaploid: A cell or individual that possesses a haploid chromosome set plus a missing single chromosome *cyto* >>> Figure 37

nulliplex type: The condition in which a >>> polyploid carries a recessive gene at a particular locus in all homologues; simplex denotes that the dominant gene is represented once, duplex twice, triplex three times, quadruplex four times, etc. *gene* >>> autotetraploid >>> Table 3

nullisome: A plant lacking both members of one specific pair of chromosomes *cyto* >>> aneuploid >>> Figure 37

nullisomic: >>> aneuploid >>> nullisome

nullisomic analysis: In nullisomic analysis, observations are made for phenotypic or other differences between the nullisomic for each chromosome and the disomic condition within the same variety; the method is applied for localization of genes within a given genome; the method can only be used in polyploids, while diploids commonly do not tolerate the loss of both homologous chromosomes *gene meth* >>> aneuploid >>> Figure 37 >>> monosomic analysis

nulli(somic)-tetrasomic line: >>> aneuploid >>> nullitetrasomic

nullisomy: >>> aneuploid >>> nullisome

nullitetrasomic: A cell or individual, usually an >>> allopolyploid, that lacks one pair of chromosomes, which is partially compensated for by a tetrasomic (four-fold dosage) of another, usually homoeologous, chromosome; a whole series of nulli-tetrasomics has been produced in hexaploid wheat, and successfully used in numerous genetic and molecular studies *cyto* >>> aneuploid >>> Figure 37

null mutation: A mutation which eliminates all enzymatic activity; usually deletion mutations *biot*

numeric aberration: The variation of the number of genomes or chromosomes, for example, ploidy variation, aneuploids (nullisomics, monosomics, trisomics, tetrasomics, etc.), substitutions, or additions *cyto* >>> Figure 37

numerical aperture (NA): Relationship between the objective light collection angle (a) and the refractive index (h) of the medium between the objective and specimen; it is calculated using a mathematical formula devised by E. ABBE for the direct comparison of the resolving power of dry and all types of immersion objectives ($NA = h \times \sin q$), where h = the refractive index of the medium between the front lens of the objective and the cover slip [when a ray of light passes from a rare medium (air) to a denser medium (oil) it is bent and refracted; air has a refractive index of 1; immersion oil has a refractive index of 1.5] and q = the aperture angle defined by the optical axis and the outermost rays still covered by the objective; thus, the numerical aperture is the sine of half the angular aperture of an objective lens *micr*

numerical taxonomy: The classification of related organisms using a multitude of characteristics, each one of which is given equal weight; the degree of similarity between them is calculated using a computer, which treats the data collected for all characters and determines the similarities taking all possible pairs *tax*

numeric constancy of chromosomes: The constant inheritance of the same number of chromosomes from generation to generation, which is facilitated by the mitotic and meiotic mechanisms *evol*

nurse crop: >>> companion crop

nurse culture: A culture technique or the callus upon which a filter paper is placed separating single cells from the callus in the paper raft technique; the callus (i.e., the nurse tissue) releases growth factors and nutrients that induce growth in the single cells supported by the filter paper and sharing the communal environment *biot*

nurse plant: A plant, often a shrub, which provides shelter and shade to developing seedlings of other plants *eco*

nursery: A place where plants are grown for sale, transplanting, or experimentation, e.g., young trees or other plants are raised, either for propagation or for testing and observations *agr*

nursery effect: Gene expression can be influenced and regulated by the plants themselves depending on various stress situations, such as drought, diseases, or nutrition; in poplar, it was demonstrated that gene expression may even differ between twins or clones when planted under different growing conditions; plants from nurseries may develop differently depending on their origin as well breeding and growing conditions; this is sometimes called the "nursery effect"; therefore, chances for pre-mature stimulation of plants are given, particularly in clonal propagation *meth biot* >>> gene expression

nurseryman's tape: >>> grafting tape

nut: A dry, indehiscent, woody fruit, e.g., in hazelnut, beech, oak acorn *bot*

nutation: The turning of a plant or plant organ toward light *bot*

nutlet: A little nut (e.g., in strawberry) *bot* >>> nut

nutraceutical: A neologism combining nutritional with pharmaceutical and meaning a food product that has been determined to have a specific physiological benefit for human health; in general, a product isolated or purified from food that is generally sold in medicinal forms not usually associated with food; it has been demonstrated to have a physiological benefit or provide protection against chronic disease; the term has no regulatory definition and is primarily used in promotion and marketing *biot* >>> (SCHLEGEL 2007)

nutrient: A nutritive substance or ingredient, such as major and minor mineral elements, necessary for plant growth and development as well as the organic addenda such as sugars, vitamins, amino acids, and others employed in plant tissue culture media *phys biot* >>> Table 33

nutrient-enhanced varieties (crops): Plants that have been modified to possess novel traits that make those plants more economically valuable for nutritional uses (e.g., higher than normal protein content in feedgrains; high zinc content or high glutenin in wheat; high amylose, high lysine, high methionine, high oil in maize; high phytase in maize and soybeans;

high oleic oil, high stearate, high sucrose in soybeans) *biot seed* >>> Table 38 >>> (SCHLEGEL 2007) >>> zinc >>> www.zinc.org/

nutrient film technique: A hydroponic technique used to grow plants; it delivers a film of water or nutrient solution either continuously or through on–off cycles (e.g., on 8 min and off 7 min) *hort meth*

nutrient gradient: A diffusion gradient of nutrients and gases that develops in tissues where only a portion of the tissue is in contact with the medium; gradients are less likely to form in liquid media than in callus cultures *biot*

nutriomics: A new frontier in plant biology that can provide innovative solutions for improving plant nutrient efficiency, thus increasing crop productivity through genetic and molecular approaches *biot*

nutritional mutant: >>> nutrient-enhanced varieties

nutritive substance: >>> nutrient

nutritive value index: A measure for daily digestible amount of forage per unit of metabolic body size relative to a standard forage; it is used as a selection criterion in forage crop breeding *meth agr* >>> Table 33

nyctanthous: Flowering by night *bot*

nymph: The immature stage of certain insects whose growing young resemble the parents in body form *zoo phyt*

O

obconic: A solid figure with the shape of a cone; widest at the apex and pointed at the base *bot*

objective: The lens or combination of lenses that first receives the rays from an observed object, forming its image in an optical device, as a microscope *micr*

objective lens: The lower lens in a microscope that is closest to what is being looked at *micr*

oblate: Pollen grain with a polar axis shorter than the equatorial diameter, e.g., in >>> *Tilia cordata bot* >>> pollen

obligate: Restricted to a particular way of life *phyt*

obligate apomixis: Seed apomixis with maternal offspring of 100% *bot*

obligate parasite: An organism that cannot live in the absence of its host *phyt*

obligate weed: A weed never found in the wild stage, but growing only in association with man *phyt*

oblique: Unequal sides; slanting *bot*

oblong: Object with round ends and parallel margins, three times longer than wide *bot*

obpyriform: Pear-shaped, with the wide end toward the apex *bot*

observation: A data collection strategy in which the activities of subjects are visually examined; the observer attempts to keep his presence from interfering in or influencing any behaviors *stat meth*

observation tube: >>> eyepiece

obsolete variety: A plant variety that is no longer grown commercially *seed*

obtuse: Pointed with an angle greater than 90°; a broad pointed apex or base *bot*

ocDNA: An open circular DNA molecule that has at least one nick; it cannot be supercoiled and has the same density as linear DNA in CsCl/EB density gradients *biot*

ochre codon: UAA stop codon gene *biot* >>> non-sense mutation

ocimene: A volatile terpene produced by alfalfa flowers, which give them aroma *phys*

octopine A rare derivate of an amino acid; it is produced in some crown galls of plants; the controlling genes are part of the T-DNA of Ti plasmids *chem phys* >>> Figure 62 >>> www .pflanzengallen.de

octoploid: Having eight chromosome sets of identical or different complements *cyto*

octovalent: Chromosome configuration consisting of eight chromosomes *cyto*

ocular: >>> eyepiece

OECD: >>> Organization for Economic Cooperation and Development

of gardens: Term used after the Latin name of a plant to denote that the name is commonly but incorrectly used *bot*

off-grade: Postharvest removal of pathogen-infested or damaged fruit, seeds, or plants by screening procedures; the culled or off-graded material can later be individually analyzed or discarded *meth*

office action: In patenting biological material, a formal response by a patent examiner regarding a patent application or amendment *adm* >>> Intellectual Property Rights

offset: A young plant produced by the parent, usually as its base (offshoot) or a small bulb at the base of a mother bulb *bot*

offspring: >>> progeny

off-target effect due to genome editing: Unintended cleavage and mutations at untargeted genomic sites with similar but not identical sequences compared to the target site *meth biot* >>> genome-editing methods

off-type: An individual differing from the population norm in morphological or other traits; the term also includes escapes and contaminants (e.g., seeds that do not conform to the characteristics of a variety, uncontrolled self-pollination during production of hybrid seed, segregates from plants, etc.) *gene agr*

Ogura (cytoplasmic male sterility and fertility-restoring [cms/Rf] system): One of the most promising hybrid systems in *Brassica* species; the Ogura cms has been introduced earlier into spring turnip rape (*Brassica rapa*) and the sterility has been stable; to enable the production of F1 hybrids in turnip rape, the Kosena fertility-restoring gene (*Rfk1*), a homologue of the Ogura fertility-restoring gene (*Rfo*), was transferred from spring oilseed rape (*Brassica napus*) into spring turnip rape by the traditional backcross method *breed*

Ohio method: >>> ear-to-row selection

oidium: The generic name given to the conidial stage of all >>> powdery mildews, the Erysiphales; the conidia are consistently similar throughout this family, being unbranched and producing chains of hyaline, i.e., oval conidia *phyt* >>> Table 51

oil additives: An additive made either synthetically or extracted from seeds that enhances herbicide uptake into the plant, for example, crop oils, methylated seed oils, or crop oil concentrates *phyt meth*

oil crops: Plants that are grown for oil or oil-like products; main oil crops are castor, peanut, rapeseed, safflower, sesame, soybean, sunflower, crambe, niger, jojoba, and poppy *agr*

oil immersion: The oil that is placed between the lens of a microscope and the coverslip above a microscopical preparation *micr*

oil-immersion objective: The objective lens system used for highest resolution with the light microscope; the space between the coverslip over the object to be examined and the lens is filled with a drop of oil of the same refractive index as the glass *micr*

oil legumes: >>> oil crops

oil poppy: >>> oil crops

oilseed: >>> oil crops

Okayama–Berg procedure: A method of cDNA cloning using specialized vectors favoring the generation of full-size clones *biot*

Okazaki fragment: Small (<1 kb), discontinuous strands of DNA produced using the lagging strand as template during DNA synthesis; DNA ligase links the fragments to give rise to a continuous strand *biot*

okra-leaf: A mutant narrow-leaf type in cotton *gene*

old English rose: >>> old garden rose

old garden rose: A hybrid rose which has been in cultivation since before 1867 *hort*

oleic acid: A fatty acid that is naturally present in the fat of animals and also in oils extracted from oilseed plants, for example, soybean or rapeseed; the soybean oil produced from traditional varieties of soybeans tends to contain 24% oleic acid *chem phys*

olein: A colorless to yellowish, oily, water-insoluble liquid, $C_{57}H_{104}O_6$, the triglyceride of oleic acid, present in many vegetable oils *chem phys*

oleosomes: Storage bodies for lipids and/or fats in the seeds of certain plants *bot*

oligocarpous: Bearing less than the typical amount of fruit *bot*

oligogene: A gene that produces a pronounced phenotypic effect on characters that show normal inheritance *gene*

oligogenic: Inheritance due to a small number of genes with discernible effects *gene*

oligogenic resistance: Resistance controlled by one or a few genes *phyt*

oligomer: A protein composed of two or a few identical polypeptide subunits *chem phys*

oligonucleotide (oligos): A small piece of >>> ssDNA or >>> ssRNA; oligos are synthesized by chemically linking together a number of specific nucleotides; they are used as synthetic genes and DNA probes or in site-directed mutagenesis *biot*

oligonucleotide-directed mutagenesis (ODM): >>> site-specific mutagenesis

oligonucleotide ligation assay (OLA): A PCR-based method for >>> SNP typing; it is a ligase-mediated gene detection system which uses exact 3′ matching of a primer to one of the SNP allele; if this happens, the other labeled oligonucleotide that binds to the nucleotide immediately next to the SNP on the other side would be joined to the primer by ligase; the resulting sample can then be tested for the presence of the label *biot meth*

oligonucleotide probe: A DNA fragment consisting of nucleotides that is used to detect the presence of its complementary sequence in a DNA sample by hybridization testing *biot meth*

oligopeptide: A small protein composed of 5–20 amino acids *chem phys*

oligos: >>> oligonucleotide

oligospermous: Showing only few seeds *bot*

O-line: A structure in young embryo, which is the lower boundary of shoot apical tissues *biot*

omnipotency: >>> totipotency

omnivorous: Of parasites, or attacking a number of different hosts *phyt*

once blooming: Refers to plants or varieties (e.g., in roses) that bloom once a year *hort*

one-gene-one-enzyme (polypeptide) hypothesis: The hypothesis that a large class of structural genes exists in which each gene encodes a single polypeptide that may function either independently or as a subunit of a more complex protein; originally it was thought that each gene encoded the whole of a single enzyme, but it has since been found that some enzymes and other proteins derive from more than one polypeptide and hence from more than one gene *gene bio*

one-point crossover: In genetic algorithms, a breeding technique using one randomly chosen point, interchanging the portions of the two breeding individuals to the right of that point *stat*

one-way analysis of variance: The one-way analysis of variance allows us to compare several groups of observations, all of which are independent but possibly with a different mean for each group; a test of great importance is whether or not all the means are equal *stat* >>> two-way analysis of variance

ontogenesis The course of growth and development of an individual from zygote formation to maturity *phys* >>> www.plantontology.org

ontogeny: >>> ontogenesis

ontology: A branch of metaphysics dealing with the nature of being or a set of concepts and categories in a subject area or domain, e.g., in plant genetics, that shows their properties and the relations between them *meth* >>> PLANTEOME >>> plant ontology

oogonium: A primordial germ cell that gives rise, by mitosis, to oocytes, from which the ovum and polar bodies develop by meiosis *bot*

opal: A UAG stop codon *biot* >>> non-sense mutation

opaque: Partially pervious to light *bot*

opaque (−2) maize (mutant): A mutant form that produces proteins rich in lysine and higher in content of calcium, magnesium, iron, zinc, and manganese *gene* >>> Table 38

opacity: The measure of how opaque or see-through a plastic film is; certain pigments added to the polyethylene resin will make the plastic more opaque or less see-through *hort meth*

open continuous culture: A cell culture in which inflow of fresh medium is balanced by outflow of a corresponding volume of culture *biot* >>> closed continuous culture >>> continuous culture

open pollination: Natural, cross, or random pollination; a free gene flow *bot*

open storage: Storage with free access to normal atmospheric conditions *seed*

open-pollinated crossing group: >>> crossing groups

open-reading frame (ORF): The mRNA region between the start and stop codon *gene*

operator (gene): A region of DNA at one end of an operon that acts as the binding site for a specific repressor protein and so controls the functioning of adjacent cistrons *gene*

operon: A set of adjacent structural genes whose mRNA is synthesized in one piece, together with the adjacent regulatory genes that affect the transcription of the structural genes; it is under the control of an operator gene, lying at one end of it *gene*

opine: A group of unusual amino acid derivatives produced by plant cells transformed by Agrobacterium in order to feed the pathogen; opines include agropines, nopalines, and octopines and are characteristic for individual strains of *Agrobacterium biot*

opposite: Applied to the leaf arrangement in which leaves arise in pairs, one pair at each node *bot*

optic chiasma: It refers to a visible chiasma on meiotic chromosomes seen through a microscope *cyto*

optical density (OD): A logarithmic unit of transmission; OD = −logT (transmission); for example, a change of the optical density from 1 to 2 represents a tenfold increase in absorption *micr*

optical mapping: In general, a technique for constructing ordered, genome-wide, high-resolution restriction maps from single, stained molecules of DNA, called "optical maps"; by mapping the location of restriction enzyme sites along the unknown DNA of an organism, the spectrum of resulting DNA fragments collectively serve as a unique "fingerprint" or "barcode" for that sequence; originally developed by David C. SCHWARTZ in the 1990s, this method has since been integral to the assembly process of many large-scale sequencing projects for both microbial and eukaryotic genomes; the technology enables whole-genome analysis, which involves the capture of individual DNA molecules, obtained directly from genomic DNA, followed by digestion *in situ* by selected restriction endonucleases; the resulting fragments are then visualized directly to produce detailed optical restriction maps; this methodology allows patterns of sequence variation to be detected across entire genomes, without the need for DNA amplification, and, unlike other genome-wide scanning methods, provides detailed haplotype information by analyzing individual DNA molecules *meth biot cyto* >>> mapping >>> Table 39

optimal sampling strategy: A sampling strategy that ensures that the genetic diversity of a species is represented in the samples *stat meth*

opuntia: Within the subtribe Opuntioideae there are several species used as crop and horticultural plants; edible fruits and fleshy parts of the plant; special use for production of the stain "carmine red" by the ectoparasite >>> cochenille (*Dactylopius coccus*) *bot*

orangery: A sheltered place, especially a greenhouse, used for the cultivation of orange trees and other tropical or subtropical plants in cool climates *meth hort*

orcein: A dye used in cytology; it is dissolved in acetic acid and used for staining of squash preparations of chromosomes *cyto meth prep*

orchard: An area of land devoted to the cultivation of (fruit or nut) trees *hort*

ordinal scale: A scale for scoring quantitative data using a series of predefined intervals arranged in a logical sequence (e.g., a typical ordinal scale may involve responses of "very good", "good," "satisfactory", "poor," or "very poor") *stat*

ordinary least squares: The estimator that minimizes the sum of squared residuals *stat*

organ asymmetry: In many plants the left and right halves of their organs have distinct shapes, e.g., the leaves of *Begonia*, *Tilia* (lime tree), and *Ulmus* (elm), and petals of *Anhirrhinum* (snapdragon) or *Pisum* (pea) flowers; they can occur in two mirror-image forms, left handed and right-handed; in many cases these two forms occur in equal numbers on the plant, either being located opposite each other or alternating along the stem; asymmetry of each organ traces back to a meristem with a single plane of symmetry (bilateral symmetry), such that mirror-image organs arise from opposite halves of the meristem *bot*

organ culture: The growth in aseptic culture of plant organs, such as roots or shoots, beginning with organ primordia or segments and maintaining the characteristics of the organ *biot*

organelle: Within a cell, a persistent structure that has a specialized function; mostly separated from the rest of the cell by selective membranes *bot*

organic agriculture: A concept and practice of agricultural production that focuses on production without the use of synthetic pesticides *eco agr*

organic conservation breeding: Varieties that are of conventional origin are conserved and propagated by cultivation under organic standards; relevant to older varieties which should be conserved as a gene pool for future breeding efforts and to varieties of high importance *meth*

organic cotton: Organically grown cotton uses crop rotation, beneficial insects, compost, and other farming methods in place of chemical fertilizers and intensive farming techniques *agr meth*

organic soil: A soil that is composed predominantly of organic matter; it usually refers to peat *agr* >>> Table 44

Organization for Economic Cooperation and Development (OECD): An international agency which, among other things, has developed specifications, procedures, and standards for international seed certification among member countries *seed org* >>> www.oecd.org/home/0,2987,en_2649_201185_1_1_1_1_1,00.html

organogenesis: The initiation and growth of an organ from cells or tissue *phys* >>> Table 47

organogenetic: Refers to cells or tissue able to form organs *biot*

organogenic: >>> organogenetic

organo-metallic complex: A general term referring to compounds in which metal ions have been complexed by organic compounds, for example, chelates; while all chelates are complexes, all complexes are not chelates *agr chem* >>> siderophore

oralmat: An extract of the >>> rye plant; it contains many beneficial, naturally occurring compounds such as beta-1,3-glucan, matairesinol, genistein, squalene, and CoQ10; oralmat is an immune modulator and supports healthy respiratory function of humans by strengthening defenses against toxins in the lungs; it is also a blood vessel modulator and causes constriction or dilation as needed *foo*

ornamental plant: A plant that is grown for visual display; it describes any plant cultivated for decorative purposes; plants classified as ornamental are generally contrasted with those grown as food sources, though there are other, non-ornamental uses for plants *hort*

ornamentals: >>> ornamental plant

ornithogamy: >>> ornithophily

ornithophilous: >>> ornithophily

ornithophily: Pollination by birds *bot*

Orobanchaceae: A family of totally parasitic herbs; often, specific root parasites of particular angiosperms and/or crop plants (e.g., sunflower) *bot phyt*

orphan crops: Those crops which are typically not traded internationally but which can play an important role in regional food security; many of these crops have received little attention from crop breeders or other research institutions *agr* >>> minor crop

orphan gene: A gene identified by sequencing; its function is unknown *gene* >>> orphanized

orphanized: Orphan 7-TM receptors (orphan GPCRs) are receptors for which there are no defined ligands; traditionally, receptors are characterized by specific ligands and stimuli; the search for ligands of orphan receptors is akin to reverse-engineering this process, since the receptors have been identified but they are responding to as-yet unidentified endogenous ligands; there are approximately 150 orphan 7-TM receptors, subdivided into three classes: A, B, and C; orphan G-protein coupled receptors are usually identified by number (e.g., GPR35, GPR139) and as they are "de-orphanized" (by discovery of ligand), they sometimes maintain this prefix (e.g., GPR55); screening of putative ligands, reverse the prediction of candidate ligands using structural similarities: putative ligands for some of the orphan receptors include kynurenic acid (GPR35) and lysophosphatidic acid (GPR87) *biot*

ortet: The original plant from which a clone is started through rooted cuttings, grafting, tissue culture, or other means of vegetative propagation (e.g., the original plus tree used to start a grafted clone for inclusion in a seed orchard) *meth hort fore* >>> cutting >>> Figure 50

orthodox seed: Seed that can be dried and stored for long periods at reduced temperatures and under low humidity; there is variation in the ability to withstand drying and storage with some seeds being more sensitive than others; some seeds are considered intermediate in their storage capability while others are fully orthodox; notable example of a long-lived orthodox seed which survived accidental storage followed by controlled germination is the case of the 2,000-year-old Judean date palm (*Phoenix dactylifera*) seed which successfully sprouted in 2005 (HANSEN 2008) *seed* >>> recalcitrant seed

orthogonal design: A design where the various factors in the design are orthogonal in the sense of partitions *stat*

orthogonality: Considered to be one of the most important features for design of experiment *stat*

orthologous genes: Homologous genes that have become differentiated in different species derived from a common ancestral species; as opposed to paralogous genes *gene*

ortholome: A set of orthologous DNA sequences between two or more species *biot*

orthoploid: >>> euploid

oryzalin: An agent that is efficient for chromosome doubling of haploid apple shoots *in vitro*; also shoot tips of diploid roses can be treated *in vitro* at concentrations of 5 to 15 µM; tetraploid shoots can be obtained in highest frequencies after exposure to 5 µM oryzalin for 14 days; thin (1 mm) nodal sections can be treated with 5 µM oryzalin for only 1 day *meth cyto*

oryzinin: >>> glutenin

osmic acid (OsO$_4$; f.w. 254.20): A fixing agent commonly used to prepare tissue samples for electron microscopy *micr prep*

osmium tetroxide: >>> osmic acid

osmolality: An important parameter in the quality control of tissue culture media or in physiological experiments; it is defined as the concentration of all solutes in a given weight of water and is expressed as units of osmolality (milliosmoles of solute per kilogram of water, mOsm/kg H$_2$O) *phys* >>> osmolarity

osmolarity: The total molar concentration of the solutes affecting the osmotic potential of a solution or nutrient medium; it is expressed as milliosmoles of solute per liter of water, mOsm/l H$_2$O; total solute concentration can be estimated by adding the concentrations of all individual ions and other solutes *phys* >>> osmolality

osmolyte: Osmolytes are osmolytic active, neutral organic compounds, such as sugars (polyols), certain amino acids, and quaternary ammonium compounds; >>> proline is the most widely distributed compatible osmolyte; there is a strong correlation between increased cellular proline levels and the capacity to survive both water deficit and the effects of high environmental salinity *phys* >>> Table 38

osmophore: Floral organ, often formed from the petal, which is long, narrow, drooping, often dark, and usually foul-smelling, frequently as part of a myophyllous syndrome *bot*

osmoprotection: A plant survival strategy brought about by the synthesis of groups of proteins (dehydrins) in response to water stress; most tissues produce these proteins which protect cellular organelles and membranes from the damaging effects of water loss (>>> desiccation), a notable exception being the cells of the starchy endosperm which undergo programmed cell death *phys*

osmosis: The net movement of water or of another solvent from a region of low solute concentration to one of higher concentration through a semipermeable membrane *phys*

osmoticum: Particles that cause osmosis *phys*

osmunda fiber: Chopped, tried roots of the fern genus Osmunda, often used in orchid cultivation *hort meth*

other crop seed: One of the four components of a purity test; the total percentage (by weight) of seed of all crop species, each comprising less than 5% of the seed lot *seed*

O-type: A maintainer plant in sugarbeet breeding; it carries the same sterility genes as the male sterile plants but has the normal cytoplasm – (N)xxzz; this genotype exists at low frequencies (3–5%) in most sugarbeet populations; it can be identified only by test-crossing prospective O-types with CMS plants; if all the offspring from a test cross are male sterile, the test-crossed pollinator plant is of the O-type genotype; by repeated selfing of an identified O-type, and simultaneous repeated backcrossing to a CMS line, inbred O-type lines and their equivalent inbred CMS lines can be developed *seed*

ounce (oz): Equals 31.1030 g

outbreeding: The crossing of plants that are not closely related genetically, in contrast to inbreeding, in which the individuals are closely related *bot*

outclassed: It refers to a crop variety that is taken away from registration *seed*

outcrossing: Cross-pollination between plants of different genotypes *bot gene*; in biotechnology, the transfer of a given gene or genes (e.g., one synthesized by humans and inserted into a plant via genetic engineering) from a domesticated organism (e.g., crop plant) to wild type (relative of crop) *biot eco*

outgrades (in potato): Outgrades are tubers considered unmarketable because of size, disease, greening, second growth, or slug or mechanical damage *seed*

outgroup: Any group, used to root a phylogenetic tree in a >>> cladistic analysis, which is not a member of the taxon group being studied *bot biot meth*

outgrowing (growing out): >>> sprouting >>> dormancy

outlier: An individual that occurs naturally some distance away from the principal area in which its population is found; they are anomalous values in the data and can be due to recording errors, which may be correctable, or they may be due to the sample not being entirely from the same population *stat*

outplant: A seedling, transplant, or cutting ready to be established on a certain site *agr fore hort*

output trait: A genetic character that alters the quality of the crop product itself, e.g., by altering its starch, protein, oil, or vitamin composition *gene* >>> input trait

ovary: The part of the flower that develops into the grain in grasses; the ovary has one or more ovules, each containing an embryo sac *bot* >>> Figure 35 >>> Tables 8, 47

ovary wall: >>> pericarp

ovate: Egg-shaped; having an outline like that of an egg, with the broader end basal *bot* >>> ovoid

overall resistance: Resistance to disease expressed at all plant growth stages *phyt*

overdominance (*syn* superdominance): After HULL (1946, 1952) the phenomenon in which the character of the heterozygotes is expressed more markedly in the phenotype than in that of either homozygote; usually the heterozygote is fitter than the two homozygotes; this can give rise to monohybrid heterosis when the hybrid vigor is obtained by crossing parents differing in a single specified pair of allelic genes *gene* >>> heterosis >>> Tables 20, 21, 36

overhang: 3′ and 5′ ssDNA overhangs of dsDNA; overhangs may also be called extensions or >>> sticky ends *biot* >>> cohesive ends

overlapping code: >>> overlapping DNA (segments)

overlapping DNA (segments): A special type of gene organization; one DNA sequence may code for different proteins; it is performed by two open reading frames, which subsequently act *gene*

overlapping inversion: A part of an inverted segment being inverted a second time, together with an ancient segment that was not included in the first inversion segment *cyto*

overpotting: A cultural problem which describes a plant being grown in a pot that is too large; when overpotted, a plant will expend the greatest part of its energy growing roots, thus reducing the amount of energy which might otherwise go to growing leaves and flowers; in addition, because the larger pot size will hold more water than the plant can absorb, conditions of excessive moisture will leave the plant susceptible to potentially fatal diseases, such as crown rot, Pythium, or root rot *meth hort* >>> Figure 62

overseeding: Seeding into an existing crop stand or turf *agr*

overstocked: The situation in which trees are so closely spaced that they are competing for resources, resulting in less than full-growth potential for individual trees *fore*

overstored seeds: >>> method of overstored seeds

overstory: The canopy in a stand of trees; in contrast to the understory which is low-growing woody or herbaceous vegetation forming a layer beneath the overstory *fore*

oversummering: The survival through the summer and/or to keep alive through summer *agr*

overwintering: The survival through the winter and/or to keep alive through winter *agr*

overyielding: The production of a yield by an intercrop that is larger than the yield produced by planting the component crops in monoculture on an equivalent area of land *agr*

ovoid: Egg-shaped *bot* >>> ovate

ovule: A structure in angiosperms and gymnosperms that, after fertilization, develops into a seed *bot* >>> Table 42

ovule primordium: Meristematic tissue of the ovary wall from which the seeds of angiosperms originate *bot* >>> Table 42, 47

ovum: >>> egg

own-root: Plants growing on their own roots, generally from cuttings or seeds *hort*

oxalate oxidase: An enzyme detected in, for example, barley seedling roots soon after germination, and in the leaves of mature plants, and in response to powdery mildew infection; the enzyme contains manganese; the enzyme shows almost identical structure to the wheat protein germin *phys* >>> Table 51

oxalic acid: A white, crystalline, water-soluble, poisonous acid, $H_2C_2O_4 \cdot 2H_2O$, used for bleaching and as a laboratory reagent *chem phys*

oxidase: An enzyme that catalyzes reactions involving the oxidation of a substrate using molecular oxygen as an electron acceptor chem *phys*

oxidation: A reaction in which atoms or molecules gain oxygen or lose hydrogen or electrons chem phys

oxidative burst: An early defense response against pathogens where 10–14 mol H_2O_2 per cell per second is generated; the early oxidative burst (phase I) is common for compatible and incompatible interactions; the later, but sustained burst (phase II) is associated with a hypersensitive response/reaction *phyt*

2-oxoglutarate-dependent dioxygenase: A nonheme iron-containing oxidoreductase that introduces an atom of molecular oxygen into succinic acid, which is formed from 2-oxoglutaric acid with the release of carbon dioxide; the second oxygen atom is used to oxidize the substrate and is commonly incorporated *phys*

P

P value: Probability value; a decimal fraction showing, for example, the number of times an event will occur in a given number of trials; it was introduced by the English statistician Ronald FISHER in the 1920s; he did not mean it to be a definitive test; he intended it simply as an informal way to judge whether evidence was significant in the old-fashioned sense, i.e., worthy of a second look; the idea was to run an experiment, then see if the results were consistent with what random chance might produce; one would first set up a >>> "null hypothesis" that one wants to disprove, such as there being no correlation or no difference between two groups; next, assuming that this "null hypothesis" was in fact true, calculate the chances of getting results at least as extreme as what was actually observed; this probability was the *P* value; the smaller it was, the greater the likelihood that the "null hypothesis" was false; nevertheless, it is necessary to realize the limits of conventional statistics; one should instead bring into its analysis elements of scientific judgement about the plausibility of a hypothesis and study limitations *stat*

P1: Parental generation; P1 individuals are the parents of the F1 generation *gene meth* >>> parental generation

P1 generation: >>> parental generation

pachnema: >>> pachytene

pachytene: The state of the prophase of first meiosis, when the homologous chromosomes are completely paired; crossing-over occurs between the non-sister chromatids of homologous chromosomes; the recombination that occurs during crossing-over in a heterozygous individual is responsible for part of the segregation observed in the progeny *cyto*

packer: Compacts loose soil to help the seeds to germinate *agr*

paddock: A grazing area that is a subdivision of a grazing management unit, and is enclosed and separated from other areas by a fence or barrier *agr meth*

paddy field: Area of land with surface water (flooded) used for growing rice *agr*

paddy rice: Rice production under flooded conditions, also called lawland rice *agr*

paired spikelet(s): Accessory spikelets within a common spike, caused by deficiency in light and regulated by the transcriptional regulator TB1, which is able to alter arrangement of spikelets in a dosage-dependent manner; it is related to the teosinte branched gene *TB1* and correlated with less tillering *phys*

pairing block: A definite chromosome segment that acts as a functional unit in meiotic chromosome pairing *cyto* >>> zygomere

pairing segment: The segment of chromosomes that pairs and crosses over (pairing segment), and the remaining segments that do not pair are known as differential segments *cyto*

palatability: The characteristics of feed that will affect the intake by animals; it has to be considered in forage crop breeding *agr*

pale (palea, paleae *pl***):** The upper or inner of the two bracts of the floret, which covers the ventral surface of the caryopsis in grasses *bot*

palea: >>> pale

paleontology: The study of the fossil record of past geological periods and of the phylogenetic relationships between ancient and contemporary plant and animal species *bio*

paleopolyploid(s): Ancient polyploids; all their diploid progenitors apparently have become extinct *evol*

palindrome: Adjacent inverted DNA repeats; the identical base sequences are on the opposite strands; long (~130 bp) uninterrupted palindromes occur in eukaryotic DNA but are lethal in bacteria *biot*

palisade cells: Columnar cells of palisade parenchyma; these may be one to three layers deep depending upon species and environment; always contain chloroplasts *bot*

palisade layer: In leaves, a somewhat compacted layer of elongated cells that underlie the upper epidermis with the long axis perpendicular to the leaf surface; in seeds, the term is used interchangeably with Malpighian layer *bot*

palmate: Refers to leaves; a compound with two or more leaflets arising from the top of a stalk or rachis; the main nerves of a leaf may palmate, a situation in which several more or less equally large ones diverge along the blade from an origin at the top of the petiole *bot*

palmitic acid: A white, crystalline, water-insoluble solid, $C_{16}H_{32}O_2$; it is one of the most common saturated fatty acids found in animals and plants; as its name indicates, it is a major component of the oil from palm trees (palm oil and palm kernel oil) *chem phys*

Panama disease (of banana): A very destructive parasite in banana (*Fusarium oxysporum f. sp. cubense*) *phyt*

pan genome: A set of high-quality sequenced diverse genotypes from same species; it describes all genes and genetic variation within a species, i.e., the entire gene set of all strains of a species; it includes genes present in all strains (core genome) and genes present only in some strains (variable or accessory genome); (1) open pan genome – number of genes of the pangenome increases with the number of additionally sequenced strains; typically, species that live in multiple environments of mixed microbial communities have multiple ways of exchanging genetic material, and hence continuously extend their total set of genes; (2) closed pan genome – after some sequenced strains, additional strains do not provide new genes to the species pan genome; a closed pangenome is typical for species that live in isolated niches with limited access to the global microbial gene pool; for those species a small number of sequenced strains already cover the complete pangenome *meth biot* >>> core genome >>> pan genome concept >>> dispensable genome >>> Figure 71

pan genome concept: MORGANTE et al. (2007) found that >>> transposable elements were largely responsible for extensive variation in both intergeneric and local genic content among individuals within a species; it was concluded that a single genome sequence might not reflect the entire genomic complement of a species, and they introduced the concept of the plant pan genome; it includes core genomic features that are common to all individuals in a species and a dispensable genome composed of partially shared or non-shared DNA sequence elements *gene* >>> Figure 71

pangene(s): >>> gemmule(s)

pangenesis: Recapitulation of certain ancestral traits during embryogenesis *evol* >>> gemmule

panicle: The branched inflorescence of, for example, oats and other grasses; in oats, it can be shaped as lag panicle, as stiff or lax panicle *bot*

panicle culture: Aseptic culture of grain panicle segments to induce microspore germination and development *biot meth*

Panicum type: Apospory in which there is no initial polarization in the embryo sac; the spindle at first meiosis lies crosswise at the micropylar end and a second mitosis leads to a mature four-nucleate, monopolar, and unreduced embryo sac *bot* >>> Figures 28, 35

panmictic: A random interbreeding population; the individuals mate at random (e.g., in rye) *gene* >>> panmixis >>> Figure 51

panmictic population: >>> panmictic

panmixia: >>> panmixis

panmixis: Random mating; a mode of sexual reproduction in which male and female gametes encounter each other incidentally (e.g., each mating between two individuals of a population shows the same probability, as in a rye field) *bot*

panogen (seed) treater: A small treatment cup, operating from a rocker arm directly off the seed dump pan and out of a small reservoir, meters one cup of treatment with each dump of the seed pan; the fungicide flows through a tube to the head of the revolving drum seed mixing chamber; it flows in with seed from the dumping pan and is distributed over the seed by the rubbing action of the seed passing through the revolving drum; the desired treating rate is obtained by the size of the treatment cup and adjusting the seed dump weight; treatment cup sizes are designated by treating the rate in ounces and not by actual size, e.g., the 22 cm^3 cup applies 22.5 cm^3 of treatment of 35 l with six dumps of 35 l *seed* >>> seed treater

pantropical: Growing throughout the warmer regions of the world; used especially of the ranges of widespread weeds *agr eco*

paper-piercing test A stress test for seedling vigor utilizing sand covered by filter paper, through which the seedlings must emerge to be considered vigorous *seed* >>> www.plantstress .com

papilla: Accumulation of material at the host plant's cell wall at the site of a penetration attempt by a fungus (Oomycete); in barley, papillae contain callose, proteins, silicon, phenolics, etc. *phyt*

pappus: A tuft of delicate fibers or bristles at the tip of a tiny fruit, such as the feathery structure of the ripe dandelion seed that is easily blown from the head *bot*

paracasein: >>> casein

paracentric: Around the centromere *cyto*

paracentric inversion: >>> pericentric inversion

paraffin section: A section of tissue cut by a >>> microtome after embedding in paraffin wax *micr*

Parafilm: The brand name for a stretchable, waxed adherent used as a glassware closure or for other purposes in scientific work *prep*

parallel evolution: Evolution of roughly similar changes in two or more closely related lineages *evol gene* >>> Figures 68, 69

parallel mutation: A mutation that causes similar phenotypes but in different species or genotypes *gene*

parallel selection: >>> unconscious selection

paralogous (genes): Homologous genes that have arisen through gene duplication and that have evolved in parallel with the same organism; as opposed to orthologous genes *gene*

parameter: A numerical quantity that specifies a population with respect to some characteristics *stat*

paramutable allele: An unstable allele where the phenotypic consequences are enhanced by the presence of a paramutagenic allele in a heterozygote *gene*

paramutagenic allele: An allele possessing the ability to cause paramutation *gene*

paramutation: A mutation in which one allele in the heterozygous condition permanently changes the partner allele *gene*

paramylon: A beta-1,3-glucan reserve; a carbohydrate similar to starch; the chloroplasts found in Euglena contain chlorophyll which aids in the synthesis of carbohydrates to be stored as starch granules and paramylon; the euglenoids have chlorophylls a and b, and they store their photosynthate in an unusual form called paramylon starch, a B-1,3 polymer of glucose; the paramylon is stored in rod-like bodies throughout the cytoplasm *phys*

paranemic coiling: Chromatids are easily separated laterally *cyto*

parapatric: Applied to species whose habitats are separate but adjoining *bot* >>> allopatric >>> sympatric

parapatric speciation: Occurs as a result of two populations diverging in adjacent geographical areas *evol*

paraphyletic: A non-monophyletic group containing some, but not all representatives of a taxon, i.e., an incomplete group of descendants from one common ancestor with one or more descendants missing *bot evo*

parasexual: Genetic systems that achieve genetic recombination by ameiotic exchange mechanisms *bot*

parasite: An organism living in or on a host organism, from which it obtains food and other support *phyt*

parasitic plant: >>> parasite

parasitization *syn* **parasitize** *syn* **parasitized:** >>> parasite

paraveinal mesophyll: Layer of flattened cells between the palisade parenchyma and the spongy mesophyll tissues; it is involved in transport of assimilates between the former layers and the phloem *bot*

paravariation: Modification that is contrary to the idiovariation and mixovariation *gene*

parenchym(a): A tissue composed of specialized plant cells with a system of air spaces running between them; the cells are regarded as the basic cells from which all other cell types have evolved *bot*

parent: Pertaining to an organism, cell, or structure that produces another *gene* >>> P1

parental generation: The generation comprising the immediate parents of the F1 generation; the symbols P2 and P3 may be used to designate grandparental and great-grandparental generations, respectively *gene* >>> P1

parental ranking: Ranking of parents based on estimates of their breeding value from tests of their progeny or clonal copies *meth stat*

parental type: An association of genetic markers, found among the progeny of a cross, that is identical to an association of markers present in a parent *gene*

parent–offspring analysis: The linear regression of the performance of offspring on that of the parents was proposed as a method of estimating heritability *stat*

parent-progeny testing: >>> parent–offspring analysis

parietal (placentation): Arrangement of the ovules around the outer walls of a carpel *bot*

Paris green: A very nasty pesticide containing copper and arsenic, which was widely used until replaced with >>> DDT *phyt*

parsimony: A set of methods that assumes that the simplest solution is the most likely one; it is used to construct >>> cladograms, and assumes that minimizing the number of character state changes on a tree is the best approximation of phylogenetic history *stat meth*

parthenocarp: >>> parthenocarpy

parthenocarpy: Production of fruit without seeds, as in bananas and some grapes *bot*

parthenogenesis: The development of an individual from an egg without fertilization; it occurs in some plants *bot* >>> Figure 28

parthenote: A cell or individual resulting from parthenogenesis *cyto*

partially balanced design: A design is partially balanced with respect to an association scheme if the number of blocks containing two points depends only on which class of the partition contains the given pair of points *stat* >>> design of experiment >>> partially balanced incomplete block design

partially balanced incomplete block design (PBIB): Refers to an experimental design where the residual variance of the difference between the candidates may adopt one out of only two different values *stat meth* >>> design of experiment >>> partially balanced design

partial correlation: Displays the numerical relationship between a single Xi and Y in multiple correlation; in contrast to the simple correlation, the partial correlation accounts for interaction between the Xi variables and illustrates the true relationship between the Xi and Y *stat*

partial diallel: >>> incomplete diallel

partial dominance (*syn* **semidominance**): Incomplete dominance; the production of an intermediate phenotype in individuals heterozygous for the gene concerned; it is generally considered to be a type of incomplete dominance, with the heterozygote resembling one homozygote more than the other *gene* >>> Tables 20, 21

partial Latin square: An n × n array whose cells are either empty or contain a symbol from an alphabet of size n, such that each symbol occurs at most once in each row or column, e.g., a >>> Latin rectangle *stat* >>> design of experiment

partial pedigree: Some ancestors of a particular genotype are known, usually the female parents; partial pedigrees are most common where open-pollinated or polycross seed is used for progeny testing *gene* >>> Figure 51 >>> www.geneflowinc.com/

partial resistance: Resistance which is expressed by the slower development of fewer pustules or lesions, compared with normally susceptible varieties *phyt*

partial seeding: Seeding confined to limited areas (drills, strips, patches, or nests), generally according to a regular spatial pattern *fore*

partial self-incompatibility: Self-incompatibility in some species results in a lower percentage of seed set than occurs with cross-pollination *gene*

participatory plant breeding: Approaches that involve close collaboration between researchers and farmers, and potentially other stakeholders, to bring about plant genetic improvements within a species; participatory plant breeding covers the whole research and development cycle of activities associated with plant genetic improvement, such as identifying breeding objectives, generating genetic variability or diversity, selecting within variable populations to develop experimental materials, evaluating these materials, release of materials, diffusion, seed production, and distribution; it also may include assessing existing policy or legislative measures, or both, and designing new ones where needed *agr* >>> participatory variety selection

participatory variety selection: Selection of fixed lines (including landraces) by farmers in their target environments using their own selection criteria; it consists of four methodological steps: (1) situation analysis and identification of farmer's varietal needs, (2) search for suitable genetic materials, (3) farmer's experimentation with new crop varieties in their own fields and with their own crop-management practices, and (4) wider dissemination of farmer-preferred crop varieties *meth* >>> participatory plant breeding

particle bombardment: >>> gene transfer >>> nano-biolistic

particle gun: >>> gene transfer >>> nano-biolistic

particulate inheritance: The model proposing that genetic information is transmitted from one generation to the next in discrete units (particles) so that the character of the offspring is not a smooth blend of factors from the parents *gene* >>> MENDEL's laws of inheritance

partitioning: The physiological process by which assimilates are distributed among competing sink tissues *phys*

parts per million (ppm): Designates the quantity of a substance contained in a million parts of a mixture or solution in a carrier, such as air or water *meth*

passage number: The number of times the cells in the *in vitro* culture have been subcultured *biot*

passage time: The interval between successive subcultures or the culture period *prep*

passive immunity: The immunity against a given disease produced by conditioning *phyt*

passive resistance: Innate resistance which does not entail any defensive reactions of the host to the presence of the pathogen *phyt* >>> disease avoidance

passport data: Information about the origin of an accession and any other relevant information, including descriptors, which assist in the identification of the accession *seed*

pasteurization: Named after the Frenchman L. PASTEUR; a technique of heating wine, milk, food, or soil to about +80°C in order to destroy harmful >>> microorganisms; this temperature does not lead to a complete sterilization, for which a temperature of about +120°C is required *meth phyt*

pasture: An area of vegetation that is used for grazing and/or devoted to the production of forage for harvest primarily by grazing *agr*

pasture experiments: A special type of field design; they are different from other types because of the response to treatments, often measured by grazing of animals; since animals are used, the plots must be fairly large, big enough to support one or more animals for an extended period of time; because of the costs involved in conducting a full-scale pasture trial, the initial evaluation of prospective treatments is often done on small plots with clipping being used to simulate grazing *stat meth*

patch: A group of flowers, often of the same species, presenting a similar syndrome to pollinators, which "major" on it *bot* >>> majoring

patch clamp(ing) (technique): A widely applied electrophysiological technique for the study of ion channels – membrane proteins that regulate the flow of ions across cellular membranes and therefore influence the physiology of all cells; the method applies the measurement of conductance of cell membranes by clamping a tiny buffer-filled pipette tip to a cell surface, so that the pipette serves as an electrode for the measurements *meth phys*

patchy distribution: Close and often repeated grazing of small patches or even individual plants while adjacent similar patches or individual plants are lightly grazed or not grazed *agr*

patent: A description of an invention; patents contain one or more claims that describe the subject matter covered in sufficient detail to permit skilled experts to practice an invention, and grant the right to exclude others from practicing an invention *adm* >>> Intellectual Property Rights

patent agent: An individual with technical training who is capable of representing an inventor in patent prosecution *adm* >>> Intellectual Property Rights

patent attorney: An individual with legal training in patent law who is capable of representing an inventor in patent prosecution and litigation *adm* >>> Intellectual Property Rights

patent pool: An agreement between two or more patent owners to license one or more of their patents to one another or third parties *adm* >>> Intellectual Property Rights

pat (phosphinotricin acetyltransferase) gene: Mediates a tolerance to the herbicide >>> phosphinotricin (syn glufosinate); a number of soil bacteria naturally possess the pat gene; it enables them to degrade phosphinotricin in the soil; in 1988, the pat gene was successfully isolated from bacteria and transferred to plants by means of genetic engineering; transgenic plants expressing the pat gene are able to degrade the herbicide agent phosphinotricin *biot*

path diagram: Pedigree arrangement showing only the direct line of descent from common ancestors *gene meth*

path-coefficient analysis: A method for analyzing regular and irregular breeding systems; it is used to calculate path coefficients (direct effects) and indirect effects between independent (x) and dependent (y) variables; the path coefficient is a measure of the importance of a given path of influence from cause to effect; it is the ratio of the standard deviation of the effect to the total standard deviation *stat* >>> www.geneflowinc.com/

pathogen: A (micro)organism (e.g., bacteria, viruses, fungi) that causes disease in a particular host or range of hosts *phyt*

pathogenesis: The complete sequence of events starting with the arrival of the pathogen at the host surface to the completion of the disease cycle *phyt*

pathogenic: Producing disease or toxic symptoms *phyt*

pathogenicity: The ability of an organism to survive at the expense of its host and to incite disease *phyt*

pathologist: A specialist in the origin, nature, and course of diseases *phyt*

pathotoxin: Microbial metabolite with a causal role in disease; it induces all disease symptoms and exhibits the same host specificity as the pathogen of origin *phyt*

pathotype: A subspecific classification of a pathogen distinguished from others of the species by its pathogenicity on a specific host *phyt* >>> pathogenicity

pathovar: In plant pathology, strains of bacteria causing disease in specific plant cultivars; in general, a strain or a set of strains with the same or similar characteristics, differentiated at the infrasubspecific level from other strains of the same species or subspecies on the basis of distinctive >>> pathogenicity to one or more plant *phyt*

patroclinous inheritance: An inheritance in which all offspring have the nucleus-based phenotype of the father *gene*

PBCC: >>> Plant Breeding Coordinating Committee

pBR322 vector The most-used plasmid cloning vector; size: 4.3 kb, about 20 copies per chromosome; it carries ampicillin and tetracycline resistance genes *biot*

PCN: >>> potato cyst nematode

PCR: >>> polymerase chain reaction technique

PIERCE's disease (of grape): Caused by the bacterium *Xylella fastidiosa hort phyt*

peat: An accumulation of dead plant material often forming a very deep layer; it shows various stages of decomposition and is completely waterlogged; it has a low pH value and may fix metal ions, such as copper or iron *agr* >>> Table 44

peat pot: A preformed plant pot (of different sizes and shapes) manufactured from peat and other supplements, usually used for seedling cultivation *meth*

peck: The fourth part of a >>> bushel, a dry measure of eight quarts; e.g., for potatoes, a peck was standardized at 15 >>> pounds; four pecks comprise a bushel; once commonly used in the >>> potato trade, it is now obsolete *meth*

pectin: One of a group of homopolysaccharides, which is especially rich in galacturonic acid; they form a kind of cement, so contributing to the structure of plant cell walls *chem phys*

pectinase: An enzyme that degrades pectin (the adhesive material that cements cells together); it is used alone or with enzymes to digest the polygalacturonic acid of plant cell walls to sugar and galacturonic acid in protoplast production *chem phys biot*

ped: A unit of soil structure, formed by joining together fine sand, silt, or clay particles *agr*

pedicel: The small laterals of the panicle branch in, for example, oats, which bear the spikelets on their swollen tips *bot*

pedigree: The record of the ancestry of an individual, genetic line, or variety *gene* >>> full pedigree >>> partial pedigree

pedigree breeding: A system of breeding in which individual plants are selected in the segregating generations from a cross on the basis of their desirability judged individually and on the basis of a pedigree record; the advantages are (1) if selection is effective, inferior genotypes may be discarded before lines are further evaluated; (2) selection in generation involves a different environment, which provides good genetic variability; and (3) the genetic relationship of lines is estimated and can be used to maximize genetic variability *meth* >>> Figure 7 >>> Tables 5, 28, 35 >>> www.geneflowinc.com/

pedigreed seed: Seed of a named cultivar that is produced under the supervision of a certification agency to ensure genotype and purity *seed*

pedology: The science of soils, including their classification, formation, structure, and composition *agr* >>> Table 44

pedon: Defined to serve as the smallest classifiable soil units; pedons are intended to be large enough to contain the entire root system of an average-sized plant; they also should be of a size suitable for field examination, description, and sampling *agr* >>> Table 44

peduncle: Flower stalk; the inflorescence stalk of a plant supporting either a cluster or a solitary flower *bot*

pedunculate: Possessing or pertaining to a peduncle; stalked *bot*

peeling efficiency: In potato breeding, the ratio of the weight of potato chips or peeled potato produced from a given weight of tubers; a high peeling efficiency is in excess of 80%; many varieties achieve a good efficiency of 70–79%; peeling efficiency deteriorates in most varieties during storage *meth*

pegging: The burying of, for example, peanut fruits *agr*

pellet: A small, roundish mass of matter *biot*

pelleted seeds: Seed that are commercially prepared for precision planting by pelleting them inside a special preparation in order to make them more uniform in size; sometimes special nutrient or growth-promoting substances are placed in the pellets to aid in seed germination and growth *seed*

pelleting drum: Used for pelleting seeds of different crops *seed* >>> pelleted seeds

pellicle: The proteinaceous outer layer of the cuticle of stigma papillae *bot*

pendulous: Drooping or hanging downward *bot*

penetrance (of genes): The proportion of individuals of a specified genotype who manifest that genotype in the phenotype under a defined set of environmental conditions; penetrance, as well as >>> expressivity, often depends on both the genotype and the environment *cyto gene*

penetration: The phenotypic expression of an allele may depend on the growing condition or other factors, such as age or sex; sometimes only a fraction of individuals with a certain genotype shows the "expected" phenotype; this portion is sometimes called penetration *gene*

penetration peg: A minute protuberance from a hypha, germ tube, or appressorium that affects penetration of the host plant surface *phyt bot*

pentaploid: Having five homologous or inhomologous genomes *gene*

pentasomic: The presence of five homologous chromosomes in a complement *cyto*

pentosan: Any of a class of polysaccharides that occur in plants, and form pentoses upon hydrolysis *chem phys* >>> arabinoxylan

pentose: A monosaccharide that consists of five carbon atoms *chem phys*

pepo: A fruit that has a hard rind (e.g., >>> watermelon, >>> cantaloupe, >>> squash, >>> cucumber) but without internal separations or septa; it derives from an inferior ovary with a single locule and parietal placentation *bot*

peptidase: Any of the class of enzymes that catalyze the hydrolysis of peptides or peptones to amino acids *phys*

peptide: A linear molecule that consists of two or more amino acids linked by peptide bonds *chem phys*

peptide bond: A covalent bond formed between the NH_2 group of one amino acid and the COOH group of another, with the elimination of H_2O *chem*

perennation: The survival of plants from growing season to growing season with a period of reduced activity in between *bot*

perennial: A plant that normally lives for more than two seasons and that, after an initial period, produces flowers annually; among perennial crops are oil palm, cacao, rubber tree, rye grass, tall fescue, alfalfa, etc. *bot*

perennialism: A plant that continues to grow from year to year *bot*

perfect flower: Having both functional pistil and stamens *bot* >>> Table 18

perfect state: The state of the life cycle of a fungus in which spores are formed after nuclear fusion or by parthenogenesis *phyt*

perianth: Flower envelope; the outer covering of a flower composed of the floral leaves, usually an outer greenish calyx and an inner, brightly colored corolla *bot*

pericarp: The mature ovary wall, which is fused with the testa in the caryopsis *bot* >>> caryopsis

pericarp layer: Comprises several layers of cells; in the mature grain, the innermost layers form a solidified, dense, protective coat, which prevents water entering and molds from infecting the grain; the outer layers form a coarse coat over the grain surface; these outer layers can become infected by molds such as *Alternaria* and *Cladosporium* by harvest time, giving a dull color to the grain *bot prep*

pericentric: An intrachromosomal inversion that includes the centromere; as opposed to paracentric *cyto*

pericentric inversion: The inversion of a chromosome segment containing a block of genes that involves the centromere *cyto*

periclinal: Referring to a layer of cells running parallel to the surface of a plant part *bot*

periclinal chimera: A plant made up of two genetically different tissues, one surrounding the other *bot* >>> Figure 56

periderm: Several layers of corky cells located on the outside of the epidermis, e.g., of a potato tuber, and containing high amounts of suberin *bot*

perigynium: A scale-like bract enclosing the pistil in *Carex* ssp. *bot*

periodicity: Repetition of events at fairly regular intervals *bot*

perisperm: May derive from the nucellus in some plant species (e.g., sugarbeet, leafy spurge); it contributes substantially to the storage tissue *bot*

perithecium: A rounded or flask-like ascocarp from which ascospores are formed; spores are discharged to the exterior via a small pore *bot*

permaculture: A contraction of permanent agriculture where the inputs equal the outputs *agr*

permanent parasite: Parasite that lives its entire adult life within or on a host *phyt*

permanent preparation: Preparing a microscopic slide (e.g., chromosome preparation) for long-term storage by using special recines (>>> NUSSINOVITCH 2009) or other substances *prep micr cyto*

permanent slide: >>> permanent preparation

permanent wilting (point): >>> wilting point

permeability: A property of a membrane, or other barrier, being the ease with which a substance will diffuse or pass through it *phys*

permutation: Changing the order of a set of elements arranged in a particular way stat; in genetics, any permanent mutation in a gene without phenotypic consequences; it predisposes the carrier for further mutations *gene*

peroxidase (Per): An enzyme that catalyzes the oxidation of certain organic compounds using hydrogen peroxide as an electron acceptor *chem phys*

peroxisome: A cytoplasmic organelle characterized by the association of peroxide-generating oxidases with a catalase *cyto*

persistence: The act or fact of persisting *phyt*

persistent: >>> persistence

persistent modification: Non-heritable morphological or physiological changes over a more or less long period of generation cycles induced by varying abiotic or biotic influences *gene*

Pesek–Baker index: >>> index selection

pest: Any form of plant or animal life, or any pathogenic agent, injurious or potentially injurious to plants or plant products *phyt*

pest control: The regulation or management of a species defined as a pest, usually because it is perceived to be detrimental to a person's health, the ecology or the economy; pest control is at least as old as agriculture, as there has always been a need to keep crops free from pests; in order to maximize food production, it is advantageous to protect crops from competing species of plants, as well as from herbivores competing with humans; the conventional approach was probably the first to be employed, since it is comparatively easy to destroy weeds by burning them or plowing them under, and to kill larger competing herbivores, such as crows and other birds eating seeds; techniques such as crop rotation, companion planting (also known as intercropping or mixed cropping), and the selective breeding of pest-resistant cultivars have a long history *phyt* >>> small interfering RNAs

pest resistance: Resistance to any form of plant or animal life or any pathogenic agent *phyt*

pesticide: A chemical preparation for destroying plant, fungal, or animal pests; the latest research reveals that pesticides may negatively interfere with the physiological communication between crop plants and their surrounding rhizosphere, leading to yield depression and/or plant stress *phyt*

petal: One of the inner floral leaves, usually brightly colored and borne in a tight spiral or whorled corolla *bot*

petiole: The stalk by which a leaf is attached *bot*

PETRI dish: A flat, circular dish about 7 cm in diameter and 1 cm high, usually with a matching dish slightly larger in diameter, in which cells, organs, seeds, or microorganisms are cultured *prep*

P1 generation: >>> parental generation

PGIPs: >>> polygalacturonase-inhibiting proteins

phage: The abbreviation for bacteriophage (i.e., a virus that infects bacteria) *bio*

phalaenophily: Pollination by moths *bot*

pharm crop(s): Genetically engineered crops that produce proteins that are pharmaceuticals, vaccines, industrial enzymes, or reagents for biochemical laboratories; they are mostly grown in open fields; then after harvest the novel protein is purified for use; most pharm crops are in pre-commercial field trials, but at least two proteins used in biochemical and diagnostic procedures are already being grown in maize for commercial use – avidin and beta-glucuronidase; other proteins from pharm crops are already in clinical trials; open-field trials of pharmaceutical crops have been taking place every growing season in the U.S. since 1992; if these altered crops were released into the environment they could never be recalled and could enter the food chain; none of the companies have a commercial permit for the cultivation of these crops; in the case of avidin and beta-glucuronidase, the companies are selling the chemical in commercial channels, but hiding their actions behind a "research" permit provided by the >>> USDA; with this type of permit, the chemicals and field locations can be kept secret *biot agr* >>> biopharming

phase-contrast microscopy: Light rays passing through an object of high refractive index that will be retarded in comparison with light rays passing through a surrounding medium with a lower refractive index; the retardation or phase change for a given light ray is a function of the thickness and the index of refraction of the material through which it passes; thus, in a given unstained specimen, transparent regions of different refractive indexes retard the light rays passing through them to differing degrees; such phase variations in the light focused on the image plane of the light microscope are not visible to the observer; the phase contrast microscope is an optical system that converts such phase variations into visible variations in light intensity or contrast; it allows observation of the cell and structures (even living) without staining and degrading treatments *micr*

phasmid: A cloning vector that has the capabilities to replicate as a plasmid or as a phage; the two modes of replication are usually functional in different bacterial species *biot*

phellogen: A layer of plant tissue outside of the true cambium, giving rise to cork tissue *bot*

phene: The phenotype of the plant, which is a product of the gene (gene > DNA > transcription > RNA > processing > translation > protein) and the interaction with the environment *gene* >>> phenotype >>> Figure 60

phenetics: A branch of biology that determines overall similarity of organisms, not evolutionary relationships *bot evo meth* >>> Figures 68, 69 >>> cladistics

phenocopy: A non-hereditary phenotypic change that is environmentally induced during a limited developmental phase of an organism; it may mimic the effect of a known genetic mutation *gene*

phenogenetics: A branch of genetics that studies the interaction of the genotype and its manifestation *gene*

phenogram: A branching tree of individuals or taxa based on >>> phenetics *bot evo meth*

phenol: A chemical used to remove proteins from DNA preparations *biot*

phenolics: Plant compounds (structurally characterized by an alcohol group on an aromatic ring) that impart a variety of functions to plants, including defense mechanisms and interactions with other organisms; phenolics can also determine plant properties such as flavor and palatability *phys*

phenol test (reaction): The color produced in the grain of, for example, wheat and barley, by treatment with a 1% solution of phenol in water *seed meth*

phenology: The study of the impact of climate on the seasonal occurrence of flora *eco*

phenomics: A field of study concerned with the characterization of phenotypes, i.e., observable or detectable characteristics of plants and/or organisms that result from interactions of its genetic constitution with the environment in which it grows; it is a high-throughput analysis of phenotypes that includes detailed and systematic analysis of phenotypes in terms of a data repository and a means of structured interrogation *biot meth*

phenotype: The observable manifestation of a specific genotype (i.e., those properties on an organism, produced by the genotype in conjunction with the environment) *gene* >>> Figure 60

phenotypic expression: The manifestation of a particular gene resulting in a particular phenotype *gene* >>> Figure 60

phenotypic plasticity: The capability of a genotype to assume different phenotypes *gene* >>> Figure 60

phenotypic segregation: The phenotypic differentiation patterns of cells or individuals in segregating populations; as opposed to genetic segregation *gene* >>> Figure 60 >>> Table 2

phenotypic selection: Development of a variety based on its physical appearance without regard to its genetic constitution *gene* >>> Figure 60

phenotypic variance (VP): The total variance observed in a character; it includes experimental error, genotype × environment interaction, and the genotypic variance *stat* >>> Figure 60

phenotyping: >>> digital plant phenotyping

phenylalanine (Phe): An aromatic amino acid ($HO_2CCH(NH_2)CH_2C_6H_5$); this essential amino acid is classified as non-polar because of the hydrophobic nature of the benzyl side chain; the codons for L-phenylalanine are UUU and UUC; it is a white, powdery solid *chem phys* >>> Table 38

pheromone: A chemical exchanged between members of the same animal species that affects behavior (sex attractants, alarm substances, aggregation-promotion substances, trail substances, etc.) *phyt bio*

phomopsis (in sunflower) (*syn* **brown stem canker**): It is caused by the fungus Phomopsis helianthi (syn Diaporthe helianthi), the disease causes significant yield losses in Eastern Europe; the disease cycle begins with ascospore production and release from >>> perithecia in cool, wet weather; the ascospores infect plants through older leaves, but infections later become systemic and grow into the water-conducting tissues, and finally through the stem cortex; the disease is most severe during periods of extended high temperatures and high humidity; spores are spread among plants by splashing rain and irrigation water and wind; the pathogen survives between sunflower crops in infested crop debris *phyt* >>> Figure 62

photomorphogenesis: Anatomical development induced in response to light *phys*

phytosanitation: >>> crop hygiene

***Ph* locus:** A gene that controls homoeologous chromosome pairing in wheat and, similarly, in other allopolyploid plants; the *Ph* gene restricts pairing between homologous chromosomes in a >>> polyploidy; during evolution, a subtelomeric heterochromatin block became inserted into a group of *cdc2*-like genes on 5B; *Ph1* is involved in the regulation of chromatin condensation; it seems likely that this insertion event generated a functional and/or regulatory change at the 5B *cdc2*-like gene family gene >>> Table 43

phloem: A tissue comprising various types of cells that transports dissolved organic and inorganic materials over long distances within vascular plants *bot*

***Phoma* black stem disease (in sunflower):** Caused by *Phoma macdonaldii*; it is one of the most important diseases of sunflower, e.g., in France *phyt*

phosphatase: An enzyme that catalyzes the reaction involving the hydrolysis of esters of phosphoric acid *chem phys*

phosphorescence: Light emitted by an excited triplet molecule as part of the de-excitation process; because a significant amount of the energy in the original exciting photon is lost in the conversion from the excited singlet to the excited triplet state, the wavelength of phosphorescence is greater (i.e., has less energy) than the wavelength of the exciting photon, and greater than the wavelength of any fluorescence from the excited singlet state in that molecule *phy chem*

phosphorus (P): An element that is required by plants in the oxidized form; it is utilized in reactions in which energy is transferred, often involving >>> ATP; phosphorus deficiency in some plant species triggers the release of organic anions such as citrate and malate from roots; these anions are widely suggested to enhance the availability of phosphate for plant uptake by mobilizing sparingly soluble forms in the soil; "Carazinho" is an old >>> wheat (*Triticum aestivum*) cultivar from Brazil, which secretes citrate constitutively from its root apices, and it also produces relatively more biomass on soils with low P availability than other cultivars that lack citrate efflux; citrate efflux improves relative biomass production where growth is limited by P availability *chem phys* >>> mitochondrion >>> nitrogen use efficiency

phosphorylation: The addition of a phosphate group to a compound, involving the formation of an ester bond between the reactants *phys*

photo(activation) (activated) localization (light) microscopy (PALM): Super-resolution imaging by precise localization of many photoactivatable fluorophores; the method can be applied for optically imaging intracellular proteins at nanometer spatial resolution; numerous sparse subsets of photoactivatable fluorescent protein molecules can be activated, localized (to ~2 to 25 nanometers), and then bleached; the new form of super-resolution fluorescence microscopy has emerged in recent years, based on the high-accuracy localization of individual photo-switchable fluorescent labels; image resolution as high as 20 nm in the lateral dimensions and 50 nm in the axial direction has been attained with this concept, representing an order-of-magnitude improvement over the diffraction limit; the demonstration of multicolor imaging with molecular specificity, three-dimensional (3D) imaging of cellular structures, and time-resolved imaging of living cells further illustrates the exciting potential of this method for biological imaging at the nanoscopic scale *micr cyto*

photobleaching: Photochemical reaction of fluorophores, light, and oxygen that causes the intensity of the fluorescence emission to decrease with time *micr*

photoinhibition: The slowing or stopping of a plant process by light (e.g., the germination of some seeds) phys

photo-insensitive plants: >>> daylength insensitivity

photometry: The measurement of the intensity of light or of relative illuminating power *meth micr*

photomorphogenesis: Changes in plant growth due to light; there is a main plant regulator protein, Cop1, that suppresses genes controlling photomorphogenesis; for example, when a seedling gets exposed to light, the Cop1 protein is reduced in the nucleus and photosynthesis is initiated – the seedling becomes green; cryptochromes may interact with photoreceptor-proteins, which can recognize blue light; thus the interaction can "switch-off" the Cop1 protein *phys*

photon: A packet of light energy; the basic unit of light; although light has some properties consistent with a wave, light from the sun is in individual packages; each photon is a particle of electromagnetic radiation traveling with the speed of light (3×10^8 m/sec); the energy in each photon is dependent on the frequency with which the electromagnetic field vibrates *phy*

photonasty: >>> nasty

photoperiod: The relative length of the periods of light and darkness associated with day and night; in many species, floral induction occurs in response to daylength; species have been categorized according to their daylength requirements as short-day, long-day, intermediate-day, or day-neutral *phys*

photoperiodism: The response of a plant to periodic, often rhythmic, changes in either the intensity of light or the relative length of day *phys*

photophosphorylation: The coupling of >>> photosystems I and II with an electron transfer chain that moves electrons from water (which is oxidized to form O_2) to NADP$^+$ (which is reduced to form NADPH); the transfer of electrons between photosystem II and photosystem I releases energy, which is conserved in the form of a trans-membrane proton gradient and used to synthesize ATP *phys*

photoreceptor: A pigment that absorbs the light used in various metabolic plant processes that require light *phys*

photosensitizers: Molecules which have efficient conversions to the triplet state after absorption of a photon; these triplet state molecules either reduce other molecules to generate free radicals, or they excite O_2, normally a triplet, to the excited singlet state; singlet oxygen is very reactive and can cause significant cellular damage *phys*

photosynthesis: The series of metabolic reactions that occur in certain autotrophs, whereby organic compounds are synthesized by the reduction of carbon dioxide using energy absorbed by chlorophyll from light; by using nothing but sunlight as the energy input, plants perform massive energy conversions, turning 1,000 billion metric tons of CO_2 into organic matter, i.e., energy for animals in the form of food, every year; that is only using 3% of the sunlight that reaches earth; worldwide efforts are being made on artificial photosynthesis using the principles and mechanisms observed in nature; an artificial photosynthesis system or a photoelectrochemical cell that mimics what happens in plants could potentially create an endless, relatively inexpensive supply of clean "gas" and electricity; in 2016, by modifying gene expression, i.e., accelerating interconversion of violaxanthin and zeaxanthin in the xanthophyll cycle and by increasing amounts of a photosystem II subunit, for the first time photosynthetic efficiency could be increased by about 15% (KROMDIJK et al. 2016) *phys* >>> Table 33

photosystem: An array of pigment-protein complexes and electron transfer components that function together to harvest light energy, transfer the energy to photochemical reaction centers, and move the excited electrons in a controlled fashion to produce usable biochemical energy; each photosystem contains hundreds of chlorophyll and carotenoid molecules functioning as antennae, while only a few chlorophyll molecules are employed in the reaction centers *phys*

pH value: The negative logarithm of the hydrogen-ion activity, expressed in terms of the pH scale from 0 to 14 *chem*

phyletic: The evolution by which a race or line is progressively transformed from its ancestral form without branching or separating into related parts *evol* >>> Figures 68, 69

phyletic gradualism: A model of evolution characterized by slow and gradual modifications of biological structures leading to speciation *evol* >>> Figures 68, 69

phyletic series: >>> phyletic

phyllode: An expanded petiole resembling and having the function of a leaf, but without a true blade *bot*

phyllody: The condition in which parts of a flower are replaced by leaf-like structures; often it is a symptom of certain diseases *bot*

phyllosphere: The surface of a living leaf *bot*

phyllotaxis: The active arrangement of leaves on a stem or axis *bot*

phyllotaxy: >>> phyllotaxis

phylloxera: A fatal vine pest that destroys the soft vine roots of *Vitis vinifera* cultivars; the only remedy is to replant on phylloxera-resistant rootstocks; roots of most American hybrid vines are immune to the effects of the pest; phylloxera will generally not inhabit soils that are 80% sand; in all other soil textures, cultivars should be grafted onto phylloxera-resistant rootstocks *hort* >>> vine-louse

phylogenesis: >>> phylogeny

phylogenetic: Of, pertaining to, or based on phylogeny *evol*

phylogenetic evolution: Gradual transformation of one species into another without branching *evol* >>> Figures 68, 69

phylogenetic tree: A diagram showing evolutionary lineages of organisms, i.e., attempts to relate and show the evolutionary history of taxa from a common ancestor at the root of the tree, e.g., trees constructed from DNA sequence data where each taxon under study has provided a DNA sequence from a gene or segment of DNA common to all taxa; the branching structure of a tree is its topology and the tree is bifurcating if three edges meet at each interior vertex and two edges meet at the root; the end of an edge at a taxon is called a terminal vertex or leaf *meth gene stat evol* >>> Figures 68, 69

phylogenetics: Study of reconstructing evolutionary genealogical ties between taxa and line of descent of species or higher taxon *evol* >>> Figures 68, 69

phylogeny: Evolutionary relationships within and between taxonomic levels, particularly the patterns of lines of descent, often branching, from one organism to another *evol* >>> crop evolution >>> Figures 10, 68, 69 >>> Tables 1, 14, 17, 45, 53, 54

phylum (phyla pl): A taxonomic rank above "kingdom" and below "class"; it is equivalent to the botanical term "division" *bio*

physical map: Map of the locations of identifiable landmarks on DNA (e.g., restriction enzyme cutting sites, genes) regardless of inheritance; the distance is measured in base pairs *gene* >>> mapping >>> Figure 43 >>> Table 39

physiological maturity: Stage is identified by the dark spot on the tip of the kernel; grain will contain 25 to 35% moisture *phys* >>> Table 13

physiological races: Pathogens of the same species with similar or identical morphology but differing pathogenic capabilities *phyt* >>> biotype

physiologic specialization: The existence of a number of races or forms of one species of pathogen based on their pathogenicity to different cultivars of a host *phyt*

physiological maturity: The maturity of a seed when it reaches its maximum dry weight; this usually occurs prior to the normal harvest date *seed*

physiological race: One of a group of forms that are alike in morphology but unlike in certain cultural, physiological, biochemical, pathological, or other characteristics, e.g., biotypes of a species that differ in the ability to infect particular varieties of the susceptible plant species *phyt*

physiologist: A specialist dealing with the functions and activities of living organisms and their parts *phys*

phytase: An enzyme that catalyzes the breakdown of phytin, the source of inorganic phosphorus in seed metabolism *chem phys*

phytate: Phosphorous-containing compound (inositol-hexaphosphate) found in cereal grains which can bind zinc and reduce its availability to humans, i.e., antinutrient *phys*

phytic acid: A myo-inositol 1,2,3,4,5,6-hexakisphosphate and the major storage compound of phosphorous in plants, predominantly accumulating in seeds (up to 4–5% of dry weight) and pollen; in cereals, phytic acid is deposited in embryo and aleurone grain tissues as a mixed "phytate" salt of potassium and magnesium, although phytates contain other mineral cations, such as iron and zinc; during germination, phytates are broken down by the action of >>> phytases, releasing their phosphorous, minerals, and myo-inositol, which become available to the growing seedling; phytic acid represents an anti-nutritional factor for animals *phys* >>> zinc >>> www.zinc.org/

phytoalexin(e): An antifungal substance that is produced by a plant in response to damage or infection *phyt phys*

phytochemicals: Certain biologically active chemical compounds that occur in fruits, vegetables, grains, herbs, flowers, or bark; they act to repel or control insects, prevent plant diseases, and control fungi and adjacent weeds; sometimes they confer beneficial health effects to the animals or humans that consume the plant containing those applicable phytochemicals; for example, vitamin C in citrus fruits; beta carotene in carrots and other orange vegetables; d-limonene in orange peels; tannins in green tea; capsaicin in chili peppers; omega-3 (ω-3) fatty acids in soybean oil and fish oil; genistein, saponins, vitamin E, or phytosterols in soybeans *chem phys*

phytochrome: A photoreversible pigment that occurs in every major taxonomic group of plants; it exists in two interchangeable forms with respect to absorption, a red and a far-red form; Pr phytochrome is receptive to orange-red light (600–680 nm) and inhibits flowering; Pf-r phytochrome is receptive to far-red light (700–760 nm) and induces flowering *phys*

phytogenetics: Synonymous with plant genetics, dealing with inheritance in plants *gene*

phytohormone: Organic substance or compound that occurs in minute quantities and affects plant growth and development profoundly, e.g., auxins, gibberellins, and ethylene *phys* >>> plant hormone >>> plant growth regulator

phytoremediation: Refers to the technologies that use living plants to clean up soil, air, and water contaminated with hazardous chemicals; it is a cost-effective plant-based approach to remediation that takes advantage of the ability of plants to concentrate elements and compounds from the environment and to metabolize various molecules in their tissues; it refers to the natural ability of certain plants called hyperaccumulators to bioaccumulate, degrade, or render harmless contaminants in soils, water, or air; toxic heavy metals and organic pollutants are the major targets for phytoremediation *biot*

phytoncide: >>> herbicide

phytopathology: The study of plant diseases *phyt*

phytoremediation: The use of plants to clean up sites that have been contaminated with chemicals or petroleum products *biot agr*

phytosanitary certificate: A certificate from a recognized plant quarantine service that indicates that a sample is substantially free from diseases or pests *seed*

phytosiderophores: Non-proteinogenic amino acids developed by plants under conditions of mineral deficiency (especially under iron and zinc deficiency); the production and exudation of phytosiderophores is controlled by several genes; there are crop plants, such as rye, showing a high level of phytosiderophore production and/or exudation toward the rhizosphere *phys* >> zinc >>> www.zinc.org/

phytotoxic: Being poisonous to plants *phyt*

phytotoxicity: When plants suffer from toxic doses of chemicals *phys agr*

phytotoxin: >>> pathotoxin

phytotron: A group of rooms or a room for growing plants under controlled and reproducible environmental conditions *prep meth*

pick-up reel: A special device on some harvesters for taking up lodging straw *agr*

picotee: A type of edging on a daylily flower where the edge is a completely different color than the tepals *hort*

pie chart: A descriptive technique used to exhibit the way in which a single total quantity is apportioned to a group of categories; each category is assigned a sector of the circle according to the percentage of the total it represents *stat*

pigment: An organic compound that produces color in the tissue of the plant *bot*

pigment strand: A cylinder of cells that lies between the vascular tissues and the nucellar projection in the crease region of cereal grains *bot*

PIIMS: >>> Purdue ionomics information management system

pileorhiza: >>> root cap

piliferous layer: >>> root hair

pilose: Hairy *bot*

pilosum: Used for describing a hairy ventral furrow, for example, barley grain *bot*

pilot test: Preliminary test or study of the program or evaluation activities to try out procedures and make any needed changes or adjustments *stat meth*

Pima cotton: Long staple cotton variety *agr*

pin: The long-styled morph of a distylous heteromorph *bot*

pin flower: Flowers with long styles and short stamens *bot* >>> Table 18

pincers: A gripping tool consisting of two pivoted limbs forming a pair of jaws and a pair of handles *prep*

pinching (pinching back, pinching out): Removing the terminal bud, for example, to encourage side-shoots or to prevent flower formation *hort meth* >>> disbud

pink rot: A potato disease caused by the soil-borne fungus *Phytophthora erythroseptica*; this generally occurs during seasons of excessively wet weather; the disease is often collectively lumped with Pythium leak, a fungal disease caused by a different fungus, *Pythium ultimum*; together, they are called "water rot" *phyt*

pinna: One of the leaflets of a pinnate leaf *bot*

pinnate: Feathered, for example, a compound leaf that has leaflets arranged on either side of a stalk *bot*

pinnate leaf: >>> pinnate

pinninervate: Pinnate-veined; feather-veined *bot*

pipette: A slender, graduated tube for measuring liquids or transferring them from one container to another *prep*

PIPs: >>> plant-incorporated protectants

pistil: The gynoecium of a syncarpous flower; a pistil includes an ovary, style, and stigma; the stigma is the receptor of the pollen *bot* >>> Figure 35

pistillate flower: Designating a flower having one or more pistils and no stamens *bot*

pistillody: The conversion of any organ of a flower into carpels (e.g., stamens into pistils or pistil-like structures) *bot* >>> pistillate flower

pistillum: >>> pistil

pit: In botany, a term used for the widening in the center of the ventral furrow in some wheat and barley grains *bot*

pith: A tissue that occupies the central part of a stem (composed of parenchyma cells), for example, in *Saccharum bot*

pit sowing: >>> hole seeding

pitting: A method of storing, e.g., potato tubers over the winter in a shallow pit covered with layers of straw and dirt *meth agr*

pixel: A picture (pix) element (el); it refers to points of information used to map images; pixels exist in arrays with specific x- and y-coordinates *phy meth*

placenta: The part of the ovary wall formed from the fused margins of the carpel or carpels on which the ovules are carried *bot*

placentation: The position of the placenta within the carpel; it may be parietal (on the walls), axial (on the axis), basal (on the basis), or free-central *bot*

plagiotropic: An orientation or growth response at an oblique angle to the vertical *bot* >>> orthotropic

plan apochromatic objective lens (plan apo): A modern, high-resolution microscope objective lens designed with high degrees of corrections for various aberrations; it is corrected for (1) spherical aberration in four wavelengths (dark blue, blue, green, red), (2) for chromatic aberration in more than these four wavelengths, and (3) for flatness of field; a single plan apo may contain as many as 11 lens elements *micr*

plantation: An area under cultivation; a closely set stand of trees or special crops that has been planted by humans; a large estate or farm on which crops are raised, often by resident workers; a newly established settlement or a colony *agr hort*

plant breeder: A person or organization actively engaged in the breeding and maintenance of varieties of plants, applying a wide range of methods of different scientific disciplines >>> http://cuke.hort.ncsu.edu/gpb/publications/baenziger2006.pdf

plant breeder's rights: The intellectual property rights that are legally accorded to plant breeders by various laws or international treaties *adm*

plant breeding The application of genetic principles and practices to the development of individuals or cultivars more suited to the needs of humans; basically three approaches are included: (1) determination of a pool of genotypes of a given crop, (2) selection of the most suitable genotypes from that pool, (3) testing of most suitable genotypes under different environments, and (4) development of a variety for commercial use; plant breeding uses knowledge from agronomy, botany, genetics, cytogenetics, molecular genetics, physiology, pathology, entomology, biochemistry, or statistics; major historical steps of plant breeding were influenced by first breeding and hybridization approaches, by Mendelian genetics, by selection methods, by polyploidization, utilization of genetic resources, by quantitative genetics and statistics, heterosis and hybrid breeding, by tissue and *in vitro* techniques, genetic engineering and gene transfer, by DNA markers and genomics, or targeted mutations by CRISP/Cas genome editing *meth* >>> Table 49 >>> www.integratedbreeding.net >>> www.youtube.com/watch?v=wiMI-uGcsIk&feature=share&fbclid=IwAR1Zpo-aJyt a3BF4Si20He3E-p6B5t4D8kx3E9Rlz72SuFoYORt298hojB4

Plant Breeding Coordinating Committee (PBCC): Serves as a forum regarding issues and opportunities of national and global importance to the public and private sectors of the United States national plant breeding effort; office at Raleigh, North Carolina State University *org* >>> http://cuke.hort.ncsu.edu/gpb/pr/pbccmain.html >>> http://globalplantbreeding .ncsu.edu/

plant crafting: >>> grafting

plant density: The rate at which seed or vegetative propagules are placed in a field or experimental planting *meth agr*

PLANTEOME: A project that provides a suite of reference and species-specific >>> ontologies for plants and annotations to genes and phenotypes; ontologies serve as common standards for semantic integration of a large and growing corpus of plant genomics, phenomics, and genetics data *biot meth* >>> www.planteome.org

plant establishment: In potato breeding, it refers to the growth period from early sprouting until the initiation of new tubers occurs, and this includes development of both roots and shoots; many growers refer to this stage as vegetative growth; the mother tuber is important during early plant growth, but becomes less important as the new plant establishes its roots and shoots; the period lasts from 30 to 70 days depending on planting date, soil temperature and other environmental factors, the physiological age of the tubers, and the variety *meth agr*

plant evolution: Evolution of eukaryotes from a presumed bacteria-like ancestor is one of the major events in evolutionary history; eukaryotes show a distinct nucleus, organelles involved in energy metabolism (such as mitochondria and chloroplasts), extensive internal membranes, and a cytoskeleton of protein fibers and filaments; chloroplasts in green plants (providing photosynthesis) and algae originated as free-living bacteria related to the Cyanobacteria (their chloroplastic DNA is more similar to free-living Cyanobacteria DNA than to sequences from the plants the chloroplasts reside in); the eukaryotic mitochondria (providing ATP synthesis) are endosymbionts like chloroplasts; mitochondria were acquired when aerobic Eubacteria were engulfed by anaerobic host cells; as they conferred useful functions like aerobic respiration and photosynthesis, they were retained as endosymbionts; this must have happened after the nucleus was acquired by the eukaryotic lineage; the origin of the eukaryotic nucleus is almost certainly autogenous and not a result of endosymbiosis; mitochondria are believed to have originated from an ancestor of the present-day purple photosynthetic bacteria that had lost its capacity for photosynthesis; all land plants evolved from the green algae or Chlorophyta; in the period before the Permian (the Carboniferous), the landscape was dominated by seedless ferns and their relatives; vascular plants first appeared in Silurian (439–409 mya); after the Permian extinction, gymnosperms became more abundant; they evolved seeds and pollen; angiosperms evolved from gymnosperms during the early Cretaceous about 140–125 mya; they further diversified and dispersed during the late Cretaceous (97.5–66.5 mya); water lilies are one of the most ancient angiosperm plants; currently, over three-fourths of all living plants are angiosperms; the angiosperms developed a close contact with insects which promoted cross-pollination and resulted in more vigorous offspring; their generation time to reproduce is short, and their seeds can be dispersed by animals; the important events in the evolution of the angiosperms were the evolution of showy flowers (to attract insects and birds), the evolution of bilaterally symmetrical flowers (adaptation for specialized pollinators), and the evolution of larger and more mobile animals (to disperse fruits and seeds); the earliest land plant fossils date back to 420 mya, the earliest angiosperms fossils are from 142 mya, the oldest grass fossils from 50 to 70 mya, and the oldest rice lineage around 40 mya; the angiosperms developed between 260 and 340 mya; the monocot–dicot divergence happened between 180 and 240 mya; the grass family arose between 65 and 100 mya, the grass subfamilies between 50 and 80 mya; divergence of maize and sorghum around 15–20 mya, wheat and barley 10–14 mya, wheat and rye 7 mya, *Triticum* and *Aegilops* 8 mya, polyploid wheat 0.8 mya, and common wheat 8 tya *bot evol* >>> Figures 68, 69

plant gall: >>> cecidium

plant growth-promoting Rhizobacteria (PGPRs): A class of bacteria that enhance >>> Canola™ plant growth by preventing deleterious effects of other microorganisms that are harmful to plants *biot phyt*

plant growth regulator: As opposed to phytohormons, a plant growth regulator is an artificial compound having effects like plant hormones; it is a compound chemically related to phytohormones but synthesized by humans for the regulation of the growth of weeds, *in vitro*-growing plants, and cultivated plants *phys* >>> phytohormon

plant hair: >>> trichome

plant hormone: A compound that is synthesized by a plant, but is not a nutrient, coenzyme, or detoxification product, and which regulates growth, differentiation, or other specific physiological processes *phys*

plantibodies: Antibodies produced in transgenic plants expressing the antibody-producing genes of an animal, for example, mouse, that had been previously injected with a pathogen (usually a virus) and that infects the plants *phyt biot*

plant-incorporated protectants (PIPs): Formerly referred to as plant-pesticides, plant-incorporated protectants are substances that act like pesticides, but that are produced and used by a plant to protect it from pests such as insects, viruses, and fungi *biot phyt*

planting bar: A hand tool used in making a slit-hole in which trees are planted *fore hort*

planting cord: String of various manufacturing and length used for marking experimental plots, paths, and margins between landmarks, or applied to mark planting rows prep

plant label: Plastic, wood, or other stakes for gardens or experiments to indicate what seeds or material are planted where until they appear, which varieties are included, what sort of evaluation is carried out, etc.; in plant conservation, paper forms to include in drying plant samples, with formal printed forms as permanent labels on herbarium specimens; the minimum information includes the name of the collector, the location collected, the date collected, and the correct identification of the specimen *meth agr hort*

plantlet: A stage of *in vitro* culture; the stage after torpedo stage and usually one of the last before a whole plant is generated *biot*

plantling: >>> plantlet

plant-made pharmaceuticals (PMP): Chemicals with pharmaceutical applications produced in genetically engineered crops *biot*

plant ontology: A structured vocabulary and database resource that links plant anatomy, morphology and growth, and development to plant genomics data *meth stat* >>> ontology >>> PLANTEOME

plant passport: An official seed label used for forthcoming marketing; it shows the crop, crop class, for example, European community grade, inspections, etc. *seed* >>> passport data

Plant Patent Act (PPA): In 1930, the PPA was enacted into law in the United States; the plant patent grants the breeder the exclusive right, for 17 years, to propagate the patented plant by asexual reproduction; the purpose of the PPA was to encourage research investment in asexually reproduced plant species; since 1930, more than 6,000 plant patents have been issued by the Patent and Trademark Office, primarily for fruit trees, flowers, ornamental trees, grape, and other horticultural species *adm* >>> Intellectual Property Rights >>> www.uspto.gov

plant pathogenesis-related proteins (PPRP): Groups of proteins with different chemical properties produced in a cell within minutes or hours following inoculation, but all being more or less toxic to pathogens *phyt*

plant pathology: >>> phytopathology >>> biological control

plant phenotyping: >>> digital plant phenotyping

plant protection: >>> phytopathology >>> biological control

plant quarantine: The isolation of newly imported plants to ensure that they are not introducing any foreign parasites; the term quarantine may refer to the quarantine station itself, or to the process of testing and purifying the plant material *seed meth*

plant succession: The natural pattern of ecosystem growth and change over time for a particular environment; plant life follows established patterns of growth and change after major disruptions, like fires, floods, agricultural damage, logging, etc.; generally, smaller, fast-growing herbaceous species and grasses grow first in an open field, followed in a few years by softwood tree seedlings and larger herbaceous species; as a young forest develops into a mature forest (30 to 70 years), an understory of smaller hardwood trees develops; the final stage is a climax hardwood forest (100 plus years) *eco*

plants with novel traits (PNTs): Any genetic plant strain with characteristics that show neither familiarity nor substantial equivalent traits to plant species already cultivated *biot*

plant variety protection: >>> plant variety rights >>> Intellectual Property Rights

Plant Variety Protection Act: Enacted in 1970 in the United States to provide patentlike protection for seed plant species; prior to 1970, breeders in the industry worked primarily with maize and sorghum, with some efforts directed to alfalfa, cotton, sugarbeet, and certain vegetables; the purpose of the PVPA was to stimulate private plant breeding research and to provide better seed cultivars to farmers and gardeners *adm* >>> Intellectual Property Rights

plant variety right(s): The legal rights of a plant breeder (not necessarily as a person) who has developed a new variety *seed* >>> Intellectual Property Rights

plant with a novel trait: A plant variety possessing a characteristic that is intentionally selected or created through a specific genetic change and is either not previously associated with a distinct and stable population of the plant species or expressed outside the normal range of a similar existing characteristic in the plant species *gene*

plaque: A circular zone of lysis produced by bacteriophage in a colony of bacteria on an artificial medium *meth*

plaque lift: Impression of bacteriophage particles blotted to a filter membrane; the technique is used to screen both with radioactively labeled DNA or RNA probes, and with serological screening of eukaryotic expression libraries *biot*

plasmagene: An extranuclear hereditary determinant showing no Mendelian inheritance *gene*

plasmalemma: >>> cell membrane

plasmatic: All functions, processes, or properties of the cytoplasm *bot* >>> cytoplasm

plasmic: >>> plasmatic

plasmid: An extrachromosomal hereditary determinant of bacteria; such ring-shaped structures are intensively used for amplification of DNA segments in recent bioengineering approaches; it is capable of self-replication; it can carry genes into a host cell *gene* >>> Figure 27

plasmid vector: A plasmid or plasmid-like structure used as a carrier for alien DNA segments or genes *biot* >>> Figure 27

plasmodesm(a) (plasmodesmata *pl***):** Cytoplasmic bridges, lined with a plasma membrane, that connect adjacent cells; they provide major pathways of communication and transport between cells *bot*

plasmodium: In cellular slime molds, a vegetative structure consisting of a non-cellular, mobile mass of naked protoplasm containing many nuclei *bot*

plasmogamy: The fusion of the cytoplasm of two or more cells after karyogamy during the process of fertilization *cyto*

plasmolysis: The result of placing plant cells in a hypertonic solution so that water is drawn out of the cell; the cytoplasm shrinks and the cell membrane is pulled away from the cell wall *phys*

plasmon: All the extrachromosomal hereditary determinants *gene* >>> plasmotype >>> Tables 53, 54

plasmon mutation: A mutation that genetically changes the cytoplasm and/or its hereditary determinants *gene* >>> Tables 53, 54

plasmotype: The sum of the extrachromosomal hereditary determinants *gene* >>> Tables 53, 54

plasticulture: The use of plastics in agriculture; in the broad sense, it includes plastic film mulches, drip irrigation tape, row covers, low tunnels, high tunnels, silage bags, hay bale wraps, and plastic trays and pots used in transplant and bedding plant production; the use of plastic in horticulture crop production has increased dramatically in the last 20 years; the use of plasticulture in the production of horticultural crops (vegetables, small fruits, flowers, tree fruits, and ornamentals) helps to mitigate the sometime extreme fluctuations in weather, especially temperature, rainfall, and wind; row covers, low tunnels, and high tunnels all have the potential to minimize the effect of these extreme weather events on the crop and optimize plant growth and development in a protected environment *agr hort fore biot*

plastid: One of a group of double-membrane-bound plant-cell organelles that vary in their structure and function (e.g., chloroplasts, leucoplasts, amyloplasts) *bot*

plastid DNA: Organelle DNA that is present in a plastid *gene*

plastid inheritance: Non-Mendelian inheritance that is caused by hereditary factors present in the plastids *gene*

plastidome: >>> plastome

plastidome mutation: Mutations that genetically change the DNA of the plastids of a plant cell *gene*

plastitude(s): >>> gemmule(s)

plastogene: The hereditary determinants located in the plastid *gene*

plastome: A term usually used for the plastids of a cell or for the genetic information of the plastid DNA *bot gene*

plastome mutation: >>> plastidome mutation

plate: To place on or in special media in a culture dish *prep*

platform: Plant species chosen for the introduction and subsequent expression of a novel gene *biot*

plating efficiency: An estimate of the percentage of viable cell colonies developing on an agar plate relative to the total number of cells spread onto the plate; the plating efficiency is a function of the tissue, medium composition, plating density, and the phase of the stock culture *biot*

pleiotropic: An allele or gene that affects several traits at the same time *gene*

pleiotropy: The phenomenon of a single gene being responsible for a number of different phenotypic effects that are apparently unrelated *gene*

plesiomorphy: Ancestral state of a particular character *bot*

plicate: Pollen grain with plicae, i.e., circumferential, parallel ridge-like folds *bot* >>> pollen

ploidy: The number of complete chromosome sets in the cell nucleus (e.g., diploid, tetraploid, etc.) *cyto*

plot: In field experiments, more or less large pieces of land used for planting and evaluation; in forestry, a group of trees, all from the same entry (family, clone, provenance) planted together; a five-tree-plot row is the most common design for forest genetics experiments; in statistics, an experimental unit, to which a single treatment is applied *meth agr fore hort stat* >>> design of experiment >>> Figure 50

plot size: In field experiments, the size, number, and distribution of plots are essential elements; an efficient combination of plot size and plot number is required; larger plots offer smaller between-plot variance, long measurement time per plot, shorter walking time between plots, less edge, and less statistical error; small plots require a higher number of plots in order to achieve the same level of precision of estimates; plot size and shape vary with crops and stage of testing, and among characters under selection; plots are generally small at the initial stage of testing and reach a maximum size during the second and third years of replicated trials; unbordered plots with few replications at one or few locations are used for traits with low heritability; row spacing and plant populations are chosen to be similar to commercial production of crop; seedbed preparation, fertilization, weed control practices, etc., are generally the same as those used for commercial production; the mechanization and computerization of most plant breeding programs have greatly increased a breeder's ability to handle more plots, populations, etc. *meth stat* >>> design of experiment

plot structure: The need for experimental design, even when comparing an unstructured set of treatments, is caused by inhomogeneity in the set of plots; a common form of structure consists of one or more partitions; in the case of a single partition a >>> block design is used; for a pair of partitions >>> nested block designs are applied or a row-column design if the partitions are nested or crossed, respectively *stat* >>> design of experiment

plow pan: A pan created by plowing at the depth of tillage, largely the result of the common practice of dropping the tractor wheels of one side of the tractor into the dead furrow for steering while performing the plowing operation *agr*

plumose: Feathery in appearance, having fine hairs on each side *bot*

plum pox virus: A disease of >>> plum and related trees, symptoms are variable according to the plant species, but usually include the appearance of pale or dark rings or spots on leaves and fruits *phyt*

plumula: The undeveloped shoot consisting of unexpanded leaves and the growing point (i.e., the terminal bud of developed embryo) *bot*

plurannual: >>> perennial

plus tree: A tree phenotype judged (but not proved by testing) to be unusually superior in some qualities (e.g., growth rate relative to site, growth habit, high wood quality, resistance to disease and insect attack, or to other adverse local factors) *fore*

PMP: >>> plant-made pharmaceuticals

pneumatophore: A specialized root in certain aquatic plants that performs respiratory functions *bot*

pod: A fruit that dehisces down both sides into two separate valves that are most typically dry and somewhat woody; they are characteristic fruits of the Leguminosae *bot* >>> legume

pod drop: Losses due to the premature drop of pods *phys agr*

pod shattering: Seed losses due to the premature shattering of pods; for example, in oilseed rape it can be as great as 50% of the potential yield in some seasons; average losses are around 10–15%, equivalent to 500 kg/ha or 10 times the sowing rate *agr*

podzol: A kind of soil that is very low in humus *agr* >>> Table 44

poikiloploid: >>> mixoploid

point mutation: A mutation that can be mapped to one specific locus; it is caused by the substitution of one nucleotide for another; it may also be caused by deletion and inversion *gene*

poison: >>> toxicity

poisonous: >>> toxicity

Poisson distribution: The basis of a method whereby the distribution of a particular attribute in a population can be calculated from its mean occurrence in a random sample of the population, provided that the population is large and the probability that the attribute will occur is less than 0.1 *stat*

polarized light (microscopy): A technique used in light microscopy; a beam of polarized light is passed through thin slices of preserved tissue; the light is diffracted by some parts of the plant tissue and not others; for example, starch grains diffract the light in a very characteristic manner producing an image of the Maltese cross; this image is diagnostic of the presence of starch grains *micr meth*

polarization: Restriction of the orientation of the vibration of electromagnetic waves of light *phy micr* >>> polarization microscope

polarization microscope: A compound light microscope used for studying the anisotropic properties of objects and for rendering objects visible because of their optical anisotropy *micr* >>> polarization

polar mutation: A mutation of one gene that affects the expression of the adjacent nonmutant gene on one side, but not of the gene on the other side *gene*

polar nuclei: >>> pole nucleus

polar plate: The functional center for meiotic division of chromosomes; spindle fibers converge on it *cyto*

pole: One of the two ends of the cell spindle toward which chromosomes move during mitotic and meiotic anaphase *cyto*

pole nucleus (pole nuclei *pl***):** The two haploid nuclei present in the center of the embryo sac after division of the megaspore; they may fuse to form a diploid definitive nucleus before fusing with the male gamete to form the triploid primary endosperm nucleus prior to double fertilization *bot* >>> Figure 25

100-polished grain weight: Weight of the 100-polished grain (without hull and pericarp) weight *meth*

pollard: A tree cut back almost to the trunk in order to form a thick head of spreading branches, which are cut for basket-making and kindling (e.g., poplars and willows can be pollarded) *bot*

pollen: Collectively, the mass of microspores or pollen grains produced within the anthers of a flowering plant; it is a highly specialized tissue whose function is the production of two sperm cells and their subsequent delivery through the style and ovary to the embryo sac cells where the double fertilization takes place; it is a highly reduced structure consisting of only three cells; the processes of pollen development, germination, and fertilization involve the specific expression of a large number of genes *bot* >>> pollination bag >>> Figures 35, 58

pollen analysis: A method to study pollen grains, particularly their size, shape, and surface; since those characters are highly specific for species the method is used for taxonomic classifications *bot* >>> pollination bag

pollen barrier: In seed production, the separation of the varieties, lines, etc. (mostly in allogamous crops) in order to prevent intercrossing; it is done by different means, such as with strips of other plants (e.g., hemp or maize), which prevent the free flow of pollen, or by isolation walls of tissue *seed* >>> pollination bag

pollen contamination: Refers, often in connection with seed orchards, to the phenomenon that some of the pollen fertilizing orchard genotypes originate outside the seed orchard *gene* >>> alien pollen >>> pollen migration >>> wild pollen >>> gene flow >>> pollination bag

pollen culture: >>> microspore culture >>> pollination bag

pollen embryoids: Embryoids that derive from anther culture *biot* >>> Figure 58

pollen grain: A microspore in flowering plants that germinates to form the male gametophyte, a structure made up of the pollen grain plus a pollen tube *bot* >>> Figure 35 >>> pollen >>> pollination bag

pollen grain mitosis: In microsporogenesis, a nuclear division that occurs in the pollen grain after the formation of tetrads; it results in a smaller generative nucleus and a larger vegetative nucleus *bot*

pollen mixing: Cross-incompatibility in interspecific crosses is associated with proteins of the pistil that interact with proteins of the pollen to prevent normal pollen tube germination and growth; this unfavorable reaction may be avoided in certain combinations by mixing pollen from a compatible species with pollen from an incompatible parent *meth* >>> pollination bag

pollen mother cell (PMC): The microsporocyte, which undergoes two meiotic divisions to produce four microspores; each microspore becomes a pollen grain *bot*

pollen parent: The parent that furnishes the pollen and that fertilizes the ovules of the other parent in the production of seed *meth*

PollenPlus® technology: The increase of pollen production and shedding of >>> restorer lines in >>> hybrid seed production of >>> rye (*Secale cereale*); by marker-aided >>> introgression (<3 cM linkage) of a restorer gene (*Rf*) on chromosome 4RL from an alien germplasm (Iran) the restorer lines are improved; there is a pleiotropic effect of the gene on fertility restoration and increased pollen production; in this way the formation of >>> ergot (*Claviceps purpurea*) is significantly reduced by preferred pollen reception on the pistil instead of spores of the fungus; the method was introduced by KWS LOCHOW Ltd., Germany *breed*

pollen pore: The pore present in the pollen, through which it imports the nutrition during its development; at the time of pollen germination, the pollen tube emerges from it *bot phys*

pollen sac: A sac within the anther of a stamen within which microspores are produced *bot*

pollen-shedding: The status or process when the pollen grains are released from the anthers *bot* >>> pollination bag

pollen sterility: Pollen that is not able to fertilize an egg cell *phys* >>> pollination bag

pollen transfer: Refers to the mode by which pollen is spread from the male to female organs (e.g., mediated by wind, insects, by hand, etc.) *bot* >>> chloroplast >>> pollination bag

pollen tube: The tube formed from a germinating pollen grain and down which the two male gametes pass to the ovum *bot* >>> pollination bag >>> Figure 35

pollen tube competition: >>> certation

pollinarium (pollinaria *pl*): A functional unit in orchid pollination that consists of two or more pollinia, stalk, or stipe, and a viscidium *bot* >>> pollination bag

pollinate: To transfer pollen from the anther to the receptive surface of the stigma of the same or another flower in angiosperms, and from male to female in gymnosperms; this process usually requires a vector in outbreeding plants *meth* >>> electric-aided pollination >>> honeybee >>> pollination bag >>> Figure 35

pollination: The transfer of pollen grains from the anther to the stigma of a flowering plant; pollination can also be done in artificial culture (i.e., *in vitro* pollination) *bot* >>> electric-aided pollination >>> honeybee >>> pollination bag >>> Figure 35

pollination bag: Usually the flowers are isolated from undesirable pollen by covering them with a pollen-tight cloth, plastic, or paper bag before they are receptive; although breeders use pollination bags routinely in their crossing programs, the effect of pollination bag materials and sizes on temperature within the bag is less determined; some bag materials may trap heat, like a miniature greenhouse, allowing spikes to reach lethal temperatures on hot days; in field experiments, notable differences are detected in spike temperatures when dialysis tubing or white onion-skin typing paper covered the emasculated spikes, e.g., in wheat; measured temperature differences between these pollination bag types are 2 to 3.5°C at noon on clear days; dialysis bags consistently produce the highest spike temperatures; maximum spike temperatures were 6°C above air temperature for bagged spikes; measured spike temperature are highest with a glassine bag and lowest with a bag of white onion skin typing paper wrapped around the spike; maximum differences at noon on clear days between these pollination bags were 4 to 6°C, and maximum spike temperatures inside bags were 8°C above air temperature; such high temperatures can lead to physiological damage when air temperatures are high *meth*

pollination techniques Particularly, in interspecific crosses different methods of pollination are applied in order to overcome progamous incompatibility: (1) the grafted-ovary method, (2) placental pollination, (3) warm-water treatment, (4) the mentor pollen technique, and (5) hormone treatment *meth* >>> electric-aided pollination >>> honeybee >>> pollination bag

pollinator: That parental individual, line, variety, or species, which is used as a donor of pollen in a cross *meth* >>> pollination bag

pollinium: The male dispersal unit in the orchids, derived from an anther locule, containing a sticky mass of pollen tetrads, and with a flexible sticky stalk (>>> caudicle) *bot*

polyacrylamide gel: A gel used to separate biological molecules (proteins); it is prepared by mixing a monomer (acrylamide) with a cross-linking agent (N,N′-methylene-bisacrylamide) in the presence of a polymerizing agent; it leads to the formation of an insoluble three-dimensional network of monomer chains, which become hydrated in water *prep*

polyacrylamide gel electrophoresis (PAGE): A method for the separation of proteins and amino acids on the basis of their molecule size; the molecules move through the gel under the influence of an electric field *meth*

polyadenylation: Posttranscriptional addition of 50–200 adenine residues to the 3′ end of eukaryotic mRNA; the poly-A tail can be used to separate eukaryotic mRNA from other RNA species with oligo T cellulose *biot* >>> Table 38

polyandrous: Having many stamens *bot*

polyandry: The state of having more than one male mate at one time; in fertilization, the fusion of one female and two or more male pronuclei within an egg cell *bot*

polybrid: A hybrid with more than two parental groups *bot*

polycarpellary: Composed of several carpels *bot*

polycarpic: In general, bearing many fruits (i.e., producing fruit many times or indefinitely); in botany, having a gynoecium forming two or more distinct ovaries or carpels *bot*

polycarpous: >>> polycarpic

polycentric: A chromosome that shows more than two centromeres *cyto* >>> centromere

polycistronic RNA: mRNA that codes for more than one polypeptide *gene*

polycotyledonous embryos: Embryos having more than two cotyledons or seed leaves (e.g., in pines and conifers) *bot*

polycross: Open pollination of a group of genotypes (generally selected) in isolation from other compatible genotypes in such a way as to promote random mating inter se; it is a widely used procedure for intercrossing parents by natural hybridization *meth* >>> Figure 51 >>> Table 35

polycross test: A progeny test to assess general combining ability from crosses among selected parents; identities can be maintained only for the seed parents; a mixture of pollen is artificially applied to each female parent *meth* >>> topcross >>> Figure 51 >>> Table 37

polyculture: Cropping systems in which different crop species are grown in mixtures in the same field at the same time *agr*

polyembryonic: >>> polyembryony

polyembryony: The condition in which an ovule has more than one embryo, like in certain grasses or cereals; in the past, the phenomenon was used for haploid selection among the embryos, which often show different ploidy levels *bot meth* >>> Figure 17

polyethene: >>> polyethylene

polyethylene A plastic polymer of ethylene; there are two major types with different chain structures: the stiffer, stronger, linear material sometimes called high-density (HDPE) and the low-pressure and more flexible, lower melting, branched polyethylene known as low-density (LDPE) or high-pressure polyethylene; more recently, linear low-density polyethylene (LLDPE) has become a major product *prep chem hort*

polyethylene glycol (PEG): A polymeric substance of >>> m.w. between 1,000 and 6,000; it is used for stimulation of protoplast fusion *chem biot*

polygalacturonase-inhibiting proteins (PGIPs): Leucine-rich repeat (LRR) proteins involved in the plant defense system; a number of PGIPs have been characterized from dicot species, whereas only a few data are available from monocots *phys biot phyt* >>> Table 38

polygamodioecious: A plant that is mostly >>> dioecious but with some >>> perfect flowers *bot*

polygamomonoecious: A plant that is mostly >>> monoecious but with some >>> perfect flowers *bot*

polygamous: Plants that show male, female, and hermaphrodite flowers on the same or different plants *bot*

polygamy: Where genets with male only, female only, and gynomonoecious and/or andromonoecious and/or hermaphrodite flowers coexist *bot*

polygene(s): One of a group of genes that together control a quantitative character; individually each gene has little effect on the resulting phenotype, which instead requires the interaction of many genes *gene* >>> Table 33

polygenic: Of traits determined by many genes, each having only a slight effect on the expression of the trait *gene* >>> multifactorial >>> Table 33

polygenic resistance: Resistance to parasites based on many genes, i.e., a series of genes controlling a quantitative character *phyt*

polygeny: A trait that is controlled by many genes *gene* >>> Table 33

polygynoecial: Having a number of pistils joined together, as in aggregate fruits (e.g., raspberry) *bot*

polygyny: The state of having more than one female mate at one time; in fertilization, the fusion of one male and two or more female pronuclei within an egg cell *bot*

polyhaploid: Haploid plant derived from a polyploid individual *cyto* >>> *Imperata cylindrica* procedure

polyhybrid: Individuals that are heterozygous with respect to the alleles of many gene loci or of crosses involving parents that differ with respect to the alleles of more loci *gene*

polykaryotic: Cells showing many nuclei *cyto*

polylinker: Synthetic oligonucleotide with recognition sites for several restriction endonucleases *biot* >>> Table 38

polymerase: An enzyme that catalyzes the replication and repair of nucleic acids *gene*

polymerase chain reaction (PCR) technique: Oligonucleotide synthesis is the chemical synthesis of nucleic acids to assemble a defined sequence; in contrast to enzymatic synthesis, which in nature proceeds in a 5' to 3' direction, synthetic oligo-nucleotide synthesis predominantly takes place in the 3' to 5' orientation; the process of oligonucleotide synthesis has evolved since the early 1950s, although the basic solid-phase synthesis methods that are currently used for the assembly of oligonucleotides have been in place for nearly 30 years; these methods rely primarily on nucleoside phosphoramidite (amidite) chemistry; starting samples of low quality or inadequate volume massively increase the chance of unreliable results or even failure; recently, the field of oligonucleotide chemistry has broadened with the introduction of new monomers, activation chemistries, and solid supports; this has enabled the creation of novel oligonucleotide constructs that are suitable for many varied applications, such as construction of microarrays, sequencing, polymerase chain reaction (PCR), and quantitative real-time PCR (qPCR); it is a technique for continuous amplification of DNA and/or DNA fragments *in vitro*; the DNA sequence must be known so that oligonucleotides can be synthesized that are complementary to the extremes of the fragment that is to be amplified; heat-stable DNA polymerase (e.g., from *Thermus aquaticus*) is used for DNA synthesis; PCR's simplicity as a molecular technique is, in some ways, responsible for the huge amount of innovation that surrounds it, as researchers continually think of new ways to tweak, adapt, and re-formulate concepts and applications (>>> NOLAN and BUSTIN 2013) *gene meth* >>> competitive allele-specific PCR (CASP) >>> http://maswheat.ucdavis.edu

polymeric genes: Genes with equal effects but cumulative action *gene*

polymerize: The act or process of forming a polymer or polymeric compound *chem*

polymery: The production of a trait by cooperation among several polymeric genes *gene*

polymix breeding: An alternative method to full-sib crossing and testing; the alternative solution is to apply a pollen mix of many male parents rather than a single pollen for each cross; polymix breeding is easy to implement, ensures good estimates of breeding values of the parent being pollinated, and provides for increased genetic gain opportunity because of the significantly increased number of effective parental combinations tested; however, it has found limited use because inbreeding and pedigree control is lost *meth stat*

polymorphic: Occurring in several different forms; in biotechnology, when a marker locus is polymorphic, there are detectable differences in DNA sequence at that locus *bot gene biot*

polymorphism: The existence of two or more forms that are genetically distinct from one another but contained within the same interbreeding population; DNA sequences are mutating all the time; as two populations diverge over time, they each accumulate different mutations at different sites, e.g., when restriction sites in the vicinity of a given gene are compared from one genotype to another, one genotype may have the site, and the other will not; this also is referred to as a polymorphism; high polymorphism between two genotypes is evidence of genetic divergence *gene*

polynemic chromosome: Describes metaphase chromosome and/or chromatids with more than two DNA helices *cyto*

polynucleotide: A sequence of many nucleotides *gene*

polypeptide: A linear polymer that consists of ten or more amino acids linked by peptide bonds *chem*

polyphenol oxidase (PPO): One of the major enzymes responsible for browning of wheat food products; wheat cultivars differ in PPO activity, and plant breeders wish to select germplasm and cultivars with low PPO activities *agr phys*

polypheny: >>> pleiotropy

polyphilic: A flower that is visited by many species of pollinator *bot*

polyphore: A receptacle or torus bearing many distinct carpels, as in *Rosa bot*

polyphylesis: Originating from several lines of descent *bot* >>> monophylesis

polyphyletic: Designating a group of species arbitrarily classified together, some of the members of which have distinct evolutionary histories, not being descended from a common ancestor *evol* >>> Figure 68 >>> crop evolution

polyploid: An individual carrying more than two complete sets of homologous chromosomes *cyto*

polyploidization: The spontaneous or induced multiplication of a haploid or diploid genome of a cell or individual *cyto* >>> Figure 68 >>> crop evolution >>> *Imperata cylindrica* procedure >>> Table 35

polyploidy: The condition in which an individual possesses one or more sets of homologous chromosomes in excess of the normal two sets found in diploids; it can be produced in nature by somatic doubling due to irregular mitosis in the meristematic cell, and by unreduced gametes due to irregular reductional division during meiosis; for example, approximately 70% of grass species and 20% of legumes are polyploidy; sometimes there is a loss of duplicated genes; in soybean (*Glycine max*) two separate polyploidy events (13 and 59 million years ago) were undergone, that resulted in 75% of its genes being present in multiple copies;

based on gene ontology and expression data it was revealed that only a small proportion of the duplicated genes have been neo-functionalized or non-functionalized; in addition, duplicated genes are often found in collinear blocks, and several blocks of duplicated genes are co-regulated, suggesting some type of epigenetic or positional regulation; transcription factors and ribosomal protein genes are differentially expressed in many tissues; the main consequence of polyploidy seems to be at the regulatory level *cyto* >>> crop evolution >>> nitrous oxide gas >>> Tables 17, 35

polysaccharide: A molecule composed of chains of sugar units *phys*

polysome: A polyribosome, consisting of two or more ribosomes bound together by their simultaneous translation of a single mRNA molecule *gene phys*

polysomic: Where a chromosome or gene is present in more than twice in a genet *cyto*

polysomy: The reduplication of some but not all of the chromosomes of a set beyond the normal diploid number *cyto*

polyspermy: The entry of more than one sperm cell into an egg cell, irrespective of whether or not the additional sperm cells fertilize *bot*

polytene chromosome: A chromosome that is formed by repeated reduplication of single chromatids; sections may appear to puff or swell due to differential gene activation; it is visible through the light microscope *cyto* >>> puff >>> lampbrush chromosome >>> Figure 66

polyteny: >>> polytene chromosome

pomaceous fruit: >>> pome

pome: A fruit (e.g., >>> apple, >>> pear, >>> quince) in which the seeds are protected by a tough carpel wall, and the entire fruit is embedded in a fleshy receptacle *bot*

pomology: The science or study of growing fruit *hort*

popbean: Any legume variety whose seed both pops and becomes fully cooked and edible during the quick exposure to heat during popping *foo*

population: A community of individuals which share a common gene pool *bot gene*; in statistics, a hypothetical (often infinitely large) series of potential observations, among which observations actually made constitute a sample *stat* >>> closed population >>> Figure 38 >>> Tables 35, 37

population density: The number of individuals of a population per unit area of a particular habitat *bot eco*; in biotechnology, the cell number per unit area or volume of a medium *biot* >>> plant density >>> Table 37

population genetics: The study of inherited variation in populations and its modulation in time and space; it relates the heritable changes in populations to the underlying individual processes of inheritance and development *gene* >>> Table 37

population, Mendelian: >>> Mendelian population

population size: The number of individuals in a population that are included in reproduction during a certain generation *meth stat* >>> Table 37

population-specific single nucleotide polymorphism: Any >>> SNP that is present in one population and absent in another *biot meth*

population variance: A measure of dispersion defined as the average of the squared deviations between the observed values of the elements of a population and the corresponding mean of those values *stat*

population waves: Irregular or rhythmic changes of the number of individuals in a population that is included in reproduction during certain generations *meth stat*

porocidal: Anther dehiscence through apical pores, e.g., in Ericaceae *bot*

porosity: The proportion, as the percentage volume, of the total bulk volume of a body of rock or soil occupied by pore space *agr* >>> Table 44

position (positional) effect: The change in the expression of a gene with respect to neighboring genes *gene*

positional candidate gene: A gene known to be located in the same region as a DNA marker that has been shown to be linked to a single-locus trait or to a quantitative trait locus and whose function suggests that it could be the source of genetic variation in the trait in question *biot*

positional cloning: A process of molecular cloning of a gene with reference to its position on the genetic or physical map; high-density linkage maps of molecular markers provide an alternative gene cloning approach, positional or map-based cloning; the positional cloning consists of identifying the markers that flank and show tight linkage to the target gene, "walking" to the gene by using various genomic libraries constructed in, e.g., yeast artificial chromosome vectors, and confirming the gene effects by the comparison of the isolated gene with a wild-type allele, or complementation of the recessive phenotype by transformation; thus positional cloning permits the isolation of any gene which can be precisely mapped *biot*

positive assortative mating: A type of non-random mating in which individuals of similar phenotype mate more often than predicted under random mating conditions *gene*

positive screening: A selection technique in which the best individuals in a genetically diverse population are preserved to become the parents, either of the next screening generation, or of a new >>> cultivar *meth* >>> negative screening >>> Figure 60

postemergence herbicide: An herbicide that affects the weeds after emergence *phyt*

postemergence treatment: Any treatment made after a specified weed or crop plant emerges; application of herbicide to crop after seedlings have emerged from soil and are growing *agr meth phyt*

postgamous incompatibility: >>> crossbreeding barrier

postharvest losses: Crop losses due to parasites that occur after harvest, usually in the store; these can be reduced or prevented by ensuring that the stored product is dry, to prevent molds developing, and that the product is in an airtight container that lacks oxygen, to prevent various animal pests from eating it *agr phyt*

postharvest decay pathogen(s) (in apple): They cause blue mold and bitter rot of apples during storage by the fungi *Penicillium expansum* and *Colletotrichum acutatum*, which result in significant economic losses; resistance to these pathogens in commercial apple cultivars has not been documented in the literature; in 2011, in an apple germplasm collection, from the center of origin in Kazakhstan, resistance in four accessions to *C. acutatum* and two accessions resistant to both *P. expansum* and *C. acutatum* were reported for the first time *phyt hort*

postplanting: Applied after transplanting a crop *agr*

poststratification: An estimation method that adjusts the sampling weights so that they sum to specified population totals corresponding to the levels of a particular response variable *stat*

post-transcriptional gene silencing (PTGS): Silencing of an endogenous gene caused by the introduction of a homologous dsRNA, transgene, or virus; in PTGS, the >>> transcript of the silenced gene is synthesized but does not accumulate because it is rapidly degraded; this is a more general term than RNAi, since it can be triggered by several different means; it was initially considered a bizarre phenomenon limited to petunias and a few other plant species; in the 1990s, a surprising observation was made in petunias; while trying to deepen the purple color of these flowers, R. JORGENSEN introduced a pigment-producing gene under the control of a powerful promoter; instead of the expected deep purple color, many of the flowers appeared variegated or even white; the observed phenomenon was named "cosuppression", since the expression of both the introduced gene and the homologous endogenous gene was suppressed; meanwhile, it became one of the most important phenomena in molecular biology; it has become clear that PTGS occurs in both plants and animals and has roles in viral defense and transposon silencing mechanisms; most exciting is the emerging use of PTGS and, in particular, RNA interference (RNAi) – PTGS initiated by the introduction of double-stranded RNA (dsRNA) – as a tool to knock out expression of specific genes *biot meth* >>> Figure 46

posttranslational modifications: Such modification can be cleavage of amino terminal peptide, hydroxylation, and oxidation of amino acids in the polypeptide chain for cross-linking; covalent modifications by acetylation, phosphorylation, or glycosylation *biot*

postzygotic incompatibility: A condition where, in the case of incompatible or wide crosses, the zygote fails to develop, often for nutritional reasons; in some cases embryo culture can be used to rescue the embryo *bot*

potassium (K): An element that is required for plant growth; deficiency leads to reduced growth and to dark or blue-green coloration in the leaves *chem phys*

potassium iodide: >>> chemical desiccation

potato cyst nematode (PCN): A major pest of potato (*Globodera rostochiensis* [golden PCN], *G. pallida* [white PCN]) causing severe yield loss if populations are allowed to reach high levels *zoo phyt*

potato fork: A fork of heavy construction with flat tines, for deep digging as is needed for potato harvesting *agr hort*

potato knife: In the past, a knife used during planting season to cut seed potatoes into seed pieces or sets *meth tech*

potato leaf roll virus (PLRV): A virus spread by aphids that can cause severe reduction in yield *phyt*

potato virus Y: A virus spread by aphids that can cause severe reduction in yield *phyt*

pot-bound: Used when a plant has an overly extensive root system in a too-small container *hort*

pot feet: Supports placed under pots and planters to raise them off the ground for better drainage and air circulation *hort*

pot up: To insert a seedling or rooted cutting in potting mix in a container *hort meth*

pound (lb): Equals 373.2420 g

poverty adjustment: The increase in percentage composition of a nutrient in plant tissue from the minimum percentage up to the critical percentage required for maximum yield *agr*

power (of statistical test): The probability that a statistical test will detect a defined pattern in data and declare the extent of the pattern as showing statistical significance *stat*

PowerMarker: A comprehensive set of statistical methods for genetic marker data analysis, designed especially for >>> SSR/SNP data analysis; it builds a powerful user interface around both new and traditional statistical methods for population genetic analysis; PowerMarker is also a 2D Viewer, which is used intensively for visualizing linkage disequilibria results *stat biot* >>> Figure 63

PPA: >>> Plant Patent Act

ppm: >>> parts per million

PPO: >>> protoporphyrinogen oxidase

PPRP: >>> plant pathogenesis-related proteins

preadaptation: An adaptation evolved in one adaptive zone, which proves especially advantageous in an adjacent zone, and so allows the organism to radiate into it; in breeding, the pretreatment of plants under moderate climatic (light, temperature, or nutrient) conditions in order to gradually accustom them to stress conditions *phys evol* >>> www.plantstress.com

prebasic seed: An >>> OECD class of pedigreed seed which is considered as equivalent to breeder's seed and is used for the production of basic seed or foundation seed *agr seed* >>> breeder's seed

prebloom: The stage or period immediately preceding blooming *agr*

prebreeding: All research and screening activities before a plant material enters the directed breeding process (e.g., the development of germplasm to a state where it is viable for breeder's use); primarily, it involves the evaluation of traits from exotic material and their introduction into more cultivated backgrounds *meth*

prechilling: The practice of exposing imbibed seeds to cool (+5 to +10°C) temperature conditions for a few days prior to germination at warmer conditions *seed* >>> stratification

precipitant: A substance or process that causes precipitation *chem*

precision: The difference between a sample-based estimate and its expected value; it is measured by the sampling error (or standard error) of an estimate *stat*

precision breeding: A newly enabled approach to plant genetic improvement that transfers only specific desirable traits among sexually compatible relatives via the relatively stable mitotic cell division pathway in order to avoid the significant genetic disruption imposed upon conventional breeding by meiosis (sexual reproduction) *bree meth* >>> smart breeding

precision farming: A farming system which uses global positioning (GPS) technology involving satellites and sensors on the ground, along with intensive information management tools to understand variations in resource conditions within fields; the information is used to more precisely apply fertilizers and other inputs, and to more accurately predict crop yields *agr* >>> global positioning system

precision seed: Calibrated or pelleted sorts of seeds used for precision drilling *agr*

precleaning: The process for removing the bulk of foreign materials grossly different in size from the harvested seed or other crop products *seed*

precocious embryo development: Asexual development of the embryo before the flower opens and anthers dehisce *bot*

predecessor: Within the crop rotation, the crop before the recent cropping *agr*

predisposition: An increase in susceptibility resulting from the influence of environment on the subject *phyt*

preemergence damping off: Death of seedlings, as a result of disease, before they have emerged above the soil surface *phyt*

preferential pairing: Chromosome pairing in allopolyploids in which the most structurally similar chromosomes preferentially pair with another *cyto* >>> *Ph* locus

prefoliation: >>> vernation

preharvest losses: Crop losses from parasites that occur in the field; as opposed to >>> postharvest losses that occur in the store *agr phyt*

preharvest sprouting: >>> sprouting >>> dormancy

premium wheats: Wheat grades segregated on the basis of their specific quality characteristics to meet particular market requirements; these may include high or low protein content, combined respectively with hardness or softness of the grain; there will generally also be the specification of certain varieties known to meet or exceed the quality requirements *agr* >>> wheat

prepotency: The capacity of a parent to impress characteristics on its offspring so they are more alike than usual *gene*

prescribed burn: A fire set and controlled by humans to achieve some management objective, such as improving pasture in grazing systems *agr*

preselection: Selection of some individuals for further tests or studies *meth* >>> Figure 51

presoaking: Presoaking of seeds in water has been suggested as a means to speed up germination *meth*

pretest: Determination of performance prior to the organization or administration of a data collection activity *stat*

prevalence: The observed frequency of a trait or disease in a population, often at a particular age or time; in statistics, proportion of a population with both new and ongoing cases or events or conditions at a given point in time or over a specified period; this contrasts with incidence, which is the rate at which new cases of the condition occur *gene stat*

prezygotic incompatibility: In the case of incompatible or wide crosses, the inhibition of pollen germination or the prevention of pollen tube growth, among other possible barriers of plant fertilization; in some cases the barrier can be overcome by *in vitro* pollination *bot*

pricking off: A method of transplanting tiny seedlings; the blade of a knife or plant marker is used to remove each plant from one spot and move it to another *meth hort*

pricking out: A method of thinning seedlings by cutting them off at soil level so as not to disturb the roots of the other plants *hort meth* >>> thinning

pricking-out peg: An adjusted peg used for thinning or transplanting seedlings by cutting them off at soil level *meth hort* >>> thinning

prickle pollination: >>> tripping mechanism >>> thinning

prickly: Having thorns *bot*

primary branch: Any branch arising from the main axis; all branches that come from the central axis of a grass inflorescence *bot*

primary constriction: The centromeric region of a chromosome *cyto*

primary culture: A culture resulting from cells, tissues, or organs taken from an organism *biot*

primary endosperm nucleus: The nucleus resulting from the fusion of the male gamete and two polar nuclei in the central cell of the embryo sac *bot*

primary fluorescence: Fluorescence originating from the specimen itself *micr*

primary gene pool: Includes the cultivated species of a crop and related species from which useful genes can be most readily obtained for breeding; in general, it is the total sum of all the genetic variation in the breeding population of a species and closely related species that commonly interbreed with, or can be routinely crossed with, the species *evol*

primary host: >>> host

primary infection: The first infection of a plant by an overwintering or oversummering pathogen *phyt*

primary inoculum: Spores or fragments of a mycelium capable of initiating a disease *phyt*

primary leaf: One of the first pair of leaves to emerge above the cotyledon during the development of a seedling; it is often morphologically distinct from subsequent leaves *bot*

primary periclinal chimera: >>> chimera >>> Figure 56

primary production: The amount of light energy converted into plant biomass in a system *agr phys*

primary sex ratio: >>> sex ratio

primary unilateral branch: Any branch that originates from the central or main axis of a panicle inflorescence with spikelets along one side or what appears to be one side *bot*

primase: An enzyme that makes the RNA primer required by DNA polymerase in DNA replication *chem biot*

primed *in situ* labeling: A technique that labels specific sequences in whole chromosomes by *in situ* DNA chain elongation or >>> PCR *meth biot cyto*

primer: An >>> oligonucleotide that forms a double strand by using a complementary segment; the primer is prolonged till the double strand is completed; in biotechnology, small section of DNA nucleotides which binds to the single-stranded DNA template during PCR; these can be sequences specific for a gene or totally random, depending upon the experiment's objective *biot*

primer annealing step: The second step of >>> PCR; in this step the primers will bind to their complementary DNA sequences on the single-stranded DNA template *prep*

primer-dimers: Occur when single stranded primer oligonucleotides bind to each other rather than the DNA template *chem biot*

primer extension: Primers designed to hybridize with a target, ending one base at the >>> SNP position; single-base primer-extension technology acts on primer–target hybrids to add a single chain-ending nucleotide, often a dideoxynucleotide; the only one of four nucleotides that will extend the primer is the one that is complementary; the identity of the added nucleotide is determined in a variety of ways; it is a method of SNP detection; primer extension analysis is utilized to quantitate mRNA levels, and to detect low-abundance mRNA species; in addition, it can be utilized to map the 5′-end of transcripts to determine the exact start sites for transcription *meth biot*

primer walking: A method for sequencing long (~1 kb) cloned pieces of DNA; the initial sequencing reaction reveals the sequence of the first few hundred nucleotides of the cloned DNA; on the basis of these data, a primer containing ~20 nucleotides and complementary to a sequence near the end of sequenced DNA is synthesized, and is then used for sequencing the next few hundred nucleotides of the cloned DNA; this procedure is repeated until the complete nucleotide sequence of the cloned DNA is determined *biot meth*

priming: The treatment of seeds with an osmotic solution (e.g., polyethylene glycol, which allows controlled hydration); the seed embryo develops to the point of germination and then is dried; it is applied for more uniform and rapid germination of certain vegetable seeds; in tobacco cropping, a term to designate the harvesting of ripe tobacco leaves from the stalk as they ripen, beginning at the bottom and progressing upward *seed agr*

primitive form: In phylogeny, seedless vascular plants with underground rhizomes; they grow in tropical to subtropical areas and are terrestrial or epiphytic; in botany and breeding, as compared to cultivated crops, plants still show wild characteristics, such as brittle rachis, seed shattering, and others *bot*

primocanes: On raspberries and blackberries, new, first-year canes *hort*

primordium (primordia *pl*): The early cells that serve as the precursors of an organ to which they later give rise by mitotic development *bot*

primosome: The mobile complex of helicase and primase that is involved in DNA replication gene *biot*

principal host: >>> host

prior art: In patenting biological material, public knowledge that exists in a field; all previously issued patents, publications, public announcements, or knowledge that bear on the invention claimed in a patent application *adm* >>> Intellectual Property Rights

prism: An integrated software solution for plant researchers and agronomists; it provides user-friendly tools to manage experimental data; PRISM consists of several different book types, for example, an experiment field book; each book type is used to classify information and to execute business rules; many of the book types are configured to produce a spreadsheet-like interface; it can create and print reports and labels (including barcodes) *meth stat* >>> www.teamcssi.com/index.html

probability sample: A sample selected by a method such that each unit has a fixed and determined probability of selection *stat*

probe: A radioactively or non-radioactively labeled and defined nucleic acid sequence that can be used in molecular hybridization; usually it is used for *ex situ* or *in situ* identification of specific, complementary nucleic acid sequences *biot* >>> Figures 52, 53, 54

processed pseudogene: A copy of a functional gene which has no promoter, no introns, and which, consequently, is not itself transcribed; pseudogenes are thought to originate from the integration into the genome of >>> cDNA copies synthesized from >>> mRNA molecules by >>> reverse transcriptase; pseudogenes therefore have a poly (dA) sequence at their 5′ ends; because they are not subject to any evolutionary pressure to maintain their coding potential, >>> pseudogenes accumulate mutations and often have stop codons in all three reading frames *biot*

processing of seeds: The complex of measures in order to clean, calibrate, disinfect, store, and pack seeds *seed*

prochromosome: A heterochromatin block that is seen during the >>> interphase of cell division and which is related to the number of chromosomes per complement or less *cyto* >>> Table 43

procumbent: Trailing or lying flat on the ground *bot*

productivity index: A measure of the amount of biomass invested in the harvested product in relation to the total amount of standing biomass present in the rest of the system *agr*

proembryo: The young embryo in its early stages of development after zygote formation *bot* >>> direct embryogenesis

proembryonal complex: A mass of embryogenic cells *bot biot*

professional plant breeding: There were no professional plant breeders before about 1900; plant breeding was undertaken by farmers and amateurs *adm*

progenitor: The original, ancestral, or parental cell, individual, or species *bot gene evol*

progenitor cell: Undifferentiated cell (i.e., immature cell), which will go on to develop into any cell type *biot*

progeny: Offspring; plants grown from the seeds produced by parental plants *bot gene* >>> Figure 19 >>> Table 57

progeny selection: Selection based on progeny performance *meth* >>> pedigree breeding >>> Table 57

progeny test(ing): A test of the value of a genotype based on the performance of its offspring produced in some definite system of mating *meth* >>> pedigree breeding >>> Figures 19, 60 >>> Table 57

prokaryote: The class of organisms that does not have discrete cell nuclei in a nuclear envelope, including bacteria, but shows single, circular DNA molecules within the cytoplasm *bio*

prokaryotic: >>> prokaryote

prolamin: A protein; it is soluble in 70–90% ethyl alcohol but not in water; it is found only in cereal seeds (e.g., gliadins in wheat and rye, zein in maize); upon hydrolysis they yield proline, glutamic acid, and ammonia; it is a storage protein acting as a reserve and deposited in protein bodies *chem phys* >>> Tables 15, 38

prolate: Pollen grain with a polar axis longer than the equatorial diameter, e.g., in >>> *Daucus carota bot*

proliferation: Successive development of new parts, organs, etc. *phys*

proline (Pro): A heterocyclic, non-polar imino acid, which is present in all proteins; the major pathway for proline synthesis, which takes place in the cytoplasm, is from >>> glutamate, through gamma-glutamyl phosphate and glutamyl-gamma-semialdehyde, a two-step reaction that is catalyzed by a single enzyme, D1-pyrroline-5-carboxylate synthetase *chem phys* >>> osmolyte >>> Table 38

prometaphase: The stage in mitosis between the dissolution of the nuclear membrane and the organization of the chromosomes on the metaphase plate *cyto*

promiscuous: Heterogeneous or haphazard mixture *gene*

promiscuous DNA: >>> mitochondrial DNA

promiscuous soybean: Naturally nodulating varieties of soybean that fix nitrogen with a wider range of native Rhizobia, and do not require external inoculation *agr*

promoter: The region of DNA which binds RNA polymerase and directs this enzyme to the correct site where transcription of the gene will begin, i.e., a nucleotide sequence within an >>> operon, lying between the operator and the structural gene or genes, that serves as a recognition site and point of attachment for the RNA polymerase; it is the starting point for transcription of the structural genes in the operon, but is not itself transcribed; the promoter controls where (e.g., which portion of a plant) and when (e.g., which stage in the lifetime) the gene is expressed; for example, the promoter "Bce4" is seed-specific; promoters may be constitutive, inducible, or tissue-specific: (1) constitutive promoter – an unregulated promoter that allows for continual transcription of its associated gene, (2) inducible promoter – the activation of a promoter, and, consequently, of the transcription of the associated gene(s), in response to the presence either of a particular compound (e.g., presence of a plant hormone) or of a defined external condition (e.g., elevated temperature), (3) tissue-specific promoter – a promoter that limits the expression of the transcribed gene to certain tissues (e.g., protein expression is limited to seed and is not expressed in the leaves, stems, or roots of the plant) *gene biot* >>> cauliflower mosaic virus (CaMV)

promoter mutation: Any mutation that occurs within the promoter sequence of a gene; as a consequence, the encoded homeobox transcription factor is not functional and the gene cannot be expressed *gene*

promoter sequence: >>> promoter

promoter single nucleotide polymorphism: Any >>> SNP that occurs in the promoter sequence of a gene; if the SMP prevents the binding of a transcriptional factor to its recognition sequence in the promoter, the latter becomes partly dysfunctional *biot*

propagate: To reproduce or cause to multiply or breed *agr*

propagating bench: A stationary, shallow box (sometimes covered by a glass pan or other means); it is usually filled with fine sand or certain soil (often sterilized), which is kept moist; cuttings, slips, or shoots after *in vitro* culture are inserted into the growing medium until they form roots *meth hort*

propagating case: >>> propagating bench

propagation: Various methods by which plants are increased (e.g., seeds, division, separation, softwood cuttings, slips, grafting, budding, or layering) *meth hort agr*

propagule: Any type of plant to be used for reproduction (e.g., a seedling, a rooted or unrooted cutting, a graft, a tissue-cultured plantlet, etc.) *meth hort*

proper block design: A >>> block design is proper if all blocks have the same cardinality, i.e., contain the same number of plots; for binary designs, this just means that all blocks contain the same number of points *stat* >>> design of experiment

prophage: The non-infectious form of a temperate bacteriophage in which the phage DNA has become incorporated into the lysogenic, host bacterial DNA *biot*

prophase: The first phase of mitosis and meiosis *cyto*

prophyll(um): The first leaf or protective scale of a lateral shoot; a sheath with two veins *bot*

proplastid: A colorless, double-membrane-bound organelle with little internal structure that acts as a precursor in the development of all plastids *bot*

prosecution: In patenting biological material, the process by which an inventor engages with the patent office to obtain a patent and determine the scope of its claims *adm* >>> Intellectual Property Rights

prostrate: Lying flat on the ground *bot*

protandry: The maturation of anthers before carpels (e.g., in sugarbeet, sunflower, or carrot) *bot* >>> Table 18

protease: An enzyme that catalyzes the hydrolysis of the peptide bonds of proteins and peptides *chem phys* >>> proteinase

protease inhibitors (PIs): Proteins that accumulate in the seeds and storage organs of many plants; they protect plants against a variety of insect, bacterial, and fungal pathogens *phyt biot*

protected variety: A variety that is released and granted a certificate of plant variety protection under the legal statutes of a given country; the owner of a protected variety has the right of selling, offering, reproduction, import, export, or using for hybrid seed production *seed* >>> Figure 55

protein: A polymer that has a high relative molecular mass of amino acids; it has many functions in the living cell *chem phys* >>> Tables 15, 16, 33, 48

proteinase: An enzyme that hydrolyzes protein molecules *chem phys*

protein engineering: Production of altered proteins by site-directed mutagenesis *biot* >>> biotechnology

protein expression map: By comparing the two-dimensional electrophoresis gel patterns of samples exposed to different physiological conditions or different treatments, it is possible to identify groups of proteins with related functions or whose expression is interdependent *meth biot*

protelomere: A terminal deep-staining structure with sharp limits, normally composed of one or three dark-staining large >>> chromomeres *cyto*

proteome: The complete set of proteins detectable in a tissue; it is the functional representation of the genome that includes the types, functions, and interactions of proteins that are present in a cell; it is not a fixed characteristic of a cell, but variable depending on developmental stage, hormonal status, etc. *phys*

proteome analysis: The basis of proteome analysis is an electrophoretic separation of the proteins on a two-dimensional protein gel, followed by silver staining of proteins, and an image analysis of the stained gel; interesting protein spots identified on the gel can be excised; the critical protein can be extracted from the excised spot and further analysis of the protein (amino acid composition, partial amino acid sequence, isoelectric point, m.w.) may result in protein identification *meth phys biot*

proteomics: The large-scale study of proteins, particularly their structures and functions, i.e., a system for identifying proteins within a cell or tissue sample; it is the systematic characterization of proteins, which are found in a tissue or in a specific physiological condition; the proteins can be identified using mass spectroscopy; it allows breeders to understand how genes behave in different parts of the plant and under different growing conditions *biot* >>> functional genomics >>> genomic technology

protoclonal variation: Variability of somatic cells derived from protoplast culture *biot*

protoclone: A plant regenerated from a protoplast culture *biot*

protoderm: A primary embryonic tissue that is located on the surface of young embryos *bot biot*

protogene: A dominant allele as opposed to >>> allogene *gene*

protogyny: A condition in which the female parts develop first (e.g., in rapeseed) *bot* >>> Table 18

protoplasm: A complex, translucent, colorless, colloidal substance within each cell, including the cell membrane, but excluding the large vacuoles, masses of secretions, ingested material, etc. *gene*

protoplast: That part of a cell that is actively engaged in metabolic processes, or a cell without a cell wall; protoplasts are produced by enzymes, which digest the wall; they are used for production of hybrid cells by protoplast fusion or for injection of foreign DNA *bot biot* >>> Figure 57

protoplast culture: The isolation and culture of plant protoplasts by mechanical means or by enzymatic digestion of plant tissue, organs, or cultures derived from these; protoplasts are utilized for selection or hybridization at the cellular level and for a variety of other purposes *biot* >>> Figure 57

protoplast fusion: A technique used in somatic hybridization experiments; it is used for overcoming crossing barriers; protoplasts are placed together and induced to fuse, applying fusogenic agents, such as polyethylene glycol or physical means; subsequent regeneration of the cell wall allows the propagation and regeneration of a somatic hybrid plant *biot* >>> Figure 57

protoporphyrinogen oxidase (protox) (PPO): An enzyme utilized in photosynthesis and the target site of many herbicides in the cell membrane disrupter mode of action *phyt biot*

prototroph: A strain of organisms capable of growth on a defined minimal medium from which they can synthesize all of the more complex biological molecules they require; as opposed to auxotroph *phys*

protozoa: Unicellular eukaryotic microorganisms that lack cell walls *zoo*

protruding ends: >>> cohesive ends

provenance: The geographical source and/or place of origin of a given lot of seed, propagules, or pollen *fore*

provenance test: An experiment, usually replicated, comparing trees grown from seed or cuttings that were collected from many geographical regions of a species distribution *fore*

provisional patent: A preliminary patent application filed without a formal patent claim, oath or declaration, or any prior art statement; it provides the means to establish an early effective filing date in a subsequent non-provisional patent application and allows the term "Patent Pending" to be applied *adm* >>> Intellectual Property Rights >>> www.uspto.gov

proximal: Toward or nearer to the place of attachment *cyto* >>> centromere >>> chromosome arm

prune: >>> pruning

pruning: Trimming branches or parts of trees and shrubs in order to trim a plant or to bring the plant into a desired shape; in addition, the removal of the growing point from a plant frequently causes the initiation of tillering or branching; the onset of flowering on the new vegetative growth may be delayed compared with that of unpruned plants; the method may also be used for synchronization of flowering prior to hybridization; branches on trees strive in an upward direction; the internal struggle is all about reaching up for the light; the result can be branches that are so intertwined that they themselves create shade for each other; that affects the formation of flower buds and, thereby, fruits; the best shape a fruit tree can have is one central trunk, with branches that grow as close to a right angle as possible and that are at the most half as thick as the trunk; the tree is usually pruned to help it achieve this shape; however, once a branch is cut off, it does not bear fruit; if the branch is cracked, it still hangs on, and it will try and save the situation by forming more flowers and, thereby, more fruits; for example, in pear trees where the crack-method gave an up to 50% higher yield; however, the method can also be used on apple, cherry, and plum trees; the branch is cracked relatively close to the trunk; on plum trees, where the branches are more brittle, one can twist the branch instead of snapping it; the cracking should be done in the early spring when the sap in tree is rising; the wound will heal quickly, and there is no entry for fungal diseases; besides resulting in a greater yield, the cracked branches also provide for more comfortable fruit-picking *hort meth* >>> self-pruning

pruning shears: In plant grafting, they should be the scissors or sliding-blade type rather than the blade and anvil type; if the blade and anvil pruner is used to harvest scion wood, plant tissue will be crushed; as with knives, pruning and lopping shears should be kept sharp to give clean, close cuts *hort*

pseudanthium (*pl* **pseudanthia**): A special type of inflorescence, in which several flowers are grouped together to form a flower-like structure; the real >>> flowers are generally small and greatly reduced, but can sometimes be quite large (as in the sunflower flowerhead); pseudanthia take various forms and occur in many plant families, e.g., Apiaceae, Asteraceae, Campanulaceae, Centrolepidaceae, Cornaceae, Cyperaceae, Dipsacaceae, Euphorbiaceae, Eriocaulaceae, Hamamelidaceae, Moraceae, Poaceae, Pontederiaceae, Proteaceae, Rubiaceae, or Saururaceae *bot*

pseudoallele: Genes that behave as alleles in the allelism test but that can be separated by crossing-over *gene*

pseudoallelism: >>> pseudoallele

pseudobivalent: A bivalent-like association of two mitotic chromosomes due to reciprocal chromatid exchange *cyto*

pseudobulb: Enlarged, bulbous stem segments that often occur on the stems of epiphytic orchids; plants use this to temporarily store water and food during times of drought *bot*

pseudocarp: A fruit consisting of one or more ripened ovules attached or fused to modified bracts or other non-floral structures *bot*

pseudocereals: The grain amaranths, mainly quinoa (*Chenopodium quinoa*), and buckwheat (*Fagopyrum* spp.), which are cereal-like grains that are not of grass origin *agr*

pseudochromosome: Rod-like GOLGI bodies of the spermatocytes *cyto biot*

pseudocompatibility: The occurrence of fertilization that normally is prevented by incompatibility mechanisms; it is caused by specific environmental or genotypic conditions; for example, in rye, self-incompatibility can be broken by high temperatures (about +30°C) so that seeds are set *bot*

pseudodominance: The apparent dominance of a recessive gene (allele), owing to a deletion of the corresponding gene in the homologous chromosome *gene*

pseudodominant: >>> pseudodominance

pseudofertility: >>> pseudocompatibility

pseudogamous heterosis: Increased vigor of maternal offspring due to male parent influence on the endosperm *bot gene*

pseudogamy: A type of apomixis in which the diploid egg cell develops into the embryo without fertilization of the egg cell, although only after fertilization of the polar nuclei with one of the sperm cells from the male gamete to form a normal triploid endosperm *bot*

pseudogene: Genes that have been switched off in evolution and no longer have any function; they are entirely neutral and evolve at a constant rate; they are formed by duplication or retrotransposition, and loss of gene function by disabling mutations; most pseudogenes evolve rapidly in terms of sequence and expression levels, showing tissue- or stage-specific expression patterns; a large fraction of nontransposable element regulatory noncoding RNAs originate from transcription of pseudogenes proximal upstream regions; rapid rewiring of pseudogenes transcriptional regulatory regions is a major mechanism driving the origin of novel regulatory modules *gene*

pseudogenization: An evolutionary phenomenon whereby a gene loses its function by disruption to its regulatory or coding sequence; such loss of function is generally thought to be detrimental to an organism and selectively disadvantageous; since gene loss leads to immediate loss of gene function, it probably affects organisms to a greater extent than do most amino acid replacements; however, if a nonessential gene is pseudogenized, there may be no selective disadvantage and such pseudogenization can be selectively neutral; OLSON (1999) proposed the "less-is-more" hypothesis, suggesting that gene loss may serve as an engine of evolutionary change; currently in humans, over 16,000 pseudogenes are estimated to be present; many of these are unprocessed pseudogenes, meaning that they contain remnants of the exon–intron structure of a functional gene but have insertion, deletion, or substitution mutations that eliminate their ability to produce a complete functional protein; these pseudogenes represent previously functional genes in the genome that have been lost during evolution *biot*

pseudoheterosis: Luxuriance; it designates hybrids between species, varieties, or lines that exceed the parents in some traits, however, neither by sheltering deleterious genes nor by balanced gene combinations *gene* >>> heterosis

pseudoisochromosome: A chromosome that shows only equal ends as a result of >>> interchanges; chromosomes pair during meiotic >>> metaphase I like an >>> isochromosome; but the interstitial segments (proximal to the >>> centromere) are non-homologous *cyto*

***Pseudomonas syringae*:** A bacteria that attacks several crop plants (for example, rye, *Secale cereale*); in genetic engineering, a genetically modified strain of this bacteria lacks a cell-surface protein that helps ice crystals to form; spraying these bacteria on crops may prevent freezing-related damages *biot phyt*

pseudo-overdominance: It is due to epistasis (*a1b1* complementary) and tight repulsion linkage *gene* >>> heterosis >>> overdominance

pseudoresistance: Apparent resistance in a potentially susceptible host resulting from chance, transitory (non-heritable) traits, or environmental conditions *phyt*

pseudostem: A false stem; for example, bananas have pseudostems which look like tree trunks but are not; each banana stem consists of layers of leaf sheaths, with the flower peduncle growing up through the center and emerging at the center of the crown *bot*

pseudovivipary: Vegetative proliferation of plantlets in the inflorescence axes *bot*

psychophily: Pollination by butterflies *bot*

PTGS: >>> post-transcriptional gene silencing

Pto: A specific resistance gene in tomato, which controls resistance against the bacterium *Pseudomonas syringae pv. tomato*; the gene encodes a cytosolic protein kinase; the *Pto* system remains the best characterized specific resistance gene system *phyt*

pubescent: Covered with soft hairs, downy *bot*

pUC18 vector A plasmid cloning vector; size: 2.7 kb; ~100 copies per chromosome; it shows ampicillin resistance for selection and alpha-complementing fragment of beta-galactosidase with in-frame polylinker for cloning *biot*

puff: A structural modified region of a polytene chromosome; it originates from the despiralization of deoxyribonucleoprotein *cyto* >>> polytene chromosome >>> lampbrush chromosome >>> Figure 66

puffing: >>> puff >>> Figure 66

pulp: The soft, succulent part of a fruit, usually composed of mesocarp and/or the pith of a stem *bot*

pulse: The edible seeds of any leguminous crop *bot* >>> legumes

pulsed-field gel electrophoresis: An electrophoretic technique in which the gel is subjected to electrical fields alternating between different angles, allowing very large DNA fragments to move through the gel, and hence permitting efficient separation of mixtures of such large fragments *meth biot*

pulvini: A swelling at the base of a leaf or a branch of the inflorescence *bot*

PUNNETT square: A diagrammatic representation of a particular cross used to predict the progeny of the cross; a grid is used as a graphic representation of the progeny zygotes resulting from different gamete fusions in a specific cross, or a checkerboard used in finding all possible zygotes produced by a fusion of parental gametes; this results in determining the genotypic and phenotypic ratios *meth*

pupa: The non-feeding stage between the larva and adult; organism undergoes complete transformation within a protective cocoon or hardened case *zoo phyt*

purebred: Derived from a line subjected to inbreeding; offspring that are the result of mating between genetically similar kinds of parents; it is the opposite of >>> hybrid; purebred is the same as >>> true breeding *gene*

pure line: A number of individuals of a successive, self-pollinated crop, which derives from a single plant; a strain homozygous at all loci *gene meth*

pure line breeding: >>> true breeding

pure-live seed: The percentage of the content of a seed lot that is pure and viable; it is determined by multiplying the percentage of pure seed by the percentage of viable seed (germination percentage) and dividing by 100 *seed*

purity testing: Determination of the degree of contamination of seed lots with genetically non-identical, damaged, or pest-infected seeds; the quality of a seed lot is judged by the relative percentage of various components; the quality is considered superior if pure seed percentage is above 98, and the other-species seeds and inert matter percentage is as low as possible; the percentage of seeds of other species should be almost negligible, or nil; genetic purity of seeds refers to the trueness to type; if the seed possesses all the genetic qualities that the breeder has placed in the variety, it is said to be genetically pure; the genetic purity has a direct effect on ultimate yields; if there is any deterioration in the genetic makeup of the variety during the seed multiplication and distribution cycle, there will definitely be a proportionate decrease in its performance (e.g., yield, disease resistance, etc.); genetic purity of a variety can deteriorate due to several factors during production cycles, such as (1) development variations (when the seed crops are grown in a difficult environment, under different soil and fertility conditions, or different climatic conditions, or under different photoperiods, or at different elevations for several consecutive generations, the development variations may arise sometimes as differential growth response; to minimize the opportunity for such shifts to occur in varieties, it is advisable to grow them in their areas of adaptation and growing seasons); (2) mechanical mixtures (this is the most important source of variety deterioration during seed production; they may often take place at the time of sowing, if more than one variety is sown with the same seed drill, through volunteer plants of the same crop in the seed field, or through different varieties grown in adjacent fields; two varieties growing alongside each other in the field are often mixed somewhat during harvesting and threshing operations; often the seed produce of all the varieties are kept on same threshing floor, resulting in considerable varietal mixtures; to avoid this sort of mechanical contamination it would be necessary to rogue the seed fields, and practice the utmost care during seed production, harvesting, threshing, and further handling); (3) mutations (this is not a serious factor of varietal deterioration; in the majority of the cases it is difficult to identify or detect minor mutations; in the case of vegetatively propagated crops, periodic increase of true-to-type stock would eliminate the deteriorating effects of mutations); (4) natural crossing (in sexually propagated crops, natural crossing is an important source of varietal deterioration due to introgression to genes from unrelated stocks, which can only be solved by prevention); (5) minor genetic variations (these may exist even in the varieties appearing phenotypically uniform and homogeneous at the time of their release; during later production cycles some of these variations may be lost because of selective elimination by the environment); (6) selective influence of diseases (new crop varieties often become susceptible to new races of diseases, often caused by obligate parasites, and are out of seed programs; similarly the vegetatively propagated stocks deteriorate fast if infected by viral, fungal, or bacterial diseases; during seed production it is, therefore, very important to produce disease-free seeds/ stocks); and (7) techniques of plant breeder (in certain instances, serious instabilities may occur in varieties due to cytogenetic irregularities not properly assessed in the new varieties prior to their release; premature release of varieties, still segregating for resistance and susceptibility to diseases or other factors, which may also be important in the deterioration of varieties; other factors, such as breakdown in male sterility, certain environmental conditions, and other heritable variations, may considerably lower the genetic purity); steps for maintaining genetic purity are: (1) providing adequate isolation to prevent contamination by natural crossing or mechanical mixtures; (2) rogueing of seed fields prior to the stage at which they could contaminate the seed crop; (3) periodic testing of varieties for genetic purity; (4) avoiding genetic shifts by growing crops in areas of their adaptation only; (5) certification of seed crops to maintain genetic purity and quality of seed; (6) adopting the generation system; and (7) grow out tests *seed*

puroindoline proteins: Small, basic, cysteine-rich proteins found in bread wheat (*Triticum aestivum*); by engineered introduction of a puroindoline gene, the hard-textured grain of durum wheat (*Triticum durum*) and other cereals, including maize and barley, can be converted into soft texture (MORRIS and GEROUX 2000); puroindoline genes, puroindoline a and puroindoline b, represent the grain hardness gene on chromosome 5DS; in hexaploid wheat, soft kernel texture is the wild type, whereas hard texture (hardness) results from any one of several mutations in the puroindolines *phys gene*

push–pull system: A strategy for controlling agricultural pests by using repellent "push" plants and trap "pull" plants, e.g., cereal crops like maize or sorghum are often infested by stem borers; grasses planted around the perimeter of the crop attract and trap the pests, whereas other plants, like *Desmodium*, planted between the rows of maize, repel the pests and control the parasitic plant *Striga*; the technology was developed at the International Centre of Insect Physiology and Ecology (ICIPE) in Kenya in collaboration with Rothamsted Research (U.K.) and national partners *agr meth*

pustule: A blister-like spot or spore mass developing below the epidermis, which usually breaks through at maturity *phyt*

P value: Probability value; a decimal fraction showing, for example, the number of times an event will occur in a given number of trials *stat*

PVPA: >>> Plant Variety Protection Act

pycnidium (pycnidia *pl*): A flask-shaped fungal receptacle bearing asexual spores, that is, pyc-
    niospores *bot*

pycniospore: A spore from a pycnidium *bot phyt*

pycnotic: The concentration of the nucleus into a compact, strongly stained mass, taking place as
    the cell dies *cyto*

pyramidal: Triangular in outline; shaped like a pyramid *bot*

pyramiding of genes: The process of bringing together several disease-resistance or agronom-
    ically important genes from different sources into one genetic background (genotype);
    increasing genetic values relies on increasing the frequency of favorable genes controlling
    a given trait; to create a superior genotype many genes and/or alleles have to be combined
    and assembled; this process is called pyramiding *biot meth gene*

pyrenoid: Refers to a proteinaceous region of chloroplast, sometimes the site of accumulation of
    carbohydrate reserves *bot phys*

pyricularia leaf spot disease (of pearl millet): One of the two most destructive diseases to pearl
    millet in the U.S., caused by *Pyricularia grisea phyt* >>> Figure 62

pyriform: Pear-shaped *bot*

pyrimidine: A basic, six-membered heterocyclic compound; the principal pyrimidines, uracil,
    thymine, and >>> cytosine, are important constituents of nucleic acids *chem phys* >>>
    Table 38

***Pythium* damping-off and root rot (of soybean):** A disease causing poor stands and consequently
    reduced yields; resistance to seedling diseases caused by *Pythium* spp. was reported in
    the soybean cultivar "Archer" and was described to be associated with the *Rps1k* gene for
    resistance to *Phytophthora sojae phyt* >>> Figure 62

Q

Q band(s): Bands on chromosomes stained by >>> quinacrine mustard exhibiting bright and dark
    bands under >>> UV light; Q bands are A-T rich regions of the chromosome *cyto* >>> Q
    banding

Q banding: A chromosomal staining technique using the fluorescence dye quinacrine mustard;
    under UV light a characteristic light and dark banding is induced *cyto meth* >>> Q band

qPCR: >>> polymerase chain reaction (PCR) technique

QTL: >>> quantitative trait locus

QTL mapping: A statistical study of the alleles that occur in a locus and the phenotypes (physical
    forms or traits) that they produce; because most traits of interest are governed by more than
    one gene, defining and studying the entire locus of genes related to a trait gives hope of
    understanding what effect the genotype of an individual might have; the statistical analysis
    is required to demonstrate that different genes interact with one another and to determine
    whether they produce a significant effect on the phenotype; QTLs identify a particular
    region of the genome as containing a gene that is associated with the trait being assayed or
    measured; they are shown as intervals across a chromosome, where the probability of asso-
    ciation is plotted for each marker used in the mapping experiment; the QTL techniques
    were developed in the late 1980s and can be performed on inbred lines of any species; to
    begin, a set of genetic markers must be developed for the species in question; a >>> marker
    is an identifiable region of variable DNA; the aim is to find a marker that is significantly
    more likely to co-occur with the trait than expected by chance, that is, a marker that has a
    statistical association with the trait; ideally, they would be able to find the specific gene or
    genes in question, but this is time-consuming and difficult; instead, they can more readily
    find regions of DNA that are very close to the genes in question; when a QTL is found,

it is often not the actual gene underlying the phenotypic trait, but rather a region of DNA that is closely linked with the gene; for organisms whose genomes are already known, one might now try to exclude genes in the identified region whose function is known with some certainty not to be connected with the trait; if the genome is not available, it may be an option to sequence the identified region and determine the putative functions of genes by their similarity to genes with known function, usually in other genomes; this can be done using >>> BLAST, an online tool that allows users to enter a primary sequence and search for similar sequences within the BLAST database of genes from various organisms; QTL mapping is also used to determine the complexity of the genetic architecture underlying a phenotypic trait, e.g., they may be interested in knowing whether a phenotype is shaped by many independent loci, or by a few loci, and whether those loci interact; this can provide information on how the phenotype may be evolving *biot meth* >>> interval mapping >>> composite interval mapping

quadriduplex type: >>> autotetraploid >>> nulliplex type

quadripartitioning: During meiosis, >>> cytokinesis does not occur after meiosis I and it is postponed until after second division *cyto*

quadrivalent: A chromosome association of four members *cyto* >>> Figure 15

quadruplex type: >>> autotetraploid

qualitative character: A character in which variation is discontinuous *gene* >>> Table 33

qualitative inheritance: An inheritance of a character that differs markedly in its expression amongst individuals of a species; variation is discontinuous; such characters are usually under the control of major genes *gene* >>> Table 33

qualitative resistance: >>> vertical resistance

quality-declared seed: A terminology introduced by the FAO for a seed system in which a proposed 10% of the seed produced and distributed is checked by an autonomous seed control agency, and the rest by the seed-producing organization *seed*

quantitative character: A character in which variation is continuous so that classification into discrete categories is arbitrary *gene* >>> metric character >>> Figure 43 >>> Table 33 >>> QTL mapping

quantitative genetics: A branch of genetics that deals with the inheritance of quantitative traits; sometimes it is also called biometrical or statistical genetics *gene* >>> QTL mapping

quantitative inheritance: An inheritance of a character that depends upon the cumulative action of many genes, each of which produces only a small effect; the character shows continuous variation (i.e., a graduation from one extreme to the other) *gene* >>> QTL mapping >>> Figure 43 >>> Table 33

quantitative resistance: >>> horizontal resistance

quantitative trait locus (QTL): A locus or DNA segment that carries more genes coding for agronomic or other traits, i.e., a region of DNA that is associated with a particular phenotypic trait – these QTLs are often found on different chromosomes; knowing the number of QTLs that explains variation in the phenotypic trait tells us about the genetic architecture of a trait, e.g., plant height is controlled by many genes of small effect, or by a few genes of large effect; another use of QTLs is to identify candidate genes underlying a trait; once a region of DNA is identified as contributing to a phenotype, it can be sequenced; the DNA sequence of any genes in this region can then be compared to a database of DNA for genes whose function is already known *gene* >>> QTL mapping >>> Figure 43 >>> Tables 33, 58

quantitative vertical resistance: A rare type of resistance; it occurs in cereals, e.g., Hessian fly resistance (*Mayetiola destructor*) in wheat or rice blast; it is a confusing term because its inheritance is qualitative while its effects are quantitative; it can easily be confused with horizontal resistance, and the best way to avoid it in a breeding program is by choosing only parents that exhibit the normal, qualitative, vertical resistance *phyt* >>> quantitative resistance >>> horizontal resistance

quantum speciation: The rapid rise of a new species, usually in small isolates, with the "founder effect" and >>> random genetic drift; it is also called saltational speciation *tax evol*

quarantine: The official confinement of plants subject to phytosanitary regulations for observation and research or for further inspection and/or testing; more general, a legal ban on the export or import of certain noxious weeds or insects that may be attached to the plants *meth seed agr*

quarternary hybrid: Considered >>> a hybrid derived from four different grandparental individuals *meth*

quartet: The four nuclei and/or cells produced during meiosis *bot* >>> tetrad

quasibivalent: >>> pseudobivalent

quasidominance: Direct transmission, generation to generation, of a recessive trait giving the impression of dominance; it happens if the recessive gene is frequent or inbreeding is intense *gene*

quasi-complete Latin square: Considered a >>> Latin square because every entry appears once in each row and once in each column; it is quasi-complete because every entry is a horizontal neighbor to every other entry exactly twice, and every entry is a vertical neighbor to every other entry exactly twice *stat* >>> design of experiment >>> Figure 49

quassia: An organic pesticide sometimes recommended for controlling aphids, caterpillars, or whiteflies; reputed to be non-toxic to humans; a quassia spray is mixed from crushed Quassia bark and boiling water (about 1/4 cup of bark to 1 quart of water); after being allowed to cool, the mixture is strained, leaving the liquid ready to spray *phyt meth hort*

quelling: Posttranscriptional gene silencing in *Neurospora crassa* induced by the introduction of a transgene *biot* >>> posttranscriptional gene silencing

quenching: The process of causing the deexcitation of a molecule from an excited singlet or triplet state back down to the ground state; quenching of chlorophyll is an important protective role of carotenoids; if a triplet molecule is not rapidly quenched, it can react with oxygen to generate singlet oxygen; singlet oxygen can cause much cellular damage *phys*

quercitron: >>> glucoside

quick test (of seed testing): A type of test for evaluating seed quality, usually germination, more rapidly than standard laboratory tests *seed*

quiescence: The absence of growth, usually implying the absence of environmental conditions favoring growth; although dormant seed are quiescent, quiescence is distinguished from dormancy, which implies the inability to germinate even in the presence of environmental conditions favoring growth *seed meth* >>> dormancy

quilled: Describes a flowerhead consisting of narrow, tubular rye flowers *bot*

quincuncial: Of the arrangement of corolla lobes in a bud; a variant of imbricate aestivation *bot*

quincunx planting: Planting four young plants to form the corners of a square with a fifth plant at its center *meth* >>> design of experiments

quinine: An alkaloid, found in the bark of the Cinchona tree, that is used to treat malaria, a tropical parasitic disease carried by mosquitoes; it is an essential bitter ingredient of certain drinks (e.g., Indian tonic water), where it can be recognized by its blue fluorescence under black UV light illumination *chem hort*

quinone: >>> growth inhibitor

quota sampling: The non-probability equivalent of stratified sampling; like >>> stratified sampling, the researcher first identifies the stratums and their proportions as they are represented in the population; then, convenience or judgment sampling is used to select the required number of subjects from each stratum; this differs from stratified sampling, where the stratums are filled by >>> random sampling *stat*

R

R1, R2, R3, etc.: The first, second, third, etc. generation following any type of irradiation in mutation breeding *meth*

RABL configuration: In plants and insects, refers to chromosomes that are spatially organized within the nucleus, with centromeres clustered on the nuclear membrane at one pole and telomeres attached to the nuclear membrane of the other hemisphere *cyto*

race: A genetically and, as a general rule, geographically distinct interbreeding division of a species; in other contexts, a population or group of populations distinguishable from other such populations of the same species by the frequencies of genes, chromosomal rearrangements, or hereditary phenotypic characteristics; a race that has received a taxonomic name is a subspecies *tax*

RACE: Rapid amplification of cDNA ends *biot*

raceme: An inflorescence in which the main axis continues to grow, producing flowers laterally, such that the youngest ones are apical or at the center *bot*

racemose: An indeterminate, unbranched inflorescence in which the flowers are borne on pedicels of about equal length along an elongated axis *bot*

race-non-specific type: Host-plant resistance that is operational against all races of a pathogen species; it is variable, sensitive to environmental changes, and usually polygenetically controlled *phyt*

race-specific resistance (*syn* **qualitative resistance**): Individually determined by single or very few genes in the plant – the specific resistance genes – and these are most usually effective against a limited proportion of the pathogen population; the proportion of the pathogen population capable of causing disease on a resistant host is termed virulent on the host; the portion incapable of causing disease as avirulent on that host; correspondingly, the host is susceptible or resistant, and the interaction compatible or incompatible, respectively; where the inheritance of virulence/avirulence has been studied, it is generally found that a specific avirulence gene in the pathogen corresponds to a specific resistance gene in the plant; this relationship is termed the "gene-for-gene hypothesis" *phyt* >>> vertical resistance

race-specific type: Host-plant resistance that is operational against one or a few races of a pathogen species; generally produces an immune or hypersensitive reaction and is controlled by one or a few genes *phyt*

rachilla: The spikelet axis; also applied to the segment of the rachilla that remains attached to the oat grain *bot*

rachis: The main axis of the ear and of the panicle of grasses *bot*

rad: An abbreviation for "radiation absorbed dose"; a measure of the amount of any ionizing radiation that is absorbed by the tissue; one rad is equivalent to 100 ergs of energy absorbed per gram tissue *phy meth* >>> radioactive

radiation absorbed dose: >>> rad

radiation use efficiency: The amount of dry weight of a crop that is produced from a unit amount of radiation on the crop; numerical values must have precise definitions of the parts of the crop measured, the part of solar radiation considered, and whether the radiation was absorbed or intercepted by the crop *phys*

radicle: The root of the embryo, which develops into the primary root of the seedling *bot*

radicule: >>> radicle

radioactive: Pertaining to a substance when a constituent chemical element is undergoing the process of changing into another element through the emission of radiant energy; radioactivity is used as a tool in research to tag or trace the movement of compounds; the presence of a compound containing a radioactive element is revealed by instruments that measure the radiant energy emitted, or by radio-sensitive films *phy meth*

radioactive tracer: >>> isotopic tracer

radiocarbon dating (*syn* **14C dating**): A dating method for organic material that is applicable to about the last 70,000 years; it relies on the assumed constancy over time of atmospheric 14C:12C ratios and the known rate of decay of radioactive carbon, of which half is lost in a period of about every 5,730 years *meth*

radio frequency impedance spectroscopy (RFIS): >>> biomass

radioisotope: In cytology, radioactive labeling of nucleic acids is usually achieved by enzymatically incorporating nucleotides containing 32P, 35S, 125I, or 3H; the labeled nucleotide is essentially identical to its unlabeled counterpart; the labeled probe is not susceptible to steric hindrance during hybridization, and this may be one reason why radioactive probes can be more sensitive than non-radioactive probes; the particular isotope that is chosen will depend upon the application, since there is an inverse relationship between sensitivity and resolution, e.g., the high emission energy (E_{max}) of 32P makes it suitable for detecting sequences in less than 7 days, but gives a wide scattering of silver grains in photographic emulsions during autoradiography and low resolution of the signal; in contrast, the weak beta-emission of 3H results in a high resolution of signal, but it can take at least 2 weeks to expose the photographic emulsion; advantages of radioactive detection are higher sensitivity, the fact that non-radioactive methods require elaborate detection methods (e.g., the supernatant and pellet cannot just be tested with a GEIGER counter to see where the sample went), and that radioactivity allows for highly accurate quantitation over many orders of magnitude *cyto meth* >>> Table 37

raffinose: The raffinose family of oligosaccharides; believed to play an important role in the resistance of plants to environmental stress by protecting membrane-bound proteins; it is relevant when seeds mature because they rapidly lose moisture as they dry out *phys* >>> www.plantstress.com

raingrown cotton: Cotton grown using mainly water provided by the natural cycle of rainfall rather than artificial irrigation, also known as dryland cotton *agr* >>> barbadense

rainfed: Farming practices that rely on rainfall for water; it provides much of the food consumed by poor communities in developing countries, e.g., rainfed agriculture accounts for more than 95% of farmed land in sub-Saharan Africa, 90% in Latin America, 75% in the Near East and North Africa, 65% in East Asia, and 60% in South Asia *agr eco*

rainfed agroecosystem: A farming system in which crop water needs are met by natural precipitation *agr*

rain forest: A tropical forest of tall, densely growing, broad-leaved evergreen trees in an area of high annual rainfall *eco*

rain gauge: An instrument for measuring rainfall *prep*

ramet: An individual that belongs to a clone *bot* >>> cutting

ramification: A physiologically independent individual that belongs to a clone and/or makes up a genet *bot* >>> genet >>> cutting

ramify: >>> ramification

rancidity: Off-flavor and aroma that occurs when foods or feeds containing oil deteriorate as the result of oxidation of lipids (oils) to form volatile compounds, including aldehydes and fatty acids; another form of rancidity is caused by hydrolysis of the glycerides (oils and fats) to form free fatty acids; this is the major cause of rancid butter *chem*

random amplified polymorphic DNA (RAPD) technique: A comparative study (among individuals, populations, or species) of the DNA fragment length produced in controlled DNA synthesis reactions started with short sequences of DNA (primers); as a genetic mapping methodology, it utilizes as its basis the fact that specific DNA sequences (polymorphic DNA) are repeated (i.e., appear in sequence) with a gene of interest; thus, the polymorphic DNA sequences are linked to that specific gene; their linked presence serves to facilitate genetic mapping within a genome *biot meth*

random effects: Effects of a treatment in an experiment in which the treatments are considered in relation to a whole range of possible treatment effects *stat*

random genetic drift: Variation in gene frequency from one generation to another due to chance fluctuations *gene evol*

randomization: The process of making assignments at random; in field trials, randomization of entries is required to obtain a valid estimate of experimental error; each entry must have an equal chance of being assigned to any plot in a replication, and an independent randomization is required for each replication *stat meth* >>> design of experiment >>> Tables 25, 40 >>> www.randomizer.org/

randomized block: The entries to be tested are assigned at random to the plots within the block *stat meth* >>> design of experiment >>> Tables 25, 40 >>> www.randomizer.org/

randomized block design: A randomized block analysis of variance design (e.g., one-way blocked ANOVA) is created by first grouping the experimental subjects into blocks; the subjects in each block are as similar as possible (e.g., littermates); there are as many subjects in each block as there are levels of the factor of interest; randomly assigning a different level of the factor to each member of the block, such that each level occurs once and only once per block; the blocks are assumed not to interact with the factor *stat* >>> design of experiment >>> randomized block >>> randomized complete block >>> Tables 25, 26, 40 >>> www.randomizer.org/

randomized complete block (RCB): Widely used in field experimentation; it is an extension of the paired t-test; this design is appropriate when quantitative data are collected, such as yield, and rigorous comparison between treatments is required; two cornerstones of the design are replication (i.e., repetition) and randomization; allows accommodation of any variability in the local environment and determination of the probability of the differences in results between treatments being real or simply due to chance; three replications are the minimum; more are better, but the statistical advantage gained is successively smaller with each added replicate; each treatment must be included once in each block of replications; the treatment locations must be randomly assigned to plots within the block; the purpose of randomizing the locations is to avoid biasing the results; if field plots are the basic experimental unit, the individual plots should be three to five times long as wide, and should be sized to comfortably handle one or two passes of the field equipment being used; proper plot location is important to reduce bias in the results; if the plots are on sloping land, run the long axis of the plots up and down the slope; with this layout, each plot will contain a portion of each slope position; similarly, if soil characteristics gradually change across the study site, run the long axis of the plots parallel to the gradient of soil variability; each block contains a plot for each candidate to be tested; in that case, the classification of the data according to the blocks and the classification according the candidates are orthogonal; two standard statistical tests are used together in analysis of data from this design; an F-test, commonly called "analysis of variance" (ANOVA), is used to determine if there are significant differences between some of the treatments; if the F-test shows that significant treatment differences do exist, then a "means comparison test", such as the Dunn's test or Duncan's multiple range test, is used to determine which "treatment means" (i.e., treatment averages) are actually different from one another; there are two advantages to using this design: one is the ability to compare numerous treatments using one analysis (F-test); the second is the ability to separate out differences between replicates caused by environmental gradients *stat meth* >>> design of experiment >>> replicated measurement t-test design >>> Tables 25, 40 >>> www.randomizer.org/

random lines: >>> isogenic lines

random mating: For a given population, where an individual of one sex has an equal probability of mating with any individual of the opposite sex, or insofar as the genotypes with respect to given genes are concerned *gene* >>> Table 37

random polycross: A >>> polycross scheme in which the pollination is random; this is possible with >>> allogamous >>> species, and with an autogamous species, which responds to a male gametocide, or which has an easily controlled genetic male sterility; the advantage is that it produces very large numbers of crosses with very little labor; the disadvantage is that there is no control over the pollination and some parents may be more widely represented than others *meth* >>> design of experiment >>> Figure 51

random priming: A method of DNA labeling; it was introduced by FEINBERG and VOGELSTEIN and depends on the ability of the >>> KLENOW fragment of DNA polymerase to copy single-stranded DNA templates primed with random hexanucleotide mixtures *biot meth*

random sample: A sample of a population selected so that all items in the population are equally likely to be included in the sample *stat*

random sampling: A sample drawn from a population in such a way that every individual of the population has an equal chance of appearing in the sample; it ensures that the sample is representative and provides the necessary basis for virtually all forms of inference from sample to population, including the informal inference, which is characteristic of rerandomization statistics *stat*

rangeland: Land on which the native vegetation is predominantly grasses, grass-like plants, forbs, or shrubs *eco*

range of reaction: The range of all possible phenotypes that may develop by interaction with various environments from a given genotype; it is also called "norm of reaction" *gene*

range pole: >>> landmark

RAPD: >>> random amplified polymorphic DNA technique

raphe: A ridge, sometimes visible on the seed surface, which is the axis along which the ovule stalk joins the ovule *bot*

rapid multiplication: Techniques to accelerate production of a new plant cultivar; for example, green >>> cuttings of a >>> potato clone can be rooted in a mist propagator and, when planted out, the cuttings will themselves produce more cuttings *meth* >>> Figure 50

rare single nucleotide polymorphism: SNPs that occur at a frequency of less than 1% in a population *biot*

rate of response to selection: >>> selection progress

ratoon cotton: Crop that is cut back or cropped and is left to grow again for another season *agr*

ratooning: A sprout or shoot from the root of a plant (e.g., in sugarcane) after it has been cropped *bot*

ratooning ability: Regeneration of the main stem and tillers after harvesting *phys*

ratooning crop: Obtaining a second crop from the same plant (e.g., in sugarcane) *agr*

ray: A pedicel in an umbellate inflorescence *bot*

ray flower: A flower head with outer ray flowers forming petallike structures surrounding the inner disc flowers (e.g., in the Asteraceae or in sunflower) *bot*

RCB: >>> randomized complete block

rDNA: >>> ribosomal DNA

rDNA-ITS: >>> ribosomal DNA internal transcribed spacers

reading frame: The mechanism that moves a >>> ribosome, one codon at a time, from a designated start sequence during genetic translation; a shift in the reading frame by any number of nucleotides other than three, or multiples of three, will cause an entirely new sequence of codons *gene*

reading mistake: The placement of an incorrect amino acid into a >>> polypeptide chain during protein synthesis *gene* >>> reading frame

reagent: A substance that, because of the reactions it causes, is used in analysis and synthesis *meth chem*

realized genetic gain: Refers to the observed difference between the mean phenotypic value of the offspring of the selected parents and the phenotypic value of the parental generation before selection *meth*

realized heritability: Refers to heritability measured by a response to selection; it is the ratio of the single-generation progress of selection to the selection differential of the parents *stat*

reaper: A machine for cutting standing grain *agr*

reaper-binder: An implement that cuts and ties hay into bundles *agr*

rearrangement: All chromosome mutations that result in modified karyotypes *cyto* >>> translocation >>> interchange >>> chromosome mutation

reassociation: >>> anneal

rebloom: In ornamental plants, a valuable characteristic in which a plant blooms at its normal period and then, after a period of rest, produces a second set of flowers *hort*

recalcitrant seed: Seed that does not survive drying and freezing, e.g., avocado, mango, lychee; in particular, seed that cannot withstand either drying or temperatures of less than +10°C and, therefore, cannot be stored for long periods, as compared to >>> orthodox seeds *seed*

receptacle: The part of the stem from which all parts of the flower arise; in Compositae, the flattened tip of the stem that bears the bracts and florets *bot*

receptaculum: >>> receptacle

receptor site: In molecular genetics, a set of reactive chemical groups in the cell wall of a bacterium, which are complementary to a similar set in the tailpiece of a bacteriophage *gene*

recessive: Describes a gene and/or allele whose phenotypic effect is expressed in the homozygous state but masked in the presence of the dominant allele; usually the dominant gene and/or allele produces a functional product, whereas the recessive gene and/or allele does not; both one and two doses per nucleus of the dominant allele may lead to expression of its phenotypes, whereas the recessive allele is observed only in the complete absence of the dominant allele *gene* >>> Table 6

recessive allele: >>> recessive

recessive epistasis: The effect of a recessive allele of a gene suppressing the phenotype manifestation of another gene *gene* >>> Table 6

recessiveness: >>> recessive

recipient: One that receives; receiver *gene meth*

recipient cell: >>> recipient

reciprocal cross: One of a pair of crosses in which the two opposite mating types are each coupled with each of two different genotypes and mated with the reciprocal combination; for example, male of genotype A × female of genotype B (first cross) and male of genotype B × female of genotype A (reciprocal cross); such crosses are used to detect (1) sex linkage, (2) maternal inheritance, or (3) cytoplasmic inheritance *meth*

reciprocal full-sib selection: A method of interpopulation improvement for species in which the commercial product is hybrid seed; a cycle of selection is completed in the fewest number of seasons by use of plants from which both selfed and hybrid seed can be obtained *meth* >>> Figure 55 >>> Table 37

reciprocal genes: Non-allelic genes that reciprocate or complement one another *gene*

reciprocal half-sib selection: >>> reciprocal recurrent selection

reciprocal recurrent selection: A breeding method used to achieve an accumulation of genes that are valuable for specific traits but also for combining ability; in practice, two populations form the basis of selection; reciprocally, one population serves as a tester for the investigation of the selections deriving from the other population; from the first population, usually a greater number of plants is selfed and at the same time crossed with several plants of the second population; the same procedure is realized with the second population; during the second year, the progenies of the test-crosses are subjected to performance trials; the progenies of the cross of one female parent, derived from population 1, are combined with several male parents from population 2; based on the performance testing, progenies from selfed seed, produced by the best plants, are grown during the third year; in the same year, the best progenies of selfings are crossed in many combinations; seeds obtained from those crosses form the improved populations for growing during the fourth year; within such populations, individual plants may be selfed again and crossed for utilization in a new cycle *meth* >>> recurrent selection >>> Figures 21, 51

reciprocal translocation: A translocation that involves an exchange of chromosomes segments between two non-homologous chromosomes *cyto* >>> translocation >>> Robertsonian translocation >>> chromosome mutation

recognition sequence (site): A nucleotide sequence composed typically of four, six, or eight nucleotides; it is recognized by a restriction endonuclease; so-called type II enzymes cut (and their corresponding modification enzymes methylate) within or very near the recognition sequence *biot gene* >>> Table 38

recombinant: An individual or cell with a genotype produced by recombination (i.e., with combinations of genes other than those carried in the parents); they result from independent assortment or crossing over *gene*

recombinant DNA molecule: DNA molecule created by ligating together two not normally contiguous DNA molecules *biot*

recombinant DNA technology: DNA molecules constructed by joining, outside the cell, natural or synthetic DNA segments to DNA molecules capable of replication in living cells *biot* >>> biotechnology

recombinant inbred lines (RILs): A population of fully homozygous individuals that is obtained through the repeated selfing of an F1 hybrid, and that comprises 50% of each parental genome in different combinations; they derive from multiple inbred strains (either by selfing or by full-sib mating) and can serve as a powerful resource for the genetic dissection of complex traits; however, the use of such multiple-strain inbred lines requires a detailed knowledge of the haplotype structure in such lines *gene meth stat biot* >>> mapping population >>> Table 61

recombinant protein: A protein synthesized from a cloned gene *biot*

recombinant type: An association of genetic markers, found among the progeny of a cross, that is different from any association of markers present in the parents *gene*

recombinase: A group of enzymes that catalyze the joining of two DNA molecules after recognizing the recombination sites *phys gene biot*

recombination: Meiotic recombination is initiated by double-strand break formation in chromosomal DNA at a large number of sites in the genome; however, few breaks, generally one to two per chromosome, result in the formation of crossovers; despite evolutionary conservation of recombination mechanisms, there are substantial differences in recombination landscapes among species as well as among genotypes and sexes within species; they often differ between male and female meioses with regard to the number of recombination events, their overall distribution along chromosomes, and fine-scale location relative to genes and chromatin features; meiotic recombination events occur in open chromatin regions, but specific characteristics of chromatin at recombination sites vary among species; the process whereby new combinations of parental characters may arise in the progeny, caused by exchange of genetic material of different parental lines; it is a requirement for response to selection, but researchers still debate whether increasing recombination beyond normal levels will result in significant gains in short-term selection; the hypothesis was tested, in the context of plant breeding, through a series of simulation experiments comparing short-term selection response ($\leq$20 cycles) between populations with normal levels of recombination and similar populations with unconstrained recombination (i.e., free recombination); additive and epistatic models were considered by examination of a wide range of values for key design variables: selection cycles, QTL number, heritability, linkage phase, selection intensity, and population size; with few exceptions, going from normal to unconstrained levels of recombination produced only modest gains in response to selection (~11% on average); breeders might capture some of this theoretical gain by increasing recombination through either (1) extra rounds of mating or (2) selection of highly recombinant individuals via use of molecular markers/maps; all methods tested captured less than half of the potential gain; another analysis indicates that the most effective method is to select for increased recombination and the trait simultaneously; this recommendation is based on evidence of a favorable interaction between trait selection and the impact of recombination on selection gains; when the relative contributions of the two components of meiotic recombination, chromosome assortment, and crossing over is examined for short-term selection gain, then, depending primarily on the presence of trait selection pressure, chromosome assortment alone accounted for 40–75% of gain in response to short-term selection; in 2018, a single *recq4* mutation of >>> *Arabidopsis thaliana* was identified that significantly increases >>> crossing over and recombination in several crops plants species *gene* >>> mapping population >>> gametocide >>> Figure 24 >>> Table 22

recombination counting and ordering (RECORD) method: A method for the ordering of loci on genetic linkage maps; it minimizes the total number of recombination events; the search algorithm is a heuristic procedure, combining elements of branch-and-bound with local reshuffling; it does not require intensive calculations; the algorithm rapidly produces an optimal ordering as well as a series of near-optimal ones; the latter provides insight into the local certainty of ordering along the genetic map; compared to "JoinMap", RECORD is much faster and less sensitive to missing observations and scoring errors, since the optimization criterion is less dependent on the position of the erroneous markers; RECORD performs better in regions of the map with high marker density *stat meth*

recombination frequency: The number of recombinants divided by the total number of progeny, expressed as a percentage or fraction; such frequencies indicate relative distances between loci on a genetic map *gene* >>> mapping >>> Table 22

recombination nodule: Swellings along DNA strands and associated proteins during prophase pairing of chromosomes; the nodules can be identified under the electron microscope; it is suggested that recombination nodules are the sites of genetic crossing over; when crossing over occurs there is a breakage and reunion of chromatin strands; a small amount of DNA is synthesized at the breaks in short pieces of ~200 bp but repeated 700–2,000 times per haploid genome; DNA synthesis inhibitors cause an increase in chromosome breakage *cyto*

recombination system: All factors that mediate and control the process of genetic recombination *gene*

recombination unit: >>> Morgan unit

recon: The smallest unit of DNA capable of recombination *gene*

record: >>> recombination counting and ordering method

recurrent full-sib selection: A method of intrapopulation improvement that involves the testing of paired-plant crosses; it is the only method of recurrent selection in which the seeds from two individuals, rather than one, are used for testing and to form the new population *meth* >>> Table 37

recurrent half-sib selection: A method of intrapopulation improvement that includes the evaluation of individuals through the use of their half-sib progeny; the general procedure for a cycle of selection is to (1) cross the plants being evaluated to a common tester, (2) evaluate the half-sib progeny from each plant, and (3) intercross the elected individuals to form a new population *meth* >>> Table 37

recurrent mass selection: A breeding method designed to increase the levels of desirable qualities, which are quantitative variables, by changing the frequency of polygenes; in each screening generation, the best individuals are selected, and they become the parents of the next screening generation; the process is repeated for as many generations as necessary, but the rate of progress declines dramatically after a few generations *meth*

recurrent (backcross) parent: The parent to which a hybrid is crossed in a backcross; it replaces the dragged alleles step-by-step with the alleles of the original variety *meth*

recurrent reciprocal selection: A recurrent selection breeding system in which genetically different groups are maintained and, in each selection cycle, individuals are mated from the different groups to test for combining ability *meth* >>> recurrent selection >>> Figures 21, 51

recurrent selection: A method designed to concentrate favorable genes scattered among a number of individuals; it is performed by repeated selection in each generation among the progeny produced by matings inter se of the selected individuals of the previous generation; in practice, plants from a population are selfed and, after the yield of the selfed seeds, the progenies of the phenotypically best individuals are grown in the second year; the best progenies are then crossed in as many combinations as possible and the seeds received thereby are grown in the third year as a population; within the already improved population, selection and selfing can be carried out again; with this population, a second cycle of recurrent selection can be started *meth* >>> facilitated recurrent selection >>> Figures 4, 20, 21 >>> Table 37

recurved: Used to describe how the tips of some flower petals curl under *hort*

rediploidization: In anther culture, the haploidization of the genome by culturing pollen grains to haploid plantlets and its rediploidization after spontaneous or induced doubling of the chromosome set *biot* >>> *Imperata cylindrica* procedure >>> Figures 17, 26, 58 >>> Table 7

red rust: Uredospore state of rusts, particularly of cereals *phyt* >>> Table 52

reduced floret: A floret that is either staminate or neuter; if it is highly reduced (e.g., awnlike structures) then it is sometimes called a rudimentary floret *bot*

reducers: One of three basic types of organisms called producers, reducers, and consumers; reducers break down the organic chemicals of dead organisms, and they make these nutrients available for reuse by other organisms *evol*

reductase: An enzyme responsible for reduction in an oxidation–reduction reaction *chem phys*

reduction division: The two nuclear divisions in meiosis that produce daughter nuclei, each of which has half as many chromosomes as the parental nucleus *bot* >>> meiosis

reductional division: >>> reduction division >>> meiosis

redundant DNA: DNA that does not appear to be genetically active and hence is not translated or transcribed; it often consists of repeated sequences *gene*

redundant gene: A gene that is present in many functional copies, so that one copy can complement the loss of another copy *gene*

reduplication: Doubling of the genetic matter of a haploid chromosome set *cyto* >>> *Imperata cylindrica* procedure >>> rediploidization

reed: The straight stalk of any of various tall grasses growing in marshy places *agr eco*

reel (at a harvester): Draws the cut crop into the intake auger, which carries it to the center of the cutting table *agr*

reencounter parasite: When a >>> crop >>> host is taken to another part of the world, some of its parasites may be left behind in the area of origin (e.g., tropical rust, when >>> maize was taken from the New World to Africa); if the parasite arrives in the new area at a later date, it is described as a reencounter parasite, usually very damaging because the crop host tends to lose >>> horizontal resistance during the absence of that parasite *phyt* >>> Table 52

referee plant breeder: A plant breeder recognized by an official organization to make decisions on varietal identification of crops *agr seed*

reference single nucleotide polymorphism (refSNP *syn* **rsSNP** *syn* **rsID** *syn* **SNP ID):** Any >>> SNP at a specific site of a genome that serves as reference point for the definition of other SNPs in its neighborhood *biot meth*

reflexed: Bent downward or backward from the apex *bot*

reforest: To replant trees on land denuded by cutting or fire *eco fore*

refraction: The change of direction of a ray of light in passing obliquely from one medium into another in which its wave velocity is different *micr* >>> refraction index

refraction index: A number indicating the speed of light in a given medium, as the ratio of the speed of light in a vacuum or in air to that in the given medium; for example, distilled water 1.336, liquid paraffin 1.343, glycerine 1.473, Euparal 1.483, xylene 1.497, cedarwood oil 1.520, or balsam 1.524 *micr* >>> refraction

refractometer: An instrument for determining the refractive index of a substance *prep* >>> refraction >>> refraction index

regenerable: >>> regeneration

regenerant: An entire plant grown from a single cell *phys*

regenerate: >>> regenerant >>> regeneration

regeneration: The replacement by a plant of tissue or organs that have been lost *hort*; in biotechnology, forming a new, entire plant from a clump of cells and/or from a single cell *biot* >>> Figure 27

regeneration cut: A timber harvest designed to promote and enhance natural establishment of trees; even-aged stands are perpetuated by seed tree, shelterwood, and clearcuts; >>> uneven-aged stands are perpetuated by selection of individuals or small groups of trees *fore*

regenerative agriculture: A broad term increasingly used to describe farming and ranching systems that are meant to restore the soil health and biological balance of an area in production through use of organic, permaculture, and other biologically integrated farming systems *agr*

reginned cotton: Cotton that has passed through the ginning process more than once, and has also already been baled; it may go through the ginning again for additional cleaning, blending, or the removal of foreign material *agr meth*

regiospecificity: The ability of an enzyme to target a particular bond within a molecule *phys chem*

registered seed: A class of certified seed that is produced from breeder seed; a progeny of breeder, select, or foundation seed; it is handled under procedures acceptable to the certifying agency to maintain satisfactory genetic purity and identity *seed* >>> Table 28

registration: >>> release of variety

RGEN-ISL: >>> RNA-guided endonuclease – *in situ* labeling

regression analysis: The process of fitting a regression equation to a set of data by using the method of least squares; it includes the various statistical tests and estimates associated with the use of the fitted equation *stat*

regression coefficient: The rate change of the dependent variable with respect to the independent variable *stat*

regression line: A line that defines how much an increase or decrease in one factor may be expected from a >>> unit increase in another *stat*

regulator(y) gene: In the operon theory of gene regulation, a gene that is involved in switching on or off the transcription of structural genes; when transcribed, the regulator gene produces a repressor protein, which switches off an operator gene and hence the operon that this controls; the regulator gene is not part of the operon and may even be on a different chromosome *gene*

regulatory region: Stretches of the DNA sequence which control the activity of genes *biot* >>> regulator(y) *gene*

regulatory sequences: DNA sequences which direct transcription to occur in the desired tissues and at the correct stages of plant development *biot*

regulatory single nucleotide polymorphism: A relatively rare >>> SNP that affects the expression of a gene; it is usually located in the promoter region *biot*

regulatory system: >>> regulator(y) gene

reiterated (DNA): Nucleotide sequences that occur many times within a genome *gene* >>> redundant DNA

rejuvenation: Synonymous with dedifferentiation or treatment that leads to culture invigoration or revival biot; in horticulture (e.g., in fruit tree planting), reversion from adult to juvenile by restoration of juvenile vigor (growth) on a mature entity *hort*

rejuvenation (of seed samples): The restoration of viability of seeds by new propagation of the material *seed*

relay cropping: Cropping systems in which two or more crops are grown in sequence in the same field in the same year, with little or no overlap in time *agr*

release: To free trees from competition by cutting or otherwise removing or killing nearby vegetation and branches; usually applied to young stands *fore*

release factor: One of a set of proteins that recognize stop codons on mRNA at the A site on the ribosome, which leads to the release of the completed protein from the tRNA in the P site of the ribosome *phys gene*

release of variety: A crop variety or germplasm that is released and designated to be reproduced, marketed, and made available as seed for public use *seed*

reliability: The ability of a measure to yield consistent results each time it is applied *stat*

REMAP: >>> retrotransposon-microsatellite amplified polymorphism

REMI: >>> restriction-enzyme-mediated integration

REML: >>> restricted maximum likelihood

remotant: Reblooming or repeat bloomer; a plant that blooms continuously or more than once yearly *hort*

remote hybridization: >>> distant hybridization

renewal pruning: >>> renovation pruning

RENNER complex: A specific gametic chromosome combination in evening primrose (*Oenothera* spp.) *cyto*

renovation: Usually refers to the mechanical removal of plants from a very dense, unproductive, or sodbound stand for the purpose of revitalizing its productivity *seed*

renovation pruning: Hard pruning to rejuvenate an old or overgrown shrub *hort meth*

repair grafting: >>> bridge grafting

repeat: Small tandem duplication or a nucleotide sequence that occurs many times within a DNA molecule *gene* >>> wheat.pw.usda.gov/ITMI/Repeats/index.shtml

repeated blocks: An incidence structure of points and >>> blocks has repeated blocks if there are two blocks incident with exactly the same points *stat* >>> design of experiment

repeated sequence A DNA sequence that occurs in many copies *gene* >>> repetitive DNA >>> wheat.pw.usda.gov/ITMI/Repeats/index.shtml

repellent: A material or substance that animals try to avoid *phyt*

repetitive (repetitious) DNA: A type of DNA that constitutes a significant fraction of the total DNA; sequence motifs are repeated hundreds or thousands of times in the genome; some may be redundant DNA of unknown function, but >>> mRNA, >>> tRNA, >>> 5S-RNA, and histones are coded by repetitive sequences; the significance of repetitive DNA in the genome is not completely understood, and it has been considered to have both structural and functional roles, or perhaps even no essential role; it makes up the major proportion of all the nuclear DNA in most eukaryotic genomes; after BISCOTTI et al. (2015) several types of repetitive DNA can be differentiated (cf. Table 59) *gene* >>> wheat.pw.usda.gov/ITMI/Repeats/index.shtml

replant disease: Replant disease, also described as soil decline, has been known for decades; the causes, however, are not understood; poor vegetative development, stunted growth, and reduced yield are visible plant reactions; in annual plants, crop rotation and partially changing the cultivation sites are measures to overcome replant disease; these possibilities are usually not available for woody species being produced in nursery and fruit production centers; replant disease is of special importance for members of the Rosaceae family, but has also been reported for some other crops, e.g., vine; the replant diseased soils cannot be used for up to 20–30 years, unless soil disinfection is applied; the used substances are ecologically harmful, and the development of other approaches is absolutely essential to maintain a sustained soil productivity *phyt agr*

replicate: A more or less exact duplication or repetition of a test, an experiment, or an experimental single plot to assure or to increase confidence in the resulting data *stat* >>> randomization

replicated control design: A very simple design type; it requires a minimum of land, labor, and statistical analysis; the results are useful for comparison based on general trends but not absolute values; each treatment is included only once; each experimental treatment is near to a control treatment; the replicated control design is not a true statistical design for an applied research experiment; it will not conclusively show whether there are differences between treatments; accepted statistical methods must be used to reliably show if and when treatment differences truly exist *agr meth stat* >>> design of experiment

replicated measurement t-test design: A well-suited design to field experiments when there is one treatment and a control condition, or two treatments; when the measured data have been determined, the data can be analyzed using an unpaired t-test; it is best to have between 10 and 20 observations for each treatment; taking fewer than ten observations makes it difficult to detect the systematic effects of the treatment; a modification to the design that would improve the precision is the paired t-test; the advantage of the method is that it detects differences between varieties A and B; the disadvantage is that it may involve more inconvenience to the breeder; the paired t-test is used when the design has been set up to link pairs of observations; data may be taken as paired when it can be assumed that sources of experimental error (factors that influence the outcome of the experiment that are not part of the treatment) are the same for each of the pairs *agr meth stat* >>> design of experiment

replicate estimate: An estimate of the population quantity based on the replicate subsample using the same estimation methods used to compute the full sample estimate *stat*

replicate sample: One of a set of subsamples, each obtained by deleting a number of observations in the original sample for the purpose of computing the appropriate variance based on the complex design of the survey *stat*

replication: In cytology, the synthesis of new daughter molecules of nucleic acid from a parent molecule, which acts as a template *cyto*; in an experimental field design, it allows not only estimation of the error variance, and consequently application of statistical tests, but it also promotes the accuracy of the estimation of genotypic values of the entries tested *stat meth* >>> randomization >>> Figures 5, 9 >>> Table 25

replication error: Any modification that prevents or disturbs the DNA replication process *gene*

replication number: The replication number of a treatment in a >>> block design is the number of times the treatment occurs in the design; in the case of a binary design, this is equal to the number of blocks in which the treatment occurs; in a resolvable design, it is equal to the number of replicates or parallel classes in a resolution *stat* >>> design of experiment

replication unit: >>> replicon

replicon: A structural gene that controls the synthesis of a specific initiator along with a replicator locus upon which the corresponding initiator acts *gene*

replum: Resistant septum after dehiscence of fruits, as in the Brassicaceae *bot*

reporter gene: In DNA or gene transfer experiments, the linkage of a gene for which (transient) expression can easily be detected with a target DNA sequence or gene, e.g., β-glucuronidase or luciferase *biot*

representational error: >>> cryptic error

representative sample: A sample that accurately reflects the distribution of relevant variables in the target population *stat*

repressor: A protein produced by a regulatory gene that inhibits the activity of an operator gene, and hence switches off an operon *gene*

repressor gene: >>> repressor

reproduce: To create another individual of the parental type that will, in turn, produce another *meth* >>> reproduction >>> Table 35

reproducible: >>> reproduction

reproduction: The process of forming new individuals of a species by sexual or asexual methods *bot gene*; in forestry, the process by which the forest is replaced or renewed, either through artificial reproduction, by means of seeding or planting, or by natural reproduction, from natural seeding or sprouting *fore* >>> Table 35

reproductive isolating mechanism: Any biological property of an organism that interferes with its interbreeding with organisms of other species *gene eco* >>> Table 35

reproductive isolation: The absence of interbreeding between members of different species *eco*

reproductive material: Seeds and vegetative parts of plants intended for the production of plants, as well as plants raised by means of seeds or vegetative parts; also includes natural regeneration *meth hort seed*

reproductive meristem: >>> generative meristem

reproductive organ: Usually refers to the sexual organ *bot*

reproductive shoot: The aerial portion of a plant bearing the flowers and associated parts *bot*

reproductive success: Number of successful gametes, thus number of offspring, of a parent *gene*

repulsion: The linkage phase of a double heterozygote for two linked gene pairs, which has received one dominant factor from each parent and the alternative recessive factor from each parent (e.g., for genes and/or alleles A, a and B, b the repulsion heterozygote receives Ab from one parent and aB from the other, where A and B are dominant, and a and b are recessive) *gene* >>> gametocide

research: Diligent and systematic inquiry into a subject in order to discover or revise facts, theories, etc. *meth*

reselection: >>> backward selection

resequencing by hybridization: A variant sequencing technique that allows the detection of an >>> SNP in a target region of a genome; an SNP is first detected in a genome-wide SNP mapping; its linkage disequilibrium with the target region is determined and the region is amplified in a >>> PCR with bracketing primers; the amplification product or oligonucleotides together with the complementary sequence are then spotted onto a microarray; with thousands of other targets they are hybridzed; finally, hybridizing probes are resequenced and the SNP is verified *biot meth*

residue: A compound such as an amino acid or a nucleotide when it is part of a larger molecule *chem*

residue seed method: >>> method of overstored seeds

resilience: The ability of a population to persist in a given environment despite disturbance or reduced population size; based upon the ability of individuals within the population to survive (fitness) and reproduce (fecundity) in a changed environment *gene eco*

resin: An exudate of tree wood or bark, liquid but becoming solid on exposure to air, consisting of a complex of terpenes and similar compounds *bot*

resistance Inherent capacity of a host plant to prevent or retard the development of an infectious disease; there are different types of resistance: (1) hypersensitivity (infection by the pathogen is prevented by the plant), (2) specific resistance (specific races of the pathogen cannot infect the plant), (3) non-uniform resistance (the host prevents the establishment of certain races), (4) major gene resistance (races of the pathogen are controlled by major genes in the host), (5) vertical resistance (host resistance controls one or a certain number of races), (6) field resistance (severe injury in the laboratory, but resistance under normal field conditions), (7) general resistance (the host is able to resist the development of all races of the pathogen), (8) non-specific resistance (host resistance is not limited to specific races of the pathogen), (9) uniform resistance (host resistance is comparable for all races of the pathogen, rather than being good for some races), (10) minor gene resistance (host resistance is controlled by a number of genes with small effects), (11) horizontal resistance (variation in host resistance is primarily due to differences between varieties and between isolates, rather than to specific variety × isolate interactions) *phyt* >>> biological control >>> systemic acquired resistance >>> Table 33

resistance breeding: Special crossing and selection methods in order to improve the inherent capacity of a crop plant to prevent or retard the development of an infectious disease; usually approaches of multiple selection (when qualitative or quantitative resistance is already present in breeding material that should just be adjusted by other suitable traits), recurrent selection (when qualitative or quantitative resistance is already present in pre-breeding material; however an improvement is needed before transferring to adapted varieties), or subsequent backcrossing (when qualitative or quantitative resistance of pre-breeding lines have to be improved by exotic or alien sources then many backcrosses are needed) are applied *phyt meth* >>> resistance >>> Table 33

resistance gene analogues (RGA): With the recent molecular cloning of several plant disease-resistance genes it became apparent that resistance genes share certain homologies in the conserved amino acid domain; >>> PCR amplification of genomic DNA using degenerate primers on the basis of these conserved amino acid domains identified sequences with homologies to plant disease-resistance genes, i.e., resistance gene analogues (RGAs); RGAs exist in large numbers in plant genomes and provide new possibilities for the investigation of resistance genetics in general and also for the analysis of certain plant disease resistances *biot meth*

resistance gene homologue polymorphism (RGHP): A group of molecular marker techniques that target groups of resistance genes by >>> PCR using primers aimed at conserved domains of resistance genes *biot meth*

resistance, induced systemic: Disease resistance in plants due to a physiological change in the plant that is induced by nonpathogenic organisms *pyth*

resistant: >>> resistance

resistant rootstocks: Some vegetatively propagated tree crops have superb agricultural or horticultural characteristics but are susceptible to various soil-borne parasites; they are then grafted on to resistant rootstocks, e.g., the grafting of classic, European, wine grapes on to American rootstocks to control >>> *Phylloxera*; many fruit trees (e.g., stone and pome fruits, citrus) and other high-yielding clones (e.g., rubber) are grafted on to resistant rootstocks for this reason *phyt hort meth* >>> Figure 50

resolution: The smallest distance by which two objects can be separated and still be resolved as separate objects; the resolving power can never be greater than the wavelength of light used; the image which is seen through the eyepiece is the aerial image formed by the microscope objective in the tube; it has a limit, where useful magnification ends and the empty magnification begins; the performance limit of the microscope is determined by the numerical aperture (NA), so that the total magnification of the microscope objective magnification multiplied by the eyepiece magnification cannot exceed 1,000 × NA *micr*

resolving power: >>> resolution

respiration: Oxidative reactions in cellular metabolism involving the sequential degradation of food substances and the use of molecular oxygen as a final hydrogen acceptor *phys*

responsible plant breeder: The plant breeder or breeding organization that is officially recognized as the maintainer of breeder seed reference samples and production for a variety *seed*

rest: A condition of a plant in which growth cannot occur, even though temperatures and other environmental factors are favorable for growth *phys* >>> dormancy

rest period: >>> rest >>> dormancy

resting bud: >>> hibernaculum

resting spore: A spore germinating after a resting period (frequently after overwintering), as does an oospore or a teliospore *phyt*

restitution nucleus: A nucleus with an unreduced chromosome number *cyto* >>> nitrous oxide gas

restorer: An inbred line that permits restoration of fertility to the progeny of male sterile lines to which it is crossed *seed* >>> Figure 23

restorer gene: A gene and/or allele that is able to restore fertility of a sterile genotype; while genes for sterility frequently belong to the mitochondrial genome (i.e., cytoplasmic), the restorer genes are very often found to belong to the nuclear complement; they are used in hybrid variety production *gene* >>> Figure 2

restorer line (R line): A pollen parent line; it contains the restorer gene or genes, which restore cytoplasmic male sterile plants to pollen fertility; it is crossed with an A line in the production of hybrid seeds *seed* >>> Figures 2, 23, 55

restoring gene: >>> restorer gene

restricted maximum likelihood (REML): A multi-trait (co)variance estimation *stat*

restriction analysis: Determination of the number and size of the DNA fragments produced when a particular DNA molecule is cut with restriction endonucleases *biot* >>> restriction enzyme >>> Table 39

restriction endonuclease: >>> restriction enzyme

restriction enzyme: An enzyme that functions in a bacterial modification-restriction system and recognizes specific nucleotide sequences and breaks the DNA chain at these sites; there is a great number of them, each with different recognition and/or cutting sites; they are intensively used as a tool in molecular genetics and also in producing chromosomal banding patterns in cytogenetics *gene* >>> restriction analysis >>> Figure 59 >>> Table 39

restriction-enzyme-mediated integration (REMI): A method of transformation that generates tagged mutations *biot*

restriction fragment length polymorphism (RFLP): A comparative study (in individuals, populations, or species) of the DNA fragment lengths produced by particular restriction enzymes; by using a DNA hybridization technique, restriction fragments can be identified if they are complementary to a specific DNA probe; each mutation that produces or eliminates a restriction site in a homologous region leads to a change of length of the restriction fragment, which has to be detected; it is used to infer genomic relationships; RFLPs represent an important tool in detecting variability; they are free of secondary effects due to pleiotropic action, and they are frequently associated with the segregation of alleles affecting morpho-physiological traits; the advantages are as follows: (1) they are present everywhere in the genome and in living organisms; (2) they show Mendelian inheritance (e.g., each band seen in a >>> SOUTHERN blot indicates the presence of one or more restriction sites in a sequence; the sequence containing a restriction site is one allele, while the corresponding sequence missing the restriction site is the other allele; the "phenotypes" of these alleles are the differences in banding patterns, due to presence or absence of bands); (3) they show codominant expression; (4) they have no pleiotropic effects; (5) they are independent of environmental effects; (6) they are available at each developmental and/ or physiological stage; (7) different loci within the genome can be identified by one DNA probe; (8) heterologous genes may also be used as probes; (9) any number of DNA probes can be established; (10) probes are available for coding and silent genes (DNA sequences); (11) probes also show the variability of flanking DNA sequences; and (12) several traits can be screened in the same experimental sample *meth* >>> Tables 29, 39 >>> Figures 48, 59

restriction map: Representation of DNA with the position of restriction sites indicated *gene* >>> restriction analysis >>> Table 39

restriction site: A certain nucleotide sequence within the double-stranded DNA; it is recognized by a restriction endonuclease; the enzyme cuts the double strand within the recognition sequence; the restriction sites are usually composed of four to six base pairs and are bilaterally symmetric; both strands are cut either on exactly opposite positions (blunt ends) or alternated ones (sticky ends); the type of cutting depends on the enzyme used *gene* >>> Figure 59 >>> Table 39

restrictive pruning: Periodic pruning to limit growth of trees or shrubs *hort meth*

resynthesis: The artificial production of autopolyploids or allopolyploids of naturally occurring autopolyploid or allopolyploid plants by utilization of the presumable parental species (e.g., it was done in wheat and rapeseed) *meth*

reticulate: In the form of a network, like some types of netted venation *bot*

retrotransposon: Retrotransposons are a ubiquitous and major component of plant genomes; those with long terminal DNA repeats (LTRs, Ty1-copia-like family) are widely distributed over the chromosomes of many plant species *gene* >>> Figure 46

retrotransposon-based insertional polymorphism (RBIP): A molecular marker technique that targets variation in >>> retrotransposon insertion sites; the codominant marker system uses PCR primers designed from the retrotransposon and its flanking DNA to examine insertional polymorphisms for individual retrotransposons; the presence or absence of insertion is investigated by two PCRs, the first using one primer from the retrotransposon and one from the flanking DNA, the second using primers designed from both flanking regions; polymorphisms are detected by simple agarose gel-electrophoresis or by dot hybridization assays; a drawback of the method is that sequence data of the flanking regions are required for primer design; a major advantage is that RBIP does not necessarily require a gel-based detection system but can easily be adapted to automated, gel-free procedures, such as >>> TaqMan or DNA chip technology in order to increase sample throughput *biot meth* >>> Table 29

retrotransposon-microsatellite amplified polymorphism (REMAP): A molecular marker technique that targets variation in >>> retrotransposon insertion sites; REMAP fragments between retrotransposons and microsatellites are generated by PCR, using one primer based on an LTR target sequence and one based on a simple sequence repeat motif. Fragments are separated by high-resolution agarose gel-electrophoresis *biot meth* >>> Table 29

reverse breeding: Reverse breeding allows the opposite to >>> hybrid breeding (by selecting and crossing parental lines to evaluate hybrid performance), i.e., selecting uncharacterized >>> heterozygotes and generating parental lines from them; with these, the selected heterozygotes can be recreated as F_1 hybrids, greatly increasing the number of hybrids that can be screened in breeding programs; key to reverse breeding is the suppression of meiotic crossovers in a hybrid plant to ensure the transmission of nonrecombinant chromosomes to haploid gametes; these gametes are subsequently regenerated as doubled-haploid (DH) offspring; each DH carries combinations of its parental chromosomes, and complementing pairs can be crossed to reconstitute the initial hybrid; >>> achiasmatic meiosis and haploid generation result in uncommon phenotypes among offspring owing to >>> chromosome number variation; these features can be dealt with during a reverse-breeding experiment, which can be completed in six generations *bree meth* >>> breeding

reverse genetics: Using linkage analysis and polymorphic markers to isolate a disease gene in the absence of a known metabolic defect, then using the DNA sequence of the cloned gene to predict the amino acid sequence of its encoded protein; in general, a technology aiming at isolating mutants of a given sequence; it is also applied for identification of gene function *gene biot*

reverse mutation: The production, by further mutation, of a premutation gene from a mutant gene; it restores the ability of the gene to produce a functional protein; strictly, reversion is the correction of a mutation (i.e., it occurs at the same site) *gene*

reverse transcriptase: An enzyme from retroviruses for the synthesis of a DNA complementary to an RNA molecule (i.e., cDNA); it is used (1) for filling-in reactions, (2) for DNA sequencing, and (3) for cDNA synthesis *biot*

reversion: >>> reverse mutation

revertant: An allele that undergoes reverse mutation or a plant bearing such an allele *gene* >>> reverse mutation

revolute: Turned under along the margins toward the abaxial surface *bot*

rewilding: Since >>> genetically modified organisms (GMO) are under strong discussion among consumers and some scientists, alternatives are sought; one of them is to give back to plants genes that had long ago been bred out of them; it is called "rewilding"; the idea was floated by a group at the University of Copenhagen, which proposes the name for the process that would result if scientists took a gene or two from an ancient plant variety and melded it with more modern species to promote greater resistant to drought, for example *bree*

RFLP: >>> restriction fragment length polymorphism

RGHP: >>> resistance gene homologue polymorphism

Rhd2 **gene:** One of a large family of genes involved in free radical production; variation in the activities of different genes in the family could cause differences in plant height and leaf size, in fact anything affected by cell expansion; the *Rhd2* gene encodes an NADPH oxidase, a protein that transfers electrons from NADPH to oxygen molecules and thus forms reactive oxygen species (ROS)-free radicals; in mutations where *Rhd2* is inactive, the accumulation of ROS is much lower than in the wildtype (non-mutant) plants; these plants accumulate less calcium in their root and root hair cells and the growth of roots and root hairs is stunted *gene phys biot* >>> free radicals

rheogameon: Refers to species composed of segments with marked morphological divergence but gene exchange takes place between them *tax gene*

rhizoid: A hair-like filamentous anchorage or absorbing organ *bot*

rhizomania (in beets): One of the most devastating diseases in sugarbeet, caused by beet necrotic yellow vein virus (BNYVV) belonging to the genus Benyvirus; the use of sugarbeet varieties with resistance to BNYVV is generally considered as the only way to maintain a profitable yield on rhizomania-infested fields; as an alternative to natural resistance, the transgenic expression of viral dsRNA for engineering resistance to rhizomania can be considered; transgenic plants expressing an inverted repeat of a 0.4 kb fragment derived from the BNYVV replicase gene display high levels of resistance against different genetic strains of BNYVV when inoculated using the natural vector, *Polymyxa betae*; the resistance is maintained under high infection pressures and over prolonged growing periods in the greenhouse as well as in the field; resistant plants accumulate extremely low amounts of transgene mRNA and high amounts of the corresponding >>> siRNA in the roots, illustrative of RNA silencing as the underlying mechanism; the transgenic resistance compares very favorably with natural sources of resistance to rhizomania, and thus offers an attractive alternative for breeding resistant sugarbeet varieties *phyt meth biot*

rhizome: A horizontally creeping underground stem that bears roots and leaves and usually persists from season to season *bot* >>> root stock

rhizosheath: A sheath of soil particles bound together with >>> polysaccharides that surrounds the roots of some desert plants; rhizosheaths provide the roots with protection against desiccation *phys agr*

rhizosphere: The soil near a living root *agr*

rhodanese: An enzyme defined biochemically by its ability to transfer sulfur from thiosulfate to cyanide, yielding thiocyanate; it is found in plants, animals, and bacteria *phys*

Rht gene: >>> short-straw mutant

Rhynchosporium leaf blotch (of barley and rye): Frequently occurs in wet seasons and in high humidity; symptoms first appear as irregular or diamond-shaped blue-gray water-soaked lesions on the leaves and leaf sheaths; as the lesions mature, they become pale brown with a dark purple margin and coalesce to form large areas of dead tissue; ears may also be infected *phyt*

rhytidome: >>> bark

rhytmicity: >>> periodicity

rib: A primary or prominent vein of a leaf *bot*

ribonuclease: >>> RNase

ribonucleic acid (RNA): A polymer composed of nucleotides that contain the sugar ribose and one of the four bases >>> adenine, >>> cytosine, >>> guanine, and >>> uracil; RNAs are defined regions of the genome, termed transcription units, that serve as templates for the synthesis of another nucleic acid; there are three main types of RNA: (1) ribosomal RNA (rRNA), which is a structural component of ribosomes, (2) transfer RNA (tRNA), which is a class of small RNAs involved in protein synthesis (or translation), (3) messenger RNA (mRNA), which carries the protein-coding information from DNA and is subsequently translated during protein synthesis; recently other classes of RNA that are extremely important in a diverse range of processes have been discovered; although rRNA and tRNA are not translated to give proteins, the DNA regions that code for these RNAs are still termed genes *gene* >>> Table 38

ribonucleoprotein: A protein composed of pre-rRNAs and ribosomal as well as non-ribosomal protein components; one of the non-ribosomal proteins, the nucleolin, is considered to play a key role in regulation of rDNA transcription, perisomal synthesis, ribosomal assembly, and maturation *phys*

ribosomal DNA internal transcribed spacers (rDNA-ITS): A multigene family with nuclear copies in eukaryotes that are arranged in tandem arrays in nucleolar organizer regions (>>> NORs), generally at more than one chromosomal location; each unit within a single array consists of genes coding for small (18S) and large (28S) rRNA subunits; the 5.8S nuclear rDNA gene lies embedded between these genes but separated by two internal transcribed spacers, ITS1 and ITS2; for example, the copy numbers of 18S–5.8S–28S rRNA genes in diploid genomes of *Quercus cerris*, *Q. ilex*, *Q. petraea*, *Q. pubescens*, and *Q. robur* are estimated to be in the range of 1,300–4,000; the small subunit is highly conserved and has been used for the relationship between Archaebacteria and Eubacteria, while more conserved domains within the 28S region have been used to cover evolutionary time through the Paleozoic and Mesozoic eras; the faster evolving ITS regions have been employed for population and congeneric phylogenies; internal transcribed spacers represent biparental nuclear mode of inheritance through several loci with several alleles per locus, are codominant, exhibit high levels of variability even within a single individual, and are composed of non-coding sequences *biot*

ribosomal RNA: The RNA molecules that are structural parts of ribosomes (i.e., 5S, 16S, and 23S RNAs in prokaryotes, and 5S, 18S, and 28S RNAs in eukaryotes) *gene*

ribosome: One of the ribonucleoprotein particles, which are the sites of translation; it consists of two unequal units bound together by magnesium ions *gene*

riboswitch: A part of an >>> mRNA molecule that can directly bind a small target molecule, and whose binding of the target affects the gene's activity; an mRNA that contains a riboswitch is directly involved in regulating its own activity, in response to the concentrations of its target molecule; the discovery that modern organisms use >>> RNA to bind small molecules, and discriminate against closely related analogs, significantly expanded the known natural repertoire of RNA beyond its ability to code for proteins or to bind other RNA or protein macromolecules; most known riboswitches occur in bacteria, but functional riboswitches of one type (the TPP riboswitch) have been discovered in plants and certain fungi; the original definition of the term "riboswitch" specified that they directly sense small-molecule metabolite concentrations; although this definition remains in common use, sometimes a broader definition that includes other cis-regulatory RNAs is applied *biot meth*

rice–*Azolla*–fish culture: A farming system originally developed in China in which rice, *Azolla*, and fish grow in the same field; *Azolla* serves as the feed for fish; this system improves both fish and rice yield *agr*

rice grassy stunt 1 (RGSV1) and 2 (RGSV2) virus: The symptoms of RGSV1 are severe stunting, excessive tillering, and pale green to yellow and narrow leaves with small rusty spots; RGSV2 causes severe stunting, excessive tillering, and yellow to orange and narrow leaves with small rusty spots *phyt* >>> Table 52 >>> www.tau.ac.il/lifesci/units/ICCI//genebank2.html

rice hoja blanca disease: Symptoms are cream-colored to yellow spots, elongating and coalescing to form longitudinal yellowish green to pale green striations; streaks may coalesce to cover the whole leaf; brown and sterile glumes with typical "parrot beak" shape of deformation *phyt*

rice ragged stunt virus (RRSV): Symptoms are stunted plants, but which remain dark green; leaves are ragged and twisted; vein swelling on leaf collar, leaf blades, and leaf sheaths *phyt*

rice tungro virus: Causal agent is rice tungro bacilliform virus (RTBV) and rice tungro spherical virus (RTSV); the disease is transmitted by the green leafhopper *Nephotettix* spp.; symptoms are yellow to yellow orange leaves, stunting, and slightly reduced tillering *phyt*

rice whorl maggot disease: Causal agent is *Hydrellia philippina*; symptoms are leaf margin feeding which causes conspicuous damage and sometimes stunting of plants *phyt*

rick: >>> stack

ridge tillage: A type of soil-conserving tillage in which the soil is formed into ridges and the seeds are planted on the tops of the ridges; the soil and the crop residue between the rows remain largely undisturbed; the practice offers opportunities to reduce crop production costs by banding fertilizers and pesticides, and reducing the need for field trips *agr*

rifamycins: A group of antibiotics that inhibit initiation of transcription in bacteria *biot phys*

rill: Small, intermittent water course with steep sides; usually only several centimeters deep *agr*

rind: A thick and firm outer coat or covering (e.g., in watermelon, orange, etc. or the bark of a tree) *bot*

ring bivalent: An association of two chromosomes with terminal chiasmata on both arms *cyto* >>> Figure 15

ring chromosome: A (sometimes aberrant) chromosome with no ends (e.g., the chromosome of bacteria); an isochromosome may also form a ring in MI of meiosis *cyto*

ringer solution: A physiological saline containing sodium, potassium, and calcium chlorides used in physiological experiments for temporarily maintaining live cells or organs *in vitro phys*

ringspot disease: A circular area of chlorosis with a green center; a symptom of many virus diseases *phyt*

ripe: >>> mature

ripewood cutting: Cutting of a mature shoot taken from an evergreen plant, from late summer to early winter *hort meth*

RNA: >>> ribonucleic acid

RNA-guided endonuclease – *in situ* labeling-tool (RGEN-ISL): A method for conducting genome sequence analysis *in situ* to study the sequence without disrupting chromosome structure; it uses >>> CRISPR/Cas9-mediated techniques; the standard for examining *in situ* DNA sequences at a chromosomal level was >>> fluorescence *in situ* hybridization (FISH), which requires the denaturation of chromatin, therefore disrupting the structure of the chromosome; combining CRISPR/Cas9 with fluorescence labeling technique bypasses the denaturation step *meth cyto biot*

RNAi: >>> RNA interference

RNA interference (RNAi): The use of double-stranded RNA to interfere with gene expression; RNAi is usually mediated by approximately 21-nt small interfering RNAs *meth biot*

RNA interference (RNAi) technology: A mechanism that inhibits gene expression by inhibiting >>> translation, causing the degradation of specific RNA molecules or hindering the >>> transcription of specific genes; RNAi targets include RNA from viruses and transposons, and also plays a role in regulating development and genome maintenance; small interfering RNA strands are key to the RNAi process, and have complementary nucleotide sequences to the targeted RNA strand; specific RNAi pathway proteins are guided by the small interfering RNA to the targeted messenger RNA, where they cleave the target, breaking it down into smaller portions that can no longer be translated into protein; the RNAi pathway is initiated by the enzyme dicer, which cleaves long, double-stranded RNA molecules into short fragments of 20–25 bp; one of the two strands of each fragment, known as the guide strand, is then incorporated into the RNA-induced silencing complex and pairs with complementary sequences; the outcome of this recognition event is posttranscriptional >>> gene silencing *biot meth*

RNA machine: >>> ribosome

RNA polymerase: An enzyme that transcribes an RNA molecule from the template strand of a DNA molecule; it adds to the 3′ end of the growing RNA molecule one nucleotide at a time using ribonucleotide triphosphates (rNTPs) as substrates; RNA polymerase I is dedicated to the synthesis of only one type of RNA molecule (pre-rRNA); RNA polymerase II is required for general transcription reactions; RNA polymerase III produces small RNAs such as tRNAs and 5S rRNA *phys biot*

RNA transcriptase: The enzyme responsible for transcribing the information encoded in DNA into RNA; it is also called transcriptase or RNA polymerase *biot*

RNase: An enzyme hydrolyzing RNA *gene*

Robertsonian translocation: A chromosomal mutation due to centric fusion or centric fission (i.e., a reciprocal translocation with breakpoints within the centromeric regions); in wheat, Robertsonian translocations arise from centric misdivision of univalents at anaphase/telophase I, followed by segregation of the derived telocentric chromosomes to the same nucleus, and fusion of the broken ends during the ensuing interkinesis *cyto* >>> translocation

rod (rd): Equals 5.03 meters

rogue: To remove and destroy individual plants that are diseased, infested by insects, or otherwise undesirable *agr meth*

rolled paper toweling: Adjusted filter paper or paper towels are used for this method in order to germinate seeds inside and/or between the layers of paper; after germination and growth, the viability and/or germability are determined *seed*

roller: A device that compacts the soil to produce a firm seedbed, like a packer *agr* >>> packer

rolling: In the U.S., a method instead of mowing cover crops, e.g., rye; many farmers flatten crop stand out by attaching a rolling, paddlewheel-like cylinder with metal slats to a tractor and barreling over the rye, tamping and crimping it into a mat; when used as cover, rye is planted in the fall, killed in the spring, and left to decompose in the same fields where soybeans and other cash crops are later planted; rolling with a roller-crimper uses less energy than mowing, is faster, and only needs to be done once a season; unlike mowing, it also leaves residue intact in the field, forming a thick mat that can provide better weed suppression *meth agr*

root: The lower part of a plant, usually underground, by which the plant is anchored and through which water and minerals enter the plant; a set of proteins plays a broad role in guiding this tissue formation; the factors, known as the BIRDs, help a root maintain its organization as it grows, guiding several distinct steps in the development of two interior layers of tissue *bot*

root ball: The roots and soil or soil mix that they are growing in when lifted from the open ground *hort*

rootbound: A condition in which the roots of a plant have grown entangled in a tight mass, or completely filled a container *hort*

root cap: A cap of cells covering the apex of the growing point of a root and protecting it as it is forced through soil *bot*

root crop: A crop, such as beets, turnips, or sweet potatoes, grown for its large, edible roots *agr*

root culture: The *in vitro* growth of roots (e.g., root tips or root meristem on a synthetic medium) *biot*

root cutting(s): Root cuttings are made by cutting off pieces of root and planting them under suitable conditions; in this way some plant species or varieties can be easily propagated *meth agr hort* >>> greenwood cutting >>> hardwood cutting >>> semiripe cutting

root exudate(s): Chemicals released by roots including sugars, amino acids, organic acids, proteins, cell debris, and carbon dioxide *phys* >>> siderophore

root graft: The natural growing together or joining of the roots of nearby plants; in horticulture, the process of grafting a shoot or stem of one plant onto the root section of another *hort meth* >>> root grafting

root grafting: The process of grafting scions (shoots) directly on a small part of the root of some appropriate stock, the grafted root then being potted *hort*

root hair: A tabular outgrowth of an epidermal cell of a root, which functions to absorb water and nutrients from the soil *bot*

rooting: The natural or induced process of root formation *phys bot*

rooting compound: A powdery substance into which fresh cuttings are dipped before inserting in soil or medium, containing hormones, such as kinetins, to encourage root growth *meth prep hort*

root knot nematode A nematode (*Meloidogyne naasi*) that induces small, gall-like growth on the roots of certain types of plants *phyt* >>> www.pflanzengallen.de

root nodule A small, gall-like growth on the roots of certain types of plants (legumes); the nodules develop as a result of infection of the root by bacteria *bio agr* >>> www.pflanzengallen.de

root pruning: Cutting the roots of large plants, mainly trees and shrubs, to force more vigorous growth or to prepare the plant for transplantation or transportation *meth hort*

rootstock (*syn* rhizome): In horticulture, the bottom or supporting root used to receive a scion in grafting; most temperate-zone fruit trees are propagated by asexual methods of grafting or budding in order to preserve the characteristics of the aerial portion, or scion, of the plant; in some cases, the scion cultivar of the plant cannot be reproduced by seed from adventitious roots on cuttings, and so propagation by grafting onto a rootstock is necessary; rootstocks are also used for other purposes, such as size control, disease resistance, or winter hardiness; early horticulturists initiated programs of selection, improvement, and hybridization for superior rootstocks; they established comparative trials and systematic experiments to determine edaphic, adaphic, and biotic adaptability, as well as growth and productivity capabilities *hort* >>> rhizome

rootstock variety: Special (fruit) tree varieties (often of wild-type character) that serve as rootstock for graftings; usually they show good root formation, resistance traits, and compatibility with the scion *hort*

root sucker: A shoot arising adventitiously from a root of a plant; mostly at some distance from the main trunk *bot*

rosarium: Since Roman times, a rose garden and breeding site of roses *hort*

rosel(l)ate: >>> rosette

rosette: An arrangement of leaves radiating from a root crown near the earth *bot* >>> Table 42

rosette plant: >>> rosette

rosette stage: An early stage of plant growth; many plants, when developing, develop leaves first to support photosynthesis, and then develop a stem; during this stage, the plants look like rosettes *bot* >>> Table 42

rosular: >>> rosette

rot: To deteriorate, disintegrate, fall, or become weak due to decay *agr*

rotary hoe: An implement that breaks the soil with a circular motion *agr*

rotary mower: A mower with a blade that spins in a horizontal plane from a central rod; its advantages are the ability to cut tall grass, versatility of movement, a less expensive purchase price, and blades that can be easily sharpened *agr*

rotation: >>> crop rotation

rotation age: The age at which the stand is considered ready for harvesting under the adopted plan of management *fore*

rotation of crops: >>> crop rotation

rotenone: A natural >>> insecticide extracted from the roots of Derris elliptica of Southeast Asia, where it is used to control body lice, and from Lonchocarpus in South America; these plants are cultivated in a number of tropical countries, and improved cultivars are available; no >>> resistance to rotenone has ever been known to develop in any species of insect *phyt*

rouge: A noun referring to an off-type plant; when used as a verb it refers to the act of removing, to uproot or destroy such plants that do not conform to a desired standard or are diseased seed; to remove individuals that have an undesirable phenotype, or that have been shown through progeny tests to have a less desirable genotype from a seed orchard, seed production area, or nursery bed *meth seed* >>> genetic thinning

round-dance neighbor design: A >>> neighbor design in which blocks are complete (each treatment occurs once in each block) and have a circular structure so that each pair of elements occurs as neighbors in a circle exactly once *stat* >>> design of experiment

Roundup™: An herbicide brand that provides non-selective control of several annual and perennial weeds; Roundup™ will also damage crops, such as maize or soybeans, that are not resistant *phyt*

row-column design: An experimental design in which the set of plots has the structure of a rectangle, in which two plots may be in the same row, or in the same column, or neither *stat* >>> design of experiment

row-complete Latin square: A >>> Latin square is row-complete if each ordered pair of distinct symbols occurs precisely once in consecutive positions in a row of the square *stat* >>> design of experiment

row cover(s): Flexible, transparent coverings made from polyester or polypropylene that are installed over single or multiple rows of horticultural crops for the purpose of enhancing plant growth by warming the air around the plants in the field *hort*

row-quasi-complete Latin square: A >>> Latin square is row-quasi-complete if each unordered pair of distinct symbols occurs precisely twice in consecutive positions in a row of the square *stat* >>> design of experiment

row spacing: The distance between rows of crop plants; it depends on needs for optimal plant growth, plant density, weed control, and harvest technology *agr* >>> design of experiment

row sprigging: Planting of sprigs in rows or furrows *agr*

royalty: The payment of a percentage of sales as compensation to breeders, product developers, patent licensors, or even investors *adm* >>> Intellectual Property Rights

rRNA: >>> ribosomal RNA

rub: >>> rubbing

rubbed seeds: >>> rubbing

rubbing: Smoothing the surface of multigerm seed of sugarbeet *seed*

rubisco: A CO_2-fixing enzyme; the key enzyme in photosynthesis; it is the most frequent protein on earth; it has a unique double function of being both a carboxylase and an oxygenase; when acting as an oxygenase, it catalyzes the light-dependent uptake of O_2 and the formation of CO_2 in a complicated process (photorespiration), which takes place concomitantly in three organelles – chloroplasts, peroxisomes, and mitochondria; there are projects to manipulate the enzyme in order to create an artificial plant that can contribute to reduction of CO_2 content in the atmosphere (i.e., decreasing the so-called greenhouse effect) *phys biot*

ruderal plant: A plant that is associated with human dwellings or agriculture, or one that colonizes waste ground *eco*

rudiment: >>> rudimentary

rudimentary: Incompletely developed *bot*

rugose: Wrinkled or folded; having horizontal folds in the surface *bot*

runoff: That portion of precipitation on a drainage area that is discharged from the area in stream channels; types include surface runoff, ground water runoff, or seepage *hort agr*

run out: Separation of nucleic acid or protein molecules by gel electrophoresis *prep*

runner: A procumbent shoot that takes root, forming a new plant that eventually is freed from connection with the parent by decay of the runner; it serves as a vegetative propagule (e.g., in strawberry) *bot* >>> stolon

running wild: Establishment of a cultivated plant in a natural ecosystem or in a biocenosis of anthropogenic origin; cultivated plants are specialized for producing higher yield, which generally makes them uncompetitive in natural environments; sometimes, cultivated plants may cross with wild relatives or be planted in new regions, which can make them more competitive in the wild *eco biot*

rush: >>> reed

russet: A brownish roughened area on the skin of fruits as a result of cork formation *phyt*

russet potato: Refers to any potato variety with a cylindrical or oblong shape with rough brown skin characterized by corky, dark brown checks (>>> russets) and numerous "eyes"; the standard russet potato is the "Russet Burbank" though it is slowly being supplanted by newer varieties that are easier to grow and which exhibit better flavor; russet potatoes generally have low moisture and high starch content which make them good for baking and frying; they are often called "Idaho potatoes" or "Idaho bakers", even when they are not grown in Idaho (U.S.) *agr*

rust A plant disease caused by a fungus of the class Urediniomycetes; the characteristic symptom is the development of spots or pustules bearing masses of powdery spores that are usually rust-colored, yellow, or brown *phyt* >>> Table 52 >>> www.tau.ac.il/lifesci/units/ICCI// genebank2.html

rye bodies: In an embryological study, M. BENNETT (1974) demonstrated that the presence of rye chromosomes in >>> antipodal cells can be detected with the help of at least eight to nine >>> heteropycnotic bodies found in addition to seven densely stained centromeric regions; they are grouped around one pole; those bodies are rather spherical or subspherical and >>> FEULGEN-positive; the number of rye bodies exceeded the number of seven rye chromosomes, because each rye chromosome has a single body except chromosome 1R which shows two, one located at the end of each chromosome arm; in chromosome 2R, 4R, 5R, and 6R, 7R the body is found on the short arm, while chromosome 3R shows it on the long arm; BENNETT (1974) suggested that the bodies represent the >>> telomeric heterochromatin, and each of the bodies constituted 4% of the total haploid genome *cyto*

S

S0: A symbol used to designate the original selfed plant *meth*

S1, S2, S3, etc.: The representation for continued selfing (self-fertilization) of plants; S1 designates the generation obtained by selfing the parent plant, S2 the generation obtained by selfing the S1 plant, etc. *meth*

saccate: Swollen or sac-shaped; in >>> pollen grains, having one or more air sacs, i.e., exinous expansions forming an air sac *bot* >>> plicate

saccharide: An alternative term for sugar *chem phys*

saccharose: A sweet, crystalline substance, $C_{12}H_{22}O_{11}$, obtained from the juice or sap of many plants (e.g., from sugarcane and sugarbeet) *chem phys*

sacrificial crop: Crop planted to distract pests safely *agr*

safener (*syn* **antidotes** *syn* **antagonists** *syn* **crop protectants**): A chemical agent that reduces the toxicity of herbicides to crop plants by a physiological or molecular mechanism *phys biot*

safranin: >>> GRAM's stain

safety duplication: A duplicate of a base collection stored under similar conditions for long-term conservation, but at a different location to insure against accidental loss of material from the base collection *seed meth*

SAGE: >>> serial analysis of gene expression

S-allele: An allele of a gene controlling incompatibility in many allogamous plants; alleles present in both style and pollen are referred to as matching S-alleles; S-alleles usually belong to a series of multiple alleles *gene*

sagittate: Arrow-shaped *bot*

salicin: >>> glucoside

salicylic acid: A phenolic substance, which is accumulated in many plants as part of the defense response; it is shown, in some plants, to be necessary for the accumulation of pathogenesis-related proteins; treatment of plants with salicylic acid can induce resistance *biot*

saline soil: A soil containing enough salts to reduce plant fertility *agr* >>> crop rotation >>> Table 44

salinization: The accumulation of soluble salts at the surface or at some point below the surface of the soil profile to levels that have negative effects on plant growth and/or on soils; this occurs due to water evaporation leaving behind salts that were dissolved in soil water *agr eco* >>> Table 44

Salmon procedure: A method for producing haploids in hexaploid wheat; "Salmon" is a name of an alloplasmic wheat variety carrying a 1RS.1BL chromosome translocation together with cytoplasm of Aegilops kotchyi; the interaction of cytoplasmically genetic determinants with, possibly, a parthenogenesis-inducing gene on chromosome arm 1RS of rye results in haploid progeny *meth* >>> Imperata cylindrica procedure >>> Figures 17, 26 >>> Table 7

saltation: A mutation occurring in the asexual state of fungal growth, especially one occurring in *in vitro* culture *phyt*

saltational speciation: >>> quantum speciation

SAM: >>> ACC synthase

samara: A fruit similar to an achene except that the entire seed coat is tightly fused with the pericarp (e.g., ash, elm, maple key, tree of heaven, etc.) *bot* >>> winged fruit

sample: A finite series of observations taken from a population *meth stat*

sample size: The number of experimental units on which observations are considered; it may be less than the number of observations in a data set, due to the possible multiplying effects of multiple variables and/or repeated measures within the experimental design; for quantitative trait loci (QTL) the following has to be considered: population size, structure, and the number of molecular markers genotyped can have significant impacts on QTL detection; the equation to calculate minimum sample size for an F2 population is given as follows: $N_{F2} = [1 - r^2_{F2}/r^2_{F2}] \times \{[z_{(1-(\alpha/2))}/(1 - r^2_{F2})^{1/2}] + z_{(1-\beta)}\}^2 \times [1 + (k^2/2)]$, a BC1 population, $N_{BC1} = [1 - r^2_{BC1}/r^2_{BC1}] \times \{[z_{(1-(\alpha/2))}/(1 - r^2_{BC1})^{1/2}] + z_{(1-\beta)}\}^2 \times [1 + (k^2/2)]$, where r^2 is the fraction of phenotypic variance explained by the QTL (r^2 is valid only for simple models; mean squares provide an alternative method to estimate fraction of phenotypic variation when models have replication and location), k is the dominance coefficient ($k = 0$ for completely additive trait, $k = 1$ for a completely dominant trait, and $k = -1$ for a completely recessive trait; this equation assumes that the marker and QTL are coincidental), α is the type I error (the probability of incorrectly identifying an association that does not exist), β is the type II error (the probability of failing to identify a true association; $z_{(1-(\alpha/2))} = 1.96$ at $\alpha = 0.05$, $z_{(1-\beta)} = 1.28$ at $\beta = 0.10$); for example, in an F2 population, the estimated sample size required to detect a dominant QTL that is responsible for 25% of the phenotypic variation is 12 *stat*

sampling: The method by which a representative sample is taken from a seed lot or something else *meth*

sampling accuracy: Nearness to true value *stat*

sampling design: A method for selecting a sample *stat*

sampling distribution: The distribution of a statistic based on all possible samples that can be drawn according to a specific sampling plan; the term almost always refers to random sampling, and the statistic is usually a function of the n sample observations *stat*

sampling error: Variability due to the limited size of the sample *stat*

sampling precision: Nearness to a given standard *stat*

sampling units: Non-overlapping collections of elements from the population *stat*

sandblasting: Damage caused to plants (young plants especially) by blowing sand; moderate to extreme sandblasting can damage stems and kill plants *hort agr*

sanitation: Plant disease control involving removal and burning of infected plant parts and decontamination of tools, equipment, etc. *phyt*

sanitation cut: A cutting made to remove trees killed or injured by fire, insects, fungi, or other harmful agencies, for the purpose of preventing the spread of insects or disease *phyt fore*

sap: The exudate (>>> NUSSINOVITCH 2009) from ruptured tissues emanating from the vascular system or parenchyma *bot*

sapling: Young tree *hort fore*

saponin: Any member of a class of glycosides that form colloidal solutions in water and foam when shaken; it occurs in many different plant species; in cereals, only oats are known to produce these compounds; in oats, the resistance to infection by the take-all fungus, *Gaeumannomyces graminis* var. *tritici*, has been attributed to the family of antifungal saponins known as avenacins, which are present in roots *chem phys*

saprophyte: An organism that lives on or in dead or decaying organic matter *bot*

sapwood: The living, softer part of the wood between the inner bark and the heartwood *bot fore*

sarment: A slender running stem *bot* >>> runner

sarmentous plant: >>> runner

SAT chromosome: >>> satellite(d) chromosome

satellite: A distal segment of a chromosome that is separated from the rest of the chromosome by a chromatic filament *cyto*

satellite(d) chromosome: Each chromosome with a secondary constriction; this constriction divides a satellite part from the rest of the chromosome (arm) *cyto*

satellite DNA: A highly repetitive DNA, composed of repeated hepta- to deca-nucleotide sequences; DNA of different buoyant density; a minor DNA fraction that has sufficiently different base composition from the bulk of the DNA in order to separate distinctly during cesium chloride density gradient centrifugation; it can derive from nucleus, plastid, or mitochondrial DNA *biot* >>> http://wheat.pw.usda.gov/ ITMI/Repeats/index.shtml

satellite virus: Part of a small virus associated with tobacco necrosis virus (TNV), which is dependent upon the TNV genome for its own replication; or part of a defective virus associated functionally with another virus which it depends on for replication *phyt*

satellite virus-induced silencing system (SVISS): A proprietary viral RNA-based technology developed by Bayer CropScience AG, which allows gene function discovery to be carried out, as well as validation research for genes expressed in different tissues of tobacco plants; for example, transgenes producing dsRNA are introduced into several crops with the aim of obtaining tolerance against abiotic stresses; the enzyme PARP [poly-(ADP-ribose)-polymerase] is well-known to play a key role in regulating cellular stress response in humans and animals; the down-regulation of the PARP genes in plants results in enhanced tolerance against abiotic stresses, like drought, heat, and ozone in model plants; this technology is applied to various crops including maize and canola *biot meth*

saturation mapping: The enrichment of specific regions of an established genetic map with molecular markers such that marker density becomes extremely high; a saturated map is a prerequisite for >>> map-based cloning of a gene; the method produces high-resolution maps and incorporates markers generated by different approaches *biot meth*

saturation mutagenesis: Induction and recovery of large numbers of mutations in one area of a genome, or in one biological function, in order to identify all the genes in that area, or affecting that function *biot*

SAT-zone: The secondary constriction of a satellite chromosome *cyto* >>> nucleolar zone

savanna: A plain characterized by coarse grasses and scattered tree growth *eco*

sawflies: Larvae of various insects that feed on leaves *phyt*

scab: A general term for any unrelated plant disease in which the symptoms include the formation of dry, corky scabs, e.g., potato common scab is caused by the bacterium *Streptomyces scabies*, and potato powdery scab is caused by the fungus *Spongospora subterranea phyt*

scabrous: Rough to the touch; caused by short, stiff, angled hairs on the surface *bot*

scaffold: The eukaryotic >>> chromosome structure remaining when DNA and histones have been removed; made from non-histone proteins; the central framework of a chromosome to which the DNA solenoid is attached as loops; composed largely of topoisomerase *cyto*

scald: A necrotic condition in which tissue is usually bleached and has the appearance of having been exposed to high temperatures *phyt*

scale: Any thin, scarious body, usually a degenerated leaf *bot*

scale insect: A small flat insect pest, with a waxy dorsal shield, that adheres to the plant epidermis and feeds from the plant sap *phyt*

scale stick: A flat stick, similar to a yardstick, which is calibrated so log volumes can be read directly when the stick is placed on the small end of the log of known length *meth fore*

scalping: The removal of material larger than the crop seed during the processing of seeds *seed*

scanning electron microscope: A microscope used to examine the surface structure of biological specimens; a three-dimensional screen image is acquired through focusing secondary electrons emitted from a sample surface bombarded by an electron beam *micr*

scape: A leafless flower stalk arising from the ground *bot hort*

SCAR: >>> sequence-characterized amplified region

scarification: The process of mechanically abrading a seed coat to make it more permeable to water; this process may also be accomplished by brief exposure to strong acids (sulfuric acid); it may enhance germination *seed*

scarious: Thin, dry, and membranous; not green (e.g., the margins of a Poa lemma) bot

scatter diagram: A diagram in which observations are plotted as points on a grid of x- and y-coordinates to see if there is any correlation *stat*

SCE: >>> sister chromatid exchange

"Schiave": Collectively refers to grapevine cultivars presently grown on the southern and northern slopes of the eastern Alps and bearing different names, such as "Schiava", "Trollinger",

"Rossara", "Rossola", "Geschlafene", "Gansfüsser", "Urban", etc.; their common origin has been suggested by historic, linguistic, and ampelographic considerations; however, a dendrogram constructed from an AFLP analysis of the 33 Schiave cultivars shows different, and in some cases relevant, degrees of genomic dissimilarity; the analyzed cultivars cluster into at least five taxonomic groups with specific geographic distribution along the valleys of Valtellina, Bergamo, and Brescia, and those of South Tyrol and Swabia; it is concluded that the common definition "Schiave'" refers to a similar cultivation practice in contiguous regions rather than to a common genetic background (FOSSATI et al. 2001) *hort*

Schiff's reagent: A reagent consisting of fuchsine bleached by sulfurous acid that produces a red color upon reaction with an aldehyde; it is used for chromosome staining *micr*

schizocarp: A dry, two-seeded fruit of some plants that separates at maturity along a midline into two mericarps; each mericarp has a dry, indehiscent pericarp enclosing a loose-fitting ovule (e.g., carrot) *bot*

scion: A portion of a shoot or a bud on one plant that is grafted onto a stock of another *hort*

scion rooting: Covering a low graft with soil so that the plant develops roots directly from both the rootstock and the scion *meth hort*

scission: >>> fission

sclerenchyma: Tissue composed of cells with thickened and hardened walls *bot*

sclerophyllous: Having tough, leathery, usually evergreen leaves *bot*

Sclerotinia **(of rape or clover):** A soil-borne disease (*Sclerotinia sclerotiorum* in rape, *S. trifoliorum* in clover) that infects a wide range of crops; symptoms appear from May onward as bleached areas of the stem with black sclerotia within the infected stem *phyt*

sclerotium (sclerotia *pl***):** A dense, compact mycelial mass capable of remaining dormant for extended periods *bot*

SCN: >>> soybean cyst nematode

scorch: "Burning" of leaf margins as a result of infection or unfavorable environmental conditions *phyt*

scorpiod cyme: A determinate inflorescence in which the lateral buds on one side are suppressed during growth, resulting in a curved or coiled arrangement *bot*

scrambling legumes: Legumes with a climbing or trailing habit *agr*

screening: Examining the properties and performance responses of individuals, lines, genotype, or other taxa under an assortment of conditions in order to evaluate the individuals or groups; a routine testing for particular properties *meth*

screening overkill: When screening a large population for horizontal resistance, there is a danger, in the early breeding cycles, that every individual will be killed and the entire breeding population lost; it can be prevented by using natural crop protection late in the season, to ensure that the least susceptible plants produce at least a few seeds *phyt*

scutellum: A shield-shaped organ of the embryo of grasses; it is often viewed as a highly modified cotyledon in monocots *bot*

scutiform: Platter-shaped *bot*

SDR: >>> second division restitution

SDS: >>> sodium dodecyl sulfate-polyacrylamide

SDS gel electrophoresis (sodium dodecyl sulfate-polyacrylamide) (SDS-PAGE): In SDS-PAGE, SDS masks protein charge and separations depending only on size as compared to common gel electrophoresis *meth* >>> gel electrophoresis

SE: >>> standard error

Sea Island cotton: Fine, long staple cotton grown in the West Indies *agr*

secalin: A prolamin glyco-protein found in the grain of >>> rye, *Secale* ssp., and one of the forms of >>> gluten proteins that people with coeliac disease cannot tolerate, and thus rye should be avoided by people with this disease; it is generally recommended that such people follow a gluten-free diet; in bread-making with rye flour, this protein requires exposure to an acid such as lactic acid so that the bread will rise; this is usually achieved with a sourdough ferment; 62 DNA sequences are known for the coding regions of omega-secalin (ω-secalin) genes; only 19 out of the 62 ω-secalin gene sequences were full-length open reading frames (ORFs), which can be expressed into functional proteins; the other 43 DNA sequences were pseudogenes, as their ORFs are interrupted by one or a few stop codons or frameshift mutations; the 19 ω-secalin genes have a typical primary structure, which is different from wheat gliadins; there is no cysteine residue in ω-secalin proteins, and the potential celiac disease (CD) toxic epitope (PQQP) was identified to appear frequently in the repetitive domains; the ω-secalin genes from various cereal species share high homology in their gene sequences; the ω-secalin gene family has involved fewer variations after the integration of the rye chromosome or whole genome into the wheat or triticale genome; the higher Ka/Ks ratio (i.e., non-synonymous to synonymous substitutions per site) in ω-secalin pseudogenes than in ω-secalin ORFs indicates that the pseudogenes can be subject to a reduced selection pressure; based on the conserved sequences of ω-secalin genes, it is possible to manipulate the expression of this gene family in rye, triticale, or wheat 1BL/1RS translocation lines, to reduce its negative effects on grain quality *chem*

secalotricum: A cross combination of >>> rye (*Secale*) and >>> wheat (*Triticum*) in which rye serves as the donor of the cytoplasm (mother plant); as opposed to triticale *bot agr* >>> triticale

secondary constriction: >>> SAT-zone

secondary crop: A crop that originated as a weed of a primary crop (e.g., rye) *evol*; in agronomy, a crop grown after a primary crop *agr*

secondary embryogenesis: The development of embryos from young embryos *bot biot*

secondary gene pool: Species in the secondary gene pool include those from which genes can be transferred to the cultivated species, however, with more difficulties as compared to species of a primary gene pool *evol*

second(ary) glume: The uppermost of the two glumes; an odd-veined, empty bract of the spikelet *bot* >>> glume

secondary host: In heteroecious aphids, plant on which only parthenogenetic reproduction takes place *phyt*

secondary infection: Any infection caused by inoculum produced as a result of a primary or a subsequent infection and/or an infection caused by secondary inoculum *phyt*

secondary noxious weeds: Annual and biennial weeds that are difficult to control and which have been designated by the state as secondary noxious *phyt agr*

secondary pairing: The association of bivalents in polyploids due to genetic, evolutionary, or structural factors; by any reason those bivalents appear in groups; sometimes it seems that the bivalents of a certain genome are closer together than at random, or the bivalents of a certain genome occupy certain domains (spatial order) within the meiotic cell (prometaphase and metaphase) *cyto*

secondary periclinal chimera: >>> chimera >>> mutation >>> Figure 56

secondary root: >>> lateral root

secondary tiller: Arises from the prophyll node and leaf node of the primary >>> tillers in cereals; in the same manner, tertiary tillers may occasionally be produced by secondary tillers; the primary tillers are usually the smallest of the tillers that emerge *bot*

second division: Second meiotic division, which is a mitotic division of chromosomes *cyto*

second division restitution (SDR): Due to premature cytokinesis before the second meiotic division takes place; sister chromatids end up in the same nucleus *cyto*

second genetic code: Refers sometimes to the nature of the amino acid residues of a protein which determine its secondary and tertiary structures, and sometimes to the features of a tRNA molecule that make it recognizable by one amino acid synthetase but not by others *gene*

second polar event: The fusion of a male gamete with two haploid female nuclei at the time of the fertilization of the ovule; the resulting triploid nucleus divides to form the endosperm *bot*

section cutting: Sections are cut from a block of wax around a plant material, usually by microtome *cyto prep*

sectorial chimera: A chimera in which the distinct meristem is cross-sectional present, like sectors of a circle *bot* >>> chimera >>> Figure 56

sedimentation test: A test for evaluating wheat protein quality where flour is suspended in an aqueous solution of lactic acid and held for a time under specified conditions, then the volume occupied by sediment is measured *meth* >>> ZELENY test

seed: A mature ovule consisting of an embryonic plant together with a store of food, all surrounded by a protective coat *bot* >>> Tables 13, 47

seed bank: A place or storage in which seeds of rare plants or obsolete varieties are kept, usually vacuum-packed and under cold conditions in order to prolong their viability *meth fore agr hort* >>> genebank

seed breeder's rights: National and international rules and laws that provide plant breeders a legal means to apply for proprietary rights to cultivated plant varieties they have bred; "breeder" means the person who bred or discovered and developed a variety; "variety" means a plant grouping within a single botanical taxon of the lowest known rank, whose grouping, irrespective of whether the conditions for the grant of a breeder's right are fully met, can be defined by the expression of the characteristics resulting from a given genotype or combination of genotypes, distinguished from any other plant grouping by the expression of at least one of said characteristics, and considered as a unit with regard to its suitability for being propagated unchanged *agr seed*

seedbed: A plot of ground prepared for seeds or seedlings *agr*

seedborne: Carried on or in seeds *phyt*

seedborne pathogens: Pathogens carried on or in seeds; for example, in wheat, the streak mosaic virus of barley, the fungi, such as snow mold (*Fusarium nivale*), Septoria spike blotch (*Septoria nodorum*), Helminthosporum leaf blotch or spot blotch (*Cochiobolus sativus, Helminthosporum sativum, syn Bipolaris sorokiana, Drechslera sorokiana*), loose smut (*Ustilago nuda* or *U. tritici*), common smut or stinking smut (*Tiletia caries*), and dwarf smut (*Tiletia controversa*); in barley, the stripe mosaic virus, the fungi, such as snow mold (*Fusarium nivale*), leaf stripe disease (*Pyrenophora graminea*), Helminthosporum leaf blotch (*Helminthosporum gramineum, syn Drechslera graminea*), net blotch disease (*Pyrenophora teres, syn Helminthosporum teres,* syn *Drechslera teres*), loose smut (*Ustilago nuda*), black smut (*Ustilago nigra*), and hard smut (*Ustilago hordei*); in rye, the streak mosaic virus of barley, the fungi, such as snow mold (*Fusarium nivale, syn Griphosphaeria nivalis*), Septoria spike blotch (*Septoria nodorum*), stalk bunt (*Urocystis occulata syn Tuburcinia occulata*), and ergot (*Claviceps purpurea*); in oats, the streak mosaic virus of barley, and the fungi, such as loose smut (*Ustilago avenea*); and in maize, the fungi, such as common smut (*Ustilago maydis*), seed rots (*Fusarium* spp., *Penicillium* spp.), and seedling rots (*Pythium* spp., *Fusarium* spp., *Helminthosporum* spp., *Penicillium* spp., *Rhizopus* spp., *Rhizoctonia* spp., *Deploida* spp.) *phyt* >>> Table 47

seed certification: A procedure developed as a means of assuring that seeds have a high standard of purity and quality; seed of an approved variety can only be marketed if it meets strict quality criteria; seed quality standards are laid down in national or international laws, and policed by agencies appointed by governments *seed meth*

seed cleaning: Done to eliminate weed seeds and various insects; grain that is intended for milling, or marketing as food, must be cleaned of all foreign matter, such as chaff, soil, stones, etc. *meth phyt*

seed coat: The protective covering of a seed, usually composed of the inner and outer integuments *bot*

seed coat permeability test: >>> tetrazolium test

seed conditioning: For marketing, seeds are usually cleaned, sized, treated with fungicides, insecticides, or inoculant, and finally bagged *seed* >>> Table 11

seed contamination: The intrusion of weeds, insects, or pathogens in seeds having an adverse effect on germination or subsequent plant growth *phyt*

seed cotton: Unginned picked cotton *agr*

seed counting: Important when determining the >>> hundred-seed (grain) weight or the >>> thousand-seed (grain) weight to ensure that high yields of grain crops are due to many large seeds, rather than to very many small seeds; the manufacturers of seed testing equipment have various designs of equipment for counting and weighing seeds *seed*

seed divider: A device that divides a seed lot and puts subsamples directly into a various number of planting envelopes *seed* >>> Table 11

seed dormancy: >>> dormancy >>> Table 13

seed dressing: A coating (either dry or wet) of protectant pesticide (fungicide, rodenticide, or bird repellent), with or without fertilizer, applied to seeds usually before planting; dry seed dressings are often physically stuck to the testa of the seed by a sticker such as methyl cellulose *agr meth*

seed drill: >>> drill

seed flat: >>> flat

seed flow: Movement of seed through a threshing or cleaning machine *agr*

seed grower (pedigreed): An individual or institution who applies for the inspection of a crop offered for pedigree, grows the crop in accordance with the official regulations, and accepts full responsibility for the production and management of the seed crop and all related financial obligations *agr seed*

seed health test: Specific tests to determine the absence or presence of certain microorganisms known to cause economic loss to crop yield are carried out in numerous forms; seeds are generally surface-sterilized and placed onto agar plates or other substrates which are known to promote the growth of the disease; the percentage of seeds exhibiting the disease is used as a measure of seed health *seed*

seed incompatibility: A postgamous sterility due to failure of tissue development involved in the formation of the seed *seed* >>> Table 47

seed increase: >>> increase

seed index: The 100-g-weight of seeds *seed* >>> thousand-grain weight >>> hundred-seed weight

seeding lath: Commonly, a wooden device for obtaining uniformly spaced drills in seedbeds and aiding the even distribution of hand-sown seed in them *meth*

seeding machine: >>> drill

seedling: A young plant grown from seed *bot*

seedling guard: A row cover to protect seeds indoors or out *meth hort*

seedling resistance: Resistance detectable at the seedling stage *phyt*

seed leaf: >>> cotyledon

seed lot: Usually a batch of seed that has all come from one farm or one crop; the whole of one seed lot can be covered by one seed certificate, and it can be expected to behave uniformly *seed*

seed maturation: >>> maturation

seed mixture: Either seed of more than one kind of cultivar, or a combination of seed of two or more species *seed*

seed mottling: A discoloration of the soybean seed that is not fungal in origin; mottling of seed can often be caused by viruses such as bean pod mottle virus (BPMV) and soybean mosaic virus (SMV) *phyt*

seed multiplication: All methods required to grow plants to maturity and produce seeds, including those practices necessary for harvesting, processing, and preparing seeds for subsequent plantings *seed* >>> Table 49

seed orchard: Plantation of fruit or forest trees, assumed or proven genetically to be superior; it is isolated in order to reduce pollination from genetically inferior outside sources; it is managed to improve the plants and produce frequent, abundant, and easily harvestable seeds *hort fore*

seed parent: The strain from which seed is harvested in the hybrid seed field; also commonly used to designate the female parent in any cross-fertilization *meth* >>> Figure 55 >>> Table 49

seed plant: An individual plant that is or was used for seed production and/or maintaining the genotype *seed* >>> seed >>> Table 49

seed plants: >>> sperma(to)phyta >>> seed

seed potato: A potato tuber that is used for the next growing season in order to produce the next generation for selection and experimental testing *seed* >>> Table 47

seed processing: The operations involved in preparing harvested seed for multiplication or marketing *seed* >>> Table 11

seed production area: In forestry and horticulture, a stand or plantation designated for collection of seeds for reforestation purposes; it may be rogued of inferior trees and treated in such a manner as to produce large quantities of seed; the wood harvest is usually also an important consideration, and the establishment and management is similar to commercial stands; usually seed production was not an initial consideration at establishment *meth fore*

seed proteins (in wheat): Consist of glutenin and gliadin, which belong to the prolamin superfamily; the prolamin superfamily also includes rice >>> prolamins, rye >>> secalins, maize zeins, and barley >>> hordeins; among these prolamin proteins, only wheat prolamins can form a gluten macropolymer; based on a DNA sequence comparison, it was shown that glutenin, gliadin, hordein, and secalin (but not rice prolamins) contained glutamine-rich tandem repetitive sequences; this structure might be involved in protein elasticity; the number of cysteine residues in these proteins involved in inter- and intra-disulfide bonds varies even within the same species; fractionation of seed proteins by aqueous alcohol with DTT (insoluble polymeric proteins) or without DTT (soluble polymeric and monomeric proteins) shows that barley hordeins do not form insoluble polymers, probably due to a very low amount of D-hordein, which shares a similar structure with wheat high-molecular-weight glutenin subunits (HMW-GSs); since HMW-GSs are known to be important to forming a structural framework of gluten polymers, the amount of D-hordein might be critical for hordein polymerization *phys*

seed quality control: Control of physiological, sanitary, and genetic seed quality characteristics *seed*

seed regulation: The total set of rules and protocols related to variety development and release, seed production, quality control, and delivery *seed*

seed set: The process of producing seeds after flowering *bot agr*

seed source: The location where a seed lot was collected; usually defined on an eco-geographic basis by distance, elevation, precipitation, latitude, etc. *fore* >>> provenance

seed spacing: >>> population density

seed stack: The erect stem on a plant that produces flowers and seed; it is particularly applied to root crops and leafy vegetable crops that produce seed after the desired product (root, head, leaves) has fully developed *agr*

seed stand: Any stand used as a source of seed *fore hort*

seed stock: Seed used as a source of germplasm for maintaining and increasing seed of crop varieties *seed* >>> stock seeds

seed testing: Seed offered for sale is usually tested in a seed-testing laboratory; the main test is for germination percentage, but other tests can include seed health, freedom from weed seeds, identity and purity of cultivar, etc. *seed* >>> Table 49

seed trap: A device for catching the seeds falling on a small area of ground from trees or shrubs; it is set for determining the amount of seedfall and the time, period, rate, and distance of dissemination *fore hort*

SeedTrayApp: User-friendly application for packaging seed in a head-row type tray for planting; the workflow takes advantage of barcoded seed packages and a visual tray map to reduce errors in laying out trays and to reduce the time for packing seed; the computer software is written in Java and can be used on any operating system with a Java runtime environment (Java Virtual Machine) including Windows, macOS, and Linux; SeedTrayApp is distributed as a runnable Java archive (.jar); this file can be saved in any directory and launched by double clicking on the executable file; it was developed in 2012 by J. POLAND, Project Leader, USDA-ARS of Kansas State University, U.S., Al WILKINSON, and Trevor RIFE, Project Programmer; SeedTrayApp is free software released under the GNU/GPL license *stat meth seed*

seed treater Designed to apply accurately measured quantities of pesticides to a given weight of seed; basically, there are three types of commercial seed treaters on the market: dust treaters, slurry treaters, and direct treaters; the Panogen and mist-o-matic treaters are examples of direct treaters *seed* >>> dust treater >>> slurry treater >>> direct treater

seed tree: A tree left standing for providing seed *fore meth* >>> seed-tree method

seed-tree method: A method of regenerating a forest stand in which all trees are removed from the area except for a small number of seed-bearing trees that are left singly or in small groups *fore*

seed vessel: The pericarp (wall of the ripened ovary), which contains the seeds *bot*

seed viability testing All methods to determine the potential for rapid uniform emergence and development of normal seedlings under both favorable and stress conditions *meth seed* >>> www.plantstress.com

seed vigor Seed properties that determine the potential for rapid uniform emergence and development of normal seedlings under both favorable and stress conditions *seed* >>> Table 47 >>> www.plantstress.com

100-seed weight: Measurements in grams of 100 well-developed whole grains (seeds with the hull), dried to 13% moisture content, weighed on a precision balance *meth*

segmental allopolyploids: A partial homology or so-called homoeology of chromosome sets combined in an allopolyploid *cyto*

segregate: >>> segregation

segregating population: A population, generally the progeny of a cross, in which genetic differences are detectable, thus permitting identification of individuals having a desired trait and their selection for further breeding *gene* >>> Table 37

segregation: The separation of alleles during meiosis so that each gamete contains only one member of each pair of alleles *gene* >>> Figure 6 >>> Tables 2, 3, 4, 6, 7, 8, 19, 20, 21

segregation distortion: The distortion of the 1:1 segregation ratio produced by a heterozygote; it can arise because of abnormalities of meiosis, which results in an Aa individual producing an unequal number of A- and a-bearing gametes, or it may arise from A- and a-bearing gametes being unequally effective in producing zygotes *gene*

select (synthetic): A category for a specific combination of seed lots from an inspected breeder, or foundation of crop used in the production of a certified seed crop *seed meth* >>> Figure 51

selectable marker: A physiological or morphological character, which may easily be determined as a marker for its own selection or for selection of other traits closely linked to that marker *gene*

selected seed: Selected seed is a class of tree seed that is the progeny of rigidly selected trees or stands of untested parentage that have promise but not proof of genetic superiority, and for which geographic source and elevation shall be stated on the certification label *fore hort seed*

selection: The process determining the relative share allotted to individuals of different genotypes in the propagation of a population; natural selection occurs if zygotic genotypes differ with regard to fitness *meth evol* >>> clonal selection >>> differential selection >>> index

selection >>> mass selection >>> recurrent selection >>> tandem selection >>> Figures
38, 39, 60 >>> Tables 7, 37, 49

selection after flowering: A selection that is only possible after flowering since the critical selective characters are expressed after flowering, seeds, and/or fruit formation periods (e.g., grain size, spike length, fruit color, etc.) *meth*

selection coefficient: A measure of the disadvantage of a given genotype in a population *stat* >>> Figures 38, 39, 60

selection criteria: The specific characters and plant reactions on which the selection is focused during the breeding cycles *meth*

selection differential: In artificial selection, the difference in mean phenotypic value between the individuals selected as parents of the following generation and the whole population *meth stat* >>> Figures 38, 39, 60

selection gain: In artificial selection, the difference in mean phenotypic value between the progeny of the selected parents and the parental generation *gene meth* >>> Figure 38

selection index (indices *pl***):** Selection indices help select the best individuals for the next breeding cycle on the basis of observed phenotypic values for several traits of each candidate individual; they assign subjective economic weights to each trait and are relatively simple to analyze; their disadvantages are that they require large amounts of information, economic weights are difficult to assign, sampling error can be large, and the statistical sampling properties of selection indices of the selection response are unknown except in the case of two traits *stat meth* >>> Figures 38, 39, 60

selection intensity: The ratio of the number of genotypes selected divided by the number of genotypes tested *meth gene stat* >>> Figures 38, 39, 60

selection limit: The exhaustion of genetic variance in a population, so that no further selection response can be expected *gene* >>> Figures 38, 39, 60

selection pressure: The effectiveness of natural selection in altering the genetic composition of a population over a series of generations *stat meth* >>> Figures 38, 39, 60

selection prior to flowering: A selection that is possible before flowering since the critical selective characters are already expressed (e.g., seedling resistance, tillering capacity, head size in cabbage, etc.) *meth*

selection response: The difference between the mean of the individuals selected to be parents and the mean of their offspring; it is expressed by the formula $R = h^2 \times S$ (h = heritability, S = selection coefficient, R = phenotypic difference between the mean of all selected fractions and the mean of total population) *gene stat* >>> Figures 38, 39, 60

selective advantage: An advantage for survival of a genotype in a population, and for production of viable progeny as compared to other genotypes, which may show a selective disadvantage with respect to fitness and viability *meth stat* >>> Figures 38, 39, 60

selective agent An environmental or chemical agent that imposes a lethal or sublethal stress on growing plants, or portions thereof in culture, enabling selection of resistant or tolerant individuals *biot* >>> www.plantstress.com

selective breeding: Selecting and mating species with characteristics that were chosen from past performance *meth* >>> Figures 38, 39, 60

selective culture medium: >>> selective agent

selective disadvantage: Inferior fitness of one genotype compared to others in the population *meth stat* >>> selective advantage

selective fertilization: The non-random participation of male or female gametes or different genotypes in the formation of zygotes and/or hybrids *bot*

selective gametocide: A treatment that inactivates certain gametes, such as one that produces male sterility but does not affect the female gametes *meth seed*

selective grazing: Using certain plant species, individual plants, or plant parts to the exclusion of others *agr meth*

selective herbicide: An herbicide that acts against either monocots or dicots, against weeds and not against crop plants, or even against species weeds *phyt agr*

selective medium: >>> selective agent

selective neutrality: The situation in which different alleles of a certain gene confer equal fitness *gene*

selective system: Any experimental method that enhances the recovery of specific genotypes *meth*

selective target evaluation in microbes (STEM): Refers to the way microbial biologists and >>> bioinformaticists use various databases and industrial applications to prioritize the genes to study using functional genomics tools *biot*

selective value: A measure of the fitness of a gene within a genotype or of a genotype within a population; it is proportional to the probability of that gene or genotype surviving, and it is also a function of gene or genotype frequencies *gene stat* >>> Figure 38

select plot grower: A seed grower who has been approved by an official organization for the production of select seed crops; this person or institution has completed a (for example) three-year probationary period of plot production with three recent years of pedigreed seed crop production *seed agr* >>> seed grower (pedigreed)

select seed: The approved progeny of breeder or select seed produced in a manner by seed growers authorized by an official organization to maintain its varietal identity and purity; select seed may be produced from select seed for a maximum of five multiplications from breeder seed *seed meth*

self: An individual plant produced by self-fertilization, as opposed to cross-bred *gene*

self-compatible: A plant that can be self-fertilized *bot* >>> Table 35

self-fertile: Capable of producing seed upon self-fertilization *bot* >>> Table 35

self-fertility: >>> autogamy

self-fertilizing: The fusion of male and female gametes from the same individual *bot* >>> Table 35

self-fertilizing crop: >>> self-fertilizing

self-incompatibility: Controlled physiological hindrance to self-fertilization; inability to set seed from application of pollen produced on the same plant; among the results obtained from the numerous hybridization experiments of the late 18th and early 19th centuries was the discovery of self-incompatibility; self-incompatibility is a barrier against inbreeding and the homozygosis caused by it; it displays its effect during the so-called progamic stage of development, i.e., before fertilization occurs, so that the chances of the egg-cell being fertilized by a foreign pollen remain unimpaired; it became possible to search for the genetic basis of self-incompatibility after the Mendelian laws of heredity were rediscovered; C. CORRENS performed the first studies on this topic; the breakthrough was achieved by E. M. EAST and A. J. MANGELSDORF with their studies on *Nicotiana* in 1925; self-incompatibility is caused by a gene (*SI*) with numerous alleles: *SI1, SI2, SI3, SI4 ... SIn*; A. LUNDQUIST and D. L. HAYMAN discovered, independent of each other, a second self-incompatibility locus (*Z*) in grasses; in such cases, the pollen is only rejected if both plants contain the same alleles in both loci (*SI* and *Z*); such bifactorial systems were later found in numerous monocot and dicot families; finally, A. LUNDQUIST and his collaborators discovered plant species with three or more incompatibility loci (*Ranunculus acris* 3, *Beta vulgaris* 4), each of which may have numerous alleles; the situation becomes completely confused in the case of polyploidy; accordingly, the models proposed by LUNDQUIST were questioned by D. L. MULCAHY and G. BERGAMINI-MULCAHY in 1983; systems with several genes are weaker self-incompatibility systems than those with just one gene; there are several mechanisms responsible for self-incompatibility in higher plants: (1) pollen may fail to germinate on the stigma; (2) pollen tube growth in the style may be inhibited to the extent that pollen fails to reach the ovary; (3) pollen tubes of sufficient length may fail to penetrate the ovule; or (4) a male gamete that enters the embryo sac may fail to unite with the egg cell; the relative length of stamens and style in a bisexual flower is associated with incompatibility; most plant species have homomorphic flowers in which the stamens and styles attain comparable length; homomorphic flowers can be gametophytic or sporophytic; some self-incompatibility species exhibit heteromorphic flowers in which the stamens and styles attain different heights; the presence of either pin or thrum flowers is termed >>> distyly; J. L. BREWBAKER discovered, in 1957, several remarkable correlations between a self-incompatibility system and pollen features: (1) gametophytic self-incompatibility occurs nearly always in pollen with two nuclei; the stigma of the respective species is usually wet; (2) sporophytic self-incompatibility occurs in pollen with three nuclei; such pollen germinates badly under natural conditions and its life expectancy is short; the respective stigmata are dry; for the controlling mechanism, it was possible to identify glycoproteins at the surfaces of pollen and stigma; the addition of ConA drastically reduces pollen adhesion in *Galanthus nivalis*, an indication that lectin– lectin receptor interactions play a substantial part in the stabilization of the cells' linkages; H. F. LINSKENS (1960) produced antibodies against the self-incompatibility antigens of petunia pollen that reacted with antigen determinants of the surface of non-compatible stigmas; the proteins of the stigma surface are subject to a strong turn-over; a large amount of new material is produced after the pollen grain has been bound; the recognition seems only to be an elicitor of the stigma tissue's actual defensive reaction; A. CLARKE cloned numerous self-incompatibility alleles and detected that the produced glycoproteins display RNase activity; plants with sporophytic self-incompatibility, like Brassica species, have glycoproteins, too, but they belong to other protein families than that of the self-incompatibility products of gametophytic self-incompatibility, and they lack RNase activity; the analyzed gene products occur at the stigma surface in large quantities, but they have not been identified at pollen surfaces; compatibility is not enough to explain the penetration of the stigma and style tissue by the pollen tube; numerous enzymes of the pollen tube surface are activated and begin to disintegrate the transfusion tissue; the orientation of the directed growth is maintained in every stage of the process; the sequence of reactions of pollen tube and style tissue has to be coordinated; this coordination is impossible with pollen of foreign species; the consequence is a rejection called incongruent reaction, which usually causes the lack of one or several steps in a chain of events that starts with the encounter of pollen and style cells; there are possibilities to avoid or break self-incompatibility (e.g., by high temperatures in rye) *gene* >>> Table 35 >>> SCHLEGEL (2007)

selfing: When a pistil is fertilized with pollen from the same plant that bears the pistil; also applied to seed resulting from such fertilization *meth* >>> self-pollination >>> Table 35

selfish DNA: A segment of the genome with no apparent function other than to ensure its own replication *gene*

self-pollination: The transfer of pollen from anther to stigma of the same plant (e.g., in barley, chickpea, clover cowpea, crambe, crotelaria, guar, field bean, field pea, linseed, jute lentil, lespedeza, millet, mungbean, oats, peanut, potato, rice, sesame, soybean, tobacco, tomato, vetch, wheat, or wheatgrass) *bot* >>> Table 35

self-pruning: Pruning is a horticultural and silvicultural practice involving the selective removal of parts of a plant, such as branches, buds, or roots; reasons to prune plants include deadwood removal, shaping (by controlling or directing growth), improving or maintaining health, reducing risk from falling branches, preparing nursery specimens for transplanting, and both harvesting and increasing the yield or quality of flowers and fruits; in nature, meteorological conditions such as wind, ice and snow, and salinity can cause plants to self-prune; a plant may also induce part of itself to die off; there is even a gene (*Sp*) on chromosome arm 6L for self-pruning in tomato that recessively regulates vegetative to reproductive switching of sympodial meristems (CARMEL-GOREN et al. 2003) *hort fore meth* >>> pruning

self-rooted: >>> true-rooted

self-seed: A plant that releases ripe seeds that sprout into new plants without human help *agr hort*

self-seeding: >>> self-seed

self-sow: When plants produce seeds and germinate from seed without assistance; most wild flowers and several types of annuals will produce seeds, release them into the soil, and produce new plants the following year *bot*

self-sown cereals: >>> self-seed

self-sterility: The inability of some hermaphrodites to form viable offspring by self-fertilization *bot* >>> Table 35

semiallele: Mutant alleles that are allelic with respect to the function but not the structure of the allele (i.e., they are not present on the same chromosome position) *gene*

semiarid: Of a region or land characterized by very little annual rainfall, usually from 250 to 5,000 mm *eco*

semiconservative replication: Replication of DNA in which the molecule divides longitudinally, each half being conserved and acting as a template for the formation of a new strand *gene*

semidesert: An extremely dry area characterized by sparse vegetation *eco*

semidomesticated: The incomplete breeding of species in order to accommodate human needs *evol*

semidominance: >>> partial dominance

semidominant: >>> partial dominance

semidwarf: A common term used in wheat breeding; it designates individuals or a variety showing intermediate stem length compared to very short mutant lines; during the 1970s, semidwarf wheat genotypes triggered the so-called >>> "Green Revolution" in agriculture, mainly of Third-World countries; several alleles of reduced height (*Rht*) loci contributed to shorter straw length and thus better lodging resistance, as well as to higher spikelet fertility and thus yield; several dwarf (*Rht*) mutants show a decrease of the amount of certain gibberellins within the meristematic cells, which reduce plant height, not only in wheat but also in rice, maize, and thale cress ssp.; the genes modifying those gibberellic acids produce several proteins regulating different cell activities; some of the *Rht* genes that show a specific gibberellic acid insensitivity are sequenced and available for use in genetic engineering *agr* >>> CIMMYT >>> gibberellic acid (GA) >>> short-straw mutant >>> *Rht* gene

semigamy: >>> hemigamy

semi-Latin squares: An (n × n) row-and-column design for kn treatments where every row and every column of the design contains a complete set of treatments, and every row-by-column intersection contains a block of k plots; in 1937, >>> YATES gave an example of a 7 × 7 semi-Latin square for 14 varieties, where the 14 varieties were divided into seven pairs and each pair was assigned to a different letter of a 7 × 7 Latin square; each row-by-column intersection of the Latin square contained a block of two plots, and the two varieties of each pair were randomly allocated to pairs of plots within individual blocks; this design was of the split-plot type, and some treatment contrasts were estimated entirely within blocks, whereas other treatment contrasts were estimated entirely between blocks; often, precision is greater for plots-within-blocks comparisons than for plots-between-blocks comparisons, and YATES suggested that a better arrangement would be to assign the varieties to the two sets of letters of a 7 × 7 >>> Graeco-Latin square where the row-by-column intersections of the superimposed squares would represent blocks of size two; each of the seven varieties assigned to one set of letters would occur once in the same block with each of the seven varieties assigned to the other set of letters; YATES showed that, where all variety comparisons are of equal importance, this is the best choice of design for comparisons in the plots-within-blocks stratum; finally, he showed that hyper-Graeco-Latin square designs with row-by-column intersections split into three or more plots could be constructed from three or more orthogonal squares *stat* >>> design of experiment

seminoferous: seed-bearing (e.g., a cone consisting of seed-bearing, overlapping scales surrounding a central axis; a seed-bearing spike of a grass; a pistil as a seed-bearing organ of a flower) *bot*

semiripe cutting: A cutting taken from wood that has begun to mature *hort meth* >>> greenwood cutting >>> hardwood cutting >>> root cutting

semispecies: >>> incipient species

semisterile: >>> semisterility

semisterility: A situation in which half or more of all zygotes are inviable *gene bot*

semolina: A granular, milled product of durum wheat, used in the making of pasta *agr*

senescence: The phase of plant growth that extends from full maturity to death; senescence or programmed cell death is a process that interacts with many biochemical and physiological changes in living organism and is generally induced by aging; many environmental stresses that accelerate the production of activated oxygen can also induce senescence artificially; one of the important aspects of senescence is possibly the degradation of macromolecules such as DNA; the random amplification of polymorphic DNA (RAPD) technique is a good method to compare the DNA quality of juvenile and senescence samples in which oxidative stress is induced; when genomes are influenced by senescence and environmental stresses they undergo genome diversity *bot*

sense mutation: A mutation that changes a termination (stop) codon into one that codes for an amino acid, usually resulting in an elongated protein *gene*

sense strand DNA: The DNA strand with the same sequence as mRNA *gene biot*

sensitivity analysis: A method used in modeling that identifies critical variables that significantly affect the system under study, and then changing their values to determine the type and extent of their effect *meth stat*

sepal: A floral part of the outer whorl, referred to collectively as the calyx *bot*

Septoria seedling disease (of wheat): The same pathogen (*Septoria nodorum, syn Leptosphaeria nodorum*) that causes Septoria glume blotch in wheat *phyt*

septum (septa *pl*): A cross-wall or partition *bot*

sequence-characterized amplified region (SCAR): A molecular marker; the DNA amplified as a PCR-derived marker (e.g., RAPD) is sequenced; the sequence is used to develop a new PCR marker comprising a single DNA species, unique to the target organism or genotype *biot*

sequence divergence: Changes in the DNA or protein sequences of homologous genes in different species due to the independent accumulation of mutations and natural selection since these species shared a common ancestor *evol gene*

sequence motif: A short conserved amino acid or nucleotide sequence pattern that represents a specific functional site of a molecule *gene phys*

sequence-tagged-site (STS) marker: A unique (single-copy) DNA sequence used as a mapping landmark on a chromosome *biot gene*

sequence-tagged-microsatellites (STMs) marker: Primers constructed from the flanking regions of >>> microsatellite DNA which can be used in >>> PCR reactions to amplify the repeat region, i.e., a locus-specific, codominant molecular marker, used for, e.g., >>> DNA fingerprinting, or genetic and physical mapping; it is generated by sequencing microsatellite-flanking sequences; specific PCR primers towards the flanks are designed and used to amplify genomic DNA *biot meth* >>> sequence-tagged-site >>> primers

sequencing: The determination of nucleotides and their order along a DNA or RNA molecule, or the determination of the amino acids and their order in a protein molecule; it includes any method or technology that is used to determine the order of the four bases – adenine, guanine, cytosine, and thymine – in a strand of DNA; the advent of rapid DNA sequencing methods has greatly accelerated biological research *gene biot* >>> next-generation sequencing

sequential cropping: A multiple cropping pattern characterized by two or more crops growing in sequence on the same field within a year; the succeeding crop is planted after the preceding one has been harvested; there is no intercrop competition *agr*

sequential sampling: A survey method without fixed sample size, in which sampling is continued until a predetermined, approximate level of objects or events is found *stat*

sere: A succession of plant communities in a given habitat leading to a climax association *eco*

serial analysis of gene expression (SAGE): A technique applied to study transcriptomes; most useful for fully sequenced genomes or for organisms having a quasi-complete collection of >>> ESTs; it is based on cloning of concatemers of very small tags (9–11 bases), each representing a transcript; sequencing of the concatemers results in a list of tags present in a population; the frequency of each tag is proportional to the abundance of the transcript it represents *biot*

serology: The study of the nature, production, and interactions of antibodies and antigens *phyt sero*

seroprotein: Protein produced in response to an antigen in an immunological reaction *phys chem*

serotype: A subdivision of virus strains distinguished by protein, or a protein component that determines its antigenic specificity *phyt*

serrate: With sharp teeth pointed toward apex or forward *bot*

sessile: Without stalk of any kind and/or without a petiole (leaf) or pedicel (stalk of a flower) *bot*

set(t): A short piece of a stem or a plantlet used for propagation, for example, in >>> sugarcane it is used for >>> vegetative propagation; each set(t) usually has three nodes, and the cut ends are often dipped in a mixture of insecticide and fungicide; the first >>> crop from these set(t)s is called the "plant crop", while all subsequent crops are called >>> ratoon crops, until replanting becomes necessary *hort agr*

set (of seeds, etc.): The development of fruit and/or seed following pollination; refers also to transplant as seedlings, to apply as a graft, a young bulb, tuber, or other type of vegetative propagule ready for planting *meth hort*

seta (setae pl): A bristle *bot*

setting: >>> set

severe mosaic virus of potato: >>> potato virus Y

sex: Contrasting and complementary traits shown by female and male individuals within a given species *bot*

sex chromatin: A condensed mass of chromatin representing an inactivated X chromosome *cyto* >>> Figure 65

sex chromosome: A chromosome whose presence or absence is linked with the sex of the bearer; it plays a role in sex determination (e.g., in asparagus, hops, hemp, etc.); sex chromosomes are more widespread than previously thought; if sex chromosome is defined as having two closely linked sex determination genes, then all established >>> dioecious species have sex chromosomes, but not all of them have heteromorphic sex chromosomes; the primary function of sex chromosomes is to reinforce >>> dioecy, i.e., a sexual system that may increase >>> outcrossing in species lacking >>> self-incompatibility; sex chromosomes in land plants can evolve as a consequence of close linkage between the two sex determination genes with complementary dominance required to establish stable dioecious populations, and they are found in at least 48 species across 20 genera; the sex chromosomes in hepatics, mosses, and gymnosperms are morphologically heteromorphic; in angiosperms, heteromorphic sex chromosomes are found in at least 19 species from 4 genera, while homomorphic sex chromosomes occur in 20 species from 13 genera; the prevalence of the XY system found in 44 out of 48 species may reflect the predominance of the evolutionary pathway from >>> gynodioecy towards >>> dioecy; all dioecious species have the potential to evolve sex chromosomes, and reversions back from dioecy to various forms of monoecy, gynodioecy, or androdioecy have also occurred; such reversals may occur especially during the early stages of sex chromosome evolution before the lethality of the

YY (or WW) genotype is established; about six stages of sex chromosome evolution can be summarized: (1) unisexual mutation of two sex determination genes with complementary dominance, (2) suppression of recombination between the two sex determination genes and YY genotype is viable, (3) suppression of recombination spread to neighboring regions and a small male-specific region of the Y chromosome region evolved (YY genotype is not viable), (4) the male-specific region of the Y chromosome expands in size and degenerates in gene content via accumulation of transposable element insertions and intrachromosomal rearrangements (the X and Y chromosomes become heteromorphic), (5) severe degeneration of the Y chromosome happens (deletion of nonfunctional DNA sequences results in reduction of Y-chromosome size), (6) suppression of recombination spreads to the entire Y chromosome (the Y chromosome is lost and X-to-autosome ratio sex determination system has evolved) *cyto* >>> heterochromosome >>> Figure 65

sex determination: The mechanism by which the sex is determined *bot*

sex dimorphism: The different morphology of individuals within a species caused by their sexual constitution (e.g., in Melandrium alba, hops, hemp, etc.) *bot*

sex expression: >>> sex-controlled

sex inheritance: >>> sex linkage

sex linkage: Genes located and inherited on a sex chromosome *gene*

sex ratio: The number of males divided by the number of females (also given in percent) at fertilization (equals primary sex ratio) *gene*

sex-controlled: A trait whose appearance is controlled by the type of sex of the individual (e.g., in asparagus, hops, hemp, etc.) *gene*

sexfoil: A leaf with six leaflets *bot*

sex-influenced: >>> sex-controlled

sex-limited: Pertaining to genetically controlled characters that are phenotypically expressed in only one sex *gene*

sexual: Processes in which meiosis and fertilization are included leading to genetic recombination *gene*

sexual dimorphism: >>> sex dimorphism

sexual partner: >>> mate

sexual reproduction: Reproduction involving the union of gametes that are haploid and derive from two sexes *gene*

sexual selection: Contributes to the sex dimorphism; it is based on male competition or on female choice *gene*

sexuality: Sexual character; possession of the structural and functional traits of sex *gene*

SGE: >>> starch gel electrophoresis >>> gel electrophoresis

shade cloth: Any of various fabrics used in the summer to lower soil temperatures, accelerate germination of cool-season autumn crops, prevent bolting, or protect against drying *meth agr hort*

shade tolerance: The degree to which a tree or common plant can grow or establish under a forest canopy or within a stand *agr*

shattercane: A summer annual that resembles grain >>> sorghum and >>> maize; it derived from wild or cultivated sorghum varieties (*Sorghum bicolor*); it became a major weed problem in maize, sorghum, and >>> soybean fields *agr phyt*

shattering: The opening or disintegrating of the seed coat, fruit, or husk before harvesting; the consequence is loss of seeds in crop plants (e.g., seeds drop to the ground prior to harvest) *agr*; in viticulture, the physiological stage following bloom when impotent flowers and small green berries begin to fall from the cluster *hort*

sheath: Of leaves, the base of a blade, or stalk that encloses the stem *bot* >>> Table 30

shedding: The release of seeds from seed-bearing organs, such as spikes in cereals, usually before harvest *agr*

sheepnosing: A physiological disorder that commonly occurs in grapefruits, e.g., found in Navelina navel oranges grown under arid conditions; the stem end of the fruit has a high collar with a depressed button that result in a snout-like appearance; sheepnosed fruits often fail to meet the demanded quality parameters; it is attributed to many factors; heat stress during flower-bud initiation mainly causes this disorder where the ovary become elongated at the stem end where fruit set and the tree load are low; the orange variety "Washington Navel" is resistant under the same conditions *phys*

shelf life: The term or period during which a stored commodity (vegetables or food) remains effective, useful, or suitable for consumption *agr*

shifting cultivation: A type of farming in which fields are used for a few years, and are then left to grow in a wild state for many years; this allows the soil to recover and become rich and fertile again, as well as discouraging the increase of certain >>> parasites; the method is used in areas of low population density *agr eco* >>> slash-and-burn farming

shifting dominance: >>> alternating dominance >>> dominance

shikimate pathway: An important chemical pathway in plants, responsible for producing amino acids and other aromatic products *phys*

shikimic acid: A crystalline acid, $C_6H_6(OH)_3COOH$, formed in plants as a precursor in the biosynthesis of aromatic amino acids and of lignin *phys*

shoot: A stem or branch and its leaves; a new young growth *bot*

shoot apex: The 0.5–2 mm tip of a shoot; the apical or lateral shoot meristematic dome together with the leaf primordia, from which emerge the leaves and subadjacent stem tissue *bot*

shoot bud: >>> bud

shoot culture: >>> shoot-tip culture

shooter mutant: A mutant of *Agrobacterium tumefaciens* in which the auxin genes have been partially or completely deleted; such mutants induce an enhanced cytokinin level in transformed plant tissue; in some species, transformation with the shooter mutant leads to a "shooty" phenotype, i.e., with many shoots; the mutants are then used to induce regeneration of species with low regeneration capacity *biot meth*

shoot meristem: A meristem located at the apex of shoots, and from which the aerial parts of a plant are derived *bot*

shoot regeneration: In tissue culture, the phenomenon of *in vitro* shoot formation from callus *biot* >>> Figure 58

shoot tip: >>> shoot apex

shoot-tip culture: The *in vitro* culture of shoot apical meristem plus one to several primordial leaves *biot*

short-day conditions: Short days and corresponding long nights *phys*

short-day plant: A plant in which flowering is favored by short days and corresponding long nights (e.g., *Chrysanthemum* spp.) *phys*

short interspersed nuclear elements (SINEs): Families of short (150 to 300 bp), moderately repetitive elements of eukaryotes, occurring about 100,000 times in the genome; these interspersed nuclear elements are ancient non-LTR retrotransposons which are highly abundant, heterogeneous, and in contrast to more conserved LTR-retrotransposons, only poorly annotated in plant genomes; they can be used as molecular markers, e.g., as in potato variety identification *biot* >>> long interspersed nuclear elements

short-lived perennial: For example, several grasses normally expected to live only two to four years *bot agr*

short-straw mutant: A mutant genotype of monocots showing reduced plant height by reduced length of internodes; several genes are known to cause the shortening of plants; some of those genes (*Rht* and *Gai* genes) have attained great attention in cereal breeding *gene* >>> CIMMYT >>> semidwarf >>> gibberellic acid >>> gibberellic acid (response) testing >>> *Rht* gene

shotgun collection: Cloning of an entire genome in the form of randomly generated fragments *biot*

shotgun method: A way of preparing DNA for genetic studies; random-cut fragments of DNA are cloned into a vector; the fragments resemble a clone library; using this collection of clones, several molecular techniques are applicable *biot* >>> sibling technique

shot hole: A leaf spot disease characterized by holes made by the dead parts dropping out; for example, caused by *Heteropatella antirrhini* in snapdragon, by *Stigmina carpophila* in peach, or by *Pseudomonas mors-prunorum* in plum *phyt* >>> Figure 62

shoulder: The upper edge of the broad wing of the glume (e.g., in wheat) *bot*

shrub: A perennial woody plant, less than 10 m tall, which branches below or near ground level into several main stems *bot*

shrubby: Shrub-like, bushy, with many stems rather than a single trunk; a form of lichen that appears bushy or hair-like *bot*

shuck: An outer covering (e.g., the husk of maize, the shell of a walnut, etc.) *bot*

shuttle breeding: A program that shuttles seed between two (or more) locations to be grown at each; this method was introduced by N. BORLAUG and is a proven contributor to the success of CIMMYT (Mexico); shuttle breeding allows two breeding cycles per year instead of one – for example, a winter cycle in the northern desert of Sonora, and a summer crop in the central Mexican highlands; this not only fast-forwards selection, but also exposes test varieties to radically different day lengths, temperatures, altitudes, and diseases; resulting plants are broadly adapted; they grow well in numerous environments *meth*

shuttle vector: A vector (e.g., a >>> plasmid) constructed in such a way that it can replicate in at least two different host species (e.g., a >>> prokaryote and a >>> eukaryote); >>> DNA recombined into such a vector can be tested or manipulated in several cell types *biot*

sib: Short for >>> sibling; one of two or more offspring with both parents in common but deriving from different gametes *tax*

sib mating: A form of inbreeding in which progeny of the same parents (>>> siblings) are crossed *meth*

sibbed: Mated individuals having the same parentage *gene*

sibing: The transfer of pollen between different plants of the same variety *meth*

sibling chromatid: >>> sister chromatid

sibling(s): One of two or more individuals having one or both parents in common; a male sibling is called a "brother", and a female sibling is called a "sister" >>> full sibs

sibling species: Morphologically similar or identical populations that are reproductively isolated *gene eco* >>> Table 37

sibling technique: A method for the isolation of bacteria containing a specific cloned gene (e.g., in shotgun cloning experiments) whose phenotype cannot be easily recognized on individual colonies, but can be detected with great sensitivity in a large population where only a small minority of the cells contains the gene wanted *biot*

side grafting: >>> inarching >>> graft

siderophore: Chelating protein, produced by both plants and microorganisms, with a high affinity for metal ions; low-molecular-weight, virtually Fe(III)-specific ligands produced as scavenging agents in order to combat low iron stress; it is also produced and expelled from the cell of some species of bacteria under conditions of iron deficiency; siderophores complex to iron in the soil solution and are then reabsorbed and processed, providing the organisms with an efficient mechanism for obtaining a scarce resource *phys* >>> root exudates

sieve cell: Long, slender, tapering cells that form part of the sieve tube; they lack nuclei but retain cytoplasm; each sieve cell ends in a sieve plate *bot*

sieve plate: Perforated wall area between two phloem cells through which their protoplasts are connected *bot*

sigmoid (curve): S-shaped (e.g., the growth curve of a plant) *stat*

sign: A visible manifestation of a causal agent of plant disease *phyt*

signal sequence: A specific sequence of 15 to 20 amino acids that is involved in the movement of a sector protein through the plasma membrane *gene*

signal transduction pathway: The process whereby a >>> receptor initiates one or more sequences of biochemical reactions that connect the stimulus to a cellular response *phys*

sign test: Designed to test a hypothesis about the location of a population distribution; it is most often used to test the hypothesis about a population median; the sign test does not require the assumption that the population is normally distributed; in many applications, this test is used instead of the one-sample t-test when the normality assumption is questionable; it is a less powerful alternative to the WILCOXON-signed ranks test, but does not assume that the population probability distribution is symmetric; this test can also be applied when the observations in a sample of data are ranks (i.e., ordinal data rather than direct measurements) *stat*

signal peptide: A short segment of about 15–30 amino acids, which is found on the N-terminal of secreted proteins; during processing, the protein of the cell metabolites recognizes the signal sequence and allows the secretion or penetration through the membranes of organelles; later in mature proteins, the sequence disappears; it is removed by a protease *phys*

signal sequence: A stretch of 13–36 hydrophobic amino acids at the amino-terminal of the nascent >>> polypeptide chain that guides polypeptide translocation through the rough endoplasmic reticulum; it helps the polypeptide to pass through the membrane via interaction with its receptor on the membrane, and is usually cleaved off at the other side of the membrane by an endopeptidase *gene biot* >>> leader sequence

signal transduction: The biochemical events that conduct the signal of a hormone or growth factor from the cell exterior, through the cell membrane, and into the cytoplasm; this involves a number of molecules, including receptors, proteins, and messengers *phys*

signed-rank test: A non-parametric statistical test of the difference between two treatments using paired observations; the differences are ranked according to their absolute magnitude, and each rank is given the sign of the original difference; the sum of the positive or negative ranks provides the test statistic developed by WILCOXON *stat* >>> sign test

significance level: The significance level of a statistical test is the preselected probability of (incorrectly) rejecting the null hypothesis when it is in fact true; usually a small value, such as 0.05, is chosen; if the P value (probability value) calculated for a statistical test is smaller than the significance level, the null hypothesis is rejected *stat*

significance test: Statistical test designed to distinguish differences due to sampling error from differences due to discrepancy between observation and hypothesis *stat*

signpost: >>> landmark

silage: A type of foodstuff for livestock, prepared from green crops and by fermentation; it contains ~65% moisture *agr*

silage crops: Crops grown for harvest and storage while in the green and high-moisture conditions; maize, grass, and sorghum are used for this purpose and cured while in storage *agr*

silencer sequence: A DNA sequence, usually located distant from the core promoter, that may inhibit gene transcription *biot*

silent mutation: A mutation in a gene that causes no detectable change in the biological characteristics and/or gene product *gene*

silica cell: Cell filled with silica as in epidermis of grasses *bot*

silica gel: A substance which absorbs water vapor and is put into air-tight containers to keep the contents dry; it is particularly valuable for the long-term storage of seeds in genetic conservation *seed meth*

siliceous: Composed of or abounding in silicle *bot*

silicious: >>> siliceous

silicle: A short, broad silique, e.g., in shepherd's purse *bot*

silicula: >>> silicle

siliqua: A dry, dehiscent fruit that has a central partition; it is elongated; it is produced by many members of Brassicaceae, e.g., in radish *bot*

silique: >>> siliqua

silk: In maize, the stigma and style of the female flower, through which the pollen tube grows to reach the embryo sac *bot*

silking: The female flowering in maize (i.e., the moment when silks emerge from the husk) *agr*

silky: Covered with close-pressed soft and straight pubescence *bot*

silver staining: Using silver nitrate ($AgNO_3$) and an appropriate cell pretreatment, interphase nuclei can be made visible without phase contrast microscopy; they appear yellow to light brown and the nucleoli appear brown; at metaphase, silver nitrate stains nucleolus organizer regions that have been active in the preceding >>> interphase; although the amount of silver nitrate is in some relation to the amount of activity or number of active rDNA copies, it cannot be used for an accurate quantitative study *cyto*

silvics: The study of the life history and general characteristics of forest trees and stands with particular reference to locality factors, as a basis for the practice of silviculture *fore* >>> silviculture

silviculture: The theory and practice of controlling the establishment, composition, growth, and quality of forest stands to achieve the objectives of its management *fore*

simple fruit: Derives from a single pistil *bot*

simple incomplete Trojan squares: With simple factorial treatment structure, they can be constructed by omitting the complete first row from a standardized Trojan square; assuming a Trojan square is constructed by using standardized orthogonal Latin squares, the first row of the square will comprise blocks that each contain only a single level of the letters factor; omitting the first row of a standardized Trojan square will give a design that has full factorial balance for the letters and squares factors in both the blocks-within-columns and the plots-within-blocks strata; simple incomplete Trojan designs have a simple factorial analysis of variance in every stratum of the analysis and, if the factorial effects of the design are associated with real factorial treatment effects, the design has a balanced factorial treatment structure *stat* >>> design of experiment >>> Trojan squares

simple linear regression: A regression analysis in which the mean value of a dependent variable y is assumed to be related to a single independent variable x by the expression $y = a + \beta x$ where a = intercept and β the slope of the regression stat

simple random sampling: A sampling technique wherein the target population is treated as a unitary whole and each element has an equal probability of being selected for the sample *stat meth*

simple sequence repeat(s) (SSR): Genetic DNA marker that can be identified through biochemical assays, also known as >>> microsatellites; they usually consist of two or three nucleotide repeats; the genetic variation occurs in the number of repeats, and therefore the difference in length can be measured; it is a co-dominant marker system, and particularly informative when heterozygote genotypes are involved; SSR markers are used during the pre-breeding and breeding phases to find intra-species genotypic variations; SSR markers can also be used for >>> bulk segregant analysis as well as seed lot testing at low costs *biot* >>> Figure 63

simplex type: >>> autotetraploid >>> nulliplex type

simulation: A way of predicting the result of an action based on modeling with a computer; the simulation may be deterministic (thus the same action always gives the same result) or Monte-Carlo (different runs can give different results because random sampling is involved in the simulation program) *stat meth*

simulation experiment: A designed experiment to investigate some phenomenon of interest about the model, e.g., effect of perturbations, parameter sensitivity *stat*

simultaneous selection: In a selection process, when several traits are considered in the same generation *meth*

sinapate ester: In rapeseed meal a high content of phenolic acid esters is common, mainly sinapate esters, which have been shown to cause a dark color and a bitter taste of meal and derived protein products; therefore, increasing the meal and protein quality of winter rapeseed (*Brassica napus*) is of importance in rapeseed breeding for food and feed purposes *chem*

SINEs: >>> short interspersed nuclear elements

single-base extension: A technique for the detection of >>> SNPs that uses primers ending directly adjacent to the SP mismatch; it is used to incorporate the complementary, fluorescently labeled ddNTP in a >>> PCR *biot*

single copy: a gene or DNA sequence is only present once in a haploid genome; most of the structural genes (coding for proteins) represent single-copy genes *gene*

single-copy probe: A radioactive-labeled DNA or RNA sequence with highly specific activity; it is used to identify complementary regions applying the >>> Northern and >>> SOUTHERN hybridization techniques or colony hybridization *meth biot* >>> Figure 48

single-copy region: A unique or non-repeated DNA region occurring only once in the haploid genome; most of the structural genes (protein-coding) are single-copy genes *gene*

single cross: A cross between two parental plants and/or lines *meth* >>> Figures 22, 31

single-cross hybrid: The first generation of a cross between two specified inbred lines *seed meth*

single-cross parent: The F1 offspring of two inbred parents, which in turn is used as a parent, usually with another single-cross parent to produce a double-cross hybrid (e.g., in maize) *meth* >>> Figures 22, 31

single-feature polymorphisms (SFPs): Identified in transcript profiling data by visualizing differences in hybridization signals in different plant varieties; the polymorphisms present in DNA are transcribed into the messenger RNA and can potentially affect hybridization to the microarrays or GeneChip probes if present in a region complementary to the probe; polymorphisms generated during mRNA processing, such as alternative splicing and polyadenylation, could also affect hybridization of the target RNA *biot meth*

single-gene character: Any genetic trait whose inheritance is controlled by a single gene; the inheritance of a single-gene character follows MENDEL's laws of inheritance *gene*

single germ: >>> monogerm(ous)

single-hit single nucleotide polymorphism: Any >>> SNP for which each allele is present only one sample from a distinct population *biot*

single hybrid: >>> double cross

single-locus association model: The simplest and most used model to identify associations between >>> SNPs and traits in crop breeding; however, hidden populations structure and sample structure, i.e., cryptic relatedness, may lead to inflated test statistics and, thus, to false-positive and false-negative associations between marker and trait; several correction methods have been proposed, particularly for single-marker association testing, where phenotypic relevance of a single putative gene position is tested in a time in isolation from other putative loci; the mixed model method including a random polygenic term is the most widely used approach *meth stat* >>> multilocus association model

single node culture: Culture of separate lateral buds, each carrying a piece of stem tissue *biot*

single-nucleotide polymorphisms (SNPs): Variations (in individual nucleotides) that occur within DNA at the rate of approximately 1 in every 1,300 bp in most organisms; SNPs usually occur in the same genomic location in different individuals; it is based on a single nucleotide exchange, i.e., deletion or insertion; SNPs are the most frequent mutations in the genome; their relatively good genome coverage and genome-wide distribution in both coding and non-coding regions recommend them as highly informative markers for mapping procedures; they are most efficient and inexpensive to use when once developed;

SNPs are based on the allelic variation of single nucleotides and are co-dominant markers; high-density SNP array data in plant breeding populations are increasingly valuable for genomic selection and for identifying regions of the genome that underlie traits of interest in genome-wide association studies; the accuracy of genomic selection and power of association studies increases with the number of individuals and with the density of SNP markers; however, the cost of genotyping many individuals at high density is high; this high cost is a barrier to the adoption of genomic selection in plant-breeding programs where the number of selection candidates in each cycle can be very large; an effective strategy to overcome this cost barrier is to genotype a proportion of the population at high density, phase their genotypes, and use these data for imputation of large numbers of individuals genotyped at low density *biot meth* >>> simple sequence repeat(s) >>> Figure 63

single-nucleotide polymorphism cluster: The accumulation of >>> SNPs in relatively small genomic regions; the cluster may extend over 1 kb; they are often rare but very old; they originate from ancestral chromosome fragments inherited from extant species *biot gene* >>> Figure 63

single-nucleotide polymorphism map: The linear arrangement of >>> SNPs along a specific region in two homologous genomes *biot meth*

single-nucleotide polymorphism scanning: The *in silico* search for >>> SNPs in a sequenced stretch of DNA (BAC clone, DNA fragment, genome) *biot meth* >>> *in silico* mapping

single-plant comparisons: A method to estimate crop losses by comparing yields of individual plants or components differentially affected by yield constraints *agr meth*

single-plant plot: Used for the replicated evaluation of experimental lines or cultivars by the, for example, honeycomb design; the number of plants evaluated is equal to the number of replications in the experiment; the plots are organized in a systematic manner to permit comparison of a plant of one line with adjacent plants of other lines *meth* >>> honeycomb design >>> Figure 32

single-polymorphic amplified test (SPLAT): If sequence-tagged-site (>>> STS) does not reveal polymorphism, it is usually converted into SPLAT; individual STS products from different genotypes are themselves sequenced; any difference revealed can be sequences from nuclear ribosomal DNA; they can be exploited in the production of internal primers; in general, single-polymorphic amplified sequences are dominant markers *biot*

single random sampling: A survey method in which every sample point has an equal chance of being sampled *stat*

single replication: An experimental plot design in which each variant is present in one replication, either randomized or non-randomized *stat*

single-row plot: An experimental field design in which a single plant row of different length represents one plot *meth* >>> design of experiment

single-seed descent (SSD): Derivation of plants by a selection procedure in which F2 plants and their progeny are advanced by single seeds until genetic purity is achieved; single-seed descent methods (single-seed, single-hill, multiple-seed) are easy ways to maintain populations during inbreeding; natural selection cannot influence the population, unless genotypes differ in their ability to produce viable seeds; artificial selection is based on the phenotype of individual plants, not on the progeny performance *meth* >>> Figure 16

single-sequence repeat (SSR) DNA marker technique: A genetic mapping technique that utilizes the fact that microsatellite sequences repeat (appear repeatedly in sequence within the DNA molecule) in a manner enabling them to be used as markers *biot* >>> simple sequence repeat >>> Figure 63 >>> http://wheat.pw.usda.gov/ITMI/Repeats/index.shtml

single-strand conformational polymorphism (SSCP): This feature relies on secondary and tertiary structural differences between denatured and rapidly cooled amplified DNA fragments that differ slightly in their DNA sequences; different SSCP alleles are resolved on non-denaturing acrylamide gels, usually at low temperatures; the ability to resolve alleles depends on the conditions of electrophoresis and this requires DNA sequence data *biot*

single-tiller comparisons: >>> single-plant comparisons

singling peg: >>> pricking-out peg

sinigrin: >>> glucoside

sink: A term used for storage organs of a plant into which products of photosynthesis are translocated and stored (grain, fruit, tubers, roots as in carrot, stem as in sugarcane) *phys*

sinuate: Having a wavy margin *bot*

siRNAs: >>> small interfering RNAs

sister chromatid: Derives from replication of one chromosome, as opposed to non-sister chromatids, which derive from the other homologous or non-homologous chromosomes *cyto*

sister chromatid exchange (SCE): An event, similar to crossing-over, that can occur between sister chromatids at mitosis and meiosis; it may be detected in "harlequin" chromosomes [sister chromatids that stain differentially so that one appears dark (usually red-violet) and the other light] *cyto*

site-determined selection: A method to select varieties (also population varieties) that are optimally adapted to specific environmental conditions; the parent lines are also selected with the final growing site of the target crop; crosses are carried out at a central location where the F1 (up to F4) generations are also sown; the F5 to Fn plants are assessed and selected at different locations by means of pedigree selection, that is, a combination of natural and artificial selection; thus environmental factors determine which characteristics are expressed and which are not; this may influence the breeder's selection of promising phenotypes; the method is often used in cereal breeding, but is also applicable to other crops *meth* >>> selection

site-directed mutagenesis (SDM): The process of introducing specific base pair mutations into a gene; a technique that can be used to make a protein that differs slightly in its structure from the protein that is normally produced; single mutation is caused by hybridizing the region in a codon to be mutated with a short, synthetic oligonucleotide; this causes the codon to code for a different specific amino acid in the protein gene product; site-directed mutagenesis holds the potential to create modified (engineered) proteins that have desirable properties not currently available in the proteins produced by the plant *meth gene biot*

site of action: The exact binding spot of an herbicide *phys phyt*

site of uptake: The location where an herbicide is taken up by the plant *phys phyt* >>> site of action

site selection: Variation in the production of the soil is commonly referred to as soil heterogeneity, caused by variation of the soil type, availability of nutrients, and moisture; the variation cannot be completely eliminated for breeding experiments, but it can be minimized by careful selection of the area in a field where plots will be placed *agr stat*

site-specific mutagenesis: The use of recombinant DNA technology to create specific deletions, insertions, or substitutions *in vitro* in a particular gene; the technique allows the production of proteins having any desired amino acid at any position; it is one of the most important techniques for introducing a mutation into a DNA sequence; there are numerous methods for achieving site-directed mutagenesis, but with decreasing costs of oligonucleotide synthesis, artificial gene synthesis is now occasionally used as an alternative to site-directed mutagenesis; since 2013, the development of the >>> CRISPR/Cas9 technology, based on a prokaryotic viral defense system, has also allowed for the editing of the genome, and mutagenesis may be performed *in vivo* with relative ease *biot gene*

site-specific recombination: A crossover event that requires homology of only a very short region and uses an enzyme specific for that recombination; recombination occurring between two specific sequences that need not be homologous, mediated by a specific recombination system *gene cyto*

skewed distribution: A frequency distribution that is not symmetrical about its mean *stat*

skiophilous: Shade-loving *bot*

skiophyte: >>> skiophilous

slant: Sterile agar medium congealed in test tubes in a slanting position *biot meth*

slapper: Blueberry, raspberry, aronia-berry, or grape-berry harvester with a slapper-type picking mechanism; it has double banks of flexible horizontal rods that shake the plants and remove the berry clusters; a pulsator type of harvester strikes the posts rapidly so that the berries fall off; fruit is immediately taken to the factory, where it is crushed and made into sap *agr*

slash-and-burn farming: A destructive type of agriculture in which the farmer burns down a new portion of the rainforest every few years in order to cultivate a crop; a type of >>> shifting cultivation *agr*

slice seed: A technique used to sow seed; a machine cuts or slices grooves into the lawn or soil and drops seeds directly into the grooves; it is used to fill in a thinning lawn without disturbing the existing grass excessively *agr meth* >>> thinning

slip: A cutting from a mother plant *hort meth*

slit planting: Prying open a cut made by a spade, mattock, or planting bar; inserting a young tree; and closing the cut by pressure *fore hort*

slow rusting (genotype): A genotype in which rust develops slowly but never reaches a high degree of severity; this is a type of partial or incomplete resistance; it is quantitatively inherited *phyt* >>> Table 52 >>> https://en-lifesci.tau.ac.il/icci

slurry (seed) treater: The principle involves suspension of wettable powder treatment material in water; the treatment material applied as a slurry is accurately metered through a sample mechanism composed of a slurry cup and seed dump pan; the cup introduces a given amount of slurry with each dump of seed into a mixing chamber where seeds are blended; the metering principle is the same in direct, ready-mix, or fully automatic treaters, that is, the introduction of a fixed amount of slurry to a given weight of seed; to obtain a given dump weight, slurry treaters are equipped with a seed gate that controls seed flow to the dump pan; with the proper seed gate setting, a constant dump weight for a given seed can be obtained; the amount of treatment material applied is adjusted by the slurry concentration and the size of the slurry cup or bucket; as the dump pan fills, a point is reached where it overbalances the counterweight and dumps into the mixing chamber; this brings the alternate weighing pan in position to receive the inflow of seed and activates a mechanism to add a cup of slurry to the mixing chamber; thus, one cup of slurry is added with each dump of seed; the mixing chamber is fitted with an auger-type agitator that mixes and moves seed to the bagging end of the chamber; the speed of the auger is important because at slow speeds more uniform distribution is obtained; slurry tanks have 15- to 35-gallon capacities, depending on the size of the treater; they are equipped with agitators that mix the slurry in the tank and keep it suspended during operation; it is important that the powder be thoroughly suspended in water before treating; if the treater has been idle for any period of time, sediment in the bottom of the slurry cups must be cleaned out; the proper size slurry cup must be used; modern machines have cups with ports and rubber plugs for 15, 23, and 46 cm³ quantities; some users prefer to mix the slurry in an auxiliary tank and then transfer to the slurry chamber as needed *seed* >>> seed treater

small-area comparison: A method to appraise crop losses based on comparative yield measurements in areas differentially affected by yield constraints *agr meth*

small grains: Small-grain cereals that are monocotyledonous plants belonging to the order of Poales and to the family of Gramineae; they are characterized by specific morphological traits such as stubble, spikelets, scutellum, etc., and by grains that are rich in carbohydrates; there are six main groups cultivated – wheat, barley, rye, triticale, oats, and rice *agr*

small interfering RNAs (siRNAs): Models of PTGS indicate that 21–23 nucleotide dsRNAs mediate PTGS; introduction of siRNAs can induce PTGS, e.g., in mammalian cells; siRNAs are apparently produced *in vivo* by cleavage of dsRNA introduced directly or via a transgene or virus; amplification by an RNA-dependent RNA polymerase (RdRP) may occur in some organisms; siRNAs are incorporated into the RNA-induced silencing complex (RISC), guiding the complex to the homologous endogenous mRNA where the complex cleaves the transcript; siRNA technology can be a good alternative to protein-based >>> GMOs for imparting pest resistance to plants; siRNA does not result in production of a "toxic" protein; breeding for lines with natural small RNAs against pests, recombinant contracts, and simple spraying of stabilized siRNAs might the future of pest control *biot* >>> posttranscriptional gene silencing >>> pest control

small nuclear RNA (snRNA): Short RNA transcripts of 100 to 300 bp that associate with proteins to form small nuclear ribonucleoprotein particles (snRNPs); most snRNPs are components of spliceosomes that excise introns from pre-mRNAs in RNA processing *biot*

small-plot intensive farming (SPIN farming): A part of the urban agriculture phenomenon that's sweeping the globe; many cities are attempting to make their communities green by addressing a variety of concerns including food security, climate change, waning economies, water conservation, and a litany of other issues coming down the pike; urban agriculture is a central component of this green city movement and SPIN farming is one model of urban agriculture that is working for farmers and eaters alike; a key principle of SPIN farming is its focus on economic viability; e.g., Canadian farmers practiced rural farming technology and expertised and miniaturized it to urban backyards; they shrank the bed sizes, intensified the plantings, and moved closer to their markets; they have succeeded in growing more in less space with fewer inputs, thus reducing their environmental impact *agr hort*

small supernumerary marker chromosomes (sSMC): >>> marker chromosomes

SMART breeding (Selection with Markers and Advanced Reproductive Technologies): A breeding technique (also sometimes called "precision breeding") which uses molecular biological methods; it refers also to an organic farming technique of reproducing a species' members together to retain desirable traits and so produce a stronger hybrid; it works in a similar way to traditional breeding, however, the gene or gene variant responsible for a specific trait can be accurately identified using molecular biological procedures (DNA sequencing, PCR); it is then possible to test the offspring of a cross for the presence of the crossed gene, even before the actual trait is signaled by a changed external appearance; only those plants which contain the desired gene are then grown on; the purpose of this is to introduce into crop plants genes from, e.g., wild populations which confer characteristics of interest to breeders (disease resistance, fruit color, sugar content, etc.); the technique was successfully used by N. KEDAR, an Israeli scientist, who perfected the technique using beefsteak tomatoes to produce a fruit that would ripen on the vine and remain firm in transit *biot* >>> breeding >>> breeding method

smear: Direct spreading of cells in a semifluid tissue over the surface of a microscopic slide with a flat-honed scalpel or other tools, and the immediate inversion of the slide over a dish of fixative *prep cyto*

Smith–Hazel index: >>> index selection

Smith model (for quantitative genetic analysis): Selection and random genetic drift are the main forces affecting selection response in recurrent selection programs; the correct assessment of both forces allows a better comparison of the efficiency of different recurrent selection schemes; the population diallel analysis proposed by HAMMOND and GARDNER (1974), and the model proposed by SMITH (1979) with full consideration of inbreeding depression due to random genetic drift, can also be applied to quantitative analysis; the effect of random genetic drift is expected to be large, particularly in studies with many selection cycles and/or high rates of inbreeding; therefore, the extension of the population diallel allows a better assessment of the selection response in recurrent selection *stat meth*

smooth: A statistical method to remove genotyping errors from genetic linkage data during the mapping process; the program SMOOTH calculates the difference between the observed and predicted values of data points based on data points of neighboring loci in a given marker order; highly improbable data points are removed by the program in an iterative process with a mapping algorithm that recalculates the map after cleaning; the method is able to detect a high amount of scoring errors and demonstrates that the program enables mapping software to successfully construct a very accurate high-density map *stat meth*

smothering crop: Refers to a crop that suppresses the growth of weeds by its heavy growth and shadowing effect *agr meth*

smudging: >>> fumigation

smut: A plant disease caused by a fungus of the order Ustilaginales; the symptoms include the formation of masses of black, shoot-like spores; infected plants often show some degree of distortion *phyt* >>> bunt

snag: A standing dead tree from which the leaves and some of the branches have fallen *fore hort*

snowball sampling: A special non-probability method used when the desired sample characteristic is rare; it may be extremely difficult or cost-prohibitive to locate respondents in these situations; snowball sampling relies on referrals from initial subjects to generate additional subjects; while this technique can dramatically lower search costs, it comes at the expense of introducing bias because the technique itself reduces the likelihood that the sample will represent a good cross section from the population *stat* >>> simple random sampling

snow line: The line, as on mountains, above which there is perpetual snow *eco*

SNPs: >>> single-nucleotide polymorphisms

snRNA: >>> small nuclear RNA

S1 nuclease: A nuclease that cuts single-stranded DNA and RNA; used for S1 protection experiments in transcript mapping *biot* >>> Table 39

soap insecticide: A solution of soft soap, that is, potassium soap, traditionally used to control aphids and similar insect pests; also water with a small content of dish detergent can be used; the affected insects are unable to breathe; these insecticides are stable *phyt*

sobole: A shoot, stolon, or sucker; i.e., a shoot arising from underground stem tissue *bot*

sod: The upper stratum of grassland, containing the roots of grass and any other herbs, or a piece of this grassy layer pared or pulled off (turf, divot, fail) *agr*

sodium (Na): A soft, silver-white, chemically active metallic element that occurs naturally only in combination *chem*

sodium chlorate: A colorless water-soluble solid, $NaClO_3$, used as a mordant, oxidizing, or bleaching agent *chem* >>> chemical desiccation

sodium dodecyl sulfate-polyacrylamide (SDS-PAA): An ionic detergent, irritant *chem biot*

sod seeding: Sowing directly into a sward without previous cultivation *agr*

soft dough (stadium): Description of the cereal grain at maximum fresh weight; when punctured, the grain contents are still wet but starting to become floury *phys bot*

soft endosperm: >>> chalky

soft red winter wheat: Seeded in the fall, has low to medium protein content, with a soft endosperm; used in making cakes, pastries, flat breads, and crackers; it is grown in the eastern third of the U.S., east of the Mississippi River; it shows a high yield, but relatively low protein *agr seed* >>> wheat

soft rot: A decomposition of plant tissue (fruits, roots, stem, etc.) by fungi or bacteria resulting in the tissue becoming soft (e.g., caused by *Erwinia carotovora* in potato) *phyt*

soft-tip cutting: >>> stem-tip cutting

soft white wheat: Used in much the same way as soft red winter (for bakery products other than bread); it is grown mainly in the Pacific Northwest, and to a lesser extent in California, Michigan, Wisconsin, and New York; low protein, but high yield; it produces flour for cakes, crackers, cookies, pastries, quick breads, muffins, and snack foods; subclasses are soft white, white club, and western white wheats *agr seed* >>> wheat

softwood: One of the botanical groups of trees that in most cases have needles or scale-like leaves; the conifers; also, the wood produced by such trees *bot fore*

soil: The natural space–time continuum occurring at the surface of the earth and supporting plant life *agr* >>> crop rotation >>> Table 44

soil aggregation: The process whereby primary soil particles (sand, silt, clay) are bound together, usually by biological activity and substances derived from root exudates and microbes *agr* >>> Table 44

soil conditioner: Poor or damaged soils impact crop growth and yield; products, known as soil conditioners, can be added to soil to improve its physical qualities, for example, by providing nutrition for plants; a wide range of materials have been described as soil conditioners due to their ability to improve soil quality; these include substances such as bone meal, peat, compost, manure, coir, and straw; many soil conditioners are available in the form of "certified organic products"; the addition of organic material can improve the water retention of sandy soils, thus stimulating microbial activity, and indirectly increasing nutrient levels to improve plant growth; some conditioners, such as calcium, magnesium, potassium, ammonium, and sodium are used to increase the cation exchange capacity of soil; this helps plants to grow by providing the required nutrition; minerals may also be used to adjust the pH value of either acidic or alkaline soils, thus increasing their suitability for certain crops *agr*

soil heterogeneity: Variation in the production of the soil is commonly referred to as soil heterogeneity, caused by variation of the soil type, availability of nutrients, and moisture *agr* >>> crop rotation >>> Table 44

soil pasteurization: A sterilization of soil that destroys unacceptable organisms without chemically altering the soil *meth hort* >>> solarization

solanine: A glycoside neurotoxin (glycoalkyloid) that is naturally present at low levels within potatoes or other Solanaceae; as a result, solanine can be found at detectable levels in the blood of consumers; it acts as a plasma cholinesterase inhibitor; the U.S. Food and Drug Administration prohibits the sale of potatoes that contain more than allowed; the level can increase in harvested potatoes that are exposed to direct sunlight *chem phys*

solar furnace: Flower that focuses solar radiation so that the temperature at the center of the flower is above the ambient temperature *bot phys*

solarization: The process of sterilizing the soil and killing soil pests by covering the moist planting area with clear plastic; it is a simple non-chemical technique that captures radiant heat energy from the sun; this energy causes physical, chemical, and biological changes in the soil; these changes lead to control or suppression of soilborne plant pathogens, such as fungi, bacteria, nematodes, and pests, along with weed seed and seedlings *meth hort*

solid medium: As opposed to liquid medium, any *in vitro* culture medium solidified with agar or other jellying agents; it is widely used in *in vitro* propagation *prep biot*

solum: The upper part of a soil profile; roots, plants, and animal-life characteristics of the soil are mainly confined to the solum *agr*

solvent: A substance that dissolves another to form a solution *chem*

somaclonal variation: Somatic (vegetative non-sexual) plant cells can be propagated *in vitro* in an appropriate nutrient medium; according to the composition and conditions, the cells may proliferate in an undifferentiated (disorganized) pattern to form a callus, or in a differentiated (organized) manner to form a plant with a shoot and root; the cells, which multiply by division of the parent somatic cells, are called somaclones and, theoretically, should be genetically identical with the parent; in fact, *in vitro* cell culture of somatic cells, whether from a leaf, a stem, a root, a shoot, or a cotyledon, frequently generates cells significantly different, genetically, from the parent; during culture, the DNA breaks up and is reassembled in different sequences which give rise to plants different in identifiable characters from the parent; such progeny are called "somaclonal variants" and provide a useful source of genetic variation *biot*

somaclone: A plant regenerated from a tissue culture of somatic cells *biot*

somatic: Body cells *bot* >>> somatic cell

somatic apospory: >>> apospory

somatic cell: A diploid, body cell, or cells other than the germ cells *bot*

somatic cell fusion: The fusion of somatic cells (usually protoplasts) by different means *in vitro biot* >>> protoplast

somatic crossing-over: Crossing-over during mitosis of somatic cells such that parent cells heterozygous for a given allele, instead of giving rise to two identical heterozygous daughter cells, give rise to non-identical daughter cells, one of which is homozygous for one of these alleles, the other being homozygous for the other allele *cyto gene*

somatic embryo: An organized embryonic structure morphologically similar to a zygotic embryo but initiated from somatic cells; somatic embryos develop into plantlets *in vitro* through developmental processes that are similar to those of zygotic embryos *biot*

somatic embryogenesis: >>> somatic embryo

somatic fusion: >>> somatic cell fusion

somatic hybridization: The fusion of genetically different somatic cells by different means, usually to overcome natural crossing (incompatibility) barriers *biot*

somatic mutation: A mutation occurring in a somatic cell; if the mutated cell continues to divide, the individual will develop a patch of tissue with a genotype different from the cells of the rest of the body *gene*

somatic regeneration: >>> regeneration

somatoplastic sterility: The collapse of zygotes during embryonic stages due to disturbances in embryo–endosperm relationships *bot* >>> cytoplasmic male sterility

sorting: A process utilized to sort and/or separate different cells, particles, chromosomes, etc.; some automated means of cell sorting in use for different applications include (1) utilizing controlled electrical fields to collect specific cell types onto electrodes in biochips, (2) fluorescence-activated cell (FAC) sorter machines, or (3) magnetic particles (e.g., attached to antibodies); the ploidy level can also be determined by cell-sorting equipment *biot meth*

source community: A community from which a local variety or a seed lot originated *seed*

source-identified seed: A class of true seed defined as seed from natural stands with known geographic source and elevation, or a plantation of known geographic location, as specified in the standards of the various certifying agencies *seed meth*

source plant: A donor plant from which an explant is taken in order to initiate an *in vitro* or *in vivo* culture *meth*

source population: The initial population where trees are considered non-inbred and non-related, which the concepts inbreeding and coancestry require *meth fore hort* >>> base population >>> Table 37

source identified: The seed certification class of prevariety germplasm which provides third-party assurance of geographic origin, usually for perennial native forage grasses, legumes, and forbs produced from parent populations which have not been selected; source-identified class seed labels identify the original geographic location, of the collection or production, that has been declared by the responsible plant breeder *seed meth*

South American leaf blight (SALB): A disease of the rubber tree caused by the fungus *Microcyclus ulei phyt* >>> Figure 62

Southern root-knot nematode: Can cause severe yield loss of soybean in the southern production region of the United States; planting root-knot nematode-resistant cultivars is the most effective method of preventing yield loss, *Meloidogyne* spp. *phyt*

Southern transfer: A method in biochemistry named after E. SOUTHERN; this technique is also called a SOUTHERN blot; as the result of a restriction analysis, an agarose gel will contain fragments of DNA; the DNA fragments in the gel are transferred to a cellulose nitrate filter (or other suitable medium) during the SOUTHERN transfer; the moistened receiving material is placed on top of the gel; absorbent materials are placed on top of the filter; DNA is eluted onto the filter because solvent is drawn up through the bed by capillary action; the filter is used for complementary hybridization tests; for example, if the filter is placed in a solution of suitable radioactively labeled RNA or denatured DNA probe, strands bound to the filter will, in turn, bind labeled probe that is complementary; after washing to remove unbound probe, autoradiography may be used for detection of positive DNA fragments; the method is used to map restriction sites for a single gene within a complex genome *meth biot* >>> Table 38 >>> Figure 48

sowing brick: A prepared block or ball of loam, peat, plastic, or foam into which one or more seeds are pressed; when planting out, the emergent seedling can have a better start in an unfavorable environment *fore*

sowing time: The optimum dates for seeding; temperature and moisture are the major determinants of sowing time *agr*

soybean cyst nematode (SCN): Small roundworms, *Heterodera glycines*, that cause root damage and subsequent above-ground symptoms to soybeans; this nematode is found in 27 states in the United States and causes the greatest yield losses of any pathogen of soybeans; 16 physiological races of the nematode have been identified; management of SCN is achieved by planting varieties resistant to a given race or races, and by rotating soybeans with a non-host crop such as field maize; nematicides can also be used to reduce populations *phyt*

sp.: Abbreviation of "species"; the expression follows the name of a genus when the single species indicated is unknown or for any other reason not specified *tax*

space planting: In contrast to broadcast seeding, the spreading of seeds in a more or less defined distance from each other; it is facilitated by automatized seed drills *meth* >>> broadcast seeding

spacer: Non-coding and/or non-genic DNA sequences that may occur between genes *gene* >>> spacer sequence

spacer sequence: A portion of DNA sequences that does not code any RNA; it is rich in >>> adenine and >>> thymine, and found between the transcriptional DNA segments *gene* >>> Table 38

spade leaf: The first true leaf of a seedling clover, lucerne, or other legume plant; the leaf is simple in contrast to the subsequent trifoliate leaf *bot*

span length: The distance spanned by a specified percentage of the fibers in a fibrograph test beard used for quality determination in cotton, e.g., span length 50% = the length that 50% of the cotton fibers in the sample will equal or exceed *meth*

spathe: A modified leaf sheath that subtends and often encloses some of the inflorescence *bot*

spatial order: Chromosomes or genomes may occupy certain domains within the nucleus and/or dividing cells *cyto*

spatial variation: Spatial variation is common in field trials, and accounting for it increases the accuracy of estimated genetic effects; there are spatial models that show possibilities for flexible modeling with respect to field trial design and joint modeling over multiple years and locations; the MATÉRN model can accommodate flexible field trial designs and yields interpretable parameters *stat meth*

spawn: A common term applied to a mixture of fungal mycelium or a nutritive organic material for the artificial propagation of mushrooms *hort*

Spearman rank correlation coefficient: Usually calculated on occasions when it is not convenient, economical, or even possible to give actual values to variables, but only to assign a rank order to instances of each variable; it may also be a better indicator that a relationship exists between two variables when the relationship is non-linear *stat*

speciation: The splitting of an originally uniform species into daughter species that coexist in time *evol*

species: The individuals of one or more populations that can interbreed, but which in nature cannot exchange genes with members belonging to other species *tax* >>> Table 12 >>> List of Crop Plants

species hybridization: The mating of two different species; based on species hybridization, several wild species are represented in the ancestry of the cultivated crop plants (e.g., of wheat); induced species hybridization is also used for the introgression of desirable genes into the cultivated crop plants *meth* >>> synthetic amphiploid >>> triticale >>> Figure 2

specific combining ability: >>> combining ability

specific resistance: >>> vertical resistance

spectral karyotype: A visualization of all chromosomes of a complement, each labeled with a different fluorescence color; this technique is useful for identifying structural chromosome abnormalities or alien chromosomes *cyto* >>> FISH

spectral reflectance: A potentially rapid technique that can assess biomass at the genotypic level without destructive sampling; spectral reflectance indices have proven to be useful in the assessment of early biomass and vigor of different plant genotypes; spectral indices as a selection tool in plant breeding improve genetic gains for different important traits; progress in grain yield is mainly attributed to better partitioning of photosynthetic products; the systematic increase in the partitioning of assimilates (harvest index) has a theoretical upper limit of approximately 60%; further yield increases in cereals through improvement in harvest index will be limited without a further increase in total crop biomass; therefore, a breeding approach is applied that can select genotypes with higher biomass capacity, while maintaining the high partitioning rate of photosynthetic products; direct estimation of biomass is a time- and labor-intensive undertaking; destructive in-season sampling involves large sampling errors; canopy light reflectance properties based mainly on the absorption of light at a specific wavelength are associated with specific plant characteristics; the spectral reflectance in the visible wavelengths (400–700 nm) depends on the absorption of light by leaf chlorophyll and associated pigments such as carotenoid and anthocyanins; the reflectance of the visible wavelengths is relatively low because of the high absorption of light energy by these pigments *meth phys*

spectrometry: The measurement of the absorption or emission of light by a substance at a specific wavelength *phy meth*

spectrophotometer: An optical system used in biology to compare the intensity of a beam of light of specified wave length before and after it passes through a light-absorbing medium *meth*

speed breeding: The growing human population and a changing environment have raised significant concern for global food security, with the current improvement rate of several important crops inadequate to meet future demand; the slow improvement rate is attributed partly to the long generation times of crop plants; a method called "speed breeding" may greatly shorten generation time and accelerate breeding; it can be used to achieve up to six generations per year for spring wheat and durum wheat, barley, chickpea, and pea, and four generations for rapeseed, instead of two to three under normal glasshouse conditions *meth*

spelling: Resting a pasture from grazing *agr*

spelt forms: >>> speltoid

speltoid: A mutation that arises spontaneously (e.g., in wheat); it resembles spelt wheat in certain features *bot* >>> wheat >>> Table 1

sperm: A male gamete *bot*

sperm nucleus (sperm nuclei *pl***):** In angiosperms, the two male gametes that are formed by division of the generative cell; they migrate down the pollen tube behind the vegetative nucleus; when the pollen tube enters the embryo sac, the tip of the tube breaks down to release the generative nuclei; one fuses with the egg nucleus to form the zygote, and the other usually fuses with the polar nuclei or definitive nucleus to form the primary endosperm nucleus *bot* >>> Figure 25

spermatogenesis: The process of differentiation of a mature sperm cell from an undifferentiated germ-line cell, including the process of meiosis *bot*

sperma(to)phyta: Seed-bearing plants; gnetophytes, a group that consists mostly of shrubs and woody vines, are the most primitive living non-flowering seed plants – present since the late >>> Mesozoic era; they are situated at the base of the evolutionary tree of seed plants *bot*

Sphagnum moss: A bog moss belonging to the genus Sphagnum; it has been frequently used as a rooting medium for plants *bot hort meth*

spherical aberration: Inaccurate focusing of light due to curved surface of lens *micr*

spheroidal: Pollen grain where the polar axis and the equatorial diameter are more or less equal in length, e.g., in *Daphne alpina bot* >>> pollen

spheroplast: A cell with a partially degraded cell wall *biot*

spherosome: Oil-storage bodies of plant cells *bot*

spica: >>> spike

spice: The fragrant vegetable condiments used for the seasoning of food; a wide range of plants, such as pepper, allspice, nutmeg, ginger, cinnamon, cloves, marjoram, oregano, etc., are bred for spice production *hort*

spiciform: Spike-shaped *bot*

spicula: A small or secondary spike *bot*

spicules: Minute teeth occurring on the nerves and other parts of the lemma and glume of, for example, wheat, barley, oats, grasses, etc. *bot*

spike: A "flower" head on which the spikelets are borne without a stalk as, for example, in wheat, barley, rye or other grasses *bot* >>> Figure 34

spike-bed method: A spike-bed is a seedbed in which the seeds from one spike are sown in the same sequence in which they were arranged in the spike; thus the plants in the >>> seedbed reflect the quality of the original spike; the method is applicable not only for cereals but also for other crops *meth* >>> selection method

spike density: The number of spikelets per unit of spike length; it is a distinctive character of certain cereal varieties *bot agr* >>> Figure 34

spikelet: The unit of inflorescence in the cereals and in some grasses consisting of a pair of glumes and one or more florets *bot*

spikelet fork: It is composed of the glume base and a part of the rachis *bot*

spikelet glume: A chafflike bract; specifically one of the two empty chaffy bracts at the base of the spikelet in the grasses *bot*

spindle: The set of microtubular fibers that appear to move the chromosomes of eukaryotes during mitotic and meiotic cell division *cyto*

spindle attachment: >>> centromere

spindle poison: Any poison affecting the correct formation or function of the spindle of the dividing cell *cyto*

spindle pole: The regions to which the spindle fibers pull the chromosomes or chromatids during anaphase *cyto*

spine: A sharp woody or rigid outgrowth from a stem, leaf, or other plant part; cacti spines that are in the center of or form a circle around the >>> areole are called central spines *bot*

SPIN farming: >>> small-plot intensive farming

spinous: Having narrow, sharply pointed processes (spines) *bot*

spiny: >>> spinous

spiral separator: A type of seed separator with no moving parts; the seeds enter at the top and slide or roll down an inclined spiral runway; the speed of seed movement and centrifugal force allows separation of the heavier, round, fast-moving seeds from those that move slower *seed*

spiraperturate: Pollen grain with one or more spiral aperture(s), e.g., in *Berberis vulgaris bot* >>> saccate

SPLAT: >>> single-polymorphic amplified test

splicing: The removal of the intron (non-coding sequence) from an mRNA gene sequence and the fusion of the exons (coding sequence) during the mRNA processing *biot* >>> gametocide

split grafting: Used for top-working larger trees to change to another variety or introduce a new pollen source; to top-work a tree to another variety, half of the main limbs are cut back to where they have a diameter of 5–7 cm; a cleft graft is performed in these stubs at this time; during the following year, the remaining limbs are cut back and split grafted; in order to make the graft, the stub is split and hatched with a special grafting tool or with a hammer; the split is kept open with a certain tool and beveled scions are positioned in the split so their cambium layers align with the stock cambium; the spreading tool is removed, and the tissue is sealed with grafting wax; this sort of graft may be made during the late winter or early spring while the wood is dormant *meth hort*

split-plot design (*syn* **nested block design**): Allows the testing of two factors in combination; one factor (the main effect) serves as a replication for the second factor (the split effect); there are many split-plot design options, but the basic principle involves assigning one set of treatments to the main plots that are arranged in >>> randomized complete blocks; the second set of treatments is assigned to subplots within each main plot; the statistical analysis is similar to that used with the randomized complete block design, i.e., an ANOVA set up for a split plot is followed by a means comparison test such as the DUNN's test or DUNCAN's multiple range test *agr meth stat* >>> design of experiment

splitting: >>> fission

spontaneous haploids: The spontaneous occurrence of haploids, usually by asexual processes *bot* >>> apomixis >>> Imperata cylindrica procedure >>> parthenogenesis >>> polyembryony >>> Figures 17, 26 >>> Table 7

spontaneous mutation: A naturally occurring mutation, as opposed to one artificially induced by chemicals or irradiation; usually such mutations are due to errors in the normal functioning of cellular enzymes *gene*

spore: A minute reproductive unit in fungi and lower plant forms *bot*

sporophyte: The diploid generation of plants; it is the conspicuous generation in the higher plants *bot* >>> Figure 28

sporophytic self-incompatibility: Self-incompatibility is based on the genotypic and phenotypic relationship between the female and male reproductive system; alleles in cells of the pistil determine its receptivity to pollen; the phenotype of the pollen, expressed as its inability to effect fertilization, may be determined by the maternal plant, referred to as sporophytic incompatibility *gene* >>> self-incompatibility

sport: A sudden deviation from type; a somatic mutation *gene* >>> bud sport

sporulation: The period of active spore production *phyt*

spreader: A substance added to fungicide or bactericide preparations to improve contact between the spray and the sprayed surface; in resistance testing, the variety, line, or genotype infected with the pathogen; from the spreader the pathogen is distributed by wind, rain, or insects to the tester genotypes; the spreader and the testers can be arranged in different experimental and field designs *phyt* >>> Figure 30

spreader bed: >>> spreader

spreader row: >>> spreader

spreader strip: >>> spreader

spreader surrounds: Differ from spreader rows in that they surround the screening population rather than run through it; they are used when parasite gradients are not a problem *phyt meth*

sprig: A small part of a plant, such as stolons, used for propagations, twigs, bearing flowers, etc. *meth hort*

sprigging: >>> sprig

sprout: A shoot of a plant, as from a germinating seed, a rootstock, tuber, runner, etc., or from the root (a sucker), stump, or trunk of a tree *bot hort* >>> shoot

sprouting (*syn* out growing *syn* unprincipled): Germination of seeds on mature spikes of grasses and/or cereals, usually before harvest in wet years, or when harvest is delayed; it can cause severe loss of quality in rye, wheat, barley, or oats *bot agr* >>> dormancy

sprout inhibitor: Chemicals applied, e.g., to potato tubers to help prevent shrinkage, blackening, nutrient loss, and susceptibility to bruising, and to reduce accumulation of some natural toxic chemicals that accompany sprouting; the most widely used and least offensive sprout inhibitor is chlorpropham (isopropyl N-(3-Chlorophenyl)-carbamate) or CIPC, which is now used on 90–95% of stored potatoes; other sprout inhibitors include maleic hydrazide (MH), dimethyl naphthalene (DMN), carvone ("Talent"), jasmonates, ethylene, and hydrogen peroxide; non-chemical inhibition of sprouting includes storage at low temperatures, shaded huts, and rustic storage *meth*

spur: A short or stunted branch or shoot, as of a tree *hort*

spur pruning: A method of pruning (fruit) trees, by which one or two eyes of the previous year's wood are left and the rest cut off, so as to leave spurs or short rods *hort*

square: An unopened cotton flower with its enclosing bracts *agr*

square design: A design if it has equally many points and blocks *stat*

square lattice design: >>> lattice design

squarrose: Spreading rigidly at a right angle; usually the shape of bracts, e.g., *Aegilops squarrosa bot*

SSCP: >>> single-strand conformational polymorphism

SSD: >>> single-seed descent

ssDNA: single-stranded DNA *biot*

sSMC >>> small supernumerary marker chromosomes >>> www.uniklinikum-jena.de/fish/sSMC /sSMC+by+chromosome/sSMC+McClintock.html

ssp.: >>> subspecies >>> sp.

SSR: >>> simple sequence repeat

stabilizing selection: A type of selection that removes individuals from a population (i.e., from both ends of a phenotypic distribution divided by deviation of a sample of means) *meth stat*

stachyose: An unfavorable sugar in soybean meal that causes flatulence for non-ruminant animals; two QTL are located on chromosome 10 and 11 for stachyose content *phys*

stack: A well-built pile of grain, forage, or straw that usually has a square or circular base *agr*

stacked genes: >>> gene stacking

stage: The small platform of a microscope on which the object to be examined is placed *micr*

stage of maturity: >>> maturation

stage-wise selection: Often used as a synonym for stepwise selection; selection of individuals in stages; individuals are preselected (the first stage) followed by reselection (the second stage) of the best individuals based on another testing method; for example, phenotypic selection followed by clonal test of those preselected based on phenotype *meth fore hort*

staggered cuts: The cleavage of two opposite strands of duplex DNA at points near one another *biot*

***Stagonospora* blotch:** A fungal disease (*Stagonospora nodorum*) that affects all above-ground parts of wheat plants; it causes serious losses in grain yield and seed quality of wheat grown in the southeastern U.S. *phyt*

stale seedbed: A seedbed which is prepared and left untouched to encourage weed germination, and into which the crop is later sown with minimum soil disturbance, before or after the weeds are killed with an herbicide *phyt agr*

stalk: A stem-like supporting structure such as a peduncle or pedicel *bot*

stalk diameter: The diameter of a stalk, usually at a designated node or internode *prep agr*

stalk rot: One type of stalk rot (Anthracnose stalk rot) is caused by a fungus (*Colletotrichum graminicola*) that enters the maize plant, for example, above ground; the disease causes the inside of the stalk to degrade from bright white to gray; it weakens the stalk, making it more likely for plants to lodge *phyt*

stalk tunneling: The longitudinal tunnels in plant stalks produced by different insects *phyt*

stamen (stamina *pl*): The part of the flower that produces the pollen *bot*

staminal (staminate) flower: A flower bearing stamens but not functional pistils (i.e., a male flower) *bot*

staminode: A sterile stamen that does not produces pollen *bot hort*

stand: Term used for an established field crop or tree plantation *agr* >>> seed production area

standard: A criterion for evaluating performance and results; it may be a quantity or quality of output to be produced, a rule of conduct to be observed, a model of operation to be adhered to, or a degree of progress toward a goal; in general, agreeing upon rules for the specification, design, and development of entities within a given class *stat*

standard deviation (SD): A measure of the variability in a population of items, i.e., a set of n measurements x1, x2, x3, ..., xn is equal to the positive square root of the variance of the measurements *stat*

standard error (SE): A measure of variation of a population of means *stat*

stand canopy: >>> canopy

standing corn: >>> standing crop

standing crop: The mass of the individuals of a crop in a field *agr*

standing power: A trait of a crop, usually of cereals, that describes the stalk, which resists lodging; it is frequently scored on a scale of 1–9 *agr*

staple: The main item of a diet, hence "staple crop" and "staple food"; staple crops can be divided into >>> major and >>> minor staples *agr*

starch: A homopolysaccharide, consisting of glucose molecules, which is the major storage carbohydrate of plants and the major nutritionally important carbohydrate in the human diet; because of its unique physical properties, it is a valuable raw material in many food and non-food industries; it is synthesized in large amounts in leaves during the day, and degraded during the subsequent night; its utilization depends on the granular structure; starch granules are supramolecular organized structures, made up of ordered and disordered, or amorphous, regions; the ordered parts consist of double helices formed from short branches of amylopectin molecules; most of the double helices are further organized into crystalline lamellae, making the granule into a so-called "semicrystalline" structure; two crystalline forms are found in starch molecules, A and B, the helices in the A-form being more densely packed than in the B-form *chem phys*

starch gel electrophoresis: A method for high-resolution separation of soluble proteins using starch as a matrix; it was introduced by O. SMITHIES in 1955; M. POULIK (a collaborator of SMITHIES') introduced discontinuous buffer systems and enzyme-histochemical staining of electrophoretically separated enzymes, which were later called "zymograms";

the problem with the method is the fact that, due to the presence of various anionic residues, such as glucuronic acid in the purest starch fractions, the measured mobilities are always biased by electroosmotic flow of water in the gels; since the gels have net negative charges, the water moves relative to the stationary gels to compensate; in addition, the concentration of glucuronic acid residues varies in starch preparations from batch to batch; therefore, reproducibility is often severely compromised *chem*

starch plants: >>> nutrient-enhanced varieties >>> potato

StarLinkTM: An insect-resistant variety of maize that was not labeled for human consumption *agr*

start codon: The codon AUG, in RNA, which represents the start of a new protein in DNA *biot gene*

starter: >>> start codon >>> primer

starter culture: Pure culture or mixture of microorganisms which is used for starting fermentation during commercial production of a biofertilizer *biot*

starter dose: A small dose of fertilizer given to a crop before planting so as to give it a head start even in fertile soil, e.g., application of 20–25 kg N/ha to a grain legume until the N fixation system becomes operational, or a small early application in a fertile soil or under adverse growth conditions *agr*

stationary phase: The plateau of the growth curve after log growth, during which the cell number remains constant; new cells are produced at the same rate as older cells die *biot*

statistic: A quantity that is calculated from a sample of data in order to give information about unknown values in a corresponding population *stat*

statistical analysis: Analyzing collected data for the purposes of summarizing information to make it more usable and/or making generalizations about a population based on a sample drawn from that population *stat*

statistical genetics: >>> quantitative genetics

statistical significance: The degree to which a value is greater or smaller than would be expected by chance; typically, a relationship is considered statistically significant when the probability of obtaining that result by chance is less than 5% if there were, in fact, no relationship in the population *stat*

statistical test: A type of statistical procedure that is applied to data to determine whether the results are statistically significant *stat*

statistics: The scientific discipline concerned with the collection, analysis, and presentation of data *stat*

statocyte: A cell that is present in pulvinus (thickened leaf sheath base) parenchyma; it contains starch grains and/or calcium oxalate crystals, which are thought to act as statoliths (i.e., balancing stones); it may be gravity-sensing; in cereals, if a culm lodges, pulvinus cells on the lower side will elongate but not divide; this helps to straighten up the culm *bot*

statolith: Any of the granules of lime, sand, calcium oxalate, etc., contained within a statocyte *bot* >>> statocyte

stearic acid: A colorless, wax-like, sparingly water-soluble fatty acid, $C_{18}H_{36}O_2$, occurring in some vegetable oils *chem phys*

steckling: The plantlet in the first year of the biennial sugarbeet stored over winter and planted for the production of seeds; it is grown from unthinned plantlets *agr seed*

steckling bed: A sugarbeet seedling bed *agr*

steep: Liquid formulation of seed protectant or disinfectant in which grain is immersed *seed*

steeping: The soaking of barley grains in the production of malt prior to germination *meth*

stele: The central cylinder, inside the cortex, of roots and stems of vascular plants *bot*

stem: In vascular plants, the part of the plant that bears buds, leaves, and flowers; it forms the central axis of the plant and often provides mechanical support *bot*

stem apex: The top or tip of a stem *bot*

stem cell: Any of reproductive and/or generative cells in an organism, as opposed to somatic cells *bot*; in biotechnology, immature cells capable of developing into various (specialized) cell types *biot*

stem cutting: A cutting taken from a portion of stem *hort meth*

stem pitting: A symptom of some viral diseases characterized by depressions on the stem of the plant *phyt*

stem rust (in wheat): A dangerous fungus (*Puccinia graminis f. sp. tritici*) causing yield loss in many countries of the world; recently, a new and virulent race, previously found in East Africa and Yemen, has moved to major wheat-growing areas in Iran; countries like Afghanistan, India, Pakistan, Turkmenistan, Uzbekistan, and Kazakhstan are most threatened by this fungus; it is estimated that as much as 80% of all wheat varieties planted in Asia and Africa are susceptible to the wheat stem rust; the spores of wheat rust are mostly carried by wind over long distances; the fungus is spreading rapidly and could seriously lower wheat production in countries at direct risk; the new race first emerged in Uganda in 1999 and is therefore called Ug99; subsequently it spread to Kenya and Ethiopia; in 2007, Ug99 had affected wheat fields in Yemen; disease surveillance and wheat breeding is already underway to monitor the fungus and to develop Ug99 resistant varieties *phyt* >>> Table 52 >>> www.tau.ac.il/lifesci/units/ICCI//genebank2.html

stem tip: >>> stem apex

stem-tip cutting: Cutting taken from the soft tip of a non-flowering stem, usually from spring to autumn *hort meth*

stenospermocarpy: Seedlessness in some fruits, for example, in some melons, squashes, table grapes, and citrus fruits; normal pollination and fertilization is required to set fruit in stenospermocarpic fruits; the near-seedless condition results from subsequent abortion of the embryo that began growing following fertilization; the remains of the undeveloped seed are often visible in the fruit *bot hort*

step-wise regression: A method of selecting the "best" set of independent variables for a regression equation; variables are introduced one at a time, with the criterion for accepting a variable based on the correlation of that variable and y in the presence of the other variables already in the model *stat*

step-wise selection: >>> stage-wise selection

stereomicroscope: A microscope having a set of optics for each eye so arranged that the sets view the object from slightly different directions and make it appear in three dimensions *micr*

sterile: Unable to produce reproductive structures (i.e., unable to reproduce) *gene*

sterile spikelet: The non-fertile lateral spikelet in two-row barley *bot*

sterility: Failure to produce functional gametes or viable zygotes *bot gene*

sterility gene: A gene that causes sterility *gene*

sticker: Substance used in inoculation of bacteria to ensure that, for example, the rhizobia adhere to the seed of legume crops during planting; it can range from milk or sugar solutions used in simple seed inoculation, to stronger adhesives (40% gum Arabic, 5% methyl ethyl cellulose) used in pelleting seed *agr seed* >>> NUSSINOVITCH 2009

stickiness: The genetically or artificially induced agglutination of chromosomes; usually it leads to different sorts of chromosome aberrations *cyto*

sticky ends: Single-stranded ends of DNA fragments, usually produced by restriction enzymes of type II *gene biot* >>> blunt ends >>> Table 38

stigma: The part of the female reproductive organs on which pollen grains germinate; it is the receptor of the pollen *bot* >>> wet stigma >>> Figure 35

stigmatic hairs: Small tiny hairs and/or hair-like structures on a stigma *bot*

stimulant: A chemical or other substance that excites an organ or tissue to a specific activity (e.g., the application of a plant regulator to a stem to induce root formation) *phys*

sting: A sharp, hollow, glandular hair that secretes an irritating or poisonous fluid (e.g., in nettle) *bot*

stinging hair: >>> sting

stipe: A stalk, especially the petiole of a fern or palm frond, or the caudicle in an orchid flower and a stem of mushrooms *bot*

stipule: A leafy or linear appendage, found usually in pairs, at or near the base of the petiole of a leaf (e.g., in pea) *bot*

stirps: A race or permanent variety of plants *tax gene*

stochastic: A process with an indeterminate or random element as opposed to a deterministic process that has no random element *stat*

stock: That part of a plant, usually consisting of the root system together with part of the stem, onto which is grafted a scion bot; in genetics, an artificial and/or experimental mating group *gene* >>> graft >>> rootstock

stock collection: >>> stock seeds

stock culture: A culture maintained in a culture collection for future study or reference *meth phyt biot*

stock plant: A mother plant kept for cuttings to reproduce the plant and/or variety *hort fore meth*

stock seeds (parent ~): The supply of seeds, tubers, or other propagules reserved for planting, multiplication, or as a source of germplasm for maintaining and increasing seed of crop varieties or genetic tester lines *seed meth*

stolon: A stem that grows horizontally, which roots at its tip to produce a new plant, as opposed to a runner *bot*

stolon nursery: An area used for producing stolons for propagation *seed meth*

stoma (stomata pl): A small opening, many of which are found in the epidermal layers of plants, allowing access for carbon dioxide and egress for water; they are surrounded by guard cells, which control the pore size *bot* >>> carbon dioxide

stomatal closure response: The response by plants to environmental stress such as drought, that reduces stomatal pore size; this leads to reduced water transpiration rates and restricted access of CO_2 to the leaves *phys*

stone fruit: Fruits of the botanical family Rosaceae that contain a single hard seed, called a stone, pit, or pip; the term includes >>> plums, >>> cherries, greengages, peaches, apricots, almonds, and sloes *bot hort* >>> drupe

stoner: A modification of the gravity separator, especially constructed to separate stones from crop seeds; the machine separates seed on the same principle as conventional specific gravity machines, however it discharges only at each end; the desirable seeds flow to the lower end and are discharged, while stones and heavier concreted earthen material drop down to the upper end of the deck and are discharged; it is useful for conditioning seed of field beans *seed* >>> Table 11

stool: The root or stump of a tree, bush, grass, or cane that produces shoots each year, or the mother plant from which young plants are propagated by the process of layering *hort agr bot* >>> tillering

stooling out: New shoots appearing on the edge of the crown *bot*

stop codon: A codon for which there is no corresponding tRNA molecule to insert an amino acid into the polypeptide chain; the protein synthesis is hence terminated and the completed polypeptide released from the ribosome; there are three stop codons (UAA, UAG, UGA) *gene*

stopping: >>> disbud

storability of seeds: The longevity of seeds after storage; it represents a trait important for the conservation of seed resources; seed oligosaccharides and especially raffinose series oligosaccharides are hypothesized to play an important role in the acquisition of desiccation tolerance and consequently in seed storability; in many plants species seed maturation is accompanied by the accumulation of soluble oligosaccharides; there are even genes controlling storability, e.g., in rice a quantitative trait locus of seed storability, qSS-9, was mapped on chromosome 9 *seed* >>> aging test

storage organ: >>> storage proteins

storage proteins: Proteins that are found in specialized storage organs, such as endosperm tissue in cereals or others; for example, the endosperm of wheat contains a great number of non-enzymatic storage proteins that are the components of gluten, one of the most intricate naturally occurring protein complexes; gluten may be subdivided, on the basis of differential solubility, into gliadin and glutenin; in wheat breeding, these proteins may provide markers for specific quality characteristics *phys* >>> Table 15

storage protein genes: Structural genes coding the synthesis of storage proteins; for example, in wheat, each homoeologous chromosome group contains structural genes having similar positions within each genome; the differences in the strength and elasticity of dough is under control of several storage protein loci [the high- and low-molecular-weight (HMW, LMW)] >>> glutenins (Glu-loci) and the >>> gliadins (Gli-loci); the HMW glutenins appear responsible for dough strength, the LMW glutenins and gliadins for dough extensibility and elasticity; by using electrophoretic methods, it has been possible to identify five loci controlling these proteins; they are located on the homoeologous chromosomes 1 and 6; alleles at these loci are highly variable and correlate with differences between good and poor bread- or biscuit-making quality *gene prep* >>> Table 15

storage tissue: >>> storage proteins

storm-proof: A type of cotton in which the fibers are held tightly in the boll; generally harvested by stripping the boll from the plant *agr*

stover: Coarse roughage used as feed for livestock; in breeding, stalks and leaves, not including grain, of such forages as corn and sorghum, i.e., stalks after removal of the spikes, usually cured and used as feed; in the U.K., fodder minus the grain portion of the plant; the use of maize stover as a feedstock for cellulosic biofuels production has become a task of recent breeding approaches *agr*

straddle: A foundation of the trunk and main branches for grafting, either of rootstock or stem builder *meth hort*

straggling: Breakage of a few plant stems, especially cereals, resulting in a few tillers falling down *agr*

strain: A term sometimes used to designate an improved selection within a variety; a group of individuals from a common origin; generally, a more narrowly defined group than a variety *tax* >>> variety

strain building: Improvement of cross-fertilizing plants by any one of a number of methods of selection *meth*

stratification: The placing of seeds between layers of moist peat or sand and exposing them to low temperatures in order to encourage germination and/or breaking dormancy; the method can also be realized in a refrigerator between +4–10°C *seed* >>> prechilling

stratified mass selection: Mass selection in which the population is split into subpopulations that are grown under different environmental conditions (i.e., in different fields or in different parts of a field); plants for next-generation seed are selected from the different subpopulations *meth*

stratified random sample: A sample obtained by separating the population elements into non-overlapping groups, called strata, and then selecting a simple random sample within each stratum *stat*

stratified sampling: A sampling procedure in which the sample space is partitioned into groups or strata and each stratum is sampled randomly; it helps to assure that the sample has values over the full range of the sample space, for example, Latin hypercube sampling *stat*

straw: The dried remains of fine-stemmed plants from which the seed has been removed in threshing *agr*

strawbreaker of cereals: >>> eyespot disease

straw pith: In cross section at middle of internode below the neck of culm; it can be hollow, thick-walled, or solid *bot*

straw walker: The part of a combine that moves the straw to the rear of the machine *agr* >>> coarse shaker

streptavidin: A protein produced by Streptomyces avidinii; it shows affinity to the vitamin "biotin"; it is used for detection of biotinylated hybrid DNA *meth micr prep* >>> FISH >>> GISH

***Streptomyces coelicolor*:** A species of the Streptomyces family, a commonly occurring family of bacteria that are highly adapted to living in soil; these bacteria are very productive natural factories, used to make over half of the naturally derived antibiotics in current use; Streptomyces can make over 6,000 different chemical products that include antibacterial agents (e.g., tetracyclines, erythromycin, rifamycin, kanamycin, and vancomycin), antifungals (candicidin and nystatin), anti-cancer drugs (bleomycin, daunorubicin, and doxorubicin), antihelminthics (avermectin), and immunosuppressants (FK506 and rapamycin); the *Streptomyces coelicolor* genome contains ca. 8,000 genes *biot*

streptomycin: An antibiotic; it inhibits the elongation of protein biosynthesis; it is produced by some strains of *Streptomyces*; in genetic experiments, streptomycin resistance is used as a selection marker *meth*

stress A specific response by the plant to a stimulus that disturbs or interferes with the normal physiological equilibrium *bio* >>> www.plantstress.com

stress proteins: Proteins made by plant cells (and other organisms); when those cells are stressed by environmental conditions (chemicals, pathogens, heat) (e.g., when maize is stressed during its growing season by high nighttime temperatures), the plant switches from its normal production of (immune system defense) chitinase to the production of heat-shock (i.e., stress) proteins *biot phys* >>> www.plantstress.com

stress tolerance: >>> biological control

striate: Displaying narrow parallel streaks or bands *bot*

***Striga hermonthica*:** The most widespread and destructive obligate root parasite infecting maize and other cereals in Africa *phyt*

strigolactone (SL): Under environmental stresses such as drought and salinity, plants may experience restricted growth and productivity – stress responses that are mediated by complex molecular signaling networks; a previously unknown signaling pathway that plays a key role in stress tolerance was recently identified; although various phytohormones are known to be involved in the regulation of plant stress responses, the role of strigolactone (SL) in this important process remains elusive; by using different molecular and physiological approaches in *Arabidopsis*, SL acts as positive regulator of plant responses to drought and salt stress, which is associated with shoot- rather than root-related traits; comparative transcriptome analysis suggests that plants integrate multiple hormone-response pathways – at least SL, abscisic acid, and cytokinin pathways – for adaptation to environmental stress; genetic modulation of SL content/response can provide a new approach for development of crops with improved stress tolerance *phys*

strigose: Bearing straight, stiff, sharp, appressed hairs *bot*

strike: To take root (e.g., of a slip of a plant) *hort meth*

string bean: >>> stringiness

stringency: Reaction conditions (e.g., temperature, salt, and pH) that dictate the annealing of single-stranded DNA/DNA, DNA/RNA, or RNA/RNA hybrids; at high stringency, duplexes form only between strands with perfect one-to-one complementarity; lower stringency allows annealing between strands with some degree of mismatch between bases *biot meth*

stringiness: Green beans, peas, or other legumes with strings that must be removed before eating or preparing for cooking; it is a wild characteristic; since several legume species are grown for their young, edible, fleshy pods, stringiness became a quality trait, however, it has largely been bred out of modern varieties *hort*

string tags: Waterproof tags used to mark collected specimens for herbarium study *meth tax*

strip cropping: The practice of growing crops in strips or bands along the contour in an attempt to reduce run-off, thereby preventing erosion or conserving moisture *agr*

strip cutting: A term used to denote the practice of harvesting alternate borders of a crop, so that some partially grown crop is maintained in the field at all times *agr*

strip tillage: Planting and tillage operations that are limited to a strip not to exceed one-third of the distance between rows; the area between is left untilled with a protective cover of crop residue on the surface for erosion control *agr*

"strong" flour: Strong flour contains strong gluten with good elastic properties (e.g., in wheat) suitable for bread-making; as opposed to "weak" flour, which contains weak gluten, which is less elastic but more extensible than that contained in strong flour; suitable for biscuit-making *prep*

structural aberration: Structural changes of the entire chromosome (e.g., deletions, duplications, translocations, inversions, telocentrics, or isosomics) *cyto* >>> Figure 37

structural gene: A gene that codes for the amino acid sequence of a protein *gene*

structural genomics: A study of the structure of all gene sequences in a fully sequenced genome *biot*

structural heterozygosity: Heterozygosity for chromosome mutations *cyto* >>> translocation >>> Figure 37

structural homozygosity: Homozygosity for chromosome mutations *cyto*

struggle for existence: The phrase used by C. DARWIN (>>> SCHLEGEL 2007) to describe the competition between organisms for environmental resources such as food or a place to live, hide, or breed *evol*

STS: >>> sequence-tagged-site

stubble: The bases of cereal stems that remain after harvest; in general, fields or areas of fields containing the cut stems of a combinable crop left undisturbed after harvest *agr*

stubble crop: A crop plant from which stubble remains after combine harvesting (e.g., rapeseed, wheat, barley, and sugarcane) *agr*

Student's test: A statistical method used to determine the significance of the difference between means of two samples *stat* >>> SCHLEGEL 2007

stunt: To retard or stop plant growth (e.g., through exposure to harsh weather or lack of water or nutrients) *agr hort*

style: An extension of the carpel, which supports the stigma *bot*

style grafting: A crossing method applied when pollen fails to germinate on the stigma of a female parent; pollen is first applied to the stigma of a plant of the same genotype (species), so that it effectively germinates and the pollen tube grows into the style; the style is then cut off, just below the point reached by the pollen germ tubes, and grafted on a cut style of a female target plant; when the two parts of style are joined, the pollen tube continues to grow down to the ovary for fertilization; the method is practicable only in plants with fairly long and thick styles, e.g., in lilies *meth* >>> crossing method >>> incompatibility

subbing (of microscopic slides): Slides are subbed so that cells or tissue adhere and are retained during subsequent steps of preparation of microscopic slides, for example, during the procedure of *in situ* hybridization; many different methods are suitable, including the use of gelatin, polylysine, Tespa, Denhardt solution, or glue *cyto meth*

subcloning: Transplantation of a piece of DNA from one vector to another *biot*

subculture: The aseptic transfer of a part of a stock culture to a fresh medium *biot*

suber: >>> cork

suberiferication: The process by which the cut surface of a stem forms a protective, corky layer, especially in conditions of high temperature and high humidity *bot*

suberin: A complex fatty substance found especially in the cell walls of cork *bot*

subirrigation: Applying irrigation water below the ground surface either by raising the water table within or near the root zone, or by using a buried perforated or porous pipe system that discharges water directly into the root zone *hort meth*

sublimation: The transition of a substance directly from the solid to the vapor phase or vice versa, e.g., vaporizing sulfur in a greenhouse in order to reduce growth of fungi *chem phyt*

sublining: Dividing a breeding population into several smaller populations; all controlled crosses for forward selection are made within a subline, leading to inbreeding within sublines *meth fore* >>> Table 37

submarine patent: A patent that emerges after it has unknowingly been infringed upon *adm* >>> Intellectual Property Rights

submedian: Not quite in the middle *cyto* >>> Figure 11

submedian centromere: >>> submedian

submicroscopic: Objects that are too small to be resolved and made visible by an ordinary light microscope *micr*

subpopulation: Either an identifiable fraction or subdivision of a population or specific biotype within a natural population *tax* >>> Tables 12, 37

subsexual reproduction: A partially asyndetic meiosis that leads to a restitution nucleus with limited crossing-over, which in turn leads to genetic variability *bot*

subshrub: >>> suffrutex

subsinuate: Somewhat wavy-margined *bot*

subsoil: The soil that lies beneath that which is cultivated; it is usually hard and infertile *agr* >>> Table 44

subspecies: Abbreviation "ssp."; used of a species forming a population or group of populations that are distinguishable from other members of the species and often partially reproductively isolated from them *tax* >>> race >>> Table 12 >>> List of Important Crop Plants

substitution line: A line in which one or more chromosomes are replaced by one or more chromosomes of a donor variety or species; it is used for genetic analysis and gene transfer *cyto*

substrate: A substance that is acted upon, as by an enzyme; also used for a culture medium *prep meth biot*

subtractive hybridization: A form of differential hybridization, where a probe is prepared in which the signal representing sequences common to the two transcript pools of interest are removed; the product is a probe population enriched for sequences preferentially present in one of the pools; it is a more sensitive technique than differential hybridization *biot*

succession: The gradual supplanting of one community of plants by another, the sequence of communities being termed a "sere" and each stage "seral" *agr hort fore*

succession cropping: The method of growing new plants in the space left by the harvested ones *agr*

succession of crops: >>> crop rotation

succession planting: Planting portions of a crop over a period of time to get a continuous harvest over a long period of time *agr hort*

succulence: Having watery or juicy tissue *bot*

succulent plant: >>> succulence

sucker: A vegetative shoot of subterranean origin *bot* >>> root sucker

sucrose: >>> saccharose

sucrose gradient centrifugation: Used for separation of DNA, RNA, or proteins according to size and conformation *meth biot*

sudden-death syndrome (of soybean): A disease caused by *Fusarium solani f. sp. glycines*; it is controlled by a number of quantitatively inherited loci (QTLs); for example, the variety

"Forrest" shows a strong field resistance to the syndrome, while the variety "Essex" is susceptible *phyt* >>> QTL mapping >>> Table 58

suffrutex (*syn* **subshrub**): Low-growing woody shrub or perennial with woody base *bot*

sugar: Any sweet, soluble, crystalline, lower-molecular-weight carbohydrate *phys chem* >>> saccharose

sugarbeet leaf curl: Caused by the sugarbeet leaf curl virus *phyt*

sugarbeet root maggot (SBRM): A major insect pest of sugarbeet throughout much of North America; recently, two insecticides with the same mode of action have been used almost exclusively for SBRM control; alternative control strategies would be required if insecticide-resistant SBRM developed or the insecticides were no longer available due to regulatory actions; germplasm lines with SBRM resistance are available, but information on the SBRM resistance of hybrid cultivars created by crossing these lines with a SBRM-susceptible cms parental line is still lacking; *Tetanops myopaeformis phyt*

sugarcane rust: Of interest because no vertical resistance occurs against it; this is because sugarcane is derived from a continuous wild pathosystem; the disease caused by Puccinia melanocephela has occasionally been damaging when it appeared in an area of susceptible cane, as happened recently in Cuba *phyt* >>> www.tau.ac.il/lifesci/units/ICCI//genebank2.html

sui generis: Literally, "of its own kind"; an intellectual property rights system on plant varieties *seed agr*

sulfonamide: A chemical having a bacteriostatic effect *chem phyt* \

sulfur (S): A yellow mineral; an element that is needed by the plant; it is found covalently bound, especially in proteins, where it stabilizes plant structures; deficiency leads to chlorosis and etiolation *chem phys*; it is also used as a dust or wettable powder for control of various diseases and insects, including powdery mildew, rust, and mites *phyt* >>> Tables 51, 52

sulfuration: A treatment with sulfur, with fumes of burning sulfur, sulfur dioxide, or with sulfites as in fumigation, bleaching, or preserving; it is used in greenhouses in order to reduce and/or prevent leaf diseases (e.g., in cereals) *phyt meth*

supercoiled: Natural conformation of DNA molecules *biot*

superdominance: Interaction of the dominant and recessive alleles of a single gene locus *gene* >>> overdominance

super-elite: Very high-grade seed used primarily for further seed production *seed*

superfemale: A female individual that carries higher doses of female determiners *gene*

supergene: A group of neighboring genes on a chromosome that tend to be inherited together and sometimes are functionally related; this chromosome segment is protected from crossing-over and so is transmitted intact from generation to generation, like a recon; if coadapted, such genes, even though not necessarily functionally related, may cooperate to produce some adaptive characteristic *gene*

superior: Refers to an ovary when the other organs of the flower are inserted below it *bot*

superior pelea: >>> upper palea >>> pale

supermale: A male individual that carries higher doses of male determiners *gene*

supernumerary chromosomes: Chromosomes present, often in varying numbers, in addition to the characteristic invariable complement of chromosomes *cyto*

supernumerary spikelet: A rudimentary or partially developed spikelet in wheat, usually borne immediately below a normal spikelet at the same rachis node *bot*

superweeds: Weeds that have resistance to herbicides *phyt biot*

suppressive soil: A soil in which certain diseases fail to develop because of the presence in the soil of microorganisms antagonistic to the pathogen *phyt* >>> Table 44

suppressor mutation: A second mutation that masks the phenotypic effects of an earlier mutation; it occurs in a different site in the genome *gene*

suppressor-sensitive mutation: A mutation whose phenotype is suppressed in a genotype that also carries an intergenic suppressor of that mutation, for example, amber, ochre, or opal *gene*

surface tension: Tension exerted by a liquid surface due to molecular cohesion and apparent at liquid boundaries *phy*

surfactant: >>> spreader

survival index: The degree of effectiveness of a given phenotype in promoting the ability of that organism to contribute offspring to the future population *stat*

survival of the fittest: The corollary of Darwin's theory of natural selection, namely that as a result of the elimination by natural selection of those individuals least adapted to the environment, those that ultimately remain are the fittest *evol*

susceptibility: The inability of a host plant to suppress or retard invasion by a pathogen or pest, or to withstand adverse environmental influences *phyt*

susceptible: Being subject to infection or injury by a pathogen *phyt*

suspension culture: Cells and groups of cells dispersed in an aerated, usually agitated, liquid culture medium *biot*

suspensor: The group or chain of cells produced from the zygote that pushes the developing pro-embryo toward the center of the ovule in contact with the nutrient supply *bot*

sustainability: The capability of entities to survive without human intervention *meth*

sustainable agriculture: A systematic approach to agriculture that focuses on ensuring the long-term productivity of human and natural resources for meeting food and industrial needs; it is an integrated system of plant and animal production practices having a site-specific application that will, over the long term, (1) satisfy human food or industrial needs; (2) enhance environmental quality and the natural resource base upon which the agricultural economy depends; (3) make the most efficient use of non-renewable resources and on-farm resources and integrate, where appropriate, natural biological cycles and controls; (4) sustain the economic viability of farm operations; and (5) enhance the quality of life for farmers and society as a whole *agr*

sustainable use: The careful exploitation of natural resources without destroying them *eco* >>> sustainable agriculture

sustainable crops: >>> sustainable agriculture

sustainable forestry: Management of forested area in order to provide wood products in perpetuity, soil and watershed integrity, persistence of most native species, and maintenance of highly sensitive species or suitable conditions for continued evolution of species *fore* >>> sustainable agriculture

sustained yield: An output of renewable resources that does not impair the productivity of the resource; it implies a balance between harvesting and incremental growth or replenishment *agr*

Svedberg unit (S): A unit of sedimentation rate during centrifugation *chem*

SVISS: >>> satellite virus-induced silencing system

sward: A stand of forage grasses or legumes *agr* >>> turf

sward plot: A stand of forage grasses or legumes in plot *meth*

swath: A wind-row; a row of cut or pulled crop usually waiting for drying or curing before further harvesting *agr*

sweep: Tree defect resulting from a gradual curve in the main stem of the tree *hort fore*

sweet potato whitefly: An insect pest of cotton, fruit, vegetable, and greenhouse crops *phyt*

symbiont: An organism living in a state of symbiosis *eco* >>> symbiosis

symbiosis: The living together for mutual benefit of two organisms belonging to different species *eco*

symmetrical distribution: A frequency distribution for which the values of the distribution that are equidistant from the mean occur with equal frequency *stat*

sympatric: Of populations or species that inhabit, at least in part, the same geographic region; as opposed to allopatric *eco* >>> parapatric

sympetalous: >>> gamopetalous >>> gametogamy

symplastic coupling: A special type of cell–cell interaction in which two or more plant cells, connected by plasmodesmata, can exchange macromolecules such as introduced dyes, proteins, or RNA; cells interconnected in this way can form a network of cells that behave as a supracellular domain *phys*

symptom: A visible response of a host plant to a pathogenic organism *phyt*

Syn-1: >>> first-generation synthetic varieties

synapsis: The side-by-side pairing of homologous chromosomes during the zygotene stage of meiotic prophase *cyto*

synaptene: >>> zygotene

synaptonemal complex (SC): Ribbonlike structures observed in electron micrographs of nuclei in the synaptic stages of meiosis; the ribbon represents a system that promotes synapsis of homologous chromosomes; the SC seems to be the essential prerequisite for homologous chromosome pairing and crossing-over *cyto*

syncarp: A structure consisting of several united fruits, usually fleshy *bot* >>> aggregate fruit

syncarpy: Plants with flowers having two or more carpels, all fused together *bot*

synchronous culture: An *in vitro* culture in which a large proportion of the cells are in the same phase of the cell cycle at the same time *biot*

syncytial endosperm: A mass of cytoplasm with several nuclei, not divided into separate cells *bot*

syndesis: >>> synapsis

synergid (synergidae pl): One of two haploid cells that lie inside the embryo sac, beside the ovum; they nourish the ovum *bot* >>> Figures 25, 35

syngamy: Sexual fusion of the sperm and egg cell *bot* >>> fertilization

syngraft: >>> isograft

synkaryon: A nucleus resulting from the fusion of two genetically different nuclei, sometimes from two different species *bot biot*

synonymous single nucleotide polymorphism: Any >>> SNP that occurs in an exon but does not change the amino acid composition of the encoded protein *biot*

synteny: Chromosomal association of genes (linkage) established in somatic-cell culture *phys gene*; in biotechnology, the conservation of the gene order on the chromosomes; it allows understanding of the evolutionary relationships between species; until the 1980s, it was imagined that each crop plant had its own genetic map; by using the first >>> RFLP markers, it turned out that related species had remarkably similar gene maps; this demonstrated conservation over a few million years of evolution in syntenous relationships between potato and tomato in the broad-leafed plants, and between the three genomes of bread wheat in the grasses; later it was shown that the same similarities held over the rice, wheat, and maize genomes, which were separated by some 60 million years of evolution; the discovery of synteny has had an enormous impact on the way we think about plant genetics; there are great opportunities to predict the presence and location of a gene in one species from what is known from another; for practical plant breeding, knowledge of synteny allows breeders access to all alleles in, for example, all cereals rather than just the species on which they are working; a key first example of this is the transfer to rice of the wheat dwarfing genes that made the >>> Green Revolution possible; in these experiments, the gene was located in rice by synteny, and then isolated and engineered with the alteration in DNA sequence that characterized the wheat genes, before replacing the engineered gene in rice; this approach can be applied to any gene in any cereal, including the so-called >>> orphan crops that have not attracted the research interest that wheat, rice, or maize have over the past century *biot evol* >>> www.rye-gene-map.de/

synteny mapping: The localization of DNA sequences (chromosome arms, genes, etc.) in a target genome that are syntenic to regions in a reference genome *biot meth*

synthetic (variety): A variety produced by crossing inter se a number of genotypes selected for good combining ability in all possible hybrid combinations, with subsequent maintenance of the variety by open pollination; usually the first generation of a synthetic variety is obtained by a polycross involving a certain number of components with a good general combining ability; the components are maintained by identical reproduction, either by vegetative propagation (clones) or by continued sib mating (inbred populations) *seed meth* >>> Figures 50, 51 >>> Tables 5, 37

synthetic amphiploid: An artificially produced amphiploid *cyto*

synthetic germplasm: Man-made genetic resources, such as transgenic or genetically engineered plants, induced mutations, induced somaclonal variation, artificial hybrids (e.g., tetra-, hexa- or octoploid triticale), induced polyploids, or experimental lines (NILs, RILs, DHs) *breed seed meth*

synthetic seed(s): It refers to alginate encapsulated somatic embryos, vegetative buds, or any other micropropagules that can be used as seeds and converted into plantlets after propagating under *in vitro* or *in vivo* conditions; moreover, synthetic seeds retain their potential for regeneration even after low-temperature storage; the production of synthetic seeds opens up new vistas in agricultural biotechnology; encapsulated propagules are used for *in vitro* regeneration and mass multiplication at reasonable cost; in addition, these propagules are used for germplasm preservation of elite plant species and the exchange of plant materials between national and international laboratories *biot seed* >>> artificial seeds

systematics: The science of describing, naming, and systematically classifying organisms *tax*

systematic sampling: A method of selecting a sample by a systematic method as opposed to random sampling, for example, select each third plant from a sample *stat*

systemic: Absorbed into the sap stream and passed to other parts of the plant *phys* >>> systemic pesticide

systemic acquired resistance (SAR): The induction of disease resistance in areas of a plant remote from the site of initial infection; it is an important element in plants' strategies to protect themselves from disease attack; the gene DIR1 is essential to the SAR response *phyt* >>> resistance >>> DIR1 gene

systemic pesticide: An agent that is distributed throughout the plant; it protects the entire host against pests for a certain time *phyt* >>> disease protection

systemic sampling: A sampling procedure in which the sample space is sampled at a priori determined points *stat*

systems biology: The simultaneous study of complex interactions of multiple levels of biological information including DNA, RNA, proteins, and biochemicals *biot bio*

T

T1 (generation): A term used in plant genetics; it refers to the progeny resulting from self-pollination of the primary transformant regenerated from tissue culture *gene*

tandem base mutation(s) (TBM): A TBM is defined as a substitution at adjacent nucleotides (primarily substitutions at two adjacent nucleotides), but substitutions at three adjacent nucleotides have been observed *biot gene* >>> InDel >>> mutation

T chromosome: A chromosome in which a terminal (T) region shows neocentric activity *cyto* >>> neocentric activity

T4 phage: A type of a bacteriophage used as a source of commonly used ligase, DNA polymerase, and polynucleotide kinase *biot*

tabular root: The main, downward-growing root of a plant, which grows deeply and produces lateral roots along its length *bot*

tail: Single-stranded DNA extension added by terminal deoxynucleotidyl transferase *biot*

tailing: The *in vitro* addition of same nucleotide by the enzyme terminal transferase, to the 3'-hydroxyl ends of a duplex DNA molecule, i.e., homopolymeric tailing *biot*

tailings: Partly threshed material that has passed through the coarse shakers or straw walkers and is eliminated at the rear of a threshing machine *seed*

take-all: >>> saponin >>> take-all disease

take-all decline: The decline in the cereal disease take-all after three or four successive cereal crops *agr*

take-all disease: A fungal disease (*Gaeumannomyces graminis* syn *Ophiobolus graminis*) that attacks wheat plant roots; it causes dry rot and premature death of the plant; certain strains of Brassica plants and *Pseudomonas* bacteria act as natural antifungal agents against the fungus *phyt* >>> biological control

TAL effector nuclease(s) (TALENs): DNA-binding domains of transcription activator-like (TAL) effectors can be engineered; the TAL effector DNA-binding domain creates chromosomal breaks at specific DNA sites; so TALENs may create targeted mutations in a variety of eukaryotes; the recent efficiency is one mutation among 100,000 cells treated *biot*

TALENs: >>> TAL effector nuclease

tandem array: The existence of two or more identical DNA sequences in a series, i.e., end to end *biot*

tandem duplication: >>> tandem repeat

tandem repeat: A chromosomal mutation in which two identical chromosome segments lie adjacent to each other, with the same gene order; the DNA, which codes for the rRNA, contains many tandem repeats *gene* >>> http://wheat.pw.usda.gov/ITMI/Repeats/index.shtml

tandem selection: In the case of successive multiple selection, the selection concerns other traits in the first few generations than in later generations *meth*

tannic acid: >>> tannin

tannin: A generic term for complex, non-nitrogenous compounds containing phenols, glycosides, or hydroxy acids, which occur widely in plants (e.g., in the testa of cocoa and beans) *chem phys*

tan spot (of wheat): A disease of wheat that is caused by the fungus *Pyrenophora tritici-repentis*; on susceptible hosts, *P. tritici-repentis* induces two phenotypically distinct symptoms, tan necrosis and chlorosis; this fungus produces several toxins that induce tan necrosis and chlorosis symptoms in susceptible cultivars; there is a wheat variety "Salamouni" that is resistant to tan spot disease, controlled by a single recessive gene tsn4 located on chromosome 3A *phyt*

***Tapesia yallundae*:** Sexual stage of *Pseudocercosporella herpotrichoides phyt* >>> white leaf spot >>> eyespot disease

tapetal cell: >>> tapetum

tapetal layer: >>> tapetum

tapetum: A layer of cells, rich in food, which surrounds the spore mother cells *bot*

tapping: Driving spouts into the trunks of maple or pine trees to let the sap out for food and industrial use *fore*

taproot: A large, descending, central root *bot* >>> tabular root

Taq (DNA) polymerase: A DNA-dependent RNA polymerase from phage T7, which recognizes a very specific promoter sequence; it is used in many expression vectors *biot*

TaqMan™: A trademark term for a high-throughput, closed-tube assay to detect specific sequences in PCR products; TaqMan indicates the probe used to detect specific sequences in >>> PCR products by employing the 5'→3' exonuclease activity of >>> Taq DNA polymerase; the TaqMan probe (20–30 bp), disabled from extension at the 3' end, and consists of a site-specific sequence labeled with a fluorescent reporter dye and a fluorescent quencher dye; during PCR the TaqMan probe hybridizes to its complementary single-strand DNA sequence within the PCR target; when amplification occurs the TaqMan probe is degraded due to the 5' → 3' exonuclease activity of Taq DNA polymerase, thereby separating the quencher from the reporter during extension; due to the release of the quenching effect on the reporter, the fluorescence intensity of the reporter dye increases; during the entire amplification process this light emission increases exponentially, the final level being measured by spectrophotometry after termination of the PCR; because increase of the fluorescence intensity of the reporter dye is only achieved when probe hybridization and amplification of the target sequence have occurred, the TaqMan assay offers a sensitive method to determine the presence or absence of specific sequences; the technique is particularly useful in diagnostic applications, such as the screening of samples for the presence or incorporation of favorable traits and the detection of pathogens and diseases; the TaqMan assay allows high sample throughput because no gel electrophoresis is required for detection; when different probes are used which are able to discriminate between allelic variants, TaqMan behaves as a >>> codominant marker *meth biot*

***Taraxum* type:** Diplospory where the spore mother cell enters the meiotic prophase, but because of synapsis there is no pairing and the univalents remain scattered over the whole spindle; the first meiosis results in a restitution nucleus; the second meiosis results in an unreduced dyad *bot*

targeting-induced local lesions in genomes (TILLING): A technique to screen a population that has either been deliberately mutagenized or possesses natural biodiversity (EcoTILLING) to identify those plants with a change in a specific gene of interest; the technique uses high-throughput methods useful to plant breeders, including a reverse molecular method that combines random chemical mutagenesis with PCR-based screening of gene regions of interest; it provides a range of allele types, including mis-sense and knockout mutations; by comparing the phenotypes of isogenic genotypes differing in single sequence motifs, TILLING provides direct proof of function of both induced and natural polymorphisms without the use of transgenic modifications *biot meth* >>> EcoTILLING >>> TAL effector nuclease

target population: The target population is the entire group of individuals a breeder is interested in; the group about which he or she wishes to draw conclusions by several means *stat* >>> Table 37

target region amplification polymorphism (TRAP) marker: A novel marker technique for plant genotyping; this large-scale DNA sequencing technology has generated a tremendous amount of sequence information for many important organisms; a rapid and efficient PCR-based technique was developed, which uses bioinformatics tools and >>> expressed sequence tag (EST) database information to generate polymorphic markers around targeted candidate gene sequences; this technique uses 2 primers of 18 nucleotides to generate markers; one of the primers, the fixed primer, is designed from the targeted EST sequence in the database; the second primer, the arbitrary primer, is an arbitrary sequence with either an AT- or GC-rich core to anneal with an intron or exon, respectively; PCR amplification is run for the first 5 cycles with an annealing temperature of 35°C, followed by 35 cycles with an annealing temperature of 50°C; for different plant species, each PCR reaction can generate as many as 50 traceable fragments with sizes ranging from 50 to 900 bp when separated on a 6.5% polyacrylamide sequencing gel; the TRAP technique is useful in genotyping germplasm collections and in tagging genes governing desirable agronomic traits of crop plants, originally developed for sunflower studies (HU and VICK, 2003) *biot meth*

target site duplication: A sequence of DNA that is duplicated when a >>> transposable element inserts, usually found at each end of insertion *biot*

target validation: Blocking the activity or preventing the synthesis of a selected protein to predict if developing a chemical targeted against that protein would likely produce the desired therapeutic effect; the demand for fast, reliable, and cost-effective target validation is rapidly growing because the >>> proteomics research, and increasing understanding of many plant diseases at the molecular level, are leading to identification of a host of new proteins *biot* >>> gene silencing

tassel: As at the top of a stalk of maize *bot*

tasseling (of maize): The time or process when maize tassels emerge *agr* >>> tassel >>> detasseling

TATA box: A canonic DNA sequence; part of a plant promoter; promotes the transcription of DNA *gene*

tautomeric shift: The transfer of a hydrogen atom from one position in an organic molecule to another position *chem*

taxis: A change of direction of locomotion in a motile cell, made in response to certain types of external stimulus, such as temperature, light, nutrients, etc. *bot*

taxon (taxa pl): A group of organisms of any taxonomic rank (i.e., family, genus, species, etc.) *tax* >>> Tables 12, 17

taxonomy: The scientific classification of organisms *tax bio* >>> Tables 1, 12, 14, 17

TBC: >>> transcript-based cloning

t-distribution: >>> STUDENT's test

T-DNA: >>> Ti plasmid

T4 DNA ligase: An enzyme from bacteriophage T4-infected cells that catalyzes the joining of duplex DNA molecules and repairs nicks in DNA molecules; it requires that one of the DNA molecules shows a 5' phosphate group and the other DNA molecule a 3' hydroxyl group *biot*

TE: Tris-EDTA buffer *prep chem*

technology transfer: The transfer of discoveries made by basic research institutions, such as universities and government laboratories, to the commercial sector for development into useful products and services *meth adm*

tectum: The outer layer of the pollen grain exine in angiosperms *bot*

teeth: The jagged edges on some leaves, like those of thistles and hawthorne *bot*

teleutospore: >>> teliospore

teliospore A thick-walled resting spore produced by rust and smut fungi *bot* >>> Table 52 >>> www.tau.ac.il/lifesci/units/ICCI//genebank2.html

telium (telia pl): Pustule containing teliospores *bot*

telocentric (chromosome): Refers to the chromosomal centromere which lies at the end *cyto* >>> aneuploid >>> Figures 11, 37

telochromomere: A chromomere that is terminally located *cyto* >>> chromomere

telochromosome: >>> aneuploid >>> telosome

telomerase: A reverse transcriptase (hTERT) containing an RNA molecule (hTR) that functions as the template for the tandem repeat at the telomere; it synthesizes the telomere to maintain its length after each cell division; it is active in young cells and gametes, inactive in differentiated somatic cells, and reactivated in malignant cells; telomerase can add one base at a time to the telomeric end of a chromosome; this maintenance work is required for cells to escape from replicative senescence *biot*

telomere: One of the two terminal chromomeres of a chromosome; telomeres contain tandemly repeated nucleotide sequences, which promote the replication of the DNA double strand; a telomere also plays a role in the spatial orientation of chromosomes within the >>> interphase nucleus; chromosomes lose ~100 bp from the telomere every time the cell divides; the enzyme telomerase can add the lost bases *cyto* >>> telomerase

telomere mapping: The localization of telomeres at the end of chromosomes by >>> FISH and the identification of other sequence elements within the telomeric region *biot cyto meth*

telophase: The fourth and final phase of mitosis and the two divisions of meiosis, during which (1) the spindle disappears, (2) nucleoli reappear, (3) the nuclear membrane starts to develop around the two groups of daughter chromosomes and/or chromatids, and (4) the chromosomes return to their extended state, in which they are no longer visible; the nuclei then enter a resting stage as they were before division occurred *cyto*

telosome: Shorthand of telochromosome; a chromosome with a terminally located centromere *cyto* >>> aneuploid >>> Figures 11, 37

telosomic: >>> aneuploid >>> telosome

telosyndesis: >>> acrosyndesis

telotrisomic: In allopolyploids, such as hexaploid wheat, a cell or individual with one missing chromosome but having a telocentric and an isochromosome for the same arm of the missing one *cyto* >>> aneuploid >>> Figures 11, 37

temperate grasses: The most important sown temperate grasses are species of the genera *Lolium*, *Festuca*, *Dactylis*, *Phleum*, and *Bromus agr bot*

temperature gradient capillary electrophoresis (TGCE): A molecular method that can be used to distinguish heteroduplex from homoduplex DNA molecules and can thus be applied to the detection of various types of DNA polymorphisms; unlike most >>> single-nucleotide polymorphism (SNP) detection technologies, TGCE can be used even in the absence of prior knowledge of the sequences of the underlying polymorphisms; TGCE is both sensitive and reliable in detecting SNPs, small >>> InDel (insertion/deletion) polymorphisms (IDPs), and simple sequence repeats, and using this technique it is possible to detect a single SNP in amplicons of over 800 bp and 1-bp IDPs in amplicons of approximately 500 bp; genotyping data obtained via TGCE are consistent with data obtained via gel-based detection technologies; for genetic mapping experiments, TGCE has a number of advantages over alternative heteroduplex-detection technologies, such as >>> celery endonuclease (CELI) and denaturing high-performance liquid chromatography (dHPLC); multiplexing can increase TGCE's throughput to 12 markers on 94 recombinant inbreds per day; given its ability to efficiently and reliably detect a variety of subtle DNA polymorphisms that occur at high frequency in genes, TGCE shows great promise for discovering polymorphisms and conducting genetic mapping and genotyping experiments *biot meth* >>> Table 39

temperature treatment (of plant sexual organs): Exposing the plant or certain sexual organs, such as the style, to higher temperatures over a certain period of time in order to break down the incompatibility mechanism in the style; after treatment, pollen can grow into the style toward the ovary; the method is used in crossing ornamental plants (lily) or cereals (rye) *meth*

template: The DNA single strand, complementary to a nascent RNA or DNA strand, which serves to specify the nucleotide sequence of the nascent strand *gene*

temporary resistance: Vertical resistances are temporary in that they stop functioning on the appearance of a matching vertical >>> pathotype; they are within the capacity for micro-evolutionary change of the parasite *phyt*

T-end: A chromosome showing a terminal centromere *cyto*

tendril: Part of a stem, leaf, or petiole that is modified as a delicate, commonly twisted, thread-like appendage; it is an aid to climbing (e.g., in pea) *bot*

teosinte: The original wild grass, native to Mexico, from which cultivated maize derived; it is now classified as part of the same species as maize, *Zea mays bot*

tepal: One of the perianth members in those flowers where there is no distinction between calyx and corolla *bot*

teratology: The science of malformation *bio*

terete: Cylindrical and slender, as the normal culm of a grass plant *bot*

terminal association: In meiosis, the achiasmatic association of homologous chromosomes just by unspecific end-to-end attachments *cyto*

terminal bud: A bud located at the apex (tip) of the stem; this type of bud is the dominant bud, since it can cause all the lateral (side) buds below it to remain dormant at all times of the year; terminal buds have special tissue cells, called apical meristem, that can divide indefinitely *bot*

terminal centromere: >>> telosome

terminalization: A progressive shift of chiasmata from their sites of origin to more terminal positions *cyto*

terminal spikelet: Marks the completion of the spikelet initiation phase in cereals, such as wheat and barley; the terminal spikelet has its plane of symmetry at right angles to the other spikelets *bot phys*

terminase: An enzyme of phage lamda, which generates the staggered cuts at the "cos" sites during packaging *biot* >>> lamda phage

termination: The incorporation of the final amino acid into a polypeptide chain and the release of the complete chain from the ribosomes during protein biosynthesis *gene*

termination codon: >>> stop codon

terminator: A nucleotide sequence that acts as a signal for the termination of transcription *biot*

terminator genes: A set of transgenes designed to arrest the final stages of seed development in a crop plant; the resulting trait is the production of seeds that will not germinate; it is also called the technology protection system *biot*

terminator seeds: A descriptive term used for seeds that have been genetically engineered to produce a crop whose first generation produces sterile seeds, thus preventing a second generation from being grown from seeds saved from the first; it might be a way to build patent protection directly into a high-value, genetically engineered crop variety and thus recoup high research investment costs *biot* >>> Intellectual Property Rights

terminator technology Plants are genetically engineered so when the crops are harvested, all new seeds produced from these crops are sterile *biot* >>> www.victoryseeds.com/news/terminator_gene.html

terpene: A hydrocarbon that is composed of two or more isoprene units; they may be linear or cyclic molecules, or combinations of both, and include important biological compounds, such as vitamins A, E, and K *chem phys* >>> growth inhibitor

terracing: Converting hillsides into terraces that follow the contour as a method of soil conservation; it is an expensive and laborious process but, once completed, it is both effective and easily maintained; it is common in mountain regions of the world where soil for cropping is rare *agr*

terracotta pots: Flower pots made of terracotta (baked clay) were common in the past; they are more expensive than plastic pots, but they provide superior aeration to plant roots *hort*

tertiary chromosome: A chromosome formed by interchange between non-homologous chromosomes *cyto*

tertiary gene pool: Gene transfer from a species of the tertiary gene pool to the cultivated species, since the primary gene pool usually requires special crossing and embryo rescue techniques in order to get viable hybrids *evol* >>> primary gene pool >>> secondary gene pool

tertiary tiller: >>> secondary tiller

tertiary trisome: A chromosome present in addition to the normal diploid complement but as a result of a reciprocal interchange between two standard chromosomes; frequently it occurs in the progeny of a translocation heterozygote *cyto* >>> balanced tertiary trisomic >>> Figure 14

testa: The seed coat; it derives from secondary outgrowths of the nucellus, and later the ovule, which are called collars or integuments; the inner and outer integuments become the testa of the mature ovule; it is commonly composed of cuticle, palisade layer, sandwich cells, and parenchym cells *bot* >>> caryopsis >>> seed coat

test cross: A cross between a heterozygote of unknown genotype and an individual homozygous for the recessive genes in question; if the unknown is heterozygous, then approximately 50% of the offspring should display the recessive phenotype; e.g., *AbC/aBc* × *abc/abc*, where *A/a*, *B/b*, and *C/c* represent three pairs of alleles, and the slashes separate the contributions from the two parents of each individual; in general, each cross that contributes to the solution of an experimental question by using more or less defined crossing partners; in other test crosses, promising parental genotypes are crossed with a number of other known genotypes; the progeny is grown separately and assessed for the desired traits; the measurements can be used for determination of >>> "general combining ability" and >>> "specific combining ability", and these are needed if effective crosses are to be made; in inbreeding, it is not enough merely to select the best varieties, but parents with good crossing ability must also be available, so that a number of important traits are combined and passed on to progeny; it is especially crucial for asexually propagated plant or hybrids; test crosses require extensive testing facilities and are time-consuming *meth* >>> Figure 51

tester (plant): Plants of like kind and similar physiological condition used in experiments to measure performance or quality characters *meth* >>> Figure 30

testing plant variety(ies): Before any new crop variety can be placed on the market, it must undergo statutory testing under a process known as national listing (single plant testing, small-plot testing, double-row testing, performance-testing field); successful varieties are placed on a register of varieties approved for marketing; national rules are determined on a European, American, or Asian basis, and apply to all the major agricultural and vegetable crop species; official trials are conducted, in most cases for a minimum of two years, to test each new variety for a range of characteristics which together determine its uniqueness, its genetic >>> uniformity, and its value to growers and the rest of the food chain; national listing is extremely rigorous – the majority of varieties entered do not complete the process *seed meth* >>> variety listing >>> UPOV

test mating: >>> test cross

test of significance: >>> significance test

test tube: A hollow cylinder of thin glass with one end closed; used to hold chemicals or specimens in laboratory experimentation and analysis *prep*

tetra-allelic: In tetraploids, when multiple allele loci all have different alleles (e.g., four alleles, *A1A2A3A4*) *gene*

tetra-allelic single-nucleotide polymorphism (tetra-allelic SNP): Any >>> SNP of which four alleles are present in the population *biot*

tetracycline: An antibacterial antibiotic from *Streptomyces* spp. *chem phys* >>> achromycin

tetrad: Four homologous chromatids in a bundle during the first meiotic prophase and metaphase *cyto*; in meiosis, the four haploid cells resulting from a single diploid cell during gametogenesis *bot* >>> polyad >>> tetrad analysis

tetrad analysis: The use of tetrads to study the behavior of chromosomes and genes in crossing-over during meiosis; particularly used in studies of fungi *gene* >>> tetrad

tetraploid: Having four sets of chromosomes in the nucleus *gene cyto*

tetraploidization: The mitotic or meiotic procedure which produces tetraploids *meth cyto*

tetraploidy: >>> tetraploid

tetrasome: A chromosome present four times *cyto* >>> aneuploid >>> tetrasomic

tetrasomic: Having one or more chromosomes of a complement represented four times in each nucleus *cyto* >>> aneuploid >>> Figure 37

tetrasomy: The state of having one or more chromosomes as four copies *cyto* >>> aneuploid >>> Figure 37

tetrazolium penetration assay: >>> tetrazolium test

tetrazolium test: Seed coat permeability is important to study as it plays significant roles in seed dormancy, germination, and protection from pathogens; tetrazolium red is a cationic dye that is widely used in seed viability testing; tetrazolium salts are amphipathic cations, which, after penetrating the dead cells of the seed coat, are reduced to red-colored insoluble precipitates made up of formazans by active dehydrogenases (NADH-dependent reductases) in the embryo of seeds; the intensity of red coloration is directly proportional to the permeability of the seeds; the quantification involves extraction of formazans from the incubated seeds and spectrophotometric determination of absorbance of formazan extracts at 485 nm; a tetrazolium test is a quick test to determine seed viability; tetrazolium is a class of chemicals that have the ability to accept hydrogen atoms from dehydrogenase enzymes during the respiration process in viable seeds; the method was developed in Germany in the early 1940s by G. LAKON; the test is used throughout the world as a highly regarded method of estimating seed viability; it can be completed in only a few hours *seed*

TGCE: >>> temperature gradient capillary electrophoresis

TGS: >>> transcriptional gene silencing

TGW: >>> thousand-grain weight

THC: >>> hemp

theca: Usually referring to the pollen sac in flowering plants or the capsule in bryophytes *bot* >>> Figure 35

theobromine: A mutagenically active purine analogue *chem phys* >>> TAL effector nuclease

thermoperiodism: In some plants (e.g., *Chrysanthemum*, tomato), the floral induction is accomplished by repeated exposure to low night temperatures, separated by periods of higher temperatures *bot*

thermophil(ic) (plants): Plants preferring moderate temperatures and/or those that cannot cope with low temperatures or frosts *phys*

thermotaxis: >>> taxis

thiamine: Contributes to the formation of the important coenzyme thiamine pyrophosphate, which is involved in the oxidative decarboxylation of alpha-keto acids and transketolase reactions *phys*

thin: In a cultivated crop, to remove some plants in order to increase the area available to others; recently, Blue River Technology (Mountain View, CA, U.S.) has developed a system for thinning lettuce plants for commercial production; one currently can thin to about 25 cm spacing to improve uniformity of lettuce head size; the system uses image analysis to identify off-size plants and plants spaced too close; seedlings are thinned by spraying concentrated fertilizer that causes a lethal burn in seedlings; four rows can be can processed at a time at a speed of 1.3 m per second; the system can be applied in many breeding situations; uniform spacing improves >>> phenotyping whether done on individual plants, head rows, or plots; on the other hand, over-sowing may not be cost effective *agr*; in forestry, removal of trees in an overstocked stand to give the remaining trees adequate room for growth *fore* >>> thinning

thinning: >>> thin

thiol: Any of a class of odiferous, sulfur-containing compounds *chem phys*

thorn: A hard, sharp outgrowth on a plant (e.g., a sharp-pointed aborted branch) *bot*

thousand-grain weight (TGW): Equals 1,000-grain weight given in grams; it refers to a measure for seed weight, and thus indirectly for seed size; from a seed lot, 1,000 seeds are randomly taken and the weight is determined in grams *meth*

three-parent cross: >>> three-way cross

three-point cross: A series of crosses designed to determine the order of three, non-allelic, linked genes upon a single chromosome on the basis of their crossing-over behavior *gene* >>> three-point test cross

three-point test cross: Cross involving one parent with three heterozygous gene pairs and another (tester) with three homozygous recessive gene pairs *gene meth* >>> three-point cross >>> Figure 51

three-way cross: A first-generation hybrid between a single-cross and an inbred line or pure line variety *meth* >>> Figure 31 >>> Table 37

three-way hybrid: A hybrid between an inbred line and single cross hybrid *seed meth* >>> Figure 31 >>> Table 37

threonine (Thr): An aliphatic, polar alpha-amino acid ($HO_2CCH(NH_2)CH(OH)CH_3$); its codons are ACU, ACA, ACC, and ACG; together with serine and tyrosine, threonine is one of three proteinogenic amino acids bearing an alcohol group; the threonine residue is susceptible to numerous posttranslational modifications *chem phys* >>> Table 38

threshing: Breaking the seeds free from the seedpods and other fibrous material or the separation of seed from chaff *agr*

threshing machine: A device that breaks the seeds free from the seedpods and other fibrous material or the separation of seed from chaff *agr*

threshold effect: A term usually applied to traits with a polygenic basis that develop when the dosage of contributory alleles exceeds a critical value in certain environments; sometimes used to explain all-or-none phenomena based on polygenically inherited characters, such as resistance *versus* susceptibility to some diseases *stat*

threshold value: A critical value on an underlying scale of liability above which individuals manifest a trait or disease *stat*

thrips: Any of several minute insects of the order Thysanoptera that have long, narrow wings fringed with hairs and that infest and feed on a wide variety of weeds and crop plants; for example, *Thrips tabaci* is a major problem in the cultivation of cabbage for storage, as this pest causes symptoms that necessitate the removal of affected leaves from the product *zoo phyt*

throat: The center of a daylily where the pistil and stamens join, often contrasting in color to the base blossom color; breeding is done for green throats that remain sunfast or non-fading throughout the day, as green is desired in order to improve overall aesthetic appeal of a cultivar; older varieties of daylilies, many now regarded as obsolete, have gold, yellow, or melon-colored throats *hort*

throw-back: >>> atavism

thrum morph: A flower morph with long stamens and short >>> pistils *versa* the pin morph *bot*

thylakoid: One of the membranaceous discs or sacs that form the principal subunit of a granum in chloroplasts *bot*

thymidine: A nucleoside with thymine as its base; it acts as an essential growth factor for microorganisms *chem phys*

thymine (T): The pyrimidine base that occurs in DNA ($C_5H_6N_2O_2$); it is also known as 5-methyluracil, a pyrimidine nucleobase; in RNA, thymine is replaced with uracil in most cases; in DNA, it binds to adenine (A) via two hydrogen bonds to assist in stabilizing the nucleic acid structures *chem gene*

TI: >>> transcript imaging

tiered: A breeding population may be tiered in different compartments, which cannot be considered as sublines, but rather elite and mainline *meth*

tigella: A short stem *bot*

tiling path: The ordered arrangement of >>> BAC clones using sequence overlaps of neighboring clones so that they completely cover the corresponding region of the chromosome *biot*

tillage: The preparation of soil for seeding; it includes manuring, plowing, harrowing, and rolling land, or whatever is done to bring it to a proper state *agr*

tillers: Shoots, some of which will eventually bear spikes, which arise from the base of the stem in the grasses *bot* >>> Figure 28

tillering: >>> tillers

tillering node: A node on the base of the stem in grasses from which shoots arise *bot*

TILLING: >>> targeting induced local lesions in genomes

tilth: The condition of the soil after preparation for seeding *agr*

timberline U.S.: >>> tree limit

time components: The time taken by basic operations (e.g., recombination, time before the test, testing time, and time after the test); breaking of cycling time into components can be used in optimization of breeding strategies (e.g., choice of the most efficient testing or selection method) *meth*

time isolation: An isolation practice to prevent crossing; it is used when distance isolation is a problem, caging is too costly or troublesome, or when only growing a couple of varieties in a season; it works with any two varieties or species that shed pollen over a limited time and have sufficiently different rates of maturation; to use time isolation, planting dates of two similar varieties are staggered so that by the time the later of the two varieties is flowering, the earlier variety has already finished flowering and is no longer producing or receptive to pollen; the earlier, faster-maturing crop is planted two or three weeks before the later, slower-maturing one *meth*

Ti (tumor-inducing) plasmid: Established tumors in plants contain only a part of the total plasmid (ca 10%); this part is called T-DNA; this part of the pathogen is integrated into the chromosomes of the host cell *gene*

tissue: A group of cells with similar origin and structurally organized into a functional unit; the organs of multicellular organisms are made up of combinations of tissues, and of one or more types of cells *bot*

tissue culture: The maintenance or growth of tissue *in vitro* in a way that may allow further differentiation, preservation, or regeneration; it can be applied to (1) clonal propagation or rapid and large-scale multiplication of genetically identical plants from a single superior stock plant, (2) establishment of disease-free *in vitro* stock plants in culture, (3) >>> germplasm storage and >>> long-term storage of stock plants, (4) selection of mutants from spontaneous or induced mutations, (5) production of rooted microcuttings in recalcitrant woody ornamental species, (6) recovery of hybrids from incompatible species through either embryo or ovule culture, (7) production of >>> haploids through >>> anther culture *biot* >>> Figure 57

titer: The amount of a standard reagent necessary to produce a certain result in a titration *chem*

tk: Abbreviation for thymidine kinase *biot*

tocopherol (vitamin E): A fat-soluble vitamin that is an important antioxidant; it is often used in skin creams; it is claimed to play a role in encouraging skin healing and reducing scarring after injuries, such as burns; vitamin E exists in eight different forms or isomers, four tocopherols and four tocotrienols; all isomers have a chromanol ring, with a hydroxyl group, which can donate a hydrogen atom to reduce free radicals and a hydrophobic side chain which allows for penetration into biological membranes; there is an alpha, beta, gamma, and delta form of both the tocopherols and tocotrienols, determined by the number of methyl groups on the chromanol ring; each form has its own biological activity; wildtype sunflower (*Helianthus annuus*) seeds are a rich source of alpha-tocopherol; an advanced breeding line IAST-413, developed by pedigree selection, shows enhanced alpha-tocopherol content, as well *phys*

tolerance: The ability of a plant to endure attack by a pathogen without severe loss of yield *phyt*

tolerance index: Refers to the characteristic production (grain yield, biomass, root or shoot length, etc.) of a genotype in a given stress environment (e.g., acid soil) relative to a nonstress environment (e.g., improved or limed acid soil); it is used in order to identify the tolerance level of tested plant entries; the simple formula is given as follows Ti_{GY} (e.g., grain yield) = $ALRL_{-L} / ALRL_{+L}$, where, Ti = tolerance index for a certain genotype, $ALRL_{-L}$ =

calculated mean longest root length of a genotype in unlimed (–L) acid soil (production in stress environment), $ALRL_{+L}$ = calculated mean longest root length of a genotype in limed (+L) acid soil (production in nonstress environment) *stat*

tolerant: The ability of a host plant to develop and reproduce fairly efficiently while sustaining disease *phyt*

tollocan: A Mexican >>> short-day >>> potato, which has an exceptionally high level of >>> horizontal resistance to >>> blight *agr phyt*

tomatinase: An extracellular enzyme, which hydrolyzes glucose from a-tomatine to form b2-tomatine; b2-tomatine has significantly less antifungal activity than a-tomatine; it is produced by the tomato leaf spot fungus *Septoria lycoperscii phys* >>> a-tomatine >>> Figure 62

tomato hornworm: The larva of the hawkmoth, *Manduca quinquemaculata*, which feeds on plants of the nightshade family, especially tomatoes *phyt zoo*

tomentose: Vesticure of dense, short, soft, matted hairs *bot*

tonoplast: A membrane that borders the vacuole of a cell *bot*

topcross: A cross between a selection, line, clone, etc., and a common pollen parent, which may be a variety, inbred line, single cross, etc.; the common pollen parent is called the "topcross" or "tester" parent; in maize, a topcross is commonly called an "inbred-variety cross"; usually, it is used in order to test the "general combining ability" *meth* >>> Figures 19, 50, 51 >>> Table 37

topcross hybrid: The first generation of a cross between an inbred line and an open pollinated variety *seed meth* >>> Figure 51 >>> Table 37

topcross progeny: Progeny from outcrossed seed of selections, clones, or lines to a common pollen parent *meth* >>> topcross >>> Figures 19, 50, 51 >>> Table 37

topcross test: >>> topcross

top-dressing: Fertilization when crop plants are already developed, usually before flowering (i.e., an additional fertilization to the basic dressing before sowing or close behind); material, such as compost or manure, that is applied to the surface around the plant to aid in drainage, decrease erosion, prevent moisture loss, and keep weeds down *agr hort*

topiary: Training, cutting, and trimming of plants into ornamental shapes *hort*

top-kill: To kill aboveground plant tissues without killing underground parts from which the plant can produce new stems and leaves *hort fore meth*

topoisomerase: Enzyme that catalyzes the reduction of supercoiling of DNA; it is important during DNA replication to relieve the negative supercoiling that occurs when the two strands of the DNA double helix are separated *biot*

topsoil: The fertile, upper part of the soil *agr* >>> Table 44

top-watering: A method whereby water is simply poured into the soil from the top; when employing a top-watering method with some ornamental plants, it is crucial to avoid overwatering and to minimize the amount of water which might splash on the leaves; at the very least, improper watering can cause >>> watering spots; in the worst cases, improper watering can encourage the growth of deadly fungi, such as those which cause crown rot, root rot, and *Pythium hort meth* >>> Figure 62

torpedo stage: During this stage, the embryo elongates into the cellular endosperm and the internal layers of the hypocotyl and radicle differentiate to form the vascular tissue; lipid deposition into the cotyledons begins at this stage; organelle differentiation occurs, leading to greening of the embryo *phys*

torus: The receptacle of a flower *bot*

total mean square: >>> variance

totipotency: The potential ability of a cell to express all its genetic information under appropriate conditions and to proceed through all the stages of development to produce a fully differentiated adult; A. HANSEN first postulated in 1879 that plants are totipotent, referring to the power of regeneration of Begonia that can develop new plantules of leaf cuttings, a feature still used by gardeners today; HANSEN's statement was generalized by G. HABERLAND *bio gene hort* >>> dedifferentiation >>> competency

totipotent(ial): >>> totipotency

tough rachis: A non-brittle rachis of a spike; for example, in wheat, the non-brittleness is determined by two recessive genes on the short arms of chromosomes 3A and 3B; the consequence of threshing is spikelets, rather than grains *gene bot agr*

trans-domestication: When a wild plant species brought from an exotic region is domesticated elsewhere in a foreign region *bot*

two-stage analysis: Series of plant breeding trials are often unbalanced and have a complex genetic structure; to reduce computing cost, it is common practice to employ a two-stage approach, where adjusted means per location are estimated and then a mixed model analysis of these adjusted means is performed; an important topic is how means from the first step should be weighted in the second step; different weighting methods in the analysis of trials using mixed models with fixed or random genetic effects require different approaches; when genetic effects are taken as random in stage two, in three of four data sets the two-stage analysis gives acceptable results; in both cases differences between weighting methods were small, and the best weighting method depended on the data set but not on the evaluation criteria; a two-stage analysis without weighting also produces acceptable results, but weighting mostly performs better; in the fourth data set the missing data pattern is informative, resulting in violation of the missing-at-random (MAR) assumption in one- and two-stage analysis; in this case both analyses are not strictly valid *stat meth*

toxicity: The quality, relative degree, or specific degree of being toxic or poisonous *meth prep* >>> Table 46

toxigenic: >>> toxicity

trabant: >>> satellite

traceability: Measures covering feed, food, and their ingredients; includes the obligation for feed and food businesses to ensure that adequate procedures are in place to withdraw feed and food from the market where a risk to the health of the consumer is posed; operators have to keep adequate records of suppliers of raw materials and ingredients so that the source of the problem can be identified *agr*

trace element: An element required only in minute amounts by an organism for its normal growth *phys* >>> micronutrient

trachea: Wood vessel *bot*

tracking dye: Dye molecule visible in room light that is mixed with a laboratory sample; this allows tracking of the loading of the sample and indicates its movement through an electrophoresis gel *prep*

trade: A pair of collections of blocks of a >>> block design, or rows of an array, such that some property of the design or array is preserved if the first set of blocks is removed and replaced by the second *stat*

trade-related intellectual property right agreement: Treaty administered by the World Trade Organization that implements minimum standards for many forms of intellectual property regulation; it contains requirements that laws of nations must meet for copyright laws, patents, and monopolies for the developers of new plant varieties; the treaty specifies enforcements procedures, remedies, and resolutions procedures; thus it is the most comprehensive international agreement on intellectual property *org* >>> Intellectual Property Rights

traditional breeding: Modification of plants through selective breeding; practices used in traditional plant breeding may include aspects of biotechnology such as tissue culture and mutation breeding *meth*

training: The operation of forming (young) tree plants to a wall or espalier, or causing them to grow in a desired shape *hort meth*

trait: A recognizable quality or attribute resulting from the interaction of a gene or group of genes with the environment *meth gene* >>> character

tramlines: Unseeded, equispaced tracks established in a field at seeding time to provide a pass in the field for tractors to use to aid in the application of chemicals and fertilizer *agr meth*

trans-acting gene: A gene acting on or cooperating with another gene on a different chromosome *gene biot*

transaminase (aminotransferase): An enzyme that catalyzes a transamination reaction *chem phys*

transamination: The transfer of an amino group from an amino acid to a keto acid in a reaction catalyzed by a transaminase *chem*

transcapsidation: >>> heterologous encapsidation

transcript: The RNA product of a gene *gene*

transcript-based cloning (TBC): An innovation of gene identification and isolation based on DNA microarrays and the instability of RNA transcripts from genes damaged by mutations; the technique works in both model systems, with relatively simple genomes, and in crop species (e.g., barley) that have a much larger and more complex genome; favorite genes can be isolated in one-tenth of the time, and for one-tenth of the cost of existing methods; TBC relies on taking "snapshots" of gene activity in normal plants and plants with genetic mutations; by comparing "snapshots", it is possible to identify and isolate the gene affected by the mutation *biot*

transcript imaging (TI): A technique based on >>> AFLP analysis applied to >>> cDNA; it provides a quantified view of all the transcripts in a sample on an electronic image where approximately 20,000 AFLP bands, each representing a transcript, can be detected and quantitated; differences between strains, samples, or treatment can be detected, and the differential bands can be sequenced; the sequence of the band leads to the identification of the transcript *meth biot*

transcription: The polymerization of ribonucleotides into a strand of RNA in a sequence complementary to that of a single strand of DNA; by this means, the genetic information contained in the latter is faithfully matched in the former; the process is mediated by a DNA-dependent RNA polymerase *gene*

transcriptional gene silencing (TGS): Gene expression is reduced by a blockade at the transcriptional level; transcriptional repression is caused by chromatin modification or DNA methylation; this was generally observed in plants but has also been seen in animals; generally in plants, transcriptional inactivation requires homology between promoters *biot meth* >>> target validation

transcription factors: Proteins that are directly involved in regulation of transcription initiation by binding to the control elements and allowing RNA polymerase to act; there are ubiquitous transcription factors as well as cell- and tissue-specific ones; several families have been identified including helix-loop-helix proteins, helix-turn-helix proteins, leucine zipper proteins, and zinc finger proteins *gene biot* >>> Table 38

transcription start site: The position in a gene where the mRNA synthesis starts; the segment from this point downstream to the translation initiation site is called the 5′ untranslated region (5′ UTR) *phys gene biot*

transcriptome: DNA sequences of the expressed genome and/or the set of all mRNAs produced by an organism, plant, tissue, or cell type; the term can be applied to the total set of transcripts or to a specific subset of transcripts *gene biot*

transcriptomics: The study of global gene expression and changes caused by genetics, environmental effects, drugs or chemicals, or tissue location; it is the application of micro- or macroarrays and sequence-based methods to conduct expression profiling to determine gene expression at a global (genome-wide) level *biot meth*

transcriptome mapping: The procedure to establish a genetic map with cDNA fragments that differ in size in the respective parents and segregate in the progeny; since the cDNA fragments are displayed on Northern plots, fragments can be isolated and sequenced; such sequences can be annotated so that the underlying gene can be identified *biot meth*

transduce: >>> transduction

transduction: The transfer of bacterial genetic material from one bacterium to another via phage *gene*

transfectant: A genetic transformation of a cell by free (naked) DNA *gene*

transfection: Introduction of pure (naked) phage DNA into a cell *biot*

transferase: An enzyme that catalyzes the transfer of a functional group from one substance to another *chem phys*

transfer RNA (tRNA): A generic term for a group of small RNA molecules, each composed of 70–80 nucleotides arranged in a clover-leaf pattern stabilized by hydrogen bonding; they are responsible for binding amino acids and transferring these to the ribosomes during the synthesis of a polypeptide *gene*

transformant: The cell or individual that was transformed during a transformation procedure *biot*

transformation: In general, the heritable modification of the properties of a plant; in biotechnology, the transfer of genetic information to a recipient strain of bacteria by DNA extracted from a donor strain, and recombination of that DNA with the DNA of the recipient *biot* >>> chloroplast

transformation efficiency: The number of bacterial and/or plant cells that uptake and express plasmid DNA and/or a foreign gene, respectively, divided by the mass of plasmid used *biot*

transgene: A gene introduced into a host genome by transfection or other similar means *biot*

transgene stacking: Multiple gene transfer to plants is important for sophisticated genetic manipulations by, e.g., the stacking of transgenes specifying different agronomic traits, the expression of different polypeptide subunits making up a multimeric protein, the introduction of several enzymes acting sequentially in a metapolic pathway, or the expression of a target protein *biot meth*

transgenesis: The act of creating organisms containing foreign genes through DNA transfer by transformation *biot* >>> cisgenesis

transgenic: >>> transgenic plants

transgenic genotype: >>> transgenic plants

transgenic plants: A plant that contains an alien or modified DNA (gene) introduced by biotechnological means, and which is more or less stably inherited *biot*

transgression: Segregants in a segregating population that fall outside the variation limit of parental lines *gene*

transgressive: >>> transgenic plants

transgressive segregation: The segregation of individuals in the F2 or a later generation of a cross that shows a more extreme development of a character than either parent *gene*

transient expression: The temporary expression of a gene or genes shortly after the transformation of a host cell [i.e., cells in which the transgene has not been physically incorporated into the genome (stable transformation), but is carried as an episome that can be lost] *biot*

transition: A type of mutation that involves the replacement in DNA or RNA of one purine with another or of one pyrimidine with another *gene*

transition rate: The rate transitions between character states in a lineage, e.g., the rate at which species switch from self-incompatibility to self-compatibility *gene*

transition stage: Occurs when the ovoid, radially symmetrical proembryo elongates to form a cone-shaped structure, establishing the future shoot–root meristem axis, about 10–12 days after pollination *phys meth* >>> Table 13

translation: The polymerization of amino acids into a polypeptide chain whose structure is determined genetically *gene*

translocation: A change in the arrangement of genetic material, altering the location of a chromosome segment; the most common forms of translocation are reciprocal, involving the exchange of chromosome segments between two non-homologous chromosomes *cyto*

translocation tester set: A series of more or less defined homozygously reciprocal translocations; they may be utilized in test crosses to identify accessory or unknown chromosomes by homologous pairing in MI of meiosis (e.g., this sort of tester set was produced in diploid rye and barley) *meth cyto*

transmission: The spread of a disease agent among individual hosts *phyt*

transmission genetics: The study of the mechanisms involved in the passage of genes from one generation to the next *gene meth*

transmitting tract: The central tissue of >>> stylus through which pollen tubes grow on their pathway to the ovary *bot*

transpiration: The loss of water vapor from a plant to the outside atmosphere; it takes place mainly through the stomata of leaves and the lenticels of stems *phys*

transplant: To relocate or remove to a new growing place meth; in biotechnology, the cultured tissue or explant relocated or transferred to a new site *in vitro biot*

transplanting board: A simple device having regularly spaced slots for the individual plants in order to ensure proper spacing and lining out in the new bed *fore hort*

transplastomic plants: Plants with genetic modifications in the DNA of their chloroplasts; plant cells contain chloroplasts, cell organelles responsible for photosynthesis; these chloroplasts contain their own DNA which is separated from the cell's nuclear DNA; in transplastomic plants, the DNA in the chloroplasts has been genetically engineered; since only the nuclear DNA is inherited, the genetic modification in the chloroplasts will not be passed on to the next generation; therefore, transplastomic plants may be a solution for ensuring biocontainment *biot meth*

transposable element: A chromosomal locus that may be transposed from one spot to another within and among the chromosomes of the genome; it happens through breakage on either side of these loci and their subsequent insertion into a new position either on the same or a different chromosome *gene* >>> Figure 46

transposition: In molecular biology, the process of moving a transposon or other inserts from one position to another within a genome *cyto* >>> translocation >>> Figure 46

transposon: Chromosomal loci capable of being transposed from one spot to another within and among the chromosomes of a complement *gene* >>> transposable element >>> transposition >>> Figure 46

transposon-arrayed gene knock out (TAG-KO): A high-throughput system for gene function identification *biot*

transposon tagging: The blocking activity of functional genes by insertion of foreign DNA *biot* >>> Figure 46

transverison: A mutation in which a purine is replaced with a pyrimidine or vice versa biot

transversal design: A >>> group-divisible design in which every block is a transversal to the groups *stat*

TRAP marker: >>> target region amplification polymorphism

trap nursery: Sets of plant genotypes are assembled that carry specific resistance to the pathogen in question; these sets are grown in different geographic locations; they provide information on pathogen populations and may also provide resistant genotypes for local breeding *phyt*

trap plants: >>> trap nursery

treatment: A unique experimental practice or effect in the experiment, e.g., in a trial comparing seven crop varieties, each variety strip is a treatment; in a trial comparing four levels of herbicide treatments, each of the four levels is a treatment; in a demonstration strip design, each treatment is included only once; a control treatment is usually either a common practice or no practice, e.g., in a crop variety trial the usual variety grown in the area might be used as the control; in a herbicide-testing trial, the control treatment might be a strip with no herbicide applied *agr meth stat* >>> demonstration strip design >>> replicated control design

tree: A woody plant, which may grow >10 m tall; trees developed during the Devonian ~385 mya; they were like palm or tree ferns in habit *bot*

tree injector: Certain types of equipment specially designed to inject chemicals, usually phytocides, into the trunk of a tree *fore phyt*

tree limit: The altitude above sea level at which timber ceases to grow *eco*

tree line: >>> tree limit

tree ring: >>> annual ring

tree shelter: A plastic tube or other tissue that can be wrapped around the stem of hardwood seedlings to increase survival and growth *fore meth*

trench planting: Setting out young trees in a shallow trench or a continuous slit *meth fore hort*

triad: A group of three, as applied to spikelets of Chrysopogon or Hordeum; consists of a sessile spikelet and two pedicellate spikelets; in Chrysopogon this represents a reduced racemose branch *bot*

triangle stage: Transient stage between the late globular and heart stage when the first cell divisions of the cotyledon primordia begin and cell elongation starts in the procambium *phys*

triazine: Any of a group of three compounds containing three nitrogen and three carbon atoms arranged in a six-membered ring and having the formula $C_3H_3N_3$; some of these compounds are used as herbicides *chem phys phyt*

tribe: A rank between family and genus, comprising genera whose shared features serve to distinguish them from other genera within the family *tax* >>> Table 12

trichasium: A >>> cyme with three branches *bot*

trichome: A hairy outgrowth on a plant's surface, as a prickle; the capitate glandular trichomes of wild sunflowers (*Helianthus* spp.) are considered to be effective defense components that act against some herbivorous insects, but cultivated sunflowers are reportedly deficient in glandular trichomes; evaluation revealed that capitate glandular trichomes are often abundant in cultivated sunflowers; relative to wild *H. annuus*, inbred maintainer lines have similar numbers of glandular trichomes per floret, while commercial hybrids have only ~20% fewer trichomes when compared with wild sunflowers; in laboratory assay, it was found that glandular trichome extracts increased the mortality rates of sunflower moth, *Homoeosoma electellum*, larvae exposed from the neonatal stage to 9 days *bot*

trier: A hand-manipulated probe for sampling seeds *seed*

trieur: A device used for separating round seeds and broken pieces; it also enables the removal of small impurities such as crushed pieces, herb seeds in cereal seed lots; it is successfully used in grain-cleaning sections of flourmills; there simple trieurs, double trieurs, or cylindrical trieurs *seed meth*

trifluralin: Used as a polyploidizing agent in order to induce tetraploids in roses *meth* >>> mutagen

trigeneric hybrid: A spontaneous or experimental hybrid consisting of three genomes of different genera *bot cyto gene*

trihybrid: Progeny resulting from a cross of parents differing in three genes *gene*

triisosomic: In allopolyploids, such as hexaploid wheat, when a cell or individual lacks one chromosome pair while three homologous isosomes for the same arm are present *cyto* >>> Figure 37

Trinitario (cocoa tree): Believed to be a natural cross from strains of the other two types (*Criollo cocoa* and *Forastero cocoa*); Trinitario has a great variety of characteristics but generally possesses good, aromatic flavor; Trinitario trees are particularly suitable for cultivation >>> cocoa tree

trioecious: Species having male, female, and hermaphroditic flowers on different individuals *bot*

triple cropping: Production of three different crops in a given area in the same year *agr hort*

triple superphosphate: A fertilizer obtained by treating phosphate rock with phosphoric acid and containing ~46% P_2O_5, mainly in water-soluble form; unlike single superphosphate, it contains very little sulfur *chem agr*

triplet: A unit of three successive bases in DNA and RNA, which code for a specific amino acid *gene*

triplex (type): >>> autotetraploid >>> nulliplex type

triploid: Applied to a cell or individual with three sets of chromosomes in its nucleus *cyto*

triploid fusion nucleus: The result of the fusion of one male gamete (pollen) with two haploid female nuclei (polar cells) in the second polar event; this nucleus divides to form the cells of the endosperm *bot* >>> second polar event

triploidy: A state in which three chromosome sets are present *cyto* >>> sugarbeet

tripping mechanism: A pollen dispersal mechanism of some legumes, in which the staminal column is sprung free of the keel and exposed; it can also be initiated by hand in order to imitate insect activity and to stimulate self-pollination, for example, in >>> broad bean (*Vicia faba*) *bot*

triradial (chromosome configuration): A chromosomal pairing configuration of three in which the homologous chromosomes are arranged like a star *cyto* >>> Figure 15

trisome: >>> aneuploid >>> trisomic

trisomic: A genome that is diploid but that contains an extra chromosome, homologous with one of the existing pairs, so that one kind of chromosome is present in triplicate *cyto* >>> aneuploid >>> Figures 14, 15, 37

trisomic analysis: A method for mapping gene loci on individual chromosomes by comparing disomic and trisomic segregation patterns of a series of individuals *gene* >>> aneuploid >>> Figure 37 >>> Table 4

trisomic series: A complete set of trisomics of a given plant species, in which all different chromosomes of the complement are available as trisomics in appropriate individuals (e.g., in barley, rye, tomato, maize, etc.) *cyto* >>> aneuploid >>> Figures 14, 15, 37.

tristeza (in Spanish = sadness): A virus disease of citrus; the diagnostic symptom is a flattening of the branches and, when the bark is peeled off, there are pits in the wood, with corresponding projections in the bark; diseased trees usually die, following severe dieback; it is a graft incompatibility disease, and it is serious mainly on trees grafted onto sour orange rootstocks; resistant scion–stock combinations have rootstocks of sweet orange, rough lemon, and "Cleopatra mandarin"; the disease is also known as "stem pitting" *hort phyt*

tristyly: Three lengths of the style relative to the anthers (short, medium, long) *bot*

trivalent: An association of three homologous chromosomes in meiosis *cyto* >>> Figures 14, 15

tRNA: >>> transfer RNA

Trojan squares: Semi-Latin square designs based on sets of mutually orthogonal superimposed Latin squares, which have been shown to be maximally efficient >>> semi-Latin square designs for pairwise treatment comparisons in the plots-within-blocks stratum; Trojan squares of size (n × n)/k are based on k sets of n × n orthogonal >>> Latin squares and have a natural factorial treatment structure; the k squares and n letters of the design represent either pseudo-factors for an unstructured treatment set, or real factors for a factorial treatment set; Trojan squares have generalized balance and have a particularly simple analysis of variance *stat* >>> design of experiment >>> semi-Latin squares

trophic: >>> tropism

tropical grasses: The most important cultivated tropical grasses are species of the genera Digitaria, Eragrostis, Chloris, Cenchrus, Melinis, Panicum, Pennisetum, Cynodon, or Paspalum *agr bot* >>> temperate grasses

tropism: A directional response by a plant to a stimulus *bot*

truck crop U.S.: Vegetables and fruits grown in large quantities for the market *hort*

true breeding: A situation in which a group of identical individuals always produce offspring of the same phenotype when intercrossed; a true-breeding plant is that which, when self-fertilized, only produces offspring with the same traits; the alleles for this type of plants are homozygous *meth gene*

true-rooted: Non-grafted plants *hort*

truncated: Appearing as though abruptly cut across toward the apex *bot*

truncation selection: A breeding method in which individuals in whom quantitative expression of a phenotype is above or below a certain desired value (i.e., truncation point) are selected as parents for the next generation *meth*

trunk: >>> stem

trunkless: >>> boleless

truss: Compact cluster of flowers or fruit, particularly of rhododendrons *bot*

truthfully labeled seed: Seed labeled by the producer with information on the seed quality *seed*

tryphine: >>> entomophilous

trypsin: An enzyme of the pancreatic juice, capable of converting proteins into peptone *chem phys*

tryptophan(e) (Trp): A heterocyclic, non-polar, alpha-amino acid *chem phys* >>> Table 38

tube nucleus: >>> vegetative nucleus

tube planting: Setting out young plants in narrow, open-ended cylinders of various materials *meth hort fore*

tuber: A swollen stem or root that functions as an underground storage organ *bot*

tuber blight: >>> late blight

tubercle(s): Small tuber(s) produced in leaf axils of leaf bud cuttings, e.g., in potato *bot*

tuberization: *In vitro* induction of (micro)tuber formation of, e.g., potato, Jerusalem artichoke, or other tuberizing species, mostly from neoformed roots derived from single-node stem segments from *in vitro* plantlets and/or callus cultures; it occurs in the presence of kinetin and auxin *biot* >>> microtuber

tuberous: >>> bulbous

tuberous roots: Resembling tubers, but actually swollen, nutrient-storing root tissue (e.g., in dahlias); during the growing season, they put out fibrous roots to take up moisture and nutrients; new growth buds, or eyes, form at the base of the stem; this area is called the crown *bot*

tubiform floret: A small flower in a flower head or other cluster showing a tube-like shape *bot*

tubular floret: >>> tubiform floret

tuft: Many stems in a close cluster at ground level; not spreading (e.g., as in some grasses) *bot*

tumbleweeds: Annual plants that have a rounded shape; they dry out in the fall, and the stem breaks off near the ground; the ball-shaped, withered plant tumbles around in the wind (hence the name), scattering its seeds; there are many different types of tumbleweed; they live in the prairie and plains of the United States; one common type of tumbleweed is Russian thistle (*Salsola kali*) *phyt*

tunic: A loose, outer covering or skin surrounding some corms and >>> bulbs (e.g., in onion and tulip) *bot*

turf: The surface of grassy land, consisting of soil or mold filled with the roots of grass and other plants *agr*

turgescent: >>> turgid

turgid: The crisp, fresh condition found when the cells of the plant are amply supplied with water to the extent that they are fully extended, as opposed to wilted *bot*

turgor: The rigidity of a plant and its cells and organs resulting from hydrostatic pressure exerted on the cell walls *phys*

turgor pressure: >>> turgor

turion: An underground bud or shoot from which an aerial stem arises *bot* >>> sucker

tussock: A clump or tuft, especially of a >>> graminoid *bot*

TWEEN® 20 (*syn*** polyethylene glycol sorbitan monolaurate, polyoxyethylenesorbitan monolaurate, polysorbate 20):** A non-ionic detergent and emulsifier, useful for protein extraction, isolation in cell lysis mixtures, and micropropagation *meth biot*

twig: A shoot or small branch of a tree or shrub *bot*

twin: A pair of individuals produced at one birth *gene*

twining: A plant coiling around objects as a measure of support *phys*

twin scaling: A method of propagating bulbous plants *hort*

twin seedling: A common feature of plants; frequently, one of the seedlings is diploid, while the other is haploid via apomictic development (e.g., in asparagus, rye, etc.); in the past, twin seedlings were used for haploid selection *bot meth* >>> haploid

twin spot: A pair of mutant sectors within wild-type tissue produced by a mitotic crossover in an individual of appropriate heterozygous genotype *gene*

two-celled stage: The embryo after the first mitotic division that produces a smaller apical cell and larger basal cell; the basal cell gives rise to the radicle and the suspensor; the apical cell gives rise to all of the embryo except the root *phys*

two-point cross: A cross involving two loci *gene*

two-rowed: >>> barley

two-way analysis of variance: A way of studying the effects of two factors separately (their main effects) and (sometimes) together (their interaction effect) *stat* >>> one-way analysis of variance

tyrosinase: An enzyme that converts >>> tyrosine to dopa and oxidizes this to dopa quinone *chem phys* >>> Table 38

tyrosine (Tyr): An aromatic, polar alpha-amino acid *chem phys* >>> Table 38

U

UAG stop codon: >>> amber codon

ubiquitin: A small protein that becomes covalently linked to a protein targeted for degradation *phys chem*

UGMS: >>> unigene-derived microsatellite markers

ultracentrifugation: Centrifugation carried out at high rotor speeds (<100,000 rpm) and therefore under high centrifugal forces (<750,000 g) *meth*

ultracentrifuge: A setup for high-speed centrifugation between 65,000 and 100,000 rounds per minute *prep* >>> ultracentrifugation

umbel: An inflorescence in which all the pedicels arise at the apex of an axis *bot*

umbellate: >>> umbelliferous plants

umbelliferous plants: Tap-rooted plants with minute flowers aggregated into flat or umbrella-shaped heads (e.g., carrot, parsnip, celery, dill, parsley, etc.) *bot*

unavailable nutrients: Plant nutrients that are present in the soil but cannot be taken up by the plant roots because they have not been released, either from the rock by weathering, or from organic matter *agr*

unavailable water: Water that is present in the soil but cannot be taken up by the roots because it is strongly adsorbed on to the surface of particles *agr*

unbalanced diallelic: A genotype involving a multiple allelic locus in autotetraploids where two alleles are represented an unequal number of times *gene* >>> design of experiments

unbalanced experimental design: An experiment or set of data in which all treatments or treatment combinations are not equally represented; a common cause of unbalanced experiments is unequal mortality among entries in a test *meth* >>> randomized-block design >>> design of experiments

unbalanced translocation: A type of chromosome translocation in which a loss of chromosomal segments results in a deleterious genetic effect *cyto*

uncomplete block design: >>> incomplete block design

unconfined release: A release into an environment of a plant with novel trait(s) that is not isolated either reproductively or physically from managed or natural environments, but may be subject to certain conditions *biot meth eco*

unconscious selection: Indirect selection by breeders (sometimes called "parallel selection") when, in addition to a target characteristic, another trait is taken to the next generation; it can be a morphological, biochemical, or physiological trait genetically linked to the target characteristic *meth*

underdominance: A condition in which the phenotypic expression of the heterozygote is less than that of either homozygote *gene* >>> dominance

underplant crop: In horticulture, adding one or more complementary, low-growing plants beneath and around taller plants *hort*; in agriculture, main crops can be underplanted with vegetables; it helps these plants to have some shade in the heat of the summer, and one can sustain a crop longer than otherwise; wide ranges of combinations between main and underplant crops are possible and are applied for several purposes *agr*

underreplication: Certain heterochromatic chromosome regions and ribosomal DNA that show a slower replication as compared to the remaining genomic DNA *gene cyto*

understock: The bottom or supporting part of a graft composed of either root or stem tissue, or both *hort* >>> rootstock >>> graft

understocked: A stand of trees so widely spaced that, even with full growth potential realized, crown closure will not occur; understocking indicates a waste of resources, as the site is not fully occupied *fore*

unequal crossing-over: A crossing-over after improper pairing between chromosome homologues that are not perfectly aligned; the result is, for example, one crossing-over chromatid with one copy of the segment and another with three copies *cyto* >>> Figure 24

uneven-aged stand: A group of trees of a variety of ages and sizes growing together on a uniform site *fore*

unicellular: An organism consisting of one cell, for example, yeast *bot bio*

unifactorial: >>> monogenic

uniflorous: Showing one flower only *bot*

uniform crossover: In genetic algorithms, a breeding technique in which it is randomly decided for each element of a breeding pair of individuals whether they should be switched *stat*

uniformity: Describes the state of a population or group in which all the individuals are genetically identical; it is a typical feature of clonal varieties (e.g., potato); in general, lack of diversity within and between plant species is apparent in modern cultivars *gene seed* >>> Table 57

uniformity trial: An experiment in which a single variety is grown under uniform cultural conditions in the whole experimental area in order to determine the pattern of variation for soil fertility and/or conditions *agr meth stat* >>> Table 57

uniform mutant: >>> chimera >>> Figure 56

unigenes: A non-redundant set of genes that is defined after clustering (computational) analysis of sequences generated through an >>> EST or a genome-sequencing project *biot meth*

unigene-derived microsatellite (UGMS) markers: A specific type of molecular markers having the advantage of assaying variation in the expressed component of the genome with unique identity and positions *biot*

unilateral: The type of panicle (e.g., in oats) where the branches are all turned to one side like a pennant *bot*

unilateral inheritance: Inheritance that is associated with linkage in sex chromosomes *gene*

unilateral sexual polyploidization (*syn* **half-meiotic polyploidization**): Formation of a polyploid plant where unreduced gametes are produced by one parent *bot* >>> bilateral sexual polyploidization

uninemic chromosome: A chromosome consisting of one double helix of DNA *cyto*

uninemy: Single-strandedness of DNA in the chromosome *cyto*

union: Point where the scion and rootstock are joined *hort*

Union for Protection of New Varieties of Plants [Union pour la Protection des Obtentions Vegetales (UPOV)]: An intergovernmental organization with headquarters in Geneva, Switzerland; it is based on the International Convention for the Protection of New Varieties of Plants, as revised since its signature in Paris on December 2, 1961; on April 16, 1993, the union consisted of 23 member states; the objective of the convention is the protection of new varieties of plants by an intellectual property right *org* >>> www.upov.int

uniparental disomy: Inheritance of both homologues of a chromosome from one parent, with loss of the corresponding homologue from the other parent *gene*

uniparental inheritance: A pattern of inheritance in which only one parent provides genes to the progeny; a phenomenon, usually exhibited by extranuclear genes, in which all progeny have the phenotype of only one parent *gene* >>> maternal inheritance >>> biparental inheritance

unipolar (spindle): A cell spindle with only one pole *cyto*

unique DNA sequence: A DNA sequence that is present only once per genome, with no repetitive nucleotide sequence *gene biot* >>> single-copy DNA

unique event polymorphisms (UEPs): Single-nucleotide polymorphisms (snips or SNPs) in which a particular nucleotide is changed; since they are very rare, they also are known as unique event polymorphisms (UEPs) *biot meth*

unique identifier: An international >>> GMO identification code; every genetically modified organism that receives authorization is given a unique identifier consisting of nine letters and/or numbers; the first two or three characters indicate the company submitting the application, while the final five or six characters specify the respective transformation event; the last digit serves as a verifier; the identification code, known internationally as the unique identifier, was established by the >>> OECD; the OECD is now overseen by the European Union and forms the basis for the legally mandated system of traceability *biot org*

unisexual (flower): A flower that possesses either stamens or carpels but not both (i.e., a plant possessing only male or female flowers) *bot*

United Nations Food and Agriculture Organization (FAO): Founded in 1945, the FAO's mission is to lead international efforts to defeat hunger; in the 1970s and 1980s, the FAO seemed to take a real interest in the concerns and needs of small farmers, and was the only international forum to seriously take on the issue of farmers' rights; but more recently, it has lost any credibility it had amongst farmers' groups around the world for its public backing of the agricultural industry as a force to overcome hunger; it has recently come under serious attack for coming out in favor of genetic engineering as a useful tool to combat hunger around the world *org* >>> www.fao.org

unit inheritance: Gregor MENDEL's idea that the characteristics of parents are passed on to descendants unchanged as units, i.e., the hereditary material of any organism is made up of discrete units (later called >>> genes) *gene*

U.S. Department of Agriculture (USDA): The agency responsible for regulation of biotechnology products in plants and animals in the U.S.; the major laws under which the agency has regulatory powers include the Federal Plant Pest Act (PPA), the Federal Seed Act, and the Plant Variety Act (PVA); in addition, the Science and Education (S and E) division has non-regulatory oversight of research activities that the agency funds *agr org* >>> www .usda.gov

univalent: A single chromosome observed during meiosis when bivalents are also present; it has no pairing mate *cyto* >>> Figure 15

univalent shift: A spontaneous change in monosomy from one chromosome to another; it is caused by partial >>> asynapsis or >>> desynapsis during >>> meiosis *cyto*

univalent shift: A meiotic irregularity of monosomics, i.e., a spontaneous change in monosomy from one chromosome to another, not that from the originally deficient parent; it is caused by partial >>> asynapsis or >>> desynapsis during >>> meiosis *cyto*

univalent switch: >>> univalent shift

Universal Cotton Standard(s): Refers to American Upland Cotton; established in 1924 as an aid to promoting domestic and foreign trade; recognized by 18 countries in Europe, South America, and Asia *adm*

universe: >>> population

unloader: A device on a combine to transport the grains from a bunker to a transporter *agr*

unorthodox seeds: >>> recalcitrant seed

unreduced gametes: Gametes not resulting from common meiosis, and so showing the number of chromosomes per cell that is characteristic of a sporophyte; they spontaneously arise as a consequence of irregular division in anaphase I of meiosis; they may contribute to spontaneous (meiotic) polyploidization; in rye, they were used for production of tetraploids via valence crosses; besides spontaneous 2n gametes, they can also be produced by nitrous oxide gas in tulips or by high temperatures (+32°C) in roses *cyto* >>> nitrous oxide gas >>> unilateral sexual polyploidization >>> bilateral sexual polyploidization

unsaturated fatty acid: A fatty acid that has a double bond between the carbon atoms at one or more places in the carbon chain; hydrogen can be added at the site of the double bond *chem phys*

unspecific resistance: >>> horizontal resistance

upgrading: The reprocessing of a seed lot to remove low-quality seeds or other materials; the remaining seeds are of higher quality than the original *seed meth*

upland cotton: Originally used to refer to cotton grown on raised lands not prone to flooding; now refers to short and medium staple cottons *agr* >>> barbadense

UPOV: >>> Union for Protection of New Varieties of Plants

upper glume: The glume bract placed higher in a glume *bot*

upper palea: Upper glume >>> pale

upper tier: In an eight-celled embryo, the upper four cells formed by mitotic division of the apical cell; these cells will give rise to the apical meristem and cotyledons *bot*

up-regulated: Adjustment of metabolism upwards or increased in some way in response to change in an environmental factor *phys*

upstream: A term used for description of the position of a DNA sequence within a DNA or protein molecule; it means that the position of the sequence lies away from the direction of the synthesis of a DNA or protein molecule *gene*

uracil (U): A pyrimidine base that occurs in RNA and constituent of nucleotides and, as such, one member of the base pair A-U (adenine-uracil) *chem gene* >>> Table 38

urban forestry: The management of vegetation, particularly trees and forests, to improve the urban environment and the quality of life of people who live, work, and spend their leisure time in urban and urbanizing landscapes *fore* >>> forestry

urea: A compound, $CO(NH_2)_2$, occurring in urine and other body fluids as a product of protein metabolism; an important plant fertilizer *chem agr*

uredospore: A sexual spore of the rust fungi *bot* >>> Table 52

ureides (allantoin and allantoic acid): Important storage and translocatory forms of nitrogen in nodulated legumes; these are synthesized within the nodules; many tropical legumes export nitrogen from the nodule in the form of ureides *phys*

USDA: >>> U.S. Department of Agriculture

USDA-ARS: >>> U.S. Department of Agriculture, Agricultural Research Service

USDA hardiness zones: Planting zones established by >>> USDA, defined by minimum winter temperatures *meth agr*

utricle: A small, thin-walled, one-seeded, bladder-like fruit, like an >>> achene, but it has a compound ovary, rather than a simple one; its fruit ovary becomes bladdery or corky, e.g., in beet *bot*

V

vacilin: >>> globulin

vacuole: A transparent vesicle that is usually large and singular in mature cells but small in some meristematic cells; it is filled with a dilute solution that is isotonic with the >>> cytoplasm *bot*

valence cross: Crossing of individuals of different ploidy level; a method, for example, used in production of tetraploid rye varieties by crossing a tetraploid genotype as female and a diploid as male, respectively; tetraploid F1 seeds could be selected by green-grained xenia on a pale-grained mother plant, caused by fusion of reduced gametes of the mother plant (carrying a recessive allele for pale seed color) with unreduced gametes of the male parent (carrying the dominant allele for green seed color) *meth*

validated target: A gene or gene product that is known to be the target of a drug or chemical *biot*

validity: The degree to which a measure accurately reflects the theoretical meaning of a variable *stat*

valine (Val): An essential, aliphatic, non-polar amino acid [$O_2CCH(NH_2)CH(CH_3)_2$] *chem phys* >>> Table 38

Valsa canker: One of the most destructive diseases in apple (*Malus domestica*), especially in eastern Asia, *Valsa ceratosperma phyt* >>> Figure 62

value-added grains: >>> nutrient-enhanced varieties

value-enhanced grains: >>> nutrient-enhanced varieties

variability: The sum of different genetic or phenotypic characters within different taxa *gene* >>> Table 49

variable expression: A variation in phenotype between affected members of the same family (i.e., individuals carrying identical mutations); it occurs in many dominant conditions and may be associated with reduced penetrance *gene* >>> Figure 60

variable genome: >>> dispensable genome

variable number tandem repeat (VNTR): Genetic markers that consist of DNA segments that are duplicated end to end; the number of copies present at a locus can vary, giving rise to a large number of alleles; these genetic markers are commonly used for DNA fingerprinting *biot meth*

variables: Operationally defined concepts that can take on more than one value *stat*

variance: When all values in a population are expressed as plus and minus deviations from the population mean, the variance is the mean of the squared deviations; it is a measure of variation of a population; it can be divided into phenotypic variance, genotypic variance, and environmental variance; most plant breeding trials involve a layout of plots in rows and columns; resolvable row–column designs have proven effective in obtaining efficient estimates of treatment effects; further improvement may be possible by postblocking or by inclusion of spatial model components *stat* >>> variance components

variance balance: A >>> block design is variance-balanced if the variances of the best estimators of all the normalized treatment contrasts are equal *stat* >>> design of experiment

variance components: Important genetic variance components are additive variance, dominance variance, epistatic variance, and environmental variance *stat* >>> variance

variant: Any seed or plant which (1) is distinct within the variety but occurs naturally within the variety; (2) is stable and predictable with a degree of reliability compared to other varieties of the same kind, within known tolerances; and (3) is described as a variation of the official variety; it is not an off-type, and only considered an impurity if reported in excess of the acceptable level specified by the responsible breeder *seed*

variate: A single observation or measurement *stat*

variation: Differences in form or function between individuals or other taxa *gene*

variegate: >>> variegation

variegation: The phenomenon in some plants in which patches of two or more different colors occur on the leaves or flowers; it may be an inherited characteristic or may be due to virus infection *bot*

varietal hybrid: The product resulting from the mating of two varieties *meth*

varietal protection: >>> protected variety

varietal purity: Trueness to type or variety *agr meth* >>> testing plant varieties

variety: A plant differing from other members of the species to which it belongs by the possession of some hereditary traits; breeding varieties can be classified according to the manner of propagation, such as clone varieties (maintained by vegetative propagation), line varieties (maintained by self-fertilization), panmictic (population) varieties (propagated by cross-fertilization), or hybrid varieties (produced by directed crosses); the heterogeneity and heterozygosity increases from line to population and/or hybrid varieties *meth* >>> testing plant varieties >>> Figures 50, 51

variety blend: Mechanical mixture of seed of two or more varieties *seed* >>> blend

variety description: Document in which the responsible plant breeder specifies the distinguishing characteristics of a variety *agr meth* >>> testing plant varieties

variety listing: A list of new varieties recommended for agriculture, forestry, and horticulture; after official performance tests, only those varieties that meet the standards are accepted for a national or international variety list; important for variety listing is the testing for value and use, as well as distinctness, homogeneity, and stability of the candidates *seed agr* >>> testing plant varieties

variety machine: A term introduced by N. F. JENSEN illustrating his breeding program; it refers to a segmented time process composed of from 10 to 15 annual input segments of homozygous selections, each of which decreases in number with attrition over time until all of the selections in a segment disappear, either through discard or elevation to parent or variety status; the "machine" is fed with ~10,000 new selections annually; not much time needs to be spent on the so-called "annual procedural flow"; most time should be spent on quality of the material fed into the "machine"; the production of superior fixed lines can be more or less automatic *meth* >>> Figure 42

variety maintainer: A special status elite parent seed grower or select/foundation plot seed grower recognized by an official organization (e.g., Canadian Seed Growers' Association) as eligible to produce breeder, inbreds, or hybrid seed under the supervision of a plant breeder recognized by the organization *agr meth*

variety mixture: A composite population made up of a random mixture of different varieties; it may exhibit considerable phenotypic variation in one or more characters, but which have one or more desirable agronomic traits in common *agr* >>> blend

variety release: The official approval of a variety for multiplication and distribution *seed* >>> testing plant varieties

variety trial: It forms the backbone of plant breeding programs; variety trials range from the initial studies involving small plots, i.e., single plants and single rows, to replicated yield trials involving fairly large plots (2.5 × 5 m); in the initial studies, plant material is selected by a subjective judgment based on the experience of the breeder; in the replicated yield trials one or more check varieties are included to serve as standards against which the performance of the new selections can be compared; such trials should include a long-term check, a variety that stays the same from year to year, to serve as a baseline against which to measure the progress of the breeding program *stat meth* >>> testing plant varieties

vascular: Furnished with vessels or ducts *bot*

vascular bundle: A discrete, longitudinal strand that consists principally of vascular tissue *bot*

Vavilovian mimicry: Describes a weed that has evolved to look more like a crop species; it occurs where a weed comes to share one or more characteristics with a domesticated plant through generations of artificial selection; it is an adaptive response to hand weeding as herbicide resistance is to chemical control of weeds; selection against the weed may occur by killing a young or adult weed, separating its seeds from those of the crop (winnowing), or both; this has been done manually since Neolithic times, and in more recent years by agricultural machinery; Vavilovian >>> mimicry is a good illustration of unintentional selection by humans; it is named after N. I. W. VAVILOV (1887–1943), a prominent Russian plant geneticist who identified the >>> centers of origin of cultivated plants; Vavilovian mimicry can be classified as reproductive, aggressive (parasitic), and, in the case of secondary crops, mutualistic; it is a form of disjunct mimicry with the model agreeable to the dupe; in disjunct mimicry complexes, three different species are involved as model, mimic, and dupe – the weed, mimicking a protected crop model, with humans as signal receivers; Vavilovian mimicry bears considerable similarity to >>> Batesian mimicry in that the weed does not share the properties that give the model its protection, and both the model and the dupe (in this case people) are negatively affected by it *evol bree*

vector: An organism capable of transmitting inoculum phyt; in genetics, a vehicle, such as a plasmid or virus, for carrying recombinant DNA into a living cell *gene*

vegetation: A collection of plants of the same or diverse species *eco*

vegetational analysis: Any of various methods of studying small (sample) areas of constituent plants, often counting the numbers of plants of communities to make extrapolations to a larger area (e.g., estimating the yield) *meth agr*

vegetative: Applied to a stage or structure that is concerned with feeding and growth rather than with sexual reproduction *bot gene*

vegetative cone: In flowering plants, the vegetative cone consists of three different layers; they are called dermatogen (L1), subdermatogen (L2), and corpus (L3); from the dermatogen the epidermis is formed, from the subdermatogen the mesophyll and the gametes, and from the corpus the vascular bundle, the flesh, the pith, and the adventitious roots; the cell division of the dermatogen and the subdermatogen is usually anticline – in other words, vertical to the layer (in this way the layers show surface expansion); the cell division of the corpus is usually periclinal (i.e., the cells divide parallel to the layers) *bot*

vegetative lag phase: The period from the end of active tillering to the beginning of the reproductive stage; tiller number decreases, height and stem diameter continue to increase, but at a slower rate *phys meth* >>> Table 13

vegetative meristem: Gives rise to parts, such as stems, leaves, roots, etc. *bot*

vegetative nucleus: The tube-nucleus of a pollen grain in a flowering plant *bot* >>> Figures 25, 35

vegetative propagation: A reproductive process that is asexual and so does not involve a recombination of genetic material (e.g., cloning of potato or strawberry) *meth* >>> Figure 28

vegetative reproduction: >>> vegetative propagation

vegetative shoot: The aerial portion of a plant, composing the stem and leaves; new or young growth that arises from some portion of a plant *bot*

vegetative stage: >>> whorl stage

vein: >>> veinure

veinure: Threads of fibrovascular tissue in a leaf or other organ, especially those that branch *bot*

velogenetics: The combined use of marker-assisted selection and embryo technologies in order to increase the rate of genetic improvement in populations *biot*

velvety: A vestiture of thick medium-length hairs *bot*

venation: >>> nervature

venom: >>> toxicity

venomous: Toxic phys

ventral: The inner side, furrowed in the grain of, for example, wheat, and barley, and in the caryopsis of oats *bot*

ventral furrow: The groove running along the length of the ventral side of the caryopsis *bot*

ventral groove: >>> ventral furrow

venturi injector: This type of injector operates by generating a differential pressure or vacuum across a venturi device; this draws the chemical into the drip irrigation system *hort meth* >>> injector

verification: Matching an unknown but keyed specimen with known material in the herbarium *bot meth*; in statistics, the process of subjecting hypotheses to empirical tests to determine whether a theory is supported or refuted *stat*

vermiculite: A porous form of mica, a mineral that makes good rooting media for seed germination because of its capacity to retain moisture and permit aeration *meth*

vernal: Growth form for spring and early summer *bot*

vernalin: A hypothetical hormonelike substance found in plant meristematic regions, produced by >>> vernalization; this substance is apparently graft transmissible, but has not yet been identified; different cold-requiring species may form different substances during vernalization *phys*

vernalization: The treatment of germinating seeds with low temperatures to induce flowering at a particular preferred time; in the genetic model plant, >>> thale cress (*Arabidopsis thaliana*), a gene, *Flc*, suppresses the formation of flowers during cold periods; another gene, *Vrn2*, triggers that suppression; however, in spring the suppression is raised in a way that the gene "remembers" the previous cold period, because the gene *Vrn2* itself is cold-insensitive; there is a similar gene in *Drosophila melanogaster*, which also serves as "chemical memory"; thus vernalization is the programmed physiological process in which prolonged cold-exposure provides competency to flower in plants; widely found in winter and biennial species, such as *Arabidopsis*, fruit trees, vegetables, and cereals; the phenomenon is regulated by diverse genetic networks, and memory of vernalization in a life cycle mainly depends on epigenetic mechanisms *meth phys*

vernation: The arrangement of bud scales or young leaves in a shoot bud *bot*

verrucose: Covered with warty protuberances *bot*

vertical farming: A concept to conduct large-scale agriculture in urban high-rises or "farmscrapers"; using recycled resources and greenhouse methods such as >>> hydroponics in order to produce fruit, vegetables, edible mushrooms, and algae year-round; its proponents argue that, by allowing traditional outdoor farms to revert to a natural state and reducing the energy costs needed to transport foods to consumers, vertical farms could significantly alleviate climate change produced by excess atmospheric carbon; Dickson DESPOMMIER, a professor of environmental health sciences and microbiology at Columbia University in New York City, developed the idea of vertical farming in 1999; architectural designs have been produced by Chris JACOBS of United Future, Andrew KRANIS at Columbia University, and Gordon GRAFF at the University of Waterloo, U.S.; the concept of indoor farming is not new, since glasshouse production of vegetables and other crops has been in vogue for some time; what is new is the urgent need to scale up this technology to accommodate another 3 billion people; an entirely new approach to indoor farming must be invented, employing cutting-edge technologies; by the year 2050, nearly 80% of the earth's population will reside in urban centers; applying the most conservative estimates to current demographic trends, the human population will increase by about 3 billion people during the interim; an estimated 10^9 ha of new land (about 20% more land than is represented by the country of Brazil) will be needed to grow enough food to feed them, if traditional farming practices continue as they are practiced today; at present, throughout the world, over 80% of the land that is suitable for raising crops is in use; historically, some 15% of that has been laid waste to by poor management practices; this is the background for the concept of vertical faming *agr hort* >>> www.verticalfarm.com

vertical gene transfer: >>> outcrossing

vertical resistance: The existence of differential levels of resistance to different races of a given pathogen conditioned by one or a few qualitative genes *phyt* >>> resistance

viruliferous: A vector (usually insect) organism that carries virions and spreads the virus from host to host by mechanical means *phyt*

verticilate: Whorled *bot*

"Vertifolia" effect: Named after the potato variety "Vertifolia" where erosion of polygenic >>> horizontal resistance occurred while selecting for race-specific >>> vertical resistance; the effect of vertical resistance genes is so strong that evaluation of horizontal resistance is not possible; even causing genes to be lost *phyt*

viability: The probability that a fertilized egg will survive and develop into an adult organism; the term is also often applied to plant germination experiments with comparisons across phenotypic classes under standard specific environmental conditions *phys seed*

viable: Capable of germinating, living, and growing, or sufficiently developed physically as to be capable of living *phys seed*

vibrator separator: A machine utilizing a vibrating deck for separating seeds on the basis of their shape and differing surface textures *seed*

vicilin: A protein common in broad bean *phys*

video microscopy: Microscopy that takes advantage of video as an imaging, image-processing, or controlling device *micr*

vignin: >>> globulin

vigor test: Vigor tests of seed determine the potential ability to develop into normal, healthy plants under a wide range of field conditions; stress tests, accelerated aging, and other techniques are used to provide a more sensitive index of seed quality; poor emergence, dormancy, seed deterioration, and other quality factors can be identified through vigor testing *seed* >>> www.plantstress.com

VIGS: >>> virus-induced gene silencing

vine: Climbing plants with woody or herbaceous stems that climb, twist, adhere, or scramble over other taller objects; they can climb by tendrils, aerial roots, twining stems, twining leafstalks, adhesive disks, or hooks *bot*

vine-growing: >>> viticulture

vine-louse: A plant louse (Phylloxera vitifoliae) that injures the grapevine *phyt*

viniculture: >>> viticulture

virescence: Greening of tissue that is normally devoid of chlorophyll (e.g., the abnormal development of flowers in which all organs are green and partly or wholly transformed into structures like small leaves) *phyt*

virion: An individual virus particle *gene*

viroid: A piece of infectious nucleic acid; in plant pathology, any of a class of plant pathogenic agents consisting of an infectious, single-stranded, free RNA molecule *gene*

virulence: The relative ability of a microorganism to overcome the resistance of a host *phyt* >>> aggressiveness .

virulent: >>> virulence

viruliferous aphids: Aphids acting as vector for plant viruses *phyt zoo*

virus: A type of non-cellular "organism", that has no metabolism of its own; a nucleoprotein entity that can replicate within living cells; it passes through bacterium-retaining filters *bot gene*

virus-free plant: A plant that shows no sign of viral particles or symptoms *phyt*

virus-induced gene silencing (VIGS): Silencing that is induced by the presence of viral genomic RNA; only replication-competent viruses cause silencing, indicating that >>> dsRNA molecules are the inducing agents; restriction of virus growth in plants is mediated by >>> PTGS, which can be initiated by production of dsRNA replicative intermediates; this silencing of gene expression is gene-specific *biot meth*

viscid: Sticky *bot*

vital coloring test: >>> vital staining

vital dye: >>> vital staining

vital staining: A stain that is capable of entering and staining a living cell without causing an injury *meth cyto seed* >>> fluorescin diacetate (FDA) staining

vitamin: An organic compound produced by plants (often functions as a coenzyme) that is required in relatively small amounts in the diet for the normal growth of animal organisms *phys*

vitamin B$_1$: >>> thiamine

vitamin E: >>> tocopherol

vitamin H: >>> biotin(e)

viticulture: The science, culture, or cultivation of grapes and grapevines *hort*

vitreous grain: Characterizing slightly translucent kernels *agr*

vitrification: >>> vitrified

vitrified (*syn* **water-soaked**): Cultured tissue having leaves and sometimes stems with a glassy, transparent, or wet and often swollen appearance; the process of vitrification is a general term for a variety of physiological disorders that lead to shoot tip and leaf necrosis *phys biot*

viviparity: >>> viviparous

viviparous: A process promoting the germination of embryos while still attached to the mother plant; applied to a plant whose seeds germinate within and obtain nourishment from the fruit; it refers also to a plant that reproduces vegetatively from shoots rather than an inflorescence *bot*

vivipary: >>> viviparity

vivisection: The action of cutting into or dissecting a living body *meth*

vivotoxin: >>> pathotoxin

VNTR: >>> variable number tandem repeat

volunteer: >>> self-sown cereals >>> volunteer plants

volunteer plant(s): Plants that have resulted from natural propagation, as opposed to having been deliberately planted by humans *seed*

volunteer potato(s): Potatoes that have been left or derived from a previous potato crop and have become a weed in subsequent crops *seed*

vomitoxin (*syn* **deoxynivalenol, DON**): A naturally occurring >>> mycotoxin produced by several species of *Fusarium fungi*; wet and cool weather from flowering time to maturity promotes infection, resulting in scab or head blight in barley, wheat, oats, and rye; wheat infected with scab has a tendency to have lighter weight kernels, some of which are removed during normal harvesting and cleaning operations *phyt*

voucher specimen: A herbarium specimen that documents the morphology and identification for chromosome, ecological and anatomical studies, plant breeding (parents and progeny), and taxonomy *meth bot tax*

vulnerability: Crop vulnerability is defined as susceptibility to an absent, foreign parasite that has epidemiological competence in the area in question; some crop vulnerabilities are particularly severe, e.g., the vulnerability of potatoes in Ireland to blight *phyt* >>> genetic vulnerability

v/v: May indicate simple proportion (e.g., 3:1 v/v) or percent volume in volume *meth*

vybrid: The first- and subsequent-generation progenies of crosses of heterozygous facultative apomicts *bot*

W

Wahlund effect: The frequency of >>> homozygotes decreases in the progeny after matings among individuals of two previously isolated subpopulations *stat gene meth* >>> Table 36

WARDA: >>> Africa Rice Center

warren hoe: A lightweight hoe with a triangular blade, fastened to the handle on the flat side *agr*

wart (of potato): It is a quarantined disease of cultivated potato (*Solanum tuberosum*); since its discovery by SCHILBERSZKY in 1896, the management of wart disease was enabled by research efforts focusing on understanding and classifying the causative agent, its mode of infection, pathogenesis, geographical distribution, detection and chemical control, on developing screening methods for host resistance, and on genetic analyses, which led to the development of resistant cultivars; the causal agent of potato wart (*Synchytrium endobioticum*) is an obligate parasitic chytrid fungus; it is included as a quarantine pathogen in 55 countries, with losses in susceptible cultivars reaching 50–100%; highly intraspecific variation of wart diseases resistance allows the selection of extremely resistant and susceptible genotypes available for future genetic and breeding studies; however, the early successes are currently challenged by new *S. endobioticum* pathotypes evolving and the increased risk of dissemination by potato tuber trade *phyt*

water-absorbing capacity: In breadmaking, a high capacity to absorb water is required; this is associated with hard milling texture, high protein content, and the degree of starch damage during the milling process *prep*

waterlogged: Soil saturated with water *agr*

watermark: The area between the >>> throat and the eye, band, or halo on a daylily blossom *hort*

water ripe: A description of the newly pollinated cereal grain; when punctured the grain contents are very watery *phys bot* >>> premilk stadium

water-soaked: >>> vitrified

water spots: Leaf spots which have specifically been caused by water splashing on the leaves; it may denote either a change in tissue color or the white residue which sometimes remains when the water has evaporated *hort* >>> Figure 62

water sprout: A shoot arising from a bud located on wood that is not older than one year *hort*

water table: The highest point in a soil profile where water saturates the soil on a seasonal or permanent basis *agr*

water used efficiency: A measure of the proportion of water used by a crop relative to wheat arrived as rainfall or was supplied as irrigation or both; the numerical value is strongly dependent on the scale of measurement and time period, whether of a leaf, a plant, a crop, or field *phys* >>> nitrogen use efficiency

wax: A diverse group of naturally occurring greasy, hard, or soft and moldable high-molecular-weight solid substances of plant or mineral origin; it may contain saturated hydrocarbons, esters of higher fatty acids, and higher alcohols; most natural waxes contain a wide range of molecular sizes and therefore soften over a wide temperature range instead of melting at a specific temperature, as do substances of a uniform molecular size *chem phys*

wax coating: A thin layer covering the stem, leaves, flowers, and fruits of most plants; waxes are manufactured as oily droplets in epidermal cells, from which they migrate to the outer surface of the plant via tiny canaliculi in cell walls, and crystallize as rods and platelets; their pattern of deposition is sometimes used as a micromorphological character below the genus level; the wax coating reduces the water transpiration of the plant and is involved in water balance and resistance mechanisms against diseases *bot*

waxiness: The phenomenon of whitish, powdery, or waxy covering of plant leaves, stems, or flowers *bot*

waxy hull-less barley (WHB): A barley mutant that is rich in soluble fiber and low in fat content; these characteristics make it a nutritionally valuable ingredient for food products; it has been shown that the soluble fiber, beta-glucan, reduces cholesterol and lowers blood glucose and insulin response following a meal *gene*

waxy maize: Maize that produces kernels in which the starch that is contained within those kernels is at least 99% amylopectin, versus the average of 72–76% amylopectin in common starch *seed gene*

waxy wheat: Varieties of wheat which produce higher amylopectin content, and thus a lower amylose content in the starch within their seeds, e.g., bread flour made from waxy seeds would contain 0–3% amylose versus 24–27% amylose in bread flour made from traditional varieties *seed*

W chromosome: A sex chromosome that is limited to the female sex *cyto*

"weak" flour: >>> "strong" flour

weathering: All the physical, chemical, and biological processes that cause the disintegration of rocks at or near the surface *agr*

weed: A plant that occurs opportunistically on land that has been disturbed by human activity or under cultivated land where it competes for nutrients, water, sunlight, or other resources with cultivated plants *phyt*

weed allelopathy: The damage to crops by weed toxic exudates *phyt*

weed beet: A weed such as *Beta vulgaris* of arable fields, usually those that have grown a beet crop; it is an annual and germinates, grows, bolts, and sets seed within the same season *agr phyt*

weed killer: >>> herbicide

weediness: Unwanted effects of a plant, i.e., the ability of a plant to colonize a disturbed habitat and compete with cultivated species *agr*

weeding: Removing weeds from the crop stand *agr*

weed mimicry: >>> Vavilovian mimicry

Weismannism: A concept of the German biologist F. L. A. WEISMANN (1834–1914) that proposes that acquired traits are not inherited and only changes in the germplasm transmitted from generation to generation *evo*

well stocked: The situation in which a forest stand contains trees spaced widely enough to prevent competition yet closely enough to utilize the entire site *fore*

Western blot: >>> Western blotting

Western blotting: A technique similar to >>> SOUTHERN blotting but for the analysis of proteins instead of >>> DNA; a technique in which proteins are separated by gel electrophoresis and transferred by capillary action to a nylon membrane or nitrocellulose sheet; a specific protein can be identified through hybridization to a labeled >>> antibody *biot meth* >>> Figure 48

wet milling: Process in which feed material is steeped in water, with or without sulfur dioxide, to soften the grains in order to help separate the various components of the kernels *meth agr*

wet rot: Any rot in which the tissue is rapidly and completely disintegrated, with release of water from the lysed cells *phyt*

wet stigma: A stigma with a surface secretion, either lipidic or aqueous *bot* >>> stigma >>> dry stigma

wettable powder: Refers to a pesticide which has been processed into a powder to be mixed with water and sprayed *meth phyt*

wetting agent: A substance that improves surface contact by reducing the surface tension of a liquid, e.g., Triton X-10™ added to disinfection solutions promotes the disinfection process *meth*

wheat root nodules: A paranodule in wheat roots, induced chemically by >>> 2-4-D; it differs from the naturally occurring legume nodule *agr*

wheat stem sawfly (WSS): Several species of wheat stem sawflies are pests of wheat, including *Cephus cinctus* Norton in North America; larvae feed inside the stems and cut the stem near plant maturity; the primary means of control is resistance due to solid stems, largely controlled by a locus on chromosome 3B (*Qss.msub-3BL*); the American cultivar "Scholar" is susceptible, while "Conan" shows resistance; "Scholar" and "Conan" possess different alleles at *Qss.msub-3BL*; both alleles confer solidness, yet the "Conan" allele confers higher WSS resistance; an allele from "Conan" on chromosome 4A also decreases infestation and stem cutting; the 3B and 4A alleles from "Conan" act in an additive fashion to provide increased WSS resistance without increasing stem solidness; stem solidness has long been used by breeders as a proxy for WSS resistance because of its simplicity *phyt*

whitefly: Insects whose adults resemble tiny moths but are related to aphids; nymphs that suck sap and damage leaves of several Brassicaceae and other horticultural plants *phyt*

whitehead: A bleached cereal spike containing little or no grain; usually a result of attack by stem base or root pathogens, particularly *Gaeumannomyces graminis* (take-all disease) *phyt*

white leaf disease (of oats): Caused by copper deficiency, often on peaty soils; can be compensated for by application of copper sulfate (2 kg per 400 l water/ha) to leaves *phyt*

white leaf spot (of rape, *Pseudocercosporella capsellae*): Characterized by large white spots with dark margins and gray-black centers in older lesions on leaves; pods can be similarly infected *phyt* >>> *Tapesia yallundae* >>> Figure 62

white mold (of bean): A major disease of common bean caused by the fungus *Sclerotinia sclerotiorum phyt*

white rot: Rotting of wood in trees invaded by lignin-destroying fungi, leaving a white cellulose residue *phyt*

white rust disease (of Indian mustard) Caused by Albugo candida, a serious disease resulting in considerable yield loss every year in India *phyt* >>> www.tau.ac.il/ lifesci/units/ICCI// genebank2.html

whole grains: Grain that has not been processed except to remove the seed coat *agr*

whorl: An arrangement of leaves, branches, etc., in a circle around a stem or node *bot* >>> verticilate

whorl stage: The developmental stage of a grass plant prior to the emergence of the inflorescence *phys*

wick: A length of specially woven string or similar item with properties that allow capillary action to work; they are often used in self-watering devices *hort meth*

wide cross: >>> wide hybridization

wide hybridization (*syn* wide cross): Cross combinations between taxonomically remote species or genera *meth*

wide hybrids: >>> wide hybridization

wide-row planting: The method of wide-spaced sowing of seeds in multiple rows (e.g., for better selection of young breeding material) or several rows with wide parallel channels for irrigation *meth agr*

widger: A tool to lift seedlings *hort meth*

wildlife plantings: Agricultural crops specifically planted for wildlife in fields or small forest openings; sometimes referred to as food plots *eco*

wild type: The most frequently observed phenotype, or the one arbitrarily designated as "normal" *gene* >>> Table 35

wilt: A type of disease in which wilting is a principal symptom *phyt* >>> Figure 62

wilting point: The percentage of water remaining in the soil when the plants wilt permanently *phys*

winch plow: A plow pulled across the field by a cable attached to a winch (of steam engines); common in the 19th century in preparing heavy soils *agr*

windbreak: A wind barrier of living trees and shrubs maintained for the purpose of protecting the farm home, other buildings, garden, and orchard or feedlots *agr eco*

windrower: A machine to lift potatoes out of their rows and lay them in adjacent undue rows; it is essentially an updated version of the old-fashioned potato digger with a rear cross-conveyor to drop the potatoes on the adjacent row instead of directly behind *meth tech agr*

wind pollination: Pollination by wind-borne pollen *bot* >>> allogamy

wind-row: A loose, continuous row of cut or uprooted plants placed on the surface of the ground for drying to facilitate harvest *agr*

windthrow: A tree pushed over by wind; windthrows or blowdowns are more common among shallow-rooted species and in areas where cutting has reduced the density of a stand so that individual trees remain unprotected from the force of the wind *fore*

wings: The two expanded parts of the glume in, for example, wheat, which lie on each side of the keel; in general, a membranous or thin and dry expansion or appendage of a seed or fruit *bot*

winged fruit: >>> wings

winnower: A simple device for cleaning seed from weeds and chaff using an airflow *seed*

winnowing mill: >>> winnower

winter annuals: Plants from autumn-sown seed that bloom and fruit in the following spring, then die *phys* >>> winter-type

winterburn: Foliar necrosis, often marginal, of plants that retain their leaves in winter; it is caused by water deficiency because of frozen soil *phys*

winter killing: >>> killing frost

winter spore: >>> teleutospore >>> resting spore

winter-and-spring wheat: Facultative growth habit *bot*

winter-type (of growth habit): Plants germinating in autumn, requiring vernalization during the wintertime for flower induction during the following year *bot* >>> vernalization

whip grafting: Scions and stocks of approximately the same diameter are used; this method of grafting is applied primarily on year-old seedlings that are being converted to named varieties by removing their tops and replacing them with scion wood; it can be done during the winter in the root cellar where the fall dug stock is stored; it can be an advantage because yard chores are not as numerous in the winter; in order to make this graft, the top from the stock is removed with a diagonal cut; the cut splits the surface vertically; a piece of scion with two or three buds is selected and a matching diagonal cut is made; the pieces are slipped together, wrapped with rubber grafting tape, and coated with grafting wax *hort meth*

witch's broom: Massed outgrowth of branches of woody plants caused by fungi (e.g., rusts) *phyt* >>> Indian bread >>> Table 52 >>> www.tau.ac.il/lifesci/units/ICCI//genebank2.html

withertip: Death of the leaf beginning at the tip, usually in young leaves *phys*

wobble hypothesis: Formulated by F. CRICK in 1966; an explanation of how one >>> tRNA may recognize more than one >>> codon; the first two bases of the >>> mRNA codon and anticodon pair properly, but the third base in the >>> anticodon has some flexibility that permits it to pair with either the expected base or an alternative; a wobble base pair is a non-WATSON–CRICK base pairing between two >>> nucleotides in >>> RNA molecules; the four main wobble base pairs are guanine–uracil, inosine–uracil, inosine–adenine, and inosine-cytosine (G–U, I–U, I–A, and I–C); wobble base pairs are fundamental in RNA secondary structure and are critical for the proper translation of the genetic code *gene biot*

wolf tree: A tree that occupies more space in the forest than its value justifies; usually a tree which is older, larger, or branchier than other trees in the stand *fore*

wood-boring emerald ash borer: A prolific invasion of the wood-boring emerald ash borer (*Agrilus planipennis*) is particularly recognized in Northern America; this beetle is native to Asia; it completes its life cycle on ash trees and results in nearly complete mortality of all infested trees; among green ash (*Fraxinus pennsylvanica*) population there is a certain level of genetic diversity that was revealed by molecular markers; selection for resistant trees can be a useful approach *phyt*

working collection: A collection of germplasm kept under short-term storage conditions, commonly used by breeders or researchers *meth*

world collection (of crop plants): A global collection of samples of a species or genera; it is a coordinated activity of several countries and institutions under the >>> IPGRI *meth* >>> genebank

w/v: Weight in volume, as the number of grams of constituent in 100 ml solution *meth* >>> v/v

X

x: Designates the basic number of chromosome sets *cyto* >>> basic number

X1, X2, X3, etc.: Symbols denoting first, second, third, etc. generations from an irradiated ancestral plant (X0) >>> M1, M2, M3, etc. >>> Figure 1

xanthophyll: Yellowish-brownish (oxygen-containing) carotenoids occurring in the chloroplasts, e.g., the lutein of leaves; lutein is an important antioxidant, and its intake is associated with lower risk of macular degeneration (blindness); it can be increased for staple food fortification as a rational and economic way to improve the health status of low-income consumers, e.g., in >>> chickpea breeding *bot*

X chromosome: A sex chromosome found in a double dose in the homogametic sex and in a single dose in the heterogametic sex *cyto*

xenia: A situation in which the genotype of the pollen influences the developing embryo of the maternal tissue (endosperm) of the fruit to produce an observable effect on the seed *bot*

xenogamy: Intercrossing between flowers of different individuals, as opposed to geitonogamy *bot* >>> cross-pollination

xenograft: A type of tissue graft in which the donor and recipient are of different species; also called >>> heterograft *bree meth*

xenosis: A term coined from the word "xenozoonosis"; it describes the transfer of infections by transplantation of xenogeneic tissues or organs; it potentially poses unique epidemiological hazards due to the efficiency of transmission of >>> pathogens, particularly viruses, with viable, cellular grafts *phyt*

xenotransplantation: Transplanting a foreign tissue into another species *meth*

xenotransplant: The implantation of an organ or limb from one species to another organism in a different species *hort* >>> heterograft

xenozoonosis: >>> xenosis

xeric: Dry *bot*

xerograft: >>> heterograft

xerophyte: A drought-resistant plant or plants that grow in extremely dry areas *bot*

xerophytic: Growing in dry conditions *bot*

X-rays: Electromagnetic radiation having wavelengths in the range of approximately 0.1–10 nm, between ultraviolet radiation and gamma-rays, and capable of penetrating solids *meth*

Xta: >>> chiasma

xylan: A polysaccharide of xylose and a component of hemicellulose *phys*

xylem: A plant tissue, consisting of various types of cells, that transports water and dissolved substances toward the leaves *bot*

xylene: >>> xylol

xylol: A liquid solvent *prep* >>> refraction index

xylose: An aldopentose sugar ($C_5H_{10}O_5$) that is commonly found in plants and especially in woody tissue *chem phys*

Y

YAC: >>> yeast artificial chromosomes

YAC clone: >>> yeast artificial chromosomes

Y chromosome: Plays a role in sex determination *cyto*

yearling: A one-year-old seedling and/or plantlet *agr fore hort*

yeast: A general term for a fungus that can exist in the form of single cells, reproducing by fission or by budding *bot*

yeast artificial chromosomes (YAC): A yeast artificial chromosome is used to clone very large DNA fragments in yeast; although most bacterial vectors cannot carry DNA pieces that are larger than 50 bp, YACs can typically carry DNA pieces that are as large as several hundred base pairs *gene biot*

yeast two-hybrid: An interaction-based screening technique carried out in yeast where the interaction between two proteins leads to activation of selectable markers or reporter genes *meth biot*

yellows: A plant disease characterized by yellowing and stunting of the host plant *phyt*

yellow sigatoka: >>> black sigatoka

yellow spot disease (in sugarcane): Caused by the fungus *Mycovellosiella koepkei phyt*

yield: Commonly, the aggregate of products resulting from the growth or cultivation of a crop and usually expressed in quantity per area *agr* >>> Table 33

yield appraisal: >>> vegetational analysis

yield drag: A negative effect on grain yield associated with crop plants that have a specific gene or a specific trait *agr*

yield lag: A relative reduction in yield observed in some hybrids or varieties compared with the yield observed in the most recently produced hybrids or varieties *agr*

yield monitoring: Collecting data on the amount of production at regular intervals and by certain means (e.g., GPS readings); the resulting yield map is basic to decisions about fertilization, pest control, and other adjustments in a system of precision farming *agr*

yield potential: The highest yield a plant (hybrid, variety, etc.) is capable of producing when grown in ideal conditions *agr*

yield structure: An analysis used to determine the numerous morphological and physiological components of a plant contributing to the final yield (given in different measures) *agr* >>> vegetational analysis >>> Table 33

yield trial: A nursery or experimental design in order to determine the yield capacity of a crop or yield components *meth* >>> Table 33

YOUDEN rectangle: >>> incomplete Latin squares

Z

Zade method: >>> long-plot design

Zadok scale: >>> physiological maturity >>> Table 13

Z chromosome: A sex chromosome that is limited to the male sex *cyto*

zDNA: A left-handed, alternate form of DNA in which the backbone phosphates zigzag *gene*

zeatin ($C_{10}H_{13}N_5O$): A mitogen isolated from maize kernels; *syn* 6-(4-hydroxy-3-methylbut-2-eny lamino)purine; N6-(4-hydroxy-3-methyl-2-buten-1-yl)adenine; it is a cytokinin growth regulator and promotes cell division; high concentrations produce adventitious shoot formation; other uses include prevention of leaf senescence, reversal of toxin effects on shoots, and inhibiting excessive root formation *chem phys* >>> Table 15

zeaxanthin: A carotenoid, i.e., light-harvesting compound utilized in photosynthesis that is naturally produced in Brussels sprouts, summer squash, maize, avocado, green beans, and dark green leafy vegetables *phys chem*

zebra chromosome: A chromosome named because of its striped genomic *in situ* hybridization pattern; it was isolated from an *Elymus trachycaulus* × *Triticum aestivum* backcross derivative; the origin was traced to nonhomologous chromosomes 5A of wheat and 1H^t of *Elymus*; four chromatin segments were derived from chromosome 1H^t and five including the centromere from 5A *cyto*

zebularine: A synthetic >>> cytidine analogue and a cytidine deaminase inhibitor with anticancer activity; following metabolic activation by >>> phosphorylation and incorporation into >>> DNA, zebularine inhibits DNA >>> methyltransferase through covalent complex formation between the enzyme and zebularine-substituted DNA, hence resulting in non-specific, genome-wide induction of >>> demethylation including the removal of aberrant methylation of promoter regions of genes critical for normal cellular functions *biot*

zein: A storage protein of maize found in the endosperm; consisting of two major proteins, with m.w. of 19,000 and 21,000 Daltons *chem phys* >>> Table 15

ZELENY test: A test to measure the protein quality; the grain is milled to form a white flour and mixed with a suspension agent; the resulting suspension volume is then measured in millimeters; for example, wheats with a ZELENY volume between 20 and 30 are acceptable (<19 low; 25 medium; 35 high; 45 very high, >50 extremely high) *meth*

zenia: >>> metaxenia

Zeocin™: A copper-chelated glycopeptide antibiotic produced by *Streptomyces* CL990; it causes cell death by intercalating into DNA and cleaving it; this antibiotic is effective on most aerobic cells and is therefore useful for selection in bacteria, eukaryotic microorganisms, and plant and animal cells; resistance to zeocin is conferred by the *Shble* gene product which inactivates zeocin by binding to the antibiotic; zeocin is used at a concentration of 50–300 µg/ml for selection in mammalian cells and 25 µg/ml for bacterial selection *biot*

zero tillage: System to improve soil conservation where the new crop is planted in stubble of the previous crop with even less soil disturbance than with minimum tillage *agr meth*

zinc (Zn): A blue-white metallic element that is a trace element (essential element) that is required by plants; it is found in various enzymes; it functions as the prosthetic group of a number of enzymes; deficiency prevents expansion of leaves and internodes, giving a rosette style of plant; it plays a key role as a structural constituent or regulatory co-factor of a wide range of different enzymes in many important biochemical pathways, and these are mainly concerned with (1) carbohydrate metabolism, both in photosynthesis and in the conversion of sugars to starch, (2) protein metabolism, (3) auxin (growth regulator) metabolism, (4) pollen formation, (5) the maintenance of the integrity of biological membranes, or (6) the resistance to infection by certain pathogens *chem phys* >>> www.zinc.org/

zinc chelate: Zinc deficiency in plants can be controlled either by applying zinc sulfate to the soil or by spraying the foliage with relatively small amounts of an organic compound to which zinc has been chelated or bound *phys* >>> chelating agent

zinc fertilizers: Compounds of zinc used to provide Zn for crop nutrition, e.g., zinc sulphates $ZnSO_4 \cdot 7H_2O$, (21% Zn) and $ZnSO_4 \cdot H_2O$ (33% Zn), Zn-EDTA chelate (12% Zn), zinc oxide ZnO (78% Zn), zinc frits (variable Zn content), and natural Zn chelates *chem agr* >>> zinc >>> www.zinc.org/

zinc finger (protein) A protein motif involved in the recognition of DNA; the structure contains a complex zinc ion and consists of an antiparallel protein-chain hairpin and loop followed by a helix *gene* >>> zinc >>> www.zinc.org/

zone of hybridization: A geographical area in which different plant taxa have the same habitat, allowing spontaneous intercrossing if some sexual and reproductive prerequisites are given (e.g., in the evolution of hexaploid wheat, such zones of hybridization played an important role, for example, Aegilops species intercrossed with primitive wheats in Asia Minor) *eco*

zoo blot: Hybridization of cloned DNA from one species to DNA from other organisms to determine the extent to which the cloned DNA is evolutionarily conserved *biot evol*

zoom: To control, by magnifying or reducing, the size of an image, either optically or electronically *micr*

zoospore: Temporarily mobile sex organs of some lower plants (e.g., in *Vaucheria sessilis*) *bot*

Z-type: In sugarbeet breeding, varieties with high sugar content (Z = Zucker = sugar) but normal yielding capacity *agr*

zucchini yellow mosaic virus (ZYMV): The virus causes significant disease, which leads to fruit yield loss in cucurbit crops; in cucumber, plant resistance is often recessively inherited in cucumber; therefore, marker-assisted selection (MAS) is a useful tool for the development of resistant cucumber cultivars; ZYMV resistance is conferred by a single recessive locus *zym* $^{A192-18}$; the locus is mapped to chromosome 6, at genetic distances of 0.9 and 1.3 cM from two closely linked SSR markers; high-resolution genetic mapping narrowed down the *zym* $^{A192-18}$ locus to a <50 kb genomic region flanked by two SSR markers, which included six candidate genes *phyt gene*

zygomere: Hypothetical pairing sites along the chromosomes; for example, it is suggested that in barley the zygomeres are located proximal, as opposed to rye in which the zygomeres are located terminal *cyto*

zygomorphic: The condition of having only one plane of symmetry (e.g., orchid flowers) *bot*

zygomorphic corolla: Dorsiventral corolla; a flower composed by one symmetric axis (e.g., *Antirrhinum majus*) *bot*

zygonema: >>> zygotene

zygote: The fertilized ovum formed from the fusion of male and female gametes; a diploid zygote is formed by the union of a sperm nucleus of the pollen with the egg cell of the embryo sac; the zygote divides mitotically to form the embryo of the seed; union of a sperm nucleus with the two polar nuclei of the embryo sac results in a >>> triploid (3x) cell that divides mitotically to form the endosperm of the seed *bot* >>> Figure 25

zygotene: The stage of first meiosis when the homologous chromosomes are associating side by side (synapse); for example, a diploid crop species with 2n = 20 chromosomes would form ten pairs of homologous chromosomes *cyto*

zygotic: >>> zygote

zygotic embryo: An embryo that derives from fusion of male and female gametes *bot*

ZYMV: >>> zucchini yellow mosaic virus

zymogens: The enzymatically inactive precursors of certain proteolytic enzymes; the enzymes are inactive because they contain an extra piece of peptide chain; when this peptide is hydrolyzed by another proteolytic enzyme, the >>> zymogen is converted into the normal, active enzyme *phys*

zymogram: Diagrammatic presentation of enzymes separated by means of >>> electrophoresis *meth*

Important Crop Plants, Weeds, Ornamental, Herb, Industrial, Woody, and Other Plants of the World

Flowering plants (angiosperms) comprise ~90% of the plant kingdom. The total number of described species exceeds 230,000, and many tropical species are as yet unnamed. During the past 130 million years, flowering plants have colonized practically every conceivable habitat on earth, from deserts and alpine summits, to fertile grasslands, freshwater marshes, dense forests, and lush mountain meadows. The total number of plant species, which are cultivated as agricultural, forest, or horticultural crops, can be estimated to be close to 7,000 botanical species. Nevertheless, it is often stated that only 30 species "feed the world", because the major crops are made up by a very limited number of species. Distinguished by FAO, tropical and subtropical fruit and nut crop species represent 30% of the 150 primary food crops, and together provide more than 15% of the world's human food supply. The major commodity crops, wheat, rice and maize, are critical to providing energy in diets. Their evolution dates back more than 9,000 years (cf. Tab. 56).

The three largest flowering plant families containing the greatest number of species are the sunflower family (Asteraceae) with ~24,000 species, the orchid family (Orchidaceae; *Ophioglossum reticulatum* has the highest known diploid chromosome number 2n = 1,260) with ~20,000 species, and the legume or pea family (Fabaceae) with 18,000 species. The total number of species for these three enormous families alone is approximately 62,000, roughly 25% of all the flowering plant species on earth.

A

abaca: The abaca plant is of Philippine origin and a close relative of "pakol", which is a close relative of the banana; while the banana is valued for its fruit, the abaca is valued for its leaf sheaths; fibers removed from the leaf sheaths make ropes, clothing, and paper-based materials; the fibers are retted out of the outer sheaths of the petioles that form the pseudostem; its fruits are sweet but practically inedible due to the presence of millions of seeds; the abaca is believed to have evolved in the Bicol area from some primordial succulent plants; *Musa textilis syn Cannabis gigantea* (Moraceae), 2n = 2x = 20; 1C DNA content ~0.63 pg *agr*

abaca banana: >>> abaca

abaca-fiber: >>> abaca

absinthe: >>> wormwood

absinthe wormwood: >>> wormwood

absinthium: >>> wormwood

acacia (*syn* **acacia gum** *syn* **chaar gund** *syn* **char goond** *syn* **meska**): A wild or semi-cultivated tree in the Sudan in areas with a rainfall as low as 300 mm/a; this spiny tree grows up to 12 m tall with creamy-white legume flowers; from cuts in the bark of the tree, gum Arabic is obtained, which is used as a food additive, in crafts, and as a cosmetic; the Sudan has exported the gum to Europe and the Middle East for over 2,000 years; gum Arabic is a natural gum made of hardened sap taken mainly from two species of the acacia tree, *Acacia senegal* and *A. seyal*; the gum is harvested commercially from wild trees throughout the Sahel from Senegal and Sudan to Somalia; gum Arabic, a complex mixture of polysaccharides and glycoproteins, is used primarily in the food industry as a stabilizer; it is edible and has E number E414; it is a key ingredient in traditional lithography and is used in printing, paint production, glue, cosmetics, and various industrial applications, including viscosity control in inks and in textile industries, although less expensive materials compete with it for many of these roles; *Acacia senegal* syn *A. circummarginata* syn *A. oxyosprion* syn *A. rupestris* syn *A. senegal* ssp. *senegalensis* syn *A. spinosa* syn *A. trispniosa* syn *A. verek* syn *A. volkii* syn *Mimosa senegal* (Leguminosae) 2n = 2x = 26; 1C DNA content ~1.50 pg *fore*

acai berry: It is a palm tree, widely distributed in northern South America; its name is based on the Brazil Tupi word "iwasai"; the "palm hearts" have been collected for many years; in recent times, the species has been used primarily as a source of berries, *Euterpe oleracea* (Arecaceae), 2n = 2x = 36 *hort*

acerola: A large bushy evergreen shrub or small tree 2 to 6 m in height; it is widely cultivated in the tropics for fresh fruit and juice; it has one of the highest vitamin C content of all fruits; *Malpighia emarginata* (Malpighiaceae), 2n = 2x = 20 *hort*

acha: >>> hungry rice

achiote: A tropical American shrub which is cultivated as a food coloring; popular red dye (bixin) used for coloring butter and cheeses; dye derived from seeds of spiny red fruits; also used for body paint by South American Indians; chemically similar to beta-carotene and may protect skin from UV light; *Bixa orellana* (Bixaceae), 2n = 2x = 14; 1C DNA content ~0.20 pg *hort*

achita: >>> jataco

acidanthera: A native of Ethiopia to Burundi and Mozambique regions; *Acidanthera bicolor* var. *murielae* (Iridaceae), 2n = 2x = 30 *hort*

acid lime: >>> lime

adlay (*syn* Job's Tears): This plant is usually grown for its pearl- or stone-looking seeds; they have been used for centuries as beads for jewelry ("Good Luck" necklaces) and rosaries; the seeds, when ripe, can be any color from pearly gray to pure black; the mature seeds grow with a premade hole through the center and can be stained with common wood stains; the plant itself is often grown as an ornamental grass that somewhat resembles maize; in the Orient, the seeds are eaten as a cereal called "adlay"; this plant is a perennial there, but is grown elsewhere as an annual; in warmer climates, adlay can easily reach 3 m tall; *Coix lacryma-jobi* (Gramineae), 2n = 2x = 20; 1C DNA content ~1.63 pg *hort agr*

adley: >>> adlay

adzuki bean: A small, dried, russet-colored bean with a sweet, nutty flavor; adzuki beans can be purchased whole or powdered at Asian markets; they are particularly popular in Japanese cooking where they are used in confections such as the popular "yokan", made with adzuki-bean paste and agar; the seeds are small, reddish-brown beans, rounded in shape with a point at one end; the plant probably originated from China, and is currently imported from China and Thailand where the beans are harvested in November and December; in the Orient, adzuki beans are usually cooked to a soft consistency, red in color, and served with such ingredients as coconut milk; they are also cooked with rice, their bright color tinting the rice an attractive pink; *Phaseolus angularis* syn *Vigna aconitifolia* (Leguminosae), 2n = 2x = 22; 1C DNA content ~1.40 pg *hort* >>> pulse

Aegilops: A genus of grasses including several species of breeding and genetic interest; some of them contributed to the evolution of bread >>> wheat; they are also used as donors of disease resistance and other genes for cultivated wheat; *Aegilops* spp. (Gramineae), x = 7 *bot* >>> crop evolution >>> Figure 10 >>> Table 60

aegle marmelos: >>> bael

aerial yam (*syn* **potato yam**): Of minor importance as a food crop but was probably important in ancient times; it is the only species that occurs wild in both Africa and Asia; *Dioscorea bulbifera (Dioscoreaceae)*, x = 10 (Old World species), x = 9 (New World species), most yams being tetraploids or hexaploids (2n = 36, 54); within Old World yams, high degrees of polyploidy occur (with chromosome numbers 2n = 50, 60, 70, 80, 90, 100, 120, 140; 1C DNA content ~1.20 pg *hort*

African daisy (*syn* **osteospermum** *syn* **Cape daisy**): The plant originates in South Africa; osteospermums are half-hardy perennials and will therefore not survive frost conditions; however, they can be propagated by cuttings or may be over-wintered in a greenhouse or conservatory; the flower heads have a central disc and petals in the form of ray florets; the central disc comes in several colors, such as blue, yellow, and purple; the colors of the petals vary from white, cream, pink, purple, and mauve to yellow; they sometimes have a different shade at the tips or towards the end of the petal; some varieties have spoon-shaped petals; *Osteospermum jucundum* (2n = 18), *O. hybrida, O. fruticosum* (2n = 18), *O. grandiflorum, O. ecklonis* (after Dr. Christian Friedrich ECKLON, 1795–1868, described alternatively as a German botanist or a Danish botanical collector and apothecary) (Calenduleae), x = 9; 1C DNA content >1.1 pg *hort*

African millet: >>> finger millet

African rice: Independently domesticated >3.000 years ago; it is believed that African rice was even the first rice the European settlers in the New World cultivated; however, higher-yielding Asian varieties displaced it later; unlike its Asian counterpart, *Oryza sativa*, it copes well under adverse conditions such as drought or acid soils; a look at the genes of *O. glaberrima* confirmed that the origins of the cultivated plant are located in the Delta of the River Niger; from there, the African rice spread at first primarily along the coasts of Senegal and Gambia as well as in the Guinea Highlands; analyses indicate that at that time farmers already selected plants with specific traits for further propagation; they promoted those plants with genetic traits that promised robust plants and successful harvest, to achieve a genetic change in the plant population; therefore, *O. glaberrima* has a much lower genetic diversity than its wild predecessors; it is the result of efforts to breed a rice to withstand the harsh climatic conditions of West Africa; some genetic characteristics of African rice are by no means specific to Africa; it was shown that farmers independently chose similar characteristics both in Africa and in Asia, selecting for similar rice properties, such as an improved yield; on both continents, the rice was also changed so that the ripe seeds stayed longer in the panicles and did not shatter and disperse so quickly; in 2014, the genome of *O. glaberrima*, was completely sequenced; *Oryza glaberrima* (Gramineae), 2n = 2x = 24 (A'A'); 1C DNA content ~0.83 pg *agr* >>> rice >>> Table 45

African violet (*syn* **violet**): A flowering house plant of the gesneriad family; the most common species is *Saintpaulia ionantha*, though altogether there are 20 confirmed species; *S. ionantha* has violet-blue flowers and quilted, dark green leaves with a serrated edge; this is the species from which almost all of today's varieties have descended; while African violets are generally recognized for their distinctive rosette growth habit, trailing forms are also common; the first recorded discovery of African violets was in 1892 in the Usambara Mountains, now a part of Tanzania (Africa); the discovery was recorded by BARON von SAINTPAUL; E. BENARY was the first commercial grower of African violets; the BENARY greenhouse was located in Erfurt, Germany; he provided the seed for two of the first ten commercial hybrids introduced in 1927, "Blue Boy" and "Sailor Boy"; *Saintpaulia* spp. (Gesneriaceae), 2n = 2x = 28; 1C DNA content ~0.75 pg (*Saintpaulia ionantha*) *hort*

African walnut (*syn* **ngag** *syn* **ngak** *syn* **nut vine** *syn* **owusa nut** *syn* **awusa**): A perennial climbing shrub found in the moist forest zones of sub-Saharan Africa; the conophor plant is cultivated principally for the nuts which are cooked and consumed as snacks, along with boiled corn; a bitter after-taste is usually observed upon drinking water immediately after eating them, and this is attributed to the presence of alkaloids and other antinutritional and toxic factors; *Plukenetia conophora syn Tetracarpidium conophorum syn Angostylidium conophorum syn Cleidion preussii* (Euphorbiaceae), 2n = ?x = 60 *hort fore*

African yam: There are more than 600 species of >>> yam; several species of African yam are cultivated for the consumption of their starchy tubers in Africa, Asia, Latin America, and Oceania; they are used in a similar fashion to potatoes and sweet potatoes; there are hundreds of varieties; the word "yam" comes from Portuguese "inhame" or Spanish "ñame," which both ultimately derive from the Wolof word "nyam," meaning "to sample or taste"; yam tubers can grow up to two meters in length and weigh up to 70 kg; the yam has a rough skin that is difficult to peel, but which softens after heating; yam skins vary in color from dark brown to light pink; the majority of the yam is composed of a much softer substance known as the "meat"; this substance ranges in color from white to bright orange in ripe yams; yams are a primary agricultural commodity in West Africa and New Guinea; they were first cultivated in Africa and Asia ~8,000 BC; to this day, yams are important for survival in these regions; yam tubers can be stored for 4–6 months without refrigeration, which makes them a valuable resource for the yearly period of food scarcity at the beginning of the wet seasons; *Dioscorea rotundata* (Dioscoreaceae), 2n = 2x = 40; 1C DNA content ~0.71 pg *hort agr*

agave: A semi-woody perennial native of the American continent; it is used for the production of pulque, aquamiel, and mescal; by federal law in Mexico, *Agave tequilana* var. Azul is the only variety of agave permitted for the production of any tequila; *Agave americana,* 2n = 2x, 4x, 6x = 60, 120, 180; 1C DNA content ~3.05 pg (Agavaceae) *agr*

agbayun: >>> miracle fruit

Agropyron **(wheatgrass):** A hermaphroditic, cespitose, and annual or perennial plant; culms are erect, ascending, or decumbent, and glabrous; internodes are hollow or terete; leaves are basal and cauline, not distinctly distichous; sheaths are terete, margins open; auricles are present; ligules are membranous; blades are flat, folded, or involute, linear, and lax; spikes are bilateral and distichous (~3 mm or less between spikelets), the rachis is persistent, pedicels are absent; the spikelets are solitary at inflorescence nodes, spreading ~40° or more from rachis; they are laterally compressed, with disarticulation above the glumes, mucronate to short awned, sessile; usually four to nine florets per spikelet; a reduced floret at the apex; the callus is glabrous or scabrous; the rachilla does not extend beyond the upper floret; it has two glumes, which are two- to five-veined, opposite relative to the floret position, slightly unequal, not subulate, shorter than the first floret, and short awned; lemmas are five-veined, chartaceous or coriaceous, glabrous or ciliate, apex entire, and mucronate; awns are apical and straight; paleas are two-veined, awnless, and glabrous; there are three stamens; anthers are yellow; caryopses are adunate to the lemma and/or the palea, dorsiventrally compressed; the genus contains about six species; these species are common in China and Russia and occur on dry, frequently stony soils, in full sunlight; >>> crested wheatgrass is an important range revegetation species, several species are used for introgression and alien gene transfer experiments with wheat; *Agropyron* sp. (Poaceae), x = 7; 2C DNA content 13.9 pg (PP) *bot* >>> Figure 43

akee (apple): A West African tree with poisonous fruits; however, the white arils from naturally matured fruits are edible; now common in the West Indies; *Blighia sapida* (Sapindaceae), 2n = 2x = 32; 1C DNA content ~0.90 pg *hort*

alexanders (*syn* **alisander** *syn* **maceron**): The plant, and especially the leaves, have a smell and flavor similar to myrrh; its use as a medicinal plant is very old (2,500 years); it is used to season food in a similar way to parsley, giving flavor to soups and stews, and to prepare sauces accompanying meat and fish; its commonest use has been as a fresh vegetable, with a preference being shown for its leaves, young shoots and leaf stalks, which impart a pleasant flavor similar to celery, although somewhat sharper; it has also been eaten cooked; the roots are used preserved in a sweet-and-sour pickle; the fruit contains an essential oil, cuminal, which is reminiscent of cumin; *Smyrnium olusatrum* (Apiaceae), 2n = 2x = 14; 1C DNA content ~2,82 pg (*Smyrnium perfoliatum*) *hort*

Alexandrinian clover: An annual clover of the multicut group; *Trifolium alexandrinum* (Leguminosae), 2n = 2x = 16; 2C DNA content 2.1 pg *agr* >>> clover

alfalfa (*syn* **lucerne**):   A forage and hay crop; also used as a nitrogen-fixing cover crop or rotation crop; the usual flower color is purple, but white-flowered ones show up regularly; each plant grows up to ~90 cm tall, looking like a small bush; it is an autotetraploid species; because of its high *in vitro* culturability, it became an early subject of biotechnological approaches; it is a cultigen species derived mainly from *Medicago caerulea*, which is indigenous to southwestern Iran, the Caucasus, and eastern Anatolia; domestication appears to have started in the Bronze Age, probably sometime between 1,000 and 2,000 BC; in the Near East, the initial cultivation of alfalfa is thought to have been stimulated by the need to feed horses; horses started being domesticated in Central Asia about 2,500 BC and were brought into the Near East by invaders from Central Asia; by 400 BC, alfalfa was being grown in Europe; there are a wide variety of *M. sativa* cultivars, some of which are the product of hybridization with other wild *Medicago* species from Europe and Asia; there are different cultivars for handling different climatic extremes; the wide climatic tolerance of alfalfa is also due to its extensive root system, which in some soils can extend 7 m below the ground surface, thus making the most of available soil moisture at a wide range of depths; alfalfa grows best in unleached, nonacid soils, particularly those rich in calcium carbonate; it grows particularly well in limestone-derived soils; seed production by alfalfa can be detrimentally affected by poor pollination, often because honeybees (*Apis mellifera*), when extracting nectar, learn to avoid triggering the anthers and releasing pollen; solitary bees, such as bumblebees, *Bombus* species, are the best pollinators; *Medicago sativa* (Leguminosae), 2n = 2x, 4x = 16, 32 (SS, SSSS); 1C DNA content ~1.75 pg *agr* >>> Figure 43

alisander: >>> alexanders

alkanet (root): >>> alkanna

alkanna: This plant is a weak, hairy herb, ~30 cm high, with alternate, oblong, entire, bicuspid leaves; the flowers are small, and carried in terminal racemes, usually in pairs, which unroll as the flowers expand; the calyx is five-lobed, and the corollas funnel-shaped, with a red tube about the length of the calyx lobes, and a blue, five-parted limb; the fruit consists of four distinct nutlets, which are contracted, and not hollowed, at the base; it is indigenous to and cultivated in the southern part of Europe; the roots of cultivated plants are not so rich in the red coloring matter as are those grown in their native soil; the genus *Alkanna* is closely related to *Anchusa* and *Lithospermum*, and the roots of all three genera yield red coloring matter; alkanna root contains the red coloring matter anchusin or alkanet-red; *Alkanna tuberculata* syn *A. tinctoria* syn *Anchusa officinalis* (Boraginaceae), 2n = 2x = 14; 1C DNA content ~1.75 pg *hort*

alligator pear: >>> avocado

allspice: >>> pimento

almond: Closely related to the peach; originated in central and western Asia; it is self-incompatible, and cross-pollination is essential for fruit formation; almonds are cultivated for the seeds, known as nuts, mainly in Turkey and the Mediterranean, as well as in California; in the Old World, almonds are normally grown from seed, while in North America, they are propagated vegetatively; the tree is 3–4 m tall, with a pale-brown, rugged bark, dividing into many spreading branches; the leaves, which are borne on glandular petioles, are 5–7 cm long; the flowers are moderately large, pink or white, sessile, and in pairs, appearing before the leaves; the calyx is reddish, with blunt segments; the petals are variable in size, always much larger than the calyx, ovate, concave, and irregularly notched; stamens are spreading, about half the length of the petals; the fruit is a leathery, hoary drupe, with the sarcocarp spontaneously cracking and dropping off the putamen; the stone is oblong, or ovate, acute, hard to various degrees, always rugged, and pitted with irregular holes; it is indigenous to most of the southern parts of Asia, and is cultivated in many parts of southern Europe; the varieties of sweet almond found in commerce are the 'Valencia', 'Italian', 'Barbary', and 'Jordan'; the last-named, which are the finest of the sweet almonds, come from Malaga; they are hard-shelled, though they are generally removed from the shell before being put on the market; *Prunus amygdalus* var. *fragilis, Amygdalus communis syn Prunus amygdalus* (Rosaceae), 2n = 2x = 16; 1C DNA content ~0.28 pg *hort*

almorta: >>> chickling pea

aloe: A large genus of plants, consisting of about 450 species, from sub-Saharan Africa, Madagascar, and parts of the Middle East; many species are widespread in warm or tropical semi-arid regions, yet the distribution of others is limited to a few living in deserts or wet mountainous regions; while some species have been adopted as medical plants (aloesin, 7-*O*-methylaloesin, 7-*O*-glycosylaloesin) since ancient times, and others are used locally in folk medicine; there are still biologically active compounds to discover; *Aloe vera, Aloe* spp. (Asphodelaceae), 2n = 2x = 14; 1C DNA content ~16,40 pg *hort*

alsike (clover):   A perennial which was introduced into agriculture early in the 19th century; the flowers are white or rosy, and resemble those of the last species; it is of little practical value; *Trifolium hybridum* (Leguminosae), 2n = 2x, 4x (?) = 16, 32; 2C DNA content 1.6–3.1 pg *agr* >>> clover

alstroemeria: A relatively new horticultural crop that is grown for cut flower production; more than 150 cultivars have been registered in the Netherlands alone; most of these have originated through hybridization of the diploid species that are endemic to either Chile or Brazil; although the first commercial hybrid was a diploid interspecific hybrid, this gave rise to superior polyploid cultivars that originated spontaneously in the breeder's nursery; this trend of spontaneous polyploidization appears to be a common phenomenon in alstroemeria, and it has contributed substantially to the development of the present-day cultivars; in addition to diploids, there are triploid and tetraploid cultivars that have originated from the interspecific hybrids; mainly two types of interspecific hybrids of alstroemeria have been used for the development of cultivars: (a) inter-Chilean species hybrids and (b) hybrids between the Chilean and Brazilian species; although more than 60 *Alstroemeria* species have been recognized, only six to seven Chilean and three to four Brazilian species have so far been utilized in breeding; *Alstroemeria inodora* (Alstroemeriaceae/Liliaceae), 2n = 2x = 16; 1C DNA content ~22.10 pg (*A. pelegrina) hort*

alverjón: >>> chickling pea

amaranth: Grain amaranth is an underutilized crop with high nutritional quality, from the Americas. Emerging genomic and biotechnological tools are becoming available that allow the integration of novel breeding techniques for rapid improvement of amaranth and other underutilized crops; amaranth is a pseudocereal that is of particular interest because of its balanced amino acid and micronutrient profiles; additionally, its C_4 photosynthetic pathway and its ability to withstand environmental stress make the crop a suitable choice for future agricultural systems; the annual plant is common in southern Europe and Asia; because of the morphological diversity, amaranth can be used as an ornamental plant, a vegetable or a grain crop, and even as a weedy plant; as a vegetable, it has been known since the Roman times when it was used similar to spinach; it shows C4 photosynthesis; the Asian amaranths are *Amaranthus tricolor* and *A. lividis*; they were selected as potherbs; they do not develop large inflorescences and exhibit only low seed yields; they are widely grown as a vegetable in India, the East Indies, Southeast Asia, and the Far East; the American amaranths were selected for increased grain production and large compound inflorescences; three species are used: (a) *A. cruentus* (eastern North America, tropical highlands of Mexico, Central America, and South America); (b) *A. hypochondria-cus,* which is mainly grown in the western Sierra Madre of Mexico; and (c) *A. caudatus,* grown in Ecuador, Peru, and Bolivia; *Amaranthus* spp. (Amaranthaceae); grain amaranth: *Amaranthus caudatus* or *A. hypochondriacus syn A. frumentaceus syn A. leucocarpus* (Amaranthaceae), 2n = 2x = 32; 2C DNA content 1.9pg *agr* >>> dye amaranth *hort* >>> http://newcrop.hort.purdue.edu/newcrop

amarelle: One of several cultivated sour cherry varieties bearing pale red fruit with colorless juice; 'Montmorency' is a famous variety and has bright red cherries with a sweet-sour flavor; amarelle are the most popular sour cherries in Canada and have given their name to a range of dishes, which include the fruit, from duck to gâteau to ice creams; English cherries are small, bright orange-red fruit with soft translucent flesh and are mainly used for preserves; *Cerasus vulgaris* var. *caproniana* (Rosaceae), 2n = 2x = 32 *hort* >>> cherry

amaryllis: >>> hippie

American chestnut: With about three to four billion trees during the 19th century, it was native to eastern U.S.; by 1940, most of the trees had been wiped out by chestnut blight; the pathogenic fungus *Cryphonectria parasitica* (*syn Endothia parasitica*) is a member of the Ascomycota phylum; the blight was accidentally introduced to North America around 1900, possibly on imported Japanese chestnut nursery stock; in 1905, American mycolo-gist William Murrill isolated and described the fungus responsible and demonstrated, by inoculation onto healthy plants, that the fungus caused the disease; resistant hybrid chest-nut trees were bred that retained the main characteristics of the American chestnut tree; in the early 1950s, James Carpenter discovered a large living American chestnut in a grove of dead and dying trees in Salem, Ohio that showed no evidence of blight infection; Carpenter sent budwood to R. T. Dunstan, a plant breeder in Greensboro, North Carolina; Dunstan grafted the scions onto chestnut rootstock and the trees grew well; he cross-pollinated one with a mixture of three Chinese chestnut selections: 'Kuling','Meiling', and 'Nanking'; the resulting fruit-producing hybrid was named the Dunstan Chestnut; the trade-off for resis-tance to the chestnut blight was that the Dunstan hybrid grew to a height of only 7.6 m; one of the most successful methods of breeding was to create a backcross of a resistant species from China or Japan to American chestnut; the two species are first hybridized to create a 50/50 hybrid; after three backcrosses to American chestnut, the remaining genome is approximately 1/16 that of the resistant tree, and 15/16 American chestnut; the strategy is to select for blight-resistance genes during the backcrossing, while preserving the more wild-type traits of American chestnut as the dominant phenotype; thus, the newly bred hybrid chestnut trees should reach the same heights as the original American chestnut; research is also being conducted to insert resistance genes from wheat into American chestnut; in 2013, the SUNY ESF company had over 100 individual transformants being tested, with more than 400 slated to be in the field or in the lab for various assay tests in the next several years, and more than 1,000 trees growing in several field sites in 2014; *Castanea dentata* (Fagaceae), 2n = 2x = 24; 1C DNA content ~0.98 pg (*Castanea sativa*) *hort*

American hazel: *Corylus americana* (Corylaceae), 2n = 2x = 22; 1C DNA content ~0.48 pg (*C. avellana*) *hort*

American persimmon (*syn* **common persimmon** *syn* **eastern persimmon** *syn* **simmon** *syn* **possumwood** *syn* **sugar-plum**): The tree grows wild but has been cultivated for its fruit and wood since prehistoric times by Native Americans; its range runs from southern Connecticut/Long Island to Florida, and west to Texas, Louisiana, Oklahoma, Kansas, and Iowa; it grows up to 20 m in well-drained soil; in summer, this species produces fragrant flowers; trees are dioecious, so one must have both male and female plants to obtain fruit; most cultivars are parthenocarpic; the flowers are pollinated by insects and the wind; fruiting typically begins when the tree is about six years old; the fruit is round or oval and usually orange-yellow, sometimes bluish, and from 2 to 6 cm in diameter; commercial varieties include the very productive 'Early Golden', the productive 'John Rick', 'Miller', 'Woolbright' and 'Ennis' as well as 'Meader', which are seedless varieties; another name for the American persimmon, 'date-plum' also refers to a persimmon species found in South Asia, *Diospyros lotus*; *Diospyros virginiana* (Ebenaceae), 2n =4x, 6x = 60, 90

American plum: The North American plum is a diploid, compared with the European plum, which is a hexaploid; *Prunus americana* (Rosaceae), 2n = 2x = 16 *hort*

American sarsaparilla: >>> sarsaparilla

American yam: Despite the economic and cultural importance of the indigenous "Amerindian" yam, very little is known about its origin, phylogeny, diversity, or genetics; consequently, conventional breeding efforts for the selection of genotypes resistant to potyviruses, which are directly involved in the regression of this species, have been seriously limited; *Dioscorea trifida* (Dioscoreaceae), 2n = 4x = 36–40 (autotetraploid) *agr hort* >>> aerial yam

Amochi: An herbaceous root crop; it belongs to the subfamily Aroideae; for more than 20 years, amochi has been grown as a food crop for its edible tubers in two subzones of southern Ethiopia; it is peculiar in that it remains dormant in the soil during the main rainy season when most crops grow; it emerges after the harvest of other crops and grows during the off-season, using residual moisture; amochi and other root crops that grow on residual moisture are gaining special attention in southern Ethiopia, as the rainfed agriculture is getting less dependable due to increased variability of rainfall; amochi yields up to 12 t/ha, which is comparable to the yields of the main-season crops of the area; it is irritating in contact with the skin and to the mouth when eating; therefore, amochi processing comprises mashing and prolonged storage before cooking; mashing of hard tubers demands considerable time and energy; prolonged storage on the other hand is in conflict with the immediate food needs of amochi growers because it is harvested at the end of the dry season when the stored food items from the preceding harvest are depleted; breeding for easily mashed and less irritating types with higher yield is a main target; *Arisaema schimperianum syn A. enneaphyllum* (Araceae*)*, 2n = 28, 56; 1C DNA content ~5.06 pg (*A. flavum*) *agr hort*

amorpha: >>> bastard indigo

Amur grape: *Vitis amurensis* (Vitaceae), 2n = 2x = 38; 1C DNA content ~0.50 pg *hort* >>> European grape

anchote (ivy gourd *syn* **scarlet-fruited** *syn* **gourd** *syn* **tindori** *syn* **courge écarlate):** A mostly prostrate or climbing herbaceous annual; there are about 90 genera and 700 species; commonly has 5-angled stems and coiled tendrils; the leaves are alternate and usually palmately 5-lobed or divided; stipules are absent; the flowers are actinomorphic and nearly always unisexual; the >>> perianth has a short to prolonged epigynous zone that bears a calyx of 3–6 segments or lobes and 3–6 petals or more frequently a 3- to 6-lobed sympetalous corolla; the >>> androecium is highly variable, consisting of basically five distinct to completely connate stamens that frequently are twisted, folded or reduced in number; the fruit is a type of berry called a pepo; it occurs wild from Senegal east to Somalia and south to Tanzania; also in Saudi Arabia, Yemen and India; it is locally naturalized in Mozambique and Mauritius, and has been introduced in many other tropical and subtropical regions; it is considered to be an invasive, obnoxious weed in Australia, Florida, and the Pacific Islands, e.g. Hawaii; it is cultivated in Asia from India to Indonesia; in East Africa, especially in Kenya, it is cultivated mainly for consumers of Indian origin; ripe red fruits of sweet-tasting cultivars are eaten raw, or they are peeled and cut into pieces and prepared as a stew with onions and tomatoes, *Coccinia abyssinica* (Cucurbitaceae*), 2n = 2x = 24; 1C DNA content ~2.08 pg (*C. indica*) *agr*

Andean arracacha (*syn* **apio** *syn* **apio criollo, Creole celery** in Venezuela, *syn* **zanahoria blanca** in Ecuador, *syn* **virraca** in Peru *syn* **mandioquinha, little cassava,** *syn* **batata-baroa, baroness potato,** in Brazil, *syn* **white carrot**): A root vegetable, originally from the Andes, somewhat intermediate between the carrot and celery; the leaves are similar to parsley, and vary from dark green to purple; the roots resemble short fat carrots, with lustrous off-white skin; the interior may be white, yellow, or purple; the starchy taproot is a popular food item in South America where it is a major commercial crop; it is vegetatively propagated; the origin is still unresolved; wild tuberous forms are present in the presumed areas of domestication and have a perennial or monocarpic life history; the name *arracacha* or *racacha* was borrowed into Spanish from Quechua *raqacha; Arracacia xanthorrhiza* (Apiaceae), 2n = 4x = 44 *hort*

Andean common bean: >>> French bean

anemone: The genus *Anemone* includes many species cultivated for ornamental purposes; most cut-flower cultivars belong to *A. coronaria* and are propagated by seed and sold for cultivation as 1-year-old tubers; as cultivars represent a population of hybrid individuals derived from crosses between heterozygous parents, the use of a true F1 hybrid became of interest in order to improve the uniformity and quality of the product; recently, somatic embryos and plantlets were regenerated from elite cultivars; the shortening to 15 months for the time required to produce homozygous lines would be of great benefit for seed companies to invest in F_1 hybrid breeding; *Anemone coronaria* (Ranunculaceae), 2n = 2x = 16; 1C DNA content ~11.16 pg *hort bot biot*

anise: An herb whose seed-like fruit (also called *"aniseed"*) is used in sweet baking as well as in anise-flavored liqueurs; all above-ground parts of the young anise plant are also eaten as a vegetable; the stems resemble those of celery in texture but are much milder in flavor than the fruits; the >>> Chinese star anise is botanically unrelated to anise, but has a very similar taste and aroma and recently has come into use in the West as a (cheaper) substitute for anise in baking, as well as in liquor production; *Pimpinella anisum* (Umbelliferae), 2n = 2x = 18, 20; 1C DNA content ~3.23 pg (*P. major*) *hort*

anise hyssop (*syn* **Korean mint** *syn* **licorice mint**): The tightly clustered whorls of tubular flowers are borne all summer long atop a three foot standard; one of the best plants for attracting pollinating insects; native to North America; recently discovered as a new herb plant; *Agastache anisata* (Labiatae), 2n = 2x = 18; 1C DNA content ~3.23 pg *hort*

annatto: >>> achiote

annual meadow grass: *Poa annua* (Gramineae), 2n = 4x = 28; 2C DNA content 2.9–5.2 pg *agr*

ape: >>> giant taro

apple: Deciduous, infrequently evergreen, branching tree or shrub; leaves are folded or twisted in buds, ovate, elliptic, lanceolate, or oblong, lobed, serrate, or serrulate; buds are ovoid with a few overlapping scales; flowers are white to pink or crimson, epigynous, and in cymes; stamens 15 to 50; styles two to five; ovary 3- to 5-celled; fruit is a pome, oblong, oblate, conic, or oblique, diameter 2 to 13 cm; various hues of green to yellow to red, varying russet and lenticel characteristics; flesh lacking stone cells; there are ~30 species known; domestic apple derived mainly from *Malus pumila* (*syn M. domestica*); domestic crab apple are hybrids of *M. pumila* and *M. baccata*; the origin of the apple is either southwest Asia,

---

* The predominant chromosome base number is x = 12, varying from 7 to 24; *Cucurbitacae* show a so-called fixed polyploidy with a base genome of n = 20; polyploidy and aneuploidy is quite common; the family is characterized by 35 bp long inversion between the genes of leucine-tRNA and phenylalanine-tRNA, not present in other families of the order *Cucurbitales*; in *Cucurbita digitata* and *Neoalsomitra ssp.* the inversion is again turned around; the genus of *Cucumis* has also the largest known mitochondrial genome, e. g., of about 1,500 kb in cucumber and 2,400 kb in muskmelon (*Cucumis melo*); the mitochondrial inheritance happens paternal, not as otherwise usual maternal.

in the Caucasus Mountains, or south-central Asia, on the slopes of the enormous mountain range that separates China, Kazakhstan, and Kyrgyzstan; Russian scientists and a team of horticulturists from Cornell University (Ithaca, New York) have studied the genetic makeup of apples; thus far, their trek has led them to identify the lower slopes of the Tian Shan mountains as a possible location of the original apple; this area to the northeast of Alma-Ata is called the Dzungarian Alps and is referred to as the "original wild apple forest"; within this wild forest, many varieties of apples have grown disease free for centuries; apples, as the Europeans knew them, were not native to America; explorers, Jesuits and Franciscan missionaries, and early European settlers brought seeds and occasionally small trees with them to plant orchards around their new homes; it is believed that J. ENDICOTT, an early governor of Massachusetts, brought the first apple tree to Massachusetts in the early 1600s; although apples were grown in Europe and other parts of the world, the fruition of the apple came when it arrived in North America; the warm summers and cold winters in America have helped perfect the fruit unlike anywhere else in the world; many new varieties were worthy of naming and are still grown; the first orchard in Massachusetts was planted around 1625 by a clergyman named W. BLAXTON who owned a farm on Beacon Hill in Boston; he later moved to Pawtucket, Rhode Island, and planted the first Rhode Island orchard in 1635; Blaxton is credited with having grown the first named apple in America; apples are frequently named after the owner or location of origin; he named his apple "Blaxton's Yellow Sweeting" but it was later referred to as "Sweet Rhode Island Greening"; branchless trees became interesting during the last decade, this columnar genotype being caused by one gene reducing the internode length, *Malus pumila* (Rosaceae), 2n = 2x, 3x = 34, 51; the basic chromosome number was 9, by recombination derived 17; 1C DNA content ~0.77 pg, genome size 742.3 Mb, about 44,105 genes *hort*

apricot: Yellow-fleshed fruit of the apricot tree, which is closely related to the almond, peach, plum, and cherry; although native to the Far East, it has long been cultivated in Armenia, from where it was introduced into Europe and the U.S.; recent varieties are not only grafted but also seed-propagated, e.g. in the Maghreb region (Tunisia and Morocco); *Armeniaca mandshurica syn Prunus armeniaca* (Rosaceae), 2n = 2x = 16; 1C DNA content ~0.30 pg *hort*

aralia: Hardy shrubs, trees, and herbaceous plants of the ivy family Araliaceae, native to North America, Asia, and Australia; *Aralia cachemirica*, Kashmir, *A. cordata* (2n = 4x = 48), Japan, and *A. nudicaulis* (2n = 2x = 24), North America, are herbaceous perennials, grown for their effective foliage; the houseplant known as aralia belongs to the related genus *Dizygotheca*; *Aralia* spp. (Araliaceae) *hort*

arandi: >>> Barbados nut

areca palm: >>> betel nut

argan tree: Endemic to Morocco, it is valued for its nutritive, cosmetic, and numerous medicinal properties; the smooth pits from the tree contain one to three kernels, which look like sliced almonds and are rich in oil; the kernels are then removed and gently roasted; this roasting accounts for part of the oil's distinctive, nutty flavor; it takes several days and about 32 kg of fruit – roughly one season's produce from a single tree – to make only 1 l of oil; the cosmetic oil, rich in vitamin E and essential fatty acids, is used for massage, facials, and as an ingredient in anti-aging cream; the edible oil is extracted from roasted kernels; most of the oil is bottled pure for cooking, as a dressing on salads, meat or fish or simply as a dip for bread; the tree, a relict species from the Tertiary Period, is extremely well adapted to the drought and other environmentally difficult conditions of southwestern Morocco; it grows wild in semi-desert soil; its deep root system helps to protect against soil erosion and the northern advance of the Sahara Desert; as a result of the recent interest in argan oil, programs of breeding, multiplication, and reforestation were initiated; *Argania spinosa* (Sapotaceae), 2n = 2x = 20 *fore hort*

Armenian cherry: >>> amarelle

arnautka: Strains of >>> durum wheat [*Triticum durum,* $2n = 4x = 28$ (BBA^uA^u); 1C DNA content ~12.66 pg] *agr* >>> Table 1

aroids, edible: *Alocasia* spp., *Colocasia* spp. (dashen or eddoe, $2n = 28$; 1C DNA content ~10.15 pg), *Cyrtsperma* spp., *Xanthsoma* spp. (Araceae) *hort*

aromatic ginger: >>> galangal

arrowhead: A species of water plant of the family Alismataceae, native to Europe; it has arrow-shaped leaves and flowers which appear above the surface; there are also leaves below the water, and these are very thin and ribbon-like; the stem is a rhizome; *Sagittaria sagittifolia* (Alismataceae), $2n = 2x = 22$; 1C DNA content ~21.25 pg *hort*

arrowroot canna: *Canna glauca* or *C. edulis* (Cannaceae) *hort* >>> Queensland arrowroot

arsorossie: >>> bitter gourd

artichoke: This Mediterranean crop is a perennial thistle and is vegetatively propagated, because true seedlings are very variable; immature flower heads are cooked and eaten; the tender receptacle and "meaty" phyllaries are dipped in butter; *Cynara cardunculus syn Cynara cardunculus* var. *scolymus syn C. scolymus* (Compositae), $2n = 2x = 34$; 1C DNA content ~1.10 pg *hort*

arugula: >>> rocket salad

arvi: >>> taro

asafetida: A herb with about 100 perennial, thick-rooted species from the Mediterranean region and central Asia; cultivated since ancient Greek and Roman times; the massive fleshy tap-roots of mature plants release milky resinous juice that is manufactured into pills, powder, or paste, and used as alternatives to onion and garlic in mushroom, meat and other dishes, *Ferula assafoetida, F. foetida, F. narthex* etc. (Apiaceae), $2n = 2x = 22$ *hort*

asaa: >>> miracle fruit

ash, white: *Fraxinus americana* (Oleaceae), $2n = 2x, 4x, 6x = 46, 92, 138$; 1C DNA content ~0.89 pg (*F. angustifolia*) *fore hort*

ash gourd: >>> wax gourd

ashoka: *Saraca indica* (Amherstieae), $2n = 2x = 24$; 1C DNA content ~2.00 pg *hort*

Asiatic pennywort: >>> centella

asp: >>> aspen

aspen: Any one of several species of poplar tree; the European quaking aspen (*Populus tremula*) has flattened leafstalks that cause the leaves to flutter in the slightest breeze; the soft, light-colored wood is used for matches and paper pulp; it is grown on plantations; *Populus* spp. (Salicaceae), $2n = 2x = 38$; 2C DNA content 1.2 pg; haploid genome size ~480 Mb *fore*

asparagus: A dioecious, perennial vegetable plant, with male and female flowers on separate individuals, that is cultivated for its young succulent shoots; it derived its name from the ancient Greeks, who used the word to refer to all tender shoots picked and savored while very young; as a dioecious vegetable, it has been used since 3,000 years ago in Egypt; as a medicinal plant against coughs and urinary problems, it has been used for more than 3,000 years in China; it has been grown in Europe since the 15th century; Germany, France, Spain, Greece, and Belgium are the European countries with the highest production; Peru has become the biggest exporter of asparagus in the world; it is grown from seeds or rhizomes for its young, succulent shoots; breeding is a challenge since this dioecious crop is inevitably out-pollinating; useful genotypes can be maintained by vegetative propagation; it is also possible to produce diploid homozygous hermaphrodite asparagus by diploidization of haploidsobtained from twin seedlings; selfed seeds from those doubled haploids are highly uniform and available as seed-propagated asparagus; the sexual dimorphism is controlled by the gene *M* located on a pair of homomorphic sex chromosomes that do not differ in morphology; in breeding and cultivation, male plants are preferred due to their higher yield and greater longevity; for breeding of male plants, homozygous males ("supermales") are important; these supermales cannot be differentiated from the heterozygous males even at the flowering stage; time-consuming testcrosses have to be carried out to identify them; *Asparagus officinalis* (Liliaceae), 2n = 2x, 4x = 20, 40; 2C DNA content 4.2 pg *hort*

asparagus bean (*syn* **bora** *syn* **Chinese long bean** *syn* **garter bean** *syn* **snake bean, yard-long bean** *syn* **long-podded cowpea**): The plant is subtropical/tropical and most widely grown in the warmer parts of Southeastern Asia, Thailand, and Southern China; the pods are actually only about half a yard long; it is a vigorous climbing annual vine; it is grown primarily for its strikingly long (35–75 cm) immature pods and has uses very similar to those of the green bean; the pods, which begin to form just 60 days after sowing, hang in pairs; they are best for vegetable use if picked before they reach full maturity; the many varieties are usually distinguished by the different colors of their mature seeds; a traditional food plant in Africa, this little-known vegetable has potential to improve nutrition, boost food security, foster rural development and support sustainable landcare; *Vigna unguiculata* ssp. *unguiculata syn Phaseolus unguiculata* ssp. *sesquipedalis* (Leguminosae), 2n = 2x = 14; 1C DNA content ~0.60 pg *hort*; polytene chromosomes have been observed in ovary, anthers, or immature seed tissues

asparagus broccoli: *Brassica oleracea* convar. *botrytis* var. *italica* (Brassicaceae), 2n = 2x = 18 (CC); 1C DNA content ~0.78 pg *hort* >>> *Brassica* >>> Figure 8

asparagus pea: In northern Europe, this is a rarely grown vegetable; however, in southern Europe, it is still common; although the habit is pea-like, it is related to clover; *Tetragonolobus purpureus syn Lotus tetragonolobus* (Leguminosae), 2n = 2x = 14 *hort*; polytene chromosomes have been observed in ovary, anthers, or immature seed tissues *hort* >>> Figure 57, 58

Assam tea: A black tea named after the region of its production, Assam, in India; it is manufactured specifically from the plant *Camellia sinensis* var. *assamica*; this tea, most of which is grown at or near sea level, is known for its body, briskness, malty flavor, and strong, bright color; Assam teas, or blends containing Assam tea, are often sold as "breakfast" teas; the state of Assam is the world's largest tea-growing region, lying on either side of the Brahmaputra River, and bordering Bangladesh and Burma (Myanmar); this part of India experiences high precipitation; during the monsoon period, as much as 250–300 mm of rain falls per day; the daytime temperature rises to about +40 °C, creating greenhouse-like conditions of extreme humidity and heat; this tropical climate contributes to Assam tea's unique malty taste, a feature for which this tea is well known. Historically, Assam has been the region producing the second-highest levels of commercial tea production, after southern China; southern China and Assam are the only two regions in the world with native tea plants, *Camellia sinensis* ssp. *assamica* (Theaceae), 2n = 2x, 3x = 30, 45 *hort*

aster: A large group of plants belonging to the same subfamily as the daisy; all asters have star-like flowers with yellow centers and outer rays (not petals) varying from blue and purple to white; many are cultivated as garden flowers, including the Michaelmas daisy (*Aster novi-belgii*); the China aster (*Callistephus chinensis*) belongs to a closely related genus; it was introduced to Europe and the U.S. from China in the early 18th century; *Aster* spp. (Compositae), 2n = 2x = 10 *hort*

astragalus: *Astragalus cicer* (Leguminosae), 2n = 4x = 64; 1C DNA content ~0.73 pg (*A. angustifolius*) *agr*

aubergine: >>> eggplant

aucuba: One of the most useful shrubs for shaded areas; it can grow, even thrive, beneath the dense canopy of shade-bearing trees, including beech or lime, and the dry conditions that are prevalent in such situations for most months of the year do not seem to have an adverse effect; it is not unattractive, not excessively prone to any dreadful pests or diseases, and fits into the usually highly saleable category of evergreen; at the time of year when deciduous trees and shrubs are casting their leaves to nature's recycling system, aucuba maintains a proud stance with glossy, healthy, and often variegated foliage; it is very tolerant of a wide range of soil conditions, although it is best suited to a well-drained, woodland environment; when selecting aucuba, it is important to remember that each plant will have either male or female flowers; this does not affect the standard of foliage, but, if interested in berries as well as foliage, then one with female flowers must be chosen; most well-known cultivars produce bright red berries, but there are a few uncommon cultivars that have pink or white berries; however, regardless of color, to be certain of a good show of berries, it is best to plant both male and female cultivars; *Aucuba japonica* (Cornaceae), 2n = 2x = 16; 1C DNA content ~12.85 pg *hort*

auricula: A species of primrose, a plant whose leaves are said to resemble a bear's ears; it grows wild in the Alps but is popular in cool-climate areas and often cultivated in gardens; *Primula auricula* (Primulaceae), 2n = 8x = 64; 1C DNA content ~0.49 pg (*P. veris*) *hort*

Australian finger lime: A thorny understorey shrub or small tree of lowland subtropical rainforest and dry rainforest in the coastal border region of Queensland and New South Wales (Australia); the plant is 2–7 m in height; leaves are small, 1–6 cm long and 3–25 mm wide, glabrous, with a notched tip and crenate towards the apex; flowers are white with petals 6–9 mm long; the fruit is cylindrical, 4–8 cm long, sometimes slightly curved, coming in different colors, including pink and green; early non-indigenous settlers consumed the fruit and retained the trees when clearing for agriculture; the finger lime has been recently popularized as a gourmet bushfood; the fruit juice is acidic and similar to that of a lime; marmalade and pickles are also made from finger lime; there is a wide range of different-colored variants of finger lime fruit, including green, yellow, orange, red, purple, black and brown; commercial use of finger lime fruit started in the mid-1990s; by 2000, the finger lime was being sold in restaurants, and fresh fruits were being exported; there is an increasing range of genetic selections which are budded onto *Citrus* rootstock; with the sudden high market demand for the fruit, the primary source of genetic material for propagation has been selections from wild stock, *Citrus australasica syn Microcitrus australasica* (Rutaceae), 2n = 2x = 18; 1C DNA content ~0.38 pg (*C. aurantium*) *hort* >>> www.users.kymp.net/citruspages/citrons.html

Australian nut: >>> Macadamia nut

Australian rice: *Oryza australiensis* (Gramineae), 2n = 2x = 24 (EE); 2C DNA content 2.50 pg *agr* >>> rice >>> Table 45

autumn crocus: >>> meadow saffron

avocado (pear): Originated in Central America (Mexico–Guatemala) but is now grown in most tropical and subtropical countries; the seed is highly heterozygous, and selected clones must be propagated vegetatively as grafts on seedling stocks; it is a highly nutritious salad fruit containing up to 30% oil that has a composition similar to olive oil; the tree belongs to the laurel family; its dark-green, thick-skinned, pear-shaped fruit has buttery-textured flesh; in 1998, three compounds were identified in unripe avocados that are effective against cancer cells and also act as pesticides; *Persea americana syn P. gratissima* (Lauraceae), 2n = 2x = 24; 1C DNA content ~0.93 pg *hort*

awnless brome: >>> smooth brome

awusa nut: >>> African walnut

azalea: A group of largely deciduous flowering shrubs belonging to the heath family; several species are native to Asia and North America, and many cultivated varieties have been derived from these; azaleas are closely related to the mostly evergreen rhododendrons; the Japanese azalea varieties make particularly fine evergreen ornamental shrubs; *Rhododendron simsii* (Ericaceae) *hort* >>> Table 46

azia cucumber: A type of giant cucumber derived from Asia (= azia); rarely grown in some European countries; it ripens with a deep yellow-orange color; it is used mainly for pickling, *Cucumis sativus* (Cucurbitaceae), 2n = 2x = 14; 1C DNA content 0.9–1.0 pg; ~35,000,000 bp per chromosome *hort*

B

babul: *Acacia arabica* (Acaieae), 2n = 2x = 52 *hort fore*

bael (*syn* **aegle marmelos** *syn* **bilva** *syn* **Bengal quince** *syn* **stone apple** *syn* **wood apple**): A monotypic genus in Rutaceae; it is a mid-sized, slender, aromatic, thorned, gum-bearing tree growing up to 18 m tall; it has a leaf with three leaflets; is native to India and present throughout Southeast Asia as a naturalized species; the tree is considered to be sacred by Hindus; its fruits are used in traditional medicine and as a food throughout its range; ripe fruits show antimutagenic potential, *Aegle marmelos* (Rutaceae), 2n = 2x = 18; 1C DNA content ~0.53 pg *hort*

Bahia grass: Not mentioned as a turfgrass until 1938; it was first found on a sodded sand bank in Florida and called "Pensacola"; this excellent low-growing grass may have arrived as a stowaway on a fruit boat from Central or South America; as a crop, it was released in 1944; in 1946, it was discovered to be a diploid plant (2n = 20) that reproduces sexually, based on the variability within the progenies; native to eastern Argentina, the Bahia grass genotype has become one of the major forage grasses of the southeastern U.S.; it is more cold tolerant than other grass cultivars; the tetraploid Bahia grass cytotype *P. notatum* var. *latiflorum* is the most common botanical variety in tropical and subtropical America, *Paspalum notatum* (Gramineae), 2n = 2x, 4x = 20, 40; 1C DNA content ~0.72 pg *agr*

bajori: *Pennisetum echinurus syn P. typhoideum* (Gramineae), 2n = 2x = 14 *agr* >>> pearl millet

bajra: >>> pearl millet

Balansa clover: A winter-growing annual, aerial seeding pasture legume species; it is a herbaceous, self-regenerating legume in Mediterranean climates; the habit is prostrate as a single plant, becoming semi-erect in dense swards; stems are predominantly green with occasional red coloration, glabrous, and hollow when elongated; leaves are trifoliate, alternate, green, glabrous; it is suitable for southern Australian farming systems because of its hard-seeded nature and time to flowering; late-maturing varieties with increased rate of hardseed breakdown, particularly in the late summer–early autumn period, are desirable to ensure reliable regeneration in permanent pastures; *Trifolium michelianum* (Leguminosae), 2n = 2x = 14; 2C DNA content 1.7 pg *agr* >>> clover

balsam: >>> garden balsam

balsamapfel: >>> bitter gourd

balsamina: >>> bitter gourd

balsamini longa: >>> bitter gourd

balsam pear: >>> bitter gourd

bambara groundnut: In Indonesia, it is known as "Kacang Bogor"; this name relates to the city of Bogor; "Kacang" means "bean" or more generally "pulse" in the Indonesian language; it seems to be grown in only a restricted area around Bogor and in limited amount; it is greatly appreciated by people in this region, not as a staple food, but sometimes offered boiled for direct consumption in local markets; *Vigna subterranea syn Voandzeia subterranea* (Leguminosae), 2n = 2x = 22; 1C DNA content ~0.90 pg *hort* >>> pulse >>> gynophore

bamboo (cane): A large group of giant grass plants, found mainly in tropical and subtropical regions; some species grow as tall as >30 m; the stems are hollow and jointed and can be used in furniture, house, and boat construction; the young shoots are edible; paper is made from the stems; bamboos flower and seed only once before the plant dies, sometimes after growing for as long as 120 years; *Bambusa* spp. (Gramineae), 2n = 2x = 72 *agr hort*

banana: Banana is one of the most important of all crops; it is a large monocotyledonous herb that originated in Southeast Asia; virtually all of the cultivars that are grown are thought to have been selected as naturally occurring hybrids in this region by the earliest farmers; in fact, banana was one of the first crops to be domesticated by man; recent studies show that banana has been grown in Indonesia for 6,500 years, in Uganda for 4,500 years, and in Cameroon (Africa) for 2,500 years; despite the current, clear understanding of its ancestry, the edible bananas' origins are often confused in the literature; almost all of the 300 or more cultivars that are known arose from two seeded, diploid species, *Musa acuminata* and *M. balbisiana*; there are diploid, triploid, and tetraploid hybrids among subspecies of *M. acuminata*, and between *M. acuminata* and *M. balbisiana*; conventionally, the haploid genome contributions of the respective species to the cultivars are noted with an A (*acuminata*) and B (*balbisiana*), e.g. the Cavendish cultivars, that are the mainstays of the export trades, are pure triploid *M. acuminata* and, thus, AAA; the species *M. paradisiaca* (the AAB plantains) and *M. sapientum* (the sweet dessert bananas, of which "Silk" AAB is the type cultivar) are invalid and no longer used; banana is now one of the most popular of all fruits; after rice, wheat, and milk, it is the fourth most valuable food; in export terms, it ranks fourth among all agricultural commodities and is the most significant of all fruits, with world trade totaling $2.5 billion annually; yet, only 10% of the annual global output of 86 million tons enters international commerce; much of the remaining harvest is consumed by poor subsistence farmers in tropical Africa, America, and Asia; for most of the latter producers, banana and plantain (which is a type of banana) are staple foods that represent major dietary sources of carbohydrates, fiber, vitamins A, B6, and C, and potassium, phosphorus, and calcium; in 2014, after 20 years of research, the Ugandan National Agricultural Research Organization (NARO) and the International Institute of Tropical Agriculture (IITA) released the first-ever hybrid varieties of the East Africa highland banana for food and juice; *Musa acuminate* (AA genomes) and *M. balbisiana* (BB genomes) are the diploid ancestors of modern bananas, that are mostly diploid or triploid cultivars with various combinations of the A and B genomes, including AA, AAA, BB, AAB, and ABB, *Musa* spp. (Musaceae), $2n = 2x = 22$ (AA); $2n = 3x = 33$ (AAA, AAB, ABB); 1C DNA content ~0.58 pg (*M. balbisiana*); the full sequence of the banana genomic DNA was published in 2012, giving access to the sequences of all the ~36,000 genes in this important crop; the international consortium was led by Angelique D'HONT and Patrick WINCKER from the French CIRAD and CEA-Genoscope research organizations, within the framework of the Global Musa Genomics; sequencing of plant genomes has often identified the *hAT* superfamily as the largest group of DNA transposons; by *in silico* analyses of the reference genome assembly and bacterial artificial chromosome (BAC) sequences, the classification and molecular characterization of *hAT* transposon families in *M. acuminata* have been performed; *Musa hAT* transposons are organized into three families, namely MuhAT I, MuhAT II and MuhAT III; in total, 70 complete autonomous elements of the MuhAT I and MuhAT II families are detected, while no autonomous MuhAT III transposons are found; based on the terminal inverted repeat (TIR)-specific sequence information of the autonomous transposons, 1722 MuhAT I- and MuhAT II-specific miniature inverted repeat transposable elements (MuhMITEs) can be identified; autonomous MuhAT I and MuhAT II elements are moderately abundant in the sections of the genus *Musa*, whereas the corresponding MITEs exhibit an amplification in *Musa* genomes; by fluorescent *in situ* hybridization, autonomous MuhAT transposons, as well as MuhMITEs, are localized in subtelomeric, most likely gene-rich regions of *M. acuminata* chromosomes; a comparison of homeologous regions of *M. acuminata* and *Musa balbisiana* BACs revealed the species-specific mobility of MuhMITEs; in particular, the activity of MuhMITEs II showed that transduplications of genomic sequences might indicate the presence of active MuhAT transposons, thus suggesting a potential role of MuhMITEs as modulators of genome evolution of *Musa*; enhanced host plant resistance to *Xanthomonas campestris* pv. *musacearum*, which causes the devastating banana *Xanthomonas* wilt in the Great Lakes Region of Africa, was achieved by plant genetic engineering; the genus *Musa* was previously separated into five sections (*Eumusa, Rhodochlamys, Callimusa,*

Australimusa and *Ingentimusa*), based on basic chromosome numbers and morphological characters; molecular analyses currently support restructuring of *Musa* species into two sections, *Musa* and *Callimusa*; simple sequence repeats pointed to a clade including species of sections *Australimusa* (x = 10) and *Callimusa* (x = 9); most other species of sections *Eumusa* and *Rhodochlamys* (x = 11) form the other clade *hort* >>> http://www.banana.go .ug/index.php/about-us/press-releases

banyan tree: *Ficus benghalensis* (Moraceae), 2n = 2x = 26; 1C DNA content ~0.73 pg *hort*

barbadense: Cotton is the most important natural fiber in the world, and its seeds are also used as a food source; breeding cotton for traits of interest, such as production and processability of fibers, will ensure that this natural product is as competitive as renewable synthetic fibers derived from petroleum; It belongs to a genus with 33 diploid and six allotetraploid species, native to the tropical and subtropical regions of the world; it is also known as Pima or Long-Staple cottons, Extra Long Staple, South American, Creole, Sea Island, and Egyptian, Algodon pais, and West Indische katoen; it is a species of cotton plant which is widely cultivated, though it originated in Peru; it is a tropical perennial plant that produces yellow flowers and has black seeds; cultivated cotton is dominated by two allotetraploid species grown in the warmer regions, *Gossypium hirsutum* and *G. barbadense*, thought to have formed about 1–2 million years ago (mya) by hybridization between a maternal Old World "A" genome taxon resembling *Gossypium herbaceum* (2n = 2x = 26) and a paternal New World "D" genome taxon resembling *Gossypium raimondii* or *G. gossypioides* (both 2n = 2 x = 26); tetraploid cottons have two major leaf types: normal and "okra" leaf; normal leaf is also called "broad leaf," and is predominant among cultivated cottons; "okra" leaf is also called "narrow leaf", and usually has a deeply-cut leaf edge; "okra" leaf has been associated with production advantages such as early maturity, reduced boll rot, reduced leaf area index and higher canopy CO_2-uptake per unit leaf area, higher light-saturated, single-leaf photosynthesis rate per unit leaf area, a shorter sympodial plastochron, increased numbers of flowers per season, better pesticide penetration, and moderate levels of pink bollworm resistance; because of these advantages, breeding has placed priority on high-yielding okra-leaf genotypes; in some areas of Australia, okra-leaf cotton now represents 50% of the cotton acreage; *Gossypium barbadense* (Malvaceae), 2n = 4x = 52 (AADD); 1C DNA content ~3.00 pg; *G. hirsutum*, 1C DNA content ~2.40 pg >>> short American staple cotton; *Gossypium arboreum* is an extant representative of the cotton A-genome lineage, which is paired with the D-genome lineage, making up present-day cultivated cottons; the A-genome species gave rise to spinnable fibers *agr* >>> agroenergy crop >>> chloroplast

Barbados nut: A perennial drought-resistant shrub that grows up to 5 m tall under favorable conditions, with spreading branches; the bark is smooth gray; it is native to Central America (Chiapas, Mexico) but has been introduced to many African countries and India; there are male and female plants; it has yellow-green flowers and large (pale) green leaves; the black thin-shelled seeds are considered to be toxic due to the toxalbumin curcin; roasting the seeds seems to denature the toxic protein; they also contain a high percentage (up to 37%) of clean oil, used for candles, soap, and recently, biodiesel production; each fruit contains two or three large black, oily seeds; it has insecticidal and fungicidal properties; its latex contains an alkaloid (jatrophine), which shows anticancer properties; it holds much promise for producing biodiesel, since it is cultivated in many tropical countries; *Jatropha curcas* (Euphorbiaceae), 2n = 2x = 22; 1C DNA content ~0.43 pg *agr* >>> agroenergy crop

barberry: A perennial plant used as an ornamental plant; it also serves as an alternate host plant of wheat rust fungi (*Puccinia graminis*), cultivation must be avoided in wheat-growing areas; *Berberis vulgaris* (Berberidaceae), 2n = 2x = 28; 2C DNA content 3.5 pg *bot hort phyt* >>> calafate >>> barberry, Iranian seedless

barberry, Iranian seedless: Iranian seedless barberry or zereshk has long been cultivated for its fruit in South Khorasan, Iran; there are more than 11,000 ha under production, producing more than 9,200 tonnes of dried fruit annually; all orchards are established by sucker propagation but grafting is possible; barberry is cold hardy and drought tolerant; however, water deficits during fruit set, growth, and maturation cause yield reduction; biennial bearing is a problem and sometimes no fruit is harvested in the "off" year; the barberry plant and its fruit have been used in traditional medicine; they are well documented for their antidiabetic, anticancer, and antimicrobial activities, with berberine as the main active constituent; the shrub also has ornamental uses in the landscape and the fruit is used as a food additive; in addition, the anthocyanin in the fruit is useful as a natural coloring agent in food industries; *Berberis integerrima* var. Bidaneh (Berberidaceae), 2n = 2x = 28 *hort* >>> barberry

barley: A common diploid cereal crop; it is as ancient as the origins of agriculture itself; barley grain is used as feed for animals, malt, and human food; barley was a staple food as far back as 18,000 years ago; it was the energy food of the masses as well; its use as human food was very popular during the Roman Empire (hordeari) and it continued to be the main food cereal of northern Europe until the 16th century; barley is still an important staple food in several developing countries; in the highlands of Tibet, Nepal, Ethiopia, the Andean countries, some areas of North Africa, Turkey, Iran, Afghanistan, India, and Russia, barley is used as human food either for bread making (usually mixed with bread wheat) or for specific recipes; the largest use of food barley is found in regions where other cereals do not grow well due to high altitude, low rainfall, or soil salinity; it evolved from wild forms of *Hordeum spontaneum*; there are types of two- or six-rowed ears; it is the fourth most commonly grown cereal in the world, with a global production of ~200 million tons; it is the cereal which has the most widespread natural distribution; *Hordeum* spp.: two-rowed barley = *H. distichon*, six-rowed barley = *H. vulgare* (Gramineae), 2n = 2x = 14 (VV); 2C DNA content 8.7–9.8 pg = 4,900–5,300 Mb; ~696,142,857 bp per chromosome; 'Bere', pronounced "bear", is an ancient six-rowed barley variety, currently cultivated mainly on 5–15 hectares of land in Orkney, Scotland; it is also grown on Shetland, Caithness and on a very small scale by a few crofters on some of the Western Isles, i.e. North Uist, Benbecula, South Uist, Islay and Barra; it is probably Britain's oldest cereal in continuous commercial cultivation; 'Bere' is a landrace adapted to growing on soils of a low pH and to a short growing season with long hours of daylight, as found in the high latitudes of northern Scotland; it is sown in the spring and harvested in the summer; because of its very rapid growth rate, it is sown late but is often the first crop to be harvested; it is known locally as "the 90-day barley"; winter genotypes are more amenable to androgenesis, presumably by better adaptation to stress conditions *agr* >>> crop evolution >>> waxy hull-less barley >>> Tables 15, 16, 30, 32, 35, 41, 48, 54

barnyard grass: An annual grass; *Echinochloa crus-galli* (Gramineae), 2n = 4x = 36; 2C DNA content 2.7 pg *bot agr*

basil: Plant with aromatic leaves, belonging to the mint family; a native of the tropics, it is cultivated in Europe as a herb and used to flavor food; its small white flowers appear on spikes; *Ocimum basilicum* (Labiatae), 2n = 4x = 48–52, *O. basilicum* var. *citriodorum* (lemon basil), 2n = 6x = 72 *hort* >>> Greek basil

basket willow: *Salix viminalis, S. dasyclados, S. amygdalina, S. americana, S. purpurea, S. daphnoides, S. pentandra* (Salicaceae), 2n = 2x = 38; 2C DNA content 0.7–0.9 pg *fore agr*

bastard indigo (*syn* **false indigo**): A native ornamental plant of the U.S., with several species occurring naturally; many hybrids now exist for home gardeners to make a selection in terms of best color and growth habit; most native false indigos have blue flowers; however, new varieties and cultivars are enhancing the gardener's choices; it generally blooms late spring through early summer; *Amorpha fruticosa* (Leguminosae), 2n = 2x = 40 *hort agr*

batata: >>> sweet potato

bay-tree: >>> laurel (-tree)

beach grass: *Ammophila arenaria* (Gramineae), 2n = 2x, 4x, 8x = 14, 28, 56; 1C DNA content ~3.88 pg *bot agr*

beach mulberry: >>> noni

bean: Seed of a large number of leguminous plants; they are rich in nitrogenous compounds, including proteins, and are grown both for human consumption and as feed for cattle and horses; varieties of bean are grown throughout Europe, the U.S., South America, China, Japan, Southeast Asia, and Australia; in 2018, common bean was grown on about 30 million hectares globally and on 7.6 million ha in Africa annually, where it is consumed and traded by more than 100 million households; the >>> broad bean has been cultivated in Europe since prehistoric times; the >>> French bean, >>> kidney bean, or >>> haricot is probably of South American origin; the >>> runner bean is closely allied to it, but differs in its climbing habit; among beans of warmer countries are the >>> lima or butter bean of South America; the >>> soya bean, extensively used in China and Japan; and the >>> winged bean of Southeast Asia; the tuberous root of the winged bean has potential as a main crop in tropical areas where protein deficiency is common; the Asian >>> mung bean produces the bean sprouts used in Chinese cookery; canned baked beans are usually a variety of *Phaseolus vulgaris*; *Phaseolus* spp. (Leguminosae), 2n = x = 22; polytene chromosomes have been observed in ovary and immature seed tissues; 2C DNA content 1.0 pg *hort*

bearded wheatgrass: *Elymus caninus* (Gramineae), 2n = 4x = 28 (SSHH), the H genome derives from different *Hordeum* species; 2C DNA content 9.3 pg (YY) *bot agr* >>> couches

beardgrass: *Andropogon ischaemum syn Bothriochloa ischaemum* (Gramineae), 2n = 4x = 40; 1C DNA content ~3.55–5.18 (*A. gerardii*) pg *bot agr*

beech: A hardwood deciduous tree used in plantation forests; native to Europe; *the* tree has been in cultivation for a long time; *Fagus sylvatica* (Fagaceae), *2n = 2x = 24*; 1C DNA content ~0.56 pg *fore*

beet: Several plants belonging to the goosefoot family, used as food crops; one variety of the common beet (*Beta vulgaris*) is used to produce sugar and another, the >>> mangelwurzel, is grown as a cattle feed; the >>> beetroot, or >>> red beet (*Beta rubra*), is a salad plant; the family also includes >>> spinach; spinach beet, used as a spinach substitute, is *B. vulgaris cicle*, commonly known as goosefoot; *Beta* spp. (Chenopodiaceae), 2n = 2x = 18; 1C DNA content ~1.25 pg *hort* >>> beet

beet rape: *Brassica rapa* ssp. *rapifera* (Brassicaceae), 2n = 2x = 20; 1C DNA content ~0.80 pg *agr* >>> beet >>> *Brassica* >>> Figure 8

beetroot: >>> red beet

begonia: A group of tropical and subtropical plants; they have fleshy, succulent leaves, and some have large, brilliant flowers; there are numerous species in the tropics, especially in South America and India, for example, *Begonia rex*; *Begonia* spp. (Begoniaceae), 2n = 2x = 32, 33, 34, 42, 43, 44; 1C DNA content ~0.25–1.46 pg *hort*

bell pepper: >>> paprika >>> sweet pepper

Bengal gram: >>> chickpea

Bengal quince: >>> bael

bent: >>> bentgrass

bentgrass: A perennial grass; creeping bent grass, also known as >>> fiorin, is common in northern North America, Asia, and Europe, including lowland Britain; it spreads by runners and has large attractive clusters (panicles) of yellow or purple flowers on thin stalks, like oats; it is often used on lawns and golf courses; it is a high-maintenance grass used on

putting greens; it requires frequent cutting with a reel mower, as well as frequent fertilization and watering; usually, it is highly susceptible to several diseases; *Agrostis stolonifera*
(Gramineae), 2n = 4x = 28 ($A^2A^2A^3A^3$); 2C DNA content ~7.0 pg; *A. capillaris*, 2n = 4x
= 28, $A^1A^1A^2A^2$); *A. canina* (2n = 2x = 14, A^1A^1); 1C DNA content ~3.50 pg *hort bot agr*

bere barley: >>> barley

bergamot: A fruit tree deriving from a cross between lemon (*Citrus limon*) and bitter orange (*C.
aurantium*), produced during the early 18th century; the name is also used for perfume
from fruit rinds; essential oil from peel is used as a flavoring in hard candy, baked goods,
desserts, and "Earl Grey" tea; in Calabria, Italy, where bergamot has been successfully
cultivated since the eighteenth century, it is commonly defined as "the prince of the Citrus
genus"; although production of bergamot and its derivatives is comparatively small, its
chemical composition and biological properties have been of great scientific interest and
the oil is considered essential in many high-quality perfumes; there is also an increased
demand for bergamot oil for food flavorings, gastronomy, and crafts; bergamot tea comes
from leaves of *Monarda didyma* and *M. citriodora* (Lamiaceae, 2n = 2x = 32, two satellite
chromosomes); also called Oswego tea or bee balm; *Citrus bergamia* (Rutaceae), 2n = 2x
= 18; *hort* >>> Figure 61 >>> http://users.kymp.net/citruspages/citrons.html

Berlandieri grape: *Vitis berlandieri* (Vitaceae), 2n = 2x = 38; 1C DNA content ~0.50 pg, but 2n =
2x = 40 in subgenus Muscadinia *hort* >>> European grape

Bermuda grass: Common Bermuda grass, *Cynodon dactylon*, and its interspecific hybrids with *C.
transvaalensis* are the most popular turfgrasses for golf courses, sports turfs, as well as for
lawns and roadsides, throughout the southern U.S.; first recorded release of an improved
cultivar was in the early 1940s; most cultivars are vegetatively propagated by plugs, sod,
or sprigs; first improved seeded Bermuda grass was the 'Guyman' cultivar released in
1982; until that time, tetraploid Bermuda grass (2n = 36) was the only seeded Bermuda
grass, being sold in the trade as 'Arizona common' or simply "common" Bermuda grass;
Cynodon dactylon (Gramineae), 2n = 2x = 18 (AA); 1C DNA content ~0.80–1.47 pg *agr*

Berseem clover: >>> Egyptian clover

betel nut (palm): Fruit of the areca palm, which is chewed together with lime and betel pepper as
a stimulant by peoples of East and Papua New Guinea; chewing it blackens the teeth and
stains the mouth deep red; *Areca catechu* (Palmae), 2n = 4x = 32; 1C DNA content ~0.58
pg (*A. minuta*) *hort fore*

bhindi: >>> okra

bilberry: >>> blueberry

bilimbi: *Averrhoa bilimbi* (Oxalidaceae), 2n = 2x = 24; 1C DNA content ~0.24 pg (*A. carambola*)
hort

billbergia: A species of the genus Billbergia native to Brazil, including about 66 species; was
already described and the name validly published by William HERBERT; in 1827, John
LINDLEY reclassified it into today's valid botanical systematics; there are several varieties on the market: 'Ambiorix', 'Astro Pink'. 'Astronaut', 'Bam', 'Charles Dewey', 'Dancing
Waters', 'E. Thomas Witte', 'El Capitan' etc. *Billbergia zebrina* (Bromelioideae); 1C DNA
content ~0.38 pg (*B. nutans*) *hort*

billion-dollar grass: >>> barnyard grass

bilva: >>> bael

birch: Broad-leafed, deciduous trees and shrubs with paper-like bark; *Betula* spp. (Betulaceae), *B.
pobulifolia*, *B. pubescens*, 2n = 2x = 28; 2C DNA content 0.4–1.5 pg *fore hort*

bird rape: *Brassica rapa* var. *silvestris* f. *oleifera* (Brassicaceae), 2n = 2x = 20; 1C DNA content
~0.80 pg *agr* >>> *Brassica* >>> Figure 8

birdsfoot: >>> finger millet

birdsfoot trefoil: *Lotus corniculatus* (Leguminosae), 2n = 2x, 4x = 12, 24; polytene chromosomes have been observed in ovary, anthers, or immature seed tissues; 2C DNA content 2.2 pg *agr*

biserrula: A deep-rooted, hard-seeded, aerially seeded, self-regenerating annual pasture legume; it can grow on a wide range of soil pH (4 to 8) and textures; it is similar in appearance to >>> seradella (*Ornithopus compressus*) but is distinguished by its heart-shaped leaves and its blue to mauve flowers; it has papery pods with serrated edges and small seeds; it produces high-quality forage in terms of dry matter digestibility, metabolizable energy, and crude protein content; it shows good early vigor in regeneration stands and has a prostrate habit when grazed heavily, but is semi-erect if ungrazed; *Biserrula pelecinus syn Astragalus pelecinus* (Leguminosae), 2n = 2x = 16; 1C DNA content ~0.43 pg (*B. pectinatus*) *agr*

bitter apple: >>> colocynth

bitter gourd: This tropical vine ("pare") is a tender perennial; the fruit is edible when harvested green and cooked; the taste is bitter; it contains twice the potassium of bananas and is also rich in vitamin A and C; it is used as functional food plant; pare is a monoecious climber with dark green, deeply lobed leaves with hairs on it; the dioecious flowers are yellow and the fruits are oblong and lumpy with a light green to greenish-white, waxy skin; it is a favorite in the Surinam kitchen; there is another, smaller variety, Balsam apple (*M. balaminal*), which has seeds surrounded by a bright red pulp; these seeds are small and black; the juice of this plant appears to be an abortifacient; in traditional Chinese medicine, the vegetable is used as an appetite stimulant and as a treatment for gastrointestinal infection and against cancer; *Momordica charantia* (Cucurbitaceae), 2n = 2x = 22, ploidy levels ranging from diploid (2n= 22) to hexaploid (2n= 66); hexaploidy in *M. charantia* is reported; chromosome length ranged from 0.8 to 2.9 μm; hexaploid accessions show submetacentric and subtelocentric chromosomes; many accessions harbour median and submedian chromosomes; only one hexaploid accession has chromosomes with subterminal centromeres; there is considerable variation in chromosome length and chromosome arm ratio; 1C DNA content ~2.05 pg *hort*

bitter melon: >>> bitter gourd

black bean: >>> lablab

black bent (*syn* **giant bentgrass** *syn* **black bentgrass** *syn* **redtop**): Native to Europe; it was widely used as a pasture grass until the 1940s in the cooler areas of North America; *Agrostis gigantea syn A. alba syn A. dispar syn A. nigra syn A. stolonifera* ssp. *gigantea syn A. stolonifera* var. *major* (Gramineae), 2n = 6x = 42; 1C DNA content ~2.68 pg *bot agr*

blackberry: A prickly shrub, closely related to raspberries and dewberries; a large genus of flowering plants in the rose family; most of these plants have woody stems with thorns, like roses; spines, bristles, and gland-tipped hairs are also common; most species are >>> apomictic hermaphrodites; the fruit, sometimes called a bramble, is an aggregate of >>> drupelets; native to northern parts of Europe, it produces pink or white blossoms and edible black compound fruits; hybrids are commonly grown with supports such as wires or canes, as raspberries, blackberries; blackberry primocanes, fruiting on first-year canes, have the potential to expand blackberry production both seasonally and geographically; the incorporation of the primocane-fruiting trait into cultivars with desirable horticultural attributes is challenging due to its recessive nature and tetrasomic inheritance; molecular marker-assisted selection has great potential to facilitate its incorporation, because breeders already use morphological marker-assisted selection of seedlings without marginal cotyledonary hairs to identify progeny that will be thornless when mature; *Rubus rosa* (Rosaceae), 2n = 2x, 4x = 14, 28; 1C DNA content ~0.24–1.23 pg (2x) *hort*

black chokeberry: *Aronia melanocarpa* (Rosaceae), 2n = 2x = 34; 1C DNA content ~1.28 pg (*A. arbutifolia*) *hort*

blackcurrant: *Ribes nigrum* (Grossulariaceae), 2n = 2x = 16; 1C DNA content ~0.97 pg (*R. rubrum*) *hort*

black-eyed pea: >>> cow bean

blackgram: >>> urd bean

blackgrass: *Alopecurus myosuroides* (Gramineae), 2n = 2x, 4x = 14, 28; 1C DNA content ~4.33 pg *bot agr*

black locust: One of the most important stand-forming tree species; e.g. covering ~20% of the forest area and providing ~18% of the annual timber output in Hungary; *Robinia pseudo-acacia* (Leguminosae), 2n = 2x = 20; 2C DNA content 1.3 pg *fore*

black medick: *Medicago lupulina* (Leguminosae), 2n = 2x = 16; 2C DNA content 1.8 pg *agr*

black mustard: Seeds used for the condiment mustard; *Brassica nigra* (Brassicaceae), 2n = 2x = 16 (BB); 1C DNA content ~0.78 pg; polytene chromosomes have been observed in ovary, anthers, or immature seed tissues *agr* >>> *Brassica* >>> Figure 8

black pepper: A vegetatively propagated, tropical crop that is difficult to breed; this species is native to India (Malabar Coast) and is a good example of >>> ancient clones that demonstrate the value and durability of >>> horizontal resistance; *Piper nigrum* (Piperaceae), 2n = 4x = 52; 1C DNA content ~1.03 pg *hort*

black poplar: A pioneer tree species of riparian ecosystems that is threatened with extinction because of the loss of its natural habitat; genetic diversity varies from region to region in Europe; the most unique alleles are identified in the Danube region (Austria), the Rhône region (France), Italy, the Rijn region (The Netherlands), and the Ebro region (Spain); *Populus nigra* (Salicaceae), 2n = 2x = 38; 1C DNA content ~0.54 pg *fore*

black salsify (Spanish salsify *syn* **black oyster plant** *syn* **viper's grass):** Has diuretic and depurative properties; the root has restorative and sudorific properties, and is an ingredient of many infusions; it is very rich in carbohydrates (18–20% fresh weight), with a high proportion of inulin and laevulin, which makes it very suitable for a diabetic diet; *Scorzonera hispanica* (Compositae), 2n = 2x = 12; 1C DNA content ~2.45 pg (*S. mollis*) *hort*

black spruce: *Picea mariana* (Pinaceae), 2n = 2x = 24; 1C DNA content ~17.45 pg *fore*

black walnut: >>> mansonia

blackwood acacia: *Acacia melanoxylon* (Leguminosae), 2n = 2x = 26; 1C DNA content ~0.75 pg *fore*

bleeding heart: *Dicentra spectabilis* (Fumariaceae), 2n = 2x = 16; 1C DNA content ~0.60 pg; shows polytene chromosomes up to 16n in antipodal cells *hort*

blueberry: Various North American shrubs belonging to the heath family, growing in acid soil; the genus also includes huckleberries, bilberries, deerberries, and cranberries, many of which resemble each other, and are difficult to distinguish from blueberries; all show small oval short-stalked leaves, slender green or reddish twigs, and whitish bell-like blossoms; only true blueberries, however, have tiny granular speckles on their twigs; true blueberries have black or blue edible fruits, often covered with a white bloom; since the beginning of the 20th century, more than 100 varieties have been developed; the cultivated blueberries do not derive from the European species *Vaccinium myrtillus* but from the North American species, which do not stain the tongue blue, as compared with the European types; inhibition of α-glucosidase activity is considered to be an effective means for controlling diabetes by regulating glucose uptake, and blueberries have been shown to possess high levels of inhibitory activity; peel tissue has, on average, about four times the levels of α-glucosidase inhibitory activity than the pulp; the first lady breeder was Elizabeth Coleman WHITE (1871–1954), starting with a selection program around the 1900s; in 1911, she collaborated with the botanist Frederick V. COVILLE (1867–1937), who also started selection experiments in 1906; the farmer family WHITE also grew cranberries and successfully marketed both blueberries and cranberries, *Vaccinium myrtillus, V. erythrocarpum* (Ericaceae), 2n = 2x = 48; 1C DNA content ~0.63 pg *hort* >>> highbush blueberry >>> lowbush blueberry

blue grama grass: A perennial forage grass native to northern America; it grows in and often dominates dry prairies, generally on rocky or clayey soils; the native range of this species extends across the central U.S. from Canada into central Mexico; in Iowa, blue grama is mainly found in the Loess Hills and northwestern counties, but it may be encountered elsewhere as it is sometimes included in native lawn mixes or planted as an ornamental; it is well suited to these uses because it is a short, mat-forming grass, and the attractive flowering branches often take on a bluish tint as they dry in the fall; the flowering heads, with one to three usually curved, densely flowered, one-sided branches that terminate in a spikelet, are distinctive; blue grama is palatable and nutritious for livestock and wildlife, providing high-quality forage in both the summer and winter; it is frequently planted as a part of rangeland reclamation efforts and is used in roadside plantings and erosion control projects as well; *Bouteloua gracilis* (Poaceae), 2n = 2x, 4x, 6x = 20, 40, 60; 1C DNA content ~19.70 pg (sometimes 2n = 84), basic number possibly x = 10 *agr*

blue grass: >>> Kentucky grass

blue lupine (*syn* **narrowleaf lupine** *syn* **European blue lupine**): It has been cultivated for over 6000 years as a food crop for its edible legume seeds, as a fodder for livestock and as a green manure; it has a protein content of 35–40% in the seeds, so that providing protein in the human diet is of great interest; currently lupin is grown in Australia and sold under the name "Australian sweet lupin"; the wild western Mediterranean population can be accepted as the founder of domesticated narrow-leafed lupin worldwide, *Lupinus angustifolius* (Leguminosae), 2n = 2x = 48; 2C DNA content ~4 pg; this species of lupin had its genome sequenced in 2013; it was sequenced due to the interest in developing low-alkaloid mutants as a food crop *agr* >>> Table 46

blue pea: >>> green pea

blue sisal: *Agave amaniensis* (Agavaceae), 2n = 2x = 60; 1C DNA content ~3.23 pg *agr*

blue sweet pea: >>> chickling pea

bok choi: >>> pak choi

Bombay hemp: >>> cantala

bora: >>> asparagus bean

borage: An annual plant, growing to 90–120 cm in height; it has hairy leaves and bright blue, star-shaped flowers; it originated in Europe, but is now found throughout most of Europe and North America; it is a species traditionally defined as allogamous; recently, it was revealed to have a high selfing rate, although a mechanism of protandry has been confirmed in this plant; studies investigating flower behavior show that several flowers open every day and that others are also receptive at the same time within a plant; moreover, pollinator behavior, mainly by bees, contributes to the selfing rate because it is demonstrated that these insects visit several flowers in a given plant before flying to other plants; ecological studies reveal the contribution of geitonogamous pollination to the high selfing rate; *Borago officinalis* (Boraginaceae), 2n = 2x = 16; 1C DNA content ~1.75 pg *hort* >>>

borassus (*syn* **Palmyra palm**): A genus of six species of fan palms, native to tropical regions of Africa, Asia and New Guinea, e.g. *Borassus aethiopum* (African Palmyra palm; tropical Africa), *B. akeassii* (Ake Assi's Palmyra palm; West Africa), *B. flabellifer* (Asian Palmyra palm, Lontar palm, Doub palm, sea coconut; southern Asia and southeast Asia), *B. heineanus* (New Guinea Palmyra palm; New Guinea), *B. madagascariensis* (Madagascar Palmyra palm; Madagascar), *B. sambiranensis* (Sambirano Palmyra palm; Madagascar); they are tall palms, capable of growing up to 30 m high; the leaves are long, fan-shaped, 2–3 m in length; the flowers are small, in densely clustered spikes, followed by large, brown, roundish fruits; Palmyra palms are widely cultivated in tropical regions; it has long been one of the most important trees of Cambodia and India, where it has over 800 uses; the leaves are used for thatching, mats, baskets, fans, hats, umbrellas, and as writing material; young plants are cooked as a vegetable or roasted and pounded to make meal; the fruits are eaten roasted or raw, and the young, jellylike seeds are also eaten; a sugary sap, called toddy, can be obtained from the young inflorescence, either male or female ones, *Borassus* spp. (Arecaceae), 2n = 2x = 18; 1C DNA content ~8.60 pg; in *B. flabellifer,* the sex chromosomes can be detected as a heteromorphic bivalent at meiosis; by C-banding, the heterochromatic sex chromosomes can also be revealed, the sex ration is about 1:1 in wild populations *hort*

bottle gourd: The only species of squash grown in Europe before the discovery of America; it is one of the oldest crop plants of the world; its origin seems to be Africa, but it is now grown in all tropical and subtropical regions; fossil seeds were found in Peru (6,000–12,000 BC) and Thailand (8,000 BC); *Lagenaria siceraria* ssp. *siceraria* (Cucurbitaceae), 2n = 2x = 22; 1C DNA content ~0.34 pg *hort*

bougainvillea: An ornamental plant; member of a tropical genus native to South America and much used throughout the tropics and subtropics as an ornamental; the plant is a woody, climbing shrub with many prominent "flowers" that are really bracts concealing the very small true flowers; these bracts vary in color from bright red, through orange and yellow, to white; A. de BOUGAINVILLE (1729–1811) was the first Frenchman to circumnavigate the world; the island of Bougainville, largest of the Solomon Islands in the South Pacific, is named after him; only three species, *B. spectabilis*, *B. glabra*, and *B. peruviana*, are of ornamental value; the three species and their varieties have given rise to a wide array of cultivated forms by spontaneous hybridization and somatic mutation; *Bougainvillea spectabilis* (Nyctaginaceae), 2n = 2x = 34; 1C DNA content ~4.40 pg *hort*

bow-string hemp: *Sansevieria* spp. (Agavaceae), 2n = 2x = 40; 1C DNA content ~1.30 pg (*S. trifasciata*) *hort*

box tree: Several small evergreen trees and shrubs, with small, leathery leaves; some species are used as hedging plants and for shaping into garden ornaments; the common box is slow growing and ideal for hedges; *Buxus sempervirens* (Buxaceae), 2n = 2x = 28; 1C DNA content ~0.81 pg *hort*

brahria: Native to the African tropical savannas; many species and accessions are polyploid and apomictic, which complicates the improvement of breeding stocks through hybridization; nevertheless, it became an important forage grass in Brazil; breeding lines are mostly related to differences in ploidy among the accessions, and to apomixis, an asexual mode of seed reproduction; usually, sexual accessions are diploid, while apomicts are polyploid; induced tetraploids have been successfully obtained; *Brachiaria brizantha* (Gramineae), 2n = 2x, 4x, 5x, 6x = 18, 36, 45, 54 *agr* >>> Congo signal grass

brahmi: A prostrate, profusely branched herb with white- to violet-colored flowers, populations of which spread both vegetatively and by seeds near river banks, ditches, or other water bodies; it is used as a medicinal plant; species are widely distributed in warmer parts of the world and are reported to occur in Asia, Australia, and North and South America; in India, it is an important component of the traditional as well as modern systems of medicine; it is a major constituent of several commercial herbal formulations available on the market; it has been indicated in the treatment of disorders like epilepsy and mental retardation, in asthma, and as a cardiotonic and diuretic; *Bacopa monnieri* (Scrophulariaceae), 2n = 2x = 64; 1C DNA content ~1.20 (*B. caroliniana*) pg *hort*

bramble: >>> blackberry

***Brassica oleracea*:** Includes the following varieties: >>> cabbage (leafy head), >>> kale (non-heading leafy sprout), >>> collards (non-heading leafy sprout), >>> broccoli (immature inflorescence and stalk or peduncle), >>> cauliflower (immature inflorescence), >>> Brussels sprouts (tall-stemmed cabbage with small edible heads or buds along stem), >>> kohlrabi (enlarged, edible, basal stem above the ground); broccoflower, a hybrid between broccoli and cauliflower; common cabbage-like vegetables provide an excellent example of remarkable crop improvements that were accomplished by simple long-term selection with no real goal in mind; in the wild, the *B. oleracea* plant is native to the Mediterranean region of Europe, and is somewhat similar in appearance to a leafy rapeseed plant; soon after the domestication of plants began, people in the Mediterranean region began growing this first ancient "cabbage" plant as a leafy vegetable; because leaves were the part of the plant which were consumed, it was natural that those plants with the largest leaves would be selectively propagated for next year's crop; this resulted in large and larger-leafed plants slowly being developed as the seed from the largest-leafed plants was favored; by the 5th century BC, continued preference for ever-larger leaves had led to the development of the vegetable we now know as kale; kale is known botanically by the name *B. oleracea* var. *acephala*, 2n = 2x = 18 (CC); kale continued to be grown as a leafy vegetable for thousands of years, and is still grown today; later, people began to express a preference for those plants with a tight cluster of tender young leaves in the center of the plant at the top of the stem; because of this preference for plants in which there were a large number of tender leaves closely packed into the terminal bud at the top of the stem, these plants were selected and propagated more frequently; a continued preference for these plants for hundreds of successive generations resulted in the gradual formation of an increasingly dense cluster of leaves at the top of the plant; eventually, the cluster of leaves became so large, it tended to dominate the whole plant, and the cabbage "head" was available; this progression is thought to have been complete by the 1st century AD; this plant was named *B. oleracea* var. *capitata*, 2n = 2x = 18 (CC); at about the same time, in a part of Europe near modern Germany, kale plants with short fleshy stems were being selected, resulting in fatter and fatter stems; selection on this basis eventually led to the ancestral "cabbage" plant developing into the vegetable known as kohlrabi, *B. oleracea* var. *caulorapam*, 2n = 2x = 18 (CC); both cabbage and kohlrabi have been cultivated for ~2,000 years; some time in the past thousand years, a preference developed in southern Europe for eating the immature flower buds of these plants; selection pressure favoring production of plants with large tender flowering heads was imposed by some growers; by the 15th century, the modern vegetable known as cauliflower [*B. oleracea* var. *botrytis*, 2n = 2x = 18 (CC)] had developed; ~100 years later, broccoli had been developed in Italy; because broccoli was developed in Italy, it was named *B. oleracea* var. *italica*, 2n = 2x = 18 (CC); in the 18th century, selections of cabbage plants, which produced a large number of large, tightly packed leafy axillary buds along the main stem, were made in Belgium; these became known as Brussels sprouts, *B. oleracea* var. *gemmifera*, 2n = 2x = 18 (CC); 1C DNA content ~0.78 pg (Brassicaceae) *hort* >>> *Brassica* >>> Figure 8

Brazil nut: A giant tree of the Amazon rainforest in South America; its distribution is Guyana and Amazonian Colombia, Venezuela, Peru, Brazil, and Bolivia; it may also occur naturally in Surinam and French Guiana but all collections from these two countries are most likely from cultivated trees; in 1825, POITEAU stated that Brazil nut trees had been cultivated in French Guiana for a long time before his visit, and that many of the home gardens in Cayenne had Brazil nut trees; he noticed that large quantities of Brazil nuts were brought from Pará to Cayenne under the name of "touka"; a young seedling of this tree takes at least 20 years to bear its first fruit, and may take as long as 80 years; the fruits take a year to ripen; the hard brown seeds are produced in large, thick-walled capsules weighing up to 2.5 kg; seeds contain 65 to 70% unsaturated fat and literally burn like a candle; Brazil nuts are harvested for their edible seeds; *Bertholletia excelsa* (Lecythidaceae), 2n = 2x = 34 *hort fore*

breadfruit: Fruit of two tropical trees belonging to the mulberry family; it is highly nutritious and, when baked, is said to taste like bread; native to many South Pacific islands; *Artocarpus altilis syn A. communis* (Moraceae), 2n = 2x = 56; 1C DNA content ~1.15 pg (*Artocarpus heterophylla*) *hort*

bread wheat: >>> wheat

brewing yeast: *Saccharomyces cerevisae* (Saccharomycetaceae), x = 16, genome size 12,156,677
bp, containing about 5,800 functional genes, at least 31% of yeast genes having homologs
in the human genome *biot*

brinjal: >>> eggplant

broad bean: From the Mediterranean region and Asia Minor; it was introduced by the Romans into
northern Europe and known as *faba majores*; the small-seeded type (var. *minua*) is used
as animal feed, while the large-seeded type is used for human consumption (var. *major*);
moreover, there are hybrids between them resulting in a third type (var. *equina*); the latter
is even grown as a winter crop in northern France; field beans contain large concentrations
of hemagglutinins; these substances are glycoproteins and are characterized by the property of agglutinating erythrocytes; they are located almost exclusively in the cotyledons
of the seed; by selection, for example, from the black-seeded Finnish cultivar Pirhonen,
after only three generations, the haemagglutinin concentration had been reduced by ~50%;
Vicia faba (Leguminosae), 2n = 2x = 12; polytene chromosomes have been observed in
ovary, anthers, or immature seed tissues; 2C DNA content 26.1–27.1, genome size, var.
Inovec, 13,154 Mbp pg *agr hort* >>> Tables 16, 35

broad red clover: A perennial clover; *Trifolium pratense* (Leguminosae), 2n = 2x = 14; 2C DNA
content 1.3–1.5 pg *agr* >>> clover >>> Table 35

broccoflower: >>> *Brassica oleracea*

broccoli: A variety of cabbage; it contains high levels of the glucosinolate compound glucora-
phanin; a breakdown product of this was found to neutralize damage to cells and so help
to prevent cancer; synthetic varieties are under development, *Brassica oleracea* var. *italica*
(Brassicaceae), 2n = 2x = 18, 2C DNA content 1.6 pg *hort* >>> *Brassica* >>> Figure 8

bromegrass: Annual grasses found in temperate regions; some are used as food for horses and
cattle, but many are weeds; *Bromus* spp. (Gramineae), 2n = 2x, 4x, 6x, 8x, 10x = 14, 28, 42,
56, 70; 1C DNA content 1.88–16.33 pg *bot agr*

broomcorn: >>> durra

broomcorn millet: >>> common millet

broomrape: A parasitic plant, having purplish or yellowish flowers and small scale-like leaves that
lack chlorophyll, and that grows on the roots of other plants (legumes, tobacco, sunflower,
etc.); branched broomrape (*Orobanche ramosa*) is a parasitic weed recently spreading in
Central Europe and threatening the production of several crops including tobacco, rapeseed, potato, carrot, and tomato; in contrast with other weeds that compete with crops for
resources, *O. ramosa* is directly attached to the host root and takes up all necessary water
and nutrients, assimilating directly from its host; this leads to significant yield and quality
losses; because the parasitic weed is attached to the crop, and, because it spends ~90% of
its life underground, it is very difficult to control; *Orobanche* spp. (Orobanchaceae), 2n =
2x, 4x, 6x = 38, 76, 114; 1C DNA content 1.45–5.81 pg *phyt*

brown mustard: Also known as Indian mustard; it originated in India and it has secondary centers
of origin in China and southern Russia; this species has the advantage that it can be com-
bine-harvested and, for this reason, has become a major crop in Canada and parts of the
northern U.S.; this area of North America now produces the bulk of the world's mustard;
the plant is self-pollinating and is cultivated as pure lines; *Brassica juncea* (Brassicaceae),
2n = 4x = 36; 1C DNA content ~1.53 pg *agr* >>> *Brassica* >>> Figure 8

brownseed paspalum: A wild forage grass species; the common races are tetraploid and apos-
porous apomictic, while sexual diploid representatives have been reported sporadically,
Paspalum plicatulum (Poaceae), 2n = 4x = 40; 1C DNA content ~1.20 pg (*P. dilatatum*) *agr*

brunching onion: >>> Welsh onion

Brussels sprouts: Was first known in its modern form in the middle of the 18th century, from
Belgium; similar types were already cultivated in the Middle Ages; in the Brussels sprout,
the stem initially grows to 30–80 cm high, then the sprouts develop from the leaf axils;

in modern agriculture, hybrid varieties have almost completely supplanted the mostly less uniform, free-flowering varieties; they are hardy, slow-growing, long-season vegetables, belonging to the cabbage family; in the appropriate season of the year, it can be grown with fair success; in mild areas, or where there is deep snow cover, the sprouts may overwinter; the "sprouts" (small heads that resemble miniature cabbages) are produced in the leaf axils, starting at the base of the stem and working upward; sprouts improve in quality and taste best during cool or even lightly frosty weather; Brussels sprouts require a long growing period, though newer hybrids have greatly reduced this requirement; in all but the most northern countries, summers are usually too warm for completely satisfactory production from spring plantings; plants set out in late spring to early summer grow satisfactorily and mature high-quality sprouts when the fall weather begins to cool; annual flowering in Brussels sprouts originally was found in one plant in 1953; it is genetically dominant over biennial flowering and the trait is monogenic; *Brassica oleracea* var. *gemmifera* (Brassicaceae), $2n = 2x = 18$; 2C DNA content 1.6 pg *hort* >>> *Brassica* >>> Figure 8

Brutian pine (*syn* **Turkish red pine**): Has the largest distribution area among forest tree species in Turkey, and occupies ~3.7 million ha of the total 13.8 million ha of high forest area in Turkey; it appears mainly (about half of its total distribution area) in the southern part of Turkey; it occurs from sea level up to 1,200 m above sea level in this area; this species is one of the most important forest tree species in Turkish forestry and the National Tree Breeding and Seed Production Program; *Pinus brutia* (Pinaceae), $2n = 2x = 24$; 1C DNA content ~20.15 pg (*P. mugo*) *fore*

buckbean: >>> marsh trefoil

buckwheat: Buckwheat seeds are often considered to be grains, though, unlike most grains, they are not true grasses; common buckwheat was probably first cultivated in China; besides the seeds, from which buckwheat flour is produced, buckwheat is also a good honey plant; the flour is made into noodles and into buckwheat grouts, often known as "kasha"; buckwheat contains rutin, a medicinal chemical; buckwheat pancakes, raised with yeast, were a common food in American pioneer days; the name "buckwheat" comes from its triangular seeds, which resemble the much larger seeds of beech; *Fagopyrum esculentum* (Polygonaceae), $2n = 2x = 16$; 1C DNA content ~1.44 pg (*F. sagittatum*) *agr* >>> Tatar buckwheat >>> notch-seeded buckwheat

Buddha's fruit: A small Asian gourd with an extremely sweet pulp; a glycoside in the fruit is 150 times sweeter than sucrose and may have economic potential as a noncaloric sugar substitute; it is a perennial, dioecious, herbaceous climbing vine, 2–5 m in length; roots are tuberous, fusiform when young, finally subglobose, 10–15 cm in diameter; it is found in warm foggy climates; *Siraitia grosvenorii syn Thladiantha grosvenorii* (Cucurbitaceae), $2n = 2x = 28$ *hort*

Buddha's hand: A citron showing large fingerlike protrusions on Its fruits, which can develop as open fingers or a closed hand, depending on the time of year and environmental conditions; the medium-sized fruits consist of a solid, heavy, sweet rind with no pulp or seeds inside; first cultivated in Middle East and Persia, *Citrus medica* (Rutaceae), $2n = 2x = 18$; 1C DNA content ~0.40 pg *hort*

buffalo grass: *Cenchrus ciliaris, Buchloe dactyloides* (Gramineae), $2n = 2x, 4x, 5x, 6x = 20, 40, 50, 60$; 2C DNA contents are 0.93, 1.80, 2.15, and 2.6 pg for diploid, tetraploid, pentaploid, and hexaploid, respectively *agr*

bulb barley: *Hordeum bulbosum* (Gramineae), $2n = 2x, 4x = 14, 28$ ($H^{bul}H^{bul}$); 2C DNA content 9.5 pg (HH) *bot* >>> haploidization >>> Tables 41, 54

bulb onion: >>> onion

busy lizzie: The leading bedding plant in the U.S.; *Impatiens walleriana* (Balsaminaceae), $2n = 2x = 18$; 2C DNA content 2.3 pg *hort*

butter bean: >>> Lima bean

butter cabbage: *Brassica napus* ssp. *arvensis* (Brassicaceae), 2n = 4x = 38 (AACC); 1C DNA
  content ~1.15 pg *hort* >>> *Brassica* >>> Figure 8
butterhead (lettuce): *Lactuca sativa* (Compositae), 2n = 2x = 18; 1C DNA content ~2.65 pg *hort*
  >>> lettuce >>> Table 46
butternut squash: Creeping plant whose large round fruit has a thick orange rind, pulpy flesh, and
  many seeds; a species originating in either Central America or northern South America;
  pumpkins are used in cookery (pies and soups); there are several varieties: aehobak –
  a summer squash, also called "Korean zucchini"; butternut squash – a popular winter
  squash in much of North America; calabaza – a commonly grown winter squash in the
  Caribbean, tropical America, and the Philippines; crookneck, Dickinson pumpkin – a proprietary strain of Dickinson is used for its canned pumpkin; giromon – a large, green
  cultivar, grown primarily in the Caribbean, where Haitians use it to make the traditional
  "soupe giromon"; golden cushaw – similar in shape but a different species than the common *Cucurbita argyrosperma* "cushaw" type; loche – a landrace of squashes from Peru;
  Long Island cheese pumpkin – the exterior resembles a wheel of cheese in shape, color,
  and texture; Musquée de Provence or Moscata di Provenza, Naples long squash, Seminole
  pumpkin – an heirloom variety originally cultivated by the Seminole Indians of Florida;
  or Tromboncino – a summer squash, also known as "Zucchetta"; *Cucurbita moschata*
  (Cucurbitaceae), 2n = 2x = 40; 1C DNA content 0.43 pg *hort* >>> vegetable marrow

C

cabbage: Any of several cultivated varieties of the plant, *B. oleracea* ssp. *capitata*, 2n = 2x =
  18 (CC), of the mustard family, having a short stem and leaves, formed into an edible
  head; cabbages are one of the most ancient of vegetables that are still grown today; they
  were cultivated as far back as 4,000 years ago; today, the common cabbage (also known
  as *Brassica oleracea* Capitata Group), so-called because of its "capita" or "head" shape,
  is classified into three types: green leafed, with smooth green leaves; >>> red cabbage,
  with purplish red leaves; and >>> Savoy cabbage, with crinkled leaves; Galician cabbage
  (traditional in Portugal's Caldo Verde, is classified as >>> *B. oleracea* Tronchuda Group;
  it stands apart from the main Capitata Group; different varieties of cabbage were also
  grown and cultivated in Asia from the earliest times: >>> pak choi (*B. rapa* Chinensis
  Group); >>> Chinese or napa cabbage (*B. rapa* Pekinensis Group); flowering cabbage
  (*B. parachinensis*); flat cabbage [(Chinese Tai goo choy and the Japanese Tatsoi (*B. rosu-
  laris*)]; >>> rape (*Brassica rapa* chinensis group), whose seed is crushed to produce canola
  oil; Chinese broccoli (*B. oleracea* Alboglabra Group); mizuna (*B. rapa* Japonica Group);
  mustard greens (*B. juncea* and *B. campestris*); collards (*B. oleracea* Acephala Group);
  >>> kale (*B. oleracea* Acephala Group) – the latter being the closest relative of the wild
  cabbage, from which all "coles" developed, its curly leaves also growing in rosettes of
  blue-green leaves; >>> kohlrabi (*B. oleracea* Gongylodes Group); >>> Brussels sprouts
  (*B. oleracea* Gemmifera Group); >>> broccoli or calabrese (*B. oleracea* Botrytis Group);
  >>> cauliflower (*B. oleracea* Botrytis Group); *Brassica oleracea* (Brassicaceae), 2n = 2x =
  18 (C'C'); 1C DNA content ~0.78 pg; 1C DNA content ~20.15 pg *hort* >>> *Brassica* >>>
  chloroplast >>> Figure 8
cacao: The tree is often called cacao, while the product is called cocoa, from which chocolate is
  manufactured; the cacao plant is an evergreen flowering tree, native to wet, warm forests
  of South and Central America; the center of origin of cacao is on the eastern equatorial
  slopes of the Andes, and it occurs throughout the Amazon Valley, where it provides an
  interesting example of a cline; all the wild trees in the center of origin are self-incompatible; as one moves down the Amazon, self-compatible types become increasingly common and, at the river mouth, they are all self-compatible; all the cacao in West Africa is

self-compatible and very uniform, with a very narrow genetic base; this tree grows to 15 m; after flowering, 25–30 cm long red fruit pods develop; in each pod are almond-shaped cacao beans and pulp; in 2013, a rare aberrant fruit phenotype was found in St. Augustine, Trinidad, where double fructification instead of single fructification on a single pedicel was observed; the aberration occurs as a result of the formation of two pistils in a single flower; fruits matured as normal, and fruit morphology and seed number were within the range of that reported for normal fruits; chocolate is made from the beans in the pods of the cacao plant; *Theobroma cacao* (Sterculiaceae), 2n = 2x = 20; 1C DNA content ~0.43 pg *hort* >>> Criollo cocoa bean >>> Forastero cocoa bean

cactus fig: Edible fruits (cactus pear) are used for juices, liqueurs, and jams; also used as an herb plant; *Opuntia stricta* var. *dilleni* (Cactaceae), 2n = 2x, 6x = 22, 66; 1C DNA content ~22.8 pg *hort agr*

cactus pear: >>> cactus fig

calabash (*syn* **gourd** *syn* **calabash tree**): A tropical South American evergreen tree with gourd-like fruits 50 cm across, whose dried skins are used as water containers; the Old World tropical vine >>> bottle gourd is sometimes also called a calabash, and it produces equally large gourds; *Crescentia cujete* (Bignoniaceae), 2n = 2x = 36; 1C DNA content ~1.60 pg *hort*

calabrese: >>> sprouting broccoli

calafate: An evergreen shrub, with shiny box-like leaves; it is native to the south of Argentina and Chile and is a symbol of Patagonia; the bush grows to a height of 1–1.5 m and has many arching branches, each covered in many tripartite spines; it has many small yellow flowers in summer; its edible blue-black berries are harvested for jams, but are also eaten fresh; the cultivar Nana is widely available as a garden shrub, and is also used in commercial plantings as a low spiny hedge to discourage intruders, but it does not fruit; *Berberis buxifolia* (Berberidaceae), 2n = 4x = 56; 1C DNA content ~3.03 pg *hort*

Calamansi orange: >>> Calamondin orange

calamint (mill mountain *syn* **mountain balm** *syn* **basil thyme** *syn* **mountain mint**): An erect, bushy plant with square stems, rarely more than 30 cm high, bearing pairs of opposite leaves, which, like the stems, are downy with soft hairs; the flowers bloom in July and August; the plant grows by waysides and in hedges, and is not uncommon, especially in dry places; it may be cultivated as a hardy perennial, propagated by seeds sown outdoors in April, by cuttings of side shoots in cold frames in spring, or by division of roots in October and April; it contains a camphoraceous, volatile, stimulating oil in common with the other mints; *Calamintha alpina* ssp. *hungarica* (Lamiaceae), 2n = 2x = 18 *hort*

Calamondin orange: A fruit tree that was developed in and is very popular throughout Southeast Asia, especially the Philippines, where it is most commonly used for cooking; in the West, it is variously known as acid orange, Chinese Orange or Panama orange; it is a shrub or small tree growing to 3–6 m, and bears small citrus fruit used to flavor foods and drinks; although sometimes described as a native of the Philippines or other areas of Southeast Asia, the tree is in fact the result of a hybrid between species in the genus *Citrus* and unknown in the wild; it is generally held that most species in cultivation are ancient apomictic hybrids and selected cultivars of these hybrids, including crosses with segregate genera, such as *Fortunella* and *Poncirus;* the calamondin is usually described as a cross between *Citrus reticulata* (>>> tangerine or >>> Mandarin orange) and *Citrus japonica* Oval Kumquat Group; if the segregate genus *Fortunella* is recognized, the calamondin should be treated as an intergeneric hybrid in the nothogenus ×*Citrofortunella*; in North America, the calamondin is grown mainly as an ornamental plant; it can be especially attractive when the fruit are present; it is frost sensitive and therefore limited to warm climates; *Citrus microcarpa* (Rutaceae), 2n = 2x = 18; 1C DNA content ~0.48 pg *hort* >>> http://users.kymp.net/citruspages/citrons.html

calamus: *Acorus calamus* (Acoraceae), 2n = 2x,4x = 18,36; 1C DNA content ~0.65 pg *hort*

calpa: >>> pongamia

calceolaria: *Calceolaria* spp. (Scrophulariaceae), 2n = 2x, 4x = 36, 72; 1C DNA content ~1.36 pg (*C. gracilis*) *hort*

California bromegrass: >>> bromegrass

camomile: >>> chamomile

Canada thistle: *Cirsium arvense* (Asteraceae), 2n = 2x = 34; 1C DNA content 1.55 pg *bot agr*

Canadian serviceberry (*syn* **juneberry** *syn* **shadblow** *syn* **serviceberry** *syn* **shadblow** *syn* **shadbush** *syn* **shadbush serviceberry** *syn* **sugarplum** *syn* **thicket serviceberry**): It became an ornamental plant in Europe, *Amelanchier canadensis syn Amelanchier canadensis* var. *subintegra* (Rosaceae), 2n = 4x = 68; 1C DNA content ~1.15 pg *hort*

canahua: *Chenopodium pallidicaule* (Chenopodiaceae), 2n = 2x = 18; 2C DNA content 1.0 pg *hort*

Canary grass: *Phalaris canariensis* (Gramineae), 2n = 2x = 12; 1C DNA content ~3.83 pg *agr*

candlenut: A tropical tree of Southwest Asia, usually 9-12 m in height; fresh fruits (candlenuts) contain a toxin, so that they are not eaten raw; nuts are roasted before being cracked open; taste is similar to macadamia nuts and mostly used as a flavoring agent in the cuisines of Hawai, Indonesia, and Malaysia, *Aleurites moluccana* (Euphorpiaceae), 2n = 2x, 4x = 22, 44 *hort*

candytuft: *Iberis* spp. (Brassicaceae), 2n = 2x = 14; 1C DNA content ~0.57 pg (*I. gibraltaria*) *hort* >>> *Brassica*

canna (*syn* **achira**): Grown especially for its edible rootstock, from which arrowroot starch is obtained; the name is derived from the Quechua word achira; other names for it in Spanish are capacho, sagu, tasca, chisqua, adura, or luano; in Portuguese it is araruta gigante; in Vietnamese it is dong; *Canna edulis syn C. indica* (Cannaceae), 2n = 2x,3x = 18,27; 1C DNA content ~0.72 pg *hort* >>> Queensland arrowroot

cantala: *Agave cantala syn A. teqilana* (Agavaceae), 2n = 3x = 90; 1C DNA content ~4.40 pg *agr*

cantaloup: >>> cantaloupe

cantaloupe: Several small varieties of muskmelon distinguished by their round, ribbed fruits with orange-colored flesh; *C. melo cantalupensis* (Cucurbitaceae), 2n = 2x = 24; 2C DNA content 2.0–2.5 pg *hort*

Cape daisy: >>> African daisy

caper: A shrub confined to the Mediterranean region or Africa; it is a straggling, spreading vine-like, deep-rooted, deciduous shrub, about 60 cm high; the capers of commerce are the unopened flower buds; they are handpicked daily, typically in the morning; the buds are pickled in strong vinegar, brine, oil, or wine, or simply preserved in granular salt, *Capparis spinosa* (Capparaceae), 2n = 2x = 24 *hort*

caper (spurge): Often known as the "mole plant" because of its reputation for deterring the activity of this creature; a native of Mediterranean areas, it has spread into Northern Europe, and has also been introduced into the U.S.; the seeds have in the past been used for medicinal purposes and for the production of lamp oil; the oil content of the seeds is high, ~50% by weight, of which some 85% is in the form of oleic acid, used in the preparation of some soaps, detergents, lubricants, paints, and cosmetics; the oil has also been suggested as a possible diesel fuel substitute; research on the crop has been conducted in America, but especially in Germany, where yields of over 2t seed per ha have been achieved from experimental plots; small trials in the United Kingdom have indicated that similarly high yields may be obtainable elsewhere; *Euphorbia lathyris* (Euphorbiaceae), 2n = 2x = 20 *hort* >>> Table 46

capulin: A true cherry cultivated since early times in the cooler mountainous regions of Central and South America; the dark red fruits contain a pale green, but sweet and juicy pulp which can be eaten raw or cooked; there is also an unrelated fruit called the Jamaican cherry (*Muntingia calabura*, Elaeocarpaceae), known as capulin or capuli in Latin America; it is indigenous to Central and South American tropics, but is now widely grown in the Philippines, India, and Malaysia, where it is known as the Japanese or Chinese cherry; its small red or yellow fruits have a light brown, soft, juicy pulp filled with tiny yellowish seeds that are too small to notice when eating; its flavor is somewhat like the sweet fig; *Prunus salicifolia syn P. capollin syn P. capuli syn Prunus serotina salicifolia* (Rosaceae), 2n = 4x = 16; 1C DNA content ~0.50 pg *hort*

carambola: An elongated, angular fruit composed of five carpels with a star-shaped cross section; the tartness is due to calcium oxalate crystals in the flesh which dissolve in the saliva, forming oxalic acid; the species is related to taling pling (*Averrhoa bilimbi*); *Averrhoa carambola* (Oxalidaceae), 2n = 2x = 24; 1C DNA content ~0.24 pg *hort*

caraway (seed): A seed used for flavoring breads, liquors, casseroles, and other foods; the slender seed, member of the Apiaceae or parsley family, has a pungent, anise-like flavor; caraway is usually used whole, *Carum carvi* (Umbelliferae), 2n = 2x = 20; 1C DNA content ~4.78 pg *hort* >>> anise

cardamom: This genus is native to Southeast Asia and is a member of the ginger family; the fruits are widely used as a spice, and are particularly prized in Arab countries for adding to coffee; the plants are open pollinated; a highly aromatic spice is derived from the seeds and dried fruits; used in curry powder, seasoning for sausages, incenses, perfumes, and medicines; hybrid breeding has been discussed; there are reports of heterosis in cardamom for yield and yield-associated characteristics; *Amomum maximum syn Elettaria cardamomum* (Zingiberaceae), 2n = 4x = 48 *hort*

cardoon (*syn* spotted golden thistle): Has occasionally been cultivated, but generally the wild plant has been used, with harvesting being limited to only the leaves in spring; at present, its cultivation is very restricted and is tending to disappear; the plant is native to eastern Mediterranean regions and closely related to artichoke; because of its large blue flowers, it is grown not only as an ornamental but also as a vegetable plant; the fleshy leaf base and lower parts of rips, which taste similar to black salsify, are used for salad; *Cynara cardunculus* **var.** *scolymus syn Scolymus maculates* (**Compositae**), 2n = 2x = 34; 1C DNA content ~1.10 pg *hort*

Caribbean pine: *Pinus caribaea* (Pinaceae), 2n = 2x = 24; 1C DNA content ~22.65 pg *fore*

carnation: A large number of double-flowered cultivated varieties of a plant belonging to the pink family; the flowers smell like cloves; they are divided into flake, bizarre, and picotees, according to whether the petals have one or more colors on their white base, have the color appearing in stripes, or have a colored border to the petals, respectively; *Dianthus caryophyllus* (Caryophyllaceae), 2n = 2x, 6x = 30, 90; 1C DNA content ~0.63 pg *hort*

carob: An evergreen tree with pinnately compound leaves (with two to six pairs of oval leaflets), which can grow to a height of 15 m; this species is grown around the world, primarily as a food crop for its sweet and nutritious fruits; the fruit is a pod, technically a legume pod, 15 to 30 cm in length and fairly thick and broad; pods are borne on the old stems of the plant on short flower stalks; most carob trees are monoecious, with separate male and female flowers on the same tree; the dark brown pods are not only edible, but also rich in sucrose (~40% plus other sugars) and protein (up to 8%); the pod contains vitamin A and B vitamins, and several important minerals; they can be eaten directly by livestock, but carob is mostly known because the pods are ground into a flour that is a cocoa substitute; although this product has a taste slightly different from that of chocolate, it has only one-third the calories, is virtually fat-free, is rich in pectin, is nonallergenic, has abundant protein, and has no oxalic acid, which interferes with absorption of calcium; carob flour is widely used in health foods for its chocolate-like flavoring; it is native to the eastern Mediterranean, probably the Middle East, where it has been in cultivation for at least 4,000 years; the plant was well known to the ancient Greeks, who planted seeds of this plant in Greece and Italy; seeds were used to weigh gold, hence the word "carat"; the seeds of various plants were often used as weights because their mass reputedly varied so little; carob seeds are unusually constant in weight; however, the variability of seeds sampled from a collection of carob trees (CV = 23%) was close to the average of 63 species reviewed from the literature (CV = 25%); in a perception experiment, observers could discriminate differences in carob seed weight of around 5% by eye, demonstrating the potential for humans to greatly reduce natural variation; interestingly, the variability of premetrication carat weight standards is also around 5%, suggesting that human rather than natural selection gave rise to the carob myth; MOHAMMED's army ate "kharoub," and Arabs planted the crop in northern Africa and Spain, along with citrus and olives; Spaniards carried carob to Mexico and South America, and the British took carob to South Africa, India, and Australia; records show that carob was intentionally introduced into the U.S. in 1854, and the first seedlings were apparently planted in California in 1873; for commercial production, cultivars with the finest quality fruits are bud grafted onto common stock; carob grows well anywhere that citrus is grown, and it prefers moderately dry climates that receive more than 300 mm of rainfall; ideal is a Mediterranean type of climate; *Ceratonia siliqua* (Leguminosae), 2n = 2x = 24; 1C DNA content ~0.65 pg hort

carpet bent: *Agrostis stolonifera syn A. alba* (Gramineae), 2n = 4x = 28; 2C DNA content 7.0 pg *bot agr*

carrot: Occurs in Central Europe as a wild plant with a thin, whitish, woody taproot; it is assumed that the garden carrot arose from a cross between the wild carrot and the giant carrot (*Daucus maximus*), found in the Mediterranean area; the fleshy, fattened yellow-red root was a new characteristic; through breeding, varieties arose which varied from ~20 cm long, through medium-sized conical shapes to varieties with short, round roots; there are early, medium to late, and perennial varieties; yield and quality characteristics are at the forefront as breeding aims; tenderness, a high amount of desirable components (carotene, sugar), uniform color of skin and flesh, shape and size of the root, and smooth peeling are important criteria; *Daucus carota* (Umbelliferae), 2n = 2x = 18; 1C DNA content ~1.00 pg *hort* >>> chloroplast >>> http://www.carrotmuseum.co.uk/history3.html

casbanan: >>> cassabanana

cashew (~ nut): Originating from the rainforest of Brazil, it is frost susceptible; cashew is one of the hardiest of trees and, in warm countries, will grow on poor soils that are unsuitable for other crops; the nuts fetch a high price and the crop is about as valuable as arabica coffee; each nut is borne externally on the end of a fairly large fruit; the fruit is edible, but very astringent, and it can be utilized for the manufacture of alcohol; there is a negative correlation between total yield and quality – the highest-yielding trees producing small nuts of low commercial quality; there is great variation among trees, and there is scope for selection within existing orchards; the cashew nut is attached to a swollen, fleshy stalk (pedicel) called the cashew "apple"; the outer shell of the "nut" contains the allergen urushiol and can cause a dermatitis reaction similar to that of poison oak and poison ivy; *Anacardium occidentale* (Anacardiaceae), 2n = 2x = 42 *fore hort* >>> agroenergy crop

cassabanana (*syn* sikana *syn* musk cucumber *syn* melocotonero *syn* calabaza de olor *syn* calabaza melón *syn* pérsico *syn* alberchigo in Mexico *syn* melocotón or melón de olor in El Salvador and Guatemala *syn* calabaza de chila in Costa Rica *syn* cojombro in Nicaragua *syn* chila in Panama *syn* pavi in Bolivia *syn* padea, olerero, secana or upe in Peru *syn* calabaza de Paraguay, curuba, or pepino melocoton in Colombia *syn* cura, coróa, curua, curuba, cruatina, melão caboclo or melão macã in Brazil *syn* cajú cajuba, cajua, cagua, calabaza de Guinea in Venezuela *syn* pepino, pepino angolo or pepino socato in Puerto Rico *syn* cohombro in Cuba): The vine is perennial, herbaceous, fast-growing, heavy, requiring a strong support, such as a trellis, or by climbing trees to 15 m or more by means of four-part tendrils equipped with adhesive discs that can adhere tightly to the smoothest surface; young stems are hairy; native to Brazil but it has been spread throughout tropical America;

the fruit is ellipsoid or nearly cylindrical, sometimes slightly curved; 30–60 cm in length, 7–12 cm thick, hard-shelled, orange-red, maroon, dark-purple with tinges of violet, or entirely jet-black, *Sicana odorifera* (Cucurbitaceae), 2n = 2x = 24 *hort*

cassabanana (*syn* **casbanan** *syn* **sikana** *syn* **musk cucumber**): A large, herbaceous perennial vine, native to tropical South America, grown as an ornamental plant and for its sweet edible fruit; the fast-growing, fleshy vine can reach 15 m or more in height, climbing with four-part adhesive tendrils; the large, hairy, palmately lobed leaves grow to 30 cm in width; the fruit is large, up to 60 cm long, with skin of variable color; the sweet, aromatic, yellow-to-orange flesh of the mature fruit is eaten raw or made into preserves; the immature fruit is cooked as a vegetable; it is grown widely in the warm parts of Latin America, as well as by the Cajun people of the southern United States; *Sicana odorifera* (Cucurbitaceae), 2n = 2x = 40 *hort*

cassava: A perennial shrub producing a high yield of tuberous roots in six months to three years after planting; originating in Central and South America, cassava spread rapidly and arrived on the west coast of Africa via the Gulf of Benin and the River Congo at the end of the 16th century, and on the east coast via Reunion Island, Madagascar, and Zanzibar at the end of the 18th century; by the beginning of the 19th century, cassava had arrived in India, but controlled breeding did not begin until the 1920s; for many farmers, it is the primary staple, but is also used as a cash crop to produce industrial starches, tapioca, and livestock feeds; world production in 1995, all from developing countries, was about 165.3 million t from ~16.2 million ha; currently, Nigeria, Brazil, Congo, Thailand, and Indonesia are the world's largest producers; *Manihot esculenta* (Euphorbiaceae), 2n = 2x = 72; 1C DNA content ~0.83 pg *hort* >>> agroenergy crop

castor (bean): Castor oil is prepared from seeds; seeds also contain the protein ricin which is more poisonous, g for g, than cyanide or rattlesnake venom; it grows wild in the western U.S.; *Ricinus communis* (Euphorbiaceae), 2n = 2x = 20; 1C DNA content ~0.52 pg *hort* >>> agroenergy crop >>> Tables 16, 46

cat's claw (climber): Has lianas up to 15 m, often rooting at the nodes, glabrous or nearly so; leaves dry dark green to nearly black; dimorphic, juvenile plants with small leaflets 1–2 cm long, 0.4–0.8 cm wide; mature leaflets are narrowly ovate to lanceolate, 5–16 cm long, 1–7 cm wide, both surfaces sparsely lepidote; tendrils are deciduous, three-forked, 0.1–3 cm long, each fork bearing a small horny hook; flowers usually are in axillary clusters of one to three; seeds are 1–1.8 cm long, 4.2–5.8 cm wide, the wings membranous, not sharply demarcated from the seed body; it prefers fertile, well-drained soils, but appears to tolerate most soil types, particularly alluvial soils; root tubers and stolons form in the plant's second year and can subsequently form at each leaf node while the vine is prostrate; as such, the plant can form a dense mat which carpets the forest floor; the vine climbs standing vegetation and can smother native trees and shrubs; *Dolichandra unguis-cati* (Bignoniaceae), x = 7 *bot biot*

cat's-claw creeper: >>> cat's claw

catclaw trumpet: >>> cat's claw

Caucasian clover: >>> Kura clover

Caucasian persimmon: *Diospyros lotus* (Ebenaceae), 2n = 2x = 30; 1C DNA content ~1.20 pg (*D. discolor*) *hort* >>> persimmon

cauliflower: A form of cabbage, with a large edible head of fleshy, cream-colored flowers which do not fully mature; it is similar to broccoli but less hardy; it is native to the Mediterranean and Middle East regions, and has been cultivated since at least 600 BC; it was moved to Central Europe from Cyprus and was therefore known as "Cyprus coleworts"; the cauliflower head itself is a degenerate, sterile flowering structure whose buds are kept white by carefully covering them to prevent the formation of chlorophyll that sunlight would cause;

Brassica oleracea var. *botrytis*, (Brassicaceae), 2n = 2x = 18; 1C DNA content ~0.78 pg *hort* >>> *Brassica* >>> chloroplast >>> Figure 8

celeriac: >>> celery

celery: A biennial umbelliferous plant, native of the seacoast of Europe and Asia; when deprived of its acrid and even poisonous properties by cultivation, it became celery; it is a plant of the parsley family, and its blanched leafstalks are used as a salad; however, the tuberous form is not as popular; *Apium graveolens* (Umbelliferae), 2n = 2x, 3x = 22, 33; 2C DNA content 2.1 pg *hort* >>> tuberous celery

centella (gotu kola, Asiatic pennywort, Indian pennywort): An important herbaceous medicinal plant with a worldwide distribution, but native to wetlands in Asia; the herb possesses medicinal value and is used extensively in traditional systems of medicine; the medicinal properties of the herb are attributed to the presence of characteristic triterpenoids and their saponins in the leaves; the major triterpenoids are asiaticoside, madecassoside and their aglycones asiatic acid and madecassic acid, respectively, among others; *Centella asiatica* (Apiaceae), 2n = 2x = 18 (sometimes B chromosomes) *hort*

ceriman: Of the many aroids that are cultivated as ornamental plants, only this one has also been grown for its fruit; it is often called monstera and, inappropriately, false breadfruit; the plant is a fast-growing, stout, herbaceous vine, spreading over the ground and forming extensive mats if unsupported, but climbing trees to a height of 9 m or more; ceriman is native to wet forests of southern Mexico, Guatemala and parts of Costa Rica and Panama; fully ripe pulp is like a blend of pineapple and banana; it may be served as dessert with a little light cream, or may be added to fruit cups, salads or ice cream; *Monstera deliciosa syn Philodendron pertusum* (Araceae), 2n = 2x, 4x = 60, 120; 2n = 2x = 50; 1C DNA content ~4.80 pg *hort*

chamomile: It is an important medical plant with a long history of usage and a wide range of medical applications; wild forms are diploid, whereas cultivated ones are diploid or tetraploid; both naturally occurring triploids and those induced through directed crosses between diploid and tetraploid parents could facilitate the exploitation of triploidy in chamomile as in other crop plants (fruit and ornamental plants); *Matricaria chamomilla syn M. recutita* (Compositae), 2n = 2x = 18; 1C DNA content ~3.88 pg *hort*

chard: >>> beet

charlock: Annual plant belonging to the cress family, found in Europe and Asia; it has hairy stems and leaves and yellow flowers; *Sinapis arvensis* (Brassicaceae), 2n = 2x, 3x = 18, 24 ($S^{ar}S^{ar}$); 1C DNA content ~0.50 pg *bot agr* >>> *Brassica* >>> Figure 8

chaya: A large, fast-growing leafy perennial shrub that is believed to have originated in the Yucatán Peninsula of Mexico; the specific >>> epithet, *"aconitifolius"*, means *Aconitum*-like leaves; it has succulent stems which exude a milky sap when cut; it can grow to be 6 m tall, but is usually pruned to about 2 m for easier leaf harvest; it is a popular leaf vegetable in Mexican and Central American cuisines, similar to spinach; the leaves must be cooked before being eaten, as the raw leaves are toxic; chaya is a good source of protein, vitamins, calcium, and iron; and is also a rich source of antioxidants; it is easy to grow, a tender perennial in the U.S., and suffers little insect damage; It is tolerant of heavy rain and has some drought tolerance; propagation is normally by woody stem cuttings about 25 cm long, as seeds are produced only rarely; early growth is slow as roots are slow to develop on the cuttings, so leaves are not harvested until the second year; chaya leaves can be harvested continuously as long as no more than 50% of the leaves are removed from the plant, which guarantees healthy new plant growth; some varieties have stinging hairs and require gloves for harvesting; cooking destroys the stinging hairs, *Cnidoscolus aconitifolius* (Euphorbiaceae), 2n = 2x = 36 *hort* >>> tree spinach

chayote: A native to Guatemala; it is popular throughout tropical regions, where it is known by several names, including vegetable pear, mirliton, and mango squash; it is a tender, perennial-rooted cucurbit, with climbing vines and leaves resembling those of the cucumber; the light green, pear-shaped fruit, which contains a single, flat edible seed, may weigh as much as 0.5–1 kg, but most often is from 150 to 500 g; fruits may be slightly grooved and prickly; *Sechium edule* (Cucurbitaceae), 2n = 2x = 28; ~35,000,000 bp per chromosome *hort*

cheat: >>> rye brome

cheese fruit: >>> noni

cherimoya (tree): A small fruit tree that bears a commercially valuable fruit; a native tree of Ecuador and Peru, it is now cultivated in several areas of the world, including California, Chile, Ecuador, Israel, Peru, and Spain; *Annona cherimola* (Annonaceae), 2n = 2x = 14; 1C DNA content ~1.33 (*A. reticulata*) pg *hort* >>> pawpaw >>> sweetsop >>> soursop >>> ylang-ylang (related species)

cherry: A group of fruit-bearing trees distinguished from plums and apricots by their fruits, which are round and smooth and not covered with a bloom; they are cultivated in temperate regions with warm summers and grow best in deep fertile soil; cultivated cherries are derived from two species, the sour cherry (*Prunus cerasus,* 2n = 4x = 32; 1C DNA content ~0.63 pg) and the gean (*P. avium,* 2n = 2x = 16; 1C DNA content ~0.35 pg), which grow wild in Europe; the former is the ancestor of morello, duke, and Kentish cherries, whereas the latter is the ancestor of the sweet cherries, namely hearts, mazzards, and bigarreaus; in addition to those varieties that are grown for their fruit, others are planted as ornamental trees; *Prunus* spp. (Rosaceae) *hort* >>> Japanese cherry >>> sour cherry >>> sweet cherry

cherry plum: A species of plum native to Central and eastern Europe, Southwest and Central Asia; it is a large shrub or small tree reaching 6–15 m tall, with deciduous leaves 4–6 cm long; the fruit is a drupe 2–3 cm in diameter and yellow or red in color; it is edible, and reaches maturity from mid-August to mid-September; it is also a very popular ornamental tree in gardens, grown for its very early flowering; numerous cultivars have been developed, many of them selected for purple foliage and pink flowers, such as 'Pissardii' and 'Nigra'; cherry plum is different from mirabelle; *Prunus cerasifera syn P. divaricata* (Rosaceae), 2n = 2x, 3x, 4x, 6x = 16, 24, 32, 48 (CC); 1C DNA content ~0.63 pg *hort*

chervil: A delicate annual herb, usually used to flavor mild-flavored foods, such as poultry, some seafoods, and young vegetables; it is a constituent of the French herb mixture fines herbes; after about 10 years of selection, H. de VILMORIN (1843–1899) domesticated this European wildflower with forked roots to smooth roots; *Anthriscus cerefolium* (Umbelliferae), 2n = 2x = 16; 1C DNA content ~2.25 pg (*A. sylvestris*) *bot hort*

chess: >>> rye brome

chestnut: There are two types of chestnut tree: (a) the sweet or Spanish chestnut and (b) the horse chestnut; the sweet chestnut (*Castanea sativa,* 2n = 2x = 24; 1C DNA content ~0.98 pg), native to South Europe, Asia, and North America, has toothed leaves and edible seeds, and can grow to a height of 21 m; *Castanea* spp. (Fagaceae) *hort fore* >>> American chestnut >>> Chinese chestnut >>> Japanese chestnut >>> sweet chestnut

chestnut rose: A deciduous shrub growing to 1.2 m tall; it is in flower in June, and the seeds ripen in August; the scented flowers are hermaphrodite and are pollinated by bees; the plant prefers light (sandy), medium (loamy), and heavy (clay) soils, requires a well-drained soil, and can grow in nutritionally poor soils; *Rosa roxburghii* (Rosaceae), 2n = 2x = 14; 1C DNA content ~0.48 pg *hort*

chia: An annual herb of the genus *Salvia*, native to Mexico; it grows to 1 m, with opposite leaves 4–8 cm long and 3–5 cm broad; its flowers are purple or white and are produced in numerous clusters in a spike at the end of each stem; it was cultivated by the Aztecs in pre-Columbian times, and was so valued that it was given as an annual tribute by the people to the rulers; it is grown commercially for its seed, a food that is very rich in omega-3 fatty acids, since the seeds yield 25–30% extractable oil, mostly α-linolenic acid; it is still widely used in Mexico and South America, with the seeds ground for nutritious drinks and as a food source; today, chia is grown commercially in its native Mexico, and in Bolivia, Argentina, Ecuador, and Guatemala; in 2008, Australia was the world's largest producer; *Salvia hispanica* (Lamiaceae), 2n = 2x = 12 *hort* >>> golden chia

Ch'iao: >>> rakkyo

chícharos: >>> chickling pea

chichinga: >>> serpent gourd

chickling pea [*syn* **blue sweet pea** *syn* **chickling vetch** *syn* **grass pea** *syn* **Indian pea** *syn* **Indian vetch** *syn* **white vetch** *syn* **almorta** *syn* **alverjón** (Spain) *syn* **guixa** (Catalonia) *syn* **chícharos** (Portugal) *syn* **cicerchia** (Italy) *syn* **guaya** (Ethiopia) *syn* **khesari** (India)]: A legume commonly grown for human consumption and livestock feed in Asia and East Africa; an important legume food crop for resource-poor farmers in the developing world; it is a particularly important crop in areas that are prone to drought and famine, and is thought of as an "insurance crop" as it produces reliable yields when all other crops fail; *Lathyrus sativus* (Leguminosae), 2n = 2x = 14; polytene chromosomes have been observed in ovary, anthers, or immature seed tissues; 2C DNA content 14.4–16.8 pg *hort*

chickling vetch: >>> chickling pea

chickpea: Also called garbanzo (Spanish), pois chiche (French), Kicher-Erbse (German), chana (Hindi); in Turkey, Romania, Bulgaria, Afghanistan, and adjacent parts of Russia, chickpea is called nakhut or nohut; it known as gram or Bengal gram (English), and is the most important pulse in India, particularly in the semi-arid areas; it is self-pollinating and a herbaceous annual plant which branches from the base; it is almost a small bush with diffused, spreading branches; the plant is mostly covered with hairs, glandular or non-glandular, but some genotypes do not possess hair; based on seed size and color, cultivated chickpeas are of two types: (a) macrosperma (kabuli type); the seeds of this type are large (100-seed mass >25 g), round or ramhead, and cream-colored; the plant is medium to tall in height, with large leaflets and white flowers, and contains no anthocyanin, (b) microsperma (desi type); the seeds of this type are small and angular in shape; the seed color varies from cream, black, brown, yellow to green; there are 2–3 ovules/pod but on an average 1–2 seeds/pod are produced; the plants are short with small leaflets and purplish flowers, and contain anthocyanin; the only common cultivated annual species is *C. arientinum*, although *C. soongaricum* is also cultivated as a food plant in some parts of Afghanistan, eastern Himalaya and Tibet; the center of diversity lies in western Asia (Southeastern Turkey), probably in the Caucasus region/ Asia Minor; biofortification is a target in chickpea breeding; high lutein concentration in chickpea grains is recessive to low lutein concentration; high lutein concentration is often associated with low grain weight; breeding large-seeded chickpea with enhanced lutein concentration seems to be difficult due to the association of lutein-promoting alleles with small-grain alleles; the chickpea grain lutein concentration trait is relatively stable, due to strong genetic control; high-lutein cultivars are likely to maintain their performance even when grown under diverse field conditions; *Cicer arietinum* (Leguminosae), 2n = 2x = 16; 1C DNA content ~0.95 pg *agr* >>> lutein >>> xanthophyll

chic(k)ory *syn* **chicorée** *syn* **chicorée de Bruxelles** *syn* **witloof** *syn* **Belgium endive** *syn* **endive witloof** *syn* **achicoria de Bruselas:** Most grown and consumed in Belgium; native to Europe and West Asia, with large, usually blue, flowers; a garden vegetable; grown under cover, its blanched leaves are used in salads; it was discovered by accident, in Belgium; the rungs are large, up to 20 cm, and its leaves are closely attached to each other and have the shape of a piston; the color of chicory changes from white to green from downwards to upwards; the darker the top of the chicoree is, the more bitter it tastes; common chicory is a bushy perennial herb; root chicory (*Cichorium intybus* var. *sativum*) has been in cultivation in Europe as a coffee substitute for a long time; around 1970, it was found that the root contains up to 20% inulin; new strains have been created, giving root chicory an >>> inulin content comparable to that of >>> sugar beet (~600 dt/ha); chicory, with sugar beet and rye, was used as an ingredient of the East German "Mischkaffee" (mixed coffee), introduced during the "coffee crisis" of 1976–1979; chicory's leaves are still used today in typical Roman recipes; it is common in Roman restaurants to eat dishes with boiled chicory leaves, olive oil and lemon juice (fried with garlic and red pepper); the plant is very common in the Roman countryside and is often picked by farmers; the plant is one of the earliest cited in recorded literature; the Roman HORACE (65–8 bc) mentions it in reference to his own diet, which he describes as very simple: "Me pascunt olivae, me cichorea, me malvae" ("As for me, olives, endives, and mallows provide sustenance."); the Scottish judge and philosopher, Lord J. B. MONBODDO (1714–1799), describes the plant in 1779 as the "chicoree", which the French cultivate as a "pot herb"; in the Napoleonic Era in France; chicory frequently appeared as either an adulterant in coffee or as a coffee substitute; this practice also became common in Germany, U.S. and U.K.; there was an attempt in 1788 to introduce its cultivation as fodder, it being grown largely for that purpose in France, especially for sheep, *Cichorium intybus* (Compositae), $2n = 2x = 18$ *hort*

chico sapote: >>> sapodilla

Chilean strawberry: *Fragaria chiloensis* (Rosaceae), $2n = 8x = 56$; 1C DNA content ~0.61 pg *hort*

chili: Pod or powder made from the pod of a small, hot, red pepper; it is widely used in cooking; the hot ingredient of chili is capsaicin; it causes a burning sensation in the mouth by triggering nerve branches in the eyes, nose, tongue, and mouth; capsaicin does not activate the taste buds and therefore has no flavor; there are several subdivisions, such as *C. frutescens* var. *fasciculatum,* which is known as the "red cluster pepper," *C. frutescens abbreviatum,* known as the "short pepper," *C. frutescens longum,* known as the "long pepper," *C. frutescens longum* var. *conoides* (*syn Capsicum conoides*), known as the "cone pepper," *C. frutescens longum* var. *abbreviatum* known as the "hot, short pepper," *C. frutescens longum* var. *cerasiforme,* known as the "hot cherry pepper," or *C. frutescens longum* var. *baccatum* known as "Peruvian pepper," "bird pepper" or "hot cherry pepper"; *Capsicum frutescens* (Solanaceae), $2n = 2x = 24$; 1C DNA content ~3.40 pg *hort*

China aster: *Callistephus hortensis syn C. chinensis* (Compositae), $2n = 2x = 18$ *hort* >>> aster

China grass: >>> ramie

Chinese bitter melon: >>> bitter gourd

Chinese bitter orange: closely related to *Citrus* and sometimes included in that genus, being sufficiently closely related to allow it to be used as a rootstock for *Citrus*; *Poncirus trifoliata syn Citrus trifoliata* (Rutaceae); it differs from *Citrus* in having deciduous, compound leaves, and pubescent (downy) fruit; It is native to northern China and Korea; the fruits are green, ripening to yellow, and 3–4 cm in diameter, resembling a small orange; they are very bitter, not edible fresh, but can be made into marmalade; when dried and powdered, they can be used as a condiment; the cultivar 'Flying Dragon' has highly twisted, contorted stems *hort* >>> http://users.kymp.net/citruspages/citrons.html

Chinese cabbage: Downy mildew, caused by the oomycete *Hyaloperonospora parasitica*, is a serious threat to members of the Brassicaceae family, in general; a major quantitative trait locus (>>> QTL) for seedling resistance (*BrDW*) and its flanking markers, K14-1030 and phosphoglucomutase, were identified in *Brassica rapa* ssp. *pekinensis*; K14-1030 was successfully converted into a sequence-characterized amplified region marker SCK14-825, and a bacterial artificial chromosome (BAC), with sequence homology to K14-1030, was identified; on the basis of the homologous and linked BAC sequences, two microsatellite simple sequence repeat markers, kbrb058m10-1 and kbrb006c05-2, were designed and mapped on the confidence intervals of *BrDW*; these three markers could explain the

QTL effect to a considerable extent and exhibited relatively high selection accuracy, which would be helpful in marker-assisted selection (MAS) for breeding downy mildew-resistant *Brassica oleracea* var. *chinensis (syn* var. *pekinensis)* (Brassicaceae), 2n = 2x = 20 (AA); 1C DNA content ~0.78 pg *hort* >>> *Brassica* >>> Figure 8

Chinese chestnut: *Castanea mollissima* (Fagaceae), 2n = 2x = 24 *hort*

Chinese chive: *Allium tuberosum* (Alliaceae), 2n = 4x = 32; 1C DNA content ~32.09 pg *hort*

Chinese garlic: *Allium macrostemon* (Alliaceae), 2n = 4x = 32; 1C DNA content ~21.63 pg *hort*

Chinese ginger: >>> galangal

Chinese gooseberry: *Actinidia chinensis* (Actinidiaceae), 2n = 2x = 58; 1C DNA content ~0.78 pg

China-grass: >>> ramie

Chinese gooseberry: >>> kiwi

Chinese jute: *Abutilon avicennae, A. theophrasti* (Malvaceae), 2n = 6x = 42; 1C DNA content ~1.40 pg *agr*

Chinese lantern(s) *(syn* **abutilon):** Species of the shrubby abutilons are described as tender, but, with wall protection, some are perfectly capable of surviving European winters; throughout the summer months, their exotic blooms can bring a touch of the unexpected to the garden; abutilons are natives of Brazil and tropical areas of India and Asia, therefore, full sun is necessary for them to produce the profusion of bell- or lantern-shaped flowers; *A. megapotanicum* is particularly attractive, with pendulous red and yellow flowers and leaves that are heart shaped at the base; the closely related *A. megapotanicum* var. *variegatum* has leaves that are richly patterned with yellow and white variegation; *Abutilon megapotanicum* (Malvaceae), 2n = 6x = 42; 1C DNA content ~1.40 pg (*A. theophrasti*) *hort*

Chinese-lantern lily: >>> Christmas-bells

Chinese leek: *Allium ramosum* (Alliaceae), 2n = 4x = 32; 1C DNA content ~31.50 pg *hort*

Chinese long bean: >>> asparagus bean

Chinese mustard: >>> pak choi

Chinese plum: Commonly called the plum; a small deciduous tree native to China; it is now also grown in fruit orchards in Korea, Japan, Europe, the United States, and Australia; it should not be confused with *Prunus mume*, a related species of plum also grown in East Asia; another tree, the Korean cherry (*P. japonica*), is a separate species despite having a Latin name similar to *P. salicina*'s common name; plant breeder >>> Luther BURBANK devoted a lot of work to hybridizing this species with the Japanese/Chinese plum (*Prunus salicina*) and developed a number of cultivars, *Prunus salicina* (*syn P. triflora syn P. thibetica*) *hort*

Chinese plume grass: >>> Chinese silvergrass

Chinese silvergrass: A tall (<3 m), densely bunched grass that invades forest edges, old fields, and other disturbed areas throughout the U.S.; the leaves are long and slender, upright to arcing, with silvery midribs; the leaves have sharp tips and rough margins; the terminal panicle is fan-shaped, long, and silvery to pink in color; it escaped from ornamental plantings, and it forms large clumps along disturbed areas, displacing native vegetation; the grass is also extremely flammable and increases fire risks of invaded areas; Chinese silvergrass is native to Asia and was introduced into the U.S. and Europe for ornamental purposes; cultivars of *Miscanthus* spp. are also used as bioenergy crops, one of the most productive taxa is the sterile triploid *M.* × *giganteus*; on it, little domestication or breeding has been undertaken, i.e., there is huge potential to utilize the extensive genetic resources of the genus for crop improvement; *Miscanthus sinensis* (Poaceae), 2n = 2x = 38; 1C DNA content ~2.65 pg *hort bot*

Chinese star anise (*syn* **star anise** *syn* **star aniseed** *syn* **badianeo):** A spice that closely resembles >>> anise in flavor, obtained from the star-shaped pericarp; a small native evergreen tree of southwest China; the fruits are harvested just before ripening; it is widely used in Chinese, Malay/Indonesian, and Indian cuisine where it is a major component of garam masala; it is widely grown for commercial use in China, India, and most other countries in Asia; it is an ingredient of the traditional five-spice powder of Chinese cooking, and is also one of the ingredients used to make the broth for the Vietnamese noodle soup called "pho"; *Illicium verum* (Illiciaceae), 2n = 2x = 28; 1C DNA content ~3.35 pg *hort*

Chinese sugar cane: >>> sweet sorghum

Chinese tallow tree: *Sapium sebiferum* (Euphorbiaceae), 2n = 2x = 36 *fore*

Chinese water chestnut: >>> water chestnut

Chinese wisteria: *Wisteria sinensis* (Leguminosae), 2n = 2x = 16 *fore hort*

chireweed: >>> stitchwort

chive: A wild plant throughout most of the northern hemisphere, forming perennial clumps ~12 cm high; it is the mildest member of the onion family, without the bitterness of raw onions or the pungency of garlic; the leaves are used as a garnish; this species is an outbreeder and is easy to breed; chives can be propagated either vegetatively or from true seed; chives do not have well-developed bulbs but they do form tillers to produce dense clumps of the plant; *Allium schoenoprasum* (Alliaceae), 2n = 4x = 32; 2C DNA content 16.6 pg *hort*

chokeberry, black: This is represented by two or three related species of deciduous shrubs in the family Rosaceae, native to eastern North America, and most commonly found in wet woods and swamps; chokeberries are cultivated as ornamental plants and also because they are very high in antioxidant pigment compounds, like anthocyanins; the name comes from the astringency of the fruits, which are inedible when raw; the berries are used to make wine, jam, syrup, juice, soft spreads, tea and tinctures; fruits are eaten by birds which then disperse the seeds in their droppings; chokeberries are often mistakenly called choke-cherries, which is the common name for *Prunus virginiana*; further adding to the ambiguity, there is a cultivar of *Prunus virginiana* named "Melanocarpa", easily confused with *Aronia melanocarpa*, one of the chokeberries; chokeberries are naturally understory and woodland edge plants, and grow well when planted under trees; they are tolerant to drought and pollution, and resistant to insects and disease; several cultivars have been developed for garden planting, including *A. arbutifolia* 'Brilliant' (2n = 2x, 4x = 34, 68), selected for its striking fall leaf color; *A. melanocarpa* 'Viking', 'Aron' (tetraploid), 'Königshof', and 'Nero' were selected for larger fruit suitable for jam-making; because they are self-fertile, only one plant is needed to produce fruits; hybrids of *Sorbus aucuparia* or *S. aria* × *Aronia melanocarpa* or *A. arbutifolia* have been produced for different end-uses; *Aronia melanocarpa* (Rosaceae), 2n = 2x = 34; 1C DNA content ~128 pg *hort*

Christmas-bells (*syn* **Chinese lantern lily**): A tuberous plant native to South Africa and grown in New Zealand as a cut-flower crop; the flowers are 2–2.5 cm long and 1.3 cm in width, are golden orange, lantern-shaped, and hang down from curved, wiry flower-stalks originating at the leaf axils; mature plants are produced from fork-shaped tubers where the stems arise from the single growing points on each tip; during the growing season, the original tuber withers and is lost but is replaced by daughter tubers that develop at the base of each stem; it is a monospecific genus in which little morphological variation has been observed; variants have been developed for the New Zealand horticulture industry by hybridization with species in related genera, for example, *Littonia modesta; Sandersonia aurantiaca* (Liliceae), 2n = 2x = 24; 2C DNA content 6.9 pg *hort*

Christmas rose: The genus *Helleborus* comprises approximately 20 species of herbaceous perennial flowering plants in the family Ranunculaceae; many species are poisonous; the flowers have five sepals surrounding a ring of small, cup-like nectaries (petals modified to hold nectar); the sepals do not fall as petals would, but remain on the plant, sometimes for many months; Christmas roses are widely grown in gardens for decorative purposes, as well as for their purported medicinal abilities and uses in witchcraft; they are particularly valued by gardeners for their winter and early spring flowering period; the plants are frost tolerant and many are evergreen; recent breeding programs have used artificial polyploidization in order to increase genetic variability for selection; *Helleborus niger* (Christmas rose), *H. Orientalis, H.* × *nigercors* (Ranunculaceae), 2n = 4x = 32; 1C DNA content ~14.73 pg *hort* >>> Table 46

christophine: >>> chayote

chumberas: >>> pencas

cibol: >>> Welsh onion

cigar plants: >>> cuphea

cicerchia: >>> chickling pea

ciku: >>> sapodilla

cilantro: >>> coriander

cinchona tree: Tropical tree that is the primary source of the antimalarial drug quinine; quinine is found in the bark of the cinchona tree; it is a chemical that cures malaria, a deadly tropical disease carried by mosquitoes; there are many species of cinchona; they range from about 15 to 20 m tall; the cinchona tree is native to rainforests of the eastern slope of the Amazonian Andes of South America, where it is called the "fever tree"; *Cinchona officinalis, C. ledgeriana, C. uccirubra, C. calisaya, C. pubescens* (*syn* common names: quinine tree, quinine, quinoa, red cinchona, fever tree) (Rubiaceae), 2n = 2x = 34 *hort*

cinnamon: The spice cinnamon consists of the dried green bark (called "quills") of an open-pollinated, evergreen tree, which is indigenous to Sri Lanka (Ceylon), and is propagated by seed; selection within existing crops should lead to improved clones which can be maintained by vegetative propagation; cinnamon and cassia, the so-called "spices of life", together constitute one of the most widely used groups of species; cinnamon is a very ancient crop and was shipped by Austronesian people to Madagascar, and taken from there to Africa and, eventually, to ancient Rome; the Portuguese conquered Ceylon in 1536 and gained a monopoly in the cinnamon trade; the Dutch conquered them, and the monopoly, in 1656; then the British conquered the Dutch, and won the monopoly, in 1796; in the 19th century, commercial production commenced in various parts of the world, and the monopoly was broken, but the Sri Lankan cinnamon remains the best; distillation of the wood of *C. camphora* produces camphor; *Cinnamomum camphora* (Lauraceae), 2n = 2x = 14; 1C DNA content ~0.60 pg *hort fore*

citron *Citrus* sp. (Rutaceae), 2n = 2x = 18; 2C DNA content 1.2 pg *hort* >>> hesperidia >>> lemon for candied peel >>> Figure 61 >>> http://users.kymp.net/citruspages/citrons.html

citrus: >>> citron

clove: Dried, unopened flower bud of the clove tree; a member of the >>> myrtle family, the tree is a native of the Maluku Islands, Indonesia; cloves are used for flavoring in cookery and confectionery; oil of cloves, which has tonic qualities and relieves wind, is used in medicine; the aroma of cloves is also shared by the leaves, bark, and fruit of the clove tree; *Eugenia caryophyllus syn Syzygium zeylanicum* (Myrtaceae), 2n = 2x = 22; 1C DNA content ~1.83 pg *hort* >>>carnation

clover Important fodder and forage crops, usually sown in mixtures with grasses; clovers occur commonly in natural grasslands in humid temperate areas, and in highland areas in the tropics; there are ten clover species that are considered to be agriculturally important: >>> Egyptian or Berseem clover (annual, *T. alexandrinum,* 2n = 2x = 16), >>> Caucasian or Kura clover (perennial, *T. ambiguum,* 2n = 2x, 4x, 6x = 16, 32, 48), >>> yellow suckling clover (annual, *T. dubium,* 2n = 4x = 32), >>> strawberry clover (perennial, *T. fragiferum,* 2n = 2x = 16), >>> Alsike clover (perennial, *T. hybridum, 2n = 2x = 16*), >>> crimson clover (annual, *T. incarnatum,* 2n = 2x = 16), >>> red clover (perennial, *T. pratense 2n = 2x, 4x, 8x = 14, 28, 56*), >>> white clover (perennial, *T. repens,* 2n = 2x = 16), >>> Persian clover (annual, *T. resupinatum,* 2n = 2x, 4x = 16, 32); >>> subterranean clover (annual, *T. subterraneum,* 2n = 2x = 16); the clovers are important because of their ability to perform nitrogen fixation with the aid of *Rhizobium* in root nodules; with only minor exceptions, the annual species are self-compatible while the perennial species are self-incompatible; pollination is commonly done by insects; *Trifolium* spp. (*Trifolium occidentale,* 2n = 2x = 16; *Trifolium nigrescens,* 2n = 2x = 16; 1C DNA content ~0.39 pg; *Trifolium uniflorum,* 2n = 4x = 32; 1C DNA content ~0.95 pg; *Trifolium isthmocarpum,* 2n = 2x = 16; *Trifolium petrisavii,* 2n = 2x = 16) (Leguminoseae), x = 8 *agr*

Clove pink: >>> carnation

club wheat: Varieties may be of either winter or spring type; stems vary in height but are generally stiff; spikes are short, usually under 5–6 cm in length, very compact and flattened; each spikelet usually contains five flowers and is spread almost at right angles to the rachis or stem; spikelets are generally awnless, but sometimes awned; kernels are small, flattened, have a very shallow, narrow crease, and a short brush; about five varieties were grown on nearly 800,000 ha in 1959, mostly in the Pacific Northwest of the U.S.; the principal use is flour manufacture; *Triticum compactum* or *Triticum compactum* ssp. *compactum* (Gramineae), 2n = 6x = 42 (BBA^{u}A^{u}DD); 1C DNA content ~17.33 pg (*T. aestivum*) >>> wheat

clubbed wheat: >>> club wheat

cluster bean: >>> guar

clustered clover: *Trifolium glomeratum* (Leguminosae), 2n = 2x = 16; 1C DNA content ~0.39 pg *agr* >>> clover

cobnut (*syn* **hazel nut** *syn* **cobnut** *syn* **filbert):** This species shows considerable diversity and some taxonomists have suggested additional specific names; *Corylus* spp. (Corylaceae), 2n = 2x = 22; 2C DNA content 1.0 pg *hort*

coca: A South American shrub belonging to the coca family, whose dried leaves are the source of the drug cocaine; it was used as a holy drug by the Andean Indians; *Erythroxylum coca* (Erythroxylaceae), 2n = 2x = 24 *hort fore*

cocksfoot Three levels of polyploidy occur in this grass: diploid, tetraploid, and hexaploid; tetraploids are the most common and generally distributed forms; it is distributed from Europe to North America, Australia, and New Zealand; *Dactylis glomerata* (Gramineae), 2n = 4x = 28; 2C DNA content 12.4 pg *agr*

cockspur: >>> barnyard grass

cocoa: >>> cacao

coco de mer (*syn* **coconut of sea** *syn* **double coconut):** The sole member of the genus *Lodoicea* is a palm endemic to the islands of Praslin and Curieuse in the Seychelles; the tree grows to 25–34 m tall; the leaves are fan-shaped, 7–10 m long and 4.5 m wide with a 4 m petiole; it is dioecious, with separate male and female plants; the male flowers are catkin-like, up to 1 m long; the mature fruit is 40–50 cm in diameter and weighs 15–30 kg, and contains the largest seed in the plant kingdom; the fruit requires 6–7 years to mature and a further two years to germinate; it is grown as an ornamental tree in many areas in the tropics, and subsidiary populations have been established on Mahé and Silhouette Islands in the Seychelles to help conserve the species; the fruit is used in Ayurvedic medicine and also in traditional Chinese medicine; the endosperm is edible; *Lodoicea maldivica* (Arecaceae), 2n = 2x = 36 *fore hort*

coconut: The nutritious meat or "copra" within the seed is endosperm tissue, and the coconut milk is liquid endosperm; the "coconut apple" is a spongy, sweet mass of cotyledon tissue inside the seed cavity that dissolves and absorbs the endosperm; the "coir" fibers come from the fibrous husk or mesocarp; there are two main types or varieties of coconuts; the "niu kafa" types have an elongate, angular fruit, up to 15 cm in diameter, with a small egg-shaped nut surrounded by an unusually thick husk; the "niu vai" types have a larger, more spherical fruit, up to 25 cm in diameter, with a large, spherical nut inside a thin husk; the "niu kafa" type represents the ancestral, naturally evolved, wild-type coconut, disseminated by floating; the "niu vai" type was derived by domestic selection for increased endosperm ("meat" and "milk") and is widely dispersed and cultivated by humans; both types of fruit can float, but the thicker, angular husk adapts the "niu kafa" type particularly well to remote atoll conditions where it can be found today; *Cocos nucifera* (Palmae), 2n = 2x = 32; 2C DNA content 3.2 pg *fore hort* >>> agroenergy crop

cocozelle: *Cucurbita pepo* (Cucurbitaceae), 2n = 2x = 40; 1C DNA content 0.55 pg *hort* >>> tanier

coffee (Arabian): The coffee plant is a small tree, but is pruned into a large bush to make harvesting easier; it produces sweet-smelling white flowers; these are followed by green berries which turn red when ripe; each berry contains two seeds, which are processed to make coffee for drinking, Robusta coffee (*Coffea canephora*); *Coffea arabica* (Rubiaceae), 2n = 4x = 44 (AABB); 1C DNA content 1.20 pg, *C. robusta* (2n = 2x = 22) *hort*

cola: >>> kola

colocynth: *Colocynthis syn Citrullus vulgaris* (Cucurbitaceae), 2n = 2x = 22; 1C DNA content 0.45 pg; endopolyploidy of 24–384n was observed *hort* >>> Table 46

Colorado spruce (*syn* **blue spruce**): *Picea pungens* (Pinaceae), 2n = 2x = 24; 1C DNA content 18.15 pg *fore*

columbine: A group of plants belonging to the buttercup family; all are perennial herbs with divided leaves and hanging flower heads with spurred petals; the wild columbine (*Aquilegia vulgaris*), with blue flowers, has been developed by repeated crossing to produce modern garden species (*A. hybrida*), with larger flowers and a wider range of colors; the eastern columbine (*A. canadensis*), with red flowers, is native to eastern North America; the key to the different lengths of columbine spurs is the shape of the cells inside; previously, it was thought that the different lengths were due to the number of cells; columbines are a living example of evolutionary diversity – they come in over 70 species, each with flowers tailored to the length of their pollinators' tongues; the short, curled spurs of *A. vulgaris* attract bees, while the longer tails of the appropriately named *A. longissima* appeal to hawkmoths; *Aquilegia* spp. (Ranunculaceae), 2n = 2x = 14; 1C DNA content ~0.51 pg (*A. vulgaris*) *hort*

common bean: >>> bean

common bugloss: >>> alkanet root >>> alkanna

common bulb onion: >>> onion

common chickling: >>> chickling pea

common chicory: >>> chicory

common corncockle: *Agrostemma githago* (Caryophyllaceae), 2n = 48, 2C DNA content 4.2 pg *bot agr* >>> Table 46

common grape: >>> European grape

common horsetail: *Equisetum arvense* (Equisetaceae), 2n = 2x = 216; 1C DNA content ~14.65 pg *bot agr* >>> Table 46

common madder *syn* **dyer's madder:** A herbaceous perennial plant species belonging to the bedstraw and coffee family; it can grow up to 1.5 m in height; the evergreen leaves are approximately 5–10 cm long and 2–3 cm broad, produced in whorls of 4–7 starlike around the central stem; it climbs with the aid of tiny hooks on the leaves and stems; the roots can be over 1 m long, up to 12 mm thick and the source of red dyes known as rose madder and Turkey red; it has been used since ancient times as a vegetable red dye for leather, wool, cotton and silk; for dye production, the roots are harvested after two years; the outer red layer gives the common form of the dye, whereas the inner yellow layer provides the refined form; early evidence of dyeing comes from India where a piece of cotton dyed with madder has been recovered from the archaeological site at Mohenjo-daro, Pakistan (3rd millennium BC); in Egypt, it has been known since the 18th dynasty (1552–1306 BC), *Rubia tinctorum syn R. peregrina* (Rubiaceae), 2n = 4x = 44

common milkweed: *Asclepias syriaca* (Asclepiadaceae), 2n = 2x = 22; 1C DNA content ~0.42 pg *bot agr*

common millet: An allotetraploid cereal which was among the first domesticated crops, but is now a minor crop, despite its high water-use efficiency; the maternal ancestor is *P. capillare* (or a close relative) and the other genome being shared with *P. repens, Panicum miliaceum* (Gramineae), 2n = 4x = 36; 1C DNA content ~1.04 pg *agr*

common onion: >>> onion

common osier: >>> basket willow

common petunia: A popular bedding plant that has a long history as a genetic model system; the genomes reveal that the *Petunia* lineage has experienced at least two rounds of hexa-ploidization, the older gamma event, which is shared with most Eudicots, and a more recent Solanaceae event, which is shared with tomato and other solanaceous species; tran-scription factors involved in the shift from bee- to moth-pollination reside in particularly dynamic regions of the genome, which may have been key to the remarkable diversity of floral color patterns and pollination systems; the high-quality genome sequences enhanced the value of *Petunia* as a model system for research on unique biological phenomena such as small RNAs, symbiosis, self-incompatibility and circadian rhythms; two wild parents, *P. axillaris* N and *P. inflate* are used in breeding; *Petunia hybrida* (Solanaceae), 2n = 2x = 14; 1C DNA content ~1.68 pg; ~1 Gb genome size containing 32,928 and 36,697 protein-coding genes, respectively *hort*

common velvet grass: A perennial grass; *Holcus lanatus* (Gramineae), 2n = 2x = 14; 2C DNA content 4 pg *bot agr*

common vetch: *Vicia sativa* ssp. *sativa* (Leguminosae), 2n = 2x = 12; polytene chromosomes have been observed in ovary, anthers, or immature seed tissues; 2C DNA content 4.5 pg *agr*

common wheat: >>> wheat

common wormwood: >>> wormwood

cone wheat: >>> rivet wheat >>> wheat

Congo pea: >>> pigeonpea

Congo signal grass (common names - Congo grass, Congo signal, Congo signal grass, Chinese cabbage, Kennedy ruzi, Kennedy ruzigrass, prostrate signal grass, ruzi, ruzigrass, ruzi grass; Spanish: Congo, Congo señal, gambutera, Kenia, pasto Congo, pasto ruzi, ruzi; Portuguese: ruzisiensis, capim Congo; French: herbe à Bengali; Thai: ya ruzi)**:** A forage crop that is grown throughout the humid tropics, native to Burundi, Rwanda and Eastern Congo; four *Brachiaria* species now cover as much as 85% of the cultivated pastures of Brazil, *Brachiaria ruziziensis syn B. eminii syn Urochloa ruziziensis* (Poaceae), 2n = 4x = 36 *agr* >>> brachiaria

Cook pine: Timber tree native to New Caledonia with beautiful grain (knots) produced by whorls of limbs along the main trunk; *Araucaria columnaris* (Araucariaceae), 2n = 2x = 26 *fore*

coracan millet: >>> finger millet

coral (cockscomb) tree: Several tropical trees with bright red or orange flowers, which produce a very lightweight wood; *Erythrina crista-galli* (Leguminosae), 2n = 42 *hort*

core watermelon: >>> watermelon

coriander: An annual herb commonly used in Latin American, Indian, and Southeast Asian cook-ing; all parts of the plant are edible, but the fresh leaves and the dried seedpods are the parts most commonly used in cooking; it is believed to have originated in the Mediterranean area, and in Southwest Europe; some believe its use began as far back as 5,000 BC, and there is evidence for its use by the Egyptians; coriander was brought to the U.S. in 1670 and was one of the first spices cultivated by early settlers; the fruits are known as "*coriander seeds*" and have a lemon citrusy flavor when crushed; it is also described as warm, nutty, spicy, and orange-flavored; fruits are usually dried but can be eaten green; ground corian-der is a major ingredient in curry powder and other aromatic dishes; *Coriandrum sativum* (Umbelliferae), 2n = 2x = 22; 2C DNA content 4.1 pg *hort*

cork oak: Cork is a light, waterproof, outer layers of the bark covering the branches and roots of almost all trees and shrubs; the cork oak is a native of southern Europe and North Africa; it is cultivated in Spain and Portugal; the exceptionally thick outer layers of its bark provide the cork that is used commercially; *Quercus suber* (Fagaceae), 2n = 2x = 24; 2C DNA content 1.0–1.6 pg *fore*

cork tree: *Phellodendron* spp. (Rutaceae), 2n = 78 *fore agr*

corn: >>> maize

corn bindweed: *Convolvulus tricolor syn C. arvensis* (Convolvulaceae), 2n = 48; 1C DNA content ~1.78 pg *bot agr*

cornelian cherry: >>> European cornel

corn flower: *Centaurea cyanus* (Asteraceae), 2n = 2x = 24; 2C DNA content 2.8–3.6 pg; *C. jacea* (2n = 4x = 44), *C. oxylepis* (2n = 4x = 44), *C. phrygia* (2n = 2x = 22) *bot agr* >>> sunflower

corn salad: *Valerianella olitoria syn V. locusta* (Valerianaceae), 2n = 2x = 16; 1C DNA content ~0.22 pg *hort*

corn thistle: >>> Canada thistle >>> creeping thistle >> cursed thistle

corn toadflax (common ~): *Linaria arvensis syn L. vulgaris* (Scrophulariaceae), 2n = 2x = 12; 1C DNA content ~0.87 pg *bot agr*

cos lettuce: A type of salad showing a more or less dense head and bag-like habit; often, the spoon-like leaves tend to grow lax; therefore the leaves were bound together by strings in order to produce more soft and leached leaves; beside a variety with pale green leaves grown in some regions of Germany and named 'Kasseler Strünkchen', there are also varieties with dark red leaves ('Romaine rouge') or with red spotted leaves (in Austria, called 'Forellensalat', = salmon salad); *Lactuca capitata* var. *romana* (Compositae), 2n = 2x = 18; 1C DNA content ~2.65 pg (*L. sativa*) *hort* >>> lettuce >>> Table 46

costmary (*syn alecost*): A herb which has been grown for many centuries and is possibly native to eastern Europe and Central Asia; it has soft, slightly hairy, fine-toothed, pale green leaves up to 30 cm long, which have a scent of balsam or mint; it bears clusters of small, white, rayed flowers with yellow centers; *syn Balsamita major syn Balsamita major* var. *tanacetoides syn Chrysanthemum balsamita syn Chrysanthemum balsamita* var. *tanacetoides syn Pyrethrum majus syn Tanacetum balsamita* (Compositae), 2n = 2x = 18; 1C DNA content ~4.90 pg *hort*

cotton: >>> barbadense

cottonwood: >>> poplar

couches (*syn elymus, wildrye, wheatgrass*): Couch plants are hermaphroditic, cespitose or rhizomatous, perennial; culms are erect, ascending, or decumbent, glabrous or scabrous; the internodes are hollow, or solid, terete; leaves are basal and cauline, not distinctly distichous; caryopses are adnate to lemma and/or palea or free from lemma and palea, longitudinally grooved or not grooved, dorsiventrally compressed; the base chromosome number is x = 7; the genus consists of ~120 species; these grassland and shrubland species are adapted to temperate climates with many important forage and weedy taxa; the genus *Elymus* is closely related to some important cereal crops and serves as potential alien genetic resources for the improvement of these crops; *Elymus* spp. (Gramineae), 2n = 4x = 28; all *Elymus* species are allopolyploid, including five basic genomes (St, H, Y, P, W); the St genome was donated by *Pseudoroegneria*, which is the pitoval genome found in all alloploid species; the H genome was donated from *Hordeum* species; the P and W genomes came from *Agropyron* and *Australopyrum* species, respectively; the origin of the Y genome remains uncertain, although it might have originated from a common progenitor of St and Y, via St^Y genomes; St^H- and St^Y-species are the most common tetraploid *Elymus* spp.; 2C DNA content 9.3 (YY) pg *bot agr*

couchgrass: >>> *Agropyron* >>> couches

courgettes: >>> marrow

cow bean: An important grain legume grown in the tropics, where it constitutes a valuable source of protein in the diets of millions of people; some abiotic and biotic stresses adversely affect its productivity; recent progress in the development of a consensus genetic map containing 37,372 single nucleotide polymorphisms (SNPs), mapped to 3,280 bins, strengthened the cowpea trait discovery pipeline; several informative markers associated with quantitative trait loci (QTLs) related to desirable attributes of cowpea were generated; genetic improvement activities aim at the development of drought tolerant, phosphorus use efficient, bacterial blight and virus resistant lines through exploiting available genetic resources, as well as deployment of modern breeding tools, that will enhance genetic gain when grown by sub-Saharan Africa farmers; *Phaseolus unguiculata* ssp. *sinensis syn Vigna unguiculata* (Leguminosae), 2n = 2x = 22; 1C DNA content ~0.60 pg; polytene chromosomes have been observed in ovary and immature seed tissues *hort* >>> pulse

cowpea: >>> cow bean

crambe: An annual herb that is sometimes grown as an oilseed crop, owing to its high content (55%–60%) of erucic acid (C22:1) in the seed oil; >>> erucic acid is a heat-stable, very-long-chain fatty acid used for production of lubricants and plastics; *Crambe abyssinica* also possesses other valuable characteristics, such as a short growing period, wide adaptation, and insect resistance, which make it a potential gene donor for the modification of rapeseed; *Crambe abyssinica* (Brassicaceae), 2n = 6x = 90; 1C DNA content ~0.90 pg *agr* >>> *Brassica* >>> Figure 8

creeping bentgrass: One of the best-adapted bentgrass species for use on golf course fairways and putting greens because of its high tolerance to low mowing height; it is a highly outcrossing allotetraploid species; *Agrostis stolonifera* (Poaceae), 2n = 4x = 28 ($A^2A^2A^3A^3$); 1C DNA content ~3.50 pg *bot agr*

creeping thistle: >>> Canada thistle

Creole bean: *Vigna reflexopilosa* var. *glabra* (Fabaceae), 2n = 4x = 44 (AABB) *hort* >>> Adzuki bean >>> black gram >>> cowpea >>> mung bean

crested dogstail: *Cynosurus cristatus* (Gramineae), 2n = 2x = 14; 1C DNA content ~3.05 pg *bot agr*

crested wheatgrass: A grass native to central Europe and from the eastern Mediterranean to Mongolia and China; *Agropyron cristatum* (Gramineae), 2n = 2x, 4x, 5x = 14, 28, 42; most common cytotype is 2n = 4x = 28 (PPPP); 2C DNA content 13.9 pg (PP) *bot agr*

cress: Several plants of the cress family, characterized by a pungent taste; the common European garden cress is cultivated worldwide; the young plants are grown along with white mustard to be eaten; *Lepidium sativum* (Brassicaceae), 2n = 2x, 4x = 16, 32; 2C DNA content 1.0 pg *hort* >>> *Brassica* >>> Figure 8

crimson clover: *Trifolium incarnatum* (Leguminosae), 2n = 2x = 14; 2C DNA content 1.7 pg *agr* >>> clover

crisphead (lettuce): Wild lettuce, also known as lactucarium, green endive, and lettuce opium, is the precursor to the common forms of cultivated lettuce; wild lettuce originated in Central Asia, but it is now grown throughout the world; lettuce was cultivated as a food crop by the Ancient Greeks, Romans, and Chinese; it was brought to the New World very early; *Lactuca sativa* (Compositae), 2n = 2x = 18; 1C DNA content ~2.65 pg *hort* >>> Table 46

crowfoot: Several white-flowered aquatic plants belonging to the buttercup family, with a touch of yellow at the base of the petals, the divided leaves are said to resemble the feet of a crow; *Ranunculus* spp. (Ranunculaceae), 2n = 4x = 32; 2C DNA content 9.8 pg *bot agr*

crown imperial: *Fritillaria imperialis* (Liliaceae), 2n = 2x = 24; 1C DNA content ~42.95 pg (often with chromosome fragments) *hort*

cucumber: Trailing annual plant belonging to the gourd family, producing long, green-skinned fruit with crisp, translucent, edible flesh; small cucumbers, called gherkins, usually the fruit of *Cucumis anguria*, are often pickled; there are ~735 species belonging to the cucumber family; the South African *C. humifructus* is unusual in that it fruits 15–30 cm underground; *Cucumis sativus* (Cucurbitaceae[1]), 2n = 2x = 14; 2C DNA content 1.4–2.5 pg; ~35,000,000 bp per chromosome *hort*

cucumber tree: >>> bilimbi

cucurbits: *Cucumis* spp., *Citrullus* spp., *Cucurbita* spp., *Lagenaria* spp. (Cucurbitaceae); endopolyploidy of 24–384n was observed, 2n = 4x = 48; 1C DNA content ~1.95 pg (*C. figarei*) *hort* >>> Table 46

cucuzzi: >>> bottle gourd

cumin: The seed comes from a plant of the parsley family; a small, delicate annual, it grows to a height of ~25 cm; the spicy seeds are boat-shaped, ridged, and brownish green in color; cumin is similar in appearance to caraway seeds, but cumin seeds do not have the licorice flavor of the former spice; it is used in curries and occasionally in snack foods; *Cuminum cyminum* (Umbelliferae), 2n = 2x = 14; 1C DNA content ~3.50 pg *hort*

cuphea: It is a genus containing about 260 species of annual and perennial flowering plants, native to warm temperate to tropical regions of the Americas; the species range from low-growing herbaceous plants to semi-woody shrubs up to 2 m tall; commonly, they are known as cupheas, or, in the case of some species, cigar plants; several species are popular ornamental plants or honey plants, e.g. *C. ignea* 'David Verity' and *C. micropetalia*; others are used to produce cuphea oil, of interest as a source of medium-length-chain triglycerides; for most purposes, cuphea oil is identical to coconut oil and palm oil; cuphea may thus produce a valuable source of income for farmers in temperate regions, by supplementing coconut and palm oil to satisfy the growing demand (e.g. for biodiesel production); early attempts at commercial production have focused on an interspecific hybrid population derived from *C. lanceolata* and clammy cuphea (*C. viscosissima*); sometimes, cuphea is introduced into the crop rotation of maize and wheat, where it has positive effects on the following harvest, including a higher yield of crop and crops that are higher in protein; *Cuphea lanceolata* (Lythraceae), 2n = 2x = 12 *agr hort*

cup plant: This perennial, herbaceous plant is up to 3–4 m tall and native to prairies of North America and Canada; the leaf pairs form cups around a square stem; there are about 33 species of *Silphium* recognized, which are divided into five sections (Composita, Dentata, Integrifolia, Laciniata, and Perfoliata), based upon leaf, stem, capitulum, and phyllary morphology; it attracts birds, butterflies, hummingbirds and honeybees; the plant might hold the promise of being not only a new source crop for cellulosic ethanol, but also a way to help capture and store carbon, not only in the U.S. but also in Europe and other countries; breeding activities have been initiated in order to develop the wild species into a bioenergy crop plant; *Silphium perfoliatum* var. *connatum* and var. *perfoliatum* (Asteraceae), 2n = 2x = 14 *hort agr*

curled lettuce: >>> lettuce >>> cutting lettuce

curly kale: >>> kale

cursed thistle: >>> Canada thistle

cuscuta (*syn* **dodder** *syn* **strangle-weed** *syn* **hellbind** *syn* **hailweed** *syn* **devil's hair**): These species belong to a monogeneric family in which all members are parasitic on other plants; cuscuta consists mainly of yellow, orange, or red, slender, vine-like stems with vestigial leaves, and the plants lack chlorophyll entirely; it can occasionally be an agricultural nuisance; *Cuscuta* spp. (*C. campestris* 2n = 8x = 56, *C. europaea* 2n = 2x = 14; 1C DNA content ~1.08 pg (Cuscutaceae) *phyt agr*

cush-cush yam: The only cultivated yam that is indigenous to the New World, *Dioscorea trifida* (Dioscoreaceae), 2n = 4x = 40; 1C DNA content ~1.05 (*D. esculenta*) pg *hort* >>> yam >>> aerial yam

custard apple: Large, edible, heart-shaped fruits are produced by a group of tropical trees and shrubs which are often cultivated; Bullock's heart (*Annona reticulata*) produces a large dark-brown fruit containing a sweet reddish-yellow pulp; it is a native of the West Indies; *Annona cherimolia* (Annonaceae), 2n = 2x = 14; 1C DNA content ~1.33 pg *hort* >>> cherimoya >>> pawpaw >>> sweetsop >>> soursop >>> ylang-ylang (related species)

cutting lettuce: *Lactuca sativa* var. *crispa* or *L. sativa* var. *secalina* (Compositae), 2n = 2x = 18; 1C DNA content 2.65 pg *hort* >>> lettuce >>> Table 46

cycad: Cycads are seed plants typically characterized by a stout and woody trunk with a crown of large, hard, and stiff, evergreen leaves; they usually have pinnate leaves; the individual plants are either all male or all female, i.e. dioecious; they vary in size from having a trunk that is only a few centimeters tall to trunks up to several meters tall; they typically grow very slowly and live for a very long time, with some specimens known to be as much as 1,000 years old; because of their superficial resemblance, they are sometimes confused with and mistaken for palms or ferns, but are only distantly related to either; they are found across much of the subtropical and tropical parts of the world; cycads are >>> gymnosperms ("naked seeded"); they have very specialized pollinators, usually a specific kind of beetle; they have been reported to fix nitrogen in association with a cyanobacterium living in the roots; these blue-green algae produce a neurotoxin, called β-methylamino-L-alanine, that is found in the seeds; a neurotoxin, cycling through the food chain from the cyanobacterium through cycads to its seeds, to bats eating the seeds, to humans consuming the bats, is hypothesized to be a source for some neurological diseases in humans, such as mad cow disease; the starch obtained from the stems of certain species is still used as food by some indigenous tribes; tribal people also grind and soak the seeds to remove the neurotoxins that may be present, making the food source generally safe to eat, although often not all the toxin is removed; in addition, consumers of bush meat may face a health threat as the meat comes from game which may have eaten cycad seeds and carry traces of the toxin in body fat; there are about 305 described cycad species, in 10–12 genera and two to three families of cycads (Cycadaceae, Stangeriaceae, and Zamiaceae); most cycads have only a few large chromosomes because of a lack of whole-genome duplication or, possibly in the genus *Zamia*, as a result of chromosomal fission; large chromosomes and extremely small effective population sizes result in substantial linkage disequilibrium, genetic hitch-hiking and genetic drift in cycads; by contrast, other seed plants have higher incidences of polyploidy and may therefore have been more prone to radiation; *Cycas elephantipes, C. siamensis, C. tansachana*, 2n = 2x = 22; 1C DNA content ~12.55 pg (*C. petraea*); *Zamia integrifolia, Z. pumila; Z. pygmaea* (2n = 2x = 16; 1C DNA content ~12.05 pg (*Z. angustifolia*) *fore*

D

daffodil: A common English name, sometimes used now for all varieties of narcissus, and the chief common name of horticultural prevalence used by the American Daffodil Society; the range of forms in cultivation has been heavily modified and extended, with new variations available from specialists almost every year; the name "daffodil" is derived from an earlier "Affodell", a variant of Asphodel; the reason for the introduction of the initial "d" is not known, although a probable source is an etymological merging from the Dutch article "de," as in "De affodil"; from at least the 16th century, "Daffadown Dilly", "daffadown dilly", and "daffydowndilly" have appeared as playful synonyms of the name; the name "jonquil" is sometimes used in North America, particularly in the South, but strictly speaking that name belongs to only the rush-leaved *Narcissus jonquilla* (Amaryllidaceae), 2n = 2x = 14, 1C DNA content ~16.40 pg, and cultivars derived from it *hort* >>> narcissus

dahlia: A group of perennial plants belonging to the daisy family, comprising 20 species and many cultivated forms; dahlias are stocky plants with tuberous roots and showy flowers that come in a wide range of colors; they are native to Mexico and Central America, e.g. *Dahlia pinnata* (Asteraceae), 2n = 8x = 64; 1C DNA content ~4.04 pg (*Dahlia coccinea*) *hort*

daikon: *Raphanus sativus* convar. *sativus* syn *Raphanus raphanistroides* (Brassicaceae), 2n = 2x = 18 (R^{r}R^r); 1C DNA content 0.555 pg, var. Saxa, genome size 543 Mbp *hort* >>> *Brassica* >>> Figure 8

daimyo oak: *Quercus dentata* (Fagaceae), 2n = 2x = 24; 1C DNA content ~0.95 pg (*Quercus suber*) *fore*

daisy fleabane: *Erigeron* spp. (*Erigeron annuus* 2n = 3x = 42; 1C DNA content ~2.17 pg) (Asteraceae) *bot agr*

Damask rose: Damask roses are grown in several European and Asiatic countries for rose oil production; *Rosa damascena* (Rosaceae), 2n = 4x = 28; 1C DNA content ~1.08 pg *hort*

Damson plum: A cultivated variety of plum tree, distinguished by its small oval edible fruits, which are dark purple or blue-black in color; *Prunus domestica* var. *insititia* (Rosaceae), 2n = 2x = 16; 1C DNA content ~0.33 pg *hort*

dandelion (*syn* **common dandelion** *syn* **lion's tooth** *syn* **priest's crown** *syn* **pu gong ying** *syn* **swine's snout** *syn* **dent de lion**): A perennial herb introduced, it is thought, from Europe and Asia to America; it is now naturalized throughout the northern and even southern hemisphere; it is a very easily grown plant, which succeeds in most soils; it becomes quite large when cultivated, the leaves reaching a foot or more in length; dandelion is often cultivated as an edible salad crop and as a medicinal herb plant; it is found growing in pastures, lawns, waste ground, sand, rocks, even cracks in concrete; from a thick, long, tap root, dark brown outside, white and milky white inside, grow long jaggedly toothed leaves, shiny, dark to light green and growing in the shape of a rosette close to the ground; a purplish flower-stalk rises straight from the center, which is leafless, smooth, hollow, and bears a single, bright golden yellow, furry- looking flower which blooms almost any time of the year; when mature, the seed in the flowers heads are round and fuzzy, carried by the wind to germinate wherever they land; it is used for medicinal purposes and is edible; the dandelion is very nutritious, having more vitamins and minerals than most vegetables; it has a long history of use as a food in many countries; the young leaves are less bitter, and flowers are eaten raw in salads; all leaves are also cooked or boiled as a pot herb, and flowers are often dipped in batter and fried; dried roots are used as a coffee substitute; herbal wine is made from fermented flowers; recent increase of prices for natural rubber from rubberwood again made the production of dandelion latex attractive; it was actually cultivated in Poland during World War II for its latex; dandelions are still cultivated today for their latex in various eastern Europe and western Asian countries, *Taraxacum officinale* (Asteraceae), 2n = 3x = 24; 1C DNA content ~1.28 pg (the latter three polyploids are probably all obligate apomictic) *hort bot agr* >>> Russian dandelion

dari: >>> durra

darnel (ryegrass): A species of ryegrass, formerly a common weed among cereal crops; often associated with weed wheat and barley; for example, in Southwestern Ethiopia, where people maintain traditional ways of subsistence by growing various kinds of crops and livestock, and by conducting sustainable shifting cultivation, man's impacts, especially cereal cultivation, on the diversity of darnel was surveyed; the grains of darnel are either awned or awnless, and the awnless is dominant over the awned; the awned form was found in emmer wheat, and the awnless one was generally associated with bread, macaroni, and rivet wheat; grain cleaning is done by winnowing and subsequent hand removal of contaminants; emmer wheat has non-free-threshing grains, and the other three crops have free-threshing grains; awned darnel's grain morphology is similar to that of emmer wheat grain, and the awnless darnel grain resembles the free-threshing grains of bread, macaroni, and rivet wheat; separating awned darnel grains from emmer wheat grains is difficult, as is separating awnless grains from free-threshing wheat grains; free-threshing wheat grains contaminated with darnel grains are sown in the emmer wheat field because the boundaries between the two fields are unclear; crop seed exchange and contamination of crop grains with darnel grains during storage or seeding of crops led to unintended artificial gene flow of darnel and consequently conserved the genetic diversity of darnel; *Lolium temulentum* (Poaceae), 2n = 2x = 14; 1C DNA content ~2.86 pg *bot agr* >>> Table 46

dasheen: >>> taro

date palm: A palm tree; it has a subtropical, semiarid origin in the Middle East; this is possibly the oldest plant domestication in the world; the plant is dioecious and breeding is exceptionally difficult; propagation by seeds is a waste of time, because of the loss of fruit quality, and vegetative propagation with basal suckers is essential; the quality of the date fruit is affected by metaxenia; a new-encounter killer disease, "Bayoud disease" (*Fusarium oxysporum* f. sp. *albidinis*) is spreading inexorably from Morocco eastwards; the female tree produces the brown oblong fruit, dates, in bunches weighing 9–11 kg; dates are an important source of food in the Middle East, being rich in sugar; they are dried for export; the tree also supplies timber and materials for baskets, rope, and animal feed; the most important species is native to northern Africa, Southwest Asia, and parts of India, it grows up to 25 m high; a single bunch can contain as many as 1,000 dates; their juice is made into a kind of wine; of the many varieties of date palms, the species *Phoenix dactylifera* is cultivated extensively and traded and consumed worldwide, *Phoenix dactylifera* (Arecaceae), 2n = 2x = 36; 1C DNA content 0.95 pg *fore hort*

date plum: >>> American persimmon

daylily: *Hemerocallis* sp. (Asphodelaceae), 2n = 2x, 3x = 22, 33 *bot hort*

deadnettle: *Lamium* sp. (Lamiaceae), 2n = 2x, 4x = 18, 36; 1C DNA content ~1.10 pg (*Lamium album*) *bot agr*

derris: A climbing leguminous plant of Southeast Asia; its roots contain rotenone, a strong insecticide; *Derris elliptica* (Fabaceae), 2n = 2x = 22 *agr*

desi chickpea: >>> chickpea

dill: A short-lived, European, annual herb; seeds are used to flavor pickles; the fernlike leaves are used to flavor many foods, such as borscht; the name "dill" is derived from a Norse word which means to soothe, the plant having the carminative property of allaying pain; the common dill grows wild in the cornfields of Spain and Portugal and the south of Europe generally; there is also a species of dill cultivated in Eastern countries known by the name of shubit; it was this species of garden plant of which the Pharisees were in the habit of using to pay tithes; the Talmud requires that the seeds, leaves, and stem of dill shall pay tithes; it is an umbelliferous plant, very like the caraway; *Anethum graveolens* (Apiaceae), 2n = 2x = 22; 1C DNA content 1.2 pg *hort* >>> caraway

Dinkel wheat: *Triticum spelta* (Poaceae), 2n = 6x = 42 (BBA^{u}A^{u}DD); 1C DNA content ~17.33 pg (*Triticum aestivum*) >>> wheat

dog dumpling: >>> noni

dogwood (flowering ~): A shrub up to 4 m high, it is a native to eastern North America; it is used as an ornamental tree that plays an important role in the U.S. nursery industry; over 100 cultivars have been released during the past decade; 'Appalachian Spring' and 'Cherokee Brave' are popular varieties; it is predominantly cross pollinated and highly self-incompatible; most cultivars are propagated from >>> axillary buds grafted onto rootstock grown from wild seedlings; the old names now rarely used include American dogwood, Florida dogwood, Indian arrowwood, cornelian tree, white cornel, false box, and false boxwood; this species has in the past been used in the production of inks, scarlet dyes, and as a >>> quinine substitute; the hard, dense wood has been used for products such as golf club heads, mallets, wooden rake teeth, tool handles, jeweler's boxes, and butcher's blocks,

Cornus florida syn Benthamidia florida (Cornaceae), 2n = 2x = 22; 1C DNA content ~1.16 (*C. sanguinea*) pg *hort*

Dolichos lablab: >>> lablab

Douglas fir: One of six species of evergreen coniferous tree belonging to the pine family; the most common is *Pseudotsuga menziesii*, native to western North America and East Asia; it grows to 60–90 m in height, has long, flat, spirally-arranged needles and hanging cones, and produces hard, strong timber; *P. glauca* has shorter, bluish needles and grows to 30 m in mountainous areas; *Pseudotsuga* spp. (Pinaceae), 2n = 2x = 26; 1C DNA content ~19.05 pg *fore hort*

doum palm: A native palm tree of Upper Egypt, Sudan, Kenya, and Tanzania; it was considered sacred by ancient Egyptians; seeds of doum nuts have been found in the pharaohs' tombs; the doum palm, also known as the gingerbread palm, grows a red-orange, apple-sized fruit that tastes like gingerbread; the fruit's hard, white nut is used to make buttons; rind from doum nuts is used to make molasses; the palm's leaves are used to make mats and writing paper and to bind parcels; the doum palm can reach heights of 6–9 m; *Hyphaene thebaica* (Palmae), 2n = 2x = 32(?); 1C DNA content ~3.40 pg (*H. benguelensis*) *hort*

doura: >>> durra

drooping brome: *Bromus tectorum syn Anisantha tectorum* (Gramineae), 2n = 2x = 14; 1C DNA content ~3.25 pg *bot agr*

duckweed: Duckweeds are the fastest-growing angiosperms that are adapted to the aquatic environment, and have the potential to become a new generation of sustainable crops; although a seed plant, *Spirodela polyrhiza* clones rarely flower and multiply mainly through vegetative propagation; most duckweeds reproduce to form the next generations by vegetative budding during spring and summer, whereas they become natural starch repositories, when they switch into the dormant stage during winter time; *Spirodela polyrhiza* (Lemnaceae), 2n = 2x = 20, genome size from 150 Mb in *S. polyrhiza* to 1,881 Mb in *Wolffia arrhizal aqu*

durian: The fruit of several tree species belonging to the genus *Durio* and the Malvaceae family; the durian has been known and consumed in southeastern Asia since prehistoric times, but has only been known to the western world for about 600 years; the earliest known European reference to the durian is the record of Niccolò da CONTI, who travelled to Southeastern Asia in the 15th century; the fruit is used to flavor a wide variety of sweet edibles such as traditional Malay candy, ice kacang, dodol, rose biscuits, and, with a touch of modern innovation, ice cream, milkshakes, mooncakes, yule logs and cappuccino; durian trees are large, growing to 25–50 m in height, depending on the species; the leaves are evergreen, elliptic to oblong and 10–18 cm long; the flowers are produced in three to thirty clusters together on large branches and directly on the trunk, with each flower having a calyx and five petals; the trees have one or two flowering and fruiting periods per year, though the timing varies depending on the species, cultivars, and localities; a typical durian tree can bear fruit after four or five years; the fruit can hang from any branch and matures roughly three months after pollination; it can grow up to 30 cm long and 15 cm in diameter, and typically weighs 1–3 kg; its shape ranges from oblong to round, the color of its husk from green to brown, and that of its flesh from pale-yellow to red, depending on the species; the durian is somewhat similar in appearance to the >>> jackfruit, an unrelated species; over the centuries, numerous durian cultivars have been maintained as vegetative clones; they used to be grown with mixed results from seeds of trees bearing superior-quality fruit, but are now propagated by layering, marcotting, or, more commonly, by grafting, including bud, veneer, wedge, whip or U-grafting onto seedlings of randomly selected rootstocks; different cultivars can be distinguished to some extent by variations in the fruit shape, such as the shape of the spines; consumers express preferences for specific cultivars, which fetch higher prices in the market; *Durio zibethinus* (Bombacaceae), 2n = 2x = 14, with median, submedian and almost subterminal primary constrictions, and four chromosomes with secondary constrictions *fore*

durra: Also known as milo, kafir, feterita, kaoliang, mtata, sorgo, jola, jawa, guinea corn, broom-corn, and cholam; it is grown mainly in Africa, India, China, and the U.S.; it is an open-pollinated, short-day plant; hybrid varieties have been produced in America; a grass, whose seeds are used to make a flour and as cattle feed; it is an important food crop in Africa, Central America, and southern Asia, and is the fifth major cereal crop grown in the world; the largest producer is the U.S.; it originated in eastern Africa and first diverged from the wild varieties in Ethiopia 5,000 years ago; it is well adapted to grow in hot arid or semiarid areas; the many subspecies are divided into four groups: (a) grain sorghums, (b) grass sorghums (for pasture and hay), (c) sweet sorghums (to produce sorghum syrups), and (d) broom corn (for brooms and brushes); *Sorghum vulgare* or *S. bicolor* (Gramineae), $2n = 2x = 20$; 1C DNA content ~0.75 pg *agr*

durrha: >>> durra

durum wheat: An ancient wheat grown since Egyptian times; a wheat used to make bread and other bakery products; the hard, flinty kernels are specially ground and refined to obtain semolina, a granular product used in making pasta items such as macaroni and spaghetti; most durum wheats are grown in Mediterranean countries, Russia, North America, and Argentina; *Triticum durum* (Gramineae), $2n = 4x = 28$ (BBA^uA^u); 1C DNA content ~12.66 pg *agr* >>> wheat

Dutch turnip: >>> stubble turnip

dwarf bean: >>> French bean

dwarf French bean: *Phaseolus vulgaris* var. *nanus* (Leguminosae), $2n = 2x = 22$; 1C DNA content 0.6 pg, polytene chromosomes have been observed in ovary and immature seed tissues *hort*

dwarf sisal: *Agave angustifolia* (Agavaceae), $2n = 4x = 120$; 1C DNA content ~7.45 pg *agr*

dye amaranth: *Amaranthus cruentus* (Amaranthaceae), $2n = 2x = 32$; 2C DNA content 1.1 pg *hort*

dyer's alcanet (*syn* **dyer's alkanet**): >>> alkanna

dyer's bugloss: >>> alkanna

dyer's chamomile: *Anthemis tinctoria* (Compositae), $2n = 2x = 18$; 1C DNA content ~3.75 *hort*

dyer's madder: >>> common madder

E

earth-smoke: >>> fumitory

Eastern gamagrass (*syn* **gamagrass** *syn* **bullgrass** *syn* **sesame grass**): A grass native to America, with a gene pool that has a lot to offer breeders of maize, including tolerance to cold and resistance to insects, as well as tolerance to drought and flooding; it can be hybridized with maize and it is used for introgression experiments; *Tripsacum dactyloides* (Poaceae), $2n = 2x, 4x = 36, 72$ (cytotypes with 36, 45, 54, 72, and 108 chromosomes are found); 1C DNA content 3.875 pg *bot agr*

Eastern white pine (*syn* **Weymouth pine**): *Pinus strobus* (Pinaceae), $2n = 2x = 24$; 1C DNA content ~25.65 *fore*

East-India arrowroot: *Curcuma* spp. (Zingiberaceae) $2n = 6x, 9x, 11x, 12x, 15x = 42, 63, 77, 84, 105$; 1C DNA content varied from 1.7 pg in *Curcuma vamana* to 4.8 pg in *C. oligantha hort*

eddo: >>> taro

eddoe: >>> taro

eggapple: >>> eggplant

eggplant: Also known as >>> aubergine or brinjal, it originated in India and is the only important Old World cultivated species of the family Solanaceae; a good source of minerals and vitamins, it has medicinal properties; the crop is cultivated for its fruit which is eaten as a vegetable; it is open-pollinated and hybrid varieties are useful; transgenic *Bacillus* *thuringiensis* (*Bt*)-eggplants have been bred to reduce pesticide use; they are in use by farmers in Bangladesh; *Solanum melongena* (Solanaceae), 2n = 2x = 24; 1C DNA content 0.95 pg *hort* >>> chloroplast

Egyptian clover: >>> Alexandrinian clover

Egyptian tree onion: >>> tree onion

einkorn: One-grained wheat, so-called because it has a single seed per spikelet; a wheat cultivated since Stone Age times, both winter and spring forms occur; spikes are awned, slender, narrow, flattened, and fragile; spikelets contain only a single fertile flower and thus produce only one seed; seeds are pale red, slender, flattened, almost without crease, and remain in the spikelets after threshing; *T. monococcum, T. boeoticum* (Gramineae), 2n = 2x = 14 (A^{m}A^m); 1C DNA content ~6.23 pg *agr* >>> wheat >>> Table 53

elephant garlic: >>> Levant garlic

elephant grass: Also known as Napier grass, it is so-called because it grows tall enough to hide an African elephant; it occurs wild in the general area of Uganda; it is a highly productive fodder, and it provides an excellent mulch; it is usually propagated by stem cuttings of 3–4 nodes; seed is produced abundantly but is difficult to collect; *Pennisetum purpureum* (Gramineae), 2n = 4x = 28; 1C DNA content ~2.90 *agr*

eleusine: >>> finger millet

elm (tree): The elms, with three or four other genera, are separated off from the great group of catkin-bearing forest trees to form a distinct natural order, the Ulmacea; they are confined to the north temperate zone, and, of the genus *Ulmus*, there are rather more than a dozen forms admitted to rank as species; these are similar in having their leaves "oblique," that is, unequally lobed at the base, one side being larger than the other; in their tufted flowers, which are not in drooping catkins, each contain both stamens and pistils; and in the enclosed ovary, are two chambers, though the >>> winged fruit which results has commonly only one chamber with one seed in it; the position of this seed-chamber in the elliptical fruit furnishes the distinguishing characters of British elms; in the common elm (*Ulmus campestris*), the seed chamber is above the center, and near the little notch at the top of the >>> samara, whereas, in the witch elm (*U. montana*), it is below the center; there is a recent study that shows that all English elm trees (*Ulmus procera*) derived from one single plant introduced by the Romans ~2,000 years ago; the closest relationship was found to recent populations in Latium, Italy; since this type of elm does not produce seeds, it was propagated by clones; Romans used elm trees as climbing supports for grape plants; the seedless plants could not be distributed among the vineyards without control; *Ulmus* spp. (Ulmaceae), 2n = 2x = 28; 1C DNA content ~1.08 pg (*U. glabra*) *fore hort*

emmer wheat: Grown since Neolithic times; one of the most ancient of cultivated cereals; it may be either winter or spring in habit; leaves generally are pubescent; spikes are very dense and flattened laterally; spikelets generally contain two flowers and generally are awned; the red or white kernels remain enclosed in the glumes after threshing; they are slender and acute at both ends; *Triticum dicoccum, T. dicoccoides* (Gramineae), 2n = 4x = 28 (BBA^{u}A^u), 2C DNA content 25 pg *agr* >>> wheat >>> Table 53

endive: Leaves used as a garnish or herb, *C. intybus* is >>> chic(k)ory, whose dried and roasted roots are used for blending with coffee; the young shoots of *C. endiva* are endives, and are used as a vegetable, mainly in salads; there are broad-leaved types: escarole, broad-leaved endive (English), scarole, chicorée scarole (French) or chicória escarola (Portuguese); and arrow-leaved types: endive, curly endive (English), frisée, chicorée frisée (French) or chicória frisada (Portuguese); it is only known from cultivation and was probably first brought into cultivation in the Mediterranean region, where its wild relatives, *Cichorium pumilum* and *C. calvum*, occur; both species can be considered as the progenitor of escarole and endive; it was probably known to the ancient Egyptians, but no archaeological evidence has been found; it spread to central Europe in the 16th century and is now grown throughout Europe and North America; it is grown throughout the tropics including tropical Africa, but is mostly of minor importance there; *Cichorium endivia* et ssp. (Compositae), 2n = 2x, 3x = 18, 27 *hort* >>> chickory >>> http://database.prota.org/PROTAhtml/Cichorium%20endivia_En.htm

Engelmann spruce: *Picea engelmannii* (Pinaceae), 2n = 2x = 24; 1C DNA content ~19.45 pg *fore*

English marigold (*syn* **pot marigold**): *Calendula officinalis* (Compositae), 2n = 28, 32(?) *hort bot*

English ryegrass: *Lolium perenne* (Gramineae), 2n = 2x = 14; 2C DNA content 6.7 pg *agr*

enset(e): This member of the banana family is grown for food in Ethiopia and was domesticated there; whereas wild enset grows over much of East and southern Africa and the genus extends across Asia to China, it has only ever been domesticated in the Ethiopian Highlands; enset contributes to improved food security for approximately 20 million Ethiopians and, according to Ethiopian researchers, there is potential for expanding consumption of the crop, the starchy stem of the plant is consumed; the most important species is *E. ventricosa*; it is a staple food crop in parts of southern Ethiopia at altitudes of 1,500 to 3,000 m; the edible starch is extracted from the corm and pseudostem and fermented to make "kocho", an important food for several million people in the region; fiber, which is made into cordage and sacking, is also obtained from the pseudostems; *Ensete ventricosa syn Musa ensete syn Ensete edule* (Musaceae), *2n = 2x = 18; DNA* sequence data suggest a genome size of approximately 547 megabases, similar to the 523-megabase genome of the closely related banana (*Musa acuminata*); at least 1.8% of the annotated *M. acuminata* genes are not conserved in *E. ventricosum*; furthermore, enset contains genes not present in banana, including reverse transcriptases and virus-like sequences as well as a homolog of the *RPP8-like* resistance gene *hort*

epazote (**wormseed** *syn* **Jesuit's tea** *syn* **Mexican tea** *syn* **paico** *syn* **herba Sancti Mariæ**): An herb native to Central America, South America, and southern Mexico; it is an annual or short-lived perennial plant, growing to 1.2 m tall, irregularly branched, with oblong-lanceolate leaves up to 12 cm long; the flowers are small and green, produced in a branched panicle at the apex of the stem; epazote is used as a leaf vegetable and herb for its pungent flavor; raw, it has a resinous, medicinal pungency, similar to anise, fennel, or even tarragon, but stronger; *Dysphania ambrosioides syn Chenopodium ambrosioides* (Chenopodioideae), 2n = 2x, 4x, 8x = 16, 32, 64 *hort*

esparsette: >>> sainfoin

estragon (*syn* **tarragon** *syn* **French tarragon** *syn* **cooking tarragon**): Tarragon is a green, glabrous perennial shrub; its branched root system with runners produces erect, bushy-branched stems to 60–70 cm high; the lower leaves are ternate, the upper leaves lanceolate to linear and small-toothed or entire, the small, drooping whitish-green or yellow flowers are almost globular and bloom from May to June in terminal panicles; it contains components common to many herbs that are routinely consumed without reported adverse effects; *Artemisia dracunculus* var. *sativa* (Compositae), 2n = 2x = 18; 1C DNA content ~2.97 pg *hort*

Ethiopian mustard: Seed oil of zero erucic acid germplasm of Ethiopian mustard is characterized by a low concentration of oleic acid and high concentrations of linoleic and linolenic acids; this crop is confined to the highlands of Northeast Africa where it is grown for oil, which is locally known as "noug oil"; *Brassica carinata* (Brassicaceae), 2n = 4x = 34; 1C DNA content ~1.58 pg *agr* >>> *Brassica* >>> Figure 8

eucalypt(s): Eucalypts are the world's most widely planted hardwood trees; their outstanding diversity, adaptability and growth have made them a global renewable resource of fiber and energy; in 2014, its genome was sequenced; it contains about 640 Mb; of 36,376 predicted protein-coding genes, 34% occur in tandem duplications, the largest proportion thus far reported in a plant genome; *Eucalyptus* also shows the highest diversity of genes for specialized metabolites, such as terpenes that act as chemical defenses and provide unique pharmaceutical oils; genome sequencing of the *E. grandis* sister species *E. globulus* and a set of inbred *E. grandis* tree genomes revealed dynamic genome evolution and hotspots of inbreeding depression; *Eucalyptus grandis* (Myrtales), 2n = 2x = 22, 1C DNA content ~0.35 to 0.75 pg for different species *fore*

European cornel: A species of dogwood native to southern Europe and Southwest Asia; in North America, the plant is known by the common name of "cornelian cherry"; the fruit is edible, but the unripe fruit is astringent; when ripe, the fruit is dark ruby red; it has an acidic flavor which is best described as a mixture of cranberry and sour cherry; it is mainly used for making jam, makes an excellent sauce, similar to cranberry sauce, when pitted and then boiled with sugar and orange, but also can be eaten dried; in Azerbaijan and Armenia, the fruit is used for distilling vodka; in Turkey and Iran it is eaten with salt as a snack in summer, and traditionally drunk as a cold drink called sherbet; cultivars selected for fruit production in the Ukraine have fruit up to 4 cm long; when ripe on the plant, the berries bear a resemblance to coffee berries, and ripen in mid to late summer; there are varieties, such as 'Alaba' (almost white fruits), 'Aurea' (yellow leaves and red fruits), 'Elegantissima' (broad leaves with yellowish margin), 'Flava' (yellow fruits), 'Macrocarpa' (pear-shaped fruits), 'Nana' (dwarf habit; from the Balkans), 'Pyramidalis' (erect habit), 'Sphaerocarpa Cretzoiu' (round fruits), 'Variegata' (leaves with white margin), 'Violacea' (with pink fruits; from Romania), 'Schumener' and 'Jolico' (with long fruits; from Austria), 'Kasanlak' (high-yielding, large fruits; from Bulgaria), 'Cormas' (with large fruits; from Denmark), varieties 'Helen', 'Pioneer', 'Red Star' or 'Elegant' from the U.S.; *Cornus mas* (Cornaceae), 2n = 2x = 18; 1C DNA content ~2.40 pg *hort fore*

European elymus: >>> lyme grass >>> couches

European grape: Grapes (*Vitis vinifera* spp. *vinifera*) have been a source of food and wine since their domestication from the wild progenitor (*Vitis vinifera* ssp. *silvestris*) around 8,000 years ago, and they are now the world's most valuable horticultural crop; in addition to being economically important, *V. vinifera* is also a model organism for the study of perennial fruit crops for two reasons: firstly, its ability to be transformed and micropropagated via somatic embryogenesis, and secondly, its relatively small genome size of 500 Mb; all 60 wild species of grapevine grow in the northern hemisphere; in North America there 33 species, such as *Vitis liparia*, >>> *V. berlandieri*, *V. rupestris*, or *V. labrusca*; in Europe and the Near East, there was only one species identified, *V. vinifera*, with subspecies *silvestris* and *caucasica*; from Europe are derived all American, South American, African, Australian, and New Zealand varieties; in Asia, 26 species are described, such as >>> *V. amurensis*, *V. coignetiae*, *V. thumbergii*, or *V. flexuosa*; most of the 7,000 grapevine cultivars can be divided into two groups, red and white, based on the presence or absence of anthocyanin in the berry skin (50% of the varieties are white ones); it has been found, from genetic experiments, that the presence of red color is controlled by a single locus; a regulatory gene, *VvMYBA1*, which can activate anthocyanin biosynthesis in a transient assay, was shown not to be transcribed in white berries due to the presence of a retrotransposon in the promoter; the berry color locus comprises two very similar genes, *VvMYBA1* and *VvMYBA2*, located on a single bacterial artificial chromosome; either gene can regulate color in the grape berry; the white berry allele of *VvMYBA2* is inactivated by two nonconservative mutations, one leading to an amino acid substitution and the other to a frameshift, resulting in a smaller protein; transient assays showed that either mutation removed the ability of the regulator to switch on anthocyanin biosynthesis; *VvMYBA2* sequence analyses, together with marker information, confirm that 55 white cultivars all contained the white berry alleles, but not red berry alleles; it seems that all extant white cultivars of grape vines have a common origin, and that rare mutational events occurring in two adjacent genes were essential for the genesis of the white grapes used to produce the white wines and white table grapes; *ex situ* conservation is applied for wild grape populations

(*V. vinifera* ssp. *silvestris*) in Iran; *in situ* preservation is only performed in the case of the Sardasht-Ghasmarash population to ensure that evolutionary dynamic forces continue to influence plant adaptation and survival in response to environmental changes; recently, an all-season seedless grape "Early Sweet" was bred that can also be harvested during the winter time, *Vitis vinifera* ssp. *sativa* (Vitaceae), 2n = 2x = 38; 1C DNA content ~0.43 pg *hort* >>> Schiave

European plum: *Prunus domestica* (Rosaceae), 2n = 6x = 48; 1C DNA content ~0.93 pg *hort*

evening primrose: A typical ruderal plant in Europe, although its origin is Peru; closely related is the species *O. missouriensis*, native to southern regions of North America; in addition to the nice smell, in the past the roots were used as a vegetable; they are edible from autumn until May; later they become woody; it has, in recent years, made the transition from being a wild flower and cottage garden plant to an established agricultural crop; its value lies in the seed oil which contains, at maturity, approximately 7–10% gamma linolenic acid (GLA), an essential fatty acid with proven value as a nutrient and pharmaceutical in humans; there are breeding activities to develop it as an oil seed crop; this seed may contain up to 30% oil; *Oenothera biennis* (Oenotheraceae), 2n = 2x = 14; 1C DNA content ~1.22 pg *hort*

F

fall rose: >>> China aster

false: >>> bastard indigo

false acacia: >>> locust tree

false goat's beard (*syn* **tall false buck's beard** *syn* **red false buck's beard**): *Astilbe* spp. (Saxifragaceae), 2n = 2x, 4x = 14, 28 *hort*

false flax (*syn* **gold-of-pleasure**): An emerging biofuel crop with potential applications in industry, medicine, cosmetics and human nutrition; sequencing studies generated a FastQ file size of 2.97 Gb with 10.83 million reads, having a maximum read length of 101 nucleotides; the number of contigs generated was 53,854 with maximum and minimum lengths of 10,086 and 200 nucleotides respectively; genes involved in lipid metabolism were identified and transcription factors are known; bioinformatics analysis revealed the presence of a total of 19,379 simple sequence repeats; *Camelina sativa* (Brassicaceae), 2n = 40 *agr* >>> agroenergy crop >>> *Brassica*

false indigo: >>> bastard indigo

false saffron: >>> safflower

false sarsaparilla: >>> sarsaparilla

farro: >>> spelt

feijoa (*syn* **pineapple guava**): Native to South America, it is a small, dense, attractive tree which is often used for windbreaks; it produces attractive flowers with bright red stamens and a small round or oval fruit, green in color; seedling populations exhibit high levels of heterozygosity; it grows in subtropical and temperate climates and is cultivated for its fruit in Australia and New Zealand; acetonic plant extracts show anticancer activities on solid and hematological cancer cells; extracts show toxic effects on normal myeloid progenitors, thus displaying a tumor-selective activity; flavones are the active components; they induce apoptosis which is accompanied by caspase activation; *Feijoa sellowiana* (Myrtaceae), 2n = 2x = 22 *hort*

fennel: A plant which produces edible seeds and leaves; a perennial herb, erect, glaucous, to 2 m tall, highly aromatic; leaves to 40 cm long, finely dissected, ultimate segments filiform; umbels terminal, 5–15 cm wide; umbellets with 20–50 tiny flowers on filiform pedicels; fruit 4–9 mm long, half as wide or less, grooved; the cultivar 'Florence fennel' has inflated leaf bases which form a sort of bulb; it comes mainly from India and Egypt and it has an anise-like flavor, but is more aromatic, and sweeter; it is used traditionally as a herb in cooking, particularly with fish. It is also used as a diuretic and to improve milk supply of breastfeeding mothers; *Foeniculum vulgare* (Apiaceae), 2n = 2x = 22; 1C DNA content ~4.55 pg *hort*

fenugreek: An erect annual plant to 50 cm, which may be branched; the leaves are trifoliate and the leaflets oblong-lanceolate, to 5 cm long; it was a fodder of very ancient cultivation in Mediterranean countries; it is also a highly aromatic plant used as a pot-herb or spice; it is widely grown in India and neighboring countries as a flavoring, and in North Africa and western Asia as a fodder and spice; its seeds are used as a component of curry powder in India; *Trigonella foenum-graecum* (Leguminosae), 2n = 2x = 16; 1C DNA content ~0.70 pg (*T. stellata*) *hort*

field bean: >>> broad bean

field brome: A winter annual with an extensive fibrous root system; *Bromus arvensis* (Gramineae), 2n = 2x = 14; 1C DNA content ~6.25 pg *bot agr*

field horsetail: >>> common horsetail

field pea: *Pisum sativum* ssp. *arvense* (Leguminosae); 2n = 2x = 14; polytene chromosomes have been observed in ovary, anthers, or immature seed tissues; 2C DNA content 7.6–11.9 pg *agr*

field pepperweed: There are approaches to domesticate it as novel oilseed crop; it is an annual plant of the mustard family, native to Europe, but commonly found in North America as an invasive weed; the most notable characteristic is the raceme of flowers which forks off the stem; these racemes are made up firstly of small white flowers and later of green, flat and oval seedpods each about 6 mm long and 4 mm wide; recently, studies have been undertaken in order to modify the fatty acid profile of the seed oil to obtain a suitable industrial oil quality, *Lepidium campestre* (Brassicaceae), 2n = 2x = 16 >>> cress

field pink: *Dianthus campestris* (Caryophyllaceae), 2n = 2x = 30; 1C DNA content ~0.61 pg (*D. sylvestris*) *bot agr*

field salad: >>> corn salad

field scabious: *Knautia arvensis* (Dipsacaceae), 2n = 2x = 20; 1C DNA content ~3.7 pg *bot agr*

field speedwell: *Veronica agrestis* (Scrophulariaceae), 2n = 2x = 18; 2C DNA content 1.6–1.7 pg *bot agr*

fig: One of the 750 species of the genus *Ficus;* plants of the fig family may range in size from small shrubs to trees reaching 40 m or more; they are distributed throughout the tropics; the edible fig originated in West Asia and has been cultivated for at least 6,000 years; it is the oldest fruit crop known as a gynodioecious and insect-pollinated (figs are pollinated by parasitic wasps) species; its utilization consists of ecotypes called common figs (unisexual female trees) and caprifigs (bisexual with functional male trees), occurring in similar frequencies in wild populations; it is now grown mainly in Italy, Turkey, Greece, and California; the fruit may be eaten fresh or preserved and dried; with a history as ancient as any cultivated fruit, many researchers believe the fig has been with humans even longer than the pomegranate; the *Ficus* constitutes one of the largest and hardiest genera of flowering plants, featuring as many as 750 species; although the extraordinary mutualism between figs and their pollinating wasps has received much attention, the phylogeny of both partners is only beginning to be reconstructed; the fig plant does have a long history of traditional use as a medicine and has been a subject of significant modern research; due to particular features in its floral structure, classical plant breeding procedures that involve hybridization are not possible in fig; thus, genetic improvement of figs by using mutagens is an important line of research; mostly epigenetic modifications are observed after mutagenesis; irradiation is capable of altering epigenetic patterns; a commercial cultivar in Brazil is 'Roxo-de-Valinhos'; *Ficus carica* (Moraceae), 2n = 2x = 26; 1C DNA content ~0.37 pg *hort* >>> SCHLEGEL (2007)

filbert: The cultivated species of hazel; the edible fruits are oval, elongated nuts with a mild, oily taste; the nuts are borne on a large shrub resembling the hazel; filbert nuts differ from hazels in having an outer husk rather longer than the nut; both types are sometimes called cobnuts; *Corylus maxima* (Corylaceae), 2n = 2x = 22; 2C DNA content 1.0 pg *hort*

fine-stem stylo: An autogamous, semi-prostrate perennial with a strong >>> taproot; one of the most important tropical forage legumes; native to South and Central America and Africa; anthracnose, caused by the fungus *Colletotrichum gloeosporioides*, is a major constraint to the extensive use of this legume as a tropical forage; *Stylosanthes guianensis syn S. montevidensis* var. *intermedia* (Leguminosae), 2n = 2x = 20 *agr*

finger millet: Also known as African millet, as well as wimbi, bullo, telebun, and other vernacular names; an important crop in the drier areas of Africa and India, although sorghum and bulrush millet are more drought-resistant; it has a wide range of uses, as flour, as an additive to various dishes, and for brewing. In a dry climate, it stores well for up to ten years; it is self-pollinated and there are innumerable cultivars in both Africa and India; *Eleusine coracana* ssp. *coracana* (Gramineae), 2n = 4x = 40; 1C DNA content ~1.63 pg *agr*

fingerroot: >>> galangal

firebush *syn* **fire bush** *syn* **firecracker plant** *syn* **firecracker shrub** *syn* **hummingbird bush** *syn* **Mexican firebush** *syn* **redhead** *syn* **scarlet bush** *syn* **scarletbush** *syn* **Texas firecracker bush** *syn* **notro:** There are approximately 27 species in this genus; native to subtropical and tropical America, from Florida in the southern U.S. to as far south as Argentina; it is a small, perennial, semi-woody bush that grows to a height of 3 m; leaves are red-tinged and deeply veined, 10–20 cm long; flowers are a mass of tubular, bright reddish-orange; fruit is an edible and juicy berry, turning from green to yellow to red, and finally, black when ripe; the plant is rich in active phytochemicals, including flavonoids and alkaloids; it contains oxindole alkaloids as found in >>> cat's claw (*Uncaria tomentosa*); it also contains apigenin, ephedrine, flavanones, isomaruquine, isopteropodine, maruquine, narirutins, oxindole alkaloids, palmirine, pteropodoine, rosmarinic acid, rumberine, rutin, senociophyulline, speciophylline, and tannins; *Hamelia patens syn Duhamelia patens syn D. sphaerocarpa syn H. brittoniana syn H. sphaerocarpa syn H. latifolia syn H. lanuginosa syn H. intermedia syn H. erecta* (Rubiaceae), 2n = 2x = 24; 1C DNA content ~1.53 pg *hort*

flax (linseed): An annual dicot plant, 40–80 cm in height; the fruit is a capsule containing less than 10 seeds, whose oil content is in the range 35–45%; it is cultivated either as a textile plant, for the fibers contained in the stem, or for its oleo-proteinaceous seeds; winter flax varieties, with their procumbent growth at the beginning of development, are differentiated from spring flax varieties, that grow erect and are sensitive to cold; textile flax has been cultivated in Europe since the Middle Ages, but has declined since the appearance of cotton and synthetic fibers; the long stem is slightly branched at the top and is rich in fibers; planting occurs in spring; harvesting occurs by uprooting the plant when the capsules are yellow-green; retting permits decomposition of "cements", largely pectin, that bind the fibers; flax seeds produce an oil used for industrial purposes and are also used as animal feed; sown in March, the oil-yielding flax ("linseed") is harvested when the seeds are mature, and drying may be necessary; this crop is grown in France, Germany, and the U.K.; flax is historically interesting in that H.H. FLOR, in 1940, discovered the gene-for-gene relationship for pathogen–host plant interaction, while working on flax rust (*Melampsora lini*) in Illinois; *Linum usitatissimum* (Linaceae), 2n = 2x = 30; 1C DNA content 0.7 pg *agr*

fleur-de-lys: >>> iris

florist's daisy: A species of perennial garden plant that have been grown since 500 BC in China; in 1630, more than 500 varieties were already mentioned there; the plant is 30–90 cm high and wide, which grows as a perennial herbaceous or slightly woody plant; the stems stand upright; in the complex inflorescences are some to many individual cup-shaped partial inflorescences; tens of thousands of different cultivars have been obtained, with flower heads of very different shapes, sizes and colors; edible and ornamental chrysanthemums are grown; Japan is by far the largest producer with two billion stalks, followed by the Netherlands (800 million), Colombia (600 million), and Italy (500 million), *Chrysanthemum morifolium* (Asteraceae), 2n = 2x = 53 to 66, B chromosomes have been recognized *hort*

fodder beet: *Beta vulgaris* var. *crassa* (Chenopodiaceae), 2n = 2x = 18; 1C DNA content ~1.25 pg *agr* >>> beet

fodder radish: *Raphanus sativus* var. *oleiformis* (Brassicaceae), 2n = 2x, 4x = 18, 36; 1C DNA content ~0.55–1.45 pg *agr* >>> *Brassica* >>> radish >>> Figure 8

fonio: >>> hungry rice

foxglove: A pharmaceutical and ornamental plant; *Digitalis purpurea* (Scrophulariaceae), 2n = 2x = 56; 2C DNA content 2.5 pg *hort*

foxtail barley: *Hordeum jubatum* (Gramineae), 2n = 2x = 14; 2C DNA content 9.5 pg (HH) *bot agr* >>> Tables 41, 54

foxtail lily: *Eremurus* spp. (Liliaceae), 2n = 2x = 14 *hort bot*

foxtail millet: This cereal is used as human food in subtropical Europe and northern Africa; it is also important in India, Japan, and China; in Russia, it is used for brewing beer, in the United Kingdom, as birdseed, and in the U.S., it is grown for hay and silage; both self-pollination and cross-pollination occur; *Setaria italica* (Gramineae), 2n = 2x = 18 (AA); 2C DNA content 5.35 pg *bot agr*

freesia: *Freesia* sp. (Iridaceae), *F. alba*, 2n = 2x = 22; 2C DNA content 3.2 pg *hort*

French bean: *Phaseolus vulgaris* (Leguminosae), 2n = 2x = 22; 2C DNA content 1.0 pg *hort* >>> pulse

fuchsia: The French botanist C. PLUMIER, first described a plant called *Fuchsia triphylla florecoccinea*; the name derives from the German medical doctor L. FUCHS (1501–1566), who was a university teacher at Tübingen (Germany); in 1788, a British sailor, Cpt. FIRTH, brought a wild fuchsia, *F. coccinea*, from Brazil to Kew Gardens, London; in the same year the breeder, J. LEE, propagated the first plants and sold them on the market; because of its attraction, further wild species were collected in the New World, e.g. *F. magellanica*, *F. fulgens*, *F. arborescens*, or *F. lycoides*; Central America seems to be the center of diversity; the fuchsia later became one of the 21 genera of the family Onagraceae; the genus contains very small species (*F. procumens*), but also one as tall as a 9-m tree (*F. excorticata*, 2n = 2x = 22; 1C DNA content ~0.73 pg); however, the majority of the species grow as bushes; even one epiphytic species is known; serious breeding began from 1840, when a British gardener found among seedlings, a segregant, very different from common plants; it was named 'Venus Victrix'; around 1900, the French breeder LEMOINE introduced more than 400 varieties; by the end of the 19th century, breeding of fuchsia had reached a peak; *Fuchsia* spp. (Onagraceae), x = 11 *hort*

fumitory: *Fumaria* spp. [*F. luteus syn Pseudofumaria lutea* (Papaveraceae), 2n = 4x = 28; 1C DNA content ~19.45 pg (*F. muralis*)] *bot agr*

fuzzy melon: >>> wax gourd

G

galangal (*syn* **galanga** *syn* **blue ginger** *syn* **laos**): A rhizome of plants of the genus *Alpinia* or *Kaempferia*, with culinary and medicinal uses, originating from Indonesia; it is used in various Asian cuisines, e.g. in Thai and Lao tom yum and tom kha gai soups, Vietnamese Hué cuisine, and throughout Indonesian cuisine, as soto; though it is related to and resembles ginger, there is little similarity in taste; *Alpinia galanga* (greater galangal, 2n = 4x =

48), *A. officinarum* (lesser galangal, 2n = 4x = 48), *Kaempferia galanga* (kencur *syn* aromatic ginger *syn* sand ginger, 2n = 2x = 22), *Boesenbergia pandurata* (Chinese ginger *syn* fingerroot, 2n = 2x, 3x = 20, 24, 36) (Zingiberaceae) *hort*

gambier (*syn* gambir *syn* pale catechu *syn* zong er cha): *Uncaria gambir syn Nauclea gambir syn Ourouparia gambir* (Rubiaceae), 2n = 8x = 88; 1C DNA content ~1.78 pg (*N. orientalis*) *hort*

garden aster: >>> China aster

garden balsam: *Impatiens balsamina* (Balsaminaceae), 2n = 2x = 14; 2C DNA content 2.3 pg *hort*

garden beet: >>> red beet

garden cress: >>> cress

garden hydrangea: *Hydrangea hortensis* (Saxifragaceae), 2n = 2x = 36; 1C DNA content ~1.93–3.33 pg (*H. macrophylla*) *hort*

gardenia: A group of subtropical and tropical trees and shrubs, found in Africa and Asia, belonging to the madder family, with evergreen foliage and flattened rosettes of fragrant waxen-looking flowers, often white in color; most cultivated varieties have double flowers and are propagated from cuttings; hardy in the tropics and subtropics, they are grown indoors in temperate regions; *Gardenia* spp. (Rubiaceae), 2n = 22; 1C DNA content ~1.30 pg (*G. resiniflua*) *hort* >>> common madder

garden leek: Leeks are tetraploids and set seed freely, while "elephant garlic" is a hexaploid and is sterile; both are open-pollinated crops; *Allium porrum* or *A. ampeloprasum* (Alliaceae), 2n = 4x = 32; 2C DNA content 63.2–65.3 pg *hort*

garden lettuce: >>> butterhead >>> lettuce

garden orach: >>> mountain spinach

garden pansy: *Viola macrocarpa* (Violaceae), 2n = 2x = 12; 2C DNA content 5.4 pg *hort*

garden pea: Despite its importance as a traditional crop plant, the pea has a history as an organism used for genetic studies going as far back as MENDEL (1866); the pea later became a subject of intensive genetic studies, and thus became one of the most genetically investigated plants; it has led to the identification and symbolization of more than 600 classical genes, in addition to ~2,500 genes identified and preserved in collections; particularly, pea mutants were used in breeding, *Pisum sativum* ssp. *hortense* (Leguminosae), 2n = 2x = 14; polytene chromosomes have been observed in ovary, anthers, or immature seed tissues; 2C DNA content 3.8–5.9 pg *hort* >>> fasciata mutant >>> Table 16

garden thyme: *Thymus vulgaris* (Labiatae), 2n = 48; 1C DNA content ~0.78 pg *hort*

garlic: A perennial bulb of the onion family; it grows ~30 cm high, and has pale spherical flowers; it has been used since ancient Egyptian times as a herb; it never sets seed, and thus can be propagated only vegetatively; the flowers sometimes produce small bulbils, which can be used for propagation, but these are also vegetative and are not the result of pollination; the formation of flowers and seeds is a major physiological sink that severely reduces the yield of vegetative parts of the plant; *Allium sativum* (Alliaceae), 2n = 2x = 16; 2C DNA content 35.7 pg *hort*

geranium: >>> pelargonium

gerbera: *Gerbera jamesonii* (Compositae), 2n = 50; 1C DNA content ~2.55 (*G. hybrida*) pg *hort*

German chamomile: >>> chamomile

germander speedwell: *Veronica chamaedrys* (Scrophulariaceae), 2n = 2x = 18; 2C DNA content 1.7 pg *bot agr*

garter bean: >>> asparagus bean

gherkin: >>> cucumber

giant swamp taro: *Cyrtosperma chamissonis* syn *C. edule, C. merkusii* (Araceae), 2n = 2x = 26 *hort*

giant taro: *Alocasia macrorhiza* syn *A. indica* (Araceae), 2n = 2x = 28 *hort*

gingelly: >>> sesame

gillyflower: Old name for the carnation and related plants; *Matthiola* spp. (Brassicaceae), 2n = 2x = 14; 1C DNA content ~2.11 pg (*M. incana*) *hort* >>> *Brassica*

ginger: Seems to have originated from southern China; today, it is cultivated all over tropical and subtropical Asia (50% of the world's harvest is produced in India), Brazil, Jamaica (exporter of the highest-quality ginger), and Nigeria, whose ginger is rather pungent, but lacks the fine aroma of other provenances; German *Ingwer*, English *ginger*, French *gingembre*, and practically all other names of ginger in European languages can, at first, be traced back to Latin *zingiber*, which was, in turn, a loan from Greek (*zingiberis*); the Modern Greek name is not related to that extensive series; instead, *piperoriza* is just a descriptive compound, referring to the pungent peppery taste; *Zingiber officinale* (Zingiberaceae), 2n = 2x = 22; 1C DNA content ~6.03 pg *hort*

gingerbread palm: >>> doum palm

ginkgo: A primitive seed-bearing tree (a gymnosperm) that was common during the Mesozoic Era, but has only one existing species now; ginkgos peaked during the Jurassic and Cretaceous periods; this deciduous tree has fan-shaped leaves, divided into two lobes; it is a dioecious gymnosperm species with both male and female plants; *Ginkgo biloba* (Ginkgoaceae), 2n = 2x = 24; 1C DNA content ~11.75 pg (ZW sex chromosomes, with the female-specific gene region being on the W) *hort*

ginseng, American ~: A herbaceous perennial plant in the ivy family, commonly used in Chinese or traditional medicine; it is native to eastern North America, though it is also cultivated in China; since the 18th century, American ginseng has been primarily exported to Asia, where it is highly valued for perceived superior quality and sweet taste; *Panax quinquefolius syn Panacis quinquefolis* (Araliaceae); 2n = 4x = 48; nuclear genome size of 4,91 Gbp >>> ginseng, Asian ~ >>> ginseng, Korean ~

ginseng, Asian ~: Ginseng is the root of plants in the genus *Panax*, such as >>> Korean ginseng, (South) China ginseng (*P. notoginseng*, 2n = 2x = 24), and >>> American ginseng, typically characterized by the presence of ginsenosides and gintonin; Araliaceae; 2n = 2x = 24; nuclear genome size of >3.5 Gbp >>> ginseng, American ~ >>> ginseng, Korean ~

ginseng, Korean ~: *Panax ginseng* (Araliaceae); 2n = 4x = 48; nuclear genome size of 4,90 Gbp >>> ginseng, American >>> ginseng, Asian ~

gladiolus: Plants of southern European and African origins; cultivated perennials belonging to the iris family, with brightly colored funnel-shaped flowers borne on a spike; the sword-like leaves spring from a corm; *Gladiolus communis* and other species (Iridaceae), 2n = 2x, 4x = 90, 180 (*G. verensis*, 2n = 2x = 30; 1C DNA content ~1.38 pg) *hort*

globe artichoke: >>> artichoke

Goa bean (*syn* **winged bean**): *Psophocarpus tetragonolobus* (Leguminosae), 2n = 2x = 18; 1C DNA content ~0.80 pg; polytene chromosomes have been observed in ovary, anthers, or immature seed tissues *hort*

goat grass: >>> *Aegilops*

goji: >>> wolfberry

goldenberry: >>> Peruvian groundcherry

golden chia: An annual plant that is commonly called chia, chia sage, golden chia, or desert chia, because its seeds are used in the same manner as >>> *Salvia hispanica* (chia); it grows in California, Nevada, Arizona, New Mexico, Sonora, and Baja California, and was an important food for native Americans; some native names, including *pashí* from Tongva and *it'epeš* from Ventureño are still in use, *Salvia columbariae* (Lamiaceae), 2n = 2x = 32 >>> chia

golden chamomile: >>> dyer's chamomile

golden gram: >>> mung bean

golden oatgrass: A perennial grass; *Trisetum flavescens syn Avena flavescens* (Gramineae), 2n = 4x = 28; 2C DNA content 5.1 pg *bot agr* >>> Table 14

gold-of-pleasure: An annual or biennial forb/herb utilized for oil production; classified as a secondary crop plant; *Camelina sativa* (Brassicaceae), 2n = 40 *agr* >>> *Brassica*

gombo: >>> okra

good King Henry: A herbaceous perennial plant native to Europe; it shows high winter hardiness and grows even in alpine regions; the arrow-shaped leaves have been used similarly to spinach; in England, it is grown on rare occasions as a substitute for asparagus, particularly in early spring when young sprouts can be prepared like asparagus; *Chenopodium bonus-henricus* (Chenopodiaceae), 2n = 4x = 36; 1C DNA content ~1.48 pg *bot agr*

gooseberry: Edible fruit of a low-growing bush found in Europe and Asia, related to the currant; it is straggling in its growth, and has straight sharp spines in groups of three, and rounded, lobed leaves; the flowers are green and hang on short stalks; the sharp-tasting fruits are round, hairy, and generally green, but there are reddish and white varieties; *Ribes uva-crispa* (Grossulariaceae), 2n = 2x = 16; 1C DNA content ~0.94 pg *hort*

goosy grass: >>> drooping brome

gotu kola: >>> centella

granadilla: >>> passionfruit

grand wormwood: >>> wormwood

grape: *Vitis* spp. (2n = 2x = 38; 1C DNA content ~0.43 pg), *Muscadinia* spp. (2n = 2x = 40) (Vitaceae), *hort* >>> Amur grape (2n = 2x = 38; 1C DNA content ~0.50 pg) >>> Berlandieri grape (2n = 2x = 38; 1C DNA content ~0.50 pg) >>> European grape >>> muscat plant

grapefruit: Round, yellow, juicy, sharp-tasting fruit of the evergreen grapefruit tree; the tree grows up to 10 m and has dark, shiny leaves and large, white flowers; the large fruits grow in grape-like clusters (hence the name); grapefruits were first established in the West Indies and subsequently cultivated in Florida by the 1880s; they are now also grown in Israel and South Africa; some varieties have pink flesh; it is of relatively recent origin and is thought to be a chance hybrid between two other *Citrus sp.*; the name "grapefruit" was apparently used for the first time in Jamaica in 1814, but its etymology is obscure; *Citrus paradisi* (Rutaceae), 2n = 2x = 18; 1C DNA content ~0.40 pg *hort* >>> hesperidia >>> Figure 61 >>> http://users.kymp.net/citruspages/citrons.html

grape hyacinth: *Muscari* spp. (Liliaceae), 2n = 2x, 4x = 18, 36; 1C DNA content ~6.23 pg (*M. moschata*) *hort*

grapevine: >>> European grape

grass: many species of the family *Gramineae bot agr* >>> Table 35

grasscloth: >>> ramie

grass pea: >>> chickling pea

greater birdsfoot trefoil: *Lotus uliginosus* (Leguminosae), 2n = 2x = 12; polytene chromosomes have been observed in ovary, anthers, or immature seed tissues; 2C DNA content 1.1 pg *bot agr*

great-headed garlic: >>> Levant garlic

great morinda: >>> noni

Greek basil: A common local cultivar in Greece, also known as "fine-leaved" basil, because of its small, narrow leaves; it is used mainly as ornamental plant in pots and gardens; its variation in terms of the time of flowering and the shape of the vegetative development (compact or loose ball, reverse cone, other shapes) led to the improvement of the Greek basil population, by selecting for plants with late blooming, compact ball shape and large size, potentially suited for ornamental uses; *Ocimum basilicum* (Lamiaceae), 2n = 4x = 52 *hort* >>> basil

Greek mountain tea (*syn* **malotira**)**:** Made using the dried leaves and flowers of ironwort plants; these plants are hardy flowering perennials that have adapted to survive with little water and little soil; only one type of this plant is cultivated, and only in Greece; the other types are gathered in the wild; the dried inflorescence with some leaves contains two flavonoids and a very high percentage of iron, 52.5 mg per 100 g; hence the name ironwort (*Sideritis* is derived from the Greek word sideros = iron); in breeding, interspecific hybrids are used for higher yield and higher quality; *Sideritis raeseri* (Lamiaceae), 2n = 2x = 32 *hort*

green cabbage: >>> kale

green ginger: >>> wormwood

greengram: >>> mung bean

green pea: *Pisum sativum* convar. *sativum* or convar. *vulgare* (Leguminosae), 2n = 2x = 14; 1C DNA content ~4.88 pg *hort* >>> field pea >>> garden pea

green pepper: >>> sweet pepper

gray mangrove: The most widely distributed mangrove species in Australia, due mainly to its tolerance of cool conditions; on the East coast of Australia, it occurs as far south as Corner Inlet in Victoria, whereas, on the West coast, its most southerly occurrence is Bunbury; it grows to 10 m tall; pencil-sized peg type above-ground roots; light green leaves are approximately 10 cm long with a silvery-grey undersurface; the underside of the leaf has special glands for secreting excess salt; the gray mangrove, *A. marina*, can survive in highly saline conditions; its genome may serve as gene pool for salinity resistance; salinization poses an increasingly serious problem in coastal and agricultural areas with negative effects on plant productivity and yield; *Avicennia marina* (Avicenniaceae), 2n = 4x, 6x = 64, 96 *eco biot*

groundcherry: The groundcherry is a wild relative of the >>> tomatillo and, much like the tomatillo, its fruits are encased within a papery husk that protects the fruit from spoiling, the berry inside the husk is small – marble-sized – but delivers a big citrusy flavor; a source of antioxidants, vitamins A, B and C, and other nutrients, these small berries are exclusively grown on small-scale farms and home gardens; commercial production does not yet occur, a void that can at least partially be attributed to the plant's unruly growth; with its long sprawling branches, it requires extensive management to tame its growth; its branches are adorned with husk-covered fruits that fall to the ground, often before ripening; this makes harvesting the fruits a labor-intensive process, and raises food safety concerns if the fruits come in contact with soil microorganisms that can cause food-borne illnesses; by use of CRISPR/Cas9 technology a natural mutation in the *self-pruning* (SP) gene of tomato, which represses flowering, resulting in plants that have more manageable growth, was successfully transferred to groundcherry; the CRISPR-edited groundcherries show shorter braches; this more diminutive growth habit is preferable for larger-scale agricultural settings, because more compact plants can be grown and harvested more easily, as a step forward to a more domesticated habit; *Physalis pruinosa* (Solanaceae), 2n = 2x = 24 *hort* >>> groundcherry, Peruvian

groundcherry, Peruvian (*syn* **Cape gooseberry** *syn* **groundcherry** *syn* **Inca berry** *syn* **husk tomato**): A species indigenous to Central America and western South America, but grows well in Africa; the fruit is a small, round berry, about the size of a marble, full of small seeds; it is bright yellow when ripe, and very sweet, making it ideal for baking into pies and making jam; the most notable feature is the single lantern-type pod that covers each berry; native to Peru and Chile, where the fruits are casually eaten and occasionally sold in markets; whereas the plant is still not an important cash crop, it has been widely introduced into cultivation in other tropical, subtropical, and even temperate areas; the plant was grown by early settlers at the Cape of Good Hope before 1807; in South Africa, it is commercially cultivated and is common as an escape, and the jam and canned whole fruits are staple commodities, often exported; it is cultivated and naturalized on a small scale in Gabon and other parts of Central Africa; in 2017, a fossilized fruit was found in Patagonia that dated back to 52 million years ago; *Physalis peruviana* (Solanaceae), 2n = 4x = 48 *hort*

groundnut: >>> peanut

guar: A member of the Leguminosae; its wild progenitors are extinct but it is thought to have been a native of Africa, taken at an early date to Southeast Asia, where it now has many uses; cluster bean, one of the most important cash legume crops, e.g. in Texas and Oklahoma in the U.S., it has played an increasingly important role in a wide range of industries; *Cyamopsis tetragonoloba* (Leguminosae), 2n = 2x = 14 *hort*

guaraná: A climbing plant in the maple family, native to the Amazon basin and especially common in Brazil; native Americans of the Maué and Andirá tribes cultivated guaraná 'Sorbilis' in the Central Amazon, and the Barés cultivated the 'Typica' variety in the upper Negro River (Brazil); it features large leaves and clusters of flowers, and is best known for its fruit, which is about the size of a coffee bean; as a dietary supplement, guaraná is an effective stimulant; it contains about twice the caffeine found in coffee beans (about 2–4.5%); the high concentration of caffeine is a defensive toxin that repels pathogens from the berry and its seeds; the plant was introduced to European colonizers and to Europe in the 16th century by OVIEDO, HERNÁNDEZ, COBO and other Spaniard chroniclers; by 1958, guaraná was commercialized; *Paullinia cupana* (Sapindaceae), 2n = 210; two cytomorphological groups can be distinguished: (a) a metacentric and submetacentric group showing 25 sets of three pairs of chromosomes (2–76), (b) a group containing only acrocentric chromosomes showing 12 sets of two pairs of chromosomes (82–105), a homologous submetacentric pair (1) and an acrocentric pair (81); 2C DNA content = 22.8 pg *fore*

guava: Tropical American tree belonging to the >>> myrtle family; the astringent yellow pear-shaped fruit is used to make guava jelly, or it can be stewed or canned; it has a high vitamin C content; *Psidium guayava* (Myrtaceae), 2n = 2x, 3x = 22, 33; 2C DNA content 0.7 pg *hort*

guaya: >>> chickling pea

Guinea grass: *Panicum maximum* (Gramineae), 2n = 4x = 32 *agr*

guixa: >>> chickling pea

gum Arabic: >>> acacia

gumbo: >>> musk okra

gum ghatti: A natural gum from the sap of a tree native to dry, deciduous forests of India and Sri Lanka; the common name "ghatti" is derived from the word "ghat" or mountain pass; this gum was originally carried by people over mountain passes or "ghats" to ports in India; the gum has properties intermediate between gum Arabic and karaya gum; because it is a superior oil emulsifier with a higher viscosity, it is used in liquid and paste waxes and for fat soluble vitamins; *Anogeissus latifolia* (Combretaceae), 2n = 2x = 24; 1C DNA content ~1.25 pg *fore*

H

hairy vetch: *Vicia villosa* (Leguminosae), 2n = 2x = 14; polytene chromosomes have been observed in ovary, anthers, or immature seed tissues; 2C DNA content 4.7 pg *bot agr*

hairy vetchling: *Lathyrus hirsutus* (Leguminosae), 2n = 2x = 14; polytene chromosomes have been observed in ovary, anthers, or immature seed tissues; 2C DNA content 22.5 pg *bot agr*

hana-daikon (*syn* **hana-daikon oh-araseitô** *syn* **hiroha-hana-daikon**): Native to the Asian temperate zone, particularly China; the plant produces a high-quality oil with high concentrations of palmitic (14.3%) and oleic (20.3%) acids, and lower concentrations of linolenic (4.8%) and erucic (0.9%) acids; it produces a large number of branches, pods per plant, and seeds per pod, which contributes to the high yield potential of this plant; it is domesticated as a new oil crop in China; *Orychophragmus violaceus syn Raphanus chanetii syn Moricandra sonchifolia syn Raphanus courtoisii, Orychophragmus sonchifolius syn Brassica violacea* (Brassicaceae), 2n = 2x = 24 (OO); 1C DNA content 1.47 pg *hort agr* >>> *Brassica*

hardy garden mum: >>> florist's daisy

harlequin flower: A genus in the family Iridaceae with ~13 species endemic to Cape Province, South Africa; all species are perennials that grow during the wet winter season, flower in spring, and survive underground as dormant corms over summer; their conspicuous flowers have six tepals, which, in most species, are equal in size and shape; *S. bulbifera* is the most common species in cultivation, with flowers from cream to yellow or purple; *S. grandiflora* is a similar but larger plant; *S. tricolor* has bright red flowers with yellow and black centers; many named hybrid cultivars were bred from *S. bulbifera* and *S. tricolor*; *Sparaxis* spp. (Iridaceae), $2n = 2x = 20$ *hort*

haricot: >>> French bean

hawkbit: >>> dandelion

Haynald wheat: Named after the Hungarian botanist, Archbishop L. HAYNALD, has been attached with further synonyms to almost every genus of the section Triticeae; it is a widespread annual and it is present in the Mediterranean region (as a common weed in Greece, Italy, Bulgaria, Albania, Serbia, Macedonia, Croatia), in South Russia, Romania, and Hungary, reaching even the north of Europe as a sporadic weed; it is an outcrossing species and its genome was designated as "V" by SEARS (1953); it is considered to be an important donor of genes for wheat breeding programs; *Dasypyrum hordeaceum* (*syn Dasypyrum breviaristatum, Haynaldia hordeacea, Triticum hordeaceum, Hordeum breviaristatum*) is a perennial species and it is present only in North Africa and Greece; it is a tetraploid species ($2n = 4x = 28$) whose genome was designated as "$V^bV^bV^bV^b$" by LÖVE (1984), however, diploids were found as well; *Dasypyrum villosum syn Haynaldia villosa* (Gramineae), $2n = 2x = 14$ (V^vV^v); 2C DNA content 10.70 pg *bot agr*

hazel(nut): Shrubs or trees that includes the European common hazel or cob (*Corylus avellana*, $2n = 2x = 22$; 1C DNA content 0.48 pg), of which the filbert is the cultivated variety; European hazelnut (*Corylus avellana*) is an important crop in Turkey, Georgia and Azerbaijan, where cultivars were selected from the native vegetation; accessions from Turkey have been assigned to the Black Sea group, and cultivars from Georgia and Azerbaijan have a similar phenotype; North American species include the American hazel (*C. americana*, $2n = 2x = 28$); *Corylus* spp. (Corylaceae), $2n = 2x = 22$ *hort* >>> Siberian hazel >>> American hazel >>> Turkish cobnut >>> cobnut >>> filbert

head cabbage: *Brassica oleracea* convar. *capitata* (Brassicaceae), $2n = 2x = 18$ (CC); 1C DNA content 0.78 pg *hort* >>> *Brassica* >>> Figure 8

head lettuce: *Lactuca sativa* var. *capitata* (Compositae), $2n = 2x = 18$; 1C DNA content 2.65 pg *hort* >>> lettuce >>> Table 46

hedge hyssop: A garden herb of the mint family, the flowers and evergreen leaves of which have long been used as a flavoring for foods and beverages, and as a folk medicine; the plant has a sweet scent and a warm, bitter taste; in the European Middle Ages, hyssop was a stewing herb; its modern uses are for flavoring meats, fish, vegetables, salads, sweets, and such liqueurs as >>> absinthe; *Gratiola officinalis syn Hyssopus officinalis* (Scrophulariaceae), $2n = 2x = 12$; 1C DNA content 0.50 pg *bot agr* >>> hyssop

hellebores: >>> Christmas rose

hemp: Cultivation and use of hemp for fiber can be traced back to 2,800 BC in China; for many centuries, hemp has been cultivated as a source of strong stem fibers, seed oil, and psychoactive drugs in its leaves and flowers; environmental concerns and recent shortages of wood fiber have renewed interest in hemp as a raw material for a wide range of industrial products; hemp is an herbaceous annual that develops a rigid woody stem ranging in height from 1 to 5 m; the stalks have a woody core surrounded by a bark layer containing long fibers that extend nearly the entire length of the stem; breeding has developed hemp varieties with increased stem fiber content and very low levels of delta-9-tetrahydrocannabinol (THC), the psychoactive ingredient of marijuana; although hemp is well adapted to the temperate climatic zone and will grow under different environmental conditions, it grows best under warm growing conditions, an extended frost-free season, highly productive agricultural soils, and abundant moisture throughout the growing season; hemp yields are in the range 1.0–3.5 t of dry stems/ha; hemp is a dioecious plant, having both staminate (male) and pistillate (female) plants, each with distinctive growth characteristics; staminate plants are tall and slender with few leaves surrounding the flowers, while pistillate plants are short and stocky with many leaves at each terminal inflorescence; staminate plants senesce and die soon after their pollen is shed, while pistillate plants remain alive until the seeds mature; quite stable monoecious varieties have been developed; hemp is also a rich source of a variety of compounds, including cannabinoids, terpenoids, and flavonoids; their content depends upon plant genetics, growth conditions, time of harvest, and drying conditions; 80-100 different cannabinoids have been identified; it has been used medicinally for 4,000 years and remained in the global pharmacopeia until 1932, and in the British Pharmaceutical Codex until 1949; medical use has been prohibited since 1973; the principal cannabinoid, THC, was first isolated in 1964; the first cannabinoid pharmaceutical product Marinol® (a synthetic THC product) was approved in the U.S. in 1985; the discovery of specific cannabinoid receptors in the human brain in the early 1990s and subsequent identification of the endocannabinoids anandamide and 2-arachadonoylglycerol, led to a resurgence of interest in the field of cannabinoid medicine, especially within the pharmaceutical industry; cannabidiol (CBD), as a non-psychoactive cannabinoid, is currently a cannabinoid of significant interest, showing a wide range of pharmacological activities; the other classes of compounds present in cannabis also have their own pharmacology (e.g. terpenoids, flavonoids); the potential for interaction and synergy between compounds within the plant may play a role in the therapeutic potential of cannabis; *Cannabis sativa* (Moraceae); 2n = 2x = 20 (XY male, XX female); 1C DNA content 0.84 pg *agr*

henbane: Poisonous plant, belonging to the nightshade family, found on waste ground throughout most of Europe and western Asia; it is a branching plant, up to 80 cm high, with hairy leaves and a sickening smell; the yellow flowers are bell-shaped; it is used in medicine as a source of the drugs hyoscyamine and scopolamine; *Hyoscyamus niger* (Solanaceae), 2n = 2x = 34; 2C DNA content 2.3 pg *bot agr*

henequen agave: A monocotyledonous plant, which has its center of origin in Mexico, where some species have been domesticated and are of economic importance as sources of fiber, steroids, spirits, and other products; the only cultivated species in the Yucatan Peninsula is henequén (*Agave fourcroydes*), that is descended from the wild ancestor, *Agave angustifolia*; both of these agaves are monocarpic perennials, that produce flowers only once toward the end of their life cycle of ~20 years, after which they die; throughout its life span, henequén propagates mainly by means of its rhizomes, the apical meristems of which emerge at a distance from the parent plant, giving rise to new individuals; the flowers develop at the top of large inflorescences that can reach 3–8 m high and, after flowering, bulbils originate from buds beneath bracteoles on the inflorescence; however, in spite of their potential utility, bulbils and seeds are not usually used for commercial propagation and have not been used for breeding processes; the traditional agricultural practice of cutting the henequén inflorescences soon after they begin to develop, in order to preserve the plant a little longer, has also limited the supply of seeds; fruits develop abnormally because carpels remain empty, probably due to insufficient pollination, and the few seeds that mature show a low germination rate of less than 10%; both phenomena could be due to the pentaploid ploidy level, which may be responsible for the low fertility of the species; in the wild species *A. angustifolia*, which is a fertile hexaploid (2n = 6x = 180), seed germination is as high as 73%; *Agave fourcroydes* (Agavaceae), 2n = 2x = 60; 1C DNA content 2.98 pg *agr*

henna (*syn* **hina**): Native to tropical and subtropical regions of Africa, southern Asia, and northern Australasia in semi-arid zones; small shrub belonging to the loosestrife family, found in

Iran, India, Egypt, and North Africa; the leaves and young twigs are ground to a powder, mixed to a paste with hot water, and applied to the fingernails and hair to give an orange-red hue; the color may then be changed to black by applying a preparation of indigo; henna produces a red-orange dye molecule, lawsone; this molecule has an affinity for bonding with protein, and thus has been used to dye skin, hair, fingernails, leather, silk, and wool; *Lawsonia inermis* syn *L. alba* (Lythraceae), 2n = 4x = 30–34; 4C DNA content 1.4 pg *hort*

herba Sancti Mariæ: >>> epazote

hevea (rubber): >>> rubber

hickory: A tree belonging to the walnut family; the genus includes 17–19 species of deciduous trees with pinnately compound leaves and large nuts; ~12 species are native to North America (11–12 in the U.S., one in Mexico), with five to six species native to China and Indochina; it provides a valuable timber, and all species produce nuts, though some are inedible; the >>> pecan (*Carya illinoensis*) is widely cultivated in the southern U.S., and the shagbark (*C. ovata*) in the North; *Carya* spp., *Hicoria* spp. (Juglandaceae), 2n = 2x = 32; 1C DNA content 0.83 pg *fore hort*

highbush blueberry: *Vaccinium corymbosum* (Ericaceae), 2n = 2x = 24; 1C DNA content 0.67 pg *hort*

hippie: Many tropical lily-like plants also belong to the amaryllis family, such as those of the genera *Haemanthus* (Cape tulip, or blood lily), *Alstroemeria* (Peruvian lily), and *Hippeastrum*; the hippeastrums, grown for their large, showy flowers, are commonly known as amaryllis; an ornamental Eurasian plant known as winter daffodil (*Sternbergia lutea*, 1C DNA content 23.96 pg) belongs to the same family; hippies are perennial herbs from a bulb with contractile roots, comprising 50 genera and 870 species; the leaves are alternate and more or less basal, simple, usually linear or lorate, flat, entire, parallel-veined, sheathing at base; stipules are absent; the flowers are bisexual, often showy, actinomorphic to zygomorphic, usually in umbelloid >>> cymes; the >>> perianth consists of six distinct or connate petaloid tepals, sometimes with an adnate corona; the androecium consists of six stamens attached to the receptacle or adnate to the perianth tube; filaments are free or connate, sometimes appendaged and forming a staminal corona; the gynoecium consists of a single compound pistil of three carpels, a single style, one capitate or three-lobed stigma, and an inferior ovary with three locules, each containing several to numerous axile ovules; the fruit is a loculicidal capsule or sometimes a berry; the seed coat usually has a black or blue crust; many hybrid varieties are grown as ornamental plants; *Hippeastrum hybridum* (Amaryllidaceae), 2n = 2x = 22; 1C DNA content 14.98 pg *hort* >>> Table 46

hippeastrum: >>> hippie

Holcus: >>> common velvet grass

hollyhock: Tall flowering plant belonging to the mallow family; originally a native of Asia; it produces spikes of large white, yellow, pink, or red flowers, 3 m high when cultivated as a biennial; it is a popular cottage garden plant; *Alcea rosea* ssp. *plena* (Malvaceae), 2n = 26, 40, 42, 56 *hort*

honeysuckle (syn honeyberry syn haskap berry syn blue-berried honeysuckle syn deepblue honeysuckle syn sweetberry): Haskap is native throughout the cool temperate northern Hemisphere in countries such as Canada, Japan, Russia, and Poland; it is a deciduous shrub growing to 1.5–2 m tall; the fruit is an edible, blue berry, somewhat rectangular in shape weighing 1.3 to 2.2 g, and about 1 cm in diameter; haskap *Lonicera caerulea* variety *edulis* has been used frequently in breeding efforts, but other varieties have been bred with it to increase productivity and flavor; in several haskap breeding programs, the variety *emphyllocalyx* has been the dominant one used; plants of many haskap cultivars grow to be 1.5 to 2 m tall and wide, can survive a wide range of soil acidities, pH 3.9–7.7, requiring high organic matter, well-drained soils, and plentiful sunlight for optimum productivity; *Lonicera caerulea* plants are more tolerant of wet conditions than most fruit species;

cultivated varieties include 'Tundra', 'Borealis', 'Indigo Treat', 'Indigo Gem', 'Indigo Yum', 'Honeybee', 'Aurora', 'Wojtek', 'Atlaj', 'Nimfa', 'Berry Blue', and 'Polar Jewel'; *L. caerulea* (Caprifoliaceae), 2n = 2x, 4x = 18, 36 *hort*

hone: >>> pongamia

hop clover: An annual clover; *Trifolium campestre* (Leguminosae); 2n = 2x = 14; 2C DNA content 1.0 pg *agr* >>> clover

hops: Hops is a long-day plant; it has been cropped in central Europe for more than 1,000 years; despite the early origin of cultivation, hops never developed into a major crop because it is only used by the brewery industry to flavor fermented malt beverages, primarily beer, and ale; it is a dioecious species; male plants are used only for breeding or yield stimulation; vegetatively propagated female plants are grown for commercial production in ~30 countries worldwide; in diploids, a 1:1 sex ratio of seedling progenies is expected for an XX (female) and XY (male) sex mechanism, although males are actually less frequent; it is propagated vegetatively, and only ~8 clones dominated world production until quite recently; these included 'Fuggle' and 'Golding' in the United Kingdom, 'Hallertau' in Germany, and 'Saaz' in the Czech Republic; the clones are mostly >>> ancient, and they demonstrate the utility and durability of horizontal disease resistance; a breeding program, started in Germany in 1922, accumulated polygenic resistance to downy mildew by breeding within the European population; this was one of the earliest examples of >>> horizontal resistance being chosen over vertical resistance in crop improvement; *Humulus lupulus* (Moraceae), 2n = 2x = 20 (XY male) 1C DNA content 2.90 pg *agr hort* >>> Figure 50

horseradish: A pungent relish is obtained from the large taproot; a delicious condiment with meat and seafood; this species does not flower or set seed; there are many clones with widely varying degrees of pungency; these >>> ancient clones have few pests or diseases and they are a good example of both the effectiveness and the durability of horizontal resistance, *Armoracia rusticana* (Brassicaceae), 2n = 4x = 32, 2C DNA content 2.5 pg *hort* >>> *Brassica* >>> radish >>> Figure 50

horseradish tree (*syn* **drumstick tree** *syn* **ben oil tree**): It is the most widely cultivated species of the genus *Moringa*; it is an exceptionally nutritious vegetable tree with a variety of potential uses; the tree itself is rather slender, with drooping branches that grow to ~10 m in height; in cultivation, it is often cut back annually to 1 m or less and allowed to regrow so that pods and leaves remain within arm's reach; the taste of the roots is similar, which can serve as a rough substitute for >>> horseradish; the fruit of the tree is quite popular as a vegetable in Asia and Africa; the long, slender, triangular seed pods are long thin pod resembling a drum stick; moringa leaves are also eaten as a leaf vegetable, particularly in the Philippines, South India and Africa, *Moringa oleifera* (Moringaceae), 2n = 2x = 28, 1C DNA content 0.60 pg *fore*

horsetail: A primitive, spore-bearing plant (a sphenopsid) with rhizomes; its side branches are arranged in rings along the hollow stem; other genera of horsetail were common during the Mesozoic Era, like *Neocalamites*, *Calamites*, etc.; horsetails date from the Devonian period 408–360 Mya, but still exist today and are invasive weeds, *Equisetum arvense* (Equisetaceae), 2n = 216; 1C DNA content 14.65 pg *phyt* >>> common horsetail >>> Table 46

hot paprika: >>> hot pepper

hot pepper: *Capsicum annum* var. *acuminatum* (Solanaceae), 2n = 2x = 24; 1C DNA content 3.60 pg *hort*

huauzontle: *Chenopodium nuttalliae* (Chenopodiaceae), 2n = 4x = 36; 1C DNA content 1.48 pg *hort*

huckleberry: >>> blueberry

Hungarian vetch: *Vicia pannonica* (Leguminosae), 2n = 2x = 12; 1C DNA content 6.75 pg; polytene chromosomes have been observed in ovary, anthers, or immature seed tissues *agr*

hungry rice (*syn* **acha** *syn* **fonio** *syn* **white fonio** *syn* **petit mil**): Probably the oldest African cereal; for thousands of years, West Africans have cultivated it across the dry savanna; hungry rice was believed to have once been the major food crop in this region of the world; even though the crop has been neglected for quite some time and few people know of it, hungry rice remains important in areas scattered from Cape Verde to Lake Chad; in certain regions of Mali, Burkina Faso, Guinea, and Nigeria, it is either a staple or a major part of the diet; each year in West Africa, farmers devote approximately 300,000 ha to cultivation of the crop, which supplies food to 3–4 million people; part of the reason for the neglect of this crop was as a result of a misunderstanding by scientists and other decision makers; the crop was nicknamed "hungry rice," a misleading term originated from the explorers from Europe who knew little of the crop or the lives of those who used it; unknown to those outsiders, the locals were harvesting hungry rice not because they were hungry, but because they liked the taste; indeed, they considered the crop to be exotic, and in some places it is reserved particularly for chiefs, royalty, and special occasions; in some societies hungry rice also forms part of the traditional bride price; it does not only deserve much greater recognition, it could have a big future; it is one of the world's best-tasting cereals; in recent times, some people have made side-by-side comparisons of dishes made with hungry rice and common rice and have greatly preferred hungry rice; it is also one of the most nutritious of all grains, its seed being rich in methionine and cysteine, amino acids which are vital to human health and which are deficient in major cereals, such as wheat, rice, maize, sorghum, barley, and rye; this combination of amino acids and taste is leading to an increasing interest in the crop; some varieties of hungry rice can mature so quickly that they are ready to harvest long before all other grains; they are perhaps the world's fastest-maturing cereal, producing grains just six to eight weeks after they are planted; they provide food easily in the growing season, when many crops are still too immature to be harvested and the previous year's production has been depleted; for this reason, it is sometimes called "grain of life"; some varieties, however, are late maturing, taking 165–180 days; among the cultivated types, two species are most prominent; white acha (*Digitaria exilis*), which is the most widely cultivated, is found from Senegal to Chad, and is very popular around the upland plateau of central Nigeria; the other species, the black acha (*Digitaria iburua*) is restricted to the Jos-Bauchi Plateau of Nigeria, as well as to northern regions of the Togo and Benin Republics; the *D. iburua* plant is a little bigger than *D. exilis*, in terms of stature and even grain size; it can tolerate poor soils and will grow where little else succeeds; it is insensitive to daylength; hungry rice is used in a variety of ways: it can be made into porridge and couscous, ground and mixed with other flours to make breads, popped and even brewed for beer; it has been described as a good substitute for semolina, the wheat product used to make spaghetti and other pastas; in the Hausa region of Nigeria and Benin Republic, people prepare couscous out of both types of hungry rice; in Togo, the Lambas brew a famous beer (*tchapalo*) from it; in southern Togo, the Akposses and Akebou people prepare hungry rice with beans in a dish that is reserved for special occasions; the grains are efficiently digested by farm animals; the straw and chaff can also be fed to farm animals; *Digitaria exilis* (2n = 4x = 36; 1C DNA content 0.98 pg), *D. iburua* (black fonio *syn* iburu *syn* black acha *syn* fonio noir *syn* manne noire *syn* ibourou), 2n = 6x = 54; 1C DNA content 1.20 pg (Gramineae) *bot agr*

hyacinth: Bulb-producing plants belonging to the lily family, native to the eastern Mediterranean and Africa; the cultivated hyacinth has large, scented, cylindrical heads of pink, white, or blue flowers; the water hyacinth is unrelated, a floating plant from South America; first cultivars had already been described between 1562 and 1640; in this period, just after its introduction from Turkey, there was little change beyond the appearance of white and pink and double mutants or segregates, and of larger spikes; van Kampin's bulb catalogue of 1739 lists nearly 400 varieties of hyacinth; before 1768, lilac and yellow colors had appeared; *Hyacinthus orientalis* (Liliaceae), 2n = 2x, 3x, 4x = 16, 24, 32; 1C DNA content 21.33, 32.38, 46.60 pg, respectively *hort*

hyacinth bean: >>> lablab

hyssop: A bushy herb formerly used in medicine; it is also used, like its relative, mint, in cooking; leaves are used in soups and teas, and, with flowers, in salads; *Hyssopus officinalis* (Labiatae), 2n = 2x = 12; 2C DNA content 1.0 pg *hort* >>> hedge hyssop

I

ice plant: Native to southern Africa and related to *Mesembryanthemum criniflorum*; it is now grown in several European countries; it is used for salads, giving a salty taste; *Mesembryanthemum crystallinum, M. edule* (Hottentot fig, Hottentot's fig, sour fig) (Aizoaceae), 2n = 2x = 18; 1C DNA content 0.43 pg >>> www.anbg.gov.au/cpbr/WfHC/Mesembryanthemum/ index .html

ilang-ilang: >>> ylang-ylang

Illinois bundleflower (*syn* **prickleweed**): A native herbaceous warm-season perennial legume that has potential as a forage and grain crop; it has a wide geographic range, stretching well beyond the prairie, from Colorado to Minnesota and Florida to Texas; it is a good nitrogen-fixer, a preferred forage for livestock, and produces relatively high yields of large seeds with favorable nutritional profiles; *Desmanthus illinoensis syn Mimosa illinoensis* (Leguminosae), 2n = 2x = 28 *agr*

Indian dwarf wheat: An early-maturing spring wheat with short, stiff stems; spikes are awnless or short-awned, and dense; they appear square in cross-section; spikelets contain six or seven flowers and develop four or five kernels; kernels are short and almost spherical, uniquely so among wheats, and thresh freely; it is grown in Northwest India; *Triticum sphaerococcum* or *Triticum aestivum* ssp. *sphaerococcum* (Gramineae), 2n = 6x = 42 (BBA^{u}A^{u}DD); 1C DNA content 17.33 pg >>> wheat

Indian hemp: A special type of *C. sativa*, which is cultivated in India as a source of narcotics; *Cannabis indica* (Moraceae), 2n = 2x = 20; 1C DNA content 0.84 pg *agr*

Indian lettuce: *Lactuca indica* (Compositae), 2n = 2x = 18; 1C DNA content 5.94 pg *hort* >>> Table 46

Indian melon (*syn* **casabas** *syn* **honeydew** *syn* **Asian melons**): Several varieties are known: canary melon, a large, bright-yellow melon with pale green to white inner flesh; casaba, bright yellow, with a smooth, furrowed skin. Less flavorful than other melons, but keeps longer, Hami melon, originally from Hami, Xinjiang, China. Flesh is sweet and crisp; honeydew, with a sweet, juicy, green-colored flesh, it is grown as bailan melon in Lanzhou, Chinam there is even a second variety which has yellow skin, white flesh and tastes like a moist pear; Kolkhoznitsa melon, with smooth, yellow skin and dense, white flesh; Japanese melons (including the sprite melon); Korean melon, a yellow melon with white lines running across the fruit and white flesh inside, which can be crisp and slightly sweet or juicy when left to ripen longer; Piel de Sapo (toad skin) or Santa Claus melon, with a blotchy green skin and white sweet-tasting flesh; Sugar melon, a smooth, white, round fruit; tiger melon, an orange, yellow and black striped melon from Turkey with a soft pulp, *C. melo inodorus* (Cucurbitaceae)

Indian millet: >>> durra

Indian mulberry: >>> noni

Indian mustard: >>> brown mustard

Indian pea: >>> chickling pea

Indian pennywort: >>> centella

Indian rape: *Brassica campestris* ssp. *dichotoma* (Brassicaceae), 2n = 2x = 18 (AA); 1C DNA content 0.58 pg *hort* >>> *Brassica* >>> Figure 8

Indian sandalwood: *Santalum album* (Santalaceae), 2n = 2x = 20; 1C DNA content 0.29 pg *fore*

Indian vetch: >>> chickling pea

indigo: Violet-blue vegetable dye obtained from various tropical plants such as anil, but now replaced by a synthetic product; it was once a major export crop of India; this dye has been used for at least 4,000 years, and it is superior to the European woad (*Isatis tinctoria*, 2n = 2x = 28; 1C DNA content 0.58 pg); however, with the development of analine dyes, the world market for natural dyes collapsed; indigo was cultivated until the 19th century; the current search for alternative crops and increased interest in natural products has led to reconsidering it as a crop to be grown in marginal areas; both morphophysiological and molecular characterization allows the accessions to be distinguished into groups of European and Asian origin; future breeding work is recommended because some accessions have good agronomic potential; *Indigofera tinctoria* (Leguminosae), 2n = 2x = 16; polytene chromosomes have been observed in ovary, anthers, or immature seed tissues *hort agr* >>> indigo woad

indigo woad: This plant provides a natural blue dye called woad, which is inferior to >>> indigo; *Isatis tinctoria* (Brassicaceae), 2n = 2x = 28; 1C DNA content 0.58 pg *hort agr* >>> *Brassica* >>> indigo

Indonesian oil palm: >>> oil palm

intermediate wheatgrass: *Agropyron intermedium syn* >>> *Thinopyrum intermedium* (Gramineae), 2n = 6x = 42; 1C DNA content 12.95 pg *bot agr*

iris: A plant of northern temperate regions; most have flattened leaves with large and showy flowers with an equal number of upright and pendulous petals, called standards and falls, respectively; *Iris* spp. (Iridaceae), *I. sibirica,* 2n = 2x = 28; 2C DNA content 4.2 pg *hort*

isanu: Closely related to the garden >>> nasturtium, *Tropaeolum majus*; the tubers are an important source of food for around 9 million people living at elevations of 2,500 to 4,000 m throughout the Andes mountains; the plant grows to over 1–1.5 m in diameter and 0.5–0.8 m high with slender and cylindrical aerial stems; *Tropaeolum tuberosum* (Tropaeolaceae), 2n = 4x = 28; 1C DNA content 1.33 pg; polytene chromosomes have been observed in ovary, anthers, or immature seed tissues *hort*

Isfahan wheat: A tetraploid, domesticated, and hulled emmer wheat; it may be winter or spring in habit; stems are usually tall, thick, and solid or pithy; leaves are broad, spikes are long and dense, sometimes compound or branched; they are almost square in cross-section, with long awns; kernels are short, ovate, and humped in shape; this wheat is closely related to durum but is somewhat inferior in terms of both productivity and macaroni-making quality, so has practically disappeared from cultivation; it is grown quite extensively in Mediterranean countries; *Triticum isphanicum* (Gramineae), 2n = 4x = 28 (BBA^{u}A^u); 2C DNA content 25 pg *agr*

Italian ryegrass: *Lolium multiflorum* (Gramineae), 2n = 2x = 14; 2C DNA content 8.2 pg *agr*

ivory (palm) nut: A genus of five species of palms, native to tropical South America; the genus name means "plant elephant," which refers to the very hard, white seed endosperm, which resembles elephant ivory; plants are medium-sized to tall, reaching up to 20 m tall, with pinnate leaves; *Phytelephas macrospora syn Elephantusia* spp. (Palmae), 2n = 2x = 36; 1C DNA content 1.00 pg *hort bot*

J

jack bean: *Canavalia ensiformis* (Leguminosae), 2n = 2x = 22 *hort* >>> pulse

jackfruit (*syn* **kathal**): A species of tree of the mulberry family; it is native to parts of southern and Southeast Asia; it is the national fruit of Bangladesh; it is believed to be indigenous to the Southwestern rain forests of India; it is widely cultivated in tropical regions of the Indian subcontinent, such as Sri Lanka, Vietnam, Thailand, Malaysia, and Indonesia; jackfruit is also found in East Africa, e.g. in Uganda and Mauritius, as well as throughout Brazil and Caribbean nations like Jamaica; it is well suited to tropical lowlands, and its fruit is the largest tree-born fruit, reaching 36 kg in weight and up to 90 cm long and 50 cm in diameter; archeological findings in India have revealed that jackfruit was cultivated in India 3,000 to 6,000 years ago; the flesh of the jackfruit is starchy, fibrous and is a source of dietary fiber; the flavor is similar to that of a tart banana; varieties of jackfruit are distinguished according to the characteristics of the fruits' flesh; *Artocarpus heterophyllus* (Moraceae), 2n = 2x = 56; 1C DNA content 1.15 pg *hort*

Japanese (flowering) apricot: One of the most popular landscape plants in China and Japan; called meihua in Chinese; it has been cultivated and bred in China for thousands of years; in China alone, more than 200 cultivars are maintained; it exhibits an S-RNase-based gametophytic self-incompatibility system like other self-incompatible *Prunus* species; an *S* haplotype-specific F-box protein gene, a candidate gene for pollen-*S*, leads to the development of a molecular typing system for an *S*-haplotype in this fruit species; *Prunus mume* (Rosaceae), 2n = 2x, 4x = 16, 32 *hort biot*

Japanese aucuba: >>> aucuba

Japanese laurel: >>> aucuba

Japanese barnyard millet (*syn* **white millet** *syn* **Indian sawa millet** *syn* **billion-dollar grass**)**:** The fastest growing of any cereal, it can produce a harvest in little more than 40 days; it is grown as a minor cereal in the Orient and India, and as a fodder crop in North America where it can produce up to eight crops a year; in India, it is cultivated on marginal lands where rice and other crops will not grow well; the grains are cooked in water, like rice, or boiled with milk and sugar; sometimes it is fermented to make beer; cytogenetic data suggest that *E. frumentacea* and *E. esculenta* are domesticated derivatives of *E. crus-galli* and *E. colona*, respectively, and that *E. oryzoides* is very closely related to *E. crus-galli*; *E. crus-galli* is an allohexaploid, produced by natural hybridization between the tetraploid *E. oryzicola* and a still-to-be-discovered diploid species of *Echinochloa*, with subsequent chromosome doubling; *Echinochloa frumentacea syn Panicum frumentaceum syn Echinochloa crus-galli* (Gramineae); two species of the genus *Echinochloa*, *E. frumentacea* (Indian barnyard millet) and *E. esculenta* (Japanese barnyard millet), are cultivated for food and fodder by hill and tribal communities in Asia, particularly in India, China and Japan; the crop has wide adaptability and occupies a special place in marginal rainfed areas, because of its short life cycle; although the area under the crop has decreased drastically in the past 50 years, the crop's ability to survive under harsh conditions makes it a better choice during famine years; in the Indian Himalayan region, the crop was traditionally used as a substitute for rice; recently, it has been identified as a suitable choice for climate-resilient agriculture; high nutrient content and antioxidant effects enable it to be considered as a functional food crop; the demand for the crop has increased due to its highly nutritious grains; thus, it has the potential to provide both food and nutritional security, particularly in the hills, where nutritional deficiencies are in abundance; despite enormous potential, the crop has not gained popularity among the masses and is still considered to be a poor man's food, 2n = 4x, 5x, 6x = 36, 48, 54; mean diploid; 2C DNA content 2.65–2.7 pg *bot agr*

Japanese black pine: *Pinus thunbergii* (Pinaceae), 2n = 2x = 24 *fore*

Japanese bunching onion: >>> Welsh onion

Japanese (flowering) cherry: A decorative ornamental tree known all over the world; famous varieties are 'Ama-no-gawa', 'K(w)anzan', 'Fudan-zakura', 'Gyoiko', 'Kiku-shidare-zakura', 'Shiro-fugen', and 'Tai-haku'; *Prunus serrulata rosea* (Rosaceae), 2n = 2x = 16; 1C DNA content 0–50 pg *hort*

Japanese chestnut: *Castanea crenata* (Fagaceae), 2n = 2x = 24 *hort*

Japanese horseradish: >>> wasabi >>> radish

Japanese laurel: >>> aucuba

Japanese lawngrass: >>> zoysiagrass

Japanese mint: *Mentha arvensis* (Labiatae), 2n = 8x = 96 (R^{a}R^{a}SSJJAA); 1C DNA content 1.50 pg *hort*

Japanese (nashi) pear: It has a gametophytic self-incompatibility mechanism controlled by a single *S*-locus with multiple *S*-haplotypes, each of which contains separate genes that determine the allelic identity of pistil and pollen; the pistil *S* gene is an S-ribonuclease gene; *Pyrus pyrifolia* (Rosaceae), 2n = 2x = 34; 1C DNA content 0.58 pg *hort*

Japanese persimmon: >>> persimmon

Japanese plum: *Prunus salicina* (Rosaceae), 2n = 2x, 4x = 16, 32 *hort*

Japanese privet: *Ligustrum japonicum* (Oleaceae), 2n = 2x = 44 *hort*

Japanese quince An East Asian genus within the subfamily Maloideae comprising four diploid (2n = 34) species: *C. cathayensis* (Chinese quince), *C. japonica* (Japanese quince), *C. speciosa* (flowering quince), and *C. thibetica* (Tibetan quince); of these, flowering quince and Japanese quince are the genetically most diverse species; Japanese quince was brought to Europe by 1869; since that time, it has been appreciated as an ornamental plant because of its early, showy, long-lasting flowering; the flowers, usually reddish-orange in color, are pollinated by honeybees and bumblebees; Japanese quince is outcrossing and has a strict self-incompatibility system; plants of Japanese quince are normally less than 1.2 m in height, and the shoots and twigs of wild plant material are usually armed with thorns; the irregularly shaped fruit (a pome) contains numerous seeds, becomes yellow and fragrant when mature, and has previously only been used to a minor extent; it has now been identified as a promising fruit crop for the North European climate; *Chaenomeles japonica* (Rosaceae), 2n = 2x = 34 *hort*

Japanese radish: *Raphanus sativus* ssp. *niger* (Brassicaceae), 2n = 2x = 18 (RR); 1C DNA content 0.55 pg *hort* >>> *Brassica* >>> radish >>> Figure 8

Japanese red pine: *Pinus densiflora* (Pinaceae), 2n = 2x = 24; 1C DNA content 25.05 pg *fore*

Japanese rice: Varieties of this rice are used strictly for the brewing of sake (Japanese rice wine); they represent a unique and traditional group of rices; they are characterized by common traits, such as a large grain size, with low protein content, and a large, central white-core structure; analysis of both nuclear and chloroplast genetic polymorphisms showed that the genetic diversity in sake-rice cultivars is much smaller than the diversity found in cooking-rice cultivars; the genetic diversity within the modern sake-brewing cultivars is about two-fold higher than the diversity within the local sake-brewing cultivars, which is in contrast to the situation with cooking cultivars; it is due to introgression of the modern cooking cultivars into the modern sake-brewing cultivars through breeding practices; cluster analysis and chloroplast haplotype analysis suggested that the local sake-brewing cultivars originated monophyletically in the western regions of Japan; *Oryza sativa* (Gramineae), 2n = 2x = 24 (AA), 2C DNA content 0.8–1.0 pg = 420–460 Mbp *agr* >>> rice >>> Table 45

Japanese snake gourd: *Trichosanthes cucumerina* (Cucurbitaceae), 2n = 2x = 22; 1C DNA content 2.83 pg *hort*

Japanese zoysiagrass: A seed-propagated, turf-type grass grown in temperate east Asia; two species extend throughout Malaysia; because of its high density and rhizomatous growth habit, zoysiagrass tends to crowd out other plants; *Zoysia japonica* (Gramineae), 2n = 4x = 40; 1C DNA content 0.50 pg *agr hort*

Japan laurel: >>> aucuba

jataco (*syn* **achita** *syn* **quihuicha**): An edible herb from the Andes of South America; edible leaves are used as a pot-herb; nutritious seeds are cooked and eaten like cereal grains; *Amaranthus caudatus* (Amaranthaceae), 2n = 2x = 32; 1C DNA content 0.63 pg *hort* >>> achita

Java cantala: >>> cantala

Java cardamom: >>> cardamom

Jerusalem artichoke: A close relative of the sunflower, it is open-pollinated and amenable to recurrent mass selection for horizontal resistance; *Helianthus tuberosus* (Compositae), 2n = 6x = 102; 1C DNA content 12.55 pg *hort* >>> agroenergy crop >>> sunflower >>> artichoke

Jesuit's tea: >>> epazote

jicama Includes three closely related cultivated species: *P. tuberosus*, *P. erosus*, and *P. ahipa*; its storage root dry matter content is usually low, although genotypes with a high dry matter content have been identified (e.g. the 'Chuin' accessions); flowers are often removed through flower pruning to increase storage root fresh matter yield, *Pachyrrhizus erosus syn Phaseolus pachyrrhizoides* (Leguminosae), 2n = 2x = 22; 1C DNA content 0.63 pg *hort* >>> pulse

Job's tears: >>> adlay

jointed charlock: >>> runch

jojoba: (from the Mexican name: ho-ho-bah) The plant is a woody perennial, a dioecious bush native to the Sonora Desert of Arizona, northern Mexico, and arid parts of southern California; jojoba plants are either male or female; only the female plants will produce beans and the males are only used for pollination, >>> overplanting is necessary so there are enough females; it takes plants at least three years to flower; until that time, one cannot distinguish the sex; after flowering the excess males can be removed from plantages and at least another two years are required before the females will produce beans; the female flower becomes a hardened capsule, which contains one or more developing seeds; as the seeds grow within the capsule during the spring and summer months, the capsule wall becomes thinner until it dries; the capsule ultimately splits and the matured seed drops to the ground; the color and shape of jojoba seeds are reminiscent of coffee beans; jojoba is the only plant that produces significant quantities of liquid wax esters akin to the natural restorative esters produced by human sebaceous glands; currently all jojoba beans are from wild plants; it is grown commercially in Argentina, Israel, and Australia, in addition to the U.S. and Mexico; the plant produces beans which contain up to 50–54% of their weight in oil; the composition of the oil found in the bean is similar to that found in the sperm whale; recently, it was demonstrated in Arabia, where it is also grown in salty deserts, that the oil is extremely useful for diesel engines as well; *Simmondsia chinensis* (Simmondsiaceae), 2n = 4x = 52; 1C DNA content 0.74 pg; the sex cannot be identified until flowering stage [2–5 years age]; sex chromosomes are not distinguishable; therefore, sex of jojoba seedlings cannot be determined by cytological methods; plantations grown from seed usually result in 84% male and 16% female plants whereas only 10% male plants are necessary for optimal yield; two male-specific markers of ~525 bp and ~325 bp were identified using primer combinations *Eco*RI-GC/*Mse*I-GCG and *Eco*RI-TAC/*Mse*I-GCG, respectively; a female-specific marker of ~270 bp was identified with the primer combination *Eco*RI-TAC/*Mse*I-GCG) *hort* >>> agroenergy crop

Joshua tree: >>> yucca

josta (syn jostaberry): A hybrid of blackcurrant x gooseberry, *Ribes nigrum × uva-crispa* (Grossulariaceae), 2n = 2x = 16; 1C DNA content 0.94 pg *hort*

jujube (*syn* **Chinese jujube** *syn* **red date** *syn* **Chinese date** *syn* **Korean date** *syn* **Indian date**): It is thought to be native of southern Asia, between Lebanon, Iran, Pakistan, northern India, Bangladesh, Nepal (Bayar), the Korean peninsula, and southern and Central China, and also Southeastern Europe, though more likely introduced there; it is a small >>> deciduous tree or shrub reaching a height of 5–10 m, usually with thorny branches; the leaves are shiny-green, ovate-acute, 2–7 cm wide and 1–3 cm broad, with three conspicuous veins at the base, and a finely toothed margin; the flowers are small, ~5 mm wide, with five inconspicuous yellowish-green petals, the fruit is an edible oval drupe 1.5–3 cm deep; when immature, it is smooth-green, with the consistency and taste of an apple, maturing brown to purplish-black and eventually wrinkled, looking like a small date; there is a single hard stone similar to an olive stone; jujube was domesticated in South Asia by 9,000 BC; over 400 cultivars have been selected; *Ziziphus zizyphus syn Z. jujuba* (Rhamnaceae), 2n = 2x, 3x, 4x = 24, 36, 48; 1C DNA content 1.55 pg (*Z. glabrata*) *hort*

juniper: A conifer (gymnosperm) with blue, woody fruits and overlapping, scale-like leaves; *Juniperus* spp. (Cupressaceae), 2n = 2x, 3x, 4x = 22, 33, 44; 1C DNA content 9.84–24.10 pg *hort* >>> Table 46

jute, white: Fiber obtained from two plants of the linden family: *Corchorus capsularis* and dark jute, *C. olitorius*; jute is used for sacks and sacking, upholstery, webbing, string, and stage canvas; jute is now often replaced by synthetic polypropylene; the world's largest producer of jute is Bangladesh, *Corchorus* spp. (Tiliaceae), 2n = 2x = 14; 1C DNA content 0.25–0.40 pg *agr*

K

kabuli chickpea: >>> chickpea

kaki: >>> persimmon

kakrol: >>> teasel gourd

kale: (French: chou frisé, chou vert) *Brassica oleracea* ssp. *acephala* var. *sabellica* (Brassicaceae), 2n = 2x = 18 (CC); 1C DNA content 0.78 pg *agr hort* >>> *Brassica* >>> Figure 8

kamut wheat: "Kamut" derives from the ancient Egyptian word for wheat, said to have been derived from seed found in the Egyptian pyramids; it appeared on the market in 1980 and was marketed as a new cereal; however, it is an ancient relative of modern durum wheat (*Triticum durum*); it is thought to have evolved contemporaneously with the free-threshing tetraploid wheats; it is also claimed that it is related to *T. turgidum*, which also includes the closely related durum wheat; the identity of the correct subspecies is in dispute; it was originally identified as *T. polonicum*; some other taxonomists believe it is *T. turanicum*, commonly called "Khorasan" wheat (also called Oriental wheat); recent molecular studies identified "kamut" as probably being a (natural ?) hybrid between *T. durum* and *T. polonicum* wheats, which occurred in the >>> Fertile Crescent; although its true history and taxonomy is not yet clear, its great taste, texture, and nutritional qualities, as well as its hypo-allergenic properties are unequivocal; the grain is two to three times the size of common wheat, with 20–40% more protein, is richer in lipids, amino acids, vitamins and minerals, and is a "sweet" alternative for all products that now use common wheat; *Triticum turanicum* (Gramineae), 2n = 4x = 28 (BBA^{u}A^u); 1C DNA content 12.28 pg >>> wheat >>> Khorasan wheat

kankro: >>> teasel gourd

kantola: >>> teasel gourd *syn* kantroli

kantoli: >>> teasel gourd

kantroli: >>> teasel gourd

kapok (tree): A large, deciduous, tropical tree that is native to tropical America, Africa, and the East Indies; this fast-growing tree is generally in the range 14–30 m tall; the kapok is the tallest tree in Africa; it has pink, white, or yellow flowers in clusters; the lightweight silky down from the seed pods (sometimes called Java cotton) is used as pillow stuffing, sleeping bag stuffing, life jacket stuffing, furniture upholstery, insulation, and for other uses; the green leaves are lanceolate and palmately compound (with five to nine leaflets); the yellow-green oil from the seeds is used in foods and to manufacture soap; young leaves are also cooked and eaten; the wood is also used; the night-blooming flowers are pollinated and the seeds are spread by fruit bats; *Ceiba pentandra* (Bombacaceae), 2n = 2x = 88; 1C DNA content 1.75 pg *hort*

karanji: >>> pongamia

karela: >>> bitter gourd

kartol: >>> teasel gourd

Katjang bean: *Phaseolus unguiculata* (*Vigna unguiculata*) ssp. *cylindrica* (Leguminosae), 2n = 2x = 22; 1C DNA content 0.60 pg; polytene chromosomes have been observed in ovary and immature seed tissues *hort*

kava (*syn* **kava-kava** *syn* **awa** *syn* **puawa**): A dicot showing long, slender inflorescences, and producing a one-seeded drupe with a thin mesocarp; although the leaves and stems of *P. methysticum* have some medicinal value, the active components are mostly concentrated within the roots and rhizomes of the plant; several resins and lactones, commonly referred to as kava lactones, have been isolated and identified from the roots and rhizomes; lactones occur naturally as odor-bearing components of various plant products, but they can also be synthesized; the identified chemicals include methysticin, yangonin, dihydromethysticin, dihyrokawain, and the three most powerful kava lactones, namely kavain, dihydrokavain, and dihydromethysticin; kava has been used throughout the Polynesian Islands for thousands of years as a beverage; the Polynesian Islands include Hawaii, Fiji, Samoa, Tonga, and Papua New Guinea (the eastern portion of New Guinea island, which also includes several hundred smaller islands); the herb has been used for ceremonial, ritual, religious, social, and medicinal purposes for centuries; in the past few decades, kava has been widely marketed as a treatment for anxiety, stress, restlessness, and sleep disorders; although kava is often compared to standard sedatives, its unique action on the nervous system is not so easily categorized; the first use of kava was documented about 1775 by Captain James COOK during voyages to the South Pacific Islands; kava is mainly processed in two ways; one method is to chew the root and rhizome fragments, soak the masticated roots in cold water or coconut milk, and then filter the resulting liquid; a few hours after filtration, the frothy beverage is ready to be consumed; the second method of kava processing is to macerate the root and rhizome pieces in cold water or coconut milk and then filter the substance just before drinking it; the rootstock prepared by the chewing method, enhanced perhaps by salivary enzymes, has a greater narcotic effect than in the macerated preparation; *Piper methysticum* (Piperaceae), 2n = 10x = 130 *hort*

keladi: >>> taro

kenaf (*syn* **bimli jute** *syn* **Deccan hemp**): A 4,000-year-old crop with roots in ancient Africa; as a member of the hibiscus family, it is related to cotton and okra; it grows quickly, rising to heights of 4 m in as little as 4–5 months; it may yield 2.5–4 t of dry fiber/ha; whereas the flowering period can last 3–4 weeks, each individual flower blooms for only one day; after blooming, the flower drops off, leaving a seed pod behind; the stalk consists of two distinct fiber types; the outer fiber is called "bast" and comprises ~40% of the stalk's dry weight; the refined bast fibers measure 2.6 mm and are similar to the best softwood fibers used to make paper; the whiter, inner fiber is called "core," and comprises 60% of the dry weight; these refined fibers measure 0.6 mm and are comparable to hardwood tree fibers, which are used in a wide range of paper products; upon harvest, the whole plant is processed in a mechanical fiber separator, similar to a cotton gin; the separation of the two fibers allows independent processing and provides raw materials for a growing number of products; *Hibiscus cannabinus syn H. cannabis* (Malvaceae), 2n = 2x = 36; 1C DNA content 1.53 pg *agr*

kencur: >>> galangal

Kentucky (blue)grass: A cool-season turfgrass that spreads by rhizomes; it is the most popular species for high-quality lawns in Ohio; it is very winter hardy; *Poa pratensis* (Gramineae), 2n = 84; 1C DNA content 5.38 pg *agr* >>> Texas blue grass

Kersting's groundnut: *Macrotyloma geocarpum* (Leguminosae), 2n = 2x = 20, 22 *hort* >>> pulse >>> gynophore

khat (*syn* **qat** *syn* **gat** *syn* **miraa**): A flowering plant native to tropical East Africa and the Arabian Peninsula; it is a slow-growing shrub or tree that grows to between 1.5 m and 20 m tall, depending on region and rainfall, with evergreen leaves 5–10 cm long and 1–4 cm broad; the flowers are produced on short axillary cymes 4–8 cm long, each flower small, with five white petals; the fruit is an oblong three-valved capsule containing one to three seeds; khat has been grown for use as a stimulant for centuries in the Horn of Africa and the Arabian Peninsula; there, chewing khat predates the use of coffee and is used in a similar social context; its fresh leaves and tops are chewed or, less frequently, dried and consumed as tea, in order to achieve a state of euphoria and stimulation; it also has anorectic side-effects; the leaves or the soft part of the stem can be chewed with either chewing gum or fried peanuts to make it easier to chew; it contains the alkaloid cathinone, an amphetamine-like stimulant which is said to cause excitement, loss of appetite, and euphoria; in 1980, the World Health Organization classified khat as a drug of abuse that can produce mild to moderate psychological dependence (less than tobacco or alcohol); the plant has been targeted by anti-drug organizations like the U.S. Drug Enforcement Administration; it is a controlled or illegal substance in many countries, but is legal for sale and production in many others; *Catha edulis* (Celastraceae), 2n = 2x = 32 *hort*

khesari: >>> chickling pea

Khorasan wheat: *Triticum turanicum* (Gramineae), 2n = 4x = 28 (BBA^{u}A^u); 1C DNA content 10.05 pg *agr* >>> wheat >>> kamut wheat

kidney bean: A plant of the humid tropical uplands which originated in Central America; the young green pods are eaten sliced and boiled, and the dried seeds can also be cooked and eaten; *Phaseolus coccineus, P. multiflorus* (Leguminosae), 2n = 2x = 22; 1C DNA content 0.68 pg, polytene chromosomes have been observed in ovary and immature seed tissues *hort* >>> pulse

kidney vetch: *Anthyllis vulneraria* (Leguminosae), 2n = 2x = 12; 2C DNA content 1.0 pg *bot agr*

Kikuyu grass: A fodder grass from Kenya that is now widespread throughout the tropics; *Pennisetum clandestinum* (Gramineae), 2n = 4x = 36; 1C DNA content 1.15 pg *agr*

kiwi: Fuzzy green-brown fruit with translucent pale green flesh surrounding a narrow ring of tiny black seeds; the flavor suggests a blend of melon, strawberry and banana; it originates from China and was introduced to New Zealand (now the main producer worldwide) in 1904 by the teacher Isabel FRASER; the common plant name, "Chinese raspberry", was changed into "kiwifruit" (related to the endemic bird) by the American company "Turners and Growers" for marketing reasons in 1959; about 80% of the world's production is from the variety "Hayward Giant Kiwi Berry", bred by the New Zealand breeder Hayward WRIGHT; he also bred the >>> nectarine "Goldmine"; of more than 60 species of *Actinidia* (kiwifruit), only two have been widely cultivated so far, and there is potential for breeding new varieties; all *Actinidia* species are dioecious, with similar sex-determining regions; in 2009, it was confirmed that, among the kiwifruit plant's small chromosomes (<1 μm), there is a pair of X/Y-like chromosomes that result in its dioecism; the genome has been mapped with 644 microsatellite markers from three genetic libraries (two from the New Zealand Institute for Plant and Food Research and one from the University of Udine, Italy); these showed 29 linkage groups, representing expression of 587 genes, and revealed that sex-linked sequence characterized amplified region (SCAR) markers and the flower sex phenotype were mapping to a subtelomeric region that bore the hallmarks of an early sex-determining locus; an absence of sex chromosome pairing means that the male-specific region on the Y chromosome is inherited as a unit, maintaining sexual dimorphism; it is suggested that at least two linked genes on the putative Y chromosome are responsible for dioecy: one suppressing pistil formation and one stimulating pollen development, *Actinidia chinensis* (Actinidiaceae), 2n = 2x = 58; 1C DNA content 0.78 pg *hort*

kodo millet (*syn* **scrobic** *syn* **scrobic paspalum**): Domesticated in India some 3,000 years ago; now widespread in Africa; introduced to Australia from Zimbabwe in 1931; it is a loosely tufted, shallow-rooting, short-lived perennial or annual grass, with ascending, somewhat succulent branched stems up to 90 cm high, tufts up to 60 cm in diameter; culms with four to six nodes; leaves up to 30 cm long and 12 mm wide, flat, soft, completely hairless on mature plants; significant weed species are: *P. ciliatifolium*, *P. conjugatum*, *P. dilatatum*, *P. fimbriatum*, *P. fluitans*, *P. laeve*, *P. lividum*, *P. longifolium*, *P. notatum*, *P. paspaloides*, *P. plicatulum*, *P. scrobiculatum*, *P. thunbergii*, *P. urvillei*, *P. vaginatum*, and *P. virgatum*; cultivated fodder species are: *P. dilatatum* (Dallis), *P. notatum* (Bahia), and *P. plicatulum*; important native pasture species are: *P. auriculatum*, *P. glumaceum*, *P. notatum*, *P. paniculatum*, and *P. scrobiculatum*; grain crop species are: *P. scrobiculatum* (kodo), and *Paspalum scrobiculatum syn P. polystachyum syn P. commersonii* (Gramineae), 2n = 2x, 4x, 6x = 20, 40, 60; mean 2C DNA content per genome 1.6 pg *agr*

kohlrabi: Variety of kale, which is itself a variety of >>> cabbage; it is used for food and resembles a turnip; the leaves of the kohlrabi shoot form a round swelling on the main stem; *Brassica oleracea* var. *gongylodes* (Brassicaceae), 2n = 2x = 18 (CC); 1C DNA content 0.78 pg *hort* >>> *Brassica* >>> radish >>> Figure 8

kola (nut): Several tropical trees, especially *Cola acuminata*; in West Africa, the nuts are chewed for their high caffeine content, and, in the West, they are used to flavor soft drinks; *Cola nitida* (Sterculiaceae), 2n = 6x = 40; 1C DNA content 0.70 pg *hort*

kolomikta: *Actinidia kolomikta* (Actinidiaceae), 2n = 2x = 58

koracan: >>> coracan >>> finger millet

Korean lawngrass: >>> zoysiagrass

Korean mint: >>> anise hyssop

Korean temple grass: >>> zoysiagrass

koroniviagrass: *Brachiaria humidicola* (Poaceae), 2n = 4x = 36 *agr* >>> palisadegrass >>> brachiaria

kosena radish: *Raphanus sativus* convar. 'Kosena' (Brassicaceae), 2n = 2x = 18 (RR); 1C DNA content 0.55 pg *hort* >>> *Brassica* >>> radish >>> Figure 8

kudzu (vine): A type of legume; a flowering plant that bears its protein-rich seeds in pods and can fix nitrogen from the soil (due to the symbiotic root bacteria *Rhizobium*); its tubers are edible and the fiber is useful; the plant has a woody stem, wide leaves, and purple flowers; it is native to Japan and was introduced to the U.S. around 1876; it soon became a nuisance weed in the southern U.S.; *Pueraria thunbergiana* (Leguminosae), 2n = 2x = 22; 1C DNA content 1.10 pg *hort phyt*

ku gua foo: >>> bitter gourd

kuikui pake: >>> Barbados nut

kui ts'ai: >>> Chinese chive

kummerovia: >>> lespedeza

kumquat: A group of small fruit-bearing trees; the edible fruit closely resembles that of the >>> orange (*Citrus sinensis*), but it is much smaller and ovular, being approximately the size and shape of an olive; the plants are slow-growing evergreen shrubs or short trees, from 2.5 to 4.5 m tall, with sparse branches, sometimes bearing small thorns; the leaves are dark glossy green, and the flowers white, similar to other citrus flowers, borne singly or clustered in the leaf axils; the plant is native to South Asia and the Asia-Pacific region; the earliest historical reference to kumquats appears in literature of China in the 12th century; they have long been cultivated in Japan, Taiwan, the Philippines and Southeast Asia; they were introduced to Europe in 1846 by Robert FORTUNE, collector for the London Horticultural Society, and, shortly thereafter, into North America; kumquats are often eaten raw; as the rind is sweet and the juicy center is sour, the raw fruit is usually consumed either whole or only the rind is eaten; the fruit is considered ripe when it reaches a yellowish-orange stage and has just shed the last tint of green; *Citrus japonica syn Fortunella japonica* (Rutaceae), 2n = 2x = 18 *hort*

kumudu: >>> noni

Kura clover: *Trifolium ambiguum* (Leguminosae), 2n = 4x = 32; 2C DNA content 1.9 pg *agr* >>> clover

kurrat: *Allium kurrat* (Alliaceae), 2n = 4x = 32 *hort*

kutki: A perennial herb which is used as a substitute for Indian gentian (*Gentiana kurroo*); a high-value medicinal herb used in herbal drug formulations like picroliv, picrolax, etc; it is one of the major income-generating non-timber forest products found in the Nepalese Himalayas and one of the oldest medicinal plants traded from the Karnali zone; the rhizome has been collected recklessly from its natural habitat, resulting in endangered status for this species; mostly, it prefers out-crossing as evinced by pollination experiments and floral architecture; *Picrorhiza kurroa* (Scrophulariaceae), 2n = 2x = 34 *hort*

L

lablab (*syn* **Bonavist bean** *syn* **black bean**): *Lablab purpureus* ssp. *purpureus syn Dolichos lablab syn Lablab purpureus* (Leguminosae), 2n = 2x = 22; 1C DNA content 0.38 pg *hort* >>> pulse

lachenalia: A bulbous genus consisting of more than 100 described species; the genus is endemic to the winter rainfall areas of South Africa and Namibia; the phenotypic variability among different species is one of the main reasons why the genus was selected for development; the bulbs produce inflorescences varying from spikes to racemes; the individual flowers vary from short and spreading to long and cylindrical; the leaves vary from filiform, to broad and ovate, to heart-shaped, and may have spots, pustules, hairs and/or ring-shaped markings; the flower color ranges from yellow, red, purple, and green to white; various basic chromosome numbers are present in the genus, i.e., x = 5, 6, 7, 8, 9, 10, 11, 13; ploidy levels range from diploid to octoploid, and polyploidy is present in many species; *Lachenalia bulbifera* (Hyacinthacea) *hort*

lady's fingers: >>> okra

lambsquarter (*syn* **lamb's quarters** *syn* **tree spinach** *syn* **fat-hen**): A very large annual leafy vegetable that grows to over 2.5 m tall; it is also known as tree spinach (not to be confused with "chaya"), though native to mountainous regions of India; it is easily cultivated in the U.K. and other areas, and may be sold under the name "Tree Spinach"; it is a leafy green which tastes like very much like >>> chard or >>> spinach with a hint of >>> asparagus when cooked; the best-tasting parts of the plant are the tender growing tips, which can be harvested continuously, the plant becoming bushy; since the plant contains oxalic acid, it should be cooked in a steel pan, not in aluminum; this plant, a relative of >>> quinoa, has edible seeds which can be cooked or ground into flour; the plant contains high concentrations of vitamins A, C, and K, and calcium, iron, phosphorus, and potassium, as well as saponins, which may have health benefits, *Chenopodium album syn Chenopodium giganteum* (Amaranthaceae), 2n = 6x = 54; 1C DNA content 2.33 pg *bot agr*

lamb's lettuce: >>> corn salad >>> lettuce

lamoot: >>> sapodilla

larch: Trees belonging to the pine family; the common larch (*Larix decidua*) grows to 40 m; it is one of the few conifers to shed its leaves annually; the small needle-like leaves are replaced every year by new bright-green foliage, which later darkens; closely resembling it is the North American tamarack (*L. laricina*); the golden larch (*Pseudolarix amabilis*), a native of China, turns golden in autumn; *Larix* spp. (Pinaceae), 2n = 2x = 24; 1C DNA content 11.45 pg *fore*

lard fruit: In 1933, a party of Chinese biologists, during a field survey in Southwest China's Yunnan Province, discovered a wild creeper, which had fruits as large as watermelons; for more than two decades, nothing was done about this plant; it was not until 1958 that serious efforts were started by the botanical garden in Shishong Baanna (China) to domesticate the plant as a source of edible oil; today, many creepers are grown from wild seeds; they produce from 40 to 80 melons a year as against 10 to 20 in the case of the wild plant; domestication of this plant has also been achieved by cutting and layering, asexual methods of propagation, which ensure the maximum rate of fruiting from selected plants; each fruit contains six seeds, each a little bigger than a hen's egg, the average seed carries two or three kernels; the kernel has a 70 to 80% oil content – higher than walnuts, peanuts and sesame seeds; plants are highly resistant to drought and waterlogging; it is relatively easy to cultivate; *Hodgsonia macrocarpa* (Cucurbitaceae), 2n = 2x = 18 *hort*

large-seeded false flax (*syn* **German sesame** *syn* **gold-of-pleasure**): An annual wild crucifer that is reported to be resistant to *Alternaria* blight; it additionally possesses valuable agronomic attributes that make it attractive as an alternative spring-sown crop for tight crop rotations; it is particularly rich in polyunsaturated C18 fatty acids, making it a valuable renewable feedstock for the oleochemical industry; *Camelina sativa syn Camelina sativa* ssp. *sativa syn Myagrum sativum* (Brassicaceae), x = 6, 14; 2n = 12, 26, 40; polytene chromosomes have been observed in ovary, anthers, or immature seed tissues *agr* >>> *Brassica*

larkspur: *Delphinium consolida* (Ranunculaceae), 2n = 2x = 16; 1C DNA content 4.28 pg, or 2n = 4x = 32 (*D. elatum*) *bot agr*

laurel (~ **tree**): European evergreen trees with glossy, aromatic leaves, yellowish flowers, and black berries; the leaves of sweet bay or poet's laurel are used in cooking; several species are cultivated worldwide; ornamental shrub laurels, e.g. cherry laurel (*Prunus laurocerasus*, Rosaceae), are poisonous; two laurel species have traditionally been considered within the genus *Laurus*: *L. nobilis* and *L. azorica*; the first is characterized by the presence of glabrous twig leaves and is located in the Mediterranean region; it can be found cultivated or naturalized, and has been reported in Spain, France, Italy, and Greece; *L. azorica* is characterized by the presence of densely tomentose to hirsute twig leaves and has been described in the Azores, Madeira, and the Canary Islands; *Laurus nobilis* (Lauraceae), 2n = 2x = 48; 1C DNA content 3.05 pg *hort*

lavender: A small, woody perennial grown for its essential oil, which is steam distilled from flowers; a sweet-smelling purple-flowering plant belonging to the mint family, native to western Mediterranean countries; the bushy low-growing species has long, narrow, upright leaves of a silver-green color; the small flowers, borne on spikes, vary in color from lilac to deep purple and are covered with small fragrant oil glands; lavender oil is widely used in pharmacy and perfumes; *Lavandula angustifolia* (Labiatae), 2n = 2x, 4x = 50, 100; 1C DNA content 0.9 pg *hort*

leaf celery: *Apium graveolens* ssp. *secalinum* (Umbelliferae), 2n = 2x = 22; 1C DNA content 1.08 pg *hort*

leaf lettuce: >>> cutting lettuce >>> lettuce

leafy spurge: A member of the spurge family, native to Europe and Asia; it is characterized by plants containing a white milky sap and flower parts in threes; leafy spurge is an erect, branching, perennial herb 1 m tall, with smooth stems and showy yellow flower bracts; stems frequently occur in clusters from a vertical root that can extend many centimeters underground; the leaves are small, oval to lance-shaped, somewhat frosted, and slightly wavy along the margin; the flowers of leafy spurge are very small and are borne in greenish-yellow structures surrounded by yellow bracts; clusters of these showy, yellow bracts open in late May or early June, while the actual flowers do not develop until mid-June; it displaces native vegetation in prairie habitats and fields through shading and by competing for available water and nutrients, and through plant toxins that prevent the growth of other plants underneath the spurge; it has also become an aggressive invasive species, and, once present, can completely take over large areas of open land; it occurs also across much of the northern U.S., with the most extensive infestations reported for Montana, North Dakota, Nebraska, South Dakota, and Wyoming; it has been identified as a serious pest; it was transported to the U.S. possibly as a seed impurity in the early 1800s; first recorded in Massachusetts in 1827, leafy spurge spread quickly and reached North Dakota within ~80 years; because of its persistent nature and ability to regenerate from small pieces of root, leafy spurge is extremely difficult to eradicate; biological control offers a highly promising management tactic for leafy spurge; the >>> United States Department of Agriculture (USDA) has reported success using six natural enemies of leafy spurge imported from Europe; these include a stem and root-boring beetle (*Oberea erythrocephala*), four root-mining flea beetles (*Aphthona* spp.), and a shoot-tip gall midge (*Spurgia esulae*); large-scale field-rearing and release programs are carried out cooperatively by federal and State officials in many northern states; several systemic herbicides have been found to be effective if applied in June, when the flowers and seeds are developing, or in early to mid-September, when the plants are moving nutrients downward into the roots; *Euphorbia esula* (Euphorbiaceae), 2n = 2x, 4x = 30, 60 *phyt* >>> Table 46

ledidi: >>> miracle fruit

leek: >>> garden leek

lemon: The lemon is a small evergreen tree and fruit; the juice of the lemon fruit contains about four times as much citric acid as is found in the orange; it is a major commercial source of citric acid; this is the origin of the term "lemonade" and this yellow fruit has always been popular in temperate countries where limes were unavailable; it is usually too sour to be eaten as a fruit, but it is widely used as a flavoring and garnish in many foods and drinks; the freshly grated peel, known as zest, is also widely used as a flavoring; lemons rank the third among the citrus industry in the world, with a total annual production of about 9% in the citrus production; the exact origin of the lemon has remained a mystery, though it is widely presumed that lemons first grew in India, northern Burma, and China; the genetic origin of the lemon, however, was reported to be a hybrid between sour orange and citron; lemon cultivation is common in China, India, Iran, Brazil, Spain, Italy, Mexico, and, to some extent. in the U.S.; in China, many lemon varieties popularly cultivated are landraces, and many of them were invariably given local or vernacular names by lemon growers, *Citrus limon* (Rutaceae), 2n = 2x, 4x = 18, 36; 1C DNA content 0.40 pg *hort* >>> hesperidia >>> Figure 61 >>> http://users.kymp.net/citruspages/citrons.html

lemon balm: Perennial herb belonging to the mint family, with lemon-scented leaves; it is widely used in teas, liqueurs, and medicines; because of its proven sedative, spasmolytic and antiviral effects, it is often used in aqueous or alcoholic extracts for self-medication or pharmaceutical purposes; its therapeutic effect is due to the content of essential oils and rosmarinic acid, *Melissa officinalis* (Labiatae), 2n = 2x, 4x = 32, 64; 2C DNA content 1.6 pg *hort*

lemon basil: >>> basil

lemon for candied peel (citron): A fragrant fruit; the designation "*medica*" is apparently derived from the similar ancient names "media" or "median apple", which were influenced by THEOPHRASTUS (>>> SCHLEGEL 2007), who believed the citron was native to Media, Persia or Assyria; citron has many similar names in diverse languages e.g. cederat, or cedro; the fruit is usually ovate or oblong, narrowing to the stylar end; however, the citron's fruit shape is highly variable, due to the large quantity of albedo, which forms independently, according to the fruit's position on the tree, twig orientation, and many other factors; this could also be the reason of its being protuberant, forming a "v" shape from the end of the segments up to the stylar end; the rind is leathery, furrowed, and adherent; there are also a fingered variety, sometimes called "Buddha's hand"; the citron is used for the fragrance or zest of its outer peel, but the most important part is still the albedo, which is a fairly important item in international trade, and is widely employed in the food industry as succade; *Citrus medica* (Rutaceae), 2n = 2x = 18; 1C DNA content 0.40 pg *hort* >>> hesperidia >>> Figure 61 >>> http://users.kymp.net/citruspages/citrons.html

lemongrass: It is native to India and the nearby island of Sri Lanka; it is found growing naturally in tropical grasslands; it grows in dense clumps that can grow to 1.8 m in height and ~1.2 m in width; the strap-like leaves are 1–2.5 cm wide, ~0.9 m long, and have gracefully drooping tips; the evergreen leaves are bright bluish-green and release a citrus aroma when crushed; it is the leaves that are used as flavoring and in medicine; they are steam distilled to extract lemongrass oil, an old standby in the perfumer's palette of scents, *Cymbopogon citratus* (Poaceae), $2n = 6x = 60$ (*C. citratus*, $2n = 2x = 20$) *hort*

lemon verbena: A species of flowering plant in the verbena family; it is native to Argentina, Paraguay, Brazil, Uruguay, Chile, Bolivia, and Peru; it was brought to Europe by the Spanish in the 17th century; is a deciduous open shrub growing to 2–3 m high; the 8-cm long glossy, pointed leaves are slightly rough to the touch and emit a powerful lemon scent when bruised; sprays of tiny lilac or white flowers appear in late spring or early summer; leaves are used to add a lemony flavor to fish and poultry dishes, vegetable marinades, salad dressings, jams, puddings, and beverages; it is also used to make herbal teas, or added to standard tea in place of actual lemon, *Aloysia citrodora* (Verbenaceae), $2n = 2x = 36$; 1C DNA content 0.74 pg (*A. triphylla*) *hort*

lenten rose: >>> Christmas rose

lentil: A short-statured, annual, self-pollinating, and food legume; the crop is grown in dryland cereal-based rotations because of its nitrogen-fixing ability, its high-protein seeds for human consumption, and its straw, which is a valued livestock feed; the putative progenitor of the cultivated lentil is *Lens culinaris* ssp. *orientalis*, which is distributed from Greece in the west to Uzbekistan in the east, and from the Crimean Peninsula in the north to Jordan in the south; the lentil ranks among the oldest and the most appreciated grain legumes of the Old World (back to 8,000–7,000 BC; the oldest carbonized remains of lentil are from the Franchthi cave in Greece, dated to 11,000 BC (the landrace is 'Eglouvis'), and from Tell Mureybit in Syria, dated to 8,500–7,500 BC); it is cultivated from the Atlantic coast of Spain and Morocco in the west, to India in the east; the place of origin of the cultivated lentil is not known with certainty; the greatest variability in the cultigen is found in the Himalaya–Hindu-Kush junction region between India, Afghanistan, and Turkestan; presently, the major lentil-producing regions are Asia (58% of the area) and the West Asia–North Africa region (37% of the acreage of developing countries); it is the most important pulse in Bangladesh and Nepal, where it contributes significantly to the diet; farmers also grow lentils in India, Iran, and Turkey; other significant producers in the developing world include Argentina, China, Ethiopia, Morocco, Pakistan, and Syria; global lentil production is growing rapidly; it rose by 112% from 1.3 million t in the period 1979–1981 to 2.9 million t in the period 1993–1995, resulting from a 54% increase in area to 3.42 million ha and an increase in average productivity of 38%, from 600 kg/ha to 825 kg/ha; >>> The International Center for Agricultural Research in the Dry Areas (ICARDA) has a mandate to improve this crop; recently, aluminum (Al)-tolerant genotypes were discovered (L-7903, L-4602), caused by a monogenic dominant trait; *Lens culinaris* (Leguminosae), $2n = 2x = 14$; 1C DNA content 4.20 pg *agr*

lespedeza: *Lespedeza stipulacea* (Leguminosae), $2n = 2x = 20$ *agr*

lesser bindweed: >>> corn bindweed

lesser broomrape: The largest genus among the holoparasitic members of the Orobanchaceae; *Orobanche minor* (Orobanchaceae), $x = 12, 19, 24$ (within the genus); $2n = 2x = 38$; 1C DNA content 1.83 pg *bot agr* >>> broomrape

lettuce: Leafy compact head; many varieties, romaine lettuce with more elongate leaves; it is the world's most popular salad plant; both its common and its Latin names are based on an easily noticeable characteristic; it has a heavy, milky juice; the word "lettuce" is probably derived from the Old French "laities" (*pl.* "laitue"*)*, meaning "milky," referring to this plant; the Latin root word *"lac"* (= milk) appears in the Latin name *Lactuca*; the ancient

Greeks called lettuce "tridax"; the old Persians, "kahn"; according to HERODOTUS (>>> SCHLEGEL 2007), lettuce was served on the tables of the Persian kings of the 6th century BC; it was popular among the Romans about the beginning of the Christian Era, and had been brought to a fairly advanced state of culture and improvement; in the first century after Christ, Roman writers already described a dozen distinctly different varieties; in China, it was described in the 5th century; one form of "stem lettuce" is native to China; the "asparagus lettuce" and others with long, narrow leaves and tall, thick, succulent, edible stems are of this type; cultivated lettuce is closely related to the wild lettuce, *Lactuca scariola,* from which it was doubtless derived; *L. scariola* is widely scattered over the globe, but it originated in inner Asia Minor, the trans-Caucasus, Iran, and Turkistan; as in the development of the cabbages, the primitive forms of lettuce were loose, leafy, and sometimes "stemmy" types; the loose-heading and firm-heading forms occurred much later; >>> cos lettuce (romaine) forms an erect, compact rosette of elongated leaves, approaching the character of a head; it is relatively tolerant to heat and evidently was developed in a moderately warm climate; the old records and its name indicate an Italian origin; light-green, dark-green, and red-spotted forms of romaine were described in 1623; this type was common in Italy in the Middle Ages and is said to have been taken to France from Italy in 1537 by F. RABELAIS (1494–1553); toward the end of the 16th century, it was still rarely grown in France and Germany; firm-heading forms had become well developed in Europe by the 16th century; the oak-leaved and curled-leaf types, and various colors were all described in the 16th and 17th centuries in Europe; C. COLUMBUS evidently carried lettuce to the New World, for its culture was reported on Isabela Island (Crooked Island) in the Bahamas in 1494; it was common in Haiti in 1565; lettuce was doubtless among the first garden seeds sown in every European colony on this continent; common lettuce is related to >>> prickly lettuce *(Lactuca serriola), Lactuca sativa* (Compositae), 2n = 2x = 18; 1C DNA content 2.65 pg *hort* >>> chloroplast >>> Table 46

Levant garlic: This species has two variants, leeks and Levant garlic; the first is tetraploid and sets seed freely, while Levant garlic is a hexaploid and is sterile; the breeding procedures are those of open-pollinated crops; *Allium ampeloprasum* (Alliaceae), 2n = 4x = 32 (AAA'A''); 1C DNA content 30.0 pg *hort*

licorice: A root from which a sweet flavor can be extracted; the licorice plant is a legume native to southern Europe, India, and parts of Asia; it is not botanically related to >>> anise, >>> star anise, or >>> fennel, which are sources of similar flavoring compounds; it grows on dry and sandy soils and likes moderate moisture and temperature; it has been common as a herb for more than 2,000 years; commercially produced in Iran, Afghanistan, China, Pakistan, Iraq, Azerbaijan, Uzbekistan, Turkmenistan, and Turkey, *Glycyrrhiza glabra* (Leguminosae), 2n = 2x = 16 *hort*

licorice mint: >>> anise hyssop

lilac: Flowering Old World shrubs, with clusters (panicles) of small, sweetly scented, white or purple flowers on the main stems; the common lilac (*Syringa vulgaris*) is a popular garden ornamental; it is one of the most widely cultivated ornamental trees and shrubs in temperate regions of the world; several hundred lilac cultivars have been generated by extensive hybridization, cultivation of chance hybrid seedlings in nurseries, and artificial selection; *Syringa* spp. (Oleaceae), 2n = 2x = 46; 1C DNA content 1.20 pg *hort*

lily: Plants belonging to the lily family, of which there are ~80 species, most with showy, trumpet-shaped flowers growing from bulbs, which occur in nature (like *L. auratum, L. pumilum, L. dauricum, L. martagon, L. bulbiferum, L. candidum, L.callosum*); the lily family includes hyacinths, tulips, asparagus, and plants of the onion genus; modern lily cultivars are developed by combination breeding of Asiatic and Oriental hybrids, *Lilium* spp. (Liliaceae), 2n = 2x = 24 *hort* >>> http://www.liliumbreeding.nl

Lima bean (*syn* **Sieva bean** *syn* **butter bean**): The plant is self-pollinating, but some natural cross-pollination occurs; named after the capital of Peru, archaeological remains of this bean having been found there, dating from 6,000 BC; however, it is thought that this bean probably originated in the Guatemala area of Central America and was taken to South America by early travelers; the green-shelled beans are eaten as a vegetable, and the dried beans are also cooked and eaten; *Phaseolus lunatus* (Leguminosae), 2n = 2x = 22; 1C DNA content 0.70 pg; polytene chromosomes have been observed in ovary, anthers, or immature seed tissues *hort* >>> pulse

limber pine: *Pinus flexilis* (Pinaceae), 2n = 2x = 24; 1C DNA content 31.20 pg *fore* >>> pine

lime: Acid lime is an important commercial fruit crop, cultivated from the Terai to the high hills of Nepal; high variation in acid lime fruits is observed among existing landraces due to crossing with other citrus species; it is a sharp-tasting green or greenish-yellow citrus fruit of the small thorny lime bush, native to India; the white flowers are followed by the fruits, which resemble lemons but are more round in shape; they are rich in vitamin C; in the late 18th century, the British Admiral Nelson insisted on his sailors drinking lime juice in order to prevent scurvy, which is due to a deficiency of Vitamin C; this earned the British the nickname of "limeys"; *Citrus aurantifolia* (Rutaceae), 2n = 2x = 18; 1C DNA content 0.38 pg *hort* >>> hesperidia >>> Figure 61 >>> http://users.kymp.net/citruspages/citrons.html

lime tree: Deciduous trees native to the northern hemisphere; the leaves are heart-shaped and coarsely toothed, and the flowers are cream-colored and fragrant; the common lime (*Tilia vulgaris*) has greenish-yellow scented flowers in clusters on a winged stalk, followed by small round fruits; the small-leafed lime (*T. cordata*) is found in areas of ancient woodland in Europe; *Tilia* spp. (Tiliaceae), 2n = 6x = 84 (allopolyploid, diploid-like chromosome pairing) *fore*

linden: >>> lime tree

lingonberry: A small-fruited crop species of considerable economic importance in Northern Europe; despite the fact that lingonberries are still harvested from the wild, changes in forest management, variable fruit quality from native stands, and fluctuations in annual yield have stimulated initiatives to domesticate this crop; since the 1960s, investigations with the aim of improving the growing of lingonberry have been conducted mainly in Sweden, Finland, Germany, and the U.S.; commercial cultivation was originally introduced into Germany during the 1980s, some years later in Sweden, and, recently, the first operations have been established in the U.S.; recent research on growing techniques has been carried out in Finland and Sweden; however, production of cultivated lingonberries is insignificant compared to that of wild berries; the species is also known for its dual-purpose uses: berries, as well as ornamental value for domestic gardens and landscaping; lingonberry is a perennial, evergreen dwarf shrub; *Vaccinium vitis-idaea* (Ericaceae), 2n = 2x = 24 *hort*

linseed: >>> flax

liquorice: >>> licorice

lisianthus (*syn* **alkali chalice** *syn* **[Texas] bluebells**): A short-lived herbaceous perennial plant with >>> taproots; the genus *Eustoma* is comprised of two species; *Eustoma grandiflorum*, commonly known as prairie gentian, is native to the Midwestern prairies of the U.S. south to Mexico; it is one of the leading cut-flowers in Japan; there are market demands for cultivars with deep-yellow flowers, but they have never been bred successfully; *Eustoma exaltatum* is native to the southern U.S., Mexico, Central America, and the West Indies; the two species may represent different ecotypes of the same species, for both species are interfertile and produce fertile progeny; in cultivation, *E. grandiflorum* grows like a biennial, and first produces a rosette which bolts into a single flowering stem after cold treatment, then dies after flowering; in contrast, *E. exaltatum* grows more like a perennial in cultivation and produces additional shoots each season; both species have purple flowers that are funnel-shaped to campanulate, but corolla lobes of *E. exaltatum* range up to

2.5 cm in length, whereas those of *E. grandiflorum* are 5 to 6 cm in length; most of the plant material in cultivation is *E. grandiflorum*; this species is an important seed-propagated cut flower crop in Europe and Japan; *Eustoma grandiflorum syn Lisianthus russelianus* (Gentianaceae), 2n = 8x = 72; 1C DNA content 1.63 pg *hort*

listada: A Spanish striped eggplant; among listada, the most internationally renowned is the 'Listada de Gandía' heirloom; *Solanum melongena* (Solanaceae), 2n = 2x = 24; 2C DNA content 1.9 pg *hort*

litchi: Evergreen tree belonging to the soapberry family; the delicately flavored egg-shaped fruit has a rough brownish outer skin and a hard seed; it is native to southern China, where it has been cultivated for 2,000 years; *Litchi chinensis* (Sapindaceae), 2n = 2x = 30; 1C DNA content 0.70 pg *hort*

little bluestem (broom beardgrass *syn* **wiregrass, bunchgrass** *syn* **prairie beardgrass** *syn* **broom):** The most important dominant perennial grass of uplands in the tallgrass prairie of the U.S. and Canada; usually tufted, sometimes with short rhizomes; presently, it is a dominant species, most significantly on coarse-textured soils, in the mixed-grass prairie region; its wide geographical adaptation to dry soils suggests little bluestem has potential for biomass production in areas unsuitable for most other grass species in the North American steppe; because of its >>> C$_4$ metabolism, little bluestem develops primarily during the warm weather of summer and early fall, and it has excellent drought resistance, *Schizachyrium scoparium syn Andropogon scoparius* (Poaceae), 2n = 40 *agr hort*

little millet: *Panicum miliare* (Gramineae), 2n = 4x = 36; 1C DNA content 1.04 pg *agr*

loblolly pine The principal commercial southern pine of the U.S.; a large, resinous, and fragrant tree with rounded crown of spreading branches, height: 24–30 m; diameter: 0.6–0.9 m; needles are evergreen and 13–23 cm long; three in a bundle, stout, stiff, often twisted, and green; the bark is blackish-gray, thick, deeply furrowed into scaly ridges, exposing brown inner layers; the cones are 7.5–13 cm long, conical, dull brown, and almost stalkless; they open at maturity but remain attached; cone-scales are raised, keeled, with a short stout spine; the habitat is deep, poorly drained flood plains to well-drained slopes of rolling, hilly uplands; it forms pure stands, often on abandoned farmland; it is among the fastest-growing southern pines; it is extensively cultivated in forest plantations for pulpwood and lumber; *Pinus taeda* (Pinaceae), 2n = 2x = 24; 1C DNA content 22.10 pg *fore*

locust: >>> carob

locust tree: >>> black locust

lodgepole pine: *Pinus contorta* (Pinaceae), 2n = 2x = 24; 1C DNA content 22.10 pg *fore*

loganberry: *Rubus* × *loganobaccus* (Rosaceae), 2n = 2x = 14 *hort*

Lolium × *hybridum*: An artificial grass hybrid between *Lolium perenne* (2n = 2x = 14) *x Lolium multiflorum* (2n = 2x = 14); it is used in agriculture; *L.* × *hybridum* (Gramineae), 2n = 2x = 14 (alloploid); 1C DNA content 2.72 pg *agr*

longan: A tropical tree native to South and Southeast Asia, in the Indomalaya ecozone; grown for its edible fruit; the tree can grow up to 6–7 m in height, and the plant is very sensitive to frost; the fruit is extremely sweet, juicy and succulent in superior agricultural varieties, and, apart from being eaten fresh, is also often used in East Asian soups, snacks, desserts, and sweet-and-sour foods, either fresh or dried, sometimes canned with syrup in supermarkets, *Dimocarpus longan syn Euphoria longan syn E. longana syn Nephelium longana* (Sapindaceae), 2n = 2x = 20 *fore* >>> litchi >>> rambutan

long-podded cowpea: >>> asparagus bean

loofah: >>> luffa

loose-leafed lettuce: >>> cutting lettuce >>> lettuce

lop grass: >>> drooping brome

loquat: A small, evergreen tree or shrub belonging to the Rosaceae; it is native to China and Japan, and is also known as the Japanese medlar; it is widely cultivated, both as a decorative tree and for its edible, plum-like, yellowish fruit; loquat cultivation is very ancient in Eastern Asia but the crop's spread to Europe occurred more recently, in 1784, when it was introduced into the Botanical Gardens of Paris; it was from here that the loquat made its way to the Mediterranean and subsequently was introduced to Florida from Europe and to California from Japan; in the 20th century, the crop has spread to India, Southeastern Asia, and South Africa, as well as Central and South America; in general, it can be found in maritime climates between 20° and 35° latitude; *Eriobotrya japonica* (Rosaceae), 2n = 2x = 34; 1C DNA content 0.78 pg *hort*

lovage: *Levisticum officinale* (Umbelliferae), 2n = 2x = 22; 2C DNA content 9.9 pg *hort*

lovegrass: *Eragrostis* spp. (Gramineae), 2n = 4x = 40; 1C DNA content 0.68 pg (e.g. in >>> tef, *Eragrostis tef*, or *E. albensis*) *bot agr*

love-in-a-mist: *Nigella damascena* (Ranunculaceae), 2n = 2x = 12; 2C DNA content 21.6 pg *bot agr*

lowbush blueberry: *Vaccinium angustifolium* (Ericaceae), 2n = 2x = 24 *hort*

lucerne: >>> alfalfa

luffa: Smooth luffa is a rampant, fast-growing annual vine that produces pretty yellow flowers and strange-looking fruits that are edible when immature and used as back scrubbers or sponges when fully mature; the vine can grow to more than 80 cm long and scrambles over anything in its path; the large leaves are lobed and have silvery patches on the topsides; the flowers are showy and conspicuous, about 5–7 cm across with five petals; the fruits are green, up to 60 cm long and 8 cm in diameter; they are cylindrical and smooth, and shaped like a club, slighter wider on one end; small fruits look like >>> okra or little cucumbers; on older fruits, the outer skin eventually dries and turns brown and papery; smooth luffa is probably native to tropical Africa and Asia; it is grown throughout most of Asia for food and for pot scrubbers, and is cultivated commercially for export in Japan; *Luffa cylindrica, L. acutangula, L. aegyptiaca* (Cucurbìtaceae), 2n = 2x = 26; 1C DNA content 0.85 pg; polytene chromosomes have been observed in ovary, anthers, or immature seed tissues *hort*

lulos: *Solanum quitoense* (Solanaceae), 2n = 2x = 24 *hort*

luo han kuo: >>> buddha's fruit

lupine: Popular herbaceous garden plants; they are members of the pea family and are related to beans and clover; lupine cultivation is at least 2,000 years old and most likely began in Egypt or in the general Mediterranean region; there are over 300 species of the genus *Lupinus*; almost all have high levels of alkaloids (bitter-tasting compounds) that make the seed unpalatable and sometimes toxic; alkaloids have been removed from the seed by soaking; in the 1920s, German breeders (>>> SCHLEGEL 2007) produced the first selections of alkaloid-free or "sweet" lupine, which can be directly consumed by humans or livestock; >>> white lupine (*L. albus*), >>> yellow lupine (*L. luteus*), and >>> blue or narrow-leafed lupine (*L. angustifolius*) are cultivated as crops; lupines are currently grown as a forage and grain legume in Russia, Poland, Germany, and the Mediterranean, and as a cash crop in Australia, where it is exported to the European seed markets; both winter-hardy and non-hardy types are available; male-sterility is described in lupin crop species of *L. angustifolius* L. and *L. luteus* L. and is also characterized in the Andean lupin, *L. mutabilis*. In *L. angustifolius* and *L. luteus*, male-sterile plants were identified in artificially induced mutation populations, whereas, in *L. mutabilis*, both naturally occurring and induced male-sterile plants were selected; *Lupinus* spp. (Leguminosae), 2n = 2x = 26; 1C DNA content ~1.5 pg *bot agr hort* >>> Table 46

lychee: >>> litchi

Lyme grass: *Elymus europaeus syn Hordelymus europaeus* (Gramineae), 2n = 4x = 28; 2C DNA content 9.3 pg (YY) *bot agr* >>> couches

***Leymus racemosus* (Gramineae):** 2n = 4x = 28 (JJNN *syn* $N^sN^sX^mX^m$); 2C DNA content 2.7–7.7 pg (X^mX^m) *bot agr*

M

maca (*syn* **Peruvian ginseng** *syn* **maino** *syn* **ayuk willku** *syn* **ayak** *syn* **chichira**): A root vegetable; it grows in the mountains of Peru at altitudes of 1,000–3,000 m; native Peruvians have used the plant as a food and a medicine, since before the times of the Incas; it contains biologically active aromatic isothiocyanates, specifically *p*-methoxybenzyl isothiocyanate, which is also found in >>> magua (*Tropaeolum tuberosum*), another species reputed to increase fertility in humans; today, dried maca roots are ground to powder and sold in drug stores in capsules as a medicine and food supplement to increase stamina and fertility; root consists of 60% carbohydrates, 10% protein, 8.5% dietary fiber, and 2.2% fats; *Lepidium meyenii* *syn L. peruvianum* (Brassicaceae), 2n = 8x = 64 *hort* >>> *Brassica* >>> magua

macadamia nut: Edible nut of a group of trees native to Australia, especially *Macadamia ternifolia*, and cultivated in Hawaii, South Africa, Zimbabwe, and Malawi; the nuts are slow-growing; they are harvested when they drop; in Australia, the most important commercial species are *M. integrifolia* (Queensland only) and *M. tetraphylla*; the northern Queensland species, *M. Whelanii* has seeds containing dangerous levels of prussic acid, but Aborigines used them as a food source after grinding, washing, and cooking them; typically, orchards are established with two to four elite cultivars that have been vegetatively propagated by grafting onto seedling rootstocks or as rooted cuttings; nut production generally starts 4 years after planting and continues till the trees are at least 30 years of age; nuts are commonly harvested from the orchard floor, de-husked, dried to 10% moisture content, and then transported to a factory where the nut is cracked to retrieve the kernel; recently, a breeding program has been established to produce new improved cultivars for the Australian industry; *Macadamia* spp. (Proteaceae), 2n = 2x = 28 *fore hort*

macaroni wheat: >>> durum wheat >>> wheat

mace: >>> nutmeg

maceron: >>> alexanders

Macha wheat: A late-maturing winter wheat with tall, hollow stems; spikes vary in density from open to dense, with short awns; kernels remain in the spikelets after threshing; they are elliptical, red, and intermediate in hardness; it is grown in Transcaucasia and Russia; *Triticum macha syn Triticum aestivum* ssp. *macha* (Gramineae), 2n = 6x = 42 (BBA^{u}A^{u}DD); 1C DNA content 17.68 pg *bot agr* >>> wheat

Madagascar bean: >>> Lima bean

Magellan barberry: >>> calafate

magua (*syn* **mashua** *syn* **isanu** *syn* **cubio** *syn* **añu** *syn* **ysaño** *syn* **puel**): An annual, herbaceous climber and tuber crop indigenous to the Andean highlands; of economic value as a food and medicinal crop; date of domestication is estimated at around 5,500 BC; this root crop ranks fourth in importance in the Andean region after >>> potato, >>> oca, and >>> ulluco; of the Andean tubers, magua is one of the highest yielding, easiest to grow, and the most frost resistant; it is cultivated in the Andes of Bolivia, Peru, Ecuador, Columbia, and Venezuela; the tubers are an important source of food for around 9 million people living at elevations of 2,500 to 4,000 m throughout the Andes mountains; *Tropaeolum tuberosum* (Tropaeolaceae), 2n = 4x = 52 (autotetraploid) *hort*

maidenhair tree: >>> ginkgo

maize, corn U.S.: The world's fourth most important crop, behind only wheat, rice, and potatoes; there are more than 327 million acres of maize planted each year worldwide; the U.S. produces over 526 million U.S. tons per year; other continents and countries that produce a large amount of maize include Africa, Argentina, Brazil, China, France, India, Mexico, Romania, Russia, and South Africa; the best place to grow maize is in well-aerated, deep, warm soil containing a lot of organic matter, nitrogen, phosphorus, and potassium; moderately high summer temperatures, warm nights, and adequate, well-distributed rainfall helps even more during the growing season; the growing season and daylength also play a role in growth; maize evolved in Mexico or Central America ~6,000 years ago (>>> SCHLEGEL 2007); it developed from a small wild plant with a pod-pop cob; modern maize has its cob enclosed in the one sheath, which prevents dissemination of seed; it has imperfect flowers; it is monoecious, cross-pollinating; almost all maize plants grown in the world are F_1 hybrid maize; there are seven types of maize: (a) *flint*; >>> flint maize kernels are hard and smooth and have little soft starch; flint was probably the first maize Europeans ever laid eyes on; it is not grown in the U.S. as much as it is in Asia, Central America, Europe, and South America; in temperate zones, flint maize matures earlier, has better germination, and the plant vigor is earlier than in dent; (b) *flour*; flour maize contains a lot of soft starch, and has almost no dent; though it is not used much anymore, it is grown in the drier sections of the U.S. and in the Andean region of South America; it is an older type of maize, and was found in a lot of graves of the Aztecs and Incas; since the kernel is so soft, the American Indians could make it into flour; (c) *pop*; an extreme form of flint; it has a very small proportion of soft starch; it is a very minor crop, and is grown mostly for humans to eat; the reason it "pops" so well is because of the horny endosperm, which is a tough, stretchy material that can resist the pressure of steam generated in the hot kernel until it has enough force to explode or "pop"; (d) *sweet*; this type of maize has an almost clear, horny kernel when it is still young; the kernels become wrinkled when dry; the ears can be eaten fresh, or can be stored in cans; the only difference between sweet and dent maize is that the sweet genotype has a gene which prevents some sugar from being converted into starch; it is grown a lot as a winter crop in the southern U.S.; *Bacillus thuringiensis* (*Bt*)-sweetcorn has also proven effective for control of some lepidopteran species and continues to be accepted in the fresh market; *Bt*-fresh-market sweetcorn hybrids are released almost every year; (e) *dent*; gets its name from the dent in the crown of the seed; is grown more than any other type of maize; millions of tons of grain are produced from dent corn; it is used for human and industrial use, and for livestock feed; the starch reaches the summit of the seed, and the sides are also starchy; the denting is caused by the drying and shrinking of the starch; the dent corn grown in the corn belt of the U.S. came from a mix of New England flints and gourseed (an old variety of corn grown by the Indians in Southeastern North America; (f) *waxy*; seeds appear waxy; chemically, it has a different type of starch than normal starch; it was developed in China, and some waxy mutations have occurred in American dent strains; very little is grown, and that which is, is used for producing a starch similar to tapioca starch; (g) *pod*; pod maize is not grown commercially, but it is used a lot in studying the phylogenesis of maize; it resembles varieties of the primitive forms; every seed is enclosed in a pod and the whole ear is also enclosed in a husk; recent molecular studies demonstrated that the two main modern divisions of maize arose ~3,000 years ago, as maize arrived in what is now the Southwestern U.S. and, at about the same time, on the islands of the Caribbean; temperate maize spread further north and east across North America, while tropical maize spread south; the temperate-tropical division remains today; what maintains it are differences in disease susceptibility and photosensitivity – essentially, how daylength affects flowering time; the two maize types are now so different from each other that they do not cross well, and their hybrids are not well adapted anywhere; there are also colored maizes; some of them are bred for making colored polenta; they are rich in anthocyanins; the anthocyanin content does not alter the taste of the colored polenta; *Zea mays* (Gramineae), 2n = 2x = 20; 2C DNA content 4.9–12.6 pg = 2,400–3,200 Mb; ~300,000,000 bp per chromosome; in 2009, the genome was completely decoded with about 32,000 genes *agr* >>> agroenergy crop >>> crop evolution >>> dent maize >>> Tables 15, 16, 32, 35, 48

Malabar nightshade: The green climbing spinach is native to eastern India, the red species to China; it is a biennial plant, although grown annually; the fleshy leaves are prepared similar to spinach in Asian countries; *Basella alba, B. rubra* (Chenopodiaceae), 2n = ?x = ~48; 4C DNA content ~7 pg *hort*

malanga: >>> tania

mallow: Plants of the mallow family, including the European common mallow (*Malva sylvestris*), the tree mallow (*Lavatera arborea*), marsh mallow (*Althaea officinalis*), and hollyhock (*A. rosea*); most mallows have pink or purple flowers; *Malva* spp. (Malvaceae), 2n = 6x = 42; 1C DNA content 1.48 pg (*Malva alcea* is a dodecaploid with 2n = 12x = 84) *hort*

mandarin (*syn* **tangerine**): A small orange; often known as the "loose-skinned" oranges because of their easy peeling; these fruits are used mainly as dessert fruits; they probably originated in Vietnam and are of ancient cultivation in China and Japan; *Citrus reticulata* (Rutaceae), 2n = 2x = 18; 1C DNA content 0.48 pg *hort* >>> hesperidia >>> Figure 61 >>> http://use rs.kymp.net/citruspages/citrons.html

mangel: *Beta vulgaris* (Chenopodiaceae), 2n = 2x = 18; 1C DNA content 1.25 pg *hort agr* >>> beet >>> sugarbeet

mangel-wurzel: >>> beet

mango: A large tree with a broad, rounded canopy; cultivated orchards kept at 6–10 m tall; narrow leaves; trees are long-lived; some specimens are 300 years old and still fruiting; flowers are tiny, red-yellow; borne in panicles of up to 3,000 individuals; flowering occurs in December; plants are self-fertile; the cultivated mango is probably a natural hybrid between *M. indica* and *M. sylvatica,* occurring in Southeastern Asia to India; selection of wild types has occurred for 4,000–6,000 years, with vegetative propagation for at least 400 years in India; mangos were brought to Britain and Europe after the British occupied India in the 1800s; they were brought to Brazil and the West Indies in the 1700s with exploration of the area; hundreds of mango cultivars exist throughout the world; however, in the western hemisphere, a few cultivars derived from a breeding program in Florida are the most popular for international trade; locally, many cultivars are used, and often seedling trees are grown as a backyard food source; 'Tommy Atkins' is the most common mango in the U.S., medium sized, with a beautiful exterior but firm, finely fibrous flesh which is low in flavor, compared with others; 'Keitt' is among the largest of major cultivars, with yellow-green skin, and the yellow-gold flesh is fiberless and full-flavored; 'Kent' has red-bluish skin, is medium sized, and relatively round, with fiberless, rich-flavored flesh, which may have a turpentine-like aftertaste; 'Haden' is the old, anthracnose-prone mainstay, now replaced by others, but 90% of Hawaiian production, and grown in Ecuador (Guayaquil area) and western Mexico; small and relatively round, the red skin color is excellent when grown in hot, sunny, dry climates; firm flesh, almost fiberless; *Mangifera indica* (Anacardiaceae), 2n = 4x = 40; 1C DNA content 0.45 pg *hort*

mango squash: >>> chayote

mangold: *Beta vulgaris* var. *cicla* (Chenopodiaceae), 2n = 2x = 18; 1C DNA content 1.25 pg *agr hort* >>> beet >>> sugarbeet

mangosteen: Evergreen tropical tree in the family Guttiferae, native to Malaysia; its red flowers are followed by round brownish-purple fruits ~6 cm in diameter; inside the thick rind, the fruits consist of segments of soft, white, sweet pulp; mangosteen trees are difficult to propagate; they grow slowly and may take 15 years to produce their first fruit; it is an obligate >>> agamosperm, which has only two close relatives, *G. hombroniana* (2n = 48) and *G. malaccensis* (2n = 42?) that are facultative agamosperms; *Garcinia mangostana* (Guttoferae), 2n = 88–90 (allopolypoid) *hort*

Manila aloe: >>> cantala

Manila hemp: >>> abaca

manioc: >>> cassava

manna: *Cassia fistula* (Leguminosae), 2n = 4x = 28; 1C DNA content 1.76 pg *hort*

mansonia (*syn* **African black walnut** *syn* **African walnut** *syn* **bête**): It occurs from Guinea and Côte d'Ivoire east to the Central African Republic and northern Congo; the heartwood is yellowish brown to dark grey-brown or even dark brown, often with purple, reddish or greyish green streaks, often in alternating light and dark bands, though it fades on exposure, to a somewhat dull brown; the wood is used for general and high-class joinery, cabinet work, furniture, turnery, decorative veneer, and handicrafts; it is also used in construction for doors and windows, in railway coaches and shop fittings, and for boxes and crates; well-colored wood resembles American black walnut and is commonly used as a substitute, e.g. for gun stocks and grips, musical instruments and loudspeaker enclosures; wood waste can be used as a substrate for the edible fungus *Pleurotus tuber-regium*, observations indicating that the fermented substrate has some value as cattle feed, 2n = 2x = 48 *fore*

mara: >>> bitter gourd

marigold: Plants belonging to the daisy family, including pot marigold (*Calendula officinalis*, 2n = 2x = 14) and the tropical American *Tagetes patula* and *T. erecta* (2n = 2x = 24; 1C DNA content 1.18 pg), commonly known as French marigold; in 1993, it was revealed that marigolds give off volatile insecticides; several compounds found in marigold tissues are insecticidal, including the volatile thiophenes; these have proved toxic to disease-carrying mosquitoes; *Calendula* spp., *Tagetes* spp. (Compositae), 2n = 54 *hort*

marjoram (*syn* **sweet marjoram**): A herbaceous, perennial plant native to Cyprus and the Eastern Mediterranean, growing up to 70 cm in height; when grown in warmer countries, it has a stronger flavor, and is dried and sold as the herb oregano; sweet marjoram is also a popular herb; in the food industry, marjoram is used mainly as a spice in sausages, but also in baked goods, processed vegetables, condiments, soups, snack foods, and gravies; the main compounds of the volatile oil of marjoram are the epimeric monoterpene alcohols *trans*-sabinene hydrate, *cis*-sabinene hydrate and *cis*-sabinene hydrate acetate; *Origanum majorana syn Majorana hortensis* (Labiatae), 2n = 3x = 30; 1C DNA content 0.68 pg *hort*

marrow fat pea: >>> green pea

marrow-stem kale: *Brassica oleracea* convar. *acephala* var. *medullosa* (Brassicaceae), 2n = 2x = 18 (CC); 1C DNA content 0.78 pg *agr hort* >>> *Brassica* >>> Figure 8

mash: >>> urd bean

maritime pine: *Pinus pinaster* (Pinaceae), 2n = 2x = 24; 1C DNA content 28.90 pg *fore*

marrow: Trailing vine that produces large pulpy fruits, used as vegetables and in preserves; the young fruits of one variety *Cucurbita pepo* are known as courgettes (U.S.: zucchini); *Cucurbita pepo* (Cucurbitaceae), 2n = 2x = 40; 1C DNA content 0.55 pg [higher levels of ploidy (6x, 12x, 24x, 48x) were encountered at later stages; the euploid increase in chromosome number is the result of endopolyploidy] *hort*

marsh trefoil (*syn* **buckbean** *syn* **water trefoil** *syn* **marsh clover**): A perennial herb; it is a green, glabrous plant, with creeping rootstock and procumbent stem, varying in length according to the situation, covered by the sheaths of the leaves, which are on long, fleshy, striated petioles and three-partite; it was held to be of great value as a remedy against the once-dreaded scurvy ("Scharbock" is a German name – a corruption of the Latin "*scorbutus*," the old medical name for the disease); *Menyanthes trifoliata* (Menyanthaceae), 2n = 54 *bot agr*

marvel of Peru (*syn* **four o'clock flower**): The most commonly grown ornamental species of *Mirabilis*; it is available in a range of colors; "mirabilis" in Latin means wonderful and Jalapa is a town in Mexico; the flowers usually open from late afternoon onwards, then producing a strong, sweet-smelling fragrance; it was in this plant that C. CORRENS first discovered, in 1909, that >>> chloroplast genes determine traits, such as foliage coloration in plants where green, white, and variegated plants occur, that is, mosaics of white and green tissue (i.e. >>> extranuclear inheritance); *Mirabilis jalapa* (Nyctaginaceae), 2n = 58; 1C DNA content 1.19 pg *hort*

masterwort: A perennial herb with a stout, knotted rhizome, a rosette of basal leaves, and a tall, furrowed, hollow stem terminated by large compound umbels of whitish- or pinkish-colored flowers; the leaves are ternate or biternate, the segments broadly ovate and serrate; the fruit is a broadly winged ribbed achene; *Astrantia* ssp. *syn Peucedanum ostruthium* (Umbelliferae), x = 7, 8; 2n = 2x = 14 *bot agr*

matè: Also called yerba maté, Paraguay tea, or Brazilian tea; it is a tea-like beverage, popular in Argentina, Uruguay, Paraguay, and southern Brazil; other names are chá-dos-jesuitas (Jesuit tea), chá-das-missões (Mission tea), congonha, congonha-das-missões, erva-de-são-bartolomeu (Saint Barthelemy herb), orelha-de-burro (donkey's ear), chá-do-paraná (Paraná tea); it is brewed from the dried leaves and stemlets of the perennial tree; the name "matè" derives from the Quichua word "matí" for the >>> gourd (*Lagenaria vulgaris*) that is used to drink the infusion; it is a stimulating drink, greenish in color, containing caffeine and tannins, and is less astringent than tea; it has a characteristic mature flavor which is somewhat sweet, sour, withered leaf-like, similar to that obtained from >>> tea (*Camellia sinensis*); of the 196 volatile chemical compounds found in yerba matè, 144 are also found in tea; it is used in popular medicine and employed in commercial herbal preparations as a stimulant to the central nervous system, a diuretic, and an antirheumatic; caffeine content = 1–1.5%; tannin content = 7–11%; on average, 300,000 tons of matè are produced each year; it grows between the parallels 10° and 30° (South) in the Paraná and Paraguay River basins at altitudes varying from 500 to 1,000 m; as a tropical or subtropical plant, it needs a mild climate, with no dry season, and 15 to 21°C average temperatures and up to 1,500 mm of rain annually; in the wild, the plant needs 25 years to develop completely, reaching in that case a height of up to 15 m; the leaves are alternate, elliptical, or oval, with the border slightly serrated; it flowers between the months of October and December; the flowers are small, polygamous, dioecious, with calyx and corolla in a tetrameric disposition; the fruit resembles a pepper-like berry; among several varieties, there are three types that are the most important: 'angustifolia', 'longifolia', and 'latifolia'; *Ilex paraguariensis* (Aquifoliaceae), 2n = 4x = 40; 1C DNA content 1.15 pg *hort*

matgrass: *Nardus* spp. (Gramineae), *N. stricta* 2n = 2x = 26; 2C DNA content 4.2 pg *bot agr*

Mauritius hemp: Also known as green aloe, female karata, maguey, mayuey criollo, cocuisa, giant cabuya, and aloes vert, it is a robust "shrub" with a basal rosette about 2.5 to 3.5 m in diameter and flowering stalks 5 to 10 m in height; it has no taproot, relatively fine lateral roots, and many fine roots; the green to yellow-green leaves are linear-lanceolate to oblanceolate, pointed at the tip, and are fleshy with thread-like parallel fibers; the inflorescences (panicles) are terminal and contain many pendulous, fragrant, white, greenish-white, yellowish-green, or pale blue-green, 2.5 to 3.3 cm long by 1.8 cm wide flowers; bulbils 1–16 cm long develop abundantly on the peduncles after flower dehiscence; Mauritius hemp helps hold the soil, furnishes cover for wildlife, and adds to the aesthetics of wildlands; the species is widely, although not heavily, used as a landscaping plant for accent and curiosity; it was once widely cultivated for fiber, hence the common name; *Furcraea gigantea* var. *willemettiana syn Furcraea gigantean syn A. gigantean syn A. gigantensis* (Agavaceae), 2n = 2x = 60; 1C DNA content 4.10 pg *agr*

May beet: There are three different types grown: (1) a spherical type of beet with white, white-purple, or yellow roots; (2) a conical type of beet with a white, yellow, or purple heads (also called >>> stubble beet or autumn beet, mainly used for animal feed and as a>>> green manure); and (3) a type with small and short beet habit (also called >>> Teltow beet or >>> Navet Petit de Berlin); May beet is one of the oldest vegetable crops; it originates from the Mediterranean region or Asia Minor; the leaves can be up to 30 cm in height; the beets are eaten when they are still young, with a diameter of ~7 cm; well-known varieties are

'Goldball', 'Mailänder', or 'Dutch White'; *Brassica napus* var. *rapa* (Brassicaceae), 2n = 4x = 38 (AACC); 1C DNA content 1.15 pg *agr hort* >>> beet >>> *Brassica* >>> Figure 8

meadow saffron: Commonly known as autumn crocus or naked lady, it is a flower which resembles the true crocuses, but flowers in autumn; the plant has been mistaken by foreigners for ramsons, which it vaguely resembles, but is a deadly poison due to the presence of >>> colchicine, a useful drug with a narrow therapeutic index; the symptoms of colchicine poisoning resemble those of arsenic and there is no antidote; despite its toxicity, colchicine is an approved treatment for gout, and is also used in plant breeding to produce polyploids; *Colchicum autumnale* (Liliaceae); *Crocus sativus*, commonly known as saffron crocus, is a species of the *Crocus* genus in the Iridaceae family; it is best known for the spice saffron, which is produced from parts of the plant's flowers; human cultivation of saffron crocus and use of saffron spans more than 3,500 years and spans different cultures, continents, and civilization; the cultivated *Crocus* species is triploid; it is self-sterile and cannot be multiplied directly by seeds, 2n = 3x = 24; 1C DNA content 2.43 pg = 3.45 Gbp (*C. arenarium*) *bot agr hort* >>> saffron >>> Table 46

meadow fescue: *Festuca pratensis* (Gramineae), 2n = 2x = 14; 2C DNA content 8.9 pg *agr*

meadowfoam: A winter annual, produces novel very long-chain seed oils (C_{20} and C_{22}) with less than 2% saturated fatty acids, which are stable under diverse conditions; the fatty-acid composition makes this seed oil valuable for use in cosmetics, lubricants, rubber additives, and plastics; while a few meadowfoam cultivars have been developed, high-yielding germplasm is required for further crop improvement; it is a recently domesticated species; the first cultivar 'Ross' was developed and released in 2003 by the Oregon Agricultural Experiment Station (U.S.); it was developed by three cycles of recurrent half-sib family selection for increased seed yield and lodging resistance; it is a heterogeneous, open-pollinated population; the first and second cycles of selection were performed between 1990 and 1997, *Limnanthes alba* (Limnanthaceae), 2n = 2x = 10; 1C DNA content 1.39 pg *bot agr*

meadow foxtail: *Alopecurus pratensis* (Gramineae), 2n = 4x = 28; 2C DNA content 13.6 pg *bot agr*

meadow soft grass: >>> common velvet grass

medic(s): Annual species constitute the most numerous component of the genus *Medicago*; they are native to the Mediterranean basin from which they spread to areas of the world with a Mediterranean climate; these species are extremely heterogeneous in morphology, environmental adaptation, and chromosome number; the greater part are diploid with 2n = 16 or 2n = 14, and two are polyploid, with 2n = 30; because of some interesting characteristics such as rapid growth, large production of pods, viability of the seeds on the soil for long periods of time, and nitrogen fixation by means of bacterial activity, medics received increasing attention in the past decade; at present, they are regarded as very promising species for sustainable agricultural systems, as well as for environmental uses; numerous studies have also been carried out to understand their phylogenesis and to explore their germplasm; one of the most studied species is *Medicago truncatula*, a diploid (2n = 2x = 16) self-pollinating plant with a small genome size (2C DNA content 0.9–1.1 pg), which responds very well to regeneration and transformation methods; such characteristics make *M. truncatula* a very suitable species for basic genetic studies; *Medicago* spp. *evol agr* >>> alfafa >>> lucerne

medlar: A small, deciduous tree native to Europe and Asia Minor; the ripe, apple-shaped pomes are eaten raw and used in preserves; *Mespilus germanica* (Rosaceae), 2n = 2x = 34; 1C DNA content 0.75 pg *bot hort*

melon: >>> cantaloupe

melongene: >>> eggplant

mengkudu: >>> noni

Mexican tea: >>> epazote

Michaelmas daisy: >>> aster

mignonette: *Reseda* sp. (Resedaceae), x = 5, 6; 2n = 2x, 3x, 4x, 6x, 8x = 20, 26, 28, 30, 40, 48 *hort*

mikan (*syn* **mandarin** *syn* **Satsuma orange**): Produces sweet, seedless and easy-peeling fruits which much resemble mandarins but belong to a different citrus species; it is of Chinese origin, but was introduced to the West *via* Japan; in Japan, it is known as "unshu mikan," in China, as "wenzhou migan"; recorded cultivation of the "wenzhou migan" date back some 2,400 years; it was listed as a tribute item for Imperial consumption in the Tang Dynasty (>>> SCHLEGEL 2007); the best record of the cultivation of this variety in ancient China is from the Jijia Julu, written by Han Yan, the governor of the region and published in 1178; *Citrus unshiu* (Rutaceae), 2n = 2x = 18 *hort* >>> Figure 61 >>> http://users.kymp.net/cit ruspages/citrons.html

milfoil: >>> yarrow

milk thistle (*syn Carduus marianus syn* blessed milk thistle *syn* **Marian thistle** *syn* **Mary thistle** *syn* **Saint Mary's thistle** *syn* **Mediterranean milk thistle** *syn* **variegated thistle** *syn* **Scotch thistle**): an annual or biennial plant; this fairly typical thistle has red to pur-ple flowers and shiny pale green leaves with white veins; originally a native of southern Europe through to Asia, it is now found throughout the world; it grows to 30 to 200 cm tall, having an overall conical shape with an approx. 160 cm maximum diameter base; the stem is grooved and more or less cottony; with the largest specimens, the stem is hollow; though its efficacy in treating diseases is still unknown, *S. marianum* is sometimes pre-scribed by herbalists to help treat liver diseases (cirrhosis, jaundice and hepatitis); silibinin (*syn* silybin *syn* sylimarin I) may have hepatoprotective (antihepatotoxic) properties that protect liver cells against toxins; fucosyltransferases are a group of enzymes that catalyze the transfer of L-fucose from a donor substrate to an acceptor molecule; these compounds also act as antioxidants for scavenging free radicals and inhibiting lipid peroxidation, *S. marianum* (Asteraceae), 2n = 2x = 34 *hort*

milkvetch: >>> astragalus

milkweed: *Asclepias incarnata* (Asclepiadaceae), 2n = 2x = 22; 1C DNA content 0.42 pg *agr*

millet: A general name for species that are grown in similar regions to sorghum; millet is more drought resistant; different millets may have evolved in different parts of the world, includ-ing Africa and Asia; it has been grown in China for ~5,000 years; five types are described: (a) common millet *syn* broomcorn millet (*Panicum miliaceum*, 2n = 4x = 36); (b) finger millet (*Eleusine coracana*, 2n = 4x = 36); (c) foxtail millet (*Setaria italica*, 2n = 2x, 4x = 18, 36); (d) pearl millet (*Pennisetum americanum*, 2n = 2x = 14); and (e) Japanese barnyard *syn* Indian sawa millet (*Echinochloa frumentacea*, x = 9; 2n = ?x = 27, 36, 42, 48, 54, 72, 108) (Gramineae) *agr hort* >>> Table 35, 48

miracle berry: >>> miracle fruit

miraculous berry: >>> miracle fruit

miracle fruit: The plant is a shrub that grows up to 6 m high in its native habitat, but does not usually grow higher than 3 m in cultivation; its leaves are 5–10 cm long, 2–4 cm wide and glabrous below; they are clustered at the ends of the branchlets; the flowers are brown; it carries red, 2-cm long fruits; each fruit contains one seed; when berries are eaten, they cause sour foods (such as lemons and limes) subsequently consumed to taste sweet; this effect is due to miraculin, which is used commercially as a sugar substitute; the berry has been used in West Africa since at least the 18th century, when European explorer Chevalier des MARCHAIS, who searched for many different fruits during a 1725 excursion to West Africa, provided an account of its use there. Marchais noticed that local people picked the berry from shrubs and chewed it before meals; in Japan, miracle fruit is popular among diabetics and dieters; today, it is being cultivated in Ghana, Puerto Rico, Taiwan, and southern Florida, *Synsepalum dulcificum* (Sapotaceae), 2n = 2x = 26

mirliton: >>> chayote

miso mold: A very important fungus used in the fermentation of soybeans to make miso paste and in the fermentation of rice to make sake; *Aspergillus oryzae* (Aspergillaceae), eight chromosomes with ~12,074 genes are known, among them are 7,124 with unknown function; four telomeres were cloned and sequenced; the telomeric repeat sequence consisted of dodecanucleotides (TTAGGGTCAACA); the length of the telomeric repeat tract was 114–136 bp, which corresponds to 9–11 repeats of the dodeca-nucleotide sequence *hort biot*

moneywort (*syn* **creeping-Jenny**): *Lysimachia nummularia* (Primulaceae), 2n = 30, 32, 36, 43 *bot agr*

Mongolian oak: *Quercus mongolica* (Fagaceae), 2n = 2x = 24 *fore*

monkeyflower: *Mimulus* spp. (Scrophulariaceae), 2n = 4x = 28; 1C DNA content 0.37 pg (*Mimulus guttatus*) *hort*

moong: >>> mung bean

moth bean (*syn* **mat bean**): A very drought-resistant, self-pollinated grain legume that requires hot tropical temperatures; the green pods may be eaten as a vegetable, the seeds are eaten cooked, and the plant makes a useful forage crop; *Vigna aconitifolia* (Leguminosae), 2n = 2x = 22; 1C DNA content 1.13 pg *hort* >>> pulse

morello: *Cerasus vulgaris* var. *austera* (Rosaceae), 2n = 2x = 32 *hort* >>> morello cherry

Moricandia: >>> wild crucifer

mountain clover: *Trifolium montanum* (Leguminosae), 2n = 2x = 16; 1C DNA content 0.52 pg *bot agr* >>> clover

mountain savory (*syn* **winter savory**): *Satureja montana* (Labiatae), 2n = 6x = 30; 1C DNA content 2.78 pg *hort*

mountain spinach: *Atriplex hortensis* ssp. *viridis* (Chenopodiaceae), 2n = 2x = 18; 2C DNA content 1.5 pg *hort*

mugwort (*syn* **artemisa** *syn* **carline thistle** *syn* **chiu ts'ao** *syn* **common mugwort** *syn* **Douglas mugwort** *syn* **felon herb** *syn* **sailor's tobacco** *syn* **wormwood**): A perennial herb native to Africa, temperate Asia, and Europe, widely naturalized in most parts of the world; growing on hedge banks and waysides, uncultivated and waste land; cultivation is fairly easy; it prefers slightly alkaline, well-drained loamy soil, in a sunny position; a tall-growing shrubby plant, with angular stems, which are often purplish; the leaves are smooth and dark green above and covered with a cottony down beneath; they are alternate, pinnately lobed, and segmented; the small greenish yellow flowers are panicled spikes with a cottony appearance; blooming is from July to October; it is closely related to common wormwood (>>> absinthe); leaves and stems are gathered when in bloom, and dried for later herb use; the leaves have an antibacterial action, inhibiting the growth of *Staphylococcus aureus, Bacillus typhi, B. dysenteriae, B. streptococci, E. coli, B. subtilis,* and *Pseudomonas*; a weak tea made from the infused plant is a good all-purpose insecticide; the fresh or the dried plant repels insects; *Artemisia vulgaris* (Asteraceae), 2n = 2x = 18; 2C DNA content 6.0 pg *hort*

mulberry: A juicy, purplish-black fruit of the black mulberry; resembles a blackberry and makes a fine preserve; male and female flowers grow on separate bushes but the fruits form on the female trees whether or not they have been fertilized; the secret art of silk culture, along with mulberry seeds, is presumed to have spread from China to other parts of the world along the famous Silk Road; *Morus alba* (Moraceae), 2n = 2x = 28; 1C DNA content 0.85 pg *hort*

multiplier (onion): *Allium cepa* var. *aggregatum* (Alliaceae), 2n = 2x = 16; 1C DNA content 16.75 pg, genome size, var. Alice, 17,061 Mbp *hort* >>> onion

mung bean: Most widely cultivated throughout the southern half of Asia including India, Pakistan, Bangladesh, Sri Lanka, Laos, Cambodia, Vietnam, eastern parts of Java, eastern Malaysia, South China, and Central Asia; domestication and cultivation started in the northwest and far south of India 4,000–6,000 years ago; the differentiation of *Medicago, Lotus, Glycine,* and *Cajanus* genera began 59 million years ago, *Phaseolus radiatus syn Vigna radiata* (Leguminosae), 2n = 2x = 22; 2C DNA content 0.53 pg = 500 Mb *hort* >>> Creole bean >>> pulse

mungo bean: >>> urd bean

muop dang: >>> bitter gourd

muscat plant: Grapevine; the most highly scented variety of *Vitis*; well known as a dessert grape, it yields, in Portugal, the "Moscatel de Setubal," a good sweet white wine, and its flavor is conspicuous in the French "Muscatel de Frontignan"; *Vitis vinifera* (Vitaceae), 2n = 2x, 4x = 38, 76; 1C DNA content 0.43 pg *hort* >>> European grape

mushroom: In general, it is the macroscopic sporing body of a fungus; mushrooms usually have gills, while toadstools have pores; both edible and poisonous mushrooms occur; the cultivation of edible mushrooms is economically important, but breeding of this crop is rather difficult; the cultivated mushroom (*A. bisporus*) or any edible fungus similar to it in appearance; *Agaricus bisporus,* 13 chromosomes, total genome size is 31 Mbp; among 400 wild specimens of *A. bisporus* collected in Europe, only three were tetrasporic *hort*

musk cucumber: >>> cassabanana

muskmelon: >>> Indian melon >>> cantaloupe

musk okra (*syn* **lady's finger** *syn* **West African okra** *syn* **bhindi**): An annual herbaceous shrub; originated in Africa where it has been cultivated for many generations; the fruit, a large green erect pod, is eaten cooked and the seeds are toasted, ground, and used as a substitute for coffee; there are many selections; *Hibiscus esculentus syn Abelmoschus esculentus* (Malvaceae), 2n = 72,130–140 *hort*

mustards: These yellow-flowered members of the >>> cabbage family produce seeds that have been ground and used as a condiment for thousands of years; white, black, or Indian, both seeds and leaves are pungent; black mustard greens (often a weed in the *U.S.)* has medical applications; *Brassica nigra* [2n = 2x = 16 (BB), polytene chromosomes have been observed in ovary, anthers, or immature seed tissues], *Brassica alba* (*Sinapis alba*), and *Brassica juncea,* 2n = 4x = 36; 1C DNA content 1.53 pg (Brassicaceae) *agr* >>> *Brassica* >>> Figure 8

Myrobalan plum: >>> cherry plum

myrrh: This dioecious tree is native to Yemen, Somalia, and eastern Ethiopia; an aromatic oleoresin comes from a number of small, thorny tree species of the genus *Commiphora*, which grow in dry, stony soil; the oleoresin is a natural blend of an essential oil and a resin; myrrh resin is a natural gum; when a tree wound penetrates through the bark and into the sapwood, the tree bleeds a resin; it is waxy, and coagulates quickly; after the harvest, the gum becomes hard and glossy; the gum is yellowish, and may be either clear or opaque; it darkens deeply as it ages, and white streaks emerge; myrrh was used by the ancient Egyptians, along with natron, for the embalming of mummies; it was a part of the "Ketoret", which is used when referring to the consecrated incense described in the Hebrew Bible and Talmud; it was offered on the specialized incense altar in the time when the Tabernacle was located in the First and Second Jerusalem Temples, and it was an important component of the Temple service in Jerusalem; myrrh was traded by camel caravans overland from areas of production in southern Arabia by the Nabataeans to their capital city of Petra, from where it was distributed throughout the Mediterranean region, *Commiphora myrrha syn Balsamodendron myrrha syn Commiphora cispidata syn C. coriacea syn C. molmol syn C. myrrha* var. *molmol syn C. playfairii* var. *benadirensis* (Burseraceae), 2n = 2x = 26; 1C DNA content 0.63 pg *hort*

myrtle: Any plant of the genus *Myrtus*, e. g. *M. communis*, having evergreen leaves, fragrant white flowers, and aromatic berries; (Myrtaceae), 2n = 2x = 22 *bot hort*

N

napier grass: Used as a natural pesticide; the grass lures the corn borer whose gluttonous larva destroys a large portion of the maize cultivation of Africa every year; the moths prefer the grass to the maize, which turns into a fatal trap for them; the grass produces a poisonous sticky substance, which sticks the caterpillars to the grass, where they remain and die; it is planted by farmers around their maize fields, e.g. in Kenya; *Pennisetum purpureum* (Gramineae), 2n = 4x = 28; 1C DNA content 2.90 pg *phyt agr*

narbon vetch: *Vicia narbonensis* (Leguminosae), 2n = 2x = 14; polytene chromosomes have been observed in ovary, anthers, or immature seed tissues; 2C DNA content 16.1 pg *hort*

narcissus: A genus of mainly hardy, mostly spring-flowering, bulbs in the Amaryllis family, subfamily Amaryllidoideae, native to Europe, North Africa, and Asia; there are also several *Narcissus* species that bloom in the autumn; plants includes the daffodil, jonquil, and narcissus itself; in the true narcissus, the petals form a small bowl rather than the long trumpet of the >>> daffodil; *Narcissus pseudonarcissus, N. poeticus* (Iridaceae), 2n = 2x = 14; 1C DNA content 11.75 pg *hort*

narrow-leafed lupine: >>> blue lupine

Natal indigo (*syn* **Bengal indigo** *syn* **Java indigo** *syn* **Indigotier chessé** *syn* **indigotier** *syn* **Indigueiro** *syn* **anileira** *syn* **mnili**): *Indigofera* is a very large genus comprising ~700 species and is distributed throughout the tropics and subtropics of Africa, Asia and the Americas; Africa and the southern Himalayas are richest in species; over 300 species have been recorded for tropical Africa; Natal indigo originates from Africa; it occurs almost throughout tropical Africa, and also in northern and eastern South Africa, Swaziland and southern Arabia; its range has probably been extended by its cultivation for indigo and subsequent naturalization; it is widely planted in India and Southeastern Asia; a variety of plants have provided indigo, a dye, throughout history, but most natural indigo is obtained from those in the genus *Indigofera*, which are native to the tropics; in temperate climates indigo can also be obtained from >>> woad (*Isatis tinctoria*) and >>> dyer's knotweed (*Polygonum tinctorum*), although the *Indigofera* species yield more dye; the word "indigo" comes from the Latin *Indicum* and the Greek *indikon* meaning "blue dye from India" or more literally "Indian substance"; the primary commercial indigo species in Asia was true indigo (*Indigofera tinctoria syn Indigofera sumatrana*); in Central and South America the two species, *Indigofera suffruticosa* (anil) and *Indigofera arrecta* (Natal indigo), were the most important; *Indigofera tinctoria* was domesticated in India; indigo made its way to the Greeks and the Romans via various trade routes, and was valued as a luxury product; it is among the oldest dyes to be used for textile dyeing and printing; many Asian countries, such as India, China, Japan and South East Asian nations have used indigo as a dye (particularly a silk dye) for centuries; the dye was also known to ancient civilizations in Mesopotamia, Egypt, Greece, Rome, Britain, Mesoamerica, Peru, Iran, and Africa; dye was obtained from the processing of the plant's leaves; these were soaked in water and fermented in order to convert the glycoside indican naturally present in the plant to the blue dye indigotin; natural indigo was the only source of the dye until 1897; within a short time, however, synthetic indigo almost completely superseded natural indigo, and today nearly all indigo produced is synthetic; *Indigofera arrecta* (Fabaceae), 2n = 2x = 16 *hort*

naseberry: >>> sapodilla

nasturtium: A native plant of Peru; it was introduced as an annual ornamental plant showing a wide range of flower colors; young and green fruits are sometimes eaten after pickling; in biological horticulture, it is used to prevent insect attacks to fruit trees, when grown around the stem; *Tropaeolum majus* (Tropaeolacae), 2n = 4x = 28; 1C DNA content 1.33 pg; polytene chromosomes have been observed in the ovary, anthers, or immature seed tissues *hort*

navet petit de Berlin: Closely related to >>> May beet and autumn beet; it was mainly grown in some regions of Prussia and around Berlin (Germany); it is a selection from a Polish variety "Piedrowski," showing yellow flesh, although, in 1885, VILMORIN (a French breeder, >>> SCHLEGEL 2007) described it as a white-fleshed type; for some time, the beet was not grown; however, some restaurants started again to include it in the spectrum of old sorts of vegetable; in addition, it had to be reselected from the Polish material since original seeds had been lost; *Brassica campestris* var. *rapa* (Brassicaceae), $2n = 2x = 20$ (AA); 1C DNA content 0.80 pg *hort* >>> beet >>> *Brassica* >>> May beet >>> Figure 8

navy bean: >>> French bean

nectarine: Smooth, shiny-skinned form of >>> peach, usually smaller than other peaches and with firmer flesh; it arose from a natural >>> mutation of the original form; *Persica davidiana syn Prunus persica* (Rosaceae), $2n = 2x = 16$; 1C DNA content 0.28 pg *hort*

neem (*syn* **niem** *syn* **margosa** *syn* **nimtree** *syn* **Indian lilac**): Neem is the Indian (Hindi) name of the tree; its leaves and seeds are known for their insecticidal properties; the seeds are also known for retarding nitrification due to the action of triterpenes present in them which have nitrification inhibiting properties; the first indication that neem was being used as a medical treatment was ~4,500 years ago; it was the high point of the Indian Harappa culture, one of the great civilizations of the ancient world; in these ancient texts, neem is mentioned in almost 100 entries for treating a wide range of diseases and symptoms, most of which continue to vex humanity; neem is a tropical evergreen, native to India and Burma, and grown in Southeast Asia and western Africa; it can live up to 200 years; *Azadirachta indica syn Melia indica* (Meliaceae), $x = 12$; $2n = 2x, 3x, 4x = 24, 36, 48$; 1C DNA content 0.48 (*M.birmanica*) pg *bot*

New Caledonia pine: >>> Cook pine

New Zealand spinach: A popular vegetable in New Zealand, Australia, and the Pacific Islands in the 18th century; it occurs naturally in coastal areas in this region and also in Japan, China, and Taiwan; perhaps it is native to New Zealand; it is eaten cooked as a green leafy vegetable; it can be used in many dishes like amaranth, spinach, or other leafy vegetables with a neutral soft taste; in the U.S., the tender tips are also eaten raw in salads; *Tetragonia tetragonoides* (Tetragoniaceae), $2n = 32$ *hort*

oca: >>> New Zealand yam

ngag (*syn* **ngak**): >>> African walnut

Nigerian walnut: >>> African walnut

niger seed (*syn* **noug**): A member of the Compositae family, it is grown as an oilseed crop in Ethiopia and India; the crop grows best on poorly drained, heavy clay soils; the maturity period of the niger seed accessions ranged from 132 to 168 days; seed oil contents ranged from 39.8 to 46.9%, with linoleic acid being the major fatty acid in the oil (76.6% of total fatty acids); in Ethiopia about 50–60% of the edible oil comes from this crop, and in India its production is estimated to be ~2% of the total annual oilseed production; *Guizotia abyssinica* (Compositae), $2n = 2x = 30$; 1C DNA content 3.85 pg *hort agr*

nira: >>> Chinese chive

noble cane: >>> sugarcane

noni (*syn* **Indian mulberry** *syn* **great morinda,** *syn* **nunaakai** (Tamil Nadu, India) *syn* **dog dumpling** (Barbados) *syn* **mengkudu** (Indonesia and Malaysia) *syn* **kumudu** (Balinese) *syn* **pace** (Javanese) *syn* **beach mulberry** *syn* **cheese fruit**): A tree in the coffee family; native range extends through Southeast Asia and Australasia, and the species is now cultivated throughout the tropics and widely naturalized; the plant bears flowers and fruits all year round; the fruit is a multiple fruit that has a pungent odor when ripening, and is hence also known as "cheese fruit" or even "vomit fruit"; it is oval in shape and reaches 4–7 cm size; at first green, the fruit turns yellow then almost white as it ripens; it contains many seeds; it is sometimes called starvation fruit; despite its strong smell and bitter taste, the fruit is nevertheless eaten as a famine food, and, in some Pacific islands, even a staple food, either raw or cooked; Southeast Asians and Australian Aborigines consume the fruit raw with salt or cook it with curry; the seeds are edible when roasted; the plant reaches maturity in about 18 months and then yields between 4–8 kilograms of fruit every month throughout the year; *Morinda citrifolia* (Rubiaceae), 2n = 4x = 44; 1C DNA content 0.68 pg (*M. tinctoria*) *hort*

nopal: >>> pencas

nori: The genus includes a number of species of intertidal red algae that are collected for food in Asian countries; nori is commonly cultivated in shallow muddy bays of Japan; the dried blades are packaged and sold in Asian markets throughout the world; nori provides the tasty black wrapper around sushi, and is also wrapped around crackers and used in soups; *Porphyra* spp. (Bangiaceae), *P. columbina* (n = 20 gametophytes) showed a basic haploid chromosome number of x = 3 *hort*

Northern wheatgrass: *Agropyron dasystachyum* (Gramineae), 2n = 4x = 28; 2C DNA content 13.9 pg (PP) *bot agr* >>> crested wheatgrass

Norway spruce: *Picea abies* (Pinaceae), 2n = 2x = 24; 1C DNA content 20.01 pg *fore*

notch-seeded buckwheat: *Fagopyrum emarginatum* (Polygonaceae), 2n = 2x = 16 *agr*

noug: >>> niger seed

nunaakai: >>> noni

nutmeg (*syn* **mace**): Kernel of the hard aromatic seed of the evergreen nutmeg tree, native to the Maluku Islands, Indonesia; both the nutmeg and its secondary covering, known as "mace," are used as spices in cookery; *Myristica fragrans* (Myristicaceae), 2n = 6x = 48; 1C DNA content 1.20 pg *hort*

O

oak: The English oak is the most common oak species in northern Europe; its lobed leaves and small acorns are characteristic of the deciduous oaks of cool temperate regions; oaks from warmer regions are usually evergreen and often have unlobed leaves; *Quercus* spp. (Fagaceae), 2n = 2x = 24 *fore* >>> Mongolian oak >>> daimyo oak

oat: The cultivated species is a >>> hexaploid and the first controlled crosses were made by a Scottish farmer, Patrick Sheriff, in 1860 (SCHLEGEL 2007); oats did not become important to man as early as wheat or barley; oat probably persisted as a weed-like plant in other cereals for centuries prior to being cultivated by itself; some scientists believe that our present cultivated oats developed as a >>> mutation from wild oat that may have taken place in Asia Minor or Southeastern Europe; molecular studies using genotyping-by-sequencing technology show that all *Avena* taxa can be assigned to one of four major genetic clusters: cluster 1 = all hexaploids including cultivated oat, cluster 2 = AC genome tetraploids, cluster 3 = C genome diploids, cluster 4 = A genome diploid and tetraploids; no evidence was found for the existence of discrete B or D genomes; it was possible to deduce that hexaploid oat likely formed by the fusion of an ancestral diploid species from cluster 3 (*A. clauda*, *A. eriantha*) with an ancestral diploid species from cluster 4D (*A. longiglumis*, *A. canariensis*, *A. wiestii*) to create the ancestral tetraploid from cluster 2 (*A. magna*, *A. murphyi*, *A. insularis*); subsequently, that ancestral tetraploid fused again with another ancestral diploid from cluster 4D to create hexaploid oat; based on the geographic distribution of these species, it is hypothesized that both the tetraploidization and hexaploidization events may have occurred in the region of Northwest Africa, followed by radiation of hexaploid oat to its current worldwide distribution; latest molecular studies classify the tetraploids *A. insularis*, *A. maroccana* (syn *A. magna*), and *A. murphyi* as containing D-plus-C and not A-plus-C genomes; oldest known oat grains were found in Egypt among remains of the 12th Dynasty, which was ~2,000 BC; these probably were weeds and not actually cultivated by the Egyptians; the oldest known cultivated oats were found in caves in Switzerland that are believed to belong to the Bronze Age; the center of greatest variety of forms is in Asia Minor where almost all subspecies are in contact with each other; oats were first brought to North America with other grains in 1602 and planted on the Elizabeth Islands off the coast of Massachusetts; as early as 1786, George Washington sowed 580 acres to oats; by the 1860s and 1870s, the westward shift of oat acreage in the U.S. had moved into the middle and upper Mississippi Valley, which is its major area of production today; oats are chiefly a European and North American crop; these areas have the cool, moist climate to which oats are best adapted; Russia, Canada, the U.S., Finland, and Poland are the leading oat-producing countries; oats are adapted to a wide range of soil types, thus temperature and moisture conditions are the usual limiting factors as to where oats are grown; some winter oats are produced in the U.S., but most are spring oats produced mainly in the North Central states; oats have been used as livestock and human foods since ancient times; some have been used as pasture, hay, or silage; but most have been used as a feed grain; oat straw has been an important bedding for livestock through history; *Avena sativa* and *Avena* spp. (Gramineae), 2n = 6x = 42 (AAC^{v}C^{v}DD); 2C DNA content 25.2 pg = 11,300 Mb; repeats make up 70% of all the DNA; transposable elements are a major contributor to the genome *bot agr* >>> Tables 14, 15, 16, 30, 32, 35, 48 >>> wild oat

oca (*syn* **oka** *syn* **New Zealand yam**): An annual plant that overwinters as underground stem tubers; the plant was brought into cultivation in the central and southern Andes for its tubers, which are used as a root vegetable; the plant is not known in the wild, but populations of wild *Oxalis* species that bear smaller tubers are known from four areas of the central Andean region; introduced to Europe in 1830 as a competitor to the potato and to New Zealand as early as 1860, it has become popular in that country under the name "New Zealand yam" and is now a common table vegetable, *Oxalis tuberosa* (Oxalidaceae), 2n = 8x = 64; 1C DNA content 1.76 pg *hort*

oil palm (*syn* **dendê oil** *syn* **macaw-fat**): A remarkable crop, producing around 40% of the world's vegetable oil from around 6% of the land devoted to oil crops; conventional breeding has clearly been the major focus of genetic improvement in this crop; a mix of improved agronomy and management, coupled with breeding and selection, have quadrupled the oil yield of the crop since breeding began in earnest in the 1920s; as for all perennial crops with long breeding cycles, oil palm faces immense challenges in the coming years with increased pressure from population growth, climate change and the need to develop environmentally sustainable oil palm plantations; the African palm tree is native to West Africa; this palm has the highest yield of vegetable oil of any crop; African oil palm is the principal source of palm oil; it is native to West and Southwest Africa, occurring between Angola and Gambia; the species name *"guineensis"* refers to one of its countries of origin, Guinea; human use of oil palms may date as far back as 5,000 years in West Africa; in the late 1800s, archaeologists discovered palm oil in a tomb at Abydos dating back to 3,000 BC; it is thought that that Arab traders brought the oil palm to Egypt; the edible vegetable oil derived from the >>> mesocarp (reddish pulp) of the fruit of the oil palms, primarily the African oil palm *Elaeis guineensis*, and, to a lesser extent, from the American oil palm, *E. oleifera*, and the maripa palm, *Attalea maripa*; the oil shows a naturally reddish color because of a high beta-carotene content; it is not to be confused with palm kernel oil derived from the kernel of the same fruit, or coconut oil derived from the kernel of the >>> coconut palm (*Cocos nucifera*); the differences are in color (raw palm kernel oil lacks carotenoids and is not red), and in saturated fat content; palm mesocarp oil is 41% saturated oil, while palm kernel oil and coconut oil are 81% and 86% saturated, respectively; palm oil is a common cooking ingredient in the tropical belt of Africa, Southeast Asia and parts of Brazil: its use in the commercial food industry in other parts of the world is buoyed by its lower cost and by the high oxidative stability of the refined product when used for frying; a recent rise in the use of palm oil in the food industry has come from changed labeling requirements that have caused a switch away from using trans fats; palm oil has been found to be a reasonable replacement for trans fats; palm oil is used as food or processed into margarine, soaps, and livestock feeds; the narrow genetic base of existing commercial oil palm cultivars has prompted oil palm breeders to place increased importance on augmenting these genetic resources because the sustainable development of the crop depends largely on the availability of genetic diversity and its use; molecular data indicate that genetic diversity among the genotypes is mainly distributed within regions, suggesting that there is no isolation by geographical distance; a relatively low number of accessions (~120), that includes at least one representative of each family, would allow to efficiently collect almost the entire genetic diversity of Cameroon; recently, a single gene "shell", was identified that is responsible for determining the thickness of the seeds; this gene determines whether a palm is designated dura, tenera or pisifera, and affects oil yield of a palm tree by 30%; the gene could be used to reduce the incidence of lower-yielding dura palms in commercial plantations from a purported rate of 10–15%, thus boosting oil-per-hectare from the mature palms. *Elaeis guineensis, E. oleifera, Attalea maripa* (Palmae), 2n = 2x = 32; 2C DNA content 2.0–2.4 pg *hort* >>> agroenergy crop >>> Table 16

oil radish: *Raphanus sativus* var. *oleiformis* (Brassicaceae), 2n = 2x = 18 (RR); 1C DNA content 0.55 pg *agr* >>> *Brassica* >>> radish >>> Figure 8

oilseed rape: >>> rapeseed

oka: >>> oca

okra: This is an annual crop grown for its fruits that are cooked and eaten as a green vegetable; there has been considerable hybridization with wild species and there is much genetic variation; it derives from the Old World hibiscus family; its red-and-yellow flowers are followed by long, sticky, green fruits known as >>> "ladies' fingers" or >>> "bhindi"; the fruits are cooked in soups and stews; cultivated okra constitutes a major economic crop in West and Central Africa as a result of its vital importance as a component of various recipes in many cuisines and preparations; it has considerable area under cultivation in Africa and Asia, in particular because of its contribution to the human diet by supplying fats, proteins, carbohydrates, minerals and vitamins; its mucilage is suitable for medicinal and various industrial applications; however, its productivity is very low because of several factors such as lack of adapted genotypes, pest and disease constraints, and the narrow genetic base of existing cultivars; *Abelmoschus esculentus* (Malvaceae), 2n = 4x = 120 (T'T'YY); 1C DNA content 1.65 pg *hort*

old cocoym: >>> taro

oleander: A Mediterranean evergreen shrub found along watercourses, gravelly places, and damp slopes; it is grown widely as an ornamental for its abundant and long-lasting flowering as well as its moderate hardiness; it is used for screens, hedging along highways, planting along beaches, and in urban areas as, by removing suckers and leaving just a few stems, it can also be formed into very attractive small trees; in northern regions it may be grown as an indoor or patio plant; it has flexible branches with green, smooth bark eventually turning to dark grey; the leaves are 5–20 cm long, narrow, acuminated or acute in the apex, shortly petiolate, with a coriaceous dark-green blade; some cultivars have white or yellow variegated leaves; flowers are produced in terminal heads and their colors vary from deep to pale pink, lilac, carmine, purple, salmon, apricot, copper, orange, yellow, and white; each flower is ~5 cm in diameter with five petals, although some cultivars have double flowers; *Nerium oleander* (Apocynaceae), 2n = 2x = 22 *hort*

olibanum (*syn* **shallaki guggal** *syn* **Indian olibanum** *syn* **shallaki** *syn* **Indian frankincense =
B. serrata**): Also known as "frankincense" in its center of production in eastern Africa; the word derives from old French *"franc encens"* (= pure incense); it is an aromatic gum

obtained from trees of the genus *Boswellia* and, when thrown on to glowing charcoal, it produces an aromatic smoke; *Boswellia* spp. (Burseraceae), 2n = 2x, 4x = 22, 44; 1C DNA content 0.70 pg *fore hort*

olive: Native to the Mediterranean region; fresh olives (drupes) are extremely bitter due to oleuropein, a phenolic glucoside; olives are soaked in lye (sodium hydroxide) to remove the bitter oleuropein; olives picked green are oxidized in air to produce a black color; green olives kept submerged will retain green color; unlike most unsaturated plant oils, which come from seeds, monounsaturated olive oil is obtained from the pulp or mesocarp of the fruit; virgin olive oil is obtained from the first pressing; *Olea europaea* (Oleaceae), 2n = 2x = 46; 1C DNA content 1.95 pg *hort*

onion (common ~): The onion belongs to the family Amaryllidaceae, containing over 300 species, of which 70 have been cultivated for 4,700 years or more; the onion is one of the major vegetable crops in the world; onions have been used by man for several centuries, e.g. during the ancient civilizations of Egypt, Rome, Greece, and China; onions are widely spread throughout the temperate northern hemisphere of the world; >500 of species are common in the Old World; there are >80 species found in the New World; chives is the only species found in both the Old and the New World; in 2012, onions were grown in 170 countries with global production of 87 million tons; onions contain outstanding levels of polyphenols, vitamins, and sulfur-containing compounds, the latter being responsible for their pungency; those compounds also affect various aspects of human health, including support for bone and connective tissues, anti-inflammatory effects, diabetes prevention, digestive tract health, and anticancer protection; hence, many traits were developed or are under development for onion breeding, including bulb shape, bulb color, bulb size, flowering time, pungency, nutritional value, and disease resistance; onions are recognized by the pungent smell or taste of "onion," which they produce when their tissues are crushed or tasted; they are biennial or perennial bulbous herbs; the bulbs are formed by the swollen leaf bases attached to the base of the underground part of the stem; bolting or premature flowering before bulb maturation, is an undesirable trait strongly selected against by breeders; a quantitative trait locus (QTL) on chromosome 1, which is designated *AcBlt1*, consistently conditions bolting susceptibility; a chromosome 3 region, which coincides with a functionally characterized acid invertase, is not associated with bolting in other environments, but shows significant association with bulb sucrose content; in a few species there are very long leaf sheath bases, which are much less swollen, and others have rhizomes or storage roots; *Allium cepa* (Alliaceae), 2n = 2x = 16; 2C DNA content = 33.5 pg; 54,165 protein-coding genes were detected among 165,179 assembled transcripts from 203.0 Mb nucleotides *hort* >>> chives

opium poppy: >>> poppy

opuntia: Fruits are eaten, leaves are used as fodder for animals; within the sub-tribe Opuntioideae, there are several species used as crops and horticultural plants; there is a special use for production of the stain "carmine red" from the >>> ectoparasite >>> "cochenille" (*Dactylopius coccus*); *Opuntia* spp. (Opuntioideae), 2n = 2x, 3x, 6x = 22, 33, 66 *agr hort*

orache (*syn* **garden orache**): A former vegetable completely substituted for by >>> spinach, except in a few southern regions of Germany; a common wild plant, although a native of western Asia; there are genotypes showing light green and red leaves; it can grow up to 2 m height; it can become a weed when seed production is not controlled; *Atriplex hortensis* (Chenopodiaceae), 2n = 2x = 18; 1C DNA content 1.18 pg *hort* >>> mountain spinach

orange (~ tree): Round orange-colored juicy citrus fruit of several species of evergreen trees, which bear white blossoms and fruits at the same time; thought to have originated in Southeast Asia, orange trees are commercially cultivated in Spain, Israel, the U.S., Brazil, South Africa, etc.; the sweet orange (*Citrus sinensis*, 2n = 2x, 3x, 4x, 5x = 18, 27, 36, 45) is the one commonly eaten fresh; the Jaffa, blood, and navel oranges are varieties of this species; >>> tangerines and >>> mandarins belong to a related species (*C. reticulata*); the sour or Seville orange (*C. aurantium*, 2n = 2x = 18) is the bitter orange used in making marmalade; *Citrus* spp. (Rutaceae), 2n = 2x = 18; 2C DNA content 1.2 pg *hort* >>> hesperidia >>> Figure 61 >>> http://users.kymp.net/citruspages/citrons.html

orchard grass: *Dactylis glomerata* (Gramineae), 2n = 2x, 4x = 14, 28; 2C DNA content 8.7, 12.4 pg *agr*

oregano: *Origanum vulgare* (Labiatae), 2n = 3x = 30; 2C DNA content 1.4 pg *hort*

Orient(al) wheat: A spring wheat, early to reach >>>maturity with narrow, pubescent leaves; spikes are long, loose, and almost square in cross-section; awns are long and often black; spikelets produce two or three kernels which are long, narrow, white, and hard; it is grown in the Mediterranean area and the Near East; *Triticum turanicum* (Gramineae), 2n = 4x = 28 (BBA^{u}A^u); 1C DNA content 12.28 pg >>> wheat

oromo dinich (wolaita-dinich *syn* Ethiopian potato): A crop indigenous to Ethiopia and has been cultivated for several years as a food crop; over the past few years, the crop has been declining in production; it is a perennial plant and frost tender; The flowers are hermaphrodite (having both male and female organs); *Plectranthus edulis syn Coleus edulis* (Lamiaceae), 2n = 64, 84 *agr*

oryzopsis: *Oryzopsis miliacea* (Gramineae), 2n = 2x = 24 *bot hort*

osier: >>> basket willow

osteospermum: >>> African daisy

ox-eye chamomile: *Anthemis tinctoria* (Asteraceae), 2n = 2x = 18; 1C DNA content 3.75 pg *hort*

oyster cap fungus (*syn* white rot fungus): An edible basidiomycete of increasing agricultural and biotechnological importance; *Pleurotus ostreatus* (Basidiomycetes), x = 11 *hort*

P

pace: >>> noni

padwal: >>> serpent gourd

paeonia: Perennial plants native to Europe, Asia, and North America, remarkable for their large, round, brilliant white, pink, or red flowers; most popular in gardens are the common peony (*Paeoni officinalis*), the white peony (*P. lactiflora*), and the taller tree peony (*P. suffruticosa*); the genus includes ~35 widely distributed species and constitutes an independent family; it includes three sections: Moutan, Oneapia, and Paeonia; the first one is a woody group and endemic to China; the second and third ones are herbaceous; intersectional hybridization between Moutan and Paeonia formed a new type of group originating from Japan and developed in America; they have been called "Itoh Hybrids" commemorating the breeders; the phenotype of these intersectional hybrids combines the characters of herbaceous and >>> tree peonies; it provided more ornamental values and met the requirements of customers; therefore the breeding of intersectional hybrids has become the most promising field for traditional breeding; though the germplasm resources of peonies are abundant in China and they are playing an important role in breeding of intersectional hybrids, it is difficult to breed intersectional hybrids due to different genetic backgrounds and incompatibility between two sections; however, crossing is feasible when the proper parents from different sections are selected; *Paeonia* spp. (Paeoniaceae), 2n = 2x = 10 *hort* >>> Table 46 *hort*

paico: >>> epazote

pak choi: *Brassica campestris* ssp. *chinensis* (Brassicaceae), 2n = 2x = 20 (AA); 1C DNA content 0.58 pg >>> *Brassica* >>> Figure 8 >>> cabbage

palisadegrass: A member of a pantropical grass genus containing ~100 species, mainly from the African continent, is found in a wide range of habitats from semi-desert to swamps; while grasses from this genus have been continuously exploited by local pastoralists for millennia, the interest in species of this genus as sown and managed forage only began in the 1960s; this occurred first on a limited scale in humid, coastal, tropical Australia, followed by tropical South Africa, and later in Brazil in the early 1970s; currently, the genus *Brachiaria* is the most widely used forage grass in the South American savannas due to its physiological tolerance to low-fertility acid soils of the tropics; *Brachiaria brizantha* (palisadegrass), *B. decumbens* (signalgrass), *B. humidicola* (koroniviagrass), and >>> *B. ruziziensis* (ruzigrass) are the most commercially exploited brachiariagrasses; their economic importance is greatest in tropical America, where extensive adoption over the past three decades has had a revolutionary impact on the productivity of vast areas of previously underused, marginal soils; *Brachiaria* alone accounts for at least 85% of the cultivated pastures in Brazil, covering over 50 million ha and sustaining the largest commercial cattle herd in the world, of ~205 million head; two cultivars, *B. decumbens* convar. 'Basilisk' and *B. brizantha* convar. 'Marandu' are undoubtedly the most widely grown species, not only in the Brazilian savannas but throughout the tropics; the rapid expansion of their acreage did not occur without problems; both cultivars have significant limitations; the first one lacks resistance to a ubiquitous family of sucking insects, the spittlebugs (*Homoptera: Cercopidae*); the second, while resistant, requires higher soil fertility and does not tolerate waterlogged soils; Brazilian pastures are severely degraded because of inadequate fertilization and mismanagement; renovation of pastures and intensification of production practices demand new cultivars; eight cultivars are presently commercialized by a dynamic seed industry, dominated by Brazilian companies, and seven of these cultivars are direct selections from naturally occurring germplasm collected in Africa; all the cultivars are polyploid (2n = 4x = 36) and apomictic, which held back the initiation of brachiariagrass breeding programs until suitable sexual germplasm was developed in the mid-1980s; to increase genetic variability in the genus in hopes of generating new cultivars for pasture diversification, an extensive program, based on intra- and interspecies hybridization, was undertaken at the EMBRAPA Beef Cattle Center in 1988, with the objective of determining the inheritance of >>> apomixis and thus manipulating this character for the development of new improved hybrids; at first, hybridizations were between sexual *B. ruziziensis* and apomictic *B. brizantha* (2n = 2x, 4x, 6x = 18, 36, 54) or *B. decumbens*; a great number of hybrids was obtained and some are under agronomic evaluations; some interesting sexual hybrids were selected to be crossed to some superior ecotypes of the paternal species; the *B. ruziziensis/B. decumbens/B. brizantha* complex provides a wealth of genetic variation for the introgression of derived genes of interest, such as for spittlebug resistance and nutritive value, among others; *Brachiaria brizantha* (Poaceae), 2n = 4x = 36 *agr*

Palmyra palm: >>> borassus

panigrahi: >>> pongamia

pansy: There are over 400 species of pansy, or >>> violet, widely distributed in temperate regions; the flowers have five unequally shaped petals; *Viola wittrockiana* ssp. *hiemalis* (Violaceae), 2n = 8x = 48 *hort*

papaya: Tropical evergreen tree, native from Florida to South America; the edible fruits are like melons, with orange-colored flesh and large numbers of blackish seeds in the center; they can weigh up to 9 kg; the fruit juice and the tree sap contain papain, an enzyme used to tenderize meat and help digestion; the name "pawpaw" is also used for this tree, but in the U.S., the pawpaw is the tree *Asimina triloba*, of the custard-apple family; transgenic papaya cultivars carrying the coat-protein gene provide effective protection against *papaya ring spot virus*; the transgenic "Honey Sweet" plum cultivar provides an interesting germplasm source for *Plum pox virus* control; *Carica papaya* (Caricaceae), 2n = 2x = 18; 2C DNA content 0.8 pg *hort*

paprika: Non-pungent bell pepper lacks the cytoplasmic male-sterility (CMS) nuclear restorer allele, *Rf*, and CMS cannot be employed in its F_1 hybrid seed production; to use that, the genic male-sterility (GMS) system in non-pungent bell pepper can be converted to the CMS male-sterility system, the conversion of GMS to CMS for non-pungent bell pepper line GC3 is conducted by introgression of S-type cytoplasm and the *Rf* allele from tropical pungent donors, *Capsicum annuum* (Solanaceae), 2n = 2x = 24; 1C DNA content 3.16 pg *hort*

paradise nut: A member of the Brazil nut family; a giant rainforest tree with seeds produced in a thick, woody, pot-like capsule; *Lecythis ollaria* (Lecythidaceae), 2n = 2x = 34 *fore hort*

Para nut: >>> Brazil nut

pare: >>> bitter gourd

park red fescue: *Festuca rubra* ssp. *genuina* (Gramineae), 2n = 2x = 14; 2C DNA content 13.9 pg *agr*

parsley: The hardy biennial parsley has been cultivated for so long that its origin is uncertain; it may have originated in Southeast Europe; the mild-flavored leaves and seeds are widely used as a herb in cooking; *Petroselinum crispum syn P. sativum* (Umbelliferae), 2n = 2x = 22; 2C DNA content 4.0–5.3 pg *hort*

parsnip: A native plant of Europe; after widespread cropping in the past, it became a rarely grown vegetable; well-known varieties are 'Sutton's Student', 'White Gem', and an F₁ hybrid 'Gladiator'; during the end of the annual season, a long whitish root is developed; it tastes similar to celery; *Pastinaca sativa* (Umbelliferae), 2n = 2x = 22; 2C DNA content 3.4 pg *hort*

paspalum: >>> brownseed paspalum

passionfruit (*syn* **granadilla** *syn* **maypops** *syn* **apricot-vine** *syn* **passiflore rouge** *syn* **Passionsblume** *syn* **purple passion-flower**): Native to North America, it is a perennial vine found growing in sandy thickets and open fields, roadsides, fence rows, and waste places; it is easily cultivated through root division or by seed; transplants from the wild do well; it requires a well-drained, sandy, slightly acid soil in full sun; a trellis should be provided, since it is a tendril climbing vine; it has many beautiful large and aromatic flowers; it grows very quickly and produces edible fruit and has medicinal uses; it has large, three-lobed, serrated leaves, with beautifully intricate purple and white sweet-scented flowers that are 5–6 cm across; flowers bloom from June to August; the fruit, when ripe, is yellow-green and the size of a small hen's egg; the yellow pulp is sweet and edible; the herb is gathered above ground; after the fruit has matured, it can be dried for later use; the edible fresh, juicy fruit is gathered when soft and light yellow-green; *Passiflora edulis, P. incarnate, P. ligularis* (Passifloraceae), 2n = 2x = 18; 2C DNA content 3.2 pg *hort*

pattypan squash: *Cucurbita pepo* convar. *pattisonina* (Cucurbitaceae), 2n = 2x = 40; 1C DNA content 0.55 pg *hort*

pawpaw (**paw paw** *syn* **paw-paw** *syn* **common pawpaw**): A species of the pawpaw genus in the same plant family as the custard-apple, cherimoya, sweetsop, ylang-ylang, and >>> soursop plants; the pawpaw is native to the eastern U.S. and the adjacent southernmost Ontario, Canada, from New York west to eastern Nebraska, and south to northern Florida and eastern Texas; it is a patch-forming (clonal) understory tree found in well-drained, deep, fertile bottom-land and hilly upland habitat, with large, simple leaves and large fruits, the largest edible fruit indigenous to North America; the large shrub or small tree growing to a height of 11 m (rarely to 14 m) with a trunk 20–30 cm or more in diameter; the large leaves of pawpaw trees are clustered symmetrically at the ends of the branches, giving a distinctive imbricated appearance to the tree's foliage; flowers are perfect, about one to two inches across, rich red-purple or maroon when mature, with three sepals and six petals; they are borne singly on stout, hairy, axillary peduncles, the flowers are produced in early spring at the same time as or slightly before the new leaves appear, and have a faint fetid or yeasty smell; the fruit of the pawpaw is a large, yellowish-green to brown berry, 5–16 cm long and 3–7 cm broad, weighing 20–500 g, containing several brown seeds 15-25 mm in diameter embedded in the soft, edible fruit pulp, the conspicuous fruits beginning development after the plants flower; they are initially green, maturing by September or October to yellow or brown; when mature, the heavy fruits bend the weak branches down; fruits, ripe after the first frost, have long been a favorite treat throughout the tree's extensive native range in eastern North America, and on occasion have been sold locally at farmers' markets; they have a sweet flavor somewhat similar to >>> banana, >>> mango, and >>> cantaloupe, varying significantly by source or cultivar, with more protein than most fruits; in recent years, cultivation of pawpaws for fruit production has attracted renewed interest, particularly among organic growers, as a native fruit with few to no pests, can be successfully grown without pesticides; the commercial cultivation and harvesting of pawpaws is strong in Southeastern Ohio, and is also being explored in Kentucky and Maryland, as well as various areas outside the species' native range, including California, the Pacific Northwest, and Europe; *Asimina triloba* (Annonaceae), 2n = 2x = 18; 1C DNA content 0.80 pg *hort*

pea: >>> garden pea

peach: Yellow-reddish round edible fruit of the peach tree, which is cultivated for its fruit in temperate regions and has oval leaves and small, usually pink, flowers; the fruits have thick velvety skins; *Persica davidiana syn Prunus persica* (Rosaceae), 2n = 2x = 16; 1C DNA content 0.28 pg *hort* >>> nectarine

peach palm: Common names are peyibay(e), pejivalle (Spanish), pejibaye (Costa Rica, Nicaragua), chantaduro (Colombia, Ecuador), pijuayo (Peru), pijiguao (Venezuela), tembé (Bolivia), pibá (Panama), cachipay (Colombia), pupunha (Portuguese); a cultivated fruit tree and "heart of palm", it is an important component of agroforestry systems in the Peruvian Amazon; in general, it was the most important palm of pre-Columbian America and constituted the main crop of the Amerindians covering an extensive area of the humid tropics and even some areas of the dry tropics; because organic material easily decomposes in the archaeological sites of the humid tropics, there are few references to findings of peach palm material which could enable its past to be reconstructed; the oldest come from seeds found in various localities on the two coasts of Costa Rica and date from 2,300 to 1,700 BC, when it is assumed that it was already cultivated; when contact with Europeans took place, accounts indicate that it was the main crop and sustenance of the indigenous population of the humid tropics of Costa Rica; the importance of the peach palm also extended to numerous tribes of lower Central America and the humid tropics of South America, scattered across the basins of the Cauca, Magdalena, San Juan, Orinoco and Amazon rivers and their tributaries as well as certain other areas; *Bactris gasipaes* (Palmae), 2n = 2x = 28; 1C DNA content ~4.08 pg *fore bot* >>> http://www.hort.purdue.edu/newcrop/1492/peach-palm.html

peanut (*syn* **groundnut**): A four-foliate legume with yellow sessile flowers and subterranean fruits; it is native to South America, and it originated between southern Bolivia and northern Argentina, from where it spread throughout the New World as Spanish explorers discovered its versatility; at present, farmers in Asia and Africa also cultivate it under a wide range of environmental conditions in areas between 40° South and 40° North of the equator; the largest producers of groundnut are China and India, followed by sub-Saharan African countries and Central and South America; most of the crop is produced where average rainfall is 600 to 1,200 mm and mean daily temperatures are more than 20°C; it became a major oilseed crop of the tropics and subtropics; the seeds are rich in protein and oil; the genus *Arachis* is widely distributed in the north and central regions of South America; a special variety is the 'Spanish peanut', the kernels of which are small- to medium-sized with smooth skin, and the kernel color ranges from a pale pinkish buff to a light brown during storage; this type of peanut is used predominantly in peanut candy, although significant quantities are also used for salted peanuts and peanut butter; they have a higher oil content than other types of peanuts; the so-called 'runner peanut' is the most widely used peanut for making peanut butter, peanut candies, baked goods, and snack nuts; *Arachis hypogaea* (Leguminosae), 2n = 4x = 40 (AABB); 1C DNA content 2.87 pg; the A and B genomes are of similar size and are composed mostly of metacentric chromosomes; the A genome is characterized by a pair of small chromosomes and the presence of strong centromeric heterochromatic bands; in contrast, B chromosomes are all of similar size and have much weaker centromeric bands; the genome of peanut is estimated at about 2.8 Gbp and with a high repetitive DNA content; polytene chromosomes have been observed in ovary, anthers, or immature seed tissues; two diploid wild species (*A. duranensis* and *A. stenosperma*) show AA genome *agr* >>> agroenergy crop >>> Kersting's groundnut >>> gynophore >>> Table 16

pear: Gritty-textured edible fruit of the pear tree, native to temperate regions of Europe and Asia; white flowers precede the fruits, which have a greenish-yellow and brown skin and taper towards the stalk; pear trees are cultivated for their fruit, which are eaten fresh or canned; a wine known as "perry" is made from pear juice; pears and apples are antique fruits, and both HOMER and PLINY the Elder recorded the names of ancient cultivars of peach (>>> SCHLEGEL 2007); *Pyrus communis* (Rosaceae), 2n = 2x, 3x = 34, 51; 1C DNA content 0.55 pg *hort*

pearl millet: Also known as bulrush millet, spiked millet, and cattail millet, and as bajra in India; this is an ancient crop and the most important of all the millets; it originated in Africa but was taken to India at an early date; its value lies in its tolerance of poor soils and low rainfall; the plant is open-pollinated and exhibits extreme variation; hybrid varieties were highly successful in India until the breakdown of vertical resistance to downy mildew (*Sclerospora graminicola*); *Pennisetum americanum* and *P. glaucum syn P. typhoides* (Gramineae), 2n = 2x = 14; 1C DNA content 2.40 pg >>> millet >>> haploidization

peat moss: Every year, 30 million cubic meters of peat are consumed worldwide, nine million of them in Germany alone; peat extraction and drainage by agriculture have meant that, in this country, only about five percent of the mire landscapes are intact; researchers from Germany try to tackle this problem through targeted (smart) breeding of peat moss; *Sphagnum* is a genus of approximately 380 accepted species of mosses, commonly known as peat moss; one can resettle it on degraded moors, i.e. by sowing multiplied clonal fragments; in axenic culture medium, breeding is attempted to increase productivity by polyploidization and protoplast transfer; the regrown peat moss can then be harvested after a few years and used as peat substitute in horticulture; such use of wet moorland can increase biodiversity, reduce carbon dioxide emissions, and provide a source of income for agriculture worldwide; *Sphagnum* spp., *Sphagnum flexuosum* (Sphagnaceae), chromosomes n = 19 + micro chromosomes or 2n = 38, mean DNA content 0.45 pg *agr eco*

pecan (~ nut): Nut-producing >>> hickory tree (*Carya illinoensis* or *C. pecan*), native to the Central U.S. and northern Mexico, and now widely cultivated; the trees grow to over 45 m, and the edible nuts are smooth-shelled, the kernel resembling a smooth, oval walnut; *Carya laciniosa syn C. pecan* (Juglandaceae), 2n = 2x = 32; 1C DNA content 0.83 pg *hort*

pedunculate oak: *Quercus robur* (Fagaceae), 2n = 2x = 24 *fore* >>> oak

pencas: Edible fleshy twigs and fruits; cropping areas: ~50,000 ha (Mexico), ~ 60,000 ha (Tunisia), ~100,000 ha (Italy); *Opuntia maxima syn Ficus indica syn Opuntia ficus-indica* (Cactaceae), 2n = 4x, 8x = 44, 88 *hort* >>> opuntia

pelargonium: Shrubby, tender flowering plants belonging to the geranium family, grown extensively for their colorful white, pink, scarlet, and black-purple flowers; they are the familiar summer bedding and pot "geraniums"; ancestors of the garden hybrids came from southern Africa; *Pelargonium graveolens* and other species (Geraniaceae), 2n = 2x = 16 *hort*

pensacola: >>> Bahia grass

peony: >>> paeonia

pepino (*syn* **sweet pepino**): A species of evergreen shrub native to South America and grown for its sweet edible fruit; the fruit resembles a >>> melon (*Cucumis melo*) in color, and its flavor recalls a succulent mixture of honeydew and cucumber, and thus it is also sometimes called pepino melon or melon pear, but pepinos are only very distantly related to melons and pears; it is a domesticated native of the Andes; the fruit is common in markets in Colombia, Ecuador, Bolivia, Peru, and Chile, but less often overseas, because it is quite sensitive to handling and does not travel well; attempts to produce commercial cultivars and to export the fruit have been made in New Zealand, Turkey and Chile, *Solanum muricatum* (Solanaceae), 2n = 2x = 24 *hort*

pepper: >>> sweet pepper

peppermint: Perennial herb of the mint family, native to Europe, with oval aromatic leaves and purple flowers; oil of peppermint is used in medicine and confectionery; *Mentha piperita* (Labiatae), 2n = 2x = 36; 2C DNA content 0.7 pg *hort*

perennial ryegrass: >>> English ryegrass

perennial sweet leek: >>> garden leek

peria: >>> bitter gourd

perilla: A common annual weed of the eastern U.S. but considered a commercial crop in Asia; it is a member of the mint family and has the characteristic square stems and four stamens of most species in that family; perilla food products are available in Korean ethnic markets, and red-leafed cultivated plants are used in landscaping; the seeds of perilla contain 31–51% of a drying oil similar to tung or linseed oil; all drying oils leave a hard protective surface when dry; perilla oil has been used as a drying oil in paints, varnishes, linoleum, printing ink, lacquers, and for protective waterproof coatings on cloth; it has also been used for cooking and fuel; *Perilla frutescens* (Lamiaceae), 2n = 2x = 38 plus ±2 B chromosomes *hort*

periwinkle: Periwinkle has become one of the very extensively investigated medicinal plants after the discovery of two powerful anticancer alkaloids, vinblastine (VLB), vincristine (VCR), and an antihypertension alkaloid, ajmalicine, in its leaves more than 50 years ago; the VCR and VLB alkaloids occur in very low concentrations of 20 mg/t and 1 g/t of plant material; they were discovered in leaves of this plant ~40 years ago and are still considered to be among the most exciting anticancer chemotherapeutic agents currently available for clinical use; biosynthesis of *these* alkaloids is a complex process involving many genes, enzymes, regulators, inter- and intra-cellular transporters, cell types, organelles, and tissues, and our current understanding of the process is still considered to be incomplete to produce alkaloids through metabolic engineering/synthetic biology; until such time, breeding periwinkle varieties with higher concentrations of anticancer alkaloids for cultivation will be an alternate approach to meet the demand for these alkaloids and reduce their costs; *Catharanthus roseus* (Apocynaceae), 2n = 2x = 16; 1C DNA content 2.43 pg *hort*

Persian clover: *Trifolium resupinatum* (Leguminosae), 2n = 2x = 16; 1C DNA content 0.52 pg *agr* >>> clover

Persian wheat: Of spring habit, early maturing, and somewhat resistant to fungal diseases; it has strongly yellow to light red stems; spikes are flexible, tending to lean over; while several flowers are present in each spikelet, only three usually develop into kernels; they are free-threshing, flinty, generally red; it grows in the Eastern Mediterranean Area, including southern Russia; *Triticum carthlicum syn T. persicum* (Gramineae), 2n = 4x = 28 (BBA^{u}A^u); 1C DNA content 12.28 pg *bot agr* >>> wheat

Persimmon (*syn* **Chinese persimmon** *syn* **Japanese persimmon** *syn* **kaki persimmon** *syn* **Oriental persimmon** *syn* **kaki**): Name is given to the usually edible fruits of some species of the genus *Diospyros*; this genus includes various shrubs and trees native to the tropics and subtropical areas of the world; persimmons are cultivated widely in Japan and China as ornamentals and for their fruits, some of which also produce a useful dye; kaki trees typically do not bear until they are 3 to 6 years old; the 2–2.5 cm wide flowers appear in the spring; female flowers have a creamy yellow color and tend to grow singly, whereas male flowers have a pink tint and tend to appear in threes; on occasion, bisexual flowers occur; the kaki is among the oldest plants in cultivation, known for its use in China for more than 2,000 years; identifying sexuality at the seedling stage of persimmon and the elimination of male progeny has been regarded as an important strategy for enhancing breeding efficiency; there is a male-linked marker "*OGI*"; the *OGI* marker in agreement with the sex phenotype is obtained in 85 plants (89.5%); the *OGI* locus is used to distinguish male from female persimmon plants at an early stage; *Diospyros kaki* (Ebenaceae), 2n = 6x = 90; 1C DNA content 2.54 pg *hort*

pe-tsai: >>> Chinese cabbage

petunia: >>> common petunia

peyote: Spineless cactus of northern Mexico and the Southwestern U.S.; it has white or pink flowers; its buttonlike tops contain mescaline, which causes hallucinations and is used by American Indians in religious ceremonies; *Lophophora williamsii* (Cactaceae), 2n = 2x = 22 *hort* >>> Table 46 >>> http://www.erowid.org/plants/peyote/peyote_info1.shtml#index-4

phacelia (*syn* **California bluebell**): Named from the Greek for cluster; refers to clustering of flowers; the best known species is *Phacelia campanularia*, the harebell phacelia, which bears gentian-blue, bell-shaped flowers with contrasting white stamens on one-sided curved racemes; the plants grow 25 cm tall and are somewhat hairy; they remain in bloom for a long time, the flowers being produced in one-sided curving clusters; *P. whitlavia*, the "bluebell phacelia," grows 40 cm tall, differing from the above in having a swollen corolla tube three times as long as the lobes; in horticulture, phacelias are excellent edging plants for blue effects and admirably suited to the rockery; in agriculture, phacelia is used for green manure and bee feeding; *Phacelia* spp. (Hydrophyllaceae), 2n = 22 *bot agr hort* >>> tancy phacelia

Philippine sisal: >>> cantala

physic nut: >>> Barbados nut

picotee: >>> carnation

pigeonpea (*syn* **red gram**): A multipurpose food legume, serving as a lifeline to resource-poor farmers in tropical and subtropical regions of Asia, Africa, and Latin America; globally, pigeonpea is cultivated on an area of 6.97 Mha, with a total production and yield/ha of 5.05 Mt and 724 kg/ha, respectively (FAO STAT, 2016); the generic name *Cajanus* derived from the word "Katjang" or "Catjang" from the Malay language, meaning "pod" or "bean"; Africa is regarded as the place of origin; this tropical pulse is also called red gram, Congo pea, and no-eye pea; the crop is self-pollinating with ~20% of out-crossing, usually by bees and other insects; for controlled hybridization, the flowers must be emasculated before 9 in the morning on the day before the flower opens; they may be hand-pollinated at the time of emasculation; pigeonpeas have a wide ecological adaptability, but they do poorly in the wet tropics and they cannot tolerate frost; most cultivars are short-day plants; it is commercially important in India; varieties are classified as tree type, tall varieties, or dwarf; new hybrids are similar in height to southern peas and beans; pigeonpeas must be grown as an annual in most parts of the U.S., since plants are killed by freezing temperatures; in 2013, the International Crops Research Institute for the Semi-Arid Tropics (ICRISAT) developed the world's first cytoplasmic-nuclear male sterility (CMS)-based commercial F_1 hybrid; the CMS, in combination with natural outcrossing of the crop, was used to develop viable hybrid breeding technology; hybrid 'ICPH 2671' recorded 47% superiority for grain yield; in on-farm trials conducted in five Indian states, mean yield of this hybrid (1,396 kg/ha) was 46.5% greater than that of the popular cv. 'Maruti' (953 kg/ha); the hybrid 'ICPH 2671' also exhibited high levels of resistance to *Fusarium* wilt and sterility mosaic disease; hybrid breeding technology and high yield advantages realized in farmers' fields have given hope for a breakthrough in pigeonpea productivity, *Cajanus cajan* (Leguminosae), 2n = 2x = 22; 1C DNA content 0.86 pg; genome analysis predicted 48,680 genes and a length of 833.07 Mbp *hort* >>> pulse

pignon d'inde: >>> Barbados nut

pimento: Evergreen trees belonging to the >>> myrtle family, found in tropical parts of the New World; the fruits of this Central American, functionally dioecious tree were mistaken for >>> black pepper by Christopher Columbus; the dried berries of the species *Pimenta dioica* are used as a spice; *Pimenta dioica* (Myrtaceae), 2n = 2x = 22; 1C DNA content 0.28 pg *hort*

pimpernel: Plants belonging to the primrose family, comprising ~30 species, mostly native to Western Europe; the European scarlet pimpernel (*Anagallis arvensis*) grows in cornfields, the small star-shaped flowers opening only in full sunshine: *Anagallis* spp., *Pimpinella* spp. (Umbelliferae), *P. Major* – 2n = 2x = 18; 2C DNA content 6.4 pg; *P. saxifraga* – 2n = 4x = 36; 2C DNA content 10.3 pg *hort*

pincushion flower: >>> scabious

pine(s): Very important lumber trees, e.g. >>> Eastern white pine (*P. strobus*), lodgepole pine (*P. contorta*), and >>> Ponderosa pine (*P. ponderosa*); raw turpentines are oleoresins (liquid resins containing essential oils) exuded as pitch; "spirits" of turpentine come from distilled pitch; rosin is left after the volatile "spirits of turpentine" are removed; most raw turpentine is from longleaf pine (*P. palustris*), loblolly pine (*P. taeda*), and slash pine (*P. elliottii*); slash pine is also used in the pulpwood industry for making paper; European sources of turpentines include cluster pine (*P. pinaster*) and Scots pine (*P. sylvestris*); *Pinus* spp. (Pinaceae), 2n = 2x = 24 *fore* >>> agroenergy crop

pineapple (*syn* **ananas** *syn* **pina**): Native to the American tropics, southern Brazil, and Paraguay; the cultivated pineapples are grown mainly between latitudes 24° North and 25° South, principally at lower altitudes, in many countries where climatic conditions are favorable; Christopher COLUMBUS and his shipmates saw the pineapple for the first time on the island of Guadeloupe in 1493 and then again in Panama in 1502; Caribbean Indians placed pineapples or pineapple crowns outside the entrances to their dwellings as symbols of friendship and hospitality; Europeans adopted the motif and the fruit was represented in carvings over doorways in Spain, England, and later in New England for many years; the plant was grown in China in 1594 and in South Africa about 1655; it reached Europe in 1650 and fruits were being produced in Holland in 1686, but trials in England were not successful until 1712; pineapple is cultivated for fruit, which is used fresh, canned, frozen, or made into juices, syrups, or candied; pineapple bran, the residue after juicing, is high in vitamin A, and is used in livestock feed; from the juice, citric acid may be extracted, or on fermentation, alcohol; commercial bromelain is generally prepared from pineapple wastes; a mixture of several proteases, bromelain is used in meat tenderizers, for chill-proofing beer, in manufacturing precooked cereals, in certain cosmetics, and in preparations to treat edema and inflammation; bromelain is also nematicidal; pineapples are tolerant of a wide range of soils, providing they possess good drainage, soil aeration, and a low percentage of lime; sandy loam, mildly acid soils of medium fertility are best; *Ananas comosus* (Bromeliaceae), 2n = 2x = 50; 1C DNA content 0.55 pg *hort*

pink: >>> carnation

pinto bean: >>> tepary bean

Pinyon pine: *Pinus edulis syn Caryopitys edulis* (Pinaceae), 2n = 2x = 24; 1C DNA content 32.93 pg *fore*

pistachio (~ **nut**): Deciduous tree of the cashew family, native to Europe and Asia, whose green nuts are eaten salted or used to enhance and flavor food; *Pistacia vera, Pistacia mutica* (Anacardiaceae), 2n = 2x = 30; 1C DNA content 0.60 pg *hort*

pita: >>> Henequen agave

pitajaya (*syn* **pitahaya, pitaya, organ pipe cactus, strawberry pear,** or **dragon fruit**): A climbing or epiphytic tropical American cactus with angular stems and mostly white, very fragrant flowers; the sprawling cactus has stems up to 2 m long; the edible, colored fruits are ~10 cm long; they have a white spongy pulp with small black seeds and thick skin; they are considered to be the best-tasting cactus fruits; the plant may grow out of and over the ground or climb onto trees using aerial roots; it grows best in dry, tropical or subtropical climates; in wet, tropical zones, plants may grow well but sometimes have problems setting fruit reliably; it tolerates temperatures up to 40°C, and short periods of frost (> -1°C), but prolonged cold will damage or kill the plant; it is recommended to grow it in soil that is supplemented with large amounts of organic matter; the plant has been grown successfully in sandy soils; shade is sometimes provided in hot climates; flowers are ornate and beautiful, and many related species are propagated as ornamentals; they bloom only at night, and usually last just one night, where pollination is necessary to set fruit; in full production, pitahaya plants can have up to 4–6 fruiting cycles per year; plants can be propagated by seed or by stem cuttings; recently it has been successfully cropped in Israel; its origin is Mexico and Nicaragua to Ecuador (Central America); *Hylocereus undatus, Hylocereus polyrhizus syn Cereus polyrhizus* (Cactaceae), 2n = 2x = 22 >>> http://www.amjbot. org/ cgi/ content/full/87/7/1058?ck = nck

pita savila: >>> savila

pitera de gogo: >>> savila

plane tree: The genus *Platanus* includes nine or ten species, some doubtful; the barriers to hybridization between the two main species of the genus (*P. occidentalis* from the U.S. and *P. orientalis* from the Eastern Mediterranean Basin and Middle East) arose from ecogeographic isolation for at least 30 million years; however, man disturbed this isolation after the discovery of America and subsequent botanical material exchanges occurred; so European plane tree populations observed everywhere are often considered to be hybrid products between the American and oriental species; in most European towns, in North America and other temperate countries, plane trees represent an important ornamental or lining tree; for example, they constitute 40% of street trees in Paris, more in London, and provide thousands of trees in important towns in Europe; today, the canker stain, a serious fungal disease native to the U.S., threatens the trees in Europe with heavy damage, being already present in the Mediterranean area; *Platanus hybrida syn P. hispanica* (Platanaceae), 2n = 2x = 16; 1C DNA content 1.92 pg *fore hort*

plantain: Northern temperate plants; the great plantain (*Plantago major*) is low growing with large oval leaves carried close to the ground, grooved stalks, and spikes of green flowers with purple anthers followed by seeds, which are used in bird food; the most common introduced species is the ribwort plantain (*P. lanceolata*), native to Europe and Asia and a widespread weed in Australia, Europe, and America; many other species are troublesome weeds; there is also a type of >>> banana known as plantain (in the tropics, a starchy banana that is eaten cooked; in industrial countries, a plantain is a large sweet banana that is eaten raw); *Plantago* spp. (Plantaginaceae), *P. major* 2n = 2x = 12; 2C DNA content 1.7 pg *bot agr*

plum: Smooth-skinned, oval, reddish-purple or green edible fruit of the plum tree; there are many varieties, including 'Victoria', 'Czar', egg-plum, greengage, and >>> damson; the wild sloe (*Prunus spinosa*), which is the fruit of the blackthorn, is closely related; dried plums are known as prunes; *Prunus domestica* and other species (Rosaceae), 2n = 6x = 48 (CCSSSS); 1C DNA content 0.93 pg *hort* >>> >>> cherry plum >>> damson plum >>> European plum

poinsettia: A native plant of deciduous tropical forest in Mexico, Guatemala, and Nicaragua; it is a shrub or small tree, typically reaching a height of 0.6–4m; the plant bears dark green dentate leaves that measure 7–16cm in length; the top leaves (>>> bracts) are flaming red, pink, or white and are often mistaken as flowers; the actual flowers (>>> cyathia) are grouped within the small yellow structures found in the center of each leaf bunch; name "poinsettia" is after J. R. POINSETT (the first U.S. Ambassador to Mexico), who introduced the plant into the U.S.A. in 1828; in Nahuatl, the language of the Aztecs, the plant is called Cuitlaxochitl meaning "star flower"; the Aztecs used the plant to produce red dye and as an antipyretic medication; in both Chile and Peru, the plant became known as "Crown of the Andes" and in Europe as "Christmas Star"; until the 1990s, the P. ECKE family (3rd generation) of Encinitas, California (U.S.), had a virtual monopoly on poinsettias owing to a technological secret that made it difficult for others to compete; the ECKEs' technique, which involved grafting two varieties of poinsettia together, made it possible to get every seedling to branch, resulting in a bushier plant; in the 1990s, a university researcher discovered the method and published it; now the technique is widespread; in 1906, the German immigrant A. ECKE founded an orchard near Eeagle Rock and grew wild plants from the area in his backyard; by subsequent selection, improved cultivation and good marketing he contributed to the worldwide cultivation and distribution, *Euphorbia pulcherrima* (Euphorbiaceae), 2n = 4x = 28; 1C DNA content 1.30 pg *hort* >>> Table 46

Polish wheat: Varieties are spring wheats with tall stems; spikes are large, open or dense, awned, and square or rectangular in cross-section; kernels are very long, narrow, and hard; they thresh free of the glumes; while grown extensively in Mediterranean countries, Polish wheat has proved inferior in other regions both in yield and in quality for bread or macaroni products; for these reasons, it has substantially disappeared from commercial production; *Triticum polonicum* (Gramineae), 2n = 4x = 28 (BBA^{u}A^u); 1C DNA content 12.28 pg *bot agr* >>> wheat

pomegranate An attractive shrub or small tree, 6 or 10 m high: it is much branched, more or less spiny, and extremely long lived – some specimens at Versailles are known to have survived two centuries; it has a strong tendency to sucker from the base; the leaves are evergreen or deciduous; showy flowers are located on the branch tips singly or with as many as five in a cluster; they are 3 cm wide and characterized by the thick, tubular, red calyx having five to eight fleshy, pointed sepals forming a vase, from which emerge the three to seven crinkled, red, white, or variegated petals enclosing the numerous stamens; the fruit, 6.25–12.5 cm wide, has a tough, leathery skin or rind, basically yellow more or less overlaid with light or deep pink or rich red; the interior is separated by membranous walls and white spongy tissue (rag) into compartments packed with transparent sacs filled with tart, flavorful, fleshy, juicy, red, pink, or whitish pulp; in each sac, there is one white or red, angular, soft or hard seed; the seeds represent about 52% of the weight of the whole fruit; it is native from Iran to the Himalayas in northern India and has been cultivated since ancient times throughout the Mediterranean region of Asia, Africa, and Europe; historically, the fruit was used in many ways as it is today and was featured in Egyptian mythology and art, praised in the Old Testament of the Bible and in the Babylonian Talmud, and it was carried by desert caravans for the sake of its thirst-quenching juice; *Punica granatum* (Punicaceae), 2n = 2x = 16; 1C DNA content 0.72 pg *hort*

pomelo: A citrus fruit native to South East Asia; usually pale green to yellow when ripe, with sweet white (or, more rarely, pink or red) flesh and very thick pudgy rind ('Chandler', a Californian variety, has a smoother skin); it is the largest citrus fruit, 15–25 cm in diameter and usually weighing 1–2 kg; pomelo is sometimes referred to as a cross/backcross product, *Citrus paradisi* (grapefruit, 2n = 2x, 3x, 4x = 18, 27, 38) × *Citrus maxima* (2n = 2x = 18), developed around 1970 in Israel, *Citrus maxima syn C. grandis*, 2n = 2x = 18; 1C DNA content 0.40 pg *hort* >>> Figure 61

pomme de merveille: >>> bitter gourd

pomo balsamo: >>> bitter gourd

Ponderosa pine: *Pinus ponderosa* (Pinaceae), 2n = 2x = 24; 1C DNA content 24.20 pg *fore* >>> pine

pongamia (*syn* **honge** *syn* **panigrahi** *syn* **karanji** *syn* **calpa** *syn* **ponge**): A legume tree thought to have originated in India and found throughout Asia; it grows to about 15–25 m in height, with a large canopy which spreads equally wide; it is well suited to intense heat and sunlight and its dense network of lateral roots and its thick, long taproot make it drought tolerant; it is often used for landscaping purposes as a windbreak or for shade due to the large canopy and showy fragrant flowers; although all parts of the plant are toxic and will induce nausea and vomiting if eaten, the fruits and sprouts, along with the seeds, are used in many traditional remedies; juices from the plant, as well as the oil, are antiseptic and resistant to pests; seeds show 25–40% lipid content of which nearly half is oleic acid; the seed oil is an important asset of this tree, having been used as lamp oil, in soap making, and as a lubricant for thousands of years; recently, the seed oil has been found to be useful in diesel generators and, along with >>> jatropha, it is being explored in hundreds of projects throughout India and the Third World as feedstock for biodiesel; it is especially attractive because it grows naturally through much of arid India, having very deep roots to reach water, and is one of the few crops well-suited to commercialization by India's large population; both the chloroplast and mitochondrial genomes of pongamia have been sequenced and annotated (152,968 and 425,718 bp, respectively), with similarities to previously characterized legume organelle genomes being identified; many nuclear genes associated with oil biosynthesis and nodulation in pongamia have been characterized, *Pongamia pinnata syn Derris indica syn Pongamia glabra* (Leguminosae), 2n = 2x = 22; 1C DNA content 1.80 pg *fore eco* >>> agroenergy crop

ponge: >>> pongamia

poplar: Deciduous trees with characteristically broad leaves; the white poplar (*Populus alba*) has a smooth gray trunk and leaves with white undersides; other species are the aspen (*P. tremula*), gray poplar (*P. canescens*), and black poplar (*P. nigra*); most species are tall; they are often grown as windbreaks in commercial orchards; *Populus* spp. (*P. deltoides, P. trichocarpa, P. tremula, P. tremuloides, P. candicans, and/or P. lasiocarpa*) (Salicaceae), 2n = 2x = 38; 1C DNA content 0.48 pg *fore* >>> chloroplast

poppy: Plants belonging to the poppy family; they have brightly colored, mainly red or orange flowers, often with dark centers, and yield a milky sap; species include the crimson European field poppy (*Papaver rhoeas*) and the Asian opium poppy (*P. somniferum*), source of the drug opium; closely related are the California poppy (*Eschscholtzia californica*) and the yellow horned or sea poppy (*Glaucium flavum*); *Papaver* spp., *Eschscholtzia* spp., *Glaucium* spp. (Papaveraceae), 2n = 2x, 4x = 22, 44 *hort* >>> Table 46

Portuguese maritime pine (*syn* **maritime pine**): *Pinus pinaster* (Pinaceae), 2n = 24 = 24; 1C DNA content 28.90 pg *fore*

portulac: *Portulaca grandiflora* (Portulacaceae), 2n = 2x = 18; 1C DNA content 1.68 pg *hort*

potato: Domesticated more than 6,000 years ago in the high Andes of South America; in the 16th century, Spanish conquistadors brought the potato from Peru to Europe, where it took two centuries before potatoes were introduced into the European diet; the cultivated potatoes, first appearing in Europe and later spreading worldwide, were first recorded outside of the Americas in 1567 on the Canary Islands archipelago; remnant landraces of these early potatoes are still grown on the Canary Islands; Canary Island landraces possess both Andean and Chilean types, as well as possible hybrids of the two; it is speculated that the early European potato was selected from Chilean introductions before the 1840s, because they were better able to reproduce in long-day conditions, in contrast to Andean potatoes, that were short-day adapted; at present, potato is the fourth most important crop in developing countries after rice, wheat, and maize; more than 3 billion people consume potatoes; potato production is 330 million tons (2009) expanding at an unprecedented rate; ~30% of the world's potato crop is currently produced in developing countries, mainly by small-scale farmers; China is the largest producer; the center of origin is South America; the potato is derived from the *andigena* subspecies, which is the progenitor of the *tuberosum* subspecies; the latter originated from different related diploid populations in different locations through sexual polyploidization; recent molecular studies even demonstrate the close relationship to tomato; particularly in potato, haploids became important for breeding and genetics; recently, they have been genetically modified for the production of pharmaceuticals, industrial starch, etc.; *Solanum tuberosum* (Solanaceae), 2n = 4x = 48 (AAA'A'); 1C DNA content 1.75 pg; genome size is 727 Mbp, of which 94% are non-gapped sequences; the 17-nucleotide depth distribution suggests a genome size of 844 Mbp, consistent with estimates from flow cytometry; about 39,000 protein-coding genes are predicted, including two genome duplication events indicative of a palaeopolyploid origin *agr* >>> chloroplast >>> Tables 17, 35

potato onion: >>> multiplier (onion)

potato yam: >>> aerial yam

poulard wheat: >>> Isfahan wheat

prairie cordgrass (*syn* **freshwater cordgrass** *syn* **marshgrass** *syn* **sloughgrass** *syn* **rip gut):** A tall grass, rhizomatous, and native to marshes, drainage ways, and moist prairies in North America; one of the dominant grasses of the tall-grass prairie region, although it is also found in southern Canada and over a good portion of the eastern and Central U.S.; it is one of the tallest native grasses, sometimes reaching 3 m in height but is often shorter under less than ideal conditions, particularly in dry years; it often grows in dense, nearly pure stands near the edges of wetlands, with the vigorous rhizomes and shade from the tall stems excluding most other species; biomass production of native warm-season grasses intended for biofuel purposes in the northern Great Plains of the U.S. may be enhanced by selecting new populations of cordgrass and other species of that family; *Spartina pectinata* (Gramineae), 2n = 6x, 12x = 42, 84 *agr*

prairie grass: A winter annual or biennial grass, native of South America, widely distributed in the Pampeana area of Argentina and cultivated in temperate regions of the world; prairie grass is a hexaploid species with a facultative cleistogamous reproductive behavior; genetic variation within and between populations of this crop species is a major concern for plant breeders and population geneticists; *Bromus catharticus* (Gramineae), 2n = 6x = 42; 1C DNA content 6.80 pg *agr*

prickly burr: >>> American chestnut

prickly lettuce: The closest wild relative of cultivated lettuce; it is an autogamous species and serves as a resource for introgression experiments; *Lactuca serriola* (Compositae) 2n = 2x = 18; 1C DNA content 2.65 pg *hort* >>> chloroplast >>> lettuce >>>Table 46

prickly pear cactus: Prickly pear cacti are North American desert succulents that have flat, fleshy, leaf-shaped pads and large spines (modified leaves) growing from tubercles (small bumps on the pads); they have red, yellow, or purple flowers; their fruit are edible; they have also been introduced to southern Europe; *Opuntia* spp. (Opuntia), 2n = 2x = 22 *hort* >>> cactus fig >>> opuntia >>> pencas

primrose: Plants belonging to the primrose family, with showy, five-lobed flowers; the common primrose (*Primula vulgaris*) is a woodland plant, native to Europe, with abundant pale yellow flowers in spring; *Primula* spp. (Primulaceae), 2n = 2x = 22 *hort* >>> evening primrose

Psathyrostachys huashanica: *Psathyrostachys* is a small genus in the tribe Triticeae that contains about 10 species, which are distributed from east Turkey to Central China and Mongolia, and all species share the same N_s genome; *P. huashanica* grows only on the rocky slopes of Mount Huashan in the central area of China and it is noted for its excellent agronomic characteristics, i.e. early maturity, dwarf stature, resistance to wheat take-all fungus, stripe rust, and powdery mildew, and tolerance of drought and salinity, that can be utilized in cereal improvement; plants are laxly tufted, with long, underground rhizomes; culms are glabrous, 30–60 cm tall and two- to four-noded, *Psathyrostachys huashanica* (Gramineae), $2n = 2x = 14$ (N_sN_s) *bot*

Pseudoroegneria strigosa: (Gramineae) $2n = 2x = 14$ (SS); 2C DNA content 8.8 (S^tS^t)–9.5 pg *bot*

pulque plant (*syn* **guamiel** *syn* **centemetl** *syn* **maguey de pulque** *syn* **maguey** *syn* **manso** *syn* **metl** *syn* **savia agaves** *syn* **teometl** *syn* **tlacametl**): Pulque is the fermented juice from the base of the flower stalk; leaves of the central cone are removed and the sap is allowed to collect in the cavity; mescal and tequila are distilled pulque; *Agave atrovirens syn A. coarctata, cochlearis, compluviata, crassispina, ferox, jacobiana, latissima, lehmannii, mitriformis, potatorum, quiotifera, tehuacanensis* (Agavaceae), $x = 30$; $2n = 2x, 3x, 4x, 5x, 6x = 60, 90, 120, 150, 180$ *hort* >>> http://www.extraplicity.com/galleries/agave.html

pumpkin: >>> butter nut squash

purging nut: >>> Barbados nut

purple false brome: A grass species native to southern Europe, northern Africa, and Southwestern Asia east to India; it is related to the major cereal grain species wheat, barley, oats, maize, rice, rye, sorghum, and millet; it has many qualities that make it an excellent model organism for functional genomics research in temperate grasses, cereals, and dedicated biofuel crops, such as switchgrass; it has a small genome (~300–320 Mbp), diploid accessions, and a series of polyploid accessions, a small physical stature (approximately 20 cm at maturity), self-fertility, inbreeding, a short lifecycle (annual, less than four months), simple growth requirements, and an efficient transformation system; diploid ecotypes have five easily distinguishable chromosomes that display high levels of chiasma formation at meiosis; the nuclear genome was indistinguishable in size from that of >>> thale cress (*Arabidopsis*), making it the simplest genome described in >>> grasses to date; immature embryos exhibited a high capacity for plant regeneration *via* somatic embryogenesis; regenerated plants display very low levels of albinism and have normal fertility; a simple transformation system has been developed based on microprojectile bombardment of embryogenic callus and >>> selection for hygromycin resistance; selected ecotypes were resistant to all tested cereal-adapted *Blumeria graminis* species (powdery mildew) and cereal brown rusts (*Puccinia recondita*); in contrast, different ecotypes displayed resistance or disease symptoms following challenge with the rice blast pathogen (*Magnaporthe grisea*) and wheat/barley yellow stripe rusts (*Puccinia striiformis*); despite its small stature, *B. distachyon* has large seeds that should prove useful for studies on grain filling; *Brachypodium distachyon* (Gramineae), $2n = 2x = 10$; 2C DNA content 0.6–1.25 pg *bot biot*

purple leaf plum: >>> cherry plum

purple vetch: *Vicia benghalensis* (Leguminosae), $2n = 2x = 14$; polytene chromosomes have been observed in ovary, anthers, or immature seed tissues; 2C DNA content 7.0 pg *bot hort*

purslane (*syn* **purslave** *syn* **parsley** *syn* **pusley**): Common prostrate weed with edible, succulent leaves and stems; a >>> C_4 plant; *Portulaca oleracea* (Portulacaceae), $2n = 2x = 54$; 1C DNA content 1.68 pg *hort* >>> http://www.hort.purdue.edu/newcrop/1492/neglected.html

pyrethrum: Popular name for several cultivated chrysanthemums; the ornamental species *Chrysanthemum coccineum*, and hybrids derived from it, are commonly grown in gardens; pyrethrum powder, made from the dried flower heads of some species, is a powerful pesticide for aphids and mosquitoes, *Chrysanthemum* spp. *syn Tanacetum cinerariifolium*; *T. cineariifolium,* is also a species of daisy, from which natural pyrethrins are extracted; it originated in Dalmatia where people still put dried pyrethrum flowers in their bedding to kill fleas and bed bugs; they have apparently been doing this for centuries, without any resistance evolving in fleas or bed bugs, demonstrating that natural pyrethrins are a durable insecticide; *Chrysanthemum* spp. (Compositae), $2n = 2x = 18$ *hort phyt*

Q

quackgrass: >>> *Agropyron*

quassia: Tropical American trees with bitter bark and wood; the heartwood of *Quassia amara* is a source of quassiin, an infusion of which was formerly used as a tonic; it is now used in >>> insecticides; the quassia family includes the Asian ailanthus (*Ailanthus altissima*), also called the "tree of heaven"; *Quassia amara* (Simaroubaceae), 2n = 4x = 36 *hort fore*

Queensland arrowroot: Also known as achira in South America, where it originated; this crop is usually called Queensland >>> arrowroot, or purple arrowroot, in English; it is grown commercially in Australia for extraction of starch from the rhizomes; hybrids of wild species of *Canna* are a popular ornamental known as the canna lily; *Canna edulis* (Cannaceae), 2n = 2x, 3x = 18, 27 *hort*

quickgrass: >>> *Agropyron*

quince: Related to the japonica and other plants of the rose family; the common species has been grown in Europe since Roman times (>>> SCHLEGEL, 2007); the yellow fruits are too hard and acidic to be eaten raw but are used to make jam or jellies; *Cydonia oblonga* (Rosaceae), 2n = 2x = 34 *hort*

quinine: A tree native to Amazon rainforest vegetation; this plant, often thought to be used for the production of quinine, which is an anti-fever agent, especially useful in the prevention and treatment of malaria; there are a number of various other chemicals which are made from this tree, and they include cinchonine, cinchonidine, and quinidine; *Cinchona calisaya* is the tree most cultivated for quinine production, *Cinchona officinalis* (quinine bark), *Q. pubescens syn C. succirubra* (Rubiaceae), 2n = 2x = 34; 1C DNA content 0.73 pg *fore*

quinoa: A species of goosefoot, which is cooked like grain; during cooking, the germ comes out of the seed and dangles, curled, from it, or falls off; it comes from the Andean region of South America, where it has been an important food for 6,000 years; it is very undemanding and altitude hardy, so it can be cultivated in the Andes up to 4,000 m; it grows best in well-drained soils and requires a relatively long growing season; it is also susceptible to a leaf miner in eastern North America which may reduce crop success; the leaf miner also affects the related common weed, *Chenopodium album*, but *C. album* is much more resistant; similar *Chenopodium* species were probably grown in the North America, before maize agriculture became popular; chenopodiums were also used in Europe as greens; *Chenopodium album*, which has widespread distribution in the U.S., produces edible seeds and greens much like quinoa, but in smaller quantities; caution should be exercised in collecting this weed, however, because, when grown in heavily fertilized agricultural fields, it can accumulate dangerously high concentrations of nitrate; in colonial times, quinoa was scorned by the Spanish colonists as food for Indians, but, in more enlightened times, the grain has come to be highly appreciated for its nutritional value; unlike cereals, such as wheat or rice, quinoa contains a full complement of amino acids; *Chenopodium quinoa* (Chenopodiaceae), 2n = 4x = 36; 2C DNA content 2.6–3.4 pg, genome size 1,385 Gbp in length, with 3,486 scaffolds; the longest scaffold is 23.8 Mbp, the scaffold N_{50} is 3.84 Mbp and 90% of the genome is contained in 439 scaffolds; the scaffolds were mapped into 18 pseudo molecules, corresponding to the haploid chromosome number; the genome has been annotated using *ab initio* prediction in combination with evidence from full-length transcripts (IsoSeq) and RNA sequencing from different tissue types to give 44,776 genes; repetitive sequences comprised 64% of the genome, with a large proportion coming from long terminal repeat (LTR) transposable elements *hort*

R

rabbitroot: >>> sarsaparilla

radicchio: A red or red-variegated chicory, very extensively cultivated in northeastern Italy as a leafy vegetable; it includes different types which represent valuable high-quality crops; five major types of radicchio are cultivated in the Veneto region (Italy); *Cichorium intybus* (Compositae), 2n = 2x = 18 *hort*

radish: Annual herb native to Europe and Asia, and cultivated for its fleshy, pungent, edible root, which is usually reddish but sometimes white or black; it is eaten raw in salads; there are four basic types, and all of them belong to this one species; the small radish is the temperate zone garden vegetable, grown commercially on quite a large scale; the large radish is popular in the Far East; mougri radish is grown in Southeast Asia, solely for its leaves and young seed pods, as it has no fleshy root; fodder radish is similar to mougri; in F_1 hybrid breeding, a *Ppr-b* gene is responsible for male-fertility restoration of the Ogura-type male-sterile radish plants, and it is located in the complex *Rfo* locus in the vicinity of the similar *Ppr-a* gene and the *Ppr-c* pseudogene; there is a wide allelic variation within the *Rfo* locus, as well as high genetic complexity of the fertility restoration mechanism in radish, *Raphanus sativus* var. *radicula* (Brassicaceae), 2n = 2x = 18 (RR); 1C DNA content 0.55 pg *hort* >>> *Brassica* >>> fodder radish >>> Figure 8

ragweed: Annual plant in the Compositae family, common on roadsides and waste ground throughout North America; it grows to 1.5 cm in height and has long spikes of yellow-green flowers; the great ragweed (*A. trifida*) grows to 2.4 m; both are wind-pollinated; they produce large quantities of pollen and are a major cause of hayfever; *Ambrosia elatior syn Ambrosia artemisiifolia* (Asteraceae), 2n = 4x = 36; 1C DNA content 1.16 pg *bot agr*

ragi: >>> finger millet

rakkyo: *Allium cepa syn A. chinense syn A. bakeri* (Alliaceae), 2n = 2x, 3x, 4x = 16, 24, 32 *hort*

rambutan: A medium-sized tropical tree and fruit of this tree; it is native to Vietnam, Indonesia, the Philippines, Sri Lanka, Malaysia, and elsewhere in Southeast Asia, although its precise natural distribution is unknown; It is closely related to several other edible tropical fruits including the >>> lychee, >>> longan, and mamoncillo; The name "rambutan" is derived from the Indonesian word *rambutan*, meaning "hairy", *Nephelium lappaceum* (Sapindaceae), 2n = 2x = 22 *hort*

rampion: A native plant from Central Europe, from which young leaves are eaten as a salad during the winter season; at the end of the first year, the plant also forms fleshy roots that can be eaten raw as well; *Campanula rapunculus* (Campanulaceae), 2n = 2x = 20; 1C DNA content 3.98 pg *hort*

ramie (*syn* China grass): It is native to eastern Asia and a herbaceous perennial growing to 1–2.5 m tall; the leaves are heart-shaped, 7–15 cm long and 6–12 cm broad, and white on the underside with dense small hairs – this gives it a silvery appearance; unlike nettles, the hairs do not sting; it is one of the oldest vegetable fibers and has been used for thousands of years; the fibers are found in the bark of the stalk; the fiber is very fine and silk-like and naturally white in color; it was used for Chinese burial shrouds over 2,000 years ago, long before cotton was introduced to the Far East; it is classified chemically as a cellulose fiber, just as are cotton, linen, and rayon; leading producers of ramie are China, Taiwan, Korea, the Philippines, and Brazil; the fibers consist of individual cells of up to 55 cm in length, making them the longest cells in the plant kingdom; *Boehmeria nivea* (Urticaceae), 2n = 2x = 14 *agr*

rape: Plant species of the mustard family grown for their seeds, which yield a pungent edible oil; the common >>> turnip is a variety of *Brassica rapa* and the >>> Swede turnip is a variety of *B. napus*; *Brassica* spp. (Brassicaceae), 2n = 4x = 38; 1C DNA content 1.15 pg *agr* >>> *Brassica* >>> rapeseed >>> Figure 8

rapeseed: Oilseed rape, or >>> Canola™ in Canada, is the world's third most important oilseed crop; by breeding, it was developed from the "weed" rapeseed (*B. napus oleifera*); rape methyl ester provides a renewable replacement for diesel fuel that gives off fewer sooty particles and none of the sulfur dioxide that causes acid rain; *Brassica napus* ssp. *napus* (Brassicaceae), 2n = 4x = 38 (AACC); 1C DNA content 1.15 pg *agr* >>> *Brassica* >>> Figures 8, 45, 46, 47 >>> agroenergy crop >>> Tables 16, 35

Raphanobrassica: An intergeneric hybrid between *Raphanus* and *Brassica* species; the first hybrid was reported in 1826; ~100 years later, G.D. KARPECHENKO produced a hybrid between *Raphanus sativus* [2n = 2x = 18 (RR)] and *Brassica oleracea* [2n = 2x = 18 (CC)] >>> 2n = 4x = 36 (RRCC) >>> *Brassica* >>> Figure 8

raspberry: *Rubus* sp. (Rosaceae), 2n = 2x = 14, *Rubus ideaus* 2n = 2x = 18 *hort*

ratabaga: >>> Figure 8

red bean: >>> Adzuki bean

red beet: Red beets derived from a common origin with sugarbeet, fodder beet, or >>> mangel; the wild type, *Beta maritima*, is found along the coastlines of Europe; relicts of the original forms of *B. vulgaris* were found in the north of Holland from 2,000 BC; in the beginning, the leaves were used; the beet-like types were yellow ('Golden Beet'), white ('Albina Vereduna'), roundish or conical; the red type was developed during the 19th century, and are still available as the Swiss and Italian landraces 'Basano' or 'Chioggia'; primary pigments in red beet are the betalains, which are comprised of the red-violet betacyanins and the yellow betaxanthins; the presence of dominant alleles at two linked loci (*R* and *Y*) condition the qualitative production of betalain pigment in the beet plant; red-pigmented roots are observed only in the presence of dominant alleles at both the *R* and *Y* loci, while white roots are conditioned by recessive alleles at the *Y* locus, and yellow roots by the genotype *rrY-*; a gene "blotchy" (*bl*) conditions a blotchy or irregular pigment patterning in either red or yellow roots; *Beta vulgaris* ssp. *vulgaris* convar. *crassa* var. *conditiva* (Chenopodiaceae), 2n = 2x = 18; 1C DNA content 1.25 pg >>> beet

redbud: *Cercis canadensis* (Leguminoseae), 2n = 2x = 12 *hort*

red cabbage: *Brassica oleracea* convar. *capitata* var. *rubra* (Brassicaceae), 2n = 2x = 18 (CC); 1C DNA content 0.78 pg *hort* >>> *Brassica* >>> Figure 8

red clover: >>> broad red clover

redcurrant: *Ribes sativum, R. rubrum* (Grossulariaceae), 2n = 2x = 16; 1C DNA content 0.97 pg *hort*

red fescue: *Festuca rubra* (Gramineae), 2n = 2x = 14; 2C DNA content 13.9 pg *agr*

red gram: >>> pigeon pea

red-head cabbage: >>> red cabbage

red pepper: >>> sweet pepper

red pine: *Pinus resinosa* (Pinaceae), 2n = 2x = 24; 1C DNA content 23.18 pg *fore*

red plum: >>> American plum

red rice: >>> African rice

red spruce: *Picea rubens* (Pinaceae), 2n = 2x = 24 *fore*

red top: >>> black bent

redwood: The coast redwood is the tallest tree, growing up to 113 m tall and living for over one thousand years; one redwood tree in California is 2,200 years old; the roots of this giant conifer are shallow, but spread sideways up to 75 m from the trunk; the bark is deeply-furrowed, fibrous, thick and lacks resin; there are many species of redwood, including the giant coast redwood; *Sequoia sempervirens* (Taxodiaceae), 2n = 6x = 66; 1C DNA content 32.14 pg *fore hort* >>> http://www.na.fs.fed.us/spfo/pubs/silvics_manual/Volume_1/sequoia /sempervirens.htm

reed canary grass: A perennial grass plant, very robust, hairless, rhizomatous; stems are erect, 50–150 cm high; leaf blade rolled when young, long, finely striate, almost smooth, pale green; ligule is long, oval-obtuse; no auricles; panicle-like inflorescence, elongated, spreading at flowering then contracted, whitish-green to purple; *Phalaris arundinacea syn Phalaroides arundinacea syn Typhoides arundinacea syn Baldingera arundinacea syn Digraphis arundinacea* (Gramineae), 2n = 4x = 28; 1C DNA content 4.13 pg *agr*

rhea: >>> ramie

Rhodes grass (*syn* **chloris**): The dominant wild grass in extensive savannas in East and southern Africa; selection has produced a number of pasture cultivars, both perennial and annual; some cultivars are turf grasses and make attractive lawns; this species can be grown over a wide range of habitats, and it has been introduced to many areas; it has reasonably high yields of hay, fodder, and grazing; *Chloris gayana* (Poaceae), x = 10; 2n = 2x, 3x, 4x, 5x, 8x, 10x = 14, 20, 26, 30, 36, 40, 72, 80, 100; 2C DNA content 0.7 pg *agr* >>> http://delta -intkey.com/grass/www/chloris.htm

rhubarb: A genus of perennial plants that grows from thick short rhizomes; the plants have large leaves that are somewhat triangular shaped with long fleshy petioles; the flowers are small, greenish-white to rose-red, and grouped in large compound leafy inflorescences; a number of varieties of rhubarb have been domesticated both as medicinal plants and for human consumption; the leaves are toxic because they contain high levels of oxalates; however, the stalks are used in pies and other foods for their tart flavor; *Rheum rhaponticum* (Polygonaceae), 2n = 2x, 4x = 22, 44 *hort*

ribbon grass: >>> reed canary grass

rice: Second only to wheat in terms of area and amount of grain produced; it probably originated in India or Southwest Asia, where several wild species are found; rice culture spread to China ~5,000 years ago and to Europe ~2,500 years ago; there are more than 1,000 culti-vated varieties; it is an annual grass, 80–150 cm height; the inflorescence is a loose panicle containing ~100 single-flowered spikelets; it is normally self-pollinating; the mature ker-nels are enclosed in the palea and lemma, and their color varies from white to brown; it grows with its roots in water; rice transports oxygen to its roots from the leaves; categories of rice are based on length of grain: short (5 mm), medium (6 mm), long (7 mm); short-grain types of the *japonica* type have short straw, whereas long-grain types of the *indica* type usually have taller and weaker stems; rice is a short-day plant; rice growing was the key for the development of Asiatic civilizations and certain African cultures; Asiatic species of rice appear to have diverse origins and to be derived from a complex in which *Oryza rufipogon* and *O. nivara* played a major role; African species are believed to have been domesticated from *O. barthii*; based on the average genetic distance among all the strains of common wild rice and cultivated rice for nuclear and mitochondrial genomes, the variability of the nuclear genome was found to be higher than that of the mitochondrial genome; the global pattern based on all genomes shows much more diversity in wild rice than in cultivated rice; rice remains, along with wheat, the main staple food source for humans, especially in high-density areas in hot and wet tropical and subtropical areas; the annual production is ~500 million tons; with ~450 million nucleotides, rice carries the smallest genome among the cereals; the estimated number of genes is between 42,000 and 63,000; the mean gene size is ~4,500 nucleotides in *indica* rice; yield heterosis and com-bining ability are significantly increased in inter-group hybrids based on heterotic groups of advanced *indica* rice accessions; variations between genes mainly arose via sequence duplications (~70% of the genes); there is high gene homeology and synteny to wheat, maize, and barley genes; the complete genome was sequenced for the first time in 2002 by scientists of the Chinese Genomics Institute, Beijing in collaboration with the University of Washington; *Oryza sativa* (Gramineae), 2n = 2x = 24 (AA); 2C DNA content 0.8–1.0 pg = 420–460 Mbp; ~37,000,000 bp per chromosome; completely sequenced with genome length of ~30,900,000 bp; a wild rice, *O. grandiglumis*, shows 2n = 48 (CCDD) *agr* >>>

agroenergy crop >>> basmati rice >>> brown rice >>> heat stress >>> Japanese rice
>>> Tables 15, 32, 45, 48

rice bean: A self-pollinating Old World tropical bean that is eaten with rice, or in place of rice, in the Far East, *Vigna umbellata* (Leguminosae), 2n = 2x = 22; 1C DNA content 0.58 pg; polytene chromosomes have been observed in ovary, anthers or immature seed tissues >>> pulse

ricegrass (coba *syn* **jiao-bai** *syn* **kuw-sun** *syn* **kwo-bai** *syn* **makomo** *syn* **makomo dake** *syn* **Manchurian water-rice** *syn* **Manchurian wild rice** *syn* **Manchurian wildrice** *syn* **Manchurian zizania** *syn* **water-bamboo** *syn* **wild rice stem):** A perennial aquatic grass, which grows up to 4 m tall, with spreading rhizomes; the flowers are hermaphrodite; the swollen stem bases, infected with the smut fungus *Ustilago esculenta*, are eaten as a vegetable by the Chinese; they must be harvested before the fungus starts to produce spores, since the flesh deteriorates at this time; they are parboiled then sautéed with other vegetables and have a nutty flavor; the wild forms of this species have developed resistance to the smut, so particularly disease-susceptible cultivars are grown specifically; the seed can be cooked as well; it is used like rice in sweet or savory dishes; *Zizania latifolia syn Zizania caduciflora syn Hydropyrum latifolium* (Gramineae), 2n = 2x = 30; 2C DNA content 4.4 pg (*Zizania aquatica*) *agr*

ricegrass (*syn* **smilo grass):** Perennial mountain rice native to Mediterranean region and introduced into North America, *Oryzopsis miliacea* (Gramineae), 2n = 2x = 24 *agr*

rice-paper plant: *Tetrapanax papyriferum* (Araliaceae), 2n = 2x = 24 *hort*

rivet wheat: >>> Isfahan wheat

rocket (salad): Mustard species cultivated around the Mediterranean Sea, Asia Minor, and Northwest India; it was grown in Greek and Roman times as a pharmaceutical and salad plant; the name rocket is used for at least 20 species of herbs (corn rocket, *Bunias erucago;* cress rocket, *Carrichteria annua;* dame's rocket, *Hesperis matronalis;* dyer's rocket, *Reseda luteola;* Eastern rocket, *Sisymbrium orientale;* French rocket, *Sisymbrium erysimoides;* garden rocket *syn* rocket *syn* salad rocket, *Eruca vesicaria* ssp. *sativa*; garden rocket, *Hesperis matronalis*; hairy rocket, *Erucastrum gallicum*; London rocket, *Sisymbrium irio*; perennial rocket, *Sisymbrium strictissimum;* prairie rocket, *Erysimum asperum*; purple rocket, *Iodanthus pinnatifidus*; salad rocket, *Eruca vesicaria* ssp. *sativa*; sea rocket, *Cakile* spp.*;* scented rocket, *Hesperis matronalis;* sky rocket, *Hesperis matronalis;* small-flowered rocket, *Erysimum inconspicuum;* sweet rocket, *Hesperis matronalis*; tall rocket, *Sisymbrium altissimum*; >>> Turkish rocket, *Bunias orientalis*; wall rocket, *Diplotaxis tenuifolia*; water rocket, *Rorippa sylvestris*; white rocket, *Diplotaxis erucoides*; wild rocket, *Diplotaxis tenuifolia*; yellow rocket, *Barbarea vulgaris*), mostly members of the mustard family Brassicaceae; these species vary from well-known and widely cultivated to obscure and rarely or never grown; they share in common a distinctive zesty or sharp flavor, that is akin to mustard or horseradish; from rocket species to species, the flavor varies; one of these plants is grown mainly for its beauty *(Hesperis matronalis);* the rest are considered edible salad herbs, wild or cultivated (Brassicaceae); the common salad rocket salad or arugula is also an edible annual plant growing to 20–100 cm, not to be confused with wild rocket (*Diplotaxis tenuifolia*); a species native to the Mediterranean region, from Morocco and Portugal east to Lebanon and Turkey; the leaves are deeply pinnately lobed with four to ten small lateral lobes and a large terminal lobe; the flowers are 2–4 cm in diameter, arranged in a >>> corymb, with the typical Brassicaceae flower structure; the petals are creamy white with purple veins, and the stamens are yellow; the sepals are shed soon after the flower opens; the fruit is a >>> siliqua (pod) 12–35 mm long with an apical beak, and contains several seeds (which are edible); it is used as a leaf vegetable, which looks like a longer-leafed and open lettuce; it has been grown in the Mediterranean area since Roman times, and is considered to be an aphrodisiac; before the 1990s, it was usually collected in the wild and was not cultivated, *Eruca sativa syn Eruca vesicaria* ssp. *sativa syn Brassica eruca* (Brassicaceae), $2n = 2x = 22$; 1C DNA content 0.58 pg *hort* >>> *Brassica* >>> Figure 8

rock maple: >>> sugar maple

rocoto: *Capsicum pubescens* (Solanaceae), $2n = 2x = 24$; 1C DNA content 4.47 pg *hort*

Roegneria ciliaris: (Gramineae); $2n = 4x = 28$ ($S^cS^cY^cY^c$) is reported to be a potential source of resistance to wheat scab *bot agr*

rooibos: A broom-like member of the >>> legume family of plants growing in South Africa's fynbos; the plant is used to make a herbal tea called "rooibos tea", "bush tea", "redbush tea", "South African red tea", or "red tea"; the product has been popular in Southern Africa for generations and is now consumed in many countries; the plant has been transformed from a wild-harvested crop to a commercial viable agribusiness, *Aspalathus linearis* (Crotalarieae), $2n = 2x = 18$ *hort*

Romain salad: >>> cos lettuce >>> lettuce

roquette: >>> rocket salad

rose: *Rosa* spp. (Rosaceae), $2n = 2x, 4x = 14, 28$ *hort* >>> http://www.helpmefind.com/rose/glossary.php

rose apple (*syn* **malay apple**): A common name used for several species, e.g. *Syzygium aqueum* = watery rose apple, *S. jambos* = rose apple or jambu, *S. malaccense* = Malay rose apple, or *S. samarangense* = Java rose apple; most used is jambu; it is a species of flowering tree that is native to Malaysia, Indonesia, and southern Vietnam; the trees grows to 8–25 m tall, the bark is grayish brown, somewhat flaky, glabrous throughout; leaves are subcoriaceous, elliptic to oblong obovate, 14–38 cm long, 5–20 cm wide; the flowers are found in axillary cymes 2–5 mm long on older branches and occasionally trunk, peduncles 0.5–1 cm long, bracts 1–1.5 mm long; the fruit is a berry, pink to red, turning maroon, rarely whitish, and obovoid; the plant was introduced throughout the tropics, including many Caribbean countries such as Jamaica, Suriname, Dominican Republic, and Trinidad and Tobago; the fruit is oblong-shaped and dark red in color, although some varieties have white or pink skin; the flesh is white and surrounds a large seed; jam is prepared by stewing the flesh with brown sugar and ginger, *Syzygium malaccense syn Caryophyllus malaccensis syn Eugenia domestica syn E. malaccensis* (Myrtaceae), $2n = 2x = 22$; 1C DNA content 0.25 pg *hort*

roselle: An annual or perennial herb or woody-based subshrub, growing to 2–2.5 m tall; the leaves are deeply three- to five-lobed, 8–15 cm long, arranged alternately on the stems; the flowers are 8–10 cm in diameter, white to pale yellow with a dark red spot at the base of each petal, and have a stout fleshy calyx at the base, 1–2 cm wide, enlarging to 3–3.5 cm, fleshy and bright red as the fruit matures; it takes about six months to mature; a species native to the Old World tropics, used for the production of bast fibre and as an infusion; in the Caribbean, sorrel drink is made from sepals of the roselle; in Malaysia, roselle calyces are harvested fresh to produce a health-beneficial drink due to its high concentrations of vitamin C and anthocyanins; China and Thailand are the largest producers and control much of the world supply; Thailand invested heavily in roselle production and their product is of superior quality, whereas China's product, with less stringent quality control practices, is less reliable and reputable; the world's best roselle comes from the Sudan, but the quantity is low and poor processing hampers quality; Mexico, Egypt, Senegal, Tanzania, Mali, and Jamaica are also important producers but production in these countries is mostly used domestically, *Hibiscus sabdariffa* (Malvaceae), $2n = 2x, 4x, 6x = 18, 36, 72$ *hort*

rosemary: A typical shrub of the Mediterranean maquis; its multiple ornamental and aromatic uses, and the great interest in its cultivation is well documented; only a few cultivars or clones have been well characterized; it contains several compounds which have been proven to have antioxidative functions, *Rosmarinus officinalis* (Lamiaceae), $2n = 2x = 24$ *hort*

rosinweed: >>> cup plant

rubber (~ tree): Large trees that live in tropical areas; these trees are tapped for their latex (from which rubber is made), which is produced in their bark layers (it is not the sap); the Pará rubber tree (*Hevea brasiliensis*) is native to South American rainforests, and grows to be over 30 m tall; in 1876, H. WICKHAM brought seeds from the Pará rubber tree (taken from the lower Amazon area of Brazil) to London; seedlings were grown in London, and later sent to Ceylon and Singapore; the technique of tapping rubber trees for their latex was developed in Southeast Asia (before that, the trees were cut down to extract the rubber); commercial natural rubber production now takes place in Malaysia, Thailand, Indonesia, and Sri Lanka (but not significantly in South America); *Hevea brasiliensis* (Euphorbiaceae), 2n = 2x = 38; 1C DNA content 2.15 pg *fore*

runch (*syn* wild radish): *Raphanus raphanistrum* (Brassicaceae), 2n = 2x = 18; 1C DNA content 0.55 pg *bot agr phyt* >>> *Brassica* >>> radish >>> Figure 8

runner bean: >>> kidney bean

Russian dandelion: A perennial >>> dandelion native to Kazakhstan, cultivated for its fleshy roots that have high rubber content; recently, two plant species have received considerable attention as potential alternative sources of natural rubber and biotechnological approaches: the Mexican shrub guayule (*Parthenium argentatum*) and the Russian dandelion, *Scorzonera tau-saghyz syn Taraxacum kok-saghyz syn T. bessarabicum* (Compositae), 2n = 2x = 16; 1C DNA content 1.11 pg *hort* >>> dandelion

Russian wildrye: A salt- and drought-tolerant, cool-season forage grass used in seeded pastures in the Northern Great Plains of the U.S., where water often limits production; seedling vigor is generally poor in diploid cultivars, but tetraploid germplasm has improved seedling vigor; it is a potential source of useful genes for wheat improvement, such as those for barley yellow dwarf virus resistance and stress tolerance; *Psathyrostachys juncea* (Gramineae), 2n = 2x = 14 (NN); 2C DNA content 16.6–16.7 pg (N^sN^s) *bot agr*

rutabaga: >>> swede

ruzigrass: *Brachiaria ruziziensis* (Poaceae), 2n = 4x = 36 *agr* >>> palisadegrass

rye: A cereal that played a major role in the feeding of European populations throughout the Middle Ages owing to its considerable winter hardiness; the cultivated rye resulted from crossbreeding between *Secale vavilovii* and the perennial species, *S. anatolicum* and *S. montanum*; it was domesticated rather late, having evolved initially as a weed among the earlier-cultivated cereals; the world production amounts to ~30 million tons; it is mainly used in bakery for black bread, in confectionery for gingerbread, blinis, etc., or for the production of rye whiskey; hybrid rye is widely grown in Germany, Poland, and other European countries; *Secale cereale* (Gramineae), 2n = 2x = 14 ($R^{cer}R^{cer}$), 2C DNA content 18.9 pg, var. Dankovske, genome size (among the cultivated cereals, the DNA 1C-value of rye is one of the highest, about 34% greater than the largest haploid genome of wheat) 7,917 Mbp *agr* >>> agroenergy crop >>> crop evolution >>> ergot >>> trisomics >>> translocation tester set >>> pentosan >>> Figure 64 >>> Tables 15, 32, 35, 48 >>> SCHLEGEL (2006)

rye brome: *Bromus secalinus* (Gramineae), 2n = 4x = 28; 1C DNA content 13.95 pg *bot agr*

ryegrass: >>> perennial ryegrass

S

safflower: Originally grown for the dye extracted from its florets and as a minor oil crop; there are two types of safflower oil, high oleic (HO), with 70–75% oleic acid, and high linoleic (HL), with about 70% linoleic acid; the original HO trait in safflower, found in an introduction from India, is controlled by a partially recessive allele *ol* at a single locus; in the lipid biosynthesis pathway of developing safflower seeds, microsomal oleoyl phosphatidylcholine desaturase (FAD2) is largely responsible for the conversion of oleic acid to linoleic acid; *in vitro* microsomal assays indicated drastically reduced FAD2 enzyme activity in the HO genotype compared to conventional HL safflower; a study indicated that a single nucleotide deletion was found in the coding region of *CtFAD2-1* that causes premature termination of translation in the HO genotypes, and the expression of the mutant *CtFAD2-1Δ* was attenuated in the HO genotypes compared to conventional HL safflower; it seems that down-regulation of *CtFAD2-1* expression in the HO genotype may be explained by nonsense-mediated RNA decay (NMD); the NMD phenomenon, indicated by gene-specific RNA degradation of the defective *CtFAD2-1Δ*, was subsequently confirmed in *Arabidopsis thaliana* seed as well as in the transient expression system in *Nicotiana benthamiana* leaves; molecular markers are available corresponding to the *olol* mutation that can facilitate a rapid screening and early detection of genotypes carrying the *olol* mutation for use in marker-assisted selection for the management of the HO trait in safflower breeding programs; *Carthamus tinctorius* (Compositae), 2n = 2x = 24 (BB); 1C DNA content 1.40 pg *hort*

saffron: The genus *Crocus* has nearly 100 species, each with unique characters of color, flowering time or geographical distribution; the wild progenitors of the saffron crocus are extinct, and this is an indication of its antiquity; saffron is the dried stigmas of the flower from *Crocus sativus*, which is grown commercially in many countries, from Iran to Mediterranean Spain and Greece, with significant production also in Kashmir; it is highly prized as a spice and coloring, used in both sweet and savory dishes, teas, and also with medical applications; about 200 stigmas from 70 flowers go into a gram of the spice; flowers are picked in the early morning, and the stigmas removed on processing tables in the shade, before the drying process, which differs between producers; despite the high price, though, only a few stigmas are needed to flavor a dish; saffron does vegetatively, by separation of corms from the parent; the cultivated crocus does not set seed, and, it can be propagated only by corms; multiplication of the crop is a very slow process because only two or three new corms are formed each year at the base of the old corm; a yellowish-orange dye is made from elongate stigmas and the tips of styles; saffron contains the glycoside crocin (derived from the diterpene crocetin); 4,000 stigmas yields ~30 g of dye; saffron is the basis of French bouillabaisse, Spanish paella, English saffron buns, Jewish gilderne, Russian challah, Indian zaffrani chawal, and Persian sholezard; *Crocus sativus* (Iridaceae), 2n = 3x = 24 (autotriploid), harboring eight chromosome triplets; molecular studies revealed that two sets of the triploid genome are related to the wild species, *Crocus cartwrightianus,* while the third one is related to *C. pallasii* ssp. *pallasii*; 1C DNA content 5.90 pg *hort* >>> meadow saffron

sage: *Salvia* is the most important and largest genus of the Lamiaceae family; its strong differentiation among species serves as a rich source of germplasm, with potential for use in breeding programs; *Salvia officinalis* (Labiatae), 2n = 4x = 32; polytene chromosomes have been observed in ovary, anthers, or immature seed tissues; 2C DNA content 1.7 pg *hort*

sago palm: Several species are used in Southeast Asia and Polynesia for the production of sago; the starch is extracted from the pith of a palm stem that is ~15 years old; *Metroxylon sagu* (Palmae), 2n = 2x = 32 *hort agr fore*

sainfoin (*syn* **esparsette**): Perennial leguminous plant, native to Asia and naturalized in Europe; it has rosy-red flowers in axillary spikes; it is sometimes grown as a fodder plant on dry, chalky soil, and makes excellent hay; it has a long history of traditional culture worldwide, but its use has declined in western countries over recent decades; it suffers from low productivity and is more difficult to maintain than other legumes but is known to have valuable characteristics such as palatability and drought tolerance; recent studies suggest that it has several other highly beneficial properties due to its unique tannin and polyphenol composition; condensed tannins have been shown to confer anthelmintic properties, to increase protein utilization and prevent bloating in livestock; they may also have the potential to reduce greenhouse gas emissions; positive effects on wildlife and honey production could also be advantageous in the context of sustainable farming; *Onobrychis viciifolia* (Leguminosae), 2n = 4x = 28; 1C DNA content 1.25 pg *agr*

saltbush: A cosmopolitan genus, often associated with halophytic plants; environmental changes induce differences in sex expression; the genus *Atriplex* contains various species distinguishable by morphology, biological cycles, and ecological adaptations; because of their favorable crude protein content, many species of *Atriplex* are excellent livestock fodder during off-season periods, when grasses are low in feed value; *Atriplex*, as well as other shrub species, are important components of arid land vegetation; tolerance to salinity, drought, and high temperature are important characteristics of species; among the species of *Atriplex* in North Africa, *Atriplex halimus*, a perennial shrub, is found in semi-arid and arid environments; this species, particularly well adapted to arid and salt-affected areas, is valued as livestock forage when herbage availability is low, and is considered to be a promising forage plant for large-scale plantings; in Morocco, it is widely distributed as a wild species; since it exhibits a substantial increase in protein content when irrigated with saline water in south Morocco, this halophyte has been considered to be a valuable source of forage; *Atriplex* spp. (Chenopodiaceae), x = 9; 2n = 2x, 4x, 6x, 8x, 10x = 18, 36, 54, 72, 90 *bot agr* >>> mountain spinach >>> orache

salsepareille: >>> sarsaparilla

salsify: Plant habit is similar to black salsify, to which it is related; however, its flowers are not yellow but light red; the roots are white; because of its fine smell, which is released during cooking, it is also called vegetable oyster; roots are harvested during the first year, flowering happens during the second year; *Tragopogon porrifolius* (Compositae), 2n = 2x = 12; 1C DNA content 3.29 pg *hort*

san chi: A perennial herb of the daisy family; it grows to 1 m tall, with yellow flowers blooming in late summer; its Chinese name, jin bu huan, translates to "gold no trade" indicating the high value placed on it for its healing qualities; it is cultivated in China and Japan and is the main ingredient in "Yunnan Bai Yao" powder used to staunch heavy bleeding and promote rapid healing; the grayish-yellow roots contain bioactive glucosides; the therapeutic effects include reduction of inflammation (antiphlogistic effects), and contraction of tissues and blood vessels to arrest bleeding (astringent, styptic and hemostatic effects); *Gynura pinnatifida* (Compositae), 2n = 2x = 20 *hort*

sandalwood: >>> Indian sandalwood

sand bluestem *syn* **sand hill bluestem** *syn* **Prairie bluestem** *syn* **Hall's bluestem** *syn* **Turkey-foot** *syn* **Hall's beardgrass:** A strongly rhizomatous and sod-forming perennial species in the grass family, it is a bunchgrass which grows in tufts and can reach about 2 m in height under favorable conditions; it is native to North America and is found growing from the Mississippi River west to the Rocky Mountains and from Canada to Chihuahua (Mexico); it prefers sandy soils; sand bluestem is a high-quality forage, with good palatability for livestock, but it cannot stand up to continuous heavy grazing, *Andropogon hallii* (Poaceae), 2n = 6x, 9x = 60, 90; 1C DNA content 5.18 pg (at meiosis, regularly 30 bivalents in 6x cytotypes) *agr*

sand ginger: >>> galangal

sanwa millet: >>> barnyard grass

sapathilla: >>> sapodilla

sapodilla (*syn* **sapote** *syn* **ciku** *syn* **chico sapote** *syn* **sapoti** *syn* **naseberry sapathilla** *syn* **lamoot** *syn* **sawo**): Tropical evergreen tree native to Central America, and also cultivated in other tropical regions for its edible fruit; the tree grows to 18 m; its fruits are about the size of an apple, greenish-brown in color, and rough-skinned; inside, the pulp is sweet and soft, and has black seeds embedded in it; its sap, which becomes solid when boiled, provides a type of nonelastic rubber used for making shoe soles; it is also used as chewing gum; chicle gum was known to the Aztecs, and large amounts are exported to the U.S.; *Manilkara zapota syn M. zapotilla syn M. achras syn Achras sapota syn A. zapota syn Sapota achras* (Sapotaceae), 2n = 2x = 26 *hort*

sapote: >>> sapodilla

sapoti: >>> sapodilla

sarson: >>> yellow-seeded sarson

sarsaparilla (wild sarsaparilla *syn* **false sarsaparilla** *syn* **American sarsaparilla** *syn* **shotbush** *syn* **small spikenard** *syn* **wild licorice** *syn* **rabbitroot** *syn* **salsepareille):** A caulescent perennial herb, 0.3–0.7 m tall, highly clonal, >>> dioecious; rhizome long, branched and horizontal; upright stem 3–10 cm long; leaves 30–60 cm long and wide, ternately compound, usually single or occasionally double, arising from the tip of the upright stem, purplish at nodes; stipules absent; petiole 10–45 cm long; it is pollinated mostly by bumblebees; the female ramets have fewer flowers per umbel than the male ramets and reach peak flowering before the male ramets; it is widely distributed in North America from Newfoundland to northern Alberta; roots and rhizomes are used as an alternative, tonic, or antisyphilitic, often used by American Indians in decoctions; *Aralia nudicaulis* (Smilacaceae), 2n = 2x, 4x = 24, 48 *bot hort*

Satsuma mandarin: >>> mikan

savila: Used in many countries of Africa, the Canary Islands, and America as a herb plant, described as early as 2,500 BC; *Aloe vera* (Aloaceae), 2n = 2x = 14; 1C DNA content 16.40 pg *hort agr* >>> aloe

savory (*syn* **summer savory**): A culinary annual herb; it is used as an essential oil plant (containing rosmarinic acid, caffeic acid); *Satureja hortensis syn Calamintha hortensis* (perennial, *Satureja montana*) (Labiatae), 2n = 6x = 30; 1C DNA content 2.78 pg *hort*

Savoy cabbage: *Brassica oleracea* var. *sabauda* (Brassicaceae), 2n = 2x = 18 (CC); 1C DNA content 0.78 pg *hort* >>> *Brassica* >>> Figure 8

sawa millet: *Echinochloa colona* syn *E. colonum* (Gramineae), 2n = 2x = 54; 1C DNA content 1.35 pg *agr hort*

sawo: >>> sapodilla

scabious: *Scabiosa* sp. (Dipsacaceae), 2n = 16, 18, 20 *bot agr*

scarlet runner bean: >>> kidney beans

scorzonera: >>> black salsify

Scots pine: *Pinus sylvestris* (Pinaceae), 2n = 2x = 24; 1C DNA content 22.98 pg *fore* >>> pine

scurvy grass: A biennial to short-lived wintergreen perennial; a common plant in western and northern Europe with perennial growth habit; particularly, it grows along salt lakes; because of the high content of vitamin C, leaves have been used as a vegetable; in the past, sailors kept it on board; *Cochlearia officinalis* (Brassicaceae), 2n = 4x = 32; 1C DNA content 0.75 pg *hort* >>> *Brassica*

sea barley: *Hordeum marinum* (Gramineae), 2n = 2x = 14; 2C DNA content 9.5 pg (HH) *bot agr* >>> Tables 41, 54

sea-buckthorn: A dioecious, wind-pollinated shrub with nitrogen-fixing ability; a fascinating plant species; it is native to Europe and Asia and has been known and used by humans for centuries; it is mentioned in the writings of ancient Greek scholars such as DIOSCORID and THEROPHAST; in ancient Greece, sea-buckthorn was known as a remedy for horses; leaves and young branches were added to the fodder; this resulted in rapid weight gain and a shiny coat for the horse; this gave the name to the plant in Latin "Hippo" = horse, "phaos" = to shine; in recent years, sea-buckthorn has attracted special attention and become an important subject for domestication in many countries; the nutritional and medicinal value of sea-buckthorn is largely unknown in North America; the medicinal value of sea-buckthorn was recorded in the Tibetan medical classic, "Gyud Bzi", in the eighth century; sea-buckthorn has become an important medicinal and nutritional product, especially in Russia, where it is referred to as "Siberian pineapple" because of its taste and juiciness; since 1982, over 300,000 ha of sea-buckthorn have been planted in China; according to recent taxonomic studies, there are five species; only *H. rhamnoides* has an extremely wide distribution in Eurasia, from China, Mongolia, Russia, Kazakhstan, Turkey, Romania, Switzerland, and France to Britain, and north to Finland, Norway, and Sweden; it grows on hills and hillsides, in valleys and river beds, along sea coasts and islands, in small isolated communities or large continuous pure stands; usually it forms a shrub or small tree 3 to 4 m in height; a tree-like appearance is often formed because only the buds on the outer portions of the plant sprout and branch; it is a dioecious species with male and female flowers on separate trees; pollination is normally aided by wind; flower buds are differentiated during the previous growing season, so the number of fruit produced in any one year depends on the growing conditions of the preceding year; the fruit is tightly clustered on two-year-old branches; it is easy to propagate by seed or cuttings; it can grow in arid to very wet conditions and tolerates cold winters; though it prefers sandy and neutral soil, sea-buckthorn survives in soils with pH values from 5–9 and tolerates seawater flooding; it is extremely variable in height, from a small bush less than 50 cm to a tree more than 20 m high; sea-buckthorn berries have attractive colors, varying from yellow to orange to red; the size of berries varies from 4–60 g/100 berries among genotypes in natural populations, and exceeds 60 g/100 berries in some Russian cultivars; it shows diverse fruit shapes, from flattened spherical, cylindrical, ovate, elliptic to many irregular shapes; the combination of fruit color, shape, and size provides diverse choice and increases the ornamental value of the plant; normally, sea-buckthorn has terminal and lateral thorns; observations show large variation in terms of density, shape, and sharpness in natural populations; in Russia, Mongolia and Germany, thornless or nearly thornless cultivars have been bred; biochemical analysis of sea-buckthorn berries has revealed a wide range of variations in vitamin C, carotene, flavonoids, and vitamin E concentrations; genetic diversity in sea-buckthorn provides a good opportunity for plant breeding and selection, while clinal variation of growth rhythm, height, and hardiness provide guidelines for seed and plant transfer as well as plant introduction; domestication of sea-buckthorn started in Siberia in the 1930s; since it is dioecious, plant breeding projects aim at producing both female and male cultivars; however, breeding objectives for female and male cultivars differ and, generally, there are more quality criteria to be met in a female cultivar; therefore, the selection pressure is higher on female cultivars and as a consequence, larger seedling population size is needed to obtain a certain number of female selections than to obtain the same number of male selections; the gender is determined by distinguishable sex chromosomes; *Hippophae rhamnoides* (*Elaeagnaceae*), 2n = 2x = 24 *hort*

sea holly: A perennial plant, native to Europe, and often found on sea shores; it produces a basal rosette, from which grow flowering spikes with stiff spiny foliage and stems; these can reach around 50 cm in height; it is often grown in gardens for its metallic bluish flowers and upper foliage; the basal foliage is a gray or silvery green; for garden use, however, it is often passed over in favor of its more strongly colored and more ornamental relatives and cultivars; the roots have been used as vegetables or for sweetmeats; they are also considered to have a number of herbal uses, including aphrodisiac qualities; young shoots and leaves are sometimes used as an asparagus substitute; *Eryngium maritimum* (Apiaceae), 2n = 2x = 16; 1C DNA content 2.80 pg *hort*

seakale: Perennial European coastal plant with broad, fleshy leaves and white flowers; it is cultivated in Europe and the young shoots are eaten as a vegetable; *Crambe maritima* (Brassicaceae), 2n = 4x = 60; 1C DNA content 2.42 pg *hort* >>> Brassica

sea lavender: *Limonium sinuatum* (Plumbaginaceae), 2n = 2x = 16; 1C DNA content 3.20 pg *hort*

sea lyme grass: *Leymus arenarius* (Gramineae), 2n = 8x = 56; 2C DNA content 2.7–7.7 pg (X^mX^m) *bot agr*

sea oats: A >>> C_4 perennial grass capable of stabilizing sand dunes; it is most abundant along the Gulf of Mexico and Southeastern Atlantic coastal regions of the U.S.; the species exhibits low seed set and low rates of germination and seedling emergence, and so extensive clonal reproduction is achieved through production of rhizomes, which may contribute to a decline in genetic diversity; in general, sea oat plants exhibit a low range of genetic similarity; *Uniola paniculata* (Chloridoideae), 2n = 4x = 40 *hort agr*
>>> http://delta-intkey.com/grass/www/uniola.htm

sea sand-reed: >>> beach grass

sequoia: *Sequoia gigantia, S. sempervirens* (Taxodiaceae), 2n = 6x = 66 (AAAABB); 1C DNA content 32.14 pg; mitochondria are paternally inherited *fore*

seradella: *Ornithopus sativus* (Leguminosae), 2n = 2x = 14 *agr*

serpent gourd (snake gourd *syn* **chichinga** *syn* **padwal):** A tropical or subtropical vine, raised for its strikingly long fruit, used as a vegetable, for medicine, and, a lesser known use, crafting didgeridoos; the narrow, soft-skinned fruit can reach 150 cm long; its soft, bland, somewhat mucilaginous flesh is similar to that of the >>> luffa and the >>> calabash; it is most popular in the cuisine of South Asia and Southeast Asia; the shoots, tendrils, and leaves are also eaten as greens; it was probably domesticated in ancient times in India, from where non-bitter and large-fruited types may have migrated to other tropical areas; the fully mature fruit contains a soft, red, tomato-like pulp; *Trichosanthes cucumerina* (Cucurbitaceae), 2n = 2x = 22; 1C DNA content 2.83 pg *hort*

serpent root: >>> black salsify

sesame (*syn* **simsim, beniseed, gingelly, and till):** Cultivated sesame (edible sesame) has been differentiated into ~3,000 varieties and strains, distributed extensively from the tropical to temperate zones around the world; the crop is self-pollinated and it exhibits great variation in many characteristics; a >>> dehiscent strain is suitable for combine harvesters; *Sesamum indicum* (Pedaliaceae), 2n = 2x = 26; 1C DNA content 0.97 pg *agr* >>> agroenergy crop

sesbania: Used as a green manure crop in rice-based cropping systems; it can be grown in the field before the rice crop is sown, then ploughed back into the soil, replenishing the nitrogen levels; it is a legume native to West Africa and forms a symbiotic relationship with *Azorhizobium caulinodans*; it is renowned for its stem nodulation; both stem and root nodules fix nitrogen, with root nodules forming at the curled root hair, whereas stem nodules occur at the sites of adventitious root primordia *via* "crack" entry; the stem nodules, unlike the root nodules, contain functioning chloroplasts in the nodule cortex and are therefore capable of carbon fixation; it has a very fast growth rate, is able to grow in flooded habitats, and is very nitrogen rich; *Sesbania rostrata* (Leguminosae), 2n = 2x = 12; 1C DNA content 1.23 pg *agr*

Seville orange: >>> orange

shaddok: *Citrus maxima* (Rutaceae), 2n = 2x = 18; 1C DNA content ~40 pg *hort* >>> hesperidia >>> Figure 61 >>> http://users.kymp.net/citruspages/citrons.html

shallot: Small onion in which bulbs are clustered like garlic; it is used for cooking and in pickles; *Allium cepa* var. *ascalonicum* (Alliaceae), 2n = 2x = 16; 1C DNA content 16.75 pg *hort*

sheep('s) fescue: *Festuca ovina* (Gramineae), 2n = 2x = 14; 1C DNA content 2.41 pg *agr*

sheepgrass: A key species in the eastern part of the Eurasian steppe and widely distributed in northern China; it is highly adaptable and holds considerable value in terms of animal husbandry and ecology; over the past thirty years, it has been collected and evaluated;

methods for utilizing new varieties in different regions have been developed; *Leymus chinensis* (Poaceae), allotetraploid, 2n = 4x = 28 (NNXX)

shiitake mushroom: A filamentous eubasidiomycete fungus; *Lentinus edodes* (Tricholomataceae), x = 11 *hort*

short American staple cotton: *Gossypium herbaceum* (Malvaceae), 2n = 2x = 26 (A1A1); 2C DNA content 2.1 pg *agr*

shortleaf pine: *Pinus echinata* (Pinaceae), 2n = 2x = 24; 1C DNA content 22.75 pg *fore*

shotbush: >>> sarsaparilla

shot wheat: >>> Indian dwarf wheat

shrubby blackberry: Raspberries, blackberries, and dewberries are common and widely distributed; most of these plants have woody stems with prickles like roses; spines, bristles, and gland-tipped hairs are also common in the genus; the fruit, sometimes called a bramble fruit, is an aggregate of drupelets; *Rubus fruticosus* (Rosaceae), 2n = 2x = 20 *hort*

Siberian hazel: *Corylus heterophylla* (Corylaceae), 2n = 2x = 28 *hort*

Sieva bean: >>> Lima bean

signalgrass: *Urochloa* is a genus of plants in the grass family, native to Eurasia, Africa, Australia, Mexico, and the Pacific Islands; it includes most of the important grasses and hybrids currently used as pastures in the tropical regions; *Urochloa* spp. (Poaceae), 2n = 4x = 36 ($B^1B^1B^2B^2$) *agr* >>> Congo signal grass

silver birch: *Betula pendula* (Betulaceae), 2n = 2x = 28; 1C DNA content 0.46 pg *fore* >>> birch

silver fir: *Abies alba* (Pinaceae), 2n = 2x = 24; 1C DNA content 16.55 pg *fore*

silver vine: *Actinidia polygama* (Actinidiaceae), 2n = 6x = 58; 1C DNA content 0.79 pg *hort*

siratro: A deep-rooting and self-pollinating perennial with trailing pubescent stems, which may root anywhere along their length; especially in moist clay soils but rarely in drier sandy soils; leaves are pinnately trifoliate, dark green and slightly hairy on the upper side, silvery and very hairy on the lower surfaces; lateral leaflets are ovate, obtuse, about 4–6 cm, often asymmetrically lobed; it is grown as a cover crop and also for green manuring; it can also be intercropped with grain legumes; the growth duration is 150–180 days; it can also be used as forage as it can be cut several times before being plowed in as a green manure; *Macroptilium atropurpureum syn Phaseolus atropurpureus* (Leguminosae), 2n = 2x = 22 *agr*

sisal: Once an important bast fiber crop in its center of origin in Mexico, and also in East Africa (Kenya and Tanzania); sisal has been largely supplanted by synthetic fibers; seed-set in sisal is extremely rare and breeding this crop is difficult; *Agave sisalana* (Agavaceae), 2n = 6x = 180; 1C DNA content 11.31 pg *agr* >>> Henequen agave

Sitka spruce: *Picea sitchensis* (Pinaceae), 2n = 2x = 24 *fore*

skirret: A herbaceous perennial plant from China, first introduced to France; the plant grows up to 1.20 m height and develops roots similar to dahlia; the fleshy root tastes sweet and was used as a substitute for sugar; *Sisum sisarum* (Umbelliferae), 2n = 2x = 22 *hort*

slender foxtail: The genus *Alopecurus*, a member of the tribe Aveneae, contains 29 species and is distributed in almost all non-tropical regions of both hemispheres and, to a certain extent, in the alpine tropics with its main distribution in Southwest Asia; the genus contains important fodder plants, such as *A. pratensis, A. arundinaceus, A. aequalis, A. geniculatus, A. myosuroides*, and *A. bulbosus*; some of them are valuable pasture grasses; *Alopecurus agrestis* (Gramineae), 2n = 2x, 4x = 14, 28; 2C DNA content 11.65 pg *bot agr*

slender wheatgrass (*syn* northern wheat grass): *Agropyron dasystachyum syn Elymus lanceolatus* (Gramineae), 2n = 4x = 28 (SSHH); 2C DNA content 9.3 pg (YY) *bot agr* >>> couches

small burnet: An evergreen forb readily utilized by livestock and wildlife, that lacks persistence under heavy grazing; the forage production of some small burnet accessions can be comparable to alfalfa, and phenotypic variation is such that the germplasm has potential for

improvement through recurrent selection; *Sanguisorba minor* (Rosaceae), 2n = 4x = 28; 1C
DNA content 0.55 pg *agr*

small-leafed sweet basil: >>> sweet basil

small melilot: *Melilotus indica* (Leguminosae), 2n = 2x = 16; 2C DNA content 2.5 pg *agr*

small radish: >>> radish

small spikenard: >>> sarsaparilla

small timothy *Phleum bertolonii* (Gramineae), 2n = 2x = 14; 2C DNA content 3.4 pg *bot agr* >>>
http://delta-intkey.com/grass/www/phleum.htm

smooth brome: Has become an important cool-season forage grass in North America; in the U.S.,
~30 cultivars or experimental populations were produced during the past 50 years; *Bromus
inermis* (Gramineae), 2n = 8x = 56; 1C DNA content 12.27 *bot agr*

smooth loofah: >>> luffa

smooth stalked meadow grass: >>> Kentucky blue grass

snake bean: >>> asparagus bean

snake gourd: >>> serpent gourd

snap bean: >>> French bean

snapdragon (*Antirrhinum majus*): A common and familiar garden plant; it has a complex flower
shape in which each petal has a distinct structure, which makes it very suitable for studying how plants control flower shape and structure; snapdragons are particularly useful for
genetic and developmental studies because they also contain transposons; *Antirrhinum*
spp. (Scrophulariaceae), 2n = 2x = 16; 1C DNA content 1.65 pg (*A. majus*) *hort*

soft brome: *Bromus mollis syn Bromus interruptus syn Bromus mollis* var. *interruptus* (Gramineae),
2n = 4x = 28 *bot agr*

sopropo: >>> bitter gourd

sorghum: The fifth most important cereal crop in the world after wheat, rice, maize, and barley;
in 1992, the world planted 45.4 million ha of sorghum with an average yield of 1,532 kg/
ha; over 80% of the area devoted to sorghum lies in Africa and Asia, where average yields
were 766 and 1,171 kg/ha, respectively; it originated from Africa ~6,000 years ago; it then
spread to India, China, Europe, and America; it is an annual grass with a panicle containing two types of spikelets (pedicelled and sessile); the sessile spikelets contain perfect
flowers, the pedicelled spikelets contain flowers that are either male only or are sterile; self-
pollination is usual; sorghum exhibits the >>> C_4 photosynthetic pathway; it is a short-day
plant; the sorghum grain is more or less rounded, ~6 mm in diameter, often with colored
lines; the color varies from white to brown or black; there are four types of *Sorghum
bicolor*: (a) milo – drought resistant, many tillers, early maturing; (b) kafir – thick stalks,
large leaves, used for forage and grain; (c) sweet – sweet juice in the stalk, grows up to 3 m
height, for animal fodder; and (d) broomcorn – has branches, used for making brooms; in
2016, positive results were emerging from a three-year project with DuPont Pioneer, U.S.,
in >>> doubled haploid development; it has seen significant strides in bringing doubled
haploid technology to sorghum; two preliminary lines have been identified as possible
inducers, which is a major first step in developing doubled haploids; several thousand more
candidates are being evaluated; doubled haploid technology will greatly accelerate the
breeding process; it allows to strongly leverage a number of other technologies, such as
genomic predictions, marker-assisted breeding and precision phenotyping etc.; *Sorghum
bicolor* (Gramineae), 2n = 2x = 20; 2C DNA content 1.5–1.7 pg; 750–780 Mb *agr* >>>
agroenergy crop >>> crop evolution >>> millet >>> Tables 15, 16, 48

sorrel: Of numerous sorrel species, several are grown in Europe as vegetables; the vegetable sorrel or English winter spinach (*Rumex patientia*) is a native of southern Europe and Asia
Minor; since the 15th century, it has been grown as a vegetable; the garden sorrel (*R. acetosa* var. *hortensis*) is still used as a garden herb (convar. "Belleville"); the Roman or
French sorrel (*R. cutatus*) is only very rarely grown anymore; it was particularly used for

soups; the Alpine sorrel (*R. alpinus*) is rarely prepared as spinach or for salad, as previously; it is also an indicator plant for nitrogen (Polygonaceae); *R. acetosa* 2n = 2x = 14; 2C DNA content 3.3 pg *hort*

sour cherry: *Prunus cerasus* (Rosaceae), 2n = 4x = 32; 1C DNA content 0.63 pg *hort*

soursop: *Annona muricata* (Annonaceae), 2n = 2x = 14; 1C DNA content ~1.15 pg *hort* >>> pawpaw >>> sweetsop >>> ylang-ylang (related species)

soya: >>> soybean

soybean (U.S.), soya bean (U.K.): A species of legume native to East Asia; it is an annual plant that has been used in China for 5,000 years as a food and a component of medicinal drugs; soy contains significant amounts of all the essential amino acids for humans, and so is a good source of protein; soybeans are the primary ingredient in many processed foods, including dairy product substitutes; it derives from the wild annual progenitor, *G. soja*; most genetic diversity is found among 12 wild perennial species, which are indigenous to Australia, South Pacific Islands, Taiwan, and southern China; soybean is an economically important leguminous seed crop for feed and food products, that is rich in seed protein (~40%) and oil (~20%); it enriches the soil by fixing nitrogen in symbiosis with bacteria; genetic resources for breeding approaches are limited; therefore, latest activities are focused on molecular means to increase mutant frequency; a sequencing-based reverse genetics approach was developed in order to obtain mutants in genes, where breeders are limited by assay methods for compositional traits, *Glycine max* (Leguminosae), 2n = 2x = 40 (GG); 1C DNA content 2,5 pg, var. Polanka, genome size 1,223 Mbp *agr* >>> agroenergy crop >>> *chloroplast* >>> Table 16

speargrass [*syn* **blady grass** (Australia) *syn* **alang-alang** *syn* **lalang** (Malaysia) *syn* **gi** (Fiji) *syn* **ngi** *syn* **paille de dys, paillotte, impérata cylindrique, impérate, satintail, kasoring** (Palau) *syn* **cotton wool grass, cotranh** (Vietnam) *syn* **illuk** (Sri Lanka) *syn* **yakha** (Laos) *syn* **kunai** (New Guinea) *syn* **silver spike** (southern Africa) *syn* **cogon grass** (U.S.) *syn* **bai mao gen** *syn* **'Red Baron'** (Europe)]: A perennial rhizomatous grass native to East and Southeast Asia, India, Micronesia, and Australia; it grows from 0.6–3 m tall; the leaves are ~2 cm wide near the base of the plant and narrow to a sharp point at the top; the margins are finely toothed and are embedded with sharp silica crystals; the red leaf color in the early growth stages is the main character of the plant; the main vein is a lighter color than the rest of the leaf and tends to be nearer to one side of the leaf; roots are up to 1.2 m deep, but 0.4 m is typical in sandy soil; intergeneric hybridization between wheat (*Triticum aestivum*) and this wild weedy species resulted in the recovery of a high frequency of wheat haploids, which were obtained through the elimination of speargrass chromosomes from the hybrid; comparisons based on the efficiency of speargrass and maize (*Zea mays*) as pollen sources indicated that speargrass-mediated haploid production is equally efficient; in haploid production of >>> *Triticum durum*, it shows significantly better pollen source for haploid induction as compared to maize, e.g. in terms of pseudoseed formation, embryo formation, haploid regeneration and haploid formation efficiency; the line × tester analysis revealed that both male and female genotypes had significant effects on all haploid induction parameters except haploid formation frequency; among the pollen sources, *I. cylindrica* emerged as the best combiner based on general combining ability (GCA) values when compared with two Himalayan maize composites; *Imperata cylindrica* (Poaceae), 2n = 2x? = 20? 1C DNA content 5.43 pg *agr hort*

spearmint: Perennial herb belonging to the mint family, with aromatic leaves and spikes of purple flowers; the leaves are used for flavoring in cookery; *Mentha arvensis syn M. piperita, M. spicata* (Labiatae), 2n = 2x = 36; 2C DNA content 0.7 pg *hort* >>> peppermint

speedwell: A group of flowering plants belonging to the snapdragon family; of the many wild species, most are low growing with small bluish flowers; the creeping common speedwell (*Veronica officinalis*) grows in dry grassy places and open woods throughout Europe; it has oval leaves and spikes of lilac flowers; *Veronica* spp. (Scrophulariaceae), 2n = 4x = 36; 1C DNA content 0.90 pg *bot agr*

spelt wheat: May be either winter or spring in habit, and awned or awnless; the spike is long and narrow; spikelets are two-kernelled and upright, closely pressed to the rachis or central stem; kernels are red, long, flattened, with a sharp tip and a narrow, shallow crease; they remain enclosed in the glumes after threshing; *Triticum spelta* (Gramineae), 2n = 6x = 42 (BBA^{u}A^{u}DD); 1C DNA content 17.33 pg >>> wheat

spice saffron: >>> meadow saffron

spider plant (*syn* **shona cabbage** *syn* **African cabbage** *syn* **spiderwisp** *syn* **cat's whiskers** *syn* **chinsaga** *syn* **stinkweed**): An African leafy vegetable with a potential to improve food security and micronutrient deficiencies; it is an annual wildflower native to Africa but has become widespread in many tropical and sub-tropical parts of the world; (*Cleome gynandra* (Cleomaceae), 2n = 2x = 34; 2C DNA content 2.31 to 2.45 pg *hort*

spikenard: >>> aralia

spinach: A vegetable grown for its edible leaves; it was first cultivated in Persia; in popular folklore, spinach is supposed to be rich in iron; in reality it has about the same iron content as any other green vegetable; but spinach is a rich source of vitamin A and E, and several vital antioxidants; *Spinacia oleracea* (Chenopodiaceae), 2n = 2x = 12; 2C DNA content 1.6 pg *hort*

sponge gourd: >>> luffa

spotted medick: A procumbent to scrambling annual growing to 60 cm, and glabrous when mature; *Medicago arabica* (Leguminosae), 2n = 2x = 16; 1C DNA content ~0.86 pg *agr*

spring oilseed rape: >>> rapeseed

spring onion: >>> Welsh onion

sprouting broccoli: *Brassica oleracea* convar. *botrytis* var. *cymosa* (Brassicaceae), 2n = 2x = 18 (CC); 1C DNA content 0.78 pg *hort* >>> *Brassica* >>> Figure 8

spurry (corn ~): Considered to be an agricultural weed; in the 1950s, it was introduced to Europe as a secondary oil crop plant; its characteristics are branched stems to 40 cm, glandular-pubescent, and annual; leaves are linear, furrowed on lower side; seeds are small, flattened, without wings or with circum-equatorial wings less than one-tenth as wide as the actual seed; *Spergula arvensis* (Caryophyllaceae), 2n = 2x = 18; 1C DNA content 1.05 pg *agr*

squash: Transgenic squash cultivars resistant to *Zucchini yellow mosaic virus*, *Watermelon mosaic virus*, and *Cucumber mosaic virus*, have been deregulated and commercialized since 1996; *Cucurbita maxima* (Cucurbitaceae), 2n = 2x = 40; 1C DNA content 0.46 pg *hort*

St. Augustine grass: An important warm-season turf and pasture grass in the U.S. and the least cold-hardy turfgrass species; St. Augustine grass is well suited for lawns and commercial landscapes; while many genotypes are cross-fertile, all cultivars are propagated vegetatively in sod production; to ensure varietal purity e.g. variety 'Raleigh', development of sterile triploid hybrids by crossing tetraploid and diploid genotypes has been successfully used in other warm-season turfgrasses; it varies genetically in many morphological and agronomic traits; *Stenotaphrum secundatum* (Poaceae), 2n = 2x = 18; 1C DNA content 0.54 pg *bot agr*

star fruit: >>> carambola

star grass: Also known as Bermuda grass or Bahamas grass; one of the most widely dispersed grasses in the tropics and subtropics, extending even to Southwest England; whereas it can be a serious weed, with fast-growing rhizomes and runners, it can be useful as both a pasture grass and a turf grass; it is usually propagated vegetatively, but some forms can be sown by seed; non-rhizomatous, high-yielding strains are known and are very useful; *Cynodon dactylon* (Gramineae), 2n = 2x = 18; 1C DNA content 0.80 pg *bot agr*

starwort: >>> aster

statice: >>> sea lavender

stevia (*syn* **sweetleaf** *syn* **sweet leaf** *syn* **sugarleaf**): A perennial plant; BERTONI was the first to describe it botanically in the year 1899, and for giving it the Latin name in honor of a chemist from Paraguay named REBAUDI; in his description, BERTONI states that he was extremely surprised to find the extreme sweetness contained in the smallest of the leaves; the plant can reach heights from 40–80 cm; it is the number one sweetener in Japan, and the production of liquid stevia, stevia extracts, and stevia plants can be found throughout the world, mainly in South America, Paraguay, Central America, the U.S., China, and Israel; as a sugar substitute, stevia's taste has a slower onset and longer duration than that of sugar, although some of its extracts may have a bitter or liquorice-like aftertaste at high concentrations; with its extracts being up to 300 times the sweetness of sugar, stevia has garnered attention with the rise in demand for low-carbohydrate, low-sugar food alternatives; stevia has a negligible effect on blood glucose, even enhancing glucose tolerance; therefore, it is attractive as a natural sweetener for diabetics and others on carbohydrate-controlled diets; "rebiana" is the trade name for a stevia-derived sweetener being developed jointly by the Coca Cola Co. and Cargill Inc., with the intent of marketing in several countries and gaining regulatory approval in the U.S.; the European Union approved it in 2011; *Stevia rebaudiana* (Asteraceae), 2n = 2x, 3x, 4x = 22, 33, 44 *hort agr*

stinging nettle: *Urtica dioica* (Urticaceae), 2n = 4x = 52; 2C DNA content 3.2 pg *bot agr*

stitchwort: *Stellaria* spp. (Caryophyllaceae), 2n = 40,42,44 *bot agr*

St John's wort: Flowers used as herb to treat symptoms of mild depression and mood swings; a European wildflower that is naturalized throughout North America; since 2000, there have even been breeding activities in several European countries, such as Poland and Germany; because of facultative apomictic growth (aposporic), pseudogamy, and its tetraploid genome constitution, breeding is difficult and only clonal varieties seem to be available, *Hypericum perforatum* (Hypericaceae), 2n = 4x = 32; 1C DNA content 0.78 pg *hort*

stone apple: >>> bael

stone onion: >>> Welsh onion

strand wheat: >>> sea lyme grass

strawberry: The garden strawberry or pineapple strawberry (*Fragaria* × *ananassa*) has an octoploid genome; the origin derives from a cross between *F. chiloensis* x *F. virginiana*, combining larger fruit size of *F. chiloensis* and hermaphroditism of *F. virginiana*; it was recognized by A. N. DUCHESNE (1766); he repeated the cross and grew progeny, which matched *F. ananassa* in appearance; his theory was only accepted in the 20th century; *F. ananassa* became the main strawberry; additional diversity from California races of *F. chiloense* brought the "everbearing" type, with a greatly extended season; other crosses between *F. ananassa* and the Rocky Mountain form of *F. virginiana* (*F. ovalis*) introduced drought and low-temperature tolerances; *Fragaria ananassa* (Rosaceae), 2n = 8x = 56; 1C DNA content 0.61 pg *hort* >>> http://www.nal.usda.gov/pgdic/Strawberry/book/boksix.htm

strawberry clover: A perennial with slender, pointed trifoliate leaves; leaflet venation is distinctive, with veins meeting the edge of the leaf at right angles; hairs are present on petioles and underside of leaflets; stipules are relatively conspicuous and become gradually narrowed; from leaf axils, form a prostrate branched network radiating from an initially tap-rooted seedling; overall growth habit is similar to that of white clover, with perennation dependent on adventitious-rooted plantlets; development occurs from stolon nodes, though the tap root of strawberry clover persists for longer than that of white clover; cross-fertilized by bumblebees and honey bees; *Trifolium fragiferum* (Leguminosae), 2n = 2x = 16; 1C DNA content 0.54 pg *bot agr* >>> clover

strawberry peach: >>> Chinese gooseberry

strawberry spinach: The name describes the red, sappy, and edible fruits showing a similar shape to wild strawberry; this old vegetable was subsequently replaced by spinach because of its bigger leaves and easier harvest; *Chenopodium capitatum* (Chenopodiaceae), 2n = 2x = 18; 1C DNA content ~0.65 pg *hort* >>> spinach >>> mountain spinach

strawberry tree: An evergreen shrub or small tree, native to the Mediterranean region and western Europe north to western France and Ireland; due to its presence in South West Ireland, it is also known as the Irish strawberry tree and the Killarney strawberry tree; it grows to 5–10 m tall, rarely up to 15 m, with a trunk diameter of up to 80 cm; the fruit is a red aggregate drupe, 1–2 cm diameter; its flowering occurs in autumn, together with fruits from the previous year; *Arbutus unedo* (Ericaceae), 2n = 2x = 26 *hort*

straw flower: *Helichrysum bracteatum* (Asteraceae), 2n = 2x = 14 *hort*

string bean: >>> French bean

stubble turnip: *Brassica rapa* var. *rapifera* (Brassicaceae), 2n = 2x = 20 (AA); 1C DNA content 0.80 pg *agr* >>> *Brassica* >>> May beet >>> Figure 8

subterranean clover: *Trifolium subterraneum* (Leguminosae), 2n = 2x = 16; 1C DNA content 0.56 pg *hort* >>> clover

subterranean vetch: *Vicia sativa* ssp. *amphicarpa* (Leguminosae), 2n = 2x = 12; 1C DNA content 2.25 pg; polytene chromosomes have been observed in ovary, anthers, or immature seed tissues *bot agr*

suckling clover: *Trifolium dubium* (Leguminosae), 2n = 4x = 28; 1C DNA content 10.73 pg *bot agr* >>> clover

Sudan grass: *Sorghum sudanese* (Gramineae), 2n = 4x = 20; 1C DNA content 4.53 pg *agr*

Sudan pearl millet: >>> millet >>> pearl millet

sugar apple: >>> sweetsop

sugarbeet: Leafy beets have been cultivated since Roman times, but sugar beet is one of the most recently domesticated crops; nowadays, a crop with major economic importance, providing ~45% of the world's sugar production; wild relatives of sugarbeet are found in Europe, Asia Minor, and North Africa; >>> polyploid breeding has long been a major concern; at present, the majority of commercial hybrids are >>> triploids produced on diploid seed parents pollinated by tetraploids; in 1958, this radically new type of sugarbeet was placed on the market for the first time under the name of 'Trirave' (in continental countries) or 'Triplex' (in the U.K.); it was a substantially true triploid variety, produced by crossing a diploid male-sterile type with a tetraploid pollinator; the pollinator was cut out before harvest, so that only the female parent was harvested; the method of seed production was expensive, but the advantage of getting a high percentage of triploids is considerable, and the increased sugar production by far outweighs the extra cost of the seed; in 2013, the genome was sequenced, it comprises 567 megabases, of which 85% could be assigned to chromosomes; the assembly covers a large proportion of the repetitive sequence content that was estimated to be 63%; 27,421 protein-coding genes are predicted; phylogenetic analyses provided evidence for the separation of the order Caryophyllales before the split of asterids and rosids, and revealed lineage-specific gene family expansions and losses; *Beta vulgaris* ssp. *vulgaris* convar. *crassa* var. *altissima* (Chenopodiaceae), 2n = 2x = 18 (VV); 1C DNA content 1.25 pg *agr* >>> beet >>> *beet yellows virus* >>> agroenergy crop >>> Figure 44 >>> Table 35 >>> ACHARD >>> SCHLEGEL (2007)

sugarcane: A tropical perennial grass, thriving in humidity; it is vegetatively propagated by planting a "seed piece" – a piece of cane stalk with at least one bud; it re-sprouts annually from underground buds on basal portions of old stalks; depending on variety and growing conditions, a 0.5–1.5 kg stalk with 15% sugar will be produced in ~12 months from an original planting, or 9 to 11 months from regrowth; types of sugarcane can be placed into one of three categories according to their physical and chemical characteristics: (a) chewing cane contains fibers that stick together when chewed, making it easier to spit out the pulp once the sugar has been consumed; (b) cane for crystals must contain a high percentage of sucrose, since this is the sugar type that easily forms into crystals when concentrated;

(c) syrup canes contain less sucrose and more of other sugars, allowing the juice to be concentrated into syrup and still not form crystals; several old named varieties are still available; sugarcane cultivars are hybrid products of at least three or four *Saccharum* spp.; they are high >>> polyploids and are genetically complex; the cultivated forms are based on *S. sinense* and *S. barberi* species from India and China; the >>> center of diversity is Indonesia; sugarcane is regarded as one of the most significant and efficient sources of biomass for biofuel production; the production is under strictly rainfed conditions; Brazil produces about 95% of the total global production on more than 8 million hectares; about 13 t/ha of raw sugar is the common yield; with irrigation, about 15 t/ha can be achieved; *Saccharum officinarum* (Gramineae), $2n = 2x = 80$, 100–130; 1C DNA content 4.05 pg *agr* >>> agroenergy crop

sugar maple: Collected from sapwood during early spring; many commercial syrups contain artificial ingredients such as colorings, flavorings, and preservatives; *Acer saccharum* (Aceraceae), $2n = 2x = 26$; 1C DNA content ~0.70 pg *fore hort*

sugar kelp: A brown algae or kelp that is phylogenetically distant from both plants and animals, having diverged from green algae early in the evolution of eukaryotes; it is also known by the common name sugar kelp, sea belt and Devil's apron, due to its shape; it is found in the North East Atlantic Ocean and the Barents Sea, south to Galicia in Spain; the species is found on sheltered rocky seabeds; the sugar custard is edible; it contains sugars, which make it easy to taste sweet; it is harvested for kombu in Ireland; a regional form in the North Pacific is eaten in Japan as Karafuto kombu; there is research to explore its genome, the population history of New England sugar kelp, and to take advantage of its biphasic lifecycle to rapidly domesticate it using cutting-edge genomic selection methods; *Saccharina latissima* (Laminariaceae); there is a heteromorphic alternative life in the brown seaweed, *S. japonica*, with macroscopic monoecious sporophytes and microscopic dioecious gametophytes; female gametophytes are genetically different from males; the chromosome number of both male and female haploid gametophytes is 31, and there are 62 chromosomes in diploid sporophytes *aqu*

sugar palm: *Arenga saccharifera* (Palmae), $2n = 2x = 32$ *fore hort*

sugar pea: *Pisum sativum* var. *axiphium* (Leguminosae), $2n = 2x = 14$; 1C DNA content 4.88 pg, var. Ctirad, genome size 4,445 Mbp; polytene chromosomes have been observed in ovary, anthers, or immature seed tissues *hort*

sunflower: The genus of cultivated sunflower (*H. annuus*) includes more than 50 species; it contains one other economically important species, *H. tuberosus,* plus several ornamentals; wild species played an important role in genetic improvement of sunflower; sunflower was a common crop among American Indian tribes throughout North America; evidence suggests that the plant was cultivated by Indians in present-day Arizona and New Mexico ~3,000 BC; some archaeologists suggest that sunflower may have been domesticated before maize; sunflower was used in many ways throughout the various Indian tribes; seed was ground or pounded into flour for cakes, mush, or bread; some tribes mixed the meal with other vegetables such as beans, squash, and maize; the seed was also cracked and eaten for a snack; there are references to squeezing the oil from the seed and using the oil in making bread; non-food uses include purple dye for textiles, body painting, and other decorations; parts of the plant were used medicinally, ranging from snakebite antivenom to other body ointments; the oil of the seed was used on the skin and hair; the dried stalk was used as a building material; the plant and the seeds were widely used in ceremonies; this exotic North American plant was taken to Europe by Spanish explorers during the 16th century; the plant became widespread throughout present-day western Europe, mainly as an ornamental, but some medicinal uses were developed; by 1716, an English patent was granted for squeezing oil from sunflower seed; sunflower became very popular as a cultivated plant in the 18th century; most of the credit is given to Peter the Great (Russia);

the plant was initially used as an ornamental, but, by 1769, literature mentions sunflowers being cultivated for oil production; by 1830, the manufacture of sunflower oil was done on a commercial scale; the Russian Orthodox Church increased the popularity of sunflower by forbidding most oil foods from being consumed during Lent; however, sunflower was not on the prohibited list and therefore gained immediately in popularity as a food; by the early 19th century, Russian farmers were growing over 0.8 million ha of sunflower; during that time, two specific types had been identified: oil-type for oil production and a large variety for direct human consumption; government research programs were implemented; V. S. PUSTOVOIT developed a very successful breeding program at Krasnodar (Russia); oil contents and yields were increased significantly; today, the world's most prestigious sunflower scientific award is known as The Pustovoit Award; by the late 19th century, Russian sunflower seed found its way into the U.S.; by 1880, seed companies were advertising the "Mammoth Russian" sunflower seed in catalogues; this particular seed name was still being offered in the U.S. in 1970, nearly 100 years later (!); a likely source of this seed movement to North America may have been Russian immigrants; the first commercial use of the sunflower crop in the U.S. was silage as feed for poultry; in 1926, the Missouri Sunflower Growers' Association participated in what is likely the first processing of sunflower seed into oil in North America; Canada started the first official government sunflower breeding program in 1930; the basic plant breeding material utilized came from Mennonite immigrants from Russia; acreage spread because of oil demand; by 1946, Canadian farmers built a small crushing plant; in 1964, the Government of Canada licensed the Russian cultivar "Peredovik"; this variety produced high yields and high oil content; acreage in the U.S. escalated in the late 1970s to over 5 million ha because of strong European demand for sunflower oil; this European demand had been stimulated by Russian exports of sunflower oil in the previous decades; during this time, demand for animal fats was negatively impacted by cholesterol concerns; however, the Russians and later Bulgarians could no longer supply the growing demand of European companies; Europeans imported sunflower seed from all over the world, which was then crushed in European mills; Europe continues to be a large consumer of sunflower oil today, but depends on its own production and breeding; sunflower-like fossils were recently discovered in approximately 47.5 million-year- old Eocene sediments of Northwestern Patagonia next to the Río Pichileufú; the sunflower family (Compositae, the largest family of vascular plants; the family has more than 22,750 currently accepted species, spread across 1620 genera, and 12 subfamilies; the largest genera are *Senecio* [1,500 species], >>> *Vernonia* [1,000 species], *Cousinia* [600 species], and >>> *Centaurea* [600 species]) had already developed 50 million years ago and its place-of-origin is presumably in today's south America, when it was still part of the supercontinent Gondwana; *Helianthus* spp. (Compositae), 2n = 2x = 34; 2C DNA content 6.6–9.9 pg *agr hort* >>> agroenergy crop >>> Jerusalem artichoke >>> Tables 31, 35 >>> SCHLEGEL (2007)

sun hemp: A legume often grown for use as a green manure crop; its plants grow up to a height of 1.5–2.5 m and can add 100–120 kg N/ha in tropical and subtropical regions in 40–45 days; it can be grown in poorly drained, low-fertility soils; its alternate uses are as forage and as a source of fiber; *Crotalaria juncea* (Leguminosae), 2n = 2x = 16; 1C DNA content 1.23 pg; polytene chromosomes have been observed in ovary, anthers, or immature seed tissues *agr*

sun plant: >>>> portulac

swamp meadow grass: *Poa palustris* (Gramineae), 2n = 2x, 4x, 5x = 14, 28, 42 *bot agr*

swamp milkweed: *Asclepias incarnata syn A. syriaca* (Asclepiadaceae), 2n = 2x = 22; 1C DNA content 0.42 pg *bot agr*

Swede: In the past, it was a vegetable crop common in northern Europe, particularly Germany, Sweden, or Denmark; it was increasingly grown at the beginning of the 20th century as a substitute for potato; at that time, potato harvest was low because of strong yield lost by

Phytophthora infestans; nowadays, swede is mainly a vegetable of local importance; there are new breeds with yellow flesh and green heads ('Seefelder') or red heads ('Marian'); *Brassica napus* var. *napobrassica* (Brassicaceae), 2n = 4x = 38 (AACC); 1C DNA content 1.34 pg *agr hort* >>> *Brassica* >>> Figure 8

Swede's rape: >>> swede

Swedish clover: >>> alsike (clover)

Swedish turnip: >>> swede

sweet basil: >>> basil

sweet berry: >>> miracle fruit

sweet cherry: Sweet cherries can be grafted onto a wide range of rootstocks belonging to the genus *Prunus*; the identification of sweet cherry rootstocks, using morphological traits, is almost impossible, particularly during the dormant season, yet it is very important for the grower to know exactly the rootstock as this has a major influence on cultivar performance and agricultural practices; DNA-based molecular analysis carried out on actively growing shoot tips, leaves or dormant buds provides a good opportunity to reliably distinguish the rootstocks, e.g. using simple sequence repeat (SSR) markers, to facilitate the identification of the most popular sweet cherry rootstocks; this assay provided a flexible, cost-effective, and closed-tube microsatellite genotyping method, *Prunus avium* (Rosaceae), 2n = 2x = 16; 1C DNA content 0.35 pg *hort*

sweet chestnut: *Castanea sativa* (Fagaceae), 2n = 2x = 22,24; 1C DNA content 0.98 pg *hort fore*

sweet cicely: Plant belonging to the carrot family, native to southern Europe; the root is eaten as a vegetable, and the aniseed-flavored leaves are used in salads; *Myrrhis odorata* (Umbelliferae), 2n = 2x = 22; 2C DNA content 1.7 pg *hort*

sweet clover: Belongs to a genus with 19 species native to Eurasia from Central Europe to Tibet; it is used for forage and for soil enrichment through nitrogen fixation; *Melilotus albus* (Leguminosae), 2n = 2x = 16; 1C DNA content 1.10 pg *agr*

sweet corn: >>> maize

sweet flag: >>> calamus

sweet orange: It is the most important of the citrus fruits, in terms of acreage, and it is now used mainly as a fresh juice at breakfast in order to provide a daily dose of vitamin C; there are three main types of cultivar; "navel oranges" have a second row of carpels opening at the apex with the appearance of a "belly button" or navel; "blood oranges" have a red, or streaky red pulp; thirdly, there are cultivars with normal fruits; 'Valencia' is the most important commercial cultivar, followed by 'Washington Navel', and 'Jaffa'; *Citrus sinensis syn C. aurantium* (Rutaceae), *Citrus sinensis,* 2n = 2x = 18; 1C DNA content 0.63 pg; ~40,777,778 bp per chromosome *hort* >>> hesperidia >>> orange >>> Figure 61 >>> http://users.kymp.net/citruspages/citrons.html

sweet pea: *Lathyrus odoratus* (Leguminosae), 2n = 2x = 14; polytene chromosomes have been observed in ovary, anthers, or immature seed tissues; 2C DNA content 10.9–17.0 pg *hort*

sweet pepper: Species of *Capsicum* are grown throughout the tropics and are valuable crops under protected cultivation in many temperate countries; peppers with pungent fruits are used as a spice, either fresh, dry, or as extracted oleoresin; those with non-pungent fruits are used as a vegetable; the genus is native to the Americas, where the fruits have been used by man for over 5,000 years; the "tabasco pepper" is a large-fruited form of domesticated *C. frutescens,* while small-fruited forms of *C. frutescens* are cultivated for oleoresin extraction; most species are self-compatible and are facultative inbreeders; *Capsicum annuum* (Solanaceae), 2n = 2x = 24; 1C DNA content 3.16 pg *hort* >>> tabasco >>> pepper

sweet potato: This crop originated in tropical South America; it was taken by Polynesians to Fiji and New Zealand, where it is known by its Peruvian name "kumara"; the Portuguese took it to Africa and the Far East where it is known by its Caribbean name of "batatas," which is the origin of the English word "potato"; the Spanish took it from Acapulco to the Philippines, where it is known by its Mexican name of "camote"; it is now one of the more important tropical food crops; although it is cultivated as clones, the crop sets true seed freely, and farmers often keep self-sown seedlings as a new cultivar; the harvestable product is a tuber which, in the U.S., is often incorrectly called a >>> yam; it was domesticated in Central America more than 5,000 years ago; the crop was reportedly introduced into China in the late 16th century; because of its hardy nature and broad adaptability, and because its planting material can be rapidly multiplied from very few roots, sweet potato spread through Asia, Africa, and Latin America during the 17th and 18th centuries; sweet potato has secondary centers of genetic diversity; in Papua New Guinea and in other parts of Asia, many types of sweet potato can be found that are genetically distinct from those found in their area of origin; today it is widely grown in tropical and temperate regions of the world due to its high yield, high nutritive values, and adaptability to a wide range of soils and drought; *Ipomoea batatas* (Convolvulaceae), 2n = 6x = 90 (BBBBBB); 1C DNA content 2.25 pg *hort agr* >>> agroenergy crop

sweet root: >>> calamus

sweet sorghum: *Miscanthus sacchariflorus* (Gramineae), 2n = 4x = 76; 1C DNA content 4.20 pg *agr*

sweetsop: A small deciduous tree that rarely reaches a height of more than 6 m; the 20–25 cm leaves are usually lanceolate to oblong-lanceolate; this species is probably indigenous to the warmest part of Central America and grows best in the warm dry areas of the tropics; the fruit, which ripens in the spring, is subglobose or ovoid; it is composed of loosely cohering carpels, which are usually covered with a white or bluish bloom; the carpels separate readily when ripe, exposing the cream-colored flesh in which are embedded numerous small, brown, glossy seeds; the fruit pulp has a custardy consistency and is sweet and pleasant tasting; the pulp is usually eaten fresh as a dessert fruit, but it is also used to make a delicious sherbet; the sweetsop is of better quality than the custard apple and deserves to be more widely planted in the tropics; as with the other annonas, the sweetsop is often grown from seeds; such seedlings usually come into bearing when three to four years of age; improved selections can be propagated readily by grafting or budding; *Annona squamosa* (Annonaceae), 2n = 2x = 14; 1C DNA content 1.23 pg *hort* >>> pawpaw >>> soursop >>> ylang-ylang (related species)

Sweet William: Biennial to perennial plant belonging to the pink family, native to southern Europe; it is grown as a commercial cut flower and as a bedding plant; the species is usually biennial, and is native to mountain pastures of Central and southern Europe; plants have dark green foliage and cymose inflorescences with densely clustered florets in shades of red, pink, salmon, or white; *Dianthus barbatus* (Caryophyllaceae), 2n = 2x, 4x, 6x = 30, 60, 90 *hort*

sweet yellow lupine: *Lupinus luteolus syn L. luteus* (Leguminosae), 2n = 2x = 48; 1C DNA content 0.59 pg *agr* >>> lupine >>> Table 46

Swiss chard: The plant was selected from the wild beet, *Beta vulgaris*; there are varieties with white, green, or red petioles; it was a common vegetable in the past; leaves were prepared like >>> spinach, stalks like >>> asparagus; most breeding work is carried out in Switzerland; *Beta vulgaris* var. *cicla* (Chenopodiaceae), 2n = 2x = 18; 1C DNA content 1.25 pg *hort* >>> beet

switch grass: An ornamental grass; it is easily grown in average, medium wet to wet soils in full sun to part shade; it tolerates a wide range of soils, including dry ones, but prefers moist, sandy or clay soils; it tends to flop in rich soils; it grows primarily in clumps, but may naturalize by rhizomes as well as self-seeding to form sizable colonies; it can be grown from seed; *Panicum virgatum* (Gramineae), 2n = 6x = 54; 1C DNA content 1.35 pg *bot hort*

sword bean: *Canavalia gladiata* (Leguminosae), 2n = 2x = 22 *hort* >>> pulse

T

taami: >>> miracle fruit

ta'amu: >>> giant taro

tabasco: *Capsicum frutescens* (Solanaceae), 2n = 2x = 24; 1C DNA content 3.40 pg *hort* >>> sweet pepper

table beet: >>> red beet

table watermelon: *Citrullus vulgaris* var. *edulis* (Cucurbitaceae), 2n = 2x = 22; 1C DNA content 0.45 pg; endopolyploidy of 24–384x was observed *hort* >>> Table 46

Tahiti arrowroot (*syn* **batflower** *syn* **Indian arrowroot** *syn* **Polynesian arrowroot**): A perennial herb of East Indies to Polynesia and Australia; it is cultivated for its large edible root; *Tacca leontopetaloides syn T. involucrata syn T. pinnatifida* (Taccaceae), 2n = 28(?); 1C DNA content 1.05 pg *hort*

taling pling: >>> carambola

tall fescue: *Festuca arundinacea* (Gramineae), 2n = 6x = 42; 1C DNA content 8.49 pg *agr*

tall oatgrass: A perennial plant, robust, hairless to more or less hairy, cespitous; stems erect, 60–150 cm high; *Arrhenatherum elatius syn Avena elatior* (Gramineae), 2n = 4x = 28; 1C DNA content 7.98 pg *bot agr* >>> Table 14

tall wheatgrass: An excellent genetic source of resistance to scab and high seed protein content that has been widely used in wheat breeding for quality improvement; *Agropyron elongatum syn Lophopyrum elongatum* (Gramineae), 2n = 2x = 14; 2C DNA content 13.9 pg (PP) *bot agr*

tamarillo (*syn* **tree tomato** *syn* **tomate de árbol**): A small tree or shrub; the plant is a fast-growing tree that grows up to 5 m; peak production is reached after four years, and the life expectancy is about 12 years; the tree usually forms a single upright trunk with lateral branches; the flowers and fruits hang from the lateral branches; the leaves are large, simple and perennial, and have a strong pungent smell; it is best known as the species that bears the tamarillo, an egg-shaped edible fruit; it is native to the Andes of Peru, Chile, Ecuador, Colombia, and Bolivia; it is still cultivated in gardens and small orchards for local production; it is one of the most popular fruits in these regions, *Solanum betaceum syn Cyphomandra betacea* (Solanaceae), 2n = 2x = 24, chromosomes 1, 2, 3, and 9 are submetacentric and the rest are metacentric, chromosome 1 is the longest (5.30 μm), chromosome 12 is the shortest (2.10 μm), 2C DNA content 20.6 pg *hort*

tamarind: A large tropical tree with a short massive trunk, ferny pinnate leaves, small yellow flowers, and fat reddish brown pods; the tree can reach 28 m tall; it has drooping branches, and a domed umbrella shaped crown, which is as wide as the height of the tree; the leaves are ~25 cm long with 10–18 pairs in 2.5-cm oblong leaflets; tamarind drops its leaves in pronounced dry seasons; in climates without a dry season, it stays evergreen; the flowers are ~2.5 cm across, pale yellow with purple or red veins; they have five unequal lobes and are borne in small drooping clusters; the velvety cinnamon-brown pods are 5–15 cm long, sausage-shaped, and constricted between the seeds; the pulp that surrounds the eight to ten seeds is both sweet and extremely sour; tamarind is cultivated and has become naturalized in the tropics throughout the world, but was probably originally native to eastern Africa; it is generally propagated by seed, but it also can be started from cuttings; tamarinds are grown as ornamental shade and street trees, and for the edible pods; the pods are fed to livestock, and the pulp within the pods is used to make beverages, curries, chutneys, and sauces; pulp is also made into a soft drink known as "refresco de tamarindo" in Latin America, and "tamarinade" in Jamaica; tamarind is used extensively in Indian and Southeast Asian cuisine, and is an important ingredient in Worcestershire sauce; *Tamarindus indica* (Leguminosae), 2n = 2x = 26; 1C DNA content 0.84 pg *fore hort*

tancy phacelia: *Phacelia tanacetifolia* (Hydrophyllaceae), x = 6, 14 *agr* >>> phacelia

tangerine: >>> mandarin

tania: A vegetatively propagated root crop, known variously as >>> taro, dasheen or cocoyam; it was the basis of the agriculture in Papua New Guinea, which is amongst the oldest in the world, dating from ~7,000 BC; it became important on the Caribbean islands; in Puerto Rican cuisine, it is called "malanga" or "yautía"; it is a labor-intensive crop, and it became only a minor staple, which lacked the potential of a major staple, capable of supporting the growth of cities and the development of a sophisticated civilization; induced tetraploids has been produced and cropped, *Xanthosoma sagittifolium* syn *X. violaceum* syn *Colocasia esculenta* syn *Caladium sagittæfolium* (Araceae), 2n = 2x = 26 *hort* >>> tanier

tanier: *Xanthosoma atrovirens* (Araceae), 2n = 2x = 26; 1C DNA content 2.30 pg *hort* >>> tania

tansy (Finnish ~): *Tanacetum vulgare* (Asteraceae), 2n = 2x = 18; 2C DNA content 8.9 pg *bot hort*

tara vine: *Actinidia arguta* (Actinidiaceae), 2n = 4x = 116; 1C DNA content 1.55 pg *hort*

taro: One of the oldest known vegetables and has been grown in some regions of the world for more than 2,000 years; the starchy corms of taro are consumed throughout the tropics and are essential for food security in many developing countries; taro corms are increasingly processed into fries, chips, flours, or flakes in urban areas, and varieties with attractive corm flesh colors are now needed; recently, there has been breeding for increasing flavonoid content in taro in order to increase the crop value; *Colocasia esculenta* var. *globulifera* (Araceae), 2n = 42; 1C DNA content 4.08 pg *hort* >>> coc(o)yam >>> aerial yam

tarragon: *Artemisia dracunculus* (Compositae), 2n = 2x = 18; 1C DNA content 2.97 pg *hort*

Tartary buckwheat: *Fagopyrum tataricum* (Polygonaceae), 2n = 2x = 16; 1C DNA content 1.44 pg *agr* >>> buckwheat

tau-sahyz: >>> Russian dandelion

tea: The grade of tea depends on the age of the leaves; in "golden tips" only the youngest bud is used; in "orange pekoe" the smallest leaf; in "pekoe" the second leaf; in "pekoe souchong" the third leaf; in "souchong" the fourth leaf; and in "congou" the fifth and largest leaf is gathered; in green tea, the leaves are dried and appear dull green; in black tea, the leaves are fermented and then dried; "oolong tea" is only partially fermented and is intermediate between black and green; the various pekoes, souchongs, and congous are black teas, while gunpowder and hyson are the most important grades of green tea; *Camellia sinensis* (Camelliaceae), 2n = 2x = 30; 1C DNA content 3.90 pg *hort*

tears of Job: >>> adlay

teasel gourd (*syn* **kakrol** *syn* **kankro** *syn* **kartol** *syn* **kantoli** *syn* **kantola** *syn* **kantroli**): A relatively small oval to ovoid vegetable and a very highly prized vegetable in the Indian subcontinent though relatively rare; it is only available during the rainy months; it is only available during the rainy months *Momordica dioica*, is usually planted in the summer, it is most possibly a native plant of the Indian sub-continent; the wild ancestor is *M. cochinchinensis*; the vegetable is rich in calcium, phosphorus, iron and carotene and is in demand for export and for internal markets; the crop can be grown in different types of soil and is propagated for cultivation, through tuberous roots, between March and April; it can be harvested in 90–100 days; yield ranges from 10–12 t/ha; fruits are consumed by frying, or cooking, with or without meat or fish, or simply, like mashed potato with green chillies, mustard oil, and salt; *M. subangulata* ssp. *renigera* exhibits morphological characters found in both *M. dioica* (2n = 28) and *M. cochinchinensis* (2n = 28), i.e. of segmentally allopolyploid origin (Momordica), 2n = 8x = 56; 1C DNA content 2.05 pg (*M. charantia*, 2n = 2x = 22) *hort*

teatree (*syn* **narrowleaf paperbark** *syn* **snow-in-summer** *syn* **bottle brush** *syn* **manuka**): Shrub or small tree native to Australia and New Zealand; it is thought that some species of teatree were used by the explorer Captain Cook to brew tea; it was used for this purpose in the first years of settlement; it contains essential oils; it is the most important species for commercial production of melaleuca oil, a topical antibacterial and antifungal used in a range of products, including antiseptics, deodorants, shampoos, soaps, and lotions; *Melaleuca alternifolia* (Myrtaceae); 1C DNA content 1.13 pg (*M. leucadendra*, 2n = 22) *hort*

tef(f): A species of lovegrass native to northern Africa; it is similar to millet in nutrition and in cooking, but the seed is much smaller; tef is an important food grain in Ethiopia, where it is used to make injera; because of its small seeds (less than 1 mm diameter), one can hold enough to sow a whole field in one hand; this property makes tef particularly well suited to a semi-nomadic lifestyle; since tef is naturally gluten-free, it is well suited for people with celiac disease or on a gluten-free diets; *Eragrostis tef* (Poaceae), 2n = 4x = 40; 1C DNA content 0.35 pg *agr*

teonochtil: >>> pencas

teosinte: The tripartite hypothesis postulates that cultivated maize was derived from a wild pod-corn, which was once indigenous to the lowlands of South America, that this wild *Z. mays* is now extinct, that teosinte originated from natural hybridization of *Z. mays* with a species of *Tripsacum* after cultivated maize was introduced into Central America, and that most modern races of maize resulted from introgression of primitive maize with teosinte, *Tripsacum*, or both; the oldest known archaeological remains suggest that a wild *Z. mays* existed before teosinte came on the scene; *Z. mays* and teosinte are conspecific and only distantly related to *Tripsacum*; it is probable that maize originated from a teosinte-like ancestor under domestication; *Euchlaena mexicana syn Zea mays* ssp. *mexicana* (Gramineae), 2n = 2x = 20; 2C DNA content 5.43 pg, var. Ce-777, genome size 2,655 Mbp *bot agr*

teosinte, perennial *Zea diploperennis* (Gramineae), 2n = 2x = 20; 1C DNA content 2.6 5 pg (*Zea perennis,* 2n = 4x = 40; 1C DNA content 5.28 pg) *bot agr* >>> teosinte >>> http://en.wikipedia.org/wiki/Teosinte

tepary bean: Of very ancient domestication in Mexico, and later replaced to a large extent by *Phaseolus vulgaris, Phaseolus acutifolius* var. *latifolius* (Leguminosae), 2n = 2x = 22; 1C DNA content 0.60 pg; polytene chromosomes have been observed in ovary and immature seed tissues *hort* >>> pulse

Texas bluegrass: A vigorous sod-forming perennial, dioecious grass, tolerant to heat; it is native to the southern Great Plains of the U.S.; interspecific hybridization between >>> Texas blue grass and >>> Kentucky (blue) grass (*Poa pratensis*) has shown to be a proven method for the development of turf-type hybrid bluegrass cultivars for the southern United States; *Poa arachnifera* (Gramineae), 2n = 8x = 56; 2C DNA content 8 to 13 pg, using flow cytometry *bot agr*

thale cress: A flowering species of the family Brassicaceae; this plant became a main subject of molecular genome analysis because it has a small and simple genome; it contains almost no repetitive DNA; its genome was completely sequenced by the end of 2000; it is the model plant system of choice because of the additional advantages of short generation time (~5 weeks), high seed production (up to 40,000 seeds per plant), and natural self-pollination; *Arabidopsis thaliana* (Brassicaceae), 2n = 2x = 10; 2C DNA content 0.15 pg = 120 Mb; ~25,000,000 bp per chromosome *bot gene* >>> *Brassica*

Thinopyrum bessarabicum: (Gramineae), 2n = 2x = 14 (JJ or E^bE^b); 2C DNA content 14,9 pg *bot*

Thinopyrum distichum syn *T. junceiforme* **(Gramineae):** A highly salt-tolerant, perennial grass that is indigenous to the shoreline of Southern Africa where it grows within the spring high-tide zone; it is rhizomatous, exhibits facultative apomixis, and occupies highly saline coastal sands with low fertility, limited soil water, and high pH; due to its adaptation to adverse environmental conditions, the grass has previously been targeted for gene mining and transfer to durum and common wheat, rye and triticale; it is a segmental autotetraploid and, in a partial polyhaploid, its two genomes show a high degree of meiotic pairing; 2n = 4x = 28 ($J^{1d}J^{1d}J^{2d}J^{2d}$); 1C or ($J^1J^1J^2J^2$) DNA content ~11.80 pg *bot*

Thinopyrum elongatum: (Gramineae), 2n = 2x = 14 (EE); 2C DNA content 12.0 pg (E^eE^e) *bot*

***Thinopyrum intermedium*:** (Gramineae), 2n = 6x = 42 ($E^eE^eE^eE^eS^tS^t$); 1C DNA content 12.95 pg
  bot agr >>> intermediate wheatgrass

thistle: *Cirsium* sp. (Asteraceae), 2n = 34 *bot agr*

thistle artichokes: >>> cardoon

thyme: Herb belonging to the mint family; garden thyme is native to the Mediterranean, grows
  to 30 cm high, and has small leaves and pinkish flowers; its aromatic leaves are used for
  seasoning in cookery; *Thymus vulgaris* (Lamiaceae), 2n = 2x = 30; 1C DNA content 0.78
  pg *hort*

tien chi: >>> san chi

til: >>> sesame

***timopheevi* wheat:** A late-maturing spring wheat with leaf blades that are pubescent on both sides;
  spikes are very compact, rather short, somewhat pyramidal in shape with soft, thin, rather
  short awns; spikelets usually contain two kernels; kernels are medium-long, slender, and
  hard or flinty; it occurs in Transcaucasia and Russia; *Triticum timopheevii* (Poaceae), 2n =
  4x = 28 (GGA^uA^u); 1C DNA content 11.30 pg >>> wheat

timothy: Perennial grass native to the United Kingdom and Europe; it is widely grown as fodder for
  livestock; it was introduced to North America by Timothy HANSEN in the 18th century,
  Phleum pratense (Gramineae), 2n = 6x = 42 (AAAABB); 2C DNA content 8.3 pg *agr*

tobacco: A tall, herbaceous plant, the leaves of which are harvested, cured, and rolled into cigars,
  shredded for use in cigarettes and pipes, or processed for chewing or as snuff; the main
  source of commercial tobacco is *Nicotiana tabacum*, although *N. rustica* is also grown
  and is used in oriental tobaccos; breeders have developed a wide range of morphologically different types, from the small-leafed, aromatic tobaccos to the large, broad-leaved
  cigar tobaccos; it is the most widely grown non-food crop in the world; it is not only of
  agronomic interest, but also for its utilization in genetic, physiological, biochemical, and
  biotechnological research; *Nicotiana tabacum, N. rustica* (Solanaceae), 2n = 4x = 48; 2C
  DNA content 6.6 pg = 1,650 Mb *agr*

tomatillo: A semi-woody annual, 60 to 150 cm high, mostly grown as a sprawling plant; it is native
  to southern Baja California and Guatemala; its fruit is globose, two-celled, 2.5–7.5 cm in
  diameter, smooth and sticky; the flesh is pale-yellow, crisp or soft; it is used as vegetable,
  not as a fruit, *Physalis philadelphica* (Solanaceae), 2n = 2x = 24 *hort* >>> groundcherry

tomato: Like potato, tomato originates from South America and became an early crop plant in
  Mexico; because of the strong smell of leaves and stem, it was believed to be a poisonous
  plant; hence, the early Latin genus name *Lycopersicon* ("wolf's peach"); in the 16th century it was already being used as a vegetable in Italy (French: pomi d'oro = golden apple);
  during the 19th century, the varieties 'King Humbert', 'Rè Umberto' (both from 1880), or
  'Purpurviolet Ponderosa' became very attractive; at the beginning of the 20th century, the
  variety 'Lukullus' was famous; as with the garden pea, it became a subject of intensive
  genetic research and is still one of the best-investigated crop plants; there are more than
  10,000 varieties described; alien gene transfer and genetic engineering were successfully
  demonstrated in tomato; *Solanum syn Lycopersicon esculentum* (Solanaceae), 2n = 2x =
  24; 2C DNA content 1.96 pg, var. Stupicke polni tyckove rane, genome size 958 Mbp; by
  2012, the whole genome was for the first time sequenced, both for both the domesticated
  type and its wild ancestor, *Solanum pimpinellifolium,* revealing the order, orientation,
  types and relative positions of its 35,000 genes (The Tomato Genome Consortium started
  its work in 2003) *hort agr* >>> chloroplast >>> potato

tomate de árbol: >>> tamarillo

tonka bean: Seeds from the egg-shaped fruits of this tropical South American tree are used as a
  substitute for vanilla; the seeds contain the fragrant phenolic compound coumarin which
  is used in the perfume industry; *Dipteryx odorata* (Fabaceae) *hort fore*

toria: >>> Indian rape

t'ou: >>> rakkyo

touch-me-not: *Impatiens balsamina* (Balsaminaceae), 2n = 2x = 14; 1C DNA content 1.33 pg *hort*

tree of heaven: *Ailanthus glandulosa* (Simaroubaceae), 2n = 32; 1C DNA content 2.18 pg *bot hort*

tree onion: *Allium cepa* var. *viviparum* (Alliaceae), 2n = 2x = 16; 1C DNA content 16.75 pg *hort*

tree sorrel: >>> bilimbi

tree tomato: >>> tamarillo

tree spinach: >>> lambsquarter

trembling aspen: *Populus tremuloides* (Salicaceae), 2n = 38; 1C DNA content 0.45 pg (1 pair of heteromorphic chromosomes are responsible for the determination of the male sex) *fore*

triticale: An artificial amphiploid hybrid between wheat (*Triticum*) and rye species (*Secale*), in which wheat is the donor of the cytoplasm; it is a cereal crop created by man from wheat and rye, based on research starting at the end of the 19th century (WILSON 1876, U.K.; RIMPAU 1888, 1891, Germany); the aim was to combine the quality advantages of wheat with the stress insensitivity of rye; growth habit is similar to wheat; it differs from wheat by having a greater vigor and larger-sized spikes and grains; among three basic ploidy levels so far developed for triticale, the hexaploid type became the most important for breeding and agriculture; there is a taxonomic proposal for classification: genus *Triticum,* section *Triticale,* notospecies *Triticale krolowii* [(2n = 4x = 28 (BBR^{ce}R^{ce}, A^{u}A^{u}R^{ce}R^{ce}, or DDR^{ce}R^{ce})], *Triticale turgidocereale* [2n = 6x = 42 (BBA^{u}A^{u}R^{ce}R^{ce}); 1C DNA content 19.80 pg] and *Triticale rimpaui* [2n = 8x = 56 (BBA^{u}A^{u}DDR^{ce}R^{ce}); 1C DNA content 25.98 pg] (Gramineae) *agr* >>> Tables 1, 15, 16, 48

tritipyrum: An artificial amphiploid hybrid between wheat (*Triticum*) and *Thinopyrum* species (>>> *Thinopyrum*) in which wheat is the donor of the cytoplasm (Gramineae) *agr*

tritordeum: An artificial amphiploid hybrid between wheat (*Triticum*) and barley species (*Hordeum*), in which wheat is the donor of the cytoplasm (Gramineae), 2n = 8x = 56 (BBA^{u}A^{u}DDH^{vu}H^{vu}); in 2013, the Institute of Sustainable Agriculture (IAS) of the Spanish National Research Council (CSIC), in collaboration with AGRASYS – a spin-off of the CSIC, has launched tritordeum, a new cereal which has accreditation as a natural crop species and has now reached the market; after 30 years of breeding, tritordeum is a cereal with good nutritive and agronomic characteristics (high tolerance to drought and heat stress); the latest strain is a cross between durum wheat and a wild barley species, *Hordeum chilense*, native to Chile and Argentina; it is the second new cereal developed by man – the first one was >>> triticale, a combination of durum wheat and rye, which is used for animal feed; tritordeum is registered in the Community Plant Variety Office (CPVO) of the European Union and is a natural crop species *agr* >>> Table 41 >>> cytoplasmic male sterility (CMS)

true turnip: >>> stubble turnip

truffle: An edible fruit-body of *Tuber* spp. or other Tuberales; of all the edible fungi, truffles are perhaps the most fascinating; they are truly the *non plus ultra* of mushroom cuisine; truffles are the fruiting bodies (ascocarps) of mycorrhizal ascomycetous fungi; unlike other common forest mushrooms, truffles are subterranean and resemble small pebbles or clods of dirt beneath the soil; they emit the odor of certain mammalian steroids and are irresistible to some mammals, including female pigs; this particular steroid is found in the saliva and breath of male pigs (boars) and explains the natural lust and talent sows have for truffle hunting; pigs and dogs can detect truffles from as far away as 50 m; the fabled truffles of France and Italy retail for more than $1,000 per kilogram; *Tuber melanosporum* (black truffle), *T. magnatum* (white truffle), *T. gibbosum* (Oregon white truffle) (Tuberaceae), 2n = 2x = 8, 10 *bot hort*

tsuru reishi: >>> bitter gourd

tuba root: >>> derris

tuberous celery: *Apium graveolens* var. *rapaceum* (Umbelliferae), 2n = 2x = 22; 6x, 9x, 11x, 12x and 15x *hort*

tulip: Introduced into Europe in 1572; Dutch growers developed many new varieties, resulting in tulipmania in the 1630s, when vast sums were paid for bulbs of rare colors; *Tulipa* spp. (Liliaceae), 2n = 2x, 3x, 4x = 24, 36, 48; 1C DNA content 12.35–60.45 pg *hort* >>> http://www.tulipworld.com/about-tulip-world

tunas: >>> pencas

tung (oil tree): *Aleurites fordii, A. montana* (Euphorpiaceae), 2n = 2x = 22 *fore*

Turkish cobnut: *Corylus colurna* (Corylaceae), 2n = 2x = 22; 1C DNA content 0.48 pg *hort* >>> hazel

Turkish red pine: >>> Brutian pine

Turkish rocket: Leaves and young stems are eaten raw or cooked; young leaves have a mild cabbage flavor that goes very well in a mixed salad, though some people find them indigestible; the leaves are a bit hairy; cooked leaves make an excellent vegetable; they are available early in the year, usually towards the end of winter; flower buds and flowering stems give a pleasant mild flavor with a delicate sweetness and cabbage-like flavor; they make an excellent broccoli substitute though they are rather smaller; *Bunias orientalis* (Brassicaceae), 2n = 2x = 14; 1C DNA content 2.59 pg *hort* >>> *Brassica* >>> rocket salad

turkterebinth nut: >>> pistachio (nut)

turmeric: Mostly, a sterile, triploid, vegetatively-propagated crop cultivated mainly in Southeast Asia; when dried rhizomes are ground, the resulting yellow powder is used by the food industry as a natural food dye; many pharmacological compounds have broadened the commercial application of the crop; plants are gathered annually and re-planted from some of those rhizomes in the following season; because of triploidy, conventional breeding is difficult and, hence, improvement has been limited to germplasm selection; *Curcuma longa* (Zingiberaceae), basic genome n = 7, different ploidy levels were found: 6x, 9x, 11x, 12x and 15x, *however common* 2n = 9x = 63, 2C DNA content 2.6–2.9 pg *hort*

turnip: *Brassica campestris* (Brassicaceae), 2n = 2x = 20 (AA); 1C DNA content 0.58 pg *hort* >>> beet >>> *Brassica* >>> May beet >>> Figure 8

turnip cabbage: >>> kohlrabi

turnip rape: *Brassica campestris* ssp. *oleifera* (Brassicaceae), 2n = 2x = 20 (AA); 1C DNA content 0.58 pg *agr* >>> *Brassica* >>> Figure 8

turnip-rooted chervil: A vegetable of middle and eastern Europe; there is also a Siberian species (*C. prescottii*), which can be spring-sown; in Germany, it was a preferred vegetable spice in the 18th and 19th centuries; *Chaerophyllum bulbosum* (Umbelliferae), 2n = 2x = 22 *hort* >>> Table 46

tussock grass: A tall grass (50–60 cm tall) that is used as an ornamental grass; it grows in clumps and has fine hair-like leaves; in summer, it produces fine green to yellow inflorescences; *Deschampsia caespitosa syn Aira caespitosa* (Gramineae), 2n = 2x, 4x = 26, 52; 1C DNA content 2.93–6.03 pg *hort*

U

ulluco [*syn* **papa lisa** *syn* **melloco** (Ecuador) *syn* **olluco** (Peru) *syn* **chugua** (Colombia) *syn* **ruba** (Venezuela)]**:** A tuberous crop native to the Andes; it is the most widely grown and economically important root crop of the Andean region; red, yellow, and red-spotted accessions are known; the leaf and the tuberous root are edible, similar to spinach and potato, respectively; known to contain high levels of protein, calcium, and carotene; the major appeal of the ulluco is its crisp texture which, like the jicama, remains even when cooked; because of its high water content, the ulluco is not suitable for frying or baking but it can be cooked in many other ways, like the potato; *Ullucus tuberosus* (Basellaceae), 2n = 2x = 24 *hort*

underground onion: >>> multiplier (onion)

urd bean: A highly prized pulse in India; the flowers are self-pollinating and crosspollination is very rare; the maximum diversity for *V. mungo* exists in the upper Western Ghats and the Deccan Hills, with a second center in Bihar – that is, the center of origin of two crops, *Vigna mungo syn Ullucus tuberosus* and *V. radiata* lies in India; *Phaseolus mungo syn Vigna mungo* (Leguminosae), 2n = 2x = 22; 1C DNA content 0.55 pg; polytene chromosomes have been observed in ovary, anthers or immature seed tissues *hort* >>> pulse >>> mung bean

urn plant: An ornamental plant; corollas of *Aechmea* change color from powder blue early in the day to deep rose-red by late afternoon and to intense red as the pistil shrivels, about one day after anthesis, whether the flower is pollinated or not; it originates from tropical and subtropical rainforests of Central and South America; over this vast range, bromeliads are found in a variety of habitats; a range of techniques, such as cut style pollination, grafted style pollination, placental pollination, and the use of mentor pollen, have been applied to overcome prefertilization barriers in hybridization research; *Aechmea fasciata* (Bromeliaceae), 2n = 2x = 50 *hort* >>> www.lapshin.org/cultivar/N36/pogany-e.htm

V

valerian: Perennial plants native to the northern hemisphere, with clustered heads of fragrant tubular flowers in red, white, or pink; the root of the common valerian or garden heliotrope is used in medicine to relieve wind and to soothe or calm patients; *Valeriana officinalis* (Valerianaceae), 2n = 8x = 56; 1C DNA content 4.08 pg *hort*

vanilla: Climbing orchids native to tropical America but cultivated elsewhere, with large, fragrant, white or yellow flowers; from fermented and dried seed capsules called vanilla beans; annual world production of vanilla pods is estimated at 1,500 t; *Vanilla planifolia* (Orchidaceae; *Ophioglossum reticulatum* has the highest known diploid chromosome number 2n = 1,260), 2n = 2x = 32 [different accessions show 2n = 16 to 32, 2n = 16 to 38, and 2n = 22 to 54 chromosomes, and 2C DNA contents of 5.03 pg (for most accessions), 7.67 pg (for the "Stérile" phenotypes), and 10.00 pg (for the "Grosse Vanille" phenotypes), resp.] *hort*

varnish tree: *Rhus venicifera syn R. verniciflua* (Anacardiaceae), 2n = 2x = 30 *fore*

***vavilovi* wheat:** A winter-type wheat, midseason in maturity, with thick, strong stems; spikes are medium dense to loose, and awned; kernels remain in the spikelets after threshing; they are ovate, white and hard; it is grown somewhat in Russia, *Triticum aestivum* ssp. *vavilovi syn T. vavilovii* (Gramineae), 2n = 6x = 42 (BBA^{u}A^{u}DD); 1C DNA content 17.74 pg >>> wheat

vegetable marrow: *Cucurbita pepo* (Cucurbitaceae), 2n = 2x = 40; 1C DNA content 0.55 pg *hort*

vegetable oyster: >>> salsify

vegetable pear: >>> chayote

velvetbean: A self-pollinated species and an important legume used in tropical agricultural systems in rotation with other crops for nematode management and/or soil improvement; *Mucuna* spp. (Leguminosae), 2n = 2x = 22 *agr*

velvetleaf: A fiber plant grown in China in many varieties; the fruit consists of a circular row of about 10–15 flattened seedpods; each seedpod has a stout beak and contains about 5–15 seeds; seeds are grayish brown, somewhat flattened, and either kidney-shaped or heart-shaped; the root system consists of a stout white taproot; the plant spreads by reseeding itself; seeds are used in herbal medicine to remove heat and damp, to counteract toxicity, and to cure nebula; *Abutilon theophrastii* (Malvaceae), 2n = 6x = 42; 1C DNA content 1.40 pg *hort*

vernonia (*syn* **ironweed**): Approximately 1,000 species are found in the large genus *Vernonia*; the species *V. galamensis* is limited in distribution primarily to eastern Africa; it is a potential new industrial crop growing wild in Ethiopia; seeds from this plant contain an oil rich in epoxy fatty acids; epoxy oils are widely used in plasticizers and as additives in flexible polyvinyl chloride resins; this market is currently supplied by the epoxidation of either soybean or linseed oil; the composition of vernonia oil has superior qualities compared to these other oils; a potential market use might be as a drying agent in reformulated oil-based or alkyd resin paints; some 1.2 billion liters of paint are manufactured annually in the U.S.; the drying agents currently used are major pollutants; the annual plant is grown on small farms in Africa, where seeds are bought and locally crushed for oil; private industries have also produced this species in other countries located close to the equator; this is because the plants with the largest seeds and the best seed retention flower only under short-day conditions; hybrids were developed between short-day flowering types and an accession flowering under any daylength; the resulting day-neutral hybrids can be planted in spring and harvested in autumn, that is, under the long-day conditions of North America; they have been grown successfully in Arizona, Kentucky, Missouri, Oregon, Texas, Virginia, and northern Argentina; commercial development of these hybrids depends on further improvement of seed yields; *Vernonia galamensis syn V. pauciflora* (Compositae), $2n = 2x = 18$ *hort agr* >>> sunflower

>>> http://www.hort.purdue.edu/newcrop/CropFactSheets/vernonia.html#Taxonomy

vetch: *Vicia* spp., *Lathyrus* spp. (Leguminosae), polytene chromosomes have been observed in ovary, anthers, or immature seed tissues *agr bot* >>> woolly-pod vetch >>> common vetch >>> hairy vetch >>> hairy vetchling >>> Hungarian vetch >>> kidney vetch >>> narbon vetch >>> purple vetch >>> subterranean vetch

Vietnamese balm (*syn* **rau mong toi**): Flowers are hermaphroditic and are pollinated by insects; the plant prefers light (sandy), medium (loamy), and heavy (clay) soils; young leaves can be eaten raw or cooked, finely cut and added to salads, or used as a potherb; the leaves also can be used as an aromatic condiment for vegetable dishes; powdered seeds are used as a condiment for flavoring foodstuffs; it contains an essential oil and is antibacterial, antipyretic, antiviral, astringent, carminative, diaphoretic, diuretic, and stomachic; it is used in the treatment of common colds, fevers, headaches, diarrhea, edema, and oliguria; the plants have a broad-spectrum antibacterial action; the plant is seed propagated; *Elsholtzia ciliata* (Labiatae), $2n = 2x = 18$ *hort*

violet: >>> African violet

viper's grass: >>> black salsify

W

wavy hairgrass: *Deschampsia flexuosa* (Gramineae), $2n = 4x = 28$; 2C DNA content 11.0 pg *agr bot*

wallflower: European perennial cottage garden plant with fragrant spikes of red, orange, yellow, brown, or purple flowers in spring; *Erysimum cheiri syn Cheiranthus cheiri* (Brassicaceae), $2n = 2x = 14$; 1C DNA content 0.26 pg *hort* >>> *Brassica*

walnut: Its spontaneity has been proved in some Asian areas and its spread in Europe; as far as the cultivated tree, its culture is owed to the Romans; it is thought that the legions took the plant to the regions near the Rhine, in Germany, as well as to Spain; afterwards, it was introduced into France and England; the walnut tree was taken to America by the Spanish navigators and it was widely accepted in California; its introduction to the southern hemisphere is more recent; at present, the walnut tree is cultivated mainly in the south of Europe, where its quality is superior to the commercial product from other continents, India, northern regions of Japan, China, South America and in general, all the regions of the temperate climate world; the walnut is a tall, deciduous tree prized for its dark timber; the wrinkled nut is contained in a hard shell which is, in turn, surrounded by a fleshy green layer; technically a walnut is the seed of a drupe or drupaceous nut and thus not a true botanical nut; it is used for food after being processed, with green walnuts being used for pickled walnuts, and mature walnuts after full ripening used for its nutmeat; cultivars (Ashley, Chandler, Cisco, Eureka, Feradam, Ferbel, Ferjean, Fernette, Fernor, Ferouette, Forde, Franquette, Grandjean, Germisara, Gillet, Hansen, Hartley, Howard, Idaho, Ivanhoe, Jupanesti, Lara, Livermore, Marbot, Mayette, Meylanaise, Paradox, Parisienne, Payne, Poe, Robert Livermore, Rita, Ronde de Montignac, Royal, Serr, Sexton, Solano, Sunland, Tehama, Tulare, Valcor, Vina, Wilson's Wonder, Yolo) are propagated vegetatively as clones and it is advisable to grow a mixture of clones to improve pollination; in 2016, worldwide production of walnuts (in the shell) was 3.7 million tonnes, with China contributing 48% of the world total; other major producers were (in the order of decreasing harvest): United States, Iran, Turkey, Mexico, Ukraine, and Chile; *Juglans regia* (Juglandaceae), 2n = 2x = 32; 1C DNA content 0.62 pg *hort* >>> African walnut >>> black walnut >>> mansonia

wasabi: The fleshy rhizome is the source of the green paste called "wasabi" that is commonly served with sashimi (raw fish) in Japan; *Eutrema wasabi syn Wasabia japonica* (Brassicaceae), 2n = 4x = 28 *hort* >>> *Brassica*

water chestnut (*syn* Chinese water chestnut): A grass-like sedge grown for its edible corms; it is not a nut at all, but an aquatic vegetable that grows in marshes, underwater in the mud; it has tube-shaped, leafless green stems that grow to about 1.5 m; the water >>> caltrop, which is also referred to by the same name, is unrelated and often confused with the water chestnut; the small, rounded corms have a crisp white flesh and can be eaten raw, slightly boiled, grilled, or are often pickled or tinned; they are a popular ingredient in Chinese dishes; they are most often eaten raw, sometimes sweetened; they can also be ground into a flour form, used for making water chestnut cake, which is common as part of dim sum cuisine; *Eleocharis dulcis syn E. equisetina syn E. indica syn E. plantaginea syn E. plantaginoides syn E. tuberosa syn E. tumida* (Cyperaceae), 2n = 200; 1C DNA content 0.42–9.00 pg *bot hort*

watercress: A water plant common in Europe; the leaves have a mustard-like taste and have a high content of vitamin C; in the past, it was used as a vegetable substitute; nowadays, it is offered as an all-season vegetable; commercial productivity is adversely affected by the obligate aquatic fungal pathogen, *Spongospora subterranea* f.sp. *nasturtii*, which causes "crook-root" disease; the fungus can additionally act as a vector for detrimental watercress viruses; three species of watercress are recorded in Europe; *Rorippa nasturtium-aquaticum* (common or green watercress, 2n = 2x = 32), *Rorippa sterilis* (brown watercress, 2n = 3x = 48) and *R. microphylla* (wild watercress, 2n = 4x = 64); *Rorippa nasturtium-aquaticum* (Brassicaceae), 2n = 2x = 32 (R'R'); 2C DNA content 1.4 pg *hort* >>> *Brassica*

watermelon: Belonging to the gourd family and native to tropical Africa, with a dark green rind and reddish juicy flesh studded with a large number of black seeds; it is widely cultivated in subtropical regions; *Citrullus lanatus, C. vulgaris* (Cucurbitaceae), 2n = 2x = 22; 1C DNA content 0.45 pg; endopolyploidy of 24–384n was observed *hort* >>> Table 46

water yam: An important tuber crop and a staple food for millions of people in many tropical and subtropical countries; the plant has a long history of asexual propagation; >>> dioecy, irregular and erratic flowering, asynchrony of sexual seed development, and apparent paucity in seed set has led the species to be regarded as sterile; however, natural and artificial pollination can be achieved in *D. alata* under optimized cultural management practices; *Dioscorea alata* (Dioscoreaceae), 2n = 4x = 40; 1C DNA content 0.58 pg *hort* >>> white yam >>> yam

wattle: Species of acacia originate in Australia, where their fluffy golden flowers are the national emblem; the leathery leaves are adapted to drought conditions and avoid loss of water through transpiration by turning their edges to the direct rays of the sun; used for tanning leather and in fencing; *Acacia senegal* (Leguminosae), 2n = 2x = 26; 1C DNA content 1.50 pg *hort*

wax gourd (*syn* **white gourd** *syn* **winter melon** (winter melon is also a common name for members of the Inodorus cultivar group of the muskmelon (*Cucumis melo*), more commonly known as casaba or honeydew melons) *syn* **ash gourd** *syn* **fuzzy melon**): It is a vine grown for its very large fruit, eaten as a vegetable when mature; it is the only species of the genus *Benincasa*; the fruit is fuzzy when young, the immature melon has thick white flesh that is sweet when eaten; by maturity, the fruit loses its hairs and develops a waxy coating, providing a long shelf life; the melon may grow as large as 80 cm in length; although the fruit is referred to as a "melon," the fully grown fruit is not sweet; originally cultivated in Southeast Asia, the wax gourd is now widely grown in East Asia and South Asia as well, *Benincasa hispida* (Cucurbitaceae), 2n = 2x = 24; 1C DNA content 1.05 pg *hort*

wax tree: *Toxicodendron succedanea syn Rhus succedanea* (Anacardiaceae), 2n = 2x = 30 *fore*

Welsh onion: A type of onion common in gardens for many years; the wild type is native to Siberia and eastern Asia; because of the small bulbs, only the leaves are used as a vegetable; *Allium fistulosum* (Alliaceae), 2n = 2x = 16; 1C DNA content 12.53 pg *hort* >>> onion >>> garlic

Western Australian sandalwood: A widespread tree in the semi-arid and arid regions of Western Australia; it yields one of the woods traded as sandalwood, which commands high prices; it grows to 5–6 m; its foliage is gray in color; the fruit is spherical, ~3 cm in diameter; it has a spherical kernel; seeds are very difficult to germinate, requiring warm, moist conditions not normally found in their natural habitat; in most years they would not naturally germinate; possibly they depend on the el Niño cycle; success has been reported by placing the kernels in moist vermiculite in sealed plastic bags at room temperature; once germinated, each seedling should be planted next to another seedling, and watered adequately; two ecotypes can be identified; *S. spicatum* shows moderate levels of genetic diversity compared with other Australian tree species; the northern populations in the arid region show greater levels of diversity and less population differentiation than the southern populations in the semi-arid region, due to differences in the distribution of rare alleles; *S. spicatum* (Santalaceae), 2n = 20; 1C DNA content 0.20 pg *fore*

Western wheatgrass: A perennial, cross-pollinating native grass that is an important component of rangelands in the mixed-grass prairies throughout the Central and northern Great Plains and in some areas of the Intermountain region (parts of Utah, Nevada, Idaho, Oregon, Arizona, and California); because of its sod-forming characteristics, it is widely recommended for use in rangeland improvement and revegetation after disturbances such as mining, construction, and fire; it has low seed yields and is difficult and slow to establish from seed because of seed dormancy and poor seedling vigor; however, thick stands may result over time from extensive rhizome development; the development of new western wheatgrass cultivars with improved seed production and seedling vigor has greatly enhanced the value of this species for revegetation of frequently disturbed rangelands, military training lands, and areas with repeated wildfires; *Pascopyrum smithii syn Agropyron smithii* (Gramineae), 2n = 8x = 56 (allooctoploid); 1C DNA content 0.40 pg *agr eco*

West Indian arrowroot (*syn* **starch arrowroot**): Substance used as a thickener in cooking, produced from the clumpy and starchy roots of various tropical plants; the true arrowroot was used by native South Americans as an antidote against the effects of poisoned arrows; the West Indian island of St. Vincent is the main source of supply today; plant roots and tubers are dried, finely powdered, and filtered; because of the small size of the starch particles, the powder becomes translucent when cooked; *Maranta arundinacea* (Marantaceae), 2n = 2x = 48; 1C DNA content 0.38 pg *hort*

wheat: Refers to a family of related small grains that are descended from the natural crossing of three Middle East grasses (*Aegilops* spp.) centuries ago; it is the most grown cereal crop in the world; it comprises 14 species; the inflorescence is a spike, containing about 20–30 spikelets; each with about four to six florets; one seed is set per floret, although the smaller florets may not bear seeds; it is normally self-pollinating; four main commercial market classes are described: (a) hard red spring wheat, (b) hard red winter wheat, (c) soft red winter wheat, and (d) white wheat; wheat is the leading human food resource with an annual world production >550 million tons; it is the staple food for 40% of the world's population, providing 20% of the calories and 55% of the carbohydrates consumed; the cultivation of wheat has developed for more than 10,000 years; the areas dedicated to wheat throughout the world exceed the area for all other crops, like rice, maize, and potato; the various wheats include a group of diploid species characterized by eight genomes and groups of tetraploid and hexaploid with nine genomes; two main species are produced: (a) soft wheat (hexaploid), *Triticum aestivum* [2n = 6x = 42 (BBA^{u}A^{u}DD]; 1C DNA content 18 pg = 17,100 Mbp; in 2018, 200 scientists from 73 research institutions in 20 countries DNA sequenceordered precise location of 107,891 genes of the total genome of the bread wheat variety >>> 'Chinese Spring'), and (b) hard wheat (tetraploid) *Triticum durum* [2n = 4x = 28 (BBA^{u}A^u)]; comparative analyses of gene sequences (*Wcor15-2A, Wcor15-2B* and *Wcor15-2D*) in diploid, tetraploid and hexaploid wheats suggest that *T. urartu, Ae. speltoides* and *Ae. tauschii subsp. strangulata* are most likely the donors of the Wcor15-2A, Wcor15-2B and Wcor15-2D loci in common wheat, respectively; *Triticum* spp. (Gramineae) *agr* >>> agroenergy crop >>> bulgur >>> 'Chinese Spring' >>> crop evolution >>> farro >>> hard red winter wheat >>> hard red spring wheat >>> hard white wheat >>> premium wheats >>> soft white wheat >>> Figure 10 >>> Tables 1, 15, 16, 30, 32, 35, 48, 50, 53 >>> http://www.wheatbp.net

wheatgrass: >>> *Agropyron* >>> *Elymus* >>> couches

white (head) cabbage: *Brassica oleracea* convar. *capitata* var. *alba* (Brassicaceae), 2n = 2x = 18 (CC); 1C DNA content 0.78 pg *hort* >>> *Brassica* >>> Figure 8

white campion: One of the few members of the plant kingdom carrying sex chromosomes; it became the subject of research into the molecular genetics of sex determination; *Silene alba* (Caryophyllaceae), 2n = 2x = 24 (XY); 1C DNA content 2.85 pg *bot biot*

white clover: This tetraploid clover could have been derived from hybridization between *T. nigrescens* and *T. uniflorum*; *Trifolium repens* (Leguminosae), 2n = 4x = 32 *agr* >>> clover

white gourd: >>> wax gourd

white lupine: *Lupinus albus* (Leguminosae), 2n = 2x = 48; 1C DNA content 0.60 pg *agr* >>> Table 46

white mustard: This is a "hot" mustard, as opposed to the three species, *Brassica juncea, B. nigra* [2n = 2x = 16 (BB); polytene chromosomes have been observed in ovary, anthers, or immature seed tissues], and *B. carinata*, which are "pungent" mustards; it is a completely cross-pollinating crop requiring recurrent mass selection for breeding; it is cultivated as an oil crop in Scandinavia; *Sinapis alba syn Brassica alba* (Brassicaceae), 2n = 3x = 18?; 1C DNA content 0.50 pg *agr* >>> *Brassica* >>> Figure 8

white pine: *Pinus strobus* (Pinaceae), 2n = 2x = 24; 1C DNA content 25.65 pg *fore*

white spruce: *Picea glauca* (Pinaceae), 2n = 2x = 24; 1C DNA content 16.15 pg *fore*

white sweet clover: *Melilotus alba* (Leguminosae), 2n = 2x = 16; 1C DNA content 11.33 pg *agr*

white vetch: >>> chickling pea

white yam: Also known as the Guinea yam, and the eight-months yam; yams are an important crop throughout the tropics, particularly in West Africa; it originated in West Africa and is the most important species agriculturally; yams were domesticated in both the Old and New World tropics, some believe as early as 9,000 BC in Southeast Asia and Africa; *Dioscorea rotundata (Dioscoreaceae)*, 2n = 4x = 36; 1C DNA content 0.71 pg *hort* >>> aerial yam

whortleberry: >>> blueberry

wild chamomile: >>> German chamomile

wild crucifer: A potential source for alien gene transfer; of special interest is its C_3–C_4 intermediate photosynthetic and/or photorespiratory mechanism; *Moricandia arvensis, M. nitens* (Brassicaceae), 2n = 4x = 28; 1C DNA content 0.65 pg *bot biot* >>> *Brassica* >>> Figure 8

wild emmer wheat: *Triticum dicoccoides* (Gramineae), 2n = 4x = 28 (BBA^{u}A^u); 1C DNA content 12.28 pg *agr* >>> emmer wheat

wild licorice: >>> sarsaparilla

wild mustard: >>> charlock

wild oat: *Avena fatua* (Gramineae), 2n = 6x = 42 (AAC^{v}C^{v}DD); 2C DNA content 29.8 pg *bot agr* >>> Table 14

wild radish: >>> runch >>> radish

wild rice: A native aquatic grain of northern North America; it is not related to true rice; *Zizania palustris* (Gramineae), 2n = 2x = 30; 1C DNA content 2.18 pg *bot* >>> heat stress >>> rice

wildrye: >>> lymegrass >>> *Elymus* >>> couches

wild sarsaparilla: >>> sarsaparilla

willow: A deciduous tree with simple leaves and small erect catkins; willows are found throughout the world, except in Australia; *Salix exigua* and *S. viminalis* are used for biomass production in sustainable agriculture, *Salix* spp. (Salicaceae), 2n = 38; 1C DNA content 0.35–0.86 pg *agr fore bot* >>> basket willow

willow-leafed foxglove: *Digitalis obscura* (Scrophulariaceae), 2n = 56 *hort* >>> foxglove

winged bean (*syn* **asparagus pea** *syn* **four-angled bean** *syn* **Manila bean** *syn* **princess pea**): *Psophocarpus tetragonolobus* (Leguminosae), 2n = 2x = 18, 26; 1C DNA content 0.80 pg; polytene chromosomes have been observed in ovary, anthers, or immature seed tissues *hort* >>> pulse

winter melon: >>> wax gourd

winter grape: >>> Berlandieri grape

winter (oilseed) rape: >>> rapeseed

wolfberry (*syn* **goji** *syn* **goji berry**) It is the common name for the fruit of two very closely related species, *Lycium barbarum* and *L. chinense*, two species of boxthorn; they are deciduous woody perennial plants, growing 1–3 m high. *L. chinense* is grown in the south of China and tends to be somewhat shorter, while *L. barbarum* is grown in the north, primarily in the Ningxia Hui Autonomous Region, and tends to be somewhat taller; leaves form on the shoot in either an alternating arrangement or bundles of up to three, each having a shape that is either lanceolate or ovate; leaf dimensions are 7 cm wide by 3.5 cm broad, with blunted or round tips; the flowers grow in groups of one to three in the leaf axils; the >>> calyx consists of bell-shaped or tubular >>> sepals forming short, triangular lobes; the >>> corolla is lavender or light purple, 9–14 mm wide with five or six lobes shorter than the tube; these species produce a bright orange-red, ellipsoid berry 1–2 cm long; the number of seeds in each berry varies widely, based on cultivar and fruit size, containing anywhere between ten and 60 tiny yellow seeds, that are compressed with a curved embryo; the berries ripen from July to October in the northern hemisphere; they have been termed a superfruit because of their nutrient value and antioxidant content; *Lycium barbarum, L. chilense* (Solanaceae), 2n = 2x, 4x = 24, 48; 1C DNA content 1.78–3.25 pg *hort*

wood apple: >>> bael

wood sorrel: In addition to the white and pink flowers (it is also used as an ornamental plant), wood sorrel develops a whitish and fleshy root that can be eaten as a vegetable; in the past it was cultivated in Europe; *Oxalis deppei syn O. esculenta syn O. tetraphylla* (Oxalidaceae), 2n = 2x = 22 *hort*

woolly-pod vetch: *Vicia villosa* ssp. *dasycarpa* (Leguminosae), 2n = 2x = 14; 1C DNA content 2.28 pg; polytene chromosomes have been observed in ovary, anthers, or immature seed tissues *bot agr* >>> vetch

wormseed: >>> epazote

wormwood: Absinthe is a green, bitter liqueur primarily flavored with wormwood, a European herbaceous perennial related to the native sagebrush species of the western U.S.; absinthe also contains thujone, a terpenoid component of many essential oils, including those found in other *Artemisia* species and in the coniferous genus *Thuja*; research has shown that thujone not only fuels creativity, but also that an overdose of the compound causes yellow-tinged vision; artist Vincent van Gogh (1853–1890) suffered from epilepsy and was treated with absinthe liqueur on a regular basis; artemisinin, as one of the secondary metabolites, is produced by *Artemisia annua*; it is a sesquiterpene lactone used for removing the causes of malaria (*Plasmodium falciparum* and *P. vivax*) and for treating different types of cancer such as leukemia, breast cancer, colon cancer, and small-cell lung carcinomas; artemisinin content in wild *A. Annua* plants is very low; therefore, it causes an increase in the price of medicines made from this product, especially for people in developing countries, where malaria is widespread; chemical synthesis of artemisinin is expensive and not economically viable; hairy root cultures under nano cobalt particle elicitation is a tool for producing artemisinin; *Artemisia absinthium* (Compositae), 2n = 2x = 18; 2C DNA content 8.52 pg *hort*

wrinkled pea: *Pisum sativum* convar. *medullare* (Leguminosae), 2n = 2x = 14; 1C DNA content 4.88 pg; polytene chromosomes have been observed in ovary, anthers, or immature seed tissues *hort*

X

Xinjiang rice wheat (*syn* **daosuimai** *syn* **ricehead wheat**): A Chinese wheat landrace with long glumes similar to those of *T. turgidum* ssp. *polonicum* or *T. ispahanicum*; it was found in the agricultural areas in the west part of Talimu Basin, Xinjiang (China) in 1948; it was supposed that Xinjiang rice wheat was a mutated form of *T. polonicum*; however, its chromosome number is 2n = 42 and it was named *T. petropavlovskyi*; Arbuzova et al. (1996) gave the gene symbol "*Eg2*" for elongated glume; other authors speculated that this species was derived from natural hybridization between *T. aestivum* and *T. polonicum*; the *Eg2* (now: *P-Apet1*) gene is located ~46.8 cM from the *cn-A1* locus, which controls the "chlorine" trait, on chromosome arm 7AL, ~12.4 cM from the centromere; *Triticum petropavlovskyi* (Gramineae), 2n = 6x = 42 *bot*

xin mai cao shu: A perennial, cespitose, and loosely clumped grass with short rhizomes; butt sheaths are glabrous; its distribution is temperate Asia and China; it became a source of resistance to wheat >>> stripe rust, >>> take-all disease, and >>> powdery mildew, with tolerance to >>> salinity and >>> drought; it has been successfully hybridized as the pollen parent to bread wheat without using immature >>> embryo rescue culture; *Psathyrostachys huashanica* (Gramineae), 2n = 2x = 14 (N^{s}N^s); 2C DNA content 16.7 pg *bot phyt*

Y

yam bean: A perennial climbing or trailing plant of Central America, Northwestern Costa Rica, and southern Mexico, producing annual, twining stems 2-6 m long from a tuberous rootstock; it also has a range of local medicinal applications and can be used as a pesticide; it is sometimes grown as a green manure; whereas the roots are edible (and often eaten in quantity), the upper portions of the plant, especially the seeds, mature seedpods and the leaves, contain a poisonous glucoside, *Sphenostylis stenocarpa syn Pachyrhizus tuberosus syn Pachyrhizus erosus* (Leguminosae), 2n = 2x = 22, 2C DNA content 1.17 to 1.22 pg *hort* >>> pulse

yam(s) (Asia): This is the Asian yam, also known as the white yam, the greater yam, the winged yam, and the water yam; climbing plants cultivated in tropical regions; the starchy tubers (underground stems) are eaten as a vegetable; this yam was of major importance to the seafaring Polynesians who took it to most of the tropical islands of the Old World; it is propagated vegetatively, because most cultivars never produce fertile seed, and some are completely sterile; the Mexican yam (*Dioscorea composita*) contains a chemical that is used in the female contraceptive pill; *Dioscorea alata* (white Lisbon or water yam, 2n = 4x = 40; 1C DNA content 1.05 pg) and *D. esculenta* (Dioscoreaceae), 2n = 3x–10x = 30–100 *hort agr* >>> aerial yam, yellow yam African yam, white yam >>> cush-cush yam >>> water yam

yam (Africa): *Dioscorea rotundata* (Dioscoreaceae), 2n = 4x = 40; 1C DNA content 0.71 pg *hort agr* >>> aerial yam

yam(s) bean: >>> jicama

yard long bean: >>> asparagus bean

yarrow (*syn* **milfoil** *syn* **old man's pepper** *syn* **soldier's woundwort** *syn* **knight's milfoil** *syn* **thousand weed** *syn* **nose bleed** *syn* **carpenter's weed** *syn* **bloodwort** *syn* **staunchweed** *syn* **sanguinary** *syn* **devil's plaything** *syn* **bad man's plaything** *syn* **yarroway):** A perennial herb, native to Europe and Asia and naturalized in North America and most other countries throughout the world; it is very common along roadsides and in old fields, pastures, and meadows in the eastern and Central U.S. and Canada; it can be easily cultivated; it prefers a well-drained soil in a sunny position; a very good companion plant, it improves the health of plants growing nearby and enhances their essential oil content, thus making them more resistant to insect predation; also improves the soil fertility; the flowers are several bunches of flat-topped panicles consisting of numerous small, white flower heads; each tiny flower resembles a daisy; the whole plant is more or less hairy, with white, silky, appressed hairs; it blooms from May to August; gathered stem, leaves, and flower heads in bloom are dried for later herb use; dry herb is edible as a spice or flavoring, strong sage flavor; *Achillea millefolium* (Asteraceae), 2n = 6x = 54; 1C DNA content 7.65 pg *hort*

yautia: >>> tanier

year bean: Occurs in both truly wild and cultivated forms; wild populations are found in Guatemala, where the center of diversity lies, and the cultivated crop is grown in the highlands of Central America and northern South America; botanists sometimes recognize it as a subspecies of *Phaseolus coccineus*; *Phaseolus polyanthus* (Leguminosae), 2n = 2x = 12; 1C DNA content 0.73 pg *agr* >>> pulse

yellow gram: >>> chickpea

yellow lucerne: *Medicago falcata* (Leguminosae), 2n = 4x = 32; 1C DNA content 0.85 pg (*M. quasifalcata*) *agr*

yellow lupine: >>> sweet yellow lupine

yellow-seeded sarson: *Brassica campestris* ssp. *tricularis* (Brassicaceae), 2n = 2x = 20 (AA); 1C DNA content 0.58 pg *hort* >>> *Brassica* >>> Figure 8

yellow sweet clover: An erect or ascending biennial, up to 1.5 m high, with trifoliate leaves; allogamous; suitable for conservation but the stands are mainly grazed; *Melilotus officinalis* (Leguminosae), 2n = 2x = 16; 1C DNA content 1.13 pg *agr*

yellow sucking: >>> suckling clover

yellow trefoil: >>> black medick

yellow yam (syn 12-month yam, yellow guinea yam): In spite of its name, this is a West African species that still occurs wild; it was taken to the New World with the slave trade; *Dioscorea cayenensis* (Dioscoreaceae), 2n = 6x = 60; 1C DNA content 0.77 pg *hort* >>> aerial yam

yerba mate: >>> mate

yew: A slow-growing evergreen, the longest living of all European trees, probably reaching ages of 1,000 years old; the seeds are enclosed in a fleshy covering called an aril which is eaten by birds; the seed itself is poisonous, like the rest of the tree; *Taxus baccata* (Taxaceae), 2n = 2x = 24; 1C DNA content 11.05 pg *hort fore*

yielding maguey: >>> cantala

ylang-ylang *syn* **ilang-ilang** *syn* **annona asiatic:** A small protogynous flower of the cananga tree,
having six yellow-green petals; a strongly allogamous and fast-growing tree that exceeds 5
m per year and attains an average height of 12 m; it grows in full or partial sun, and prefers
the acidic soils of its native rainforest habitat; the flower is greenish yellow (rarely pink),
curly like a sea star, and yields a highly fragrant essential oil (canaga oil); the fragrance
is rich and deep with notes of rubber and custard, and bright with hints of jasmine and
neroli; the main aromatic components are benzyl acetate, linalool and *p*-cresyl methyl
ether and methyl benzoate; a related species is *C. fruticosa*, which is a dwarf ylang-ylang,
that grows as a small tree or compact shrub with highly scented flowers; ylang-ylang has
been cultivated in temperate climates, e.g. Madagascar, Indonesia, or Philippines, under
conservatory conditions, *Cananga odorata* (Annonaceae), 2n = 2x = 16; 1C DNA content
0.78 pg *hort fore* >>> >>> pawpaw >>> soursop

yucca: A plant in the agave family that has stiff, sword-like leaves and clusters of white or purplish,
waxy flowers; this succulent plant is pollinated by the yucca moth; the Joshua tree (*Y.
brevifolia*) is a type of yucca; *Yucca* spp. (Agavaceae), 2n = 60; 1C DNA content 2.55–6.00
pg *hort*

Z

zedoary: A perennial herb and member of the genus *Curcuma*; this fragrant plant bears yellow
flowers with red and green bracts, and the underground stem section is large and tuberous,
with numerous branches; the leaf shoots of the zedoary are long and can reach 1 m; it is
native to India and Indonesia, and was introduced to Europe by Arabs around the 6th century; its use as a spice in the West today is extremely rare, having been replaced by ginger;
the edible root has a white interior and a fragrance reminiscent of >>> mango, *Curcuma
zedoaria* (Zingiberaceae), 2n=63, including three SAT-chromosomes, 2C DNA content 3.4
pg *bot hort*

zereshk: >>> Iranian seedless barberry

zinnia: An ornamental plant; a genus of flowering annual plants that are native to Mexico; zinnias
are named for J. G. ZINN (1727–1759), a German botanist; *Zinnia elegans* (Compositae),
2n = 2x = 24 *hort*

zoysiagrass: The genus *Zoysia* consists of 16 species that are naturally distributed on sea coasts
and grasslands around the Pacific; of these, *Zoysia japonica*, *Z. matrella*, and *Z. tenuifolia* are grown extensively as turfgrasses, and *Z. japonica* (*syn Osterdamia japonica syn
Zoysia pungens syn Zoysia pungens japonica*) is also used as a forage grass in Japan and
other countries in East Asia; it is in leaf all year, and in flower from June to August, and
the seeds ripen from June to August; the flowers are hermaphroditic and are pollinated by
wind; the plant prefers light (sandy), medium (loamy), and heavy (clay) well-drained soil; it
also grows on acid, neutral, and basic (alkaline) soils; it cannot grow in the shade; *Zoysia*
spp., 2n = 4x = 40, allotetraploid (Gramineae); 1C DNA content 0.43 pg *agr*

zucchini: *Cucurbita pepo* convar. *giromontiia* (Cucurbitaceae), 2n = 2x = 40; 1C DNA content 0.55
pg *hort* >>> marrow

Figures

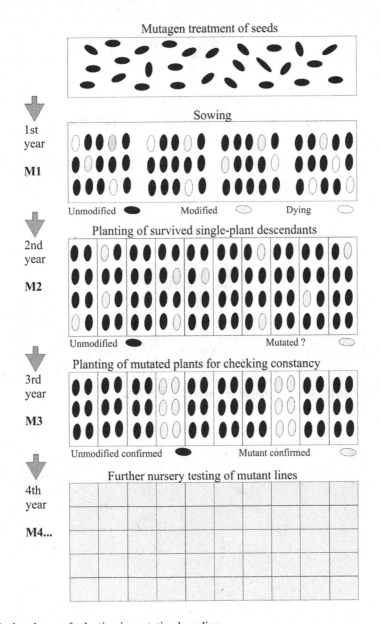

FIGURE 1   Basic scheme of selection in mutation breeding.

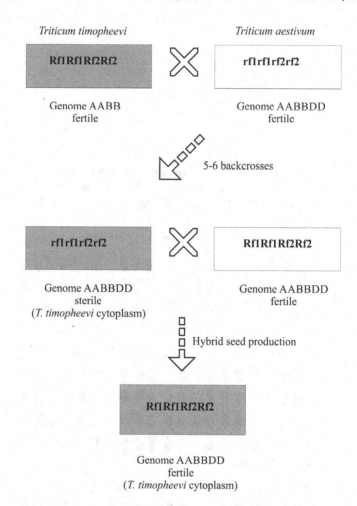

FIGURE 2   Main steps in the establishment of a hybrid variety of wheat.

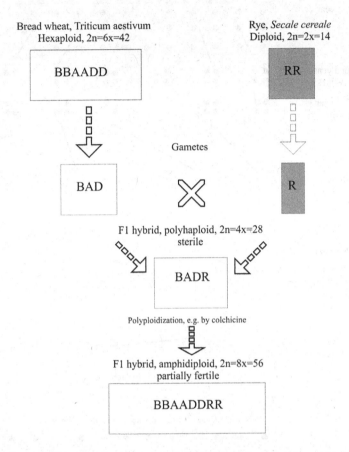

Donor species and their genome constitution

Bread wheat, Triticum aestivum
Hexaploid, 2n=6x=42

Rye, *Secale cereale*
Diploid, 2n=2x=14

BBAADD

RR

Gametes

BAD

R

F1 hybrid, polyhaploid, 2n=4x=28
sterile

BADR

Polyploidization, e.g. by colchicine

F1 hybrid, amphidiploid, 2n=8x=56
partially fertile

BBAADDRR

FIGURE 3 Development of allopolyploid (amphidiploid) hybrids, for example, a wheat–rye hybrid (octo-ploid triticale).

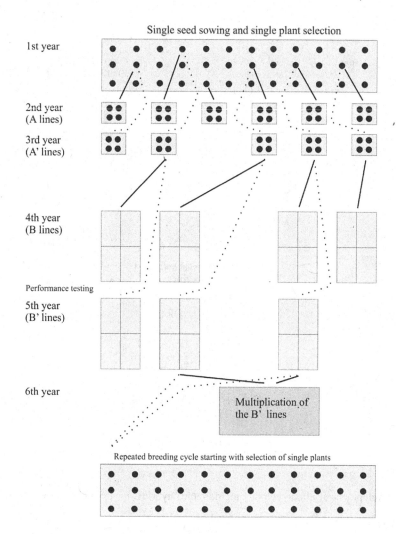

FIGURE 4 Residue seed method of breeding: half-sib progeny selection or method of overstored seeds.

Single-seed sowing and single-plant selection from a heterogeneous
population (e.g. a F2 population of a special cross of genetically different
parents) or landrace

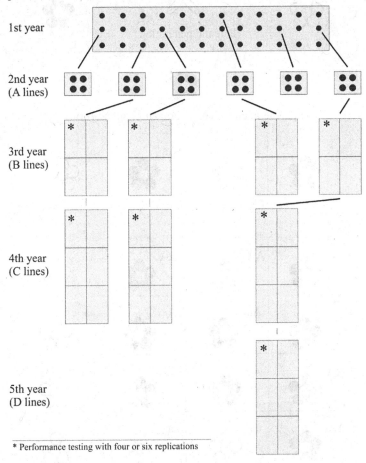

1st year

2nd year
(A lines)

3rd year
(B lines)

4th year
(C lines)

5th year
(D lines)

* Performance testing with four or six replications

FIGURE 5   Single plant selection, including testing of progeny, in autogamous plants.

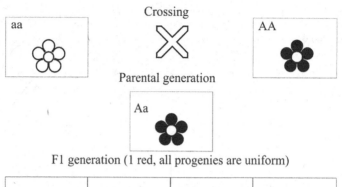

F1 generation (1 red, all progenies are uniform)

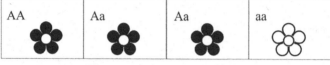

F2 generation (3 red : 1 white)

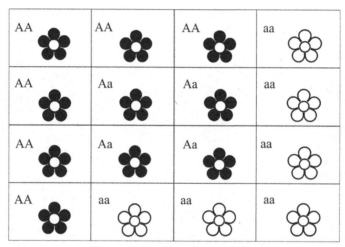

F3 generation (10 red : 6 white)

FIGURE 6 Inheritance and segregation patterns in subsequent generations of flower color from a cross of a red-flowered with a white-flowered plant, and with dominant inheritance of red flowers.

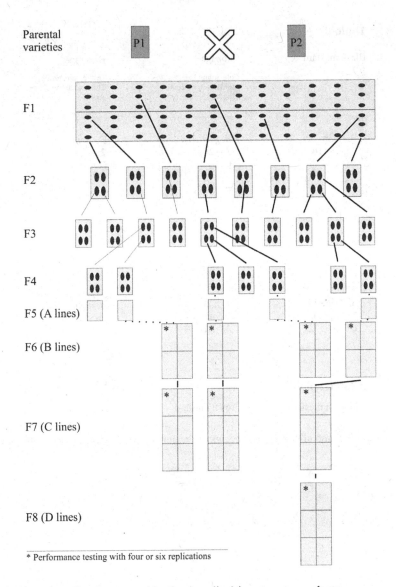

FIGURE 7 Pedigree breeding (cross-combination breeding) in autogamous plants.

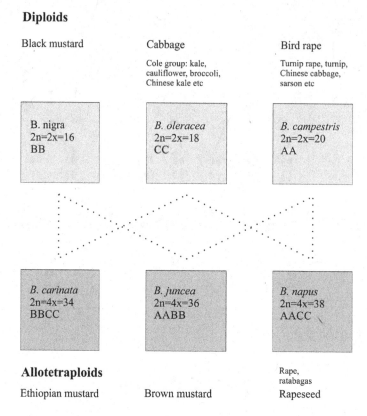

FIGURE 8 Some diploid and allopolyploid species of the genus *Brassica,* having agricultural and breeding importance.

Treatment/Variant

Replication	columns			
	I	II	III	IV

blocks	I	II	III	IV
a	1	2	3	4
b	2	3	4	1
c	3	4	1	2
d	4	1	2	3

blocks	I		II		III		IV		V	
	α	β	α	β	α	β	α	β	α	β
a	2	5	3	6	8	9	1	10	4	7
b	6	9	7	10	2	3	5	4	8	1
c	8	1	9	2	4	5	7	6	10	3
d	4	7	5	8	10	1	3	2	6	9
e	10	3	1	4	6	7	9	8	2	5

FIGURE 9 Experimental field design of a Latin square considering four variants and four replications (four blocks, four columns) and a Latin rectangle considering ten variants (five blocks, five columns).

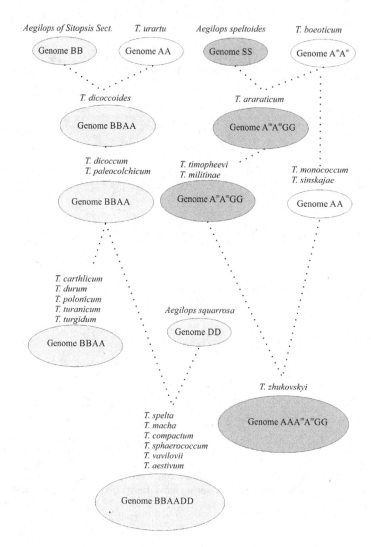

FIGURE 10   The phylogeny of wheat (*Triticum* spp.).

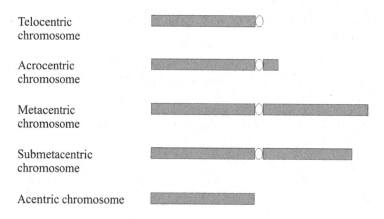

FIGURE 11   Types of chromosome and/or centromere constrictions.

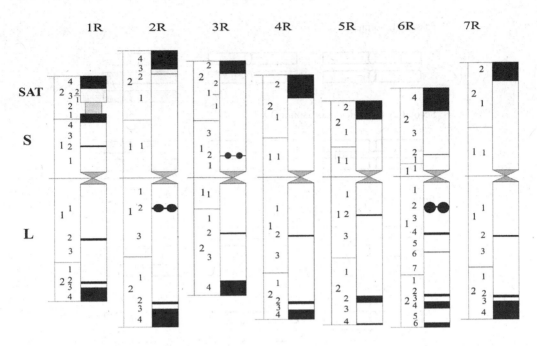

FIGURE 12 The karyogram of diploid rye, *Secale cereale* L.

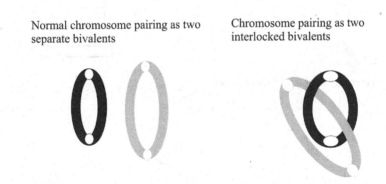

Normal chromosome pairing as two separate bivalents

Chromosome pairing as two interlocked bivalents

FIGURE 13 Pairing failure of meiotic chromosomes in an interlocked configuration.

Primary trisomics

Secondary trisomics (Isotrisomics)

Tertiary trisomics

Telotrisomics

Compensating trisomics

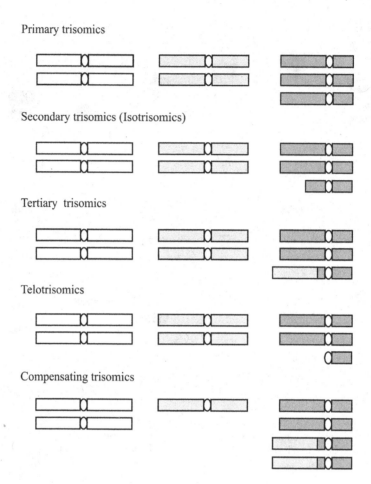

FIGURE 14   Different types of trisomics in plants.

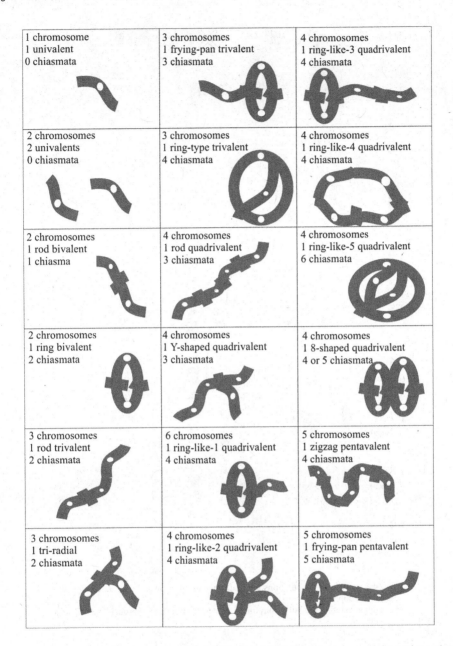

FIGURE 15 A diagrammatic representation of different meiotic chromosome configurations observed at diakinesis and metaphase, including the minimum chiasmata.

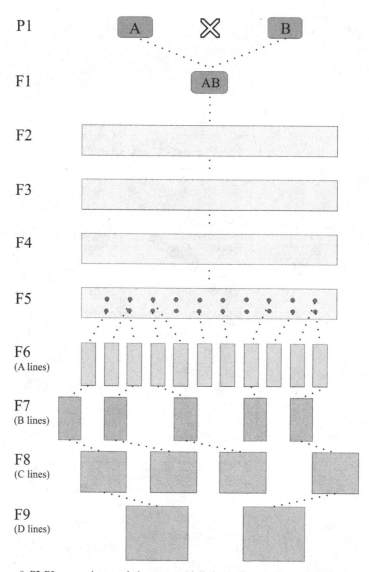

* F2-F5 segregating populations are multiplied as bulks, usually regulated by mass selection approaches
* F6-F9 separating the bulks into lines by pedigree breeding approach and progeny testing

FIGURE 16 Combination breeding, using the bulk method.

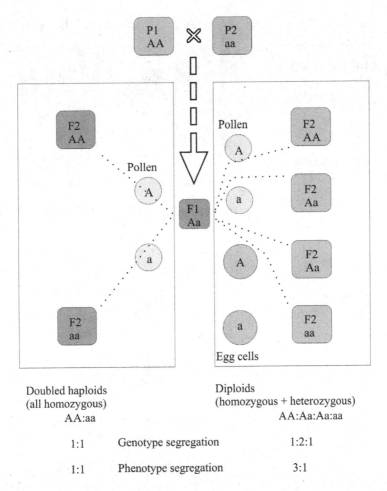

FIGURE 17 A comparison of the genetic segregation patterns of sexually derived and doubled-haploid-derived F₂ progenies from F₁ heterozygotes.

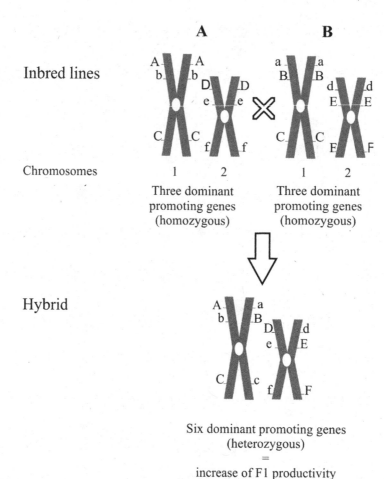

FIGURE 18   Schematic interpretation of the "dominance hypothesis," considering inbred lines and the F_1 hybrid.

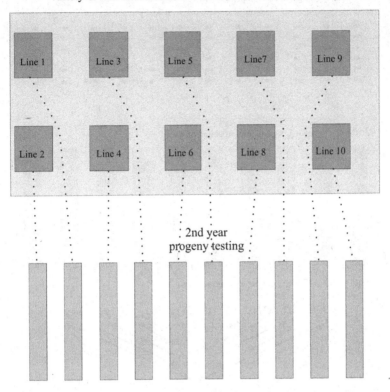

1st year
Variety or line as cross tester with different lines, clones etc.

2nd year
progeny testing

FIGURE 19 Schematic drawing of a topcross design, including progeny testing for general combining ability.

1st cycle
1st year (selfings ⌢ within population)

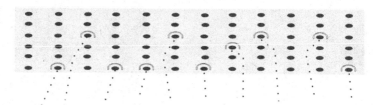

2nd year (growing of progenies of selfed plants)

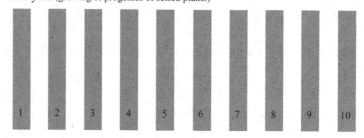

2nd year (crosses among the selfed progenies)

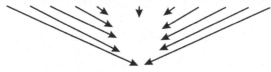

1x3 4x5 2x7 2x3 4x8 5x7 7x8 3x6 1x7 9x2 10x1 2x9 .... etc

3rd year (improved populations and repeated selfings)

2nd cycle
3rd year (progenies of selfings and crosses from first cycle)

 = isolation of plant

FIGURE 20   Schematic drawing of a recurrent selection design.

1st cycle

1st year (selfings ⌒within population and test crosses of populations 1 x population 2 or *vice versa*)

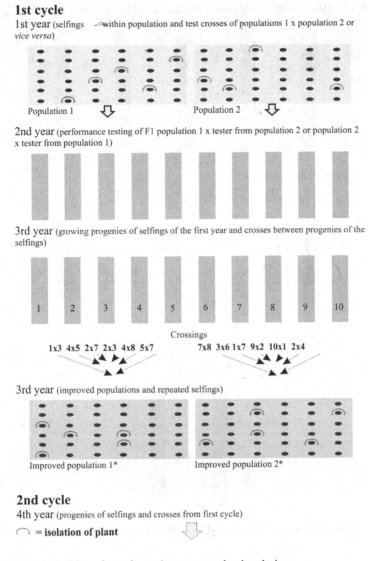

Population 1 ⇩

Population 2 ⇩

2nd year (performance testing of F1 population 1 x tester from population 2 or population 2 x tester from population 1)

3rd year (growing progenies of selfings of the first year and crosses between progenies of the selfings)

| 1 | 2 | 3 | 4 | 5 | 6 | 7 | 8 | 9 | 10 |

Crossings

1x3 4x5 2x7 2x3 4x8 5x7   7x8 3x6 1x7 9x2 10x1 2x4

3rd year (improved populations and repeated selfings)

Improved population 1*

Improved population 2*

2nd cycle

4th year (progenies of selfings and crosses from first cycle)

⌒ = **isolation of plant** ⇩

FIGURE 21 Schematic drawing of a reciprocal recurrent selection design.

(1) Single-cross hybrids (females are detasseled and pollinated by males

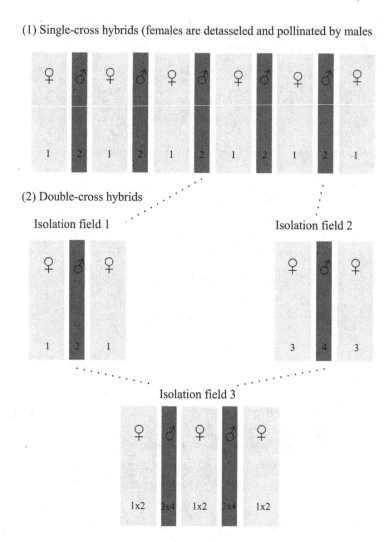

FIGURE 22 Schematic drawing of F$_1$ hybrid seed production in maize.

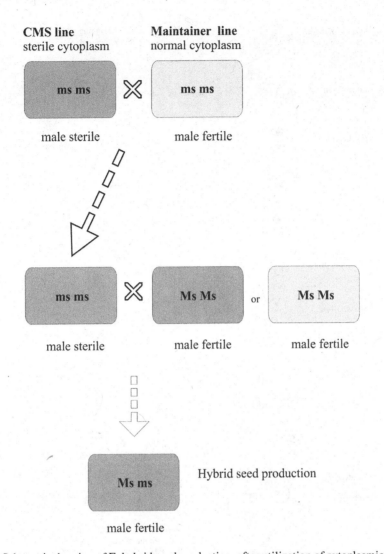

FIGURE 23 Schematic drawing of F₁ hybrid seed production, after utilization of cytoplasmic male-sterility.

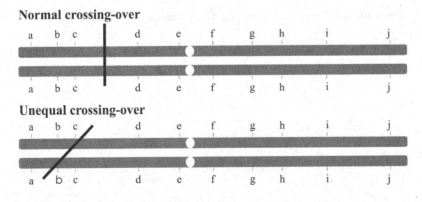

FIGURE 24 Schematic drawing of unequal crossing-over.

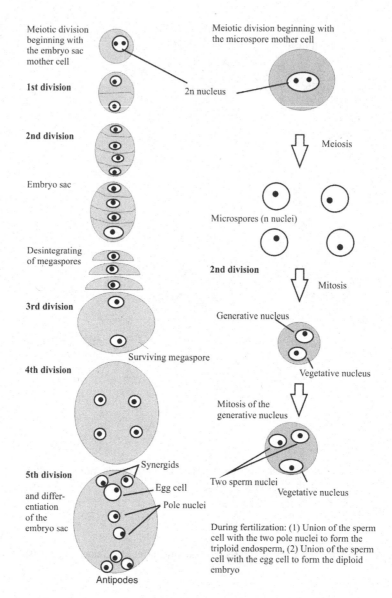

FIGURE 25 Schematic drawing of embryo sac and pollen formation.

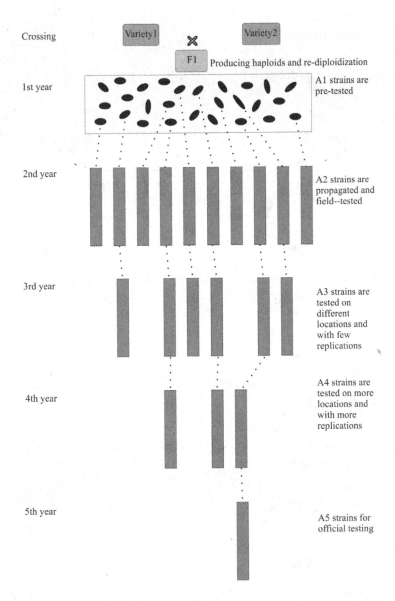

FIGURE 26   Breeding scheme using doubled haploids.

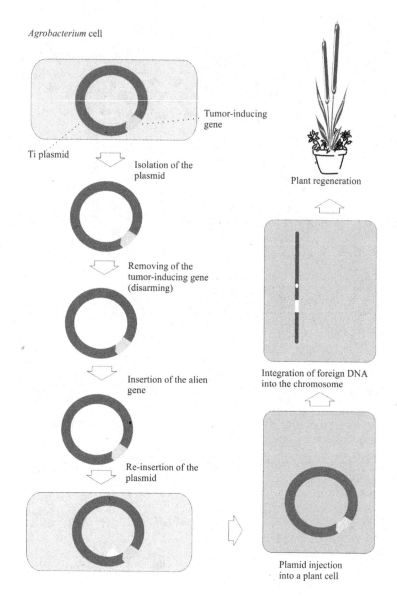

FIGURE 27 Schematic drawing of *Agrobacterium*-mediated gene transfer.

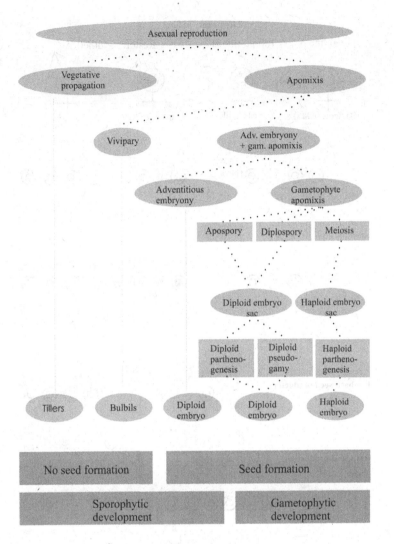

FIGURE 28   Different methods of asexual reproduction in plants.

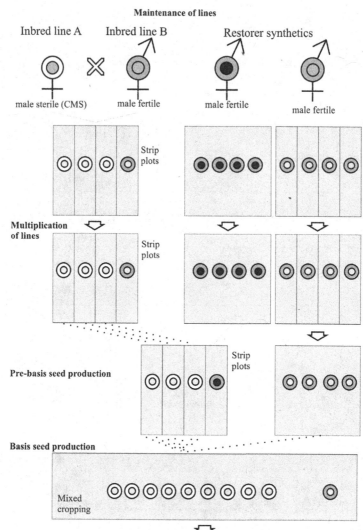

FIGURE 29 F₁ hybrid seed production in allogamous rye.

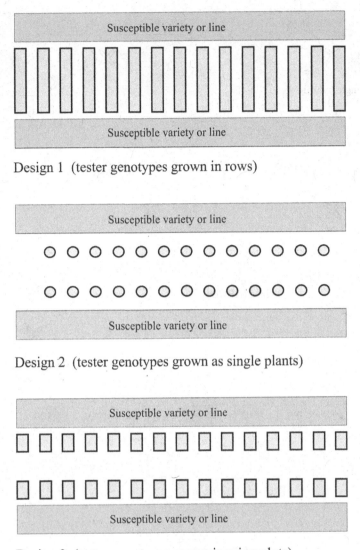

Design 1 (tester genotypes grown in rows)

Design 2 (tester genotypes grown as single plants)

Design 3 (tester genotypes grown in microplots)

FIGURE 30 Designs of spreader nurseries.

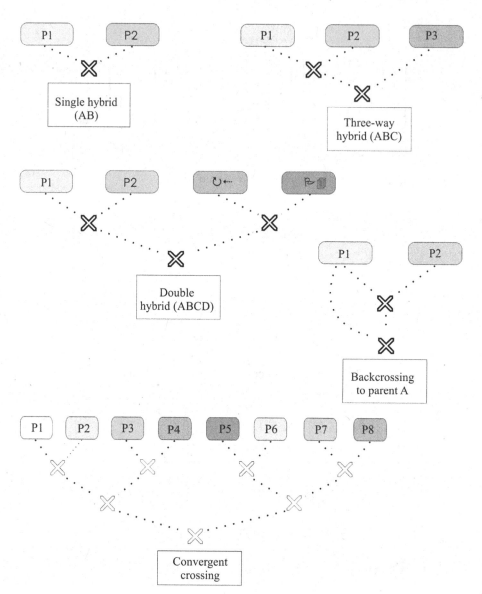

FIGURE 31 Basic crossing schemes in plant breeding.

Honeycomb design (the honeycombs outlined represent two different
selection intensities)

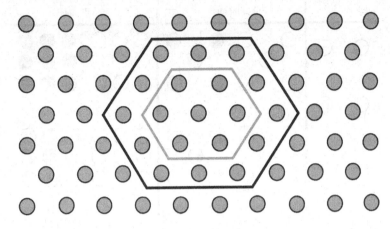

Grid design

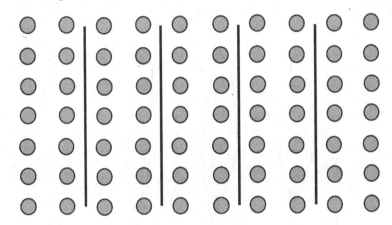

FIGURE 32 Schematic drawing of honeycomb and grid designs.

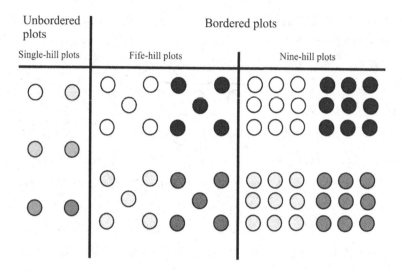

FIGURE 33   Several types of hill plots.

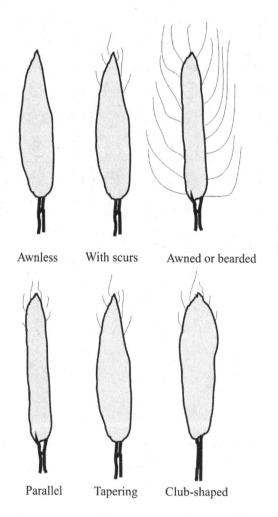

FIGURE 34   Basic shapes and types of spikes in wheat.

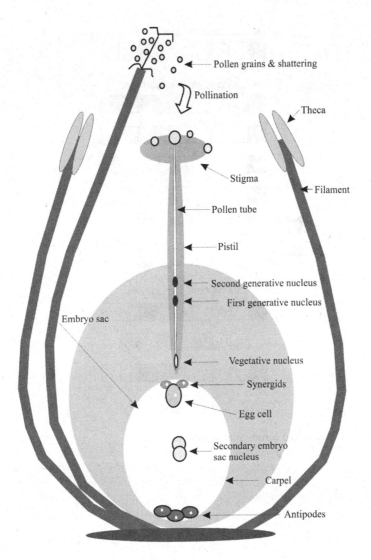

FIGURE 35 Schematic drawing of sexual organs of a plant flower.

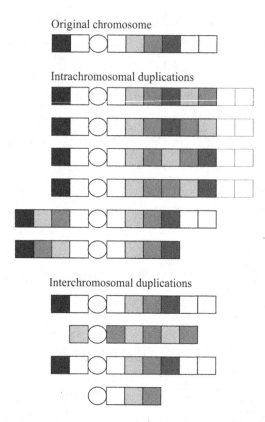

FIGURE 36   Types of duplicated chromosome segments.

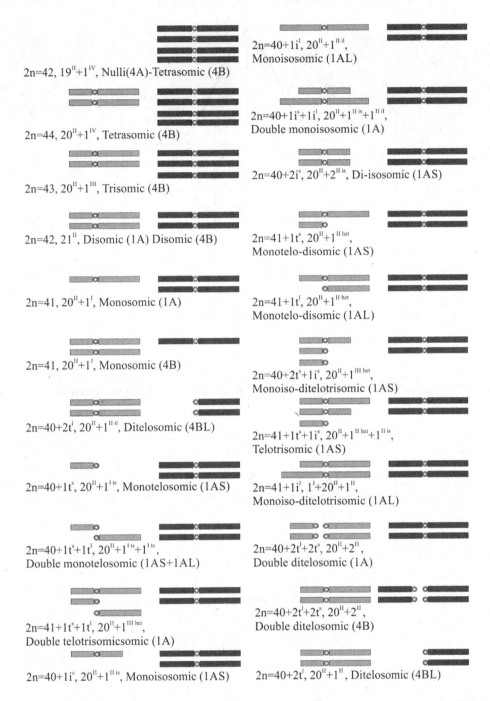

FIGURE 37 Schematic drawing of aneuploid types in wheat, applicable to other diploids and polyploids; it describes the chromosome number, meiotic configuration, and name of the aberration; in isosomics, 1^{II} means a ring-like structure).

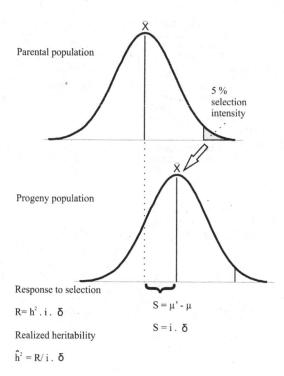

FIGURE 38 Estimation of heritability, h², based on selection advantage, "R", selection coefficient, "S", and selection intensity, "i".

Mass selection (positive)

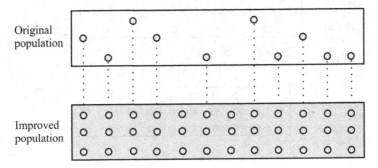

Original population

Improved population

Single-plant descent selection

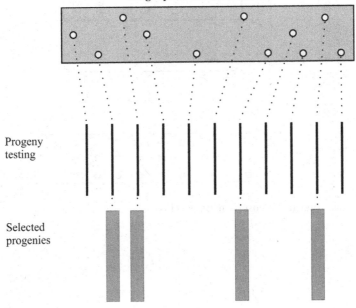

Progeny testing

Selected progenies

FIGURE 39 Schematic drawing of mass selection and single-plant selection.

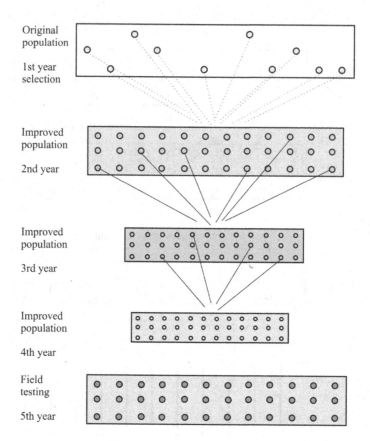

FIGURE 40   Schematic drawing of positive mass selection.

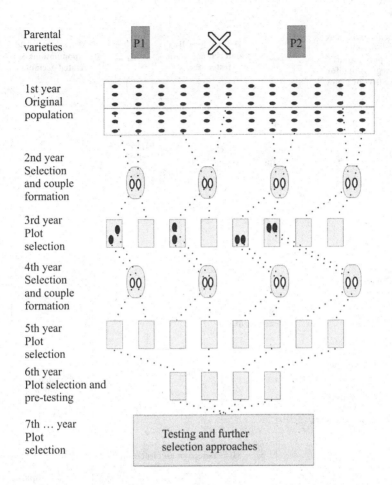

Parental varieties

1st year Original population

2nd year Selection and couple formation

3rd year Plot selection

4th year Selection and couple formation

5th year Plot selection

6th year Plot selection and pre-testing

7th ... year Plot selection

Testing and further selection approaches

FIGURE 41 Couple method of breeding in allogamous plants.

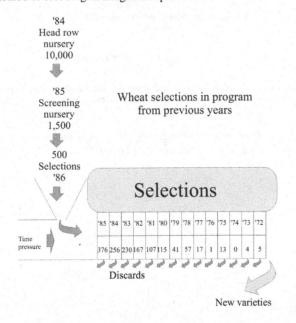

'84 Head row nursery 10,000

'85 Screening nursery 1,500

Wheat selections in program from previous years

500 Selections '86

Selections

Time pressure

'85	'84	'83	'82	'81	'80	'79	'78	'77	'76	'75	'74	'73	'72
376	256	230	167	107	115	41	57	17	1	13	0	4	5

Discards

New varieties

FIGURE 42 A general breeding scheme called "the variety machine." (Modified after N. F. Jensen, 1988. *Plant Breeding Methodology*. John Wiley & Sons, New York, pp. 676.)

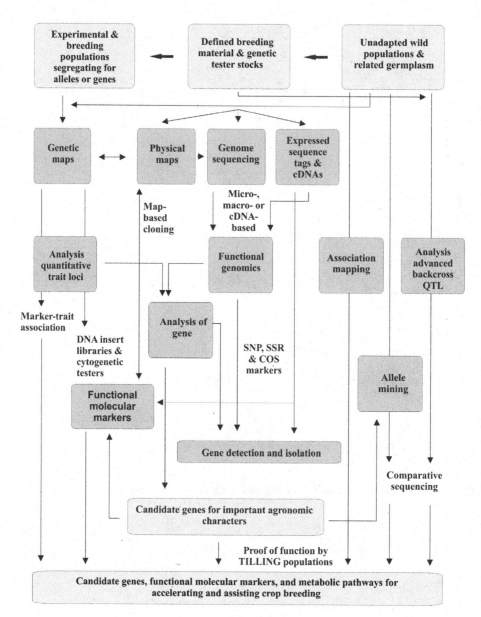

FIGURE 43 A general scheme of integration of genomic resources for crop improvement. (Modified after Varshney, R. K. et al., 2005. *Trends in Plant Science* 10: 621–630.)

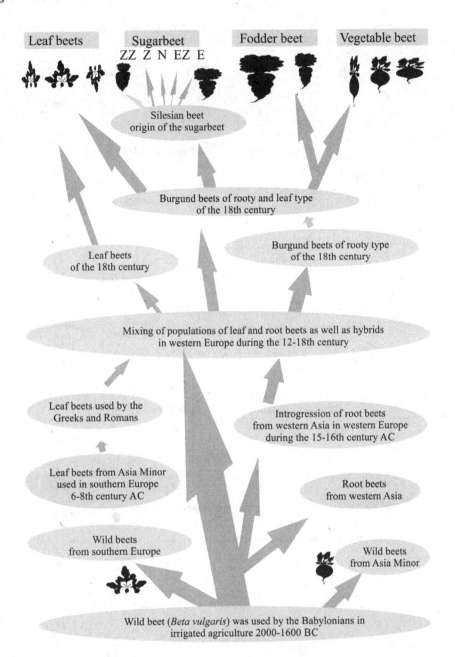

FIGURE 44 Schematic drawing of presumed pedigree of sugar beet. (Modified after Zhukovski, 1971.)

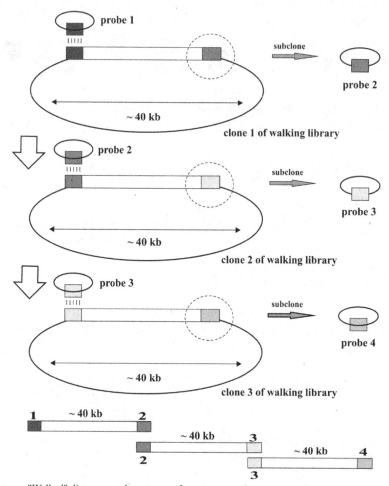

"Walked" distance on chromosome from sequence 1 to sequence 4

FIGURE 45   Molecular procedure of chromosome walking.

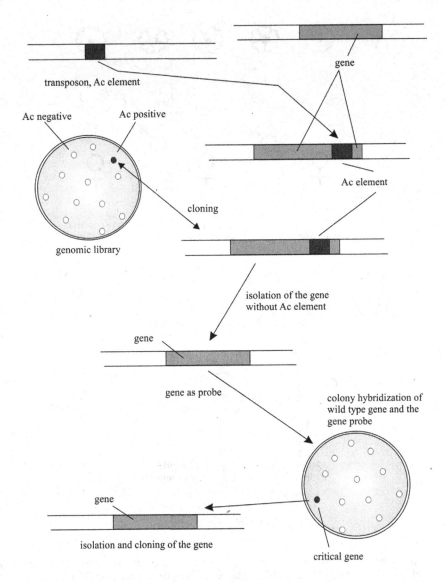

FIGURE 46 Schematic drawing of gene tagging, using transposons.

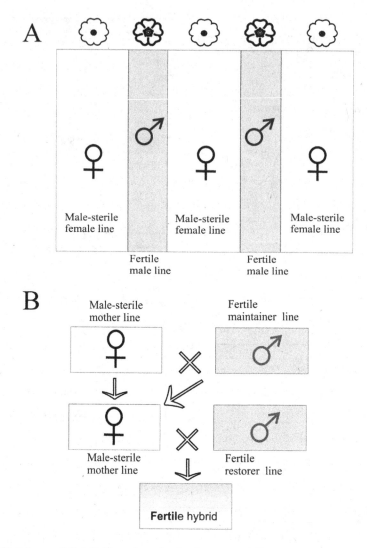

FIGURE 47 (A) Scheme of F₁ hybrid seed production in rapeseed (*Brassica napus* ssp. *napus*), strip cropping as ratio 2:1; and (B) scheme of line production by using cytoplasmic male-sterile mutants.

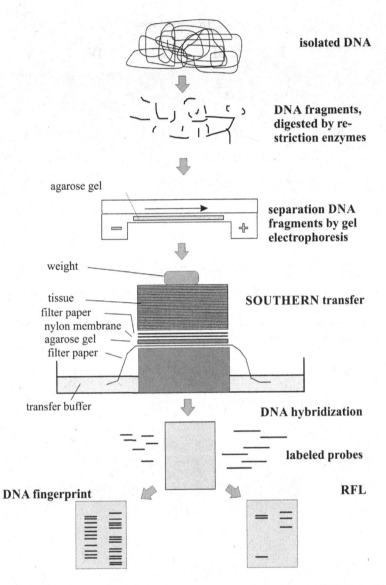

isolated DNA

DNA fragments, digested by restriction enzymes

agarose gel

separation DNA fragments by gel electrophoresis

weight

tissue
filter paper
nylon membrane
agarose gel
filter paper

SOUTHERN transfer

transfer buffer

DNA hybridization

labeled probes

DNA fingerprint

RFL

FIGURE 48 Schematic drawing of restriction fragment length polymorphism (RFLP) analysis, using gel electrophoresis and a Southern transfer technique.

Treatment/Variant

columns

Replication (blocks)	I	II	III	IV	V
a	1	2	3	4	5
b	5	3	1	2	4
c	3	4	2	5	1
d	4	1	5	3	2
e	2	5	4	1	3

FIGURE 49 Experimental field design of quasi-complete Latin square, considering five variants and five replications (five blocks, five columns).

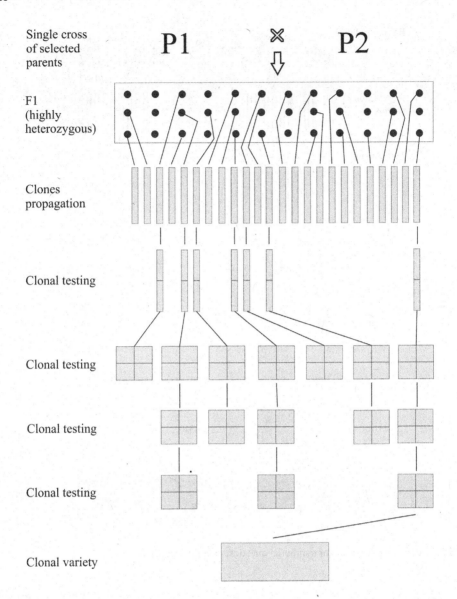

Single cross
of selected
parents

P1   ✕   P2
           ⇩

F1
(highly
heterozygous)

Clones
propagation

Clonal testing

Clonal testing

Clonal testing

Clonal testing

Clonal variety

FIGURE 50   Breeding schemes for clonal varieties.

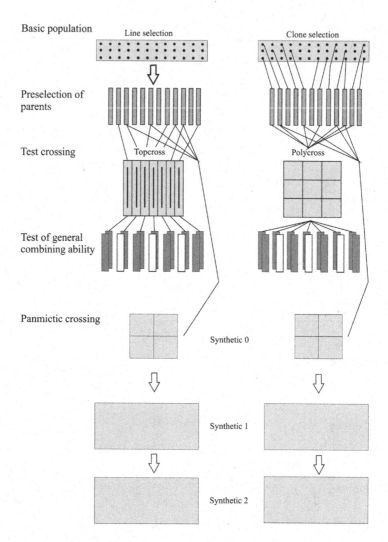

FIGURE 51   Breeding schemes for synthetic varieties, using inbred lines.

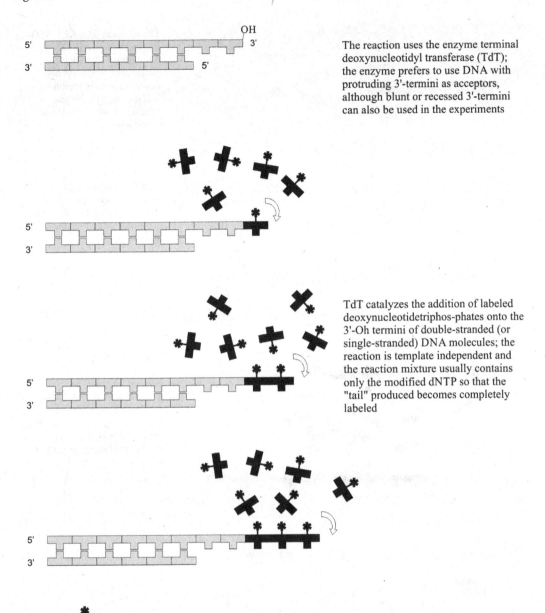

The reaction uses the enzyme terminal deoxynucleotidyl transferase (TdT); the enzyme prefers to use DNA with protruding 3'-termini as acceptors, although blunt or recessed 3'-termini can also be used in the experiments

TdT catalyzes the addition of labeled deoxynucleotidetriphos-phates onto the 3'-Oh termini of double-stranded (or single-stranded) DNA molecules; the reaction is template independent and the reaction mixture usually contains only the modified dNTP so that the "tail" produced becomes completely labeled

Labeled dNTP, e.g. digoxigenin-11-dUTP

FIGURE 52 Schematic drawing of a labeling reaction with labeled dNTP for *in situ* hybridization experiments.

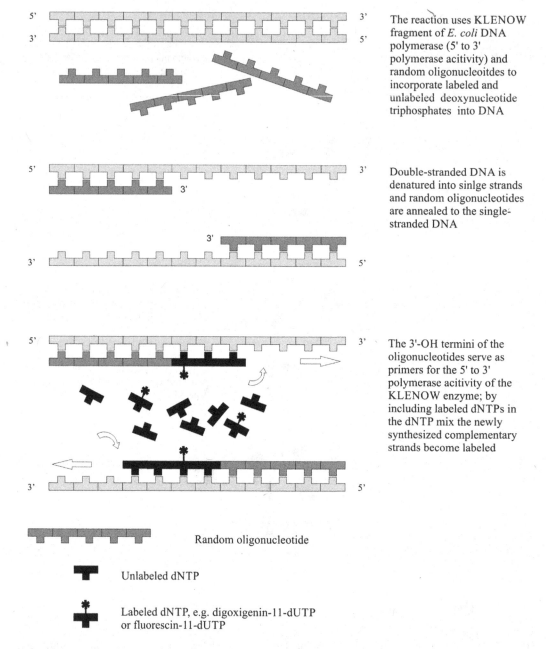

The reaction uses KLENOW fragment of *E. coli* DNA polymerase (5' to 3' polymerase acitivity) and random oligonucleoitdes to incorporate labeled and unlabeled deoxynucleotide triphosphates into DNA

Double-stranded DNA is denatured into sinlge strands and random oligonucleotides are annealed to the single-stranded DNA

The 3'-OH termini of the oligonucleotides serve as primers for the 5' to 3' polymerase acitivity of the KLENOW enzyme; by including labeled dNTPs in the dNTP mix the newly synthesized complementary strands become labeled

Random oligonucleotide

Unlabeled dNTP

Labeled dNTP, e.g. digoxigenin-11-dUTP or fluorescin-11-dUTP

FIGURE 53 Schematic drawing of oligolabeling reaction with unlabeled and labeled dNTP for *in situ* hybridization experiments.

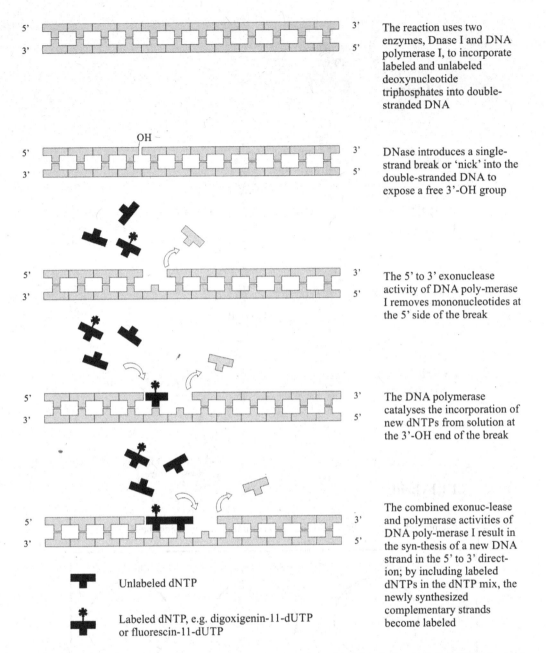

5' ⬜⬜⬜⬜⬜⬜⬜⬜⬜⬜ **3'** The reaction uses two enzymes, Dnase I and DNA polymerase I, to incorporate labeled and unlabeled deoxynucleotide triphosphates into double-stranded DNA

3' ⬜⬜⬜⬜⬜⬜⬜⬜⬜⬜ **5'**

OH

5' ⬜⬜⬜⬜⬜⬜⬜⬜⬜ **3'** DNase introduces a single-strand break or 'nick' into the double-stranded DNA to expose a free 3'-OH group

3' ⬜⬜⬜⬜⬜⬜⬜⬜⬜ **5'**

5' ⬜⬜⬜⬜⬜⬜⬜⬜⬜ **3'** The 5' to 3' exonuclease activity of DNA poly-merase I removes mononucleotides at the 5' side of the break

3' ⬜⬜⬜⬜⬜⬜⬜⬜⬜ **5'**

5' ⬜⬜⬜⬜⬜⬜⬜⬜⬜ **3'** The DNA polymerase catalyses the incorporation of new dNTPs from solution at the 3'-OH end of the break

3' ⬜⬜⬜⬜⬜⬜⬜⬜⬜ **5'**

5' ⬜⬜⬜⬜⬜⬜⬜⬜⬜ **3'** The combined exonuc-lease and polymerase activities of DNA poly-merase I result in the syn-thesis of a new DNA strand in the 5' to 3' direct-ion; by including labeled dNTPs in the dNTP mix, the newly synthesized complementary strands become labeled

3' ⬜⬜⬜⬜⬜⬜⬜⬜⬜ **5'**

⊥ Unlabeled dNTP

⊥ Labeled dNTP, e.g. digoxigenin-11-dUTP or fluorescin-11-dUTP

FIGURE 54 Schematic drawing of the nick translation reaction with unlabeled and labeled dNTP for *in situ* hybridization experiments.

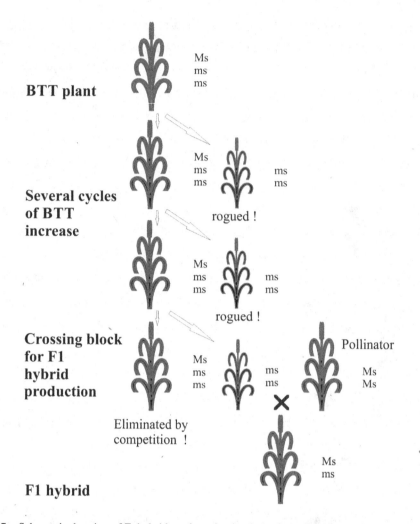

BTT plant

Ms
ms
ms

Several cycles of BTT increase

Ms
ms
ms

ms
ms

rogued !

Ms
ms
ms

ms
ms

rogued !

Crossing block for F1 hybrid production

Ms
ms
ms

ms
ms

Pollinator

Ms
Ms

Eliminated by competition !

×

F1 hybrid

Ms
ms

FIGURE 55   Schematic drawing of F$_1$ hybrid seed production in barley, using the BTT system.

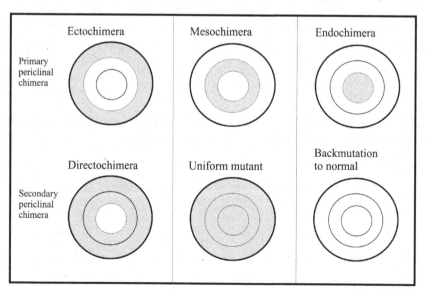

Ectochimera

Mesochimera

Endochimera

Primary periclinal chimera

Directochimera

Uniform mutant

Backmutation to normal

Secondary periclinal chimera

FIGURE 56   Types of periclinal chimera.

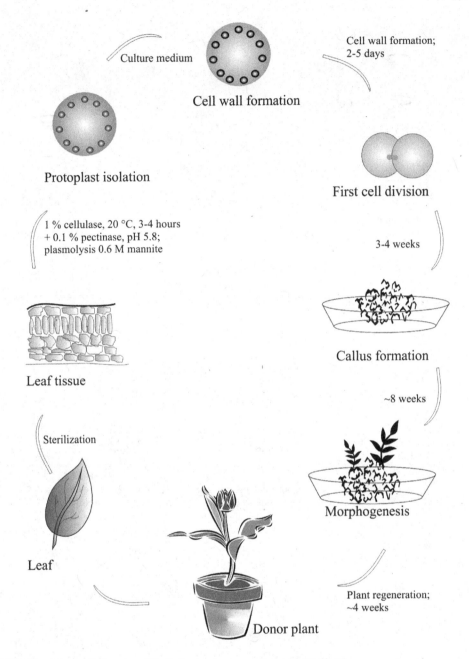

Culture medium

Cell wall formation

Cell wall formation;
2-5 days

Protoplast isolation

First cell division

1 % cellulase, 20 °C, 3-4 hours
+ 0.1 % pectinase, pH 5.8;
plasmolysis 0.6 M mannite

3-4 weeks

Leaf tissue

Callus formation

~8 weeks

Sterilization

Morphogenesis

Leaf

Plant regeneration;
~4 weeks

Donor plant

FIGURE 57   Schematic drawing of isolation and regeneration of protoplasts.

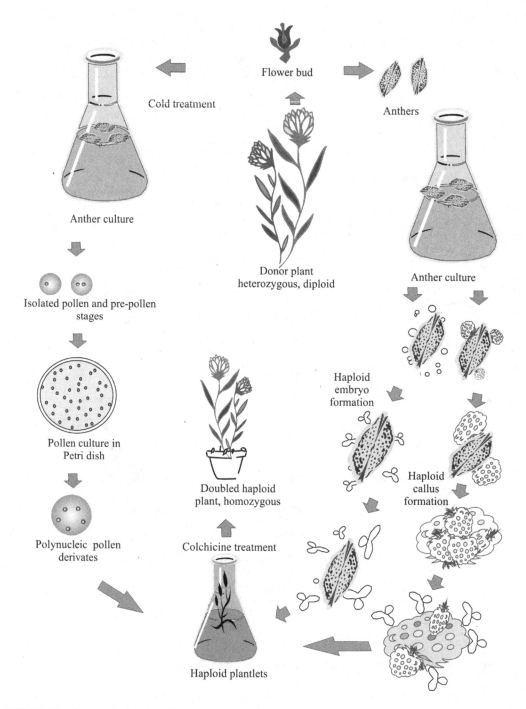

FIGURE 58 Schematic drawing of haploid production by androgenesis (anther and pollen culture).

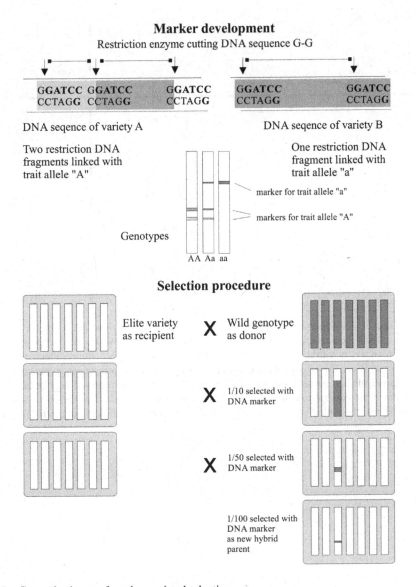

FIGURE 59 General scheme of marker-assisted selection.

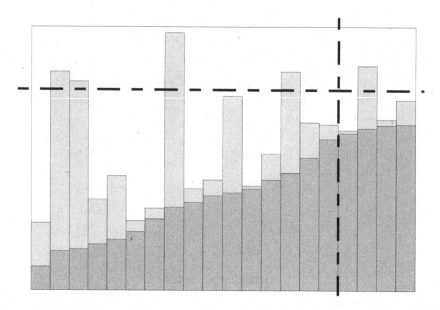

FIGURE 60 Diagramatic scheme of phenotype vs. genotype selection. The height of the lighter portion of each column reflects the variability caused by the genetic constitution, while the darker portion gives the phenotypic variation, i.e. non-inherited characters. The total height of each column reflects the total expression of the character. When 20 individuals of cross progeny are arranged by their genotypic background, then five superior individuals can be selected, based on a specific quality limit (dotted horizontal line). When it is selected by genotypic variation (as revealed by progeny testing), then just four individuals would need to be considered (beyond the dotted vertical line). Therefore, genotypic selection is more efficient when applied from the very beginning.

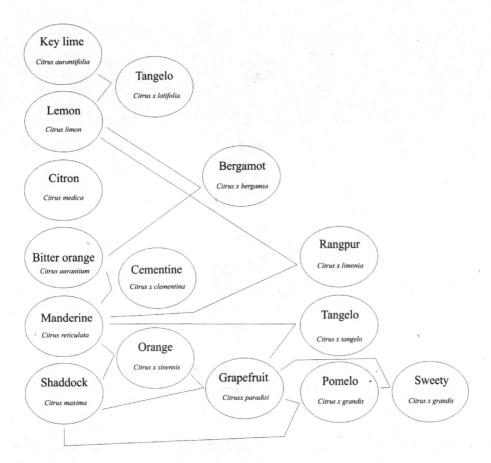

FIGURE 61 Genomic relationships among citrus fruits.

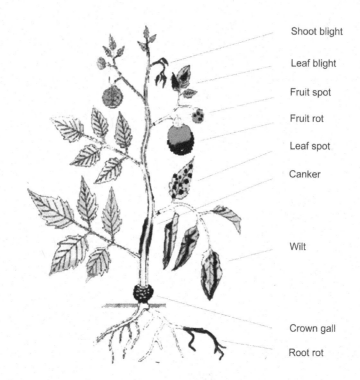

Shoot blight

Leaf blight

Fruit spot

Fruit rot

Leaf spot

Canker

Wilt

Crown gall

Root rot

FIGURE 62   Schematic drawing of common plant diseases.

SSRs

Genotype A: GTGA**GTGT**TGGC

di-nucleotide repeat

Genotype B: GTGA**GTGTGT**TGGC

di-nucleotide repeat

Genotype C: GTGA**GTGTGTGT**TGGC

di-nucleotide repeat

Electrophoretic separation
pattern of PCR fragments

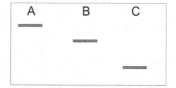

SNPs

Genotype A: CCTGTTAA C
GGTACATT

FIGURE 63   Schematic drawing of SSR and SNP markers. (Source: R. Schlegel.)

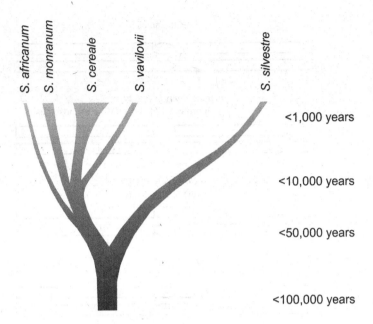

FIGURE 64   Phylogenetic relationships among rye species (*Secale* spp.) and their evolutionary development. (Source: R. Schlegel.)

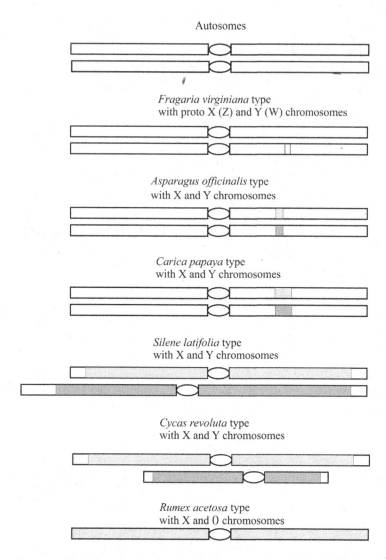

FIGURE 65 Main steps of sex chromosome evolution in plants. (Modified after Ming et al. 2011.)

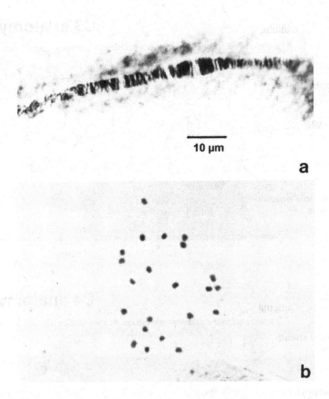

10 µm

a

b

FIGURE 66 Polytene chromosome of bean (*Phaseolus vulgaris*) antipodal cells in comparison to the standard metaphase chromosomes. (Used with permission of E. Badaeva, 2012.)

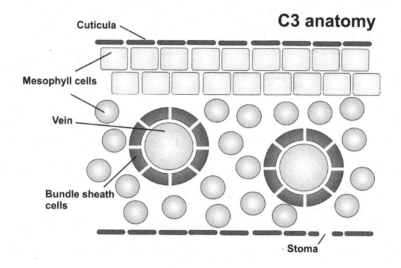

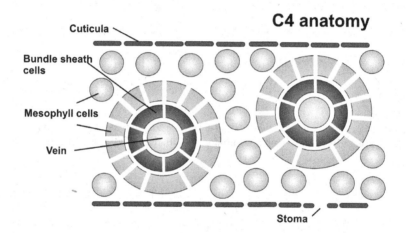

FIGURE 67 Schematic drawing of leaf anatomy in C_3 and C_4 plants. In C_4 plants, a wreath-like arrangement of cells acts to concentrate carbon dioxide. A ring of mesophyll cells captures the carbon dioxide, that is conveyed to an inner ring of bundle-sheath cells. It is called Kranz ("wreath") anatomy.

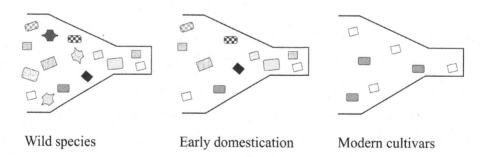

Wild species   Early domestication   Modern cultivars

FIGURE 68 Scheme of reduction of variability during the process of domestication, i.e. from wild plants to advanced varieties.

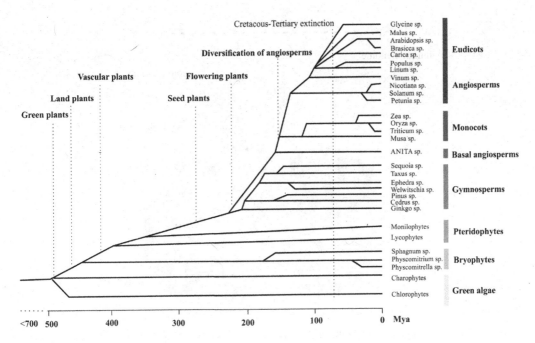

FIGURE 69 Simplified phylogeny of the green plant lineage focusing on the occurrence of events. Estimates for the age of angiosperms have suggested a range of 167–199 million years ago (Mya); then rapid radiations responsible for the extant angiosperm diversity occurred after the early diversification of the Mesangiospermae 139–156 Mya, with a burst of diversification specific to the Cretaceous Period, <125 Mya. (Modified after Alix et al. (2017), http://molcyt.org/2017/07/13/polyploidy-and-interspecific-hybridisation-partners-for-adaptation-speciation-and-evolution-in-plants/.)

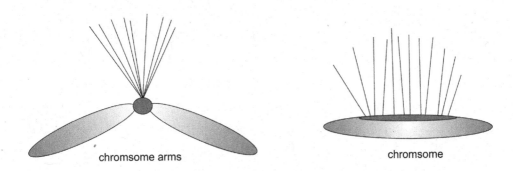

FIGURE 70 Schematic drawing of a monocentric and a holocentric chromosome. (Source: R. Schlegel.)

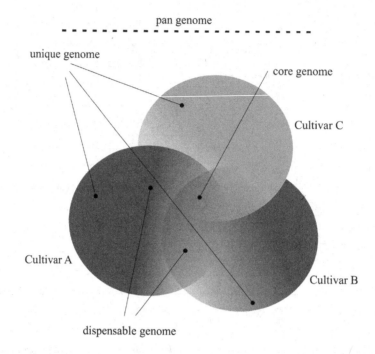

FIGURE 71 Schematic drawing of genomic compartments of a plant species. (Source: R. Schlegel.)

Tables

TABLE 1
Classification of Wheats (*Triticum* spp.)

Ploidy level	Species	Chromosome number (2n =)	Genome formula
Diploid	*urartu*	14	$A^u A^u$
	boeoticum	14	$A^m A^m$
	monococcum	14	$A^m A^m$
	ssp.	*monococcum* (cultivated form,	
	ssp.	einkorn or small spelt wheat)	
		aegilopoides (wild type)	
	sinskajae (mutant, free-threshing form of *T. monococcum*)	14	$A^m A^m$
Tetraploid	*turgidum*	28	$BBA^u A^u$
	ssp.	*turgidum* (rivet, cone, or pollard wheat)	
	ssp.	*carthlicum* (Peresian or Persian black wheat)	
	ssp.	*dicoccum* (emmer wheat)	
	ssp.	*durum* (macaroni, hard, or durum wheat)	
	ssp.	*paleocolchicum*	
	ssp.	*polonicum* (Polish wheat)	
	ssp.	*turanicum* (Khorasan wheat)	
	ssp.	*dicoccoides* (wild emmer wheat)	
	araraticum	28	$GGA^m A^m$
	timopheevi	28	$GGA^m A^m$
	militinae (mutant of *T. timopheevi*)	28	$GGA^m A^m$
Hexaploid	*aestivum*	42	$BBA^u A^u DD$
	ssp.	*aestivum* (bread wheat)	
	ssp.	*compactum* (club, dwarf, hedgehog, or cluster wheat)	
	ssp.	*macha*	
	ssp.	*spelta* (large spelt or dinkel wheat	
	ssp.	*sphaerococcum*	
	vavilovii	42	$BBA^u A^u DD$
	zhukovskyi (*T. timopheevi* × *T. monococcum*)	42	$GGA^u A^u A^m A^m$

There are five known combinations in the genus *Triticum* leading to five super species:

- A^m *T. monococcum*
- A^u *T. urartu*
- BA^u *T. turgidum*
- GA^u *T. timopheevi*
- $BA^u D$, *T. aestivum*

In the genus *Triticum*, five genomes, all originally found in diploid species, have been identified:

- A^m - present in wild einkorn (*T. boeoticum*)
- A^u - present in *T. urartu* (closely related to *T. boeoticum*)
- B - present in most tetraploid wheats; source most similar to *Aegilops speltoides*
- G - present in *T. timopheevi* group of wheats; source most similar to *Ae. speltoides.*
- D - present in *Aegilops squarrosa*, i.e. in all hexaploid wheats

TABLE 2

Number of Gametes, Genotypes, and Phenotypes, Considering One and Multifactorial Heterozygosity in F_1 and F_2 Generations

Number of heterozygous factors in F_1 generation	Number of different gametes in F_1 Number of homozygous genotypes in F_2 Number of phenotypes in F_2, considering complete dominance of all characters	Number of genotypes in F_2	Number of combinations (among these numbers, all genotypes may be realized in F_2)
1	2	3	4
2	4	9	16
3	8	27	64
4	16	81	256
5	32	243	1024
6	64	729	4096
7	128	2187	16384
8	256	6561	65536
9	512	19683	262144
10	1024	59049	1048576
n	2^n	3^n	$2^n \times 2^n = 4^n$

TABLE 3

Segregation of Recessive Nulliplex Genotypes from Triplex, Duplex, and Simplex Genotypes, Considering Selfing, Random Chromosome Distribution, and Complete Dominance

	Genotypes	
Triplex	**Duplex**	**Simplex**
AAAa	AAaa	Aaaa
Gametes		
AA	AA	Aa
AA	Aa	Aa
AA	Aa	Aa
Aa	Aa	aa
Aa	Aa	aa
Aa	aa	aa
3AA:3Aa	1AA:4Aa:1aa	3Aa:3aa
Progeny		
9AAAA	1AAAA	9AAaa
18AAAa	8AAAa	18Aaaa
9AAaa	18AAaa	9aaaa
	8Aaaa	
	1aaaa	
No nulliplex aaaa type	Segregating nulliplex aaaa type of ratio 1:35	Segregating nulliplex aaaa type of ratio 9:27

TABLE 4
Expected F$_2$ Segregations of Trisomic F$_1$ Plants from a Critical Cross of Trisomic by Disomic, Excluding Any Selection, Male Transmission of *n + 1* Gametes and Abnormal Chromosome Segregation

F$_1$ genotype	Selfing or backcrossing of disomic recessive	Expected segregation depending on the transmission of n + 1 gametes through female	
		50%	**30%**
Aa	Selfing	3:1	3.0:1.0
Aa	Backcrossing	1:1	1.0:1.0
Aaa	Selfing	17:1	11.9:1.0
AAa	Backcrossing	5:1	3.3:1
Aaa	Selfing	2:1	1.6:1.0
Aaa	Backcrossing	1:1	1.0:1.3

TABLE 5
Basic Methods of Plant Breeding

Selection breeding	Mass selection	Positive mass selection
		Negative mass selection
	Pedigree selection	Autogamous plants
		Allogamous plants
		Vegetatively grown plants
Combination breeding		Autogamous plants
		Allogamous plants
Hybrid breeding		Autogamous plants
		Allogamous plants
Synthetics		Autogamous plants

TABLE 6

Genotypic and Phenotypic Segregation in F_2 Populations, Considering Two Genes and Interacting in Different Manners

Genotypes	Phenotypes*				
	a	b	c	d	e
1 AABB					
2 AABb					
2 AaBB					
4 AaBb					
1 AAbb					
2 Aabb					
1 aaBB					
2 aaBb					
1 aabb					
Segregation	9:3:3:1	12:3:1	15:1	1:4:6:4:1	9:7

*Phenotype determination:

a = factor A is dominant over a, factor B is dominant over b, and there is no gene interaction.

b = factor A is epistatic of Bb and bb.

c = factor A is equidirectionally substituted by B, and factor B is equidirectionally substituted by A.

d = the dominant alleles A and B show additive interaction in the same direction, i.e. $AABb = AaBB$, $AaBb = AAbb, etc.$

e = factor A can be phenotypically expressed only when A and B are present due to complementary gene action.

TABLE 7

Frequencies and Ratios of Completely Recessive Plants in F_2 Progeny of Doubled Haploids, Diploids and Tetraploids

Number of alleles	Frequency of recessive plants as determined by one of the plants given below (1/x)		
	Doubled-haploid	Diploid	Tetraploid
1	2	4	36
2	4	16	1296
3	8	64	46656
4	16	256	1679616
5	32	1024	60466176
n	2^n	2^{2n}	6^{2n}

TABLE 8
Genome Relationships between Embryo, Endosperm, and Ovary, after Crossing Parents with Different Ploidy Levels

Female × Male	Embryo	Endosperm	Ovary
2x × 2x	2x	3x	2x
2x × 3x	3x	4x	2x
4x × 2x	3x	5x	4x
2x × 6x	4x	5x	2x
6x × 2x	4x	7x	6x

TABLE 9
Genetic Segregation Patterns Depending on the Number of Genes Involved

No. of different gene pairs	No. of different gametes	No. of different $F_1 \times F_1$ gametes	No. of F_2 genotypes	No. of homozygous F_2 genotypes	No. of heterozygous F_2 genotypes	No. of F_2 phenotypes*
1	2	4	3	2	1	2
2	4	16	9	4	5	4
3	8	64	27	8	19	16
4	16	256	81	16	65	32
5	32	1024	243	32	211	64
n	$2n$	4^n	3^n	2^n	$3^n - 2^n$	2^n

*when completely dominant.

TABLE 10
Frequencies of Homozygotes and Heterozygotes in a Progeny of a Heterozygous Individual after Subsequent Self-Pollinations

Number of generations	Genotypes			Relative proportion of heterozygotes per population
	AA	Aa	aa	
0	0	1	0	1
1	1/4	2/4	1/4	1/2
2	3/8	2/8	3/8	1/4
3	7/16	2/16	7/16	1/8
4	15/32	2/32	15/32	1/16
5	31/64	2/64	31/64	1/32
10	1023/2048	2/2048	1023/2048	1/1024
n				$1/2^n$

TABLE 11
Examples of Seed Conditioning in Some Crop Plants

Small-seeded grasses	Field beans	Small grains	Small-seeded legumes
Scalper	Air screen cleaner	Aspirator	Scalper
Debearder	Gravity separator	Debearder	Huller, scarifier
Air screen cleaner	Stoner	Air screen cleaner	Air screen cleaner
Disc and/or cylinder separator	Color separator	Indent cylinder	Gravity separator
Treater	Processing belts	Treater	Stoner
Bagging	Treater	Bagging	Roll mill
	Bagging		Cylinder separator
			Treater
			Bagging

TABLE 12
Taxonomic Classification System in Plants

Classification	Example
Kingdom (Regnum)	Fungi
Subkingdom	Cryptogamia
Division	Fungi
Subdivision	Eumycotina
Class	Basidiomycetes
Subclass	Heterobasidiomycetes
Order	Uredinales
Suborder	Uredineae
Family	Pucciniaceae
Subfamily	Puccinioideae
ribe	Puccineae
Subtribe	Puccinia
Genus	*Puccinia*
Subgenus	*Hetero-Puccinia*
Section	–
Subsection	*Puccinia graminis*
Species	*Puccinia graminis* var. *phlei-pratensis*
Subspecies	–
Variety	–
Subvariety	–
Form	–
Special form (*forma specialis*)	–
Physiological race	*P. graminis* var. *phlei-pratensis* Race 3

TABLE 13
Decimal Code for Plant Growth in Cereals

Code	Stage	Code	Stage
0	Germination	5	Inflorescence emergence (FEEKES scale 10.1, HAUN scale 10.2)
00	Dry seed	50-51	1st spikelet visible
01-02	Start of imbibition	52-53	1/4 of inflorescence emerged (FEEKES scale 10.2)
03-04	Imbibition completed	54-55	1/2 of inflorescence emerged
05-06	Radicle emerged	56-57	3/4 of inflorescence emerged (FEEKES scale 10.4, HAUN scale 10.7)
07-08	Coleoptile emerged	58-59	Inflorescence emerged (FEEKES scale 10.5, HAUN scale 11)
09	Leaf on coleoptile tip	6	Anthesis (FEEKES scale 10.51, HAUN scale 11.4)
1	Seedling growth	60-63	Beginning of anthesis
10	1st leaf through coleoptile (FEEKES scale 1)	64-67	1/2 of anthesis
11	1st leaf unfolded (HAUN scale 1.+)	68-69	Anthesis completed (HAUN scale 11.6)
12	2nd leaf unfolded (HAUN scale 1.+)	7	Milk development
13-19	3rd leaf-9th leaves unfolded (HAUN scale 2.+ - 7.+)	70-71	Caryopsis waterripen (FEEKES scale 10.54, HAUN scale 12.1)
2	Tillering	73-74	Early milk ripe (HAUN scale 13)
20	Main shoot only	75-76	Medium milk ripe (FEEKES scale 11.1)
21	Main shoot + 1 tiller (FEEKES scale 2)	77-79	Late milk ripe
22	Main shoot + 2 tillers	8	Dough development
23-29	Main shoot + 3-9 tillers (FEEKES scale 3)	80-84	Early dough (HAUN scale 14)
3	Stem elongation (FEEKES scale 4-5)	85-86	Soft dough
31	Pseudoerection of stem	87-89	Hard dough (HAUN scale 15)
32	1st node detectable (FEEKES scale 7)	9	Ripening
33	2nd node detectable	90	Caryopsis hard
34-36	3rd-6th node detectable	91	Caryopsis hard
37-38	Flag leaf detectable (FEEKES scale 8)	92	Caryopsis hard (FEEKES scale 11.4; HAUN scale 16)
39	Flag leaf ligula + collar visible (FEEKES scale 9)	93	Caryopsis loosening
4	Booting	94	Overripe, straw dead
40-41	Flag leaf sheath extending (HAUN scale 8 to 9)	95	Seed dormant
42-43	Boots just visibly swollen	96	Seeds giving 50% germination
44-45	Boots swollen (FEEKES scale 10, HAUN scale 9.2)	97	Seed not dormant
46-47	Flag leaf sheath opening	98	2nd dormancy
48-49	First awns visible (HAUN scale 10.1)	99	2nd dormancy lost

Source: After ZADOK et al., 1974.

TABLE 14

Genome Relationships of Oats (*Avena* spp.)

Section	Wild species	Cultivated species	Chromosome number (2n =)	Genome formula
Avenotrichon/ Ventricosa	*A. macrostachya*		28	MMMM (?)
	A. clauda		14	$C^p C^p$
	A. eriantha		14	$C^p C^p$
	A. pillosa		14	$C^p C^p$
	A. ventricosa		14	$C^v C^v$
Agraria	*A. brevis*		14	$A^s A^s$
	A. hispanica		14	$A^s A^s$
	A. nuda		14	$A^s A^s$
	A. strigosa	*A. strigosa*	14	$A^s A^s$
	A. wiesti		14	$A^s A^s$
Tenuicarpa	*A. agadiriana*		28	????
	A. atlantica		14	$A^s A^s$
	A. canariensis		14	$A^c A^c$
	A. canariensis	*A. brevis*	14	$A^c A^c$
	A. canariensis	*nudibrevis*	14	$A^c A^c$
	A. damascena		14	$A^d A^d$
	A. hirtula		14	$A^s A^s$
	A. longiglumis		14	$A^l A^l$
	A. matritensis		14	$A^s A^s$ (?)
	A. prostrata		14	$A^p A^p$
Ethiopica	*A. abyssinica*		28	AABB
	A. barbata	*abyssinica*	28	AABB
	A. vaviloviana		28	AABB
Pachycarpa	*A. insularis*		28	$DDC^v C^v$
	A. macrocana syn		28	$DDC^v C^v$
	A. magna			
	A. murphyi		28	$DDC^v C^v$
Avena	*A. atherantha*		42	$AAC^v C^v DD$
	A. byzanthina		42	$AAC^v C^v DD$
	A. fatua	*byzantina*	42	$AAC^v C^v DD$
	A. fatua	*nuda*	42	$AAC^v C^v DD$
	A. hybrida		42	$AAC^v C^v DD$
	A. occidentalis		42	$AAC^v C^v DD$
	A. sativa		42	$AAC^v C^v DD$
	A. sterilis	*sativa*	42	$AAC^v C^v DD$
	A. trichophylla		42	$AAC^v C^v DD$

TABLE 15
The Approximate Protein Composition (%) in Some Cereals

Cereal	Albumin	Globulin	Prolamin	Glutelin
Barley	13	12	(hordein) 52	(hordenin) 23
Maize	4	2	(zein) 55	39
Oats	11	56	(avenin) 9	23
Rice	5	10	(oryzin) 5	(oryzenin) 80
Sorghum	6	10	(kafirin) 46	38
Wheat	9	5	(gliadin) 40	(glutenin) 46

Source: After PAYNE And RHODES, 1982.
Note: In parentheses are the common names of some storage proteins.

TABLE 16
Food Reserves of Some Crop Plants

Crop	Average composition (%)			Storage organ
	Protein	Fat	Carbohydrate	
Barley	12	3	76	Endosperm
Maize	10	5	88	Endosperm
Oats	13	8	66	Endosperm
Rye	12	2	76	Endosperm
Wheat	12	2	75	Endosperm
Broad bean	23	1	56	Cotyledons
Pea	25	6	52	Cotyledons
Peanut	31	48	12	Cotyledons
Soybean	37	17	26	Cotyledons
Castor bean	18	64	–	Endosperm
Oil palm	9	49	28	Endosperm
Pine	35	48	6	Gametophyte
Rapeseed	21	48	19	Cotyledons

Source: After BEWLEY and BLACK, 1995.

TABLE 17
Taxonomic Relationships of Some Tuberous *Solanum* spp.

Series	Species	(2n =)	Genome
Etuberosa	brevidens, etuberosum, fernandezianum	24	E^bE^b, E^eE^e, E^fE^f
Morelliformia	morelliformae	24	A^mA^m
Bulbocastana	bulbocastanum, clarum	24	A^bA^b
Pinnatisecta	caridophyllum, jamesii, pinnatisectum	24	$A^{pi}A^{pi}$
Commersoniana	chacoense, commersonii	24	AA
Conicibacata	santolallae, chromatophillum,	24	$A^{c1}A^{c1}$, $A^{c2}A^{c2}$
	agrimonifolium, longiconicum, oxycarpum	48	$A^{c1}A^{c1}C^aC^a$, $A^{c1}A^{c1}C^lC^l$, $A^{c1}A^{c1}C^oC^o$
Piurna	piurnae	24	A^pA^p
	tuquerrense	48	A^pA^pPP
Acaulia	acaule	48	AAA^aA^a
	acaule ssp. *albicans*	72	$AAA^aA^aX^bX^b$
Demissa	brachycarpum, demissum, guerreoense,	72	$AADDD^bD^b$, $AADDD^dD^d$, $AADDD^gD^g$,
	spectabile		$AADDD^sD^s$
Longipedicellata	vallis-mexici	36	AAB
	fendleri, polytrichon, stoliferum	48	AABB
Polyadenia	polyadenium, infundibuliformae,	24	AA
	megistacrolobum, raphanifolium,		
	sanctae-rosae, toralapanum		
Ingaefolia	rachialatum	24	A^iA^i
Olmosiana	olmosense	24	A^oA^o
Tuberosa (wild)	abancayense, bukasovii, canasense,	24	AA
	gourlayi, kurtzianum, leptophyes, maglia,	48	AAA^sA^s
	microdontum, soukupii, sparsipilum,		
	speganzinii, vernei, verrucosum		
	sucrense		
Tuberosa	ajanhuiri, goniocalyx, phureja,	24	AA
(cultivated)	stenotomum	36	AAA^t, AAA^a
	chaucha, juzepszukii	48	AA A^tA^t
	tuberosum ssp. andigena	48	AA, A^tA^t
	tuberosum ssp. tuberosum	60	$AAAA^aA^t$
	cortilobum		

TABLE 18
Types of Flowers in Higher Plants

Flower type	Characteristics
Bisexual, hermaphroditic, monoclinous, perfect	Male and female in one flower
Protandry	Pollen shed before stigma is receptive
Protgyny	Stigma matures and ceases to be receptive before pollen is shed
Chaemogamy	Stigma receptive and pollen shed after flower opens
Cleistogamy	Stigma receptive and pollen shed in closed flower
Pin flower	Long styles and short stamens
Thrum flower	Short styles and long stamens
Diclinous, unisexual, imperfect	Males and females in separate flowers
Male, staminate	Male flower
Carpellate, female, pistillate	Female flower
Monoecious	Male and female flowers on one plant
Dioecious	Male and female flowers on separate plants
Mixed, polygamous	Presence of male, female, and perfect flowers
Polygamo-monoecious	Presence of male, female, and perfect flowers on the same plant
Polygamo-dioecious	Presence of male, female, and perfect flowers on separate plants

TABLE 19
Segregation of a Single Gene and/or Alleles in Subsequent Generations

Generation	Parent 1 × Parent 2		
F_1		Aa (1/1)	
F_2	AA (1/4)	Aa (1/2)	aa (1/4)
F_3	AA (3/8)	Aa (1/4)	aa (3/8)
F_4	AA (7/16)	Aa (1/8)	aa (7/16)
F_5	AA (15/32)	Aa (1/16)	aa (7/16)

TABLE 20
Phenotypic Relations of Homozygotes and Heterozygotes, Depending on Different Dominance Levels in Diploids

Genotype	Dominance level			
	Additive interaction	Partial dominance	Complete dominance	Over-dominance
AA	5	5	5	5
Aa	3	4	5	6
aa	1	1	1	1

TABLE 21

Phenotypic Ratios in the F_2 Generation for Two Unlinked Genes, Depending on the Degree of Dominance at Each Locus and Epistasis between Loci

Genetic explanation	F_2 genotypes								
	AA BB	AA Bb	Aa BB	Aa Bb	AA bb	Aa bb	aa BB	aa Bb	aa bb
Complete dominance lacking at either locus, no epistasis, phenotypic and genotypic ratios are equal	1	2	2	4	1	2	1	2	1
Complete dominance lacking in *A*, complete dominance in *B*, no epistasis	3		6		1	2	3		1
Complete dominance in *A* and *B*, no epistasis	9				3		3		1
Recessive epistasis, *aa* epistatic to *B* and *b*	9				3		4		
Dominant epistasis, *A* epistatic to *B* and *b*	12						3		1
Dominant and recessive epistasis, *A* epistatic to *B* and *b*, *bb* epistatic *A* and *a*, *A* and *bb* produce identical phenotypes	13							3	
Duplicate recessive epistasis, *aa* epistatic to *Bb*, *bb* epistatic to *A* and *a*	9					7			
Duplicate dominant epistasis, *A* epistatic to *B* and *b*, *B* eptistatic to *A* and *a*	15								1

TABLE 22

Calculation of Recombination Frequency between Two Loci from a Cross between *AaBb* x *aaaa*

	Genotypes				
	AaBb	*Aabb*	*aaBb*	*aabb*	Total
Observed	190	38	35	201	464
Expected	116	116	116	116	464
Description	parental	non	non	parental	

Recombination (%) = (sum of nonparental class/total number of individuals) × 100
Recombination = (38 + 35 / 190 + 38 + 35 + 201) × 100 = 14.4%
Recombination = (*Aabb*+*aaBb*/*AaBb*+*Aabb*+*aaBb*+*aabb*) × 100

TABLE 23
Test Crosses with Monosomics to Determine the Location of a Dominant Allele

Possible crosses	F1
1 The critical gene is not located on the missing chromosome	
Cross: AA × aa	Aa (dominant)
2 The critical gene is located on the missing chromosome	
Cross: A0 × aa	Aa (dominant)

TABLE 24
Possible Planting Arrangements for a Diallel Crossing (Six Parents, No Reciprocal Cross, No Self-Pollination)

Unpaired parents
P1 P2 P3 P4 P5 P6

Paired parents
P1 × P2 P1 × P3 P1 × P4 P1 × P5 P1 × P6
P2 × P3 P2 × P4 P2 × P5 P2 × P6
P3 × P4 P3 × P5 P3 × P6
P4 × P5 P4 × P6
P5 × P6

Semi-Latin square
P1 × P2 × P6 × P3 × P5 × P4
P2 × P3 × P1 × P4 × P6 × P5
P6 × P1 × P5 × P2 × P4 × P3

Bulk design
P1 × P1 P2 × P2 P3 × P3 P4 × P4 P5 × P5 P6 × P6

TABLE 25

A Randomized Complete-Block Design for Five Entries and Ten Replications

Block	Replication									
	1	2	3	4	5	6	7	8	9	10
I	P5	P1	P5	P4	P2	P4	P1	P5	P2	P5
II	P1	P2	P1	P1	P1	P1	P4	P3	P1	P4
III	P3	P5	P4	P3	P5	P5	P5	P2	P3	P2
IV	P2	P3	P2	P2	P4	P2	P3	P4	P5	P3
V	P4	P4	P3	P5	P3	P3	P2	P1	P4	P1

TABLE 26

Lattice Design (42 Entries, Three Replications, No Blocks within the Replication, Entries Assigned at Random to the 42 Plots)

Replication 1

1	2	3	4	5	6
7	8	9	10	11	12
13	14	15	16	17	18
19	20	21	22	23	24
25	26	27	28	29	30
31	32	33	34	35	36
37	38	39	40	41	42

Replication 2

7	13	19	37	25	31
1	14	20	32	26	38
8	2	21	27	33	39
3	9	28	15	34	40
16	4	10	22	35	41
42	29	23	17	11	5
18	12	6	24	30	36

Replication 3

38	33	28	22	17	12
13	2	24	29	40	35
42	36	25	20	9	4
6	11	16	27	37	32
18	1	7	23	34	39
3	8	14	19	41	30
26	31	21	15	10	5

TABLE 27
Recovering of Genes from the Recurrent Parent During Backcrossing

Generation	% of parentage	
	Recurrent	Nonrecurrent
F_1	50	50
BC_1	75	25
BC_2	87.5	12.5
BC_3	93.75	06.25
BC_4	96.875	03.125
BC_5	98.4375	01.5625

% homozygous individuals = $(2^m - 1/2^m)^n$

n = number of backcrosses

TABLE 28
Scheme of Seed Purification and Increase

1. Harvest of individual plants		
2. Plant individual rows, discard off-types (or rows), bulk seeds of similar type	Pedigree seed	⇩
3. Plant pedigree seeds, remove off-types, harvest breeder seeds	Breeder seed	⇩
4. Plant breeder seeds, remove off-types, harvest foundation seeds	Foundation seed	⇩
5. Plant foundation seeds, harvest registered seeds	Registered seed	⇩
6. Plant registered seeds, harvest certified seeds	Certified seed	

TABLE 29
Types and Characteristics of Several Markers in Breeding and Genetics

Characteristics	Morphological traits	Isoenzymes	RFLPs	AFLPs RAPDs	Microsatellites or Simple Sequence Repeats (SSRs)	Inter-simple sequence repeats (ISSRs)	Retrotransposon-microsatellite amplified or inter-retrotransposon amplified polymorphism (IRAPs REMAPs)
Number of loci	Limited	Limited	Almost unlimited	Unlimited	High	High	High
Inheritance	Dominant	Codominant	Codominant	Codominant or dominant	Codominant	Dominant	Codominant
Positive features	Visible	Easy to detect	Utilized before the latest techniques were available; robust; reliable; transferable across populations	PCR-based; quick assay with many markers; multiple loci; high levels of polymorphism generated; quick and simple; inexpensive; multiple loci from a single primer possible; small amounts of DNA required	PCR-based; welldistributed within the genome, many polymorphisms	PCR-based; highly polymorphic; robust in usage; can be automated	PCR-based; highly polymorphic depending on the transposon; robust in usage; can be automated; species-specific
Negative features	Possibly negative linkage to other characters	Possibly tissue-specific	Radio-activity required; time-consuming; laborious; rather expensive; large amounts of DNA required; limited polymorphism (especially in related lines)	High basic investment, patented; large amounts of DNA required; complicated methodology; problems with reproducibility; generally not transferable	Long development of the marker; expensive	Usually dominant; species-specific	Alleles cannot be detected; can be technically challenging

TABLE 30
Typical Characteristics for Identification of Wheat, Barley, or Wild Oat

Characteristic	Wheat	Barley	Wild oat
Ligule	Membranous	Membranous	Membranous
Auricle	Short + hairy or without hair	Long + clasping	Absent
Blades, collar	~ hairy	Without hair	Long hair on margin
Sheath	~ hairy	Without hair	~ without hair
Blade twist	Clockwise	Clockwise	Counterclockwise

TABLE 31
Taxonomy of the Genus *Helianthus*

Section	Series	Species	Subspecies	(2n =)
I. ANNUUI		*niveus	niveus, tephrodes, canescens	34
		*debilis	debilis, vestitus, tardilorus, silvestris, cucumerifolius	34
		*praecox	praecox, runyonii, hirtus	34
		*petiolaris	petiolaris, fallax	34
		*neglectus, *annuus, *argo-phyllus, *bolanderi, *anomalus, *paradoxus, *agrestis		34
II. CILIARES	1. Pumili	gracilenthus, pumilus, cusickii		34
	2. Ciliares	arizonensis, laciniatus		34
		ciliaris		68
III. DIVARICATI	1. Divaricati	mollis, divaricatus, decapetalus		34
		occidentalis	occidentalis, plantagineus	34
		hirsutus, strumosus		68
		eggertii, tuberosus		102
		rigidus	rigidus, subrhomboideus	102
	2. Gigantei	giganteus, grosseserratus		34
		nuttallii	nuttallii, parishii, rydbergii	34
		maximiliani, salicifolius, californicus		34
		resinosus, schweinitzii		102
	3. Microcephali	microcephalus, glaucophyllus, smithii, longifolius		34
		laevigatus		68
	4. Angus-tifolius	angustifolius, simulans, floridanus		34
	5. Atroru-bentes	silphioides, atrorubens, hetero-phyllus, radula, carnosus		34

TABLE 32
About the Evolution of Triticeae

Divergence	~ Million years ago
Earliest land plant fossils	420
Origin of angiosperms	200-340
Monocot-dicot divergence	160-240
Origin of grass family	65-100
Oldest known grass fossils	50-70
Divergence of the subfamilies	50-80
Pooideae (wheat, barley, oat)	..
Bambusoideae (rice)	..
Panicoideae (maize, sorghum)	..
Earliest fossil of the rice lineage	40
Divergence of maize and sorghum lineages	15-20
Divergence of wheat and barley lineages	10-14
Divergence of wheat and rye lineages	7

Note: Panicoideae diverged from the Pooideae-Bambusoideae first, followed shortly by divergence of the latter two subfamilies, or authors suggest that either the Pooideae or the Bambusoideae branched off first. In general, it seems that all three subfamilies diverged about the same time. Some phylogenies divide Bambusoideae into two subfamilies, Oryzoideae and Bambusoideae.

TABLE 33
Inheritance and Variation of Several Breeding Characteristics

Breeding aims	Qualitative variation	Quantitative variation
Yield traits	Dwarf growth, monogermous fruits	Yield/ha; harvest index, nutritional efficiency, rate of photosynthesis
Quality traits	Linolenic acid, large seeds	Protein content, cooking ability, digestibility
Resistance traits	Monogenic resistance, tolerance to aluminum	Polygenic resistance, drought tolerance

TABLE 34
Some Examples of Heritability of Breeding Characteristics

Heritability	Characteristic	Begin selection from: .
High	Certain resistance traits, ripening time, awns, grain color, glume shape	F_2-F_3
Medium	Seed or spike characters, straw length, lodging resistance	F_4-F_5
Low	Yield components, yield potential, physiological traits	F_7-F_8

TABLE 35
Proposed Breeding Schemes Depending on the Crop, Reproduction System and Basic Population Features (+ Applicable)

	Crops										
	1	**2**	**3**	**4**	**5**	**6**	**7**	**8**	**9**	**10**	**11**
Reproduction											
Autogamous	+	–	–	+	–	–	+	–	+	–	–
Allogamous	–	+	+	–	+	–	+	+	–	+	+
Anemophilous	–	+	+	–	+	–	–	–	–	–	+
Entomophilous	–	–	–	–	–	–	+	+	–	+	–
Self-fertile	+	–	+	+	–	–	+	+	+	–	+
Self-sterile	–	+	–	–	+	–	–	–	–	+	–
Isolation required	–	+	+	–	+	+	+	+	–	+	+
Clonable	–	+	–	–	+	+	–	–	–	+	+
Basic population derived from ...											
Crossing	+	+	+	+	+	+	+	+	+	+	+
Wild origin	–	–	–	–	–	+	–	–	–	+	+
Polyploidization	–	+	–	–	+	–	+	–	–	+	+
Induced mutation	+	–	+	+	–	–	+	–	–	–	–
Breeding method											
Mass selection	–	+	–	–	–	+	+	–	–	+	+
Pedigree method	+	–	+	+	+	+	+	+	+	+	+
Combination breeding	+	+	+	+	+	+	–	+	+	+	+

1 = Wheat, oats, barley, rice; 2 = Rye; 3 = Maize; 4 = Millet; 5 = Sugar beet; 6 = Potato; 7 = Rapeseed; 8 = Sunflower; 9 = Broad bean; 10 = Clover; 11 = Forage grasses

TABLE 36
The Size Reduction of Populations Depending on Various Numbers of Overdominance Loci and Different Selective Disadvantages for Homozygotes

	Reduction of the population at disadvantage of homozygotes		
Number of overdominant loci	**5%**	**1%**	**0.1%**
10	0.78	0.95	0.97
100	0.08	0.61	0.78
1,000	1.01×10^{-11}	0.01	0.08

TABLE 37
Types of Hybrids and/or Hybridization

Type of hybrid	Crossing scheme
Variety cross	Population I x Population II; Variety I x Variety II
Topcross hybrid	Line or single cross x variety; A x Population or (A x B x Population
Single hybrid	Line I x Line II; A x B
Modified single hybrid	Sister line cross x Sister line cross; (A' x A) x (B' x B)
Three-way hybrid	Single hybrid x Line; (A x B) x C
Modified three-way hybrid	Single hybrid x Sister line cross: (A x B) x (C' x C)
Double hybrid	Single hybrid x Single cross: (A x B) x (C x D)

TABLE 38
The Genetic Code

Position I (5' end)	Position II				Position III (3' end)
	U	C	A	G	
U	Phe	Ser	Tyr	Cys	U
	Phe	Ser	Tyr	Cys	C
	Leu	Ser	Stop	Stop	A
	Leu	Ser	Stop	Trp	G
C	Leu	Pro	His	Arg	U
	Leu	Pro	His	Arg	C
	Leu	Pro	Gln	Arg	A
	Leu	Pro	Gln	Arg	G
A	Ile	Thr	Asn	Ser	U
	Ile	Thr	Asn	Ser	C
	Ile	Thr	Lys	Arg	A
	Met	Thr	Lys	Arg	G
G	Val	Ala	Asp	Gly	U
	Val	Ala	Asp	Gly	C
	Val	Ala	Glu	Gly	A
	Val	Ala	Glu	Gly	G

C= cytosine; A= adenine; G= guanine; U = Uracil   Phe = Phenylalanine   Ser = Serine

Tyr = Tyrosine	Lys = Lysine	C = Cytosine
Leu = Leucine	Pro = Proline	His = Histidine
Asp = Aspartic acid	A = Adenine	Ile = Isoleucine
Thr = Threonine	Gln = Glutamine	Glu = Glutamate
G = Guanine	Val = Valine	Ala = Alanine
Asn = Asparagine	Cys = Cysteine	Trp = Tryptophan
Arg = Argenine	Gly = Glycine	Stop = Stop codon

TABLE 39
Cleaving Sites of Some Restriction Enzymes

Restriction enzyme	Derived from bacterium:	Cleaving sites 5'..................3' 3'..................5'	Cleaving
*Bam*HI	*Bacillus amyloliquefaciens*	GGATCC CCTAGG	G-G
*Bal*1	*Brevibacterium albidum*	TGGCCA ACCGGT	G-C
*Eco*RI	*Escherichia coli*, RY 13	GAATTC CTTAAG	G-A
*Hae*III	*Haemophilus aegyptius*	GGCC CCGG	G-C
*Hind*III	*Haemophilus influenzae*	AAGCTT TTCGAA	A-A
*Hpa*I	*Haemophilus parainfluenzae*	GTTAAC CAATTG	T-A
*Pst*I	*Providencia stuartii* 164	CTGCAG GACGTC	A-G
*Sal*I	*Streptomyces albus* G	GTCGAC GAGCTG	G-T

TABLE 40
Guidelines for Selection of an Experimental Design

Number of factors	Comparative objective	Screening objective	Response objective
1	1-factor completely randomized design		
2-4	Randomized block design	Full or fractional factorial	Central composite
5...	Randomized block design	Fractional factorial	Screen first to reduce number of factors

TABLE 41
Genome Relationships of Barley (*Hordeum* sp.)

Section	Species	Subspecies	Chromosome number (2n =)	Genome formula
Hordeum	*H. vulgare*	*vulgare, spontaneum, agriochriton*	14	II
	H. bulbosum		14, 28	II, IIII
	H. murinum	*murinum*	28	YYYY
		leporinum	14	YYYY
		leporinum	42	YYYYYY
		glaucum	14	YY
Anisolepis	*H. pusillum, H. intercedens, H. euclaston, H. flexuosum, H. muticum, H. chilense, H. cordobense, H. stenostachys*		14	HH
Critesion	*H. pubiflorum, H. halophilum, H. comosum*		14	HH
	H. jubatum		14, 28	HH, HHHH
	H. arizonicum, H. procerum, H. lechleri		42	HHHHHH
Stenostachys	*H. marinum*	*marium, gussoneanum*	14	XX
	H. secalinum, H. capense		28	HHHH
	H. bogdanii, H. roshevitzii		14	HH
	H. brevisubulatum	*brevisubulatum, nevskianum*	14, 28	HH, HHHH
		turkestanicum	28, 42	HHHH, HHHHHH
		violaceum	14, 28	HH, HHHH
		iranicum	28, 42	HHHH, HHHHHH
	H. brachyantherum	*brachyantherum*	28, 42	HHHH, HHHHHH
		californicum	14	HH
	H. depressum, H. guatemalense		28	HHHH
	H. erectifolium		14	HH
	H. tetraploidum, H. fuegianum		28	HHHH
	H. parodii		42	HHHHHH
	H. patagonicum	*patagonicum, setifolium, santacruense, mustersii, magellanicum*	14	HH

TABLE 42
Developmental Stages in Plants

Ovule stage	Description	Flower stage	Description	Rosette stage	Description
1-I	ovule primordium arise	1	emergence of floral meristem on flank of inflorescence meristem	early rosette growth	rosette is 20% of final size
1-II	ovule primordium elongates	2	floral meristem becomes demarcated from the inflorescence meristem by slight indentation	2-rosette leaf	first 2 rosette leaves have reached >1mm in length
2-I	megaspore mother cell enlarges	3	the sepal primordia arise	3-rosette leaf	first 3 rosette leaves have reached >1mm in length
2-II	inner integument initiates	4	sepals overlie floral meristem	4-rosette leaf	first 4 rosette leaves have reached >1mm in length
2-III	outer integument initiates, chalazal nucellus divides	5	petal and stamen primordia arise	5-rosette leaf	first 5 leaves have reached >1 mmlength
2-IV	megaspore meiosis occurs	6	sepals enclose floral bud	6-rosette leaf	first 6 leaves have reached >1mm in length
2-V	tetrad of megaspores formed, integuments elongate towards apex of the nucellus	7	stamen primordia are stalked at base	7-rosette leaf	first 7 leaves have reached >1mm length
3-I	non-functional megaspores degenerate, outer integument envelops the nucellus and inner integument, funiculus and nucellus begin to curve	8	locules appear on long stamens	8-rosette leaf	first 8 leaves have reached >1mm length
3-II	two nucleate megagametophyte stage, outer integument surrounds nucellus, micropylar end points >90° away from funiculus, differential growth of integuments	9	petal primordia are stalked at base	9-rosette leaf	stage at which the first 9 leaves have reached >1mm length
3-III	megagametophyte develops large central vacuole, micropylar end of ovule at 90° angle from funiculus	10	petal height is level with stamens	10-rosette leaf	stage at which the first 10 leaves have reached >1mm length

(Continued)

TABLE 42 (CONTINUED)
Developmental Stages in Plants

3-IV	4-nucleate megagametophyte, inner integument surrounds nucellus, endothelium differentiates	11	stigmatic papillae appear	11-rosette leaf	stage at which the first 11 leaves have reached >1mm length
3-V	eight nucleate megagametophyte	12	petals are level with long stamens	12-rosette leaf	stage at which the first 12 leaves have reached >1mm length
3-VI	central cell nuclei fuse, antipodal cells degenerate, additional cell layer forms in inner integument	13	floral bud opens and petals are visible to naked eye	13-rosette leaf	stage at which the first 13 leaves have reached >1mm length
		14	anthers extend above stigma	14 rosette leaf	stage at which the first 14 leaves have reached >1mm length
		15	stigma extends above long anthers	15-rosette leaf	stage at which the plant has made 16 rosette leaves
		16	petals and sepals begin to wither	complete rosette	stage at which leaf growth has ceased and the rosette no longer increases in size
		17	all of floral organs have abscised and only green silique remains	late rosette growth	rosette is 70% of final size
		18	silique color changes from green to yellow		

TABLE 43
Types of Chromatin in Chromosomes and Their Properties

Property	Constitutive heterochromatin	Intercalary heterochromatin	Euchromatin
Presence	in C-bands	in C-bands	in R-bands
Condition in interphase	condensed	condensed	dispersed
Localization	centromeric	proximal	proximal
Genetic activity	inactive	inactive (?)	active
Time of replication	late S-phase	S-phase	early S-phase
A-T content	G-C rich	A-T rich	G-C rich
Repeated DNA	highly	moderate	moderate
DNA methylation	high	moderate	methylated

TABLE 44
Scheme of Soil Classification (According to the Guidelines for Soil Description, FAO 1990)

Character	Variants
Topography	Gently undulating – Undulating – Rolling – Hilly – Steeply dissected – Mountainous – Flat – Almost flat
Landform	Hill – Upland – Plain – Mountain – Plateau – Basin – Valley
Gradient	Nearly level – Very gently sloping – Gently sloping – Sloping – Flat – Strongly sloping – Moderately steep – Steep – Very steep – Level
Crop agriculture	Shifting cultivation – Fallow system cultivation – Ley system cultivation – Annual field cropping – Rainfed arable cultivation – Wet rice cultivation – Irrigated cultivation
Parent material	Lacustrine deposits – Fluvial deposits – Alluvial deposits – Unconsolidated – Volcanic ash – Loess – Aeolian deposits – Aeolian sand – Marine deposits – Pyroclastic deposits – Glacial deposits – Organic deposits – Colluvial deposits – *In situ* weathered – Saprolite – Littoral deposits – Lagoonal deposits
Rock type	Acid igneous rock – Granite – Gneiss – Granite/Gneiss – Quarzite – Schist – Andesite – Diorite – Basic igneous rock – Ultra basic rock – Gabbro – Basalt – Dolerite – Volcanic rock – Quarzite sandstone – Shale – Marl – Travertine – Conglomerate – Siltstone – Tuff – Pyroclastic rock – Evaporite – Gypsum rock – Sedimentary rock – Limestone – Dolomite – Sandstone
Depth	Shallow 30–50 cm – Moderately deep 50–100 cm – Deep 100–150 cm – Very deep >150 cm – Very shallow <30 cm
Size classes	Medium gravel 0.6–2.0 cm – Coarse gravel 2–6 cm – Stones 6–20 cm – Boulder 20–60 cm – Large boulder 60–200 cm – Fine gravel 0.2–0.6 cm
Flooding	Daily – Weekly – Monthly – Annually – Biennially – None
Groundwater depth	Very shallow – Shallow – Moderately deep – Deep – Very deep – Not observed
Groundwater quality	Brackish – Fresh – Polluted – Oxygenated – Stagnating – Saline
Color	White – Red – Reddish – Yellowish – Brown – Brownish – Reddish-brown – Yellowish-brown – Yellow – Reddish yellow – Greenish, green – Gray – Grayish – Blue – Bluish-black – Black
Texture	Clay < 2 mm – Fine silt 2–20 mm – Coarse silt 20–63 mm – Very fine sand 63–125 mm – Fine sand 125–200 mm – Medium sand 200–630 mm – Coarse sand 630–1250 mm – Very coarse sand 1250–2000 mm
Texture classes	Clay – Loam – Clay loam – Silt – Silty clay – Silty clay loam – Silt loam – Sandy clay – Sandy clay loam – Sandy loam – Fine sandy loam – Coarse sandy loam – Loamy sand – Loamy very fine sand – Loamy fine sand – Loamy coarse sand – Very fine sand – Fine sand – Medium sand – Coarse sand – Sand, unsorted
Weathering	Weathered – Strongly weathered – Fresh or slightly weathered

TABLE 45

Species of Rice (*Oryza* spp.), Their Karyological Composition and Distribution

Species	Genome	Chromosome number 2n=	Regional distribution
Sativa complex			
Oryza sativa	AA	24	Worldwide
Oryza glaberrima	A^gA^g	24	West Africa
Oryza nivara	AA	24	Tropical and subtropical Asia
Oryza rufipogon	AA	24	Tropical and subtropical Asia; tropical Australia
Oryza breviligulata	A^gA^g	24	Africa
Oryza longistaminata	A^lA^l	24	Africa
Oryza meridionalis	A^mA^m	24	Tropical Australia
Oryza glumaepatula	$A^{gp}A^{gp}$	24	South and Central America
Officinalis complex			
Oryza punctata	BB, BBCC	24, 48	Africa
Oryza minuta	BBCC	48	Philippines and Papua New Guinea
Oryza officinalis	CC	24	Tropical and subtropical Asia; tropical Australia
Oryza rhizomatis	CC	24	Sri Lanka
Oryza eichingeri	CC	24	South Asia and East Africa
Oryza latifolia	CCDD	48	South Asia and East Africa
Oryza alta	CCDD	48	South Asia and East Africa
Oryza grandiglumis	CCDD	48	South and Central America
Oryza australiensis	EE	24	Tropical Australia
Meyeriana complex			
Oryza granulata	GG	24	South and Southeast Asia
Oryza meyeriana	GG	24	Southeast Asia
Ridleyi complex			
Oryza longiglumis	HHJJ	48	Irian Jaya, Indonesia, Papua New Guinea
Oryza ridleyi	HHJJ	48	South Asia
Uncertain classification			
Oryza brachyantha	FF	24	Africa

TABLE 46
Toxic Plants and Their Toxins

Botanical family	Species	Toxic substance	Toxic organ	Origin
Amaryllidaceae	*Amaryllis bella-donna*	bellamarine	all organs	Southern Africa
Apiaceae	*Chaerophyllum temulum*	chaerophylline	shoot & fruits	Europe
	Cicuta virosa	cicutoxin	all organs	Europe-Northern Asia
	Conium maculatum	coniine	all organs	Europe, Asia, Northern America
	Nerium oleander	oleandrine	all organs	Minor Asia
Apocynaceae	*Apocynum androsaemifolium*	cymarine	sap	Northern America
	Vinca minor	vincamine	all organs	Southern Europe
	Arum maculatum	aronine, saponine	all organs	Europe, Minor Asia, Northern America
	Calla palustris		all organs	Northern Europe
Araceae	*Aristolochia clematitis*	aristolochic acids	shoot & seeds	Minor Asia, Southern Europe
	Asarum europaeum	asaron	all organs	Europe-Siberia
Aristolochiaceae	*Vincetoxicum hirundinaria*	vincetoxin	all organs	Europe, Asia
	Dryopteris filix-mas	filicine	rhizome	Northern hemisphere
Asclepiadaceae	*Lactuca virosa*	lactucine	sap	Europe, Afrika, Minor Asia
Aspidiaceae	*Senecio jacobaea*	jacobine	all organs	Europe, Asia
Asteraceae	*Podophyllum letatum*	peltatine	rhizome	Northern America
	Lophophora williamsii	mescaline	all organs	Central America
Berberidaceae	*Lobelia inflata*	lobeline	all organs	Northern America
Cactaceae	*Lonicera nigra*	xylosteine	berries	Europe, Siberia
Campanulaceae	*Agrostemma githago*	githagine	seeds	Minor Asia
Caprifoliaceae	*Euonymus europaeus*	glycosides	all organs	Europe, Siberia
Caryophyllaceae	*Sedum acre*	sedamine	all organs	Europe, Asia
Celastraceae	*Bryonia dioica*	ryresine, bryonine	all organs	Southern Europe
Crassulaceae	*Citrullus colocynthis*	cucurbitacine, saponine	fruits	Tropical Africa
Cucurbitaceae	*Juniperus sabina*	sabine	all organs	Europe, Asia
	Equisetum sylvaticum	palustrine	all organs	Northern hemisphere
Cypressaceae	*Ledum palustre*	ledol	all organs	Northern hemisphere
Equisetaceae	*Rhododendron ferrugineum*	andromedotoxin	flowers, leaves	Alp Mountains
Ericaceae	*Erythroxylum coca*	cocaine	leaves	Peru, Bolivia
	Croton tiglium	crotine	seeds	Tropical Asia
Erythroxylaceae	*Euphorbia cyparissias*	phytotoxin	all organs	Europe
Euphorbiaceae	*Ricinus communis*	ricine	seeds	Northern Africa
	Abrus precatorius	abrine	seeds	Tropical regions
	Coronilla varia	coronilline	leaves & flowers	Europe

(Continued)

TABLE 46 (CONTINUED)
Toxic Plants and Their Toxins

Fabaceae	*Cytisus scoparius*	sparteine	seeds, leaves	Western Europe
	Laburnum anagyroides	cytisine	all organs	Southern Europe
	Lupinus polyphyllus	d-lupanine	seeds	Northern America
	Physostigma venenosum	physostigmine	seeds	Western Africa
	Illicium anisatum	anisatine	fruits	China
	Amianthium muscaetoxicum	jervine, amianthine	all organs	Northern America
Illiciaceae	*Colchicum autumnale*	colchicine	flower, seeds, bulb	Minor Asia
Liliaceae	*Convallaria majalis*	convallarine	all organs	Northern hemisphere
	Paris quadrifolia	paridine	all organs	Europe, Siberia
	Polygonatum odoratum	saponine	rhizome, berries	Europe, Asia
	Urginea maritima	scilliroside	all organs	Southern Europe
	Veratrum album	protoveratrine	all organs	Europe, Siberia
	Gelsemium sempervirens	gelsemine	all organs	America
	Strychnos nux-vomica	strychnine	all organs	India
Loganiaceae	*Viscum album*	vincetoxin	leaves, sap	Europe, Asia
	Antiaris toxicaria	antiarine, antiosidine	sap	South-east Asia
Loranthaceae	*Paeonia officinalis*	paeonine	rhizome, seeds, flowers	Southern Europe
Moraceae	*Chelidonium majus*	chelidonine	sap	Europe-Northern Asia
Paeoniaceae	*Corydalis cava*	corydaline, corycavin	all organs	Europe
Papaveraceae	*Dicentra cucullaria*	cucullarine	roots	Northern America, China
	Papaver somniferum	morphine	sap	Minor Asia
	Piper betle	piperine	leaves	Asia
	Lolium temulentum	temuline	seeds	Europe, Asia
Piperaceae	*Primula obconica*	primine	hairs	Central Asia
Poaceae	*Aconitum nabellus*	aconitine	all organs	Europe
Primulaceae	*Actaea spicata*	protoanemonine	?	Europe-Siberia

(*Continued*)

TABLE 46 (CONTINUED)
Toxic Plants and Their Toxins

Ranunculaceae	*Adonis annua, A. vernalis*	adonitoxin	shoot	Southern Europe-Near East; South-eastern Europe-Western Siberia
	Anemone nemorosa	anemonine	all organs	Europe, Northern America, Northern Asia
	Aquilegia vulgaris	cyano-glycoside	shoot	Europe-Asia
	Caltha palustris	saponine	all organs	Northern Europe
	Consolida regalis	delsoline, delcosine	all organs	Souterhn Europe-Minor Asia
	Helleborus niger	hellebrine	all organs	Europe
	Hydrastis canadensis	berberine	all organs	Northern America
	Ranunculus acris; R. ficaria; R. repens	protoanemonine	all organs	Europe, Asia; Europe, Northern Africa
	Trollius europaeus	magnoflorine	all organs	Europe
	Rhamnus catharticus	glycosides	fruits, seeds	Northern hemisphere
	Prunus laurocerasus; P. padus	prunasine; amygdaline	all organs; seeds, leaves	Southern Europe; Europe, Siberia
Rhamnaceae	*Dictamnus albus*	dictamnine	leaves, fruits, flowers	Europe
Rutaceae	*Ruta graveolens*	furokumarine	all organs	Europe
	Digitalis purpurea	digitoxin	all organs	Western Europe
	Gratiola officinalis	gratioline	all organs	Europe, Siberia
Scrophulariaceae	*Atropa bella-donna*	hyoscyamine, atropine	all organs	Europe
	Datura arborea, D. stramonium	Scopolamine; L-hyoscyamine	all organs	Central America; Northern America
Solanaceae	*Duboisia myoporoides*	scopolamine	leaves	Australia
	Hyoscyamus niger	l-hyoscyamine	all organs	Europe, Asia, Africa
	Nicotiana tabacum	nicotine	all organs	America
	Physalis alkekengi	hygrine	leaves, rhizome	Europe
	Scopolia carniolica	L-hyoscyamine	all organs	Central Europe
	Solanum dulcamara	saponine	all organs	Europe, Asia
	Taxus baccata	taxine	all organs	Europe, Asia
	Daphne mezereum	mezereine	fruits, stem	Europe, Asia, Northern America

TABLE 47

Different Techniques of Plant Tissue Culture and Their Applications in Plant Improvement

Type of technique	Application
Seed culture	• Increasing efficiency of germination and germ production • Precocious germination by application of plant growth regulators • Induction of multiple shoot formation and organogenesis • Elimination of viruses as seeds do not carry viruses
Embryo culture	• Overcoming embryo abortion due to incompatibility barriers • Overcoming seed dormancy and self-sterility of seeds • Embryo rescue in interspecific or intergeneric hybridization • Production of haploids • Shortening of breeding cycle • Development of callus cultures
Ovary or ovule culture	• Production of haploids • Recovery of hybrid embryos overcoming embryo abortion • Achievement of *in vitro* fertilization
Anther and microspore culture	• Production of haploid plants • Production of homozygous diploid • Genetic transformation of microspores • Production of useful gametoclonal variations • Mutation studies with single set of chromosomes • Fixation genetic characters from heterozygous source materials
In vitro pollination	• Production of hybrids difficult to produce by embryo rescue
In vitro fertilization	• Production of distant hybrids incompatibility • Production of transgenic plants by injecting exogenous DNA
Organ culture	• Mass production of plants of elite and rare germplasm • Production of calli, shoots and roots for secondary metabolites • Development of germplasm banks
Shoot apical meristem culture	• Production of virus free germplasm • Mass production of desirable genotypes • Cryopreservation or *in vitro* conservation of germplasm • Phytosanitary transport
Somatic embryogenesis	• Mass multiplication of elite germplasm • Production of artificial seeds • As source material for embryogenic protoplasts • Genetic transformation • Production of primary metabolites • Amenable to mechanization and for bioreactors
Organogenesis and enhanced axillary budding	• Mass multiplication of elite germplasm • Material for protoplast work, genetic transformation and micrografting • Conservation of endangered genotypes
Callus cultures	• Plantlets through somatic embryogenesis or organogenesis • Obtaining virus-free plants • Generation of useful somaclonal and gametoclonal variants • Source of protoplasts and suspension cultures • Production of useful secondary metabolites • Biotransformation studies • Selection of cell lines with valuable properties
In vitro production of secondary metabolites	• Production of drugs, aromatic substances, pigments, or flavors • Production of novel metabolites • Biotransformation and elicitor studies

(*Continued*)

TABLE 47 (CONTINUED)
Different Techniques of Plant Tissue Culture and Their Applications in Plant Improvement

Cell culture and _in vitro_ selection at cellular level	• Production of somatic embryos, nodules and entire plantlets • Over-production of primary and secondary metabolites • Induction and selection of useful mutants or somaclones
Genetic or epigenetic somaclonal variations	• Isolation of useful variants • Creation of additional genetic variation without hybridization
In vitro mutagenesis	• Induction of polyploidy • Introduction of genetic variability • Elucidation of biochemical processes
Protoplast isolation, culture and fusion	• Combining distant genomes to produce somatic or asymmetric hybrids • Production of organelle recombinants • Transfer of cytoplasmic male sterility into elite lines • Source material for genetic transformation • Creation of genetic variants
Genetic transformation	• Introduction of foreign DNA • Study of structure and function of genes. • Induction of hairy roots or shooty terratomas
In vitro flowering	• Reduction in long life cycle in perennials (bamboo or conifers) • Continuous supply of flowers, fruits and seeds irrespective of season
Micrografting	• Overcoming graft incompatibility • Rapid mass propagation of elite scions • Multiplication and survival of difficult to root species or transformants • Development of virus-free plants
Cryopreservation or storage at low temperature	• Long term preservation
Culture of hairy roots	• Studying and manipulating root-specific metabolism • Co-culture with arbuscular mycorrhizal fungi to increase metabolite production • Co-culture with insects to study pathogenesis

(Modified after A. KULKARNI, 2005).

TABLE 48

Total World Area of Major Cereal Crops and World Production in 2005, As Well As Comparative Values of Energy Provided, Total Content Of Protein, and Lipid in 100 g of Cereal Grains

Cereal crop	Area (million hectares)	Production (million tons)	Energy (kJ)	Protein (g)	Lipid (g)
Wheat	213	561	1,420	12.0	2.0
Rice	158	596	1,296	8.0	2.0
Maize	147	642	1,471	10.0	4.0
Barley	61	147	-	-	-
Sorghum	48	63	1,455	10.0	5.0
Millet	37	32	-	-	-
Oats	14	28	-	-	-
Rye	8	17	-	-	-
Triticale	7	11	-	-	-

Source: http://www.faostat.fao.org; http://www.cix.co.uk

TABLE 49

General Issues of Variety Development and Maintenance

Inducing variability
by using other varieties, species, genera, mutagenesis, or molecular means

Selection of parental genotypes
with desired properties

Crossing of parental genotype
to combine suitable properties

Selection of genotypes
with suitable traits

Testing (official) of new genotypes, lines, clones, or varieties
in field and greenhouse tests

Maintenance of new varieties and breeding material
for purity

Multiplication
for marketable, seed, clones, or transplants

TABLE 50
Examples of Monosomic Series Developed in Europe

Country	Wheat variety
Belarus	Opal
Bulgaria	Bezostaya 1, Roussalka, Sadovo 1
Czech Republic	Zlatka
France	Courtot
Germany	Caribo, Carola, Kranich, Poros, *T. spelta saharense*
Great Britain	Bersee, Cappelle Desprez, Glennson 81, Hobbit sib, Holdfast, Koga II, Mercia
Hungary	Mironovskaya 808, Rannyaya 12
Italy	Mara
Kazakhstan	Kazakstanskaya 126
Netherlands	Starke
Poland	Grana, Luna
Romania	Bezostaya 1, Favorites
Russia/USSR	Aurora, Bezostaya 2, Diamant II, Kavkaz, Milturum 533, Saratovskaya 29, Skorospelka 35
Spain	Aragon 3, Ariana 8, Pawe 247
Switzerland	Probus
Ukraine	Novostepniachka, Priboy
Yugoslavia	Novosadska Rana, Sava

Source: modified after http://www.ewac.eu/tables.htm

TABLE 51

Genetically Localized Genes of Mildew Resistance (*Blumeria graminis* f. sp. *tritici*) in Wheat

Gene	Chromosomal location	Variety/strain	Source
Pm1a	7AL	Axminster	*Triticum aestivum*
Pm1b	7AL	MocZlatka	*Triticum monococcum*
Pm1c	7AL	Weihenstephan M1N	*Triticum aestivum*
Pm1d	7AL	TRI2258	*Triticum spelta*
Pm1e	7AL	Virest	*Triticum aestivum*
Pm2	5DS	Ulka	*Triticum aestivum*
Pm3a	1AS	Asosan	*Triticum aestivum*
Pm3b	1AS	Chul	*Triticum aestivum*
Pm3c	1AS	Indian	*Triticum aestivum*
Pm3d	1AS	Kolibri	*Triticum aestivum*
Pm3e	1AS	W150	*Triticum aestivum*
Pm3f	1AS	Michigan Amber	*Triticum aestivum*
Pm3g	1AS	Aristide	*Triticum aestivum*
Pm4a	2AL	Khapli	*Triticum durum*
Pm4b	2AL	Solo	*Triticum carthlicum*
Pm5a	7BL	Hope, Selpek	*Triticum aestivum*
Pm5b	7BL	Ibis, Kormoran	*Triticum aestivum*
Pm5c	7BL	Kolandi	*Triticum sphaerococcum*
Pm5d	7BL	IGV 1-455	*Triticum aestivum*
Pm6	2BL	TP144	*Triticum timopheevii*
Pm7	T4BS.4BL-2RL	Transec	*Secale cereale* cv. Rosen
Pm8	T1RS.1BL	Neuzucht	*Secale cereale* cv. Petkus
Pm9	7AL	Normandie	*Triticum aestivum*
*Pm10**	1D	Norin 26	*Triticum aestivum*
*Pm11**	6BS	Chinese Spring	*Triticum aestivum*
Pm12	6BS-6SS.6SL	Wembley	*Aegilops speltoides*
Pm13	3SS-3BL+3DL	Chinese Spring	*Aegilops longissimum*
*Pm14**	6BS	Norin 10	*Triticum aestivum*
*Pm15**	7DS	Norin 26	*Triticum aestivum*
Pm16	4A	Maris Nimrod	*Triticum dicoccoides*
Pm17	T1RS.1AL	Amigo	*Secale cereale* cv. Insave
*Pm18***			
Pm19	7D	XX186	*Aegilops tauschii*
Pm20	T6BS.6RL	KS93WGRC28	*Secale cereale*
Pm21	T6VS.6AL	Yangmai 5	*Desypyrum villosa*
*Pm22****			
Pm23	5A	82-7241	*Triticum aestivum*
Pm24	1DS	Chiyacao	*Triticum aestivum*
Pm25	1A	NC96BGTA5	*Triticum monococcum*
Pm26	2BS	TTD140	*Triticum dicoccoides*
Pm27	6B	K-38555	*Triticum timopheevii*
Pm28	1B	Meri	*Triticum aestivum*
Pm29	7DL	Pova	*Aegilops ovata*
Pm30	5BS	C20	*Triticum dicoccoides*

* resistant against *Blumeria graminis* f.sp. *agropyri*

** renamed as *Pm1c*

*** renamed as *Pm1e*

(Modified after SINGRÜN 2002)

TABLE 52
Genetically Localized Genes of Brown Rust (*Puccinia triticina*) in Wheat

Gene	Chromosomal location	Variety/strain	Source
Lr1	5DL	Malakof	*Triticum aestivum*
Lr2a	2DS	Webster	*Triticum aestivum*
Lr2b	2DS	Carina	*Triticum aestivum*
Lr2c	2DS	Brevit	*Triticum aestivum*
Lr3a	6BL	Democrat	*Triticum aestivum*
Lr3bg	6BL	Bage	*Triticum aestivum*
Lr3ka	6BL	Klein Aniversario	*Triticum aestivum*
Lr9	6BL	-	*Triticum umbellulatum*
Lr10	1AS	Lee	*Triticum aestivum*
Lr11	2A	Hussar	*Triticum aestivum*
Lr12	4BS	Exchange	*Triticum aestivum*
Lr13	2BS	Frontana	*Triticum aestivum*
Lr14a	7BL	Hope	*Triticum aestivum*
Lr14b	7BL	Bowie	*Triticum aestivum*
Lr15	2DS	Kenya 1-12 E-19-J	*Triticum aestivum*
Lr16	2BS	Exchange	*Triticum aestivum*
Lr17a	2AS	Klein Lucero	*Triticum aestivum*
Lr17b	2AS	Harrier	*Triticum aestivum*
Lr18	5BL	-	*Triticum timopheevii*
Lr19	7DL	-	*Agropyron elongatum*
Lr20	7AL	Thew	*Triticum aestivum*
Lr21	1DL	-	*Aegilops tauschii*
Lr22a	2DS	Thatcher	*Triticum aestivum*
Lr22b	2DS	Norin 26	*Aegilops tauschii*
Lr23	2BS	Gabo	*Triticum aestivum*
Lr24	3DL	-	*Aegilops elongatum*
Lr25	4A	-	*Secale cereale* cv. Rosen
Lr26	1BL	-	*Secale cereale* cv. Imperial
Lr27	3BS	Gatcher	*Triticum aestivum*
Lr28	4AL	-	*Triticum speltoides*
Lr29	7DS	-	*Aegilops elongatum*
Lr30	4BL	Terenzio	*Triticum aestivum*
Lr31	4BS	Gatcher	*Triticum aestivum*
Lr32	3D	-	*Aegilops tauschii*
Lr33	1BL	PI58458	*Triticum aestivum*
Lr34	7D	Terenzio	*Triticum aestivum*
Lr35	2B	-	*Triticum speltoides*
Lr36	6BS	-	*Triticum speltoides*
Lr37	2AS	-	*Triticum ventricosa*
Lr38	2AL	-	*Agropyron intermedium*
Lr39*	2DS	-	*Aegilops tauschii*
Lr40*	1D	-	*Aegilops tauschii*
Lr41	1D	-	*Aegilops tauschii*
Lr42	1D	-	*Aegilops tauschii*
Lr43	7D	-	*Aegilops tauschii*
Lr44	1BL	-	*Triticum spelta*
Lr45	2AS	-	*Secale cereale*
Lr46	1BL	Pavon 76	*Triticum aestivum*
Lr47	7AS	-	*Triticum speltoides*
Lr48	-	CSP44	*Triticum aestivum*
Lr49	-	VL404	*Triticum aestivum*
Lr50	-	-	*Triticum armeniacum*

* possibly *Lr39* and *Lr40* are allelic or identical with *Lr21*

(Modified after SINGRÜN 2002)

TABLE 53

Phylogenetic Relationships in Wheat Considering Genome-Plasmon Constitutions

Species diploid (genome-plasmon)	Species tetraploid (genome-plasmon)	Species heaxaploid (genome-plasmon)
T. monococcum $A - A, A^2$		
Ae. speltoides $S - S$	*T. dicoccum, T. dicoccoides* $AB - B$	*T. aestivum* $ABD - B$
$G - S$	*T. timopheevi* $AG - G$	*T. zhukovskyi* $AAG - G$
Ae. searsii $S^v - S^s$	*Ae. kotchyi, Ae. variabilis* $US - S^v$	
Ae. bicornis $S^b - S^b$	*Ae. columnaris* $UM - U^*$	
Ae. longissima, Ae. sharonensis $S^l, S^{l*} - S^l$	*Ae. biuncalis, Ae. triaristata* $UM - U$	*Ae. triaristata* $UMN - U$
Ae. umbellulata $U - U$	*Ae. triuncialis* $UC - U$	
Ae. caudata $C - C$	*Ae. triuncialis* $UC - C^*$	
Ae. squarrosa $D - D$	*Ae. cylindrica* $CD - C^*$	
Ae. comosa $M - M$	*Ae. ventricosa* $DN - D$	
Ae. heldreichii $M^h - M$	*Ae. crassa* $DM - D^2$	*Ae. juvenalis* $DMU - D^2$
Ae. uniaristata $N - N$	*Ae. ovata* $UM - M^0$	*Ae. crassa* $DMD - D^2$
Ae. mutica $T^u - T$		*Ae. vavlovii* $DMS - D^2$
$T - T$		

(Modified after TSUNEWAKI 2009)

TABLE 54

Phylogenetic Relationships in Barley Considering Genome-Plasmon Constitutions

Species diploid	Species tetraploid	Species heaxaploid
H. intercedens	H. depressum (intercedens x californicum)	
H. euclaston		
H. pusillum		H. arizonicum (pusillum x jubatum)
H. erectifolium		
H. stenostachis		
H. cordobense		H. procerum (cordobense x fuegianum)
H. muticum		
H. pubiflorum		
H. patagonicum	H. fuegianum, H. tetraploidum (patagonicum x roshevitzii)	H. parodii (fuegianum x comosum)
H. comosum		H. lecheri (pubiflorum x jubatum)
H. flexuosum	H. guatemalense (intercedens x flexuosum)	
H. chilense		
??	H. jubatum, H. brachyantherum (?? x roshevitzii)	H. brachyantherum (brachyantherum x gussoneanum)
H. californicum		
H. roshevitzii		
H. bogdanii		
H. brevisubulatum	H. brevisubulatum (brevisubulatum x gussoneanum)	H. brevisubulatum
H. marinum	H. secalinum (pubiflorum x marinum) >>> H. capense	
H. gussoneanum	H. gussoneanum	
??		H. murinum leporinum (?? x H. murinum leporinum)
??	H. murinum leporinum (?? x marinum glaucum), H. murinum murinum (?? x marinum glaucum)	
H. murinum glaucum		
H. bulbosum	H. bulbosum	
H. vulgare		

(Modified after BLATTNER 2009)

TABLE 55
Major Food Crops of the World, Ranked According to Total Tonnage Produced Annually

Number	Crop
1	Wheat
2	Rice
3	Maize
4	Potato
5	Barley
6	Sweet potato
7	Cassava
8	Grape
9	Soybean

Number	Crop
10	Oat
11	Sorghum
12	Sugarcane
13	Millet
14	Banana
15	Tomato
16	Sugarbeet
17	Rye

Number	Crop
18	Orange
19	Coconut
20	Cotttonseed
21	Apple
22	Yam
23	Peanut
24	Watermelon
25	Cabbage

TABLE 56

Estimated Domestication of Main Crop Plants and Their Origin

Domestication (~ years)	Crop	Origin
9,000	Barley	Middle East
	Oil plam	Central Africa
	Pea	Middle East
	Orange	East Asia
8,500	Wheat	Middle East
8,000	Sorghum	Central Africa
7,000	Maize	Central America
	Bean	Central America
	Lentil	Middle East
	Potato	Peru
	Grape	Eastern Asia
6,000	Sweet potato	Central America
5,000	Cassava	Brazil, Mexico
4,500	Rice	Southern China
	Onion	Hindu Kush
	Banana	South-east Asia
	Cotton	Central America
4,000	Lucerne	Hindu Kush
3,000	Sunflower	Western-north America
	Tomato	Western-south America
	Cabbage	Mediterranean Europe
	Apple	Asia Minor
2,000	Soybean	Northern China
500	Rapeseed	Mediterranean Europe
	Coffee	West Ethiopia
300	Sugarbeet	Mediterranean Europe, France, Germany
200	Rubber	Amazonas Region
100	Coconut palm	Southern Asia
50	Triticale	Europe

TABLE 57

Scheme of a Conventional Breeding Program in Cereals, Including Subsequent Performance Testing

Plant generation	Parent selection	Stage of trial	Selection criteria		Seed stock	Marketed seed
			Crop performance	End-use quality		
F1	~500 crosses					
F2	~2 M. plants		Disease			
F3	~350.000 plants	Ear row	Disease, maturity, field			
F4	~10.000 lines	Progeny plots	traits, lodging, uniformity	Greenhouse testing		
F5	~1.000 lines	Greenhouse trials, one location				
F6	~200 lines	Greenhouse trials, three locations		End-use testing by processors		
F7	~50 lines	Greenhouse trials, sixlocations				
F8	~5 lines	National trials		Quality testing	0.5 ha breeders' seed	
F9	~3 lines				1.5 ha breeders' seed	1 ha prebasic seed
F10	2 lines				8 ha breeders' seed	40 ha prebasic seed
F11	1 line	Commercial evaluation			Stock maintenance	First multiplication and/or commercial growing

TABLE 58
Comparision of Association and QTL Mapping

Activity & requirements	QTL mapping	Association mapping
Goal of detection	Quantitative trait loci, i.e., wide region within specific pedigrees within a QTL is located	Quantitative trait nucleotides, i.e., physically as close as to causative DNA sequence
Resolution	Low; only moderate density linkage maps are required	High; disequilibrium within small physical regions require many markers
Experimental population	Defined pedigrees (backcrosses, F_2s, RILs, two and three generation pedigrees and/or families, half sibs)	Unrelated individuals, large number of small unrelated families
Cost	Moderate	Moderate for few traits moderate; expensive, however, for many traits
Interference	Pedigree-specific, except where species has extent LD	Species- and subspecies-wide
Number of markers required	$10^2 \ldots 10^3$ per genome	10^5 for small, 10^9 for large genomes

TABLE 59
Types of Repetitive DNA

Nuclear genome	Genes, regulatory and non-coding single copy sequences	Whole genome duplication
		Chromosomal duplications
		45S and 5S rRNA genes
		Other repeated genes
Repetitive DNA sequences	Dispersed repeats: Transposable elements	Class I Retrotransposons amplifying via an RNA intermediate
		Class II
		DNA transposons copied and moved via DNA
	Structural components of chromosomes	Centromeric repeats
		Telomeric repeats
	Tandem repeats	Blocks of tandem repeats at discrete chromosomal loci
		Subtelomeric repeats
		Simple sequence repeats or microsatellites

(Modified after BISCOTTI et al. 2015)

TABLE 60
Aegilops Species Used in Introgression Experiments of Wheat and Their Genomic Designations

Ploidy level (x)	Species	Genome symbols
Diploids (2x)	Ae. caudata	CC
	Ae. tauschii	DD
	Ae. comosa	MM
	Ae. uniaristata	NN
	Ae. speltoides	SS
	Ae. bicornis	$S^b S^b$
	Ae. longissimi	$S^l S^l$
	Ae. searsii	$S^s S^s$
	Ae. sharonensis	$S^{sh} S^{sh}$
	Ae. mutica	TT
	Ae. umbellulata	UU
Tetraploids (4x)	Ae. geniculata	MMUU
	Ae. peregrina	SSUU
	Ae kotschyi	SSUU
	Ae. triuncialis	UUCC
	Ae. biuncialis	UUMM
	Ae. columnaris	UUMM
	Ae. neglecta	UUMM
	Ae. cylindrical	CCDD
	Ae. ventricosa	DDNN
	Ae. crassa	DDMM
Hexaploids (6x)	Ae. neglecta	UUMMNN
	Ae. crassa	DDDDMM
	Ae. vavilovii	DDMMSS
	Ae. juvenalis	DDMMUU

TABLE 61
Genotypes Derived from a Single-Locus Heterozygote and Their Frequencies in Subsequent Selfing Generations

Generation	Genotypes			Frequency of heterozygotes	Frequency of homozygotes (%)
	AA	Aa	aa		
0	0	1	0	1	0
1	1/4	2/4	1/4	1/2	50.0
2	3/8	2/8	3/8	3/8	75.0
3	7/16	2/16	7/16	1/8	87.5
4	15/32	2/32	15/32	1/16	93.8
5	31/64	2/64	31/64	1/32	96.9
...					
10	1023/2048	2/2048	1023/2048	1/2048	99.9
n	$(2^n-1)/2^{n+1}$	$2/2^{n+1}=1/2^n$	$(2^n-1)/2^{n+1}$	$1/2^n$	$1-1/2^n$

TABLE 62

Applied Methods of Micropropagation, from a Stock Plant to Final Plantlet Regeneration

Stock plants							
Explants from leaves, roots, anthers etc.					Nodal sections	Shoot apex	Meristem
Indirect morphogenesis			Direct morphogenesis				
Callus growth on explant			Somatic embryos	Adventitious shoots	Nodal growth	Meristem or shoot apex growth	
Suspension culture	Callus					Shoot tip culture/ auxiliary branching	Meristem culture
Synth. Embryos from single cells	Synth. Embryos from callus	Adventitious shoots on callus					
Indirect somatic embryogenesis		Indirect shoot formation	Direct embryogenesis	Direct shoot formation	Nodal culture	Shoot tip culture	
Regeneration of plantlets							

Bibliography

Agrios, G. N. 1997. *Plant pathology.* San Diego, CA: Academic Press.

Alix, K., Gérard, P. R., Schwarzacher, T., and Heslop-Harrison, J. S. 2017. Polyploidy and interspecific hybridisation: Partners for adaptation, speciation and evolution in plants. *Annals of Botany* 120(2): 183–194.

Allard, R. W. 1999. *Principles of plant breeding.* 2nd ed. New York: John Wiley & Sons, Inc., 254, ISBN 0471023094.

Anonymous. 1952. *Manual for testing agricultural and vegetable seeds.* Agricultural Handbook 30. Washington, DC: USDA.

Anonymous. 1989. *Intellectual property rights associated with plants.* Madison, WI: Crop Science Society.

Anonymous. 1990a. *Detailed description of varieties of wheat, barley, oats, rye and triticale.* Cambridge, United Kingdom: National Institute of Botany.

Anonymous. 1990b. *Guidelines for soil description.* Rome: Food and Agriculture Organization of United Nations.

Barigozzi, C. 1981. *The origin and domestication of cultivated plants.* Amsterdam: Elsevier Science Publisher.

Becker, H. 1993. *Pflanzenzüchtung* (Plant breeding). Stuttgart: Verlag Eugen Ulmer.

Bennett, W. F. 1994. *Nutrient deficiencies and toxicities in crop plants.* St. Paul, MN: APS Press.

Berg, S. O. 1946. Är vetemjölets örighetsgrad huvudsakligen en sortegenskap? *Agri Hortique Genetica* 4: 1–14.

Bewley, J. D. and Black, M. 1995. *Seeds–physiology of development and germination.* New York: Plenum Press.

Biscotti, M. A., Olmo, E., and Heslop-Harrison, J. S. 2015. Repetitive DNA in eukaryotic genomes. *Chromosome Research.* doi:10.1007/s10577-015-9499-z.

Blattner, F. R. 2009. Progress in phylogenetic analysis and a new infrageneric classification of the barley genus *Hordeum* (Poaceae: Triticeae). *Breeding Science* 59(5): 471–480.

Borojevic, S. 1990. *Principles and methods of plant breeding. Developments in crop science 17.* New York: Elsevier Science Publishing Company, Inc.

Bos, I. and Caligari, P. 1995. *Selection methods in plant breeding.* London: Chapman and Hall.

Bos, I. and Caligari, P. 2007. *Selection methods in plant breeding.* Berlin, Heidelberg, Germany: Springer Science & Business Media B.V.

Briggs, F. N. and Knowles, P. F. 1967. *Introduction to plant breeding.* New York: Reinhold Publishing Corporation.

Brown, J. and Caligari, P. 2008. *An introduction to plant breeding.* Oxford, UK: Blackwell Publ., 209.

Carmel-Goren, L., Liu, Y. S., Lifschitz, E., and Zamir, D. 2003. The self-pruning gene family in tomato. *Plant Molecular Biology* 52(6): 1215–1222.

Chapman, G. P., Mantell, S. H., and Daniels, R. W. 1985. *Experimental manipulation of ovule tissues.* New York: Longman.

Clement, S. L. and Quisenberry, S. S. 1999. *Global plant genetic resources for insect-resistant crops.* Boca Raton, FL: CRC Press LLC.

Cobley, L. S. 1976. *An introduction of the botany of tropical plants.* London, England: Longman.

Cooke, D. A. and Scott, R. K. 1993. *The sugarbeet crop.* London: Chapman and Hall Ltd.

Copeland, L. O. and McDonald, M. B. 1985. *Principles of seed science and technology.* Minneapolis, MN: Burgess Publishing Company.

Dahlgreen, R. M. T., Clifford, H. T., and Yeo, P. F. 1985. *The families of the monocotyledons: Structure, evolution and taxonomy.* Stuttgart: Springer Verlag.

Enríquez, J. 1998. Genomics and the world's economy. *Science* 281(5379): 925–926.

Evers, K. H. 1924. *Die Organisation der Pflanzenzucht und des Saatbaus in der Deutschen Landwirtschaft.* Berlin: W. de Gruyter & co.

Exbrayat, J.-M. 2013. *Histochemical and cytochemical methods of visualization.* Boca Raton, FL: CRC Press/ Taylor & Francis Group.

Falconer, D. S. 1989. *Introduction to quantitative genetics.* New York: Longman Scientific Technology.

Fehr, W. R. and Hadley, H. H. 1980. *Hybridization of crop plants.* Madison, WI: American Society of Agronomy, Inc.

Fehr, W. R. 1987. *Principles of cultivar development. Volume 1: Theory and technique.* New York: Macmillan Publishing Company.

Frey, K. J. 1981. *Plant breeding 11.* Ames, IA: Iowa State University Press.

Fu, Y.-B. and Somers, D. J. 2009. Genome-wide reduction of genetic diversity in wheat breeding. *Crop Science* 49(1): 161–168.

Gabriel, K. R. 1971. The biplot graphic display of matrices with application to principal component analysis. *Biometrika* 58(3): 453–467.

Gassen, H., Martin, G., and Bertram, S. 1985. *Gentechnik.* Jena, Germany: G. Fischer verlag.

Gauch, H. G. 1992. *Statistical analysis of regional yield trials: AMMI analysis of factorial designs.* Amsterdam, The Netherlands: Elsevier Publ.

Gilad, Y., Rifkin, S. A., and Pritchard, J. K. 2008. Revealing the architecture of gene regulation: The promise of eQTL studies. *Trends in Genetics* 24(8): 408–415.

Graham, R. D. 1984. *Breeding for nutritional characteristics in cereals.* Advances in plant research. New York: Praeger.

Gressel, J. 2007. *Crop ferality and volunteerism.* Boca Raton, FL: CRC Press.

Gu, W. K., Weeden, N. F., Yu, J., and Wallace, D. H. 1995. Large-scale, cost-effective screening of PCR products in marker-assisted selection applications. *Theoretical and Applied Genetics* 91(3): 465–470.

Hansen, W. 2008. Seed from Masada is the oldest to germinate. *Los Angeles Times*, June 13, 2008.

Heffner, E. L., Sorrells, M. E., and Jannink, J.-L. 2009. Genomic selection for crop improvement. *Crop Science* 49(1): 1–12.

Heinz, D. J. 1987. *Sugarcane improvement through breeding.* Amsterdam: Elsevier.

Heyne, E. G. 1987. *Wheat and wheat improvement.* Madison, WI: American Society of Agronomy Publisher.

Hillmann, P. 1910. *Die deutsche landwirtschaftliche Pflanzenzucht.* Berlin: Deutsche Landwirtschafts-Gesellschaft.

Hillmann, P. 1911. *Die Bestimmung der Sortenreinheit und Sortenechtheit bei Beurteilung von Saatgutfeldern unter Zuhilfenahme variationsstatistischer Untersuchungen.* Berlin, Germany: Publ. Genr. Unger.

Hu, J. and Vick, B. A. 2003. Target region amplification polymorphism: A novel marker technique for plant genotyping. Plant Molecular Biology Reporter 21: 289–294.

Jensen, N. F. 1988. *Plant breeding methodology.* New York: Wiley & Sons Publ., 676.

John, J. A. and Williams, E. R. 1995. *Cyclic and computer generated designs.* Monographs on Statistics and Applied Probability 38. London: Chapman and Hall.

Kamenetsky, R. and Okubo, H. 2012. *Ornamental geophytes: From basic science to sustainable production.* Boca Raton, FL: CRC Press/Taylor & Francis Group.

Karp, A., Isaac, P. G., and Ingram, D. S. 1998. *Molecular tools for screening biodiversity.* London: Chapman and Hall.

Kosambi, D. D. 1944. The estimation of map distance from recombination values. *Annals of Eugenics* 12(1): 172–175.

Kromdijk, J., Głowacka, K., Leonelli, L., Gabilly, S., Iwai, M., Niyogi, K. K., and Long, S. P. 2016. Improving photosynthesis and crop productivity by accelerating recovery from photoprotection. *Science* 354(6314): 857–861.

Kuckuck, H., Kobabe, G., and Wenzel, G. 1991. *Fundamentals of plant breeding.* Berlin: Springer Verlag.

Lampeter, W. 1982. *Saat- und Pflanzgutproduktion. Dt.* (Seed and plant production). Berlin, Germany: Deutscher Landwirt Schaftsverlag.

Leitch, A. R., Schwarzacher, T., Jackson, D., and Leitch, I. J. 1994. *In situ hybridization.* Oxford, United Kingdom: Bios Scientific Publishers.

Love, A. 1984. Conspectus of the Triticeae. *Feddes Repetorium* 95(7–8): 425–521.

Lupton, F. G. H. 1987. *Wheat breeding.* London: Chapman and Hall Ltd.

Martin, J. M., Leonard, W. H., and Stamp, D. L. 1976. *Principles of field crop production.* New York: Macmillan.

Mayo, O. 1987. *The theory of plant breeding.* Oxford: Clarendon Press.

McDonald, M. B. and Copeland, L. O. 1989. *Seed science and technology laboratory manual.* Ames, IA: Iowa State University Press.

McNerney, R. 1999. *Glossary of terms commonly used in molecular diagnostics.* UK: Department of Infectious and Tropical Diseases, London School of Hygiene & Tropical Diseases.

Mendel, G. 1901. *Versuche über Pflanzenhybriden. Zwei Abhandlungen (1865 und 1869).* Leipzig: Verlag W. Engelmann.

Metcalfe, D. S. and Elkins, D. M. 1980. *Crop production.* New York: Macmillan.

Miller, T. E. 1983. Preferential transmission of alien chromosome in wheat. In P. E. Brandham and M. D. Bennett (eds.), *Kew chromosome conference II*. London: George Allen & Unwin, 173–182.

Ming, R., Bendahmane, A., and Renner, S. S. 2011. Sex chromosomes in land plants. *Annual Review of Plant Biology* 62: 485–514.

Morris, C. F. and Geroux, M. J. 2000. Modification of cereal grain hardness via expression of puroindoline proteins. Beltsville, MD: USDA, Office of Technology Transfer, 1998, Patent No. 61580.

Nakajima, G. 1979. Cytogenetics of hypo-autotetraploid rye with having 2n=27 chromosomes and its progenies. *Japanese Journal of Breeding* 4: 153–157.

Newbury, H. J. 2003. *Plant molecular breeding*. Birmingham, UK: The University of Birmingham, 280, ISBN 184127321X.

Ohta, Y. and Chuong, P. V. 1975. Hereditary changes in *Capsicum annuum* L. I. Induced by ordinary grafting. *Euphytica* 24(2): 355–368.

Oka, H. I. 1988. *The origin of cultivated rice*. Amsterdam: Elsevier Science Publisher.

Oliver, S. G. and Ward, J. M. 1988. *Wörterbuch der Gentechnik*. Jena, Germany: G. Fischer verlag.

Olson, M. V. 1999. When less is more: Gene loss as an engine of evolutionary change. *American Journal of Human Genetics* 64(1): 18–23.

Payne, P. I. and Rhodes, A. P. 1982. *Encyclopedia of plant physiology 14*. Berlin, Heidelberg: Springer Verlag.

Peters, J. H. and Baumgarten, H. 1992. *Monoclonal antibodies*. Berlin: Springer Verlag.

Poehlman, J. M. 1966. *Breeding field crops*. New York: Holt H. and Company Inc.

Poehlman, J. M. and Sleper, D. A. 1995. *Breeding field crops*. Ames, IA: Iowa State University Press.

Raghavarao, D. 1971. *Constructions and combinatorial problems in design of experiments*. New York: John Wiley & Sons.

Reed, C. D. 1977. *The origin of agriculture*. The Hague, The Netherlands: Mouton Publisher.

Reidie, G. P. 1982. *Genetics*. New York: Macmillan.

Rimpau, W. 1891. *Kreuzungsprodukte landwirthschaftlicher Kulturpflanzen*. Berlin: P. Parey Verl.

Rümker, K. von. 1889. *Anleitung zur Getreidezüchtung auf wissenschaftlicher und praktischer Grundlage*. Berlin: P. Parey Verl.

Rümker, K. von. 1909. *Über die Organisation der Pflanzenzüchtung*. Berlin: P. Parey Verl.

Rümker, K. von. 1924. *Die Leistung In- und Ausländischer Getreidezuchten im Lichte der Sortenprüfungen von 1905–1923*. Berlin: P. Parey Verl.

Rusk, N. 2009. Grafting as a potent molecular tool. *Nature Metabolism* 6(7): 484.

Sauer, J. D. 1993. *Historical geography of crop plants – A select roster*. Boca Raton, FL: CRC Press.

Schena, M., Shalon, D., Davis, R. W., and Brown, P. O. 1995. Quantitative monitoring of gene expression patterns with a complementary DNA microarray. *Science* 270(5235): 467–470.

Schlegel, R. 1990. Effektivität sowie Stabilität Des interspezifischen Chromosomen- und Gentransfers beim hexaploiden Weizen (Efficiency and stability of interspecific chromosome and gene transfer in hexaploid wheat), *Triticum aestivum* L. *Die Kulturpflanze* 38(1): 67–78.

Schlegel, R. 1996. Triticale–Today and tomorrow. In H. Guedes-Pinto, N. Darvey, and V. Carnide (eds.), *Triticale today and tomorrow, Volume 5: Developments in plant breeding*. Dordrecht, The Netherlands: Kluwer Academic Publishers, 21–32.

Schlegel, R. and Cakmak, I. 1997. Micronutritional efficiency in crop plants – A new challenge for cytogenetic research. In T. Lelley (ed.), *Current topics in plant cytogenetics related to plant improvement*. Wien, Austria: Wiener Universitätsverlag Facultas, 91–102.

Schlegel, R., Melz, G., and Korzun, V. 1997. Genes, marker and linkage data of rye (*Secale cereale* L.). Fifth updated inventory. *Euphytica* 101: 23–67.

Schlegel, R., Cakmak, I., Ekiz, H., Kalayci, M., and Braun, H. J. 1998. Screening for zinc efficiency among wheat relatives and their utilisation for an alien gene transfer. *Euphytica* 100(1/3): 281–286.

Schlegel, R. 2006. Rye (*Secale cereale* L.) – A younger crop plant with bright future. In R. J. Sing and P. Jauher (eds.), *Genetic resources, chromosome engineering, and crop improvement: Volume II: Cereals*. Boca Raton, FL: CRC Press, 365–394.

Schlegel, R. 2007. *Concise encyclopedia of crop improvement. Institutions, persons, theories, methods, and histories*. New York/London/Oxford, USA: Haworth Press, 1–331.

Schnell, F. W. 1961. Heterosis and inbreeding effect. Max Plack Institute für Tierzucht und Tierernährung. *Sonderband* 1961: 251–272.

Schoonhoven, A. V. and Voysest, O. 1993. *Common beans – Research for crop improvement*. Wallingford, United Kingdom: CAB International.

Sears, E. R. 1953. Addition of the genome of *Haynaldia villosa* to *Triticum aestivum*. *American Journal of Botany* 40(3): 168–174.

Shah, P., Jogani, V., Bagchi, T., and Misra, A. 2006. Role of Caco-2 cell monolayers in prediction of intestinal drug absorption. *Biotechnology Progress* 22(1): 186–198.

Sharp, P. J. and Lagudah, E. S. 2003. Identification and validation of markers linked to broad-spectrum stem rust resistance gene *Sr2* in wheat (*Triticum aestivum* L.). *Crop Science* 43: 333–336.

Siegel, A. F. 1988. *Statistics and data analysis*. Toronto, ON: John Wiley & Sons.

Simmonds, N. W. 1976. *Evolution of crop plants*. London: Longman, Green and Co. Ltd.

Simmonds, N. W. 1981. *Principles of crop improvement*. New York: Longman Publisher.

Singrün, C. 2002. Untersuchungen zur Lokalisierung und Kartierung von Genen für Resistenz Gegen Mehltau und Braunrost in Saatweizen (*Triticum aestivum* L.) und Dinkel (*Triticum spelta* L.). Ph.D. thesis, Techn. Univ., München (Germany), 1–150.

Snedecor, W. G. and Cochran, W. G. 1980. *Statistical methods*. Ames, IA: Iowa State University Press.

Stanek, V. und Pavlas, P. 1934/1935. Über eine schnelle, informative Methode zur Bestimmung des schädlichen Stickstoffs der Amide und der Aminosäuren in der Rübe. *Zeitschrift für die Zuckerindustrie* 59: 129–142.

Steel, R. G. D. and Torrie, J. H. 1980. *Principles and procedures of statistics: A biometrical approach*. Tokyo, Japan: McGraw-Hill Inc.

Stegemann, S. and Bock, R. 2009. Exchange of genetic material between cells in plant tissue grafts. *Science* 324(5927): 649–651.

Strickberger, M. W. 1988. *Genetik*. München/Wien: C. Hanser Verlag.

Taller, J., Hirata, Y., Yagishita, N., Kita, M., and Ogata, S. 1998. Graft-induced changes and the inheritance of several characteristics in pepper (*Capsicum annuum* L.). *Theoretical and Applied Genetics* 97(5–6): 705–713.

Tan, K. H. 2005. *Soil sampling, preparation, and analysis*. Boca Raton, FL: CRC Press/Taylor & Francis Group.

Tanksley, S. D. and McCouch, S. R. 1997. Seed banks and molecular maps: Unlocking genetic potential from the wild. *Science* 277(5329): 1063–1066.

Tsunewaki, K. 2009. Plasmon analysis *Triticum-Aegilops* complex. *Breeding Science (Japan)* 59: 455–470.

van Harten, A. M. 1998. *Mutation breeding, theory and practical application*. Cambridge, United Kingdom: University Press.

Varshney, R. K., Graner, A., and Sorrells, M. E. 2005. Genomics-assisted breeding for crop improvement. *Trends in Plant Science* 10(12): 621–630.

Vavilov, N. I. 1928. Geographische Zentren unserer Kulturpflanzen (Geographic centers of our crop plants). *Zeitschrift für induklive Abstammungs und vererbungslehre* 45: 342–369.

Webster, C. C. and Wilson, P. N. 1980. *Agriculture in the tropics*. London: Longman.

Weising, K., Nybom, H., Wolff, K., and Kahl, G. 2005. *DNA fingerprinting in plants: Principles, methods, and applications*. Boca Raton, FL: CRC Press, 472.

Yan, W. and Kang, M. S. 2002. *GGE biplot analysis: A graphical tool for breeders, geneticists, and agronomists*. Boca Raton, FL: CRC Press.

Yang, R.-C., Crossa, J., Cornelius, P. L., and Burgueno, J. 2009. Biplot analysis of genotype × environment interaction: Proceed with caution. *Crop Science* 49(5): 1564–1156.

Yates, F. 1937. *The design and analysis of factorial experiments*. Technical Communication 35. Harpenden, England: Imperial Bureau of Soil Science.

Young, N. D. and Tanksley, S. D. 1989. RFLP analysis of the size of chromosomal segments retained around the Tm-2 locus of tomato during backcross breeding. *Theoretical and Applied Genetics* 77(3): 353–359.

Zadoks, J. C., Chang, T. T., and Konzak, C. F. 1974. A decimal code for the growth stages of cereals. *Weed Research* 4(6): 415–421.

Zeven, A. C. and deWet, J. M. J. 1982. *Dictionary of cultivated plants and their regions of diversity*. Wageningen: Centre for Agricultural Publications.

Zhukovski, P. M. 1971. *Cultivated plants and their wild relatives*. Leningrad: Kolos.